T0190365

Graduate Texts in Physics

Graduate Texts in Physics

Graduate Texts in Physics publishes core learning/teaching material for graduate- and advanced-level undergraduate courses on topics of current and emerging fields within physics, both pure and applied. These textbooks serve students at the MS- or PhD-level and their instructors as comprehensive sources of principles, definitions, derivations, experiments and applications (as relevant) for their mastery and teaching, respectively. International in scope and relevance, the textbooks correspond to course syllabi sufficiently to serve as required reading. Their didactic style, comprehensiveness and coverage of fundamental material also make them suitable as introductions or references for scientists entering, or requiring timely knowledge of, a research field.

More information about this series at http://www.springer.com/series/8431

Masahito Hayashi

Quantum Information Theory

Mathematical Foundation

Second Edition

Masahito Hayashi
Graduate School of Mathematics
Nagoya University
Nagoya, Aichi
Japan

Original Japanese version published by Saiensu-sha Company, Ltd., Tokyo, Japan, 2004

ISSN 1868-4513 ISSN 1868-4521 (electronic)
Graduate Texts in Physics
ISBN 978-3-662-57032-6 ISBN 978-3-662-49725-8 (eBook)
DOI 10.1007/978-3-662-49725-8

Printed on acid-free paper

This Springer imprint is published by Springer Nature
The registered company is Springer-Verlag GmbH Germany
The registered company address is: Heidelberger Platz 3, 14197 Berlin, Germany

The original version of the book frontmatter was revised: The additional bibliographic information on the copyright page was included. The erratum PDF is available at DOI:10.1007/978-3-662-49725-8_11

Preface

This book was originally published in Japanese by Saiensu-sha, Tokyo, Japan in May 2003. Then, the first English edition was published by Springer in 2006 with some revision from Japanese version. It has been 10 years since the publication of the first English edition. During this decade, so many remarkable progresses have been made in the area of quantum information theory. So, I decided to publish the second English edition with considerable revision to include these latest progresses.

I believe that the most important progress among this decade is the resolution of the additivity problem. This problem was proposed as a problem equivalent to various kinds of additivity problems in entanglement theory and channel capacity. To include this progress, I have added Sect. 8.13: Violation of superadditivity of entanglement formation. Since this topic needs special knowledge for classical information, I have also added Sect. 2.6: Large Deviation on Sphere. Although Sect. 2.6 is important for the understanding of quantum information, it seems that its content is partially misunderstood among quantum information researchers. So, this section is also helpful for understanding quantum information. Further, since this topic affects the channel capacity, I rewrote Sect. 9.2: C-Q Channel Coding with Entangled Inputs.

The second most important progress is considerable progress on quantum hypothesis testing. This progress has been made by so many authors. Although Chernoff bound, Hoeffding bound, and Han–Kobayashi bound in this topic had not been exactly solved at that time, these exact forms have been completely solved during this decade. To reflect this progress, I completely rewrote Chap. 3. Since the quantum Han–Kobayashi bound is closely related to the new type of quantum Rényi relative entropy, the new Chap. 3 also discusses it. Further, to make Chap. 3 more self-contained, I have moved Section: Information Quantities in Quantum Systems to Chap. 3. The content of this chapter is employed in Chap. 4 because the hypothesis testing is closely related to channel coding. Hence, I also rewrote Chap. 4 partially. Recently, this relation has been of interest for many researchers. In this revision, I emphasize this relation more strongly while this relation was discussed in the first edition. I also summarize its history in Sect. 4.10.

The third most important progress is considerable progress in conditional Rényi entropy. To discuss this issue, I newly added Sect 2.1.5: Conditional Rényi entropy and Sect 5.6: Conditional Rényi Entropy and Duality. This progress has made notable influence on entropic uncertainty relation, secure random number generation, entanglement measure, and the duality relation between coherence and information leakage. Since these four areas also have greatly progressed during this decade, I summarize them in the following new sections: Sect. 7.3: Entropic Uncertainty Relation, Sect. 8.14: Secure Random Number Generation, Sect. 8.8: Maximally Correlated State, and Sect. 8.15: Duality Between Two Conditional Entropies. Further, using the contents of Sect. 8.15, I simplified the proof in Sect. 9.6: Channel Capacity for Quantum-State Transmission. Also, based on the previously gained knowledge, I newly added Subsect. 9.6.3: Decoder with assistance by local operations.

Other topics have advanced recently, and we can list discord, Bregman divergence, and matrix convex function, among them. The first edition discussed discord, however, its treatment is not perfect. So, in the second edition, I have completed it in new Sect. 8.10: Discord. To deal with recent progress of Bregman divergence, I added Sect. 2.2.2: Bregman divergence. Recently, extremal point decomposition of matrix convex functions was completed. This decomposition brings us a more detailed analysis of quantum f-relative entropy. So, to include the decomposition, I rewrote Appendix A.4: Convex Functions and Matrix Convex Functions. Then, I have newly added Sect. 6.7: Relative Modular Operator and Quantum f-Relative Entropy.

As one of the features of this book, I have discussed the axiomatic approach, while the first edition emphasizes mainly in entanglement measure. However, this approach is also important in the entropy theory. To clarify this relation, I have newly added Sect. 2.5: Continuity and Axiomatic Approach. Also, I have added several descriptions related to this approach.

In this edition, I additionally have included around 120 new exercises, so that this edition totally has 450 exercises. I also have completed solutions for all exercises for readers' convenience. Since each chapter can be understood separately, I have organized the second edition so that each chapter contains solutions for exercises and proofs of theorems in that chapter. In particular, since Chap. 2 is composed of knowledge from classical information and has a distinguished description from existing textbooks, this chapter might be useful for readers interested only in classical information. Recently, I have published another book "Introduction to Quantum Information Science" with S. Ishizaka, A. Kawachi, G. Kimura, and T. Ogawa, which is more introductory. Since this book is more mathematically oriented, I changed the title to "Quantum Information Theory: A Mathematical Foundation."

I am grateful to Prof. Fumio Hiai, Prof. Francesco Buscemi, Dr. Motohisa Fukuda, Mr. Kengo Takeuchi, and Mr. Kosuke Ito for their comments. I would like to express my appreciation for their cooperation. I would also like to thank Dr. Claus E. Ascheron of Springer Science+Business Media for his encouragement and patience during the preparation of the manuscript.

Nagoya, Japan Masahito Hayashi

Preface to the First English Edition

This book is the revised and English edition of the Japanese book *Introduction to Quantum Information Theory*, which systematically describes quantum information theory and was originally published by Saiensu-sha, Tokyo, Japan in May 2003. The study of information processing based on the physical principles of quantum mechanics began in the 1960s. Recently, some protocols of quantum information processing have been demonstrated experimentally, and their theoretical aspects have been examined more deeply and mathematically. In particular, the field that is concerned with their theoretical bounds is called quantum information theory and has been studied by many researchers from diverse viewpoints.

However, only Holevo's book *Probabilistic and Statistical Aspects of Quantum Theory*, which was published back in 1980 (English version in 1982), places a heavy emphasis on the mathematical foundation of quantum information theory. Several books concerning quantum information science have been published since the late 1990s. However, they treat quantum computation, the physical aspects of quantum information, or the whole of quantum information science and are not mainly concerned with quantum information theory. Therefore, it seemed to me that many researchers would benefit from an English book on quantum information theory, and so I decided to publish the English version of my book. I hope that it will make a contribution to the field of quantum information theory.

This book was written as follows. First, the author translated the original Japanese version in cooperation with Dr. Tim Barnes. Next, the book was revised through the addition of many new results to Chaps. 8–10 and a historical note to every chapter. Several exercises were also added, so that the English version has more than 330 exercises. Hence, I take full responsibility for the content of this English version. In this version, theorems and lemmas are displayed along with the names of the researchers who contributed them. However, when the history of the theorems and lemmas is not so simple, they are displayed without the contributing researchers' names and their histories are explained in a historical note at the end of the given chapter.

I am indebeted to Prof. Masanao Ozawa and Dr. Tohya Hiroshima for their feedback on the Japanese version, which been incorporated into the English version. I am also grateful to (in alphabetical order) Dr. Giulio Chiribella, Mr. Motohisa Fukuda, Prof. Richard Gill, Dr. Michael Horodecki, Dr. Satoshi Ishizaka, Dr. Paolo Perinotti, Dr. Toshiyuki Shimono, and Dr. Andreas Winter, for reviewing the technical aspects of the English version. Further, Dr. Tomohisa Hayakawa, Mr. Daichi Isami, Mr. Takashi Okajima, Mr. Tomotake Sasaki, Mr. Taiji Suzuki, Mr. Fuyuhiko Tanaka, and Mr. Ken'ichiro Tanaka used the draft of the English version in their seminar and verified its contents. Miss Rika Abe commented on the nontechnical parts of the book. Further, Mr. Motohisa Fukuda helped me in compiling the references. I would like to express my appreciation for their cooperation.

I also would like to thank Prof. Hiroshi Imai of the University of Tokyo and the people associated with the ERATO Quantum Computation and Information Project for providing the research environments for this English version. I would like to express my gratitude to Dr. Glenn Corey and editorial staffs of Springer for good excellent editing process. I would also like to thank Dr. Claus E. Ascheron of Springer Science+Business Media for his encouragement and patience during the preparation of the manuscript.

Hongo, Tokyo, Japan
November 2005

Masahito Hayashi

Preface to the Japanese Edition

This textbook attempts to describe quantum information theory, which is a presently evolving field. It is organized so that the reader can understand its contents with very elementary prior knowledge. This research field has been developed by many researchers from various backgrounds and has matured rapidly in the last 5 years.

Recently, many people have considered that more interdisciplinary activities are needed in the academic world. Hence, education and research must be performed and evaluated on a wide scope. However, since the extreme segmentation of each research area has increased the difficulty of interdisciplinary activities. On the other hand, quantum information theory can in some sense form a bridge between several fields because it deals with topics in a variety of disciplines including physics and information science. Hence, it can be expected to contribute in some way to removing the segmentation of its parent fields. In fact, information science consists of subfields such as computer science, mathematical statistics, and Shannon's information theory. These subfields are studied in separate contexts.

However, in quantum information theory, we must return to the fundamentals of the topic, and there are fewer boundaries among the different fields. Therefore, many researchers now transcend these boundaries.

Given such a starting point, the book was written to enable the reader to efficiently attain the interdisciplinary knowledge necessary for understanding quantum information theory. This book assumes only that the reader has knowledge of linear algebra, differential and integral calculus, and probability/statistics at the undergraduate level. No knowledge of quantum mechanics is assumed.

Some of the exercises given in the text are rather difficult. It is recommended that they be solved in order to acquire the skills necessary for tackling research problems. Parts of the text contain original material that does not appear elsewhere. Comments will be given for such parts.

The author would like to thank Prof. Hiroshi Imai of the University of Tokyo, Prof. Shun-ichi Amari of the Brain Science Institute at RIKEN, Prof. Kenji Ueno of Kyoto University, and the people associated with the ERATO Quantum Computation and Information Project, the Brain Science Institute at RIKEN, and the Department of Mathematics at Kyoto University for providing me with the means to continue my research. The author also wishes to thank Prof. Hiroshi Nagaoka of the University of Electro-Communications, Prof. Akio Fujiwara of Osaka University, Prof. Keiji Matsumoto of the National Institute of Informatics, and Dr. Tomohiro Ogawa of the University of Tokyo for helpful discussions and advice. This text would not have been possible without their enlightening discussions.

I also received valuable comments from Prof. Alexander Holevo of the Steklov Mathematical Institute, Prof. Masanao Ozawa of Tohoku University, Dr. Ryutaroh Matsumoto of the Tokyo Institute of Technology, Dr. Fumiaki Morikoshi of NTT, Dr. Yodai Watanabe of RIKEN, and Dr. Mitsuru Hamada, Dr. Yoshiyuki Tsuda, Dr. Heng Fan, Dr. Xiangbin Wang, and Mr. Toshiyuki Shimono of the ERATO Quantum Computation and Information Project regarding the contents of this text. They have also earned a debt of gratitude. I would also like to thank Mr. Kousuke Hirase of Saiensu-sha for his encouragement and patience during the preparation of the manuscript

Hongo, Tokyo, Japan
December 2003

Masahito Hayashi

Contents

Notations

Basic Notations

$\|M\|$	Number of POVM elements, p. 5
$\bar{x}$	Complex conjugate of the given number x, p. 2
$\mathcal{S}(\mathcal{H})$	Set of density matrices of given Hilbert space $\mathcal{H}$, p. 6
A^T	Transpose of a matrix A, p. 2
$\overline{A}$	Complex conjugate matrix of a matrix A, p. 2
A^*	Adjoint of a matrix A, p. 2
$[X, Y]$	Commutator of matrices A and B, p. 4
$X \circ Y$	Symmetrized product of matrices A and B, p. 4
P_ρ^M	Probability distribution when measurement is $\boldsymbol{M}$ and state is ρ, p. 6
ρ_{mix}	Completely mixed state, p. 7
Tr_A	Partial trace concerning system $\mathcal{H}_A$, p. 13
$\{X \geq 0\}$	Projection defined by (1.37), p. 16
κ_M	Pinching of PVM $\boldsymbol{M}$ (1.13), p. 8
κ_X	Pinching of Hermitian matrix X (1.14), p. 8
κ_M	Pinching of POVM $\boldsymbol{M}$ (1.15), p. 8
S_i	Pauli matrix (1.16), p. 9
ρ_x	Stokes parameterization (1.17), p. 9
$\mathcal{T}(\mathcal{H})$	Set of Hermitian matrices on $\mathcal{H}$, p. 98
$\mathcal{M}(\mathcal{H})$	Set of matrices on $\mathcal{H}$, p. 98
$\eta(x)$	$-x \log x$ (Theorem 5.12), p. 223
$\eta_0(x)$	See (5.91), p. 223

Information Quantities in Classical System

$D(p\|q)$	Relative entropy (2.12), p. 28
$D_f(p\|q)$	f-relative entropy $\sum_i p_i f\left(\frac{q_i}{p_i}\right)$ (Theorem 2.1), p. 29
$\phi(s\|p\|q)$	$\log(\sum_i p_i^{1-s} q_i^s)$, p. 30

$D_{1-s}(p\|q)$	Relative Rényi entropy $-\frac{\phi(s\|p\|q)}{s}$ (2.19), p. 30
$D_{\min}(p\|q)$	Minimum relative entropy (2.20), p. 30
$D_{\max}(p\|q)$	Maximum relative entropy (2.20), p. 30
$d_2(p,q)$	Hellinger distance (2.17), p. 29
$d_1(p,q)$	Variational distance $\frac{1}{2}\sum_i \|p_i - q_i\|$(2.23), p. 31
$I(X:Y)$	Mutual information $D(\mathrm{P}_{X,Y}\|\mathrm{P}_X \times \mathrm{P}_Y)$(2.30), p. 34
$I(X:Y\|Z)$	Conditional mutual information (2.31), p. 34
$I(p,Q)$	Transmission information (2.34), p. 35
$H(p)$	Entropy of distribution p (2.2), p. 26
$H(X)$	Entropy of random variable X, p. 26
$h(x)$	Binary entropy, p. 26
$\psi(s\|p)$	$\log \sum_i p_i^{1-s}$ (2.38), p. 36
$H_{1-s}(p)$	Rényi entropy $\frac{\psi(s\|p)}{s}$ (2.38), p. 36
$H_{\min}(p)$	Minimum entropy $-\log \max_i p_i$ (2.39), p. 36
$H_{\max}(p)$	Maximum entropy $\log \|\{i\|p_i > 0\}\|$ (2.39), p. 36
$H(X\|Y)$	Conditional entropy (2.5), p. 26
$H_{1+s}(X\|Y)$	Conditional Rényi entropy $\log \|\chi\| - D_{1+s}(\mathrm{P}_{XY}\|p_{\mathrm{mix},\chi} \times \mathrm{P}_Y)$ (2.74), p. 42
$H^{\uparrow}_{1+s}(X\|Y)$	Conditional Rényi entropy $\log \|\chi\| - \min_{Q_Y} D_{1+s}(\mathrm{P}_{XY}\|p_{\mathrm{mix},\chi} \times Q_Y)$ (2.75), p. 42
$H_{\min}(X\|Y)$	Conditional minimum entropy $\lim_{s\to\infty} H_{1+s}(X\|Y)$ (2.77), p. 42
$H^{\uparrow}_{\min}(X\|Y)$	Conditional minimum entropy $\lim_{s\to\infty} H^{\uparrow}_{1+s}(X\|Y)$ (2.77), p. 42
$H_{\max}(X\|Y)$	Conditional maximum entropy $\lim_{s\to-1} H_{1+s}(X\|Y)$ (2.78), p. 42
$H^{\uparrow}_{\max}(X\|Y)$	Conditional maximum entropy $\lim_{s\to-1} H^{\uparrow}_{1+s}(X\|Y)(X\|Y)$ (2.78), p. 42

Notations for Information Geometry

J_θ	Fisher information (2.103), p. 47
$l_\theta(\omega)$	Logarithmic derivative, p. 47
$\boldsymbol{J}_\theta$	Fisher information matrix, p. 47
$D^\mu(\bar{\theta}\|\theta)$	Bregman divergence (2.111), (2.116), p. 50, 51
$\nu(\eta)$	Legendre transform of μ (2.112), (2.119), p. 50, 51
$\eta(\theta)$	Expectation parameter (2.116), (2.131), p. 51, 55
$\mu(\theta)$	Potential function (Cumulant generating function) (2.128), (2.130), p. 53, 54

Notation Related to Probability

$\mathcal{P}(\Omega)$	Set of probability distributions on the probability space Ω, p. 26
$p_{\mathrm{mix},\Omega}$	Uniform distribution on Ω, p. 26
$p_{\mathrm{mix},k}$	Uniform distribution on Ω when $\lvert\Omega\rvert = k$, p. 26
p_{mix}	Uniform distribution (Simplification of the above), p. 26
$\mathrm{E}_p(X)$	Expectation of X under distribution p (2.1), p. 25
$\mathrm{V}_p(X)$	Variance of X under distribution p (2.94), p. 46
$\mathrm{Cov}_p(X, Y)$	Covariance between X and Y under distribution p (2.93), p. 45
$\kappa_p(X)$	Conditional expectation of X under distribution p (2.107), p. 48
$\kappa_{\mathfrak{U},p}(X)$	Conditional expectation of X with respect to the subspace $\mathcal{U}$ (2.110), p. 49
$p_i^{\downarrow}$	Element of $\{p_i\}$ that is reordered according to size, p. 38
$P(p, L)$	$\sum_{i=1}^{L} p_i^{\downarrow}$ (2.48), p. 38
$\hat{\mathrm{V}}_\theta(\hat{\theta})$	Mean square error of estimator $\hat{\theta}$ (2.137), p. 56
$\mathrm{Med}_p(X)$	Median of X (2.218), p. 78
$\mu_{\mathcal{H}}$	Haar measure on Hilbert space $\mathcal{H}$ (2.210), p. 77
μ_{S^n}	Haar measure on the n-dimensional sphere S^n (2.211), p. 77
$\mathrm{Med}_{S^{2l-1}}(f)$	Median of f under the Haar measure $\mu_{S^{2l-1}}$ on $S^{2l-1}\mathrm{Med}_{\mu_{S^{2l-1}}}(f)$, p. 81
$\mathrm{E}_{S^{2l-1}}$	Expectation under the Haar measure $\mu_{S^{2l-1}}$ on S^{2l-1} $\left(\mathrm{E}_{\mu_{S^{2l-1}}}\right)$, p. 82

Notations for Large Deviation

$\mathbb{N}_d$	$\{1, \ldots, d\}$, p. 61
T_n	Set of empirical distributions on $\mathbb{N}_d$ (Set of types on $\mathbb{N}_d$), p. 61
T_q^n	Set of data with the empirical distribution q, p. 61
$\beta(\{\hat{\theta}_n\}, \theta, \epsilon)$	Rate function of error probability (2.173), p. 66
$\alpha(\{\hat{\theta}_n\}, \theta)$	First-order coefficient of rate function (2.174), p. 66

Fundamental Information Quantities and Related Notations in Quantum Systems

$H(\rho)$	von Neumann entropy $-\mathrm{Tr}\,\rho\log\rho$ (3.1), p. 98
$\psi(s\lvert\rho)$	$\log\mathrm{Tr}\,\rho^{1-s}$, p. 98
$H_{1-s}(\rho)$	Rényi entropy $\frac{\psi(s\lvert\rho)}{s}$, p. 98
$H_{\min}(\rho)$	Minimum entropy $-\log\lVert\rho\rVert$, p. 98
$H_{\max}(\rho)$	Maximum entropy $\log\mathrm{Tr}\{\rho > 0\}$, p. 98

$D(\rho\|\sigma)$	Quantum relative entropy $\operatorname{Tr}\rho(\log\rho - \log\sigma)$ (3.7), p. 99
$I_\rho(A:B)$	Mutual information (5.89) (8.34), p. 223, 369
$I_\rho(A:B\|C)$	Conditional mutual information (5.90), p. 223
$\phi(s\|\rho\|\sigma)$	$\log \operatorname{Tr}\rho^{1+s}\sigma^{-s}$, p. 99
$\phi(s)$	Abbreviation of $\phi(s\|\rho\|\sigma)$, p. 99
$D_{1+s}(\rho\|\sigma)$	relative Rényi entropy $\frac{\phi(-s\|\rho\|\sigma)}{s}$ (3.9), p. 99
$D_{\max}(\rho\|\sigma)$	Maximum relative entropy $\log \|\sigma^{-\frac{1}{2}}\rho\sigma^{-\frac{1}{2}}$ (3.10), p. 99
$D_{\min}(\rho\|\sigma)$	Minimum relative entropy $-\log \operatorname{Tr}\sigma\{\rho>0\}$ (3.10), p. 99
$\tilde{\phi}(s\|\rho\|\sigma)$	$\log \operatorname{Tr}(\sigma^{\frac{s}{2(1-s)}}\rho\sigma^{\frac{s}{2(1-s)}})^{1-s} = \lim \frac{1}{n}\phi(s\|\kappa_{\sigma^{\otimes n}}(\rho^{\otimes n})\|\sigma^{\otimes n})$, p. 99
$\underline{D}_{1+s}(\rho\|\sigma)$	Sandwiched relative Rényi entropy $\frac{\tilde{\phi}(-s\|\rho\|\sigma)}{s}$ (3.13), p. 99
$b(\rho\|\sigma)$	Bures distance (3.42), p. 103
$F(\rho,\sigma)$	Fidelity $\operatorname{Tr}\|\sqrt{\rho}\sqrt{\sigma}\|$, p. 103
$d_1(\rho,\sigma)$	Trace norm distance (3.45), p. 104
$\beta_\epsilon^n(\rho\|\sigma)$	Minimum value of second error probability (3.92), p. 118
p_{guess}	Guessing probability, p. 113
$B(\rho\|\sigma)$	Maximum decreasing rate of second error probability when first error probability goes to 0 (Theorem 3.3), p. 119
$\tilde{B}(\rho\|\sigma)$	Maximum decreasing rate of second error probability when first error probability goes to 0 and measurement is separable (Theorem 3.5), p. 123
$B^\dagger(\rho\|\sigma)$	Maximum decreasing rate of second error probability when first error probability does not go to 1 (Theorem 3.3), p. 119
$B(r\|\rho\|\sigma)$	Maximum decreasing rate of first error probability when second error probability goes to 0 at rate r (3.98), p. 120
$B^*(r\|\rho\|\sigma)$	Minimum decreasing rate of first error probability when second error probability goes to 0 at rate r (3.99), p. 120
$P_{(\rho\|\sigma)}$	A distribution defined by ρ and σ, p. 110
$Q_{(\rho\|\sigma)}$	Another distribution defined by ρ and σ, p. 110

Information Quantities of c-q Channel W

$I(p,W)$	Transmission information (4.1), p. 159
$I(\boldsymbol{M},p,W)$	Classical transmission information with measurement $\boldsymbol{M}$, p. 164
$I_{1-s}(p,W)$	$-\frac{1}{s}\log\sum_x p(x)\operatorname{Tr}W_x^{1-s}W_p^s$ (4.13), p. 161
$I_{1-s}^{\downarrow}(p,W)$	$-\frac{(1-s)}{s}\log\operatorname{Tr}(\sum_x p(x)W_x^{1-s})^{\frac{1}{1-s}}$ (4.14), p. 162
$J(p,\sigma,W)$	$\sum_{x\in\chi}p(x)D(W_x\|\sigma)$ (Exercise 4.14), p. 176
$J_{1-s}(p,\sigma,W)$	$-\frac{1}{s}\log\sum_x p(x)\operatorname{Tr}W_x^{1-s}\sigma^s$ (4.3), p. 159
$C_c(W)$	C-q channel capacity (4.9), p. 161
$C_c^\dagger(W)$	Strong converse c-q channel capacity (4.10), p. 161
$\tilde{C}_c(W)$	C-q channel capacity with adaptive decoding and feedback (4.27), p. 164

$C_{c\|c\leq K}(W)$	C-q channel capacity with a cost function, p. 164
$C^{\downarrow}_{1-s}(W)$	$\sup_{p\in P_f(X)} I^{\downarrow}_{1-s}(p,W)$ (4.15), p. 162
$C_c(W^1,\ldots,W^M)$	C-q channel capacity for multiple receivers (4.20), p. 163
$B(R\|W)$	Reliability function (4.54), p. 172

Notations Related to c-q Channel W

Φ	Code (N,φ,Y) (Sect. 4.1.2), p. 160
$\tilde{\Phi}^{(n)}$	Feedback-allowing coding (Sect. 4.2), p. 164
$\|\Phi\|$	Size of code (4.8), p. 160
$\varepsilon[\Phi]$	Average error probability of code (4.8), p. 160
$\mathcal{P}_f(\mathcal{X})$	Set of probability distributions with a finite support in $\mathcal{X}$ (Theorem 4.1), p. 161
$\mathcal{P}_{c\leq K}(\mathcal{X})$	$\{p\in\mathcal{P}_f(\mathcal{X})\|\sum_x p(x)c(x)\leq K\}$ (Theorem 4.3), p. 167
p_{1-s}	$\operatorname{argmax}_{p\in\mathcal{P}_f(\mathcal{X})} I^{\downarrow}_{1-s}(p,W)$ (4.62), p. 174
p'_{1-s}	$\operatorname{argmax}_{p\in\mathcal{P}_{c\leq K}(\mathcal{X})} I^{\downarrow}_{1-s}(p,W)$ (Proof of Lemma 4.4), p. 176
W_p	Average state $\sum_x p(x)W_x$ (4.2), p. 159
$\sigma_{1-s\|p}$	$(\sum_x p(x)W_x^{1-s})^{\frac{1}{1-s}}/\operatorname{Tr}[(\sum_{x'} p(x')W_{x'}^{1-s})^{\frac{1}{1-s}}]$ (4.23), p. 163
$W^A\otimes W^B$	Product channel of W^A and W^B (4.4), p. 159
$W^{(n)}$	n-fold stationary memoryless channel, p. 172
$p\times W$	Correlated state (4.45), p. 170
$p\otimes\sigma$	Independent state (4.45), p. 170

Conditional Entropies in Quantum System

$H_\rho(A\|B)$	Conditional entropy $H_\rho(AB)-H_\rho(A)$ (5.88), p. 223
$H_{1+s\|\rho}(A\|B)$	Conditional Rényi entropy $-D_{1+s}(\rho\|I_A\otimes\rho_B)$ (5.112), p. 228
$\widetilde{H}_{1+s\|\rho}(A\|B)$	Conditional Rényi entropy $-\underline{D}_{1+s}(\rho\|I_A\otimes\rho_B)$ (5.113), p. 228
$H^{\uparrow}_{1+s\|\rho}(A\|B)$	Conditional Rényi entropy $\max_{\sigma_B} -D_{1+s}(\rho\|I_A\otimes\sigma_B)$ (5.114), p. 228
$\widetilde{H}^{\uparrow}_{1+s\|\rho}(A\|B)$	Conditional Rényi entropy $\max_{\sigma_B} -\underline{D}_{1+s}(\rho\|I_A\otimes\sigma_B)$ (5.115), p. 228
$H_{\min\|\rho}(A\|B)$	Conditional minimum entropy $\lim_{s\to\infty} H_{1+s\|\rho}(A\|B)$ (5.123), p. 229
$H^{\uparrow}_{\min\|\rho}(A\|B)$	Conditional minimum entropy $\lim_{s\to\infty} H^{\uparrow}_{1+s\|\rho}(A\|B)$ (5.123), p. 229
$\widetilde{H}_{\min\|\rho}(A\|B)$	Conditional minimum entropy $\lim_{s\to\infty} \widetilde{H}_{1+s\|\rho}(A\|B)$ (5.124), p. 229

$\widetilde{H}^{\uparrow}_{\min\|\rho}(A\|B)$	Conditional minimum entropy $\lim_{s\to\infty} \widetilde{H}^{\uparrow}_{1+s\|\rho}(A\|B)$ (5.124), p. 229
$H_{\max\|\rho}(A\|B)$	Conditional maximum entropy $\lim_{s\to-1} H_{1+s\|\rho}(A\|B))$ (5.125), p. 229
$H^{\uparrow}_{\max\|\rho}(A\|B)$	Conditional maximum entropy$\lim_{s\to-1} H^{\uparrow}_{1+s\|\rho}(A\|B)$ (5.125), p. 229
$\widetilde{H}_{\max\|\rho}(A\|B)$	Conditional maximum entropy $\lim_{s\to-1} \widetilde{H}_{1+s\|\rho}(A\|B)$ (5.126), p. 229
$\widetilde{H}^{\uparrow}_{\max\|\rho}(A\|B)$	Conditional maximum entropy $\lim_{s\to-1} \widetilde{H}^{\uparrow}_{1+s\|\rho}(A\|B)$ (5.126), p. 229

Notations of q-q Channels

$K(\kappa)$	Matrix representation of κ (5.4), p. 200
κ^E	TP-CP map to environment (5.7), p. 201
$\kappa_{M,W}$	Entanglement-breaking channel (Example 5.4), p. 206
τ	Transpose (5.14), p. 207
$\kappa_{d,\lambda}$	Depolarizing channel (5.12), p. 206
$\kappa^T_{d,\lambda}$	Transpose depolarizing channel (5.18), p. 208
κ^{GP}_p	Generalized Pauli channel (5.16), p. 208
κ^{PD}_D	Phase-damping channel (5.19), p. 208
$\kappa^{\mathrm{pns}}_{d,n\to m}$	PNS channel (5.21), p. 209
$\kappa^{\mathrm{era}}_{d,p}$	Erasure channel (5.22), p. 209

Quantum Fisher Information

$J_{\theta,x}$	Fisher metric based on inner product x, p. 260
$J_{\theta,s}$	SLD Fisher metric, p. 260
$J_{\theta,b}$	Bogoljubov Fisher metric, p. 260
$J_{\theta,r}$	RLD Fisher metric, p. 260
$\boldsymbol{J}_{\theta,x}$	Fisher information matrix based on inner product x (6.39), p. 262
$\boldsymbol{J}_{\theta,s}$	SLD Fisher information matrix (6.39), p. 262
$\boldsymbol{J}_{\theta,b}$	Bogoljubov Fisher information matrix (6.39), p. 262
$\boldsymbol{J}_{\theta,r}$	RLD Fisher information matrix (6.39), p. 262

Variants of Quantum Relative Entropy

$D_{c,p}(\rho\|\sigma)$	Maximum relative entropy with projective measurement $\max_{M:PVM} D(\mathrm{P}^M_\rho\|\mathrm{P}^M_\sigma)$ (5.42), p. 218
$D_c(\rho\|\sigma)$	Maximum relative entropy with measurement $\max_{M:POVM} D(\mathrm{P}^M_\rho\|\mathrm{P}^M_\sigma)$ (5.42), p. 218

$D_x^{(e)}(\rho\|\sigma)$	x-e-divergence (6.52), p. 267
$D_s^{(e)}(\rho\|\sigma)$	SLD e-divergence $2\operatorname{Tr}\rho\log\sigma^{-\frac{1}{2}}(\sigma^{\frac{1}{2}}\rho\sigma^{\frac{1}{2}})^{\frac{1}{2}}\sigma^{-\frac{1}{2}}$ (6.57), p. 268
$D_b^{(e)}(\rho\|\sigma)$	Bogoljubov e-divergence ($= D(\rho\|\sigma)$) (6.58), p. 268
$D_r^{(e)}(\rho\|\sigma)$	RLD e-divergence $\operatorname{Tr}\rho\log(\rho^{\frac{1}{2}}\sigma^{-1}\rho^{\frac{1}{2}})$(6.59), p. 268
$D_{\frac{1}{2}}^{(e)}(\rho\|\sigma)$	e-divergence with $x = \frac{1}{2}$ (6.60), p. 268
$D_x^{(m)}(\rho\|\sigma)$	x-m-divergence (6.63), p. 268
$D_b^{(m)}(\rho\|\sigma)$	Bogoljubov m-divergence ($= D(\rho\|\sigma)$) (6.66), p. 269
$D_r^{(m)}(\rho\|\sigma)$	RLD m-divergence $\operatorname{Tr}\rho\log(\sqrt{\rho}\sigma^{-1}\sqrt{\rho})$ (6.67), p. 269
$D_f(\rho\|\sigma)$	quantum f-relative entropy (6.116), p. 290

Notations Related to Quantum Information Geometry

$E_{\rho,s}(X)$	See (6.8), p. 254
$E_{\rho,b}(X)$	See (6.9), p. 254
$E_{\rho,r}(X)$	See (6.10), p. 254
$E_{\rho,p}(X)$	See (6.11), p. 254
$E_{\rho,\lambda}(X)$	See (6.12), p. 254
$\mathcal{K}_{\rho,x}(\mathcal{H})$	Kernel of $E_{\rho,x}$, p. 256
$\mathcal{M}_{\rho,x}(\mathcal{H})$	Quotient matrix space $\mathcal{M}(\mathcal{H})/\mathcal{K}_{\rho,x}(\mathcal{H})$, p. 256
$\mathcal{M}_{\rho,x}^{(m)}(\mathcal{H})$	Image of the map $E_{\rho,x}$ $(\{X \in \mathcal{M}(\mathcal{H}) \| P_\rho X = X\})$, p. 257
P_ρ	Projection to the range of ρ, p. 257
$\langle Y, X\rangle_{\rho,x}^{(e)}$	See (6.13), p. 254
$\|X\|_{\rho,x}^{(e)}$	See (6.14), p. 255
$\langle A, B\rangle_{\rho,x}^{(m)}$	See (6.18), p. 255
$\|A\|_{\rho,x}^{(m)}$	See (6.19), p. 255
$\kappa_{\rho,x}$	See (6.21), p. 256
$L_{\theta,x}$	e representation based on inner product x (6.30), p. 260
$L_{\theta,s}$	SLD e representation (6.31), p. 260
$L_{\theta,b}$	Bogoljubov e representation, p. 261
$L_{\theta,r}$	RLD e representation (6.31), p. 260
$\Pi_{L,s}^{\theta}\rho_0$	SLD e parallel transport, p. 266
$\Pi_{L,b}^{\theta}\rho_0$	Bogoljubov e parallel transport, p. 266
$\Pi_{L,r}^{\theta}\rho_0$	RLD e parallel transport, p. 266
$\mathbf{Re}X$	Real part of matrix X, p. 3
$\mathbf{Im}\,X$	Imaginary part of matrix X, p. 3
$\mathbf{V}_\theta(X)$	Matrix with components $(\operatorname{Tr}\rho_\theta X^i X^j)$, p. 284
$\boldsymbol{\Delta}_{\rho,\sigma}$	Relative modular operator, p. 290

Error Criteria

$\hat{\mathrm{V}}_\theta(\boldsymbol{M}^n, \hat{\theta}_n)$	Mean square error (MSE) of estimator $(\boldsymbol{M}^n, \hat{\theta}_n)$ (6.71), p. 273
$\hat{\boldsymbol{V}}_\theta(\boldsymbol{M}^n, \hat{\theta}_n)$	Mean square error matrix (6.94), p. 281
$\hat{\boldsymbol{V}}_\theta(\{\boldsymbol{M}^n, \hat{\theta}_n\})$	Matrix of components $\hat{\mathrm{V}}_\theta^{i,j}(\{\boldsymbol{M}^n, \hat{\theta}_n\}) \stackrel{\mathrm{def}}{=} \lim n \hat{\mathrm{V}}_\theta^{i,j}(\boldsymbol{M}^n, \hat{\theta}_n)$, p. 281
$\beta(\{(\boldsymbol{M}^n, \hat{\theta}_n)\}, \theta, \epsilon)$	Rate function of error probability (6.84), p. 276
$\alpha(\{(\boldsymbol{M}^n, \hat{\theta}_n)\}, \theta)$	First-order coefficient of rate function (6.86), p. 278

Disturbances and Uncertainties

$\boldsymbol{\Delta}_1(X, \rho)$	Uncertainty of an observable (7.12), p. 330
$\boldsymbol{\Delta}_2(\boldsymbol{M}, \rho)$	Uncertainty of a measurement (7.13), p. 330
$\boldsymbol{\Delta}_3(\boldsymbol{M}, X, \rho)$	Deviation of POVM $\boldsymbol{M}$ from observable X (7.17), p. 330
$\boldsymbol{\Delta}_4(\kappa, X, \rho)$	Disturbance of X caused by κ (7.23), p. 332
$\boldsymbol{\Delta}_4(\boldsymbol{\kappa}, \boldsymbol{X}, \boldsymbol{\rho})$	Disturbance of X caused by $\boldsymbol{\kappa}$ (7.25), p. 332
$\varepsilon(\rho, \boldsymbol{k})$	Amount of state reduction by $\boldsymbol{\kappa}$ (7.59), p. 342

Information Quantities of q-q Channel κ and State ρ

$F_e(\rho, \kappa)$	Entanglement fidelity for TP-CP κ (8.18), p. 365
$F_e(\rho, \boldsymbol{\kappa})$	Entanglement fidelity for an instrument (8.29), p. 366
$I(\rho, \kappa)$	Transmission information of q-q channel κ (8.35), p. 370
$I_c(\rho, \kappa)$	Coherent information (8.37), p. 370
$\tilde{I}_c(\rho, \kappa)$	Pseudocoherent information (8.48), p. 372
$H_e(\kappa, \rho)$	Entropy exchange $H((\kappa \otimes \iota_R)(\lvert x\rangle\langle x\rvert))$, p. 372
$\chi_\kappa(\rho)$	Holevo information (9.6), p. 494
$H_\kappa(\rho)$	Minimum average output entropy (9.7), p. 494

Class of Local Operations ($C =$)

$\emptyset$	Only local operations, p. 360
	Local operations and zero-rate classical communications from A to B, p. 403
$\rightarrow$	Local operations and classical communications from A to B, p. 360
$\leftarrow$	Local operations and classical communications from B to A, p. 360
$\leftrightarrow$	Local operations and two-way classical communications between A and B, p. 360
S	Separable operations, p. 361
PPT	Positive partial transpose (PPT) operations, p. 418

Entanglement Measures

$E_{sq}(\rho)$	Squashed entanglement (8.127), p. 394
$E_c(\rho)$	Entanglement of cost with zero-rate communication (8.161), p. 403
$E_f(\rho)$	Entanglement of formation (8.97), p. 387
$E_{r,S}(\rho)$	Entanglement of relative entropy with separable states $\min_{\sigma \in S} D(\rho\|\sigma)$ (8.77), p. 383
$E_{r,S}^{\infty}(\rho)$	Asymptotic entanglement of relative entropy with separable states (8.82), p. 383
$E_{1+s\|S}(\rho)$	Entanglement of relative Rényi entropy with separable states $\min_{\sigma \in S} D_{1+s}(\rho\|\sigma)$ (8.133), p. 396
$\tilde{E}_{1+s\|S}(\rho)$	Entanglement of relative Rényi entropy with separable states $\min_{\sigma \in S} \underline{D}_{1+s}(\rho\|\sigma)$ (8.134), p. 396
$E_{r,\mathrm{PPT}}(\rho)$	Entanglement of relative entropy with PPT states $\min_{\sigma:PPT} D(\rho\|\sigma)$, p. 418
$E_{\mathrm{SDP}}(\rho)$	SDP bound $\min_{\sigma} D(\rho\|\sigma) + \log \|\tau^A(\sigma)\|_1$, p. 418
$E_{1+s\|\,\mathrm{SDP}}(\rho)$	SDP bound with relative Rényi entropy (8.244) $\min_{\sigma} D_{1+s}(\rho\|\sigma) + \log \|\tau^A(\sigma)\|_1$, p. 425
$\tilde{E}_{1+s\|\,\mathrm{SDP}}(\rho)$	SDP bound with relative Rényi entropy (8.245) $\min_{\sigma} \underline{D}_{1+s}(\rho\|\sigma) + \log \|\tau^A(\sigma)\|_1$, p. 425
$E_p(\rho)$	Entanglement of purification (8.164), p. 404
$E_{sr}(\rho)$	Logarithm of Schmidt rank (8.113), p. 391
$C_o(\rho)$	Concurrence (8.317), p. 444

Operational Entanglement Measure with Class C

$E_{d,1}^{C}(\rho)$	Entanglement of distillation (8.72), p. 382
$E_{d,1}^{C,\dagger}(\rho)$	Strong converse entanglement of distillation (8.73), p. 382
$E_{d,2}^{C}(\rho)$	ntanglement of distillation (8.75), p. 382
$E_{d,2}^{C,\dagger}(\rho)$	Strong converse entanglement of distillation (8.76), p. 382
$E_{d,e}^{C,\infty}(\rho)$	Asymptotic entanglement of exact distillation (8.89), p. 386
$E_{d,e}^{C}(\rho)$	E ntanglement of exact distillation (8.89), p. 386
$E_{d,i}^{C}(r\|\rho)$	Exponential decreasing rate for entanglement of distillation (8.90), p. 386
$E_{c}^{C}(\rho)$	Entanglement of cost (8.107), p. 390
$E_{c,e}^{C,\infty}(\rho)$	Asymptotic entanglement of exact cost (8.112), p. 391
$E_{c,e}^{C}(\rho)$	Entanglement of exact cost (8.112), p. 391
$E_{d,i}^{C}(r\|\rho)$	Exponential decreasing rate for entanglement of cost (8.91), p. 386
$E_{m}^{C}(\rho)$	Maximum of negative conditional entropy (8.119), p. 393

$E^C_{1+s\|m}(\rho)$	Maximum of negative conditional Rényi entropy (8.131), p. 396
$\tilde{E}^C_{1+s\|m}(\rho)$	Maximum of negative conditional Rényi entropy (8.132), p. 396
$E^C_{d,L}(\rho)$	Conclusive teleportation fidelity (8.88), p. 385

Security Measures

$d_1(A:E\|\rho)$	Measure for independence $\\|\rho-\rho_A\otimes\rho_E\\|_1$ (8.283), p. 433
$F(A:E\|\rho)$	Measure for independence $F(\rho,\rho_A\otimes\rho_E)$ (8.284), p. 433
$I'_\rho(A:E)$	Measure for independence and uniformity $D(\rho\\|\rho_{\mathrm{mix},A}\otimes\rho_E)$ (8.285), p. 434
$d_{1'}(A:E\|\rho)$	Measure for independence and uniformity $\\|\rho-\rho_{\mathrm{mix},A}\otimes\rho_E\\|_1$ (8.287), p. 434
$F'(A:E\|\rho)$	Measure for independence and uniformity $F(\rho,\rho_{\mathrm{mix},A}\otimes\rho_E)$ (8.288), p. 434

Other Types of Correlation

$C_d^{A\to B}(\rho)$	Measure of classical correlation (8.170), p. 407
$D(B\|A)_\rho$	Discord $I_\rho(A:B)-C_d^{A\to B}(\rho)$ (8.177), p. 408
$C_c(\rho)$	See (8.198), p. 413
$C(\rho,\delta)$	See (8.200), p. 413
$\tilde{C}(\rho,\delta)$	See (8.201), p. 413
$C(\rho)$	$C(\rho,0)=\tilde{C}(\rho,0)$ (8.203), p. 413
$C_k^{A\to B-E}(\rho)$	Optimal generation rate of secret key with one-way communication (9.82), p. 521
$C_d^{A\to B-E}(\rho)$	See (9.83), p. 521

Notations for Bipartite System

$\mathcal{H}_s$	Symmetric space, p. 408
$\mathcal{H}_a$	Antisymmetric space, p. 408
F	Flip operator P_s-P_a, p. 408

Entangled States

$\|\Phi_L\rangle\langle\Phi_L\|$	Maximally entangled state of size L, p. 360
σ_α	Maximally correlated state (8.142), p. 398
$\rho_{W,p}$	Werner state (8.323), p. 445
$\rho_{I,p}$	Isotropic state (8.328), p. 447

Channel Capacities

$C_c(\kappa)$	Classical capacity without entangled input states (9.1), p. 493
$C_c^e(\kappa)$	Classical capacity with entangled input states (9.2), p. 493
$C_a(\rho^{A,B})$	Amount of assistance for sending information by state $\rho^{A,B}$ (9.37), p. 502
$C_{c,e}^e(\kappa)$	Entanglement-assisted classical capacity (9.42), p. 505
$C_r(W,\sigma)$	Quantum-channel resolvability capacity (9.57), p. 511
$C_c^{B,E}(W)$	Wiretap channel capacity (9.73), p. 517
$C_{q,1}$	Quantum capacity in worst case (9.101), p. 527
$C_{q,2}$	Quantum capacity with entanglement fidelity (9.101), p. 527
$C_{q,C}^{\dagger}(\kappa)$	Strong converse quantum capacity (9.122), p. 535
$C_{\mathrm{SDP}}(\kappa)$	SDP bound (9.127), p. 536
$C_{c,r}(W)$	Channel capacity for sending classical information with shared randomness (10.83)
$C_{c,r}^R(W)$	Reverse channel capacity for sending classical information with shared randomness (10.82), p. 594
$C_{c,e}(W)$	Channel capacity for sending classical information with shared entanglement, p. 596
$C_{c,e}^R(W)$	Reverse channel capacity for sending classical information with shared entanglement, p. 596
$C_{c,e}^e(\kappa)$	Channel capacity for sending classical information with shared entanglement and entangled input, p. 505
$C_{c,e}^{e,R}(\kappa)$	Reverse channel capacity for sending classical information with shared entanglement and entangled input, p. 596
$C_{c,r}^e(\kappa)$	Channel capacity for sending classical information with shared randomness and entangled input, p. 505
$C_{c,r}^{e,R}(\kappa)$	Reverse channel capacity for sending classical information with shared randomness and entangled input, p. 597
$C_{q,e}(\kappa)$	Channel capacity for sending quantum states with shared entanglement and entangled input, p. 597
$C_{q,e}^R(\kappa)$	Reverse channel capacity for sending quantum states with shared entanglement and entangled input, p. 597
$C_{q,r}(\kappa)$	Channel capacity for sending quantum states with shared randomness and entangled input, p. 597
$C_{q,r}^R(\kappa)$	Reverse channel capacity for sending quantum states with shared randomness and entangled input, p. 597

Minimum Compression Rates

$R_{B,q}(p,W)$	Minimum compression rate in blind and ensemble setting (10.4), p. 572
$R_{V,q}(p,W)$	Minimum compression rate in visible and ensemble setting (10.5), p. 572

$R_{P,q}(\rho)$	Minimum compression rate in purification setting (10.15), p. 573
$R^{\dagger}_{B,q}(p, W)$	Strong converse compression rate in blind and ensemble setting (10.6), p. 572
$R^{\dagger}_{V,q}(p, W)$	Strong converse compression rate in visible and ensemble setting (10.7), p. 572
$R^{\dagger}_{P,q}(\rho)$	Strong converse compression rate in purification setting (10.16), p. 573
$R_{V,c}(p, W)$	Minimum visible compression rate with classical memory (10.60), p. 587
$R_{V,q,r}(p, W)$	Minimum visible compression rate with quantum memory and shared randomness (10.72), p. 591
$R_{V,c,r}(p, W)$	Minimum visible compression rate with classical memory and shared randomness (10.73), p. 591

Codes for Quantum Source Coding

$\mathit{\Psi}$	Blind code, p. 571
Ψ	Visible code, p. 571
Ψ_c	Visible code by classical memory, p. 586
Ψ_r	Visible code with common randomness, p. 590
$\Psi_{c,r}$	Visible code with common randomness by classical memory, p. 591

About the Author

Masahito Hayashi was born in Japan in 1971. He received the B.S. degree from the Faculty of Sciences in Kyoto University, Japan, in 1994 and the M.S. and Ph.D. degrees in Mathematics from Kyoto University, Japan, in 1996 and 1999, respectively.

He worked in Kyoto University as a Research Fellow of the Japan Society of the Promotion of Science (JSPS) from 1998 to 2000, and worked in the Laboratory for Mathematical Neuroscience, Brain Science Institute, RIKEN from 2000 to 2003, and worked in ERATO Quantum Computation and Information Project, Japan Science and Technology Agency (JST) as the Research Head from 2000 to 2006. He also worked in the Superrobust Computation Project Information Science and Technology Strategic Core (21st Century COE by MEXT) Graduate School of Information Science and Technology, the University of Tokyo as Adjunct Associate Professor from 2004 to 2007. He worked in the Graduate School of Information Sciences, Tohoku University as Associate Professor from 2007 to 2012. In 2012, he joined the Graduate School of Mathematics, Nagoya University as Professor. He also worked at the Centre for Quantum Technologies, National University of Singapore as Visiting Research Associate Professor from 2009 to 2012 and as Visiting Research Professor from 2012 to now. In 2011, he received Information Theory Society Paper Award (2011) for Information-Spectrum Approach to Second-Order Coding Rate in Channel Coding. In 2016, he received the Japan Academy Medal from the Japan Academy and the JSPS Prize from Japan Society for the Promotion of Science.

He is a member of the Editorial Board of International Journal of Quantum Information and International Journal On Advances in Security. His research interests include classical and quantum information theory, information-theoretic security, and classical and quantum statistical inference.

Prologue

Invitation to Quantum Information Theory

Understanding the implications of recognizing matter and extracting information from it has been a long-standing issue in philosophy and religion. However, recently this problem has become relevant to other disciplines such as cognitive science, psychology, and neuroscience. Indeed, this problem is directly relevant to quantum mechanics, which forms the foundation of modern physics. In the process of recognition, information cannot be obtained directly from matter without any media. To obtain information, we use our five senses; that is, a physical medium is always necessary to convey information to us. For example, in vision, light works as the medium for receiving information. Therefore, observations can be regarded as information processing via a physical medium. Hence, this problem can be treated by physics. Of course, to analyze this problem, the viewpoint of information science is also indispensable because the problem involves, in part, information processing.

In the early twentieth century, physicists encountered some unbelievable facts regarding observations (measurements) in the microscopic world. They discovered the contradictory properties of light, i.e., the fact that light has both wave- and particle-like properties. Indeed, light behaves like a collection of minimum energy particles called photons. In measurements using light, we observe the light after interactions with the target. For example, when we measure the position of the matter, we detect photons after interactions with them. Since photons possess momentum and energy, the speed of the object is inevitably disturbed.[1] In particular, this disturbance cannot be ignored when the mass of the measured object is small in comparison with the energy of the photon. Thus, even though we measure the velocity of an object after the measurement of its position, we cannot know the velocity of an object precisely because the original velocity has been disturbed by

[1]The disturbance of measurement is treated in more detail in the formulation of quantum mechanics in Chap. 7.

the first measurement. For the same reason, when we measure the velocity first, its position would be disturbed. Therefore, our naive concept of a "perfect measurement" cannot be applied, even in principle. In the macroscopic world, the mass of the objects is much larger than the momentum of the photons. We may therefore effectively ignore the disturbance by the collisions of the photons. Although we consider that a "perfect measurement" is possible in this macroscopic world, the same intuition cannot be applied to the microscopic world.

In addition to the impossibility of "perfect measurements" in the microscopic world, no microscopic particles have both a determined position and a determined velocity. This fact is deeply connected to the wave-particle duality in the microscopic world and can be regarded as the other side of the nonexistence of "perfect measurements."[2] Thus it is impossible to completely understand this microscopic world based on our macroscopic intuitions, but it is possible to predict probabilistically its measured value based on the mathematical formulation of quantum theory.

So far, the main emphasis of quantum mechanics has been on examining the properties of matter itself, rather than the process of *extracting information*. To discuss how the microscopic world is observed, we need a quantitative consideration from the viewpoint of "information." Thus, to formulate this problem clearly, we need various theories and techniques concerning information. Therefore, the traditional approach to quantum mechanics is insufficient. On the other hand, theories relating to information pay attention only to the data-processing rather than the extraction process of information. Therefore, in this quantum-mechanical context, we must take into account the process of obtaining information from microscopic (quantum-mechanical) particles. We must open ourselves to the new research field of *quantum information science*. This field is to be broadly divided into two parts: (1) *quantum computer science*, in which algorithms and complexity are analyzed using an approach based on computer science, and (2) *quantum information theory*, in which various protocols are examined from the viewpoint of information theory and their properties and limits are studied. Specifically, since quantum information theory focuses on the amount of accessible information, it can be regarded as the theory for quantitative evaluation of the process of *extracting information*, as mentioned above.

Since there have been only a few textbooks describing the recent developments in this field [1, 2], the present textbook attempts to provide comprehensive information ranging from the fundamentals to current research. Quantum computer science is not treated in this book because it has been addressed in many other textbooks. Since quantum information theory forms a part of the basis of quantum computer science, this textbook may be useful for not only researchers in quantum information theory but also those in quantum computer science.

---

[2]The relation between this fact and nonexistence can be mathematically formulated by (7.27) and (7.30).

History of Quantum Information Theory in Twentieth Century

Although quantum information theory has been very actively studied in the twenty first century, the root can be traced to the studies in the twentieth century. Let us briefly discuss the history of quantum information theory in the twentieth century. Quantum mechanics was first formulated by Schrödinger (wave mechanics) and Heisenberg (matrix mechanics). However, their formulations described the dynamics of microscopic systems, but they had several unsatisfactory aspects in descriptions of measurements. Later, the equivalence between both formulations were proved. To resolve this point, von Neumann [3] established the formulation of quantum theory that describes measurements as well as dynamics based on operator algebra, whose essential features will be discussed in Chap. 1. However, in studies of measurements following the above researches, the philosophical aspect has been emphasized too much, and a quantitative approach to extracting information via measurements has not been examined in detail. This is probably because approaches to mathematical engineering have not been adopted in the study of measurements.

In the latter half of the 1960s, a Russian researcher named Stratonovich, who is one of the founders of stochastic differential equations, and two American researchers, Helstrom and Gordon, proposed a formulation of optical communications using quantum mechanics. This was the first historical appearance of quantum information theory. Gordon [4, 5], Helstrom [6], and Stratonovich [7] mainly studied error probabilities and channel capacities for communications. Meanwhile, Helstrom [8] examined the detection process of optical communication as parameter estimation. Later, many American and Russian researchers such as Holevo [9, 10], Levitin [11], Belavkin [12], Yuen [13], and Kennedy [14] also examined these problems.[3] In particular, Holevo obtained the upper bound of the communication speed in the transmission of a classical message via a quantum channel in his two papers [9, 10] published in the 1970s. Further, Holevo [16, 18], Yuen [13], Belavkin, and their coworkers also analyzed many theoretically important problems in quantum estimation.

Unfortunately, the number of researchers in this field rapidly decreased in the early 1980s, and this line of research came to a standstill. Around this time, Bennett and Brassard [19] proposed a quantum cryptographic protocol (BB84) using a different approach to quantum mechanical systems. Around the same time, Ozawa [20] gave a precise mathematical formulation of the state reduction in the measurement process in quantum systems.

[3]Other researchers during this period include Grishanin, Mityugov, Kuriksha, Liu, Personick, Lax, Lebedev, Forney [15] in the United States and Russia. Many papers were published by these authors; however, an accurate review of all of them is made difficult by their lack of availability. In particular, while several Russian papers have been translated into English, some of them have been overlooked despite their high quality. For details, see [16, 17].

In the latter half of the 1980s, Nagaoka investigated quantum estimation theory as a subfield of mathematical statistics. He developed the asymptotic theory of quantum-state estimation and quantum information geometry [21]. This research was continued by many Japanese researchers, including Fujiwara, Matsumoto, and the present author in the 1990s [22–39]. For this history, see Hayashi [40].

In the 1990s, in the United States and Europe several researchers started investigating quantum information processing, e.g., quantum data compression, quantum teleportation, superdense coding, another quantum cryptographic protocol (B92), etc. [41–46]. In the second half of the 1990s, the study of quantum information picked up speed. In the first half of the 2000s, several information-theoretic approaches were developed, and research has been advancing at a rapid pace.

We see that progress in quantum information theory has been achieved by connecting various topics. This text clarifies these connections and discusses current research topics starting with the basics.

Structure of the Book

Quantum information theory has been studied by researchers from various backgrounds. Their approach can be broadly divided into two categories. The first approach is based on information theory. In this approach, existing methods for information processing are translated (and extended) into quantum systems. The second approach is based on quantum mechanics.

In this text, four chapters are dedicated to examining problems based on the first approach, i.e., establishing information-theoretic problems. These are Chap. 3, "Quantum Hypothesis Testing and Discrimination of Quantum States," Chap. 4, "Classical Quantum Channel Coding (Message Transmission)," Chap. 6, "Quantum Information Geometry and Quantum Estimation," and Chap. 10, "Source Coding in Quantum Systems." Problems based on the second approach is treated in three chapters: Chap. 5, "State evolution and Trace-Preserving Completely Positive Maps," Chap. 7, "Quantum measurements and State Reduction," and Chap. 8, "Entanglement and Locality Restrictions."

Advanced topics in quantum communication such as quantum teleportation, superdense coding, quantum-state transmission (quantum error correction), and quantum cryptography are often discussed in quantum information theory. Both approaches are necessary for understanding these topics, which are covered in Chap. 9, "Analysis of Quantum Communication Protocols."

Some quantum-mechanical information quantities are needed to handle these problems mathematically, and these problems are covered in Sects. 3.1, 5.4, 5.5, 5.6, 8.2, and 8.3. This allows us to touch upon several important information-theoretic problems using a minimum amount of mathematics. The book also includes 450 exercises together with solutions. Solving these problems should provide readers not only with knowledge of quantum information theory but also the necessary techniques for pursuing original research in the field.

Chapter 1 covers the mathematical formulation of quantum mechanics in the context of quantum information theory. It also gives a review of linear algebra. Chapter 2 summarizes classical information theory. This not only provides an introduction to the later chapters but also serves as a brief survey of classical information theory. This chapter covers entropy, Fisher information, information geometry, estimation of probability distribution, large deviation principle. Also, it discusses the axiomatic characterization of entropy. This concludes the preparatory part of the text. Section 2.6 treats the large deviation on the sphere, which is used only in Sect. 8.13. So, a reader can skip it before stating Sect. 8.13.

Chapter 3 covers quantum hypothesis testing and the discrimination of quantum states. This chapter starts with introduction of information quantities in quantum systems. Then, this chapter serves to answer the question: If there are two states, which is the true state? The importance of this question may not at first be apparent. However, this problem provides the foundation for other problems in information theory and is therefore crucially important. Also, this problem provides the basic methods for quantum algorithm theory. Many of the results of this chapter will be used in subsequent chapters. In particular, the quantum version of Stein's lemma is discussed here; it can be used a basic tool for other topics. Furthermore, many of the difficulties associated with the noncommutativity of quantum theory can be seen here in their simplest forms. This chapter can be mainly read after Chap. 1 and Sects. 2.1 and A.3.

Chapter 4 covers classical quantum channel coding (message transmission). That is, we treat the tradeoff between the transmission speed and the error probability in the transmission of classical messages via quantum states. In particular, we discuss the channel capacity, i.e., the theoretical bound of the transmission rate when the error probability is 0, as well as its associated formulas. This chapter can be read after Chap. 1 and Sects. 2.1, 3.1, 3.5, 3.7, and 3.8.

Chapter 5 discusses the trace-preserving completely positive map, which is the mathematical description of state evolution in quantum systems. Its structure will be illustrated with examples in quantum two-level systems. We also briefly discuss the relationship between the state evolution and information quantities in quantum systems (the entropy and relative entropy). In particular, the part covering the formulation of quantum mechanics (Sects. 5.1–5.3) can be read after only Chap. 1.

Chapter 6 describes the relation among quantum information geometry, quantum information quantities, and quantum estimation. First, the inner product for the space of quantum states is briefly discussed. Next, we discuss the geometric structure naturally induced from the inner product. The theory of state estimation in quantum systems is then discussed by emphasizing the Cramér–Rao inequality. Most of this chapter except for Sect. 6.7 can be read after Chaps. 1 and 2 and Sect. 5.1. Section 6.7 can be read after Chap. 1 and Sects. 5.1, 5.4, and 6.1.

Chapter 7 covers quantum measurement and state reduction. First, it is shown that the state reduction due to a quantum measurement follows naturally from the axioms of the quantum systems discussed in Chap. 1. Next, we discuss the relation between quantum measurement and two types of uncertainty relations, square error type uncertainty and entropic uncertainty. Finally, it is shown that under certain conditions it is possible, in principle, to perform a measurement such that the required information can be obtained while the state demolition is negligible. Readers who only wish to read Sects. 7.1 and 7.4 can read them after Chap. 1 and Sect. 5.1. Section 7.2 requires the additional background of Sect. 6.1. Section 7.3 can be read after Chap. 1 and Sects. 5.1, 5.4, 5.5, and 5.6.

Chapter 8 discusses the relation between locality and entanglement, which are fundamental topics in quantum mechanics. First, we examine state operations when the locality condition is imposed on quantum operations. Next, the information quantities related to entanglement are considered. The theory for distilling a perfect entangled state from a partially entangled state is discussed. Information-theoretic methods play a central role in entanglement distillation. Quantification of entanglement is discussed from various viewpoints. As opposite task, we discuss the entanglement of dilution, which evaluates the cost to generate a given partially entangled state. While this task is characterized by using the entanglement formation, we discuss the nonadditivity of this quantity. As another types of correlation, we discuss discord. Further, we consider the duality of conditional entropy, secure random number generation, and state generation from shared randomness.

Chapter 9 delves deeply into topics in quantum channels such as quantum teleportation, superdense coding, quantum-state transmission (quantum error correction), and quantum key distribution based on the theory presented in previous chapters. These topics are very simple when noise is not present. However, if noise is present in a channel, these problems require the information-theoretic methods discussed in previous chapters. The relationship among these topics is also discussed. Further, the relation between channel capacities and entanglement theory is also treated. The additivity problem for the classical-quantum channel capacity is discussed in Sects. 8.13 and 9.2.

Finally, Chap. 10 discusses source coding in quantum systems. We treat not only the theoretical bounds of quantum fixed-length source coding but also universal quantum fixed-/variable-length source coding, which does not depend on the form of the information source. The beginning part of this chapter, excepting the purification scheme, requires only the contents of Chaps. 1 and 2 (Sects. 2.1–2.4) and Sect. 5.1. In particular, in universal quantum variable-length source coding, a measurement is essential for determining the coding length. Hence this measurement causes the demolition of the state to be sent, which makes this a more serious problem. However, it can be solved by a measurement with negligible state demolition, which is described in Chap. 7. Then we treat quantum-state compression with mixed states and its several variants. The relations between these problems and entanglement theory are also treated. Further, we treat the relationships between the reverse capacities (reverse Shannon theorem) and these problems. Excluding Sects. 10.6–10.9, this chapter can be read after Chap. 1 and Sects. 2.1, 2.3, 3.1, 4.1, and 5.1.

This text thus covers a wide variety of topics in quantum information theory. Quantum hypothesis testing, quantum-state discrimination, and quantum-channel coding (message transmission) have been discussed such that only a minimal amount of mathematics is needed to convey the essence of these topics. Prior to this text, these topics required the study of advanced mathematical theories for quantum mechanics, such as those presented in Chap. 5. Further, Chaps. 5 ("State Evolution and Trace Preserving Completely Positive Maps in Quantum Systems") and 7 ("Quantum Measurement and State Reduction") have been written such that they can be understood with only the background provided in Chap. 1. Therefore, this text should also be suitable for readers who are interested in either the information-theoretic aspects of quantum mechanics or the foundations of quantum mechanics

References

1. M.M. Wilde, *Quantum Information Theory* (Cambridge University Press, 2013)
2. M. Hayashi, S. Ishizaka, A. Kawachi, G. Kimura, T. Ogawa, *Introduction to Quantum Information Science* (Graduate Texts in Physics, 2014)
3. J. von Neumann, *Mathematical Foundations of Quantum Mechanics* (Princeton University Press, Princeton, NJ, 1955). (Originally appeared in German in 1932)
4. J.P. Gordon, Proc. IRE. **50**, 1898–1908 (1962)
5. J.P. Gordon, Noise at optical frequencies; information theory, in *Quantum Electronics and Coherent Light, Proceedings of the International School Physics "Enrico Fermi," Course XXXI*, ed. by P.A. Miles. (Academic, New York, 1964), pp. 156–181
6. C.W. Helstrom, Detection theory and quantum mechanics. Inf. Contr. **10**, 254–291 (1967)
7. R.L. Stratonovich, Izvest. VUZ Radiofiz., **8**, 116–141 (1965)
8. R.L. Stratonovich, The transmission rate for certain quantum communication channels. Problemy Peredachi Informatsii, **2**, 45–57 (1966). (in Russian). English translation: Probl. Inf. Transm. **2**, 35–44 (1966.)
9. A.S. Holevo, Bounds for the quantity of information transmitted by a quantum communication channel. Problemy Peredachi Informatsii. **9**, 3–11 (1973) (in Russian). (English translation: Probl. Inf. Transm. **9**, 177–183 (1975)
10. A.S. Holevo, On the capacity of quantum communication channel. Problemly Peredachi Informatsii, **15**, 4, 3–11 (1979) (in Russian). (English translation: Probl. Inf. Transm. **15**, 247–253 (1979)
11. L.B. Levitin, On quantum measure of information, in *Proceedings of the 4th All-Union Conference on Information Transmission and Coding Theory*, pp. 111–115 (Tashkent, 1969) (in Russian). English translation: *Information, Complexity and Control in Quantum Physics*, ed. by, A. Blaquiere, S. Diner, G. Lochak (Springer, Berlin, Heidelberg, New York, 1987), pp. 15–47
12. V.P. Belavkin, Generalized uncertainty relations and efficient measurements in quantum systems. Teor. Mat. Fiz. **26**, 3, 316–329 (1976). (quant-ph/0412030, 2004)
13. H.P. Yuen, M. Lax, Multiple-parameter quantum estimation and measurement of non-selfadjoint observables. IEEE Trans. Inf. Theory **19**, 740 (1973)
14. H.P. Yuen, R.S. Kennedy, M. Lax, Optimum testing of multiple hypotheses in quantum detection theory. IEEE Trans. Inf. Theory 125–134 (1975)
15. G.D. Forney, Jr., S.M. Thesis, (MIT, 1963, unpublished)
16. A.S. Holevo, *Probabilistic and Statistical Aspects of Quantum Theory*, (North-Holland, Amsterdam, 1982); (originally published in Russian, 1980)
17. C.W. Helstrom, Minimum mean-square error estimation in quantum statistics. Phys. Lett. **25A**, 101–102 (1976)

18. A.S. Holevo, Covariant measurements and uncertainty relations. Rep. Math. Phys. **16**, 385–400 (1979)
19. C.H. Bennett, G. Brassard, Quantum cryptography: public key distribution and coin tossing, in *Proceedings of the IEEE International Conference on Computers, Systems and Signal Processing*, (Bangalore, India, 1984), pp. 175–179
20. M. Ozawa, Quantum measuring processes of continuous observables. J. Math. Phys. **25**, 79 (1984)
21. H. Nagaoka, Differential geometrical aspects of quantum state estimation and relative entropy, in *Quantum Communications and Measurement*, ed. by V.P. Belavkin, O. Hirota, R.L. Hudson (Plenum, New York, 1995), pp. 449–452
22. A. Fujiwara, *Statistical Estimation Theory for Quantum States*, master's thesis Department of Mathematical Engineering and Information Physics, Graduate School of Engineering, University of Tokyo, Japan (1993) (in Japanese)
23. A. Fujiwara, *A Geometrical Study in Quantum Information Systems*, Ph.D. thesis, Department of Mathematical Engineering and Information Physics, Graduate School of Engineering, University of Tokyo, Japan (1995)
24. A. Fujiwara, H. Nagaoka, Quantum Fisher metric and estimation for pure state models. Phys. Lett. **201A**, 119–124 (1995)
25. A. Fujiwara, H. Nagaoka, Coherency in view of quantum estimation theory, in *Quantum Coherence and Decoherence*, ed. by K. Fujikawa, Y.A. Ono. (Elsevier, Amsterdam, 1996), pp. 303–306
26. A. Fujiwara, H. Nagaoka, An estimation theoretical characterization of coherent states. J. Math. Phys. **40**, 4227–4239 (1999)
27. M. Hayashi, *Minimization of Deviation Under Quantum Local Unbiased Measurements,* master's thesis, Department of Mathematics, Graduate School of Science, Kyoto University, Japan (1996)
28. M. Hayashi, A linear programming approach to attainable cramer-rao type bound and randomness conditions. Kyoto-Math 97–08; quant-ph/9704044 (1997)
29. M. Hayashi, A linear programming approach to attainable Cramer–Rao type bound, in *Quantum Communication, Computing, and Measurement*, ed. by, O. Hirota, A.S. Holevo, C.M. Caves. (Plenum, New York, 1997), pp. 99–108. (Also appeared as Chap. 12 of *Asymptotic Theory of Quantum Statistical Inference,*, ed. by M. Hayashi)
30. M. Hayashi, Asymptotic estimation theory for a finite dimensional pure state model. J. Phys. A Math. Gen. **31**, 4633–4655 (1998). (Also appeared as Chap. 23 of *Asymptotic Theory of Quantum Statistical Inference*, ed. by, M. Hayashi)
31. M. Hayashi, Asymptotic quantum estimation theory for the thermal states family, in *Quantum Communication, Computing, and Measurement* 2, ed. by P. Kumar, G. M. D'ariano, O. Hirota. (Plenum, New York, 2000) pp. 99–104; quant-ph/9809002 (1998). (Also appeared as Chap. 14 of *Asymptotic Theory of Quantum Statistical Inference,*, ed. by M. Hayashi)
32. M. Hayashi, Asymptotic large deviation evaluation in quantum estimation theory, in *Proceedings of the Symposium on Statistical Inference and Its Information-Theoretical Aspect*, pp. 53–82 (1998) (in Japanese)
33. M. Hayashi, Quantum estimation and quantum central limit theorem. *Sugaku*. **55**, 4, 368–391 (2003) (in Japanese); English translation is in *Selected Papers on Probability and Statistics* (American Mathematical Society Translations Series 2) vol. 277, pp. 95–123 (2009)
34. M. Hayashi, K. Matsumoto, Statistical model with measurement degree of freedom and quantum physics. RIMS koukyuroku Kyoto Univiversity, **1055**, 96–110 (1998) (in Japanese). (Also appeared as Chap. 13 of *Asymptotic Theory of Quantum Statistical Inference,* ed. by M. Hayashi)
35. K. Matsumoto, *Geometry of a Quantum State*, master's thesis, Department of Mathematical Engineering and Information Physics, Graduate School of Engineering, University of Tokyo, Japan (1995) (in Japanese)

36. K. Matsumoto, A new approach to the Cramér–Rao type bound of the pure state model. J. Phys. A Math. Gen. **35**, 3111–3123 (2002)
37. K. Matsumoto, *A Geometrical Approach to Quantum Estimation Theory*, Ph.D. thesis, Graduate School of Mathematical Sciences, University of Tokyo (1997)
38. K. Matsumoto, The asymptotic efficiency of the consistent estimator, Berry-Uhlmann' curvature and quantum information geometry, in *Quantum Communication, Computing, and Measurement 2*, ed. by P. Kumar, G. M. D'ariano, O. Hirota. (Plenum, New York, 2000), pp. 105–110
39. K. Matsumoto, Seminar notes (1999)
40. M. Hayashi (eds.), *Asymptotic Theory of Quantum Statistical Inference: Selected Papers*, (World Scientific, Singapore, 2005)
41. B. Schumacher, Quantum coding. Phys. Rev. A **51**, 2738–2747 (1995)
42. R. Jozsa, B. Schumacher, A new proof of the quantum noiseless coding theorem. J. Mod. Opt. **41(12)**, 2343–2349 (1994)
43. C.H. Bennett, G. Brassard, C. Crepeau, R. Jozsa, A. Peres, W. K. Wootters, Teleporting an unknown quantum state via dual classical and Einstein-Podolsky-Rosen channels. Phys. Rev. Lett. **70**, 1895 (1993)
44. C.H. Bennett, H.J. Bernstein, S. Popescu, B. Schumacher, Concentrating partial entanglement by local operations. Phys. Rev. A, **53**, 2046 (1996)
45. C.H. Bennett, S.J. Wiesner, Communication via one- and two-particle operators on Einstein-Podolsky-Rosen states. Phys. Rev. Lett. **69**, 2881 (1992)
46. C.H. Bennett, Quantum cryptography using any two nonorthogonal states. Phys. Rev. Lett. **68**, 3121–3124 (1992)

36. K. Matsumoto, A new approach to the Cramér-Rao type bound of the pure state model. J. Phys. A: Math. Gen. 35, 3111–3123 (2002)
37. K. Matsumoto, A geometrical approach to quantum estimation theory. Doctoral Thesis, Graduate School of Mathematical Sciences, University of Tokyo (1997)
38. K. Matsumoto, The asymptotic efficiency of the consistent estimator, Berry-Uhlmann curvature and quantum information geometry, in Quantum Communication, Computing, and Measurement 2, ed. by P. Kumar, G.M. D'Ariano, O. Hirota (Plenum, New York, 2000), pp. 105–110
39. K. Matsumoto, Seminar notes (1999)
40. M. Hayashi (ed.), Asymptotic Theory of Quantum Statistical Inference: Selected Papers (World Scientific, Singapore, 2005)
41. B. Schumacher, Quantum coding. Phys. Rev. A 51, 2738–2747 (1995)
42. R. Jozsa, B. Schumacher, A new proof of the quantum noiseless coding theorem. J. Mod. Opt. 41(12), 2343–2349 (1994)
43. C.H. Bennett, G. Brassard, C. Crépeau, R. Jozsa, A. Peres, W.K. Wootters, Teleporting an unknown quantum state via dual classical and Einstein-Podolsky-Rosen channels. Phys. Rev. Lett. 70, 1895 (1993)
44. C.H. Bennett, S.J. Wiesner, Communication via one- and two-particle operators on Einstein-Podolsky-Rosen states. Phys. Rev. Lett. 69, 2881 (1992)
45. B. Schumacher, M.D. Westmoreland, Sending classical information via noisy quantum channels. Phys. Rev. A 56, 131 (1997)
46. A.S. Holevo, The capacity of the quantum channel with general signal states. IEEE Trans. Inf. Theory 44, 269 (1998)

Chapter 1
Mathematical Formulation of Quantum Systems

Abstract In this chapter, we cover the fundamentals of linear algebra and provide a mathematical formulation of quantum mechanics for use in later chapters. It is necessary to understand these topics since they form the foundation of quantum information processing discussed later. In the first section, we cover the fundamentals of linear algebra and introduce some notation. The next section describes the formulation of quantum mechanics. Further, we examine a quantum two-level system, which is the simplest example of a quantum-mechanical system. Finally, we discuss the tensor product and matrix inequalities. More advanced discussions on linear algebra are available in Appendix.

1.1 Quantum Systems and Linear Algebra

In order to treat information processing in quantum systems, it is necessary to mathematically formulate fundamental concepts such as quantum systems, measurements, and states. First, we consider the quantum system. It is described by a Hilbert space $\mathcal{H}$ (a finite- or infinite-dimensional complex vector space with a Hermitian inner product), which is called a **representation space**. Before considering other important concepts such as measurements and states, we give a simple overview of linear algebra. This will be advantageous because it is not only the underlying basis of quantum mechanics but is also as helpful in introducing the special notation used for quantum mechanics. In mathematics, a Hilbert space usually refers to an infinite-dimensional complex vector space with a Hermitian inner product. In physics, however, a Hilbert space also often includes finite-dimensional complex vector spaces with Hermitian inner products. This is because in quantum mechanics, the complex vector space with a Hermitian inner product becomes the crucial structure. Since infinite-dimensional complex vector spaces with Hermitian inner products can be dealt with analogously to the finite-dimensional case, we will consider only the finite-dimensional case in this text. Unless specified, the dimension will be labeled d.

The representation space of a given system is determined by a physical observation. For example, spin-$\frac{1}{2}$ particles such as electrons possess, an internal degree of freedom corresponding to "spin" in addition to their motional degree of freedom.

M. Hayashi, *Quantum Information Theory*, Graduate Texts in Physics,
DOI 10.1007/978-3-662-49725-8_1

The representation space of this degree of freedom is $\mathbb{C}^2$. The representation space of a one-particle system with no internal degrees of freedom is the set of all square integrable functions from $\mathbb{R}^3$ to $\mathbb{C}$. In this case, the representation space of the system is an infinite-dimensional space, which is rather difficult to handle. Such cases will not be examined in this text.

Before discussing the states and measurements, we briefly summarize some basic linear algebra with some emphasis on Hermitian matrices. This will be important particularly for later analysis. The Hermitian product of two vectors

$$u = \begin{pmatrix} u^1 \\ u^2 \\ \vdots \\ u^d \end{pmatrix}, \quad v = \begin{pmatrix} v^1 \\ v^2 \\ \vdots \\ v^d \end{pmatrix} \in \mathcal{H}$$

is given by

$$\langle u|v\rangle \stackrel{\text{def}}{=} \overline{u^1}v^1 + \overline{u^2}v^2 + \ldots + \overline{u^d}v^d \in \mathbb{C},$$

where the complex conjugate of a complex number x is denoted by $\overline{x}$. The norm of the vector is given by $\|u\| \stackrel{\text{def}}{=} \sqrt{\langle u|u\rangle}$. The inner product of the vectors satisfies the **Schwarz inequality**

$$\|u\|\|v\| \geq |\langle u|v\rangle|. \tag{1.1}$$

When a matrix

$$X = \begin{pmatrix} x^{1,1} & x^{1,2} & \ldots & x^{1,d} \\ x^{2,1} & x^{2,2} & \ldots & x^{2,d} \\ \vdots & \vdots & \ddots & \vdots \\ x^{d,1} & x^{d,2} & \ldots & x^{d,d} \end{pmatrix} \tag{1.2}$$

satisfies the following condition

$$X = X^* \stackrel{\text{def}}{=} \begin{pmatrix} \overline{x^{1,1}} & \overline{x^{2,1}} & \ldots & \overline{x^{d,1}} \\ \overline{x^{1,2}} & \overline{x^{2,2}} & \ldots & \overline{x^{d,2}} \\ \vdots & \vdots & \ddots & \vdots \\ \overline{x^{1,d}} & \overline{x^{2,d}} & \ldots & \overline{x^{d,d}} \end{pmatrix}, \tag{1.3}$$

it is called **Hermitian**. We also define the complex conjugate matrix $\overline{X}$ and its transpose matrix X^T as follows:

$$\overline{X} \stackrel{\text{def}}{=} \begin{pmatrix} \overline{x^{1,1}} & \overline{x^{1,2}} & \dots & \overline{x^{1,d}} \\ \overline{x^{2,1}} & \overline{x^{2,2}} & \dots & \overline{x^{2,d}} \\ \vdots & \vdots & \ddots & \vdots \\ \overline{x^{d,1}} & \overline{x^{d,2}} & \dots & \overline{x^{d,d}} \end{pmatrix}, \quad X^T \stackrel{\text{def}}{=} \begin{pmatrix} x^{1,1} & x^{2,1} & \dots & x^{d,1} \\ x^{1,2} & x^{2,2} & \dots & x^{d,2} \\ \vdots & \vdots & \ddots & \vdots \\ x^{1,d} & x^{2,d} & \dots & x^{d,d} \end{pmatrix}. \tag{1.4}$$

Also, we denote the real part $\frac{X+\overline{X}}{2}$ of matrix X and the imaginary part $\frac{X-\overline{X}}{2i}$ of matrix X by $\mathbf{Re}\, X$ and $\mathbf{Im}\, X$, respectively. Then, a Hermitian matrix X satisfies $X^T = \overline{X}$. If a Hermitian matrix X satisfies $\langle u|Xu\rangle \geq 0$ for an arbitrary vector $u \in \mathcal{H}$, it is called **positive semidefinite** and denoted by $X \geq 0$. If $\langle u|Xu\rangle > 0$ for nonzero vectors u, X is called **positive definite**. The condition of positive semidefiniteness is equivalent to all the eigenvalues of a diagonalized Hermitian matrix X that are either zero or positive. As shown later, the trace of the product of two positive semidefinite matrices X and Y satisfies

$$\mathrm{Tr}\, XY \geq 0 . \tag{1.5}$$

However, in general, the product XY is not a Hermitian matrix. Note that although the matrix $XY + YX$ is Hermitian, it is generally not positive semidefinite.

We can regard each element $u \in \mathcal{H}$ as an element of the dual space $\mathcal{H}^*$ according to the correspondence between $\mathcal{H}$ and $\mathcal{H}^*$ given by the inner product. We denote the corresponding element of the dual space $\mathcal{H}^*$ by $\langle u|$, in accordance with the conventional notation in physics. If we wish to emphasize that u is an element not of $\mathcal{H}^*$ but of $\mathcal{H}$, we write $|u\rangle$. That is,

$$|u\rangle = \begin{pmatrix} u^1 \\ u^2 \\ \vdots \\ u^d \end{pmatrix} \in \mathcal{H}, \quad \langle u| = \left(\overline{u^1}\ \overline{u^2} \cdots \overline{u^d}\right) \in \mathcal{H}^*.$$

The Hermitian inner product $\langle u|v\rangle$ can also be considered as the matrix product of $\langle u|$ and $|v\rangle$. Note that this notation is used in this text even if the norm of v is not equal to 1. On the other hand, the opposite matrix product $|v\rangle\langle u|$ is a $d \times d$ matrix:

$$|v\rangle\langle u| = \begin{pmatrix} v^1 \\ v^2 \\ \vdots \\ v^d \end{pmatrix} \left(\overline{u^1}\ \overline{u^2} \cdots \overline{u^d}\right) = \begin{pmatrix} v^1\overline{u^1} & v^1\overline{u^2} & \dots & v^1\overline{u^d} \\ v^2\overline{u^1} & v^2\overline{u^2} & \dots & v^2\overline{u^d} \\ \vdots & \vdots & \ddots & \vdots \\ v^d\overline{u^1} & v^d\overline{u^2} & \dots & v^d\overline{u^d} \end{pmatrix}. \tag{1.6}$$

Although $|Xv\rangle = X|v\rangle$, $\langle Xv| = \langle v|X^*$. Evidently, if matrix X is Hermitian, then $\langle u|Xv\rangle = \langle Xu|v\rangle$. This also equals $\mathrm{Tr}\, |v\rangle\langle u|X$, which is often denoted by $\langle u|X|v\rangle$. Using this notation, matrix X given by (1.2) may be written as $X = \sum_{i,j} x^{i,j}|u_i\rangle\langle u_j|$, where u_i is a unit vector whose ith element is 1 and remaining elements are 0.

A Hermitian matrix X may be transformed into the diagonal form U^*XU by choosing an appropriate unitary matrix U. Since $X = U(U^*XU)U^*$, we may write

$$X = \begin{pmatrix} u_1^1 & \dots & u_d^1 \\ \vdots & \ddots & \vdots \\ u_1^d & \dots & u_d^d \end{pmatrix} \begin{pmatrix} x^1 & & O \\ & \ddots & \\ O & & x^d \end{pmatrix} \begin{pmatrix} \overline{u_1^1} & \dots & \overline{u_1^d} \\ \vdots & \ddots & \vdots \\ \overline{u_d^1} & \dots & \overline{u_d^d} \end{pmatrix}. \tag{1.7}$$

Define d vectors $u_1, u_2, \dots, u_d$

$$u_i = \begin{pmatrix} u_i^1 \\ \vdots \\ u_i^d \end{pmatrix}.$$

Then the unitarity of U implies that $\{u_1, u_2, \dots, u_d\}$ forms an orthonormal basis, which will be simply called a **basis** latter. Using (1.6), the Hermitian matrix X may then be written as $X = \sum_i x^i |u_i\rangle\langle u_i|$. This process is called diagonalization. If X and Y commute, they may be written as $X = \sum_{i=1}^d x^i |u_i\rangle\langle u_i|$, $Y = \sum_{i=1}^d y^i |u_i\rangle\langle u_i|$ using the same orthonormal basis $\{u_1, u_2, \dots, u_d\}$. If X and Y do not commute, they cannot be diagonalized using the same orthonormal basis.

Furthermore, we can characterize positive semidefinite matrices using this notation. A matrix X is positive semidefinite if and only if $x^i \ge 0$ for arbitrary i. Thus, this equivalence yields inequality (1.5) as follows:

$$\operatorname{Tr} XY = \operatorname{Tr} \sum_{i=1} x^i |u_i\rangle\langle u_i| Y = \sum_{i=1} x^i \operatorname{Tr} |u_i\rangle\langle u_i| Y = \sum_{i=1} x^i \langle u_i|Y|u_i\rangle \ge 0 .$$

We also define the commutator $[X, Y]$ and symmetrized product $X \circ Y$ of two matrices X and Y as[1]

$$[X, Y] \stackrel{\text{def}}{=} XY - YX, \quad X \circ Y \stackrel{\text{def}}{=} \frac{1}{2}(XY + YX). \tag{1.8}$$

Exercises

1.1 Show Schwarz's inequality (1.1) noting that $\langle u + rcv | u + rcv \rangle \ge 0$ for an arbitrary real number r, where $c = \langle v|u\rangle / |\langle v|u\rangle|$.

1.2 Suppose that k vectors $u_1, \dots, u_k$ ($u_j = (u^{ij})$) satisfy $\langle u_{j'}|u_j\rangle = \sum_{i=1}^d \overline{u^{ij'}} u^{ij} = \delta_{j'j}(1 \le j, j' \le k)$, where $\delta_{j,j'}$ is defined as 1 when $j = j'$ and as 0 otherwise. Show that there exist $d - k$ vectors $u_{k+1}, \dots, u_d$ such that$\langle u_{j'}|u_j\rangle = \delta_{j'j}$, i.e., the matrix $U = (u^{ij})$ is unitary.

[1] A vector space closed under commutation $[X, Y]$ is called a Lie algebra. A vector space closed under the symmetrized product $X \circ Y$ is called a Jordan algebra.

1.3 Let $X = \sum_{i,j} x^{i,j}|u_i\rangle\langle u_j|$ be a Hermitian positive semidefinite matrix. Show that the transpose matrix X^T is also positive semidefinite.

1.4 Let X_θ, Y_θ be matrix-valued functions. Show that the derivative of the product $(X_\theta Y_\theta)'$ can be written as $(X_\theta Y_\theta)' = X_\theta Y_\theta' + X_\theta' Y_\theta$, where X_θ', Y_θ' are the derivatives of X_θ, Y_θ, respectively. Show that the derivative of $\operatorname{Tr} X_\theta$, i.e., $(\operatorname{Tr} X_\theta)'$, is equal to $\operatorname{Tr}(X_\theta')$.

1.5 Let U be a unitary matrix and X be a matrix. Show that the equation $(UXU^*)^* = UX^*U^*$ holds. Also give a counterexample of the equation $(UXU^*)^T = UX^TU^*$.

1.2 State and Measurement in Quantum Systems

To discuss information processing in quantum systems, we must first be able to determine the probability that every measurement outcome appears as an outcome of the measurement. Few standard texts on quantum mechanics give a concise and accurate description of the probability distribution of each measurement outcome. Let us discuss the fundamental framework of quantum theory so as to calculate the probability distribution of a measurement outcome.

In the spin-$\frac{1}{2}$ system discussed previously, when the direction of "spin" changes, the condition of the particle also changes. In quantum systems, a description of the current condition of a system, such as the direction of the spin, is called a **state**. Any state is described by a Hermitian matrix ρ called a **density matrix** or simply **density**:

$$\operatorname{Tr} \rho = 1, \quad \rho \geq 0. \tag{1.9}$$

Since quantum systems are too microscopic for direct observation, we must perform some measurement in order to extract information from the system. Such a measurement is described by a set of Hermitian matrices $\boldsymbol{M} \stackrel{\text{def}}{=} \{M_\omega\}_{\omega\in\Omega}$ satisfying the following conditions:

$$M_\omega \geq 0, \quad \sum_{\omega\in\Omega} M_\omega = I,$$

where I denotes the identity matrix. Here, Ω forms a set of measurement outcomes ω, and is called a probability space. The set $\boldsymbol{M} = \{M_\omega\}_{\omega\in\Omega}$ is called a **positive operator valued measure** (**POVM**). For readers who have read standard texts on quantum mechanics, note that M_ω is not restricted to projection matrices. When ω is continuous, the summation $\sum$ is replaced by an integration $\int$ on the probability space Ω. Here we denote the set of the measurement outcomes ω by the probability space Ω and omit it if there is no risk of confusion. If a probability space Ω is a discrete set, then the number of elements of the probability space Ω is denoted by $|\boldsymbol{M}|$. When the rank of M_ω is 1 for any $\omega \in \Omega$, the POVM $\boldsymbol{M}$ is called **rank-one**.

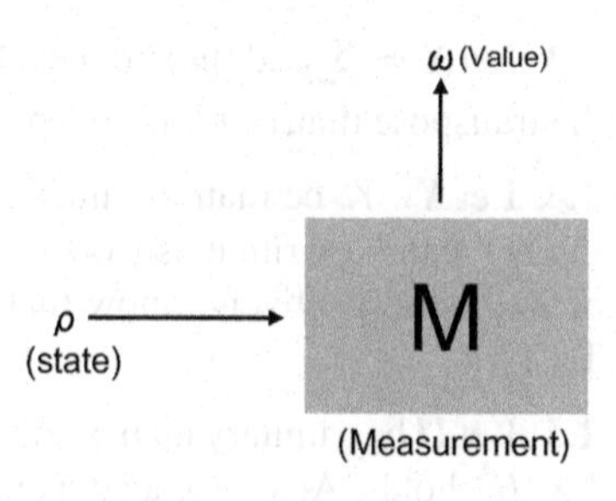

Fig. 1.1 Measurement scheme

The density matrix ρ and the POVM M form the mathematical representations of a state of the system and the measurement, respectively, in the following sense (Fig. 1.1). If a measurement corresponding to $\boldsymbol{M} = \{M_\omega\}_{\omega \in \Omega}$ is performed on a system in a state corresponding to ρ, then the probability $\mathrm{P}^{\boldsymbol{M}}_\rho(\omega)$ of obtaining ω is[2]

$$\mathrm{P}^{\boldsymbol{M}}_\rho(\omega) := \operatorname{Tr} \rho M_\omega . \tag{1.10}$$

The above definition satisfies the axioms of a probability since $\operatorname{Tr} \rho M_\omega$ is positive and its summation becomes 1, as follows. The inequality $\operatorname{Tr} \rho M_\omega \geq 0$ can be verified by the fact that ρ and M_ω are both positive semidefinite. Furthermore, since

$$\sum_{\omega \in \Omega} \operatorname{Tr} \rho M_\omega = \operatorname{Tr} \rho \sum_{\omega \in \Omega} M_\omega = \operatorname{Tr} \rho I = \operatorname{Tr} \rho = 1 ,$$

we see that this is indeed a probability distribution. In this formulation, it is implicitly assumed that both the state and the measurement are reproducible (otherwise, it would be impossible to verify experimentally (1.10)). For brevity, we shall henceforth refer to the system, state, and measurement by $\mathcal{H}$, ρ, and $\boldsymbol{M}$, respectively.

Let us now discuss the structure of a set of density matrices that shall be denoted by $\mathcal{S}(\mathcal{H})$. Consider a system that is in state ρ_1 with a probability λ and in state ρ_2 with a probability $1 - \lambda$. Now, let us perform a measurement $\boldsymbol{M} = \{M_\omega\}$ on the system. The probability of obtaining the measurement outcome ω is given by

$$\lambda \operatorname{Tr} \rho_1 M_\omega + (1 - \lambda) \operatorname{Tr} \rho_2 M_\omega = \operatorname{Tr}[(\lambda \rho_1 + (1 - \lambda)\rho_2) M_\omega] . \tag{1.11}$$

The state of the system may be considered to be given by $\rho' = \lambda \rho_1 + (1 - \lambda)\rho_2$ [Exe. 1.7]. Thus, even by using (1.10) with this state and calculating the probability distributions of the measurement outcomes, we will still be entirely consistent with the experiment. Therefore, we may believe that the state of the system is given by ρ'. This is called a probabilistic mixture (or incoherent mixture).

[2] In quantum mechanics, one often treats the state after the measurement rather than before it. The state change due to a measurement is called **state reduction**, and it requires more advanced topics than those described here. Therefore, we postpone its discussion until Sect. 7.1.

In quantum mechanics, a state $|u\rangle\langle u| \in \mathcal{S}(\mathcal{H})$ represented by a vector $u \in \mathcal{H}$ of norm 1 is called a **pure state**. This u is referred to as a state in the sense of $|u\rangle\langle u|$. The set of all vectors of norm 1 is written as $\mathcal{H}^1$. In contrast, when a state is not a pure state, it is called a **mixed state**. A pure state cannot be written as the probabilistic mixture of states except for itself[Exe. 1.8]. However, all mixed states may be written as the probabilistic mixture of other states, such as pure states. For example, if the dimensionality of $\mathcal{H}$ is d, then $\frac{1}{d}I$ is a mixed state. In fact, it is called a **completely mixed state** and is written as ρ_{mix}.

On the other hand, when $u_1, \ldots, u_d$ form an orthonormal basis of $\mathcal{H}$, the vector $|x\rangle = \sum_i x^i|u_i\rangle$ is called a quantum-mechanical superposition of $u_1, \ldots, u_d$. Note that this is different from the probabilistic mixture discussed above. The probabilistic mixture is independent of the choice of the orthonormal basis. However, the quantum-mechanical superposition depends on the choice of the basis, which depends on the physical properties of the system under consideration.

When the operators M_ω in a POVM $\boldsymbol{M} = \{M_\omega\}$ are projection matrices, i.e., $M_\omega^2 = M_\omega$, the POVM is called a **projection valued measure** (**PVM**) (only PVMs are examined in elementary courses in quantum mechanics). This is equivalent to $M_\omega M_{\omega'} = 0$ for different ω, ω'[Exe. 1.10]. Hermitian matrices are sometimes referred to as "observables" or "physical quantities." We now explain its reason.

Let the eigenvalues of a Hermitian matrix X be x^i, and the projection matrices corresponding to this eigenspace be $E_{X,i}$, i.e.,$X = \sum_i x^i E_{X,i}$. The right-hand side of this equation is called the **spectral decomposition** of X. The decomposition $\boldsymbol{E}_X = \{E_{X,i}\}$ is then a PVM. When more than one eigenvector corresponds to a single eigenvalue, the diagonalization $X = \sum_{i=1}^d x^i|u_i\rangle\langle u_i|$ is not unique, while the spectral decomposition is unique. Suppose that a measurement corresponding to a PVM $\boldsymbol{E}_X$ applied the quantum system $\mathcal{H}$ with the state ρ. Then, by using (1.10), the expectation and variance of the measurement outcome may be calculated to $\operatorname{Tr}\rho X$ and $\operatorname{Tr}\rho X^2 - (\operatorname{Tr}\rho X)^2$, respectively. Note that these are expressed completely in terms of X and ρ. Therefore, we identify the Hermitian matrix X as PVM $\boldsymbol{E}_X$ and refer to it as the measurement for the Hermitian matrix X.

When two Hermitian matrices X and Y commute each other, we can use a common orthonormal basis $u_1, \ldots, u_d$ and two sets of real numbers $\{x_i\}, \{y_i\}$ to diagonalize the matrices as

$$X = \sum_i x^i|u_i\rangle\langle u_i|, \qquad Y = \sum_i y^i|u_i\rangle\langle u_i|. \tag{1.12}$$

Then, the observables X and Y can be measured simultaneously by using the PVM $\{|u_i\rangle\langle u_i|\}_i$. Evidently, if all $X_1, \ldots, X_k$ commute, it is also possible to diagonalize them using the common basis.

In general, the elements M_ω of the PVM $\boldsymbol{M} = \{M_\omega\}$ and the state ρ do not necessarily commute. This noncommutativity often causes many difficulties in their mathematical treatment. To avoid these difficulties, we sometimes use the **pinching** map $\kappa_{\boldsymbol{M}}$ defined by[Exe. 1.11]

$$\kappa_{M}(\rho) \stackrel{\text{def}}{=} \sum_{\omega} M_{\omega} \rho M_{\omega} . \tag{1.13}$$

This is because the pinching map κ_{M} modifies the state ρ such that the state becomes commutative with M_{ω}. Hence, the pinching map is an important tool for overcoming the difficulties associated with noncommutativities. We often treat the case when PVM M is expressed as the spectral decomposition of a Hermitian matrix X. In such a case, we use the shorthand κ_{X} instead of κ_{E_X}. That is,

$$\kappa_{X} \stackrel{\text{def}}{=} \kappa_{E_X} . \tag{1.14}$$

For a general POVM $\boldsymbol{M}$, we may define[Exe. 1.11]

$$\kappa_{M}(\rho) \stackrel{\text{def}}{=} \sum_{\omega} \sqrt{M_{\omega}} \rho \sqrt{M_{\omega}} . \tag{1.15}$$

Note that this operation does not necessarily have the same effect as making the matrices commute.

Exercises

1.6 Show that when one performs a PVM $\boldsymbol{E}_X$ on a system in a state ρ, the expectation and the variance are given by $\operatorname{Tr} \rho X$ and $\operatorname{Tr} \rho X^2 - (\operatorname{Tr} \rho X)^2$, respectively.

1.7 Show that $\lambda \rho_1 + (1-\lambda)\rho_2$ is a density matrix for $\lambda \in (0, 1)$ when ρ_1 and ρ_2 are density matrices.

1.8 Suppose that any pure state ρ is written as $\rho = \lambda \rho_1 + (1-\lambda)\rho_2$ with a real number $\lambda \in (0, 1)$ and two density matrices ρ_1 and ρ_2. Show that $\rho_1 = \rho_2 = \rho$.

1.9 Let X and Y be positive semidefinite matrices. Show that $XY = 0$ if and only if $\operatorname{Tr} XY = 0$.

1.10 Let $\boldsymbol{M} = \{M_{\omega}\}$ be a POVM. Show that $\boldsymbol{M} = \{M_{\omega}\}$ is a PVM if and only if $M_{\omega} M_{\omega'} = 0$ for different ω, ω'.

1.11 Show that $\sum_{\omega} \sqrt{M_{\omega}} \rho \sqrt{M_{\omega}}$ is a density matrix for a density matrix ρ and a POVM $\boldsymbol{M}$. In particular, show that $\sum_{\omega} M_{\omega} \rho M_{\omega}$ is a density matrix when $\boldsymbol{M}$ is a PVM.

1.3 Quantum Two-Level Systems

A quantum system with a two-dimensional representation space is called a **quantum two-level system** or a **qubit**, which is the abbreviation for quantum bit. This is a particularly important special case for examining a general quantum system. The spin-$\frac{1}{2}$ system, which represents a particle with a total angular momentum of $\frac{1}{2}$, is the archetypical quantum two-level system. The electron is an example of such a spin-$\frac{1}{2}$ system. A spin-$\frac{1}{2}$ system precisely represents a specific case of angular momentum in a real system; however, it is sometimes referred to as any quantum system with two levels. In particle physics, one comes across a quantum system of isospin, which does not correspond to the motional degrees of freedom but to purely internal degrees of freedom.

Mathematically, they can be treated in the same way as spin-$\frac{1}{2}$ systems. In this text, since we are interested in the general structure of quantum systems, we will use the term quantum two-level system to refer generically to all such systems.

In particular, the Hermitian matrices S_0, S_1, S_2, and S_3 given below are called **Pauli matrices**:

$$S_0 = I, \quad S_1 = \begin{pmatrix} 0 & 1 \\ 1 & 0 \end{pmatrix}, \quad S_2 = \begin{pmatrix} 0 & -i \\ i & 0 \end{pmatrix}, \quad S_3 = \begin{pmatrix} 1 & 0 \\ 0 & -1 \end{pmatrix}. \tag{1.16}$$

They will help to simplify the expressions of matrices. The density matrix can be parameterized by using the Pauli matrices:

$$\rho_{\boldsymbol{x}} = \frac{1}{2} \begin{pmatrix} 1 + x^3 & x^1 - x^2 i \\ x^1 + x^2 i & 1 - x^3 \end{pmatrix} = \frac{1}{2} \left(S_0 + \sum_{i=1}^{3} x^i S_i \right), \tag{1.17}$$

which is called Stokes parameterization. The range of $\boldsymbol{x} = (x^1, x^2, x^3)$ is the unit sphere $\{\boldsymbol{x} | \sum_{i=1}^{3} (x^i)^2 \leq 1\}$, which is called Bloch sphere (Fig. 1.2)$^{\text{Exe. 1.12}}$.

We often focus on the basis

$$e_0 \stackrel{\text{def}}{=} \begin{pmatrix} 1 \\ 0 \end{pmatrix}, \quad e_1 \stackrel{\text{def}}{=} \begin{pmatrix} 0 \\ 1 \end{pmatrix}$$

in the space $\mathbb{C}^2$. If there are several representation spaces $\mathcal{H}_A, \mathcal{H}_B$, etc. equivalent to $\mathbb{C}^2$, and we wish to specify the space of the basis e_0, e_1, we will write e_0^A, e_1^A, etc. In

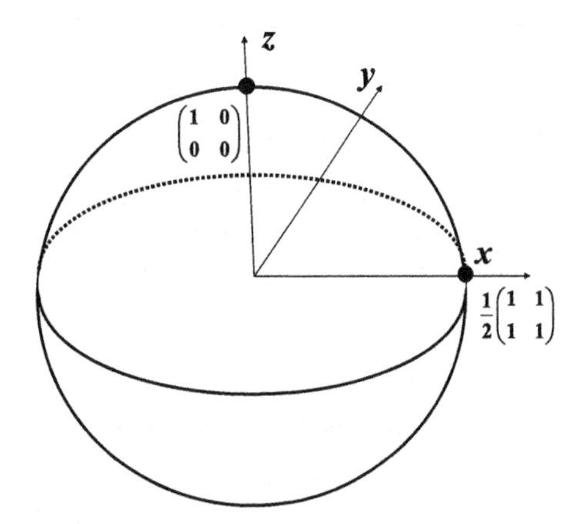

Fig. 1.2 Bloch sphere

this case, the Pauli matrix S_i and the identity matrix will be denoted by S_i^A and I_A, respectively.

Next, we consider the measurements in a quantum two-level system. The measurement of the observable S_1 is given by the PVM $\boldsymbol{E}_{S_1} = \{E_1, E_{-1}\}$, where

$$E_1 = \frac{1}{2}\begin{pmatrix} 1 & 1 \\ 1 & 1 \end{pmatrix}, \quad E_{-1} = \frac{1}{2}\begin{pmatrix} 1 & -1 \\ -1 & 1 \end{pmatrix}.$$

Given a density matrix ρ_x, the probability $\mathrm{P}_{\rho_x}^{\boldsymbol{E}_{S_1}}(1)$ of obtaining the measurement outcome 1 is $\mathrm{Tr}\,\rho E_1 = \dfrac{1+x_1}{2}$. Similarly, the probability $\mathrm{P}_{\rho_x}^{\boldsymbol{E}_{S_1}}(-1)$ of obtaining the measurement outcome -1 is $\dfrac{1-x_1}{2}$. A more detailed treatment of quantum two-level systems will be deferred until Sect. 5.3.

Exercises

1.12 Verify that the set $\{\boldsymbol{x} = (x^1, x^2, x^3) | \rho_x \geq 0\}$ is equal to $\{\boldsymbol{x} | \sum_{i=1}^3 (x^i)^2 \leq 1\}$ by showing $\det \rho_x = \frac{1-\|\boldsymbol{x}\|^2}{4}$.

1.13 Verify that ρ_x is a pure state if and only if $\|\boldsymbol{x}\| = 1$.

1.14 Show that all 2×2 Hermitian matrices with trace 0 can be written as a linear combination of S^1, S^2, and S^3.

1.4 Composite Systems and Tensor Products

A combined system composed of two systems $\mathcal{H}_A$ and $\mathcal{H}_B$ is called the **composite system** of $\mathcal{H}_A$ and $\mathcal{H}_B$. When the system $\mathcal{H}_A$ ($\mathcal{H}_B$) has an orthonormal basis $\{u_1^A, \ldots, u_{d_A}^A\}$ ($\{u_1^B, \ldots, u_{d_B}^B\}$), respectively, the representation space of the composite system is given by the Hilbert space $\mathcal{H}_A \otimes \mathcal{H}_B$ with the orthonormal basis $\{u_1^A \otimes u_1^B, \ldots, u_1^A \otimes u_{d_B}^B, u_2^A \otimes u_1^B, \ldots, u_2^A \otimes u_{d_B}^B, \ldots, u_{d_A}^A \otimes u_1^B, \ldots, u_{d_A}^A \otimes u_{d_B}^B\}$. The space $\mathcal{H}_A \otimes \mathcal{H}_B$ is called the **tensor product space** of $\mathcal{H}_A$ and $\mathcal{H}_B$; its dimension is $d_A \times d_B$. Using $d_A \times d_B$ complex numbers $(z^{i,j})$, the elements of $\mathcal{H}_A \otimes \mathcal{H}_B$ may be written as $\sum_{i,j} z^{i,j} u_i^A \otimes u_j^B$. The tensor product of two vectors $u^A = \sum_k x^k u_k^A$ and $u^B = \sum_j y^j u_j^B$ is defined as $u^A \otimes u^B \stackrel{\mathrm{def}}{=} \sum_k \sum_j x^k y^j u_k^A \otimes u_j^B$. We simplify this notation by writing $|u^A \otimes u^B\rangle$ as $|u^A, u^B\rangle$.

The tensor product $X_A \otimes X_B$ of a matrix X_A on $\mathcal{H}_A$ and a matrix X_B on $\mathcal{H}_B$ is defined as a matrix on $\mathcal{H}_A \otimes \mathcal{H}_B$ by

$$X_A \otimes X_B(u_i \otimes v_j) \stackrel{\mathrm{def}}{=} X_A(u_i) \otimes X_B(v_j).$$

The trace of this tensor product satisfies the relation[Exe. 1.15]

$$\mathrm{Tr}\, X_A \otimes X_B = \mathrm{Tr}\, X_A \cdot \mathrm{Tr}\, X_B. \tag{1.18}$$

Two matrices X_A and Y_A on $\mathcal{H}_A$ and two matrices X_B and Y_B on $\mathcal{H}_B$ satisfy

$$(X_A \otimes X_B)(Y_A \otimes Y_B) = (X_A Y_A) \otimes (X_B Y_B).$$

Hence it follows that

$$\mathrm{Tr}(X_A \otimes X_B)(Y_A \otimes Y_B) = \mathrm{Tr}(X_A Y_A) \cdot \mathrm{Tr}(X_B Y_B).$$

If the systems $\mathcal{H}_A$ and $\mathcal{H}_B$ are independent and their states are represented by the densities ρ_A and ρ_B, respectively, then the state of the composite system may be represented by the tensor product of the density matrices $\rho_A \otimes \rho_B$[Exe. 1.19]. Such a state is called a **tensor product state**.

When the density matrix ρ on the composite system $\mathcal{H}_A \otimes \mathcal{H}_B$ can be written as a probabilistic mixture of tensor product states, it is called **separable**:

$$\rho = \sum_i p_i \rho_A^i \otimes \rho_B^i, \quad p_i \geq 0, \quad \sum_i p_i = 1, \quad \rho_A^i \in \mathcal{S}(\mathcal{H}_A), \rho_B^i \in \mathcal{S}(\mathcal{H}_B). \tag{1.19}$$

Such separable states do not have a typical quantum-mechanical correlation (entanglement)[3]. When ρ is a pure state $|x\rangle\langle x|$, it is separable if and only if the vector $|x\rangle$ has a tensor product form $|u^A, u^B\rangle$. When a state ρ does not have the form (1.19), it is called an **entangled state**.

When all the n systems are identical to $\mathcal{H}$, their composite system is written as $\underbrace{\mathcal{H} \otimes \cdots \otimes \mathcal{H}}_{n}$; this will be denoted by $\mathcal{H}^{\otimes n}$ for brevity. In particular, if all the quantum systems are independent, and the state in each system is given by ρ, the composite state on $\mathcal{H}^{\otimes n}$ is $\underbrace{\rho \otimes \cdots \otimes \rho}_{n}$, which is denoted by $\rho^{\otimes n}$. Such states can be regarded as quantum versions of independent and identical distributions (discussed later).

Let us now focus on the composite state of the quantum two-level systems $\mathcal{H}_A$ and $\mathcal{H}_B$. By defining $e_0^{A,B} \stackrel{\text{def}}{=} \frac{1}{\sqrt{2}} \left(e_0^A \otimes e_0^B + e_1^A \otimes e_1^B \right)$, we see that $|e_0^{A,B}\rangle\langle e_0^{A,B}|$ is not separable, i.e., it is an entangled state. Other entangled states include

$$\begin{aligned}
e_1^{A,B} &\stackrel{\text{def}}{=} (S_1^A \otimes I_B) e_0^{A,B} = \frac{1}{\sqrt{2}} \left(e_1^A \otimes e_0^B + e_0^A \otimes e_1^B \right), \\
e_2^{A,B} &\stackrel{\text{def}}{=} (S_2^A \otimes I_B) e_0^{A,B} = \frac{i}{\sqrt{2}} \left(e_1^A \otimes e_0^B - e_0^A \otimes e_1^B \right), \\
e_3^{A,B} &\stackrel{\text{def}}{=} (S_3^A \otimes I_B) e_0^{A,B} = \frac{1}{\sqrt{2}} \left(e_0^A \otimes e_0^B - e_1^A \otimes e_1^B \right).
\end{aligned} \tag{1.20}$$

[3]In fact, even a separable state has a kind of quantum-mechanical correlation, which is called discord and is discussed in Sect. 8.10. This kind of correlation is measured by the quantity defined in (8.177).

They are mutually orthogonal, i.e.,

$$\langle e_k^{A,B} | e_l^{A,B} \rangle = \delta_{k,l} . \tag{1.21}$$

In general, any vector $|x\rangle$ on $\mathcal{H}_A \otimes \mathcal{H}_B$ can be expressed as $|x\rangle = |\sum_{i,j} x^{i,j} u_i^A \otimes u_j^B\rangle = (X \otimes I_B)|\sum_{i=1}^{d_B} u_i^B \otimes u_i^B\rangle$ by a linear map $X = \sum_{i,j} x^{i,j} |u_i^A\rangle\langle u_j^B|$ from $\mathcal{H}_B$ to $\mathcal{H}_A$. Since we can identify the vector $|x\rangle$ by using the linear map X, we denote it by $|X\rangle$.[4] So, we obtain the following properties$^{\text{Exe. 1.17, 1.18}}$:

$$(Y \otimes Z^T)|X\rangle = |YXZ\rangle, \tag{1.22}$$

$$\langle Y|X\rangle = \operatorname{Tr} Y^* X . \tag{1.23}$$

In particular, when $\sqrt{d}X$ is a unitary matrix, $|X\rangle\langle X|$ is called a **maximally entangled state** of size d. Also, an entangled state is called a partially entangled state when it is not a maximally entangled state. In this book, we denote the vector $|\frac{1}{\sqrt{d}} I\rangle$ by $|\Phi_d\rangle$. Then, we have $|u^A, u^B\rangle = |(|u^A\rangle\langle u^B|)\rangle$. So, we find that the vector $|X\rangle$ has a tensor product form if and only if the matrix X is written with the form $|u^A\rangle\langle u^B|$. This condition is equivalent with the condition that the matrix X is a rank-one matrix.

Next, let us consider the independent applications of the measurements $\boldsymbol{M}_A = \{M_{A,\omega_A}\}_{\omega_A \in \Omega_A}$ and $\boldsymbol{M}_B = \{M_{B,\omega_B}\}_{\omega_B \in \Omega_B}$ on systems $\mathcal{H}_A$ and $\mathcal{H}_B$, respectively. This is equivalent to performing a measurement $\boldsymbol{M}_A \otimes \boldsymbol{M}_B \stackrel{\text{def}}{=} \{M_{A,\omega_A} \otimes M_{B,\omega_B}\}_{(\omega_A,\omega_B) \in \Omega_A \times \Omega_B}$ on the composite system. Such a measurement is called an **independent** measurement. If a measurement $\boldsymbol{M} = \{M_\omega\}_{\omega \in \Omega}$ on the composite system $\mathcal{H}_A \otimes \mathcal{H}_B$ has the form

$$M_\omega = M_{A,\omega} \otimes M_{B,\omega}, \quad M_{A,\omega} \geq 0, M_{B,\omega} \geq 0 \tag{1.24}$$

or the form

$$M_\omega = \sum_i M_{A,\omega,i} \otimes M_{B,\omega,i}, \quad M_{A,\omega,i} \geq 0, M_{B,\omega,i} \geq 0 ,$$

the measurement $\boldsymbol{M}$ is said to be **separable**. Otherwise, it is called collective. Of course, independent measurements are always separable, but the converse is not always true.

Since the vectors $e_0^{A,B}, \ldots, e_3^{A,B}$ defined previously form an orthonormal basis in the composite system $\mathbb{C}^2 \otimes \mathbb{C}^2$, the set $\{|e_0^{A,B}\rangle\langle e_0^{A,B}|, \ldots, |e_3^{A,B}\rangle\langle e_3^{A,B}|\}$ is a PVM. This measurement is a collective measurement because it does not have the separable form (1.24).

---

[4] A notation similar to $|X\rangle$ was introduced in [1]. However, the relations (1.22) and (1.23) were essentially pointed out in [2].

On the other hand, adaptive measurements are known as a class of separable POVMs, and their definition is given as follows.[5] Suppose that we perform a measurement $\boldsymbol{M}_A = \{M_{A,\omega_A}\}_{\omega_A \in \Omega_A}$ on system $\mathcal{H}_A$ and then another measurement $\boldsymbol{M}_B^{\omega_A} = \{M_{B,\omega_B}^{\omega_A}\}_{\omega_B \in \Omega_B}$ on system $\mathcal{H}_B$ according to the measurement outcome ω_A. The POVM of this measurement on the composite system $\mathcal{H}_A \otimes \mathcal{H}_B$ is given as

$$\{M_{A,\omega_A} \otimes M_{B,\omega_B}^{\omega_A}\}_{(\omega_A,\omega_B) \in \Omega_A \times \Omega_B} . \tag{1.25}$$

Such a measurement is called **adaptive**, and it satisfies the separable condition (1.24).

Presently, it is not clear how different the adaptive condition (1.25) is from the separable condition (1.24). In Chaps. 3 and 4, we focus on the restriction of our measurements to separable or adaptive measurements and discuss the extent of its effects on the performance of information processing.

Similarly, a separable measurement $\boldsymbol{M} = \{M_\omega\}_{\omega \in \Omega}$ in the composite system $\mathcal{H}_1 \otimes \ldots \otimes \mathcal{H}_n$ of n systems $\mathcal{H}_1, \ldots, \mathcal{H}_n$ is given by

$$M_\omega = M_{1,\omega} \otimes \cdots \otimes M_{n,\omega}, \quad M_{1,\omega} \geq 0, \ldots, M_{n,\omega} \geq 0 .$$

An adaptive measurement may be written in terms of a POVM as

$$\{M_{1,\omega_1} \otimes \cdots \otimes M_{n,\omega_n}^{\omega_1,\ldots,\omega_{n-1}}\}_{(\omega_1,\ldots,\omega_n) \in \Omega_1 \times \cdots \times \Omega_n} .$$

We also denote n applications of the POVM $\boldsymbol{M}$ on the composite system $\mathcal{H}^{\otimes n}$ by $\boldsymbol{M}^{\otimes n}$.

Consider a composite system $\mathcal{H}_A \otimes \mathcal{H}_B$ in a state $\rho \in \mathcal{S}(\mathcal{H}_A \otimes \mathcal{H}_B)$. Assume that we can directly access only system $\mathcal{H}_A$ for performing measurements. In this case, we would only be interested in the state of system $\mathcal{H}_A$, and the density matrix on $\mathcal{H}_A$ is given by the **reduced density** matrix $\mathrm{Tr}_{\mathcal{H}_B} \rho \in \mathcal{S}(\mathcal{H}_A)$, which is defined to satisfy[6]

$$\mathrm{Tr}(\mathrm{Tr}_{\mathcal{H}_B} \rho) X = \mathrm{Tr}(X \otimes I_{\mathcal{H}_B}) \rho . \tag{1.26}$$

We often abbreviate $(\mathrm{Tr}_{\mathcal{H}_B} \rho)$ to ρ_A. Then, $\mathrm{Tr}_{\mathcal{H}_B}$ can be regarded as a map from the density on the composite system to the reduced density matrix and called a **partial trace**, often abbreviated to Tr_B. To specify the space on which the trace is acting, we denote the trace by Tr_A even if it is a full trace. The partial trace can be calculated according to

[5] Adaptive measurements are often called **one-way LOCC** measurements in entanglement theory. See Sect. 8.1.

[6] The uniqueness of this definition can be shown as follows. Consider the linear map $X \mapsto \mathrm{Tr}(X \otimes I_{\mathcal{H}_B})\rho$ on the set of Hermitian matrices on $\mathcal{H}_A$. Since the inner product $(X, Y) \mapsto \mathrm{Tr}\, XY$ is non-degenerate on the set of Hermitian matrices on $\mathcal{H}_A$, there uniquely exists a Hermitian matrix Y satisfying (1.26).

$$\rho^{i,j} = \sum_{k=1}^{d'} \langle u_i^A \otimes u_k^B | \rho | u_j^A \otimes u_k^B \rangle, \quad \mathrm{Tr}_B\, \rho = \sum_{i,j} \rho^{i,j} |u_i^A\rangle\langle u_j^A| , \tag{1.27}$$

where the orthonormal basis of $\mathcal{H}_A$ ($\mathcal{H}_B$) is $u_1^A, \ldots, u_d^A$ ($u_1^B, \ldots, u_{d'}^B$). This may also be written as

$$\mathrm{Tr}_B\, \rho = \sum_{i,i'} \Big(\sum_j \rho^{(i,j),(i'j)} \Big) |u_i^A\rangle\langle u_{i'}^A| , \tag{1.28}$$

where $\rho = \sum_{i,j,i',j'} \rho^{(i,j),(i'j')} |u_i^A, u_j^A\rangle\langle u_{i'}^A, u_{j'}^B|$. We may also write

$$\mathrm{Tr}_B\, \rho = \sum_{k=1}^{d'} \mathrm{Tr}_B\, P_k \rho P_k , \tag{1.29}$$

where P_k is a projection from $\mathcal{H}_A \otimes \mathcal{H}_B$ to $\mathcal{H}_A \otimes u_k^B$, where we denote the linear space spanned by the vector u_k^B by u_k^B. Further, for a given vector $|u\rangle \in \mathcal{H}_B$, we use the notation

$$\langle u|\rho|u\rangle := \mathrm{Tr}_B\, \rho |u\rangle\langle u| \otimes I_A . \tag{1.30}$$

Exercises

1.15 Show (1.18).

1.16 Show that $(X \otimes I_B)(I_A \otimes Y) = (I_A \otimes Y)(X \otimes I_B)$.

1.17 Show (1.22).

1.18 Show (1.23).

1.19 Show that the following conditions for a state ρ on the composite system $\mathcal{H}_A \otimes \mathcal{H}_B$ are equivalent.

① The state ρ has the tensor product form $\rho_A \otimes \rho_B$.
② Any normalized vector $|u\rangle \in \mathcal{H}_B$ satisfies that $\frac{\langle u|\rho|u\rangle}{\mathrm{Tr}\langle u|\rho|u\rangle} = \mathrm{Tr}_B\, \rho$.
③ Assume that the independent measurement $\boldsymbol{M}_A \otimes \boldsymbol{M}_A$ of arbitrary POVMs $\boldsymbol{M}_A = \{M_{A,\omega_A}\}$ and $\boldsymbol{M}_B = \{M_{B,\omega_B}\}$ is applied to the composite system $\mathcal{H}_A \otimes \mathcal{H}_B$ with the state ρ. The measurement outcomes ω_A and ω_B are independent of each other.

1.20 Suppose that the spaces $\mathcal{H}_A$ and $\mathcal{H}_B$ also have other bases $\{v_1^A, \ldots, v_{d_A}^A\}$ and $\{v_1^B, \ldots, v_{d_B}^B\}$ and that the unitary matrices $V_A = (v_A^{ij})$ and $V_B = (v_B^{ij})$ satisfy $v_j^A = \sum_i v_A^{ij} u_i^A$ and $v_j^B = \sum_i v_B^{ij} u_i^B$. Show that $v_j^A \otimes v_k^B = \sum_{i,l} v_A^{ij} v_B^{lk} u_i^A \otimes u_l^B$. Hence, the definition of the tensor product is independent of the choice of the bases on $\mathcal{H}_A$ and $\mathcal{H}_B$.

1.21 Prove (1.21).

1.22 Prove formulas (1.27)–(1.29), which calculate the partial trace.

1.23 Consider two Hermitian matrices $\rho_A \geq 0$ and $\sigma_A \geq 0$ on $\mathcal{H}_A$ and other two Hermitian matrices $\rho_B \geq 0$ and $\sigma_B \geq 0$ on $\mathcal{H}_B$. Show that the following two conditions are equivalent, following the steps below.

$$\circ \quad [\rho_A \otimes \rho_B, \sigma_A \otimes \sigma_B] = 0. \tag{1.31}$$

$$\circ \quad (\mathrm{Tr}\, \sigma_B \rho_B)[\rho_A, \sigma_A] = 0 \text{ and } (\mathrm{Tr}\, \sigma_A \rho_A)[\rho_B, \sigma_B] = 0. \tag{1.32}$$

(**a**) Show that (1.31) holds when $[\rho_A, \sigma_A] = [\rho_B, \sigma_B] = 0$.
(**b**) Show that (1.31) holds when $\mathrm{Tr}\, \rho_A \sigma_A = 0$.
(**c**) Show that (1.32) $\Rightarrow$ (1.31).
(**d**) Show that (1.31) $\Rightarrow$ (1.32).

1.24 Show that $\mathrm{Tr}_B X(I_A \otimes Y) = \mathrm{Tr}_B (I_A \otimes Y)X$, where X is a matrix on $\mathcal{H}_A \otimes \mathcal{H}_B$ and Y is a matrix on $\mathcal{H}_B$.

1.25 Further, show the following formula when ρ and ρ_0 are states on $\mathcal{H}_A$ and $\mathcal{H}_B$:

$$\begin{aligned} &\mathrm{Tr}_B \sqrt{\rho \otimes \rho_0}[X, Y \otimes I_B]\sqrt{\rho \otimes \rho_0} \\ =&\sqrt{\rho}[\mathrm{Tr}_B \left(I_A \otimes \sqrt{\rho_0}\right) X \left(I_A \otimes \sqrt{\rho_0}\right), Y]\sqrt{\rho}. \end{aligned}$$

1.26 Let P be a projection from $\mathcal{H}_A \otimes \mathcal{H}_B$ to the subspace $\{u^A \otimes u^B | u^B \in \mathcal{H}_B\}$ for any element $u^A \in \mathcal{H}_A$. Show that

$$\mathrm{Tr}_A(|u^A\rangle\langle u^A| \otimes I_B)X = \mathrm{Tr}_A\, PXP.$$

1.5 Matrix Inequalities and Matrix Monotone Functions

In later chapters, we will encounter quantities such as error probabilities that require us to handle inequalities in various situations. Of course, probabilities such as error probabilities are real numbers. However, in quantum systems these probabilities are expressed in terms of matrices, as we show in (1.10). Therefore, it is often helpful to use inequalities involving matrices when evaluating probabilities. By using the definition of positive semidefiniteness defined in Sect. 1.1, we may define the order (matrix inequality)

$$X \geq Y \overset{\text{def}}{\Longleftrightarrow} X - Y \geq 0 \tag{1.33}$$

for two Hermitian matrices X and Y[Exe. 1.27]. Such an order requires some care as it may involve some unexpected pitfalls arising from the noncommutativity of X and Y. In order to examine this order in greater detail, let us first analyze the properties of positive semidefiniteness again. Let X be a $d \times d$ positive semidefinite (≥ 0) Hermitian matrix and Y be a $d \times d'$ matrix. It follows that Y^*XY is a $d' \times d'$ positive semidefinite Hermitian matrix. This can be verified from

$$\langle v|Y^*XY|v\rangle = \langle Yv|X|Yv\rangle \geq 0 ,$$

where v is a vector of length d'. Furthermore, if X_1 and X_2 are two $d \times d$ Hermitian matrices satisfying $X_1 \geq X_2$, it follows that

$$Y^*X_1Y \geq Y^*X_2Y . \tag{1.34}$$

Now, we define another type of product

$$X \circ Y := \frac{1}{2}(XY + YX) .$$

If the matrices commute, then some additional types of matrix inequalities hold. For example, if $d \times d$ positive semidefinite Hermitian matrices X and Y commute, then[Exe. 1.28]

$$X \circ Y \geq 0 . \tag{1.35}$$

Inequality (1.35) does not hold unless X and Y commute. A simple counterexample exists for the noncommuting case[Exe. 1.30].

Let X_1 and X_2 be two $d \times d$ Hermitian matrices satisfying $X_1 \geq X_2 \geq 0$, and Y be a $d \times d$ positive semidefinite Hermitian matrix. When Y is commutative with X_1 and X_2, we have[Exe. 1.29]

$$X_1YX_1 \geq X_2YX_2 . \tag{1.36}$$

Inequality (1.36) does not hold unless all matrices commute[Exe. 1.31]. In general, when noncommutativity is involved, matrix inequalities are more difficult to handle and should therefore be treated with care.

Let us now define the projection $\{X \geq 0\}$ with respect to a Hermitian matrix X with a spectral decomposition $X = \sum_i x_i E_{X,i}$ [3]:

$$\{X \geq 0\} \stackrel{\text{def}}{=} \sum_{x_i \geq 0} E_{X,i} . \tag{1.37}$$

Consider the probability of the set $\{x_i \geq 0\}$ containing the measurement outcome for a measurement corresponding to the spectral decomposition $\{E_{X,i}\}$ of X. This probability is $\sum_{x_i \geq 0} \operatorname{Tr} \rho E_{X,i} = \operatorname{Tr} \rho\{X \geq 0\}$ when the state is given as a density ρ.

Therefore, this notation generalizes the concept of the subset to the noncommuting case. In other words, the probability $\mathrm{Tr}\,\rho\{X \geq 0\}$ can be regarded as a generalization of the probability $p\{\omega \in \Omega | X(\omega) \geq 0\}$, where p is a probability distribution and X is a random variable. Then, we define X_+ as $X\{X \geq 0\}$. Similarly, we may also define $\{X > 0\}$, $\{X \leq 0\}$, $\{X < 0\}$, and $\{X \neq 0\}$. Further, given two Hermitian matrices X and Y, we define the projections $\{X > Y\}$, $\{X \leq Y\}$, $\{X < Y\}$, and $\{X \neq Y\}$ as $\{X - Y > 0\}$, $\{X - Y \leq 0\}$, $\{X - Y < 0\}$, and $\{X - Y \neq 0\}$. Further, we define the matrix $(X)_+ := X\{X \geq 0\}$.

If two Hermitian matrices X and Y commute, we obtain the matrix inequality[Exe. 1.28]

$$\{X \geq 0\} + \{Y \geq 0\} \geq \{X + Y \geq 0\} \tag{1.38}$$

in the sense defined above. The range of the projection $\{X \neq 0\}$ is called the **support** of X. If the projection $\{X \neq 0\}$ is not equal to I, then the matrix X does not have its inverse matrix. In this case, the Hermitian matrix Y satisfying $XY = YX = \{X \neq 0\}$ is called the **generalized inverse matrix** of a Hermitian matrix X. It should be noted that this is not generally true unless X and Y commute. It is known that two noncummutative Hermitian matrices X and Y cannot be diagonalized simultaneously. This fact often causes many technical difficulties in the above method.

We now examine **matrix monotone functions**, which are useful for dealing with matrix inequalities. Given a function f, which maps a real number to a real number, we denote the Hermitian matrix $\sum_i f(x_i)E_{X,i}$ by $f(X)$ with respect to a Hermitian matrix $X = \sum_i x_i E_{X,i}$.

f is called a matrix monotone function in $S \subset \mathbb{R}$ if $f(X) \geq f(Y)$ for two Hermitian matrices X and Y satisfying $X \geq Y$ with eigenvalues S. Some known matrix monotone functions in $[0, \infty)$ are, for example, $f(x) = x^s$ $(0 < s \leq 1)$, and those in $(0, \infty)$ are $f(x) = \log x$ and $f(x) = -1/x$ [4]. See Exercise A.7 for the $s = 1/2$ case. Since the function $f(x) = -x^{-s}$ $(0 < s \leq 1)$ is the composite function of $-1/x$ and x^s, it is also a matrix monotone function. Note that the function $f(x) = x^s$ $(s > 1)$ ($f(x) = x^2$, etc.) is not matrix monotone[Exe. 1.32].

Exercises

1.27 Show that the order $\geq$ defined in (1.33) satisfies the axiom of order, which is equivalent with the following conditions.
(**a**) When $X \geq Y$ and $Y \geq Z$, then $X \geq Z$.
(**b**) When $X \geq Y$ and $Y \geq X$, then $X = Y$.

1.28 Suppose that X and Y commute. Show inequalities (1.35) and (1.38) using (1.12).

1.29 Verify inequality (1.36) when Y is commutative with X_1 and X_2

1.30 Show that $X = \begin{pmatrix} 1 & 0 \\ 0 & 0 \end{pmatrix}$, $Y = \frac{1}{2}\begin{pmatrix} 1 & 1 \\ 1 & 1 \end{pmatrix}$ form a counterexample to (1.35).

1.31 Show that $X_1 = I$, $X_2 = \begin{pmatrix} 1 & 0 \\ 0 & 0 \end{pmatrix}$, $Y = \frac{1}{2}\begin{pmatrix} 1 & 1 \\ 1 & 1 \end{pmatrix}$ form a counterexample to (1.36).

1.32 Verify that the following X and Y provide a counterexample to $f(x) = x^2$ as a matrix monotone function:

$$X = \begin{pmatrix} 1 & 1 \\ 1 & 1 \end{pmatrix}, \quad Y = \begin{pmatrix} 2 & 1 \\ 1 & 1 \end{pmatrix}.$$

1.33 Show that $\operatorname{rank}\{X - xI \geq 0\} \geq \operatorname{rank} P$ when a Hermitian matrix X, a projection P, and real number x satisfy $X \geq xP$.

1.34 Show that $\operatorname{Tr} X \geq \operatorname{Tr} |Y|$ for Hermitian matrices X and Y when $X \geq Y$ and $X \geq -Y$.

1.35 Show that $\operatorname{Tr} f(X) > f(\langle u_1|X|u_1\rangle) + f(\langle u_2|X|u_2\rangle)$ for a strictly convex function f when X is a positive matrix on $\mathbb{C}^2$ and X is not commutative the PVM $\{|u_1\rangle\langle u_1|, |u_2\rangle\langle u_2|\}$.

1.36 Let $\rho = |\psi\rangle\langle\psi|$ be a pure state on $\mathcal{H}_A \otimes \mathcal{H}_B$. Show the following relation for a function f.

$$f(\rho_A) \otimes I_B|\psi\rangle = I_A \otimes f(\rho_B)|\psi\rangle \tag{1.39}$$

1.37 Let $\rho = |\psi\rangle\langle\psi|$ be a pure state on $\mathcal{H}_A \otimes \mathcal{H}_B$ such that

$$|\psi\rangle = \sum_i \lambda_i |u_i^A, u_i^B\rangle.$$

Let V be the isometry $\sum_i |u_i^A\rangle\langle u_i^B|$. Show the relation for a function f, a matrix X on $\mathcal{H}_A$, and a Hermitian matrix Y on $\mathcal{H}_B$.

$$\operatorname{Tr} \rho_A^{\frac{1}{2}} X \rho_A^{\frac{1}{2}} f((VYB^{\dagger})^T) = \langle\psi|X \otimes f(Y)|\psi\rangle, \tag{1.40}$$

where ^{T} is the transpose under the basis $\{|u_i^A\rangle\}$.

1.6 Solutions of Exercises

Exercise 1.1 Use the fact that the discriminant of $\langle u + rcv|u + rcv\rangle$ concerning r is negative.

Exercise 1.2 Consider the matrix $A := \sum_{i=1}^k |e_i\rangle\langle u_i|$, where $|e_i\rangle$ is the vector that has the non-zero value 1 only in the i-th entry. The kernel of A has dimension $d - k$. So, we can choose such desired $d - k$ vectors in the Kernel of A.

Exercise 1.3 Since $X^T = \sum_{i,j} \overline{x^{i,j}}|u_i\rangle\langle u_j|$, we have $\overline{\langle x|Xx\rangle} = \langle \overline{x}|X^T\overline{x}\rangle \geq 0$, where $x = \sum_i x^i u_i$ and $\overline{x} = \sum_i \overline{x^i} u_i$.

Exercise 1.4 Let $(x_{\theta;i,j})$ and $(y_{\theta;i,j})$ be the elements of X_θ and Y_θ. Since the elements of $X_\theta Y_\theta$ are $(\sum_k x_{\theta;i,k}y_{\theta;k,j})$, their derivatives are $(\sum_k x'_{\theta;i,k}y_{\theta;k,j} + x_{\theta;i,k}y'_{\theta;k,j})$. Also, we have $(\operatorname{Tr} X_\theta)' = (\sum_i x_{\theta;i,i})' = (\sum_i x'_{\theta;i,i}) \operatorname{Tr}(X'_\theta)$.

Exercise 1.5 Consider the unitary $U = \begin{pmatrix} 0 & 1 \\ i & 0 \end{pmatrix}$ and the matrix $X = \begin{pmatrix} 0 & c \\ d & 0 \end{pmatrix}$. Then, $(UXU^*)^T = \begin{pmatrix} 0 & -id \\ ic & 0 \end{pmatrix}^T = \begin{pmatrix} 0 & ic \\ -id & 0 \end{pmatrix}$. However, $UX^TU^* = \begin{pmatrix} 0 & -ic \\ id & 0 \end{pmatrix}$.

Exercise 1.6 The expectation is $\sum_i x_i \operatorname{Tr} E_{X,i}\rho = \operatorname{Tr} \rho X$. The variance is $\sum_i x_i^2 \operatorname{Tr} E_{X,i}\rho - \operatorname{Tr} \rho X^2 = \operatorname{Tr} \rho X^2 - (\operatorname{Tr} \rho X)^2$.

Exercise 1.7 Since $\rho_1, \rho_2 \geq 0$, $\lambda\rho_1 + (1-\lambda)\rho_2 \geq 0$. Also, $\operatorname{Tr} \lambda\rho_1 + (1-\lambda)\rho_2 = \lambda \operatorname{Tr} \rho_1 + (1-\lambda) \operatorname{Tr} \rho_2 = 1$.

Exercise 1.8 Let $\rho = |x\rangle\langle x|$. Consider the projection $P := I = \rho$. Then, $0 = P\rho P = \lambda P\rho_1 P + (1-\lambda)P\rho_2 P$. So, $P\rho_1 P = P\rho_2 P = 0$. Hence, $0 = \operatorname{Tr} P\rho_1 P = \operatorname{Tr} P\rho_1$. So, $1 = \operatorname{Tr} \rho_1 = \operatorname{Tr} \rho\rho_1 = \langle x|\rho|x\rangle$, which implies $\rho_1 = \rho$. Similarly $\rho_2 = \rho$.

Exercise 1.9 Choose the diagonalizations $X = \sum_i x^i|u_i\rangle\langle u_i|$ and $Y = \sum_j y^j|v_j\rangle\langle v_j|$ with $x^i \geq 0, y^j \geq 0$. Then, $\operatorname{Tr} XY = \sum_{i,j} x^i y^j |\langle v_j|u_i\rangle|^2$.

Now, we assume that $\operatorname{Tr} XY = 0$. Then, $|\langle v_j|u_i\rangle|^2 = 0$ for non-zero x^i and y^j. This relation implies that $XY = 0$. The opposite direction is trivial.

Exercise 1.10 Assume that $\boldsymbol{M} = \{M_\omega\}$ is a PVM. $M_\omega = M_\omega I = M_\omega(M_\omega + I - M_\omega) = M_\omega + M_\omega(I - M_\omega)$. So, we have $M_\omega(I - M_\omega) = 0$. Hence, $0 = \operatorname{Tr} M_\omega(I - M_\omega) = \operatorname{Tr} M_\omega M_{\omega'}$ because $\operatorname{Tr} M_\omega(I - M_\omega - M_{\omega'}) \geq 0$. Due to Exercise 1.9, we have $M_\omega M_{\omega'} = 0$.

Conversely, we assume that $M_\omega M_{\omega'} = 0$ for different ω, ω'. Then, $M_\omega = M_\omega I = M_\omega(M_\omega + \sum_{\omega' \neq \omega} M_{\omega'} = M_\omega^2$, which implies that $\boldsymbol{M} = \{M_\omega\}$ is a PVM.

Exercise 1.11 Since ρ is positive semidefinite, $\sqrt{M_\omega}\rho\sqrt{M_\omega}$ is also positive semidefinite. $\operatorname{Tr} \sum_\omega \sqrt{M_\omega}\rho\sqrt{M_\omega} = \sum_\omega \operatorname{Tr} \sqrt{M_\omega}\rho\sqrt{M_\omega} = \sum_\omega \operatorname{Tr} \sqrt{M_\omega}\sqrt{M_\omega}\rho = \sum_\omega \operatorname{Tr} M_\omega\rho = \operatorname{Tr} \sum_\omega M_\omega\rho = \operatorname{Tr} I\rho = \operatorname{Tr} \rho = 1$. When $\boldsymbol{M}$ is a PVM, we have $\sqrt{M_\omega} = M_\omega$ So, $\operatorname{Tr} \sum_\omega M_\omega\rho M_\omega = \operatorname{Tr} \sum_\omega \sqrt{M_\omega}\rho\sqrt{M_\omega} = 1$.

Exercise 1.12 The relation $\rho_{\boldsymbol{x}} \geq 0$ holds if and only if $\det \rho_{\boldsymbol{x}} \geq 0$ and $1 \geq x^3$. Since $\det \rho_{\boldsymbol{x}} = \frac{1}{4}((1 + (x^3)^2) + (x^1 - x^2 i)(x^1 + x^2 i)) = \frac{1 - \|\boldsymbol{x}\|^2}{4}$, the above conditions are equivalent with $\sum_{i=1}^3 (x^i)^2 \leq 1$.

Exercise 1.13 If and only if $\rho_{\boldsymbol{x}}$ is a pure state, the relation $\det \rho_{\boldsymbol{x}} = 0$ holds. This condition is equivalent with the condition $\|\boldsymbol{x}\| = 1$.

Exercise 1.14 Any 2×2 Hermitian matrix A can be written as $\sum_{i=0}^3 a_i S^i$. Since $\operatorname{Tr} A = a^0$, the condition $\operatorname{Tr} A = 0$ is equivalent with $a^0 = 0$.

Exercise 1.15 $\operatorname{Tr} X_A \otimes X_B = \sum_{i,j} \langle u_i^A, u_j^B | X_A \otimes X_B | u_i^A, u_j^B \rangle$
$= \sum_{i,j} \langle u_i^A | X_A | u_i^A \rangle \langle u_j^B | X_B | u_j^B \rangle = \operatorname{Tr} X_A \cdot \operatorname{Tr} X_B$.

Exercise 1.16 We can show that $(X \otimes I_B)(I_A \otimes Y) = (X \otimes Y)$. Similarly, we can show that $(I_A \otimes Y)(X \otimes I_B) = (X \otimes Y)$. So, we obtain the desired argument.

Exercise 1.17 Since $(X \otimes I_B)|I\rangle = (X \otimes I_B)| \sum_{i=1}^{d_B} u_i^B \otimes u_i^B \rangle = | \sum_{i,j} x^{i,j} u_i^A \otimes u_j^B \rangle =$ $(I_A \otimes X^T)| \sum_{i=1}^{d_B} u_i^B \otimes u_i^B \rangle = (I_A \otimes X^T)|I\rangle$, we have $|YXZ\rangle = (YXZ \otimes I_B)|I\rangle = (YX \otimes I_B)(Z \otimes I_B)|I\rangle = (Y \otimes I_B)(X \otimes I_B)(I_A \otimes Z^T)|I\rangle = (Y \otimes I_B)(I_A \otimes Z^T)(X \otimes I_B)|I\rangle$ $= (Y \otimes Z^T)|Z\rangle$.

Exercise 1.18 $\langle Y|X\rangle = \sum_{i',j'} \sum_{i,j} \overline{y^{i',j'}} x^{i,j} \langle u_{i'}^A, u_{j'}^B | u_i^A, u_j^B \rangle = \sum_{i,j} \overline{y^{i,j}} x^{i,j} = \operatorname{Tr} Y^* X$.

Exercise 1.19 Assume ①. $\operatorname{Tr}(M_{A,\omega_A} \otimes M_{B,\omega_B})\rho_A \otimes \rho_B$ $= (\operatorname{Tr} M_{A,\omega_A}\rho_A)(\operatorname{Tr} M_{B,\omega_B}\rho_B)$. The measurement outcomes ω_A and ω_B are independent of each other. So, we obtain ③.

Assume ③. We fix $\boldsymbol{M}_B = \{|u\rangle\langle u|, I - |u\rangle\langle u|\}$, i.e., $M_{B,0} = |u\rangle\langle u|$, $M_{B,1} = I - |u\rangle\langle u|$. We choose an arbitrary POVM $\boldsymbol{M}_A = \{M_{A,\omega_A}\}$ on $\mathcal{H}_A$. Assume that we apply independent measurement $\boldsymbol{M}_A \otimes \boldsymbol{M}_A$. The marginal distribution of ω_A is $\sum_\omega \operatorname{Tr}(M_{A,\omega_A} \otimes M_{B,\omega_B})\rho = \operatorname{Tr}(M_{A,\omega_A} \otimes I_B)\rho = \operatorname{Tr} M_{A,\omega_A}(\operatorname{Tr}_B \rho)$. Due to the condition ③, when the outcome of ω_B is 0, the conditional distribution is also $\operatorname{Tr} M_{A,\omega_A}$ $(\operatorname{Tr}_B \rho)$. So, we have $\operatorname{Tr} M_{A,\omega_A} \langle u|\rho|u\rangle = \operatorname{Tr} M_{A,\omega_A} \operatorname{Tr}_B (I \otimes |u\rangle\langle u|)\rho = \operatorname{Tr}(M_{A,\omega_A} \otimes |u\rangle\langle u|)\rho = (\operatorname{Tr} M_{A,\omega_A}(\operatorname{Tr}_B \rho))(\operatorname{Tr}(I_A \otimes |u\rangle\langle u|)\rho) = (\operatorname{Tr} M_{A,\omega_A}(\operatorname{Tr}_B \rho))\langle u| \operatorname{Tr}_A \rho |u\rangle =$ $(\operatorname{Tr} M_{A,\omega_A}(\operatorname{Tr}_B \rho)) \operatorname{Tr}\langle u|\rho|u\rangle$. Since the above equation holds for any POVM $\boldsymbol{M}_A$ on $\mathcal{H}_A$, we have ②.

Assume ②. For any two vectors $|u_A\rangle \in \mathcal{H}_A$ and $|u_B\rangle \in \mathcal{H}_B$, we have $\operatorname{Tr} |u_A, u_B\rangle\langle u_A, u_B|\rho = \langle u_A| \operatorname{Tr}_B \rho |u_A\rangle \langle u_B| \operatorname{Tr}_A \rho |u_B\rangle = \operatorname{Tr} |u_A, u_B\rangle\langle u_A, u_B|(\operatorname{Tr}_B \rho) \otimes (\operatorname{Tr}_A \rho)$. Since any Hermitian matrices on $\mathcal{H}_A \otimes \mathcal{H}_B$ can be written as linear combinations of $|u_A, u_B\rangle\langle u_A, u_B|$, $\rho = (\operatorname{Tr}_B \rho) \otimes (\operatorname{Tr}_A \rho)$, which implies ①.

Exercise 1.20 We have $v_j^A \otimes v_k^B = \sum_i v_A^{ij} u_i^A \otimes \sum_l v_B^{lk} u_l^B = \sum_{i,l} v_A^{ij} v_B^{lk} u_i^A \otimes u_l^B$.

Exercise 1.21 Since $\operatorname{Tr} S_i^A S_j^A = 2\delta_{i,j}$, we have

$$\langle e_k^{A,B} | e_l^{A,B} \rangle = \frac{1}{2} \operatorname{Tr} S_i^A S_j^A = \delta_{k,l}.$$

Exercise 1.22 To show (1.27), it is sufficient to show that

$$\operatorname{Tr} \rho (X \otimes I_B) = \operatorname{Tr} \sum_{i,j} \rho^{i,j} |u_i^A\rangle\langle u_j^A| X.$$

When $X = \sum_{i,j} x_{j,i} |u_j^A\rangle\langle u_i^A|$, we have

$$\begin{aligned}\operatorname{Tr}\rho(X\otimes I_B) &= \sum_{i,j,k} x_{j,i}|u_j^A\otimes u_k^B\rangle\langle u_i^A\otimes u_k^B|\rho \\ &=\sum_{i,j} x_{i,j}\rho^{i,j} = \operatorname{Tr}\sum_{i,j}\rho^{i,j}|u_i^A\rangle\langle u_j^A|X.\end{aligned}$$

Since $\rho^{i,j} = \sum_j \rho^{(i,j),(i'j)}$, (1.27) implies (1.28). Since

$$\operatorname{Tr}_A \operatorname{Tr}_B \rho X = \operatorname{Tr}\rho X\otimes I_B = \operatorname{Tr}\rho\sum_{k=1}^{d'} P_k(X\otimes I_B) = \sum_{k=1}^{d'}\operatorname{Tr}\rho P_k(X\otimes I_B)$$
$$=\sum_{k=1}^{d'}\operatorname{Tr}\rho P_k(X\otimes I_B)P_k = \sum_{k=1}^{d'}\operatorname{Tr}P_k\rho P_k(X\otimes I_B) = \operatorname{Tr}_A\sum_{k=1}^{d'}(\operatorname{Tr}_B P_k\rho P_k)X,$$

we have (1.29).

Exercise 1.23

(**a**) When $[\rho_A,\sigma_A]=[\rho_B,\sigma_B]=0$,

$$\begin{aligned}&\rho_A\otimes\rho_B\sigma_A\otimes\sigma_B = (\rho_A\otimes I_B)(I_A\otimes\rho_B)(\sigma_A\otimes I_A)(I_A\otimes\sigma_B)\\ =&(\sigma_A\otimes I_A)(I_A\otimes\sigma_B)(\rho_A\otimes I_B)(I_A\otimes\rho_B) = \sigma_A\otimes\sigma_B\rho_A\otimes\rho_B.\end{aligned}$$

(**b**) When $\operatorname{Tr}\rho_A\sigma_A=0$, we have $\rho_A\sigma_A=0$. Hence,

$$\rho_A\otimes\rho_B\sigma_A\otimes\sigma_B = 0 = \sigma_A\otimes\sigma_B\rho_A\otimes\rho_B.$$

(bf c) Assume that the relations (1.32) hold. When $\operatorname{Tr}\rho_A\sigma_A=0$ or $\operatorname{Tr}\rho_B\sigma_B=0$, (**b**) implies (1.31). When neither $\operatorname{Tr}\rho_A\sigma_A=0$ or $\operatorname{Tr}\rho_B\sigma_B=0$ does not holds, $[\rho_A,\sigma_A]=[\rho_B,\sigma_B]=0$. Then, (**a**) implies (1.31).
(d) Assume that the relations (1.31) hold. Take the partial trace on A.

$$(\operatorname{Tr}\rho_A\sigma_A)\rho_B\sigma_B = (\operatorname{Tr}\sigma_A\rho_A)\sigma_B\rho_B,$$

which implies the first condition of (1.32). Similarly, taking the partial trace on B, we obtain the second condition of (1.32).

Exercise 1.24 It is sufficient to show

$$\operatorname{Tr}X(I_A\otimes Y)(Z\otimes I_B) = \operatorname{Tr}(I_A\otimes Y)X(Z\otimes I_B)$$

for an matrix Z on $\mathcal{H}_A$. Since $(I_A\otimes Y)$ is commutative with $(Z\otimes I_B)$,

$$\operatorname{Tr}X(I_A\otimes Y)(Z\otimes I_B) = \operatorname{Tr}X(Z\otimes I_B)(I_A\otimes Y) = \operatorname{Tr}(I_A\otimes Y)X(Z\otimes I_B).$$

Exercise 1.25

$$
\begin{aligned}
&\mathrm{Tr}_B \sqrt{\rho \otimes \rho_0}[X, Y \otimes I_B]\sqrt{\rho \otimes \rho_0} \\
=&\mathrm{Tr}_B(\sqrt{\rho} \otimes I_B)(I_A \otimes \sqrt{\rho_0})(X(Y \otimes I_B) - (Y \otimes I_B)X)(\sqrt{\rho} \otimes I_B)(I_A \otimes \sqrt{\rho_0}) \\
=&\mathrm{Tr}_B(\sqrt{\rho} \otimes I_B)[(I_A \otimes \sqrt{\rho_0})X(I_A \otimes \sqrt{\rho_0})(Y \otimes I_B) \\
&- (Y \otimes I_B)(I_A \otimes \sqrt{\rho_0})X(I_A \otimes \sqrt{\rho_0})](\sqrt{\rho} \otimes I_B) \\
=& \sqrt{\rho}(\mathrm{Tr}_B[(I_A \otimes \sqrt{\rho_0})X(I_A \otimes \sqrt{\rho_0})]Y - Y\,\mathrm{Tr}_B[(I_A \otimes \sqrt{\rho_0})X(I_A \otimes \sqrt{\rho_0})])\sqrt{\rho} \\
=&\sqrt{\rho}[\mathrm{Tr}_B \left(I_A \otimes \sqrt{\rho_0}\right) X \left(I_A \otimes \sqrt{\rho_0}\right), Y]\sqrt{\rho}.
\end{aligned}
$$

Exercise 1.26 It is sufficient to show that

$$
\mathrm{Tr}(|u^A\rangle\langle u^A| \otimes I_B)X(I_A \otimes Z) = \mathrm{Tr}\, PXP(I_A \otimes Z)
$$

for a matrix Z on $\mathcal{H}_B$. This can be shown as follows.

$$
\begin{aligned}
&\mathrm{Tr}(|u^A\rangle\langle u^A| \otimes I_B)X(I_A \otimes Z) = \mathrm{Tr}\, PX(I_A \otimes Z) \\
=&\mathrm{Tr}\, PX(I_A \otimes Z)P = \mathrm{Tr}\, PXP(I_A \otimes Z).
\end{aligned}
$$

Exercise 1.27

(**a**) Since $X - Y \ge 0$ and $Y - Z \ge 0$, we have $X - Z = X - Y + Y - Z \ge 0$.
(**b**) Since $X - Y \ge 0$ and $-(X - Y) \ge 0$, $X - Y = 0$, which implies $X = Y$.

Exercise 1.28 Since X and $\sqrt{Y}$ commute, we have

$$
X \circ Y = \frac{1}{2}(XY + YX) = \frac{1}{2}\left(\sqrt{Y}X\sqrt{Y} + \sqrt{Y}X\sqrt{Y}\right) = \sqrt{Y}X\sqrt{Y} \ge 0.
$$

Take a common diagonal basis $\{|u_i\rangle\}$. Then, X and Y are written as $X = \sum_i x_i|u_i\rangle\langle u_i|$ and $Y = \sum_i y_i|u_i\rangle\langle u_i|$. So,

$$
\begin{aligned}
&\{X \ge 0\} + \{Y \ge 0\} = \sum_{i:x_i \ge 0} |u_i\rangle\langle u_i| + \sum_{j:y_j \ge 0} |u_j\rangle\langle u_j| \\
\ge& \sum_{i:x_i+y_i \ge 0} |u_i\rangle\langle u_i| = \{X + Y \ge 0\}
\end{aligned}
$$

Exercise 1.29 Since $\sqrt{Y}$ is commutative with X_1 and X_2, we have

$$
X_1 Y X_1 = \sqrt{Y}X_1^2\sqrt{Y} \ge \sqrt{Y}X_2^2\sqrt{Y} = X_2 Y X_2.
$$

Exercise 1.30 Since $X \circ Y = \frac{1}{4}\begin{pmatrix} 2 & 1 \\ 1 & 0 \end{pmatrix}$, we have $\det(X \circ Y) = -\frac{1}{16}$. Since a matrix cannot be positive semidefinite if its determinant is negative, the matrix $X \circ Y$ is not positive semidefinite.

Exercise 1.31 Since $X_1 Y X_1 = \frac{1}{2}\begin{pmatrix} 1 & 1 \\ 1 & 1 \end{pmatrix}$ and $X_2 Y X_2 = \frac{1}{2}\begin{pmatrix} 1 & 0 \\ 0 & 0 \end{pmatrix}$, we have

$$X_1 Y X_1 - X_2 Y X_2 = \frac{1}{2}\begin{pmatrix} 0 & 1 \\ 1 & 1 \end{pmatrix}.$$

Since $\det \frac{1}{2}\begin{pmatrix} 0 & 1 \\ 1 & 1 \end{pmatrix} = -\frac{1}{4}$, the relation $X_1 Y X_1 \ge X_2 Y X_2$ does not hold.

Exercise 1.32 The relation

$$Y - X = \begin{pmatrix} 1 & 0 \\ 0 & 0 \end{pmatrix} \ge 0$$

holds. Since

$$Y^2 - X^2 = \begin{pmatrix} 5-2 & 3-2 \\ 3-2 & 2-2 \end{pmatrix} = \begin{pmatrix} 3 & 1 \\ 1 & 0 \end{pmatrix},$$

$\det(Y^2 - X^2) = -1$, which implies the matrix $Y^2 - X^2$ is not positive semi-definite.

Exercise 1.33 Let x_0 be the maximum eigenvalue of X among eigenvalues strictly smaller than x. Assume that $X = \sum_i x_i |u_i\rangle\langle u_i$. Then, let X_0 be the Hermitian matrix $\sum_{i:x_i \ge x} x_i |u_i\rangle\langle u_i + \sum_{i:x_i<x} x_0 |u_i\rangle\langle u_i$. Then, we have $X_0 \ge X \ge xP$ and $\operatorname{rank}\{X - xI \ge 0\} = \operatorname{rank}\{X_0 - xI \ge 0\}$. The relation $X_0 \ge xP$ implies that $P(X_0 - x_0 I)P \ge (x - x_0)P$. Hence, $\operatorname{rank}\{X_0 - xI \ge 0\} \ge \operatorname{rank}\{X_0 - x_0 I > 0\} \ge \operatorname{rank} P(X_0 - x_0 I) P = \operatorname{rank} P$.

Exercise 1.34 Since $\{Y \ge 0\}X\{Y \ge 0\} \ge \{Y \ge 0\}Y\{Y \ge 0\}$ and $\{Y < 0\}X\{Y < 0\} \ge -\{Y < 0\}Y\{Y < 0\}$, we have $\operatorname{Tr}\{Y \ge 0\}X\{Y \ge 0\} \ge \operatorname{Tr}\{Y \ge 0\}Y\{Y \ge 0\}$ and $\operatorname{Tr}\{Y < 0\}X\{Y < 0\} \ge -\operatorname{Tr}\{Y < 0\}Y\{Y < 0\}$. Hence, we have $\operatorname{Tr} X = \operatorname{Tr}\{Y \ge 0\} X\{Y \ge 0\} + \operatorname{Tr}\{Y < 0\}X\{Y < 0\} \ge \operatorname{Tr}\{Y \ge 0\}Y\{Y \ge 0\} - \operatorname{Tr}\{Y < 0\}Y\{Y < 0\} = \operatorname{Tr}|Y|$.

Exercise 1.35 Assume that a and b are eigenvalues of X and $\langle u_1|X|u_1\rangle = ap + b(1-p)$ with $0 < p < 1$. Then, we have $\langle u_2|X|u_2\rangle = bp + a(1-p)$. Since f is strictly convex, $f(a)p + f(b)(1-p) > f(ap + b(1-p))$ and $f(b)p + f(a)(1-p) > f(bp + a(1-p))$. Thus, $\operatorname{Tr} f(X) = f(a)p + f(b)(1-p) + f(b)p + f(a)(1-p) > f(ap + b(1-p)) + f(bp + a(1-p)) = f(\langle u_1|X|u_1\rangle) + f(\langle u_2|X|u_2\rangle)$.

Exercise 1.36 According to (A.10), we choose the bases $\{|u_i^A\rangle\}$ and $\{|u_i^B\rangle\}$ of $\mathcal{H}_A$ and $\mathcal{H}_B$ such that

$$|\psi\rangle = \sum_i \lambda_i |u_i^A, u_i^B\rangle,$$

where $\lambda_i \ge 0$. Then, $\rho_A = \sum_i \lambda_i^2 |u_i^A\rangle\langle u_i^A|$ and $\rho_B = \sum_i \lambda_i^2 |u_i^B\rangle\langle u_i^B|$. Thus,

$$f(\rho_A) \otimes I_B|\psi\rangle = \sum_i f(\lambda_i^2)|u_i^A\rangle\langle u_i^A| \otimes I_B|\psi\rangle$$
$$= \sum_i f(\lambda_i^2)\lambda_i|u_i^A, u_i^B\rangle = I_A \otimes f(\rho_B)|\psi\rangle.$$

Exercise 1.37 We have

$$\langle\psi|X \otimes f(Y)|\psi\rangle = \sum_{i,j} \lambda_i\lambda_j\langle u_i^A|X|u_j^A\rangle\langle u_i^B|f(Y)|u_j^B\rangle$$
$$= \sum_{i,j} \lambda_i\lambda_j\langle u_i^A|X|u_j^A\rangle\langle u_i^A|Vf(Y)V^\dagger|u_j^A\rangle$$
$$= \sum_{i,j} \lambda_i\lambda_j\langle u_i^A|X|u_j^A\rangle\langle u_i^A|f(VYV^\dagger)|u_j^A\rangle$$
$$= \sum_{i,j} \lambda_i\lambda_j\langle u_i^A|X|u_j^A\rangle\langle u_j^A|f((VYV^\dagger)^T)|u_i^A\rangle = \mathrm{Tr}\, \rho_A^{\frac{1}{2}} X \rho_A^{\frac{1}{2}} f((VYB^\dagger)^T).$$

References

1. G.M. D'Ariano, P. Lo, Presti, M.F. Sacchi, Bell measurements and observables. Phys. Lett. A **272**, 32 (2000)
2. M. Horodecki, P. Horodecki, R. Horodecki, General teleportation channel, singlet fraction and quasi-distillation. Phys. Rev. A **60**, 1888 (1999)
3. H. Nagaoka, Information spectrum theory in quantum hypothesis testing, in *Proceedings 22th Symposium on Information Theory and Its Applications (SITA)*, (1999), pp. 245–247 (in Japanese)
4. R. Bhatia, *Matrix Analysis* (Springer, Berlin, 1997)

Chapter 2
Information Quantities and Parameter Estimation in Classical Systems

Abstract For the study of quantum information theory, mathematical statistics, and information geometry, which are mainly examined in a nonquantum context. This chapter briefly summarizes the fundamentals of these topics from a unified viewpoint. Since these topics are usually treated individually, this chapter will be useful even for nonquantum applications.

2.1 Information Quantities in Classical Systems

When all the given density matrices $\rho_1, \ldots, \rho_n$ commute, they may be simultaneously diagonalized using a common orthonormal basis $\{u^1, \ldots, u^d\}$ according to $\rho_1 = \sum_i p_{1,i}|u^i\rangle\langle u^i|, \ldots, \rho_n = \sum_i p_{n,i}|u^i\rangle\langle u^i|$. In this case, it is sufficient to treat only the diagonal elements, i.e., we discuss only the probability distributions $p_1, \ldots, p_n$. Henceforth we will refer to such cases as **classical** because they do not exhibit any quantum properties. Let us now examine various information quantities with respect to probability distributions.

2.1.1 *Entropy*

Before proceeding to the definition of information quantities, we prepare the notations for basic probability theory. For a given probability distribution $p = \{p_x\}_{x\in\Omega}$ of the real-valued random variable X, we define the expectation $\mathrm{E}_p(X)$ as

$$\mathrm{E}_p(X) \stackrel{\text{def}}{=} \sum_{x\in\Omega} x p_x. \tag{2.1}$$

When the number $-\log p_x$ is regarded as a real-valued random variable, the **Shannon entropy** is defined as the expectation of the real-valued random variable under the probability distribution p, i.e.,[1]

[1]In this case, we consider $0 \log 0$ to be 0 here.

M. Hayashi, *Quantum Information Theory*, Graduate Texts in Physics,
DOI 10.1007/978-3-662-49725-8_2

$$H(p) \stackrel{\text{def}}{=} \sum_{x \in \Omega} -p_x \log p_x. \tag{2.2}$$

It is often simply called **entropy**. That is, when $\mathcal{P}(\Omega)$ denotes the set of probability distributions on the probability space Ω, H is a real-valued function on $\mathcal{P}(\Omega)$. Sometimes, we denote the probability distribution of a random variable X by P_X. In this case, we write the entropy of P_X as $H(X)$. For $\Omega = \{0, 1\}$, the probability distribution is written as $(a, 1-a)$ and the entropy is called a **binary entropy**, which is given by $h(a) \stackrel{\text{def}}{=} -a \log a - (1-a) \log(1-a)$.

When the number of elements of Ω is a finite number k, it is possible to choose the distribution so that all probabilities p_i have the same value. Such a probability distribution $p = (p_i)$ is called a **uniform distribution** and is denoted by $p_{\mathrm{mix},\Omega}$. It is simplified to p_{mix} for simplicity. If it is necessary to denote the number of supports k explicitly, we write $p_{\mathrm{mix},k}$. As shown later, any distribution p on Ω satisfies the relation

$$H(p) \le \log k = H(p_{\mathrm{mix},\Omega}). \tag{2.3}$$

The entropy $H(\mathrm{P}_{X,Y}(x, y))$ of the joint distribution $\mathrm{P}_{X,Y}$ for two random variables X and Y is denoted by $H(X, Y)$. In particular, if Y can be expressed as $f(X)$, where f is a function, then[Exe. 2.1]

$$H(X, Y) = H(X, f(X)) = H(X). \tag{2.4}$$

Given a conditional probability $\mathrm{P}_{X|Y=y} = \{\mathrm{P}_{X|Y}(x|y)\}_x$, the entropy of X is given by $H(X|Y = y) \stackrel{\text{def}}{=} H(\mathrm{P}_{X|Y=y})$ when the random variable Y is known to be y. The expectation of this entropy with respect to the probability distribution of Y is called the **conditional entropy** denoted by $H(X|Y)$. We may write it as

$$\begin{aligned}
H(X|Y) &\stackrel{\text{def}}{=} \sum_y \sum_x -\mathrm{P}_Y(y)\mathrm{P}_{X|Y}(x|y) \log \mathrm{P}_{X|Y}(x|y) \\
&= -\sum_{x,y} \mathrm{P}_{X,Y}(x, y) \log \frac{\mathrm{P}_{X,Y}(x, y)}{\mathrm{P}_Y(y)} \\
&= -\sum \mathrm{P}_{X,Y}(x, y) \log \mathrm{P}_{X,Y}(x, y) + \sum_y \mathrm{P}_Y(y) \log \mathrm{P}_Y(y) \\
&= H(X, Y) - H(Y).
\end{aligned} \tag{2.5}$$

The final equation in (2.5) is called **chain rule**. Using chain rule (2.5) and (2.4), we have

$$H(X) = H(f(X)) + H(X|f(X)) \ge H(f(X)), \tag{2.6}$$

which is called **monotonicity**.

Applying (2.4) to the distribution $\mathrm{P}_{X|Y=y}$, we have

$$H(X, f(X, Y)|Y) = \sum_y \mathrm{P}_Y(y) H(X, f(X, y)|Y = y)$$
$$= \sum_y \mathrm{P}_Y(y) H(X|Y = y) = H(X|Y). \tag{2.7}$$

Since (as will be shown later)

$$H(X) + H(Y) - H(X, Y) \geq 0, \tag{2.8}$$

we have

$$H(X) \geq H(X|Y). \tag{2.9}$$

If Y takes values in $\{0, 1\}$, (2.9) is equivalent to the concavity of the entropy$^{\text{Exe. 2.2}}$:

$$\lambda H(p) + (1 - \lambda) H(p') \leq H(\lambda p + (1 - \lambda) p'), \quad 0 < \forall \lambda < 1. \tag{2.10}$$

Exercises

2.1 Verify (2.4) if the variable Y can be written $f(X)$ for a function f.

2.2 Verify that (2.9) and (2.10) are equivalent.

2.3 Given a distribution $p = \{p_x\}$ on $\{1, \ldots, k\}$. Assume that the maximum probability p_x is larger than a. Verify that $H(p) \leq h(a) + (1 - a) \log(k - 1)$.

2.4 Define $p_A \times p_B(\omega_A, \omega_B) = p_A(\omega_A) p_B(\omega_B)$ in $\Omega_A \times \Omega_B$ for probability distributions p_A in Ω_A, p_B in Ω_B. Show that

$$H(p_A) + H(p_B) = H(p_A \times p_B). \tag{2.11}$$

2.1.2 Relative Entropy

We now consider a quantity that expresses the closeness between two probability distributions $p = \{p_i\}_{i \in \Omega}$ and $q = \{q_i\}_{i \in \Omega}$. It is called an information quantity because our access to information is closely related to the difference between the distributions reflecting the information of our interest. A typical example is the **relative entropy**[2] $D(p\|q)$, which is defined as

[2]The term relative entropy is commonly used in statistical physics. In information theory, it is generally known as the **Kullback–Leibler divergence**, while in statistics it is known as the **Kullback–Leibler information**.

$$D(p\|q) \stackrel{\text{def}}{=} \sum_{i\in\Omega} p_i \log \frac{p_i}{q_i} . \tag{2.12}$$

This quantity is always no less than 0, and it is equal to 0 if and only if $p = q$. This can be shown by applying the **logarithmic inequality**[Exe. 2.6] "$\log x \le x - 1$ for $x > 0$" to (2.12):

$$0 - D(p\|q) = \sum_{i=1}^{k} p_i \left(-\frac{q_i}{p_i} + 1 + \log \frac{q_i}{p_i} \right) \le \sum_{i=1}^{k} p_i \, 0 = 0 .$$

Note that the equality of $\log x \le x - 1$ holds only when $x = 1$. We may obtain (2.3) by using the positivity of the relative entropy for the case $q = \{1/k\}$.

Let us now consider possible information processes. For simplicity, we assume that the probability space Ω is given as the set $\mathbb{N}_k \stackrel{\text{def}}{=} \{1, \ldots, k\}$. When an information process converts a set $\mathbb{N}_k \stackrel{\text{def}}{=} \{1, \ldots, k\}$ to another set $\mathbb{N}_l$ deterministically, we may denote the information processing by a function from $\mathbb{N}_k$ to $\mathbb{N}_l$. If it converts probabilistically, it is denoted by a real-valued matrix $\{Q^i_j\}$ in which every element Q^i_j represents the probability of the output data $j \in \mathbb{N}_l$ when the input data are $i \in \mathbb{N}_k$. This matrix $Q = (Q^i_j)$ satisfies $\sum_{j=1}^{l} Q^i_j = 1$ for each i. Such a matrix Q is called a **stochastic transition matrix**. In this notation, Q^i expresses the distribution $(Q^i_1, \ldots, Q^i_k)$ on the output system with the input i. When the input signal is generated according to the probability distribution p, the output signal is generated according to the probability distribution $Q(p)_j \stackrel{\text{def}}{=} \sum_{i=1}^{k} Q^i_j p_i$. The stochastic transition matrix Q represents not only such probabilistic information processes but also probabilistic fluctuations in the data due to noise. Furthermore, since it expresses the probability distribution of the output system for each input signal, we can also use it to model a channel transmitting information.

A fundamental property of a stochastic transition matrix Q is the inequality

$$D(p\|q) \ge D(Q(p)\|Q(q)) , \tag{2.13}$$

which is called an **information-processing inequality**. This property is often called **monotonicity**.[3] The inequality implies that the amount of information should not increase via any information processing. This inequality will be proved for the general case in Theorem 2.1. It may also be shown using a logarithmic inequality.

For example, consider the stochastic transition matrix $Q = (Q^i_j)$ from $\mathbb{N}_{2k}$ to $\mathbb{N}_k$, where Q^i_j is 1 when $i = j, j + k$ and 0 otherwise. Given two probability distributions p, p' in $\mathbb{N}_k$, we define the probability distribution $\tilde{p}$ for $\mathbb{N}_{2k}$ as

$$\tilde{p}_i = \lambda p_i , \ \tilde{p}_{i+k} = (1 - \lambda) p'_i , \quad 1 \le \forall i \le k$$

[3]In this book, monotonicity refers to only the monotonicity regarding the change in probability distributions or density matrices.

with a real number $\lambda \in (0, 1)$. Similarly, we define $\tilde{q}$ for two probability distributions q, q' in $\mathbb{N}_k$. Then,

$$D(\tilde{p}\|\tilde{q}) = \lambda D(p\|q) + (1-\lambda)D(p'\|q') .$$

Since $Q(\tilde{p}) = \lambda p + (1-\lambda)p'$ and $Q(\tilde{q}) = \lambda q + (1-\lambda)q'$, the information-processing inequality (2.13) yields the **joint convexity** of the relative entropy

$$\lambda D(p\|q) + (1-\lambda)D(p'\|q') \geq D(\lambda p + (1-\lambda)p' \| \lambda q + (1-\lambda)q'). \quad (2.14)$$

Next, let us consider other information quantities that express the difference between the two probability distributions p and q. In order to express the amount of information, these quantities should satisfy the property given by (2.13). This property can be satisfied by constructing the information quantity in the following manner. First, we define convex functions. When a function f satisfies

$$f(\lambda x_1 + (1-\lambda)x_2) \leq \lambda f(x_1) + (1-\lambda)f(x_2), \quad 0 \leq \forall\lambda \leq 1, \forall x_1, x_2 \in \mathbb{R} ,$$

it is called a **convex function**. For a probability distribution $p = \{p_i\}$, a convex function f satisfies **Jensen's inequality**:

$$\sum_i p_i f(x_i) \geq f\left(\sum_i p_i x_i\right) . \quad (2.15)$$

Theorem 2.1 (Csiszár [1]) *Let f be a convex function. The information quantity $D_f(p\|q) \stackrel{\text{def}}{=} \sum_i q_i f\left(\frac{p_i}{q_i}\right)$ then satisfies the* **monotonicity** *condition*

$$D_f(p\|q) \geq D_f(Q(p)\|Q(q)). \quad (2.16)$$

*Henceforth, $D_f(p\|q)$ will be called an f-***relative entropy**.[4]

For example, for $f(x) = x \log x$ we obtain the relative entropy. For $f(x) = 1 - \sqrt{x}$,

$$D_f(p\|q) = 1 - \sum_i \sqrt{p_i}\sqrt{q_i} = \frac{1}{2}\sum_i \left(\sqrt{p_i} - \sqrt{q_i}\right)^2 . \quad (2.17)$$

Its square root is called the **Hellinger distance** and is denoted by $d_2(p, q)$. This satisfies the axioms of a distance$^{\text{Exe. 2.14}}$. When $f(x) = \frac{4}{1-\alpha^2}(1 - x^{(1+\alpha)/2})(-1 < \alpha < 1)$, $D_f(p\|q)$ is equal to the α-divergence $\frac{4}{1-\alpha^2}\left(1 - \sum_i p_i^{(1+\alpha)/2} q_i^{(1-\alpha)/2}\right)$ according

[4]This quantity is more commonly used in information theory, where it is called f-divergence [1]. In this text, we prefer to use the term "relative entropy" for all relative-entropy-like quantities.

to Amari and Nagaoka [2]. By applying inequality (2.16) to the concave function $x \to x^s$ $(0 \le s \le 1)$ and the convex function $x \to x^s$ $(s \le 0)$, we obtain the inequalities

$$\sum_i p_i^{1-s} q_i^s \le \sum_j Q(p)_j^{1-s} Q(q)_j^s \text{ for } 0 \le s \le 1\,,$$
$$\sum_i p_i^{1-s} q_i^s \ge \sum_j Q(p)_j^{1-s} Q(q)_j^s \text{ for } s \le 0\,.$$

Hence, the quantity $\phi(s|p\|q) \stackrel{\text{def}}{=} \log(\sum_i p_i^{1-s} q_i^s)$ satisfies the **monotonicity**

$$\phi(s|p\|q) \le \phi(s|Q(p)\|Q(q)) \text{ for } 0 \le s \le 1\,,$$
$$\phi(s|p\|q) \ge \phi(s|Q(p)\|Q(q)) \text{ for } s \le 0\,.$$

The relative entropy can be expressed as

$$\phi'(0|p\|q) = -D(p\|q), \quad \phi'(1|p\|q) = D(q\|p). \tag{2.18}$$

Since $\phi(s|p\|q)$ is a convex function of s[Exe. 2.16], the **relative Rényi entropy** [3]

$$D_{1-s}(p\|q) \stackrel{\text{def}}{=} -\frac{\phi(s|p\|q)}{s} = -\frac{\phi(s|p\|q) - \phi(0|p\|q)}{s} = -\frac{1}{s}\log \sum_i p_i^{1-s} q_i^s \tag{2.19}$$

is monotone decreasing for s[Exe. 2.17]. More precise analyses for these quantities are given in Exercises 3.45, 3.52, and 3.53.

We will abbreviate it to $\phi(s)$ if it is not necessary to specify p and q explicitly. Hence, we define the minimum and the maximum relative entropies as

$$D_{\max}(p\|q) \stackrel{\text{def}}{=} -\log \max_i \frac{p_i}{q_i}, \quad D_{\min}(p\|q) \stackrel{\text{def}}{=} -\log \sum_{i:p_i>0} q_i\,. \tag{2.20}$$

Hence, we obtain the relations[Exe. 2.18, 2.19]

$$\lim_{s\to-\infty} D_{1-s}(p\|q) = D_{\max}(p\|q), \quad \lim_{s\to 1} D_{1-s}(p\|q) = D_{\min}(p\|q), \tag{2.21}$$
$$\lim_{s\to 0} D_{1-s}(p\|q) = D(p\|q). \tag{2.22}$$

That is, $D_{\max}(p\|q)$ and $D_{\min}(p\|q)$ give the maximum and the minimum values of $D_{1-s}(p\|q)$, respectively.

Proof of Theorem 2.1 Since f is a convex function, Jensen's inequality ensures that

$$\sum_i \frac{Q^i_j q_i}{\sum_{i'} Q^{i'}_j q_{i'}} f\left(\frac{p_i}{q_i}\right) \geq f\left(\sum_i \frac{Q^i_j q_i}{\sum_{i'} Q^{i'}_j q_{i'}} \frac{p_i}{q_i}\right) = f\left(\frac{\sum_i Q^i_j p_i}{\sum_{i'} Q^{i'}_j q_{i'}}\right).$$

Therefore,

$$\begin{aligned}
D_f(Q(p)\|Q(q)) &= \sum_j \sum_{i''} Q^{i''}_j q_{i''} f\left(\frac{\left(\sum_i Q^i_j p_i\right)}{\left(\sum_{i'} Q^{i'}_j q_{i'}\right)}\right) \\
&\leq \sum_j \sum_{i''} Q^{i''}_j q_{i''} \sum_i \frac{Q^i_j q_i}{\sum_{i'} Q^{i'}_j q_{i'}} f\left(\frac{p_i}{q_i}\right) \\
&= \sum_j \sum_i Q^i_j q_i f\left(\frac{p_i}{q_i}\right) = \sum_i q_i f\left(\frac{p_i}{q_i}\right) = D_f(p\|q).
\end{aligned}$$

■

We consider the **variational distance** as another information quantity. It is defined as

$$d_1(p,q) \stackrel{\text{def}}{=} \frac{1}{2} \sum_i |p_i - q_i| . \tag{2.23}$$

It is the f-relative entropy when $f(x)$ is chosen to be $\frac{1}{2}|1-x|$. However, it satisfies the **monotonicity** property[Exe. 2.9]

$$d_1(Q(p), Q(q)) \leq d_1(p,q) . \tag{2.24}$$

The variational distance, Hellinger distance, and relative entropy are related by the following formulas:

$$d_1(p,q) \geq d_2^2(p,q) \geq \frac{1}{2} d_1^2(p,q) , \tag{2.25}$$

$$D(p\|q) \geq -2\log\left(\sum_i \sqrt{p_i}\sqrt{q_i}\right) \geq 2 d_2^2(p,q) . \tag{2.26}$$

The last inequality may be deduced from the logarithmic inequality. The combination of (2.25) and (2.26) is called **Pinsker inequality**.

When a stochastic transition matrix $Q = (Q^i_j)$ satisfies $\sum_i Q^i_j = 1$, i.e., its transpose is also a stochastic transition matrix, the stochastic transition matrix $Q = (Q^i_j)$ is called a **double stochastic transition matrix**. Now, we assume that the input symbol i and the output symbol j take the values in $1, \ldots, k_1$ and $1, \ldots, k_2$, respectively. When the stochastic transition matrix $Q = (Q^i_j)$ is double stochastic, we have $k_2 = \sum_{j=1}^{k_2} 1 = \sum_{j=1}^{k_2} \sum_{i=1}^{k_1} Q^i_j = \sum_{i=1}^{k_1} \sum_{j=1}^{k_2} Q^i_j = \sum_{i=1}^{k_1} 1 = k_1$. That is, any double stochastic matrix is a square matrix.

A stochastic transition square matrix Q is a double stochastic transition matrix if and only if the output distribution $Q(p_{\mathrm{mix}})$ is a uniform distribution because $Q(p_{\mathrm{mix}})_j = \sum_i Q^i_j \frac{1}{k} = \frac{1}{k}$. The double stochastic transition matrix Q and the probability distribution p satisfy

$$\log k - H(Q(p)) = D(Q(p)\|p_{\mathrm{mix},k}) \geq D(p\|p_{\mathrm{mix},k}) = \log k - H(p),$$

which implies that

$$H(Q(p)) \geq H(p). \tag{2.27}$$

Exercises

2.5 Show that

$$D(p_A\|q_A) + D(p_B\|q_B) = D(p_A \times p_B\|q_A \times q_B) \tag{2.28}$$

for probability distributions p_A, q_A in Ω_A and p_B, q_B in Ω_B.

2.6 Show the logarithmic inequality, i.e., the inequality $\log x \leq x - 1$, holds for $x > 0$ and the equality holds only for $x = 1$.

2.7 Show that the f-relative entropy $D_f(p\|q)$ of a convex function f satisfies $D_f(p\|q) \geq f(1)$.

2.8 Prove (2.17).

2.9 Show that the variational distance satisfies the monotonicity condition (2.24).

2.10 Show that $d_1(p,q) \geq d_2^2(p,q)$ by first proving the inequality $|x - y| \geq (\sqrt{x} - \sqrt{y})^2$.

2.11 Show that $d_2^2(p,q) \geq \frac{1}{2} d_1^2(p,q)$ following the steps below.
(a) Prove

$$\left(\sum_i |p_i - q_i|\right)^2 \leq \left(\sum_i |\sqrt{p_i} - \sqrt{q_i}|^2\right)\left(\sum_i |\sqrt{p_i} + \sqrt{q_i}|^2\right)$$

using the Schwarz inequality.
(b) Show that $\sum_i |\sqrt{p_i} + \sqrt{q_i}|^2 \leq 4$.
(c) Show that $d_2^2(p,q) \geq \frac{1}{2} d_1^2(p,q)$ using the above results.

2.12 Show that $d_1(p,q) \leq \sum_{x \neq x_0} |p_x - q_x|$ for any x_0.

2.13 Show that $D(p\|q) \geq -2\log\left(\sum_i \sqrt{p_i}\sqrt{q_i}\right)$.

2.14 Verify that the Hellinger distance satisfies the axioms of a distance by following the steps below.
(**a**) Prove the following for arbitrary vectors x and y

$$(\|x\| + \|y\|)^2 \geq \|x\|^2 + \langle x, y\rangle + \langle y, x\rangle + \|y\|^2 .$$

(**b**) Prove the following for arbitrary vectors x and y:

$$\|x\| + \|y\| \geq \|x + y\| .$$

(**c**) Show the following for the three probability distributions p, q, and r:

$$\sqrt{\sum_i \left(\sqrt{p_i} - \sqrt{q_i}\right)^2} \leq \sqrt{\sum_i \left(\sqrt{p_i} - \sqrt{r_i}\right)^2} + \sqrt{\sum_i \left(\sqrt{r_i} - \sqrt{q_i}\right)^2} .$$

Note that this formula is equivalent to the axiom of a distance $d_2(p, q) \leq d_2(p, r) + d_2(r, q)$ for the Hellinger distance.

2.15 Show (2.18).

2.16 Show that $\phi(s|p\|q)$ is convex for s.

2.17 Show that $\frac{f(s)}{s}$ is (strictly) monotone increasing for s when $f(0) = 0$ and $f(s)$ is (strictly) convex for s.

2.18 Show that $\lim_{s\to-\infty} D_{1-s}(p\|q) = D_{\max}(p\|q)$ by following the steps below.
(**a**) Show that $\frac{1}{t}\log(\sum_{i=1}^k a_i b_i^t) \to \log\max(b_1, \ldots, b_k)$ as $t \to \infty$ for $a_i, b_i \geq 0$.
(**b**) Show the desired equation.

2.19 Show that $\lim_{s\to 1} D_{1-s}(p\|q) = D_{\min}(p\|q)$.

2.20 Show that

$$D(p\|q) = \max_{\lambda=(\lambda_1,\ldots,\lambda_k)\in\mathbb{R}^k} \sum_{i=1}^k p_i\lambda_i - \log\sum_{i=1}^k q_i e^{\lambda_i} \tag{2.29}$$

for two probability distributions p and q on $\{1, \ldots, k\}$.

2.1.3 Mutual Information

Given the joint probability distribution $\mathrm{P}_{X,Y}$ of two random variables X and Y, the **marginal distributions** P_X and P_Y of $\mathrm{P}_{X,Y}$ are defined as

$$\mathrm{P}_X(x) \stackrel{\mathrm{def}}{=} \sum_y \mathrm{P}_{X,Y}(x,y) \quad \text{and} \quad \mathrm{P}_Y(y) \stackrel{\mathrm{def}}{=} \sum_x \mathrm{P}_{X,Y}(x,y) .$$

Then, the conditional distribution is calculated as

$$\mathrm{P}_{X|Y}(x|y) = \frac{\mathrm{P}_{X,Y}(x,y)}{\mathrm{P}_Y(y)} .$$

When $\mathrm{P}_X(x) = \mathrm{P}_{X|Y}(x|y)$, two random variables X and Y are **independent**. In this case, the joint distribution $\mathrm{P}_{X,Y}(x,y)$ is equal to the product of marginal distributions $\mathrm{P}_X \times \mathrm{P}_Y(x,y) := \mathrm{P}_X(x)\mathrm{P}_Y(y)$. That is, the relative entropy $D(\mathrm{P}_{X,Y}\|\mathrm{P}_X \times \mathrm{P}_Y)$ is equal to zero. We now introduce **mutual information** $I(X:Y)$, which expresses how different the joint distribution $\mathrm{P}_{X,Y}(x,y)$ is from the product of marginal distributions $\mathrm{P}_X(x)\mathrm{P}_Y(y)$. This quantity satisfies the following relation:

$$\begin{aligned} I(X:Y) &\stackrel{\mathrm{def}}{=} D(\mathrm{P}_{X,Y}\|\mathrm{P}_X\mathrm{P}_Y) = \sum_{x,y} \mathrm{P}_{X,Y}(x,y) \log \frac{\mathrm{P}_{X,Y}(x,y)}{\mathrm{P}_X(x)\mathrm{P}_Y(y)} \\ &= H(X) - H(X|Y) = H(Y) - H(Y|X) = H(X) + H(Y) - H(X,Y) . \end{aligned} \tag{2.30}$$

Hence, inequality (2.8) may be obtained from the above formula and the positivity of $I(X:Y)$. Further, we can define a **conditional mutual information** in a manner similar to that of the entropy. This quantity involves another random variable Z (in addition to X and Y) and is defined as

$$\begin{aligned} I(X:Y|Z) &\stackrel{\mathrm{def}}{=} \sum_z \mathrm{P}_Z(z) I(X:Y|Z=z) \qquad (2.31) \\ &= \sum_{x,y,z} \mathrm{P}_{X,Y,Z}(x,y,z) \log \frac{\mathrm{P}_{XY|Z}(x,y|z)}{\mathrm{P}_{X|Z}(x|z)\mathrm{P}_{Y|Z}(y|z)} \ge 0 , \end{aligned}$$

where $I(X:Y|Z=z)$ is the mutual information of X and Y assuming that $Z=z$ is known. By applying (2.5) and (2.30) to the case $Z=z$, we obtain

$$\begin{aligned} I(X:Y|Z) &= H(X|Z) + H(Y|Z) - H(XY|Z) = H(X|Z) - H(X|YZ) \\ &= -(H(X) - H(X|Z)) + (H(X) - H(X|YZ)) \\ &= -I(X:Z) + I(X:YZ) . \end{aligned}$$

This equation is called the **chain rule** of mutual information, which may also be written as

$$I(X:YZ) = I(X:Z) + I(X:Y|Z) . \tag{2.32}$$

Hence, it follows that

$$I(X:YZ) \geq I(X:Z).$$

Note that (2.32) can be generalized as

$$I(X:YZ|U) = I(X:Z|U) + I(X:Y|ZU). \tag{2.33}$$

Next, we apply the above argument to the case where the information channel is given by a stochastic transition matrix $Q = (Q_y^x)$ and the input distribution is given by p. Let X and Y be, respectively, the random variables of the input system and output system. That is, their joint distribution is given as $\mathrm{P}_{X,Y}(x, y) = Q_y^x p_x$. Then, the mutual information $I(X:Y)$ can be regarded as the amount of information transmitted via channel Q when the input signal is generated with the distribution p. This is called **transmission information**, and it is denoted by $I(p, Q)$. Therefore, we can define the transmission information by

$$I(p, Q) \stackrel{\text{def}}{=} H(Q(p)) - \sum_x p_x H(Q^x). \tag{2.34}$$

We will now discuss **Fano's inequality**, which is given by the following theorem.

Theorem 2.2 (Fano [4]) *Let X and Y be random variables that take values in the same data set $\mathbb{N}_k = \{1, \ldots, k\}$. Then, the following inequality holds:*

$$\begin{aligned} H(X|Y) &\leq \mathrm{P}\{X \neq Y\} \log(k-1) + h(\mathrm{P}\{X \neq Y\}) \\ &\leq \mathrm{P}\{X \neq Y\} \log k + \log 2. \end{aligned} \tag{2.35}$$

Proof We define the random variable $Z \stackrel{\text{def}}{=} \begin{cases} 0 & X = Y \\ 1 & X \neq Y \end{cases}$. Applying (2.5) to X and Z under the condition $Y = y$, we obtain

$$\begin{aligned} H(X|Y=y) &= H(X, Z|Y=y) \\ &= \sum_z \mathrm{P}_{Z|Y}(z|y) H(X|Z=z, Y=y) + H(Z|Y=y). \end{aligned}$$

The first equality follows from the fact that the random variable Z can be uniquely obtained from X. Taking the expectation with respect to y, we get

$$\begin{aligned} H(X|Y) &= H(X|Z, Y) + H(Z|Y) \leq H(X|Z, Y) + H(Z) \\ &= H(X|Z, Y) + h(\mathrm{P}\{X \neq Y\}). \end{aligned} \tag{2.36}$$

Applying (2.3), we have

$$H(X|Y=y, Z=0) = 0, \quad H(X|Y=y, Z=1) \leq \log(k-1).$$

Therefore,

$$H(X|Y,Z) \le \mathrm{P}\{X \ne Y\} \log(k-1). \tag{2.37}$$

Finally, combining (2.36) and (2.37), we obtain (2.35). ■

Exercise

2.21 Show the chain rule of conditional mutual information (2.33) based on (2.32).

2.1.4 The Independent and Identical Condition and Rényi Entropy

Given a probability distribution $p = \{p_i\}_{i=1}^k$, we define the **Rényi entropy** $H_{1-s}(p)$ of order $1-s$ as

$$H_{1-s}(p) \stackrel{\mathrm{def}}{=} \frac{\psi(s|p)}{s}, \quad \psi(s|p) \stackrel{\mathrm{def}}{=} \log \sum_i p_i^{1-s} \tag{2.38}$$

for a real number s in addition to the entropy $H(p)$. We will abbreviate the quantity $\psi(s|p)$ to $\psi(s)$ when there is no risk of ambiguity. When $0 < s < 1$, the quantity $\psi(s)$ is a positive quantity that is larger when the probability distribution is closer to the uniform distribution. When $s < 0$, the quantity $\psi(s)$ is a negative quantity that is smaller when the probability distribution is closer to the uniform distribution. Finally, when $s = 0$, the quantity $\psi(s)$ is equal to 0. The derivative $\psi'(0)$ of $\psi(s)$ at $s = 0$ is equal to $H(p)$.

Hence, Rényi entropy $H_{1-s}(p)$ is always positive, and the limit $\lim_{s\to 0} H_{1-s}(p)$ equals $H(p)$. Further, since $\psi(s)$ is convex, Rényi entropy $H_{1-s}(p)$ is monotone increasing for s. In particular, Rényi entropy $H_{1-s}(p_{\mathrm{mix},k})$ is equal to $\log k$. Hence, Rényi entropy $H_{1-s}(p)$ expresses the amount of the uncertainty of the distribution of p. We also define the **minimum entropy** $H_{\min}(p)$ and the **maximum entropy** $H_{\max}(p)$ as

$$H_{\min}(p) \stackrel{\mathrm{def}}{=} -\log \max_i p_i, \quad H_{\max}(p) \stackrel{\mathrm{def}}{=} \log |\{i | p_i > 0\}|. \tag{2.39}$$

Then, we obtain

$$\lim_{s\to -\infty} H_{1-s}(p) = H_{\min}(p), \quad \lim_{s\to 1} H_{1-s}(p) = H_{\max}(p). \tag{2.40}$$

These give the minimum and the maximum of Rényi entropies $H_{1-s}(p)$.

Now consider n data $i_1, \dots, i_n$ that are generated independently with the same probability distribution $p = \{p_i\}_{i=1}^k$. The probability of obtaining a particular data sequence $i^n = (i_1, \dots, i_n)$ is given by $p_{i_1} \cdot \dots \cdot p_{i_n}$. This probability distribution is called an n-fold **independent and identical distribution** (abbreviated as n-i.i.d.) and denoted by p^n. Then, we have $\psi(s|p^n) = n\psi(s|p)$, i.e., $H_{1-s}(p^n) = nH_{1-s}(p)$[Exe. 2.22]. When a sufficiently large number n of data are generated according to the independent and identical condition, the behavior of the distribution may be characterized by the entropy and the Rényi entropy.

The probability of the likelihood being less than $a \geq 0$ under the probability distribution p, i.e., the probability that $\{p_i \leq a\}$, is

$$p\{p_i \leq a\} = \sum_{i:p_i \leq a} p_i \leq \sum_{i:1 \leq \frac{a}{p_i}} \left(\frac{a}{p_i}\right)^s p_i \leq \sum_{i=1}^k p_i^{1-s} a^s = e^{\psi(s)+s \log a} \tag{2.41}$$

if $0 \leq s \leq 1$. Accordingly,

$$p^n\{p_{i^n}^n \leq e^{-nR}\} \leq e^{n \min_{0 \leq s \leq 1}(\psi(s)-sR)}. \tag{2.42}$$

Conversely, the probability of the likelihood being greater than a, i.e., the probability that $\{p_i > a\}$, is

$$p\{p_i > a\} \leq \sum_{i:1 > \frac{a}{p_i}} \left(\frac{a}{p_i}\right)^s p_i \leq \sum_{i=1}^k p_i^{1-s} a^s = e^{\psi(s)+s \log a} \tag{2.43}$$

if $s \leq 0$. Similarly, we obtain

$$p^n\{p_{i^n}^n > e^{-nR}\} \leq e^{n \min_{s \leq 0}(\psi(s)-sR)}. \tag{2.44}$$

The exponential decreasing rate (exponent) on the right-hand side (RHS) of (2.42) is negative when $R > H(p)$. Hence, the probability $p^n\{p_{i^n}^n \leq e^{-nR}\}$ approaches 0 exponentially. This fact can be shown as follows. Choosing a small $s_1 > 0$, we have $H_{1-s_1}(p) - R < 0$. Hence, we have

$$\min_{0 \leq s \leq 1}(\psi(s) - sR) = \min_{0 \leq s \leq 1} s(H_{1-s}(p) - R) \leq s_1(H_{1-s_1}(p) - R) < 0. \tag{2.45}$$

Hence, we see that the exponent on the RHS of (2.42) is negative. Conversely, the exponent on the RHS of (2.44) is negative when $R < H(p)$, and the probability $p^n\{p_{i^n}^n \leq e^{-nR}\}$ approaches 0 exponentially. This can be verified from (2.45) by choosing $s_2 < 0$ with a sufficiently small absolute value.

We may generalize this argument for the likelihood $q_{i^n}^n$ of a different probability distribution q as follows. Defining $\tilde{\psi}(s) \stackrel{\text{def}}{=} \log \sum_i p_i q_i^{-s}$, we can show that

$$p^n\{q^n_{i^n} \le e^{-nR}\} \le e^{n \min_{0\le s}(\tilde{\psi}(s)-sR)}, \tag{2.46}$$

$$p^n\{q^n_{i^n} > e^{-nR}\} \le e^{n \min_{s\le 0}(\tilde{\psi}(s)-sR)}. \tag{2.47}$$

The Rényi entropy $H_{1-s}(p)$ and the entropy $H(p)$ express the concentration of probability under independent and identical distributions with a sufficiently large number of data. To investigate the concentration, let us consider the probability $P(p, L)$ of the most frequent L outcomes for a given probability distribution $p = (p_i)$.[5] This can be written as

$$P(p, L) = \sum_{i=1}^{L} p_i^{\downarrow}, \tag{2.48}$$

where $p_i^{\downarrow}$ are the elements of p_i that are reordered according to size. Let us analyze this by reexamining the set $\{p_i > a\}$. The number of elements of the set $|\{p_i > a\}|$ is evaluated as

$$|\{p_i > a\}| \le \sum_{i:p_i>a} \left(\frac{p_i}{a}\right)^{1-s} \le \sum_{i=1}^{k} p_i^{1-s} a^{-1+s} = e^{\psi(s)-(1-s)\log a} \tag{2.49}$$

when $0 < s < 1$. By using (2.41) and defining $b(s, R) \stackrel{\text{def}}{=} \frac{\psi(s)-R}{1-s}$ for R and $0 \le s < 1$, we have

$$|\{p_i > e^{b(s,R)}\}| \le e^R, \quad p\{p_i \le e^{b(s,R)}\} \le e^{\frac{\psi(s)-sR}{1-s}}.$$

We choose $s_0 \stackrel{\text{def}}{=} \operatorname{argmin}_{0\le s\le 1} \frac{\psi(s)-sR}{1-s}$[6] and define $P^c(p, e^R) \stackrel{\text{def}}{=} 1 - P(p, e^R)$; hence,

$$P^c(p, e^R) \le e^{\frac{\psi(s_0)-s_0 R}{1-s_0}} = e^{\min_{0\le s\le 1} \frac{\psi(s)-sR}{1-s}}. \tag{2.50}$$

Applying this argument to the n-i.i.d p^n, we have

$$P^c(p^n, e^{nR}) \le e^{n\frac{\psi(s_0)-s_0 R}{1-s_0}} = e^{n \min_{0\le s\le 1} \frac{\psi(s)-sR}{1-s}}. \tag{2.51}$$

Now, we let $R > H(p)$ and choose a sufficiently small number $0 < s_1 < 1$. Then, inequality (2.45) yields

$$\min_{0\le s<1} \frac{\psi(s)-sR}{1-s} = \min_{0\le s<1} \frac{s(H_{1-s}(p)-R)}{1-s} \le \frac{s_1(H_{1-s_1}(p)-R)}{1-s_1} < 0.$$

[5] If L is not an integer, we consider the largest integer that does not exceed L.

[6] $\operatorname{argmin}_{0\le s\le 1} f(s)$ returns the value of s that yields $\min_{0\le s\le 1} f(s)$. argmax is similarly defined.

Hence, the probability $P^c(p^n, e^{nR})$ approaches 0 exponentially. That implies that the probabilities are almost concentrated on the most frequent e^{nR} elements because $1 - P^c(p^n, e^{nR})$ equals the probability on the most frequent e^{nR} elements. Since this holds when $R > H(p)$, most of the probabilities are concentrated on $e^{nH(p)}$ elements. Therefore, this can be interpreted as meaning that the entropy $H(p)$ asymptotically expresses the degree of concentration. This will play an important role in problems such as source coding, which will be discussed later.

On the other hand, when $H(p) > R$, $P(p^n, e^{nR})$ approaches 0. To prove this, let us consider the following inequality for an arbitrary subset A:

$$pA \le a|A| + p\{p_i > a\}. \tag{2.52}$$

We can prove this inequality by considering the set $A = (A \cap \{p_i \le a\}) \cup (A \cap \{p_i > a\})$. Defining $R \stackrel{\text{def}}{=} \log |A|$ and $a \stackrel{\text{def}}{=} e^{b(s,R)}$ and using (2.43), we obtain $pA \le 2e^{\frac{\psi(s)-sR}{1-s}}$. Therefore,

$$P(p, e^R) \le 2e^{\min_{s\le 0} \frac{\psi(s)-sR}{1-s}}, \tag{2.53}$$

and we obtain

$$P(p^n, e^{nR}) \le 2e^{n \min_{s\le 0} \frac{\psi(s)-sR}{1-s}}. \tag{2.54}$$

We also note that in order to avoid $P(p^n, e^{nR}) \to 0$, we require $R \ge H(p)$ according to the condition $\min_{s\le 0} \frac{\psi(s)-sR}{1-s} < 0$.

Exercises

2.22 Show that $\psi(s|p_A \times p_B) = \psi(s|p_A) + \psi(s|p_B)$.

2.23 Define the distribution $p_s(x) := p(x)^{1-s}e^{-\psi(s)}$ and assume that a distribution q satisfies $H(q) = H(p_s)$. Show that $D(p_s\|p) \le D(q\|p)$ for $s \le 1$ by following steps below.
(**a**) Show that $\frac{1}{1-s}D(q\|p_s) = \frac{1}{1-s}\sum_x q(x)\log q(x) - \sum_x q(x)\log p(x) + \frac{\psi(s)}{1-s}$.
(**b**) Show $D(q\|p) - \frac{1}{1-s}D(q\|p_s) = D(p_s\|p)$.
(**c**) Show the desired inequality.

2.24 Show the equation

$$\sup_{0\le s\le 1} \frac{sR - \psi(s)}{1-s} = \min_{q:H(q)\ge R} D(q\|p) \tag{2.55}$$

following the steps below.
(**a**) Show that $\frac{sR-\psi(s)}{1-s} \le 0$ for $R \le H(p)$ and $s \in [0, 1]$.
(**b**) Show that both side of (2.55) are zero when $R \le H(p)$.
(**c**) Show that

$$H(p_s) = (1-s)\psi'(s) + \psi(s), \tag{2.56}$$
$$D(p_s\|p) = s\psi'(s) - \psi(s). \tag{2.57}$$

(**d**) Show that

$$\frac{d}{ds}(1-s)\psi'(s) + \psi(s) = (1-s)\psi''(s) < 0, \tag{2.58}$$
$$\frac{d}{ds}s\psi'(s) - \psi(s) = s\phi''(s) > 0 \tag{2.59}$$

for $s \in (0, 1)$.
(**e**) In the following, we consider the case $R > H(p)$. Show that there uniquely exists $s_R \in (0, 1)$ such that $H(p_{s_R}) = R$.
(**f**) Show that

$$\min_{q:H(q)=R} D(q\|p) = D(p_{s_R}\|p). \tag{2.60}$$

(**g**) Show that

$$\min_{q:H(q)\geq R} D(q\|p) = D(p_{s_R}\|p). \tag{2.61}$$

(**h**) Show that

$$D(p_{s_R}\|p) = \frac{s_R R - \psi(s_R)}{1-s_R}. \tag{2.62}$$

(**i**) Show that

$$\frac{d}{ds}\frac{sR - \psi(s)}{1-s} = \frac{R + (s-1)\psi'(s) - \psi(s)}{(1-s)^2}. \tag{2.63}$$

(**j**) Show that

$$\sup_{0\leq s\leq 1} \frac{sR - \psi(s)}{1-s} = \frac{s_R R - \psi(s_R)}{1-s_R}. \tag{2.64}$$

(**k**) Show (2.55).

2.25 Show that

$$\sup_{s\leq 0}\frac{sR - \psi(s|p)}{1-s} = \min_{q:H(q)\leq R} D(q\|p). \tag{2.65}$$

(**a**) Show that there uniquely exists $s_R \leq 0$ such that $H(p_{s_R}) = R$.
(**b**) Show that

$$\min_{q:H(q)=r} D(q\|p) = D(p_{s_R}\|p). \tag{2.66}$$

(**c**) Show that

$$\min_{q:H(q)\le R} D(q\|p) = D(p_{s_R}\|p). \tag{2.67}$$

(**d**) Show that

$$D(p_{s_R}\|p) = \frac{s_R R - \psi(s_R)}{1-s_R}. \tag{2.68}$$

(**e**) Show that

$$\sup_{s\le 0}\frac{sR-\psi(s)}{1-s} = \frac{s_R R - \psi(s_R)}{1-s_R}. \tag{2.69}$$

(**f**) Show (2.65).

2.26 Assume that $R \le H_{\min}(p)$. Show that

$$\sup_{s\le 0}\frac{-\psi(s)}{1-s} = \min_{q:H(q)=0} D(q\|p) = H_{\min}(p) \tag{2.70}$$

2.27 Show that

$$-\log\max_i p_i \le H_\alpha(p) \le -\log\min_i p_i \tag{2.71}$$

for $\alpha \ge 0$.

2.1.5 Conditional Rényi Entropy

Next, we consider the conditional extension of Rényi entropy. For this purpose, we focus on the following relation between the conditional entropy and the relative entropy. For a given joint distribution P_{XY} on $\mathcal{X}\times\mathcal{Y}$, we have two characterization for the conditional entropy[Exe. 2.28]

$$H(X|Y) = \log|\mathcal{X}| - D(P_{XY}\|p_{\mathrm{mix},\mathcal{X}}\times P_Y) \tag{2.72}$$

$$H(X|Y) = \log|\mathcal{X}| - \min_{Q_Y} D(P_{XY}\|p_{\mathrm{mix},\mathcal{X}}\times Q_Y). \tag{2.73}$$

Based on the above relations, we define two kinds of **conditional Rényi entropies** for $s\in(-1,\infty)\setminus\{0\}$ as follows.

$$H_{1+s}(X|Y) \stackrel{\text{def}}{=} \log |\mathcal{X}| - D_{1+s}(\mathrm{P}_{XY} \| p_{\text{mix},\mathcal{X}} \times \mathrm{P}_Y)$$
$$= -\frac{1}{s} \log \sum_y \mathrm{P}_Y(y) \sum_x \mathrm{P}_{X|Y=y}(x)^{1+s} \tag{2.74}$$

$$H^{\uparrow}_{1+s}(X|Y) \stackrel{\text{def}}{=} \log |\mathcal{X}| - \min_{Q_Y} D_{1+s}(\mathrm{P}_{XY} \| p_{\text{mix},\mathcal{X}} \times Q_Y),$$
$$= \max_{Q_Y} -\frac{1}{s} \log \sum_{x,y} \mathrm{P}_{X,Y}(x,y)^{1+s} Q_Y(y)^{-s} \tag{2.75}$$

where Q_Y is an arbitrary distribution on $\mathcal{Y}$. In the case of $s = 0$, they are defined as $H(X|Y)$ because[Exe. 2.29]

$$\lim_{s \to 0} H_{1+s}(X|Y) = \lim_{s \to 0} H^{\uparrow}_{1+s}(X|Y) = H(X|Y). \tag{2.76}$$

According to the relations (2.40), **conditional minimum entropies** $H_{\min}(X|Y)$ and $H^{\uparrow}_{\min}(X|Y)$ and **conditional maximum entropies** $H_{\max}(X|Y)$ and $H^{\uparrow}_{\max}(X|Y)$ are defined as

$$H_{\min}(X|Y) \stackrel{\text{def}}{=} \lim_{s \to \infty} H_{1+s}(X|Y), \quad H^{\uparrow}_{\min}(X|Y) \stackrel{\text{def}}{=} \lim_{s \to \infty} H^{\uparrow}_{1+s}(X|Y), \tag{2.77}$$

$$H_{\max}(X|Y) \stackrel{\text{def}}{=} \lim_{s \to -1} H_{1+s}(X|Y), \quad H^{\uparrow}_{\max}(X|Y) \stackrel{\text{def}}{=} \lim_{s \to -1} H^{\uparrow}_{1+s}(X|Y). \tag{2.78}$$

From the definition, we find the relation

$$H_{1+s}(X|Y) \leq H^{\uparrow}_{1+s}(X|Y). \tag{2.79}$$

Unfortunately, these two conditional Rényi entropies are not the same in general. Thanks to the property of the relative Rényi entropy, we have the following lemma[Exe. 2.31].

Lemma 2.1 *The functions* $s \mapsto sH_{1+s}(X|Y)$ *and* $sH^{\uparrow}_{1+s}(X|Y)$ *are concave for* $s \in (-1, \infty)$. *The functions* $s \mapsto H_{1+s}(X|Y)$ *and* $H^{\uparrow}_{1+s}(X|Y)$ *are monotonically decreasing.*

Lemma 2.2 *The quantity* $H^{\uparrow}_{1+s}(X|Y)$ *has the following form.*

$$H^{\uparrow}_{1+s}(X|Y) = \log |\mathcal{X}| - D_{1+s}\left(\mathrm{P}_{XY} \| p_{\text{mix},\mathcal{X}} \times \mathrm{P}_Y^{(1+s)}\right) \tag{2.80}$$

$$= -\frac{1+s}{s} \log \sum_y \mathrm{P}_Y(y) \left(\sum_x \mathrm{P}_{X|Y}(x|y)^{1+s}\right)^{\frac{1}{1+s}} \tag{2.81}$$

$$= -\frac{1+s}{s} \log \sum_y \left(\sum_x \mathrm{P}_{XY}(x,y)^{1+s}\right)^{\frac{1}{1+s}}, \tag{2.82}$$

where $P_Y^{(1+s)}(y) := \frac{(\sum_x P_{XY}(x,y)^{1+s})^{\frac{1}{1+s}}}{\sum_{y'}(\sum_x P_{XY}(x,y')^{1+s})^{\frac{1}{1+s}}}$.

Proof Substituting $\sum_x P_{XY}(x,y)^{1+s}$ and $Q_Y(y)^{-s}$ to f and g in the reverse Hölder inequality (A.27) with $p = \frac{1}{1+s}$ and $q = -\frac{1}{s}$, we obtain

$$
\begin{aligned}
& e^{-s(\log|\mathcal{X}| - D_{1+s}(P_{XY}\|p_{\text{mix},\mathcal{X}} \times Q_Y))} \\
= & \sum_y \sum_x P_{XY}(x,y)^{1+s} Q_Y(y)^{-s} \\
\ge & \left(\sum_y \left(\sum_x P_{XY}(x,y)^{1+s} \right)^{1/(1+s)} \right)^{1+s} \left(\sum_y Q_Y(y)^{-s\cdot -1/s} \right)^{-s} \\
= & \left(\sum_y \left(\sum_x P_{XY}(x,y)^{1+s} \right)^{\frac{1}{1+s}} \right)^{1+s}
\end{aligned}
$$

for $s \in (0,\infty]$. Since the equality holds when $Q_Y(y) = P_Y^{(1+s)}(y)$, we obtain

$$
e^{-sH^{\uparrow}_{1+s}(X|Y)} = \left(\sum_y \left(\sum_x P_{XY}(x,y)^{1+s} \right)^{\frac{1}{1+s}} \right)^{1+s},
$$

which implies (2.81) with $s \in (0,\infty]$.

The same substitution to the Hölder inequality (A.25) yields

$$
e^{-s(\log|\mathcal{X}| - D_{1+s}(P_{XY}\|p_{\text{mix},\mathcal{X}} \times Q_Y))} \le \left(\sum_y \left(\sum_x P_{XY}(x,y)^{1+s} \right)^{\frac{1}{1+s}} \right)^{1+s}
$$

for $s \in (-1,0)$. Since the equality holds when $Q_Y(y) = P_Y^{(1+s)}(y)$, we obtain (2.81) with $s \in (-1,0)$.

Finally, (2.82) follows from a simple calculation. ■

Taking the limits $s \to -1$ and $s \to \infty$ in Lemma 2.2, we obtain the following lemma[Exe. 2.30].

Lemma 2.3 *The quantities* $H_{\min}(X|Y)$, $H^{\uparrow}_{\min}(X|Y)$, $H_{\max}(X|Y)$, *and* $H^{\uparrow}_{\max}(X|Y)$ *are characterized as*

$$
H_{\min}(X|Y) = -\log \max_{x,y:P_Y(y)>0} P_{X|Y=y}(x), \tag{2.83}
$$

$$
H^{\uparrow}_{\min}(X|Y) = -\log \sum_y P_Y(y) \max_x P_{X|Y=y}(x), \tag{2.84}
$$

$$H_{\max}(X|Y) = -\log \sum_y P_Y(y)|\{x|P_{X|Y=y}(x) > 0\}|, \tag{2.85}$$

$$H^{\uparrow}_{\max}(X|Y) = -\log \max_{y:P_Y(y)>0} |\{x|P_{X|Y=y}(x) > 0\}|. \tag{2.86}$$

Further, as an inequality opposite to (2.79), we have

Lemma 2.4 ([5, Lemma 5]) *For $s \in (-1, 1)\setminus\{0\}$, we have*

$$H_{1+s}(X|Y) \geq H^{\uparrow}_{\frac{1}{1-s}}(X|Y). \tag{2.87}$$

Proof Next, we consider the case with $s \in (0, 1)$. Substituting $P_{XY}(x, y)$ and $(\frac{P_{XY}(x,y)}{P_Y(y)})^s$ to f and g in the Hölder inequality (A.25) with $p = \frac{1}{1-s}$ and $q = \frac{1}{s}$, we obtain

$$\begin{aligned} e^{-sH_{1+s}(X|Y)} &= \sum_y \sum_x P_{XY}(x, y) \left(\frac{P_{XY}(x, y)}{P_Y(y)}\right)^s \\ &\leq \sum_y \left(\sum_x P_{XY}(x, y)^{1/(1-s)}\right)^{1-s} \left(\sum_{x'} \frac{P_{XY}(x', y)}{P_Y(y)}\right)^s \\ &= \sum_y \left(\sum_x P_{XY}(x, y)^{1/(1-s)}\right)^{1-s} = e^{-sH^{\uparrow}_{\frac{1}{1-s}}(X|Y)} \end{aligned} \tag{2.88}$$

for $s \in (0, 1)$ because $\sum_x \frac{P_{XY}(x,y)}{P_Y(y)} = \frac{P_Y(y)}{P_Y(y)} = 1$.

Next, we consider the case with $s \in (-1, 0)$. The same substitution to the reverse Hölder inequality (A.27) with $p = 1/(1-s)$ and $q = \frac{1}{s}$ yields

$$e^{-sH_{1+s}(X|Y)} \geq e^{-sH^{\uparrow}_{\frac{1}{1-s}}(X|Y)}$$

because $(\sum_x \frac{P_{X,Y}(x,y)}{P_Y(y)})^s = (\frac{P_Y(y)}{P_Y(y)})^s = 1$. ■

Now, we consider the meaning of two kinds of conditional Rényi entropies. For this purpose, we discuss the case when $P_{X^nY^n}$ is the independent and identical distribution of P_{XY}. Applying (2.42) and (2.44) to the distribution $P_{X^n|Y^n=y}$ and taking the average with respect to y under the distribution P_{Y^n}, we have

$$P_{X^nY^n}\{(x, y)|P_{X^n|Y^n}(x|y) \leq e^{-nR}\} \leq e^{n \min_{-1\leq s\leq 0} s(R-H_{1+s}(X|Y))} \tag{2.89}$$

$$P_{X^nY^n}\{(x, y)|P_{X^n|Y^n}(x|y) > e^{-nR}\} \leq e^{n \min_{s\geq 0} s(R-H_{1+s}(X|Y))}, \tag{2.90}$$

which gives an operational meaning of the conditional Rényi entropy $H_{1+s}(X|Y)$. Similarly, applying (2.50) and (2.53) to the distribution $P_{X^n|Y^n=y}$ and taking the average with respect to y under the distribution P_{Y^n}, we have

$$\sum_y \mathrm{P}_{Y^n}(y) P(\mathrm{P}_{X^n|Y^n=y}, e^{nR}) \le e^{n \min_{-1\le s\le 0} \frac{s}{1+s}(R - H^{\uparrow}_{1+s}(X|Y))} \tag{2.91}$$

$$\sum_y \mathrm{P}_{Y^n}(y) P^c(\mathrm{P}_{X^n|Y^n=y}, e^{nR}) \le e^{n \min_{s\ge 0} \frac{s}{1+s}(R - H^{\uparrow}_{1+s}(X|Y))}, \tag{2.92}$$

which gives an operational meaning of the conditional Rényi entropy $H^{\uparrow}_{1+s}(X|Y)$. These inequalities clarify the difference between two kinds of conditional Rényi entropies.

Exercises

2.28 Show (2.72) and (2.73).

2.29 Show (2.76).

2.30 Show Lemma 2.3.

2.31 Show Lemma 2.1.

2.32 Show that the equality in (2.87) holds for a real $s \in (-1, 1)\setminus\{0\}$ if and only if $\mathrm{P}_{XY}(x, y) = \frac{1}{|\mathcal{X}|}\mathrm{P}_Y(y)$.

2.2 Geometry of Probability Distribution Family

2.2.1 Inner Product for Random Variables and Fisher Information

In Sect. 2.1, we introduced the mutual information $I(X : Y)$ as a quantity that expresses the correlation between two random variables X and Y. However, for calculating this quantity, one must calculate the logarithm of each probability, which is a rather tedious calculation amount. We now introduce the **covariance** $\mathrm{Cov}_p(X, Y)$ as a quantity that expresses the correlation between two real-valued random variables X and Y. Generally, calculations involving the covariance are less tedious than those of mutual information. Given a probability distribution p in a probability space Ω, the covariance is defined as

$$\mathrm{Cov}_p(X, Y) \stackrel{\mathrm{def}}{=} \sum_{\omega\in\Omega} (X(\omega) - \mathrm{E}_p(X))(Y(\omega) - \mathrm{E}_p(Y))p(\omega). \tag{2.93}$$

If X and Y are independent, the covariance $\mathrm{Cov}_p(X, Y)$ is equal to 0[Exe. 2.33]. Thus far it has not been necessary to specify the probability distribution, and therefore we had no difficulties in using notations such as $H(X)$ and $I(X : Y)$. However, since it is important to emphasize the probability distribution treated in our discussion, we will use the above notation without their abbreviation. If X and Y are the same, the covariance $\mathrm{Cov}_p(X, Y)$ coincides with the **variance** $\mathrm{V}_p(X)$ of X:

$$\mathrm{Cov}_p(X, Y) \stackrel{\text{def}}{=} \sum_{\omega \in \Omega} (X(\omega) - \mathrm{E}_p(X))^2 p(\omega). \tag{2.94}$$

Given real-valued random variables $X_1, \ldots, X_d$, the matrix $\mathrm{Cov}_p(X_k, X_j)$ is called a **covariance matrix**. Now, starting from a given probability distribution p, we define the inner product in the space of real-valued random variables as[7]

$$\langle A, B \rangle_p^{(e)} \stackrel{\text{def}}{=} \sum_{\omega} A(\omega) B(\omega) p(\omega) . \tag{2.95}$$

Then, the covariance $\mathrm{Cov}_p(X, Y)$ is equal to the above inner product between the two real-valued random variables $(X(\omega) - \mathrm{E}_p(X))$ and $(Y(\omega) - \mathrm{E}_p(Y))$ with a zero expectation. That is, the inner product (2.95) implies the correlation between the two real-valued random variables with zero expectation in classical systems. This inner product is also deeply related to statistical inference in another sense, as discussed below.

When we observe n independent real-valued random variables $X_1, \ldots, X_n$ identical to real-valued random variable X, the average value

$$\boldsymbol{X}^n \stackrel{\text{def}}{=} \frac{X_1 + \ldots + X_n}{n} \tag{2.96}$$

converges to the expectation $\mathrm{E}_p(X)$ in probability. That is,

$$p^n\{|\boldsymbol{X}^n - \mathrm{E}_p(X)| > \epsilon\} \to 0, \quad \forall \epsilon > 0 , \tag{2.97}$$

which is called the **law of large numbers**. Further, the distribution of the real-valued random variable

$$\sqrt{n}(\boldsymbol{X}^n - \mathrm{E}_p(X)) \tag{2.98}$$

goes to the Gaussian distribution with the variance $V = \mathrm{V}_p(X)$:

$$P_{G,V}(x) = \frac{1}{\sqrt{2\pi V}} e^{-\frac{x^2}{2V}} , \tag{2.99}$$

i.e.,

$$p^n\{a \le \sqrt{n}(\boldsymbol{X}^n - \mathrm{E}_p(X)) \le b\} \to \int_a^b P_{G,V}(x) dx , \tag{2.100}$$

[7]The superscript (e) means "exponential." This is because A corresponds to the exponential representation, as discussed later.

which is called the **central limit theorem**. Hence, the asymptotic behavior is almost characterized by the expectation $\mathrm{E}(X)$ and the variance $\mathrm{V}(X)$.

For l real-valued random variables $X_1, \ldots, X_l$, we can similarly define the real-valued random variables $\boldsymbol{X}_1^n, \ldots, \boldsymbol{X}_l^n$. These converge to their expectation in probability. The distribution of the real-valued random variables

$$(\sqrt{n}(\boldsymbol{X}_1^n - \mathrm{E}_p(X)), \ldots, \sqrt{n}(\boldsymbol{X}_k^n - \mathrm{E}_p(X))) \tag{2.101}$$

converges the k-multirate Gaussian distribution and the covariance matrix $\boldsymbol{V} = \mathrm{Cov}_p(X_k, X_j)$:

$$P_{G,\mathrm{V}}(x) \stackrel{\text{def}}{=} \frac{1}{\sqrt{(2\pi)^l \det \mathrm{V}}} e^{\langle x|\mathrm{V}^{-1}|x\rangle} . \tag{2.102}$$

Therefore, the asymptotic behavior is almost described by the expectation and the covariance matrix.

Consider the set of probability distributions p_θ parameterized by a single real number θ. For example, we can parameterize a binomial distribution with the probability space $\{0, 1\}$ by $p_\theta(0) = \theta$, $p_\theta(1) = 1 - \theta$. When the set of probability distributions is parameterized by a single parameter, it is called a **probability distribution family** and is represented by $\{p_\theta | \theta \in \Theta \subset \mathbb{R}\}$. Based on a probability distribution family, we can define the **logarithmic derivative** as $l_{\theta_0}(\omega) \stackrel{\text{def}}{=} \frac{d \log p_\theta(\omega)}{d\theta}|_{\theta=\theta_0} = \left.\frac{dp_\theta(\omega)}{d\theta}\right|_{\theta=\theta_0} / p_{\theta_0}(\omega)$. Since it is a real-valued function of the probability space, it can be regarded as a real-valued random variable. We can consider that this quantity expresses the sensitivity of the probability distribution to the variations in the parameter θ around θ_0. The **Fisher metric** (**Fisher information**) is defined as the variance of the logarithmic derivative l_{θ_0}. Since the expectation of l_{θ_0} with respect to p_{θ_0} is 0, the Fisher information can also be defined as

$$J_\theta \stackrel{\text{def}}{=} \langle l_\theta, l_\theta \rangle_{p_\theta}^{(e)} . \tag{2.103}$$

Therefore, this quantity represents the amount of variation in the probability distribution due to the variations in the parameter. Alternatively, it can indicate how much the probability distribution family represents the information related to the parameter. As discussed later, these ideas will be further refined from the viewpoint of statistical inference. The Fisher information J_θ may also be expressed as the limits of relative entropy and Hellinger distance[Exe. 2.35, 2.36]:

$$\frac{J_\theta}{2} = 4 \lim_{\epsilon \to 0} \frac{d_2^2(p_\theta, p_{\theta+\epsilon})}{\epsilon^2} \tag{2.104}$$

$$= \lim_{\epsilon \to 0} \frac{D(p_\theta \| p_{\theta+\epsilon})}{\epsilon^2} = \lim_{\epsilon \to 0} \frac{D(p_{\theta+\epsilon} \| p_\theta)}{\epsilon^2} . \tag{2.105}$$

The Fisher information J_θ is also characterized by the limit of relative Rényi entropy[Exe. 2.37]:

$$\frac{J_\theta}{2} = \lim_{\epsilon \to 0} \frac{-\phi(s|p_\theta \| p_{\theta+\epsilon})}{\epsilon^2 s(1-s)} . \tag{2.106}$$

Next, let us consider the probability distribution family $\{p_\theta | \theta \in \Theta \subset \mathbb{R}^d\}$ with multiple parameters. For each parameter, we define the logarithmic derivative $l_{\theta:k}(\omega)$ as

$$l_{\theta:k}(\omega) \stackrel{\text{def}}{=} \frac{\partial \log p_\theta(\omega)}{\partial \theta^k} = \frac{\partial p_\theta(\omega)}{\partial \theta^k} \Big/ p_\theta(\omega).$$

We use the covariance matrix $\langle l_{\theta:k}, l_{\theta:j} \rangle^{(e)}_{p_\theta}$ for the logarithmic derivatives $l_{\theta:1}, \ldots, l_{\theta:d}$ instead of the Fisher information. This matrix is called the **Fisher information matrix** and will be denoted by $\boldsymbol{J}_\theta = (J_{\theta:k,j})$. This matrix takes the role of the Fisher information when there are multiple parameters; we discuss this in greater detail below.

This inner product is closely related to the conditional expectation as follows. Suppose that we observe only the subsystem Ω_1, although the total system is given as $\Omega_1 \times \Omega_2$. Let us consider the real-valued random variable X of the total system. We denote the random variable describing the outcome in the probability space Ω_j by Z_j for $j = 1, 2$. Then, dependently of the distribution p of the total system, the **conditional expectation** $\kappa_p(X)$ of X is defined as a function of $\omega_1 \in \Omega_1$ by

$$\kappa_p(X)(\omega_1) := \sum_{\omega_2 \in \Omega_2} p(Z_2 = \omega_2 | Z_1 = \omega_1) X(\omega_1, \omega_2). \tag{2.107}$$

Then, we define the inclusion map i from the set of real-valued random variables on Ω_1 to the set of real-valued random variables on $\Omega_1 \times \Omega_2$. That is, for a random variable Y on Ω_1, the real-valued random variable $i(Y)$ on $\Omega_1 \times \Omega_2$ is defined as

$$i(Y)(\omega_1, \omega_2) = Y(\omega_1), \quad \forall(\omega_1, \omega_2) \in \Omega_1 \times \Omega_2 . \tag{2.108}$$

To see the relation with the above defined inner product, we focus on an arbitrary real-valued random variable Y on Ω_1, which given as a function of Z_1. Then, the **conditional expectation** $\kappa_p(X)$ of X satisfies

$$\begin{aligned} \langle Y, \kappa_p(X) \rangle^{(e)}_p &= \sum_{\omega_1} p(Z_1 = \omega_1) Y(\omega_1) \sum_{\omega_2 \in \Omega_2} p(Z_2 = \omega_2 | Z_1 = \omega_1) X(\omega_1, \omega_2) \\ &= \sum_{\omega_1, \omega_2} Y(\omega_1) X(\omega_1, \omega_2) p(Z_1 = \omega_1, Z_2 = \omega_2) = \langle i_p(Y), X \rangle^{(e)}_p . \end{aligned} \tag{2.109}$$

In fact, when a real-valued random variable $\kappa_p(X)$ satisfies the condition (2.109) for an arbitrary real-valued random variable Y on Ω_1, it is uniquely determined because the condition (2.109) guarantees that $\kappa_p(X)$ is the image of X for the dual map of i with respect to the inner product $\langle Y, X\rangle_p^{(e)}$. That is, when the linear space of random variables on Ω_1 is regarded as a subspace of the linear space of random variables on $\Omega_1 \times \Omega_2$ via the inclusion map i, the map $\kappa_p(X)$ is the projection from the linear space of random variables on $\Omega_1 \times \Omega_2$ to the sub linear space of random variables on Ω_1. So, we can regard the condition (2.109) as another definition of the conditional expectation $\kappa_p(X)$ of X. That is, the conditional expectation $\kappa_p(X)$ of X is the real-valued random variable describing the behavior of the random variable X of the total system $\Omega_1 \times \Omega_2$ in the subsystem Ω_1.

Generally, when we focus on a subspace $\mathfrak{U}$ of real-valued random variables for an arbitrary random variable X, we can define the **conditional expectation** $\kappa_{\mathfrak{U},p}(X) \in \mathfrak{U}$ as

$$\langle Y, \kappa_{\mathfrak{U},p}(X)\rangle_p^{(e)} = \langle Y, X\rangle_p^{(e)}, \quad \forall Y \in \mathfrak{U}. \tag{2.110}$$

This implies that the map $\kappa_{\mathfrak{U},p}(\,)$ is the projection from the space of all real-valued random variables to the subspace $\mathfrak{U}$ with respect to the inner product $\langle , \rangle_p$.

Exercises

2.33 Show that $\mathrm{Cov}_p(X, Y) = 0$ for real-valued random variables X and Y if they are independent.

2.34 Let J_θ be the Fisher information of a probability distribution family $\{p_\theta|\theta \in \Theta\}$. Let p_θ^n be the n-fold independent and identical distribution of p_θ. Show that the Fisher information of the probability distribution family $\{p_\theta^n|\theta \in \Theta\}$ at p_θ is nJ_θ.

2.35 Prove (2.104) using the second equality in (2.17), and noting that $\sqrt{1+x} \cong 1 + \frac{1}{2}x - \frac{1}{8}x^2$ for small x.

2.36 Prove (2.105) following the steps below.
(a) Show the following approximation with the limit $\epsilon \to 0$.

$$\log p_{\theta+\epsilon}(\omega) - \log p_\theta(\omega) \cong \frac{d \log p_\theta(\omega)}{d\theta}\epsilon + \frac{1}{2}\frac{d^2 \log p_\theta(\omega)}{d^2\theta}\epsilon^2.$$

(b) Prove the first equality in (2.105) using **(a)**.
(c) Show the following approximation with the limit $\epsilon \to 0$.

$$p_{\theta+\epsilon}(\omega) \cong p_\theta(\omega) + \frac{dp_\theta(\omega)}{d\theta}\epsilon + \frac{1}{2}\frac{d^2 p_\theta(\omega)}{d^2\theta}\epsilon^2.$$

(d) Prove the second equality in (2.105) using **(a)** and **(c)**.

2.37 Prove (2.106) using the approximation $(1+x)^s \cong 1 + sx + \frac{s(s-1)}{2}x^2$ for small x.

2.2.2 *Bregman Divergence*

To discuss divergence from a more general viewpoint, we formulate Bregman divergence based on a general strictly convex function $\mu(\theta)$ on $\mathbb{R}$. Assume that the strictly convex function $\mu(\theta)$ is twice-differentiable. Then, we define the **Bregman divergence (canonical divergence)** of $\mu(\theta)$ as

$$D^{\mu}(\bar{\theta}\|\theta) := \mu'(\bar{\theta})(\bar{\theta}-\theta) - \mu(\bar{\theta}) + \mu(\theta)$$
$$\stackrel{(a)}{=} \max_{\tilde{\theta}} \mu'(\bar{\theta})(\tilde{\theta}-\theta) - \mu(\tilde{\theta}) + \mu(\theta) = \int_{\theta}^{\bar{\theta}} \mu''(\tilde{\theta})(\tilde{\theta}-\theta)d\tilde{\theta}. \tag{2.111}$$

Here (a) can be derived as follows. Since the inside function of the maximum is concave for $\tilde{\theta}$, the maximum is realized when the derivative is zero, which implies that $\tilde{\theta} = \theta$. Hence, we obtain (a). In this case, the convex function $\mu(\theta)$ is called the **potential** of the Bregman divergence. Further, when $\bar{\theta} > \theta$, the above maximum is replaced by $\max_{\tilde{\theta}:\tilde{\theta}\geq\bar{\theta}}$.

Since the function μ is strictly convex, the correspondence $\theta \leftrightarrow \eta = \frac{d\mu}{d\theta}$ is one-to-one. Hence, the divergence $D^{\mu}(\bar{\theta}\|\theta)$ can be expressed with the parameter η. For this purpose, we define the **Legendre transform** ν of μ

$$\nu(\eta) \stackrel{\text{def}}{=} \max_{\tilde{\theta}} \eta\tilde{\theta} - \mu(\tilde{\theta}). \tag{2.112}$$

Then, the function ν is a convex function[Exe. 2.38], and we can recover the functions μ and θ as

$$\mu(\theta) = \max_{\tilde{\eta}} \theta\tilde{\eta} - \nu(\tilde{\eta}), \quad \theta = \frac{d\nu}{d\eta}.$$

Due to the inverse function theorem, the second derivative ($\frac{d^2\nu}{d\eta^2}$ of ν is calculated to $\frac{d\theta}{d\eta} = \frac{d\eta}{d\theta}^{-1} = \frac{d^2\mu}{d\theta^2}^{-1}$.

In particular, when $\eta = \frac{d\mu}{d\theta}(\theta)$,

$$\nu(\eta) = \theta\eta - \mu(\theta) = D^{\mu}(\theta\|0) - \mu(0), \tag{2.113}$$
$$\mu(\theta) = \theta\eta - \nu(\eta) = D^{\nu}(\eta\|0) - \nu(0). \tag{2.114}$$

Using these relations, we can obtain

$$D^{\mu}(\bar{\theta}\|\theta) = D^{\nu}(\eta\|\bar{\eta}) = \theta(\eta-\bar{\eta}) - \nu(\eta) + \nu(\bar{\eta}). \tag{2.115}$$

That is, the Bregman divergence of μ can be written by the Bregman divergence of the Legendre transform of μ.

Now, we extend Bregman to the multi-parametric case. Let $\mu(\theta)$ be a twice-differentiable and strictly convex function defined on a subset Θ of the d-dimensional real vector space $\mathbb{R}^d$. The Bregman divergence concerning the convex function μ is defined by

$$D^{\mu}(\bar{\theta}\|\theta) \stackrel{\text{def}}{=} \sum_k \eta_k(\bar{\theta})(\bar{\theta}^k - \theta^k) - \mu(\bar{\theta}) + \mu(\theta), \quad \eta_k(\theta) \stackrel{\text{def}}{=} \frac{\partial \mu}{\partial \theta^k}(\theta) . \tag{2.116}$$

This quantity has the following two characterizations:

$$D^{\mu}(\bar{\theta}\|\theta) = \max_{\tilde{\theta}} \sum_k \frac{\partial \mu}{\partial \theta^k}(\tilde{\theta})(\tilde{\theta}^k - \theta^k) - \mu(\tilde{\theta}) + \mu(\theta) \tag{2.117}$$

$$= \int_0^1 \sum_{k,j} (\bar{\theta}^k - \theta^k)(\bar{\theta}^j - \theta^j) \frac{\partial^2 \mu}{\partial \theta^k \partial \theta^j}(\theta + (\bar{\theta} - \theta)t) t dt . \tag{2.118}$$

Since the strict positivity of μ implies the strict positivity of inside of the above integral, $D^{\mu}(\bar{\theta}\|\theta)$ is strictly positive unless $\bar{\theta} = \theta$. The strict positivity of μ is also guarantees that the correspondence $\theta^k \leftrightarrow \eta_k = \frac{\partial \mu}{\partial \theta^k}$ is one-to-one. Hence, the Bregman divergence $D^{\mu}(\bar{\theta}\|\theta)$ can be expressed with the parameter η. For this purpose, we define the Legendre transform ν of μ

$$\nu(\eta) \stackrel{\text{def}}{=} \max_{\tilde{\theta}} \sum_k \eta_k \tilde{\theta}^k - \mu(\tilde{\theta}). \tag{2.119}$$

Then, the function ν is a convex function[Exe. 2.38], and we can recover the functions μ and θ as

$$\mu(\theta) = \max_{\tilde{\eta}} \sum_k \theta^k \tilde{\eta}_k - \nu(\tilde{\eta}), \quad \theta^k = \frac{\partial \nu}{\partial \eta_k} . \tag{2.120}$$

Due to the inverse function theorem, the second derivative matrix $(\frac{\partial^2 \nu}{\partial \eta_k \partial \eta_j})_{k,j}$ of ν is calculated to $(\frac{\partial \theta^k}{\partial \eta_j})_{k,j} = ((\frac{\partial \eta^k}{\partial \theta_j})_{k,j})^{-1} = ((\frac{\partial^2 \mu}{\partial \theta^k \partial \theta^j})_{k,j})^{-1}$, which is the inverse of the matrix $\frac{\partial^2 \mu}{\partial \theta^k \partial \theta^j}$.

In particular, when $\eta_k = \frac{\partial \mu}{\partial \theta^k}(\theta)$,

$$\nu(\eta) = \sum_k \eta_k \theta^k - \mu(\theta) = D^{\mu}(\theta\|0) - \mu(0), \tag{2.121}$$

$$\mu(\theta) = \sum_k \theta^k \eta_k - \nu(\eta) = D^{\nu}(\eta\|0) - \nu(0). \tag{2.122}$$

Using these relations, we can characterize the Bregman divergence concerning the convex function μ by the Bregman divergence concerning the convex function ν as

$$D^{\mu}(\bar{\theta}\|\theta) = D^{\nu}(\eta\|\bar{\eta}) = \sum_k \theta^k(\eta_k - \bar{\eta}_k) - \nu(\eta) + \nu(\bar{\eta}) \tag{2.123}$$

$$= \int_0^1 \sum_{k,j} (\eta_k(\theta) - \eta_k(\bar{\theta}))(\eta_j(\theta) - \eta_j(\bar{\theta})) \frac{\partial^2 \nu}{\partial \eta_k \partial \eta_j}(\eta(\bar{\theta}) + (\eta(\bar{\theta}) - \eta(\theta))t) t dt, \tag{2.124}$$

where (2.124) follows from (2.118) for the Bregman divergence with respect to ν.

A subset $\mathcal{E}$ of Θ is called an exponential subfamily of Θ when there exist an element $\theta' \in \Theta$ and l independent vectors $v_1, \ldots, v_l \in \mathbb{R}^d$ such that $\mathcal{E} = \{\theta \in \Theta | \theta = \theta' + \sum_{j=1}^k a^j v_j \exists (a^1, \ldots, a^l) \in \mathbb{R}^l\}$. A subset $\mathcal{M}$ of Θ is called a mixture subfamily of Θ when there exist a l-dimensional vector $(b_1, \ldots, b_l)$ and l independent vectors $v_1, \ldots, v_l \in \mathbb{R}^d$ such that $\mathcal{M} = \{\theta \in \Theta | b_k = \sum_{j=1}^d v_k^j \eta_j(\theta)\}$. In particular, the set of vectors $\{v_1, \ldots, v_l\}$ is called a generator of $\mathcal{E}$ and $\mathcal{M}$, respectively.

Now, we focus on two points $\theta' = (\theta'^1, \ldots, \theta'^d)$ and $\theta'' = (\theta''^1, \ldots, \theta''^d)$. We choose the exponential subfamily $\mathcal{E}$ of Θ whose natural parameters $\theta^{l+1}, \ldots, \theta^d$ are fixed to $\theta''^{l+1}, \ldots, \theta''^d$, and the mixture subfamily $\mathcal{M}$ of Θ whose expectation parameters $\eta^1, \ldots, \eta^l$ are fixed to $\eta(\theta')^1, \ldots, \eta(\theta')^l$. Let $\tilde{\theta} = (\tilde{\theta}^1, \ldots, \tilde{\theta}^d)$ be an element of the intersection of these two subfamily of Θ. That is, $\tilde{\theta}^j = \theta''^j$ for $j = l+1, \ldots, d$ and $\eta_j(\tilde{\theta}) = \eta_j(\theta')$ for $j = 1, \ldots, l$.

Then, since

$$\theta'^j - \theta''^j = \begin{cases} (\theta'^j - \tilde{\theta}^j) & \text{if } j \geq l+1 \\ (\theta'^j - \tilde{\theta}^j) + (\tilde{\theta}^j - \theta''^j) & \text{if } j \leq l, \end{cases} \tag{2.125}$$

the definition (2.116) implies that

$$\begin{aligned} D^{\mu}(\theta'\|\theta'') &= \sum_{j=1}^d (\theta'^j - \theta''^j)\eta_j(\theta') - \mu(\theta') + \mu(\theta'') \\ &= \sum_{j=1}^d (\theta'^j - \tilde{\theta}^j)\eta_j(\theta') - \mu(\theta') + \mu(\tilde{\theta}) + \sum_{j=1}^l (\tilde{\theta}^j - \theta''^j)\eta_j(\tilde{\theta}) - \mu(\tilde{\theta}) + \mu(\theta'') \\ &= D^{\mu}(\theta'\|\tilde{\theta}) + D^{\mu}(\tilde{\theta}\|\theta''). \end{aligned} \tag{2.126}$$

Using (2.126), we obtain the **Pythagorean theorem** [2] as follows.

Theorem 2.3 (Amari [6]) *Given an element $\theta \in \Theta$ and a mixture subfamily $\mathcal{M}$ of Θ with the generator $\{v_1, \ldots, v_l\}$, we define $\theta^* := \mathrm{argmin}_{\theta' \in \mathcal{M}} D^{\mu}(\theta'\|\theta)$. Then, we obtain the following two items as Fig. 2.1.*

(1) Any element $\theta' \in \mathcal{M}$ satisfies $D^{\mu}(\theta'\|\theta) = D^{\mu}(\theta'\|\theta^*) + D^{\mu}(\theta^*\|\theta)$.

(2) The element θ^* is the unique element of the intersection of the mixture subfamily $\mathcal{M}$ and the exponential subfamily $\mathcal{E}$ containing θ with the generator $\{v_1, \ldots, v_l\}$.

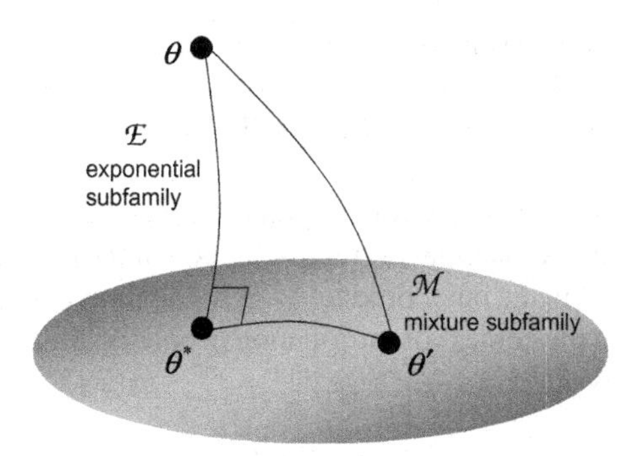

Fig. 2.1 Pythagorean theorem

Proof Choose an element $\tilde{\theta}$ in the intersection of the mixture subfamily $\mathcal{M}$ and the exponential subfamily $\mathcal{E}$ containing θ with the generator $\{v_1, \ldots, v_l\}$. Now, we choose additional vectors $\{v_{l+1}, \ldots, v_d\}$ such that the set $\{v_1, \ldots, v_d\}$ forms a basis. Then, we introduce another coordinate a^j such that $\sum_{j=1}^d a^j v_j^l = \theta^l$. Now, we apply the new coordinate a^j to the relation (2.126). Thus, any element $\theta' \in \mathcal{M}$ satisfies that $D^\mu(\theta'\|\theta) = D^\mu(\theta'\|\tilde{\theta}) + D^\mu(\tilde{\theta}\|\theta)$. Since $D^\mu(\theta'\|\tilde{\theta}) > 0$ except for $\theta' = \tilde{\theta}$, we have $\min_{\theta' \in \mathcal{M}} D^\mu(\theta'\|\theta) = D^\mu(\tilde{\theta}\|\theta)$, which implies that $\theta^* = \tilde{\theta}$, i.e., (2). Hence, we obtain (1). ■

We also have another version of the Pythagorean theorem as follows.

Theorem 2.4 (Amari [6]) *Given an element $\theta' \in \Theta$ and an exponential subfamily $\mathcal{E}$ of Θ with the generator $\{v_1, \ldots, v_l\}$, we define $\theta'_* := \operatorname{argmin}_{\theta \in \mathcal{E}} D^\mu(\theta'\|\theta)$.*

(1) Any element $\theta \in \mathcal{E}$ satisfies $D^\mu(\theta'\|\theta) = D^\mu(\theta'\|\theta'_*) + D^\mu(\theta'_*\|\theta)$.

(2) The element θ'_* is the unique element of the intersection of the exponential subfamily $\mathcal{E}$ and the mixture subfamily containing θ' with the generator $\{v_1, \ldots, v_l\}$.

Exercises

2.38 Show that $\nu(\eta)$ is a convex function.

2.39 Solve Exercise 2.23 by using Theorem 2.3.

2.2.3 Exponential Family and Divergence

In Sect. 2.1, relative entropy $D(p\|q)$ is defined. In this subsection, we characterize it as Bregman divergence.

Let $p(\omega)$ be a probability distribution and $X(\omega)$ be a real-valued random variable. When the family $\{p_\theta | \theta \in \Theta\}$ has the form

$$p_\theta(\omega) = p(\omega) e^{\theta X(\omega) - \mu(\theta)}, \tag{2.127}$$

$$\mu(\theta) \stackrel{\text{def}}{=} \log \sum_\omega p(\omega) e^{\theta X(\omega)}, \tag{2.128}$$

the logarithmic derivative at respective points equals the logarithmic derivative at a fixed point with the addition of a constant. In this case, the family, X, and $\mu(\theta)$ are called an **exponential family**, the generator, and the **cumulant generating function** of X, respectively. In particular, in an exponential family, the logarithmic derivative does not depend on the point θ except for constant differences. Hence, it is often called the exponential (e) representation of the derivative. Therefore, we use the superscript (e) in the inner product $\langle\, ,\, \rangle_p^{(e)}$. The function $\mu(\theta)$ is often called a **potential function** in the context of information geometry. Since the first derivative of $\mu(\theta)$ is calculated as $\mu'(\theta) = (\frac{d}{d\theta} e^{\mu(\theta)}) e^{-\mu(\theta)} = \sum_\omega p_\theta(\omega) X(\omega)$, the second derivative is as

$$\begin{aligned}\mu''(\theta) &= \left(\frac{d^2}{d\theta^2} e^{\mu(\theta)}\right) e^{-\mu(\theta)} - \left(\left(\frac{d}{d\theta} e^{\mu(\theta)}\right) e^{-\mu(\theta)}\right)^2 \\ &= \sum_\omega p_\theta(\omega) X(\omega)^2 - \left(\sum_\omega p_\theta(\omega) X(\omega)\right)^2 = J_\theta > 0,\end{aligned}$$

is the Fisher information. So, the cumulant generating function $\mu(\theta)$ is a strictly convex function. Therefore, the first derivative $\mu'(\theta) = \sum_\omega p_\theta(\omega) X(\omega)$ is monotone increasing. That is, we may regard it as another parameter identifying the distribution p_θ, and denote it by η. The original parameter θ is called a **natural parameter**, and the other parameter η is an **expectation parameter**. When the distribution is parametrized by the expectation parameter η, it is written as $\hat{p}_\eta$. Hence, we have $\hat{p}_{\eta(\theta)} = p_\theta$.

For example, in the one-trial binomial distribution, the generator X is given as $X(i) = i$, and the distribution p_0 is given as $p_0(i) = \frac{1}{2}$, for $i = 0, 1$. Then, the cumulant generating function μ is calculated to be $\mu(\theta) = \log \frac{1+e^\theta}{2}$. The distribution is written as $p_\theta(0) = 1/(1+e^\theta)$, $p_\theta(1) = e^\theta/(1+e^\theta)$ in the natural parameter θ. Hence, the binomial distribution is an exponential family. The expectation parameter is $\eta(\theta) = e^\theta/(1+e^\theta)$. That is, the distribution is written as $\hat{p}_\eta(1) = \eta$, $\hat{p}_\eta(0) = 1-\eta$ in the expectation parameter η.

Since $\mu(\theta)$ is twice-differentiable and strictly convex, we can consider the Bregman divergence of $\mu(\theta)$. Then, the divergence $D(p_{\bar{\theta}} \| p_\theta)$ can be written by using the Bregman divergence of $\mu(\theta)$ as follows.

$$\begin{aligned}&D(p_{\bar{\theta}} \| p_\theta) = D(\hat{p}_{\eta(\bar{\theta})} \| \hat{p}_{\eta(\theta)}) = (\bar{\theta} - \theta)\eta(\bar{\theta}) - \mu(\bar{\theta}) + \mu(\theta) \\ &= D^\mu(\bar{\theta} \| \theta) = \int_\theta^{\bar{\theta}} J_{\tilde{\theta}} (\tilde{\theta} - \theta) d\tilde{\theta} = \max_{\tilde{\theta}} (\tilde{\theta} - \theta)\eta(\bar{\theta}) - \mu(\tilde{\theta}) + \mu(\theta).\end{aligned} \tag{2.129}$$

where equations in (2.129) follow from (2.111). When $\bar{\theta} > \theta$, the above maximum is replaced by $\max_{\tilde{\theta}:\tilde{\theta} \ge \theta}$.

Next, we consider the multi-parameter case. Let $X_1(\omega), \ldots, X_d(\omega)$ be d real-valued random variables. We can define a d-parameter exponential family

$$p_\theta(\omega) \stackrel{\text{def}}{=} p(\omega) e^{\sum_k \theta^k X_k(\omega) - \mu(\theta)}, \quad \mu(\theta) \stackrel{\text{def}}{=} \log \sum_\omega p(\omega) e^{\sum_k \theta^k X_k(\omega)}. \tag{2.130}$$

The parameters θ^k are natural parameters, and the other parameters

$$\eta_k(\theta) \stackrel{\text{def}}{=} \frac{\partial \mu}{\partial \theta^k} = \sum_\omega p_\theta(\omega) X_k(\omega) \tag{2.131}$$

are expectation parameters. Since the second derivative $\frac{\partial^2 \mu(\theta)}{\partial \theta^j \partial \theta^k}$ is equal to the Fisher information matrix $J_{\theta:k,j}$, the cumulant generating function $\mu(\theta)$ is a convex function. Using (2.118), we obtain

$$D(p_{\bar{\theta}} \| p_\theta) = \int_0^1 \sum_{k,j} (\bar{\theta}^k - \theta^k)(\bar{\theta}^j - \theta^j) J_{\theta+(\bar{\theta}-\theta)t:k,j} t dt \tag{2.132}$$

similar to (2.129). Since the second derivative matrix $(\frac{\partial^2 \nu}{\partial \eta_k \partial \eta_j})_{k,j}$ of ν appearing in (2.124) is the inverse of the matrix $\frac{\partial^2 \mu}{\partial \theta^k \partial \theta^j}$, the application of (2.124) yields that

$$D(p_{\bar{\theta}} \| p_\theta) = \int_0^1 \sum_{k,j} (\eta_k(\theta) - \eta_k(\bar{\theta}))(\eta_j(\theta) - \eta_j(\bar{\theta}))(J_{\theta(t)}^{-1})^{k,j} t dt, \tag{2.133}$$

where $\theta(t)$ is defined as $\eta(\theta(t)) = \eta(\bar{\theta}) + (\eta(\bar{\theta}) - \eta(\theta))t$. Note that the inverse matrix $J_{\theta(t)}^{-1}$ is the Fisher information matrix with respect to the parameter η.

In what follows, we consider the case where p is the uniform distribution p_{mix}. Let the real-valued random variables $X_1(\omega), \ldots, X_d(\omega)$ be a basis of the space $\mathcal{R}_0(\Omega)$ of random variables that have expectation 0 under the uniform distribution p_{mix}. We also choose the dual basis $Y^1(\omega), \ldots, Y^k(\omega)$ of the space $\mathcal{R}_0(\Omega)$ satisfying $\sum_\omega Y^k(\omega) X_j(\omega) = \delta_j^k$. Then, any distribution p can be parameterized by the expectation parameter as

$$p(\omega) = \hat{p}_{\eta(\theta)}(\omega) := p_{\text{mix}}(\omega) + \sum_i \eta_k(\theta) Y^k(\omega)$$

because $p - p_{\text{mix}}$ can be regarded as an element of $\mathcal{R}_0(\Omega)$.

From (2.123) and (2.120),

$$D(\hat{p}_{\bar{\eta}} \| \hat{p}_\eta) = D^\nu(\eta \| \bar{\eta}) = \sum_k \frac{\partial \nu}{\partial \eta_k}(\eta_k - \bar{\eta}_k) - \nu(\eta) + \nu(\bar{\eta}), \tag{2.134}$$

$$\nu(\eta) = D(\hat{p}_\eta \| p_{\text{mix}}) = -H(\hat{p}_\eta) + H(p_{\text{mix}}) \tag{2.135}$$

because $\mu(0) = 0$. The second derivative matrix of ν is the inverse of the second derivative matrix of μ, i.e., the Fisher information matrix concerning the natural parameter θ. That is, the second derivative matrix of ν coincides with the Fisher information matrix concerning the expectation parameter η.

Now, for given distributions p and q, we consider the case when $Y^1(\omega) = q(\omega) - p(\omega)$. In this case, the distribution $\hat{p}_t := (1-t)p + tq$ $(0 \le t \le 1)$ depends on the first expectation parameter η_1. Other expectation parameters η_k are constants for the distribution p_t. Hence, $\eta_1(\hat{p}_t) - \eta_1(\hat{p}_{t'}) = t - t'$ and $\eta_k(\hat{p}_t) - \eta_k(\hat{p}_{t'}) = 0$ for $k \geq 2$. Thus, as a special case of (2.133), we have

$$D(p\|q) = \int_0^1 J_t t dt\,, \tag{2.136}$$

where J_t is the Fisher information for the parameter t.

2.3 Estimation in Classical Systems

An important problem in mathematical statistics is the estimation of the parameter θ from some given data $\omega \in \Omega$ for a probability distribution that generates the data. To solve this problem, a mapping $\hat{\theta}$ called an estimator from the probability space Ω to the parameter space $\Theta \subset \mathbb{R}$ is required. The accuracy of the estimator is most commonly evaluated by the **mean square error**, which is the expectation of the square of the difference $\hat{\theta} - \theta$:

$$\mathrm{V}_\theta(\hat{\theta}) \stackrel{\text{def}}{=} \mathrm{E}_{p_\theta}((\hat{\theta} - \theta)^2), \tag{2.137}$$

where θ is the true parameter. Note that sometimes the mean square error is not the same as the variance $\mathrm{V}_{p_\theta}(X)$. The estimator

$$\mathrm{E}_\theta(\hat{\theta}) \stackrel{\text{def}}{=} \mathrm{E}_{p_\theta}(\hat{\theta}) = \theta, \quad \forall \theta \in \Theta \tag{2.138}$$

is called an **unbiased estimator**, and such estimators form an important class of estimators. The mean square error of the unbiased estimator $\hat{\theta}$ satisfies the **Cramér–Rao inequality**

$$\mathrm{V}_\theta(\hat{\theta}) \geq J_\theta^{-1}\,. \tag{2.139}$$

When an unbiased estimator attains the RHS of (2.139), it is called **efficient**. This inequality can be proved from the relations

$$\langle(\hat{\theta} - \theta), l_{\theta_0}\rangle_p^{(e)} = \left.\frac{d\mathrm{E}_\theta(\hat{\theta} - \theta_0)}{d\theta}\right|_{\theta=\theta_0} = 1$$

and

$$\langle(\hat{\theta} - \theta_0), (\hat{\theta} - \theta_0)\rangle_{p_{\theta_0}}^{(e)} \langle l_{\theta_0}, l_{\theta_0}\rangle_{p_{\theta_0}}^{(e)} \geq \left|\langle(\hat{\theta} - \theta), l_{\theta_0}\rangle_{p_{\theta_0}}^{(e)}\right|^2 = 1\,, \tag{2.140}$$

which follows from Schwarz's inequality. The equality of (2.139) holds for every value of θ if and only if the probability distribution family is a one-parameter exponential family (2.127) and the expectation parameter $\eta(\theta) = \sum_\omega X(\omega) p_\theta(\omega)$ is to be estimated. In this case, the efficient estimator for the expected parameter is given as $\hat{\eta}(\omega) := X(\omega)$ (Exercise 2.40). Even in the estimation for an exponential family, there is necessarily no estimator for the natural parameter θ in (2.127) such that the equality of (2.139) holds for all θ.

Let n data $\omega^n = (\omega_1, \ldots, \omega_n) \in \Omega^n$ be generated with the n-i.i.d. of the probability distribution p_θ. The estimator may then be given by the mapping $\hat{\theta}^n$ from Ω^n to $\Theta \subset \mathbb{R}$. In this case, the Fisher information of the probability distribution family is nJ_θ, and the unbiased estimator $\hat{\theta}^n$ satisfies the Cramér–Rao inequality

$$\mathrm{V}_\theta(\hat{\theta}^n) \geq \frac{1}{n} J_\theta^{-1} .$$

However, in general, it is not necessary to restrict our estimator to unbiased estimators. In fact, rare estimators satisfy such conditions for finite n.

Therefore, in mathematical statistics, we often study problems in the asymptotic limit $n \to \infty$ rather than those with a finite number of data elements. For this purpose, let us apply the asymptotic unbiasedness conditions

$$\lim_{n\to\infty} \mathrm{E}_\theta(\hat{\theta}_n) = \theta, \quad \lim_{n\to\infty} \frac{d}{d\theta} \mathrm{E}_\theta(\hat{\theta}_n) = 1, \quad \forall \theta \in \Theta \tag{2.141}$$

to a sequence of estimators $\{\hat{\theta}^n\}$. Evaluating the accuracy with $\underline{\lim}\, n \mathrm{V}_\theta(\hat{\theta}_n)$, we have the **asymptotic Cramér–Rao inequality**[8]:

$$\underline{\lim}\, n \mathrm{V}_\theta(\hat{\theta}_n) \geq J_\theta^{-1}, \tag{2.142}$$

which is shown as follows. Based on a derivation similar to (2.139), we obtain

$$n J_\theta \mathrm{V}_\theta(\hat{\theta}_n) \geq \left| \frac{d}{d\theta} \mathrm{E}_\theta(\hat{\theta}_n) \right|^2 . \tag{2.143}$$

Combination of (2.141) and (2.143) derives Inequality (2.142).

Now, we consider what estimator attains the lower bound of (2.142). The **maximum likelihood estimator** $\hat{\theta}_{n,ML}(\omega^n)$

$$\hat{\theta}_{n,ML}(\omega^n) = \operatorname*{argmax}_{\theta \in \Theta} p_\theta^n(\omega^n) \tag{2.144}$$

[8]This inequality still holds even if the asymptotic unbiasedness condition is replaced by another weak condition. Indeed, it is a problem to choose a suitable condition to be assumed for the inequality (2.142). For details, see van der Vaart [7].

achieves this lower bound, and the limit of its mean squared error is equal to J_θ^{-1} [7]. Indeed, in an exponential family with the expectation parameter, the maximum likelihood estimator is equal to the efficient estimator[Exe. 2.41]. Hence, the maximum likelihood estimator plays an important role in statistical inference.[9]

Indeed, we choose the mean square error as the criterion of estimation error because (1) its mathematical treatment is easy and (2) in the i.i.d. case, the sample mean can be characterized by a Gaussian distribution. Hence, we can expect that a suitable estimator will also approach a Gaussian distribution asymptotically. That is, we can expect that its asymptotic behavior will be characterizable by the variance. In particular, the maximum likelihood estimator $\hat{\theta}_{n,ML}$ obeys the Gaussian distribution asymptotically:

$$p_\theta^n\{a \le \sqrt{n}(\hat{\theta}_{n,ML} - \theta) \le b\} \to \int_a^b P_{G,1/J_\theta}(x)dx, \quad \forall a, b .$$

Let us now consider the probability distribution family $\{p_\theta | \theta \in \Theta \subset \mathbb{R}^d\}$ with multiple parameters. We focus on the Fisher information matrix $\boldsymbol{J}_\theta = (J_{\theta:k,j})$, which was defined at the end of Sect. 2.2.1, instead of the Fisher information. The estimator is given by the map $\hat{\theta} = (\hat{\theta}^1, \ldots, \hat{\theta}^d)$ from the probability space Ω to the parameter space Θ, similar to the one-parameter case. The unbiasedness conditions are

$$\mathrm{E}_\theta^k(\hat{\theta}) \stackrel{\text{def}}{=} \mathrm{E}_{p_\theta}(\hat{\theta}^k) = \theta^k, \quad \forall \theta \in \Theta, 1 \le \forall k \le d .$$

The error can be calculated using the **mean square error matrix** $\boldsymbol{V}_\theta(\hat{\theta}) = (\mathrm{V}_\theta^{k,j}(\hat{\theta}))$:

$$\mathrm{V}_\theta^{k,j}(\hat{\theta}) \stackrel{\text{def}}{=} \mathrm{E}_{p_\theta}((\hat{\theta}^k - \theta^k)(\hat{\theta}^j - \theta^j)).$$

Then, we obtain the **multiparameter Cramér–Rao inequality**

$$\boldsymbol{V}_\theta(\hat{\theta}) \ge \boldsymbol{J}_\theta^{-1} . \tag{2.145}$$

Proof of (2.145) For the proof, let us assume that any vectors $|b\rangle = (b_1, \ldots, b_d)^T \in \mathbb{C}^d$ and $|a\rangle \in \mathbb{C}^d$ satisfy

$$\langle b|\boldsymbol{V}_\theta(\hat{\theta})b\rangle\langle a|\boldsymbol{J}_\theta|a\rangle \ge |\langle b|a\rangle|^2 . \tag{2.146}$$

By substituting $a = (\boldsymbol{J}_\theta)^{-1}b$, inequality (2.146) becomes

$$\langle b|\boldsymbol{V}_\theta(\hat{\theta})|b\rangle \ge \langle b|(\boldsymbol{J}_\theta)^{-1}|b\rangle$$

[9]This is generally true for all probability distribution families, although some regularity conditions must be imposed. For example, consider the case in which Ω consists of finite elements. These regularity conditions are satisfied when the first and second derivatives with respect to θ are continuous. Generally, the central limit theorem is used in the proof [7].

since $(\boldsymbol{J}_\theta)^{-1}$ is a symmetric matrix. Therefore, we obtain (2.145) if (2.146) holds. Now, we prove (2.146) as follows. Since

$$\delta_k^j = \left.\frac{\partial \mathrm{E}_\theta^j(\hat{\theta}) - \theta_0^j}{\partial \theta^k}\right|_{\theta=\theta_0} = \left\langle l_{\theta_0:k}, (\hat{\theta}^j - \theta_0^j)\right\rangle_{\theta_0}^{(e)},$$

similarly to the proof of (2.139), the Schwarz inequality yields

$$\langle b|\boldsymbol{V}_{\theta_0}(\hat{\theta})b\rangle = \left\langle \left(\sum_{k=1}^d (\hat{\theta}^k - \theta_0^k)b_k\right), \left(\sum_{k=1}^d (\hat{\theta}^k - \theta_0^k)b_k\right)\right\rangle_{p_{\theta_0}}^{(e)}$$

$$\geq \frac{\left|\left\langle \left(\sum_{k=1}^d l_{\theta_0:k}a_k\right), \left(\sum_{k=1}^d (\hat{\theta}^k(\omega) - \theta_0^k)b_k\right)\right\rangle_{p_{\theta_0}}^{(e)}\right|^2}{\left\langle \left(\sum_{k=1}^d l_{\theta_0:k}a_k\right), \left(\sum_{k=1}^d l_{\theta_0:k}a_k\right)\right\rangle_{p_{\theta_0}}^{(e)}} = \frac{|\langle a|b\rangle|^2}{\langle a|\boldsymbol{J}_{\theta_0}|a\rangle}.$$

■

Moreover, since the sequence of estimators $\{\hat{\theta}_n = (\hat{\theta}_n^1, \ldots, \hat{\theta}_n^d)\}$ satisfies the asymptotic unbiasedness condition

$$\lim_{n\to\infty} \mathrm{E}_\theta^k(\hat{\theta}_n) = \theta^k, \quad \lim_{n\to\infty} \frac{\partial}{\partial \theta^j}\mathrm{E}_\theta^k(\hat{\theta}_n) = \delta_j^k, \quad \forall \theta \in \Theta, \tag{2.147}$$

the asymptotic Cramér–Rao inequality for the multiparameter case

$$\boldsymbol{V}_\theta(\{\hat{\theta}_n\}) \geq \boldsymbol{J}_\theta^{-1} \tag{2.148}$$

holds if the limit $\boldsymbol{V}_\theta(\{\hat{\theta}_n\}) \stackrel{\text{def}}{=} \lim_{n\to\infty} n\boldsymbol{V}_\theta(\hat{\theta}_n)$ exists. Next, we prove (2.148). Defining $A_{n,i}^j \stackrel{\text{def}}{=} \frac{\partial}{\partial \theta^j}\mathrm{E}_\theta^k(\hat{\theta}_n)$, we have

$$n\langle a|\boldsymbol{J}_\theta|a\rangle\langle b|\boldsymbol{V}_\theta(\hat{\theta}_n)|b\rangle \geq |\langle a|\mathbf{A}_n|b\rangle|^2$$

instead of (2.146). We then obtain

$$\langle a|\boldsymbol{J}_\theta|a\rangle\langle b|\boldsymbol{V}_\theta(\{\hat{\theta}_n\})|b\rangle \geq |\langle a|b\rangle|^2,$$

from which (2.148) may be obtained in a manner similar to (2.145).

Similarly to the one-parameter case, the equality of (2.145) holds if and only if the following conditions hold: (1) The probability distribution family is a multiparameter exponential family. (2) The expectation parameter η is to be estimated. (3) The estimator for η is given by

$$\hat{\eta}_k(\omega) = X_k(\omega). \tag{2.149}$$

In this case, this estimator (2.149) equals the maximum likelihood estimator $\hat{\theta}_{n,ML} = (\hat{\theta}^1_{n,ML}, \ldots \hat{\theta}^d_{n,ML})$ defined by (2.144)[Exe. 2.41], i.e.,

$$\max_{\eta} \hat{p}_{\eta}(\omega) = \hat{p}_{X_k(\omega)}(\omega) . \tag{2.150}$$

A probability distribution family does not necessarily have such an estimator; however, a maximum likelihood estimator $\hat{\theta}_{n,ML}$ can be defined by (2.144). This satisfies the asymptotic unbiasedness property (2.147) in a similar way to (2.144), and it satisfies the equality of (2.148). Moreover, it is known that the maximum likelihood estimator $\hat{\theta}_{n,ML}$ satisfies [7]

$$\boldsymbol{V}_{\theta}(\{\hat{\theta}_n\}) = \boldsymbol{J}_{\theta}^{-1} .$$

Note that this inequality holds independently of the choice of coordinate. Hence, for a large amount of data, it is best to use the maximum likelihood estimator. Its mean square error matrix is almost in inverse proportion to the number of observations n. This coefficient of the optimal case is given by the Fisher information matrix. Therefore, the Fisher information matrix can be considered to yield the best accuracy of an estimator.

Indeed, usually any statistical decision with the given probability distribution family $\{q_{\gamma}|\gamma \in \Gamma\}$ is based on the likelihood ratio $\log q_{\gamma}(\omega) - \log q_{\gamma'}(\omega)$. For example, the maximum likelihood estimator depends only on the likelihood ratio. A probability distribution family $\{q_{\gamma}|\gamma \in \Gamma\}$ is called a **curved exponential family** when it belongs to a larger multiparameter exponential family $\{p_{\theta}|\theta \in \Theta\}$, i.e., q_{γ} is given as $p_{\theta(\gamma)}$ with use of a function $\theta(\gamma)$. When $p_{\theta}(\omega)$ is given by (2.130), the likelihood ratio can be expressed by the relative entropy

$$\begin{aligned}
&\log q_{\gamma}(\omega) - \log q_{\gamma'}(\omega) = \log p_{\theta(\gamma)}(\omega) - \log p_{\theta(\gamma')}(\omega) \\
=&\sum_k (\theta(\gamma)^k - \theta(\gamma')^k) X_k(\omega) - \mu(\theta(\gamma)) + \mu(\theta(\gamma')) \\
=&\sum_k X_k(\omega)(\theta''^k - \theta(\gamma')^k) + \mu(\theta(\gamma')) - \mu(\theta'') \\
&- \left(\sum_k X_k(\omega)(\theta''^k - \theta(\gamma)^k) + \mu(\theta(\gamma)) - \mu(\theta'') \right) \\
=&D(\hat{p}_{X(\omega)} \| q_{\gamma'}) - D(\hat{p}_{X(\omega)} \| q_{\gamma}),
\end{aligned} \tag{2.151}$$

where θ'' is chosen as $\eta_k(\theta'') = X_k(\omega)$. That is, our estimation procedure can be treated from the viewpoint of the relative entropy geometry.

Exercises

2.40 Show that the following two conditions are equivalent for a probability distribution family $\{p_{\theta}|\theta \in \mathbb{R}\}$ and its estimator X by following the steps below.

① There exists a parameter η such that the estimator X is an unbiased estimator for the parameter η and the equality of (2.139) holds at all points.

② The probability distribution family $\{p_\theta|\theta \in \mathbb{R}\}$ is an exponential family, $p_\theta(\omega)$ is given by (2.127) using X, and the parameter to be estimated is the expectation parameter $\eta(\theta)$.

(**a**) Show that the estimator X is an unbiased estimator of the expectation parameter under the exponential family (2.127).
(**b**) Show that ① may be deduced from ②.
(**c**) For the exponential family (2.127), show that the natural parameter θ is given as a function of the expectation parameter η with the form $\theta = \int_0^\eta J_{\eta'}\, d\eta'$.
(**d**) Show that $\mu(\theta(\eta)) = \int_0^\eta \eta' J_{\eta'}\, d\eta'$.
(**e**) Show that $\frac{l_\eta}{J_\eta} = X - \eta$ if ① is true.
(**f**) Show that $\frac{dp_\eta}{d\eta} = J_\eta(X - \eta)p_\eta$ if ① is true.
(**g**) Show that ② is true if ① is true.

2.41 Show equation (2.150) from (2.151).

2.42 Consider the probability distribution family $\{p_\theta|\theta \in \mathbb{R}\}$ in the probability space $\{1, \ldots, l\}$ and the stochastic transition matrix $Q = (Q^i_j)$. Let the Fisher information of p_{θ_0} in the probability distribution family $\{p_\theta|\theta \in \mathbb{R}\}$ be J_{θ_0}. Let J'_{θ_0} be the Fisher information of $Q(p_{\theta_0})$ in the probability distribution family $\{Q(p_\theta)|\theta \in \mathbb{R}\}$. Show then that $J_{\theta_0} \geq J'_{\theta_0}$. This inequality is called the **monotonicity** of the Fisher information. Similarly, define $\boldsymbol{J}_{\theta_0}$, $\boldsymbol{J}'_{\theta_0}$ for the multiple variable case, and show that the matrix inequality $\boldsymbol{J}_{\theta_0} \geq \boldsymbol{J}'_{\theta_0}$ holds.

2.4 Type Method and Large Deviation Evaluation

In this section, we analyze the case of a sufficiently large number of data by using the following two methods. The first method involves an analysis based on empirical distributions, and it is called the **type method**. In the second method, we consider a particular random variable and examine its exponential behavior.

2.4.1 Type Method and Sanov's Theorem

Let n data be generated according to a probability distribution in a finite set of events $\mathbb{N}_d = \{1, \ldots, d\}$. Then, we can perform the following analysis by examining the empirical distribution of the data [8]. Let T_n be the set of empirical distributions obtained from n observations. We call each element of this set a **type**. For each type $q \in T_n$, let the subset $T^n_q \subset \mathbb{N}^n_d$ be a set of data with the empirical distribution q. Since the probability $p^n(\boldsymbol{i})$ depends only on the type q for each $\boldsymbol{i} \in T^n_q$, we can denote this probability by $p^n(q)$. Then, when the n data are generated according to the probability distribution p^n, the empirical distribution matches $q \in T_n$ with the probability $p^n(T^n_q) \left(\stackrel{\text{def}}{=} \sum_{\boldsymbol{i} \in T^n_q} p^n(\boldsymbol{i})\right)$.

Theorem 2.5 *Any type $p \in T_n$ and any data $\boldsymbol{i} \in T_q^n$ satisfy the following:*

$$p^n(T_q^n) \le p^n(T_p^n), \tag{2.152}$$

$$p^n(\boldsymbol{i}) = e^{-n(H(q)+D(q\|p))}. \tag{2.153}$$

Denoting the number of elements of T_n and T_q^n by $|T_n|$ and $|T_q^n|$, respectively, we obtain the relations

$$|T_n| = \frac{n!}{n_1! \cdots n_d!} \le (n+1)^{d-1}, \tag{2.154}$$

$$\frac{1}{(n+1)^d} e^{nH(q)} \le |T_q^n| \le e^{nH(q)}, \tag{2.155}$$

$$\frac{1}{(n+1)^d} e^{-nD(q\|p)} \le p^n(T_q^n) \le e^{-nD(q\|p)}. \tag{2.156}$$

Proof Let $p(i) = \frac{n_i}{n}$ and $q(i) = \frac{n_i'}{n}$. Then,

$$p^n(T_p^n) = |T_p^n| \prod_{i=1}^d p(i)^{n_i} = \frac{n!}{n_1! \cdots n_d!} \prod_{i=1}^d p(i)^{n_i},$$

$$p^n(T_q^n) = |T_q^n| \prod_{i=1}^d p(i)^{n_i'} = \frac{n!}{n_1'! \cdots n_d'!} \prod_{i=1}^d p(i)^{n_i'}.$$

Using the inequality[Exe. 2.43]

$$\frac{n!}{m!} \le n^{n-m}, \tag{2.157}$$

we have

$$\frac{p^n(T_q^n)}{p^n(T_p^n)} = \prod_{i=1}^d \left(\frac{n_i!}{n_i'!} p(i)^{n_i'-n_i} \right) \le \prod_{i=1}^d \left(n_i^{n_i-n_i'} \left(\frac{n_i}{n}\right)^{n_i'-n_i} \right)$$

$$= \prod_{i=1}^d \left(\frac{1}{n}\right)^{n_i'-n_i} = \left(\frac{1}{n}\right)^{\sum_{i=1}^d (n_i'-n_i)} = 1.$$

Therefore, inequality (2.152) holds. For $\boldsymbol{i} \in T_q^n$, we have

$$p^n(\boldsymbol{i}) = \prod_{i=1}^d p(i)^{n_i'} = \prod_{i=1}^d p(i)^{n\left(\frac{n_i'}{n}\right)}$$

$$= \prod_{i=1}^d e^{n \log p(i) \left(\frac{n_i'}{n}\right)} = e^{n \sum_{i=1}^d q(i) \log p(i)} = e^{-n(H(q)+D(q\|p))},$$

which implies (2.153).

Each element q of T_n may be written as a d-dimensional vector. Each component of the vector then assumes one of the following $n+1$ values: $0, 1/n, \ldots, n/n$. Since $\sum_{i=1}^d q_i = 1$, the dth element is decided by the other $d-1$ elements. Therefore, inequality (2.154) follows from a combinatorial observation. Applying inequality (2.153) to the case $p = q$, we have the relation $p^n(T_q^n) = e^{-nH(p)}|T_p^n|$. Since $1 = \sum_{q \in T_n} p^n(T_q^n) \geq p^n(T_q^n)$ for $p \in T_n$, we obtain the inequality on the RHS of (2.155). Conversely, inequality (2.152) yields that $1 = \sum_{q \in T_n} p^n(T_q^n) \leq \sum_{q \in T_n} p^n(T_p^n) = e^{-nH(p)}|T_p^n||T_n|$. Combining this relation with (2.154), we obtain the inequality on the LHS of (2.155). Inequality (2.156) may be obtained by combining (2.153) and (2.155). ∎

We obtain **Sanov's Theorem** using these inequalities.

Theorem 2.6 (Sanov [9]) *The following holds for a subset $\mathcal{R}$ of distributions on $\mathbb{N}_d$:*

$$\frac{1}{(n+1)^d} \exp(-n \min_{q \in \mathcal{R} \cap T_n} D(q \| p)) \leq p^n(\cup_{q \in \mathcal{R} \cap T_n} T_q^n)$$
$$\leq (n+1)^d \exp(-n \inf_{q \in \mathcal{R}} D(q \| p)).$$

In particular, when the closure of the interior of $\mathcal{R}$ coincides with the closure of $\mathcal{R}$,[10]

$$\lim_{n \to \infty} -\frac{1}{n} \log p^n \left(\cup_{q \in \mathcal{R} \cap T_n} T_q^n \right) = \inf_{q \in \mathcal{R}} D(q \| p)$$

in the limit $n \to \infty$.

Based on this theorem, we can analyze how different the true distribution is from the empirical distribution. More precisely, the empirical distribution belongs to the neighborhood of the true distribution with a sufficiently large probability, i.e., the probability of its complementary event approaches 0 exponentially. This exponent is then given by the relative entropy. The discussion of this exponent is called a large deviation evaluation.

However, it is difficult to consider a quantum extension of Sanov's theorem. This is because we cannot necessarily take the common eigenvectors for plural densities. That is, this problem must be treated independently of the choice of basis. One possible way to fulfill this requirement is the group representation method. If we use this method, it is possible to treat the eigenvalues of density of the system instead of the classical probabilities [10, 11]. Since eigenvalues do not identify the density matrix, they cannot be regarded as the complete quantum extension of Sanov's theorem. Indeed, a quantum extension is available if we focus only on two densities; however, it should be regarded as the quantum extension of Stein's lemma given in

[10]The set is called the interior of a set X when it consists of the elements of X without its boundary. For example, for a one-dimensional set, the interior of $[0, 0.5] \cup \{0.7\}$ is $(0, 0.5)$ and the closure of the interior is $[0, 0.5]$. Therefore, the condition is not satisfied in this case.

Sect. 3.5. Since the data are not given without our operation in the quantum case, it is impossible to directly extend Sanov's theorem to the quantum case.

In fact, the advantage of using the type method is the universality in information theory [8]. However, if we apply the type method to quantum systems independently of the basis, the universality is not available in the quantum case. A group representation method is very effective for a treatment independent of basis [10, 12–17]. Indeed, several universal protocols have been obtained by this method.

Exercise

2.43 Prove (2.157) by considering the cases $n \geq m$ and $n < m$ separately.

2.4.2 Cramér Theorem and Its Application to Estimation

Next, we consider the asymptotic behavior of a random variable in the case of independent and identical trials of the probability distribution p.

For this purpose, we first introduce two fundamental inequalities[Exe. 2.44]. The **Markov inequality** states that for a real-valued random variable X where $X \geq 0$,

$$\frac{\mathrm{E}_p(X)}{c} \geq p\{X \geq c\}. \tag{2.158}$$

Applying the Markov inequality to the variable $|X - \mathrm{E}_p(X)|$, we obtain the **Chebyshev inequality**:

$$p\{|X - \mathrm{E}_p(X)| \geq a\} \leq \frac{\mathrm{V}_p(X)}{a^2}. \tag{2.159}$$

Now, consider the real-valued random variable

$$X^n \stackrel{\text{def}}{=} \sum_{i=1}^{n} \frac{1}{n} X_i, \tag{2.160}$$

where $X_1, \ldots, X_n$ are n independent random variables that are identical to the real-valued random variable X subject to the distribution p. When the variable X^n obeys the independent and identical distribution p^n of p, the expectation of X^n coincides with the expectation $\mathrm{E}_p(X)$. Let $\mathrm{V}_p(X)$ be the variance of X. Then, its variance with n observations equals $\mathrm{V}_p(X)/n$.

Applying Chebyshev's inequality (2.159), we have

$$p^n\{|X^n - \mathrm{E}_p(X)| \geq \epsilon\} \leq \frac{\mathrm{V}_p(X)}{n\epsilon^2}$$

for arbitrary $\epsilon > 0$. This inequality yields the **(weak) law of large numbers**

$$p^n\{|X^n - \mathrm{E}_p(X)| \geq \epsilon\} \to 0, \quad \forall \epsilon > 0\,. \tag{2.161}$$

In general, if a sequence of pairs $\{(X^n, p_n)\}$ of a real-valued random variable and a probability distribution satisfies

$$p_n\{|X^n - x| \geq \epsilon\} \to 0, \quad \forall \epsilon > 0 \tag{2.162}$$

for a real number x, then the real-valued random variable X^n is said to **converge in probability** to x.

Since the left-hand side (LHS) of (2.161) converges to 0, the next focus is the speed of this convergence. Usually, this convergence is exponential. The exponent of this convergence is characterized by **Cramér's Theorem** below.

Theorem 2.7 (Cramér [18]) *Define the* **cumulant generating function** $\mu(\theta) \stackrel{\mathrm{def}}{=} \log\left(\sum_\omega p(\omega)e^{\theta X(\omega)}\right)$. *Then*

$$\underline{\lim} -\frac{1}{n}\log p^n\{X^n \geq x\} \geq \max_{\theta\geq 0}\left(\theta x - \mu(\theta)\right), \tag{2.163}$$

$$\overline{\lim} -\frac{1}{n}\log p^n\{X^n \geq x\} \leq \lim_{x'\to x+0}\max_{\theta\geq 0}\left(\theta x' - \mu(\theta)\right), \tag{2.164}$$

$$\underline{\lim} -\frac{1}{n}\log p^n\{X^n \leq x\} \geq \max_{\theta\leq 0}\left(\theta x - \mu(\theta)\right) \tag{2.165}$$

$$\overline{\lim} -\frac{1}{n}\log p^n\{X^n \leq x\} \leq \lim_{x'\to x-0}\max_{\theta\leq 0}\left(\theta x' - \mu(\theta)\right). \tag{2.166}$$

If we replace $\{X^n \geq x\}$ *and* $\{X^n \leq x\}$ *with* $\{X^n > x\}$ *and* $\{X^n < x\}$, *respectively, the same inequalities hold.*

When the probability space consists of finite elements, the function $\max_{\theta\geq 0}(\theta x - \mu(\theta))$ is continuous, i.e., $\lim_{x'\to x+0}\max_{\theta\geq 0}\left(\theta x' - \mu(\theta)\right) = \max_{\theta\geq 0}(\theta x - \mu(\theta))$. Hence, the equality of (2.163) holds. Conversely, if the probability space contains an infinite number of elements as the set of real numbers $\mathbb{R}$, we should treat the difference between the RHS and LHS more carefully. Further, the inequality of (2.163) holds without limit, and is equivalent to (2.46) when we replace the real-valued random variable $X(\omega)$ with $-\log q(\omega)$. The same argument holds for (2.165).

Proof Inequality (2.165) is obtained by considering $-X$ in (2.163). Therefore, we prove only (2.163). Inequality (2.166) is also obtained by considering $-X$ in (2.164). Here we prove only inequality (2.163). Inequality (2.164) will be proved at the end of this section.

For a real-valued random variable X with $X(\omega)$ for each ω,

$$\mathrm{E}_{p^n}(e^{n\theta X^n}) = \mathrm{E}_{p^n}\left(\prod_{i=1}^n e^{\theta X_i}\right) = (\mathrm{E}_p e^{\theta X})^n = e^{n\mu(\theta)}\,. \tag{2.167}$$

Using the Markov inequality (2.158), we obtain

$$p^n\{X^n \geq x\} = p^n\{e^{n\theta X^n} \geq e^{n\theta x}\} \leq \frac{e^{n\mu(\theta)}}{e^{n\theta x}} = e^{n(\mu(\theta)-\theta x)} \text{ for } \theta \geq 0. \tag{2.168}$$

Taking the logarithm of both sides, we have

$$-\frac{1}{n} \log p^n\{X^n \geq x\} \geq \theta x - \mu(\theta).$$

Let us take the maximum on the RHS with respect to $\theta \geq 0$ and then take the limit on the LHS. We obtain inequality (2.163). ■

This theorem can be extended to the non-i.i.d. case as the **Gärtner–Ellis theorem.**

Theorem 2.8 (Gärtner [19], Ellis [20]) *Let $\{p_n\}$ be a general sequence of the probabilities with the real-valued random variables X_n. Define the* **cumulant generating functions** $\mu_n(\theta) \stackrel{\text{def}}{=} \frac{1}{n} \log \left(\sum_\omega p_n(\omega) e^{\theta n X_n(\omega)}\right)$ *and* $\mu(\theta) \stackrel{\text{def}}{=} \lim_{n\to\infty} \mu_n(\theta)$ *and the set* $G \stackrel{\text{def}}{=} \{\mu'(\theta)|\theta\}$. *Then*

$$\underline{\lim} -\frac{1}{n} \log p_n\{X_n \geq x\} \geq \max_{\theta \geq 0} (\theta x - \mu(\theta)), \tag{2.169}$$

$$\overline{\lim} -\frac{1}{n} \log p_n\{X_n \geq x\} \leq \inf_{\bar{x} \in G: \bar{x} > x} \max_{\theta \geq 0} (\theta \bar{x} - \mu(\theta)), \tag{2.170}$$

$$\underline{\lim} -\frac{1}{n} \log p_n\{X_n \leq x\} \geq \max_{\theta \leq 0} (\theta x - \mu(\theta)), \tag{2.171}$$

$$\overline{\lim} -\frac{1}{n} \log p_n\{X_n \leq x\} \leq \inf_{\bar{x} \in G: \bar{x} < x} \max_{\theta \leq 0} (\theta \bar{x} - \mu(\theta)). \tag{2.172}$$

If we replace $\{X_n \geq x\}$ and $\{X_n \leq x\}$ by $\{X_n > x\}$ and $\{X_n < x\}$, respectively, the same inequalities hold.

Inequalities (2.169) and (2.171) can be proved in a similar way to Theorem 2.7.

Next, we apply large deviation arguments to estimation theory. Our arguments will focus not on the mean square error but on the decreasing rate of the probability that the estimated parameter does not belong to the ϵ-neighborhood of the true parameter. To treat the accuracy of a sequence of estimators $\{\hat{\theta}_n\}$ with a one-parameter probability distribution family $\{p_\theta|\theta \in \mathbb{R}\}$ from the viewpoint of a large deviation, we define

$$\beta(\{\hat{\theta}_n\}, \theta, \epsilon) \stackrel{\text{def}}{=} \underline{\lim} -\frac{1}{n} \log p_\theta^n\{|\hat{\theta}_n - \theta| > \epsilon\}, \tag{2.173}$$

$$\alpha(\{\hat{\theta}_n\}, \theta) \stackrel{\text{def}}{=} \underline{\lim}_{\epsilon \to 0} \frac{\beta(\{\hat{\theta}_n\}, \theta, \epsilon)}{\epsilon^2}. \tag{2.174}$$

As an approximation, we have

$$p_\theta^n\{|\hat{\theta}_n - \theta| > \epsilon\} \cong e^{-n\epsilon^2 \alpha(\{\hat{\theta}_n\},\theta)}.$$

Hence, an estimator functions better when it has larger values of $\beta(\{\hat{\theta}_n\}, \theta, \epsilon)$ and $\alpha(\{\hat{\theta}_n\}, \theta)$.

Theorem 2.9 (Bahadur [21–23]) *Let a sequence of estimators* $\{\hat{\theta}_n\}$ *satisfy the weak consistency condition*

$$p_\theta^n\{|\hat{\theta}_n - \theta| > \epsilon\} \to 0, \quad \forall \epsilon > 0, \quad \forall \theta \in \mathbb{R}. \tag{2.175}$$

Then, it follows that

$$\beta(\{\hat{\theta}_n\}, \theta, \epsilon) \le \inf_{\theta' : |\theta' - \theta| > \epsilon} D(p_{\theta'} \| p_\theta). \tag{2.176}$$

Further, if

$$D(p_{\theta'} \| p_\theta) = \lim_{\bar{\theta} \to \theta'} D(p_{\bar{\theta}} \| p_\theta), \tag{2.177}$$

the following also holds:

$$\alpha(\{\hat{\theta}_n\}, \theta) \le \frac{1}{2} J_\theta. \tag{2.178}$$

If the probability space consists of finite elements, condition (2.177) holds.

Proof of Theorem 2.9 Inequality (2.178) is obtained by combining (2.176) with (2.105). Inequality (2.176) may be derived from monotonicity (2.13) as follows. From the consistency condition (2.175), the sequence $a_n \stackrel{\text{def}}{=} p_\theta^n\{|\hat{\theta}_n - \theta| > \epsilon\}$ satisfies $a_n \to 0$. Assume that $\epsilon' \stackrel{\text{def}}{=} |\theta - \theta'| > \epsilon$. Then, when $|\hat{\theta}_n - \theta'| < \epsilon' - \epsilon$, we have $|\hat{\theta}_n - \theta| > \epsilon$. Hence, the other sequence $b_n \stackrel{\text{def}}{=} p_{\theta'}^n\{|\hat{\theta}_n - \theta| > \epsilon\} \ge p_{\theta'}^n\{|\hat{\theta}_n - \theta'| < \epsilon' - \epsilon\}$ satisfies $b_n \to 1$ because of the consistency condition (2.175). Thus, monotonicity (2.13) implies that

$$D(p_{\theta'}^n \| p_\theta^n) \ge b_n(\log b_n - \log a_n) + (1 - b_n)(\log(1 - b_n) - \log(1 - a_n)).$$

Since $nD(p_{\theta'} \| p_\theta) = D(p_{\theta'}^n \| p_\theta^n)$ follows from (2.28) and $-(1 - b_n)\log(1 - a_n) \ge 0$, we have $nD(p_{\theta'} \| p_\theta) \ge -h(b_n) - b_n \log a_n$, and therefore

$$-\frac{1}{n} \log a_n \le \frac{D(p_{\theta'} \| p_\theta)}{b_n} + \frac{h(b_n)}{nb_n}. \tag{2.179}$$

As the convergence $h(b_n) \to 0$ follows from the convergence $b_n \to 1$, we have

$$\beta(\{\hat{\theta}_n\}, \theta, \epsilon) \le D(p_{\theta'} \| p_\theta).$$

Considering $\inf_{\theta':|\theta'-\theta|>\epsilon}$, we obtain (2.176). In addition, this proof is valid even if we replace $\{|\hat{\theta}_n - \theta| > \epsilon\}$ in (2.173) by $\{|\hat{\theta}_n - \theta| \geq \epsilon\}$. ■

If no estimator satisfies the equalities in inequalities (2.176) and (2.178), these inequalities are not sufficiently useful. The following proposition gives a sufficient condition for the equalities of (2.176) and (2.178).

Proposition 2.1 *Suppose that the probability distribution family (2.127) is exponential, and the parameter to be estimated is an expectation parameter. If a sequence of estimators is given by $X^n(\omega^n)$ (see (2.160)), then the equality of (2.176) holds. The equality of (2.178) also holds.*

It is known that the maximum likelihood estimator $\hat{\theta}_{n,ML}$ satisfies (2.178) if the probability distribution family satisfies some regularity conditions [23, 24].

Proof of Proposition 2.1 *and* (2.164) *and* (2.166) *in Theorem* 2.7 Now, we prove Proposition 2.1 and its related formulas ((2.163) and (2.164) in Theorem 2.7) as follows. Because (2.129) implies $\max_{\theta'\geq\theta}(\theta'-\theta)(\eta(\theta)+\epsilon) - (\mu(\theta')-\mu(\theta)) = D(\hat{p}_{\eta(\theta)+\epsilon}\|\hat{p}_{\eta(\theta)})$, Proposition 2.1 follows from the inequalities

$$\begin{aligned}
&\underline{\lim} -\frac{1}{n}\log \hat{p}^n_{\eta(\theta)}\{X^n(\omega^n) > \eta(\theta)+\epsilon\} \\
&\geq \max_{\theta'\geq\theta}(\theta'-\theta)(\eta(\theta)+\epsilon) - (\mu(\theta')-\mu(\theta)),
\end{aligned} \tag{2.180}$$

$$\overline{\lim} -\frac{1}{n}\log \hat{p}^n_{\eta(\theta)}\{X^n(\omega^n) > \eta(\theta)+\epsilon\} \leq \lim_{\epsilon'\to\epsilon+0} D(\hat{p}_{\eta(\theta)+\epsilon'}\|\hat{p}_{\eta(\theta)}) \tag{2.181}$$

for the expectation parameter η of the exponential family (2.127) and arbitrary $\epsilon > 0$. When $x = \eta(\theta)+\epsilon = \eta(\tilde{\theta}) \geq 0$ and $\theta = 0$, the formula (2.181) is the same as (2.164) in Theorem 2.7 with replacing $\geq$ by $>$ in the LHS because $D(\hat{p}_{\eta(\theta)+\epsilon}\|\hat{p}_{\eta(\theta)}) = \tilde{\theta}\eta(\tilde{\theta}) - \mu(\tilde{\theta}) = \max_\theta \theta\eta(\tilde{\theta}) - \mu(\theta)$. Since the LHS of (2.181) is not smaller than the LHS of (2.164) in this correspondence, (2.181) yields (2.164). Considering $-X$ instead of X, (2.164) implies (2.166).

To show (2.180), we choose arbitrary $\bar{\epsilon} > \epsilon$ and $\bar{\theta}$ such that $\mu'(\bar{\theta}) = \eta(\theta)+\bar{\epsilon}$. Based on the proof of (2.163) in Theorem 2.7, since the expectation of $e^{n(\theta'-\theta)X^n(\omega^n)}$ under the distribution p^n_θ is $e^{\mu(\theta')-\mu(\theta)}$, we can show that

$$\begin{aligned}
&-\frac{1}{n}\log p^n_\theta\{X^n(\omega^n) > \eta(\theta)+\epsilon\} \\
&\geq \max_{\theta':\theta'\geq\theta}(\theta'-\theta)(\eta(\theta)+\epsilon) - (\mu(\theta')-\mu(\theta)),
\end{aligned} \tag{2.182}$$

$$\begin{aligned}
&-\frac{1}{n}\log p^n_{\bar{\theta}}\{X^n(\omega^n) \leq \eta(\theta)+\epsilon\} \\
&\geq \max_{\theta':\theta'\leq\bar{\theta}}(\theta'-\bar{\theta})(\eta(\theta)+\epsilon) - (\mu(\theta')-\mu(\bar{\theta})) = D(\hat{p}_{\eta(\theta)+\epsilon}\|\hat{p}_{\eta(\theta)+\bar{\epsilon}}) > 0.
\end{aligned} \tag{2.183}$$

Then, (2.182) implies (2.180).

Next, using (2.183), we show (2.181) as follows. According to a discussion similar to the proof of (2.176) in Theorem 2.9, we have

$$-\frac{1}{n}\log \hat{p}^n_{\eta(\theta)}\{X^n(\omega^n) > \eta(\theta)+\epsilon\} \le \frac{D(\hat{p}_{\eta(\theta)+\epsilon'}\|\hat{p}_{\eta(\theta)})}{b_n} + \frac{h(b_n)}{nb_n} \tag{2.184}$$

for $\epsilon' > \epsilon$, where $b_n \stackrel{\text{def}}{=} \hat{p}^n_{\eta(\theta)+\epsilon'}\{X^n(\omega^n) > \eta(\theta)+\epsilon\}$. From (2.183), $b_n \to 1$. Hence, we obtain the last inequality in (2.181). ∎

Proof of (2.170) *and* (2.172) *in Theorem* 2.8 Finally, we will prove inequality (2.170) in Theorem 2.8, i.e., we will prove that

$$\overline{\lim} -\frac{1}{n}\log p_n\{X_n(\omega) \ge x\} \le \max_{\theta\ge 0}\left(\theta\mu'(\bar{\theta}) - \mu(\theta)\right) \tag{2.185}$$

for any $\bar{\theta}$ satisfying $\mu'(\bar{\theta}) > x$. Inequality (2.172) can be shown in the same way. Define the exponential family $p_{n,\theta}(\omega) \stackrel{\text{def}}{=} p_n(\omega)e^{n\theta X_n(\omega)-n\mu_n(\theta)}$. Similarly to (2.184), we have

$$-\frac{1}{n}\log p_{n,0}\{X_n(\omega) > x\} \le \frac{D(p_{n,\bar{\theta}}\|p_{n,0})}{nb_n} + \frac{h(b_n)}{nb_n},$$

where $b_n \stackrel{\text{def}}{=} p_{n,\bar{\theta}}\{X_n(\omega) > x\}$. From (2.129), $\frac{D(p_{n,\bar{\theta}}\|p_{n,0})}{n} = \max_{\theta\ge 0}\left(\theta\mu_n'(\bar{\theta}) - \mu_n(\theta)\right)$. Hence, if we show that $b_n \to 1$, we obtain (2.185). To show that $b_n \to 1$, similarly to (2.183), the inequality

$$-\frac{1}{n}\log p_{n,\bar{\theta}}\{X_n(\omega) \le x\} \ge \max_{\theta\le\bar{\theta}}(\theta-\bar{\theta})x - \mu(\theta) + \mu(\bar{\theta})$$

holds. Since the set of differentiable points of μ is open and μ' is monotone increasing and continuous in this set, there exists a point θ' in this set such that

$$\theta' < \bar{\theta}, \quad x < \mu'(\theta').$$

Since μ' is monotone increasing, we obtain

$$\begin{aligned}&\max_{\theta\le\bar{\theta}}(\theta-\bar{\theta})x - \mu(\theta) + \mu(\bar{\theta}) \ge (\theta'-\bar{\theta})x - \mu(\theta') + \mu(\bar{\theta})\\ \ge&(\mu'(\theta') - x)(\bar{\theta}-\theta') > 0,\end{aligned}$$

which implies that $b_n \to 1$. ∎

Exercises

2.44 Prove Markov's inequality by using the inequality $\sum_{i:x_i\ge c} p_i x_i \ge c\sum_{i:x_i\ge c} p_i$.

2.45 Using Cramér's theorem and (2.42) and (2.44), show the following equations below. Show analogous formulas for (2.46), (2.47), (3.5), and (3.6).

$$\lim_{n\to\infty} -\frac{1}{n} \log p^n\{p^n_{i^n} \le e^{-nR}\} = -\min_{0\le s}(\psi(s) - sR), \tag{2.186}$$

$$\lim_{n\to\infty} -\frac{1}{n} \log p^n\{p^n_{i^n} > e^{-nR}\} = -\min_{s\le 0}(\psi(s) - sR). \tag{2.187}$$

2.46 Show that

$$\lim_{n\to\infty} -\frac{1}{n} \log P^c(p^n, e^{nR}) = -\min_{0\le s\le 1} \frac{\psi(s) - sR}{1-s} \tag{2.188}$$

by first proving (2.189) and then combining this with (2.55). The $\ge$ part may be obtained directly from (2.51)

$$\begin{aligned} P^c(p^n, e^{nR}) \ge & \max_{q\in T_n:|T^n_q|>e^{nR}} (|T^n_q| - e^{nR}) e^{-n(H(p)+H(p\|q))} \\ \ge & \max_{q\in T_n:\frac{e^{nH(q)}}{(n+1)^d}>e^{nR}} \left(\frac{e^{nH(q)}}{(n+1)^d} - e^{nR}\right) e^{-n(H(p)+H(p\|q))} \\ = & \max_{q\in T_n:\frac{e^{nH(q)}}{(n+1)^d}>e^{nR}} e^{-nD(p\|q)} \left(1 - \frac{(n+1)^d e^{nR}}{e^{nH(q)}}\right). \end{aligned} \tag{2.189}$$

2.47 Show that

$$\lim_{n\to\infty} -\frac{1}{n} \log P(p^n, e^{nR}) = -\min_{s\le 0} \frac{\psi(s) - sR}{1-s} \tag{2.190}$$

by first proving (2.191) and then combining this with (2.55). The inequality $\ge$ may be obtained directly from (2.54)

$$P(p^n, e^{nR}) \ge \max_{q\in T_n:|T^n_q|\le e^{nR}} p^n(T^n_q) \ge \max_{q\in T_n:H(q)\le R} \frac{e^{-nD(q\|p)}}{(n+1)^d}. \tag{2.191}$$

2.48 Consider the case where $\Omega_n = \{0, 1\}$, $p_n(0) = e^{-na}$, $p_n(1) = 1 - e^{-na}$, $X_n(0) = a$, $X_n(1) = -b$ with $a, b > 0$. Show that $\mu(\theta) = -\min\{(1-\theta)a, \theta b\}$ and the following for $-b < x < a$:

$$\max_{\theta>0}(x\theta - \mu(\theta)) = \frac{a(x+b)}{a+b} < a, \quad \lim_{n\to\infty} \frac{1}{n} \log p_n\{X_n \ge x\} = a.$$

It gives a counterexample of Gärtner–Ellis Theorem in the nondifferentiable case.

2.5 Continuity and Axiomatic Approach

In this section, we consider how to characterize the entropy $H(p)$ by axioms. Indeed, when a real-value function S satisfies several axiomatic rules, the function S must be the entropy $H(p)$ given in (2.2). Here, we consider the following five axioms for a real-value function S for distribution, which is close to the axioms by Khinchin [25].

K1 (Normalization)

$$S(p_{\mathrm{mix},\{0,1\}}) = \log 2. \tag{2.192}$$

K2 (Continuity) S is continuous on $\mathcal{P}(\{0,1\})$.
K3 (Nonnegativity) S is nonnegative.
K4 (Expandability) For any function f, we have

$$S(\mathrm{P}_X) = S(\mathrm{P}_{f(X)X}). \tag{2.193}$$

K5 (Chain rule) When P_{XY} is a joint distribution for X and Y, the marginal distribution P_X and the conditional distribution $\mathrm{P}_{Y|X=e}$ satisfies that

$$S(\mathrm{P}_{XY}) = S(\mathrm{P}_X) + \sum_x \mathrm{P}_X(x) S(\mathrm{P}_{Y|X=x}). \tag{2.194}$$

Here, we consider another set of axioms as follows.

A1 (Normalization)

$$S(p_{\mathrm{mix},\{0,1\}}) = \log 2. \tag{2.195}$$

A2 (Weak additivity)

$$S(p^n) = nS(p) \tag{2.196}$$

A3 (Monotonicity) For any function f, we have

$$S(\mathrm{P}_X) \ge S(\mathrm{P}_{f(X)}). \tag{2.197}$$

A4 (Asymptotic continuity) Let p_n and q_n be distributions on the set $\{0,1\}^n$. When $d_1(p_n, q_n) \to 0$, we have

$$\frac{|S(p_n) - S(q_n)|}{n} \to 0. \tag{2.198}$$

Then, the following theorem shows the uniqueness of a function satisfying one of the above sets of axioms.

Theorem 2.10 *For a function S defined on the set of distributions, the following three conditions are equivalent.*

(1) S satisfies Axioms **K1-K5**.
(2) S satisfies Axioms **A1-A4**.
(3) $S(p) = -\sum_i p_i \log p_i$.

Before proceeding to the proof of Theorem 2.10, we consider the asymptotic convertibility for the independent and identical distribution.

Lemma 2.5 *For a distribution p on* Ω *and an arbitrary real number* $\epsilon > 0$*, there exists a sequence of maps* f_n *from* Ω^n *to* $\Omega_n := \{0,1\}^{\lfloor (H(p)-\epsilon)n/\log 2 \rfloor}$ *such that* $d_1(p^n \circ f_n^{-1}, p_{\mathrm{mix},\Omega_n}) \to 0$.

Lemma 2.6 *For a distribution p on* Ω *and an arbitrary real number* $\epsilon > 0$*, there exists a sequence of maps* f_n *from* $\Omega'_n := \{0,1\}^{\lfloor (H(p)+\epsilon)n/\log 2 \rfloor}$ *to* Ω^n *such that* $d_1(p^n, p_{\mathrm{mix},\Omega'_n} \circ f_n^{-1}) \to 0$.

These two lemmas show that the entropy $H(p)$ gives the asymptotic conversion rate between the independent and identical distribution and the uniform distribution. Rényi entropy $H_{1+s}(p)$ also satisfies Axioms **K1-K4** and **A1-A3**. However, it does not satisfies K5 (Chain rule) or A4 (Asymptotic continuity)[Exe. 2.49, 2.50]. Indeed, although the quantity $e^{-H_2(p)}$ satisfies A4 (Asymptotic continuity)[Exe. 2.51] as well as A3 (Monotonicity), it does not satisfy A2 (Weak additivity). Only the information quantity satisfying Axioms **K1-K5** or **A1-A4** gives the asymptotic conversion between the independent and identical distribution and the uniform distribution. Hence, we can conclude that K5 (Chain rule) and A4 (Asymptotic continuity) are crucial for the asymptotic conversion.

Proof of Theorem 2.10 First, we show (1) $\Rightarrow$ (2). A2 (Weak additivity) follows from K5 (Chain rule). A3 (Monotonicity) follows from K3 (Nonnegativity), K4 (Expandability), and K5 (Chain rule) by the same discussion as (2.6).

Now, we start to show A4 (Asymptotic continuity). Since the set $\mathcal{P}(\{0,1\})$ is compact, due to K2 (Continuity), S is uniformly continuous on $\mathcal{P}(\{0,1\})$. So, there exists the maximum value $R := \max_{p \in \mathcal{P}(\{0,1\})} S(p)$. For any $\epsilon > 0$, we choose $\delta > 0$ such that $|S(p) - S(q)| \le \epsilon$ for any $d_1(p,q) \le \delta$. Consider two distributions $\mathrm{P}^n_{X_n}$ and $\overline{\mathrm{P}}^n_{X_n}$ on the set $\{0,1\}^n$ such that $\delta_n := 2 d_1(\mathrm{P}^n_{X_n}, \overline{\mathrm{P}}^n_{X_n})$ goes to zero as $n \to \infty$. Then, we can choose a sufficiently large integer N such that $\delta_n \le \frac{\epsilon\delta}{2R}$ for $n \ge N$.

Here, X_i denotes the random variable on the i-th set $\{0,1\}$ in $\{0,1\}^n$ and $\boldsymbol{X}_n := (X_1, \ldots, X_n)$. For any integer $i \le n$, we have

$$\sum_{\boldsymbol{x}_{i-1}} \left| \mathrm{P}^n_{\boldsymbol{X}_{i-1}}(\boldsymbol{x}_{i-1}) - \overline{\mathrm{P}}^n_{\boldsymbol{X}_{i-1}}(\boldsymbol{x}_{i-1}) \right| \le \delta_n .$$

Also, for any value $x'_i \in \{0,1\}$, we have

$$
\begin{aligned}
&\sum_{\boldsymbol{x}_{i-1}} \mathrm{P}^n_{\boldsymbol{X}_{i-1}}(\boldsymbol{x}_{i-1}) \left| \mathrm{P}^n_{X_i|\boldsymbol{X}_{i-1}=\boldsymbol{x}_{i-1}}(x_i') - \overline{\mathrm{P}}^n_{X_i|\boldsymbol{X}_{i-1}=\boldsymbol{x}_{i-1}}(x_i') \right| \\
\le& \sum_{\boldsymbol{x}_{i-1}} \left| \mathrm{P}^n_{\boldsymbol{X}_{i-1}}(\boldsymbol{x}_{i-1}) \mathrm{P}^n_{X_i|\boldsymbol{X}_{i-1}=\boldsymbol{x}_{i-1}}(x_i') - \overline{\mathrm{P}}^n_{\boldsymbol{X}_{i-1}}(\boldsymbol{x}_{i-1}) \overline{\mathrm{P}}^n_{X_i|\boldsymbol{X}_{i-1}=\boldsymbol{x}_{i-1}}(x_i') \right| \\
&+ \sum_{\boldsymbol{x}_{i-1}} \left| \mathrm{P}^n_{\boldsymbol{X}_{i-1}}(\boldsymbol{x}_{i-1}) - \overline{\mathrm{P}}^n_{\boldsymbol{X}_{i-1}}(\boldsymbol{x}_{i-1}) \right| \overline{\mathrm{P}}^n_{X_i|\boldsymbol{X}_{i-1}=\boldsymbol{x}_{i-1}}(x_i') \\
\le& \sum_{\boldsymbol{x}_i} \left| \mathrm{P}^n_{\boldsymbol{X}_i}(\boldsymbol{x}_i) - \overline{\mathrm{P}}^n_{\boldsymbol{X}_i}(\boldsymbol{x}_i) \right| + \sum_{\boldsymbol{x}_{i-1}} \left| \mathrm{P}^n_{\boldsymbol{X}_{i-1}}(\boldsymbol{x}_{i-1}) - \overline{\mathrm{P}}^n_{\boldsymbol{X}_{i-1}}(\boldsymbol{x}_{i-1}) \right| \\
\le& \delta_n + \delta_n = 2\delta_n .
\end{aligned} \tag{2.199}
$$

We define the function $Y_{x_i'}(\boldsymbol{x}_{i-1}) := |\mathrm{P}^n_{X_i|\boldsymbol{X}_{i-1}=\boldsymbol{x}_{i-1}}(x_i') - \overline{\mathrm{P}}^n_{X_i|\boldsymbol{X}_{i-1}=\boldsymbol{x}_{i-1}}(x_i')|$. Applying Markov inequality to the random variable $Y_{x_i'}(\boldsymbol{X}_{i-1})$, from (2.199), we have the inequality

$$
\mathrm{P}^n_{\boldsymbol{X}_{i-1}}(\{\boldsymbol{x}_{i-1} | \, |\mathrm{P}^n_{X_i|\boldsymbol{X}_{i-1}=\boldsymbol{x}_{i-1}}(x_i') - \overline{\mathrm{P}}^n_{X_i|\boldsymbol{X}_{i-1}=\boldsymbol{x}_{i-1}}(x_i')| \le \delta\}) \ge 1 - \frac{2\delta_n}{\delta}. \tag{2.200}
$$

Let Ω_i be the set of $\boldsymbol{x}_{i-1} = (x_1, \dots x_{i-1})$ satisfying the condition inside of the parenthesis in the LHS of (2.200). Then, K3 (Nonnegativity) implies that

$$
\begin{aligned}
&\sum_{\boldsymbol{x}_{i-1}} \mathrm{P}^n_{\boldsymbol{X}_{i-1}}(\boldsymbol{x}_{i-1}) \left| S(\mathrm{P}^n_{X_i|\boldsymbol{X}_{i-1}=\boldsymbol{x}_{i-1}}) - S(\overline{\mathrm{P}}^n_{X_i|\boldsymbol{X}_{i-1}=\boldsymbol{x}_{i-1}}) \right| \\
=& \sum_{\boldsymbol{x}_{i-1} \in \Omega_i} \mathrm{P}^n_{\boldsymbol{X}_{i-1}}(\boldsymbol{x}_{i-1}) \left| S(\mathrm{P}^n_{X_i|\boldsymbol{X}_{i-1}=\boldsymbol{x}_{i-1}}) - S(\overline{\mathrm{P}}^n_{X_i|\boldsymbol{X}_{i-1}=\boldsymbol{x}_{i-1}}) \right| \\
&+ \sum_{\boldsymbol{x}_{i-1} \in \Omega_i^c} \mathrm{P}^n_{\boldsymbol{X}_{i-1}}(\boldsymbol{x}_{i-1}) \left| S(\mathrm{P}^n_{X_i|\boldsymbol{X}_{i-1}=\boldsymbol{x}_{i-1}}) - S(\overline{\mathrm{P}}^n_{X_i|\boldsymbol{X}_{i-1}=\boldsymbol{x}_{i-1}}) \right| \\
=& \sum_{\boldsymbol{x}_{i-1} \in \Omega_i} \mathrm{P}^n_{\boldsymbol{X}_{i-1}}(\boldsymbol{x}_{i-1}) \epsilon + \sum_{\boldsymbol{x}_{i-1} \in \Omega_i^c} \mathrm{P}^n_{\boldsymbol{X}_{i-1}}(\boldsymbol{x}_{i-1}) R \\
\le& \epsilon + 2\frac{\delta_n}{\delta} R \le \epsilon + \epsilon = 2\epsilon.
\end{aligned} \tag{2.201}
$$

Also, K3 (Nonnegativity) implies that

$$
\begin{aligned}
&\sum_{\boldsymbol{x}_{i-1}} \left| \mathrm{P}^n_{\boldsymbol{X}_{i-1}}(\boldsymbol{x}_{i-1}) - \overline{\mathrm{P}}^n_{\boldsymbol{X}_{i-1}}(\boldsymbol{x}_{i-1}) \right| S(\overline{\mathrm{P}}_{X_i|X_1=x_1,\dots,X_{i-1}=x_{i-1}}) \\
\le& \sum_{\boldsymbol{x}_{i-1}} \left| \mathrm{P}^n_{\boldsymbol{X}_{i-1}}(\boldsymbol{x}_{i-1}) - \overline{\mathrm{P}}^n_{\boldsymbol{X}_{i-1}}(\boldsymbol{x}_{i-1}) \right| R \le \delta_n R \le \frac{\epsilon\delta}{2}.
\end{aligned} \tag{2.202}
$$

On the other hand, K5 (Chain rule) implies that

$$S(\mathrm{P}^n_{X_n}) = \sum_{i=1}^{n} \sum_{\boldsymbol{x}_{i-1}} \mathrm{P}^n_{\boldsymbol{X}_{i-1}}(\boldsymbol{x}_{i-1}) S(\mathrm{P}^n_{X_i|\boldsymbol{X}_{i-1}=\boldsymbol{x}_{i-1}}). \tag{2.203}$$

Thus, we have

$$\begin{aligned}
&\left| S(\mathrm{P}_{X_1,\ldots,X_n}) - S(\overline{\mathrm{P}}_{X_1,\ldots,X_n}) \right| \\
\overset{(a)}{\le} & \sum_{i=1}^{n} \sum_{\boldsymbol{x}_{i-1}} \left| \mathrm{P}^n_{\boldsymbol{X}_{i-1}}(\boldsymbol{x}_{i-1}) S(\mathrm{P}^n_{X_i|\boldsymbol{X}_{i-1}=\boldsymbol{x}_{i-1}}) - \overline{\mathrm{P}}^n_{\boldsymbol{X}_{i-1}}(\boldsymbol{x}_{i-1}) S(\overline{\mathrm{P}}^n_{X_i|\boldsymbol{X}_{i-1}=\boldsymbol{x}_{i-1}}) \right| \\
\le & \sum_{i=1}^{n} \sum_{\boldsymbol{x}_{i-1}} \left| \mathrm{P}^n_{\boldsymbol{X}_{i-1}}(\boldsymbol{x}_{i-1}) S(\mathrm{P}^n_{X_i|\boldsymbol{X}_{i-1}=\boldsymbol{x}_{i-1}}) - \mathrm{P}^n_{\boldsymbol{X}_{i-1}}(\boldsymbol{x}_{i-1}) S(\overline{\mathrm{P}}^n_{X_i|\boldsymbol{X}_{i-1}=\boldsymbol{x}_{i-1}}) \right| \\
& + \left| \mathrm{P}^n_{\boldsymbol{X}_{i-1}}(\boldsymbol{x}_{i-1}) S(\overline{\mathrm{P}}^n_{X_i|\boldsymbol{X}_{i-1}=\boldsymbol{x}_{i-1}}) - \overline{\mathrm{P}}^n_{\boldsymbol{X}_{i-1}}(\boldsymbol{x}_{i-1}) S(\overline{\mathrm{P}}^n_{X_i|\boldsymbol{X}_{i-1}=\boldsymbol{x}_{i-1}}) \right| \\
= & \sum_{i=1}^{n} \sum_{\boldsymbol{x}_{i-1}} \mathrm{P}^n_{\boldsymbol{X}_{i-1}}(\boldsymbol{x}_{i-1}) \left| S(\mathrm{P}^n_{X_i|\boldsymbol{X}_{i-1}=\boldsymbol{x}_{i-1}}) - S(\overline{\mathrm{P}}^n_{X_i|\boldsymbol{X}_{i-1}=\boldsymbol{x}_{i-1}}) \right| \\
& + \left| \mathrm{P}^n_{\boldsymbol{X}_{i-1}}(\boldsymbol{x}_{i-1}) - \overline{\mathrm{P}}^n_{\boldsymbol{X}_{i-1}}(\boldsymbol{x}_{i-1}) \right| S(\overline{\mathrm{P}}^n_{X_i|\boldsymbol{X}_{i-1}=\boldsymbol{x}_{i-1}}) \\
\overset{(b)}{\le} & \sum_{i=1}^{n} 2\epsilon + \frac{\epsilon\delta}{2} = n(2\epsilon + \frac{\epsilon\delta}{2}),
\end{aligned}$$

where (a) follows from (2.203), and (b) follows from (2.201) and (2.202). Hence, A4 (Asymptotic continuity) holds.

Next, we show (2) $\Rightarrow$ (3). For a distribution p and $\epsilon > 0$, according to Lemma 2.5, we choose a sequence of maps f_n. A1 (Normalization) and A2 (Weak additivity) imply that $S(p_{\mathrm{mix},\Omega_n}) = \lfloor (H(p) - \epsilon)n/\log 2 \rfloor \log 2$. A2 (Weak additivity) and (Monotonicity) imply that $S(p^n \circ f_n^{-1}) \le S(p^n) \le nS(p)$. By using these relations, A4 (Asymptotic continuity) implies that $H(p) - \epsilon \le S(p)$. Since ϵ is arbitrary, we have $H(p) \le S(p)$. Similarly, using Lemma 2.6, we can show that $H(p) \ge S(p)$. Thus, we obtain $H(p) = S(p)$.

Now, we show (3) $\Rightarrow$ (1). K1 (Normalization), K2 (Continuity), and K3 (Non-negativity) are oblivious from the definition (2.2). K4 (Expandability) and K5 (Chain rule) follow from (2.4) and (2.5), respectively. ■

To show Lemmas 2.5 and 2.6, we prepare another lemma as follows.

Lemma 2.7 (Han [26, Lemma 2.1.1.]) *For any two distributions* P_X *on* $\mathcal{X}$ *and* P_Y *on* $\mathcal{Y}$, *there exists a function* f *from* $\mathcal{X}$ *to* $\mathcal{Y}$ *such that*

$$d_1(\mathrm{P}_{f(X)}, \mathrm{P}_Y) \le e^{-\gamma} + \max(\mathrm{P}_X(S(a+\gamma)^c), \mathrm{P}_Y(T(a)^c)), \tag{2.204}$$

where

$$S(a) := \{x \in \mathcal{X} | \mathrm{P}_X(x) \le e^{-a}\}, \quad T(a) := \{y \in \mathcal{Y} | \mathrm{P}_Y(y) \ge e^{-a}\}.$$

Proof We define a map f from $\mathcal{X}$ to $\mathcal{Y}$ as follows. We number all of elements of $T(a)$ as $T(a) = \{y_1, \ldots, y_n\}$. So, we have

$$n = |T(a)| \le e^a. \tag{2.205}$$

For this purpose, we define n disjoint subsets $f^{-1}(y_1), \ldots, f^{-1}(y_n)$ as subsets of $\mathcal{X}$. First, we choose a subset $f^{-1}(y_1) \subset S(a+\gamma)$ such that

$$\sum_{x \in f^{-1}(y_1)} P_X(x) \le P_Y(y_1) < \sum_{x \in f^{-1}(y_1)} P_X(x) + e^{-a-\gamma}.$$

for any $x' \in S(a+\gamma) \setminus f^{-1}(y_1)$. Next, we choose a subset $f^{-1}(y_2) \subset S(a+\gamma) \setminus f^{-1}(y_1)$ such that

$$\sum_{x \in f^{-1}(y_2)} P_X(x) \le P_Y(y_2) < \sum_{x \in f^{-1}(y_2)} P_X(x) + e^{-a-\gamma}.$$

We repeat this selection as long as possible. Let y_l be the final element y whose inverse set $f^{-1}(y)$ can be defined in this way.

Consider the case $l = n$. We reselect $f^{-1}(y_n)$ to be $(\cup_{i=1}^{n-1} f^{-1}(y_i))^c$. Then, the set $f^{-1}(y)$ is empty for $y \in T(a)^c$. Due to Exercise 2.12, we have

$$\begin{aligned} d_1(P_{f(X)}, P_Y) \le & \sum_{i=1}^{n-1} |P_{f(X)}(y_i) - P_Y(y_i)| + \sum_{y \in T(a)^c} |P_{f(X)}(y) - P_Y(y)| \\ \le & \sum_{i=1}^{n-1} e^{-a-\gamma} + \sum_{y \in T(a)^c} P_Y(y) \overset{(a)}{\le} e^{-\gamma} + P_Y(T(a)^c), \end{aligned}$$

where (a) follows from (2.205).

Next, we consider the case $l < n$. We define $f^{-1}(y_{l+1}) := \mathcal{X} \setminus (\cup_{i=1}^{l} f^{-1}(y_i))^c$. Then, for $y \in \{y_1, \ldots y_{l+1}\}^c$, $f^{-1}(y)$ is empty. Since

$$\sum_{i=1}^{l+1} P_Y(y_i) \ge \sum_{x \in S(a+\gamma)} P_X(x),$$

we have

$$\sum_{y \in \{y_1, \ldots y_{l+1}\}^c} P_Y(y) \le \sum_{x \in S(a+\gamma)^c} P_X(x). \tag{2.206}$$

Hence, due to Exercise 2.12, we have

$$
\begin{aligned}
d_1(\mathrm{P}_{f(X)}, \mathrm{P}_Y) \le & \sum_{i=1}^{l} |\mathrm{P}_{f(X)}(y_i) - \mathrm{P}_Y(y_i)| + \sum_{y \in \{y_1,\ldots y_{l+1}\}^c} |\mathrm{P}_{f(X)}(y) - \mathrm{P}_Y(y)| \\
\le & \sum_{i=1}^{l} e^{-a-\gamma} + \sum_{y \in \{y_1,\ldots y_{l+1}\}^c} \mathrm{P}_Y(y) \\
\overset{(a)}{\le} & e^{-\gamma} + \sum_{x \in S(a+\gamma)^c} \mathrm{P}_X(x) = e^{-\gamma} + \mathrm{P}_X(S(a+\gamma)^c),
\end{aligned}
$$

where (a) follows from (2.205) and (2.206). ■

Now, using Lemma 2.7, we show Lemmas 2.5 and 2.6.

Proof of Lemma 2.5 We apply Lemma 2.7 to the case when $a = (H(p) - \epsilon)n$, $\gamma = n\frac{\epsilon}{2}$, and P_X and P_Y are p^n and the uniform distribution $p_{\mathrm{mix},\Omega_n}$ on the set $\Omega_n = \{0,1\}^{\lfloor (H(p)-\epsilon)n/\log 2 \rfloor}$, respectively. Then, $\mathrm{P}_Y(T(a)^c) = 0$ and $e^{-\gamma} \to 0$. Since RHS of (2.44) goes to zero with $R < H(p)$, we have $\mathrm{P}_X(S(a+\gamma)^c) \to 0$. Therefore, we obtain the desired argument. ■

Proof of Lemma 2.6 We apply Lemma 2.7 to the case when $a + \gamma = (H(p) + \epsilon)n$, $\gamma = n\frac{\epsilon}{2}$, and P_Y and P_X are p^n and the uniform distribution $p_{\mathrm{mix},\Omega_n'}$ on the set $\Omega_n' = \{0,1\}^{\lfloor (H(p)+\epsilon)n/\log 2 \rfloor}$, respectively. Then, $\mathrm{P}_X(S(a+\gamma)^c) = 0$ and $e^{-\gamma} \to 0$. Since RHS of (2.42) goes to zero with $R > H(p)$, we have $\mathrm{P}_Y(T(a)^c) \to 0$. Therefore, we obtain the desired argument. ■

Exercises

2.49 Show that the Rényi entropy $H_{1+s}(p)$ and the min entropy $H_{\min}(p)$ do not satisfy A4 (Asymptotic continuity) for $s > 0$ as follows.
(**a**) Define the distribution $p_{d,\epsilon}$ on $\{0, 1, \ldots, d-1\}$ by

$$
p_{d,\epsilon}(i) := \begin{cases} \frac{1}{d} + \epsilon & \text{if } i = 0 \\ \frac{1}{d} - \frac{\epsilon}{d-1} & \text{if } i > 0. \end{cases} \tag{2.207}
$$

Show that $d_1(p_{d,\epsilon}, p_{\mathrm{mix},d}) = \epsilon$.
(**b**) Show that $H_{\min}(p_{d,\epsilon}) = \log d - \log(1 + d\epsilon)$.
(**c**) Assume that $d\epsilon \to \infty$ as $d \to \infty$. Show that $\frac{H_{\min}(p_{\mathrm{mix},d}) - H_{\min}(p_{d,\epsilon})}{\log d} = 1 + \frac{\log \epsilon}{\log d} + O(\frac{1}{d\epsilon \log d})$ as $d \to \infty$.
(**d**) Show that $H_{1+s}(p_{d,\epsilon}) = \log d - \frac{1}{s}\log(\frac{1}{d}(1 + d\epsilon)^{1+s} + \frac{d-1}{d}(1 - \frac{d\epsilon}{d-1})^{1+s})$.
(**e**) Assume that $\frac{1}{d}(d\epsilon)^{1+s} \to \infty$ as $d \to \infty$. Show that $\frac{H_{1+s}(p_{\mathrm{mix},d}) - H_{1+s}(p_{d,\epsilon})}{\log d} = 1 + \frac{(1+s)\log \epsilon}{s \log d} + O((d\epsilon)^{-(1+s)}\frac{d}{\log d})$ as $d \to \infty$.

2.50 Show that the Rényi entropy $H_{1-s}(p)$ and the max entropy $H_{\max}(p)$ do not satisfy A4 (Asymptotic continuity) for $s \in (0, 1)$ as follows.
(**a**) Define the distribution $p'_{d,\epsilon}$ on $\{0, 1, \ldots, d-1\}$ by

$$p'_{d,\epsilon}(i) := \begin{cases} 1-\epsilon & \text{if } i = 0 \\ \frac{\epsilon}{d-1} & \text{if } i > 0. \end{cases} \tag{2.208}$$

Show that $d_1(p'_{d,\epsilon}, p'_{d,0}) = \epsilon$.
(**b**) Show that $\frac{H_{\max}(p'_{\mathrm{mix},d}) - H_{\max}(p'_{d,0})}{\log d} = 1$ for $\epsilon > 0$.
(**c**) Show that $H_{1-s}(p'_{d,\epsilon}) = -\frac{1}{s}\log((1-\epsilon)^{1-s} + (d-1)(\frac{\epsilon}{d-1})^{1-s})$.
(**d**) Show that $\frac{H_{1+s}(p_{\mathrm{mix},d}) - H_{1+s}(p_{d,\epsilon})}{\log d} = \frac{1-s}{s \log d}(d-1)^s \epsilon^{1-s} + O(\frac{\epsilon}{\log d}) + O(\frac{\epsilon^{2(1-s)}}{\log d})$ as $\epsilon \to 0$.

2.51 Show that $e^{-2H_2(p)}$ satisfies A4 (Asymptotic continuity) for $s > 0$ by showing the following inequality. That is, show that the continuity of $e^{-2H_2(p)}$ does not depend on the cardinality of the supports of p and q.

$$|e^{-H_2(p)} - e^{-H_2(q)}| \le 2d_1(p, q). \tag{2.209}$$

2.6 Large Deviation on Sphere

Next, we consider a probability distribution on the set of pure states. In quantum information, if we have no information on the given system $\mathcal{H} = \mathbb{C}^l$, it is natural to assume that the probability distribution is invariant with respect to the action of the unitary group $\mathrm{U}(l)$ on the set of pure states. Such a distribution is unique and is called the Haar measure, which is denoted by $\mu_{\mathcal{H}}$. Since the normalized vector is given as $|\phi\rangle \in \mathbb{C}^l$ satisfying $\|\phi\| = 1$, the distribution $\mu_{\mathcal{H}}$ is given as a distribution on the set of pure states satisfying that

$$\int_B \mu_{\mathcal{H}}(d\phi) = \int_B \mu_{\mathcal{H}}(dU\phi) \text{ for } U \in \mathrm{U}(l). \tag{2.210}$$

That is, the Haar measure is defined as the unique distribution satisfying (2.210). When the pure state is regarded as an element of the $2l-1$-dimensional sphere S^{2l-1}, the distribution $\mu_{\mathcal{H}}$ is given as a distribution on the $2l-1$-dimensional sphere. More generally, the Haar measure μ_{S^n} on n-dimensional sphere S^n is given as the distribution satisfying that

$$\int_B \mu_{S^n}(dx) = \int_B \mu_{S^n}(dgx) \text{ for } g \in \mathrm{O}(n+1). \tag{2.211}$$

The Haar measure has several useful properties. For example, the invariance guarantees that

$$\int |\phi\rangle\langle\phi|\mu(d\phi) = \frac{1}{l} I. \tag{2.212}$$

Further, when $\mathcal{H} = \mathbb{C}^l$ is spanned by the basis $\{|e_i\rangle\}_{i=1}^l$, for n-th permutation π, we define the unitary U_π on $\mathcal{H}^{\otimes n}$ as

$$U_\pi(|v_1, \ldots, v_n\rangle) := |v_{\pi(1)}, \ldots, v_{\pi(n)}\rangle. \tag{2.213}$$

Then, we define the n-th symmetric subspace $\mathcal{H}_{s,n} \subset \mathcal{H}^{\otimes n}$ as the space spanned by $\{\sum_\pi U_\pi(|e_1, \ldots, e_1, e_2, \ldots, e_2, \ldots, e_l, \ldots, e_l\rangle)\}$. The dimension of $\mathcal{H}_{s,n}$ is $\binom{l+n-1}{l-1}$, and the invariance implies that

$$\int |\phi\rangle\langle\phi|^{\otimes n} \mu_{\mathcal{H}}(d\phi) = \frac{1}{\binom{l+n-1}{l-1}} P_{\mathcal{H}_{s,n}}, \tag{2.214}$$

where $P_{\mathcal{H}_{s,n}}$ is the projection to $\mathcal{H}_{s,n}$. When a pure state ρ on $\mathcal{H}^{\otimes n}$ is invariant for U_π with an arbitrary n-th permutation π, the pure state ρ is a state on $\mathcal{H}_{s,n}$. Hence, we have

$$\rho \leq \binom{l+n-1}{l-1} \int |\phi\rangle\langle\phi|^{\otimes n} \mu_{\mathcal{H}}(d\phi). \tag{2.215}$$

Here, $\binom{l+n-1}{l-1}$ is upper bounded by $(n+1)^{d-1}$.

In quantum information, we often consider the stochastic behavior of a function of a pure state under the Haar measure $\mu_{\mathcal{H}}$. In order to discuss this issue, we need the following preparation. First, we define the median of a real-valued random variable X as

$$\mathrm{Med}_p(X) \stackrel{\text{def}}{=} \frac{\overline{\mathrm{Med}_p(X)} + \underline{\mathrm{Med}_p(X)}}{2} \tag{2.216}$$

$$\overline{\mathrm{Med}_p(X)} \stackrel{\text{def}}{=} \inf\{r | p\{x | x \geq r\} < 1/2\} \tag{2.217}$$

$$\underline{\mathrm{Med}_p(X)} \stackrel{\text{def}}{=} \sup\{r | p\{x | x \leq r\} < 1/2\}. \tag{2.218}$$

The cumulative distribution function of the real-valued random variable X is defined as

$$F_{X,p}(a) := p\{x | x \leq a\}, \tag{2.219}$$

where $p(\Omega)$ is defined for a subset $S \subset \Omega$ as

$$p(S) := \sum_{x \in S} p_x. \tag{2.220}$$

Then, we have the following lemma.

Lemma 2.8 *When given two real-valued random variables X and Y satisfies $F_{X,p} \le F_{Y,p}$, we have* $\mathrm{E}_p X \ge \mathrm{E}_p Y$.

Then, we define the metric $d(x, y)$ between two wave functions x and y in S^{2l-1} as

$$d(x, y) := \cos^{-1} \mathbf{Re}\langle x, y\rangle \in [0, \pi]. \tag{2.221}$$

Then, for a wave function $y \in S^{2l-1}$, we define the subset $D(y, r)$ as

$$D(y, r) := \{x \in S^{2l-1} | d(x, y) \le r\}. \tag{2.222}$$

Then, the probability $\mu_{S^{2l-1}}(D(y, r))$ depends only on r. For a given probability $p \in (0, 1)$, we define $r(p)$ as $\mu_{S^{2l-1}}(D(y, r(p))) = p$. For a given subset $\Omega \subset S^{2l-1}$, we define the subset Ω_ϵ for $\epsilon > 0$ as

$$\Omega_\epsilon := \{x \in S^{2l-1} | d(x, y) \le \epsilon, \exists y \in \Omega\}. \tag{2.223}$$

Then, we prepare the following fundamental lemma.

Lemma 2.9 ([27, Theorem 2.1]) *For a given $p \in (0, 1)$ and $\epsilon > 0$, we have*

$$\min\{\mu_{S^{2l-1}}(\Omega_\epsilon) | \mu_{S^{2l-1}}(\Omega) = p\} = \mu_{S^{2l-1}}(D(y, r(p))_\epsilon), \tag{2.224}$$

where the set $D(y, r(p))_\epsilon$ is illustrated as Fig. 2.2.

Proof We give only an intuitive proof. First, we consider an infinitesimal $\epsilon > 0$. In this case, it is enough to consider the boundary of Ω because the size of boundary

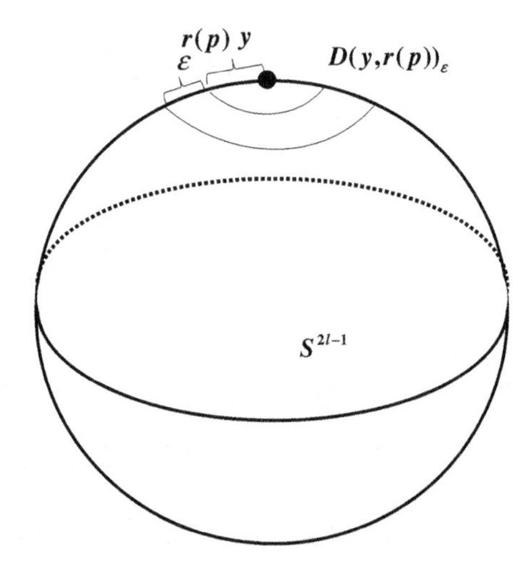

Fig. 2.2 Set $D(y, r(p))_\epsilon$

of Ω is proportional to $\frac{d\mu_{S^{2l-1}}(\Omega_\epsilon)}{d\epsilon}|_{\epsilon=0}$. We can intuitively find that the set $D(y, r(p))$ has the minimum boundary among the subsets Ω satisfying $\mu_{S^{2l-1}}(\Omega) = p$. That is, we obtain $\frac{d\mu_{S^{2l-1}}(\Omega_\epsilon)}{d\epsilon}|_{\epsilon=0} \geq \frac{d\mu_{S^{2l-1}}(D(y,r(p))_\epsilon)}{d\epsilon}|_{\epsilon=0}$.

Next, for $p' > p$ and a subset Ω satisfying $\mu_{S^{2l-1}}(\Omega) = p$, we define the function $f(p', \Omega)$ as $\mu_{S^{2l-1}}(\Omega_{f(p',\Omega)}) = p'$. Then, we have

$$\frac{df(p', \Omega)}{dp'} = \frac{1}{\frac{d\mu_{S^{2l-1}}(\Omega_{f(p',\Omega)+\epsilon})}{d\epsilon}|_{\epsilon=0}} \leq \frac{1}{\frac{d\mu_{S^{2l-1}}(D(y,r(p))_{f(p',D(y,r(p)))+\epsilon})}{d\epsilon}|_{\epsilon=0}}, \tag{2.225}$$

which implies

$$f(p', \Omega) \leq f(p', D(y, r(p))). \tag{2.226}$$

Hence, we obtain

$$\mu_{S^{2l-1}}(\Omega_{f(p',D(y,r(p)))}) \geq \mu_{S^{2l-1}}(\Omega_{f(p',\Omega)}) = \mu_{S^{2l-1}}(D(y, r(p))_{f(p',D(y,r(p)))}).$$

■

Using the above lemma, we obtain the following lemma.

Lemma 2.10 ([27, Corollary 2.2]) *When a subset $\Omega \subset S^{2l-1}$ satisfies $\mu_{S^{2l-1}}(\omega) \geq \frac{1}{2}$, we have*

$$\mu_{S^{2l-1}}(\Omega_\epsilon) \geq 1 - e^{-\epsilon^2(l-1)}/2. \tag{2.227}$$

Proof Thanks to Lemma 2.9, since $D(y, \frac{\pi}{2}) = \frac{1}{2}$, it is enough to show that $D(y, \frac{\pi}{2})_\epsilon = D(y, \frac{\pi}{2} + \epsilon) \geq 1 - e^{\epsilon^2(l-1)}/2$. The size of the boundary of $D(y, \theta)$ is proportional to $\sin^{2l-2}\theta = \cos^{2l-2}(\theta - \frac{\pi}{2})$ for $\theta \in [0, \pi]$. Hence, choosing $\theta' := \theta - \frac{\pi}{2}$, we have

$$D\left(y, \frac{\pi}{2} + \epsilon\right) = \frac{\int_{-\frac{\pi}{2}}^{\epsilon} \cos^{2l-2}\theta' d\theta'}{I_{l-1}}, \tag{2.228}$$

where

$$I_{l-1} := \int_{-\frac{\pi}{2}}^{\frac{\pi}{2}} \cos^{2l-2}\theta' d\theta' = B\left(l - \frac{1}{2}, \frac{1}{2}\right) = \frac{\Gamma(l - \frac{1}{2})\Gamma(\frac{1}{2})}{\Gamma(l)} = \frac{2l-3}{2l-2} I_{l-2}. \tag{2.229}$$

Since $\frac{2l-3}{\sqrt{2l-2}\sqrt{2l-4}} \geq 1$, we have

$$\sqrt{2l-2} I_{l-1} \geq \sqrt{2l-4} I_{l-2}, \tag{2.230}$$

which implies $\sqrt{2l-2} I_{l-1} \geq \sqrt{2} I_1 = \sqrt{2} B\left(\frac{3}{2}, \frac{1}{2}\right) = \frac{\pi}{\sqrt{2}}$.

For $t \in [0, \frac{\pi}{2}]$, the inequality $\cos t \le e^{-\frac{t^2}{2}}$ holds. Using the parameter $u := \sqrt{l-1}\theta$, we have

$$1 - D\left(y, \frac{\pi}{2} + \epsilon\right) = \frac{\int_{\epsilon}^{\frac{\pi}{2}} \cos^{2l-2}\theta' d\theta'}{I_{l-1}} = \frac{1}{\sqrt{l-1}} \frac{\int_{\epsilon\sqrt{l-1}}^{\frac{\pi}{2}\sqrt{l-1}} \cos^{2l-2} \frac{u}{\sqrt{l-1}} du}{I_{l-1}}$$

$$\le \frac{\int_{\epsilon\sqrt{l-1}}^{\frac{\pi}{2}\sqrt{l-1}} \left(e^{-\frac{u^2}{2(l-1)}}\right)^{2l-2} du}{\frac{\pi}{\sqrt{2}}} = \frac{\int_{\epsilon\sqrt{l-1}}^{\frac{\pi}{2}\sqrt{l-1}} e^{-u^2} du}{\frac{\pi}{\sqrt{2}}} \le e^{-\epsilon^2(l-1)}/2,$$

where the final inequality follows from Exercise 2.56. ■

A real-valued continuous function f of S^{2l-1} can be regarded as a real-valued random variable on S^{2l-1}. Then, we define the set Ω_f as

$$\Omega_f := \{x \in S^{2l-1} | f(x) \le \mathrm{Med}_{S^{2l-1}}(f)\}, \tag{2.231}$$

where $\mathrm{Med}_{S^{2l-1}}(f)$ is the abbreviation of the median $\mathrm{Med}_{\mu_{S^{2l-1}}}(f)$ under the Haar measure $\mu_{S^{2l-1}}$ on S^{2l-1}. Using Lemma 2.9, we obtain the inequality

$$\mu_{S^{2l-1}}((\Omega_f)_\epsilon) \ge 1 - e^{-\epsilon^2(l-1)}/2. \tag{2.232}$$

Now, we say that the function f is Lipschitz continuous with the Lipschitz constant C_0 with respect to the metric d in subset $\Omega \subset S^{2l-1}$ when

$$\frac{|f(x) - f(y)|}{d(x, y)} \le C_0, \quad \forall x, y \in \Omega. \tag{2.233}$$

In particular, when $\Omega = S^{2l-1}$, we simply say that the function f is Lipschitz continuous with the Lipschitz constant C_0 with respect to the metric d, which is assumed in the following. Since $(\Omega_f)_\epsilon \subset \{x \in S^{2l-1} | f(x) \ge \mathrm{Med}_{S^{2l-1}}(f) + C_0\epsilon\}^c$, (2.232) implies that

$$\mu_{S^{2l-1}}\{x \in S^{2l-1} | f(x) \ge \mathrm{Med}_{S^{2l-1}}(f) + C_0\epsilon\} \le \mu_{S^{2l-1}}((\Omega_f)_\epsilon^c) \le \frac{e^{-\epsilon^2(l-1)}}{2}. \tag{2.234}$$

Similarly, we can show that

$$\mu_{S^{2l-1}}\{x \in S^{2l-1} | f(x) \le \mathrm{Med}_{S^{2l-1}}(f) - C_0\epsilon\} \le \frac{e^{-\epsilon^2(l-1)}}{2}. \tag{2.235}$$

Hence, we obtain

$$\mu_{S^{2l-1}}\{x \in S^{2l-1} | \, |f(x) - \mathrm{Med}_{S^{2l-1}}(f)| \ge \epsilon\} \le e^{-\frac{\epsilon^2(l-1)}{C_0^2}}, \tag{2.236}$$

which implies that the cumulative distribution function of the real-valued random variable $|f(x) - \mathrm{Med}_{S^{2l-1}}(f)|$ is less than $F(x) := 1 - e^{-\frac{x^2(l-1)}{C_0^2}}$. Now, we simplify the expectation $\mathrm{E}_{\mu_{S^{2l-1}}}$ under the Haar measure $\mu_{S^{2l-1}}$ on S^{2l-1} to $\mathrm{E}_{S^{2l-1}}$. Thus, Lemma 2.8 guarantees that

$$\mathrm{E}_{S^{2l-1}}|f(X) - \mathrm{Med}_{S^{2l-1}}(f)| \le \int_0^\infty x \frac{dF(x)}{dx} dx$$
$$= \int_0^\infty \frac{2(l-1)}{C_0^2} x^2 e^{-\frac{x^2(l-1)}{C_0^2}} dx = \frac{C_0}{2}\sqrt{\frac{\pi}{l-1}},$$

where we used the relation in Exercise 2.55. Thus, we obtain

$$|\mathrm{E}_{S^{2l-1}} f(X) - \mathrm{Med}_{S^{2l-1}}(f)| \le \mathrm{E}_{S^{2l-1}}|f(X) - \mathrm{Med}_{S^{2l-1}}(f)| \le \frac{C_0}{2}\sqrt{\frac{\pi}{l-1}}. \tag{2.237}$$

Finally, given positive numbers δ and C_1, we define the sets

$$\begin{aligned}
\Omega_{\delta,C_1} :=& \left\{x \in S^{2l-1} \middle| f(x) \ge \mathrm{E}_{S^{2l-1}} f(X) + \frac{C_0}{2}\sqrt{\frac{\pi}{l-1}} + C_1\delta \right\} \\
&\subset \{x \in S^{2l-1} | f(x) \ge \mathrm{Med}_{S^{2l-1}}(f) + C_1\delta\}, \\
\tilde{\Omega}_{\delta,C_1} :=& \left\{x \in S^{2l-1} \middle| \mathrm{E}_{S^{2l-1}} f(X) - \frac{C_0}{2}\sqrt{\frac{\pi}{l-1}} < f(x) \right. \\
&\left. < \mathrm{E}_{S^{2l-1}} f(X) + \frac{C_0}{2}\sqrt{\frac{\pi}{l-1}} + C_1\delta \right\} \\
&\supset \{x \in S^{2l-1} | \mathrm{Med}_{S^{2l-1}}(f) < f(x) < \mathrm{Med}_{S^{2l-1}}(f) + C_1\delta\}.
\end{aligned}$$

Then, we obtain the large deviation type bound with respect to the Haar measure on the $2l-1$-dimensional sphere as follows.

Theorem 2.11 *When the function $f(x)$ has the Lipschitz constant C_1 on the subset $\tilde{\Omega}_{\delta,C_1}$, we have*

$$\mu_{S^{2l-1}}(\Omega_{\delta,C_1}) \le e^{-\delta^2(l-1)}/2. \tag{2.238}$$

Here, C_0 is the Lipschitz constant for the whole set, and C_1 is the Lipschitz constant for the specific subset $\tilde{\Omega}_{\delta,C_1}$.

Next, we apply the Haar measure to construct a proper subset of S^{2l-1}. A subset Ω of S^{2l-1} is called an ϵ net of S^{2l-1} when for any element $x \in S^{2l-1}$, there exists an element $y \in S^{2l-1}$ such that $d(x, y) \le \epsilon$.

Lemma 2.11 *There exists an ϵ net Ω of S^{2l-1} whose cardinality is less than $\frac{\sqrt{(2l-1)\pi}}{\sin^{2l-1}\frac{\epsilon}{2}} < \sqrt{(2l-1)\pi}(\frac{2}{\sin\epsilon})^{2l-1}$.*

Proof We choose a subset Ω of S^{2l-1} satisfying the condition that $d(x, y) > \epsilon$ for any two distinct elements $x, y \in \Omega$. We choose the subset Ω so that no subset Ω' strictly larger than Ω satisfies the required condition. Here, a set Ω' is called strictly larger than Ω when Ω' contains Ω and there is at least an element of Ω' that is not included in Ω. A rigorous proof of the existence of such a subset can be given by using Zorn's lemma.

Hence, for any element $x \in S^{2l-1}$, there exists an element $y \in S^{2l-1}$ such that $d(x, y) \le \epsilon$. That is, the set Ω is an ϵ net of S^{2l-1}. Due to the construction, $D(x, \epsilon/2) \cap D(y, \epsilon/2) = \emptyset$ for any two distinct elements $x, y \in \Omega$. Thus, $|\Omega| \mu_{S^{2l-1}} (D(x, \epsilon/2)) = \sum_{x\in\Omega} \mu_{S^{2l-1}}(D(x, \epsilon/2)) \le 1$. That is, $|\Omega| \le \frac{1}{\mu_{S^{2l-1}}(D(x,\epsilon/2))}$. The probability $\mu_{S^{2l-1}}(D(x, \epsilon/2))$ is evaluated by using Exercise 2.57 as

$$\begin{aligned}
&\mu_{S^{2l-1}}(D(x, \epsilon/2)) = \int_0^{\epsilon/2} \sin^{2l-2}\theta d\theta/I_{l-1} \ge \int_0^{\epsilon/2} \left(\frac{\sin\frac{\epsilon}{2}}{\epsilon/2}\theta\right)^{2l-2} d\theta/I_{l-1} \\
=&\frac{\sin^{2l-2}\frac{\epsilon}{2}}{(\epsilon/2)^{2l-2}}[\theta^{2l-1}/(2l-1)I_{l-1}]_0^{\epsilon/2} = \frac{\sin^{2l-2}\frac{\epsilon}{2}}{(\epsilon/2)^{2l-2}}(\frac{\epsilon}{2})^{2l-1}/(2l-1)I_{l-1} \\
=&\frac{\frac{\epsilon}{2}\sin^{2l-2}\frac{\epsilon}{2}}{(2l-1)I_{l-1}} \ge \frac{\sin^{2l-1}\frac{\epsilon}{2}}{\sqrt{(2l-1)\pi}}.
\end{aligned}$$

where the relation $\frac{\epsilon}{2} \ge \sin\frac{\epsilon}{2}$ is used. ∎

Exercises

2.52 Show that $\| |x\rangle\langle x| - |y\rangle\langle y| \|_1 \le 2\sin\epsilon$ when $d(x, y) = \epsilon \le \frac{\pi}{2}$ and $x, y \in S^{2l-1}$.

2.53 Show that $\| |x\rangle\langle x| - |y\rangle\langle y| \|_2 \le \sqrt{2}d(x, y)$.

2.54 Show that $\| |x\rangle - |y\rangle \| \le 2\sin\frac{d(x,y)}{2} \le d(x, y)$.

2.55 Show that $\int_0^\infty 2cx^2 e^{-cx^2} dx = \frac{1}{2}\sqrt{\frac{\pi}{c}}$.

2.56 Show $\frac{\int_{\epsilon\sqrt{l-1}}^{\frac{\pi}{2}\sqrt{l-1}} e^{-u^2} du}{\frac{\pi}{\sqrt{2}}} \le e^{-\epsilon^2(l-1)}/2$ when $u \ge 0$ and $\epsilon > 0$.

2.57 Show that

$$(2l-1)B\left(l-\frac{1}{2},\frac{1}{2}\right) \le \sqrt{(2l-1)\pi} \tag{2.239}$$

by following the steps below.

(a) Show the equation $B(l - \frac{1}{2}, \frac{1}{2}) = \pi \cdot \prod_{k=1}^{l-1} \frac{2k-1}{2k}$.
(b) Show the inequality $\sum_{k=1}^{l-1} \log\frac{2k}{2k-1} \ge \frac{1}{2}\log(2l-1)$.
(c) Show the inequality (2.239).

2.7 Related Books

In this chapter, we treat several important topics in information science from the probabilistic viewpoint. In Sect. 2.1, information quantities e.g., entropy, relative entropy, mutual information, Rényi entropy, and conditional Rényi entropy are discussed. Its discussion and its historical notes except for Rényi entropy and Conditional Rényi entropy appear in Chap. 2 of Cover and Thomas [28]. Conditional Rényi entropy is recently introduced and discussed by several papers [29–31] from various viewpoints. This quantity will be investigated much more deeply in future.

Section 2.2 focuses on information geometry. Amari and Nagaoka [2] is a textbook on this topic written by the pioneers in the field. Bregman divergence plays a central role in this section. Although their book [2] contains the Bregman divergence, it discusses information geometry from a more general viewpoint. Recent Amari's paper [6] focuses on the Bregman divergence and derives several important theorems only from the structure of Bregman divergence. This section follows his derivation.

Section 2.3 briefly treats the estimation theory of probability distribution families. Lehmann and Casella [32] is a good textbook covering all of estimation theory. For a more in-depth discussion of its asymptotic aspect, see van der Vaart [7].

Section 2.4.1 reviews the type method. It has been formulated by Csiszár and Köner [8]. Section 2.4.2 treats the large deviation theory including estimation theory. Its details are given in Dembo and Zeitouni [33] and Bucklew [34]. In this book, we give a proof of Cramér's theorem and the Gártner–Ellis theorem. In fact, (2.163), (2.165), (2.169), and (2.171) follow from Markov's inequality. However, its opposite parts are not simple. Many papers and books give their proof. In this book, we prove these inequalities by combining the estimation of the exponential theory and the Legendre transform. This proof seems to be the simplest of known proofs.

Section 2.5 explains how to derive the entropy from natural axioms. This section addresses two sets of axioms. One is close to the axioms proposed by Khinchin [25]. The other is related to asymptotic continuity, and has not been given in anywhere. The latter is related to the entropy measure discussed in Sect. 8.7.

Section 2.6 focuses on the Haar measure, which is a natural distribution on the set of pure states. Milman and Schechtman [27] discusses the asymptotic behavior of a function of the random variable subject to the Haar measure. Since this type discussion attracts much attention in quantum information recently and is applied in Sects. 8.13, 2.6 is devoted to this topic.

2.8 Solutions of Exercises

Exercise 2.1 When $y = f(x)$, $P_{X,Y}(x, y) = P_X(x)$. Hence, $H(X, f(X)) = -\sum_{x,y:y=f(x)} P_{X,Y}(x, y) \log P_{X,Y}(x, y) = -\sum_x P_X(x) \log P_X(x) = H(X)$.

Exercise 2.2 Consider the case $\mathrm{P}_Y(1) = \lambda$, $\mathrm{P}_Y(0) = 1 - \lambda$, $\mathrm{P}_{X|Y=1} = p$, $\mathrm{P}_{X|Y=0} = p'$.

Exercise 2.3 The concavity of entropy guarantees that the maximum of $H(p)$ under the above condition is realized by the distribution $(a, \frac{1-a}{k-1}, \ldots, \frac{1-a}{k-1})$, whose entropy is $h(a) + (1-a)\log(k-1)$.

Exercise 2.4 $H(p_A \times p_B) = -\sum_{\omega_A,\omega_B} p_A(\omega_A)p_B(\omega_B)\log(p_A(\omega_A)p_B(\omega_B)) = -\sum_{\omega_A} p_A(\omega_A)\log p_A(\omega_A) - \sum_{\omega_B} p_B(\omega_B)\log p_B(\omega_B) = H(p_A) + H(p_B)$.

Exercise 2.5 $D(p_A \times p_B \| q_A \times q_B) = \sum_{\omega_A,\omega_B} p_A(\omega_A)p_B(\omega_B)(\log(p_A(\omega_A)p_B(\omega_B)) - \log(q_A(\omega_A)q_B(\omega_B))) = \sum_{\omega_A} p_A(\omega_A)(\log p_A(\omega_A) - \log q_A(\omega_A)) + \sum_{\omega_B} p_B(\omega_B)(\log p_B(\omega_B) - \log q_B(\omega_B)) = D(p_A\|q_A) + D(p_B\|q_B)$.

Exercise 2.6 Define $f(x) := \log x - (x-1)$. Since $f'(x) = \frac{1}{x} - 1$, we find that the maximum of $f(x)$ is attained only when $x = 1$. That is, $f(x) < f(1) = 0$.

Exercise 2.7 Apply a stochastic transition matrix of rank 1 to Theorem 2.1.

Exercise 2.8 $D_f(p\|q) = \sum_i p_i \left(1 - \sqrt{\frac{q_i}{p_i}}\right) = 1 - \sum_i \sqrt{p_i q_i} = \frac{1}{2}\sum_i \left(\sqrt{p_i} - \sqrt{q_i}\right)^2$.

Exercise 2.9 Use the fact that $\sum_j \sum_i Q^i_j |p_i - q_i| \geq \sum_j |\sum_i Q^i_j (p_i - q_i)|$.

Exercise 2.10 Consider the $x \geq y$ and $x < y$ cases separately.

Exercise 2.11

(**a**) Use $|p_i - q_i| = |\sqrt{p_i} - \sqrt{q_i}||\sqrt{p_i} + \sqrt{q_i}|$.
(**b**) Use $p_i + q_i \geq 2\sqrt{p_i}\sqrt{q_i}$.

Exercise 2.12 We find that $p_{x_0} - q_{x_0} = -\sum_{x\neq x_0}(p_x - q_x)$. Thus, $|p_{x_0} - q_{x_0}| \leq \sum_{x\neq x_0} |p_x - q_x|$. Hence, $d_1(p,q) = \frac{1}{2}|p_{x_0} - q_{x_0}| + \frac{1}{2}\sum_{x\neq x_0}|p_x - q_x| \leq \sum_{x\neq x_0} |p_x - q_x|$.

Exercise 2.13 Assume that the datum i generates with the probability distribution p_i. Apply Jensen's inequality to the random variable $\sqrt{q_i/p_i}$ and the convex function $-\log x$.

Exercise 2.14

(**a**) Since Schwartz inequality implies that $\|x\|\|y\| \geq \langle x, y\rangle$ and $\|x\|\|y\| \geq \langle y, x\rangle$, we have

$$
\begin{aligned}
&(\|x\| + \|y\|)^2 - (\|x\|^2 + \langle x, y\rangle + \langle y, x\rangle + \|y\|^2) \\
=&2\|x\|\|y\| - \langle x, y\rangle - \langle y, x\rangle \geq 0.
\end{aligned}
$$

(**b**)

$$
(\|x\| + \|y\|)^2 \geq \|x\|^2 + \langle x, y\rangle + \langle y, x\rangle + \|y\|^2 = \|x + y\|^2.
$$

(**c**) Substitute $\sqrt{p_i} - \sqrt{r_i}$ and $\sqrt{r_i} - \sqrt{q_i}$ into x and y in the inequality given in (**b**).

Exercise 2.15 Check that $\phi'(s|p\|q) = \frac{\sum_i p_i^{1-s} q_i^s (\log q_i - \log p_i)}{\sum_i p_i^{1-s} q_i^s}$.

Exercise 2.16 Check that $\phi''(s|p\|q) =$

$$\frac{(\sum_i p_i^{1-s} q_i^s)(\sum_i p_i^{1-s} q_i^s (\log q_i - \log p_i)^2) - (\sum_i p_i^{1-s} q_i^s (\log q_i - \log p_i))^2}{(\sum_i p_i^{1-s} q_i^s)^2}.$$

Next, use Schwarz's inequality between two vectors 1 and $(-\log p_i + \log q_i)$.

Exercise 2.17 For $0 < s < s'$, we have $\frac{s}{s'} f(s') = (1 - \frac{s}{s'}) f(0) + \frac{s}{s'} f(s') \geq f((1 - \frac{s}{s'}) \cdot 0 + \frac{s}{s'} \cdot s') = f(s)$, which implies that $\frac{f(s')}{s'} \geq \frac{f(s)}{s}$. Similarly, for $0 > s > s'$, we have $\frac{f(s')}{s'} \leq \frac{f(s)}{s}$. Thus, $\frac{f(s)}{s}$ is monotone increasing When $f(s)$ is strictly convex for s, the above inequalities $\leq$ and $\geq$ can be replaced by $<$ and $>$. Hence, $\frac{f(s)}{s}$ is strictly monotone increasing

Exercise 2.18

(a) For simplicity, we denote $\max(b_1, \ldots, b_k)$ by b_M. We choose a subset $S \subset \{1, \ldots, k\}$ such that $b_M = b_i$ for $i \in S$ and $b_M > b_i$ for $i \notin S$. Thus, $\frac{1}{t} \log(\sum_{i=1}^k a_i b_i^t)$ $= \log b_M + \frac{1}{t} \log(\sum_{i \in S} a_i + \sum_{i \notin S} a_i (\frac{b_i}{b_M})^t) \to \log b_M + \frac{1}{t}$ $\log(\sum_{i \in S} a_i) \to \log b_M$ as $t \to \infty$.

Exercise 2.19 $\sum_i p_i^{1-s} q_i^s = \sum_{i:p_i>0} p_i^{1-s} q_i^s \to \sum_{i:p_i>0} q_i^s$ as $s \to 1$.

Exercise 2.20 Solving the equation that the partial derivative equals zero on the RHS. Then, we obtain $\lambda_i = p_i/q_i$. Substituting it into th RHS, we obtain the LHS.

Exercise 2.21 Apply the formula (2.32) to the conditional distribution $P_{XYZ|U=u}$. Then, we have

$$I(X : YZ|U = u) = I(X : Z|U = u) + \sum_z P_Z(z) I(X : Y|Z = z, U = u). \tag{2.240}$$

Taking the expectation for U, we obtain (2.33).

Exercise 2.22 $e^{\psi(s|P_A \times P_B)} = \sum_{a,b} P_A(a)^{1-s} P_B(b)^{1-s} = \sum_a P_A(a)^{1-s} \sum_b P_B(b)^{1-s}$ $= e^{\psi(s|P_A)} e^{\psi(s|P_B)}$.

Exercise 2.23

(b)

$$\begin{aligned} & D(q\|p) - \frac{1}{1-s} D(q\|p_s) \\ =& \sum_x q(x)(\log q(x) - \log p(x)) - \frac{1}{1-s} \sum_x q(x) \log q(x) \end{aligned}$$

$$
\begin{aligned}
&+\sum_x q(x)\log p(x) + \frac{\psi(s)}{1-s} \\
=& -\frac{s}{1-s}\sum_x q(x)\log q(x) + \frac{\psi(s)}{1-s} \\
=& -\frac{s}{1-s}H(q) + \frac{\psi(s)}{1-s} = \frac{s}{1-s}H(p_s) + \frac{\psi(s)}{1-s} \\
=& -\frac{s}{1-s}\sum_x p_s(x)(1-s)\log p(x) - \frac{s}{1-s}\psi(s) + \frac{\psi(s)}{1-s} \\
=& -s\sum_x p_s(x)\log p(x) + \psi(s) = D(p_s\|p).
\end{aligned}
$$

(**c**) The desired inequality follows from the inequality $\frac{1}{1-s}D(q\|p_s) \geq 0$ for $s \leq 1$.

Exercise 2.24

(**a**) It follows from $\psi(s) \geq H(p)$ for $s \in [0, 1]$.
(**b**) The left hand side is zero when $s = 0$.
(**c**) $\psi'(s) = -\sum_x p_s(x)\log p(x)$, $H(p_s) = -(1-s)\sum_x p_s(x)\log p(x) + \psi(s)$, and $D(p_s\|p) = \sum_x p_s(x)\log p(x) - \psi(s)$.
(**e**) It follows from the relations $\frac{d}{ds}H(p_s) < 0$ and $H(p_1) = H(p) < R$.
(**f**) It follows from Exercise 2.23.
(**g**) It follows from (**f**) and the continuity of $H(q)$ and $D(q\|p)$ for q.
(**h**) Since $\psi'(s_R) = (H(p_{s_R}) - \psi(s_R))/(1-s_R) = (R - \psi(s_R))/(1-s_R)$, we have $D(p_{s_R}\|p) = s_R\psi'(s_R) - \psi(s_R) = s_R(R - \psi(s_R))/(1-s_R) - \psi(s_R)$ $= \frac{s_R R - \psi(s_R)}{1-s_R}$.
(**j**) When $s = s_R$, $\frac{R+(s-1)\psi'(s)-\psi(s)}{(1-s)^2}. = 0$. Further, since $\frac{d}{ds}(R + (s-1)\psi'(s) - \psi(s))$ $= (s-1)\psi''(s) > 0$, $\frac{R+(s-1)\psi'(s)-\psi(s)}{(1-s)^2} > 0$ for $s > s_R$ and $\frac{R+(s-1)\psi'(s)-\psi(s)}{(1-s)^2} < 0$ for $s < s_R$. Hence, the maximum of $\frac{sR-\psi(s)}{1-s}$ can be realized with $s = s_R$.
(**k**) Combine (**g**), (**h**), and (**j**).

Exercise 2.25

(**a**) See (**e**) of Exercise 2.24.
(**b**) It follows from Exercise 2.23.
(**c**) See (**g**) of Exercise 2.24.
(**d**) See (**h**) of Exercise 2.24.
(**e**) See (**j**) of Exercise 2.24.
(**f**) Combine (**c**), (**d**), and (**e**).

Exercise 2.26 $\frac{d}{ds}\frac{-\psi(s)}{1-s} = \frac{(s-1)\psi'(s)-\psi(s)}{(1-s)^2} = \frac{-H(p_s)}{(1-s)^2} < 0$. Hence, the supremum is attained with $s \to -\infty$.

Exercise 2.27 Since $-\log \max_i p_i \le H_\alpha(p) = H_{\min}(p) \le H_\alpha(p) \le H_{\max}(p)$, it is enough to show $H_{\max}(p) \le -\log \min_i p_i$. This inequality is equivalent with $\min_i p_i \le \frac{1}{|\{i|p_i>0\}|}$.

Exercise 2.28 Equation (2.72) can be shown by a simple calculation. Equation (2.73) is shown by the following way.

$$\log |\mathcal{X}| - \min_{Q_Y} D(\mathrm{P}_{XY} \| p_{\mathrm{mix},\mathcal{X}} \times Q_Y) = H(X|Y) - \min_{Q_Y} D(\mathrm{P}_Y \| Q_Y). = H(X|Y).$$

Exercise 2.29 Due to (2.74), we have

$$\lim_{s \to 0} H_{1+s}(X|Y) = -\frac{d}{ds} \sum_y \mathrm{P}_Y(y) \sum_x \mathrm{P}_{X|Y=y}(x)^{1+s}|_{s=0} = H(X|Y).$$

Due to (2.74), we have

$$\lim_{s \to 0} H^{\uparrow}_{1+s}(X|Y) = \max_{Q_Y} -\frac{d}{ds} \sum_{x,y} \mathrm{P}_{X,Y}(x,y)^{1+s} Q_Y(y)^{-s}|_{s=0}$$
$$= \max_{Q_Y} - \sum_{x,y} \mathrm{P}_{X,Y}(x,y)(\log \mathrm{P}_{X,Y}(x,y) - \log Q_Y(y)) = H(X|Y).$$

Exercise 2.30 The second expression in (2.74) yields (2.83) and (2.85). (2.81) yields (2.84) and (2.86).

Exercise 2.31 The concavity of $s \mapsto sH_{1+s}(X|Y)$ can be shown from the convexity of $s \mapsto D_{1+s}(p\|q)$(Exercise 2.16). Since the function $s \mapsto D_{1+s}(\mathrm{P}_{XY} \| p_{\mathrm{mix},\mathcal{X}} \times Q_Y)$ is convex, the function $s \mapsto \min_{Q_Y} D_{1+s}(\mathrm{P}_{XY} \| p_{\mathrm{mix},\mathcal{X}} \times Q_Y)$ is also convex. Hence, the function $s \mapsto sH^{\uparrow}_{1+s}(X|Y)$ is concave. Similar to Exercise 2.17, we can show that the functions $s \mapsto H_{1+s}(X|Y)$ and $H^{\uparrow}_{1+s}(X|Y)$ are monotonicallly decreasing.

Exercise 2.32 Due to the equality condition of Hölder inequality, the equality in (2.88) holds if and only if there exists a function $c(y)$ such that $\mathrm{P}_{X|Y=y}(x) = c(y)\mathrm{P}_{XY}(x,y)$, which implies that $\mathrm{P}_{XY}(x,y)^{-s/(1-s)} = c(y)\mathrm{P}_Y(y)$. Hence, we obtain $\mathrm{P}_{XY}(x,y) = c(y)^{-(1-s)/s}\mathrm{P}_Y(y)^{-(1-s)/s}$. This condition is equivalent to $\mathrm{P}_{XY}(x,y) = \frac{1}{|\mathcal{X}|}\mathrm{P}_Y(y)$.

Exercise 2.33 We denote the marginal distributions of X and Y p_X and p_Y respectively. Then, $\mathrm{Cov}_p(X,Y) = \sum_{x,y} p(x,y)(X - \mathrm{E}_p X)(Y - \mathrm{E}_p Y) = \sum_{x,y} p_X(x)$ $p_Y(y)(X - \mathrm{E}_p X)(Y - \mathrm{E}_p Y) = \sum_x p_X(x)(X - \mathrm{E}_p X) \sum_y p_Y(y)(Y - \mathrm{E}_p Y) = 0$.

Exercise 2.34 For $i \ne j$, we have $\sum_{\omega_1,\ldots,\omega_n} p_\theta(\omega_1)\cdots p_\theta(\omega_n) \frac{d \log p_\theta(\omega_i)}{d\theta} \frac{d \log p_\theta(\omega_j)}{d\theta} =$ 0. Hence, $\sum_{\omega_1,\ldots,\omega_n} p^n_\theta(\omega_1,\ldots,x_n)(\frac{d \log p^n_\theta(\omega_1,\ldots,x_n)}{d\theta})^2$
$= \sum_{\omega_1,\ldots,\omega_n} p_\theta(\omega_1)\cdots p_\theta(\omega_n)(\frac{d \log p_\theta(\omega_1)+\cdots+\log p_\theta(\omega_n)}{d\theta})^2$
$= \sum_{\omega_1,\ldots,\omega_n} p_\theta(\omega_1)\cdots p_\theta(\omega_n)(\frac{d \log p_\theta(\omega_1)}{d\theta} + \cdots + \frac{d \log p_\theta(\omega_n)}{d\theta})^2$

$= \sum_{\omega_1,\ldots,\omega_n} p_\theta(\omega_1)\cdots p_\theta(\omega_n) \sum_{i=1}^n (\frac{d\log p_\theta(\omega_i)}{d\theta})^2 + \sum_{i\neq j} \frac{d\log p_\theta(\omega_i)}{d\theta}\frac{d\log p_\theta(\omega_j)}{d\theta}$
$= \sum_{\omega_1,\ldots,\omega_n} p_\theta(\omega_1)\cdots p_\theta(\omega_n) \sum_{i=1}^n (\frac{d\log p_\theta(\omega_i)}{d\theta})^2$
$= \sum_{i=1}^n \sum_{\omega_i} p_\theta(\omega_i)(\frac{d\log p_\theta(\omega_i)}{d\theta})^2 = nJ_\theta.$

Exercise 2.35 Use the approximation
$\sqrt{p_{\theta+\epsilon}(\omega)} \cong \sqrt{p_\theta(\omega)}\sqrt{1 + l_\theta(\omega)\epsilon + \frac{1}{2}\frac{d^2 p_\theta(\omega)}{d\theta^2}\epsilon^2}.$

Exercise 2.36

(**a**) It follows from the Taylor expansion of $p_{\theta+\epsilon}(\omega)$ for ϵ.
(**b**) Since $\frac{d^2\log p_\theta(\omega)}{d^2\theta} = \frac{d^2 p_\theta(\omega)}{d^2\theta}/p_\theta(\omega) - (\frac{dp_\theta(\omega)}{d\theta}/p_\theta(\omega))^2$, we have
$\sum_\omega p_\theta(\omega)\frac{d^2\log p_\theta(\omega)}{d^2\theta}\epsilon^2 = \sum_\omega p_\theta(\omega)(-(\frac{dp_\theta(\omega)}{d\theta}/p_\theta(\omega))^2 + \frac{d^2 p_\theta(\omega)}{d^2\theta}/p_\theta(\omega))$
$= -\sum_\omega p_\theta(\omega)(\frac{dp_\theta(\omega)}{d\theta}/p_\theta(\omega))^2 = -J_\theta$. Thus, $D(p_\theta\|p_{\theta+\epsilon})$
$= \sum_\omega p_\theta(\omega)(\log p_\theta(\omega) - \log p_{\theta+\epsilon}(\omega)) \cong -\sum_\omega p_\theta(\omega)(\frac{d\log p_\theta(\omega)}{d\theta}\epsilon +$
$\frac{1}{2}\frac{d^2\log p_\theta(\omega)}{d^2\theta}\epsilon^2)$
$= -\sum_\omega p_\theta(\omega)\frac{d\log p_\theta(\omega)}{d\theta}\epsilon - \frac{1}{2}\sum_\omega p_\theta(\omega)\frac{d^2\log p_\theta(\omega)}{d^2\theta}\epsilon^2$
$= -\frac{1}{2}\sum_\omega p_\theta(\omega)\frac{d^2\log p_\theta(\omega)}{d^2\theta}\epsilon^2 = \frac{1}{2}J_\theta\epsilon^2.$
(**d**) $D(p_{\theta+\epsilon}\|p_\theta) = \sum_\omega p_{\theta+\epsilon}(\omega)(\log p_{\theta+\epsilon}(\omega) - \log p_\theta(\omega))$
$\cong \sum_\omega (p_\theta(\omega) + \frac{dp_\theta(\omega)}{d\theta}\epsilon + \frac{1}{2}\frac{d^2 p_\theta(\omega)}{d^2\theta}\epsilon^2)(\frac{d\log p_\theta(\omega)}{d\theta}\epsilon + \frac{1}{2}\frac{d^2\log p_\theta(\omega)}{d^2\theta}\epsilon^2)$
$\cong \sum_\omega p_\theta(\omega)(\frac{d\log p_\theta(\omega)}{d\theta}\epsilon + \frac{1}{2}\frac{d^2\log p_\theta(\omega)}{d^2\theta}\epsilon^2) + \sum_\omega \frac{dp_\theta(\omega)}{d\theta}\epsilon\frac{d\log p_\theta(\omega)}{d\theta}\epsilon$
$= \sum_\omega p_\theta(\omega)\frac{1}{2}\frac{d^2\log p_\theta(\omega)}{d^2\theta}\epsilon^2) + \sum_\omega \frac{dp_\theta(\omega)}{d\theta}\frac{d\log p_\theta(\omega)}{d\theta}\epsilon^2 = J_\theta\epsilon^2 - \frac{1}{2}J_\theta\epsilon^2 = \frac{1}{2}J_\theta\epsilon^2.$

Exercise 2.37 $e^{\phi(s|p_\theta\|p_{\theta+\epsilon})} = \sum_x p_\theta(\omega)^{1-s}p_{\theta+\epsilon}(\omega)^s$
$\cong \sum_x p_\theta(\omega)^{1-s}(p_\theta(\omega) + \frac{dp_\theta(\omega)}{d\theta}\epsilon + \frac{1}{2}\frac{d^2 p_\theta(\omega)}{d^2\theta}\epsilon^2)^s$
$= \sum_x p_\theta(\omega)^{1-s}p_\theta(\omega)^s(1 + \frac{dp_\theta(\omega)}{d\theta}p_\theta(\omega)^{-1}\epsilon + \frac{1}{2}\frac{d^2 p_\theta(\omega)}{d^2\theta}p_\theta(\omega)^{-1}\epsilon^2)^s$
$\cong \sum_x p_\theta(\omega)(1 + s(\frac{dp_\theta(\omega)}{d\theta}p_\theta(\omega)^{-1}\epsilon + \frac{1}{2}\frac{d^2 p_\theta(\omega)}{d^2\theta}p_\theta(\omega)^{-1}\epsilon^2) + \frac{s(s-1)}{2}(\frac{dp_\theta(\omega)}{d\theta}$
$p_\theta(\omega)^{-1}\epsilon)^2)$
$= 1 + \sum_x p_\theta(\omega)s(\frac{dp_\theta(\omega)}{d\theta}p_\theta(\omega)^{-1}\epsilon + \sum_x p_\theta(\omega)\frac{1}{2}\frac{d^2 p_\theta(\omega)}{d^2\theta}p_\theta(\omega)^{-1}\epsilon^2$
$+ \frac{s(s-1)}{2}\sum_x p_\theta(\omega)(\frac{dp_\theta(\omega)}{d\theta})^2 p_\theta(\omega)^{-2}\epsilon^2$
$= 1 + \frac{s(s-1)}{2}\sum_x p_\theta(\omega)^{-1}(\frac{dp_\theta(\omega)}{d\theta})^2\epsilon^2 = 1 + \frac{s(s-1)}{2}\epsilon^2 J_\theta$. Thus, $\phi(s|p_\theta\|p_{\theta+\epsilon}) \cong \log$
$(1 + \frac{s(s-1)}{2}\epsilon^2 J_\theta) \cong \frac{s(s-1)}{2}\epsilon^2 J_\theta.$

Exercise 2.38 For arbitrary η and η', and a real number $\lambda \in (0, 1)$, we choose $\tilde{\theta}_0$ such that $\max_{\tilde{\theta}} \sum_k (\lambda\eta_k + (1-\lambda)\eta'_k)\tilde{\theta}^k - \mu(\tilde{\theta}) = \sum_k (\lambda\eta_k + (1-\lambda)\eta'_k)\tilde{\theta}_0^k - \mu(\tilde{\theta})$.
Hence,

$$\nu(\lambda\eta + (1-\lambda)\eta') = \sum_k (\lambda\eta_k + (1-\lambda)\eta'_k)\tilde{\theta}_0^k - \mu(\tilde{\theta})$$
$$= \lambda \sum_k \eta_k \tilde{\theta}_0^k - \mu(\tilde{\theta}) + (1-\lambda)\sum_k \eta_k \tilde{\theta}_0^k - \mu(\tilde{\theta})$$
$$\leq \lambda \max_{\tilde{\theta}} \sum_k \eta_k \tilde{\theta}^k - \mu(\tilde{\theta}) + (1-\lambda)\max_{\tilde{\theta}} \sum_k \eta_k \tilde{\theta}^k - \mu(\tilde{\theta}) = \lambda\nu(\eta) + (1-\lambda)\nu(\eta').$$

Exercise 2.39 Choose the generator $-\log p(x)$. Then, the set $\{p_s(x)\}$ is an exponential family generated by $-\log p(x)$. The set $\{q|H(q) = H(p_s)\}$ is a mixture family generated by $-\log p(x)$. So, Theorem 2.3 directly solves Exercise 2.23.

Exercise 2.40

(a) Since $\eta(\theta) = \sum_\omega p_\theta(\omega)X(\omega)$, X is an unbiased estimator.
(b) Since $\frac{d}{d\eta}\log p_\theta(\omega) = \frac{d\theta}{d\eta}\frac{d}{d\theta}\log p_\theta(\omega) = (\frac{d\eta}{d\theta})^{-1}\frac{d}{d\theta}\log p_\theta(\omega) = (J_\theta)^{-1}\frac{d}{d\theta}$ $\log p_\theta(\omega)$, the Fisher information for η is $J_\theta(J_\theta)^{-2} = J_\theta^{-1}$. Then, the lower bound of the variance of unbiased estimator given by Crámer-Rao inequality is J_θ. The variance of X is also J_θ.
(c) Use $\frac{d\theta}{d\eta} = J_\eta$.
(d) Since $\frac{d\mu}{d\theta} = \eta$, we have $\frac{d\mu}{d\eta} = \frac{d\theta}{d\eta}\frac{d\mu}{d\theta} = J_\eta \eta$. Taking the integral, we obtain the desired equation.
(e) Inequality (2.140) is derived by Schwartz inequality. Since $|\langle X - \eta, l_\eta\rangle_{p_\eta}| = 1$, the equality condition is $\frac{l_\eta}{J_\eta} = X - \eta$.
(f) Replace l_η by $\frac{p_\eta}{d\eta}/p_\eta$. We obtain $\frac{dp_\eta}{d\eta} = J_\eta(X-\eta)p_\eta$.
(g) Define $\theta := \int_0^\eta J_{\eta'}\,d\eta'$, and $\mu(\theta(\eta)) := \int_0^\eta \eta' J_{\eta'}\,d\eta'$.
$\frac{d\mu(\theta(\eta))}{d\theta} = \frac{d\mu(\theta(\eta))}{d\eta}\frac{d\eta}{d\theta} = \eta J_\eta(\frac{d\theta}{d\eta})^{-1} = \eta J_\eta J_\eta^{-1} = \eta$. The function $\log\sum_\omega$ $p_\eta(\omega)e^{\theta X(\omega)}$ also satisfies the same differential equation. Due to the uniqueness of the solution of the differential equation, we have $\mu(\theta(\eta)) = \log\sum_\omega p_\eta(\omega)e^{\theta X(\omega)}$.
Since $\frac{d\log p_\eta}{d\eta} = \frac{dp_\eta}{d\eta}/p_\eta = J_\eta(X-\eta) = J_\eta X - \eta J_\eta$, we have $\log p_\eta = \theta$ $X - \mu(\theta(\eta))$. Hence, we have $p_\eta = e^{\theta X - \mu(\theta)}$.

Exercise 2.41 Show that $\frac{p_\theta(\omega)}{d\theta} = 0$ if and only if $\eta(\theta) = X(\omega)$.

Exercise 2.42 Combine (2.13) and (2.105).

Exercise 2.43 The case of $n \geq m$ can be obtained from $n, n-1, \ldots, m+1 \geq m$. The $n < m$ case may be obtained from $\frac{1}{m}, \frac{1}{m-1}, \ldots, \frac{1}{n+1} \leq \frac{1}{n}$.

Exercise 2.44 $\mathrm{E}_p X = \sum_i p_i x_i \geq \sum_{i:x_i\geq c} p_i x_i \geq c\sum_{i:x_i \geq c} p_i$.

Exercise 2.45 Apply Cramér's theorem to the random variable $\log p_i$.

Exercise 2.46 Equation (2.189) implies that

$$\begin{aligned}
&\lim_{n\to\infty} -\frac{1}{n}\log P^c(p^n, e^{nR}) \\
&\le \lim_{n\to\infty} -\frac{1}{n}\log \max_{q\in T_n:\frac{e^{nH(q)}}{(n+1)^d}>e^{nR}} e^{-nD(p\|q)}\left(1-\frac{(n+1)^d e^{nR}}{e^{nH(q)}}\right) \\
&\le \min_{q:H(q)\ge R} D(p\|q).
\end{aligned}$$

Combing (2.55), we obtain the $\le$ part of (2.188).

Exercise 2.47

$$\begin{aligned}
\lim_{n\to\infty} -\frac{1}{n}\log P(p^n, e^{nR}) &\le \lim_{n\to\infty} -\frac{1}{n}\log \max_{q\in T_n:H(q)\le R} \frac{e^{-nD(q\|p)}}{(n+1)^d} \\
&\le \min_{q:H(q)\le R} D(p\|q).
\end{aligned}$$

Combing (2.65), we obtain the $\le$ part of (2.188).

Exercise 2.48 Since $p_n(0)e^{n\theta\cdot a}+p_n(1)e^{n\theta\cdot -b}=e^{-na}e^{n\theta a}+(1-e^{-na})e^{-n\theta b}$, we have $\mu(\theta)=\lim_{n\to\infty}\frac{1}{n}\log(e^{-na}e^{n\theta a}+(1-e^{-na})e^{-n\theta b})=-\theta b$ for $\theta<\frac{a}{a+b}$ and $\mu(\theta)=-a(1-\theta)$ for $\theta\ge\frac{a}{a+b}$. Hence, we obtain $\mu(\theta)=-\min\{(1-\theta)a,\theta b\}$. Since $-b<x<a$, we have $\max_{\theta>0}(x\theta-\mu(\theta))=\max(\max_{\frac{a}{a+b}>\theta>0}(x\theta+\theta b),$ $\max_{\theta\ge\frac{a}{a+b}}(x\theta+a(1-\theta)))=\max((x+b)\frac{a}{a+b},a+(x-a)\frac{a}{a+b})=\max(\frac{a(x+b)}{a+b},$ $\frac{a(x+b)}{a+b})=\frac{a(x+b)}{a+b}<a$.

On the other hand, since $a>x>-b$, $\lim_{n\to\infty}\frac{1}{n}\log p_n\{X_n\ge x\}=\lim_{n\to\infty}\frac{1}{n}$ $\log p_n(0)=\lim_{n\to\infty}\frac{1}{n}\log e^{-na}=a$.

Exercise 2.49

(**b**) Since $e^{-H_{\min}(p_{d,\epsilon})}=\frac{1}{d}+\epsilon$, we have $H_{\min}(p_{d,\epsilon})=\log d-\log(1+d\epsilon)$.
(**c**) Since $H_{\min}(p_{\mathrm{mix},d})-H_{\min}(p_{d,\epsilon})=\log(1+d\epsilon)=(\log d+\log\epsilon)+O(\frac{1}{d\epsilon})$, we have $\frac{H_{\min}(p_{\mathrm{mix},d})-H_{\min}(p_{d,\epsilon})}{\log d}=1+\frac{\log\epsilon}{\log d}+O(\frac{1}{d\epsilon\log d})$.
(**d**) Since $e^{-sH_{1+s}(p_{d,\epsilon})}=(\frac{1}{d}+\epsilon)^{1+s}+(d-1)(\frac{1}{d}-\frac{\epsilon}{d-1})^{1+s}$, we have $H_{1+s}(p_{d,\epsilon})$ $=-\frac{1}{s}\log((\frac{1}{d}+\epsilon)^{1+s}+(d-1)(\frac{1}{d}-\frac{\epsilon}{d-1})^{1+s})=\log d-\frac{1}{s}\log(\frac{1}{d}(1+d\epsilon)^{1+s}$ $+\frac{d-1}{d}(1-\frac{d\epsilon}{d-1})^{1+s})$.
(**e**) Since $H_{1+s}(p_{\mathrm{mix},d})-H_{1+s}(p_{d,\epsilon})=\frac{1}{s}\log(\frac{1}{d}(1+d\epsilon)^{1+s}+\frac{d-1}{d}(1-\frac{d\epsilon}{d-1})^{1+s})+$ $O(d(d\epsilon)^{-(1+s)})=\frac{1}{s}\log(\frac{1}{d}(1+d\epsilon)^{1+s})+O(d(d\epsilon)^{-(1+s)})=-\frac{\log d}{s}+\frac{1+s}{s}\log(d\epsilon)+$ $O(d(d\epsilon)^{-(1+s)})=\log d+\frac{1+s}{s}\log\epsilon+O(d(d\epsilon)^{-(1+s)})$, we have $\frac{H_{1+s}(p_{\mathrm{mix},d})-H_{1+s}(p_{d,\epsilon})}{\log d}$ $=1+\frac{(1+s)\log\epsilon}{s\log d}+O((d\epsilon)^{-(1+s)}\frac{d}{\log d})$ as $d\to\infty$.

Exercise 2.50

(**b**) Since the cardinality of $p'_{d,\epsilon}$ is d, we have $H_{\max}(p'_{d,\epsilon})=\log d$. Thus, $\frac{H_{\max}(p'_{\mathrm{mix},d})-H_{\max}(p'_{d,0})}{\log d}=1$ for $\epsilon>0$.

(c) Since $e^{sH_{1-s}(p'_{d,\epsilon})} = (1-\epsilon)^{1-s} + (d-1)(\frac{\epsilon}{d-1})^{1-s} = 1-(1-s)\epsilon + (d-1)^s \epsilon^{1-s} + O(\epsilon^2) = 1-(d-1)^s\epsilon^{1-s} + O(\epsilon)$, we have $H_{1-s}(p'_{d,\epsilon}) = -\frac{1}{s}\log((1-\epsilon)^{1-s} + (d-1)(\frac{\epsilon}{d-1})^{1-s})$.

(d) Since $(1-\epsilon)^{1-s} + (d-1)(\frac{\epsilon}{d-1})^{1-s} = 1-(1-s)\epsilon + (d-1)^s\epsilon^{1-s} + O(\epsilon^2) = 1-(d-1)^s\epsilon^{1-s} + O(\epsilon)$, we have $H_{1-s}(p'_{d,\epsilon}) = -\frac{1}{s}\log(1-(d-1)^s\epsilon^{1-s} + O(\epsilon)) = \frac{1-s}{s}(d-1)^s\epsilon^{1-s} + O(\epsilon) + O(\epsilon^{2(1-s)})$. Thus, $\frac{H_{1+s}(p'_{d,\epsilon})-H_{1+s}(p'_{d,0})}{\log d} = \frac{1-s}{s\log d}(d-1)^s\epsilon^{1-s} + O(\frac{\epsilon}{\log d}) + O(\frac{\epsilon^{2(1-s)}}{\log d})$ as $\epsilon \to 0$.

Exercise 2.51 We have

$$
\begin{aligned}
&|e^{-H_2(p)} - e^{-H_2(q)}| = |e^{-H_2(p)} - 2c + dc^2 - e^{-H_2(q)} + 2c - dc^2| \\
=&|\sum_i (p_i - c)^2 - (q_i - c)^2| = |\sum_i (p_i - q_i)(p_i + q_i - 2c)| \\
\le&\sum_i |p_i - q_i||p_i + q_i - 2c| \le \left(\sum_i |p_i - q_i|\right)\max_i |p_i + q_i - 2c| \\
=&2d_1(p,q)\max_i |p_i + q_i - 2c|.
\end{aligned}
$$

Since $\min_c \max_i |p_i + q_i - 2c| \le 1$, we obtain $|e^{-H_2(p)} - e^{-H_2(q)}| \le 2d_1(p,q)$, which implies (2.209).

Exercise 2.52 It is enough to show that $\| |x\rangle\langle x| - |y\rangle\langle y| \|_1 = 2\sin\epsilon$ when $|\langle x|y\rangle| = \cos\epsilon$. When the state $|x\rangle\langle x|$ is written as $\begin{pmatrix} 1 & 0 \\ 0 & 0 \end{pmatrix}$, the other state $|y\rangle\langle y|$ is written as $\begin{pmatrix} cos^2\theta & \cos\theta\sin\theta \\ \cos\theta\sin\theta & \sin^2\theta \end{pmatrix}$. Hence,

$$
|x\rangle\langle x| - |y\rangle\langle y| = \begin{pmatrix} cos^2\theta - 1 & \cos\theta\sin\theta \\ \cos\theta\sin\theta & \sin^2\theta \end{pmatrix}.
$$

Solving the characteristic equation, we obtain the eigenvalues $\pm\sin\epsilon$. Thus, we have $\| |x\rangle\langle x| - |y\rangle\langle y| \|_1 = 2\sin\epsilon$.

Exercise 2.53 It is enough to show the same case as Exercise 2.52. Since the eigenvalues of $|x\rangle\langle x| - |y\rangle\langle y|$ are $\pm\sin\epsilon$, we have $\| |x\rangle\langle x| - |y\rangle\langle y| \|_2 = \sqrt{2\sin^2\epsilon} = \sqrt{2}\sin\epsilon$.

Exercise 2.54 It is enough to show the same case as Exercise 2.52. Choose ϵ as $d(x,y) = \epsilon$. Then, $|x\rangle - |y\rangle = \begin{pmatrix} 1-\cos\epsilon \\ \sin\epsilon \end{pmatrix}$. Thus $\| |x\rangle - |y\rangle \|^2 = (1-\cos\epsilon)^2 + \sin^2\epsilon = 2(1-\cos\epsilon) = 4\sin^2\frac{\epsilon}{2}$. The second inequality follows from $\sin\frac{\epsilon}{2} \le \frac{\epsilon}{2}$.

Exercise 2.55 Use the relation $\int_0^\infty x^2 e^{-\frac{x^2}{2}} dx = \sqrt{\frac{\pi}{2}}$.

Exercise 2.56 Since $u, \epsilon \geq 0$, we have $(u + \epsilon\sqrt{l-1})^2 \geq u^2 + (\epsilon\sqrt{l-1})^2$. Thus

$$\frac{\int_{\epsilon\sqrt{l-1}}^{\frac{\pi}{2}\sqrt{l-1}} e^{-u^2} du}{\frac{\pi}{\sqrt{2}}} = \frac{\int_{\epsilon\sqrt{l-1}}^{\frac{\pi}{2}\sqrt{l-1}} e^{-u^2} du}{\frac{\pi}{\sqrt{2}}} = \frac{\int_0^{\frac{\pi}{2}\sqrt{l-1}-\epsilon\sqrt{l-1}} e^{-(u+\epsilon\sqrt{l-1})^2} du}{\frac{\pi}{\sqrt{2}}}$$

$$\leq e^{-\epsilon^2(l-1)} \frac{\int_0^{\frac{\pi}{2}\sqrt{l-1}-\epsilon\sqrt{l-1}} e^{-u^2} du}{\frac{\pi}{\sqrt{2}}} \leq e^{-\epsilon^2(l-1)} \frac{\int_0^{\infty} e^{-u^2} du}{\frac{\pi}{\sqrt{2}}}$$

$$= e^{-\epsilon^2(l-1)} \frac{\sqrt{2\pi}/4}{\frac{\pi}{\sqrt{2}}} = e^{-\epsilon^2(l-1)}/2$$

Exercise 2.57

(**a**) Use $B(k - \frac{1}{2}, \frac{1}{2}) = \frac{k-1-\frac{1}{2}}{k-1} B(k - 1 - \frac{1}{2}, \frac{1}{2})$ and $B(\frac{1}{2}, \frac{1}{2}) = \pi$.

(**b**) Since $\log(1 + x)$ is concave, we have $\log(1 + \frac{x}{2}) \geq \frac{1}{2}\log(1 + x)$. Thus $\sum_{k=1}^{l-1} \log \frac{2k}{2k-1} = \sum_{k=1}^{l-1} \log(1 + \frac{1}{2k-1}) \geq \frac{1}{2}\sum_{k=1}^{l-1} \log(1 + \frac{1}{k-1/2}) = \frac{1}{2}\sum_{k=1}^{l-1} \log \frac{k+1/2}{k-1/2} = \frac{1}{2} \log \frac{l-1/2}{1/2} = \frac{1}{2}\log(2l - 1)$.

(**c**) Due to (**b**), we have $\sum_{k=1}^{l-1} \log \frac{2k-1}{2k} \leq -\frac{1}{2}\log(2l - 1)$. Thus $(2l - 1)B(l - \frac{1}{2}, \frac{1}{2}) \leq (2l - 1)\pi(2l - 1)^{-1/2} = \sqrt{(2l - 1)}\pi$.

References

1. I. Csiszár, Information type measures of difference of probability distribution and indirect observations. Studia Scient. Math. Hungar. **2**, 299–318 (1967)
2. S. Amari, H. Nagaoka, *Methods of Information Geometry* (AMS & Oxford University Press, Oxford, 2000)
3. A. Rényi, On measures of information and entropy, in *Proceedings of the Fourth Berkeley Symposium on Mathematical Statistics and Probability* (University of California Press, Berkeley, 1961), pp. 547–561
4. R.M. Fano, *Transmission of Information: A Statistical Theory of Communication* (Wiley, New York, 1961)
5. M. Hayashi, Security analysis of ε-almost dual universal$_2$ hash functions: smoothing of min entropy vs. smoothing of Rényi entropy of order 2 (2013). arXiv:1309.1596
6. S. Amari, α-divergence Is unique, belonging to both f-divergence and Bregman divergence classes. IEEE Trans. Inform. Theory **55**(11), 4925–4931 (2009)
7. A.W. van der Vaart, *Asymptotic Statistics* (Cambridge University Press, Cambridge, 1998)
8. I. Csiszár, J. Körner, *Information Theory: Coding Theorems for Discrete Memoryless Systems* (Academic, 1981)
9. I.N. Sanov, On the probability of large deviations of random variables. Mat. Sbornik **42**, 11–44 (1957) (in Russian). English translation: Selected Translat. Math. Stat. **1**, 213–244 (1961)
10. M. Keyl, R.F. Werner, Estimating the spectrum of a density operator. Phys. Rev. A **64**, 052311 (2001)
11. K. Matsumoto, Seminar notes (1999)
12. M. Hayashi, Optimal sequence of POVMs in the sense of Stein's lemma in quantum hypothesis. J. Phys. A Math. Gen. **35**, 10759–10773 (2002)

13. M. Hayashi, Exponents of quantum fixed-length pure state source coding. Phys. Rev. A **66**, 032321 (2002)
14. M. Hayashi, K. Matsumoto, Variable length universal entanglement concentration by local operations and its application to teleportation and dense coding, quant-ph/0109028 (2001); K. Matsumoto, M. Hayashi, Universal entanglement concentration. Phys. Rev. A **75**, 062338 (2007)
15. M. Hayashi, K. Matsumoto, Quantum universal variable-length source coding. Phys. Rev. A **66**, 022311 (2002)
16. M. Hayashi, K. Matsumoto, Simple construction of quantum universal variable-length source coding. Quant. Inf. Comput. **2**, Special Issue, 519–529 (2002)
17. M. Hayashi, Asymptotics of quantum relative entropy from a representation theoretical viewpoint. J. Phys. A Math. Gen. **34**, 3413–3419 (2001)
18. H. Cramér, Sur un nouveaux theoorème-limite de la théorie des probabilités, in *Actualités Scientifiques et Industrielles*, no. 736, in *Colloque consacré à la thèorie des probabilités* (Hermann, Paris, 1938), pp. 5–23
19. J. Gärtner, On large deviations from the invariant measure. Theory Prob. Appl. **22**, 24–39 (1977)
20. R. Ellis, Large deviations for a general class of random vectors, Ann. Probab. **12**, 1, 1–12 (1984); *Entropy, Large Deviations and Statistical Mechanics* (Springer, Berlin, 1985)
21. R.R. Bahadur, On the asymptotic efficiency of tests and estimates. Sankhyā **22**, 229 (1960)
22. R.R. Bahadur, Rates of Convergence of Estimates and Test Statistics. Ann. Math. Stat. **38**, 303 (1967)
23. R.R. Bahadur, Some limit theorems in statistics, in *Regional Conference Series in Applied Mathematics*, no. 4 (SIAM, Philadelphia, 1971)
24. J.C. Fu, On a theorem of Bahadur on the rate of convergence of point estimators. Ann. Stat. **1**, 745 (1973)
25. A.I. Khinchin, *Mathematical Foundations of Information Theory* (Dover, New York, 1957)
26. T.S. Han, *Information-Spectrum Methods in Information Theory* (Springer, Berlin, 2002) (originally appeared in Japanese in 1998)
27. V.D. Milman, G. Schechtman, *Asymptotic theory of finite-dimensional normed spaces*, vol. 1200, Lecture Notes in Mathematics (Springer, Berlin, 1986)
28. T. Cover, J. Thomas, *Elements of Information Theory* (Wiley, New York, 1991)
29. M. Hayashi, Exponential decreasing rate of leaked information in universal random privacy amplification. IEEE Trans. Inf. Theory **57**, 3989–4001 (2011)
30. M. Iwamoto, J. Shikata, Information theoretic security for encryption based on conditional Rényi entropies. Inform. Theor. Secur. Lect. Notes Comput. Sci. **8317**(2014), 103–121 (2014)
31. M. Müller-Lennert, F. Dupuis, O. Szehr, S. Fehr, M. Tomamichel, On quantum Renyi entropies: a new generalization and some properties. J. Math. Phys. **54**, 122203 (2013)
32. E.L. Lehman, G. Casella, *Theory of Point Estimation* (Springer, Berlin Heidelberg New York, 1998)
33. A. Dembo, O. Zeitouni, *Large Deviation Techniques and Applications* (Springer, Berlin, 1997)
34. J.A. Bucklew, *Large Deviation Techniques in Decision, Simulation, and Estimation* (Wiley, New York, 1990)

Chapter 3
Quantum Hypothesis Testing and Discrimination of Quantum States

Abstract Various types of information processing occur in quantum systems. The most fundamental processes are state discrimination and hypothesis testing. These problems often form the basis for an analysis of other types of quantum information processes. The difficulties associated with the noncommutativity of quantum mechanics appear in the most evident way among these problems. Therefore, we examine state discrimination and hypothesis testing before examining other types of information processing in quantum systems in this text. In two-state discrimination, we discriminate between two unknown candidate states by performing a measurement and examining the measurement data. Note that in this case, the two hypotheses for the unknown state are treated symmetrically. In contrast, if the two hypotheses are treated asymmetrically, the process is called hypothesis testing rather than state discrimination. Hypothesis testing is not only interesting in itself but is also relevant to other topics in quantum information theory. In particular, the quantum version of Stein's lemma, which is the central topic of this chapter, is closely related to quantum channel coding discussed in Chap. 4. Moreover, Stein's lemma is also connected to the distillation of maximally entangled states, as discussed in Sect. 8.5, in addition to other topics discussed in Chap. 9. The importance of Stein's lemma may not be apparent at first sight since it considers the tensor product states of identical states, which rarely appear in real communications. However, the asymptotic analysis for these tensor product states provides the key to the analysis of asymptotic problems in quantum communications. For these reasons, this topic is discussed in an earlier chapter in this text.

3.1 Information Quantities in Quantum Systems

3.1.1 Quantum Entropic Information Quantities

For the preparation, we discuss the quantum extension of information quantities given in Sect. 2.1. Let us first consider the **von Neumann entropy** of the density

M. Hayashi, *Quantum Information Theory*, Graduate Texts in Physics,
DOI 10.1007/978-3-662-49725-8_3

matrix ρ with the spectral decomposition $\rho = \sum_{i=1}^{d} p_i |u^i\rangle\langle u^i|$ as its quantum extension of the entropy.[1] The von Neumann entropy is defined as the entropy of the probability distribution $p = \{p_i\}$ of the eigenvalues of the density ρ, and it is denoted by $H(\rho)$. Applying the arguments of Sect. 1.5 to $f(x) = \log(x)$, we have $\log \rho \stackrel{\text{def}}{=} \sum_{i=1}^{d} (\log p_i)|u^i\rangle\langle u^i|$, and we can write $H(\rho)$ as

$$H(\rho) = -\operatorname{Tr} \rho \log \rho . \tag{3.1}$$

The von Neumann entropy also satisfies the concavity, as proved in Sect. 5.5. Similarly, the **Rényi entropy** is defined as $H_{1-s}(\rho) \stackrel{\text{def}}{=} \frac{\psi(s|\rho)}{s}$ with $\psi(s|\rho) \stackrel{\text{def}}{=} \log \operatorname{Tr} \rho^{1-s}$. Henceforth, we will use its abbreviation $\psi(s)$ as mentioned previously. The minimum and maximum entropies are defined as $H_{\min}(\rho) \stackrel{\text{def}}{=} -\log \|\rho\|$ and $H_{\max}(\rho) \stackrel{\text{def}}{=} \log \operatorname{Tr}\{\rho > 0\}$, and satisfy the similar relations as (2.40).

Since the diagonal elements of a diagonal matrix forms a probability distribution, we can therefore interpret the tensor product $\rho^{\otimes n}$ as the quantum-mechanical analog of the independent and identical distribution. In other words, the eigenvalues of $\rho^{\otimes n}$ are equal to the n-i.i.d. of the probability distribution resulting from the eigenvalues of ρ. Since $\{\rho^s a^{-s} > 1\}(\rho^s a^{-s} - I) \geq 0$ for $s \geq 0$, the inequalities

$$\{\rho > a\} = \{\rho^s a^{-s} > 1\} \leq \{\rho^s a^{-s} > 1\}\rho^s a^{-s} \leq \rho^s a^{-s} \tag{3.2}$$

hold. Similarly,

$$\{\rho \leq a\} \leq \rho^{-s} a^s .$$

Hence, we obtain

$$\operatorname{Tr} \rho \{\rho > a\} \leq \operatorname{Tr} \rho^{1+s} a^{-s} , \tag{3.3}$$

$$\operatorname{Tr} \rho \{\rho \leq a\} \leq \operatorname{Tr} \rho^{1-s} a^{s} , \tag{3.4}$$

for $a > 0$ and $0 \leq s$. Treating the independent and identical distribution in a manner similar to (2.42) and (2.44), we obtain

$$\operatorname{Tr} \rho^{\otimes n} \left\{\rho^{\otimes n} \leq e^{-nR}\right\} \leq e^{n \min_{0 \leq s} (\psi(s) - sR)} \tag{3.5}$$

$$\operatorname{Tr} \rho^{\otimes n} \left\{\rho^{\otimes n} > e^{-nR}\right\} \leq e^{n \min_{s \leq 0} (\psi(s) - sR)} . \tag{3.6}$$

Certainly, the relationship similar to the classical system holds concerning $\frac{\psi(s) - sR}{1-s}$ and $H(\rho)$.

As an extension of the relative entropy, we define the **quantum relative entropy** $D(\rho\|\sigma)$ for two density matrices ρ and σ as

[1] Historically, the von Neumann entropy for a density matrix ρ was first defined by von Neumann [1]. Following this definition, Shannon [2] defined the entropy for a probability distribution.

$$D(\rho\|\sigma) \stackrel{\text{def}}{=} \text{Tr}\, \rho(\log \rho - \log \sigma)\,. \tag{3.7}$$

The quantum relative entropy satisfies an inequality similar to (2.13), which will be discussed in greater detail in Sect. 5.4. For the two quantum states ρ and σ, we can also define the function $\phi(s|\rho\|\sigma) \stackrel{\text{def}}{=} \log(\text{Tr}\, \rho^{1-s}\sigma^s)$ and obtain Exe. 3.4

$$\phi'(0|\rho\|\sigma) = -D(\rho\|\sigma)\,, \quad \phi'(1|\rho\|\sigma) = D(\sigma\|\rho)\,. \tag{3.8}$$

When it is not necessary to explicitly specify ρ and σ, we will abbreviate this value to $\phi(s)$. If ρ commutes with σ, the quantity $\phi(s|\rho\|\sigma)$ is equal to the quantity $\phi(s|P\|Q)$ with the probability distributions P and Q that consist of eigenvalues.

Since $\phi(s|\rho\|\sigma)$ is a convex function of s Exe. 3.5, a quantum extension of **relative Rényi entropy**

$$D_{1+s}(\rho\|\sigma) \stackrel{\text{def}}{=} \frac{\phi(-s|\rho\|\sigma)}{s} = \frac{1}{s}\log \text{Tr}\, \rho^{1+s}\sigma^{-s} \tag{3.9}$$

is monotone increasing for s. Hence, we define the minimum and the maximum relative entropies as [3]

$$D_{\max}(\rho\|\sigma) \stackrel{\text{def}}{=} \log \|\sigma^{-\frac{1}{2}}\rho\sigma^{-\frac{1}{2}}\|, \quad D_{\min}(\rho\|\sigma) \stackrel{\text{def}}{=} -\log \text{Tr}\, \sigma\{\rho > 0\}. \tag{3.10}$$

Hence, we obtain the relations

$$\lim_{s\to 1} D_{1-s}(\rho\|\sigma) = D_{\min}(\rho\|\sigma), \quad \lim_{s\to 0} D_{1-s}(\rho\|\sigma) = D(\rho\|\sigma). \tag{3.11}$$

Also, we can show the inequality Exe. 3.20

$$D_{1-s}(\rho\|\sigma) \le D_{\max}(\rho\|\sigma) \tag{3.12}$$

for $s \in [-1, 1)$. Due to the non-commutativity, we define another function $\tilde{\phi}(s|\rho\|\sigma) \stackrel{\text{def}}{=} \log \text{Tr}(\sigma^{\frac{s}{2(1-s)}}\rho\sigma^{\frac{s}{2(1-s)}})^{1-s}$ and another quantum extension of relative Rényi entropy (sandwiched relative Rényi entropy) [4, 5]:

$$\underline{D}_{1+s}(\rho\|\sigma) \stackrel{\text{def}}{=} \frac{\tilde{\phi}(-s|\rho\|\sigma)}{s} = \frac{1}{s}\log \text{Tr}(\sigma^{-\frac{s}{2(1+s)}}\rho\sigma^{-\frac{s}{2(1+s)}})^{1+s}. \tag{3.13}$$

These relative entropies satisfy the additivity Exe. 3.2. By a simple calculation Exe. 3.6, we obtain

$$\tilde{\phi}'(0|\rho\|\sigma) = -D(\rho\|\sigma), \tag{3.14}$$

which implies

$$\lim_{s\to 0} \underline{D}_{1+s}(\rho\|\sigma) = D(\rho\|\sigma). \tag{3.15}$$

Also, by a calculation[Exe. 3.7,3.59], we obtain

$$\lim_{s\to\infty} \underline{D}_{1+s}(\rho\|\sigma) = D_{\max}(\rho\|\sigma). \tag{3.16}$$

Further, as shown in Sect. 3.8, we have

$$\underline{D}_{1+s}(\rho\|\sigma) = \lim_{n\to\infty} \frac{D_{1+s}(\kappa_{\sigma^{\otimes n}}(\rho^{\otimes n})\|\sigma^{\otimes n})}{n}. \tag{3.17}$$

for $s > -1$, which is equivalent to $\tilde{\phi}(-s|\rho\|\sigma) = \lim_{n\to\infty} \frac{\phi(-s|\kappa_{\sigma^{\otimes n}}(\rho^{\otimes n})\|\sigma^{\otimes n})}{n}$ for $s > -1$.

Lemma 3.1 *The functions $s \mapsto \phi(s|\rho\|\sigma)$ and $\tilde{\phi}(s|\rho\|\sigma)$ are convex for $s \in [-1, \infty)$. The functions $s \mapsto D_{1+s}(\rho\|\sigma)$ and $\underline{D}_{1+s}(\rho\|\sigma)$ are monotone increasing with respect to $s \in [-1, \infty)$.*

Proof Since the limit of convex function is also convex, Relation (3.17) implies that the function $s \mapsto \tilde{\phi}(s|\rho\|\sigma)$ is convex. The convexity of $s \mapsto \phi(s|\rho\|\sigma)$ is shown in Exercise 3.5. Using these two facts, we can show that the functions $s \mapsto D_{1+s}(\rho\|\sigma)$ and $\underline{D}_{1+s}(\rho\|\sigma)$ are monotone increasing with respect to $s \in [-1, \infty)$.

The above information quantities satisfy the **monotonicity**[2] with respect to the measurement $\boldsymbol{M}$ as follows.

$$D(\rho\|\sigma) \ge D(\mathrm{P}_\rho^{\boldsymbol{M}}\|\mathrm{P}_\sigma^{\boldsymbol{M}}), \tag{3.18}$$

$$\phi(s|\rho\|\sigma) \le \phi(s|\mathrm{P}_\rho^{\boldsymbol{M}}\|\mathrm{P}_\sigma^{\boldsymbol{M}}) \text{ for } 0 \le s \le 1, \tag{3.19}$$

$$\phi(s|\rho\|\sigma) \ge \phi(s|\mathrm{P}_\rho^{\boldsymbol{M}}\|\mathrm{P}_\sigma^{\boldsymbol{M}}) \text{ for } s \le 0, \tag{3.20}$$

$$\tilde{\phi}(s|\rho\|\sigma) \le \phi(s|\mathrm{P}_\rho^{\boldsymbol{M}}\|\mathrm{P}_\sigma^{\boldsymbol{M}}) \text{ for } 0 \le s \le \frac{1}{2}, \tag{3.21}$$

$$\tilde{\phi}(s|\rho\|\sigma) \ge \phi(s|\mathrm{P}_\rho^{\boldsymbol{M}}\|\mathrm{P}_\sigma^{\boldsymbol{M}}) \text{ for } s \le 0. \tag{3.22}$$

Proofs of (3.18), (3.20), and (3.22) are given in Sect. 3.8. Inequality (3.19) is shown in Sect. A.4 as a more general argument (5.52). However, we omit the proof of (3.21). For the proof, see [6]. In contrast to $b(\rho, \sigma)$ and $d_1(\rho, \sigma)$, although there exists a POVM M satisfying the equalities in (3.18) and (3.22) only when ρ and σ commute, as shown in Theorem 3.6 and Exercise 3.62, there exists a sequence of POVMs attaining the equalities in (3.18) and (3.22) in an asymptotic sense, as mentioned in Exercise 5.44 and Exercise 3.62, respectively. However, the equality in (3.20) for $s \le -1$ does not necessarily hold even in an asymptotic sense, as verified from Exercises 3.58 and 5.22. The inequalities (3.19), (3.20), (3.21), and (3.22) are rewritten as the monotonicity of the quantum relative Rényi entropy with measurement

[2]Here, the monotonicity concerns only the state evolution, not parameter s.

$$D_{1-s}(\rho\|\sigma) \ge D_{1-s}(\mathrm{P}^{M}_{\rho}\|\mathrm{P}^{M}_{\sigma}) \text{ for } s \le 1 \tag{3.23}$$

$$\underline{D}_{1-s}(\rho\|\sigma) \ge D_{1-s}(\mathrm{P}^{M}_{\rho}\|\mathrm{P}^{M}_{\sigma}) \text{ for } s \le \frac{1}{2}. \tag{3.24}$$

Further, combining (3.23), (3.17), and their additivity, we have

$$D_{1+s}(\rho\|\sigma) \ge \underline{D}_{1+s}(\rho\|\sigma) \tag{3.25}$$

for $s > -1$. For another proof of (3.25), see Exercise 3.8.

Exercises

3.1 Show that the information quantities $D(p\|q)$ and $D_{1+s}(p\|q)$ between q and p are equal to their quantum versions $D(\rho\|\sigma)$ and $D_{1+s}(\rho\|\sigma)$ for commuting ρ and σ with diagonalizations $\rho = \sum_i p_i |u_i\rangle\langle u_i|$ and $\sigma = \sum_i q_i |u_i\rangle\langle u_i|$.

3.2 Choose density matrices ρ_A, σ_A in $\mathcal{H}_A$ and density matrices ρ_B, σ_B in $\mathcal{H}_B$ Show that

$$H(\rho_A) + H(\rho_B) = H(\rho_A \otimes \rho_B), \tag{3.26}$$

$$D(\rho_A\|\sigma_A) + D(\rho_B\|\sigma_B) = D(\rho_A \otimes \rho_B\|\sigma_A \otimes \sigma_B), \tag{3.27}$$

$$D_{1-s}(\rho_A\|\sigma_A) + D(\rho_B\|\sigma_B) = D_{1-s}(\rho_A \otimes \rho_B\|\sigma_A \otimes \sigma_B), \tag{3.28}$$

$$\underline{D}_{1-s}(\rho_A\|\sigma_A) + D(\rho_B\|\sigma_B) = \underline{D}_{1-s}(\rho_A \otimes \rho_B\|\sigma_A \otimes \sigma_B), \tag{3.29}$$

$$D_{\max}(\rho_A\|\sigma_A) + D(\rho_B\|\sigma_B) = D_{\max}(\rho_A \otimes \rho_B\|\sigma_A \otimes \sigma_B), \tag{3.30}$$

$$D_{\min}(\rho_A\|\sigma_A) + D(\rho_B\|\sigma_B) = D_{\min}(\rho_A \otimes \rho_B\|\sigma_A \otimes \sigma_B). \tag{3.31}$$

3.3 Show that

$$\mathrm{Tr}\, \rho f(X) \ge f(\mathrm{Tr}\, \rho X) \tag{3.32}$$

for a convex function f, a Hermitian matrix X, and a state ρ. This is a quantum version of Jensen's inequality.

3.4 Show (3.8) using Exercise 1.4.

3.5 Show that $\phi(s|\rho\|\sigma)$ is convex by following the steps below [7].

(**a**) Show that $\phi'(s|\rho\|\sigma) = \dfrac{\mathrm{Tr}\, \rho^{1-s}\sigma^s(\log\sigma - \log\rho)}{\mathrm{Tr}\, \rho^{1-s}\sigma^s}$ by using Exercise 1.4.

(**b**) Show that $\dfrac{d\,\mathrm{Tr}\, \rho^{1-s}\sigma^s(\log\sigma - \log\rho)}{ds} = \mathrm{Tr}\rho^{1-s}(\log\sigma - \log\rho)\sigma^s(\log\sigma - \log\rho)$.

(**c**) Show that $\phi''(s|\rho\|\sigma) = \dfrac{\mathrm{Tr}\, \rho^{1-s}(\log\sigma - \log\rho)\sigma^s(\log\sigma - \log\rho)}{\mathrm{Tr}\, \rho^{1-s}\sigma^s}$
$-\dfrac{(\mathrm{Tr}\, \rho^{1-s}\sigma^s(\log\sigma - \log\rho))^2}{(\mathrm{Tr}\, \rho^{1-s}\sigma^s)^2}$.

(**d**) Show the convexity of $\phi(s|\rho\|\sigma)$ using Schwarz's inequality.

3.6 Show (3.14).

3.7 Show (3.16).

3.8 Show (3.25) by using Araki-Lieb-Thirring inequalities [8, 9] $\mathrm{Tr}\, B^{\frac{r}{2}} A^r B^{\frac{r}{2}} \le \mathrm{Tr}(B^{\frac{1}{2}} A^r B^{\frac{1}{2}})^r$ with $r \in (0, 1)$. and $\mathrm{Tr}\, B^{\frac{r}{2}} A^r B^{\frac{r}{2}} \ge \mathrm{Tr}(B^{\frac{1}{2}} A^r B^{\frac{1}{2}})^r$ with $r \ge 1$.

3.9 Define the state $\rho_s := e^{-\psi(s|\rho)}\rho^{1-s}$ and assume that a state σ satisfies $H(\sigma) = H(\rho_s)$. Show that $D(\rho_s\|\sigma) \le D(\sigma\|\rho)$ for $s \le 1$ by following steps below.
(**a**) Show that $\frac{1}{1-s} D(\sigma\|\rho_s) = \frac{1}{1-s} \mathrm{Tr}\, \sigma \log \sigma - \mathrm{Tr}\, \sigma \log \rho - \frac{\psi(s|\rho)}{1-s}$.
(**b**) Show $D(\sigma\|\rho) - \frac{1}{1-s} D(\sigma\|\rho_s) = D(\rho_s\|\rho)$.
(**c**) Show the desired inequality.

3.10 Show the quantum extension of (2.55):

$$\sup_{0\le s\le 1} \frac{sR - \psi(s|\rho)}{1-s} = \min_{\sigma:H(\sigma)\ge R} D(\sigma\|\rho) \tag{3.33}$$

following the steps below.
(**a**) Define s_R as the same way as Exercise 2.24. Show that

$$\min_{\sigma:H(\sigma)=R} D(\sigma\|\rho) = D(\rho_{s_R}\|\rho). \tag{3.34}$$

(**b**) Show that

$$\min_{\sigma:H(\sigma)\ge R} D(\sigma\|\rho) = D(\rho_{s_R}\|\rho). \tag{3.35}$$

(**c**) Show that (3.33).

3.11 Show that

$$\sup_{s\le 0} \frac{sR - \psi(s|\rho)}{1-s} = \min_{\sigma:H(\sigma)\le R} D(\sigma\|\rho). \tag{3.36}$$

(**a**) Show that

$$\min_{\sigma:H(\sigma)=r} D(\sigma\|\rho) = D(\rho_{s_R}\|\rho). \tag{3.37}$$

(**b**) Show that

$$\min_{\sigma:H(\sigma)\le R} D(\sigma\|\rho) = D(\rho_{s_R}\|\rho). \tag{3.38}$$

(**c**) Show (3.36).

3.12 Show that $\underline{D}_{1+s}(\rho\|\sigma)$ has the following expressions.

$$\underline{D}_{1+s}(\rho\|\sigma) = \frac{1}{s}\log \mathrm{Tr}(\rho^{\frac{1}{2}}\sigma^{-\frac{s}{1+s}}\rho^{\frac{1}{2}})^{1+s} \tag{3.39}$$

$$= \frac{1+s}{s}\log \|\rho^{\frac{1}{2}}\sigma^{-\frac{s}{1+s}}\rho^{\frac{1}{2}}\|_{1+s} = \frac{1+s}{s}\log \|\sigma^{-\frac{s}{2(1+s)}}\rho\sigma^{-\frac{s}{2(1+s)}}\|_{1+s}. \tag{3.40}$$

3.13 Show that

$$\|\sigma^{-\frac{1}{2}}\rho\sigma^{-\frac{1}{2}}\| = \min\{x|\rho \le x\sigma\}. \tag{3.41}$$

3.1.2 Other Quantum Information Quantities

Next, we discuss another types of quantum information quantities. As the quantum version of the Hellinger distance $d_2(p,q)$, we introduce the **Bures distance** $b(\rho,\sigma)$ defined as

$$b^2(\rho,\sigma) \stackrel{\mathrm{def}}{=} \min_{U:\mathrm{unitary}} \frac{1}{2}\mathrm{Tr}(\sqrt{\rho}-\sqrt{\sigma}U)(\sqrt{\rho}-\sqrt{\sigma}U)^*. \tag{3.42}$$

The Bures distance $b(\rho,\sigma)$ also satisfies the axioms of a distance in a similar way to the Hellinger distance [Exe. 3.15]. Using (A.19), this quantity may be rewritten as

$$\begin{aligned} b^2(\rho,\sigma) &= 1 - \frac{1}{2}\max_{U:\mathrm{unitary}} \mathrm{Tr}\left(U\sqrt{\rho}\sqrt{\sigma} + U^*(\sqrt{\rho}\sqrt{\sigma})^*\right) \\ &= 1 - \max_{U:\mathrm{unitary}} \mathbf{Re}\,\mathrm{Tr}\,U\sqrt{\rho}\sqrt{\sigma} \\ &= 1 - \mathrm{Tr}\,|\sqrt{\rho}\sqrt{\sigma}| = 1 - \mathrm{Tr}\sqrt{\sqrt{\rho}\sigma\sqrt{\rho}}. \end{aligned}$$

Therefore, this value does not change when ρ and σ are interchanged. Later, we will also see that this quantity also satisfies similar information inequalities (Corollary 8.4). The quantity $\mathrm{Tr}\,|\sqrt{\rho}\sqrt{\sigma}|$ is called **fidelity** and is denoted by $F(\rho,\sigma)$, which satisfies that $\log F(\rho,\sigma) = \tilde{\phi}(\frac{1}{2}|\rho\|\sigma)$. Then, it follows that

$$b^2(\rho,\sigma) = 1 - F(\rho,\sigma). \tag{3.43}$$

If one of the states is a pure state, then

$$F(|u\rangle\langle u|,\rho) = \sqrt{\langle u|\rho|u\rangle}. \tag{3.44}$$

The square of this value corresponds to a probability. If both ρ and σ are pure states $|u\rangle\langle u|$ and $|v\rangle\langle v|$, respectively, then $\mathrm{Tr}\sqrt{\sqrt{\rho}\sigma\sqrt{\rho}} = |\langle u|v\rangle|$ and the Bures distance is given by

$$b^2(|u\rangle\langle u|, |v\rangle\langle v|) = 1 - |\langle u|v\rangle| .$$

We also define the **trace norm distance** $d_1(\rho, \sigma)$ as a quantum version of the variational distance by

$$d_1(\rho, \sigma) \stackrel{\text{def}}{=} \frac{1}{2}\|\rho - \sigma\|_1 , \tag{3.45}$$

where $\|\cdot\|_1$ denotes the trace norm (Sect. A.3). This also satisfies the monotonicity [see (5.51) and Exercise 3.29].

If the states involved in the above quantities are pure states such as $|u\rangle\langle u|$ and $|v\rangle\langle v|$, we shall abbreviate the notation to label the states, i.e., $b(\rho, \sigma)$ will be written as $b(u, v)$, and so on.

The above information quantities satisfy the **monotonicity** with respect to the measurement $\boldsymbol{M}$ as follows.

$$b(\rho, \sigma) \ge d_2(\mathrm{P}_\rho^{\boldsymbol{M}}, \mathrm{P}_\sigma^{\boldsymbol{M}}) , \tag{3.46}$$

$$d_1(\rho, \sigma) \ge d_1(\mathrm{P}_\rho^{\boldsymbol{M}}, \mathrm{P}_\sigma^{\boldsymbol{M}}). \tag{3.47}$$

Proofs of (3.46) and (3.47) are given in Exercise 3.19 and Sect. 3.4, respectively. As is discussed in Exercises 3.21–3.23, the equalities of (3.46) and (3.47) hold when the POVM $\boldsymbol{M}$ is chosen appropriately.

Further, similarly to inequalities (2.25) and (2.26), the inequalities

$$d_1(\rho, \sigma) \ge b^2(\rho, \sigma) = 1 - F(\rho, \sigma) \ge \frac{1}{2} d_1^2(\rho, \sigma) \tag{3.48}$$

$$D(\rho\|\sigma) \ge -2\log \mathrm{Tr}\, |\sqrt{\rho}\sqrt{\sigma}| \ge 2b^2(\rho, \sigma) \tag{3.49}$$

hold [Exe. 3.24-3.27]. Thus, we can show

$$F(\rho, \sigma) \ge 1 - \frac{D(\rho\|\sigma)}{2}. \tag{3.50}$$

From these inequalities we can see that the convergence of $d_1(\rho_n, \sigma_n)$ to 0 is equivalent to the convergence of $b(\rho_n, \sigma_n)$ to 0.

In order to express the difference between the two states ρ and σ, we sometimes focus on the quantity $1 - F^2(\rho, \sigma)$, which is slightly different from $b^2(\rho, \sigma)$. Their relation can be characterized as[Exe. 3.28]

$$2b^2(\rho, \sigma) \ge 1 - F^2(\rho, \sigma) \ge b^2(\rho, \sigma). \tag{3.51}$$

Also, the quantity $1 - F^2(\rho, \sigma)$ upper bounds the quantity $d_1(\rho, \sigma)$ in a way different from (3.48) as[Exe. 8.2]

$$1 - F^2(\rho, \sigma) \geq d_1^2(\rho, \sigma). \tag{3.52}$$

Also, we have the **quantum Pinsker inequality** as follows[Exe. 3.30].

$$D(\rho\|\sigma) \geq 2d_1^2(\rho, \sigma). \tag{3.53}$$

Note that this is a stronger requirement between $D(\rho\|\sigma)$ and $d_1(\rho, \sigma)$ than the combination of (3.48) and (3.49).

Exercises

3.14 Show that the information quantities $d_2(p, q)$, and $d_1(p, q)$ between q and p are equal to their quantum versions $b(\rho, \sigma)$ and $d_1(\rho, \sigma)$ for commuting ρ and σ with diagonalizations $\rho = \sum_i p_i |u_i\rangle\langle u_i|$ and $\sigma = \sum_i q_i |u_i\rangle\langle u_i|$.

3.15 Show that the Bures distance satisfies the axioms of a distance by following the steps below.
(**a**) Show the following for arbitrary matrices X and Y:

$$\sqrt{\operatorname{Tr} XX^*} + \sqrt{\operatorname{Tr} YY^*} \geq \sqrt{\operatorname{Tr}(X-Y)(X-Y)^*}\,.$$

(**b**) Show the following for density matrices ρ_1, ρ_2, ρ_3 and unitary matrices U_1, U_2:

$$\begin{aligned}&\sqrt{\operatorname{Tr}\left(\sqrt{\rho_1} - \sqrt{\rho_2}U_1\right)\left(\sqrt{\rho_1} - \sqrt{\rho_2}U_1\right)^*}\\ \leq&\sqrt{\operatorname{Tr}\left(\sqrt{\rho_1}-\sqrt{\rho_3}U_2\right)\left(\sqrt{\rho_1}-\sqrt{\rho_3}U_2\right)^*}+\sqrt{\operatorname{Tr}\left(\sqrt{\rho_3}-\sqrt{\rho_2}U_1U_2^*\right)\left(\sqrt{\rho_3}-\sqrt{\rho_2}U_1U_2^*\right)^*}\,.\end{aligned}$$

(**c**) Show that $b(\rho_1, \rho_2) \leq b(\rho_1, \rho_3) + b(\rho_3, \rho_2)$ for density matrices ρ_1, ρ_2, and ρ_3.

3.16 Show that the square of the Bures distance satisfies

$$b^2(\rho_1, \rho_2) \leq 2b^2(\rho_1, \rho_3) + 2b^2(\rho_3, \rho_2)\,.$$

3.17 Show the following regarding the Bures distance for two different orthogonal bases $\{u_k\}$ and $\{v_k\}$.
(**a**) Show that the vectors $u = \sum_{k=1}^d \sqrt{p_k} e^{i\theta_k} u_k$ and $v = \sum_{k=1}^d \sqrt{p_k} e^{i\theta_k} v_k$ satisfy

$$\langle u|v\rangle = \sum_i p_i \langle u_i|v_i\rangle \tag{3.54}$$

for $\langle u_k|v_j\rangle = 0, k \neq j$, an arbitrary real number θ_i, and a probability distribution p_i.
(**b**) Show that (3.54) still holds if θ_i is chosen appropriately, even if the above conditions do not hold.

3.18 Show that $d_1(|u\rangle\langle u|, |v\rangle\langle v|) = \sqrt{1 - |\langle u|v\rangle|^2}$ using Exercise A.3.

3.19 Show that

$$\sum_i \sqrt{\mathrm{P}_\rho^M(i)}\sqrt{\mathrm{P}_\sigma^M(i)} \geq \operatorname{Tr}|\sqrt{\rho}\sqrt{\sigma}| \tag{3.55}$$

for a POVM $\boldsymbol{M}$ and ρ, σ following the steps below [10]. This is equivalent to (3.46).
(**a**) Show that $\sqrt{\operatorname{Tr} X^* X}\sqrt{\operatorname{Tr} Y^* Y} \geq |\operatorname{Tr} X^* Y|$ for two matrices X and Y.
(**b**) Show that $\sqrt{\operatorname{Tr} U\rho^{1/2} M_i \rho^{1/2} U^*}\sqrt{\operatorname{Tr} \sigma^{1/2} M_i \sigma^{1/2}} \geq |\operatorname{Tr} U\rho^{1/2} M_i \sigma^{1/2}|$ for a unitary matrix U.
(**c**) Show (3.55).

3.20 Show (3.12) following the steps below.
(**a**) For any matrix A, show that $\|AA^\dagger\| = \|A^\dagger A\|$ by using the polar decomposition of A.
(**b**) Show that $\operatorname{Tr} \rho^2 \sigma^{-1} \leq \|\sigma^{-\frac{1}{2}} \rho \sigma^{-\frac{1}{2}}\|$.
(**c**) Show that $D_{1-s}(\rho\|\sigma)$ is monotonically decreasing with respect to s.
(**d**) Show (3.12).

3.21 Suppose that the density matrix σ possesses the inverse. Show that the equality in (3.55) holds if $\boldsymbol{M} = \{M_i\}$ is chosen to be the spectral decomposition of $\rho^{1/2} U^* \sigma^{-1/2} = \sigma^{-1/2}(\sigma^{1/2}\rho\sigma^{1/2})^{1/2}\sigma^{-1/2}$, for U satisfying $|\rho^{1/2}\sigma^{1/2}| = U\rho^{1/2}\sigma^{1/2} = \sigma^{1/2}\rho^{1/2}U^*$ [10].

3.22 Suppose that the density matrix σ does not possess the inverse. Show that there exists a POVM satisfying the equality in (3.55) by following the steps below.
(**a**) Show that the support of matrix U is included in the support $\mathcal{H}_1$ of σ when $|\rho^{1/2}\sigma^{1/2}| = U\rho^{1/2}\sigma^{1/2} = \sigma^{1/2}\rho^{1/2}U^*$.
(**b**) Let $\boldsymbol{M} = \{M_i\}$ be the spectral decomposition of the matrix $\rho^{1/2}U^*\sigma^{-1/2}$ on $\mathcal{H}_1$ and let P be a projection onto $\mathcal{H}_1$. Show that the POVM $\{M_i\} \cup \{I - P\}$ in $\mathcal{H}$ satisfies the equality in (3.55).

3.23 Show that the equality in (3.47) holds when the POVM $\boldsymbol{M}$ is the diagonalization of $\rho - \sigma$.

3.24 Show that $d_1(\rho, \sigma) \geq b^2(\rho, \sigma)$ by choosing a POVM $\boldsymbol{M}$ satisfying the equality in (3.46).

3.25 Show that $b^2(\rho, \sigma) \geq \frac{1}{2} d_1^2(\rho, \sigma)$ by choosing a POVM $\boldsymbol{M}$ satisfying the equality in (3.47).

3.26 Show that $D(\rho\|\sigma) \geq -2 \log \operatorname{Tr}|\sqrt{\rho}\sqrt{\sigma}|$ by choosing a POVM $\boldsymbol{M}$ satisfying the equality in (3.46).

3.27 Show that $-\log \operatorname{Tr}|\sqrt{\rho}\sqrt{\sigma}| \geq b^2(\rho, \sigma)$.

3.28 Show (3.51) by writing $x = F(\rho, \sigma) = 1 - b^2(\rho, \sigma)$.

3.29 Show (5.51) using (3.59).

3.30 Show the quantum Pinsker inequality (3.53) following the steps below.
(**a**) Show that binary relative entropy $h(x, y) \stackrel{\text{def}}{=} x \log \frac{x}{y} + (1-x) \log \frac{1-x}{1-y}$ satisfies $2(y-x)^2 \le h(x, y)$ for $0 \le x \le y \le 1$.
(**b**) Show that $2(\operatorname{Tr} \sigma P - \operatorname{Tr} \rho P) = \operatorname{Tr} |\sigma - \rho| = \operatorname{Tr}(\sigma - \rho)(P - (I - P)) \ge 0$ for $P = \{\sigma - \rho > 0\}$ or $\{\sigma - \rho \ge 0\}$.
(**c**) Show (3.53).

3.2 Two-State Discrimination in Quantum Systems

Consider a quantum system $\mathcal{H}$ whose state is represented by the density matrix ρ or σ. Let us consider the problem of determining the density matrix that describes the true state of the quantum system by performing a measurement. This procedure may be expressed as a Hermitian matrix T satisfying $I \ge T \ge 0$ in $\mathcal{H}$, and it is called **state discrimination** for the following reason.

Consider performing a measurement corresponding to a POVM $\boldsymbol{M} = \{M_\omega\}_{\omega \in \Omega}$ to determine whether the true state is ρ or σ. For this purpose, we must first choose subsets of Ω that correspond to ρ and σ. That is, we first choose a suitable subset A of Ω, and if $\omega \in A$, we can then determine that the state is ρ, and if $\omega \in A^c$ (where A^c is the complement of A), then the state is σ. The Hermitian matrix $T \stackrel{\text{def}}{=} \sum_{\omega \in A} M_\omega$ then satisfies $I \ge T \ge 0$. When the true state is ρ, we erroneously conclude that the state is σ with the probability:

$$\sum_{\omega \in A^c} \operatorname{Tr} \rho M_\omega = \operatorname{Tr} \rho \sum_{\omega \in A^c} M_\omega = \operatorname{Tr} \rho (I - T).$$

On the other hand, when the true state is σ, we erroneously conclude that the state is ρ with the probability:

$$\sum_{\omega \in A} \operatorname{Tr} \sigma M_\omega = \operatorname{Tr} \sigma \sum_{\omega \in A} M_\omega = \operatorname{Tr} \sigma T.$$

More generally, when we observe $\omega \in \Omega$, we decide the true state is ρ with the probability t_ω and it is σ with the probability $1 - t_\omega$. This discrimination may therefore be represented by a map t_ω from Ω to the interval $[0, 1]$. When the true state is ρ, defining the Hermitian matrix $T \stackrel{\text{def}}{=} \sum_{\omega \in \Omega} t_\omega M_\omega$, we erroneously conclude that the state is σ with the probability:

$$\sum_{\omega \in \Omega} (1 - t_\omega) \operatorname{Tr} \rho M_\omega = \operatorname{Tr} \rho \sum_{\omega \in \Omega} (1 - t_\omega) M_\omega = \operatorname{Tr} \rho (I - T).$$

On the other hand, when the true state is σ, we erroneously conclude that the state is ρ with the probability:

$$\sum_{\omega\in\Omega} t_\omega \operatorname{Tr}\sigma M_\omega = \operatorname{Tr}\sigma \sum_{\omega\in\Omega} t_\omega M_\omega = \operatorname{Tr}\sigma T.$$

Therefore, in order to treat state discrimination, it is sufficient to examine the Hermitian matrix T. The two-valued POVM $\{T, I-T\}$ for a Hermitian matrix T satisfying $I \ge T \ge 0$ allows us to perform the discrimination. That is, we obtain

$$\begin{aligned}&\min_{I\ge T\ge 0} (\operatorname{Tr}\rho(I-T) + \operatorname{Tr}\sigma T)\\ =&\min_{M:\mathrm{POVM}} \min_{t_\omega:1\ge t_\omega\ge 0} \sum_{\omega\in\Omega} \left(\mathrm{P}^{\boldsymbol{M}}_\rho(\omega)(1-t_\omega) + \mathrm{P}^{\boldsymbol{M}}_\sigma(\omega)t_\omega\right).\end{aligned} \tag{3.56}$$

Henceforth, T will be called a **test**.

The problem in state discrimination is to examine the tradeoff between the two error probabilities $\operatorname{Tr}\sigma T$ and $\operatorname{Tr}\rho(I-T)$. We then prepare the following lemma.

Lemma 3.2 (Holevo [11]; Helstrom [12]) *Any two non-negative matrices A and B satisfy that*

$$\min_{I\ge T\ge 0} (\operatorname{Tr} A(I-T) + \operatorname{Tr} BT) = \operatorname{Tr} A\{A \le B\} + \operatorname{Tr} B\{A > B\}. \tag{3.57}$$

The minimum value is attained when $T = \{A \ge B\}$.

Thus, substituting ρ and σ into A and B, Lemma 3.2 guarantees that

$$\begin{aligned}\min_{I\ge T\ge 0} (\operatorname{Tr}\rho(I-T) + \operatorname{Tr}\sigma T) &= \operatorname{Tr}\rho\{\rho-\sigma \le 0\} + \operatorname{Tr}\sigma\{\rho-\sigma > 0\} &(3.58)\\ &= 1 - \frac{1}{2}\|\rho-\sigma\|_1, &(3.59)\end{aligned}$$

the second equation follows from the following relation [Exe. 3.32].

$$\|X\|_1 = \max_{T:-I\le T\le I} \operatorname{Tr} XT = \operatorname{Tr} X(\{X>0\}-\{X\le 0\}) = \operatorname{Tr} X(I-2\{X\le 0\}). \tag{3.60}$$

For any POVM $\boldsymbol{M}$, since the RHS of (3.56) is not greater than $\min_{t_\omega:1\ge t_\omega\ge 0}\sum_{\omega\in\Omega} \left(\mathrm{P}^{\boldsymbol{M}}_\rho(\omega)(1-t_\omega) + \mathrm{P}^{\boldsymbol{M}}_\sigma(\omega)t_\omega\right)$, the combination of (3.56) and (3.59) implies the inequality $\|\rho-\sigma\|_1 \ge \|\mathrm{P}^{\boldsymbol{M}}_\rho - \mathrm{P}^{\boldsymbol{M}}_\sigma\|_1$, i.e., we obtain (3.47).

The minimization of the weighted sum $\operatorname{Tr}\rho(I-T) + c\operatorname{Tr}\sigma T$ can be treated by substituting ρ and $c\sigma$ into A and B. Therefore, the trace norm gives a measure for the discrimination of two states. Hence, for examining the tradeoff between the two error probabilities $\operatorname{Tr}\sigma T$ and $\operatorname{Tr}\rho(I-T)$, it is sufficient to discuss the test $T = \{\rho - c\sigma \ge 0\}$ alone. This kind of test is called a **likelihood test**.

The error probabilities $\operatorname{Tr}\rho\{\rho \le c\sigma\}$ and $\operatorname{Tr}\sigma\{\rho > c\sigma\}$ are monotone with respect to c as follows. Using Lemma 3.2, we can show that

$$\operatorname{Tr} \rho\{\rho \le c\sigma\} \le \operatorname{Tr} \rho\{\rho \le c'\sigma\} \tag{3.61}$$
$$\operatorname{Tr} \sigma\{\rho > c\sigma\} \ge \operatorname{Tr} \sigma\{\rho > c'\sigma\} \tag{3.62}$$

when $0 < c < c'$ as follows. Lemma 3.2 implies that

$$\operatorname{Tr} \rho\{\rho \le c\sigma\} + c \operatorname{Tr} \sigma\{\rho > c\sigma\} \le \operatorname{Tr} \rho\{\rho \le c'\sigma\} + c \operatorname{Tr} \sigma\{\rho > c'\sigma\} \tag{3.63}$$
$$\operatorname{Tr} \rho\{\rho \le c\sigma\} + c' \operatorname{Tr} \sigma\{\rho > c\sigma\} \ge \operatorname{Tr} \rho\{\rho \le c'\sigma\} + c' \operatorname{Tr} \sigma\{\rho > c'\sigma\}. \tag{3.64}$$

Hence,

$$\begin{aligned} c(\operatorname{Tr} \sigma\{\rho > c\sigma\} - \operatorname{Tr} \sigma\{\rho > c'\sigma\}) \le & \operatorname{Tr} \rho\{\rho \le c'\sigma\} - \operatorname{Tr} \rho\{\rho \le c\sigma\} \\ \le & c'(\operatorname{Tr} \sigma\{\rho > c\sigma\} - \operatorname{Tr} \sigma\{\rho > c'\sigma\}). \end{aligned}$$

The condition $c < c'$ guarantees (3.62). Since $c > 0$, (3.62) implies (3.61).

In order to consider the intuitive picture of the likelihood test $\{\rho > c\sigma\}$, we consider the case when ρ and σ are commute, in which, they may be simultaneously diagonalized as $\rho = \sum_i p_i |u^i\rangle\langle u^i|$ and $\sigma = \sum_i q_i |u^i\rangle\langle u^i|$ using a common orthonormal basis $\{u^1, \ldots, u^d\}$. Therefore, the problem reduces to the discrimination of the probability distributions $p = \{p_i\}$ and $q = \{q_i\}$, as discussed below. Henceforth, such cases wherein the states ρ and σ commute will be henceforth called "**classical**."

Now, we discriminate between the two probability distributions $p = \{p_i\}$ and $q = \{q_i\}$ by the following process. When the datum i is observed, we decide the true distribution is p with the probability t_i. This discrimination may therefore be represented by a map t_i from $\{1, \ldots, d\}$ to the interval $[0, 1]$. Defining the map $t_i \stackrel{\text{def}}{=} \langle u^i|T|u^i\rangle$ from $\{1, \ldots, d\}$ to the interval $[0, 1]$ for an arbitrary discriminator T, we obtain

$$\sum_i (1 - t_i)p_i = \operatorname{Tr} \rho(I - T), \quad \sum_i t_i q_i = \operatorname{Tr} \sigma T.$$

These are the two error probabilities for discriminating the probability distributions $p = \{p_i\}_i$ and $q = \{q_i\}_i$. If the function t_i is defined on the data set of the measurement $\{|u^i\rangle\langle u^i|\}_i$ such that it is equal to 1 on the set $\{i|p_i > cq_i\}$ and 0 on the set $\{i|p_i \le cq_i\}$, then the test T is equal to $\{\rho > c\sigma\}$. Therefore, if ρ and σ commute, $T = \{\rho > c\sigma\}$ has a correspondence with the subset of the data set. If these density matrices commute, the problem may be reduced to that of probability distributions, which simplifies the situation considerably. The notation $\{\rho > c\sigma\}$ can be regarded as a generalization of a subset of data set.

For the likelihood test, we have the following lemma.

Lemma 3.3 (Audenaert et al. [13]) *Any two non-negative matrices A and B satisfy that*

$$\operatorname{Tr} A\{A \le B\} + \operatorname{Tr} B\{A > B\} \le \operatorname{Tr} A^{1-s} B^s \tag{3.65}$$

for $s \in [0, 1]$.

Thus, substituting ρ and σ into A and B, we have a useful upper bound of the sum of the two error probabilities as follows.

$$\min_{I \ge T \ge 0} (\operatorname{Tr} \rho(I - T) + \operatorname{Tr} \sigma T) \le \min_{s \in [0,1]} e^{\phi(s)}. \tag{3.66}$$

Next, we consider a lower bound of the sum of the two error probabilities.

Lemma 3.4 (Nussbaum and Szkoła [14]) *For two non-negative matrices A and B, we make their diagonalizations as*

$$A = \sum_i a_i |u_i\rangle\langle u_i|, \quad B = \sum_j b_j |v_j\rangle\langle v_j|. \tag{3.67}$$

Then, we have

$$\min_{I \ge T \ge 0} (\operatorname{Tr} A(I - T) + \operatorname{Tr} BT) \ge \frac{1}{2} \sum_{i,j} \min\{a_i, b_j\} |\langle v_j | u_i \rangle|^2. \tag{3.68}$$

Now, we consider the case when $A = \rho$ and $B = \sigma$, and define two distributions $P_{(\rho\|\sigma)}(i, j) := a_i |\langle v_j | u_i \rangle|^2$ and $Q_{(\rho\|\sigma)}(i, j) := b_j |\langle v_j | u_i \rangle|^2$ on $\Omega := \{1, \dots, \dim \mathcal{H}\} \times \{1, \dots, \dim \mathcal{H}\}$ based on the notations in Lemma 3.4. Then, the right hand side of (3.68) $\sum_{i,j} \min\{a_i, b_j\} |\langle v_j | u_i \rangle|^2$ can be transformed to $\min_{t_{(i,j)}:1 \ge t_{(i,j)} \ge 0} \sum_{(i,j)} \big(P_{(\rho\|\sigma)}(i, j)(1 - t_{(i,j)}) + Q_{(\rho\|\sigma)}(i, j) t_{(i,j)}\big)$. That is, the minimum discrimination probability between two states ρ and σ is lower bounded by the half of that between two distributions $P_{(\rho\|\sigma)}$ and $Q_{(\rho\|\sigma)}$. The pair of distributions $P_{(\rho\|\sigma)}$ and $Q_{(\rho\|\sigma)}$ reflects the properties of the pair of two states ρ and σ as follows. That is, we can show the following relations[Exe. 3.31].

$$D(P_{(\rho\|\sigma)} \| Q_{(\rho\|\sigma)}) = D(\rho\|\sigma) \tag{3.69}$$

$$\phi(s | P_{(\rho\|\sigma)} \| Q_{(\rho\|\sigma)}) = \phi(s|\rho\|\sigma) \tag{3.70}$$

$$P_{(\rho^{\otimes n}\|\sigma^{\otimes n})} = P^n_{(\rho\|\sigma)}, \quad Q_{(\rho^{\otimes n}\|\sigma^{\otimes n})} = Q^n_{(\rho\|\sigma)}. \tag{3.71}$$

These relations play important roles in latter sections.

Proof of Lemma 3.2 The quantity to be minimized can be rewritten as

$$\operatorname{Tr} A(I - T) + \operatorname{Tr} BT = \operatorname{Tr} A + \operatorname{Tr}(B - A)T.$$

Now, we diagonalize $B - A$ as $B - A = \sum_i \lambda_i |u_i\rangle\langle u_i|$. Then,

$$\operatorname{Tr}(B - A)T = \sum_i \lambda_i \operatorname{Tr} |u_i\rangle\langle u_i| T.$$

The test T minimizing the above satisfies the following conditions: $\operatorname{Tr} |u_i\rangle\langle u_i| T = 0$ when $\lambda_i \geq 0$; $\operatorname{Tr} |u_i\rangle\langle u_i| T = 1$ when $\lambda_i < 0$. The test T satisfying these conditions is nothing other than $\{B - A < 0\}$. Accordingly, we have

$$\min_{I \geq T \geq 0} \operatorname{Tr}(B - A)T = \operatorname{Tr}(B - A)\{B - A < 0\}. \tag{3.72}$$

Equality (3.57) can be proved according to

$$\begin{aligned} \min_{I \geq T \geq 0} \operatorname{Tr} A(I - T) + \operatorname{Tr} BT &= \operatorname{Tr} A + \operatorname{Tr}(B - A)\{B - A < 0\} \\ &= \operatorname{Tr} A\{A - B \leq 0\} + \operatorname{Tr} B\{A - B > 0\}. \end{aligned}$$

Then, we can also obtain (3.58). See Exercise 3.33 for the derivation of (3.59). ■

Proof of Lemma 3.3 We employ an alternative proof by Narutaka Ozawa [15, 16]. Since $A - B \leq (A - B)_+$, we have $A \leq B + (A - B)_+$. Similarly, the inequality $B + (A - B)_+ \geq B$ holds. Hence, the matrix monotonicity of $x \mapsto x^s$ (Sect. 1.5) yields that

$$A^s \leq (B + (A - B)_+)^s \tag{3.73}$$

$$(B + (A - B)_+)^s - B^s \geq 0 \tag{3.74}$$

$$(B + (A - B)_+)^{1-s} \geq B^{1-s}. \tag{3.75}$$

Hence,

$$\operatorname{Tr} A - \operatorname{Tr} A^{1-s} B^s = \operatorname{Tr} A^{1-s}\{A^s - B^s\}$$

$$\leq \operatorname{Tr} A^{1-s} \left\{(B + (A - B)_+)^s - B^s\right\} \tag{3.76}$$

$$\leq \operatorname{Tr}(B + (A - B)_+)^{1-s} \left\{(B + (A - B)_+)^s - B^s\right\} \tag{3.77}$$

$$= \operatorname{Tr}(B + (A - B)_+) - \operatorname{Tr}(B + (A - B)_+)^{1-s} B^s$$

$$\leq \operatorname{Tr}(B + (A - B)_+) - \operatorname{Tr} B^{1-s} B^s \tag{3.78}$$

$$= \operatorname{Tr} B + \operatorname{Tr}(A - B)_+ - \operatorname{Tr} B = \operatorname{Tr}(A - B)_+, \tag{3.79}$$

where (3.76), (3.77), and (3.78), follow from (3.73), (3.74), and (3.75), respectively. Thus, (3.79) implies

$$\begin{aligned} &\operatorname{Tr} A\{A \leq B\} + \operatorname{Tr} B\{A > B\} = \operatorname{Tr} A - \operatorname{Tr}(A - B)\{A - B > 0\} \\ =&\operatorname{Tr} A - (A - B)_+ \leq \operatorname{Tr} A^{1-s} B^s. \end{aligned}$$

■

Proof of Lemma 3.4 Thanks to Lemma 3.2, we can restrict the matrix T to a projection. Then, we obtain

$$\mathrm{Tr}\, BT = \mathrm{Tr}\, BTT = \mathrm{Tr} \sum_j b_j |v_j\rangle\langle v_j| T \sum_i |u_i\rangle\langle u_i| T = \sum_{i,j} b_j |\langle u_i|T|v_j\rangle|^2.$$

Similarly, we have

$$\mathrm{Tr}\, A(I - T) = \sum_{i,j} a_i |\langle u_i|I - T|v_j\rangle|^2.$$

Since $|\langle u_i|T|v_j\rangle|^2 + |\langle u_i|I - T|v_j\rangle|^2 \geq \frac{1}{2}|\langle u_i|T|v_j\rangle + \langle u_i|I - T|v_j\rangle|^2 = \frac{1}{2}|\langle v_j|u_i\rangle|^2$, we have

$$\begin{aligned}
\mathrm{Tr}\, A(I - T) + \mathrm{Tr}\, BT &= \sum_{i,j} b_j |\langle u_i|T|v_j\rangle|^2 + a_i |\langle u_i|I - T|v_j\rangle|^2 \\
&\geq \sum_{i,j} \min\{b_j, a_i\}(|\langle u_i|T|v_j\rangle|^2 + |\langle u_i|I - T|v_j\rangle|^2) \\
&\geq \frac{1}{2} \sum_{i,j} \min\{b_j, a_i\} |\langle v_j|u_i\rangle|^2.
\end{aligned}$$

■

Exercises

3.31 Show the relations (3.69), (3.70), and (3.71).

3.32 Show (3.60) referring to the proof of (3.72).

3.33 Show (3.59) using (3.60).

3.34 Show that $\|\rho - \rho_{\mathrm{mix}}\|_1 \geq 2(1 - \frac{\mathrm{rank}\,\rho}{d})$.

3.35 Show that

$$\mathrm{Tr}\, A\{\sqrt{A} \leq \sqrt{B}\} + \mathrm{Tr}\, B\{\sqrt{A} > \sqrt{B}\} \leq \mathrm{Tr}\, \sqrt{A}\sqrt{B} \tag{3.80}$$

by following the steps below.
(**a**) Show the following inequalities.

$$\mathrm{Tr}\, A\{\sqrt{A} \leq \sqrt{B}\} \leq \mathrm{Tr}\, \sqrt{A}\sqrt{B}\{\sqrt{A} \leq \sqrt{B}\} \tag{3.81}$$

$$\mathrm{Tr}\, B\{\sqrt{A} > \sqrt{B}\} \leq \mathrm{Tr}\, \sqrt{A}\sqrt{B}\{\sqrt{A} > \sqrt{B}\}. \tag{3.82}$$

(**b**) Show the inequality (3.80).

3.3 Discrimination of Plural Quantum States

In this section, we extend the discussion of the previous section, where there were only two hypothesis states, to the case of k hypothesis states $\rho_1, \ldots, \rho_k$. The state discrimination in this case is given by a POVM $\boldsymbol{M} = \{M_i\}_{i=1}^k$ with k measurement outcomes. For a fixed i, the quality of the discrimination is given by the error probability $1 - \operatorname{Tr} \rho_i M_i$. We are then required to determine a POVM $\boldsymbol{M} = \{M_i\}_{i=1}^k$ that minimizes this error probability. However, since it is impossible to reduce the error probability for all cases, some a priori probability distribution p_i is often assumed for the k hypotheses, and the average error probability $\sum_{i=1}^k p_i(1 - \operatorname{Tr} \rho_i M_i)$ is minimized. Therefore, we maximize the linear function

$$\Lambda(M_1, \ldots, M_k) \stackrel{\text{def}}{=} \sum_{i=1}^k p_i \operatorname{Tr} \rho_i M_i = \operatorname{Tr}\left(\sum_{i=1}^k p_i \rho_i M_i\right)$$

with respect to the matrix-valued vector $(M_1, \ldots, M_k)$ under the condition

$$M_i \geq 0, \quad \sum_{i=1}^k M_i = I.$$

For this maximization problem, we have the following equation[3] [17]:

$$p_{\text{guess}} := \max\left\{\Lambda(M_1, \ldots, M_k) \middle| M_i \geq 0, \sum_{i=1}^k M_i = I\right\} \tag{3.83}$$

$$= \min\{\operatorname{Tr} F | F \geq p_i \rho_i\}. \tag{3.84}$$

When the matrix F and the POVM $(M_1, \ldots, M_k)$ satisfy this constraint condition, they satisfy $\operatorname{Tr} F - \operatorname{Tr}\left(\sum_{i=1}^k p_i \rho_i M_i\right) = \sum_i \operatorname{Tr} M_i (F - p_i \rho_i) \geq 0$. Hence, the inequality LHS $\leq$ RHS in (3.84). The direct derivation of the reverse inequality is rather difficult; however, it can be treated by generalized linear programming. Equation (3.84) can be immediately obtained from Theorem 3.1, called the generalized duality theorem for the linear programing, as explained after Theorem 3.1.

Theorem 3.1 *Consider the real vector spaces V_1, V_2. Let L be a closed convex cone of V_1 (Sect. A.4). Assume that arbitrary $a > 0$ and $x \in L$ satisfy $ax \in L$. If A is a linear map from V_1 to V_2, then the following relation is satisfied for $b \in V_2$ and $c \in V_1$:*

$$\max_{x \in V_1}\{\langle c, x\rangle | x \in L, A(x) = b\} = \min_{y \in V_2}\{\langle y, b\rangle | A^*(y) - c \in L^*\}, \tag{3.85}$$

where $L^* \stackrel{\text{def}}{=} \{x \in V_1 | \langle x, x'\rangle \geq 0, \forall x' \in L\}$.

For a proof, see Sect. 3.9.

Using the relation (3.84), we derive another characterization for p_{guess} in Sect. 5.6. Now, we explain how to derive (3.84) by using Theorem 3.1. In our current

[3] $\max\{a|b\}$ denotes the maximum value of a satisfying condition b.

problem, V_1 is the space consisting of k Hermitian matrices with an inner product $\langle (M_1, \ldots, M_k), (M'_1, \ldots, M'_k) \rangle = \sum_{i=1}^k \operatorname{Tr} M_i M'_i$, V_2 is the space of Hermitian matrices with the usual product $\langle X, Y \rangle = \operatorname{Tr} XY$, L is the subset in V_1 such that all the matrices are positive semidefinite, A is the map $(M_1, \ldots, M_k) \mapsto \sum_{i=1}^k M_i$, and b and c are I and $(p_1\rho_1, \ldots, p_k\rho_k)$, respectively. Applying Theorem 3.1, we obtain (3.84). Therefore, we have rewritten the multivariable maximization problem on the (left-hand side) LHS of (3.84) into a single-variable minimization problem involving only one Hermitian matrix on the (right-hand side) RHS of (3.84). In general it is difficult to further analyze such optimization problems except for problems involving special symmetries [18]. Due to the fundamental nature of this problem, it is possible to reuse our results here in the context of other problems, as will be discussed in later sections. In these problems, it is sufficient to evaluate only the upper and lower bounds. Therefore, although it is generally difficult to obtain the optimal values, their upper and lower bounds can be more readily obtained.

In this section, the problem was formulated in terms of generalized linear programming [19]. However, it is also possible to formulate the problem in terms of semidefinite programming (SDP) [20]. The semidefinite programming problem has been studied extensively, and many numerical packages are available for this problem. Therefore, for numerical calculations it is convenient to recast the given problem in terms of SDP [21].

The generalized duality theorem given here may also be applied to other problems such as the minimization problem appearing on the RHS of (6.106) in Sect. 6.6 [22–24] and the problem involving the size and accuracy of the maximally entangled state [25] that can be produced by class ② introduced in Sect. 8.16. Therefore, this theorem is also interesting from the viewpoint of several optimization problems in quantum information theory.[4]

Exercise

3.36 Show that the average correct probability $\sum_{i=1}^k \frac{1}{k} \operatorname{Tr} \rho_i M_i$ with the uniform distribution is less than $\frac{d}{k}$. Furthermore, show that it is less than $\frac{d}{k} \max_i \|\rho_i\|$.

3.4 Asymptotic Analysis of State Discrimination

It is generally very difficult to infer the density matrix of the state of a single quantum system. Hence, an incorrect guess might be obtained from a single measurement. To avoid this situation, one can prepare many independent systems with the same state and then perform measurements on these. In this case, we would perform individual

[4]An example of a numerical solution of the maximization problem in quantum information theory is discussed in Sect. 4.1.2, where we calculate the classical capacity $C_c(W)$. Nagaoka's quantum version of the Arimoto–Blahut algorithm [26, 27], known from classical information theory [28, 29]. The connection between these quantities and linear programming has also been discussed widely [20, 30].

measurements on each system and analyze the obtained data statistically. However, it is also possible to infer the unknown state via a single quantum measurement on the composite system. There are many methods available for the second approach as compared with the first. Therefore, it would be interesting to clarify the difference between the optimal performances of the two approaches.

Let us consider the problem of state discrimination for unknown states given by the tensor product states such as $\rho^{\otimes n}$ and $\sigma^{\otimes n}$. This may be regarded as a quantum-mechanical extension of an independent and identical distribution. If arbitrary measurements on $\mathcal{H}^{\otimes n}$ are allowed to perform this discrimination, we can identify the Hermitian matrix T satisfying $I \geq T \geq 0$ on $\mathcal{H}^{\otimes n}$ as the discriminator. If we restrict the allowable measurements performed on $\mathcal{H}^{\otimes n}$ to be separable or adaptive, the problem becomes somewhat more complicated.

Let us consider the first case. The minimum of the sum of the two error probabilities is then given by $\min_{I \geq T \geq 0} \left(\operatorname{Tr} \rho^{\otimes n}(I - T) + \operatorname{Tr} \sigma^{\otimes n} T \right)$, and it asymptotically approaches 0 as n increases. Since this quantity approaches 0 exponentially with n, our problem is then to calculate this exponent. By using Lemmas 3.3 and 3.4, the exponent can be characterized as follows.

Lemma 3.5 (Chernoff [31]) *Any two density matrices ρ and σ on the system $\mathcal{H}$ satisfy*

$$\min_{I \geq T \geq 0} \left(\operatorname{Tr} \rho^{\otimes n}(I - T) + \operatorname{Tr} \sigma^{\otimes n} T \right) \leq \exp(n \inf_{1 \geq s \geq 0} \phi(s)) \tag{3.86}$$

where $\phi(s) = \phi(s|\rho\|\sigma)$ was defined in Sect. 3.1. In the limit for, the equation

$$\lim_{n \to \infty} -\frac{1}{n} \log \min_{I \geq T \geq 0} \left(\operatorname{Tr} \rho^{\otimes n}(I - T) + \operatorname{Tr} \sigma^{\otimes n} T \right) = - \inf_{1 \geq s \geq 0} \phi(s) \tag{3.87}$$

holds.

Since the classical case of Lemma 3.5 has been shown by Chernoff [31], the bound $-\inf_{1 \geq s \geq 0} \phi(s)$ is called Chernoff bound.

Proof (3.86) follows from Lemma 3.3 with $A = \rho^{\otimes n}$ and $B = \sigma^{\otimes n}$. (3.86) shows the part "$\geq$" in (3.87). Hence, we show the part "$\leq$" in (3.87) by using Lemma 3.4. Now, we define two distributions $P := P_{(\rho\|\sigma)}$ and $Q := Q_{(\rho\|\sigma)}$. Then, we can show that $\phi(s|P\|Q) = \phi(s|\rho\|\sigma) = \phi(s)^{\text{Exe. 3.31}}$. Applying Lemma 3.4 to the case with $A = \rho^{\otimes n}$ and $B = \sigma^{\otimes n}$, we have

$$\begin{aligned} & 2 \min_{I \geq T \geq 0} \left(\operatorname{Tr} \rho^{\otimes n}(I - T) + \operatorname{Tr} \sigma^{\otimes n} T \right) \\ \geq & P^n\{\omega^n \in \Omega^n | P^n(\omega^n) \leq Q^n(\omega^n)\} + Q^n\{\omega^n \in \Omega^n | P^n(\omega^n) > Q^n(\omega^n)\}. \end{aligned}$$

Hence, it is sufficient to show that

$$\begin{aligned}&\lim_{n\to\infty} -\frac{1}{n}\log\Big(P^n\{\omega^n \in \Omega^n | P^n(\omega^n) \le Q^n(\omega^n)\}\\&\quad + Q^n\{\omega^n \in \Omega^n | P^n(\omega^n) > Q^n(\omega^n)\}\Big)\\&= \max_{s\in[0,1]} -\phi(s|P\|Q).\end{aligned} \tag{3.88}$$

The application of Cramér Theorem (Theorem 2.7) to the random variable $\log \frac{P(\omega)}{Q(\omega)}$ yields that

$$\begin{aligned}&\lim_{n\to\infty} -\frac{1}{n}\log P^n\{\omega^n \in \Omega^n | P^n(\omega^n) \le Q^n(\omega^n)\}\\&= \lim_{n\to\infty} -\frac{1}{n}\log P^n\{\omega^n \in \Omega^n | \log \frac{P^n(\omega^n)}{Q^n(\omega^n)} \le 0\}\\&= \sup_{s\in[0,\infty)} -\phi(s|P\|Q).\end{aligned} \tag{3.89}$$

Similarly,

$$\lim_{n\to\infty} -\frac{1}{n}\log Q^n\{\omega^n \in \Omega^n | P^n(\omega^n) > Q^n(\omega^n)\} = \sup_{s\in(-\infty,1]} -\phi(s|P\|Q). \tag{3.90}$$

Now, we note that $\phi(s|P\|Q) > 0$ for $s \in (-\infty, 0) \cup (1, \infty)$ and $\phi(s|P\|Q) \le 0$ for $s \in [0, 1]$. Since $\sup_{s\in[0,\infty)} -\phi(s|P\|Q)$ and $\sup_{s\in(-\infty,1]} -\phi(s|P\|Q)$ are greater than zero, we have $\max_{s\in[0,1]} -\phi(s|P\|Q) = \sup_{s\in[0,\infty)} -\phi(s|P\|Q) = \sup_{s\in(-\infty,1]} -\phi(s|P\|Q)$. Thus, combination of (3.89) and (3.90) yields (3.88).

Exercises

3.37 Show equation (3.87).

3.38 Show (3.86) when ρ is a pure state $|u\rangle\langle u|$ following the steps below.
(**a**) Show that $\inf_{1>s>0} \phi(s) = \log\langle u|\sigma|u\rangle$.
(**b**) Show (3.86) when $T = |u\rangle\langle u|^{\otimes n}$.

3.39 Check that the bound $\inf_{1>s>0} \phi(s)$ can be attained by the test based on the multiple application of the POVM $\{|u\rangle\langle u|, I - |u\rangle\langle u|\}$ on the single system.

3.40 Show (3.87) when ρ and σ are pure states $|u\rangle\langle u|$ and $|v\rangle\langle v|$ following the steps below.
(**a**) Show that $\min_{I\ge T\ge 0} \left(\mathrm{Tr}\, \rho^{\otimes n}(I - T) + \mathrm{Tr}\, \sigma^{\otimes n} T\right) = 1 - \sqrt{1 - |\langle u|v\rangle|^{2n}}$.
(**b**) Show that $\lim_{n\to\infty} \frac{1}{n} \log 1 - \sqrt{1 - |\langle u|v\rangle|^{2n}} = \log |\langle u|v\rangle|^2$.
(**c**) Show (3.87).

3.41 Show that $\inf_{1\ge s\ge 0} \phi(s) = \phi(1/2)$ when σ has the form $\sigma = U\rho U^*$ and $U^2 = 1$.

3.42 Assume the same assumption as Exercise 3.41. We also assume that $\phi''(1/2) > 0$ and $\phi(1/2|\rho\|\sigma) < \log F(\rho, \sigma)$. Show that $\phi(1/2|\rho\|\sigma) < \min_M \inf_{1 \ge s \ge 0} \phi(s|\mathrm{P}_\rho^M \|\mathrm{P}_\sigma^M)$ using (3.19) and $\log F(\rho, \sigma) \le \phi(1/2|\mathrm{P}_\rho^M \|\mathrm{P}_\sigma^M)$, which is shown in Corollary 8.4 in a more general form. Also show that $\phi(1/2|\rho\|\sigma) < \underline{\lim}_{n\to\infty} \frac{1}{n} \min_{M^n} \inf_{1\ge s\ge 0} \phi(s|\mathrm{P}_{\rho^{\otimes n}}^{M^n} \|\mathrm{P}_{\sigma^{\otimes n}}^{M^n})$.

3.5 Hypothesis Testing and Stein's Lemma

Up until now, the two hypotheses for the two unknown states have been treated equally. However, there are situations where the objective is to disprove one of the hypotheses (called the **null hypothesis**) and accept the other (called the **alternative hypothesis**). This problem in this situation is called **hypothesis testing**. In this case, our errors can be classified as follows. If the null hypothesis is rejected despite being correct, it is called the **error of the first kind**. Conversely, if the null hypothesis is accepted despite being incorrect, it is called the **error of the second kind**. Then, we make our decision only when we support the alternative hypothesis and withhold our decision when we support the null one. Hence, the probability that we make a wrong decision is equal to the **error probability of the first kind**, i.e., the probability that an error of the first kind is made (if the null hypothesis consists of more than one element, then it is defined as the maximum probability with respect to these elements). Hence, we must guarantee that the error probability of the first kind is restricted to below a particular threshold. This threshold then represents the reliability, in a statistical sense, of our decision and is called the **level of significance**. The usual procedure in hypothesis testing is to fix the level of significance and maximize the probability of accepting the alternative hypothesis when it is true; in other words, wes minimize the **error probability of the second kind**, which is defined as the probability of an error of the second kind. For simplicity, we assume that these two hypotheses consist of a single element, i.e., these are given by ρ and σ, respectively. Such hypotheses are called **simple** and are often assumed for a theoretical analysis because this assumption simplifies the mathematical treatment considerably.

As before, we denote our decision by a test T where $0 \le T \le I$, despite the asymmetry of the situation (the event of rejecting the null hypothesis then corresponds to $I - T$). The error probability of the first kind is $\mathrm{Tr}\, \rho(I - T)$, and the error probability of the second kind is $\mathrm{Tr}\, \sigma T$. The discussion in Sect. 3.2 confirms that in order to optimize our test, it is sufficient to treat only the tests of the form $T = \{\sigma - c\rho < 0\}$ [see RHS of (3.57)]. However, the analysis in Sect. 3.4 cannot be reused due to the asymmetrical treatment of the problem here and therefore another kind of formalism is required. Let us first examine the asymptotic behavior of the error probability of the second kind when the null and the alternative hypotheses are given by the tensor product states $\rho^{\otimes n}$ and $\sigma^{\otimes n}$ on the tensor product space $\mathcal{H}^{\otimes n}$ with a level of significance of $\epsilon > 0$.

Theorem 3.2 (Hiai and Petz [32], Ogawa and Nagaoka [7]) *The minimum value of the error probability of the second kind* $\beta_\epsilon^n(\rho\|\sigma)$ *satisfies*

$$\lim_{n\to\infty} -\frac{1}{n} \log \beta_\epsilon^n(\rho\|\sigma) = D(\rho\|\sigma), \quad 1 > \forall\epsilon > 0 \tag{3.91}$$

$$\beta_\epsilon^n(\rho\|\sigma) \stackrel{\text{def}}{=} \min_{I\geq T\geq 0} \{\operatorname{Tr} \sigma^{\otimes n} T \mid \operatorname{Tr} \rho^{\otimes n}(I-T) \leq \epsilon\} \tag{3.92}$$

when the error probability of the first kind is below $\epsilon > 0$ *(i.e., the level of significance is equal to* ϵ*).*

This theorem is called **quantum Stein's lemma**, which is based on its classical counterpart of Stein's lemma. Of course, if ρ and σ commute, we may treat this testing problem by classical means according to the arguments given after Lemma 3.2 in the previous section. From (3.8) the relation between the quantum relative entropy $D(\rho\|\sigma)$ and Chernoff's bound $\inf_{1\geq s\geq 0} \phi(s)$ is illustrated in Fig. 3.1. In particular, when $\sigma = U\rho U^*$ and $U^2 = I$, $D(\rho\|\sigma) \geq -2\phi(1/2) = -2\inf_{1\geq s\geq 0} \phi(s)$.

Since the proof below also holds for commuting ρ and σ, it can be regarded as a proof of the classical Stein's lemma, although it is rather elaborate. The proof of Theorem 3.2 is obtained by first showing Lemmas 3.6 and 3.7.

Lemma 3.6 (Direct Part, Hiai and Petz [32]) *There exists a sequence of Hermitian matrices* $\{T_n\}$ *on* $\mathcal{H}^{\otimes n}$ *with* $I \geq T_n \geq 0$ *such that for arbitrary* $\delta > 0$

$$\underline{\lim}_{n\to\infty} -\frac{1}{n} \log \operatorname{Tr} \sigma^{\otimes n} T_n \geq D(\rho\|\sigma) - \delta, \tag{3.93}$$

$$\lim_{n\to\infty} \operatorname{Tr} \rho^{\otimes n}(I-T_n) = 0. \tag{3.94}$$

Lemma 3.7 (Converse Part, Ogawa and Nagaoka [7]) *If a sequence of Hermitian matrices* $\{T_n\}$ $(I \geq T_n \geq 0)$ *on* $\mathcal{H}^{\otimes n}$ *satisfies*

$$\overline{\lim}_{n\to\infty} -\frac{1}{n} \log \operatorname{Tr} \sigma^{\otimes n} T_n > D(\rho\|\sigma), \tag{3.95}$$

then

$$\overline{\lim}_{n\to\infty} \operatorname{Tr} \rho^{\otimes n}(I-T_n) = 1. \tag{3.96}$$

Proof of Theorem 3.2 *using Lemmas* 3.6 *and* 3.7 For $1 > \epsilon > 0$, we take $\{T_n\}$ to satisfy (3.93) and (3.94) according to Lemma 3.6. Taking a sufficiently large N, we

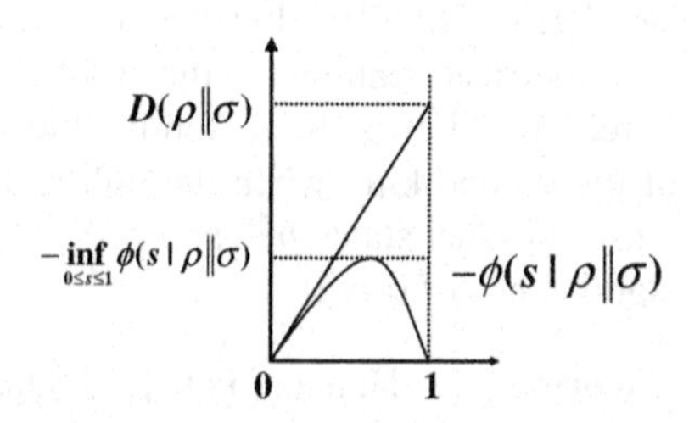

Fig. 3.1 Chernoff's bound and quantum relative entropy

have $\mathrm{Tr}\,\rho^{\otimes n}(I-T_n) \le \epsilon$ for $n \ge N$ from (3.94). Therefore, $\beta_\epsilon^n(\rho\|\sigma) \le \mathrm{Tr}\,\sigma^{\otimes n} T_n$, and we see that $\underline{\lim}_{n\to\infty} -\frac{1}{n}\log\beta_\epsilon^n(\rho\|\sigma) \ge D(\rho\|\sigma) - \delta$. Since $\delta > 0$ is arbitrary, we obtain $\underline{\lim}_{n\to\infty} -\frac{1}{n}\log\beta_\epsilon^n(\rho\|\sigma) \ge D(\rho\|\sigma)$ by taking the limit $\delta \to 0$.

Now, let $\overline{\lim}_{n\to\infty} -\frac{1}{n}\log\beta_\epsilon^n(\rho\|\sigma) > D(\rho\|\sigma)$ for a particular $1 > \epsilon > 0$. Then, we can take a sequence of Hermitian matrices $\{T_n\}$ on $\mathcal{H}^{\otimes n}$ with $I \ge T_n \ge 0$ that satisfies

$$\overline{\lim_{n\to\infty}} -\frac{1}{n}\log\mathrm{Tr}\,\sigma^{\otimes n}T_n > D(\rho\|\sigma), \quad \mathrm{Tr}\,\rho^{\otimes n}(I-T_n) \le \epsilon.$$

However, this contradicts Lemma 3.7, and hence it follows that $\overline{\lim}_{n\to\infty} -\frac{1}{n}\log\beta_\epsilon^n(\rho\|\sigma) \le D(\rho\|\sigma)$. This proves (3.91). ∎

It is rather difficult to prove the above two lemmas at this point. Hence, we will prove them after discussing several other lemmas forming the basis of the asymptotic theory described in Sect. 3.8. In fact, combining Lemmas 3.6 and 3.7, we obtain the following theorem (Theorem 3.3), which implies Theorem 3.2.

Theorem 3.3 *Define*

$$B(\rho\|\sigma) \stackrel{\mathrm{def}}{=} \sup_{\{T_n\}}\left\{ \underline{\lim_{n\to\infty}} -\frac{1}{n}\log\mathrm{Tr}\,\sigma^{\otimes n}T_n \,\middle|\, \lim_{n\to\infty}\mathrm{Tr}\,\rho^{\otimes n}(I-T_n) = 0\right\},$$

$$B^\dagger(\rho\|\sigma) \stackrel{\mathrm{def}}{=} \sup_{\{T_n\}}\left\{ \underline{\lim_{n\to\infty}} -\frac{1}{n}\log\mathrm{Tr}\,\sigma^{\otimes n}T_n \,\middle|\, \underline{\lim_{n\to\infty}}\mathrm{Tr}\,\rho^{\otimes n}(I-T_n) < 1\right\}.$$

Then,

$$B(\rho\|\sigma) = B^\dagger(\rho\|\sigma) = D(\rho\|\sigma).$$

As a corollary, we can show the following.

Corollary 3.1

$$D(\rho\|\sigma) = \lim_{n\to\infty}\frac{1}{n}\max_{\boldsymbol{M}} D(\mathrm{P}^{\boldsymbol{M}}_{\rho^{\otimes n}}\|\mathrm{P}^{\boldsymbol{M}}_{\sigma^{\otimes n}}). \tag{3.97}$$

Proof Applying the classical Stein's lemma to the case with $\mathrm{P}^{\boldsymbol{M}}_{\rho^{\otimes n}}$ and $\mathrm{P}^{\boldsymbol{M}}_{\sigma^{\otimes n}}$, we can show that $D(\mathrm{P}^{\boldsymbol{M}}_{\rho^{\otimes n}}\|\mathrm{P}^{\boldsymbol{M}}_{\sigma^{\otimes n}}) = B(\mathrm{P}^{\boldsymbol{M}}_{\rho^{\otimes n}}\|\mathrm{P}^{\boldsymbol{M}}_{\sigma^{\otimes n}}) \le B(\rho^{\otimes n}\|\sigma^{\otimes n}) = nD(\rho\|\sigma)$. Then, we obtain the $\ge$ part. Let $\{T_n\}$ be a sequence of tests achieving the optimal. Then, for any $\epsilon > 0$, we can prove that $\frac{1}{n}D(\mathrm{P}^{\{T_n,I-T_n\}}_{\rho^{\otimes n}}\|\mathrm{P}^{\{T_n,I-T_n\}}_{\sigma^{\otimes n}}) = \frac{\epsilon}{n}(\log\epsilon - \log(1-\beta_\epsilon^n(\rho\|\sigma))) + \frac{1-\epsilon}{n}(\log(1-\epsilon) - \log\beta_\epsilon^n(\rho\|\sigma)) \to (1-\epsilon)D(\rho\|\sigma)$. Hence, $\lim_{n\to\infty}\frac{1}{n}\max_{\boldsymbol{M}} D(\mathrm{P}^{\boldsymbol{M}}_{\rho^{\otimes n}}\|\mathrm{P}^{\boldsymbol{M}}_{\sigma^{\otimes n}}) \ge (1-\epsilon)D(\rho\|\sigma)$. Taking the limit $\epsilon \to 0$, we obtain the $\le$ part. ∎

For a further analysis of the direct part, we focus on the decreasing exponent of the error probability of the first kind under an exponential constraint for the error probability of the second kind. For details on the converse part, we assume an exponential constraint of the error probability of the second kind and optimize the decreasing exponent of the correct probability of the first kind. In other words, we treat the following values:

$$B(r|\rho\|\sigma) \stackrel{\text{def}}{=} \sup_{\{T_n\}} \left\{ \underline{\lim}_{n\to\infty} -\frac{1}{n} \log \operatorname{Tr} \rho^{\otimes n}(I - T_n) \middle| \underline{\lim}_{n\to\infty} -\frac{1}{n} \log \operatorname{Tr} \sigma^{\otimes n} T_n \geq r \right\}, \tag{3.98}$$

$$B^*(r|\rho\|\sigma) \stackrel{\text{def}}{=} \inf_{\{T_n\}} \left\{ \overline{\lim}_{n\to\infty} -\frac{1}{n} \log \operatorname{Tr} \rho^{\otimes n} T_n \middle| \underline{\lim}_{n\to\infty} -\frac{1}{n} \log \operatorname{Tr} \sigma^{\otimes n} T_n \geq r \right\}. \tag{3.99}$$

Then, we obtain the following theorem.

Theorem 3.4 *Then, the relations*

$$\sup_{0\leq s\leq 1} \frac{-sr - \phi(s|\rho\|\sigma)}{1-s} = B(r|\rho\|\sigma) \leq \min_{\tau:D(\tau\|\sigma)\leq r} D(\tau\|\rho), \tag{3.100}$$

$$\sup_{s\leq 0} \frac{-sr - \phi(s|\rho\|\sigma)}{1-s} \leq B^*(r|\rho\|\sigma) = \sup_{s\leq 0} \frac{-sr - \tilde{\phi}(s|\rho\|\sigma)}{1-s} \leq \min_{\tau:D(\tau\|\sigma)\geq r} D(\tau\|\rho) \tag{3.101}$$

hold.

The equation in (3.100) will be shown in Sect. 3.7. The inequality in (3.100) will be shown in Exercise 3.57. The equation in (3.101) will be shown in Sect. 3.8. The first inequality in (3.100) will be shown in Sect. 3.8. The second inequality in (3.100) will be shown in Exercise 3.179.

The commutative case i.e., the classical case is easier. For two probability distributions p and $\bar{p}$, we have equations

$$B(r|p\|\bar{p}) = \sup_{0\leq s\leq 1} \frac{-sr - \phi(s|p\|\bar{p})}{1-s} = \min_{q:D(q\|\bar{p})\leq r} D(q\|p), \tag{3.102}$$

$$B^*(r|p\|\bar{p}) = \sup_{s\leq 0} \frac{-sr - \phi(s|p\|\bar{p})}{1-s} = \min_{q:D(q\|\bar{p})\leq r} D(q\|p) + r - D(q\|\bar{p}), \tag{3.103}$$

where all equations except for the second equation in (3.103) hold with all $r \geq 0$. The second equation in (3.103) holds with $r \geq D(p\|\bar{p})$. The first equation in (3.102) will be shown jointly with the general case, i.e., the equation in (3.100). The first equation (3.103) will be shown in Exercise 3.54. The second equation in (3.102) is shown in Exercise 3.45. The second equation in (3.103) is shown in Exercises 3.52 and 3.53.

We can also characterize the asymptotic optimal performance of quantum simple hypothesis testing with a general sequence of two quantum states [33, 34]. In this general setting, the main problem is to determine the behavior of $\operatorname{Tr} \rho\{\rho - e^a\sigma \geq 0\}$ as a function of a [35].

Finally, using Corollary 3.1, we characterize other quantum versions of relative entropy because there exist many quantum versions of relative entropy even though we impose the condition that the quantity equals the relative entropy with two commutative inputs. To discuss this issue, we denote such a quantity by $\tilde{D}(\rho\|\sigma)$. Then, the condition is written as

$$\widetilde{D}(\rho\|\sigma) = D(p\|q) \tag{3.104}$$

for two commutative density matrices ρ and σ whose eigenvalues form the probability distributions p and q. Now, we impose two additional conditions for a quantity $\tilde{D}(\rho\|\sigma)$; One is the monotonicity for a measurement $\boldsymbol{M}$;

$$\widetilde{D}(\rho\|\sigma) \geq D(\mathrm{P}^{\boldsymbol{M}}_{\rho}\|\mathrm{P}^{\boldsymbol{M}}_{\sigma}). \tag{3.105}$$

The other is the additivity;

$$\widetilde{D}(\rho_1 \otimes \rho_2\|\sigma_1 \otimes \sigma_2) = \widetilde{D}(\rho_1\|\sigma_1) + \widetilde{D}(\rho_2\|\sigma_2). \tag{3.106}$$

Then, Corollary 3.1 implies that

$$\widetilde{D}(\rho\|\sigma) = \lim_{n\to\infty} \frac{1}{n}\widetilde{D}(\rho^{\otimes n}\|\sigma^{\otimes n}) \geq \lim_{n\to\infty} \frac{1}{n} \max_{\boldsymbol{M}} D(\mathrm{P}^{\boldsymbol{M}}_{\rho^{\otimes n}}\|\mathrm{P}^{\boldsymbol{M}}_{\sigma^{\otimes n}}) = D(\rho\|\sigma). \tag{3.107}$$

That is, the quantum relative entropy $D(\rho\|\sigma)$ is the minimum quantum analog of relative entropy with the monotonicity for measurement and the additivity. Note that Condition (3.105) is a weaker requirement than the monotonicity for TP-CP map (5.36), which will be explained in Chap. 5.

Exercises

In the following, we abbreviate $\phi(s|p\|\bar{p})$ to $\phi(s)$.

3.43 Define the distribution $p_s(x) := p(x)^{1-s}\bar{p}(x)^s e^{-\phi(s)}$ and assume that a distribution q satisfies $D(q\|\bar{p}) = D(p_s\|\bar{p})$. Show that $D(p_s\|p) \leq D(q\|p)$ for $s \leq 1$ by following steps below.
(**a**) Show that $\frac{1}{1-s}D(q\|p_s) = \frac{1}{1-s}\sum_x q(x)(\log q(x) - \log \bar{p}(x)) - \sum_x q(x)(\log p(x) - \log \bar{p}(x)) + \frac{\phi(s)}{1-s}$.
(**b**) Show $D(q\|p) - \frac{1}{1-s}D(q\|p_s) = D(p_s\|p)$.
(**c**) Show the desired inequality.

3.44 Show the same argument as Exercise 3.43 by the following alternative way.
(**a**) Show that $\{p_s(x)\}$ is an exponential family
(**b**) Show the desired argument by using Theorem 2.3.

3.45 Show the equation

$$\sup_{0\leq s\leq 1} \frac{-sr - \phi(s)}{1-s} = \min_{q:D(q\|\bar{p})\leq r} D(q\|p). \tag{3.108}$$

following the steps below.
(**a**) Show that $\frac{-sr-\phi(s)}{1-s} \le 0$ for $r \ge D(p\|\bar{p})$ and $s \in [0, 1]$.
(**b**) Show that both sides of (3.108) are zero when $r \ge D(p\|\bar{p})$.
(**c**) Show that

$$D(p_s\|p_1) = (s-1)\phi'(s) - \phi(s) \tag{3.109}$$
$$D(p_s\|p_0) = s\phi'(s) - \phi(s) \tag{3.110}$$

(**d**) Show that

$$\frac{d}{ds}(s-1)\phi'(s) - \phi(s) = (s-1)\phi''(s) < 0 \tag{3.111}$$
$$\frac{d}{ds}s\phi'(s) - \phi(s) = s\phi''(s) > 0 \tag{3.112}$$

for $s \in (0, 1)$.
(**e**) In the following, we consider the case $r < D(p\|\bar{p})$. Show that there uniquely exists $s_r \in (0, 1)$ such that $D(p_{s_r}\|p_1) = r$.
(**f**) Show that

$$\min_{q:D(q\|\bar{p})=r} D(q\|p) = D(p_{s_r}\|p). \tag{3.113}$$

(**g**) Show that

$$\min_{q:D(q\|\bar{p})\le r} D(q\|p) = D(p_{s_r}\|p). \tag{3.114}$$

(**h**) Show that

$$D(p_{s_r}\|p) = \frac{-s_r r - \phi(s_r)}{1-s_r}. \tag{3.115}$$

(**i**) Show that

$$\frac{d}{ds}\frac{-sr-\phi(s)}{1-s} = \frac{-r+(s-1)\phi'(s)-\phi(s)}{(1-s)^2}. \tag{3.116}$$

(**j**) Show that

$$\sup_{0\le s\le 1}\frac{-sr-\phi(s)}{1-s} = \frac{-s_r r-\phi(s_r)}{1-s_r}. \tag{3.117}$$

(**k**) Show (3.108).

3.6 Hypothesis Testing by Separable Measurements

In the previous section, we performed the optimization with no restriction on the measurements on $\mathcal{H}^{\otimes n}$. In this section, we will restrict the possible measurements to separable measurements. In other words, our test T is assumed to have the separable form:

$$T = \sum_{\omega_n} M^n_{1,\omega_n} \otimes \cdots \otimes M^n_{n,\omega_n}, \quad M^n_{1,\omega_n} \geq 0, \ldots, M^n_{n,\omega_n} \geq 0 \text{ on } \mathcal{H}^{\otimes n},$$

which is called a **separable test**. This class of tests includes cases such as making identical measurements on every system $\mathcal{H}$ and analyzing measurement data statistically. As explained in (1.25), it also includes other methods such as adaptive improvement of the measurements and statistical analysis of measurement data. The following theorem evaluates the asymptotic performance of the tests based on these measurements.

Theorem 3.5 *Defining $\tilde{B}(\rho\|\sigma)$ as*

$$\tilde{B}(\rho\|\sigma) \stackrel{\text{def}}{=} \sup_{\{T_n\}:\text{separable}} \left\{ \underline{\lim}_{n\to\infty} -\frac{1}{n} \log \operatorname{Tr} \sigma^{\otimes n} T_n \middle| \lim_{n\to\infty} \operatorname{Tr} \rho^{\otimes n}(I - T_n) = 0 \right\},$$

we have

$$\tilde{B}(\rho\|\sigma) = \max_{\boldsymbol{M}} D\left(\mathrm{P}^{\boldsymbol{M}}_{\rho} \| \mathrm{P}^{\boldsymbol{M}}_{\sigma}\right). \tag{3.118}$$

When the measurement $\boldsymbol{M}_{\max} \stackrel{\text{def}}{=} \operatorname{argmax}_{\boldsymbol{M}} D\left(\mathrm{P}^{\boldsymbol{M}}_{\rho} \| \mathrm{P}^{\boldsymbol{M}}_{\sigma}\right)$ is performed n times, the bound $\max_{\boldsymbol{M}} D\left(\mathrm{P}^{\boldsymbol{M}}_{\rho} \| \mathrm{P}^{\boldsymbol{M}}_{\sigma}\right)$ can be asymptotically attained by suitable statistical processing of the n data.

This theorem shows that in terms of quantities such as $\tilde{B}(\rho\|\sigma)$, there is no asymptotic difference between the optimal classical data processing according to identical measurements $\boldsymbol{M}_{\max}$ on each system and the optimal separable test across systems.

Therefore, at least for this problem, we cannot take advantage of the correlation between quantum systems unless a non-separable measurement is used. Since $\tilde{B}(\rho\|\sigma) \leq B(\rho\|\sigma)$, we have the **monotonicity** of quantum relative entropy for a measurement

$$D\left(\mathrm{P}^{\boldsymbol{M}}_{\rho} \| \mathrm{P}^{\boldsymbol{M}}_{\sigma}\right) \leq D(\rho\|\sigma). \tag{3.119}$$

The following theorem discusses the equality condition of the above inequality.

Theorem 3.6 (Ohya and Petz [36], Nagaoka [37], Fujiwara [38]) *The following conditions are equivalent for two states ρ and σ and a rank-one PVM $\boldsymbol{M} = \{M_i\}_{i=1}^d$.*

① *The equality in (3.119) is satisfied.*

② $[\sigma, \rho] = 0$ *and there exists a set of real numbers* $\{a_i\}_{i=1}^d$ *satisfying*

$$\rho = \sigma \left(\sum_{i=1}^{d} a_i M_i \right) = \left(\sum_{i=1}^{d} a_i M_i \right) \sigma. \tag{3.120}$$

Here, notice that a PVM $\boldsymbol{M}$ satisfying Condition ② is not limited to the simultaneous spectral decomposition of ρ and σ.[Exe. 3.46] Theorem 3.6 will be shown in Sect. 5.4. Also, another proof will be given in Exercise 6.32.

Proof of Theorem 3.5 The fact that $\max_{\boldsymbol{M}} D\left(\mathrm{P}_\rho^{\boldsymbol{M}} \| \mathrm{P}_\sigma^{\boldsymbol{M}}\right)$ can be attained, i.e., the "$\geq$" sign in (3.118), follows from the relation $B\left(\mathrm{P}_\rho^{\boldsymbol{M}} \| \mathrm{P}_\sigma^{\boldsymbol{M}}\right) = \tilde{B}\left(\mathrm{P}_\rho^{\boldsymbol{M}} \| \mathrm{P}_\sigma^{\boldsymbol{M}}\right) = D\left(\mathrm{P}_\rho^{\boldsymbol{M}} \| \mathrm{P}_\sigma^{\boldsymbol{M}}\right)$ shown by Stein's lemma with the classical case. Therefore, we show that $\tilde{B}(\rho\|\sigma)$ does not exceed this value, i.e., the "$\leq$" sign in (3.118). It is sufficient to treat $\underline{\lim}_{n\to\infty} -\frac{1}{n} \log \mathrm{P}_{\sigma^{\otimes n}}^{\boldsymbol{M}^n}(A_n)$ for the pair of separable measurements $\boldsymbol{M}^n = \{M_{\omega_n}^n\}_{\omega_n \in \Omega_n}$:

$$M_{\omega_n}^n = M_{1,\omega_n}^n \otimes \cdots \otimes M_{n,\omega_n}^n, \quad M_{1,\omega_n}^n \geq 0, \ldots, M_{n,\omega_n}^n \geq 0 \text{ on } \mathcal{H}^{\otimes n}$$

and a subset A_n of Ω_n with $\mathrm{P}_{\rho^{\otimes n}}^{\boldsymbol{M}^n}(A_n^c) \to 0$. First, we show that

$$\frac{1}{n} D\left(\mathrm{P}_{\rho^{\otimes n}}^{\boldsymbol{M}^n} \middle\| \mathrm{P}_{\sigma^{\otimes n}}^{\boldsymbol{M}^n}\right) \leq \max_{\boldsymbol{M}} D\left(\mathrm{P}_\rho^{\boldsymbol{M}} \| \mathrm{P}_\sigma^{\boldsymbol{M}}\right). \tag{3.121}$$

For this purpose, we define $a_{k,\omega_n} \stackrel{\text{def}}{=} \prod_{j\neq k} \operatorname{Tr} M_{j,\omega_n}^n \rho$ and $M_{\omega_n}^{n,k} \stackrel{\text{def}}{=} a_{k,\omega_n} M_{k,\omega_n}^n$. Since an arbitrary state ρ' on $\mathcal{H}$ satisfies

$$\begin{aligned} \operatorname{Tr} \rho' \sum_{\omega_n} M_{\omega_n}^{n,k} &= \operatorname{Tr} \sum_{\omega_n} \rho^{\otimes(k-1)} \otimes \rho' \otimes \rho^{\otimes(n-k)} M_{\omega_n}^n \\ &= \operatorname{Tr} \rho^{\otimes(k-1)} \otimes \rho' \otimes \rho^{\otimes(n-k)} = 1, \end{aligned}$$

we see that $\sum_{\omega_n} M_{\omega_n}^{n,k} = I$; hence, we can verify that $\{M_{\omega_n}^{n,k}\}$ is a POVM @. Moreover, we can show [Exe. 3.47] that

$$D\left(\mathrm{P}_{\rho^{\otimes n}}^{\boldsymbol{M}^n} \middle\| \mathrm{P}_{\sigma^{\otimes n}}^{\boldsymbol{M}^n}\right) = \sum_{k=1}^{n} D\left(\mathrm{P}_{\rho}^{\boldsymbol{M}^{n,k}} \middle\| \mathrm{P}_{\sigma}^{\boldsymbol{M}^{n,k}}\right), \tag{3.122}$$

and thus verify (3.121). Since the monotonicity of the relative entropy for a probability distribution yields

$$\begin{aligned} &\mathrm{P}_{\rho^{\otimes n}}^{\boldsymbol{M}^n}(A_n)\left(\log \mathrm{P}_{\rho^{\otimes n}}^{\boldsymbol{M}^n}(A_n) - \log \mathrm{P}_{\sigma^{\otimes n}}^{\boldsymbol{M}^n}(A_n)\right) \\ &+ \mathrm{P}_{\rho^{\otimes n}}^{\boldsymbol{M}^n}(A_n^c)\left(\log \mathrm{P}_{\rho^{\otimes n}}^{\boldsymbol{M}^n}(A_n^c) - \log \mathrm{P}_{\sigma^{\otimes n}}^{\boldsymbol{M}^n}(A_n^c)\right) \\ \leq &D\left(\mathrm{P}_{\rho^{\otimes n}}^{\boldsymbol{M}^n} \middle\| \mathrm{P}_{\sigma^{\otimes n}}^{\boldsymbol{M}^n}\right) \leq n \max_{\boldsymbol{M}} D\left(\mathrm{P}_\rho^{\boldsymbol{M}} \| \mathrm{P}_\sigma^{\boldsymbol{M}}\right), \end{aligned}$$

we obtain

$$-\frac{1}{n}\log \mathrm{P}^{\boldsymbol{M}^n}_{\sigma^{\otimes n}}(A_n) \le \frac{\max_{\boldsymbol{M}} D\left(\mathrm{P}^{\boldsymbol{M}}_{\rho}\|\mathrm{P}^{\boldsymbol{M}}_{\sigma}\right) + \frac{1}{n}h(\mathrm{P}^{\boldsymbol{M}^n}_{\rho^{\otimes n}}(A_n))}{\mathrm{P}^{\boldsymbol{M}^n}_{\rho^{\otimes n}}(A_n)}, \tag{3.123}$$

where we used the fact that $-\mathrm{P}^{\boldsymbol{M}^n}_{\rho^{\otimes n}}(A_n^c)\log \mathrm{P}^{\boldsymbol{M}^n}_{\sigma^{\otimes n}}(A_n^c) \ge 0$, and $h(x)$ is a binary entropy that is expressed as $h(x) = -x\log x - (1-x)\log(1-x)$. Noting that $h(x) \le \log 2$ and $\mathrm{P}^{\boldsymbol{M}^n}_{\rho^{\otimes n}}(A_n) \to 1$, we have

$$\overline{\lim}\, -\frac{1}{n}\log \mathrm{P}^{\boldsymbol{M}^n}_{\sigma^{\otimes n}}(A_n) \le \max_{\boldsymbol{M}} D\left(\mathrm{P}^{\boldsymbol{M}}_{\rho}\|\mathrm{P}^{\boldsymbol{M}}_{\sigma}\right),$$

from which we obtain (3.118). ∎

Exercises

3.46 Give an example of ρ, σ, and a PVM $\boldsymbol{M}$ satisfying Condition ② such that a PVM $\boldsymbol{M}$ is not the simultaneous spectral decomposition of ρ and σ.

3.47 Prove (3.122).

3.7 Proof of Direct Part of Stein's Lemma and Hoeffding Bound

In order to prove the direct part of Stein's Lemma, i.e., Lemma 3.6, firstly, we show

$$B(r|\rho\|\sigma) \ge \sup_{0\le s\le 1} \frac{-sr - \phi(s|\rho\|\sigma)}{1-s}. \tag{3.124}$$

When $r < D(\rho|\sigma)$, the right hand side of (3.124) is strictly greater than zero because $-\frac{d\phi(s|\rho\|\sigma)}{ds}|_{s=0} = D(\rho|\sigma)$. This fact proves Lemma 3.6.

In order to show (3.124), we apply Lemma 3.3 with $A = e^{-nR}\rho^{\otimes n}$ and $B = \sigma^{\otimes n}$ with an arbitrary real number R. Then, we obtain

$$\mathrm{Tr}\, e^{-nR}\rho^{\otimes n}\{e^{-nR}\rho^{\otimes n} \le \sigma^{\otimes n}\} + \mathrm{Tr}\,\sigma^{\otimes n}\{e^{-nR}\rho^{\otimes n} > \sigma^{\otimes n}\} \tag{3.125}$$

$$\le e^{-n(1-s)R}\,\mathrm{Tr}(\rho^{\otimes n})^{1-s}(\sigma^{\otimes n})^s = e^{n(-(1-s)R+\phi(s))}, \tag{3.126}$$

which implies that

$$\mathrm{Tr}\,\rho^{\otimes n}\{e^{-nR}\rho^{\otimes n} \le \sigma^{\otimes n}\} \le e^{n(sR+\phi(s))},$$
$$\mathrm{Tr}\,\sigma^{\otimes n}\{e^{-nR}\rho^{\otimes n} > \sigma^{\otimes n}\} \le e^{n(-(1-s)R+\phi(s))}.$$

Given a positive real number r and $s \in [0, 1]$, we choose $R = \frac{r+\phi(s)}{1-s}$. Then,

$$\operatorname{Tr} \rho^{\otimes n}\{\rho^{\otimes n} \le e^{nR}\sigma^{\otimes n}\} \le e^{-n\frac{-sr-\phi(s|\rho\|\sigma)}{1-s}},$$
$$\operatorname{Tr} \sigma^{\otimes n}\{\rho^{\otimes n} > e^{nR}\sigma^{\otimes n}\} \le e^{-nr},$$

which implies

$$B(r|\rho\|\sigma) \ge \frac{-sr - \phi(s|\rho\|\sigma)}{1-s}.$$

Taking the maximum with respect to $s \in [0, 1]$, we obtain (3.124).

In order to show the inequality opposite to (3.124), we prepare the following lemma.

Lemma 3.8 *When $R \in [-D(\sigma\|\rho), D(\rho\|\sigma)]$, we have*

$$\lim_{n\to\infty} -\frac{1}{n} \log \operatorname{Tr} \rho^{\otimes n}\{e^{-nR}\rho^{\otimes n} \le \sigma^{\otimes n}\} = \max_{s\in[0,1]} -sR - \phi(s), \tag{3.127}$$
$$\lim_{n\to\infty} -\frac{1}{n} \log \operatorname{Tr} \sigma^{\otimes n}\{e^{-nR}\rho^{\otimes n} > \sigma^{\otimes n}\} = \max_{s\in[0,1]} (1-s)R - \phi(s). \tag{3.128}$$

Now, we recall the relation (3.61). Then, (3.127) implies

$$\lim_{n\to\infty} -\frac{1}{n} \log \operatorname{Tr} \rho^{\otimes n}\{e^{-nD(\rho\|\sigma)}\rho^{\otimes n} \le \sigma^{\otimes n}\} \le \max_{s\in[0,1]} -sR - \phi(s) \tag{3.129}$$

for $R < D(\rho\|\sigma)$. The limit $R \to D(\rho\|\sigma)$ yields that

$$\lim_{n\to\infty} -\frac{1}{n} \log \operatorname{Tr} \rho^{\otimes n}\{e^{-nD(\rho\|\sigma)}\rho^{\otimes n} \le \sigma^{\otimes n}\} = 0 \tag{3.130}$$

Due to the relation (3.61), the relation (3.127) implies $\lim_{n\to\infty} -\frac{1}{n} \log \operatorname{Tr} \rho^{\otimes n}\{e^{-nR} \rho^{\otimes n} \le \sigma^{\otimes n}\}$ is positive if and only if $R < D(\rho\|\sigma)$. Similarly, the left hand side of (3.128) is positive if and only if $R > -D(\sigma\|\rho)$.

Since Lemma 3.2 yields that

$$\begin{aligned}&\min_{I\ge T\ge 0} \left(e^{-nR} \operatorname{Tr} \rho^{\otimes n}(I-T) + \operatorname{Tr} \sigma^{\otimes n} T\right)\\ =&\operatorname{Tr} \rho^{\otimes n}\{e^{-nR}\rho^{\otimes n} \le \sigma^{\otimes n}\} + c \operatorname{Tr} \sigma^{\otimes n}\{e^{-nR}\rho^{\otimes n} > \sigma^{\otimes n}\},\end{aligned}$$

our test can be restricted to tests with the form $\{\{e^{-nR}\rho^{\otimes n} \le \sigma^{\otimes n}\}, \{e^{-nR}\rho^{\otimes n} > \sigma^{\otimes n}\}\}$. Thanks to the above observation, using Lemma 3.8, we obtain

$$\begin{aligned}&B(r|\rho|\sigma)\\ =&\sup_{R\in(-D(\sigma\|\rho),D(\rho\|\sigma))} \left\{\lim_{n\to\infty} -\frac{1}{n} \log \operatorname{Tr} \rho^{\otimes n}\{e^{-nR}\rho^{\otimes n} \le \sigma^{\otimes n}\}\right|\end{aligned}$$

$$
\begin{aligned}
&\lim_{n\to\infty} -\frac{1}{n}\log \operatorname{Tr}\sigma^{\otimes n}\{e^{-nR}\rho^{\otimes n} > \sigma^{\otimes n}\} > r\Big\} \\
=& \sup_{R\in(-D(\sigma\|\rho),D(\rho\|\sigma))} \{\max_{s\in[0,1]} -sR-\phi(s) | \max_{s\in[0,1]}(1-s)R-\phi(s)\ge r\} \\
=& \sup_{s\in(0,1)} \{-sR_s-\phi(s)|(1-s)R_s-\phi(s)\ge r\},
\end{aligned}
$$

where $R_s = -\frac{d}{ds}\phi(s)$ (See Exercise 3.48). Now, we choose $s_0 \in (0,1)$ such that $(1-s_0)R_{s_0} - \phi(s_0) = r$. Then, we obtain

$$
B(r|\rho|\sigma) = -\, s_0 R_{s_0} - \phi(s_0) = \frac{-s_0 r - \phi(s_0)}{1-s_0}. \tag{3.131}
$$

Due to (3.124), we obtain

$$
\max_{s\in(0,1)} \frac{-sr-\phi(s)}{1-s} = \frac{-s_0 r - \phi(s_0)}{1-s_0}.
$$

Hence, we obtain the inequality opposite to (3.124).

The remaining inequality concerning the Hoeffding bound in Theorem 3.4 is

$$
B(r|\rho\|\sigma) \le \min_{\tau:D(\tau\|\sigma)\le r} D(\tau\|\rho), \tag{3.132}
$$

which is shown in Exercise 3.57.

Proof of Lemma 3.8 Similar to the proof of Lemma 3.5, we define two distributions $P := P_{(\rho\|\sigma)}$ and $Q := Q_{(\rho\|\sigma)}$. The application of Cramér Theorem (Theorem 2.7) to the random variable $\log\frac{P(\omega)}{Q(\omega)}$ yields that

$$
\begin{aligned}
&\lim_{n\to\infty} -\frac{1}{n}\log P^n\{\omega^n\in\Omega^n|e^{-nR}P^n(\omega^n)\le Q^n(\omega^n)\} \\
=& \lim_{n\to\infty} -\frac{1}{n}\log P^n\{\omega^n\in\Omega^n|\frac{1}{n}\log\frac{P^n(\omega^n)}{Q^n(\omega^n)}\le R\} \\
=& \sup_{s\in[0,\infty)} -sR-\phi(s|P\|Q).
\end{aligned} \tag{3.133}
$$

Here, we can show that

$$
\sup_{s\in[0,\infty)} -sR-\phi(s|P\|Q) = \max_{s\in(0,1)} -sR-\phi(s|P\|Q). \tag{3.134}
$$

Since the map $s\mapsto\phi(s|P\|Q)$ is convex, the value $R_s = -\frac{d}{ds}\phi(s|P\|Q)$ is monotonically decreasing with respect to s. Since $R\in(-D(Q\|P), D(P\|Q))$ and $\frac{d}{ds}\phi(s|P\|Q)|_{s=1} = D(Q\|P)$ and $\frac{d}{ds}\phi(s|P\|Q)|_{s=0} = -D(P\|Q)$, using the fact shown in Exercise 3.48, we can show that the above maximum is realized in (0, 1), i.e., (3.134) holds.

Similarly, we can show that

$$\begin{aligned} &\lim_{n\to\infty} -\frac{1}{n}\log Q^n\{\omega^n \in \Omega^n | e^{-nR}P^n(\omega^n) > Q^n(\omega^n)\} \\ =& \max_{s\in(0,1)} (1-s)R - \phi(s|P\|Q). \end{aligned} \tag{3.135}$$

Now we employ Lemma 3.4 with $A = e^{-nR}\rho^{\otimes n}$ and $B = \sigma^{\otimes n}$. Then,

$$\begin{aligned} &e^{-nR}\operatorname{Tr}\rho^{\otimes n}\{e^{-nR}\rho^{\otimes n} \le \sigma^{\otimes n}\} + \operatorname{Tr}\sigma^{\otimes n}\{e^{-nR}\rho^{\otimes n} > \sigma^{\otimes n}\} \\ \ge& \frac{1}{2}\Big(e^{-nR}P^n\{\omega^n \in \Omega^n | e^{-nR}P^n(\omega^n) \le Q^n(\omega^n)\} \\ &\quad + Q^n\{\omega^n \in \Omega^n | e^{-nR}P^n(\omega^n) > Q^n(\omega^n)\}\Big). \end{aligned}$$

Thus,

$$\begin{aligned} &\min\{R + \lim_{n\to\infty} -\frac{1}{n}\log\operatorname{Tr}\rho^{\otimes n}\{e^{-nR}\rho^{\otimes n} \le \sigma^{\otimes n}\}, \\ &\qquad \lim_{n\to\infty} -\frac{1}{n}\log\operatorname{Tr}\sigma^{\otimes n}\{e^{-nR}\rho^{\otimes n} > \sigma^{\otimes n}\}\} \\ \le& \max_{s\in(0,1)} (1-s)R - \phi(s|P\|Q). \end{aligned} \tag{3.136}$$

In fact, the opposite inequality holds due to (3.125) and (3.126). That is, the inequality (3.127) or (3.128) holds at least. Assume that the inequality (3.127) holds. We choose a sufficiently small $\epsilon > 0$ such that $R - \epsilon \in (-D(Q\|P), D(P\|Q))$. Then, Lemma 3.4 implies that

$$\begin{aligned} &e^{-n(R+\epsilon)}\operatorname{Tr}\rho^{\otimes n}\{e^{-nR}\rho^{\otimes n} \le \sigma^{\otimes n}\} + \operatorname{Tr}\sigma^{\otimes n}\{e^{-nR}\rho^{\otimes n} > \sigma^{\otimes n}\} \\ \ge& e^{-n(R+\epsilon)}\operatorname{Tr}\rho^{\otimes n}\{e^{-n(R+\epsilon)}\rho^{\otimes n} \le \sigma^{\otimes n}\} + \operatorname{Tr}\sigma^{\otimes n}\{e^{-n(R+\epsilon)}\rho^{\otimes n} > \sigma^{\otimes n}\}. \end{aligned}$$

Applying (3.136) with $R + \epsilon$, we have

$$\begin{aligned} &\min\Big\{R + \epsilon + \lim_{n\to\infty} -\frac{1}{n}\log\operatorname{Tr}\rho^{\otimes n}\{e^{-nR}\rho^{\otimes n} \le \sigma^{\otimes n}\}, \\ &\qquad \lim_{n\to\infty} -\frac{1}{n}\log\operatorname{Tr}\sigma^{\otimes n}\{e^{-nR}\rho^{\otimes n} > \sigma^{\otimes n}\}\Big\} \\ \le& \min\Big\{R + \epsilon + \lim_{n\to\infty} -\frac{1}{n}\log\operatorname{Tr}\rho^{\otimes n}\{e^{-n(R+\epsilon)}\rho^{\otimes n} \le \sigma^{\otimes n}\}, \\ &\qquad \lim_{n\to\infty} -\frac{1}{n}\log\operatorname{Tr}\sigma^{\otimes n}\{e^{-n(R+\epsilon)}\rho^{\otimes n} > \sigma^{\otimes n}\}\Big\} \\ =& \max_{s\in(0,1)} (1-s)(R+\epsilon) - \phi(s|P\|Q) \overset{(a)}{<} \epsilon + \max_{s\in(0,1)} (1-s)R - \phi(s|P\|Q) \\ =& R + \epsilon + \lim_{n\to\infty} -\frac{1}{n}\log\operatorname{Tr}\rho^{\otimes n}\{e^{-nR}\rho^{\otimes n} \le \sigma^{\otimes n}\}, \end{aligned}$$

where the strict inequality (a) follows from the fact that the maximum $\max_{s\in(0,1)}(1-s)(R+\epsilon)-\phi(s|P\|Q)$ is realized with $s>0$. Since $\min\{A,B\}<A$ implies that $\min\{A,B\}=B$, we have

$$\begin{aligned}&\lim_{n\to\infty} -\frac{1}{n}\log \mathrm{Tr}\,\sigma^{\otimes n}\{e^{-nR}\rho^{\otimes n}>\sigma^{\otimes n}\}\\ \le & \max_{s\in(0,1)}(1-s)(R+\epsilon)-\phi(s|P\|Q).\end{aligned}$$

Taking the limit $\epsilon\to 0$, we obtain (3.128).

Conversely, we assume that the inequality (3.128) holds. Replacing ϵ by $-\epsilon$, we have

$$\begin{aligned}&\min\Big\{R+\epsilon+\lim_{n\to\infty} -\frac{1}{n}\log \mathrm{Tr}\,\rho^{\otimes n}\{e^{-nR}\rho^{\otimes n}\le\sigma^{\otimes n}\},\\ &\qquad \lim_{n\to\infty} -\frac{1}{n}\log \mathrm{Tr}\,\sigma^{\otimes n}\{e^{-nR}\rho^{\otimes n}>\sigma^{\otimes n}\}\Big\}\\ \le & \max_{s\in(0,1)}(1-s)(R-\epsilon)-\phi(s|P\|Q)\overset{(a)}{<}\max_{s\in(0,1)}(1-s)R-\phi(s|P\|Q)\\ = & \lim_{n\to\infty} -\frac{1}{n}\log \mathrm{Tr}\,\sigma^{\otimes n}\{e^{-nR}\rho^{\otimes n}>\sigma^{\otimes n}\},\end{aligned}$$

where the strict inequality (a) follows from the fact that the maximum $\max_{s\in(0,1)}(1-s)(R-\epsilon)-\phi(s|P\|Q)$ is realized with $s<1$. Since $\min\{A,B\}<A$ implies that $\min\{A,B\}=B$, we have

$$R+\epsilon+\lim_{n\to\infty} -\frac{1}{n}\log \mathrm{Tr}\,\rho^{\otimes n}\{e^{-nR}\rho^{\otimes n}\le\sigma^{\otimes n}\}\le\max_{s\in(0,1)}(1-s)R-\phi(s|P\|Q),$$

which implies that

$$\lim_{n\to\infty} -\frac{1}{n}\log \mathrm{Tr}\,\rho^{\otimes n}\{e^{-nR}\rho^{\otimes n}\le\sigma^{\otimes n}\}\le -\epsilon+\max_{s\in(0,1)}-sR-\phi(s|P\|Q)$$

Taking the limit $\epsilon\to 0$, we obtain (3.127). ■

Exercise

3.48 Show that $\max_{s\in[0,1]}-sR_{s_0}-\phi(s)=-s_0R_{s_0}-\phi(s_0)$ for $s_0\in(0,1)$. Use the fact that $\phi(s)$ is convex.

3.8 Information Inequalities and Proof of Converse Part of Stein's Lemma and Han-Kobayashi Bound

In this section, we first prove the converse part of Stein's lemma based on inequality (3.20). After this proof, we show the information inequalities (3.18) and (3.20).

Proof of Lemma 3.7 Applying inequality (3.20) to the two-valued POVM $\{T_n, I - T_n\}$, we have

$$\begin{aligned}&(\mathrm{Tr}\,\rho^{\otimes n}T_n)^{1-s}(\mathrm{Tr}\,\sigma^{\otimes n}T_n)^s\\ \le&(\mathrm{Tr}\,\rho^{\otimes n}T_n)^{1-s}(\mathrm{Tr}\,\sigma^{\otimes n}T_n)^s + (\mathrm{Tr}\,\rho^{\otimes n}(I-T_n))^{1-s}(\mathrm{Tr}\,\sigma^{\otimes n}(I-T_n))^s\\ \le& e^{n\phi(s|\rho\|\sigma)}\end{aligned} \tag{3.137}$$

for $s \le 0$. Hence,

$$\frac{1-s}{n}\log(\mathrm{Tr}\,\rho^{\otimes n}T_n) + \frac{s}{n}\log(\mathrm{Tr}\,\sigma^{\otimes n}T_n) \le \phi(s|\rho\|\sigma). \tag{3.138}$$

Solving the above inequality with respect to $-\frac{1}{n}\log(\mathrm{Tr}\,\rho^{\otimes n}T_n)$, we have

$$-\frac{1}{n}\log(\mathrm{Tr}\,\rho^{\otimes n}T_n) \ge \frac{-\phi(s|\rho\|\sigma) - s - \frac{1}{n}\log \mathrm{Tr}\,\sigma^{\otimes n}T_n}{1-s}. \tag{3.139}$$

When $r = \overline{\lim} -\frac{1}{n}\log \mathrm{Tr}\,\sigma^{\otimes n}T_n$, taking the limit, we obtain

$$\overline{\lim} -\frac{1}{n}\log(\mathrm{Tr}\,\rho^{\otimes n}T_n) \ge \frac{-\phi(s|\rho\|\sigma) - sr}{1-s}. \tag{3.140}$$

When $r > D(\rho\|\sigma)$, the equation $-\phi'(0) = D(\rho\|\sigma)$ implies $\frac{\phi(s_0)}{-s_0} = \frac{\phi(s_0)-\phi(0)}{-s_0} < r$ for an appropriate $s_0 < 0$. Therefore, $\frac{-\phi(s_0)-s_0 r}{1-s_0} = \frac{s_0}{1-s_0}\left(\frac{\phi(s_0)}{-s_0} - r\right) > 0$, and we can show that $\underline{\lim}_{n\to\infty} \mathrm{Tr}\,\rho^{\otimes n}T_n = 0$. ■

Proof of Han-Kobayashi bound (3.101) In the above proof, when $r = \underline{\lim}_{n\to\infty} -\frac{1}{n}\log \mathrm{Tr}\,\sigma^{\otimes n}T_n$, we have

$$B^*(r|\rho\|\sigma) \ge \sup_{s\le 0}\frac{-sr-\phi(s)}{1-s}. \tag{3.141}$$

Since the quantity $\tilde{\phi}(s|\rho\|\sigma)$ satisfies the information processing inequality with respect to POVM as (3.22), similar to (3.141), we can show the inequality

$$B^*(r|\rho\|\sigma) \ge \sup_{s\le 0}\frac{-sr-\tilde{\phi}(s)}{1-s}. \tag{3.142}$$

As is shown in Exercise 3.54, we can show the opposite inequality in the classical case, in which, $\phi(s) = \tilde{\phi}(s)$. We choose the POVM $\boldsymbol{M}$ on the tensor product space $\mathcal{H}^{\otimes m}$ such that the element of $\boldsymbol{M}$ is commutative with $\kappa_{\sigma^{\otimes m}}(\rho^{\otimes m})$. Then, $\phi(s|\mathrm{P}^{\boldsymbol{M}}_{\rho^{\otimes m}}\|\mathrm{P}^{\boldsymbol{M}}_{\sigma^{\otimes m}}) = \phi(s|\kappa_{\sigma^{\otimes m}}(\rho^{\otimes m})\|\sigma^{\otimes m})$. We apply the classical result to the distributions $\mathrm{P}^{\boldsymbol{M}}_{\rho^{\otimes m}}$ and $\mathrm{P}^{\boldsymbol{M}}_{\sigma^{\otimes m}}$. Then, we obtain

$$B^*(r|\rho\|\sigma) \le \frac{1}{m}B^*(mr|\rho^{\otimes m}\|\sigma^{\otimes m}) \le \sup_{s\le 0}\frac{-sr-\frac{1}{n}\phi(s|\kappa_{\sigma^{\otimes n}}(\rho^{\otimes n})\|\sigma^{\otimes n})}{1-s}. \tag{3.143}$$

Hence, by taking the limit, the above relation and (3.17) yield the inequality opposite to (3.142), which implies [39]

$$B^*(r|\rho\|\sigma) = \sup_{s\le 0} \frac{-sr - \tilde{\phi}(s)}{1-s}. \tag{3.144}$$

The remaining argument concerning $B^*(r|\rho\|\sigma)$ is

$$B^*(r|\rho\|\sigma) \le \min_{\tau: D(\tau\|\sigma)\ge r} D(\tau\|\rho), \tag{3.145}$$

which is shown in Exercise 3.56. ■

Next, we prove information inequalities (3.17), (3.18), (3.20), and (3.22). For this purpose, we require two lemmas (Lemmas 3.9 and 3.10).

Lemma 3.9 *Let X be a Hermitian matrix in a d-dimensional space. The Hermitian matrix $X^{\otimes n}$ given by the tensor product of X has at most $(n+1)^{d-1}$ distinct eigenvalues, i.e., $|\boldsymbol{E}_{X^{\otimes n}}| \le (n+1)^{d-1}$.*

Proof Let $X = \sum_{i=1}^d x^i |u_i\rangle\langle u_i|$. Then, the eigenvalues of $X^{\otimes n}$ may be written as $(x_1)^{j_1} \cdots (x_d)^{j_d}$ $(n \ge j_i \ge 0)$. The possible values of $(j_1, \dots, j_d)$ are limited to at most $(n+1)^{d-1}$ values because d_j is decided from other values $j_1, \dots, j_{d-1}$.

Lemma 3.10 ([40]) *For a PVM $\boldsymbol{M}$, any positive matrix ρ and the pinching map $\kappa_{\boldsymbol{M}}$ defined in (1.13) satisfy*

$$|\boldsymbol{M}|\kappa_{\boldsymbol{M}}(\rho) \ge \rho. \tag{3.146}$$

Proof We first show (3.146) for when ρ is a pure state. Let us consider the case where $|\boldsymbol{M}| = k$, and its probability space is $\{1, \dots, k\}$. Then, the Schwarz inequality yields

$$\begin{aligned} k\langle v|\kappa_{\boldsymbol{M}}(|u\rangle\langle u|)|v\rangle &= \left(\sum_{i=1}^k 1\right)\left(\sum_{i=1}^k \langle v|M_i|u\rangle\langle u|M_i|v\rangle\right) \\ &\ge \left|\sum_{i=1}^k \langle v|M_i|u\rangle\right|^2 = \left|\left\langle v\left|\sum_{i=1}^k M_i\right|u\right\rangle\right|^2 = |\langle v\,|I|\,u\rangle|^2 = \langle v|u\rangle\langle u|v\rangle. \end{aligned}$$

Therefore, we obtain $|\boldsymbol{M}|\kappa_{\boldsymbol{M}}(|u\rangle\langle u|) \ge |u\rangle\langle u|$. Next, consider the case where $\rho = \sum_j \rho^j |u_j\rangle\langle u_j|$. Then,

$$\begin{aligned} |\boldsymbol{M}|\kappa_{\boldsymbol{M}}(\rho) - \rho =& |\boldsymbol{M}|\kappa_{\boldsymbol{M}}\left(\sum_j \rho^j |u_j\rangle\langle u_j|\right) - \sum_j \rho^j |u_j\rangle\langle u_j| \\ =& \sum_j \rho^j \left(|\boldsymbol{M}|\kappa_{\boldsymbol{M}}(|u_j\rangle\langle u_j|) - |u_j\rangle\langle u_j|\right) \ge 0, \end{aligned}$$

from which we obtain (3.146). ■

We are now ready to prove information inequalities (3.18) and (3.20). In what follows, these inequalities are proved only for densities $\rho > 0$ and $\sigma > 0$. This is sufficient for the general case due to the following reason. First, we apply these inequalities to two densities $\rho_\epsilon \stackrel{\text{def}}{=} (\rho + \epsilon I)(1 + d\epsilon)^{-1}$ and $\sigma_\epsilon \stackrel{\text{def}}{=} (\sigma + \epsilon I)(1 + d\epsilon)^{-1}$. Taking the limit $\epsilon \to 0$, we obtain these inequalities in the general case.

Proofs of (3.18) *and* (3.20) Inequality (3.18) follows from (3.20) by taking the limits $\lim_{s\to 0} \frac{\phi(s|\rho\|\sigma)}{-s}$ and $\lim_{s\to 0} \frac{\phi(s|\mathrm{P}^{M}_{\rho}\|\mathrm{P}^{M}_{\sigma})}{-s}$. Hence, we prove (3.20).

Step 1: Firstly, we show

$$\phi(s|\rho\|\sigma) \geq \phi(s|\kappa_\sigma(\rho)\|\sigma). \tag{3.147}$$

Let $\sigma = \sum_j \sigma_j E_{\sigma,j}$ be the spectral decomposition of σ and $E_{\sigma,j}\rho E_{\sigma,j} = \sum_k \rho_{k,j} E_{k,j}$ be that of $E_{\sigma,j}\rho E_{\sigma,j}$. Hence, $\kappa_\sigma(\rho) = \sum_{k,j} \rho_{k,j} E_{k,j}$. Since $E_{k,j}\rho E_{k,j} = \rho_{k,j} E_{k,j}$, it follows that $\mathrm{Tr}\, \rho \frac{E_{k,j}}{\mathrm{Tr}\, E_{k,j}} = \rho_{k,j}$. For $0 \leq s$, applying Inequality (3.32) (the quantum version of Jensen inequality) with the Hermitian matrix ρ and the density matrix $\frac{E_{k,j}}{\mathrm{Tr}\, E_{k,j}}$, we have

$$\left(\mathrm{Tr}\, \rho \frac{E_{k,j}}{\mathrm{Tr}\, E_{k,j}}\right)^{1-s} \leq \mathrm{Tr}\, \rho^{1-s} \frac{E_{k,j}}{\mathrm{Tr}\, E_{k,j}}. \tag{3.148}$$

Thus,

$$\begin{aligned}
\mathrm{Tr}\, \sigma^s \kappa_\sigma(\rho)^{1-s} &= \mathrm{Tr} \sum_{k,j} \sigma_j^s \rho_{k,j}^{1-s} E_{k,j} = \sum_{k,j} \sigma_j^s \,\mathrm{Tr}\, E_{k,j} \rho_{k,j}^{1-s} \\
&= \sum_{k,j} \sigma_j^s \,\mathrm{Tr}\, E_{k,j} \left(\mathrm{Tr}\, \rho \frac{E_{k,j}}{\mathrm{Tr}\, E_{k,j}}\right)^{1-s} \leq \sum_{k,j} \sigma_j^s \,\mathrm{Tr}\, E_{k,j} \left(\mathrm{Tr}\, \rho^{1-s} \frac{E_{k,j}}{\mathrm{Tr}\, E_{k,j}}\right) \\
&= \sum_{k,j} \sigma_j^s \,\mathrm{Tr}\, \rho^{1-s} E_{k,j} = \mathrm{Tr}\, \rho^{1-s} \sigma^s,
\end{aligned}$$

which implies (3.147).

Step 2: Next, we show

$$\phi(s|\kappa_\sigma(\rho)\|\sigma) \geq \phi(s|\mathrm{P}^{M}_{\rho}\|\mathrm{P}^{M}_{\sigma}) - (1-s)\log|\boldsymbol{E}_\sigma|. \tag{3.149}$$

For any POVM $\boldsymbol{M} = \{M_i\}$, we define the POVMs $\boldsymbol{M}' = \{M'_{i,j,k}\}$ and $\boldsymbol{M}'' = \{M''_i\}$ by $M'_{i,j,k} \stackrel{\text{def}}{=} E_{k,j} M_i E_{k,j}$ and $M''_i \stackrel{\text{def}}{=} \sum_{k,j} M'_{i,j,k}$, respectively. Then, $\mathrm{Tr}\, \sigma M'_{i,j,k} = \sigma_j \,\mathrm{Tr}\, E_{k,j} M_i E_{k,j}$ and $\mathrm{Tr}\, \rho M'_{i,j,k} = \rho_{k,j} \,\mathrm{Tr}\, E_{k,j} M_i E_{k,j}$. Thus,

$$\begin{aligned}
\mathrm{Tr}\, \rho^{1-s}\sigma^s \geq \mathrm{Tr}\, \sigma^s \kappa_\sigma(\rho)^{1-s} &= \sum_{i,j,k} (\mathrm{Tr}\, \rho M'_{i,j,k})^{1-s} (\mathrm{Tr}\, \sigma M'_{i,j,k})^s \\
&\geq \sum_i (\mathrm{Tr}\, \rho M''_i)^{1-s} (\mathrm{Tr}\, \sigma M''_i)^s,
\end{aligned}$$

where the last inequality follows from the monotonicity in the classical case. In addition, $\mathrm{Tr}\,\sigma M_i'' = \sum_{k,j} \mathrm{Tr}\,\sigma_j E_{k,j} M_i E_{k,j} = \mathrm{Tr}\,\sigma M_i$, and $\mathrm{Tr}\,\rho M_i'' = \sum_{k,j} \mathrm{Tr}\,\rho_{k,j} E_{k,j} M_i E_{k,j} = \mathrm{Tr}\,\kappa_\sigma(\rho) M_i$. Lemma 3.10 ensures that

$$|\boldsymbol{E}_\sigma|^{1-s} (\mathrm{Tr}\,\kappa_\sigma(\rho) M_i)^{1-s} \ge (\mathrm{Tr}\,\rho M_i)^{1-s},$$

which implies (3.149).
Step 3: Next, we consider the tensor product case. Applying (3.147) and (3.149) to the case with $\rho^{\otimes n}$ and $\sigma^{\otimes n}$, we have

$$\begin{aligned}\phi(s|\rho^{\otimes n}\|\sigma^{\otimes n}) \ge & \phi(s|\kappa_{\sigma^{\otimes n}}(\rho^{\otimes n})\|\sigma^{\otimes n}) \\ \ge & \phi(s|\mathrm{P}_{\rho^{\otimes n}}^{\boldsymbol{M}^{\otimes n}}\|\mathrm{P}_{\sigma^{\otimes n}}^{\boldsymbol{M}^{\otimes n}}) - (1-s)\log|\boldsymbol{E}_{\sigma^{\otimes n}}|.\end{aligned}$$

Since $\phi(s|\rho^{\otimes n}\|\sigma^{\otimes n}) = n\phi(s|\rho\|\sigma)$ and $\phi(s|\mathrm{P}_{\rho^{\otimes n}}^{\boldsymbol{M}^{\otimes n}}\|\mathrm{P}_{\sigma^{\otimes n}}^{\boldsymbol{M}^{\otimes n}}) = n\phi(s|\mathrm{P}_\rho^{\boldsymbol{M}}\|\mathrm{P}_\sigma^{\boldsymbol{M}})$, the inequality

$$\phi(s|\rho\|\sigma) \ge \frac{\phi(s|\kappa_{\sigma^{\otimes n}}(\rho^{\otimes n})\|\sigma^{\otimes n})}{n} \ge \phi(s|\mathrm{P}_\rho^{\boldsymbol{M}}\|\mathrm{P}_\sigma^{\boldsymbol{M}}) - \frac{\log|\boldsymbol{E}_{\sigma^{\otimes n}}|}{n}$$

holds. The convergence $\frac{\log|\boldsymbol{E}_{\sigma^{\otimes n}}|}{n} \to 0$ follows from Lemma 3.9. Thus, we obtain

$$\phi(s|\rho\|\sigma) \ge \lim_{n\to\infty} \frac{\phi(s|\kappa_{\sigma^{\otimes n}}(\rho^{\otimes n})\|\sigma^{\otimes n})}{n} \ge \phi(s|\mathrm{P}_\rho^{\boldsymbol{M}}\|\mathrm{P}_\sigma^{\boldsymbol{M}}). \tag{3.150}$$

Note that the convergence $\lim_{n\to\infty} \frac{\phi(s|\kappa_{\sigma^{\otimes n}}(\rho^{\otimes n})\|\sigma^{\otimes n})}{n}$ is guaranteed by Lemma A.1 because $\phi(s|\kappa_{\sigma^{\otimes (n+m)}}(\rho^{\otimes (n+m)})\|\sigma^{\otimes (n+m)}) \ge \phi(s|\kappa_{\sigma^{\otimes m}}(\rho^{\otimes m})\|\sigma^{\otimes m}) + \phi(s|\kappa_{\sigma^{\otimes n}}(\rho^{\otimes n})\|\sigma^{\otimes n}) - d\log|\boldsymbol{E}_\sigma|$. In addition, as is discussed in Exercise 5.22, the equality in $\phi(s|\rho\|\sigma) \ge \tilde{\phi}(s|\rho\|\sigma)$ does not necessarily hold for $s \le -1$. ∎

Proofs of (3.17) *and* (3.22) By using (3.150), (3.17) implies (3.22). So, we show only (3.17).
Step 1: We show (3.17) in the case with $s < 0$. We employ the notations given in the above proof. Similar to (3.148), applying Inequality (3.32) (the quantum version of Jensen inequality) with the Hermitian matrix $\sigma^{\frac{s}{2(1-s)}}\rho\sigma^{\frac{s}{2(1-s)}}$ and the density matrix $\frac{E_{k,j}}{\mathrm{Tr}\,E_{k,j}}$, we have

$$\begin{aligned}&\sigma_j^s\left(\mathrm{Tr}\,\rho\frac{E_{k,j}}{\mathrm{Tr}\,E_{k,j}}\right)^{1-s} = \left(\mathrm{Tr}\,\sigma^{\frac{s}{2(1-s)}}\rho_{k,j}\frac{E_{k,j}}{\mathrm{Tr}\,E_{k,j}}\sigma^{\frac{s}{2(1-s)}}\right)^{1-s} \\ =&\left(\mathrm{Tr}\,\sigma^{\frac{s}{2(1-s)}}\rho\sigma^{\frac{s}{2(1-s)}}\frac{E_{k,j}}{\mathrm{Tr}\,E_{k,j}}\right)^{1-s} \le \mathrm{Tr}\left(\sigma^{\frac{s}{2(1-s)}}\rho\sigma^{\frac{s}{2(1-s)}}\right)^{1-s}\frac{E_{k,j}}{\mathrm{Tr}\,E_{k,j}}.\end{aligned}$$

Thus,

$$\operatorname{Tr} \sigma^s \kappa_\sigma(\rho)^{1-s} = \sum_{k,j} \operatorname{Tr} E_{k,j} \sigma_j^s \left(\operatorname{Tr} \rho \frac{E_{k,j}}{\operatorname{Tr} E_{k,j}} \right)^{1-s}$$

$$\leq \sum_{k,j} \operatorname{Tr} E_{k,j} \operatorname{Tr} \left(\sigma^{\frac{s}{2(1-s)}} \rho \sigma^{\frac{s}{2(1-s)}} \right)^{1-s} \frac{E_{k,j}}{\operatorname{Tr} E_{k,j}}$$

$$= \sum_{k,j} \operatorname{Tr} \left(\sigma^{\frac{s}{2(1-s)}} \rho \sigma^{\frac{s}{2(1-s)}} \right)^{1-s} E_{k,j} = \operatorname{Tr} \left(\sigma^{\frac{s}{2(1-s)}} \rho \sigma^{\frac{s}{2(1-s)}} \right)^{1-s},$$

which implies

$$\tilde{\phi}(s|\rho\|\sigma) \geq \phi(s|\kappa_\sigma(\rho)\|\sigma). \tag{3.151}$$

Conversely, using (3.146), we have

$$\sigma^{\frac{s}{2(1-s)}} \rho \sigma^{\frac{s}{2(1-s)}} \leq \sigma^{\frac{s}{2(1-s)}} |\boldsymbol{E}_\sigma| \kappa_\sigma(\rho) \sigma^{\frac{s}{2(1-s)}}. \tag{3.152}$$

Thus, Lemma A.13 yields that

$$\operatorname{Tr} \left(\sigma^{\frac{s}{2(1-s)}} \rho \sigma^{\frac{s}{2(1-s)}} \right)^{1-s} \leq \operatorname{Tr} \left(\sigma^{\frac{s}{2(1-s)}} |\boldsymbol{E}_\sigma| \kappa_\sigma(\rho) \sigma^{\frac{s}{2(1-s)}} \right)^{1-s},$$

which implies

$$\phi(s|\kappa_\sigma(\rho)\|\sigma) + (1-s) \log |\boldsymbol{E}_\sigma| \geq \tilde{\phi}(s|\rho\|\sigma). \tag{3.153}$$

By considering the tensor product case, (3.151) and (3.153) imply

$$\frac{\phi(s|\kappa_{\sigma^{\otimes n}}(\rho^{\otimes n})\|\sigma^{\otimes n})}{n} + \frac{(1-s)\log|\boldsymbol{E}_\sigma^{\otimes n}|}{n} \geq \tilde{\phi}(s|\rho\|\sigma)$$
$$\geq \frac{\phi(s|\kappa_{\sigma^{\otimes n}}(\rho^{\otimes n})\|\sigma^{\otimes n})}{n}.$$

Taking the limit $n \mapsto \infty$, we obtain

$$\tilde{\phi}(s|\rho\|\sigma) = \lim_{n\to\infty} \frac{\phi(s|\kappa_{\sigma^{\otimes n}}(\rho^{\otimes n})\|\sigma^{\otimes n})}{n}, \tag{3.154}$$

which implies (3.17) for $s < 0$. Using (3.150), we obtain (3.22) for $s < 0$.
Step 2: Next, we show (3.17) in the case with $s \in (0,1)$. We notice that (3.152) holds even for $s \in (0,1)$. Now, we choose the basis $\{|e_i\rangle\}$ such that $\kappa_\sigma(\rho) = \sum_i \rho^i |e_i\rangle\langle e_i|$ and $\kappa_\sigma(|e_i\rangle\langle e_i|) = |e_i\rangle\langle e_i|$. Hence, since $x \mapsto x^{1-s}$ is concave, we have

$$e^{-s\underline{D}_{1-s}(\rho\|\sigma)} = \operatorname{Tr}(\sigma^{\frac{s}{2(1-s)}} \rho \sigma^{\frac{s}{2(1-s)}})^{1-s} = \sum_i \langle e_i|(\sigma^{\frac{s}{2(1-s)}} \rho \sigma^{\frac{s}{2(1-s)}})^{1-s}|e_i\rangle$$

$$\leq \sum_i (\langle e_i|\sigma^{\frac{s}{2(1-s)}} \rho \sigma^{\frac{s}{2(1-s)}} |e_i\rangle)^{1-s}$$

$$=\sum_i (\langle e_i|\sigma^{\frac{s}{2(1-s)}}|e_i\rangle\langle e_i|\rho|e_i\rangle\langle e_i|\sigma^{\frac{s}{2(1-s)}}|e_i\rangle)^{1-s}$$
$$=\mathrm{Tr}(\sigma^{\frac{s}{2(1-s)}}\kappa_\sigma(\rho)\sigma^{\frac{s}{2(1-s)}})^{1-s}=e^{-s\underline{D}_{1-s}(\kappa_\sigma(\rho)\|\sigma)}. \tag{3.155}$$

Since $x \mapsto -x^{-s}$ is matrix monotone, we have

$$e^{-s\underline{D}_{1-s}(\rho\|\sigma)}=\mathrm{Tr}(\sigma^{\frac{s}{2(1-s)}}\rho\sigma^{\frac{s}{2(1-s)}})(\sigma^{\frac{s}{2(1-s)}}\rho\sigma^{\frac{s}{2(1-s)}})^{-s}$$
$$\geq \mathrm{Tr}(\sigma^{\frac{s}{2(1-s)}}\rho\sigma^{\frac{s}{2(1-s)}})(\sigma^{\frac{s}{2(1-s)}}|\boldsymbol{E}_\sigma|\kappa_\sigma(\rho)\sigma^{\frac{s}{2(1-s)}})^{-s}$$
$$=|\boldsymbol{E}_\sigma|^{-s}e^{-s\underline{D}_{1-s}(\kappa_\sigma(\rho)\|\sigma)}. \tag{3.156}$$

Combining (3.155) and (3.156), we have

$$\underline{D}_{1-s}(\kappa_\sigma(\rho)\|\sigma)+\log|\boldsymbol{E}_\sigma| \geq \underline{D}_{1-s}(\rho\|\sigma) \geq \underline{D}_{1-s}(\kappa_\sigma(\rho)\|\sigma).$$

Considering the tensor product case, we obtain

$$\frac{1}{n}\underline{D}_{1-s}(\kappa_{\sigma^{\otimes n}}(\rho^{\otimes n})\|\sigma^{\otimes n})+\frac{1}{n}\log|\boldsymbol{E}_\sigma^{\otimes n}| \geq \underline{D}_{1-s}(\rho\|\sigma)$$
$$\geq \frac{1}{n}\underline{D}_{1-s}(\kappa_{\sigma^{\otimes n}}(\rho^{\otimes n})\|\sigma^{\otimes n}).$$

Taking the limit $n \mapsto \infty$, we obtain

$$\underline{D}_{1-s}(\rho\|\sigma)=\lim_{n\to\infty}\frac{1}{n}\underline{D}_{1-s}(\kappa_{\sigma^{\otimes n}}(\rho^{\otimes n})\|\sigma^{\otimes n})=\lim_{n\to\infty}\frac{1}{n}D_{1-s}(\kappa_{\sigma^{\otimes n}}(\rho^{\otimes n})\|\sigma^{\otimes n}),$$

which implies (3.17) for $s \in (0,1)$. ■

Exercises

From Exercise 3.49 to Exercise 3.55, we consider only the classical case with p and $\bar{p}$. So, we abbreviate $\phi(s|p\|\bar{p})$ by $\phi(s)$ (The results of these exercises are summarized as Table 3.1).

3.49 Show that

$$\lim_{s\to-\infty}-\phi'(s)=D_{\max}(p\|\bar{p}). \tag{3.157}$$

Table 3.1 Behaviors of $\phi(s)$, $\phi'(s)$, $s\phi'(s)-\phi(s)$, and $(s-1)\phi'(s)-\phi(s)$

s	$-\infty$		0		1
$\phi(s)$	$+\infty$	↘	0	↺	0
$\phi'(s)$	$-D_{\max}(p\|\bar{p})$	↗	$-D(p\|\bar{p})$	↗	$D(\bar{p}\|p)$
$s\phi'(s)-\phi(s)$	$-\overline{P}(p\|\bar{p})$	↘	0	↗	$D(\bar{p}\|p)$
$(s-1)\phi'(s)-\phi(s)$	$-\underline{P}(p\|\bar{p})$	↘	$D(p\|\bar{p})$	↘	0

3.50 Define $\overline{P}(p\|\bar{p}) := \sum_{x:\log p(x)-\log \bar{p}(x)} p(x)$ and $\underline{P}(p\|\bar{p}) := \sum_{x:\log p(x)-\log \bar{p}(x)} \bar{p}(x)$. Show that

$$\lim_{s\to-\infty} D(p_s\|\bar{p}) = \lim_{s\to-\infty} (s-1)\phi'(s) - \phi(s)$$

$$= -\log \underline{P}(p\|\bar{p}) = D_{\max}(p\|\bar{p}) - \log \overline{P}(p\|\bar{p}), \tag{3.158}$$

$$\lim_{s\to-\infty} D(p_s\|p) = \lim_{s\to-\infty} s\phi'(s) - \phi(s) = -\log \overline{P}(p\|\bar{p}). \tag{3.159}$$

3.51 Show that

$$\frac{ds_r}{dr} = \frac{1}{(s_r-1)\phi''(s_r)}. \tag{3.160}$$

3.52 Show that

$$\sup_{s\le 0} \frac{-sr-\phi(s)}{1-s} = \min_{q:D(q\|\bar{p})\ge r} D(q\|p) = \min_{q:D(q\|\bar{p})\le r} D(q\|p) + r - D(q\|\bar{p}) \tag{3.161}$$

for $r \in [D(p\|\bar{p}), -\log \underline{P}(p\|\bar{p})]$.
(a) Show that there uniquely exists $s_r \le 0$ such that $D(p_{s_r}\|p_1) = r$ for $r \in [D(p\|\bar{p}), -\log \underline{P}(p\|\bar{p}))$.
(b) Show that

$$\min_{q:D(q\|\bar{p})=r} D(q\|p) = D(p_{s_r}\|p). \tag{3.162}$$

(c) Show that

$$\min_{q:D(q\|\bar{p})\le r} D(q\|p) + r - D(q\|\bar{p}) = \min_{q:D(q\|\bar{p})\ge r} D(q\|p) = D(p_{s_r}\|p). \tag{3.163}$$

(d) Show that

$$D(p_{s_r}\|p) = \frac{-s_r r - \phi(s_r)}{1-s_r}. \tag{3.164}$$

(e) Show that

$$\frac{d}{ds}\frac{-sr-\phi(s)}{1-s} = \frac{-r+(s-1)\phi'(s)-\phi(s)}{(1-s)^2}. \tag{3.165}$$

(f) Show that

$$\sup_{s\le 0} \frac{-sr-\phi(s|p\|\bar{p})}{1-s} = \frac{-s_r r - \phi(s_r)}{1-s_r}. \tag{3.166}$$

(**g**) Show (3.161).

3.53 Show that

$$\sup_{s\le 0}\frac{-sr-\phi(s|p\|\bar{p})}{1-s} = \min_{q:D(q\|\bar{p})\le r} D(q\|p)+r-D(q\|\bar{p})$$
$$=\min_{q} D(q\|p)+r-D(q\|\bar{p}) = r - D_{\max}(p\|\bar{p}). \tag{3.167}$$

for $r > -\log \underline{P}(p\|\bar{p})$.
(**a**) Show that

$$\sup_{s\le 0}\frac{-sr-\phi(s|p\|\bar{p})}{1-s} = \sup_{s\to-\infty}\frac{-sr-\phi(s|p\|\bar{p})}{1-s} = r - D_{\max}(p\|\bar{p}). \tag{3.168}$$

(**b**) Show that

$$\min_{q} D(q\|p)+r-D(q\|\bar{p}) = r - D_{\max}(p\|\bar{p}). \tag{3.169}$$

(**c**) Show that

$$\min_{q:D(q\|\bar{p})\le r} D(q\|p)+r-D(q\|\bar{p}) \le r - D_{\max}(p\|\bar{p}). \tag{3.170}$$

3.54 Show the inequality opposite to (3.141) following the steps below. Therefore, we obtain the equation

$$B^*(r|p\|\bar{p}) = \sup_{s\le 0}\frac{-sr-\phi(s|p\|\bar{p})}{1-s} \tag{3.171}$$

(**a**) Show the following equations for $\phi(s)=\phi(s|p\|\bar{p})$ using Cramér's theorem.

$$\lim_{n\to\infty} -\frac{1}{n}\log p^n\left\{-\frac{1}{n}\log\left(\frac{p^n(x^n)}{\bar{p}^n(x^n)}\right)\le R\right\} = \max_{s\le 0} sR-\phi(s), \tag{3.172}$$
$$\lim_{n\to\infty} -\frac{1}{n}\log \bar{p}^n\left\{-\frac{1}{n}\log\left(\frac{p^n(x^n)}{\bar{p}^n(x^n)}\right)\le R\right\} = \max_{s\le 1} -(1-s)R-\phi(s). \tag{3.173}$$

(**b**) Show that

$$s\phi'(s)-\phi(s) = \max_{s_0\le 0}(s_0\phi'(s)-\phi(s_0)) \text{ for } s\le 0, \tag{3.174}$$
$$(s-1)\phi'(s)-\phi(s) = \max_{s_0\le 1}((s_0-1)\phi'(s)-\phi(s_0)) \text{ for } s\le 1. \tag{3.175}$$

(**c**) Show that

$$r = \max_{s\le 1}(s-1)\phi'(s_r)-\phi(s), \tag{3.176}$$

$$\sup_{s \le 0} \frac{-sr - \phi(s)}{1-s} = s_r \phi'(s_r) - \phi(s_r) = \max_{s \le 0} s\phi'(s_r) - \phi(s). \tag{3.177}$$

(d) Assume that $D(p\|\bar{p}) < r < -\log \underline{P}(p\|\bar{p})$. Show the inequality $B^*(r|p\|\bar{p}) \le \sup_{s \le 0} \frac{-sr-\phi(s|p\|\bar{p})}{1-s}$ by using (3.166), (3.176), and (3.177).
(e) Assume that $r \ge -\log \underline{P}(p\|\bar{p})$. Show the inequality $B^*(r|p\|\bar{p}) \le \sup_{s \le 0} \frac{-sr-\phi(s|p\|\bar{p})}{1-s}$ by using (3.167).

3.55 Show that

$$\frac{dB^*(r|p\|\bar{p})}{dr} = \frac{1}{s_r - 1}, \quad \frac{d^2 B^*(r|p\|\bar{p})}{dr^2} = -\frac{1}{(s_r - 1)^3 \phi''(s_r)} \ge 0, \tag{3.178}$$

which implies the convexities of $B^*(r|p\|\bar{p})$.

Now, we proceed to the quantum case with ρ and σ.

3.56 Show the following inequality by following the steps below.

$$B^*(r|\rho\|\sigma) \le \inf_{\tau: D(\tau\|\sigma) > r} D(\tau\|\rho) = \min_{\tau: D(\tau\|\sigma) \ge r} D(\tau\|\rho) \tag{3.179}$$

(a) For any state τ, show that there exists a sequence $\{T_n\}$ such that $\lim_{n\to\infty} \mathrm{Tr}\, \tau^{\otimes n} (I - T_n) = 1$ and $\lim_{n\to\infty} -\frac{1}{n} \log \mathrm{Tr}\, \sigma^{\otimes n} T_n = r$.
(b) Show that the above sequence $\{T_n\}$ satisfies $\lim_{n\to\infty} -\frac{1}{n} \log \mathrm{Tr}\, \rho^{\otimes n} T_n \le D(\tau\|\rho)$.
(c) Show (3.179).

3.57 Let a sequence of tests $\{T_n\}$ satisfy $R = \underline{\lim} -\frac{1}{n} \log \mathrm{Tr}\, \rho^{\otimes n} T_n$ and $r \le \underline{\lim} -\frac{1}{n} \log \mathrm{Tr}\, \sigma^{\otimes n} (I - T_n)$. Show that $R \le D(\tau\|\rho)$ when $D(\tau\|\sigma) < r$ using Lemma 3.7 twice. That is, show that

$$B(r|\rho\|\sigma) \le \inf_{\tau: D(\tau\|\sigma) < r} D(\tau\|\rho) = \min_{\tau: D(\tau\|\sigma) \le r} D(\tau\|\rho).$$

3.58 Show $\tilde{\phi}(s|\rho\|\sigma) = \lim_{n\to\infty} \frac{1}{n} \max_{M} \phi(s|\mathrm{P}^{M}_{\rho^{\otimes n}} \| \mathrm{P}^{M}_{\sigma^{\otimes n}})$ for $s < 0$.

3.59 To prove (3.16), show that $\lim_{s\to\infty} \frac{\tilde{\phi}(-s|\rho\|\sigma)}{s} = D_{\max}(\rho\|\sigma)$ following the steps below.
(a) Show that $D_{\max}(\rho\|\sigma) \ge D_{\max}(\kappa_\sigma(\rho)\|\sigma)$.
(b) Show that $\lim_{n\to\infty} \frac{1}{n} D_{\max}(\kappa_{\sigma^{\otimes n}}(\rho^{\otimes n})\|\sigma^{\otimes n}) \ge D_{\max}(\rho\|\sigma)$.
(c) Show that $\lim_{s\to\infty} \frac{\tilde{\phi}(-s|\rho\|\sigma)}{s} = D_{\max}(\rho\|\sigma)$.

3.60 Assume that a rank-one PVM $\boldsymbol{M} = \{M_i\}$ is not commutative with ρ and is commutative with σ. Show that

$$\phi(s|\kappa_{\sigma^{\otimes 2}}(\rho^{\otimes 2})\|\sigma^{\otimes 2}) > 2\phi(s|\mathrm{P}_{\rho}^{\boldsymbol{M}}\|\mathrm{P}_{\sigma}^{\boldsymbol{M}}) \tag{3.180}$$

for $s < 0$ by using Exercise 1.35.

3.61 Assume that a rank-one PVM $\boldsymbol{M} = \{M_i\}$ is commutative with ρ and is not commutative with ρ. Show that (3.180) for $s < 0$.

3.62 Assume that σ is not commutative with ρ. Show that

$$\tilde{\phi}(s|\rho\|\sigma) > \phi(s|\mathrm{P}_{\rho}^{\boldsymbol{M}}\|\mathrm{P}_{\sigma}^{\boldsymbol{M}}) \tag{3.181}$$

for $s < 0$ and any POVM $\boldsymbol{M}$ following the steps below. Therefore, there exists a POVM such that $\tilde{\phi}(s|\rho\|\sigma) = \phi(s|\mathrm{P}_{\rho}^{\boldsymbol{M}}\|\mathrm{P}_{\sigma}^{\boldsymbol{M}})$ if and only if σ is commutative with ρ.
(**a**) Show that it is sufficient to show (3.181) for a rank-one PVM $\boldsymbol{M}$. (Hint: Use Theorem 4.5 given in Sect. 4.7.)
(**b**) Show (3.181) in the above case by using Exercises 3.60 and 3.61.

3.9 Proof of Theorem 3.1

In this proof, we only consider the case in which there exists an element $x \in L$ such that $A(x) = b$.[5] Otherwise, since both sides are equal to $-\infty$, the theorem holds.

When $x \in L$ satisfies $A(x) = b$ and y satisfies $A^*(y) - c \in L^*$, we have $0 \le \langle A^*(y) - c, x\rangle = \langle y, A(y)\rangle - \langle c, x\rangle = \langle y, b\rangle - \langle c, x\rangle$. Hence, we can check that

$$\max_{x\in V_1}\{\langle c, x\rangle | x \in L, A(x) = b\} \le \min_{y\in V_2}\{\langle y, b\rangle | A^*(y) - c \in L^*\}. \tag{3.182}$$

Furthermore,

$$\begin{aligned}&\min_{y\in V_2}\{\langle y, b\rangle | A^*(y) - c \in L^*\}\\ =&\min_{(\mu,y)\in\mathbb{R}\times V_2}\{\mu | \exists y \in V_2, \forall x \in L, \langle y, b\rangle - \langle A^*(y) - c, x\rangle \le \mu\}.\end{aligned}$$

This equation can be checked as follows. When $y \in V_2$ satisfies $A^*(y) - c \in L^*$, the real number $\mu = \langle y, b\rangle$ satisfies the condition on the right-hand side (RHS). Hence, we obtain the $\ge$ part. Next, we consider a pair (μ, y) satisfying the condition on the RHS. Then, we can show that $\langle A^*(y) - c, x\rangle$ is greater than zero for all $x \in L$, by reduction to absurdity. Assume that there exists an element $x \in L$ such that $\langle A^*(y) - c, x\rangle$ is negative. By choosing a sufficiently large number $t > 0$, $tx \in L$, but $\langle y, b\rangle - \langle A^*(y) - c, tx\rangle \le \mu$ does not hold. It is a contradiction. This proves the $\le$ part.

[5]Our proof follows [19].

Let $\eta_0 \stackrel{\text{def}}{=} \max_{x \in V_1}\{\langle c, x\rangle | x \in L, A(x) = b\}$. Then $(\eta_0, 0)$ is a point that lies on the boundary of the convex set $\{(\langle c, x\rangle, A(x) - b)\}_{x \in L} \subset \mathbb{R} \times V_2$. Choosing an appropriate $y_0 \in V_2$ and noting that $(1, -y_0) \in \mathbb{R} \times V_2$, we have

$$\eta_0 = \eta_0 - \langle y, 0\rangle \geq \langle c, x\rangle - \langle y_0, A(x) - b\rangle, \quad \forall x \in L.$$

From this fact we have

$$\eta_0 \geq \min_{(\mu, y) \in \mathbb{R} \times V_2} \{\mu | \exists y \in V_2, \forall x \in L, \langle y, b\rangle - \langle A^*(y) - c, x\rangle \leq \mu\}.$$

This proves the reverse inequality of (3.182) and completes the proof.

3.10 Historical Note

The Rényi relative entropy $D_{1+s}(\rho\|\sigma)$ was introduced by Petz [41] as quantum f-divergence. Recently, another kind of Rényi relative entropy $\underline{D}_{1+s}(\rho\|\sigma)$ was introduced by the papers [4, 5] to connect the fidelity $F(\rho, \sigma)$ and the max relative entropy $D_{\max}(\rho\|\sigma)$, which was introduced by the paper [3]. Based on advanced knowledge, i.e., matrix convex functions (See Section A.4), Petz [41] showed the monotonicity for $D_{1+s}(\rho\|\sigma)$. A different paper [6] showed that for $\underline{D}_{1+s}(\rho\|\sigma)$ by using a more difficult method. In this text, we prove the monotonicity of the relative Rényi entropies $D_{1+s}(\rho\|\sigma)$ $(s \geq 0)$ and $\underline{D}_{1+s}(\rho\|\sigma)$ $(s \geq 0)$ for a measurement based only on elementary knowledge.

The problem of discriminating two states was treated by Holevo [11] and Helstrom [12]. Its extension to multiple states was discussed by Yuen et al. [17]. If we allowed any POVM, the possibility of perfect discrimination is trivial. That is, it is possible only when the hypothesis states are orthogonal to each other. However, if our measurement is restricted to LOCC, its possibility is not trivial. This problem is called local discrimination and has been studied by many researchers recently [42–50].

On the nonperfect discrimination, Chernoff's lemma is essential in the asymptotic setting with two commutative states. However, no results were obtained concerning the quantum case of Chernoff's lemma. Hence, Theorem 3.5 is the first attempt to obtain its quantum extension. Regarding the quantum case of Stein's lemma, many results were obtained, the first by Hiai and Petz [32]. They proved that $B(\rho\|\sigma) = D(\rho\|\sigma)$. The part $B(\rho\|\sigma) \leq D(\rho\|\sigma)$ essentially follows from the same discussion as (3.123). They proved the other part $B(\rho\|\sigma) \geq D(\rho\|\sigma)$ by showing the existence of the POVMs $\{\boldsymbol{M}^n\}$ such that

$$\lim_{n\to\infty} \frac{1}{n} D^{\boldsymbol{M}^n}(\rho^{\otimes n}\|\sigma^{\otimes n}) = D(\rho\|\sigma). \tag{3.183}$$

An impetus for this work was the first meeting between Hiai and Nagaoka in 1990 when they were at the same university (but in different departments). During their discussion, Nagaoka asked about the possibility of extending Stein's lemma to the quantum case. After their achievement, Hayashi [51] proved that there exists a sequence of POVMs $\{\boldsymbol{M}^n\}$ that satisfies (3.97) and depends only on σ. Hayashi [40] also proved that the asymptotically optimal condition for a measurement in terms of quantum hypothesis testing depends only on σ. Moreover, Ogawa and Hayashi [52] also derived a lower bound of the exponent of the second error probability. After the first edition of this book, two big breakthroughs have been done by Audenaert et al. [13], Nussbaum and Szkoła [14]. Audenaert et al. [13] showed very helpful evaluation as Lemma 3.3. Although the original their proof is rather complicated, Narutaka Ozawa [53] gave its much simpler proof, which is presented in this book. On the other hand, Nussbaum and Szkoła [14] introduced simultaneous distributions $P_{(\rho\|\sigma)}$ and $Q_{(\rho\|\sigma)}$ for two non-commutative density matrices ρ and σ. Then, they derived a lower bound of error probability. However, this kind of distribution was essentially discussed in Hayashi [40] by considering the pinched state $\kappa_\sigma(\rho)$.

Regarding the strong converse part $B^\dagger(\rho\|\sigma) \leq D(\rho\|\sigma)$, Ogawa and Nagaoka [7] proved it by deriving the lower bound of the exponent $\sup_{-1\leq s\leq 0} \frac{-sr-\phi(s)}{1-s}$, which is equal to the RHS of (3.141) when $s \leq 0$ is replaced by $-1 \leq s \leq 0$. Its behavior is slightly worse for a large value r. After this, the same exponent was obtained by Nagaoka [54] in a more simple way. However, these two approaches are based on the monotonicity of the relative Rényi entropy $\phi(s|\rho\|\sigma)$ $(-1 \leq s \leq 0)$. In this text, we apply this monotonicity to Nagaoka's proof. Hence, we derive the better bound $\sup_{s\leq 0} \frac{-sr-\phi(s)}{1-s}$, which was derived by Hayashi [55] using a different method. In addition, the second inequality in (3.100) was first proved in the first version of this book. Further, the first version of this book showed that

$$B^*(r|\rho\|\sigma) = \sup_{s\leq 0} \frac{-sr - \lim_{n\to\infty} \frac{1}{n}\phi(s|\kappa_{\sigma^{\otimes n}}(\rho^{\otimes n})\|\sigma^{\otimes n})}{1-s} \tag{3.184}$$

by showing the monotonicity for the information quantity $\lim_{n\to\infty} \frac{1}{n}\phi(s|\kappa_{\sigma^{\otimes n}}(\rho^{\otimes n})\|\sigma^{\otimes n})$. Recently, Mosonyi and Ogawa [39] showed (3.144) by showing the relation (3.154). Furthermore, Nagaoka invented a quantum version of the information spectrum method, and Nagaoka and Hayashi [34] applied it to the simple hypothesis testing of a general sequence of quantum states.

Finally, we should remark that the formulation of hypothesis testing is based on industrial demands. In particular, in order to guarantee product quality, we usually use test based on random sampling and statistically evaluate the quality. It is natural to apply this method to check the quality of produced maximally entangled states because maximally entangled states are used as resources of quantum information processing. Tsuda et al. formulated this problem using statistical hypothesis testing [56] and demonstrated its usefulness by applying it to maximally entangled states that produced spontaneous parametric down conversion [57]. Further, Hayashi [58] analyzed this problem more extensively from a theoretical viewpoint. However, concerning quantum hypothesis testing, the research on the applied side is not sufficient. Hence, such a study is strongly desired.

3.11 Solutions of Exercises

Exercise 3.1 These can be shown by simple calculations of $D(\rho\|\sigma)$ and $D_{1+s}(\rho\|\sigma)$

Exercise 3.2

$$\begin{aligned}
&H(\rho_A \otimes \rho_B) = -\operatorname{Tr} \rho_A \otimes \rho_B \log(\rho_A \otimes \rho_B) \\
=& -\operatorname{Tr} \rho_A \otimes \rho_B \log(\rho_A \otimes I_B) + \log(I_A \otimes \rho_B) = -\operatorname{Tr} \rho_A \log \rho_A - \operatorname{Tr} \rho_B \log \rho_B \\
=& H(\rho_A) + H(\rho_B).
\end{aligned}$$

(3.27) can be shown by a similar calculation.

$$\begin{aligned}
&e^{D_{1-s}(\rho_A \otimes \rho_B \| \sigma_A \otimes \sigma_B)} = \operatorname{Tr}(\rho_A \otimes \rho_B)^{1-s}(\sigma_A \otimes \sigma_B)^s \\
=& \operatorname{Tr}(\rho_A^{1-s} \otimes \rho_B^{1-s})(\sigma_A^s \otimes \sigma_B^s) = \operatorname{Tr}(\rho_A^{1-s}\sigma_A^s \otimes \rho_B^{1-s}\sigma_B^s) \\
=& e^{D_{1-s}(\rho_A \| \sigma_A)} e^{D(\rho_B \| \sigma_B)}.
\end{aligned}$$

(3.29) can be shown by a similar calculation.

$$\begin{aligned}
&e^{-D_{\max}(\rho_A \otimes \rho_B \| \sigma_A \otimes \sigma_B)} = \|(\sigma_A \otimes \sigma_B)^{-\frac{1}{2}}(\rho_A \otimes \rho_B)(\sigma_A \otimes \sigma_B)^{-\frac{1}{2}}\| \\
=& \|(\sigma_A^{-\frac{1}{2}} \otimes \sigma_B^{-\frac{1}{2}})(\rho_A \otimes \rho_B)(\sigma_A^{-\frac{1}{2}} \otimes \sigma_B^{-\frac{1}{2}})\| \\
=& \|(\sigma_A^{-\frac{1}{2}} \rho_A \sigma_A^{-\frac{1}{2}} \otimes \sigma_B^{-\frac{1}{2}} \rho_B \sigma_B^{-\frac{1}{2}} \| = e^{-D_{\max}(\rho_A \| \sigma_A)} e^{-D(\rho_B \| \sigma_B)}.
\end{aligned}$$

$$\begin{aligned}
&e^{-D_{\min}(\rho_A \otimes \rho_B \| \sigma_A \otimes \sigma_B)} = \operatorname{Tr}(\sigma_A \otimes \sigma_B)\{\rho_A \otimes \rho_B > 0\} \\
=& \operatorname{Tr}(\sigma_A \otimes \sigma_B)\{\rho_A > 0\} \otimes \{\rho_B > 0\} = \operatorname{Tr} \sigma_A\{\rho_A > 0\} \operatorname{Tr} \sigma_B\{\rho_B > 0\} \\
=& r^{-D_{\min}(\rho_A \| \sigma_A)} e^{-D(\rho_B \| \sigma_B)}.
\end{aligned}$$

Exercise 3.3 Consider the spectral decomposition $\boldsymbol{M}$ of X and apply Jensen's inequality to $\mathrm{P}_\rho^{\boldsymbol{M}}$.

Exercise 3.4 $\phi(0|\rho\|\sigma) = \frac{d}{ds} \operatorname{Tr} \rho^{1-s}\sigma^s|_{s=0} / \operatorname{Tr} \rho^{1-s}\sigma^s|_{s=0}$. $\frac{d}{ds} \operatorname{Tr} \rho^{1-s}\sigma^s|_{s=0} = -\operatorname{Tr} \rho^{1-s} \log \rho\sigma^s|_{s=0} + \operatorname{Tr} \rho^{1-s}\sigma^s \log \sigma|_{s=0} = -\operatorname{Tr} \rho \log \rho + \operatorname{Tr} \rho \log \sigma$. The other inequality can be shown in the same way.

Exercise 3.5

(a) $\phi(s|\rho\|\sigma) = \frac{d}{ds} \operatorname{Tr} \rho^{1-s}\sigma^s / \operatorname{Tr} \rho^{1-s}\sigma^s$. $\frac{d}{ds} \operatorname{Tr} \rho^{1-s}\sigma^s = -\operatorname{Tr} \rho^{1-s} \log \rho\sigma^s + \operatorname{Tr} \rho^{1-s}\sigma^s \log \sigma = -\operatorname{Tr} \log \rho\rho^{1-s}\sigma^s + \operatorname{Tr} \rho^{1-s}\sigma^s \log \sigma = \operatorname{Tr} \rho^{1-s}\sigma^s(-\log \rho + \log \sigma)$.
(d) Use Schwarz's inequality with respect to the inner product $\operatorname{Tr} XY^*$ with two vectors $\rho^{(1-s)/2}(\log \sigma - \log \rho)\sigma^{s/2}$ and $\rho^{(1-s)/2}\sigma^{s/2}$.

Exercise 3.6 $\frac{d\tilde{\phi}(s|\rho\|\sigma)}{ds} = -\operatorname{Tr}(\sigma^{\frac{s}{2(1-s)}}\rho\sigma^{\frac{s}{2(1-s)}})^{1-s}\log(\sigma^{\frac{s}{2(1-s)}}\rho\sigma^{\frac{s}{2(1-s)}}) + \operatorname{Tr}(1-s)\cdot(\sigma^{\frac{s}{2(1-s)}}\rho\sigma^{\frac{s}{2(1-s)}})^{-s}\cdot\frac{1}{(1-s)^2}\cdot\log\sigma\cdot(\sigma^{\frac{s}{2(1-s)}}\rho\sigma^{\frac{s}{2(1-s)}})$. Hence, $\frac{d\tilde{\phi}(s|\rho\|\sigma)}{ds}|_{s=0} = -\operatorname{Tr}\rho\log\rho + \operatorname{Tr}\rho\log\sigma = -D(\rho\|\sigma)$.

Exercise 3.7 $\lim_{s\to\infty}\underline{D}_{1+s}(\rho\|\sigma) = \lim_{s\to\infty}\frac{1}{s}\log\operatorname{Tr}(\sigma^{\frac{-s}{2(1+s)}}\rho\sigma^{\frac{-s}{2(1+s)}})^{1+s} = \lim_{s\to\infty}\log\|\sigma^{\frac{-s}{2(1+s)}}\rho\sigma^{\frac{-s}{2(1+s)}}\| = \log\|\sigma^{-\frac{1}{2}}\rho\sigma^{-\frac{1}{2}}\|$.

Exercise 3.8 For $s\in(0,1)$, applying Araki-Lieb-Thirring inequality to the case $r = 1-s$, we have

$$e^{-s\underline{D}_{1-s}(\rho\|\sigma)} = \operatorname{Tr}(\sigma^{\frac{s}{2(1-s)}}\rho\sigma^{\frac{s}{2(1-s)}})^{1-s} \ge \operatorname{Tr}\sigma^{\frac{s}{2}}\rho^{1-s}\sigma^{\frac{s}{2}} = \operatorname{Tr}\sigma^{s}\rho^{1-s}e^{-sD_{1-s}(\rho\|\sigma)}.$$

For $s<0$, applying Araki-Lieb-Thirring inequality to the case $r = 1-s$, we have

$$e^{-s\underline{D}_{1-s}(\rho\|\sigma)} = \operatorname{Tr}(\sigma^{\frac{s}{2(1-s)}}\rho\sigma^{\frac{s}{2(1-s)}})^{1-s} \le \operatorname{Tr}\sigma^{\frac{s}{2}}\rho^{1-s}\sigma^{\frac{s}{2}} = \operatorname{Tr}\sigma^{s}\rho^{1-s}e^{-sD_{1-s}(\rho\|\sigma)}.$$

Exercise 3.9

(**b**)

$$\begin{aligned}
&D(\sigma\|\rho) - \frac{1}{1-s}D(\sigma\|\rho_s)\\
=&\operatorname{Tr}\sigma(\log\sigma - \log\rho) - \frac{1}{1-s}\operatorname{Tr}\sigma\log\sigma + \operatorname{Tr}\sigma\log\rho + \frac{\psi(s|\rho)}{1-s}\\
=&-\frac{s}{1-s}\operatorname{Tr}\sigma\log\sigma + \frac{\psi(s|\rho)}{1-s} = -\frac{s}{1-s}H(\sigma) + \frac{\psi(s|\rho)}{1-s}\\
=&\frac{s}{1-s}H(\rho_s) + \frac{\psi(\rho)}{1-s} = -\frac{s}{1-s}\operatorname{Tr}\rho_s(1-s)\log\rho - \frac{s}{1-s}\psi(s|\rho) + \frac{\psi(s|\rho)}{1-s}\\
=&-s\operatorname{Tr}\rho_s\log\rho + \psi(s|\rho) = D(\rho_s\|\rho).
\end{aligned}$$

(**c**) The desired inequality follows from the inequality $\frac{1}{1-s}D(\sigma\|\rho_s)\ge 0$ for $s\le 1$.

Exercise 3.10

(**a**) It follows from Exercise 3.9.
(**b**) It follows from (**a**) and the continuity of $H(\sigma)$ and $D(\sigma\|\rho)$.
(**c**) (3.33) follows from (**b**) and similar relations as (2.62)–(2.64).

Exercise 3.11

(**a**) It follows from Exercise 3.9.
(**b**) It follows from (**a**) and the continuity of $H(\sigma)$ and $D(\sigma\|\rho)$.
(**c**) (3.33) follows from (**b**) and similar relations as (2.68) and (2.69).

Exercise 3.12 Since $\rho^{\frac{1}{2}}\sigma^{-\frac{s}{1+s}}\rho^{\frac{1}{2}}$ is unitary equivalent with $\sigma^{-\frac{s}{2(1+s)}}\rho\sigma^{-\frac{s}{2(1+s)}}$. So, we have the first expression. The next, expression follows from the definition of the p-norm of matrices given in (A.24). The final expression also follows from the definition of the p-norm and the original definition of $\underline{D}_{1+s}(\rho\|\sigma)$.

Exercise 3.13 We can show (3.41) as follows.

$$\min\{x|\rho \le x\sigma\} = \min\{x|\sigma^{-\frac{1}{2}}\rho\sigma^{-\frac{1}{2}} \le x\} = \|\sigma^{-\frac{1}{2}}\rho\sigma^{-\frac{1}{2}}\|.$$

Exercise 3.14 These can be shown by simple calculations of $b(\rho,\sigma)$ and $d_1(\rho,\sigma)$.

Exercise 3.15

(**a**) Schwartz inequality implies that $\sqrt{\operatorname{Tr} XX^*}\sqrt{\operatorname{Tr} YY^*} \ge \operatorname{Tr} XY^*$ and $\sqrt{\operatorname{Tr} XX^*}\sqrt{\operatorname{Tr} YY^*} \ge \operatorname{Tr} YX^*$. Hence,

$$\begin{aligned}(\sqrt{\operatorname{Tr} XX^*} + \sqrt{\operatorname{Tr} YY^*})^2 &= \operatorname{Tr} XX^* + \operatorname{Tr} YY^* + 2\sqrt{\operatorname{Tr} XX^*}\sqrt{\operatorname{Tr} YY^*} \\ \ge &\operatorname{Tr} XX^* + \operatorname{Tr} YY^* + \operatorname{Tr} XY^* + \operatorname{Tr} YX^* = \operatorname{Tr}(X-Y)(X-Y)^*.\end{aligned}$$

(**b**) Substitute $(\sqrt{\rho_1}-\sqrt{\rho_3}U_2\ \sqrt{\rho_3}U_2-\sqrt{\rho_2}U_1$ into X and Y. Since $\sqrt{\operatorname{Tr}\left(\sqrt{\rho_3}U_2-\sqrt{\rho_2}U_1\right)\left(\sqrt{\rho_3}U_2^*-\sqrt{\rho_2}U_1\right)^*} = \sqrt{\operatorname{Tr}\left(\sqrt{\rho_3}-\sqrt{\rho_2}U_1U_2^*\right)\left(\sqrt{\rho_3}-\sqrt{\rho_2}U_1U_2^*\right)^*}$, we have the desired inequality.

(**c**) Choose U_1 and U_2 such that $b(\rho_1,\rho_3) = \sqrt{\operatorname{Tr}\left(\sqrt{\rho_1}-\sqrt{\rho_3}U_2\right)\left(\sqrt{\rho_1}-\sqrt{\rho_3}U_2\right)^*}$ and $b(\rho_3,\rho_2) = \sqrt{\operatorname{Tr}\left(\sqrt{\rho_3}-\sqrt{\rho_2}U_1U_2^*\right)\left(\sqrt{\rho_3}-\sqrt{\rho_2}U_1U_2^*\right)^*}$. Then, $b(\rho_1,\rho_2) \le \sqrt{\operatorname{Tr}\left(\sqrt{\rho_1}-\sqrt{\rho_2}U_1\right)\left(\sqrt{\rho_1}-\sqrt{\rho_2}U_1\right)^*}$. Combining the inequality and (**b**), we obtain the desired inequality.

Exercise 3.16 Note that $2(x^2+y^2) \ge (x+y)^2$.

Exercise 3.17

(**b**) Take the average by integrating between $[0, 2\pi]$ for each θ_i. Note that $\langle u|v\rangle$ is continuous for each θ_i.

Exercise 3.18 Choose the orthogonal basis $\{u_1, u_2, \dots\}$ such that $u = u_1$ and $v = xu_1 + yu_2$ with $x, y \ge 0$. Then, using the matrix representation under the basis, we have $|u\rangle\rangle u| - |v\rangle\rangle v| = \begin{pmatrix} 1-x^2 & -xy \\ -xy & y^2 \end{pmatrix} = \begin{pmatrix} y^2 & -xy \\ -xy & y^2 \end{pmatrix}$. Due to Exercise A.3, its trace norm is $2\sqrt{y^4+(xy)^2} = 2y\sqrt{y^2+x^2} = 2y = 2\sqrt{1-|\langle u|v\rangle|^2}$.

Exercise 3.19

(**a**) Use the Schwarz inequality.

(**c**) Choose U such that $|\rho^{1/2}\sigma^{1/2}| = U\rho^{1/2}\sigma^{1/2}$. Note that $|\operatorname{Tr} U\rho^{1/2}M_i\sigma^{1/2}| \ge \operatorname{Tr} U\rho^{1/2}M_i\sigma^{1/2}$.

Exercise 3.20

(**a**) Make the polar decomposition $A = U|A|$. Hence, $\|AA^\dagger\| = \|U|A||A|U^\dagger\| = \||A||A|\| = \|U|A|U^\dagger U|A|\| = \|A^\dagger A\|$.

(**b**) $\operatorname{Tr}\rho^2\sigma^{-1} = \operatorname{Tr}\rho\rho^{\frac{1}{2}}\sigma^{-1}\rho^{\frac{1}{2}} \le \|\rho^{\frac{1}{2}}\sigma^{-1}\rho^{\frac{1}{2}}\| = \|\sigma^{-\frac{1}{2}}\rho\sigma^{-\frac{1}{2}}\|$.

(**c**) This fact can be shown by using Exercise 3.5.

(**d**) Using (**b**), we can show that $D_2(\rho\|\sigma) \le D_{\max}(\rho\|\sigma)$. Combing (**c**), we can show (3.12).

Exercise 3.21 Note that the spectral decomposition $\sum_i \lambda M_i$ of $\rho^{1/2}U^*\sigma^{-1/2}$ satisfies $M_i^{1/2}\sigma^{1/2} = \lambda_i M_i^{1/2}\rho^{1/2}U^*$.

Exercise 3.22

(**b**) See the hint for Exercise 3.21.

Exercise 3.23 Assume that $\rho - \sigma = \sum_i x_i M_i$, where $\{M_i\}$ is a PVM. Then, $\mathrm{Tr}(\rho - \sigma)M_i = x_i \mathrm{Tr}\, M_i$. Hence,

$$
\begin{aligned}
2d_1(\rho, \sigma) = \|\rho - \sigma\|_1 = \mathrm{Tr} \sum_i |x_i| M_i \\
= |\mathrm{Tr}\, \rho M_i - \mathrm{Tr}\, \sigma M_i| = d_1(\mathrm{P}_\rho^{\boldsymbol{M}}, \mathrm{P}_\sigma^{\boldsymbol{M}}).
\end{aligned}
$$

Exercise 3.24 Let $\boldsymbol{M}$ be a POVM that satisfies the equality in (3.46). Applying (2.25) to $\mathrm{P}_\rho^{\boldsymbol{M}}$ and $\mathrm{P}_\sigma^{\boldsymbol{M}}$, we obtain $d_1(\mathrm{P}_\rho^{\boldsymbol{M}}, \mathrm{P}_\sigma^{\boldsymbol{M}}) \ge b^2(\mathrm{P}_\rho^{\boldsymbol{M}}, \mathrm{P}_\sigma^{\boldsymbol{M}})$. Finally, adding (3.47), we obtain $d_1(\rho, \sigma) \ge b^2(\rho, \sigma)$.

Exercise 3.25 It can be shown by the similar way as Exercise 3.24.

Exercise 3.26 It can be shown by the similar way as Exercise 3.24.

Exercise 3.27 $-\log \mathrm{Tr}\, |\sqrt{\rho}\sqrt{\sigma}| = -\log 1 - b^2(\rho, \sigma) \ge b^2(\rho, \sigma)$.

Exercise 3.28 Since $0 \le x \le 1$, we have $1 - x \le 1 - x^2 \le 2(1 - x)$, which implies (3.51).

Exercise 3.29 Due to (3.59), we have

$$
\begin{aligned}
& 1 - \frac{1}{2} d_1(\kappa(\rho), \kappa(\sigma)) = \min_{I \ge T \ge 0} (\mathrm{Tr}\, \kappa(\rho)(I - T) + \mathrm{Tr}\, \kappa(\sigma)T) \\
= & \min_{I \ge T \ge 0} \left(\mathrm{Tr}\, \rho(I - \kappa^*(T)) + \mathrm{Tr}\, \sigma\kappa^*(T)\right) \\
\ge & \min_{I \ge T \ge 0} (\mathrm{Tr}\, \rho(I - T) + \mathrm{Tr}\, \sigma T) = 1 - \frac{1}{2} d_1(\rho, \sigma),
\end{aligned}
$$

which implies (5.51).

Exercise 3.30

(**a**) Define the function $g(y)$ as

$$
g(y) := h(x, y) - 2(y - x)^2 = x \log \frac{x}{y} + (1 - x) \log \frac{1 - x}{1 - y} - 2(y - x)^2.
$$

Then,

$$\begin{aligned}
\frac{dg}{dy}(y) = & -\frac{x}{y} + \frac{1-x}{1-y} - 4(y-x) \\
= & \frac{y(1-x) - x(1-y)}{y(1-y)} - 4(y-x) = \frac{y-x}{y(1-y)} - 4(y-x) \\
= & (y-x)(\frac{1}{y(1-y)} - 4) = (y-x)\frac{1-4y(1-y)}{y(1-y)} = (y-x)\frac{(2y-1)^2}{y(1-y)}.
\end{aligned}$$

Thus, $g(y)$ takes the minimum 0 at $y = x$.
(**c**) Choose a two-valued POVM $\boldsymbol{M} = \{P, I - P\}$, where P is given in (**b**). Then,

$$D(\rho\|\sigma) \geq D\left(\mathrm{P}_\rho^{\boldsymbol{M}} \| \mathrm{P}_\sigma^{\boldsymbol{M}}\right) \geq 2d_1^2\left(\mathrm{P}_\rho^{\boldsymbol{M}}, \mathrm{P}_\sigma^{\boldsymbol{M}}\right) = 2d_1^2(\rho, \sigma).$$

Exercise 3.31

$$\begin{aligned}
D(\rho\|\sigma) = & \operatorname{Tr} \sum_i a_i |u_i\rangle\langle u_i| \left(\log \sum_i a_i |u_i\rangle\langle u_i| - \log \sum_j b_j |v_j\rangle\langle v_j| \right) \\
= & \operatorname{Tr} \sum_i a_i |u_i\rangle\langle u_i| \left(\sum_i (\log a_i)|u_i\rangle\langle u_i| - \sum_j (\log b_j)|v_j\rangle\langle v_j| \right) \\
= & \sum_i a_i \log a_i - \sum_{i,j} a_i \log b_j \langle v_j|u_i\rangle\langle u_i|v_j\rangle \\
= & \sum_{i,j} a_i (\log a_i - \log b_j)|\langle u_i|v_j\rangle|^2 \\
= & \sum_{i,j} a_i |\langle u_i|v_j\rangle|^2 (\log a_i |\langle u_i|v_j\rangle|^2 - \log b_j |\langle u_i|v_j\rangle|^2) \\
= & D(P_{(\rho\|\sigma)} \| Q_{(\rho\|\sigma)}).
\end{aligned}$$

Exercise 3.32 Note that $\operatorname{Tr}|X|$ is equal to the sum of the absolute values of the eigenvalues of X. Using this fact, we can show (3.60).

Exercise 3.33 Since $\operatorname{Tr}(\rho - \sigma) = 0$, Substituting $(\rho - \sigma)$ into X of (3.60), we obtain

$$\begin{aligned}
& \operatorname{Tr} \rho\{\rho - \sigma \leq 0\} + \operatorname{Tr} \sigma\{\rho - \sigma > 0\} \\
= & \operatorname{Tr} \rho\{\rho - \sigma \leq 0\} + \operatorname{Tr} \sigma(I - \{\rho - \sigma \leq 0\}) \\
= & 1 + \frac{1}{2} \operatorname{Tr}(\rho - \sigma) 2\{\rho - \sigma \leq 0\} \\
= & 1 - \left(\frac{1}{2} \operatorname{Tr}(\rho - \sigma)(I - 2\{\rho - \sigma \leq 0\} \right) = 1 - \frac{1}{2}\|\rho - \sigma\|_1.
\end{aligned}$$

Exercise 3.34 $\|\rho_{\mathrm{mix}} - \rho\|_1 = 2\operatorname{Tr}(\rho_{\mathrm{mix}} - \rho)\{\rho_{\mathrm{mix}} - \rho \geq 0\}$. Hence, $2 - \|\rho_{\mathrm{mix}} - \rho\|_1 = 2 - 2\operatorname{Tr}(\rho_{\mathrm{mix}} - \rho)\{\rho_{\mathrm{mix}} - \rho \geq 0\} \leq 2 - 2\operatorname{Tr}\rho_{\mathrm{mix}}\{\rho_{\mathrm{mix}} - \rho \geq 0\} = 2\operatorname{Tr}\rho_{\mathrm{mix}}\{\rho_{\mathrm{mix}} - \rho < 0\} \leq 2\operatorname{Tr}\rho_{\mathrm{mix}}\{0 < \rho\} = 2\frac{\operatorname{rank}\rho}{d}$.

Exercise 3.35

(**a**) Since $(\sqrt{A}-\sqrt{B})\{\sqrt{A}\le\sqrt{B}\}\le 0$, we have $\mathrm{Tr}\,\sqrt{A}(\sqrt{A}-\sqrt{B})\{\sqrt{A}\le\sqrt{B}\}$ ≤ 0, which implies (3.81). Similarly, we obtain (3.82).
(**b**) Summing the inequalities (3.81) and (3.82), we obtain (3.80).

Exercise 3.36 $\sum_{i=1}^k \frac{1}{k}\mathrm{Tr}\,\rho_i M_i \le \sum_{i=1}^k \frac{1}{k}\mathrm{Tr}\,M_i = \frac{1}{k}\mathrm{Tr}\sum_{i=1}^k M_i = \frac{1}{k}\mathrm{Tr}\,I = \frac{d}{k}$. Further, $\sum_{i=1}^k \frac{1}{k}\mathrm{Tr}\,\rho_i M_i \le \sum_{i=1}^k \frac{1}{k}\|M_i\|_1\|\rho_i\| = \frac{1}{k}\max_{i'}\|\rho_{i'}\|\sum_{i=1}^k\|M_i\|_1 = \frac{1}{k}\max_{i'}$ $\|\rho_{i'}\|\sum_{i=1}^k \mathrm{Tr}\,M_i = \frac{d}{k}\max_i\|\rho_i\|$.

Exercise 3.37 Use Cramér's theorem with $X=-\log\frac{p(\omega)}{\bar{p}(\omega)}$, $\theta=s$, $x=0$, and show that $\lim_{n\to\infty}-\frac{1}{n}\log p^n\left\{-\frac{1}{n}\log\left(\frac{p^n(x^n)}{\bar{p}^n(x^n)}\right)\ge 0\right\}=\max_{s\ge 0}-\phi(s)$ and $\lim_{n\to\infty}$ $-\frac{1}{n}\log\bar{p}^n\left\{-\frac{1}{n}\log\left(\frac{p^n(x^n)}{\bar{p}^n(x^n)}\right)\le R\right\}=\max_{s\le 1}-\phi(s)$.

Exercise 3.38

(**a**) $\phi(s)=\log|u\rangle\langle u|^{1-s}\sigma^s=\log\langle u|\sigma^s|u\rangle$. Since $\sigma^s\ge\sigma^s$ for $s\in(0,1)$, we have $\langle u|\sigma^s|u\rangle\ge$ $\langle u|\sigma|u\rangle$. Hence, we have $\inf_{1>s>0}\phi(s)=\inf_{1>s>0}\langle u|\sigma^s|u\rangle=\langle u|\sigma|u\rangle$.
(**b**)

$$
\begin{aligned}
&\min_{I\ge T\ge 0}\left(\mathrm{Tr}\,\rho^{\otimes n}(I-T)+\mathrm{Tr}\,\sigma^{\otimes n}T\right)\\
\le&\left(\mathrm{Tr}\,|u\rangle\langle u|^{\otimes n}(I-|u\rangle\langle u|^{\otimes n})+\mathrm{Tr}\,\sigma^{\otimes n}|u\rangle\langle u|^{\otimes n}\right)\\
=&\mathrm{Tr}\,\sigma^{\otimes n}|u\rangle\langle u|^{\otimes n}=\langle u|\sigma|u\rangle^n=\exp(n\inf_{1\ge s\ge 0}\phi(s)).
\end{aligned}
$$

Exercise 3.39 When the POVM $\{|u\rangle\langle u|, I-|u\rangle\langle u|\}$ is applied, the outcome obeys the distribution P_0 or P_1, where $P_0(0):=1$, $P_0(1):=0$, $P_1(0):=\langle u|\sigma|u\rangle$, and $P_1(1):=$ $1-\langle u|\sigma|u\rangle$. Then, $\phi(s|P_0\|P_1)=\log\langle u|\sigma|u\rangle^s=s\log\langle u|\sigma|u\rangle$. Thus, $\inf_{1>s>0}\phi$ $(s|P_0\|P_1)=\inf_{1>s>0}s\log\langle u|\sigma|u\rangle=\phi(s|P_0\|P_1)$ because $\phi(s|P_0\|P_1)\le 0$.

Exercise 3.40

(**a**) Exercise 3.18 and (3.59) implies that

$$
\begin{aligned}
&\min_{I\ge T\ge 0}\left(\mathrm{Tr}\,|u\rangle\langle u|^{\otimes n}(I-T)+\mathrm{Tr}\,|v\rangle\langle v|^{\otimes n}T\right)\\
=&1-\frac{1}{2}\||u\rangle\langle u|^{\otimes n}-|v\rangle\langle v|^{\otimes n}\|_1=1-\sqrt{1-|\langle u|v\rangle|^{2n}}
\end{aligned}
$$

(**b**)

$$
\begin{aligned}
&\lim_{n\to\infty}\frac{1}{n}\log 1-\sqrt{1-|\langle u|v\rangle|^{2n}}=\lim_{n\to\infty}\frac{1}{n}\log 1-(1-\frac{1}{2}|\langle u|v\rangle|^{2n})\\
=&\lim_{n\to\infty}\frac{1}{n}\log+\frac{1}{2}|\langle u|v\rangle|^{2n})=\log|\langle u|v\rangle|^2
\end{aligned}
$$

(**c**) Since $\inf_{1>s>0}\phi(s)=\inf_{1>s>0}\log|\langle u|v\rangle|^2=\log|\langle u|v\rangle|^2$, we have (3.87).

Exercise 3.41 Since $U^* = U$, we have $\mathrm{Tr}\,\rho^{1-s}\sigma^s = \mathrm{Tr}\,\rho^{1-s}U\rho^s U^* = \mathrm{Tr}\,U\rho^{1-s}U^* \rho^s = \mathrm{Tr}\,\sigma^{1-s}\rho^s$, which implies the symmetry $\phi(s) = \phi(1-s)$. The conclusion can be derived from the convexity of $\phi(s)$ and the symmetry.

Exercise 3.42 We choose s_1 as the solution of $\phi(s) = 2s \log F(\rho, \sigma)$. We also choose s_2 as the solution of $\phi(s) = 2(1-s)\log F(\rho, \sigma)$. Assume that a convex function f on $(0, 1)$ satisfies $\phi(s) \le f(s) \le 0$ and $\log F(\rho, \sigma) \le f(1/2)$. Considering the graph of $f(s)$, we find that $f(s) \ge \phi(s_2)$ for $s \in [0, 1/2]$ and $f(s) \ge \phi(s_1)$ for $s \in [1/2, 1]$.

Remember that the assumption guarantees the symmetry $\phi(s) = \phi(1-s)$. Since $\phi''(1/2) > 0$ and $\phi(1/2|\rho\|\sigma) < \log F(\rho, \sigma)$, we have $\phi(s_2) > \phi(1/2)$ and $\phi(s_1) > \phi(1/2)$. Further, for any POVM $\boldsymbol{M}$, $\phi(s|\mathrm{P}_\rho^{\boldsymbol{M}}\|\mathrm{P}_\sigma^{\boldsymbol{M}})$ satisfies the condition for $f(s)$. Thus, $\min_{\boldsymbol{M}} \inf_{1\ge s\ge 0} \phi(s|\mathrm{P}_\rho^{\boldsymbol{M}}\|\mathrm{P}_\sigma^{\boldsymbol{M}}) \ge \min(\phi(s_2), \phi(s_1)) > \phi(1/2)$.

Similarly, for any POVM $\boldsymbol{M}^n$, $\frac{1}{n}\phi(s|\mathrm{P}_{\rho^{\otimes n}}^{\boldsymbol{M}^n}\|\mathrm{P}_{\sigma^{\otimes n}}^{\boldsymbol{M}^n})$ satisfies the condition for $f(s)$. Hence, $\frac{1}{n}\min_{\boldsymbol{M}^n} \inf_{1\ge s\ge 0} \phi(s|\mathrm{P}_{\rho^{\otimes n}}^{\boldsymbol{M}^n}\|\mathrm{P}_{\sigma^{\otimes n}}^{\boldsymbol{M}^n}) \ge \min(\phi(s_2), \phi(s_1)) > \phi(1/2)$. Therefore, $\underline{\lim}_{n\to\infty}\frac{1}{n}\min_{\boldsymbol{M}^n} \inf_{1\ge s\ge 0} \phi(s|\mathrm{P}_{\rho^{\otimes n}}^{\boldsymbol{M}^n}\|\mathrm{P}_{\sigma^{\otimes n}}^{\boldsymbol{M}^n}) \ge \min(\phi(s_2), \phi(s_1)) > \phi(1/2)$.

Exercise 3.43

(b)

$$
\begin{aligned}
& D(q\|p) - \frac{1}{1-s}D(q\|p_s) \\
=& \sum_x q(x)(\log q(x) - \log p(x)) - \frac{1}{1-s}\sum_x q(x)(\log q(x) - \log \bar{p}(x)) \\
& + \sum_x q(x)(\log p(x) - \log \bar{p}(x)) - \frac{\phi(s)}{1-s} \\
=& -\frac{s}{1-s}\sum_x q(x)(\log q(x) - \log \bar{p}(x)) - \frac{\phi(s)}{1-s} \\
=& -\frac{s}{1-s}D(q\|\bar{p}) - \frac{\phi(s)}{1-s} = -\frac{s}{1-s}D(p_s\|\bar{p}) - \frac{\phi(s)}{1-s} \\
=& -\frac{s}{1-s}\sum_x p_s(x)((1-s)\log p(x) + s\log \bar{p}(x) - \log \bar{p}(x)) \\
& + \frac{s}{1-s}\phi(s) - \frac{\phi(s)}{1-s} \\
=& -s\sum_x p_s(x)(\log p(x) - \log \bar{p}(x)) - \phi(s) \\
=& \sum_x p_s(x)(\log(p(x)^{1-s}\bar{p}(x)^s) - \log p(x)) - \phi(s) = D(p_s\|p).
\end{aligned}
$$

(c) The desired inequality follows from the inequality $\frac{1}{1-s}D(q\|p_s) \ge 0$ for $s \le 1$.

Exercise 3.44

(b) Notice that the set of all distributions forms an exponential family by adding generators $\{g_i\}_{i\ge 2}$. Here, we choose $g_1(x)$ to be $\log \bar{p}(x) - \log p(x)$. $D(q\|\bar{p})$ can be regarded as a Bregmann divergence. Next, we apply Theorem 2.3 to the following case: $\mathcal{M}$ is the mixture subfamily containing p_s with the generator g_1. $\mathcal{E}$ is the exponential subfamily $\{p_s(x)\}$, which is generated by g_1. Then, $\mathcal{M}$ contains p_s and q. Choosing the parameters θ, θ', and θ^* to indicate the distributions $\bar{p}$, q, and p_s, respectively, we obtain the desired argument.

Exercise 3.45

(**a**) Use $\frac{\phi(s)}{s} \le D(p\|\overline{p})$.
(**b**) Substitute $q = p$ and $s = 0$ in the right and left hand sides, respectively.
(**c**) $(s-1)\phi'(s) - \phi(s) = (s-1)\sum_x p_s(x)(\log \bar{p}(x) - \log p(x)) - \phi(s) = \sum_x p_s(x)(1-s)\log p(x) + s\log \bar{p}(x) - \phi(s) - \log \bar{p}(x) = D(p_s\|p_1)$.
$s\phi'(s) - \phi(s) = s\sum_x p_s(x)(\log \bar{p}(x) - \log p(x)) - \phi(s) = \sum_x p_s(x)(1-s) \log p(x) + s\log \bar{p}(x) - \phi(s) - \log p(x) = D(p_s\|p_0)$.
(**e**) The map $s \mapsto D(p_s\|\overline{p})$ is continuous and monotonically decreasing in the domain $[0, 1]$. It has the range $[0, D(p\|\overline{p})]$.
(**f**) It follows from Exercise 3.43.
(**g**) $\frac{d}{ds}D(p_s\|p) = \frac{d}{ds}s\phi'(s) - \phi(s) = s\phi''(s) > 0$. Thus, the map $s \mapsto D(p_s\|p)$ is monotonically increasing. the map $r \mapsto s_r$ is monotonically decreasing. So, the map $r \mapsto D(p_{s_r}\|p)$ is also monotonically decreasing. Thus, (3.114) follows from Exercise (**f**).
(**h**) $\frac{-s_r r - \phi(s_r)}{1-s_r} = \frac{-s_r D(p_{s_r}\|p_1) - \phi(s_r)}{1-s_r} = \frac{-s_r((s-1_r)\phi'(s_r) - \phi(s_r)) - \phi(s_r)}{1-s_r} = s_r\phi'(s_r) - \phi(s_r) = D(p_{s_r}\|p)$.
(**j**) $\frac{-r+(s_r-1)\phi'(s_r)-\phi(s_r)}{(1-s_r)^2} = 0$. Since $\frac{d}{ds}(-r + (s-1)\phi'(s) - \phi(s)) = (s-1)\phi''(s) < 0$, $\frac{-r+(s-1)\phi'(s)-\phi(s)}{(1-s)^2} < 0$ for $s > s_r$ and $\frac{-r+(s-1)\phi'(s)-\phi(s)}{(1-s)^2} > 0$ for $s < s_r$. Hence, the maximum of $\frac{-sr-\phi(s)}{1-s}$ is realized only when $s = s_r$.

(**k**) Equation (3.108) follows from the combination of (**g**), (**h**), and (**j**).

Exercise 3.46 Choose $\rho = \begin{pmatrix} 1-p & 0 & 0 \\ 0 & \frac{p}{2} & \frac{p}{2} \\ 0 & \frac{p}{2} & \frac{p}{2} \end{pmatrix}$ and $\sigma = \begin{pmatrix} 1-q & 0 & 0 \\ 0 & \frac{q}{2} & \frac{q}{2} \\ 0 & \frac{q}{2} & \frac{q}{2} \end{pmatrix}$. The PVM $\boldsymbol{M} = \{M_i\}$ is given as $M_1 = \begin{pmatrix} 1 & 0 & 0 \\ 0 & 0 & 0 \\ 0 & 0 & 0 \end{pmatrix}$, $M_2 = \begin{pmatrix} 0 & 0 & 0 \\ 0 & 1 & 0 \\ 0 & 0 & 0 \end{pmatrix}$ $M_3 = \begin{pmatrix} 0 & 0 & 0 \\ 0 & 0 & 0 \\ 0 & 0 & 1 \end{pmatrix}$. Although M_1 commutes ρ and σ, M_2 and M_3 do not commute ρ and σ. However, choosing $a_1 = \frac{1-p}{1-q}$ and $a_2 = a_3 = \frac{p}{q}$, we have $\sum_{i=1}^3 a_i M_i = \begin{pmatrix} \frac{1-p}{1-q} & 0 & 0 \\ 0 & \frac{p}{q} & 0 \\ 0 & 0 & \frac{p}{q} \end{pmatrix}$, which satisfies Condition (3.120).

Exercise 3.47

$$
\begin{aligned}
& D\left(\mathrm{P}^{\boldsymbol{M}^n}_{\rho^{\otimes n}} \middle\| \mathrm{P}^{\boldsymbol{M}^n}_{\sigma^{\otimes n}}\right) \\
= & \sum_{\omega_n} \left(\prod_{k=1}^n \operatorname{Tr} M^n_{k,\omega_n}\rho\right) \log\left(\prod_{k=1}^n \operatorname{Tr} M^n_{k,\omega_n}\rho\right) - \log\left(\prod_{k=1}^n \operatorname{Tr} M^n_{k,\omega_n}\sigma\right)
\end{aligned}
$$

$$
\begin{aligned}
&= \sum_{\omega_n} \left(\prod_{k=1}^{n} \mathrm{Tr}\, M^n_{k,\omega_n} \rho \right) \sum_{k=1}^{n} (\log \mathrm{Tr}\, M^n_{k,\omega_n} \rho - \log \mathrm{Tr}\, M^n_{k,\omega_n} \sigma) \\
&= \sum_{\omega_n} \sum_{k=1}^{n} \mathrm{Tr}\, a_{k,\omega_n} M^n_{k,\omega_n} \rho \left(\log \mathrm{Tr}\, a_{k,\omega_n} M^n_{k,\omega_n} \rho - \log \mathrm{Tr}\, a_{k,\omega_n} M^n_{k,\omega_n} \sigma \right) \\
&= \sum_{k=1}^{n} \sum_{\omega_n} \mathrm{Tr}\, a_{k,\omega_n} M^n_{k,\omega_n} \rho \left(\log \mathrm{Tr}\, a_{k,\omega_n} M^n_{k,\omega_n} \rho - \log \mathrm{Tr}\, a_{k,\omega_n} M^n_{k,\omega_n} \sigma \right) \\
&= \sum_{k=1}^{n} D\left(\mathrm{P}^{M^{n,k}}_{\rho} \middle\| \mathrm{P}^{M^{n,k}}_{\sigma} \right).
\end{aligned}
$$

Exercise 3.48 The desired equation follows from $R_{s_0} = -\frac{d\phi}{ds}(s_0)$.

Exercise 3.49 $-\phi'(s) = \sum_x p_s(x)(\log p(x) - \log \bar{p}(x)) = D_{\max}(p\|\bar{p}) + o(1)$ as $s \to -\infty$.

Exercise 3.50 From the derivation of Exercise 3.50, we find that $-\phi'(s) = D_{\max}(p\|\bar{p}) + o(1/s)$ as $s \to -\infty$.

Since $\frac{p(x)^s \bar{p}(s)^{-s}}{e^{s D_{\max}(p\|\bar{p})}} \to o(1)$ as $s \to -\infty$, we have $\phi(s) = \log \overline{P}(p\|\bar{p}) - s D_{\max}(p\|\bar{p}) + \log \sum_x \frac{p(x)}{\overline{P}(p\|\bar{p})} \frac{p(x)^s \bar{p}(s)^{-s}}{e^{s D_{\max}(p\|\bar{p})}} = \log \overline{P}(p\|\bar{p}) - s D_{\max}(p\|\bar{p}) + o(1)$. Thus, (3.109) implies that $D(p_s\|\bar{p}) = (s-1)\phi'(s) - \phi(s) = -(s-1)D_{\max}(p\|\bar{p}) + o(1) - \log \overline{P}(p\|\bar{p}) + s D_{\max}(p\|\bar{p}) + o(1) = D_{\max}(p\|\bar{p}) - \log \overline{P}(p\|\bar{p}) + o(1)$ as $s \to -\infty$.

Similarly, (3.110) implies that $D(p_s\|p) = s\phi'(s) - \phi(s) = -s D_{\max}(p\|\bar{p}) + o(1) - \overline{P}(p\|\bar{p}) + s D_{\max}(p\|\bar{p}) + o(1) = -\log \overline{P}(p\|\bar{p}) + o(1)$ as $s \to -\infty$.

Exercise 3.51 It follows from $1 = \frac{dr}{dr} = \frac{d}{dr} D(p_{s_r}\|p_1) = \frac{d}{dr}(s_r - 1)\phi'(s_r) - \phi(s_r) = \frac{ds_r}{dr}\frac{d}{ds}(s-1)\phi'(s) - \phi(s)|_{s=s_r} = \frac{ds_r}{dr}(s_r - 1)\phi''(s_r)$.

Exercise 3.52

(**a**) The map $s \mapsto D(p_s\|\overline{p})$ is continuous and monotonically decreasing in the domain $(-\infty, 0]$. It has the range $[D(p\|\bar{p}), -\log \underline{P}(p\|\bar{p}))$.
(**b**) It follows from Exercise 3.43.
(**c**) $\frac{d}{ds} D(p_s\|p) = \frac{d}{ds} s\phi'(s) - \phi(s) = s\phi''(s) < 0$. Thus, the map $s \mapsto D(p_s\|p)$ is monotonically decreasing. Since the map $r \mapsto s_r$ is also monotonically decreasing, the map $r \mapsto D(p_{s_r}\|p)$ is monotonically increasing. Thus, $\min_{q:D(q\|\bar{p})\ge r} D(q\|p) = D(p_{s_r}\|p)$ follows from (3.162).

The relations (3.110), (3.160), and (3.162) imply that $\min_{q:D(q\|\bar{p})=r'} D(q\|p) + r - D(q\|\bar{p}) = r + s_{r'}\phi'(s_{r'}) - \phi(s_{r'}) - ((s_{r'} - 1)\phi'(s_{r'}) - \phi(s_{r'})) = r + \phi'(s_{r'})$. Thus, $\frac{d}{dr'} \min_{q:D(q\|\bar{p})=r'} D(q\|p) + r - D(q\|\bar{p}) = \frac{d}{dr'} r + \phi'(s_{r'}) = \phi''(s_{r'})\frac{d}{dr'}s_{r'} = \phi''(s_{r'})\frac{1}{(s_r-1)\phi''(s_r)} = \frac{1}{(s_r-1)} < 0$. Hence, $\min_{q:D(q\|\bar{p})=r'} D(q\|p) + r - D(q\|\bar{p})$ is monotone decreasing for r'. Thus $\min_{q:D(q\|\bar{p})\le r} D(q\|p) + r - D(q\|\bar{p}) = \min_{q:D(q\|\bar{p})=r} D(q\|p) + r - D(q\|\bar{p}) = \min_{q:D(q\|\bar{p})=r} D(q\|p) = D(p_{s_r}\|p)$.
(**d**) See (**h**) of Exercise 3.45.
(**f**) See (**j**) of Exercise 3.45.
(**g**) (3.161) follows from the combination of (3.163), (3.164), and (3.166).

Exercise 3.53

(**a**) The relation (3.165) implies that $\frac{d}{ds}\frac{-sr-\phi(s)}{1-s} = \frac{-r+(s-1)\phi'(s)-\phi(s)}{(1-s)^2} = \frac{-r+D(p_s\|p_1)}{(1-s)^2}$. Since $\lim_{s\to-\infty} D(p_s\|\bar{p}) = D_{\max}(p\|\bar{p})$ and the map $s \mapsto D(p_s\|\bar{p})$ is monotonically decreasing, we have $D(p_s\|\bar{p}) \le D_{\max}(p\|\bar{p})$. Hence, $\sup_{s\le 0}\frac{-sr-\phi(s|p\|\bar{p})}{1-s} = \sup_{s\to-\infty}\frac{-sr-\phi(s|p\|\bar{p})}{1-s}$.

(**b**) $D(q\|p) - D(q\|\bar{p}) = \sum_x q(x)(\log\bar{p}(x) - \log p(x)) \ge -D_{\max}(p\|\bar{p})$. Consider the subset $\{x|\log p(x) - \log\bar{p}(x) = D_{\max}(p\|\bar{p})\}$. When the support of q is included in the subset, we have the equality.

(**c**) Consider $q = p_s$ with the limit $s \to -\infty$.

Exercise 3.54

(**a**) For the derivations of (3.172) and (3.173), substitute $\theta = s$ and $X = -\log\left(\frac{p(x)}{\bar{p}(x)}\right)$ in (2.163) and (2.165), respectively. For the derivation of (3.173), substitute $\theta = s-1$ $X = -\log\left(\frac{p(x)}{\bar{p}(x)}\right)$ in (2.165).

(**b**) They can be shown by using the convexity of $\phi(s)$.

(**c**) The definition of s_r, (3.109), and (3.175) imply $r = D(p_{s_r}\|\bar{p}) = (s_r-1)\phi'(s_r) - \phi(s_r) = \max_{s\le 1}(s-1)\phi'(s_r) - \phi(s)$. The relations (3.166), (3.164), (3.110), and (3.175) yield $\sup_{s\le 0}\frac{-sr-\phi(s)}{1-s} = \frac{-s_r r-\phi(s_r)}{1-s_r} = D(p_{s_r}\|p) = s_r\phi'(s_r) - \phi(s_r) = \max_{s\le 0} s\phi'(s_r) - \phi(s)$.

(**d**) Choose the test $\left\{-\frac{1}{n}\log\left(\frac{p^n(x^n)}{\bar{p}^n(x^n)}\right) \le \phi'(s_r)\right\}$. The relations (3.176) and (3.173) guarantee that the error exponent of first kind is r. The relations (3.177) and (3.172) show that the exponent of the correct decision when true is p is $\sup_{s\le 0}\frac{-sr-\phi(s|p\|\bar{p})}{1-s}$.

(**e**) Choose the random test as follows: When the outcome belongs to the set $\left\{-\frac{1}{n}\log\left(\frac{p^n(x^n)}{\bar{p}^n(x^n)}\right) = D_{\max}(p\|\bar{p})\right\}$, we support the hypothesis p with probability $e^{-n(r+\log \underline{P}(p\|\bar{p}))}$. The definitions of $\underline{P}(p\|\bar{p})$ and $\overline{P}(p\|\bar{p})$ implies that

$$\lim_{n\to\infty} -\frac{1}{n}\log p^n\left\{-\frac{1}{n}\log\left(\frac{p^n(x^n)}{\bar{p}^n(x^n)}\right) = D_{\max}(p\|\bar{p})\right\} = -\log\overline{P}(p\|\bar{p}), \quad (3.185)$$

$$\lim_{n\to\infty} -\frac{1}{n}\log \bar{p}^n\left\{-\frac{1}{n}\log\left(\frac{p^n(x^n)}{\bar{p}^n(x^n)}\right) = D_{\max}(p\|\bar{p})\right\} = -\log\underline{P}(p\|\bar{p}). \quad (3.186)$$

The relation (3.186) guarantees that the error exponent of first kind is $r + \log\underline{P}(p\|\bar{p}) - \log\underline{P}(p\|\bar{p}) = r$. The relation (3.185) shows that the exponent of correct decision when true is p is $r + \log\underline{P}(p\|\bar{p}) - \log\overline{P}(p\|\bar{p}) = r + D_{\max}(p\|\bar{p})$, which equals $\sup_{s\le 0}\frac{-sr-\phi(s|p\|\bar{p})}{1-s}$, as shown in (3.167).

Exercise 3.55 The relations (3.171), (3.177), and (3.160) yield that

$$\begin{aligned}\frac{dB^*(r|p\|\bar{p})}{dr} &= \frac{d}{dr}(s_r\phi'(s_r) - \phi(s_r)) = \frac{ds_r}{dr}\frac{d}{ds}(s\phi'(s) - \phi(s))|_{s=s_r} \\ &= \frac{1}{(s_r-1)\phi''(s_r)}\phi''(s_r) = \frac{1}{s_r-1}.\end{aligned}$$

Then, the relation (3.160) yields that

$$\frac{d^2 B^*(r|p\|\bar{p})}{dr^2} = \frac{d}{dr}\frac{1}{s_r - 1}$$
$$= \frac{ds_r}{dr} \cdot \frac{1}{(s_r-1)^2} = \frac{1}{(s_r-1)\phi''(s_r)} \cdot -\frac{1}{(s_r-1)^2} = -\frac{1}{(s_r-1)^3\phi''(s_r)} \geq 0.$$

Exercise 3.56

(**a**) The existence of $\{T_n\}$ follows from the direct part of quantum Stein' lemma.
(**b**) The desired inequality follows from the converse part of quantum Stein' lemma.
(**c**) The inequality $B^*(r|\rho\|\sigma) \leq \inf_{\tau:D(\tau\|\sigma)>r} D(\tau\|\rho)$ follows from (**a**) and (**b**). The equation $\inf_{\tau:D(\tau\|\sigma)>r} D(\tau\|\rho) = \min_{\tau:D(\tau\|\sigma)\geq r} D(\tau\|\rho)$ follows from the continuity of $D(\tau\|\rho)$.

Exercise 3.57 Lemma 3.7 guarantees that $\lim_{n\to\infty} \operatorname{Tr} \tau^{\otimes n}(I - T_n) = 1$ because $D(\tau\|\sigma) \leq \underline{\lim} -\frac{1}{n} \log \operatorname{Tr} \sigma^{\otimes n}(I - T_n)$. Hence, applying Lemma 3.7 again, we have $\underline{\lim} -\frac{1}{n} \log \operatorname{Tr} \rho^{\otimes n} T_n \leq D(\tau\|\rho)$.

Exercise 3.58 The desired equation follows from $\tilde{\phi}(s|\rho\|\sigma) \geq \lim_{n\to\infty} \frac{1}{n}\phi(s|\mathrm{P}^{\boldsymbol{M}}_{\rho^{\otimes n}}\|\mathrm{P}^{\boldsymbol{M}}_{\sigma^{\otimes n}})$ and $\tilde{\phi}(s|\rho\|\sigma) = \lim_{n\to\infty} \frac{1}{n}\phi(s|\kappa_{\sigma^{\otimes n}}(\rho^{\otimes n})\|\sigma^{\otimes n})$.

Exercise 3.59

(**a**) We give a spectral decomposition $\sigma = \sum_i \sigma_i E_i$. Then, $e^{D_{\max}(\kappa_\sigma(\rho)\|\sigma)} = \max_i \|E_i \sigma^{-1/2} \rho\sigma^{-1/2} E_i\| \leq \|\sigma^{-1/2}\rho\sigma^{-1/2}\| = e^{D_{\max}(\rho\|\sigma) \geq D_{\max}(\kappa_\sigma(\rho)\|\sigma)}$.
(**b**) Lemma 3.10 implies that $|\boldsymbol{E}_{\sigma^{\otimes n}}|\|(\sigma^{\otimes n})^{-1/2}\kappa_{\sigma^{\otimes n}}(\rho^{\otimes n})(\sigma^{\otimes n})^{-1/2}\| \geq \|(\sigma^{\otimes n})^{-1/2} \rho^{\otimes n}(\sigma^{\otimes n})^{-1/2}\|$. Hence, $D_{\max}(\kappa_{\sigma^{\otimes n}}(\rho^{\otimes n})\|\sigma^{\otimes n}) + \log|\boldsymbol{E}_{\sigma^{\otimes n}}| \geq D_{\max}(\rho^{\otimes n}\|\sigma^{\otimes n}) = nD_{\max}(\rho\|\sigma)$. Finally take the limit $n \to \infty$ after the dividing the both side with n.
(**c**) Due to (**a**), $\frac{1}{n}\frac{\phi(-s|\kappa_{\sigma^{\otimes n}}(\rho^{\otimes n})\|\sigma^{\otimes n})}{s} \leq \frac{1}{n} D_{\max}(\kappa_{\sigma^{\otimes n}}(\rho^{\otimes n})\|\sigma^{\otimes n}) \leq \frac{1}{n} D_{\max}(\rho^{\otimes n}\|\sigma^{\otimes n}) = D_{\max}(\rho\|\sigma)$. Taking the limit $n \to \infty$, we have $\lim_{s\to\infty} \frac{\tilde{\phi}(-s|\rho\|\sigma)}{s} \leq D_{\max}(\rho\|\sigma)$.

Since $\frac{\tilde{\phi}(-s|\rho\|\sigma)}{s} \geq \frac{1}{n}\frac{\phi(-s|\kappa_{\sigma^{\otimes n}}(\rho^{\otimes n})\|\sigma^{\otimes n})}{s}$, we have $\lim_{s\to\infty} \frac{\tilde{\phi}(-s|\rho\|\sigma)}{s} \geq \frac{1}{n} \lim_{s\to\infty} \frac{\phi(-s|\kappa_{\sigma^{\otimes n}}(\rho^{\otimes n})\|\sigma^{\otimes n})}{s} = \frac{1}{n} D_{\max}(\kappa_{\sigma^{\otimes n}}(\rho^{\otimes n})\|\sigma^{\otimes n})$.

Taking the limit $n \to \infty$, we have $\lim_{s\to\infty} \frac{\tilde{\phi}(-s|\rho\|\sigma)}{s} \geq \lim_{n\to\infty} \frac{1}{n} D_{\max}(\kappa_{\sigma^{\otimes n}}(\rho^{\otimes n}) \|\sigma^{\otimes n}) \geq D_{\max}(\rho\|\sigma)$, which follows from (**b**).

Exercise 3.60 Assume that $\sigma = \sum_i \sigma_i M_i$. Then, $\kappa_{\sigma^{\otimes 2}}(\rho^{\otimes 2}) = \sum_{i>j} (M_i \otimes M_j + M_j \otimes M_i)\rho^{\otimes 2}(M_i \otimes M_j + M_j \otimes M_i) + \sum_i (M_i \otimes M_i)\rho^{\otimes 2}(M_i \otimes M_i)$, Hence, $\operatorname{Tr} \kappa_{\sigma^{\otimes 2}}(\rho^{\otimes 2})(\sigma^{\otimes 2})^s = \sum_{i>j} \sigma_i^s \sigma_j^s \operatorname{Tr}[(M_i \otimes M_j + M_j \otimes M_i)\rho^{\otimes 2}(M_i \otimes M_j + M_j \otimes M_i)]^{1-s} + \sum_i \sigma_i^{2s} \operatorname{Tr}((M_i \otimes M_i)\rho^{\otimes 2}(M_i \otimes M_i))^{1-s}$.

On the other hand, $e^{\phi(s|\mathrm{P}^{\boldsymbol{M}}_\rho\|\mathrm{P}^{\boldsymbol{M}}_\sigma)} = \sum_{i>j} \sigma_i^s \sigma_j^s \operatorname{Tr}[(M_j \otimes M_i)\rho(M_j \otimes M_i) + (M_i \otimes M_j)\rho^{\otimes 2}(M_i \otimes M_j)]^{1-s} + \sum_i \sigma_i^{2s} \operatorname{Tr}((M_i \otimes M_i)\rho^{\otimes 2}(M_i \otimes M_i))^{1-s}$.

Hence, we have $e^{\phi(s|\kappa_{\sigma^{\otimes 2}}(\rho^{\otimes 2})\|\sigma^{\otimes 2})} - e^{2\phi(s|\mathrm{P}^{\boldsymbol{M}}_\rho\|\mathrm{P}^{\boldsymbol{M}}_\sigma)} = \sum_{i>j} \sigma_i^s \sigma_j^s \operatorname{Tr}[((M_i \otimes M_j + M_j \otimes M_i)\rho^{\otimes 2}(M_i \otimes M_j + M_j \otimes M_i))^{1-s} - ((M_j \otimes M_i)\rho^{\otimes 2}(M_j \otimes M_i))^{1-s}.$
$- ((M_i \otimes M_j)\rho^{\otimes 2}(M_i \otimes M_j)))^{1-s}]$. Since the rank-one PVM $\boldsymbol{M} = \{M_i\}$ is not

commutative with ρ, there exists a pair i, j such that $(M_j \otimes M_i)\rho^{\otimes 2}(M_i \otimes M_j) \neq 0$. Since $x \mapsto x^{1-s}$ is strictly convex, Exercise 1.35 implies $\mathrm{Tr}[((M_i \otimes M_j + M_j \otimes M_i)\rho^{\otimes 2}(M_i \otimes M_j + M_j \otimes M_i))^{1-s} - ((M_j \otimes M_i)\rho^{\otimes 2}(M_j \otimes M_i))^{1-s}. - ((M_i \otimes M_j)\rho^{\otimes 2}(M_i \otimes M_j)))^{1-s}] > 0$.

Exercise 3.61 Due to a discussion similar to Exercise 3.60, it is sufficient to show that $\mathrm{Tr}[((M_i \otimes M_j + M_j \otimes M_i)\sigma^{\otimes 2}(M_i \otimes M_j + M_j \otimes M_i))^s - ((M_j \otimes M_i)\sigma^{\otimes 2} (M_j \otimes M_i))^s. - ((M_i \otimes M_j)\sigma^{\otimes 2}(M_i \otimes M_j))^s] > 0$. This inequality follows from Exercise 1.35 and the strict convexity of $x \mapsto x^s$.

Exercise 3.62

(**a**) For any POVM $\boldsymbol{M} = \{M_i\}$, there exists a rank-one POVM $\boldsymbol{M}' := \{M'_{i,j}\}$ such that $\sum_j M'_{i,j} = M_i$. Next, we choose the Naimark extension $\boldsymbol{M}''$ of $\boldsymbol{M}'$. Then, $\boldsymbol{M}''$ is a rank-one PVM $\boldsymbol{M}$ and $\phi(s|\mathrm{P}_\rho^{\boldsymbol{M}''}\|\mathrm{P}_\sigma^{\boldsymbol{M}''}) \geq \phi(s|\mathrm{P}_\rho^{\boldsymbol{M}}\|\mathrm{P}_\sigma^{\boldsymbol{M}})$.
(**b**) Consider the case when the rank-one PVM $\boldsymbol{M} = \{M_i\}$ is not commutative with ρ. Then, apply Exercise 3.60 to the states $\kappa_{\boldsymbol{M}}(\sigma)$ and ρ. Further, Exercise 3.58 implies $2\overline{\phi}(s|\rho\|\sigma) \geq \phi(s|\kappa_{\kappa_{\boldsymbol{M}}(\sigma)^{\otimes 2}}(\rho^{\otimes 2})\|\kappa_{\boldsymbol{M}}(\sigma)^{\otimes 2})$. Therefore, we obtain (3.181) in this case.

Consider the case when the rank-one PVM $\boldsymbol{M} = \{M_i\}$ is commutative with ρ. We can show (3.181) by using Exercises 3.58 and 3.61 in a similar way.

References

1. J. von Neumann, *Mathematical Foundations of Quantum Mechanics*, (Princeton University Press, Princeton, NJ, 1955). (Originally appeared in German in 1932)
2. C.E. Shannon, A mathematical theory of communication. Bell Syst. Tech. J. **27**, 623–656 (1948)
3. N. Datta, Min- and max- relative entropies and a new entanglement monotone. IEEE Trans. Inf. Theory **55**, 2816–2826 (2009)
4. M. Müller-Lennert, F. Dupuis, O. Szehr, S. Fehr, M. Tomamichel, On quantum Renyi entropies: a new generalization and some properties. J. Math. Phys. **54**, 122203 (2013)
5. M.M. Wilde, A. Winter, D. Yang, Strong converse for the classical capacity of entanglement-breaking and Hadamard channels via a sandwiched Renyi relative entropy. Comm. Math. Phys. **331**(2), 593 (2014)
6. R.L. Frank, E.H. Lieb, Monotonicity of a relative Renyi entropy. J. Math. Phys. **54**, 122201 (2013)
7. T. Ogawa, H. Nagaoka, Strong converse and Stein's lemma in quantum hypothesis testing. IEEE Trans. Inf. Theory **46**, 2428–2433 (2000)
8. E.H. Lieb, W.E. Thirring, Inequalities for the moments of the eigenvalues of the Schrödinger Hamiltonian and their relation to sobolev inequalities, in *Studies in Mathematical Physics*, ed. by E. Lieb, B. Simon, A. Wightman (Princeton University Press, 1976), pp. 269–303
9. H. Araki, On an inequality of lieb and thirring. Lett. Math. Phys. **19**, 167–170 (1990)
10. C.A. Fuchs, Distinguishability and Accessible Information in Quantum Theory, quant-ph/9601020 (1996)
11. A.S. Holevo, An analog of the theory of statistical decisions in noncommutative theory of probability. *Trudy Moskov. Mat. Obšč.***26**, 133–149 (1972) (in Russian). (English translation: *Trans. Moscow Math. Soc.***26**, 133–149 (1972))

12. C.W. Helstrom, *Quantum Detection and Estimation Theory* (Academic, New York, 1976)
13. K.M.R. Audenaert, J. Calsamiglia, Ll. Masanes, R. Munoz-Tapia, A. Acin, E. Bagan, F. Verstraete, Discriminating states: the quantum Chernoff bound. Phys. Rev. Lett. **98**, 160501 (2007)
14. M. Nussbaum, A. Szkoła, The chernoff lower bound for symmetric quantum hypothesis testing. Ann. Stat. **37**, 1040–1057 (2009)
15. Y. Ogata, A generalization of powers-stormer inequality. Lett. Math. Phys. **97**, 339–346 (2011)
16. V. Jaksic, Y. Ogata, C.-A. Piller, R. Seiringer, Quantum hypothesis testing and non-equilibrium statistical mechanics. Rev. Math. Phys. **24**, 1230002 (2012)
17. H.P. Yuen, R.S. Kennedy, M. Lax, Optimum testing of multiple hypotheses in quantum detection theory. IEEE Trans. Inf. Theory, 125–134 (1975)
18. M. Ban, K. Kurokawa, R. Momose, O. Hirota, Optimum measurements for discrimination among symmetric quantum states and parameter estimation. Int. J. Theor. Phys. **36**, 1269 (1997)
19. R.M. Van Slyke, R.J.-B. Wets, A duality theory for abstract mathematical programs with applications to optimal control theory. J. Math. Anal. Appl. **22**, 679–706 (1968)
20. H. Imai, M. Hachimori, M. Hamada, H. Kobayashi, K. Matsumoto, "Optimization in quantum computation and information," *Proc. 2nd Japanese-Hungarian Symposium on Discrete Mathematics and Its Applications*, Budapest, Hungary (2001)
21. A. Ben-Tal, A. Nemirovski, *Lectures on Modern Convex Optimization* (SIAM/MPS, Philadelphia, 2001)
22. M. Hayashi, *Minimization of deviation under quantum local unbiased measurements, master's thesis* (Graduate School of Science, Kyoto University, Japan, Department of Mathematics, 1996)
23. M. Hayashi, A linear programming approach to attainable cramer-rao type bound and randomness conditions, Kyoto-Math 97–08; quant-ph/9704044 (1997)
24. M. Hayashi, A linear programming approach to attainable Cramer–Rao type bound, in *Quantum Communication, Computing, and Measurement*, ed. by O. Hirota, A.S. Holevo, C.M. Caves, (Plenum, New York, 1997), pp. 99–108. (Also appeared as Chap. 12 of *Asymptotic Theory of Quantum Statistical Inference,* M. Hayashi eds.)
25. E.M. Rains, A semidefinite program for distillable entanglement. IEEE Trans. Inf. Theory **47**, 2921–2933 (2001)
26. S. Arimoto, An algorithm for computing the capacity of arbitrary discrete memoryless channels. IEEE Trans. Inf. Theory **18**, 14–20 (1972)
27. R. Blahut, Computation of channel capacity and rate-distortion functions. IEEE Trans. Inf. Theory **18**, 460–473 (1972)
28. H. Nagaoka, Algorithms of Arimoto-Blahut type for computing quantum channel capacity, in *Proceedings of 1998 IEEE International Symposium on Information Theory*, 354 (1998)
29. H. Nagaoka, S. Osawa, Theoretical basis and applications of the quantum Arimoto-Blahut algorithms, in *Proceedings of 2nd Quantum Information Technology Symposium (QIT2)*, (1999), pp. 107–112
30. P.W. Shor, Capacities of quantum channels and how to find them. Math. Programm. **97**, 311–335 (2003)
31. H. Chernoff, A measure of asymptotic efficiency for tests of a hypothesis based on the sum of observations. Ann. Math. Stat. **23**, 493–507 (1952)
32. F. Hiai, D. Petz, The proper formula for relative entropy and its asymptotics in quantum probability. Comm. Math. Phys. **143**, 99–114 (1991)
33. H. Nagaoka, Information spectrum theory in quantum hypothesis testing, in *Proceedings of 22th Symposium on Information Theory and Its Applications (SITA)*, (1999), pp. 245–247 (in Japanese)
34. H. Nagaoka, M. Hayashi, An information-spectrum approach to classical and quantum hypothesis testing. IEEE Trans. Inf. Theory **53**, 534–549 (2007)
35. H. Nagaoka, Limit theorems in quantum information theory. Suurikagaku **456**, 47–55 (2001). (in Japanese)
36. M. Ohya, D. Petz, *Quantum Entropy and Its Use* (Springer, Berlin Heidelberg New York, 1993)

37. H. Nagaoka, Private communication to A. Fujiwara (1991)
38. A. Fujiwara, private communication to H. Nagaoka (1996)
39. M. Mosonyi, T. Ogawa, Quantum hypothesis testing and the operational interpretation of the quantum Renyi relative entropies. Comm. Math. Phys. **334**(3), 1617–1648 (2015)
40. M. Hayashi, Optimal sequence of POVMs in the sense of Stein's lemma in quantum hypothesis. J. Phys. A Math. Gen. **35**, 10759–10773 (2002)
41. D. Petz, Quasi-entropies for finite quantum systems. Rep. Math. Phys. **23**, 57–65 (1986)
42. J. Walgate, A.J. Short, L. Hardy, V. Vedral, Local distinguishability of multipartite orthogonal quantum states. Phys. Rev. Lett. **85**, 4972 (2000)
43. S. Virmani, M. Sacchi, M.B. Plenio, D. Markham, Optimal local discrimination of two multipartite pure states. Phys. Lett. A **288**, 62 (2001)
44. Y.-X. Chen, D. Yang, Distillable entanglement of multiple copies of Bell states. Phys. Rev. A **66**, 014303 (2002)
45. A. Chefles, Condition for unambiguous state discrimination using local operations and classical communication. Phys. Rev. A **69**, 050307(R) (2004)
46. S. Virmani, M.B. Plenio, Construction of extremal local positive-operator-valued measures under symmetry. Phys. Rev. A **67**, 062308 (2003)
47. H. Fan, Distinguishability and indistinguishability by local operations and classical communication. Phys. Rev. Lett. **92**, 177905 (2004)
48. S. Ghosh, G. Kar, A. Roy, D. Sarkar, Distinguishability of maximally entangled states. Phys. Rev. A **70**, 022304 (2004)
49. M. Owari, M. Hayashi, Local copying and local discrimination as a study for non-locality. Phys. Rev. A **74**, 032108 (2006); *Phys. Rev. A*77, 039901(E) (2008)
50. M. Hayashi, D. Markham, M. Murao, M. Owari, S. Virmani, Bounds on multipartite entangled orthogonal state discrimination using local operations and classical communication. Phys. Rev. Lett. **96**, 040501 (2006)
51. M. Hayashi, Asymptotics of quantum relative entropy from a representation theoretical viewpoint. J. Phys. A Math. Gen. **34**, 3413–3419 (2001)
52. T. Ogawa, M. Hayashi, On error exponents in quantum hypothesis testing. IEEE Trans. Inf. Theory **50**, 1368–1372 (2004); quant-ph/0206151 (2002)
53. N. Ozawa, Private communication to T. Ogawa (2010)
54. H. Nagaoka, Strong converse theorems in quantum information theory, in *Proceedings of ERATO Conference on Quantum Information Science (EQIS) 2001*, 33 (2001). (also appeared as Chap. 3 of *Asymptotic Theory of Quantum Statistical Inference*, M. Hayashi eds.)
55. M. Hayashi, Quantum hypothesis testing for the general quantum hypotheses, in *Proceedings of 24th Symposium on Information Theory and Its Applications (SITA)*, (2001), pp. 591–594
56. M. Hayashi, K. Matsumoto, Y. Tsuda, A study of LOCC-detection of a maximally entangled state using hypothesis testing. J. Phys. A: Math. and Gen. **39**, 14427–14446 (2006)
57. M. Hayashi, B.-S. Shi, A. Tomita, K. Matsumoto, Y. Tsuda, Y.-K. Jiang, Hypothesis testing for an entangled state produced by spontaneous parametric down conversion. Phys. Rev. A **74**, 062321 (2006)
58. M. Hayashi, Group theoretical study of LOCC-detection of maximally entangled state using hypothesis testing. New J. of Phys. **11**, 043028 (2009)

Chapter 4
Classical-Quantum Channel Coding (Message Transmission)

Abstract Communication systems such as the Internet have become part of our daily lives. In any data-transmission system, data are always exposed to noise, and therefore it might be expected that information will be transmitted incorrectly. In practice, however, such problems can be avoided entirely. How is this possible? For explaining this, let us say that we send some information that is either 0 or 1. Now, let us say that the sender and receiver agree that the former will send "000" instead of "0" and "111" instead of "1." If the receiver receives "010" or "100," he or she can deduce that the sender in fact sent 0. On the other hand, if the receiver receives a "110" or "101," he or she can deduce that a 1 was sent. Therefore, we can reduce the chance of error by introducing redundancies into the transmission. However, in order to further reduce the chance of an error in this method, it is necessary to indefinitely increase the redundancy. Therefore, it had been commonly believed that in order to reduce the error probability, one had to increase the redundancy indefinitely. However, in 1948, Shannon (Bell Syst Tech J 27:623–656, 1948) showed that by using a certain type of encoding scheme, it is possible to reduce the error probability indefinitely without increasing the redundancy beyond a fixed rate. This was a very surprising result since it was contrary to naive expectations at that time. The distinctive part of Shannon's method was to treat communication in the symbolic form of 0s and 1s and then to approach the problem of noise using encoding. In practical communication systems such as optical fibers and electrical wires, codes such as 0 and 1 are sent by transforming them into a physical medium. In particular, in order to achieve the theoretical optimal communication speed, we have to treat the physical medium of the communication as a microscopic object, i.e., quantum-mechanical object. In this quantum-mechanical scenario, it is most effective to treat the encoding process not as a transformation of the classical bits, e.g., 0s, 1s, and so on, but as a transformation of the message into a quantum state. Furthermore, the measurement and decoding process can be thought of as a single step wherein the outcome of the quantum-mechanical measurement directly becomes the recovered message.

M. Hayashi, *Quantum Information Theory*, Graduate Texts in Physics,
DOI 10.1007/978-3-662-49725-8_4

4.1 Formulation of the Channel Coding Process in Quantum Systems

There are two main processes involved in the transmission of classical information through a quantum channel. The first is the conversion of the classical message into a quantum state, which is called **encoding**. The second is the decoding of the message via a quantum measurement on the output system. For a reliable and economical communication, we should optimize these processes. However, it is impossible to reduce the error probability below a certain level in the single use of a quantum channel even with the optimal encoding and decoding. This is similar to a single use of a classical channel with a nonnegligible bit-flip probability. However, when we use a given channel repeatedly, it is possible in theory to reduce the error probability to almost 0 by encoding and decoding. In this case, this reduction requires that the transmission rate from the transmission bit size to the original message bit size should be less than a fixed rate. This fixed rate is the bound of the transmission rate of a reliable communication and is called the **capacity**. This argument has been mathematically proved by Shannon [1] and is called the **channel coding theorem**. Hence, it is possible to reduce the error probability without reducing the transmission rate; however, complex encoding and decoding processes are required. This implies that it is possible to reduce the error probability to almost 0 while keeping a fixed communication speed if we group an n-bit transmission and then perform the encoding and decoding on this group. More precisely, the logarithm of the decoding error can then be decreased in proportion to the number n of transmissions, and the number n can be considered as the level of complexity required by the encoding and decoding processes. These facts are known in the quantum case as well as in the classical case.

In this chapter, we first give a mathematical formulation for the single use of a quantum channel. Regarding n uses of the quantum channel as a single quantum channel, we treat the asymptotic theory in which the number n of the uses of the given channel is large.

In the transmission of classical information via a quantum channel, we may denote the channel as a map from the alphabet (set of letters) $\mathcal{X}$ to the set $\mathcal{S}(\mathcal{H})$ of quantum states on the output system $\mathcal{H}$, i.e., a **classical-quantum channel (c-q channel)** $W : \mathcal{X} \to \mathcal{S}(\mathcal{H})$.[1] For mathematical simplicity, we assume that the linear span of all of supports of W_x equals the whole Hilbert space $\mathcal{H}$. The relevance of this formulation may be verified as follows. Let us consider the state transmission channel from the input system to the output system described by the map $\Gamma : \mathcal{S}(\mathcal{H}') \to \mathcal{S}(\mathcal{H})$, where $\mathcal{H}'$ denotes the finite-dimensional Hilbert space of the input system. When the states to be produced in the input system are given by the set $\{\rho_x\}_{x\in\mathcal{X}}$, the above map W is given by $W_x = \Gamma(\rho_x)$. That is, sending classical information via the above channel reduces to the same problem as that with the c-q channel W.

[1] As discussed later, these types of channels are called c-q channels to distinguish them from channels with quantum inputs and outputs. Here, $\mathcal{X}$ is allowed to contain infinite elements with continuous cardinality.

When all the densities W_x are simultaneously diagonalizable, the problem is reduced to a channel given by a stochastic transition matrix. As in hypothesis testing, we may call such cases "classical." Then, Theorem 4.1 (to be discussed later) also gives the capacity for the classical channels given by a stochastic transition matrix.

4.1.1 Transmission Information in C-Q Channels and Its Properties

As in the classical case (2.34), the **transmission information** $I(p, W)$ and the average state W_p for the c-q channel W are defined as[2]

$$I(p, W) \stackrel{\text{def}}{=} \sum_{x\in\mathcal{X}} p(x)D(W_x\|W_p) = H(W_p) - \sum_{x\in\mathcal{X}} p(x)H(W_x), \tag{4.1}$$

$$W_p \stackrel{\text{def}}{=} \sum_{x\in\mathcal{X}} p(x)W_x . \tag{4.2}$$

The transmission information $I(p, W)$ satisfies the following two properties:

① (**Concavity**) Any two distributions p^1 and p^2 satisfy

$$I(\lambda p^1 + (1-\lambda)p^2, W) \geq \lambda I(p^1, W) + (1-\lambda)I(p^2, W). \tag{4.3}$$

See Exercise 5.27.

② (**Subadditivity**) Given two c-q channels W^A from $\mathcal{X}_A$ to $\mathcal{H}_A$ and W^B from $\mathcal{X}_B$ to $\mathcal{H}_B$, we can naturally define the c-q channel $W^A \otimes W^B$ from $\mathcal{X}_A \times \mathcal{X}_B$ to $\mathcal{H}_A \otimes \mathcal{H}_B$ as

$$\left(W^A \otimes W^B\right)_{x_A, x_B} \stackrel{\text{def}}{=} W^A_{x_A} \otimes W^B_{x_B}. \tag{4.4}$$

Let p_A, p_B be the marginal distributions in $\mathcal{X}_A, \mathcal{X}_B$ for a probability distribution p in $\mathcal{X}_A \times \mathcal{X}_B$, respectively. Then, we have the **subadditivity** for

$$I(p, W^A \otimes W^B) \leq I(p_A, W^A) + I(p_B, W^B) . \tag{4.5}$$

This inequality can be shown as follows (Exercise 4.2):

$$\begin{aligned} & I(p_A, W^A) + I(p_B, W^B) - I(p, W^A \otimes W^B) \\ =& D\left(\left(W^A \otimes W^B\right)_p \| W^A_{p_A} \otimes W^B_{p_B}\right) \geq 0 . \end{aligned} \tag{4.6}$$

[2]In many papers, the quantity $I(p, W)$ is called the quantum mutual information. In this text, it will be called the transmission information of the c-q channel, for reasons given in Sect. 5.4. Occasionally we will denote this as $I(p_x, W_x)$.

From this property (4.5), we can show that

$$\max_p I(p, W^A \otimes W^B) = \max_{p_A} I(p_A, W^A) + \max_{p_B} I(p_B, W^B),$$

which is closely connected to the additivity discussed later. Another property of the transmission information is the inequality

$$I(p, W) \stackrel{\text{def}}{=} \sum_{x \in \mathcal{X}} p(x) D(W_x \| W_p) \leq \sum_{x \in \mathcal{X}} p(x) D(W_x \| \sigma), \ \forall \sigma \in \mathcal{S}(\mathcal{H}). \tag{4.7}$$

This inequality can be verified by noting that the LHS minus the RHS equals $D(W_p \| \sigma)$.

Exercises

4.1 Show that $C_c(W) = h\left(\frac{1+|\langle v|u\rangle|}{2}\right)$ if $\{W_x\}$ is composed of the two pure states $|u\rangle\langle u|$ and $|v\rangle\langle v|$.

4.2 Show (4.6).

4.1.2 C-Q Channel Coding Theorem

Next, we consider the problem of sending a classical message using a c-q channel $W : \mathcal{X} \to \mathcal{S}(\mathcal{H})$. For this purpose, we must mathematically define a **code**, which is the combination of an **encoder** and a **decoder** as Fig. 4.1. These are given by the triplet (N, φ, Y). The number N is a natural number corresponding to the **size** of the encoder. φ is a map, $\varphi : \{1, \ldots, N\} \to \mathcal{X}$, corresponding to the encoder. The decoder is a quantum measurement taking values in the probability space $\{1, \ldots, N\}$. Mathematically, it is given by the set of N positive semi-definite Hermitian matrices $Y = \{Y_i\}_{i=1}^N$ with $\sum_i Y_i \leq I$. In this case, $I - \sum_i Y_i$ corresponds to the undecodable decision.

For an arbitrary code $\Phi = (N, \varphi, Y)$, we define the size $|\Phi|$ and the **average error probability** $\varepsilon[\Phi]$ as

$$|\Phi| \stackrel{\text{def}}{=} N, \quad \varepsilon[\Phi] \stackrel{\text{def}}{=} \frac{1}{N} \sum_{i=1}^{N} (1 - \operatorname{Tr}[W_{\varphi(i)} Y_i]). \tag{4.8}$$

If we need to identify the c-q channel W to be discussed, we denote the average error probability by $\varepsilon_W[\Phi]$. We then consider the encoding and decoding

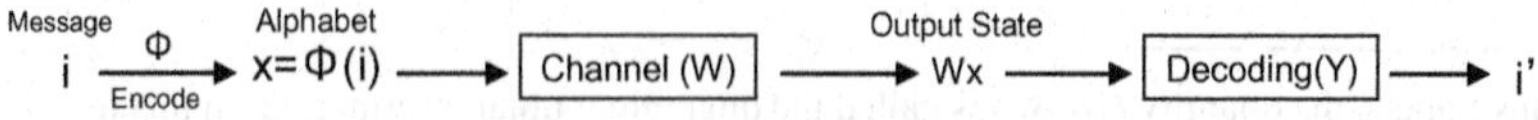

Message i —Φ (Encode)→ Alphabet x=Φ(i) → Channel (W) → Output State Wx → Decoding(Y) → i'

Fig. 4.1 Encoding and decoding

for n communications grouped into one. For simplicity, let us assume that each communication is independent and identical. That is, we discuss the case where the Hilbert space of the output system is $\mathcal{H}^{\otimes n}$, and the c-q channel is given by the map $W^{(n)} : x^n \stackrel{\rm def}{=} (x_1, \ldots, x_n) \mapsto W^{(n)}_{x^n} \stackrel{\rm def}{=} W_{x_1} \otimes \cdots \otimes W_{x_n}$ from the alphabet $\mathcal{X}^n$ to $\mathcal{S}(\mathcal{H}^{\otimes n})$. Such a channel is called **stationary memoryless**. An encoder of size N_n is given by the map $\varphi^{(n)}$ from $\{1, \ldots, N_n\}$ to $\mathcal{X}^n$, and it is written as $\varphi^{(n)}(i) = (\varphi^{(n)}_1(i), \ldots, \varphi^{(n)}_n(i))$. The decoder is also given by the POVM $Y^{(n)}$ on $\mathcal{H}^{\otimes n}$. Let us see how much information can be sent per transmission if the error probability asymptotically approaches 0. For this purpose, we look at the limit of the transmission rate $R = \underline{\lim}_{n\to\infty} \frac{1}{n} \log |\Phi^{(n)}|$ ($|\Phi^{(n)}| \cong e^{nR}$) for the sequence of reliable codes $\{\Phi^{(n)} = (N_n, \varphi^{(n)}, Y^{(n)})\}$ and discuss its bound, i.e., the **c-q channel capacity** $C_c(W)$[3]:

$$C_c(W) \stackrel{\rm def}{=} \sup_{\{\Phi^{(n)}\}} \left\{ \underline{\lim} \frac{1}{n} \log |\Phi^{(n)}| \,\middle|\, \lim_{n\to\infty} \varepsilon[\Phi^{(n)}] = 0 \right\}, \tag{4.9}$$

where $\Phi^{(n)}$ denotes a code for the quantum channel $W^{(n)}$. We may also define the **strong converse c-q channel capacity** as the dual capacity

$$C_c^{\dagger}(W) \stackrel{\rm def}{=} \sup_{\{\Phi^{(n)}\}} \left\{ \underline{\lim} \frac{1}{n} \log |\Phi^{(n)}| \,\middle|\, \underline{\lim}\, \varepsilon[\Phi^{(n)}] < 1 \right\}, \tag{4.10}$$

which clearly satisfies the inequality $C_c(W) \le C_c^{\dagger}(W)$. We then have the following theorem.

Theorem 4.1 ([2–6]) *Let $\mathcal{P}_{\rm f}(\mathcal{X})$ be the set of probability distributions with a finite support in $\mathcal{X}$. Then,*

$$C_c^{\dagger}(W) = C_c(W) = \sup_{p\in\mathcal{P}_{\rm f}(\mathcal{X})} I(p, W) = \min_{\sigma\in\mathcal{S}(\mathcal{H})} \sup_{x\in\mathcal{X}} D(W_x\|\sigma) \tag{4.11}$$

holds.

Thus, this theorem connects the c-q channel capacity $C_c(W)$ to the transmission information $I(p, W)$. Here, note that the former is operationally defined, while the latter is formally defined. The **additivity** of the c-q channel capacity

$$C_c(W^A) + C_c(W^B) = C_c(W^A \otimes W^B) \tag{4.12}$$

also follows from inequality (4.5).

For a proof of Theorem 4.1, we introduce two quantities:

$$I_{1-s}(p, W) \stackrel{\rm def}{=} -\frac{1}{s} \log \sum_x p(x) \operatorname{Tr} W_x^{1-s} W_p^s \tag{4.13}$$

[3]The subscript c of C_c indicates the sending of "classical" information.

$$I^{\downarrow}_{1-s}(p, W) \stackrel{\text{def}}{=} -\frac{(1-s)}{s} \log \operatorname{Tr} \left(\sum_x p(x) W_x^{1-s} \right)^{\frac{1}{1-s}} \tag{4.14}$$

$$C^{\downarrow}_{1-s}(W) \stackrel{\text{def}}{=} \sup_{p \in \mathcal{P}_f(\mathcal{X})} I^{\downarrow}_{1-s}(p, W), \tag{4.15}$$

where $I_1(p, W)$ and $I^{\downarrow}_1(p, W)$ are defined to be $I(p, W)$. These quantities satisfy[Exe. 4.4,4.23]:

$$\lim_{s \to 0} I_{1-s}(p, W) = I(p, W) \tag{4.16}$$

$$\lim_{s \to 0} C^{\downarrow}_{1-s}(W) = \sup_{p \in \mathcal{P}_f(\mathcal{X})} I(p, W) = \min_{\sigma \in \mathcal{S}(\mathcal{H})} \sup_{x \in \mathcal{X}} D(W_x \| \sigma). \tag{4.17}$$

Then, Theorem 4.1 may be proved using these properties and the two lemmas given below in a similar way to that of hypothesis testing.

Lemma 4.1 (Direct Part [4, 5]) *For an arbitrary real number $R > 0$ and an arbitrary distribution $p \in \mathcal{P}_f(\mathcal{X})$, there exists a code $\Phi^{(n)}$ for the stationary memoryless quantum channel $W^{(n)}$ such that*

$$\varepsilon[\Phi^{(n)}] \le 4 e^{n \min_{s \in [0,1]} s(R - I_{1-s}(p, W))}, \quad |\Phi^{(n)}| = e^{nR}. \tag{4.18}$$

Lemma 4.2 (Converse Part [6]) *When a code $\Phi^{(n)}$ for the stationary memoryless quantum channel satisfies $|\Phi^{(n)}| = e^{nR}$, the relation*

$$1 - \varepsilon[\Phi^{(n)}] \le e^{n \max_{s \in (-\infty, 0]} \frac{s(R - C^{\downarrow}_{1-s}(W))}{1-s}} \tag{4.19}$$

holds.

Proof of Theorem 4.1 Thanks to Lemma 4.1, when $R < I(p, W) \le \sup_{p' \in \mathcal{P}_f(\mathcal{X})} I(p', W)$, the relation (4.16) guarantees that the exponent $\min_{s \in [0,1]} s(R - I_{1-s}(p, W))$ is strictly negative, which implies the decoding error probability $\varepsilon[\Phi^{(n)}]$ goes to zero exponentially. On the other hand, thanks to Lemma 4.2, when $R > \sup_{p \in \mathcal{P}_f(\mathcal{X})} I(p, W)$, the relation (4.17) guarantees that the exponent $\max_{s \in (-\infty, 0]} \frac{s(R - C^{\downarrow}_{1-s}(W))}{1-s}$) is strictly negative, which implies the quantity $1 - \varepsilon[\Phi^{(n)}]$ goes to zero exponentially. These two facts and (4.17) show (4.11). ∎

Indeed, this theorem can be generalized to the case when a sender sends the message to M receivers. This case is formulated by M-output channel $W^1, \ldots, W^M$, with M output systems $\mathcal{H}_1, \ldots, \mathcal{H}_M$ and a single input system, where $W^i = (W^i_x)$. In this case, the encoder is defined in the same way, i.e., $\varphi : \{1, \ldots, N\} \to \mathcal{X}$. However, the decoder is defined by M POVMs $Y^1, \ldots, Y^M$. That is, the code is described by $\Phi \stackrel{\text{def}}{=} (N, \varphi, Y^1, \ldots, Y^M)$. In this case, the error of Φ is given as the worst decoding error probability $\varepsilon[\Phi] \stackrel{\text{def}}{=} \max_{1 \le i \le M} \frac{1}{N} \sum_{j=1}^N (1 - \operatorname{Tr} W^i_{\varphi(i)} Y^i_j)$. Further, in the same way as (4.9), the capacity $C_c(W^1, \ldots, W^M)$ is defined as

$$C_c(W^1, \ldots, W^M) \stackrel{\text{def}}{=} \sup_{\{\Phi^{(n)}\}} \left\{ \underline{\lim} \frac{1}{n} \log |\Phi^{(n)}| \,\middle|\, \lim_{n\to\infty} \varepsilon[\Phi^{(n)}] = 0 \right\}. \tag{4.20}$$

Then, we have the following proposition.

Proposition 4.1

$$C_c(W^1, \ldots, W^M) = \sup_p \min_{1\le i\le M} I(p, W^i). \tag{4.21}$$

For a proof, see Exercises 4.13 and 4.32.

Exercises

4.3 Define $J_{1-s}(p, \sigma, W) := -\frac{1}{s} \log \sum_x p(x) \operatorname{Tr} W_x^{1-s} \sigma^s$. Show that

$$I^{\downarrow}_{1-s}(p, W) = \min_\sigma J_{1-s}(p, \sigma, W) \tag{4.22}$$

and that the above minimum is attained only

$$\sigma_{1-s|p} \stackrel{\text{def}}{=} \left(\sum_x p(x) W_x^{1-s} \right)^{\frac{1}{1-s}} \Big/ \operatorname{Tr} \left[\left(\sum_{x'} p(x') W_{x'}^{1-s} \right)^{\frac{1}{1-s}} \right]. \tag{4.23}$$

Hint: Use the matrix Hölder inequality (A.26) and the reverse matrix Hölder inequality (A.28).

4.4 Show (4.16) and

$$\lim_{s\to 0} I^{\downarrow}_{1-s}(p, W) = I(p, W). \tag{4.24}$$

4.5 Show that

$$I^{\downarrow}_{1-s}(p, W) \le I_{1-s}(p, W). \tag{4.25}$$

4.6 Show the following inequality for $s \in [-1, 1] \setminus \{0\}$ as an inequality opposite to (4.25) when all of W_x are commutative with each other [7, (16)].

$$I^{\downarrow}_{\frac{1}{1+s}}(p, W) \ge I_{1-s}(p, W). \tag{4.26}$$

Inequality (4.26) can be shown by a similar way as (5.145) in the general case.

4.2 Coding Protocols with Adaptive Decoding and Feedback

In the previous section, there was no restriction on the measurements for decoding. Now, we shall restrict these measurements to adaptive decoding, and they have the following form:

$$\boldsymbol{M}^n = \{M_{1,y_1} \otimes \cdots \otimes M_{n,y_n}^{y_1,\ldots,y_{n-1}}\}_{(y_1,\ldots,y_n)\in\mathcal{Y}_1\times\cdots\times\mathcal{Y}_n}.$$

Therefore, the decoder may be written as the POVM $\boldsymbol{M}^n$ and the mapping $\tau^{(n)}$: $\mathcal{Y}_1 \times \cdots \times \mathcal{Y}_n \to \{1,\ldots,N_n\}$.

We also allow feedback during the encoding process. That is, the receiver is allowed to send his or her measurement outcomes back to the sender, who then performs the encoding based on these outcomes. In the previous section, we considered the encoder to be a map $\varphi^{(n)}(i) = (\varphi_1^{(n)}(i),\ldots,\varphi_n^{(n)}(i))$ from $\{1,\ldots,N_n\}$ to $\mathcal{X}^n$. If we allow feedback, the kth encoding element will be given by a map $\tilde{\varphi}_k^{(n)}$ from $\{1,\ldots,N_n\} \times \mathcal{Y}_1 \times \cdots \times \mathcal{Y}_{k-1}$ to $\mathcal{X}$. Therefore, in this case, we denote the encoder as $\tilde{\varphi}^{(n)} \stackrel{\text{def}}{=} (\tilde{\varphi}_1^{(n)},\ldots,\tilde{\varphi}_n^{(n)})$.

Henceforth, we call $\tilde{\Phi}^{(n)} = (N_n, \tilde{\varphi}^{(n)}, \boldsymbol{M}^n, \tau^{(n)})$ the code with adaptive decoding and feedback and denote its size N_n by $|\tilde{\Phi}^{(n)}|$. The average error probability of the code is denoted by $\varepsilon[\tilde{\Phi}^{(n)}]$. If the code has no feedback, it belongs to the restricted subclass of codes given in the previous section. However, if it has feedback, it does not belong to this subclass. That is, the class of codes given in this section is not a subclass of codes given in the previous section. This class of codes is the subject of the following theorem.

Theorem 4.2 (Fujiwara and Nagaoka [8]) *Define the c-q channel capacity with adaptive decoding and feedback* $\tilde{C}_c(W)$ *as*

$$\tilde{C}_c(W) \stackrel{\text{def}}{=} \sup_{\{\tilde{\Phi}^{(n)}\}} \left\{ \underline{\lim} \frac{1}{n} \log |\tilde{\Phi}^{(n)}| \,\middle|\, \lim_{n\to\infty} \varepsilon[\tilde{\Phi}^{(n)}] = 0 \right\}, \tag{4.27}$$

where $\tilde{\Phi}^{(n)}$ *is a code with adaptive decoding and feedback. Then,*

$$\tilde{C}_c(W) = \sup_{\boldsymbol{M}} \sup_{p\in\mathcal{P}_f(\mathcal{X})} I(\boldsymbol{M}, p, W), \tag{4.28}$$

where $I(\boldsymbol{M}, p, W) \stackrel{\text{def}}{=} \sum_{x\in\mathcal{X}} p(x) D(\mathrm{P}_{W_x}^{\boldsymbol{M}} \| \mathrm{P}_{W_p}^{\boldsymbol{M}})$.

When the maximum $\max_{\boldsymbol{M}} \sup_{p\in\mathcal{P}_f(\mathcal{X})} I(\boldsymbol{M}, p, W)$ exists, the capacity $\tilde{C}_c(W)$ can be attained by performing the optimal measurement $\boldsymbol{M}_M \stackrel{\text{def}}{=} \operatorname{argmax}_{\boldsymbol{M}} \sup_{p\in\mathcal{P}_f(\mathcal{X})} I(\boldsymbol{M}, p, W)$ on each output system. Thus, there is no improvement if we use adaptive decoding and encoding with feedback.

Proof For an arbitrary positive real number $\epsilon > 0$, we choose a POVM $\boldsymbol{M}$ such that $\sup_p I(\boldsymbol{M}, p, W) \geq \sup_{\boldsymbol{M}'} \sup_p I(\boldsymbol{M}', p, W) - \epsilon$. Then, the relation between the input letter and the output measurement data is described by the stochastic transition matrix $x \mapsto \mathrm{P}^{\boldsymbol{M}_M}_{W_x}$. Applying Theorem 4.1 to the classical channel $x \mapsto \mathrm{P}^{\boldsymbol{M}_M}_{W_x}$, we see that a code attaining $\sup_p I(\boldsymbol{M}_M, p, W)$ exists. That is, $\tilde{C}_c(W) \geq \sup_{\boldsymbol{M}'} \sup_p I(\boldsymbol{M}', p, W) - \epsilon$. Since $\epsilon > 0$ is arbitrary, we have $\tilde{C}_c(W) \geq \sup_{\boldsymbol{M}'} \sup_p I(\boldsymbol{M}', p, W)$.

Next, we show that there is no code with a rate exceeding the RHS of (4.28). Consider a sequence of codes $\{\tilde{\Phi}^{(n)} = (N_n, \tilde{\varphi}^{(n)}, \boldsymbol{M}^n, \tau^{(n)})\}$ satisfying $\lim_{n\to\infty} \varepsilon[\tilde{\Phi}^{(n)}] = 0$. Let X be a uniformly distributed random variable taking values in the input messages $\{1, \ldots, N_n\}$ and $Y^k = (Y_1, \ldots, Y_k)$ be the random variable corresponding to the outcome of the measurement $\boldsymbol{M}^n$. Since $\varepsilon[\tilde{\Phi}^{(n)}] = \mathrm{P}\{X \neq \tau^{(n)}(Y^n)\}$, Fano's inequality (2.35) yields

$$\begin{aligned} \log 2 + \varepsilon[\tilde{\Phi}^{(n)}] \log N_n &\geq H(X) - I(X : \tau^{(n)}(Y^n)) \\ &= \log N_n - I(X : \tau^{(n)}(Y^n)) \end{aligned} \tag{4.29}$$

because $H(X) = \log N_n$.

Now, to evaluate $I(X : \tau^{(n)}(Y^n))$, we define

$$\mathrm{P}_{Y_k|X,Y^{k-1}}(y_k|x, y^{k-1}) \overset{\text{def}}{=} \mathrm{P}^{\boldsymbol{M}_n^{y_1,\ldots,y_{k-1}}}_{W_{\tilde{\varphi}_k^{(n)}(x,y^{k-1})}}(y_k) = \operatorname{Tr} W_{\tilde{\varphi}_k^{(n)}(x,y^{k-1})} M_{n,y_k}^{y_1,\ldots,y_{k-1}}.$$

From the monotonicity of the classical relative entropy and the chain rule (2.32) for mutual information,

$$\begin{aligned} I(X : \tau^{(n)}(Y^n)) &\leq I(X : Y^n) = \sum_{k=1}^{n} I(X : Y_k|Y^{k-1}) \\ &= \sum_{k=1}^{n} \sum_{y^{k-1}} \mathrm{P}_{Y^{k-1}}(y^{k-1}) \sum_{x=1}^{N_n} \mathrm{P}_{X|Y^{k-1}=y^{k-1}}(x) D\left(\mathrm{P}^{\boldsymbol{M}_n^{y_1,\ldots,y_{k-1}}}_{W_{\tilde{\varphi}_k^{(n)}(x,y^{k-1})}} \middle\| \mathrm{P}^{\boldsymbol{M}_n^{y_1,\ldots,y_{k-1}}}_{W_{\mathrm{P}_{X|Y^{k-1}=y^{k-1}}}} \right) \\ &= \sum_{k=1}^{n} \sum_{y^{k-1}} \mathrm{P}_{Y^{k-1}}(y^{k-1}) I(\boldsymbol{M}_n^{y_1,\ldots,y_{k-1}}, \mathrm{P}_{X|Y^{k-1}=y^{k-1}}, W) \\ &\leq n \sup_{\boldsymbol{M}} \sup_{p \in \mathcal{P}_f(\mathcal{X})} I(\boldsymbol{M}, p, W). \end{aligned} \tag{4.30}$$

(4.29) and (4.30) yield that

$$\log 2 + \varepsilon[\tilde{\Phi}^{(n)}] \log N_n \geq \log N_n - n \sup_{\boldsymbol{M}} \sup_{p \in \mathcal{P}_f(\mathcal{X})} I(\boldsymbol{M}, p, W), \tag{4.31}$$

which can be rewritten as

$$\frac{1}{n} \log N_n \leq \frac{(\log 2)/n + \sup_{\boldsymbol{M}} \sup_{p \in \mathcal{P}_f(\mathcal{X})} I(\boldsymbol{M}, p, W)}{1 - \varepsilon[\tilde{\Phi}^{(n)}]}. \tag{4.32}$$

Since $\varepsilon[\tilde{\Phi}^{(n)}] \to 0$,

$$\overline{\lim} \frac{1}{n} \log N_n \le \sup_{M} \sup_{p \in \mathcal{P}_f(\mathcal{X})} I(M, p, W),$$

completing the proof. ■

Therefore, if the decoder uses no correlation in the measuring apparatus, the c-q channel capacity is given by $\tilde{C}_c(W)$. Next, we consider the c-q channel capacity when correlations among n systems are allowed. In this assumption, we can regard the n uses of the channel W as a single channel $W^{(n)}$ and then reuse the arguments presented in this section. Therefore, the c-q channel capacity is given by $\frac{\tilde{C}_c(W^{(n)})}{n}$. Its limiting case is

$$\lim_{n\to\infty} \frac{\tilde{C}_c(W^{(n)})}{n} = C_c(W), \tag{4.33}$$

while $\tilde{C}_c(W) < C_c(W)$, except for special cases such as those given in Sect. 4.7. An interesting question is whether it is possible to experimentally realize a transmission rate exceeding $\tilde{C}_c(W)$. This is indeed possible, and a channel W and a measurement M have been experimentally constructed with $I(M, p, W^{(2)}) > 2\tilde{C}_c(W)$ by Fujiwara et al. [9].

Exercise

4.7 Show (4.33) using Fano's inequality (2.35) in a similar way to the proof of Theorem 4.2.

4.3 Channel Capacities Under Cost Constraint

Thus far, there have been no constraints on the encoding, and we have examined only the size of the code and error probabilities. However, it is not unusual to impose a constraint that the cost, e.g., the energy required for communication, should be less than some fixed value. In this situation, we define a cost function and demand that the cost for each code should be less than some fixed value. More precisely, a cost $c(x)$ is defined for each state W_x used in the communication. In the stationary memoryless case, the cost for the states $W_{\boldsymbol{x}}^{(n)}$ is given by $c^{(n)}(\boldsymbol{x}) \stackrel{\text{def}}{=} \sum_{i=1}^{n} c(x_i)$. The states $W_{\boldsymbol{x}}^{(n)}$ used for communication are then restricted to those that satisfy $\sum_{\boldsymbol{x}} c(x_i) \le Kn$. That is, any code $\Phi^{(n)} = (N_n, \varphi^{(n)}, Y^{(n)})$ must satisfy the restriction $\max_i c^{(n)}(\varphi^{(n)}(i)) \le Kn$. The following theorem can be proved in a similar way to Theorem 4.1.

Theorem 4.3 ([10, 11]) *Define the c-q channel capacities under the cost constraint*

$$C_{c|c\le K}(W) \stackrel{\text{def}}{=} \sup_{\{\Phi^{(n)}\}} \left\{ \underline{\lim} \frac{\log|\Phi^{(n)}|}{n} \,\middle|\, \max_i \frac{c^{(n)}(\varphi^{(n)}(i))}{n} \le K,\ \lim_{n\to\infty} \varepsilon[\Phi^{(n)}] = 0 \right\},$$

$$C^{\dagger}_{c|c\le K}(W) \stackrel{\text{def}}{=} \sup_{\{\Phi^{(n)}\}} \left\{ \underline{\lim} \frac{\log|\Phi^{(n)}|}{n} \,\middle|\, \max_i \frac{c^{(n)}(\varphi^{(n)}(i))}{n} \le K, \underline{\lim}\, \varepsilon[\Phi^{(n)}] < 1 \right\}.$$

Then,

$$\begin{aligned} C_{c|c\le K}(W) &= C^{\dagger}_{c|c\le K}(W) \\ &= \sup_{p\in\mathcal{P}_{c\le K}(\mathcal{X})} I(p, W) = \min_{\sigma\in\mathcal{S}(\mathcal{H})} \sup_{p\in\mathcal{P}_{c\le K}(\mathcal{X})} \sum_x p_x D(W_x\|\sigma), \end{aligned} \tag{4.34}$$

where $\mathcal{P}_{c\le K}(\mathcal{X}) \stackrel{\text{def}}{=} \{p\in\mathcal{P}_{\mathrm{f}}(\mathcal{X}) \,|\, \sum_x p(x)c(x) \le K\}$.

For a proof of Theorem 4.3, we introduce a quantity:

$$C^{\downarrow}_{1-s|c\le K}(W) \stackrel{\text{def}}{=} \sup_{p\in\mathcal{P}_{c\le K}} I^{\downarrow}_{1-s}(p, W). \tag{4.35}$$

These quantities satisfy[Exe. 4.27]

$$\lim_{s\to 0} C^{\downarrow}_{1-s|c\le K}(W) = \sup_{p\in\mathcal{P}_{c\le K}(\mathcal{X})} I(p, W) = \min_{\sigma\in\mathcal{S}(\mathcal{H})} \sup_{p\in\mathcal{P}_{c\le K}(\mathcal{X})} \sum_x p_x D(W_x\|\sigma). \tag{4.36}$$

Then, Theorem 4.3 can be obtained from the following two lemmas in a similar way to Theorem 4.1. These lemmas will be proved later in Sects. 4.5 and 4.6.

Lemma 4.3 (Direct Part) *For arbitrary real numbers* $R > 0$ *and* K *and an arbitrary distribution* $p \in \mathcal{P}_{c\le K}(\mathcal{X})$, *there exists a code* $\Phi^{(n)} = (N_n, \varphi^{(n)}, Y^{(n)})$ *for the stationary memoryless quantum channel* $W^{(n)}$ *such that*

$$\varepsilon[\Phi^{(n)}] \le \frac{4}{C_{n,K}^2} e^{n \min_{s\in[0,1]} s(R - I_{1-s}(p,W))}, \quad |\Phi^{(n)}| = e^{nR} \tag{4.37}$$

$$\max_i \frac{c^{(n)}(\varphi^{(n)}(i))}{n} \le K, \tag{4.38}$$

where $C_{n,K} \stackrel{\text{def}}{=} p^{(n)}\{c^{(n)}(\mathbf{x}) \le nK\}$.

Lemma 4.4 (Converse Part) *When a code* $\Phi^{(n)}$ *for the stationary memoryless quantum channel satisfies* $|\Phi^{(n)}| = e^{nR}$ *and* $\max_i \frac{c^{(n)}(\varphi^{(n)}(i))}{n} \le K$, *the relation*

$$1 - \varepsilon[\Phi^{(n)}] \le e^{n \max_{s\in(-\infty,0]} \frac{s(R - C^{\downarrow}_{1-s|c\le K}(W))}{1-s}} \tag{4.39}$$

holds.

Proof of Theorem 4.3 We choose $p \in \mathcal{P}_{c \le K}(\mathcal{X})$ such that $R < I(p, W) \le \sup_{p' \in \mathcal{P}_{c \le K}(\mathcal{X})} I(p', W)$. Then, the central limit theorem guarantees that the quantity $C_{n,K}$ goes to $1/2$. The relation (4.16) guarantees that the exponent $\min_{s \in [0,1]} s(R - I_{1-s}(p, W))$ is strictly negative. The decoding error probability $\varepsilon[\Phi^{(n)}]$ in Lemma 4.3 goes to zero exponentially. On the other hand, thanks to Lemma 4.4, when $R > \sup_{p \in \mathcal{P}_{c \le K}(\mathcal{X})} I(p, W)$, the relation (4.36) guarantees that the exponent $\max_{s \in (-\infty,0]} \frac{s(R - C^{\downarrow}_{1-s|c \le K}(W))}{1-s}$ is strictly negative, which implies the quantity $1 - \varepsilon[\Phi^{(n)}]$
goes to zero exponentially. These two facts show Theorem 4.3. ■

Exercise

4.8 Let the set $\{W_x\}$ consist entirely of pure states. Let the cost function c be given by $c(x) = \mathrm{Tr}\, W_x E$, where E is a positive semidefinite Hermitian matrix on $\mathcal{H}$. Show that $C_{c;c \le K}(W) \stackrel{\mathrm{def}}{=} C_{c|c \le K}(W) = H(\rho_{E,K})$, where $\rho_{E,K} \stackrel{\mathrm{def}}{=} e^{-\beta_K E} / \mathrm{Tr}\, e^{-\beta_K E}$ and β_K satisfies $\mathrm{Tr}(e^{-\beta_K E} / \mathrm{Tr}\, e^{-\beta_K E}) E = K$.

4.4 A Fundamental Lemma

In this section, we will prove the lemma required for the proof of Theorem 4.1.

Lemma 4.5 (Hayashi and Nagaoka [11]) *When two arbitrary Hermitian matrices S and T satisfy $I \ge S \ge 0$ and $T \ge 0$, the following inequality holds:*

$$I - \sqrt{S+T}^{-1} S \sqrt{S+T}^{-1} \le 2(I - S) + 4T, \tag{4.40}$$

where $\sqrt{S+T}^{-1}$ is the generalized inverse matrix of $\sqrt{S+T}$ given in Sect. 1.5.

Proof Let P be a projection to the range of $S + T$. Since P commutes with S and T, for proving (4.40), it is sufficient to show that

$$P\left[I - \sqrt{S+T}^{-1} S \sqrt{S+T}^{-1}\right] P \le P[2(I - S) + 4T]P,$$
$$P^{\perp}\left[I - \sqrt{S+T}^{-1} S \sqrt{S+T}^{-1}\right] P^{\perp} \le P^{\perp}[2(I - S) + 4T]P^{\perp},$$

where we defined $P^{\perp} = I - P$. The second inequality follows from $P^{\perp} S = P^{\perp} T = P^{\perp} \sqrt{S+T}^{-1} = 0$. Thus, for proving (4.40), it is sufficient to show that (4.40) holds when the range of $S + T$ is equal to $\mathcal{H}$.

Since $(A - B)^*(A - B) \ge 0$ for matrices, we obtain $A^*B + B^*A \le A^*A + B^*B$. Applying this inequality to the case of $A = \sqrt{T}$ and $B = \sqrt{T}(\sqrt{S+T}^{-1} - I)$, we obtain

$$T\left(\sqrt{S+T}^{-1} - I\right) + \left(\sqrt{S+T}^{-1} - I\right)T$$

$$\leq T + \left(\sqrt{S+T}^{-1} - I\right) T \left(\sqrt{S+T}^{-1} - I\right). \tag{4.41}$$

Furthermore,

$$\sqrt{S+T} \geq \sqrt{S} \geq S \tag{4.42}$$

since $f(x) = \sqrt{x}$ is a matrix monotone function[Exe. A.7] and $0 \leq S \leq I$. Finally,

$$\begin{aligned}
I - \sqrt{S+T}^{-1} S \sqrt{S+T}^{-1} &= \sqrt{S+T}^{-1} T \sqrt{S+T}^{-1} \\
=& T + T\left(\sqrt{S+T}^{-1} - I\right) + \left(\sqrt{S+T}^{-1} - I\right) T + \left(\sqrt{S+T}^{-1} - I\right) T \left(\sqrt{S+T}^{-1} - I\right) \\
\leq& 2T + 2\left(\sqrt{S+T}^{-1} - I\right) T \left(\sqrt{S+T}^{-1} - I\right) \\
\leq& 2T + 2\left(\sqrt{S+T}^{-1} - I\right) (S+T) \left(\sqrt{S+T}^{-1} - I\right) \\
=& 2T + 2\left(I + S + T - 2\sqrt{S+T}\right) \\
\leq& 2T + 2(I + S + T - 2S) = 2(I - S) + 4T,
\end{aligned}$$

where the first inequality follows from (4.41) and the third inequality follows from (4.42). Thus, we obtain the matrix inequality (4.40). ■

Exercise

4.9 Show the generalized version of inequality (4.40) under the same conditions as Lemma 4.5 [11]:

$$I - \sqrt{S+T}^{-1} S \sqrt{S+T}^{-1} \leq (1+c)(I-S) + (2 + c + c^{-1})\, T. \tag{4.43}$$

4.5 Proof of Direct Part of C-Q Channel Coding Theorem

The arguments used for hypothesis testing in Chap. 3 may be reused for the proof of the converse theorem (Lemma 4.1) using the following lemma.

Lemma 4.6 (Hayashi and Nagaoka [11]) *Given a c-q channel $x \in \mathcal{X} \mapsto W_x$, there exists a code Φ of size N such that*

$$\begin{aligned}
\varepsilon[\Phi] \leq& \sum_{x \in \mathcal{X}} p(x) \left(2 \operatorname{Tr} W_x \{W_x - 2N W_p \leq 0\} + 4N \operatorname{Tr} W_p \{W_x - 2N W_p > 0\}\right) \\
=& 2 \operatorname{Tr}(p \times W)\{(p \times W) \leq 2Np \otimes W_p\} \\
&+ 4N \operatorname{Tr} p \otimes W_p \{(p \times W) > 2Np \otimes W_p\}
\end{aligned} \tag{4.44}$$

where p is a probability distribution in $\mathcal{X}$, and the matrices $p \times W$ and $p \otimes \sigma$ on $\mathcal{H} \otimes \mathbb{C}^{|\operatorname{supp}(p)|}$ are defined as follows.

$$p \times W \stackrel{\text{def}}{=} \begin{pmatrix} p(x_1)W_{x_1} & & 0 \\ & \ddots & \\ 0 & & p(x_k)W_{x_k} \end{pmatrix}, \quad p \otimes \sigma \stackrel{\text{def}}{=} \begin{pmatrix} p(x_1)\sigma & & 0 \\ & \ddots & \\ 0 & & p(x_k)\sigma \end{pmatrix}. \tag{4.45}$$

The RHS of (4.45) is called dependence test (DT) bound because the projection $\{(p \times W) > 2Np \otimes W_p\}$ tests the correlated state $(p \times W)$ with comparing the independent case $p \otimes W_p$.

Before proceeding to the proof, we notice that $I(p, W), I_{1-s}(p, W)$, and $I^{\downarrow}_{1-s}(p, W)$ are written as

$$I(p, W) = D(p \times W \| p \otimes W_p) \tag{4.46}$$

$$I_{1-s}(p, W) = D_{1-s}(p \times W \| p \otimes W_p) \tag{4.47}$$

$$I^{\downarrow}_{1-s}(p, W) = \min_{\sigma \in \mathcal{S}(\mathcal{H})} D_{1-s}(p \times W \| p \otimes \sigma). \tag{4.48}$$

Proof of Lemma 4.1 For the simplicity, we use notations $S_1 := p \times W$ and $S_2 := p \otimes W_p$. Applying Lemma 4.6 to the pair of channels $W^{(n)}$ and the n-fold independent and identical distribution of $p \in \mathcal{P}_{\mathrm{f}}(\mathcal{X})$, we can take a code of size N_n satisfying

$$\begin{aligned}
&\varepsilon[\Phi^{(n)}] \\
\le & \sum_{x^n \in \mathcal{X}^n} p^n(x^n)\big(2 \operatorname{Tr} W^{(n)}_{x^n}\{W^{(n)}_{x^n} - 2N_n W^{(n)}_{p^n} \le 0\} \\
& \quad + 4N_n \operatorname{Tr} W^{(n)}_{p^n}\{W^{(n)}_{x^n} - 2N_n W^{(n)}_{p^n} > 0\}\big) \\
= & 2 \operatorname{Tr} S_1^{\otimes n}\{S_1^{\otimes n} - 2N_n S_2^{\otimes n} \le 0\} + 4N_n \operatorname{Tr} S_2^{\otimes n}\{S_1^{\otimes n} - 2N_n S_2^{\otimes n} > 0\} \\
\le & 2 \operatorname{Tr}(S_1^{\otimes n})^{1-s}(2N_n S_2^{\otimes n})^s \\
= & 2^{1+s} e^{ns(R - D_{1-s}(S_1 \| S_2))} = 2^{1+s} e^{ns(R - I_{1-s}(p, W))},
\end{aligned} \tag{4.49}$$

where (a) follows from Lemma 3.3 with $A = S_1^{\otimes n}$ and $B = 2N_n S_2^{\otimes n}$. This completes the proof of Lemma 4.1. ■

Proof of Lemma 4.6 We prove this lemma by employing the **random coding method** in which we randomly generate a code (N, φ, Y) of fixed size and prove that the expectation of the average error probability is less than ϵ. Based on the above strategy, we can show that there exists a code whose average error probability is less than ϵ. For this purpose, we consider N random variables $X \stackrel{\text{def}}{=} (X_1, \ldots, X_N)$ independently obeying a probability distribution p in $\mathcal{X}$, define the encoder $\varphi_X(i)$ by $\varphi_X(i) = X_i$, and denote the expectation by E_X.

For a given encoder φ of size N, a decoder $Y(\varphi)$ is defined

$$Y(\varphi)_i \stackrel{\text{def}}{=} \left(\sum_{j=1}^{N} \pi_j\right)^{-\frac{1}{2}} \pi_i \left(\sum_{j=1}^{N} \pi_j\right)^{-\frac{1}{2}}, \tag{4.50}$$

$$\pi_i \stackrel{\text{def}}{=} \left\{ W_{\varphi(i)} - 2N W_p > 0 \right\}. \tag{4.51}$$

Then, the average error probability of the code $\Phi(\varphi) \stackrel{\text{def}}{=} (N, \varphi, Y(\varphi))$ is

$$\begin{aligned}
&\varepsilon[\Phi(\varphi)] \\
=& \frac{1}{N} \sum_{i=1}^{N} \operatorname{Tr} W_{\varphi(i)} (I - Y(\varphi)_i) \le \frac{1}{N} \sum_{i=1}^{N} \operatorname{Tr} W_{\varphi(i)} \left(2(I - \pi_i) + 4 \sum_{j:i \neq j} \pi_j \right) \\
=& \frac{1}{N} \left(\sum_{i=1}^{N} \operatorname{Tr} 2 W_{\varphi(i)} (I - \pi_i) + \operatorname{Tr} \left(4 \sum_{j:i \neq j} W_{\varphi(j)} \right) \pi_i \right).
\end{aligned} \tag{4.52}$$

For evaluating the expectation $\mathrm{E}_X[\varepsilon[\Phi(\varphi_X)]]$, let us rewrite $\mathrm{E}_X[\operatorname{Tr} W_{\varphi_X(i)}(I - \pi_i)]$ and $\mathrm{E}_X[\operatorname{Tr} W_{\varphi_X(j)} \pi_i]$ as

$$\begin{aligned}
&\mathrm{E}_X \left[\operatorname{Tr} W_{\varphi_X(i)} (I - \pi_i) \right] = \sum_{x \in \mathcal{X}} p(x) \operatorname{Tr} W_x \left\{ W_x - 2N W_p \le 0 \right\}, \\
&\mathrm{E}_X \left[\operatorname{Tr} W_{\varphi_X(j)} \pi_i \right] = \sum_{x' \in \mathcal{X}} p(x') \sum_{x \in \mathcal{X}} p(x) \operatorname{Tr} W_{x'} \left\{ W_x - 2N W_p > 0 \right\} \\
=& \sum_{x \in \mathcal{X}} p(x) \operatorname{Tr} W_p \left\{ W_x - 2N W_p > 0 \right\}.
\end{aligned}$$

$\mathrm{E}_X\,[\varepsilon[\Phi(\varphi_X)]]$ then becomes

$$\begin{aligned}
\mathrm{E}_X\,[\varepsilon[\Phi(\varphi_X)]] \le& \mathrm{E}_X \left[\frac{1}{N} \sum_{i=1}^{N} \left(\operatorname{Tr} 2 W_{\varphi_X(i)} (I - \pi_i) + \operatorname{Tr} \left(4 \sum_{i \neq j} W_{\varphi_X(j)} \right) \pi_i \right) \right] \\
=& \frac{1}{N} \sum_{i=1}^{N} \left(2 \mathrm{E}_X \left[\operatorname{Tr} W_{\varphi_X(i)} (I - \pi_i) \right] + 4 \sum_{i \neq j} \mathrm{E}_X \left[\operatorname{Tr} W_{\varphi_X(j)} \pi_i \right] \right) \\
=& \sum_{x \in \mathcal{X}} p(x) \Big(2 \operatorname{Tr} W_x \left\{ W_x - 2N W_p \le 0 \right\} \\
& + 4(N-1) \operatorname{Tr} W_p \left\{ W_x - 2N W_p > 0 \right\} \Big).
\end{aligned} \tag{4.53}$$

Since the RHS of this inequality is less than the RHS of (4.44), we see that there exists a code $\Phi(\varphi)$ that satisfies (4.44). ■

Proof of Lemma 4.3 We show this lemma using the random coding method, as in Lemma 4.6. For a channel $W_x^{(n)}$ with $x \in \mathcal{X}^n$, we consider a random coding for the probability distribution $\hat{p}(x) \stackrel{\text{def}}{=} p^n(x)/C_{n,K}$ on the subset $\hat{\mathcal{X}} \stackrel{\text{def}}{=} \{c^{(n)}(x) \le nK\}$. In this construction, we choose π_i to be $\{W_{\varphi(i)} - 2N_n W_p^{\otimes n} > 0\}$. Using the same notation as that of Lemma 4.6, we obtain

$$\mathrm{E}_{\mathcal{X}}[\varepsilon[\Phi(\varphi)]] \le \sum_{x \in \hat{\mathcal{X}}^n} \hat{p}(x) \Bigg(2 \operatorname{Tr} \left[W_x^{(n)} \left\{ W_x^{(n)} - 2N_n W_p^{\otimes n} \le 0 \right\} \right]$$
$$+ 4N_n \operatorname{Tr} \left[\left(\sum_{x' \in \hat{\mathcal{X}}} \hat{p}(x') W_{x'}^{(n)} \right) \left\{ W_x^{(n)} - 2N_n W_p^{\otimes n} > 0 \right\} \right] \Bigg).$$

By noting that $\hat{p}(x) = p^n(x)/C_{n,K}$ and $\hat{\mathcal{X}} \subset \mathcal{X}^n$, we find

$$E_{\mathcal{X}}[\varepsilon[\Phi(\varphi)]]$$
$$\le \sum_{x \in \hat{\mathcal{X}}^n} \frac{2p^n(x)}{C_{n,K}} \Bigg(\operatorname{Tr} \left[W_x^{(n)} \left\{ W_x^{(n)} - 2N_n W_p^{\otimes n} \le 0 \right\} \right]$$
$$+ 2N_n \operatorname{Tr} \left(\frac{W_p^{\otimes n}}{C_{n,K}} \right) \left\{ W_x^{(n)} - 2N_n W_p^{\otimes n} > 0 \right\} \Bigg)$$
$$\le \frac{2}{C_{n,K}^2} \sum_{x \in \mathcal{X}} p^n(x) \Bigg(\operatorname{Tr} W_x^{(n)} \left\{ W_x^{(n)} - 2N_n W_p^{\otimes n} \le 0 \right\}$$
$$+ 2N_n \operatorname{Tr} \left[W_p^{\otimes n} \left\{ W_x^{(n)} - 2N_n W_p^{\otimes n} > 0 \right\} \right] \Bigg).$$

By using the same arguments as those in the random coding method, the proof is completed. ■

Exercises

4.10 Let α and β be defined by the following. Show inequality (4.44) in Lemma 4.6 with its RHS replaced by $\alpha + 2\beta + \sqrt{\beta(\alpha + \beta)}$ using Exercise 4.9.

$$\alpha \stackrel{\text{def}}{=} \sum_{x \in \mathcal{X}} p(x) \operatorname{Tr} W_x \left\{ W_x - 2N W_p \le 0 \right\},$$
$$\beta \stackrel{\text{def}}{=} N \operatorname{Tr} W_p \left\{ W_x - 2N W_p > 0 \right\}.$$

4.11 Consider the sequence of codes $\{\Phi^{(n)}\}$ for the stationary memoryless channel of the c-q channel $x \mapsto W_x$. Let us focus on the optimal decreasing exponential rate of the average error probability when the communication rate of $\{\Phi^{(n)}\}$ is greater than R:

$$B(R|W) \stackrel{\text{def}}{=} \sup_{\{\Phi^{(n)}\}} \left\{ \underline{\lim} -\frac{1}{n} \log \varepsilon[\Phi^{(n)}] \middle| \underline{\lim} \frac{1}{n} \log |\Phi^{(n)}| \ge R \right\}. \tag{4.54}$$

This optimal rate is called the **reliability function** as a function of the communication rate R and is an important quantity in quantum information theory.
(**a**) Show that

$$B(R|W) \geq \sup_{p \in \mathcal{P}_f(\mathcal{X})} \max_{0 \leq s \leq 1} s(I_{1-s}(p, W) - R). \tag{4.55}$$

(**b**) When all states W_x are pure states, show that

$$B(R|W) \geq \sup_{p \in \mathcal{P}_f(\mathcal{X})} \max_{0 \leq s \leq 1} s(H_{1+s}(W_p) - R). \tag{4.56}$$

4.12 Define another c-q channel capacity by replacing the condition that the average error probability goes to 0 by the alternative condition that the maximum error probability goes to 0. Show that the modified c-q channel capacity is equal to the original c-q channel capacity.

4.13 Show the inequality $C_c(W^1, \ldots, W^M) \geq \sup_p \min_{1 \leq i \leq M} I(p, W^i)$ following the steps below.
(**a**) Show that there exists a code $\Phi^{(n)}$ with size e^{nR} for M-output channel $W^1, \ldots, W^M$ such that

$$\varepsilon[\Phi^{(n)}] \leq \max_{1 \leq i \leq M} M \cdot 2^{1+s} e^{ns(R - I_{1-s}(p, W^i))}$$

as an extension of (4.49).
(**b**) Show the desired inequality.

4.6 Proof of Converse Part of C-Q Channel Coding Theorem

In this section, we prove the converse parts of the c-q channel coding theorem, i.e., Lemmas 4.2 and 4.4, by using the information inequality (3.20) proved in Sect. 3.8.

Before the proof of Lemma 4.2, we prepare the following lemma.

Lemma 4.7 *Let $\Phi = (N, \varphi, Y)$ be a code with the size N, and σ be a state. Then, we have*

$$\varepsilon[\Phi] \geq \beta_{\frac{1}{N}}(S_2(\sigma) \| S_1(\varphi)), \tag{4.57}$$

where the density matrices $S_1(\varphi)$ and $S_2(\sigma)$ on $\mathcal{H}^{\otimes n} \otimes \mathbb{C}^N$ are given as

$$S_2(\sigma) \stackrel{\text{def}}{=} \frac{1}{N} \begin{pmatrix} \sigma & & 0 \\ & \ddots & \\ 0 & & \sigma \end{pmatrix}, \quad S_1(\varphi) \stackrel{\text{def}}{=} \frac{1}{N} \begin{pmatrix} W_{\varphi(1)} & & 0 \\ & \ddots & \\ 0 & & W_{\varphi(N)} \end{pmatrix}.$$

When P_X is the uniform distribution over the image of φ, we have $S_2(\sigma) = \mathrm{P}_X \otimes \sigma$ and $S_1(\varphi) = \mathrm{P}_X \times W$. Taking the infimum for P_X in both sides in (4.57), we have

$$\min_{\Phi:|\Phi|=N} \varepsilon[\Phi] \geq \inf_{\mathrm{P}_X} \beta_{\frac{1}{N}}(\mathrm{P}_X \otimes \sigma \| \mathrm{P}_X \times W) \tag{4.58}$$

for any state σ. Then, taking the supremum for σ, we have

$$\min_{\Phi:|\Phi|=N} \varepsilon[\Phi] \geq \sup_{\sigma} \inf_{\mathrm{P}_X} \beta_{\frac{1}{N}}(\mathrm{P}_X \otimes \sigma \| \mathrm{P}_X \times W). \tag{4.59}$$

The above lower bound is called the Meta converse bound, which is helpful for calculating the lower bound of the minimum decoding error probability.

Proof We choose a matrix T

$$T \stackrel{\text{def}}{=} \begin{pmatrix} Y_1 & & 0 \\ & \ddots & \\ 0 & & Y_N \end{pmatrix}.$$

Since $I \geq T \geq 0$, we have

$$\operatorname{Tr} S_1(\varphi) T = \sum_{i=1}^{N} \frac{1}{N} \operatorname{Tr} W_{\varphi(i)} Y_i = 1 - \varepsilon[\Phi]. \tag{4.60}$$

On the other hand, since $I = \sum_{i=1}^{N} Y_i$, we have

$$\operatorname{Tr} S_2(\sigma) T = \sum_{i=1}^{N} \frac{1}{N} \operatorname{Tr} \sigma Y_i = \frac{1}{N} \operatorname{Tr} \sigma \sum_{i=1}^{N} Y_i = \frac{1}{N} \operatorname{Tr} \sigma = \frac{1}{N}. \tag{4.61}$$

Combining these two relations, we obtain (4.57). ■

Now, using (4.60) and (4.61) in the proof of Lemma 4.7, we show Lemma 4.2.

Proof of Lemma 4.2 Here, we show Lemma 4.2 only when the maximum $\max_{p \in \mathcal{P}_f(\mathcal{X})} I_{1-s}^{\downarrow}(p, W)$ exists. For the general case, see Exercise 4.28. Firstly, given $s \in (-\infty, 0]$, the distribution

$$p_{1-s} \stackrel{\text{def}}{=} \operatorname*{argmax}_{p \in \mathcal{P}_f(\mathcal{X})} I_{1-s}^{\downarrow}(p, W) \tag{4.62}$$

satisfies that

$$\operatorname{Tr} W_x^{1-s} \left(\sum_{x'} p_{1-s}(x') W_{x'}^{1-s} \right)^{\frac{s}{1-s}} \leq \operatorname{Tr} \left(\sum_{x'} p_{1-s}(x') W_{x'}^{1-s} \right)^{\frac{1}{1-s}} \tag{4.63}$$

for any $x \in \mathcal{X}^{\text{Exe. 4.20}}$. That is, the state $\sigma_{1-s|p_{1-s}}$ defined in (4.23) satisfies that

$$\operatorname{Tr} W_x^{1-s} \sigma_{1-s|p_{1-s}}^s \le \left(\operatorname{Tr} \left(\sum_{x'} p_{1-s}(x') W_{x'}^{1-s} \right)^{\frac{1}{1-s}} \right)^{1-s}. \tag{4.64}$$

Assume that a code $\Phi^{(n)}$ for the stationary memoryless quantum channel satisfies $N_n \stackrel{\text{def}}{=} |\Phi^{(n)}| = e^{nR}$. Then, we have

$$\begin{aligned}
&\log \operatorname{Tr}\left[(W^{(n)}_{\varphi^{(n)}(i)})^{1-s} \sigma_{1-s|p_{1-s}}^s \right] = \sum_{l=1}^n \log \operatorname{Tr}\left[(W_{\varphi_l^{(n)}(i)})^{1-s} \sigma_{1-s|p_{1-s}}^s \right] \\
=& n \sum_{l=1}^n \frac{1}{n} \log \operatorname{Tr}\left[(W_{\varphi_l^{(n)}(i)})^{1-s} \sigma_{1-s|p_{1-s}}^s \right] \\
\stackrel{(a)}{\le}& n \log \sum_{l=1}^n \frac{1}{n} \operatorname{Tr}\left[(W_{\varphi_l^{(n)}(i)})^{1-s} \sigma_{1-s|p_{1-s}}^s \right] \\
\stackrel{(b)}{\le}& n \log \left[\sum_{l=1}^n \frac{1}{n} \left(\operatorname{Tr} \left(\sum_x p_{1-s}(x) W_x^{1-s} \right)^{\frac{1}{1-s}} \right)^{1-s} \right] \\
=& n(1-s) \log \left(\operatorname{Tr} \left(\sum_x p_{1-s}(x) W_x^{1-s} \right)^{\frac{1}{1-s}} \right),
\end{aligned} \tag{4.65}$$

(a) follows from the concavity of $x \mapsto \log x$ and (b) follows from (4.64).

Now, we apply the discussion in the proof of Lemma 4.7 to the case with $\sigma = \sigma_{1-s|p_{1-s}}^{\otimes n}$. We have

$$\begin{aligned}
&(1-\varepsilon[\Phi^{(n)}])^{1-s} N_n^{-s} \stackrel{(a)}{=} (\operatorname{Tr} S_1(\varphi^{(n)}) T)^{1-s} (\operatorname{Tr} S_2(\sigma_{1-s|p_{1-s}}^{\otimes n}) T)^s \\
\stackrel{(b)}{\le}& \operatorname{Tr} S_1(\varphi^{(n)})^{1-s} S_2(\sigma_{1-s|p_{1-s}}^{\otimes n})^s = \frac{1}{N_n} \sum_{i=1}^{N_n} \operatorname{Tr}\left[(W^{(n)}_{\varphi^{(n)}(i)})^{1-s} \sigma_{1-s|p_{1-s}}^s \right] \\
\stackrel{(c)}{\le}& \left(\operatorname{Tr} \left(\sum_x p_{1-s}(x) W_x^{1-s} \right)^{\frac{1}{1-s}} \right)^{n(1-s)}
\end{aligned} \tag{4.66}$$

for $s \le 0$, where (a) follows from (4.60) and (4.61) and (c) follows from (4.65). Here, (b) can be shown from the monotonicity (3.20) of the quantum relative Rényi entropy in the same way as (3.137).

Thus,

$$\frac{1}{n} \log(1-\varepsilon[\Phi^{(n)}]) \le \log \operatorname{Tr} \left(\sum_x p_{1-s}(x) W_x^{1-s} \right)^{\frac{1}{1-s}} + \frac{s}{1-s} R, \tag{4.67}$$

which implies (4.19). ■

Proof of Lemma 4.4 We show Lemma 4.4 only when the maximum $\max_{p\in\mathcal{P}_{c\le K}(\mathcal{X})} I^{\downarrow}_{1-s}(p,W)$ exists. For the general case, see Exercise 4.29. Firstly, for $s\in(-\infty,0]$, we choose $p'_{1-s}\stackrel{\rm def}{=}\mathop{\rm argmax}_{p\in\mathcal{P}_{c\le K}(\mathcal{X})} I^{\downarrow}_{1-s}(p,W)$. Then, similar to (4.75), the relation

$$\sum_x p(x)\,\mathrm{Tr}\, W_x^{1-s}\Big(\sum_{x'} p'_{1-s}(x')W_{x'}^{1-s}\Big)^{\frac{s}{1-s}} \le \mathrm{Tr}\Big(\sum_{x'} p'_{1-s}(x')W_{x'}^{1-s}\Big)^{\frac{1}{1-s}} \tag{4.68}$$

holds for any $p\in\mathcal{P}_{c\le K}(\mathcal{X})^{\text{Exe. 4.21}}$. That is, the state $\sigma_{1-s|p'_{1-s}}$ satisfies that

$$\sum_x p(x)\,\mathrm{Tr}\, W_x^{1-s}\sigma^s_{1-s|p'_{1-s}} \le \left(\mathrm{Tr}\Big(\sum_{x'} p'_{1-s}(x')W_{x'}^{1-s}\Big)^{\frac{1}{1-s}}\right)^{1-s} \tag{4.69}$$

for any $p\in\mathcal{P}_{c\le K}(\mathcal{X})$. Then, using (4.69), we have

$$\begin{aligned}
&\log \mathrm{Tr}\Big[(W^{(n)}_{\varphi^{(n)}(i)})^{1-s}\sigma^s_{1-s|p'_{1-s}}\Big] \le n\log\sum_{l=1}^n\frac{1}{n}\mathrm{Tr}\Big[(W_{\varphi^{(n)}_l(i)})^{1-s}\sigma^s_{1-s|p'_{1-s}}\Big]\\
\le& n\log\left[\left(\mathrm{Tr}\Big(\sum_x p'_{1-s}(x)W_x^{1-s}\Big)^{\frac{1}{1-s}}\right)^{1-s}\right]\\
=& n(1-s)\log\left(\mathrm{Tr}\Big(\sum_x p'_{1-s}(x)W_x^{1-s}\Big)^{\frac{1}{1-s}}\right).
\end{aligned} \tag{4.70}$$

Hence, similar to the proof of Lemma 4.2, we obtain (4.39). ■

Exercises

4.14 Define $J(p,\sigma,W)\stackrel{\rm def}{=}\sum_{x\in\mathcal{X}}p(x)D(W_x\|\sigma)$. Show the following relations [12, 13] including the existence of the minimums appearing the relations by following the steps below.

$$\begin{aligned}
\sup_{p\in\mathcal{P}_f(\mathcal{X})} I(p,W) =& \sup_{p\in\mathcal{P}_f(\mathcal{X})}\min_{\sigma\in\mathcal{S}(\mathcal{H})} J(p,\sigma,W)\\
=& \min_{\sigma\in\mathcal{S}(\mathcal{H})}\sup_{p\in\mathcal{P}_f(\mathcal{X})} J(p,\sigma,W) = \min_{\sigma\in\mathcal{S}(\mathcal{H})}\sup_{x\in\mathcal{X}} D(W_x\|\sigma).
\end{aligned} \tag{4.71}$$

(**a**) Show that $\min_{\sigma\in\mathcal{S}(\mathcal{H})} J(p,\sigma,W)=I(p,W)$.
(**b**) Show that $\sigma\mapsto D(W_x\|\sigma)$ is convex. (See (5.38).)
(**c**) Show the existence of the minimums $\min_{\sigma\in\mathcal{S}(\mathcal{H})}\sup_{p\in\mathcal{P}_f(\mathcal{X})} J(p,\sigma,W)$ and $\min_{\sigma\in\mathcal{S}(\mathcal{H})}\sup_{x\in\mathcal{X}} D(W_x\|\sigma)$ by using Lemma A.8.
(**d**) Show (4.71) by applying Lemma A.9.

4.15 Give an alternative proof of (4.71) by following steps below when the maximum $\max_{p \in \mathcal{P}_f(\mathcal{X})} I(p, W)$ exists.
(**a**) Choose $p_1 := \operatorname{argmax}_{p \in \mathcal{P}_f(\mathcal{X})} I(p, W)$. Show that $D(W_x \| W_{p_1}) \le I(p_1, W)$.
(**b**) Show (4.71).

4.16 Give an alternative proof of (4.12) by using (4.71).

4.17 Similarly to (4.71), show the following relations including the existence of the minimums appearing the relations by using Lemma A.9.

$$\sup_{p \in \mathcal{P}_{c \le K}(\mathcal{X})} I(p, W) = \sup_{p \in \mathcal{P}_{c \le K}(\mathcal{X})} \min_{\sigma \in \mathcal{S}(\mathcal{H})} J(p, \sigma, W)$$
$$= \min_{\sigma \in \mathcal{S}(\mathcal{H})} \sup_{p \in \mathcal{P}_{c \le K}(\mathcal{X})} J(p, \sigma, W). \tag{4.72}$$

4.18 Give an alternative proof of (4.72) similar to Exercise 4.14.

4.19 Let c_A and c_B be cost functions on $\mathcal{X}_A$ and $\mathcal{X}_B$. Show the following equation by using (4.71).

$$C_{c|c_A + c_B \le K}(W^A \otimes W^B) = \max_{K'} C_{c|c_A \le K'}(W^A) + C_{c|c_B \le K - K'}(W^B) \tag{4.73}$$

4.20 Show (4.63) by using the function $f(t) := \operatorname{Tr}(t W_x^{1-s} + (1-t) \sum_{x'} p_{1-s}(x) W_{x'}^{1-s})^{\frac{1}{1-s}}$ when the maximum $\max_{p \in \mathcal{P}_f(\mathcal{X})} I_{1-s}^{\downarrow}(p, W)$ exists.

4.21 Show (4.68) by using the function $f(t) := \operatorname{Tr}(\sum_{x'} (t p(x') + (1-t) p'_{1-s}(x)) W_{x'}^{1-s})^{\frac{1}{1-s}}$ when the maximum $\max_{p \in \mathcal{P}_{c \le K}(\mathcal{X})} I_{1-s}^{\downarrow}(p, W)$ exists.

4.22 Show the following relations including the existence of the minimums appearing the relations for $s \in [-1, 1] \setminus \{0\}$.

$$C_{1-s}^{\downarrow}(W) = \sup_{p \in \mathcal{P}_f(\mathcal{X})} \min_{\sigma \in \mathcal{S}(\mathcal{H})} J_{1-s}(p, \sigma, W)$$
$$= \min_{\sigma \in \mathcal{S}(\mathcal{H})} \sup_{p \in \mathcal{P}_f(\mathcal{X})} J_{1-s}(p, \sigma, W) = \min_{\sigma \in \mathcal{S}(\mathcal{H})} \sup_{x \in \mathcal{X}} D_{1-s}(W_x \| \sigma). \tag{4.74}$$

Hint: Use the matrix convexity of $x \mapsto x^s$ for $s \in [-1, 0)$ and the matrix concavity of $x \mapsto x^s$ for $s \in (0, 1]$.

4.23 Show (4.17) by following the steps below.
(**a**) Since the second equation of (4.17) was shown in Exercise 4.14 including the existence of the minimum $\min_{\sigma \in \mathcal{S}(\mathcal{H})} \sup_{x \in \mathcal{X}} D(W_x \| \sigma)$, it is sufficient to show the first equation. Show $\sup_p I(p, W) \le \liminf_{s \to 0} \sup_p I_{1-s}^{\downarrow}(p, W)$.
(**b**) Show that the convergence $D_{1-s}(W_x \| \sigma) \to D(W_x \| \sigma)$ is uniform for x as $s \to 0$ when the supports of W_x are included in that of σ. (Hint: Use (**c**) of Exercise 3.5.)
(**c**) Show $\sup_p I(p, W) \ge \limsup_{s \to 0} \sup_p I_{1-s}^{\downarrow}(p, W)$ by using the final expression at (4.74) and $\sigma_1 := \operatorname{argmin}_\sigma \sup_x D(W_x \| \sigma)$ (See (**c**) of Exercise 4.14).

4.24 Give an alternative proof of (4.65) for $s \in [-1, 0]$ by following the steps below.
(**a**) Show the following by using (4.74).

$$\sup_{p \in \mathcal{P}_f(\mathcal{X})} \mathrm{Tr}\Big[\sum_x p(x) W_x^{1-s}\Big] \sigma_{1-s|p_{1-s}}^s \le \sup_{p \in \mathcal{P}_f(\mathcal{X})} \left[\mathrm{Tr} \left(\sum_x p(x) W_x^{1-s} \right)^{\frac{1}{1-s}} \right]^{1-s}. \tag{4.75}$$

(**b**) Show (4.65).

4.25 Show the following for $s \in [-1, 1] \setminus \{0\}$ by using (4.74).

$$C_{1-s}^{\downarrow}(W^A \otimes W^B) = C_{1-s}^{\downarrow}(W^A) + C_{1-s}^{\downarrow}(W^B). \tag{4.76}$$

4.26 Show the following relations including the existence of the minimums appearing the relations for $s \in [-1, 1] \setminus \{0\}$.

$$\begin{aligned} C_{1-s|c \le K}^{\downarrow}(W) &= \sup_{p \in \mathcal{P}_{c \le K}(\mathcal{X})} \min_{\sigma \in \mathcal{S}(\mathcal{H})} J_{1-s}(p, \sigma, W) \\ &= \min_{\sigma \in \mathcal{S}(\mathcal{H})} \sup_{p \in \mathcal{P}_{c \le K}(\mathcal{X})} J_{1-s}(p, \sigma, W). \end{aligned} \tag{4.77}$$

4.27 Show (4.36) by following the steps below.
(**a**) Since the second equation of (4.36) was shown in Exercise 4.17 including the existence of the minimum $\min_{\sigma \in \mathcal{S}(\mathcal{H})} \sup_{p \in \mathcal{P}_{c \le K}(\mathcal{X})} J(p, \sigma, W)$, it is sufficient to show the first equation. Show $\sup_{p \in \mathcal{P}_{c \le K}(\mathcal{X})} I(p, W) \le \liminf_{s \to 0} \sup_{p \in \mathcal{P}_{c \le K}(\mathcal{X})} I_{1-s}^{\downarrow}(p, W)$.
(**b**) Show that the convergence $J_{1-s}(p, \sigma, W) \to J(p, \sigma, W)$ is uniform for p as $s \to 0$ when the support of W_x are included in that of σ for any element x in the support of p. (Hint: Use (**c**) of Exercise 3.5.)
(**c**) Show $\sup_{p \in \mathcal{P}_{c \le K}(\mathcal{X})} I(p, W) \ge \limsup_{s \to 0} \sup_{p \in \mathcal{P}_{c \le K}(\mathcal{X})} I_{1-s}^{\downarrow}(p, W)$ by using the final expression at (4.77) and $\sigma_1' := \mathrm{argmin}_\sigma \sup_{p \in \mathcal{P}_{c \le K}(\mathcal{X})} J(p, \sigma, W)$ (See Exercise 4.17).

4.28 Show Lemma 4.2 in the following way when the maximum $\max_{p \in \mathcal{P}_f(\mathcal{X})} I_{1-s}^{\downarrow}(p, W)$ does not necessarily exist.
(**a**) Choose a sequence of distributions $\{p_n\}$ such that $\lim_{n \to \infty} I_{1-s}^{\downarrow}(p_n, W) = \sup_{p \in \mathcal{P}_f(\mathcal{X})} I_{1-s}^{\downarrow}(p, W)$ and the matrix $\sum_{x'} p_n(x') W_{x'}^{1-s}$ converges as $n \to \infty$. So, we denote the limit of $\sum_{x'} p_n(x') W_{x'}^{1-s}$ by S_{1-s}, and define the function $f(t) := \mathrm{Tr}(t W_x^{1-s} + (1-t) S_{1-s})^{\frac{1}{1-s}}$ and the state $\sigma_{1-s} := S_{1-s}^{\frac{1}{1-s}} / \mathrm{Tr}\, S_{1-s}^{\frac{1}{1-s}}$. Show that

$$\mathrm{Tr}\, W_x^{1-s} S_{1-s}^{\frac{s}{1-s}} \le \mathrm{Tr}\, S_{1-s}^{\frac{1}{1-s}} \tag{4.78}$$

for any $x \in \mathcal{X}$ in th same way as Exercise 4.20.

(**b**) Show (4.19) by replacing $\sigma_{1-s|p_{1-s}}$ in the proof given in the main body by σ_{1-s}.

4.29 Show Lemma 4.4 in the following way when the maximum $\max_{p\in\mathcal{P}_{c\le K}(\mathcal{X})} I^{\downarrow}_{1-s}(p, W)$ does not necessarily exist.
(**a**) Choose a sequence of distributions $\{p_n\}$ such that $\lim_{n\to\infty} I^{\downarrow}_{1-s}(p_n, W) = \sup_{p\in\mathcal{P}_{c\le K}(\mathcal{X})} I^{\downarrow}_{1-s}(p, W)$ and the matrix $\sum_{x'} p_n(x') W_{x'}^{1-s}$ converges as $n\to\infty$. So, we denote the limit of $\sum_{x'} p_n(x') W_{x'}^{1-s}$ by S'_{1-s}, and define the function $f(t) := \mathrm{Tr}(t\sum_x p(x) W_x^{1-s} + (1-t)S'_{1-s})^{\frac{1}{1-s}}$ and the state $\sigma'_{1-s} := {S'_{1-s}}^{\frac{1}{1-s}} / \mathrm{Tr}\, {S'_{1-s}}^{\frac{1}{1-s}}$. Show that

$$\mathrm{Tr}\, W_x^{1-s} {S'_{1-s}}^{\frac{s}{1-s}} \le \mathrm{Tr}\, {S'_{1-s}}^{\frac{1}{1-s}} \tag{4.79}$$

for any $x\in\mathcal{X}$ in th same way as Exercise 4.21.
(**b**) Show (4.39) by replacing $\sigma_{1-s|p'_{1-s}}$ in the proof given in the main body by σ'_{1-s}.

4.30 Give an alternative proof of (4.70) by following the steps below when the minimum $\max_{p\in\mathcal{P}_{c\le K}(\mathcal{X})} I^{\downarrow}_{1-s}(p, W)$ exists.
(**a**) Show the following inequality for $s\in[-1,0]$ by using (4.77).

$$\begin{aligned}&\max_{p\in\mathcal{P}_{c\le K}(\mathcal{X})} \mathrm{Tr}\Big[\sum_x p(x) W_x^{1-s}\Big]\sigma^s_{1-s|p'_{1-s}}\\ \le &\max_{p\in\mathcal{P}_{c\le K}(\mathcal{X})} \left[\mathrm{Tr}\left(\sum_x p(x) W_x^{1-s}\right)^{\frac{1}{1-s}}\right]^{1-s}.\end{aligned} \tag{4.80}$$

(**b**) Show (4.70).

4.31 Define another c-q channel capacity under cost constraint by replacing the condition that the maximum cost $\max_i \frac{c^{(n)}(\varphi^{(n)}(i))}{n}$ is less than the given cost K by the alternative condition that the average cost $\frac{1}{N}\sum_i \frac{c^{(n)}(\varphi^{(n)}(i))}{n}$ is less than the given cost K. Show that the modified c-q channel capacity under cost constraint is equal to the original c-q channel capacity under cost constraint following the steps below.
(**a**) First, assume that $c(x_0)=0$ or redefine the cost as $c(x)-c(x_0)$, where $x_0 \stackrel{\mathrm{def}}{=} \mathrm{argmin}_{x\in\mathcal{X}} c(x)$. Let a code $\Phi^{(n)} = (N_n, \varphi^{(n)}, Y^{(n)})$ satisfy $\varepsilon[\Phi^{(n)}]\to 0$ and $\frac{1}{N_n}\sum_i \frac{c^{(n)}(\varphi^{(n)}(i))}{n} \le K$. For arbitrary $\delta>0$, focus a code $\tilde{\Phi}^{((1+\delta)n)} = (\tilde{N}_{(1+\delta)n}, \tilde{\varphi}^{((1+\delta)n)}, \tilde{Y}^{((1+\delta)n)})$ satisfying $\tilde{N}_{(1+\delta)n} = N_n$, $\tilde{\varphi}^{((1+\delta)n)}(i) = \varphi^{(n)}(i)\otimes W_{x_0}^{\otimes\delta n}$, and $\tilde{Y}_i^{((1+\delta)n)} = Y^{(n)}$.
Show that there exist $k \stackrel{\mathrm{def}}{=} [(1-\frac{1}{1+\delta})N_n]$ messages $i_1,\ldots,i_k$ such that $c^{((1+\delta)n)}(\tilde{\varphi}^{((1+\delta)n)}(i_k)) \le K$.
(**b**) Examine the subcode of $\tilde{\Phi}^{(n)}$ consisting of $[(1-\frac{1}{1+\delta})N_n]$ messages, and show that the rate of this subcode is asymptotically equal to $\frac{1}{1+\delta}\lim_{n\to\infty}\frac{1}{n}\log|\Phi^{(n)}|$.
(**c**) Show that the modified capacity is equal to the original capacity. (Note that this method gives the strong converse concerning the modified c-q channel capacity by combining the strong converse of the original capacity.)

4.32 Show the inequality $C_c(W^1, \ldots, W^M) \le \sup_p \min_{1\le i\le M} I(p, W^i)$ following the steps below.
(**a**) Show that there exists a distribution p for any distribution $p^{(n)}$ on $\mathcal{X}^n$ such that

$$nI(p, W^i) \ge I(p^{(n)}, (W^i)^{(n)}), \quad i = 1, \ldots, M$$

from (4.3) and (4.5).
(**b**) Show the desired inequality using Fano inequality (2.35).

4.7 Pseudoclassical Channels

Finally, we treat the capacity of a c-q channel when the quantum correlation is not allowed in the measuring apparatus, again. In Sect. 4.2 we showed that the c-q channel capacity is not improved even when feedback and adaptive decoding are allowed in encoding as long as the quantum correlation is not used in the measuring apparatus. That is, the capacity can be attained when the optimal measurement with a single transmission is performed on each system. Then, we may ask, when does the c-q channel capacity $\tilde{C}_c(W)$ with individual measurements equal the channel capacity $C_c(W)$ with the quantum correlation in the measuring apparatus? The answer to this question is the subject of the following theorem.

Theorem 4.4 (Fujiwara and Nagaoka [8]) *Suppose that* $\mathrm{Tr}\, W_x W_{x'} \neq 0$ *for any* $x, x' \in \mathcal{X}$. *Then, the following three conditions with respect to the c-q channel* W *are equivalent if* $\mathcal{X}$ *is compact.*

① *There exists a distribution* $p \in \mathcal{P}_f(\mathcal{X})$ *such that* $[W_x, W_{x'}] = 0$ *for any two elements* $x, x' \in \mathrm{supp}(p)$ *and* $I(p, W) = C_c(W)$.
② $\tilde{C}_c(W) = C_c(W)$.
③ *There exists an integer* n *such that* $\frac{\tilde{C}_c(W^{(n)})}{n} = C_c(W)$.

A quantum channel W is called **pseudoclassical** if it satisfies the above conditions.

Proof Since ①⇒② and ②⇒③ by inspection, we show that ③⇒①. The proof given below uses Theorems 3.6 and 4.5 (**Naĭmark extension** [14]). The proofs for these theorems will be given later.

Theorem 4.5 (Naĭmark [14]) *Given a POVM* $\boldsymbol{M} = \{M_\omega\}_{\omega\in\Omega}$ *on* $\mathcal{H}_A$ *with a finite probability space* Ω, *there exist a space* $\mathcal{H}_B$, *a state* ρ_0 *on* $\mathcal{H}_B$, *and a PVM* $\boldsymbol{E} = \{E_\omega\}_{\omega\in\Omega}$ *in* $\mathcal{H}_A \otimes \mathcal{H}_B$ *such that*

$$\mathrm{Tr}_A\, \rho M_\omega = \mathrm{Tr}_{A,B} (\rho \otimes \rho_0) E_\omega, \quad \forall \rho \in \mathcal{S}(\mathcal{H}_A), \forall \omega \in \Omega.$$

For the proof of Theorem 4.5, see Exercise 5.7 or the comments regarding Theorem 7.1.

Using Condition ③, we choose a measurement $\boldsymbol{M}^{(n)}$ on $\mathcal{H}^{\otimes n}$ and a distribution $p^{(n)}$ on $\mathcal{X}^n$ such that $I(p^{(n)}, \boldsymbol{M}^{(n)}, W^{(n)}) = \tilde{C}_c(W^{(n)})$. Since

$$I(p^{(n)}, \boldsymbol{M}^{(n)}, W^{(n)}) \le I(p^{(n)}, W^{(n)}), \;\; \tilde{C}_c(W^{(n)}) \ge nC_c(W) \ge I(p^{(n)}, W^{(n)}),$$

we obtain $I(p^{(n)}, \boldsymbol{M}^{(n)}, W^{(n)}) = I(p^{(n)}, W^{(n)})$. This is equivalent to

$$\sum_{\boldsymbol{x} \in \mathrm{supp}(p^{(n)})} p^{(n)}(\boldsymbol{x}) \left(D(\mathrm{P}^{\boldsymbol{M}^{(n)}}_{W^{(n)}_{\boldsymbol{x}}} \| \mathrm{P}^{\boldsymbol{M}^{(n)}}_{W^{(n)}_{p^{(n)}}}) - D(W^{(n)}_{\boldsymbol{x}} \| W^{(n)}_{p^{(n)}}) \right) = 0,$$

and we have $D(\mathrm{P}^{\boldsymbol{M}^{(n)}}_{W^{(n)}_{\boldsymbol{x}}} \| \mathrm{P}^{\boldsymbol{M}^{(n)}}_{W^{(n)}_{p^{(n)}}}) = D(W^{(n)}_{\boldsymbol{x}} \| W^{(n)}_{p^{(n)}})$ for $\boldsymbol{x} \in \mathrm{supp}(p^{(n)})$, where $\boldsymbol{x}^n$ is simplified to $\boldsymbol{x}$. Following Theorem 4.5, we choose an additional system $\mathcal{H}_A$, a pure state ρ_A on $\mathcal{H}_A$, and a PVM $\boldsymbol{E} = \{E_k\}$ on the composite system $\mathcal{H}_A \otimes \mathcal{H}$ such that

$$D\left(\mathrm{P}^{\boldsymbol{E}}_{W^{(n)}_{\boldsymbol{x}} \otimes \rho_A} \| \mathrm{P}^{\boldsymbol{E}}_{W^{(n)}_{p^{(n)}} \otimes \rho_A} \right) = D\left(W^{(n)}_{\boldsymbol{x}} \otimes \rho_A \| W^{(n)}_{p^{(n)}} \otimes \rho_A \right).$$

According to Theorem 3.6, we take a real number $a_k(\boldsymbol{x})$ for every $\boldsymbol{x} \in \mathrm{supp}(p^{(n)})$ such that the Hermitian matrix $X_{\boldsymbol{x}} = \sum_k a_k(\boldsymbol{x}) E_k$ satisfies $W^{(n)}_{\boldsymbol{x}} \otimes \rho_A = \left(W^{(n)}_{p^{(n)}} \otimes \rho_A \right) X_{\boldsymbol{x}}$. Since $X_{\boldsymbol{x}}$ is Hermitian, we obtain

$$\begin{aligned} &\left(W^{(n)}_{\boldsymbol{x}} \otimes \rho_A \right) \left(W^{(n)}_{\boldsymbol{x}'} \otimes \rho_A \right) = (W^{(n)}_{p^{(n)}} \otimes \rho_A) X_{\boldsymbol{x}} X_{\boldsymbol{x}'} \left(W^{(n)}_{p^{(n)}} \otimes \rho_A \right) \\ =& \left(W^{(n)}_{p^{(n)}} \otimes \rho_A \right) X_{\boldsymbol{x}'} X_{\boldsymbol{x}} \left(W^{(n)}_{p^{(n)}} \otimes \rho_A \right) = \left(W^{(n)}_{\boldsymbol{x}'} \otimes \rho_A \right) \left(W^{(n)}_{\boldsymbol{x}} \otimes \rho_A \right) \end{aligned}$$

for $\boldsymbol{x}, \boldsymbol{x}' \in \mathrm{supp}(p^{(n)})$. Therefore, we obtain

$$W^{(n)}_{\boldsymbol{x}} W^{(n)}_{\boldsymbol{x}'} = W^{(n)}_{\boldsymbol{x}'} W^{(n)}_{\boldsymbol{x}}.$$

Defining $p_i^{(n)}$ by $p_i^{(n)}(x) \stackrel{\mathrm{def}}{=} \sum_{\boldsymbol{x}=(x_1,\ldots,x_n):x_i=x} p^{(n)}(\boldsymbol{x})$, from Exercise 1.23 we find that W_x and W_y commute each other for any two elements $x, y \in \mathrm{supp}(p_i^{(n)})$ because $\mathrm{Tr}\, W_x W_{x'} \neq 0$ for any $x, x' \in \mathcal{X}$. Equation (4.5) yields

$$\sum_{i=1}^{n} I(p_i^{(n)}, W) \ge I(p^{(n)}, W^{(n)}) = nC_c(W).$$

Therefore, $I(p_i^{(n)}, W) = C_c(W)$, and thus we obtain ①. ∎

4.8 Historical Note

4.8.1 C-Q Channel Capacity

Here, we briefly mention the history of the c-q channel coding theorem. Since this problem was independently formulated by several researchers, it is difficult to determine who formulated it first. The first important achievement for this theorem is the inequality

$$I(p, W) \geq I(\boldsymbol{M}, p, W), \tag{4.81}$$

which was conjectured by Levitin [15] and proved by Holevo [2]. Indeed, this inequality can be proved easily from the monotonicity of the quantum relative entropy (5.36) [16, 17]; however, at that time, it had not been proved. During that period, a strong subadditivity of the von Neumann entropy (5.83) was proved by Lieb and Ruskai [18, 19]. Using the strong subadditivity, we can easily prove the above inequality[Exe. 5.45]; however, this relation between the strong subadditivity and inequality (4.81) was not known at that time. Combining inequality (4.81) with Fano's inequality, Holevo [3] showed that the weaker version of the converse part, i.e., $C_c(W) \leq \sup_{p\in\mathcal{P}_f(\mathcal{X})} I(p, W)$, held. Twenty years later, Ogawa and Nagaoka [6] proved the strong converse part $C_c^\dagger(W) \leq \sup_{p\in\mathcal{P}_f(\mathcal{X})} I(p, W)$. Moreover, Nagaoka [20] invented a more simple proof of the strong converse. His proof is based on the relation with the hypothesis testing explained in the next subsection and the monotonicity of the Rényi entropy (3.20). In this book, we prove (3.20) using elementary knowledge in Sect. 3.8 and give a proof of the strong converse part combining Nagaoka's proof and (3.20).

Regarding the direct part, in the late 1970s, Stratonovich and Vantsjan [21] treated the pure state case, i.e., the case in which all the states W_x are pure. In this case, $C_c(W)$ is equal to $\sup_p H(W_p)$, but they found the lower bound $\sup_p -\log \mathrm{Tr}\, W_p^2$ of $C_c(W)$, i.e., they proved that $\sup_p -\log \mathrm{Tr}\, W_p^2 \leq C_c(W)$. Sixteen years later, Hausladen et al. [22] proved the attainability of $\sup_p H(W_p)$ in the pure-states case. This result was presented by Jozsa, who is a coauthor of this paper, in the QCMC'96 conference held at Hakone in Japan. Holevo attended this conference and extended this proof to the mixed-state case during his stay at Tamagawa University after this conference. Later, Schumacher and Westmoreland [5] independently obtained the same result. Their method was based on the conditional typical sequence, and its classical version appeared in Cover and Thomas [23]. Therefore, we can conclude that Holevo played a central role in the formulation of the c-q channel coding theorem. Hence, some researchers call the capacity $C_c(W)$ the **Holevo capacity**, while Theorem 4.1 is called the HSW theorem. Due to this achievement, Holevo received Shannon award in 2015, which is the most prestigious award in information theory.

In the classical case, Csiszár and Körner [24] have established the type method, which is a unified method in classical information theory and is partially summarized in Sect. 2.4.1. Applying it to its classical version, the researchers obtained another proof of this theorem and examined channel coding in greater detail. Winter [25, 26] tried to apply the type method to c-q channels. He obtained another proof of the c-q channel coding theorem but could not obtain an analysis of the error exponents as precise as that by Csiszár and Körner. Since there is an ambiguity regarding the orthogonal basis in the quantum case, a simple application of the type method to the c-q channel is not as powerful as the application to the classical case. To resolve this problem, Hayashi [27] invented a different method for universal channel coding, and succeeded in giving the universal channel coding for c-q channel. Also, Bjelakovic and Boche [28] showed the same fact by another approach independently.

4.8.2 Hypothesis Testing Approach

As another unified method in classical information theory, Han [29] established the method of information spectrum. Verdú and Han [30] applied it to classical channel coding and succeeded in obtaining the capacity of a general sequence of classical channels without any assumption, e.g., stationary memoryless, etc. This result suggests the relation between channel coding and hypothesis testing. Based on the result, around 2000, Nagaoka proposed an idea to understand all of topics in information theory based on binary hypothesis testing in the classical and quantum setting. He considered that this idea holds without the independent and identical/memoryless condition because the results of information spectrum [29] hold in the general sequence of information source and channels. As an evidence of this idea, he showed Lemma 4.7 in 2000 [20], whose classical case was shown by Polyanskiy et al. [31] as the meta converse theorem latter. This method much simplifies the proof of converse part.

Further, Ogawa and Nagaoka [32] extended Verdú and Han's method to the quantum case and obtained another proof of the direct part of the c-q channel coding theorem. Their result also supports Nagaoka's idea, i.e., clarifies the relation between the c-q channel coding and the quantum hypothesis testing. However, they could not obtain the capacity of the general sequence in the quantum case. Motivated by their proof, Hayashi and Nagaoka [11] derived Lemma 4.6, which more clarifies the relation between the c-q channel coding and the quantum hypothesis testing, whose classical case was shown by Polyanskiy et al. [31] as the dependent test (DT) bound latter. In this way, several fundamental results had been shown in the quantum case firstly. Then, a decade later, the classical cases were shown independently as special cases. Recently, many researchers of classical and quantum information theory are interested in this direction because this kind of hypothesis testing approach provides an unified viewpoint for information theory. Then, they produced many results to support Nagaoka's idea, i.e., they showed many results to clarify the relation between respective topics in classical and quantum information theory and the binary hypothesis testing. In particular, the second order asymptotic analysis has been discussed in classical and quantum information theory, and this kind of hypothesis testing approach plays an essential role in the second order asymptotic analysis.

4.8.3 Other Topics

Moreover, we sometimes discuss the error exponential decreasing rate (error exponent) in channel coding. Burunashev and Holevo [33] first obtained the lower bound of the optimal error exponent in the pure-state case, which is equal to (4.56). Their method differs from the method of Exercise 4.11. In the mixed state case, combining the dependent test (DT) bound and Hoeffding bound, Hayashi [34] obtained the lower bound (4.55) of the optimal error exponent by the same method as Exercise 4.11. Dalai [35] derived an upper bound of the optimal error exponent as the quantum version of Sphere-Packing Bound. Unfortunately, the lower bound does not match the upper bound because the lower bound (4.55) is different from the tight bound even in the classical case. So, to obtain the tight bound for the optimal error exponent in the quantum case, we need to improve the lower bound (4.55).

In addition, Fujiwara and Nagaoka [8] discussed coding protocols with adaptive decoding and feedback and obtained Theorem 4.2. They also introduced pseudoclassical channels (Sect. 4.7) and obtained the equivalence of Conditions ① and ② in Theorem 4.4. This textbook slightly improves their proof and proves the equivalence among the three Conditions ①, ②, and ③. Bennett et al. [36] obtained an interesting result regarding the classical capacity with feedback and quantum correlation. The c-q channel capacity with a cost constraint was first treated by Holevo [10], and its strong converse part was shown by Hayashi and Nagaoka [11].

On the other hand, Stratonovich and Vantsjan [21] found the result of $-\log \operatorname{Tr} W_p^2$ and not $H(W_p)$ due to some weak evaluations with respect to the error probability for the pure-state case. It would be interesting to determine the difference between these two quantities. Fujiwara [37] considered an ensemble of pure states generated randomly under the asymptotic setting. He focused on two types of orthogonality relations and found that the two quantities $H(W_p)$ and $-\log \operatorname{Tr} W_p^2$ correspond to their respective orthogonality relations.

4.9 Solutions of Exercises

Exercise 4.1 Since $H(p|u\rangle\langle u| + (1-p)|v\rangle\langle v|) = H((1-p)|u\rangle\langle u| + p|v\rangle\langle v|)$, the concavity implies $H(1/2|u\rangle\langle u| + 1/2|v\rangle\langle v|) \ge H((1-p)|u\rangle\langle u| + p|v\rangle\langle v|)$. Show that the larger eigenvalue of $1/2|u\rangle\langle u| + 1/2|v\rangle\langle v|$ is $\frac{1+|\langle v|u\rangle|}{2}$.

Exercise 4.2

$$I(p_A, W^A) + I(p_B, W^B) - I(p, W^A \otimes W^B)$$

$$
\begin{aligned}
&= \sum_{x_A, x_B} p_A(x_A) p_B(x_B) D(W^A_{x_A} \otimes W^B_{x_B} \| W^A_{p_A} \otimes W^B_{p_B}) \\
&\quad - \sum_{x_A, x_B} p(x_A, x_B) D(W^A_{x_A} \otimes W^B_{x_B} \| W^A_{p_A} \otimes W^B_{p_B}) \\
&\quad + \sum_{x_A, x_B} p(x_A, x_B) D(W^A_{x_A} \otimes W^B_{x_B} \| W^A_{p_A} \otimes W^B_{p_B}) \\
&\quad - \sum_{x_A, x_B} p(x_A, x_B) D(W^A_{x_A} \otimes W^B_{x_B} \| \left(W^A \otimes W^B\right)_p) \\
&= \sum_{x_A, x_B} (p_A(x_A) p_B(x_B) - p(x_A, x_B)) \left(D(W^A_{x_A} \| W^A_{p_A}) + D(W^B_{x_B} \| W^B_{p_B})\right) \\
&\quad + \sum_{x_A, x_B} p(x_A, x_B) \Big(- \operatorname{Tr} \left(W^A_{x_A} \otimes W^B_{x_B}\right) \log \left(W^A_{p_A} \otimes W^B_{p_B}\right) \\
&\quad + \operatorname{Tr} \left(W^A_{x_A} \otimes W^B_{x_B}\right) \log \left(W^A \otimes W^B\right)_p \Big) \\
&= D\left(\left(W^A \otimes W^B\right)_p \| W^A_{p_A} \otimes W^B_{p_B}\right) \geq 0.
\end{aligned}
$$

Exercise 4.3 It is sufficient to show that

$$
\max_{\sigma \in \mathcal{S}(\mathcal{H})} \sum_x p(x) \operatorname{Tr} W_x^{1-s} \sigma^s = \left(\operatorname{Tr} \left(\sum_x p(x) W_x^{1-s} \right)^{\frac{1}{1-s}} \right)^{1-s}, \quad \forall s \in [0, 1] \tag{4.82}
$$

$$
\min_{\sigma \in \mathcal{S}(\mathcal{H})} \sum_x p(x) \operatorname{Tr} W_x^{1-s} \sigma^s = \left(\operatorname{Tr} \left(\sum_x p(x) W_x^{1-s} \right)^{\frac{1}{1-s}} \right)^{1-s}, \quad \forall s \in (-\infty, 0]. \tag{4.83}
$$

Using the matrix Hölder inequality (A.26) $|\operatorname{Tr} XY| \leq (\operatorname{Tr} X^{\frac{1}{1-s}})^{1-s} (\operatorname{Tr} Y^{\frac{1}{s}})^s$, we have

$$
\begin{aligned}
\operatorname{Tr} \sum_x p(x) W_x^{1-s} \sigma^s &\leq \left(\operatorname{Tr} \left(\sum_x p(x) W_x^{1-s} \right)^{\frac{1}{1-s}} \right)^{1-s} (\operatorname{Tr} (\sigma^s)^{\frac{1}{s}})^s \\
&= \left(\operatorname{Tr} \left(\sum_x p(x) W_x^{1-s} \right)^{\frac{1}{1-s}} \right)^{1-s}.
\end{aligned}
$$

The equality holds when $\sigma = (\sum_x p(x) W_x^{1-s})^{\frac{1}{1-s}} / \operatorname{Tr}(\sum_x p(x) W_x^{1-s})^{\frac{1}{1-s}}$. Hence, we obtain (4.82).

Equation (4.83) can be shown similarly by replacing the role of the matrix Hölder inequality (A.26) by the reverse matrix Hölder inequality (A.28).

Exercise 4.4 The relation (4.16) can be shown as follows.

$$\lim_{s\to 0} I_{1-s}(p, W) = -\frac{ds I_{1-s}(p, W)}{ds}|_{s=0}$$
$$= \frac{\sum_x p(x) \operatorname{Tr} W_x^{1-s}(\log W_x - \log W_p) W_p^s}{\sum_x p(x) \operatorname{Tr} W_x^{1-s} W_p^s}|_{s=0}$$
$$= \sum_x p(x) \operatorname{Tr} W_x(\log W_x - \log W_p) = I(p, W).$$

The relation (4.24) can be shown as follows.

$$\lim_{s\to 0} I^{\downarrow}_{1-s}(p, W) = \lim_{s\to 0} \frac{-(1-s)\log \operatorname{Tr}\left(\sum_x p(x) W_x^{1-s}\right)^{\frac{1}{1-s}}}{s}$$
$$= \lim_{s\to 0} \frac{-\log \operatorname{Tr}\left(\sum_x p(x) W_x^{1-s}\right)^{\frac{1}{1-s}}}{s} = -\frac{d \log \operatorname{Tr}\left(\sum_x p(x) W_x^{1-s}\right)^{\frac{1}{1-s}}}{ds}|_{s=0}$$
$$= \frac{-\operatorname{Tr} \frac{1}{(1-s)^2}\left(\sum_x p(x) W_x^{1-s}\right)^{\frac{1}{1-s}} \log\left(\sum_x p(x) W_x^{1-s}\right)}{\operatorname{Tr}\left(\sum_x p(x) W_x^{1-s}\right)^{\frac{1}{1-s}}}|_{s=0}$$
$$+ \frac{\operatorname{Tr}\left(\sum_x p(x) W_x^{1-s}\right)^{\frac{1}{1-s}-1} \sum_x p(x) W_x^{1-s} \log W_x}{\operatorname{Tr}\left(\sum_x p(x) W_x^{1-s}\right)^{\frac{1}{1-s}}}|_{s=0}$$
$$= -\operatorname{Tr}\left(\sum_x p(x) W_x\right) \log\left(\sum_x p(x) W_x\right) + \operatorname{Tr} \sum_x p(x) W_x \log W_x = I(p, W).$$

Exercise 4.5 Equation (4.22) implies (4.25).

Exercise 4.6 Since all of W_x are commutative with each other, we can denote W_x by $\sum_y W_x(y)|y\rangle\langle y|$. For $s \in [-1, 0)$, the Hölder inequality (A.25) implies that

$$e^{-sI_{1-s}(p,W)} = \sum_{x,y} p(x) \sum_y W_x(y)^{1-s} W_p(y)^s$$
$$= \sum_y \sum_x p(x)^{1+s} W_x(y) (p(x) W_x(y) W_p^{-1}(y))^{-s}$$
$$\le \sum_y \left(\sum_x (p(x)^{1+s} W_x(y))^{\frac{1}{1+s}}\right)^{1+s} \left(\sum_x ((p(x) W_x(y) W_p^{-1}(y))^{-s})^{\frac{1}{-s}}\right)^{-s}$$
$$= \sum_y \left(\sum_x (p(x)^{1+s} W_x(y))^{\frac{1}{1+s}}\right)^{1+s} \left(\sum_x p(x) W_x(y) W_p^{-1}(y)\right)^{-s}$$
$$= \sum_y \left(\sum_x (p(x)^{1+s} W_x(y))^{\frac{1}{1+s}}\right)^{1+s}$$

$$= \operatorname{Tr}\left(\sum_x p(x) W_x^{\frac{1}{1+s}}\right)^{1+s} = e^{-sI^{\downarrow}_{\frac{1}{1+s}}(p,W)}.$$

Replacing the Hölder inequality (A.25) by the reverse Hölder inequality (A.27), we can show the case with $s \in (0, 1]$.

Exercise 4.7 The $\le$ part in (4.33) follows from $I(\boldsymbol{M}, p, W) \le I(p, W)$. Use the Fano inequality noting the definition of $C(W)$ for the proof of the $\ge$ part.

Exercise 4.8 It is enough to show that $\min_{\sigma:\operatorname{Tr}\sigma E \le K} H(\sigma) = H(\rho_{E,K})$. Define the state $\rho := e^{-E}/\operatorname{Tr} e^{-E}$. For a given state σ, we choose the state ρ_s such that $H(\sigma) = H(\rho_s)$. Due to Exercise 3.9, $-\operatorname{Tr}\sigma E + \log \operatorname{Tr} e^{-E} - H(\sigma) = D(\sigma\|\rho) \ge D(\rho_s\|\rho) = -\operatorname{Tr}\rho_s E + \log \operatorname{Tr} e^{-E} - H(\rho_s)$, which implies that $\operatorname{Tr}\sigma E \le \operatorname{Tr}\rho_s E$. Hence, $\max_{\sigma:\operatorname{Tr}\sigma E \le K} H(\sigma) = \max_{s:\operatorname{Tr}\rho_s E \le K} H(\rho_s)$. Since $H(\rho_s) = \psi(s|\rho) - s\frac{d\psi(s|\rho)}{ds}$, $\frac{dH(\rho_s)}{ds} = -s\frac{d^2\psi(s|\rho)}{ds^2} \le 0$. Thus, $H(\rho_s)$ is monotonically decreasing for s. On the other hand, $\frac{d}{ds}\operatorname{Tr}\rho_s E = -\operatorname{Tr}\rho_s E^2 + (\operatorname{Tr}\rho_s E)^2 \le 0$. Thus, $\operatorname{Tr}\rho_s E$ is monotonically decreasing for s. Therefore, $\max_{s:\operatorname{Tr}\rho_s E \le K} H(\rho_s)$ is realized when maximum energy $\operatorname{Tr}\rho_s E$ is realized. That is, the state $\rho_{E,K}$ realizes the maximum entropy $\max_{s:\operatorname{Tr}\rho_s E \le K} H(\rho_s)$.

Exercise 4.9 From $(A - cB)^*(A - cB) \ge 0$ we have $A^*B + B^*A \le c^{-1}A^*A + cB^*B$.

Exercise 4.10 Consider the case of $c = \sqrt{\beta/(\alpha+\beta)}$.

Exercise 4.11

(**a**) Equation (4.55) follows from (4.49).
(**b**) In this case, $-sI_{1-s}(p, W) = \log \sum_x p(x) \operatorname{Tr} W_x W_p^{-s} = \log \operatorname{Tr} W_p W_p^{-s} = \log \operatorname{Tr} W_p^{1-s} = sH_{1-s}(W_p)$. Hence, (4.55) yields (4.56).

Exercise 4.12 Order the N_n signals from smallest to largest, and note that the error probability of the first $N_n/2$ signals is less than twice the average error probability.

Exercise 4.13

(**a**) Applying the Markov inequality (2.158) to (4.49) for each channel $(W^i)^{(n)}$, we have

$$\mathrm{P}_X\left\{\varepsilon[\Phi_X] > M \cdot 2^{1+s} e^{n(\phi(s|W^i,p)+sR)}\right\} < \frac{1}{M}$$

for $i = 1, \ldots, M$. Then, there exists a code $\Phi^{(n)}$ with size e^{nR} such that $\varepsilon[\Phi^{(n)}] \le M 2^{1+s} e^{n(\phi(s|W^i,p)+sR)}$ for $i = 1, \ldots, M$.
(**b**) When $R < \sup_p \min_{1\le i\le M} I(p, W^i)$ and p realizes $\sup_p \min_{1\le i\le M} I(p, W^i)$, the quantity $\max_i M 2^{1+s} e^{n(\phi(s|W^i,p)+sR)}$ goes to zero.

Exercise 4.14

(**a**) The inequality $J(p, \sigma, W) - I(p, W) = D(W_p\|\sigma) \ge 0$ holds. The equality condition is $\sigma = W_p$.
(**b**) Since $x \mapsto -\log x$ is matrix convex, $\sigma \mapsto D(W_x\|\sigma)$ is convex.

(**c**) Assume that σ is not full-rank. Since there exists an element x such that the support σ contains the support W_x, $\sup_x D(W_x\|\sigma) = \infty$. So, when we define the function $\sigma \mapsto \sup_x D(W_x\|\sigma)$ on the set of full rank densities, it satisfies the condition of Lemma A.8. So, Lemma A.8 implies the existence of $\min_{\sigma\in\mathcal{S}(\mathcal{H})}\sup_{x\in\mathcal{X}} D(W_x\|\sigma)$, Since $\sup_{p\in\mathcal{P}_f(\mathcal{X})} J(p,\sigma,W) = \sup_{x\in\mathcal{X}} D(W_x\|\sigma)$, the minimum $\min_{\sigma\in\mathcal{S}(\mathcal{H})}\sup_{p\in\mathcal{P}_f(\mathcal{X})} J(p,\sigma,W)$ also exists.
(**d**) Since $\sigma \mapsto J(p,\sigma,W)$ is convex and $p \mapsto J(p,\sigma,W)$ is linear, Lemma A.9 guarantees that $\sup_{p\in\mathcal{P}_f(\mathcal{X})}\min_{\sigma\in\mathcal{S}(\mathcal{H})} J(p,\sigma,W) = \min_{\sigma\in\mathcal{S}(\mathcal{H})}\sup_{p\in\mathcal{P}_f(\mathcal{X})} J(p,\sigma,W)$. Since $\sup_{p\in\mathcal{P}_f(\mathcal{X})} J(p,\sigma,W) = \sup_{x\in\mathcal{X}} D(W_x\|\sigma)$, we obtain (4.71).

Exercise 4.15

(**a**) Define the function $f(t) := tD(W_x\|(1-t)W_{p_1} + tW_x) + \sum x'(1-t)p_1(x')$ $D(W_{x'}\|(1-t)W_{p_1} + tW_x)$. We have $0 \le \frac{df}{dt}(0) = D(W_x\|W_{p_1}) - \sum x' p_1(x') D(W_{x'}\|$ $W_{p_1}) + \mathrm{Tr}(-W_{p_1} + W_x) = D(W_x\|W_{p_1}) - \sum x' p_1(x') D(W_{x'}\|W_{p_1})$.
(**b**)

$$\begin{aligned}
&\sup_{p\in\mathcal{P}_f(\mathcal{X})} I(p,W) = \sup_{p\in\mathcal{P}_f(\mathcal{X})}\min_{\sigma\in\mathcal{S}(\mathcal{H})} J(p,\sigma,W)\\
\le &\min_{\sigma\in\mathcal{S}(\mathcal{H})}\sup_{p\in\mathcal{P}_f(\mathcal{X})} J(p,\sigma,W) = \min_{\sigma\in\mathcal{S}(\mathcal{H})}\sup_{x\in\mathcal{X}} D(W_x\|\sigma)\\
\le &\sup_{x\in\mathcal{X}} D(W_x\|W_{p_1}) \le I(p_1,W).
\end{aligned}$$

Exercise 4.16 Since $C_c(W^A\otimes W^B) \ge C_c(W^A) + C_c(W^B)$, it is enough to show the opposite inequality. We will show the inequality only when there exist $\max_{p\in\mathcal{P}_f(\mathcal{X}_A)} I(p,W^A)$ and $\max_{p\in\mathcal{P}_f(\mathcal{X}_B)} I(p,W^B)$. However, the general case can be shown similarly. Choose $p_{A,1} \stackrel{\text{def}}{=} \mathrm{argmax}_{p\in\mathcal{P}_f(\mathcal{X}_A)} I(p,W^A)$ and $p_{B,1} \stackrel{\text{def}}{=} \mathrm{argmax}_{p\in\mathcal{P}_f(\mathcal{X}_B)} I(p,W^B)$. Then, (4.71) implies

$$\begin{aligned}
C_c(W^A\otimes W^B) \le &\sup_{(x_A,x_B)\in\mathcal{X}_A\times\mathcal{X}_B} D(W^A_{x_A}\otimes W^B_{x_B}\|W^A_{p_{A,1}}\otimes W^B_{p_{B,1}})\\
= &\sup_{(x_A,x_B)\in\mathcal{X}_A\times\mathcal{X}_B} D(W^A_{x_A}\|W^A_{p_{A,1}}) + D(W^B_{x_B}\|W^B_{p_{B,1}}) = C_c(W^A) + C_c(W^B),
\end{aligned}$$

where (a) follows from (4.84).

Exercise 4.17 We can show the existence of the minimum $\min_{\sigma\in\mathcal{S}(\mathcal{H})}\sup_{p\in\mathcal{P}_{c\le K}(\mathcal{X})}$ $J(p,\sigma,W)$ in the same way as $\min_{\sigma\in\mathcal{S}(\mathcal{H})}\sup_x D(W_x\|\sigma)$ by using Lemma A.8. Replacing $\mathcal{P}_f(\mathcal{X})$ by $\mathcal{P}_{c\le K}(\mathcal{X})$, we can show (4.72) in the same way as (4.71).

Exercise 4.18 Choose $p_1' := \mathrm{argmax}_{p\in\mathcal{P}_{c\le K}(\mathcal{X})} I(p,W)$. For a distribution $p \in \mathcal{P}_{c\le K}$ $(\mathcal{X})$, we define the function $f(t) := t\sum x(1-t)p(x)D(W_x\|(1-t)W_{p_1'} + tW_p) +$ $\sum x(1-t)p_1'(x)D(W_x\|(1-t)W_{p_1'} + tW_p)$. We have $0 \le \frac{df}{dt})(0) = \sum x p(x) D(W_x\|$ $W_{p_1'}) - \sum x p_1'(x) D(W_x\|W_{p_1'}) + \mathrm{Tr}(-W_{p_1'} + W_p) = \sum x p(x) D(W_x\|W_{p_1'}) - \sum x p_1'$ $(x) D(W_x\|W_{p_1'})$, which implies $J(p, W_{p_1'}, W) \le I(p_1', W)$.Thus,

$$\sup_{p\in\mathcal{P}_{c\le K}(\mathcal{X})} I(p,W) = \sup_{p\in\mathcal{P}_{c\le K}(\mathcal{X})} \min_{\sigma\in\mathcal{S}(\mathcal{H})} J(p,\sigma,W)$$
$$\le \min_{\sigma\in\mathcal{S}(\mathcal{H})} \sup_{p\in\mathcal{P}_{c\le K}(\mathcal{X})} J(p,\sigma,W) \le \sup_{p\in\mathcal{P}_{c\le K}(\mathcal{X})} J(p,W_{p'_1},W) \le I(p'_1,W).$$

Exercise 4.19 For a given distribution p on $\mathcal{X}_A\times\mathcal{X}_B$, we denote the marginal distributions of p by p^A and p^B. Then, we have

$$J(p^A,\sigma_A,W^A)+J(p^B,\sigma_B,W^B)-J(p,\sigma_A\otimes\sigma_B,W^A\otimes W^B)$$
$$=D((W^A\otimes W^B)_p\|W^A_{p_A}\otimes W^B_{p_B})\ge 0. \tag{4.84}$$

Since $C_{c|c_A+c_B\le K}(W^A\otimes W^B)\ge \max_{K'} C_{c|c_A\le K'}(W^A)+C_{c|c_B\le K-K'}(W^B)$, it is enough to show the opposite inequality. We will show the inequality only when there exist $\max_{p\in\mathcal{P}_{c_A\le K}(\mathcal{X}_A)} I(p,W^A)$, $\max_{p\in\mathcal{P}_{c_B\le K}(\mathcal{X}_B)} I(p,W^B)$, and $\max_{p\in\mathcal{P}_{c_A+c_B\le K}(\mathcal{X}_B)} I(p,W^A\otimes W^B)$ for any K. However, the general case can be shown similarly. Choose $p_{AB,1}\stackrel{\mathrm{def}}{=}\operatorname{argmax}_{p\in\mathcal{P}_{c_A+c_B\le K}}(\mathcal{X}_A\times\mathcal{X}_B)I(p,W^A\otimes W^B)$. Let K_A and K_B be the averages of c_A and c_B under the joint distribution $p_{AB,1}$. Choose $p_{A,1}\stackrel{\mathrm{def}}{=}\operatorname{argmax}_{p\in\mathcal{P}_{c_A=K_A}}(\mathcal{X}_A)I(p,W^A)$ and $p_{B,1}\stackrel{\mathrm{def}}{=}\operatorname{argmax}_{p\in\mathcal{P}_{c_B=K_B}(\mathcal{X}_B)}I(p,W^B)$. Then, (4.72) implies

$$C_{c|c_A+c_B\le K}(W^A\otimes W^B)$$
$$\le \sup_{p\in\mathcal{P}_{c_A+c_B\le K}(\mathcal{X}_A\times\mathcal{X}_B)} J(p,W^A_{p_{A,1}}\otimes W^B_{p_{B,1}},W^A\otimes W^B)$$
$$\stackrel{(a)}{\le} \sup_{p\in\mathcal{P}_{c_A+c_B\le K}(\mathcal{X}_A\times\mathcal{X}_B)} J(p^A,W^A_{p_{A,1}},W^A)+J(p^B,W^B_{p_{B,1}},W^B)$$
$$\le \max_{K'} C_{c|c_A\le K'}(W^A)+C_{c|c_B\le K-K'}(W^B),$$

where p^A and p^B are marginal distributions of p, and (a) follows from (4.84).

Exercise 4.20 First, notice that $\frac{df}{dt}(0)\le 0$. Using the matrix

$$\tilde{\sigma}_{1-s|p_{1-s}} := \left(\sum_{x'} p_{1-s}(x')W_{x'}^{1-s}\right)^{\frac{1}{1-s}-1} = \left(\sum_{x'} p_{1-s}(x')W_{x'}^{1-s}\right)^{\frac{s}{1-s}}$$
$$=\left[\mathrm{Tr}\left(\sum_{x'} p_{1-s}(x')W_{x'}^{1-s}\right)^{\frac{1}{1-s}}\right]^{s}\sigma^{s}_{1-s|p_{1-s}},$$

we have

$$0 \leq \frac{df}{dt}(0) = \frac{1}{1-s}\left(\operatorname{Tr} W_x^{1-s}\tilde{\sigma}_{1-s|p_{1-s}} - \sum_{x'} p_{1-s}(x') \operatorname{Tr} W_{x'}^{1-s}\tilde{\sigma}_{1-s|p_{1-s}}\right).$$

Hence,

$$\begin{aligned}
\operatorname{Tr} W_x^{1-s}\sigma_{1-s|p_{1-s}}^s &= \frac{\operatorname{Tr} W_x^{1-s}\tilde{\sigma}_{1-s|p_{1-s}}}{\left[\operatorname{Tr}\left(\sum_{x'} p_{1-s}(x') W_{x'}^{1-s}\right)^{\frac{1}{1-s}}\right]^s} \\
\leq & \frac{\sum_{x'} p_{1-s}(x') \operatorname{Tr} W_{x'}^{1-s}\tilde{\sigma}_{p_{1-s}}}{\left[\operatorname{Tr}\left(\sum_{x'} p_{1-s}(x') W_{x'}^{1-s}\right)^{\frac{1}{1-s}}\right]^s} \\
= & \frac{\operatorname{Tr}\left(\sum_{x'} p_{1-s}(x') W_{x'}^{1-s}\right)\left(\sum_{x'} p_{1-s}(x') W_{x'}^{1-s}\right)^{\frac{s}{1-s}}}{\left[\operatorname{Tr}\left(\sum_{x'} p_{1-s}(x') W_{x'}^{1-s}\right)^{\frac{1}{1-s}}\right]^s} \\
= & \frac{\operatorname{Tr}\left(\sum_{x'} p_{1-s}(x') W_{x'}^{1-s}\right)^{\frac{1}{1-s}}}{\left[\operatorname{Tr}\left(\sum_{x'} p_{1-s}(x') W_{x'}^{1-s}\right)^{\frac{1}{1-s}}\right]^s} = \left(\operatorname{Tr}\left(\sum_{x'} p_{1-s}(x') W_{x'}^{1-s}\right)^{\frac{1}{1-s}}\right)^{1-s},
\end{aligned}$$

which implies (4.63).

Exercise 4.21 Since the matrix

$$\tilde{\sigma}_{1-s|p'_{1-s}} := \left(\sum_{x'} p'_{1-s}(x') W_{x'}^{1-s}\right)^{\frac{1}{1-s}-1}$$

satisfies

$$0 \leq \frac{df}{dt}(0) = \frac{1}{1-s}\left(\sum_x p(x) \operatorname{Tr} W_x^{1-s}\tilde{\sigma}_{1-s|p'_{1-s}} - \sum_x p'_{1-s}(x) \operatorname{Tr} W_x^{1-s}\tilde{\sigma}_{1-s|p'_{1-s}}\right),$$

we have

$$\begin{aligned}
\sum_x p(x) \operatorname{Tr} W_x^{1-s}\sigma_{1-s|p'_{1-s}}^s &= \frac{\operatorname{Tr} \sum_x p(x) W_x^{1-s}\tilde{\sigma}_{1-s|p'_{1-s}}}{\left[\operatorname{Tr}\left(\sum_{x'} p'_{1-s}(x') W_{x'}^{1-s}\right)^{\frac{1}{1-s}}\right]^s} \\
\leq & \frac{\sum_{x'} p'_{1-s}(x') \operatorname{Tr} W_{x'}^{1-s}\tilde{\sigma}_{1-s|p'_{1-s}}}{\left[\operatorname{Tr}\left(\sum_{x'} p'_{1-s}(x') W_{x'}^{1-s}\right)^{\frac{1}{1-s}}\right]^s} = \left(\operatorname{Tr}\left(\sum_{x'} p'_{1-s}(x') W_{x'}^{1-s}\right)^{\frac{1}{1-s}}\right)^{1-s},
\end{aligned}$$

which implies (4.68).

Exercise 4.22 We can show the existence of the minimums $\min_{\sigma\in\mathcal{S}(\mathcal{H})}\sup_{x\in\mathcal{X}} D_{1-s}(W_x\|\sigma)$ and $\min_{\sigma\in\mathcal{S}(\mathcal{H})}\sup_{p\in\mathcal{P}_f(\mathcal{X})} J_{1-s}(p,\sigma,W)$ in the same way as $\min_{\sigma\in\mathcal{S}(\mathcal{H})}\sup_x D(W_x\|\sigma)$ by using Lemma A.8. (4.22) implies the first equation in (4.74). Also, the final equation in (4.74) follows from

$$\inf_{p\in\mathcal{P}_f(\mathcal{X})} \mathrm{Tr}\Big[\sum_x p(x)W_x^{1-s}\Big]\sigma^s = \inf_{x\in\mathcal{X}} \mathrm{Tr}\Big[W_x^{1-s}\Big]\sigma^s \text{ for } \forall s\in(0,1]$$

$$\sup_{p\in\mathcal{P}_f(\mathcal{X})} \mathrm{Tr}\Big[\sum_x p(x)W_x^{1-s}\Big]\sigma^s = \sup_{x\in\mathcal{X}} \mathrm{Tr}\Big[W_x^{1-s}\Big]\sigma^s \text{ for } \forall s\in[-1,0).$$

Hence, it is sufficient to show the following:

$$\max_\sigma \inf_{p\in\mathcal{P}_f(\mathcal{X})} \mathrm{Tr}\Big[\sum_x p(x)W_x^{1-s}\Big]\sigma^s = \inf_{p\in\mathcal{P}_f(\mathcal{X})}\max_{\sigma\in\mathcal{S}(\mathcal{H})} \mathrm{Tr}\Big[\sum_x p(x)W_x^{1-s}\Big] \quad \text{for } \forall s\in(0,1] \tag{4.85}$$

$$\min_{\sigma\in\mathcal{S}(\mathcal{H})} \sup_{p\in\mathcal{P}_f(\mathcal{X})} \mathrm{Tr}\Big[\sum_x p(x)W_x^{1-s}\Big]\sigma^s = \sup_{p\in\mathcal{P}_f(\mathcal{X})}\min_{\sigma} \mathrm{Tr}\Big[\sum_x p(x)W_x^{1-s}\Big]\sigma^s \quad \text{for } \forall s\in[-1,0). \tag{4.86}$$

Since the function $x\mapsto x^s$ is matrix concave for $s\in(0,1)$, $\sigma\mapsto\mathrm{Tr}\Big[\sum_x p(x)W_x^{1-s}\Big]\sigma^s$ is convex. Hence, Lemma A.9 yields (4.85). Similarly, since the function $x\mapsto x^s$ is matrix convex for $s\in[-1,0)$, $\sigma\mapsto\mathrm{Tr}\Big[\sum_x p(x)W_x^{1-s}\Big]\sigma^s$ is convex. Hence, Lemma A.9 yields (4.86).

Exercise 4.23

(**a**) Since $I_{1-s}(p',W)\le\sup_p I_{1-s}^{\downarrow}(p,W)$, we have $I(p',W)=\liminf_{s\to0} I_{1-s}(p',W)$ $\le\liminf_{s\to0}\sup_p I_{1-s}^{\downarrow}(p,W)$. Taking the supremum for p', we have $\sup_{p'} I(p',W)\le$ $\liminf_{s\to0}\sup_p I_{1-s}^{\downarrow}(p,W)$.
(**b**) For any s and x, there exists a parameter $\tilde{s}(s)$ between s and 0 such that $\phi(s|W_x\|\sigma)=\phi(0|W_x\|\sigma)+s\phi'(0|W_x\|\sigma)+\frac{s^2}{2}\phi''(0|W_x\|\sigma)$, i.e., $D_{1-s}(W_x\|\sigma)=D(W_x\|\sigma)-\frac{s}{2}\phi''(\tilde{s}(s)|W_x\|\sigma)$. Since (**c**) of Exercise 3.5 guarantees that

$$\phi''(\tilde{s}(s)|W_x\|\sigma)=\frac{\mathrm{Tr}\, W_x^{1-s}(\log\sigma-\log W_x)\sigma^s(\log\sigma-\log W_x)}{\mathrm{Tr}\,\rho^{1-s}\sigma^s} - \frac{(\mathrm{Tr}\, W_x^{1-s}\sigma^s(\log\sigma-\log W_x))^2}{(\mathrm{Tr}\, W_x^{1-s}\sigma^s)^2},$$

the quantity $\sup_{s\in[-\epsilon,\epsilon]}\sup_x\phi''(s|W_x\|\sigma)$ exists with sufficiently small $\epsilon>0$. So, the convergence $D_{1-s}(W_x\|\sigma)\to D(W_x\|\sigma)$ is uniform for x.
(**c**) The state σ_1 satisfies the condition in (**b**). Since $\min_\sigma\sup_x D_{1-s}(W_x\|\sigma)\le$ $\sup_x D_{1-s}(W_x\|\sigma_1)$, we have $\limsup_{s\to0}\min_\sigma\sup_x D_{1-s}(W_x\|\sigma)\le\lim_{s\to0}\sup_x D_{1-s}$

$(W_x\|\sigma_1) = \sup_x D(W_x\|\sigma_1) = \min_\sigma \sup_x D(W_x\|\sigma)$. Hence, (4.71) and (4.74) imply $\sup_p I(p, W) \geq \limsup_{s\to 0} \sup_p I_{1-s}^{\downarrow}(p, W)$.

Exercise 4.24

(**a**) Since $\sigma_{1-s|p_{1-s}}$ gives the minimum $\min_\sigma \mathrm{Tr}\Big[\sum_x p_{1-s}(x) W_x^{1-s}\Big]\sigma^s$ due to (4.83), (4.86) implies (4.75).
(**b**) The inequality (4.65) follows as

$$
\begin{aligned}
&\log \mathrm{Tr}\Big[(W^{(n)}_{\varphi^{(n)}(i)})^{1-s}\sigma^s_{1-s|p_{1-s}}\Big] = \sum_{l=1}^{n} \log \mathrm{Tr}\Big[(W_{\varphi_l^{(n)}(i)})^{1-s}\sigma^s_{1-s|p_{1-s}}\Big] \\
=&n \sum_{l=1}^{n} \frac{1}{n} \log \mathrm{Tr}\Big[(W_{\varphi_l^{(n)}(i)})^{1-s}\sigma^s_{1-s|p_{1-s}}\Big] \\
\overset{(a)}{\leq}&n \log \sum_{l=1}^{n} \frac{1}{n} \mathrm{Tr}\Big[(W_{\varphi_l^{(n)}(i)})^{1-s}\sigma^s_{1-s|p_{1-s}}\Big] \\
=&n \log \mathrm{Tr}\Big[\sum_{l=1}^{n} \frac{1}{n}(W_{\varphi_l^{(n)}(i)})^{1-s}\Big]\sigma^s_{1-s|p_{1-s}} \\
\leq&n \log \max_p \mathrm{Tr}\Big[\sum_x p(x)(W_x)^{1-s}\Big]\sigma^s_{1-s|p_{1-s}} \\
\overset{(b)}{\leq}&n(1-s)\log\left(\mathrm{Tr}\left(\sum_x p(x) W_x^{1-s}\right)^{\frac{1}{1-s}}\right),
\end{aligned}
$$

where (a) follows from the concavity of $x \mapsto \log x$ and (b) follows from (4.75).

Exercise 4.25 This exercise can be solved by the same way as Exercise 4.16 by replacing the role of (4.71) by that of (4.74).

Exercise 4.26 We can show the existence of the minimums $\min_{\sigma\in\mathcal{S}(\mathcal{H})} \sup_{p\in\mathcal{P}_{c\leq K}(\mathcal{X})} J_{1-s}(p, \sigma, W)$ in the same way as $\min_{\sigma\in\mathcal{S}(\mathcal{H})} \sup_x D(W_x\|\sigma)$ by using Lemma A.8. Due to (4.22), to show (4.74), it is sufficient to show the following:

$$
\begin{aligned}
&\max_{\sigma\in\mathcal{S}(\mathcal{H})} \inf_{p\in\mathcal{P}_{c\leq K}(\mathcal{X})} \mathrm{Tr}\Big[\sum_x p(x) W_x^{1-s}\Big]\sigma^s \\
=&\inf_{p\in\mathcal{P}_{c\leq K}(\mathcal{X})} \max_{\sigma\in\mathcal{S}(\mathcal{H})} \mathrm{Tr}\Big[\sum_x p(x) W_x^{1-s}\Big] \quad \forall s \in (0, 1],
\end{aligned}
\tag{4.87}
$$

$$
\begin{aligned}
&\min_{\sigma\in\mathcal{S}(\mathcal{H})} \sup_{p\in\mathcal{P}_{c\leq K}(\mathcal{X})} \mathrm{Tr}\Big[\sum_x p(x) W_x^{1-s}\Big]\sigma^s \\
=&\sup_{p\in\mathcal{P}_{c\leq K}(\mathcal{X})} \min_{\sigma\in\mathcal{S}(\mathcal{H})} \mathrm{Tr}\Big[\sum_x p(x) W_x^{1-s}\Big]\sigma^s \quad \forall s \in [-1, 0).
\end{aligned}
\tag{4.88}
$$

Since the function $x \mapsto x^s$ is matrix concave for $s \in (0,1)$, $\sigma \mapsto \operatorname{Tr}\left[\sum_x p(x) W_x^{1-s}\right]\sigma^s$ is convex. Hence, Lemma A.9 yields (4.87). Similarly, since the function $x \mapsto x^s$ is matrix convex for $s \in [-1,0)$, $\sigma \mapsto \operatorname{Tr}\left[\sum_x p(x) W_x^{1-s}\right]\sigma^s$ is convex. Hence, Lemma A.9 yields (4.88).

Exercise 4.27

(**a**) Since $I_{1-s}(p', W) \le \sup_{p \in \mathcal{P}_{c \le K}(\mathcal{X})} I^{\downarrow}_{1-s}(p, W)$ for $p' \in \mathcal{P}_{c \le K}(\mathcal{X})$, we have $I(p', W) = \liminf_{s \to 0} I_{1-s}(p', W) \le \liminf_{s \to 0} \sup_{p \in \mathcal{P}_{c \le K}(\mathcal{X})} I^{\downarrow}_{1-s}(p, W)$. Taking the supremum for p', we have $\sup_{p' \in \mathcal{P}_{c \le K}(\mathcal{X})} I(p', W) \le \liminf_{s \to 0} \sup_{p \in \mathcal{P}_{c \le K}(\mathcal{X})} I^{\downarrow}_{1-s}(p, W)$.
(**b**) We replace $\phi(s|W_x\|\sigma)$ by $-sJ_{1-s}(p, \sigma, W)$ in the proof of (**b**) of Exercise 4.23. Then, we can show the desired argument.
(**c**) The state σ'_1 satisfies the condition in (**b**). Since $\min_\sigma \sup_{p \in \mathcal{P}_{c \le K}(\mathcal{X})} J_{1-s}(p, \sigma, W) \le \sup_{p \in \mathcal{P}_{c \le K}(\mathcal{X})} J_{1-s}(p, \sigma'_1, W)$, we have
$\limsup_{s \to 0} \min_\sigma \sup_{p \in \mathcal{P}_{c \le K}(\mathcal{X})} J_{1-s}(p, \sigma, W) \le \lim_{s \to 0} \sup_{p \in \mathcal{P}_{c \le K}(\mathcal{X})} J_{1-s}(p, \sigma'_1, W) = \sup_{p \in \mathcal{P}_{c \le K}(\mathcal{X})} J(p, \sigma'_1, W) = \min_\sigma \sup_{p \in \mathcal{P}_{c \le K}(\mathcal{X})} J(p, \sigma, W)$. Hence, (4.72) and (4.77) imply $\sup_{p \in \mathcal{P}_{c \le K}(\mathcal{X})} I(p, W) \ge \limsup_{s \to 0} \sup_{p \in \mathcal{P}_{c \le K}(\mathcal{X})} I^{\downarrow}_{1-s}(p, W)$.

Exercise 4.28

(**a**) First, notice that $\frac{df}{dt}(0) \le 0$. Using the matrix $\tilde{\sigma}_{1-s} := S_{1-s}^{\frac{1}{1-s}-1} = S_{1-s}^{\frac{s}{1-s}} = \sigma^s_{1-s}$, we have

$$0 \le \frac{df}{dt}(0) = \frac{1}{1-s}\left(\operatorname{Tr} W_x^{1-s}\tilde{\sigma}_{1-s} - \operatorname{Tr} S_{1-s}\tilde{\sigma}_{1-s}\right).$$

Hence,

$$\begin{aligned}\operatorname{Tr} W_x^{1-s}\sigma^s_{1-s} &= \frac{\operatorname{Tr} W_x^{1-s}\tilde{\sigma}_{1-s}}{[\operatorname{Tr} S_{1-s}^{\frac{1}{1-s}}]^s} \le \frac{\operatorname{Tr} S_{1-s}\tilde{\sigma}_{p_{1-s}}}{[\operatorname{Tr} S_{1-s}^{\frac{1}{1-s}}]^s} \\ &= \frac{\operatorname{Tr} S_{1-s} S_{1-s}^{\frac{s}{1-s}}}{[\operatorname{Tr} S_{1-s}^{\frac{1}{1-s}}]^s} = \frac{\operatorname{Tr} S_{1-s}^{\frac{1}{1-s}}}{[\operatorname{Tr} S_{1-s}^{\frac{1}{1-s}}]^s} = (\operatorname{Tr} S_{1-s}^{\frac{1}{1-s}})^{1-s},\end{aligned}$$

which implies (4.78).
(**b**) Inequality (4.78) shown in (**a**) implies that

$$\operatorname{Tr} W_x^{1-s}\sigma^s_{1-s} \le (\operatorname{Tr} S_{1-s}^{\frac{1}{1-s}})^{1-s}.$$

Then, we can show that

$$\log \operatorname{Tr}\left[(W^{(n)}_{\varphi^{(n)}(i)})^{1-s}\sigma^s_{1-s}\right] \le n(1-s)\log(\operatorname{Tr} S_{1-s}^{\frac{1}{1-s}})$$

in the same way as (4.65). Hence, we can show that

$$\frac{1}{n}\log(1-\varepsilon[\Phi^{(n)}]) \le \log(\mathrm{Tr}\, S_{1-s}^{\frac{1}{1-s}}) + \frac{s}{1-s}R$$

in the same way as (4.67). So, we obtain (4.19).

Exercise 4.29

(**a**) Since the matrix $\tilde{\sigma}'_{1-s} := {S'_{1-s}}^{\frac{1}{1-s}-1}$ satisfies

$$0 \le \frac{df}{dt}(0) = \frac{1}{1-s}\left(\sum_x p(x)\,\mathrm{Tr}\, W_x^{1-s}\tilde{\sigma}'_{1-s} - \mathrm{Tr}\, S'_{1-s}\tilde{\sigma}'_{1-s}\right),$$

we have

$$\sum_x p(x)\,\mathrm{Tr}\, W_x^{1-s}{\sigma'_{1-s}}^{s} = \frac{\mathrm{Tr}\sum_x p(x) W_x^{1-s}\tilde{\sigma}'_{1-s}}{[\mathrm{Tr}\, {S'_{1-s}}^{\frac{1}{1-s}}]^s}$$
$$\le \frac{\mathrm{Tr}\, S'_{1-s}\tilde{\sigma}'_{1-s}}{[\mathrm{Tr}\, {S'_{1-s}}^{\frac{1}{1-s}}]^s} = (\mathrm{Tr}\, {S'_{1-s}}^{\frac{1}{1-s}})^{1-s},$$

which implies (4.79).
(**b**) Inequality (4.79) shown in (**a**) implies that

$$\sum_x p(x)\,\mathrm{Tr}\, W_x^{1-s}{\sigma'_{1-s}}^{s} \le (\mathrm{Tr}\, {S'_{1-s}}^{\frac{1}{1-s}})^{1-s}$$

for $p \in \mathcal{P}_{c\le K}(\mathcal{X})$. Then, we can show that

$$\log \mathrm{Tr}\left[(W^{(n)}_{\varphi^{(n)}(i)})^{1-s}{\sigma'_{1-s}}^{s}\right] \le n(1-s)\log(\mathrm{Tr}\, {S'_{1-s}}^{\frac{1}{1-s}})$$

in the same way as (4.70). Hence, we obtain (4.39).

Exercise 4.30

(**a**) Since $\sigma_{1-s|p'_{1-s}}$ gives the minimum of $\min_\sigma \mathrm{Tr}\left[\sum_x p'_{1-s}(x) W_x^{1-s}\right]\sigma^s$ due to (4.88), we have (4.80).
(**b**) The inequality (4.65) follows as

$$\log \mathrm{Tr}\left[(W^{(n)}_{\varphi^{(n)}(i)})^{1-s}\sigma^s_{1-s|p'_{1-s}}\right] \overset{(a)}{\le} n\log\sum_{l=1}^n \frac{1}{n}\mathrm{Tr}\left[(W_{\varphi_l^{(n)}(i)})^{1-s}\sigma^s_{1-s|p'_{1-s}}\right]$$
$$= n\log \mathrm{Tr}\left[\sum_{l=1}^n \frac{1}{n}(W_{\varphi_l^{(n)}(i)})^{1-s}\right]\sigma^s_{1-s|p'_{1-s}}$$
$$\le n\log \max_{p\in\mathcal{P}_{c\le K}(\mathcal{X})} \mathrm{Tr}\left[\sum_x p(x)(W_x)^{1-s}\right]\sigma^s_{1-s|p'_{1-s}}$$

$$\stackrel{(b)}{\leq} n(1-s)\log\left(\mathrm{Tr}\left(\sum_x p'_{1-s}(x)W_x^{1-s}\right)^{\frac{1}{1-s}}\right), \tag{4.89}$$

where (a) follows form the concavity of $x \mapsto \log x$ and (b) follows from (4.80).

Exercise 4.31

(**a**) Apply the Markov inequality to the uniform distribution on the message set $\{1, \ldots, N_n\}$.

Exercise 4.32

(**a**) We denote the j-th marginal distribution of $p^{(n)}$ by p_j. Then, (4.5) yields that $\sum_{j=1}^n I(p_j, W^i) \geq I(p^{(n)}, (W^i)^{(n)})$. Equation (4.3) yields that $\sum_{j=1}^n I(p_j, W^i) \leq nI(\sum_{j=1}^n \frac{1}{n}p_j, W^i)$. Hence, the distribution $\sum_{j=1}^n \frac{1}{n}p_j$ satisfies the desired condition.

(**b**) Consider an encoder $\tilde{\Phi}^{(n)}$ with the size N_n and the decoder $Y^{(n)}$. Choose the distribution $p^{(n)}$ on $\mathcal{X}^n$ as $p^{(n)}(x) = \frac{1}{N_n}$ for $x \in \mathbf{Im}\,\tilde{\Phi}^{(n)}$ and $p^{(n)}(x) = 0$ for $x \notin \mathbf{Im}\,\tilde{\Phi}^{(n)}$. Since the error is given as the maximum value for the choice of the channels $W^1, \ldots, W^M$, Fano inequality (2.35) implies that

$$\log 2 + \varepsilon[\tilde{\Phi}^{(n)}]\log N_n \geq \log N_n - I(Y^{(n)}, p^{(n)}, (W^i)^{(n)}) \tag{4.90}$$

for $i = 1, \ldots, M$. Then, choosing the distribution p on $\mathcal{X}$ according to (**a**), we have

$$I(Y^{(n)}, p^{(n)}, (W^i)^{(n)}) \leq I(p^{(n)}, (W^i)^{(n)}) \leq nI(p, W^i) \tag{4.91}$$

Thus, we obtain

$$\frac{1}{n}(\log 2 + \varepsilon[\tilde{\Phi}^{(n)}]\log N_n) \geq \frac{1}{n}\log N_n - \min_i I(p, W^i). \tag{4.92}$$

Taking the limit $n \to \infty$, we have

$$\lim_{n\to\infty} \frac{1}{n}\log N_n \leq \min_i I(p, W^i), \tag{4.93}$$

That is, we have $C_c(W^1, \ldots, W^M) \leq \sup_p \min_{1\leq i\leq M} I(p, W^i)$.

References

1. C.E. Shannon, A mathematical theory of communication. Bell Syst. Tech. J. **27**, 623–656 (1948)
2. A.S. Holevo, Bounds for the quantity of information transmitted by a quantum communication channel. Problemy Peredachi Informatsii **9**, 3–11 (1973) (in Russian). (English translation: Probl. Inf. Transm. **9**, 177–183 (1975))

3. A.S. Holevo, On the capacity of quantum communication channel. Problemly Peredachi Informatsii **15**, 4, 3–11 (1979) (in Russian). (English translation: Probl. Inf. Transm. **15**, 247–253 (1979).)
4. A.S. Holevo, The capacity of the quantum channel with general signal states. IEEE Trans. Inf. Theory **44**, 269 (1998)
5. B. Schumacher, M.D. Westmoreland, Sending classical information via noisy quantum channels. Phys. Rev. A **56**, 131 (1997)
6. T. Ogawa, H. Nagaoka, Strong converse to the quantum channel coding theorem. IEEE Trans. Inf. Theory **45**, 2486–2489 (1999)
7. M. Hayashi, Exponential decreasing rate of leaked information in universal random privacy amplification. IEEE Trans. Inf. Theory **57**, 3989–4001 (2011)
8. A. Fujiwara, H. Nagaoka, Operational capacity and pseudoclassicality of a quantum channel. IEEE Trans. Inf. Theory **44**, 1071–1086 (1998)
9. M. Fujiwara, M. Takeoka, J. Mizuno, M. Sasaki, Exceeding classical capacity limit in quantum optical channel. Phys. Rev. Lett. **90**, 167906 (2003)
10. A.S. Holevo, On quantum communication channels with constrained inputs. quant-ph/9705054 (1997)
11. M. Hayashi, H. Nagaoka, General formulas for capacity of classical-quantum channels. IEEE Trans. Inf. Theory **49**, 1753–1768 (2003)
12. M. Ohya, D. Petz, N. Watanabe, On capacities of quantum channels. Prob. Math. Stat. **17**, 179–196 (1997)
13. B. Schumacher, M.D. Westmoreland, Optimal signal ensembles. Phys. Rev. A **63**, 022308 (2001)
14. M.A. Naĭmark, Comptes Rendus (Doklady) de l'Acadenie des Sience de l'URSS, **41**, 9, 359 (1943)
15. L.B. Levitin, On quantum measure of information, in *Proceedings 4th All-Union Conference on Information Transmission and Coding Theory*, pp. 111–115 (Tashkent, 1969) (in Russian). English translation: *Information, Complexity and Control in Quantum Physics*, eds. by A. Blaquiere, S. Diner, G. Lochak (Springer, Berlin, 1987), pp. 15–47
16. G. Lindblad, Completely positive maps and entropy inequalities. Commun. Math. Phys. **40**, 147–151 (1975)
17. A. Uhlmann, Relative entropy and the Wigner-Yanase-Dyson-Lieb concavity in an interpolation theory. Commun. Math. Phys. **54**, 21–32 (1977)
18. E. Lieb, M.B. Ruskai, A fundamental property of quantum mechanical entropy. Phys. Rev. Lett. **30**, 434–436 (1973)
19. E. Lieb, M.B. Ruskai, Proof of the strong subadditivity of quantum mechanical entropy. J. Math. Phys. **14**, 1938–1941 (1973)
20. H. Nagaoka, Strong converse theorems in quantum information theory, in *Proceedings ERATO Conference on Quantum Information Science (EQIS) 2001*, vol. 33 (2001). (also appeared as Chap. 3 of *Asymptotic Theory of Quantum Statistical Inference*, ed. by M. Hayashi)
21. R.L. Stratonovich, A.G. Vantsjan, Probl. Control Inf. Theory **7**, 161–174 (1978)
22. P. Hausladen, R. Jozsa, B. Schumacher, M. Westmoreland, W. Wooters, Classical information capacity of a quantum channel. Phys. Rev. A **54**, 1869–1876 (1996)
23. T. Cover, J. Thomas, *Elements of Information Theory* (Wiley, New York, 1991)
24. I. Csiszár, J. Körner, *Information Theory: Coding Theorems for Discrete Memoryless Systems* (Academic, 1981)
25. A. Winter, *Coding Theorems of Quantum Information Theory*, Ph.D. Dissertation, Universität Bielefeld (2000); quant-ph/9907077 (1999)
26. A. Winter, Coding theorem and strong converse for quantum channels. IEEE Trans. Inf. Theory **45**, 2481–2485 (1999)
27. M. Hayashi, Universal coding for classical-quantum channel. Commun. Math. Phys. **289**(3), 1087–1098 (2009)
28. I. Bjelakovic, H. Boche, Classical capacities of averaged and compound quantum channels. IEEE Trans. Inf. Theory **55**(7), 3360–3374 (2009)

29. T.S. Han, *Information-Spectrum Methods in Information Theory* (Springer, Berlin, 2002) (originally appeared in Japanese in 1998)
30. S. Verdú, T.S. Han, A general formula for channel capacity. IEEE Trans. Inf. Theory **40**, 1147–1157 (1994)
31. Y. Polyanskiy, H.V. Poor, S. Verdu, Channel coding rate in the finite blocklength regime. IEEE Trans. Inf. Theory **56**(5), 2307–2359 (2010)
32. T. Ogawa, H. Nagaoka, Making good codes for classical-quantum channel coding via quantum hypothesis testing. IEEE Trans. Inf. Theory **53**, 2261–2266 (2007)
33. M.V. Burnashev, A.S. Holevo, On reliability function of quantum communication channel. Prob. Inf. Trans. **34**, 97–107 (1998)
34. M. Hayashi, Error exponent in asymmetric quantum hypothesis testing and its application to classical-quantum channel coding. Phys. Rev. A **76**, 062301 (2007)
35. M. Dalai, Lower bounds on the probability of error for classical and classical-quantum channels. IEEE Trans. Inf. Theory **59**(12), 8027–8056 (2013)
36. C.H. Bennett, I. Devetak, P.W. Shor, J.A. Smolin, Inequalities and separations among assisted capacities of quantum channels. Phys. Rev. Lett. **96**, 150502 (2006)
37. A. Fujiwara, Quantum birthday problems: geometrical aspects of quantum randomcoding. IEEE Trans. Inf. Theory **47**, 2644–2649 (2001)

29. T.S. Han, *Information-Spectrum Methods in Information Theory* (Springer, Berlin, 2003) (Originally appeared in Japanese in 1998)
30. S. Verdú, T.S. Han, A general formula for channel capacity. IEEE Trans. Inf. Theory 40, 1147–1157 (1994)
31. Y. Polyanskiy, H.V. Poor, S. Verdú, Channel coding rate in the finite blocklength regime. IEEE Trans. Inf. Theory 56(5), 2307–2359 (2010)
32. T. Ogawa, H. Nagaoka, Making good codes for classical-quantum channel coding via quantum hypothesis testing. IEEE Trans. Inf. Theory 53, 2261–2266 (2007)
33. M. Hayashi, H. Nagaoka, General formulas for capacity of classical-quantum channels. IEEE Trans. Inf. Theory 49, 1753–1768 (2003)
34. M. Hayashi, Error exponent in asymmetric quantum hypothesis testing and its application to classical-quantum channel coding. Phys. Rev. A 76, 062301 (2007)
35. M. Dalai, Lower bounds on the probability of error for classical and classical-quantum channels. IEEE Trans. Inf. Theory 59, 8027–8056 (2013)
36. C.H. Bennett, [illegible] quantum channels. Phys. Rev. Lett. [illegible] (1999)
37. A. [illegible], Quantum [illegible] of quantum [illegible]. IEEE Trans. Inf. Theory 57, [illegible] (2011)

Chapter 5
State Evolution and Trace-Preserving Completely Positive Maps

Abstract Until now, we have considered only quantum states and quantum measurement as quantum concepts. In order to prefer information processing with quantum systems, we should manipulate a wider class of state operations. This chapter examines what kinds of operations are allowed on quantum systems. The properties of these operations will also be examined.

5.1 Description of State Evolution in Quantum Systems

The time evolution over a time t of a closed quantum-mechanical system $\mathcal{H}$ is given by

$$\rho \mapsto e^{itH} \rho e^{-itH},$$

where H is a Hermitian matrix in $\mathcal{H}$ called the Hamiltonian. However, this is true only if there is no interaction between the system $\mathcal{H}$ and another system. The state evolution in the presence of an interaction cannot be written in the above way. Furthermore, the input system for information processing (i.e., state evolution) is not necessarily the same as its output system. In fact, in some processes it is crucial for the input and output systems to be different. Hence, we will denote the input and output system by $\mathcal{H}_A$ and $\mathcal{H}_B$, respectively, and investigate the map κ from the set $\mathcal{S}(\mathcal{H}_A)$ of densities on the system $\mathcal{H}_A$ to $\mathcal{S}(\mathcal{H}_B)$, which gives the relationship between the input and the output (state evolution). First, we require the map κ to satisfy the condition

$$\kappa(\lambda \rho_1 + (1 - \lambda)\rho_2) = \lambda \kappa(\rho_1) + (1 - \lambda)\kappa(\rho_2)$$

for $1 > \lambda > 0$ and arbitrary $\rho_1, \rho_2 \in \mathcal{S}(\mathcal{H}_A)$. Maps satisfying this property are called **affine** maps. Since the space $\mathcal{S}(\mathcal{H}_A)$ is not a linear space, we cannot claim that κ is linear; however, these two conditions are almost equivalent. In fact, we may extend the map κ to a linear map $\tilde{\kappa}$ that maps from the linear space $\mathcal{T}(\mathcal{H}_A)$ of Hermitian matrices on $\mathcal{H}_A$ to the linear space $\mathcal{T}(\mathcal{H}_B)$ of the Hermitian matrices on $\mathcal{H}_B$ as

M. Hayashi, *Quantum Information Theory*, Graduate Texts in Physics,
DOI 10.1007/978-3-662-49725-8_5

follows. Since an arbitrary matrix $X \in \mathcal{T}(\mathcal{H}_A)$ can be written as a linear sum

$$X = \sum_i a^i X_i, \quad a^i \in \mathbb{R} \tag{5.1}$$

by using elements $X_1, \ldots, X_{d^2}$ of $\mathcal{S}(\mathcal{H}_A)$, the map $\tilde{\kappa}$ may be defined as

$$\tilde{\kappa}(X) \stackrel{\text{def}}{=} \sum_i a^i \kappa(X_i).$$

The affine property guarantees that this definition does not depend on (5.1). Henceforth, we shall identify $\tilde{\kappa}$ with κ. The linear combination of the elements in $\mathcal{T}(\mathcal{H}_A)$ multiplied by complex constants gives the space $\mathcal{M}(\mathcal{H}_A)$ of the matrices on $\mathcal{H}_A$. Since any element of $\mathcal{M}(\mathcal{H}_A)$ can be written as $Z = X + iY$ with two Hermitian matrices X and Y, κ may be extended to a map from the space $\mathcal{M}(\mathcal{H}_A)$ of matrices on $\mathcal{H}_A$ to the space $\mathcal{M}(\mathcal{H}_B)$ of matrices on $\mathcal{H}_B$ as $\kappa(X + iY) := \kappa(X) + i\kappa(Y)$, which satisfies $\kappa(Z^*) = \kappa(Z)^*$. It is often more convenient to regard κ as a linear map in discussions on its properties; hence, we will often use κ as the linear map from $\mathcal{T}(\mathcal{H}_A)$ to $\mathcal{T}(\mathcal{H}_B)$. Occasionally, it is even more convenient to treat κ as a map from $\mathcal{M}(\mathcal{H}_A)$ to $\mathcal{M}(\mathcal{H}_B)$ We shall examine these cases explicitly.

In order to recover the map from $\mathcal{S}(\mathcal{H}_A)$ to $\mathcal{S}(\mathcal{H}_B)$ from the linear map κ from $\mathcal{T}(\mathcal{H}_A)$ to $\mathcal{T}(\mathcal{H}_B)$, we assume that the linear map transforms positive semidefinite matrices to positive semidefinite matrices. This map is called a **positive map**. The trace also needs to be preserved.

However, there are still more conditions that the state evolution must satisfy. In fact, we consider the state evolution κ occurring on the quantum system $\mathcal{H}_A$ whose state is entangled with another system $\mathbb{C}^n$. We also suppose that the additional system $\mathbb{C}^n$ is stationary and has no state evolution. Then, any state on the composite system of $\mathbb{C}^n$ and $\mathcal{H}_A$ obeys the state evolution from $\mathcal{H}_A \otimes \mathbb{C}^n$ to $\mathcal{H}_B \otimes \mathbb{C}^n$, which is given by the linear map $\kappa \otimes \iota_n$ from $\mathcal{T}(\mathcal{H}_A \otimes \mathbb{C}^n) = \mathcal{T}(\mathcal{H}_A) \otimes \mathcal{T}(\mathbb{C}^n)$ to $\mathcal{T}(\mathcal{H}_B \otimes \mathbb{C}^n) = \mathcal{T}(\mathcal{H}_B) \otimes \mathcal{T}(\mathbb{C}^n)$, where ι_n denotes the identity operator from $\mathcal{T}(\mathbb{C}^n)$ to itself. The map $\kappa \otimes \iota_n$ then must satisfy positivity and trace-preserving properties, as discussed above. In this case, the system $\mathbb{C}^n$ is called the **reference system**. The trace-preserving property follows from the trace-preserving property of κ:

$$\operatorname{Tr}(\kappa \otimes \iota_n)\left(\sum_i X_i \otimes Y_i\right) = \sum_i \operatorname{Tr}\left(\kappa(X_i) \otimes \iota_n(Y_i)\right)$$

$$= \sum_i \operatorname{Tr}\kappa(X_i) \cdot \operatorname{Tr}\iota_n(Y_i) = \sum_i \operatorname{Tr}X_i \cdot \operatorname{Tr}Y_i = \operatorname{Tr}\left(\sum_i X_i \otimes Y_i\right).$$

However, as shown by the counterexample given in Example 5.7 of Sect. 5.2, it is not possible to deduce that $\kappa \otimes \iota_n$ is a positive map from the fact that κ is a positive map. In a composite system involving an n-dimensional reference system $\mathbb{C}^n$, the map κ

is called an **n-positive map** if $\kappa \otimes \iota_n$ is a positive map. If κ is an n-positive map for arbitrary n, the map κ is called a **completely positive map**, which we abbreviate to **CP map**. Since the trace of a density matrix is always 1, the state evolution of a quantum system is given by a **trace-preserving completely positive map**, which is abbreviated to **TP-CP map**. It is currently believed that it is, in principle, possible to produce state evolutions corresponding to arbitrary TP-CP maps, as is shown by Theorem 5.1 discussed later. If a channel has a quantum input system as well as a quantum output system, it can be represented by a TP-CP map. Such channels are called quantum–quantum channels to distinguish them from classical-quantum channels. Strictly speaking, κ is a linear map from $\mathcal{T}(\mathcal{H}_A)$ to $\mathcal{T}(\mathcal{H}_B)$; however, for simplicity, we will call it a TP-CP map from the quantum system $\mathcal{H}_A$ to the quantum system $\mathcal{H}_B$. In particular, since the case with the pure input state is important, we abbreviate $\kappa(|x\rangle\langle x|)$ to $\kappa(x)$.

In fact, it is often convenient to discuss the adjoint map κ^* defined as

$$\operatorname{Tr} \kappa(X) Y^* = \operatorname{Tr} X \kappa^*(Y)^*, \quad \forall X \in \mathcal{M}(\mathcal{H}_A), \forall Y \in \mathcal{M}(\mathcal{H}_B). \tag{5.2}$$

That is, κ^* can be regarded as the dual map with respect to the inner product $\langle X, Y \rangle \stackrel{\text{def}}{=} \operatorname{Tr} XY^*$. Then, the trace-preserving property of κ can be translated to identity-preserving property of κ^*. That is, κ is trace-preserving if and only if κ^* is identity-preserving due to (5.2) with $Y = I$[Exe. 5.1].

A map κ is positive if and only if the adjoint map κ^* is positive because their conditions are written as $\operatorname{Tr} \kappa(X) Y = \operatorname{Tr} X \kappa^*(Y) \geq 0$ for any non-negative matrices $X \in \mathcal{T}(\mathcal{H}_A)$ and $Y \in \mathcal{T}(\mathcal{H}_B)$. Similarly, a map κ is n-positive if and only if the adjoint map κ^* is n-positive. Therefore, the completely positivity for κ is equivalent to that for κ^*. Therefore, we can discuss the adjoint map κ^* instead of the original map κ.

Before proceeding to analysis on completely positive maps, we discuss 2-positive maps. A map κ is 2-positive if and only if $\begin{pmatrix} \kappa(A) & \kappa(B^*) \\ \kappa(B) & \kappa(C) \end{pmatrix} \geq 0$ for any matrices A, B, and C satisfying the matrix inequality $\begin{pmatrix} A & B^* \\ B & C \end{pmatrix} \geq 0$ in $\mathcal{T}(\mathcal{H}_A \otimes \mathbb{C}^2)$. Now, we assume that an identity-preserving map κ^* is 2-positive. Since $\begin{pmatrix} X^*X & X^* \\ X & I \end{pmatrix} = \begin{pmatrix} X^* \\ I \end{pmatrix} (X I) \geq 0$, we find that $\begin{pmatrix} \kappa^*(X^*X) & \kappa^*(X)^* \\ \kappa^*(X) & I \end{pmatrix} = \begin{pmatrix} \kappa^*(X^*X) & \kappa^*(X^*) \\ \kappa^*(X) & \kappa^*(I) \end{pmatrix} \geq 0$, which implies[Exe. 5.2]

$$\kappa^*(X^*X) \geq \kappa^*(X)\kappa^*(X)^* = \kappa(X)^*\kappa^*(X^*). \tag{5.3}$$

We now give the matrix representation of the linear map from $\mathcal{T}(\mathcal{H}_A)$ to $\mathcal{T}(\mathcal{H}_B)$ and a necessary and sufficient conditions for it to be a TP-CP map. We denote the basis of the quantum systems $\mathcal{H}_A$ and $\mathcal{H}_B$ by $e_1^A, \ldots, e_d^A$ and $e_1^B, \ldots, e_{d'}^B$, respectively. We define $K(\kappa)$ as a matrix in $\mathcal{H}_A \otimes \mathcal{H}_B$ for κ according to

$$K(\kappa)^{(j,l),(i,k)} \stackrel{\text{def}}{=} \langle e_k^B | \kappa(|e_i^A\rangle\langle e_j^A|) | e_l^B \rangle. \tag{5.4}$$

Let $X = \sum_{i,j} x^{i,j} |e_i^A\rangle\langle e_j^A|$, $Y = \sum_{k,l} y_{k,l} |e_k^B\rangle\langle e_l^B|$. Then, we can write

$$\operatorname{Tr} \kappa(X)Y = \sum_{i,j,k,l} x^{i,j} y_{k,l} \langle e_l^B | \kappa(|e_i^A\rangle\langle e_j^A|) | e_k^B \rangle = \operatorname{Tr}(X \otimes Y^T) K(\kappa).$$

Now, let $\mathcal{H}_R$ be the space spanned by $e_1^R, \ldots, e_d^R$. Then, the maximally entangled state $|\Phi_d\rangle = \frac{1}{\sqrt{d}} \sum_i e_i^A \otimes e_i^R$ characterizes this matrix representation of κ as

$$\begin{aligned} &(\kappa \otimes \iota_R)(|\Phi_d\rangle\langle\Phi_d|) \\ =&\frac{1}{d} \sum_{i,j} \kappa(|e_i^A\rangle\langle e_j^A|) \otimes |e_i^R\rangle\langle e_j^R| = \frac{1}{d} \sum_{i,j,k,l} K(\kappa)^{(j,l),(i,k)} |e_k^B, e_i^R\rangle\langle e_l^B, e_j^R|. \end{aligned} \tag{5.5}$$

Combining these equations, we have the following representation of the output state

$$\kappa(\rho) = \operatorname{Tr}_R d(\kappa \otimes \iota_R)(|\Phi_d\rangle\langle\Phi_d|)(I_B \otimes \rho^T), \tag{5.6}$$

where ρ^T is regarded as a state on the reference system $\mathcal{H}_R$ while ρ is a state on $\mathcal{H}_A$. In the following, we omit ι_R, i.e., abbreviate $(\kappa \otimes \iota_R)$ to κ. Also, $I_B \otimes \rho^T$ is simplified to ρ^T.

The definition of the matrix $K(\kappa)$ can be naturally extended to a map from $\mathcal{M}(\mathcal{H}_A)$ to $\mathcal{M}(\mathcal{H}_B)$, as discussed above. Since the matrix $K(\kappa)$ uniquely characterizes the TP-CP map κ, as explained in the following theorem, it is called the Choi-Jamiołkowski matrix of κ.

Note that d (d') is the dimension of $\mathcal{H}_A$ ($\mathcal{H}_B$) above. $K(\kappa)$ may be used to characterize the TP-CP map as follows.

Theorem 5.1 *The following conditions are equivalent for a linear map κ from $\mathcal{T}(\mathcal{H}_A)$ to $\mathcal{T}(\mathcal{H}_B)$ [1–5].*

① *κ is a TP-CP map.*
② *The map κ^* is a completely positive map and satisfies $\kappa^*(I_B) = I_A$.*
③ *κ is a trace-preserving* $\min\{d, d'\}$*-positive map.*
④ *The matrix $K(\kappa)$ in $\mathcal{H}_A \otimes \mathcal{H}_B$ is positive semidefinite and satisfies* $\operatorname{Tr}_B K(\kappa) = I_A$.
⑤ **(Stinespring representation)** *There exists a Hilbert space $\mathcal{H}_C$ identical to $\mathcal{H}_B$, a pure state $\rho_0 \in \mathcal{S}(\mathcal{H}_B \otimes \mathcal{H}_C)$, and a unitary matrix U_κ in $\mathcal{H}_A \otimes \mathcal{H}_B \otimes \mathcal{H}_C$ such that* $\kappa(\rho) = \operatorname{Tr}_{A,C} U_\kappa(\rho \otimes \rho_0)U_\kappa^*$. *Note that the structure of $\mathcal{H}_C$ depends only on $\mathcal{H}_B$, not on κ. Only U_κ depends on κ itself.*
⑥ **(Choi–Kraus representation)** *It is possible to express κ as $\kappa(\rho) = \sum_i F_i \rho F_i^*$ using $\sum_i F_i^* F_i = I_{\mathcal{H}_A}$, where $F_1, \ldots, F_{dd'}$ are a set of dd' linear maps from $\mathcal{H}_A$ to $\mathcal{H}_B$.*

The above conditions are also equivalent to a modified Condition ⑤ (which we call Condition ⑤′), where ρ_0 is not necessarily a pure state, and the dimension of $\mathcal{H}_C$ is arbitrary. Another equivalent condition is a modification of Condition ⑥ (which we call Condition ⑥′), where the number of linear maps $\{F_i\}$ is arbitrary.

This theorem will be shown in Sect. 5.7. If the input system $\mathcal{H}_A$ and the output system $\mathcal{H}_B$ are identical to $\mathbb{C}^d$, the channel is called a **d-dimensional channel**. In this case, the Stinespring representation can be rewritten as follows.

Corollary 5.1 *The following conditions are equivalent for a linear map κ from $\mathcal{T}(\mathcal{H}_A)$ to $\mathcal{T}(\mathcal{H}_A)$.*

① *κ is a TP-CP linear map.*
② *$\kappa(\rho) = \mathrm{Tr}_E V_\kappa(\rho \otimes \rho_0)V_\kappa^*$ for a quantum system $\mathcal{H}_E$, a state $\rho_0 \in \mathcal{S}(\mathcal{H}_E)$, and an appropriate unitary matrix V_κ in $\mathcal{H}_A \otimes \mathcal{H}_E$ for κ.*

It is possible to make the dimension of $\mathcal{H}_E$ less than d^2.

The triplet $(\mathcal{H}_C, \rho_0, U_\kappa)$ in ⑤ of Theorem 5.1 is called the Stinespring representation. The equivalence to ① has been proved by Stinespring for the dual map κ^* under general conditions.

The Stinespring representation is important not only as a mathematical representation theorem, but also in terms of its physical meaning. When the input system is identical to the output system, as in Corollary 5.1, we can interpret it as a time evolution under an interaction with an external system $\mathcal{H}_E$. The system $\mathcal{H}_E$ is therefore called the **environment**.

When the map κ describes a quantum communication channel and the input and output systems are different, $\mathcal{H}_C$ can be regarded as the communication medium as in Fig. 5.1. Since the input system $\mathcal{H}_A$ and the communication medium $\mathcal{H}_C$ can be regarded as parts of the environment of $\mathcal{H}_B$, we may regard $\mathcal{H}_A \otimes \mathcal{H}_C$ as the environment, which we again denote by $\mathcal{H}_E$. We can then define the map κ_E transforming the initial state in the input system to the final state in the environment as

$$\kappa^E(\rho) \stackrel{\text{def}}{=} \mathrm{Tr}_B U_\kappa(\rho \otimes \rho_0)U_\kappa^*. \tag{5.7}$$

As shown in the following theorem, the final state in the environment of a Stinespring representation is unitarily equivalent to that of another Stinespring representation as long as the initial state of the environment ρ_0 is chosen as a pure state. That is, the state $\kappa^E(\rho)$ essentially does not depend on the Stinespring representation.

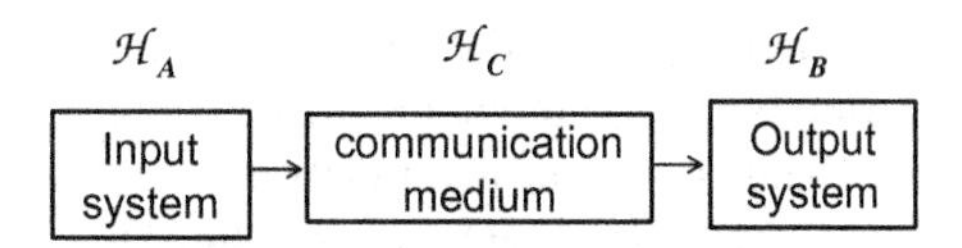

Fig. 5.1 Quantum communication channel

Theorem 5.2 *Given two Stinespring representations* $(\mathcal{H}_C, \rho_0, U_\kappa)$ *and* $(\mathcal{H}'_C, \rho'_0, U'_\kappa)$ *with the condition* $\operatorname{rank} \rho_0 = \operatorname{rank} \rho'_0 = 1$, *there exists an partial isometry* V *from* $\mathcal{H}_A \otimes \mathcal{H}_C$ *to* $\mathcal{H}_A \otimes \mathcal{H}'_C$ *such that*

$$\operatorname{Tr}_B U'_\kappa(\rho \otimes \rho'_0)U'^*_\kappa = V \operatorname{Tr}_B U_\kappa(\rho \otimes \rho_0)U^*_\kappa V^*. \tag{5.8}$$

Proof Consider the reference system $\mathcal{H}_R$ and the maximally entangled state $|\Phi\rangle$ between $\mathcal{H}_A$ and $\mathcal{H}_R$. Hence, $U_\kappa(|\Phi\rangle\langle\Phi| \otimes \rho_0)U^*_\kappa$ and $U'_\kappa(|\Phi\rangle\langle\Phi| \otimes \rho'_0)U'^*_\kappa$ are purifications of the same state $\kappa(|\Phi\rangle\langle\Phi|)$. Then, due to Lemma 8.1, there exists a partial isometry V from $\mathcal{H}_A \otimes \mathcal{H}_C$ to $\mathcal{H}_A \otimes \mathcal{H}'_C$ such that $VU_\kappa(|\Phi\rangle\langle\Phi| \otimes \rho_0)U^*_\kappa V^* = U'_\kappa(|\Phi\rangle\langle\Phi| \otimes \rho'_0)U'^*_\kappa$. Since the output state on the composite systems with the input state ρ are given as $d \operatorname{Tr}_R U_\kappa(|\Phi\rangle\langle\Phi| \otimes \rho_0)U^*_\kappa \rho^T$ and $d \operatorname{Tr}_R U'_\kappa(|\Phi\rangle\langle\Phi| \otimes \rho'_0)U'^*_\kappa \rho^T$, respectively, due to (5.6), we have (5.8). ■

As a characterization for a special TP-CP map, we have the following theorem.

Theorem 5.3 *The following conditions are equivalent for a linear map* κ *from* $\mathcal{T}(\mathcal{H}_A)$ *to* $\mathcal{T}(\mathcal{H}_B)$.

① *The system* $\mathcal{H}_B$ *can be regarded as a subspace of* $\mathcal{H}_A$ *in the following sense. There exists another system* $\mathcal{H}_E$ *such that* $\mathcal{H}_A = \mathcal{H}_B \otimes \mathcal{H}_E$ *and* $\kappa(\rho) = \operatorname{Tr}_E U_\kappa \rho U^*_\kappa$ *by choosing a unitary matrix* U_κ *in* $\mathcal{H}_A \otimes \mathcal{H}_B$.

② $\kappa^*(X)\kappa^*(Y) = \kappa^*(XY)$ *for any two matrices* $X, Y \in \mathcal{M}(\mathcal{H}_B)$.

Due to Theorem 5.1, maps κ satisfying ① forms a special class of TP-CP maps. This theorem guarantees that the dual κ^* of such a TP-CP map is a homomorphism for matrix algebras. This property is helpful for the latter discussion.

Proof Assume Condition ①. Then, $\kappa^*(X) = X \otimes I_E$ for $X \in \mathcal{M}(\mathcal{H}_B)$. Hence, $\kappa^*(X)\kappa^*(Y) = (X \otimes I_E)(Y \otimes I_E) = (XY \otimes I_E) = \kappa^*(XY)$ for any two matrices $X, Y \in \mathcal{M}(\mathcal{H}_B)$.

Next, we assume Condition ②. First, we choose a CONS $\{|u_i\rangle\}_i$ of $\mathcal{H}_B$. Since $\kappa^*(|u_i\rangle\langle u_i|)\kappa^*(|u_i\rangle\langle u_i|) = \kappa^*(|u_i\rangle\langle u_i|)$, $\kappa^*(|u_i\rangle\langle u_i|)$ is projection. We choose the basis $\{|v_{k,1}\}_k$ of the image of $\kappa^*(|u_1\rangle\langle u_1|)$. Next, we define the basis $|v_{k,i}\rangle := \kappa^*(|u_i\rangle\langle u_1|)|v_{k,1}\rangle$. Since $\kappa^*(|u_i\rangle\langle u_1|)\kappa^*(|u_i\rangle\langle u_1|)^* = \kappa^*(|u_i\rangle\langle u_i|)$, the set $\{|v_{k,i}\rangle\}_k$ forms a basis of the image of $\kappa^*(|u_i\rangle\langle u_i|)$. As $\sum_i \kappa^*(|u_i\rangle\langle u_i|) = \kappa^*(\sum_i |u_i\rangle\langle u_i|) = \kappa^*(I) = I$, the set $\{|v_{k,i}\rangle\}_{k,i}$ forms a basis of $\mathcal{H}_A$. Now, we define another system $\mathcal{H}_C$ spanned by $\{|w_k\rangle\}_k$ and a unitary $U : |u_i\rangle \otimes |w_k\rangle \mapsto |v_{k,i}\rangle$. Then, we have $\kappa^*(|u_i\rangle\langle u_j|) = U(|u_i\rangle\langle u_j| \otimes I_C)U^*$. Since $\mathcal{M}(\mathcal{H}_B)$ is spanned by matrices $|u_i\rangle\langle u_j|$, we have $\kappa^*(X) = U(X \otimes I_C)U^*$ for $X \in \mathcal{M}(\mathcal{H}_B)$, which implies ①. ■

In the next section, we will use this theorem to obtain a concrete example of a TP-CP map. In fact, the partial trace and the map $\rho \mapsto \rho \otimes \rho_0$ are TP-CP maps, as is easily verified from Theorem 5.1 above. Another representation is the output state (5.5) of the channel κ for the maximally entangled state Φ_d between the input system and the same-dimensional reference system. This representation has not only mathematical meaning but also a theoretical importance because it is possible to identify the channel

κ by identifying this final state (5.5) [6, 7]. Further, using this notation, we can describe the output state of any input pure state entangled with the reference system as follows as well as the output state $\kappa(\rho)$ as (5.6). From the discussion in (1.22) [GC1], any pure entangled state $|X\rangle$ can be described as $(I_A \otimes \sqrt{d} X^T)|\Phi_d\rangle$. Hence, we have

$$(\kappa \otimes \iota_R)(|X\rangle\langle X|) = (\kappa \otimes \iota_R)((I_A \otimes \sqrt{d} X^T)|\Phi_d\rangle\langle\Phi_d|(I_A \otimes \sqrt{d}\bar{X}))$$
$$= d(I_B \otimes X^T)(\kappa \otimes \iota_R)(|\Phi_d\rangle\langle\Phi_d|)(I_B \otimes \bar{X}). \tag{5.9}$$

In Condition ⑥, another representation $\{F_i\}$ of the completely positive map is given and is called the Choi–Kraus representation. From (1.22) the state $(\kappa \otimes \iota_R)(|\Phi_d\rangle\langle\Phi_d|)$ has the form

$$(\kappa \otimes \iota_R)(|\Phi_d\rangle\langle\Phi_d|) = \frac{1}{d_A}\sum_i |F_i\rangle\langle F_i|. \tag{5.10}$$

Hence, when $\{F_i'\}$ is another Choi–Kraus representation of κ, F_i' is represented as a linear sum of $\{F_i\}$. As is shown in Appendix 5.7, the TP-CP map κ^E can be characterized by a Choi–Kraus representation as follows.

Lemma 5.1 *When $\{F_i\}_{i=1}^d$ is a Choi–Kraus representation of κ, the environment system is described by $\mathbb{C}^d$ and the matrix elements of $\kappa^E(\rho)$ are given as*

$$\kappa^E(\rho)_{i,j} = \mathrm{Tr}\, F_j^* F_i \rho. \tag{5.11}$$

Using Choi–Kraus representation we can characterize extremal points of TP-CP maps from $\mathcal{H}_A$ to $\mathcal{H}_B$ as follows.

Lemma 5.2 (Choi [3]) *A TP-CP map κ is an extremal point of TP-CP maps from $\mathcal{H}_A$ to $\mathcal{H}_B$ if and only if κ has Choi–Kraus representation $\{F_i\}$ such that $\{F_i^* F_j\}_{i,j}$ is a linearly independent set.*

Proof Suppose that κ is an extremal point. Let $\{F_i\}$ be a Choi–Kraus representation of κ such that F_i is linearly independent (See (**a**) of Exercise 5.5). Suppose $\sum_{i,j} \lambda_{i,j} F_i^* F_j = 0$ and the matrix norm $\|(\lambda_{i,j})\|$ is less than 1. Define $\kappa_\pm$ as $\kappa_\pm(\rho) \stackrel{\text{def}}{=} \sum_i F_i \rho F_i^* \pm \sum_{i,j} \lambda_{i,j} F_i \rho F_j^*$. Since $I \pm (\lambda_{i,j}) \ge 0$, $\kappa_\pm$ is a TP-CP map. It also follows that $\kappa = \frac{1}{2}\kappa_+ + \frac{1}{2}\kappa_-$. Since κ is extremal, $\kappa_+ = \kappa$. That is, $\lambda_{i,j} = 0$. Therefore, $\{F_i^* F_j\}_{i,j}$ is a linearly independent set.

Conversely, suppose that $\{F_i^* F_j\}_{i,j}$ is a linearly independent set. We choose TP-CP maps κ_1 and κ_2 and a real number $0 < \lambda < 1$ such that $\kappa = \lambda\kappa_1 + (1-\lambda)\kappa_2$. Let $\{F_i^k\}$ be a Choi–Kraus representation of κ_k. Then, κ has Choi–Kraus representation $\{\sqrt{\lambda} F_i^1\} \cup \{\sqrt{1-\lambda} F_i^2\}$. Thus, F_i^1 is written as $\sum_j \lambda_{i,j} F_j$ (See (**b**) of Exercise 5.5). From the condition $\sum_i (F_i^1)^* F_i^1 = \sum_j F_j^* F_j$ and the assumption[GC2], $\sum_i \overline{\lambda_{i,j'}} \lambda_{i,j} = \delta_{j',j}$. Hence, we obtain $\kappa = \kappa_1$ (See Exercise 5.4).

Corollary 5.2 *When a TP-CP map κ from $\mathcal{H}_A$ to $\mathcal{H}_B$ is extremal, i.e., a Choi–Kraus representation satisfies Condition given in Lemma 5.2, the Choi–Kraus representation has only d_A elements and any image is included in the d_A^2-dimensional space of $\mathcal{H}_B$ at most. Further, when $d_A \le d_B$, we can construct Stinespring representation with $d_C = 1$.*

A Stinespring representation guarantees that the state evolution corresponding to the TP-CP map κ can be implemented by the following procedure. The initial state ρ_0 is first prepared on $\mathcal{H}_B \otimes \mathcal{H}_C$; then, the unitary evolution U_κ is performed on $\mathcal{H}_A \otimes \mathcal{H}_B \otimes \mathcal{H}_C$. It is commonly believed that in principle, state evolutions corresponding to an arbitrary unitary matrix U_κ can be implemented, and hence state evolutions corresponding to arbitrary TP-CP maps can also in principle be implemented.

Let us now consider the case where we are given two TP-CP maps κ and κ' that map from the quantum systems $\mathcal{H}_A, \mathcal{H}'_A$ to $\mathcal{H}_B, \mathcal{H}'_B$, respectively. The state evolution of the composite system from $\mathcal{H}_A \otimes \mathcal{H}'_A$ to $\mathcal{H}_B \otimes \mathcal{H}'_B$ is given by $\kappa \otimes \kappa'$. One may wonder whether the map $\kappa \otimes \kappa'$ also satisfies the condition for TP-CP maps. Indeed, this condition is guaranteed by the following corollary.

Corollary 5.3 *Given a linear map κ' from $\mathcal{T}(\mathcal{H}'_A)$ to $\mathcal{T}(\mathcal{H}'_B)$, the following two conditions are equivalent.*

① *κ' is a TP-CP map.*
② *$\kappa \otimes \kappa'$ is a TP-CP map when κ is a TP-CP map from $\mathcal{H}_A$ to $\mathcal{H}_B$.*

As another condition for positive maps, we focus on the tensor product positivity. A positive map κ is called **tensor product positive** if $\kappa^{\otimes n}$ is positive for any integer n. It follows from the above corollary that any CP map is tensor product positive.

Proof of Corollary 5.3 The proof will be shown based on Condition ④ of Theorem 5.1. Condition ② is equivalent to the condition that $K(\kappa \otimes \kappa')$ is positive semidefinite, which is equal to $K(\kappa) \otimes K(\kappa')$. Since $K(\kappa)$ is positive semidefinite, then $K(\kappa')$ is positive semidefinite. This is then equivalent to Condition ①. ∎

The fact that the dimension of the space of ρ is d'^2 is important in connection with quantum computation. One of the main issues in quantum computation theory is the classification of problems based on their computational complexity. One particularly important class is the class of problems that are solvable in polynomial time with respect to the input size. This class is called the polynomial class. The classification depends on whether operations are restricted to unitary time evolutions that use unitary gates such as C-NOT gates or if TP-CP maps are allowed. However, as confirmed by Theorem 5.1, TP-CP maps can be simulated by $d(d'^2)$-dimensional unitary evolutions. Therefore, it has been shown that the class of problems that can be solved in polynomial time is still the same [8].[1]

[1]More precisely, we can implement only a finite number of unitary matrices in a finite amount of time. For a rigorous proof, we must approximate the respective TP-CP maps by a finite number of unitary matrices and evaluate the level of these approximations.

Remark 5.1 The discussion presented here can be extended to a more general physical system, i.e., the case where the states are given as the duals of the general operator algebra, e.g., C^*-algebra, von Neumann algebra, and CCR algebra.

In the area of operator algebra, the dynamics is given as the map κ satisfying the condition of Theorem 5.3. Although such discussions in the area of operator algebra contain the infinite-dimensional case, they do not cover the case when the input system is strictly smaller than the system $\mathcal{H}_B \otimes \mathcal{H}_E$. In such a case, the state of the system $\mathcal{H}_B \otimes \mathcal{H}_E$ is given as $U_\kappa(\rho \otimes \rho_0)U_\kappa^*$, which is not invertible. Hence, if an analysis for the dynamics κ under the condition of Theorem 5.3 covers only invertible state, it cannot be extended to the case with general dynamics κ. This point is a blind spot when an analysis by operator algebra is employed.

Exercises

5.1 Show that κ is trace-preserving if and only if κ^* is identity-preserving by using (5.2) with $Y = I$.

5.2 Show (5.3) by using $\begin{pmatrix} \kappa^*(X^*X) & \kappa^*(X)^* \\ \kappa^*(X) & I \end{pmatrix} \geq 0.$

5.3 Show Corollary 5.1 using Theorem 5.1.

5.4 Let $\{F_i\}$ be a Choi–Kraus representation of the TP-CP map κ and $u_{i,j}$ be a unitary matrix. Show that $F_i' \stackrel{\text{def}}{=} \sum_j u_{i,j} F_j$ is also its Choi–Kraus representation.

5.5 Show the following items for a TP-CP map κ from $\mathcal{S}(\mathcal{H}_A)$ to $\mathcal{S}(\mathcal{H}_B)$.
(**a**) Show that there exists a Choi–Kraus representation $\{F_i\}$ of the TP-CP map κ such that matrices F_i are linearly independent. (Hint: Use Exercise 5.4.)
(**b**) Choose two Choi–Kraus representation $\{F_i\}_{i=1}^d$ and $\{F_j'\}_{j=1}^{d'}$ of the TP-CP map κ. Show that F_j' can be written as a linear combination of F_i. (Hint: Use a method similar to the proof of Theorem 5.2.)

5.2 Examples of Trace-Preserving Completely Positive Maps

In addition to the above-mentioned partial trace, the following examples of TP-CP maps exist.

Example 5.1 (*Unitary evolution*) Let U be a unitary matrix on $\mathcal{H}$. The state evolution $\kappa_U : \rho \mapsto \kappa_U(\rho) \stackrel{\text{def}}{=} U\rho U^*$ from $\mathcal{S}(\mathcal{H})$ to itself is a TP-CP map. This can be easily verified from Condition ⑥ in Theorem 5.1. Since an arbitrary unitary matrix U has the form $U = Ve^{i\theta}$, where $e^{i\theta}$ is a complex number with modulus 1 and V is a unitary matrix with determinant 1, we can write $U\rho U^* = V\rho V^*$. Therefore, we can restrict V to unitary matrices with determinant 1. Such matrices are called **special**

unitary matrices. However, there are no such unitary state evolutions when the dimension of $\mathcal{H}_A$ is smaller than the dimension of $\mathcal{H}_B$. In such a case, we consider isometric matrices (i.e., matrices satisfying $U^*U = I$) from $\mathcal{H}_A$ to $\mathcal{H}_B$. The TP-CP map $\kappa_U(\rho) \stackrel{\text{def}}{=} U\rho U^*$ is then called an **isometric state evolution**.

Example 5.2 (*Partial trace*) The partial trace $\rho \mapsto \mathrm{Tr}_{\mathcal{H}}\, \rho$ can be regarded as a state evolution from quantum system $\mathcal{H} \otimes \mathcal{H}'$ to quantum system $\mathcal{H}'$. It is also a completely positive map because it is a special case of Condition ⑤ of Theorem 5.1.

Example 5.3 (*Depolarizing channel*) For arbitrary $1 \geq \lambda \geq 0$, a map

$$\kappa_{d,\lambda}(\rho) \stackrel{\text{def}}{=} \lambda\rho + (1-\lambda)(\mathrm{Tr}\,\rho)\rho_{\mathrm{mix}} \tag{5.12}$$

is a d-dimensional TP-CP map and is called a depolarizing channel. In particular, when $d = 2$, we have[Exe. 5.6]

$$\kappa_{2,\lambda}(\rho) = \frac{3\lambda+1}{4}\rho + \frac{1-\lambda}{4}\sum_{i=1}^{3} S_i \rho S_i^*. \tag{5.13}$$

A depolarizing channel $\kappa_{d,\lambda}$ satisfies $\kappa_{d,\lambda}(U\rho U^*) = U\kappa_{d,\lambda}(\rho)U^*$ for all unitary matrices U. Conversely, when a d-dimensional channel satisfies this property, it is a depolarizing channel.

Example 5.4 (*Entanglement-breaking channel*) A TP-CP map from $\mathcal{H}_A$ to $\mathcal{H}_B$ satisfying the following conditions is called an entanglement-breaking channel. For an arbitrary reference system $\mathcal{H}_C$ and an arbitrary state $\rho \in \mathcal{S}(\mathcal{H}_A \otimes \mathcal{H}_C)$, the output state $(\kappa \otimes \iota_C)(\rho)$ on the space $\mathcal{H}_B \otimes \mathcal{H}_C$ is separable.

The entanglement-breaking channel is characterized as follows.

Theorem 5.4 (Horodecki et al. [9]) *The following two conditions are equivalent for a TP-CP map κ from $\mathcal{H}_A$ to $\mathcal{H}_B$.*

① *κ is an entanglement-breaking channel.*
② *κ can be written as*

$$\kappa(\rho) = \kappa_{\boldsymbol{M},W}(\rho) \stackrel{\text{def}}{=} \sum_{\omega\in\Omega} (\mathrm{Tr}\,\rho M_\omega) W_\omega,$$

where $\boldsymbol{M} = \{M_\omega\}_{\omega\in\Omega}$ is an arbitrary POVM on $\mathcal{H}_A$ and W is a map from Ω to $\mathcal{S}(\mathcal{H}_B)$.

For a proof, see Exercise 7.5.

If the W_ω maps are mutually orthogonal pure states, then $\kappa_{\boldsymbol{M},W}(\rho)$ can be identified with the probability distribution $\mathrm{P}^{\boldsymbol{M}}_\rho$. A POVM $\boldsymbol{M}$ is not only a map that gives the probability distribution $\mathrm{P}^{\boldsymbol{M}}_\rho$ from the quantum state ρ, but it can also be regarded as an entanglement-breaking channel (and therefore a TP-CP map).

Example 5.5 (*Unital channel*) A TP-CP map κ from $\mathcal{H}_A$ to $\mathcal{H}_B$ satisfying $\kappa\left(\rho_{\mathrm{mix}}^A\right) = \rho_{\mathrm{mix}}^B$ is called a unital channel. The depolarizing channel defined previously is a unital channel.

Example 5.6 (*Pinching*) Recall that the pinching $\kappa_{\boldsymbol{M}}$: $\rho \mapsto \sum_{\omega\in\Omega} M_\omega \rho M_\omega$ is defined with respect to the PVM $\boldsymbol{M} = \{M_\omega\}_{\omega\in\Omega}$ in Sect. 1.2. This satisfies the conditions for a TP-CP map only when $\boldsymbol{M}$ is a PVM. For a general POVM $\boldsymbol{M}$, the map $\rho \mapsto \sum_{\omega\in\Omega} \sqrt{M_\omega}\rho\sqrt{M_\omega}$ is a TP-CP map.

If all the elements M_ω are one-dimensional, the pinching $\kappa_{\boldsymbol{M}}$ is an entanglement-breaking channel. If the POVM has a non-one-dimensional element M_ω, it is not an entanglement-breaking channel.

Example 5.7 (*Transpose*) For a quantum system $\mathcal{H}_A$, we define the transpose operator τ with respect to its orthonormal basis $u_0, \ldots, u_{d-1}$ as

$$\rho = \sum_{i,j} \rho_{i,j} |u_i\rangle\langle u_j| \mapsto \tau(\rho) \stackrel{\mathrm{def}}{=} \rho^T = \sum_{i,j} \rho_{j,i} |u_i\rangle\langle u_j|. \tag{5.14}$$

Then, τ is a positive map, but not a two-positive map. Therefore, it is not a completely positive map. However, it is a tensor-product-positive map[Exe. 5.13].

According to Exercise 1.3, any tensor product state $\rho^A \otimes \rho^B$ satisfies $(\tau_A \otimes \iota_B)(\rho^A \otimes \rho^B) = \tau_A(\rho^A) \otimes \rho^B \geq 0$. Hence, any separable state $\rho \in \mathcal{S}(\mathcal{H}_A \otimes \mathcal{H}_B)$ also satisfies $(\tau \otimes \iota_B)(\rho) \geq 0$. The converse is the subject of the following theorem.

Theorem 5.5 (Horodecki [10]) *Assign orthogonal bases for $\mathcal{H}_A$ and $\mathcal{H}_B$. Let τ be the transpose with respect to $\mathcal{H}_A$ under these coordinates. If either $\mathcal{H}_A$ or $\mathcal{H}_B$ is two-dimensional and the other is three-dimensional or less, then the condition $(\tau \otimes \iota_B)(\rho) \geq 0$ is the necessary and sufficient condition for the density matrix ρ on the composite system $\mathcal{H}_A \otimes \mathcal{H}_B$ to be separable.*

Counterexamples are available for $\mathbb{C}^2 \otimes \mathbb{C}^4$ and $\mathbb{C}^3 \otimes \mathbb{C}^3$ [11]. If the input and output systems of κ are quantum two-level systems, we have the following corollary.

Corollary 5.4 *Let τ be a transpose under some set of coordinates. If κ is a channel for a quantum two-level system (i.e., a TP-CP map), the following two conditions are equivalent.*

1. *$\tau \circ \kappa$ is a CP map.*
2. *κ is an entanglement-breaking channel.*

Example 5.8 (*Generalized Pauli channel*) Define unitary matrices $\mathbf{X}_d$ and $\mathbf{Z}_d$ using the same basis as that in Example 5.7 for the quantum system $\mathcal{H}_A$ as follows:

$$\mathbf{X}_d |u_j\rangle = |u_{j-1 \bmod d}\rangle, \quad \mathbf{Z}_d |u_j\rangle = w^j |u_j\rangle, \tag{5.15}$$

where w is the dth root of 1, i.e., $e^{-2\pi i/d}$. The generalized Pauli channel κ_p^{GP} is given by

$$\kappa_p^{\mathrm{GP}}(\rho) \stackrel{\mathrm{def}}{=} \sum_{i=0}^{d-1}\sum_{j=0}^{d-1} p(i,j)(\mathbf{X}_d^i\mathbf{Z}_d^j)^*\rho(\mathbf{X}_d^i\mathbf{Z}_d^j) \tag{5.16}$$

for the probability distribution p in $\{0,\ldots,d-1\}^{\times 2}$. We often denote $\mathbf{X}_d$ and $\mathbf{Z}_d$ by $\mathbf{X}_A$ and $\mathbf{Z}_A$ respectively for indicating the space $\mathcal{H}_A$ that these act on. The above channel is also unital. For a quantum two-level system, we can write this channel as

$$\kappa(\rho) = \sum_{i=0}^{3} p_i S_i \rho S_i^*, \tag{5.17}$$

where p is a probability distribution in $\{0,1,2,3\}$, and the Pauli matrices S_i were defined in Sect. 1.3. This is called a **Pauli channel** and will be denoted by κ_p^{GP}.

Example 5.9 (*Transpose depolarizing channel (Werner–Holevo channel, Antisymmetric channel)*) If and only if a real number λ belongs to $[-\frac{1}{d-1}, \frac{1}{d+1}]$, the map

$$\kappa_{d,\lambda}^T(\rho) \stackrel{\mathrm{def}}{=} \lambda\rho^T + (1-\lambda)\rho_{\mathrm{mix}} \tag{5.18}$$

is a TP-CP map, and it is called a transpose depolarizing channel, where d is the dimension of the system [12] (see Exercise 8.84 and Theorem 5.1). In particular, when $\lambda = -\frac{1}{d-1}$, the channel $\kappa_{d,-\frac{1}{d-1}}^T$ is called an antisymmetric channel [13] or a Werner–Holevo channel [14]. This channel satisfies the anticovariance $\kappa_{d,\lambda}^T(U\rho U^*) = \overline{U}\kappa_{d,\lambda}^T(\rho)U^T$.

Example 5.10 (*Phase-damping channel*) Let $D = (d_{i,j})$ be a positive semidefinite matrix satisfying $d_{i,i} = 1$. The following channel is called a phase-damping channel:

$$\kappa_D^{\mathrm{PD}}(\rho) \stackrel{\mathrm{def}}{=} \sum_{i,j} d_{i,j}\rho_{i,j}|u_i\rangle\langle u_j|, \tag{5.19}$$

where $\rho = \sum_{i,j}\rho_{i,j}|u_i\rangle\langle u_j|$.

For example, any pinching κ_M with a PVM M is a phase-damping channel. Since

$$\kappa_D^{\mathrm{PD}}(\frac{1}{d}\sum_{k,l}|u_k, u_k^R\rangle\langle u_l, u_l^R|) = \frac{1}{d}\sum_{k,l} d_{k,l}|u_k, u_k^R\rangle\langle u_l, u_l^R|,$$

Condition ④ of Theorem 5.1 guarantees that any phase-damping channel κ_D^{PD} is a TP-CP map.

Lemma 5.3 *A phase-damping channel κ_D^{PD} is a generalized Pauli channel κ_p^{GP} satisfying that the support of p belongs to the set $\{(0,0),\ldots,(0,d-1)\}$ if and only if*

$$d_{i,j} = d_{i',0} \text{ for } i - j = i' \bmod d. \tag{5.20}$$

See Exercise 5.14.

Example 5.11 (*PNS channel*) Define the n-fold **symmetric space** $\mathcal{H}^n_{s,d}$ of $\mathbb{C}^d$ as the space spanned by $\{v^{\otimes n} | v \in \mathbb{C}^d\} \subset (\mathbb{C}^d)^{\otimes n}$. Let the input system be the n-fold symmetric space $\mathcal{H}^n_{s,d}$ and the output system be the m-fold symmetric space $\mathcal{H}^m_{s,d}$ ($n \geq m$). The PNS (photon number splitting) channel $\kappa^{\mathrm{pns}}_{d,n\to m}$ is given by

$$\kappa^{\mathrm{pns}}_{d,n\to m}(\rho) \stackrel{\mathrm{def}}{=} \mathrm{Tr}_{(\mathbb{C}^d)^{\otimes n-m}}\, \rho, \tag{5.21}$$

where we regard ρ as a state on the n-fold tensor product space. In this case, the support of the output state is contained by the m-fold symmetric space $\mathcal{H}^m_{s,d}$. Hence, we can check that it is a TP-CP map from the n-fold symmetric space $\mathcal{H}^n_{s,d}$ to the m-fold symmetric space $\mathcal{H}^m_{s,d}$. Indeed, this channel corresponds to the photon number splitting attack in the quantum key distribution.

Example 5.12 (*Erasure channel*) Let the input system be $\mathbb{C}^d$ with the basis $u_0, \ldots, u_{d-1}$ and the input system be $\mathbb{C}^d$ with the basis $u_0, \ldots, u_{d-1}, u_d$. The erasure channel $\kappa^{\mathrm{era}}_{d,p}$ with the probability is given as

$$\kappa^{\mathrm{era}}_{d,p}(\rho) \stackrel{\mathrm{def}}{=} (1-p)\rho + p|u_d\rangle\langle u_d|. \tag{5.22}$$

Exercises

5.6 Show formula (5.13) for the depolarizing channel on the quantum two-level system.

5.7 Prove Theorem 4.5 from Condition ⑤ of Theorem 5.1.

5.8 Show that $(\mathbf{X}_d^{j_1}\mathbf{Z}_d^{k_1})(\mathbf{X}_d^{j_2}\mathbf{Z}_d^{k_2}) = \omega^{j_1k_2 - k_1j_2}(\mathbf{X}_d^{j_2}\mathbf{Z}_d^{k_2})(\mathbf{X}_d^{j_1}\mathbf{Z}_d^{k_1})$ for symbols defined as in Example 5.8.

5.9 Show that

$$\frac{1}{d^2}\sum_{j=0}^{d-1}\sum_{k=0}^{d-1}(\mathbf{X}_d^j\mathbf{Z}_d^k)^* X (\mathbf{X}_d^j\mathbf{Z}_d^k) = (\mathrm{Tr}\, X)\rho_{\mathrm{mix}} \tag{5.23}$$

for an arbitrary matrix X by following the steps below.
(**a**) Show that

$$
\begin{aligned}
&(\mathbf{X}_d^{j'}\mathbf{Z}_d^{k'})^* \left(\frac{1}{d^2}\sum_{j=0}^{d-1}\sum_{k=0}^{d-1}(\mathbf{X}_d^{j}\mathbf{Z}_d^{k})^* X (\mathbf{X}_d^{j}\mathbf{Z}_d^{k}) \right) (\mathbf{X}_d^{j'}\mathbf{Z}_d^{k'}) \\
=&\frac{1}{d^2}\sum_{j=0}^{d-1}\sum_{k=0}^{d-1}(\mathbf{X}_d^{j}\mathbf{Z}_d^{k})^* X (\mathbf{X}_d^{j}\mathbf{Z}_d^{k})
\end{aligned}
$$

for arbitrary j', k'.
(**b**) Show that the matrix $A = \sum_{j,k} a^{j,k}|u_j\rangle\langle u_k|$ is diagonal if $\mathbf{Z}_d A = A\mathbf{Z}_d$.
(**c**) Show that all of the diagonal elements of A are the same if $\mathbf{X}_d A = A\mathbf{X}_d$.
(**d**) Show (5.23) using the above.

5.10 Show that

$$
\frac{1}{d^2}\sum_{j=0}^{d-1}\sum_{k=0}^{d-1}(\mathbf{X}_A^{j}\mathbf{Z}_A^{k}\otimes I_B)^* \rho (\mathbf{X}_A^{j}\mathbf{Z}_A^{k}\otimes I_B) = \rho_{\mathrm{mix}}^A \otimes \mathrm{Tr}_A \rho
$$

for a state ρ on $\mathcal{H}_A \otimes \mathcal{H}_B$ using formula (1.28).

5.11 Let $\mathcal{H}_A, \mathcal{H}_B$ be the spaces spanned by $u_0^A, \ldots, u_{d-1}^A$ and $u_0^B, \ldots, u_{d-1}^B$. Define

$$
u_{0,0}^{A,B} \stackrel{\mathrm{def}}{=} \frac{1}{\sqrt{d}}\sum_{i=1}^{d} u_i^A \otimes u_i^B, \quad u_{i,j}^{A,B} \stackrel{\mathrm{def}}{=} (\mathbf{X}_A^{i}\mathbf{Z}_A^{j}\otimes I_B) u_{0,0}^{A,B},
$$

and show that these vectors form a CONS of $\mathcal{H}_A \otimes \mathcal{H}_B$.

5.12 Suppose that the classical-quantum channel W_ρ is given by a depolarizing channel $\kappa_{d,\lambda}$ as $W_\rho \stackrel{\mathrm{def}}{=} \kappa_{d,\lambda}(\rho)$. In this case, the set of states $\mathcal{S}(\mathcal{H}_A)$ of the input system is regarded as the set of input alphabets $\mathcal{X}$. Show that the depolarizing channel $\kappa_{d,\lambda}$ is pseudoclassical and its capacity is given by

$$
\begin{aligned}
&C_c(\kappa_{d,\lambda}) = \tilde{C}_c(\kappa_{d,\lambda}) \\
=&\frac{1+(d-1)\lambda}{d}\log(1+(d-1)\lambda) + \frac{(d-1)(1-\lambda)}{d}\log(1-\lambda).
\end{aligned}
$$

5.13 Show that the transpose τ is tensor product positive.

5.14 Show Lemma 5.3 by following the steps below.
(**a**) Show that the generalized Pauli channel κ_p^{GP} satisfies (5.19) and (5.20) when the support of p belongs to the set $\{(0,0), \ldots, (0, d-1)\}$.
(**b**) Assume that a phase-damping channel κ_D^{PD} satisfies (5.20). Define $p(0,m) \stackrel{\mathrm{def}}{=} \frac{1}{d}\mathrm{Tr}\, D\mathbf{X}_d^m = \frac{1}{d}\sum_{j=0}^{d-1} d_{j,0}\omega^{-jm}$. Show that $\kappa_p^{\mathrm{GP}} = \kappa_D^{\mathrm{PD}}$.
(**c**) Show Lemma 5.3.

5.15 Show that $(\kappa^{\mathrm{era}}_{d,p})^E = \kappa^{\mathrm{era}}_{d,1-p}$.

5.16 Show that $(\kappa^{\mathrm{pns}}_{d,n\to m})^E = \kappa^{\mathrm{pns}}_{d,n\to n-m}$.

5.3 State Evolutions in Quantum Two-Level Systems

As mentioned in Sect. 1.3, the states in a quantum two-level system may be parameterized by a three-dimensional vector $\boldsymbol{x}$:

$$\rho_x = \frac{1}{2}\left(S_0 + \sum_{i=1}^{3} x^i S_i\right). \tag{5.24}$$

Let κ be an arbitrary TP-CP map from a quantum two-level system to another quantum two-level system. We shall now investigate how this map κ can be characterized under the parameterization (5.24). As discussed in Sect. 4.1, this kind of map is characterized by a linear map from the set of Hermitian matrices on $\mathbb{C}^2$ to itself. Consider a state evolution of the unitary type given in Example 5.1. The special unitary matrix V may then be diagonalized by a unitary matrix. The two eigenvalues are complex numbers with an absolute value 1 and their product yields 1. Therefore, we represent the two eigenvalues by $e^{i\theta}$ and $e^{-i\theta}$ and the two eigenvectors by u_1 and u_2. We write $V = e^{i\theta}|u_1\rangle\langle u_1| + e^{-i\theta}|u_2\rangle\langle u_2| = \exp(i(\theta|u_1\rangle\langle u_1| - \theta|u_2\rangle\langle u_2|))$. The unitary matrix V may therefore be written as $\exp(iX)$, where X is a Hermitian matrix with trace 0. We will use this description $V = \exp(iX)$ to examine the state evolution when a special unitary matrix V acts on both sides of the density matrix of a quantum two-level system.

Let us examine some algebraic properties of the Pauli matrices. Define $\epsilon_{j,k,l}$ to be 0 if any of j, k, and l are the same, $\epsilon_{1,2,3} = \epsilon_{3,1,2} = \epsilon_{2,3,1} = 1$, and $\epsilon_{3,2,1} = \epsilon_{1,3,2} = \epsilon_{2,1,3} = -1$. Then, $\left[S^j, S^k\right] = -2i\sum_{l=1}^{3}\epsilon_{j,l,k}S^l$. This is equivalent to

$$\left[\sum_{j=1}^{3} x_j S^j, \sum_{k=1}^{3} y_k S^k\right] = -2i\sum_{j=1}^{3}\sum_{k=1}^{3}\sum_{l=1}^{3} x_j y_k \epsilon_{j,l,k} S^l.$$

Defining $R_j \stackrel{\mathrm{def}}{=} [\epsilon_{j,l,k}]_{l,k}$, $S_x \stackrel{\mathrm{def}}{=} \sum_{j=1}^{3} x_j S^j$, and $R_x \stackrel{\mathrm{def}}{=} \sum_{j=1}^{3} x_j R^j$, we may rewrite the above expression as

$$\left[\frac{i}{2}S_x, S_y\right] = S_{R_x y}.$$

As shown later, this equation implies that

$$\exp\left(\frac{i}{2}S_x\right) S_y \exp\left(-\frac{i}{2}S_x\right) = S_{\exp(R_x)y}. \tag{5.25}$$

Applying this equation to states, we obtain

$$\exp\left(\frac{i}{2}S_x\right) \rho_y \exp\left(-\frac{i}{2}S_x\right) = \rho_{\exp(R_x)y}. \tag{5.26}$$

This shows that a 2×2 unitary matrix $\exp(\frac{i}{2}S_x)$ of determinant 1 corresponds to a 3×3 real orthogonal matrix $\exp(R_x)$.

Proof of (5.25) Since S_x is Hermitian, the matrix $i\frac{s}{2}S_x$ can be diagonalized by a unitary matrix with purely imaginary eigenvalues. Therefore, $\exp(i\frac{s}{2}S_x)$ is a unitary matrix. Note that $\exp(i\frac{s}{2}S_x)^* = \exp(-i\frac{s}{2}S_x)$. Since $\exp(i\frac{s}{2}S_x)S_y\exp(-i\frac{s}{2}S_x)$ is a Hermitian matrix with trace 0 like S_y, it can be rewritten as $S_{y(s)}$ according to Exercise 1.14. Let us write down the vector $y(s)$. Differentiating $\exp(i\frac{s}{2}S_x)S_y\exp(-i\frac{s}{2}S_x)$ with respect to s, we obtain

$$\begin{aligned}
S_{y'(s)} =& \left(\exp\left(i\frac{s}{2}S_x\right) S_y \exp\left(-i\frac{s}{2}S_x\right)\right)' \\
=& \left(\exp\left(i\frac{s}{2}S_x\right)\right)' S_y \exp\left(-i\frac{s}{2}S_x\right) + \exp\left(i\frac{s}{2}S_x\right) S_y \left(\exp\left(-i\frac{s}{2}S_x\right)\right)' \\
=& \frac{i}{2}S_x \exp\left(i\frac{s}{2}S_x\right) S_y \exp\left(-i\frac{s}{2}S_x\right) + \exp\left(i\frac{s}{2}S_x\right) S_y \exp\left(-i\frac{s}{2}S_x\right)\left(-\frac{i}{2}S_x\right) \\
=& \left[\frac{i}{2}S_x, \exp\left(i\frac{s}{2}S_x\right) S_y \exp\left(-i\frac{s}{2}S_x\right)\right] = \left[\frac{i}{2}S_x, S_{y(s)}\right] = S_{R_x y(s)},
\end{aligned}$$

and we find that $y(t)$ satisfies the differential equation

$$y'(s) = R_x y(s). \tag{5.27}$$

It can be verified that

$$y(s) = \exp(sR_x)y \tag{5.28}$$

satisfies this differential equation. The uniqueness of the solution of an ordinary differential equation guarantees that only the function $y(s)$ given in (5.28) satisfies $y(0) = y$ and (5.27). Applying this to when $s = 1$, we obtain (5.25). ■

Next, we consider a general TP-CP map. Define $\tilde{K}(\kappa)^{i,j} \stackrel{\text{def}}{=} \frac{1}{2}\operatorname{Tr} S_i\kappa(S_j)$. The trace-preserving property guarantees that

$$\tilde{K}(\kappa)^{0,0} = \frac{1}{2}\operatorname{Tr}\kappa(I) = 1, \quad \tilde{K}(\kappa)^{0,i} = \frac{1}{2}\operatorname{Tr}\kappa(S_i) = 0$$

for $i \neq 0$. Now define $\boldsymbol{t}$ according to $t^i \stackrel{\text{def}}{=} \tilde{K}(\kappa)^{i,0}$. Then, we have $\kappa(\rho_{\text{mix}}) = \rho_{\boldsymbol{t}}$. Let T be the 3×3 matrix $[\tilde{K}(\kappa)^{i,j}]_{1 \le i,j \le 3}$. The TP-CP map κ may then be denoted by a vector $\boldsymbol{t}$ and a matrix T as [4]

$$\kappa(\rho_x) = \rho_{Tx+\boldsymbol{t}}. \tag{5.29}$$

For example, when a unitary matrix operates on either side, then $\boldsymbol{t} = 0$ and T is an orthogonal matrix given by (5.26). To give a few more examples, let us rewrite the examples given in the previous section using $\boldsymbol{t}$ and T for the quantum two-level system. When κ is a depolarizing channel, we have $\boldsymbol{t} = 0$, and therefore $T = \lambda I$. The necessary and sufficient condition for the channel to be unital is then $\boldsymbol{t} = 0$. When κ is the transpose, we have

$$\boldsymbol{t} = 0, \quad T = \begin{pmatrix} 1 & 0 & 0 \\ 0 & -1 & 0 \\ 0 & 0 & 1 \end{pmatrix}.$$

Next, we consider the necessary and sufficient conditions for a map to be a positive map, a completely positive map, an entanglement-breaking channel, and a Pauli channel, respectively, by assuming that the channel is unital, i.e., $\boldsymbol{t} = 0$. Recall from the discussion in Sect. A.2 that special orthogonal matrices O_1, O_2 may be chosen such that $T' \stackrel{\text{def}}{=} O_1^* T O_2$ is diagonal (i.e., a singular decomposition). Taking unitary matrices U_1, U_2 corresponding to O_1, O_2 based on the correspondence (5.26), and letting κ^T be the TP-CP map corresponding to T, we have $\kappa_{U_1} \circ \kappa^T \circ \kappa_{U_2} = \kappa^{T'}$. For the analysis of the TP-CP map κ^T, it is sufficient to analyze the TP-CP map $\kappa^{T'}$. Now, using the eigenvalues λ_1, λ_2, and λ_3 of T', we give the necessary and sufficient conditions for the above types of channels as follows.

Positivity: Let us first consider a necessary and sufficient condition for a positive map. It is positive if and only if the image by T' of the unit sphere $\{\mathbf{x} | \|\mathbf{x}\| \le 1\}$ is contained by the unit sphere. Thus, its necessary and sufficient condition is

$$|\lambda_1|, |\lambda_2|, |\lambda_3| \le 1. \tag{5.30}$$

Completely positivity: Next, we use Condition ④ of Theorem 5.1 to examine the completely positive map. In order to check this condition, we calculate $K(\kappa^{T'})$:

$$K(\kappa^{T'}) = \frac{1}{2} \begin{pmatrix} 1+\lambda_3 & 0 & 0 & \lambda_1+\lambda_2 \\ 0 & 1-\lambda_3 & \lambda_1-\lambda_2 & 0 \\ 0 & \lambda_1-\lambda_2 & 1-\lambda_3 & 0 \\ \lambda_1+\lambda_2 & 0 & 0 & 1+\lambda_3 \end{pmatrix}.$$

Swapping the second and fourth coordinates, we have

$$\begin{pmatrix} 1+\lambda_3 & \lambda_1+\lambda_2 & 0 & 0 \\ \lambda_1+\lambda_2 & 1+\lambda_3 & 0 & 0 \\ 0 & 0 & 1-\lambda_3 & \lambda_1-\lambda_2 \\ 0 & 0 & \lambda_1-\lambda_2 & 1-\lambda_3 \end{pmatrix}.$$

Thus, the necessary and sufficient condition for $K(\kappa^{T'}) \geq 0$ is [4, 15]

$$(1+\lambda_3)^2 \geq (\lambda_1+\lambda_2)^2, \quad (1-\lambda_3)^2 \geq (\lambda_1-\lambda_2)^2. \tag{5.31}$$

This condition can be rewritten as

$$1 \geq \lambda_1+\lambda_2-\lambda_3, \lambda_1-\lambda_2+\lambda_3, -\lambda_1+\lambda_2+\lambda_3 \geq -1 \tag{5.32}$$

from Condition (5.30).

Entanglement-breaking: Due to Corollary 5.4, the channel κ^T is entanglement-breaking if and only if the eigenvalues of $T\begin{pmatrix} 1 & 0 & 0 \\ 0 & -1 & 0 \\ 0 & 0 & 1 \end{pmatrix}$ satisfy (5.32). Since these eigenvalues are $\lambda_1, -\lambda_2, \lambda_3$, the following is a necessary and sufficient condition for a channel to be an entanglement-breaking channel [16]:

$$1 \geq |\lambda_1| + |\lambda_2| + |\lambda_3|. \tag{5.33}$$

Pauli channel: Next, we treat the necessary and sufficient condition for a channel to be a Pauli channel. When the channel is a Pauli channel, firstly, we have $t = 0$. When the state evolution is given by the unitaries S_1, S_2, and S_3, the matrix T is given, respectively, as

$$\begin{pmatrix} 1 & 0 & 0 \\ 0 & -1 & 0 \\ 0 & 0 & -1 \end{pmatrix}, \begin{pmatrix} -1 & 0 & 0 \\ 0 & 1 & 0 \\ 0 & 0 & -1 \end{pmatrix}, \begin{pmatrix} -1 & 0 & 0 \\ 0 & -1 & 0 \\ 0 & 0 & 1 \end{pmatrix}.$$

Using p_i in (5.17), the matrix T is given by

$$\begin{pmatrix} p_0+p_1-p_2-p_3 & 0 & 0 \\ 0 & p_0-p_1+p_2-p_3 & 0 \\ 0 & 0 & p_0-p_1-p_2+p_3 \end{pmatrix}$$
$$= \begin{pmatrix} \lambda_1 & 0 & 0 \\ 0 & \lambda_2 & 0 \\ 0 & 0 & \lambda_3 \end{pmatrix}. \tag{5.34}$$

That is, the real numbers λ_1, λ_2, and λ_3 are characterized as Fig. 5.2.

Finally, the following theorem holds regarding the pseudoclassical property of channels examined in Sect. 4.7.

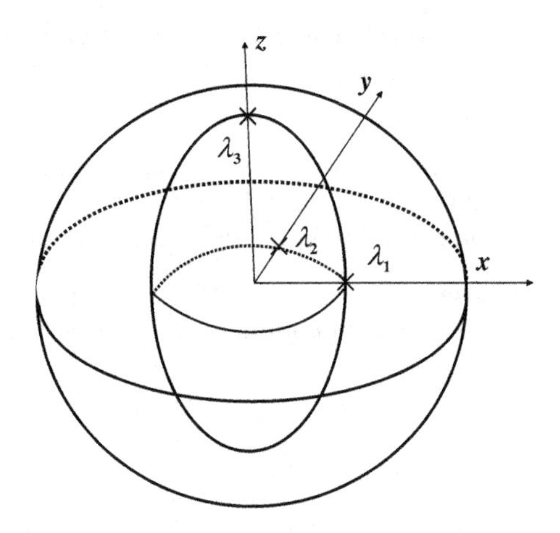

Fig. 5.2 Pauli channel

Theorem 5.6 (Fujiwara and Nagaoka [17]) *Let $\mathcal{H}$ be two-dimensional, $\mathcal{X}$ be given by $\mathcal{S}(\mathbb{C}^2)$, and W be given by the trace-preserving positive map κ from $\mathbb{C}^2$ to $\mathbb{C}^2$. A necessary and sufficient condition for a channel to be pseudoclassical is that one of the conditions given below should be satisfied.*

1. $\boldsymbol{t} = 0$.
2. *Let $\boldsymbol{t}$ be an eigenvector of TT^*. Let r be one of its eigenvalues and r_0 be the larger of the other two eigenvalues. Then,*

$$r_0 \le r^2 - \|\boldsymbol{t}\| r + \frac{(\|\boldsymbol{t}\| - r)\left(h\left(\frac{1+\|\boldsymbol{t}\|+r}{2}\right) - h\left(\frac{1+\|\boldsymbol{t}\|-r}{2}\right)\right)}{h'\left(\frac{1+\|\boldsymbol{t}\|-r}{2}\right)}.$$

Exercises

5.17 Check Condition (5.32) in the Pauli channel case (5.34).

5.18 Show that the Pauli channel given by (5.34) is entanglement-breaking if and only if $p_i \le \frac{1}{2}$ for $i = 1, 2, 3$.

5.19 Show that the positive map $\mathrm{Inv}_\lambda : \begin{pmatrix} a & b \\ c & d \end{pmatrix} \mapsto \lambda \begin{pmatrix} d & -b \\ -c & a \end{pmatrix} + (1-\lambda) \begin{pmatrix} a & b \\ c & d \end{pmatrix}$ is completely positive if and only if $\frac{2}{3} \ge \lambda \ge 0$.

5.20 Show that

$$F(\rho_{\boldsymbol{x}}, \rho_{\boldsymbol{y}})^2 = \frac{1 + \sqrt{1 - |\boldsymbol{x}|^2}\sqrt{1 - |\boldsymbol{y}|^2} + \langle \boldsymbol{x}, \boldsymbol{y} \rangle}{2}. \tag{5.35}$$

5.4 Information-Processing Inequalities in Quantum Systems

In this section, we will show that the quantum versions of the information quantities introduced in Sect. 3.1 satisfy the information-processing inequalities (i.e., the monotonicity) under the state evolutions given previously.

Theorem 5.7 (Lindblad [18], Uhlmann [19]) *Let κ be a TP-CP map from $\mathcal{H}_A$ to $\mathcal{H}_B$. Then, the* **monotonicity** *of the quantum relative entropy*

$$D(\rho\|\sigma) \geq D(\kappa(\rho)\|\kappa(\sigma)) \tag{5.36}$$

holds.

This theorem may be used to show many properties of the quantum relative entropy and the von Neumann entropy. For example, let $\rho_1, \ldots, \rho_k$ and $\sigma_1, \ldots, \sigma_k$ be density matrices on $\mathcal{H}$ and let p_i be a probability distribution in $\{1, \ldots, k\}$. Consider now the density matrix

$$R \stackrel{\text{def}}{=} \begin{pmatrix} p_1\rho_1 & & O \\ & \ddots & \\ O & & p_k\rho_k \end{pmatrix}, \quad S \stackrel{\text{def}}{=} \begin{pmatrix} p_1\sigma_1 & & O \\ & \ddots & \\ O & & p_k\sigma_k \end{pmatrix} \tag{5.37}$$

on $\mathcal{H} \otimes \mathbb{C}^k$. Since the partial trace $\mathrm{Tr}_{\mathbb{C}^k}$ is a TP-CP map, the inequality

$$D\left(\sum_{i=1}^k p_i\rho_i \middle\| \sum_{i=1}^k p_i\sigma_i\right) \leq D(R\|S) = \sum_{i=1}^k p_i D(\rho_i\|\sigma_i) \tag{5.38}$$

holds [20]. This inequality is called the **joint convexity** of the quantum relative entropy.

Proof of Theorem 5.7 Examine the connection with hypothesis testing. Let κ be a TP-CP map from $\mathcal{H}_A$ to $\mathcal{H}_B$. If a Hermitian matrix T on $\mathcal{H}_B^{\otimes n}$ satisfies $I \geq T \geq 0$, then $(\kappa^{\otimes n})^*(T)$ must also satisfy $I \geq (\kappa^{\otimes n})^*(T) \geq 0$. Therefore, from Condition ② of Theorem 5.1 and Corollary 5.3, we deduce that $(\kappa^{\otimes n})^*(T) \geq 0$. On the other hand, we see that $I \geq (\kappa^{\otimes n})^*(T)$ from

$$I - (\kappa^{\otimes n})^*(T) = (\kappa^{\otimes n})^*(I) - (\kappa^{\otimes n})^*(T) = (\kappa^{\otimes n})^*(I - T) \geq 0.$$

Since a state $\rho \in \mathcal{S}(\mathcal{H}_A)$ satisfies

$$\mathrm{Tr}(\kappa(\rho))^{\otimes n} T = \mathrm{Tr}\, \rho^{\otimes n} (\kappa^{\otimes n})^*(T),$$

the test $(\kappa^{\otimes n})^*(T)$ with the hypotheses $\rho^{\otimes n}$ and $\sigma^{\otimes n}$ has the same accuracy as the test T with the hypotheses $\kappa(\rho)^{\otimes n}$ and $\kappa(\sigma)^{\otimes n}$. That is, any test with the hypotheses

$\kappa(\rho)^{\otimes n}$ and $\kappa(\sigma)^{\otimes n}$ can be simulated by a test with the hypotheses $\rho^{\otimes n}$ and $\sigma^{\otimes n}$ with the same performance. We therefore have

$$B(\rho\|\sigma) \geq B(\kappa(\rho)\|\kappa(\sigma)).$$

Note that $B(\rho\|\sigma)$ is defined in Theorem 3.3. Hence, applying Theorem 3.3 then completes the proof. ■

Indeed, this proof requires only the tensor product positivity. Hence, since the transpose τ is tensor product positive, inequality (5.36) holds when κ is the transpose τ. Uhlmann [19] showed this inequality only with the two-positivity. This argument will be shown with a more general form in Theorem 6.12 in Sect. 6.7.1. Further, the equality condition of (5.36) can be characterized as follows.

Theorem 5.8 *For a TP-CP map κ, we assume that $D(\rho\|\sigma) < \infty$. Then, the ranges of σ and $\kappa(\sigma)$ contain those of ρ and $\kappa(\rho)$, respectively. Then, the following conditions are equivalent.*

① *The equality of (5.36) holds for a state ρ.*
② *The relation following relation holds.*

$$\rho = \sqrt{\sigma}\kappa^*(\sqrt{\kappa(\sigma)}^{-1}\kappa(\rho)\sqrt{\kappa(\sigma)}^{-1})\sqrt{\sigma}. \tag{5.39}$$

③ *The relation $P_\sigma\kappa^*(\kappa(\sigma)^{-t}\kappa(\rho)^t)) = \sigma^{-t}\rho^t$ holds for any $t > 0$.*

Here, we use the generalized inverse.

Theorem 5.8 will be shown with a more general form as Corollary 6.1 in Sect. 6.7. Now, using Theorem 5.8, we show Theorem 3.6.

Proof of Theorem 3.6 First, we show ①⇒②. Since $D(\mathrm{P}_\rho^{\boldsymbol{M}}\|\mathrm{P}_\sigma^{\boldsymbol{M}}) = D(\kappa_{\boldsymbol{M}}(\rho)\|\kappa_{\boldsymbol{M}}(\sigma))$ and $\kappa_{\boldsymbol{M}}^*(\rho') = \rho'$, Condition ③ of Theorem 5.8 with $t = 1$ implies that

$$\sigma^{-1}\rho = \kappa_{\boldsymbol{M}}^*(\kappa_{\boldsymbol{M}}(\sigma)^{-1}P_{\kappa_{\boldsymbol{M}}(\sigma)}\kappa_{\boldsymbol{M}}(\rho)) = \sum_{i:\mathrm{Tr}\, M_i\sigma>0} \frac{\mathrm{Tr}\, M_i\rho}{\mathrm{Tr}\, M_i\sigma} M_i. \tag{5.40}$$

Taking the adjoint, we have

$$\rho\sigma^{-1} = \sum_{i:\mathrm{Tr}\, M_i\sigma>0} \frac{\mathrm{Tr}\, M_i\rho}{\mathrm{Tr}\, M_i\sigma} M_i. \tag{5.41}$$

So, we have (3.120). Equation (3.120) implies the commutativity, i.e., $[\rho, \sigma] = 0$. Thus, we obtain ②.

Next, we show ②⇒①. Equation (3.120) and the commutativity imply that $D(\rho\|\sigma) = \mathrm{Tr}\,\rho\log(\sum_i a_i M_i) = \mathrm{Tr}\,\rho(\sum_i \log a_i M_i) = \sum_i \log a_i \mathrm{Tr}\,\rho M_i$, which equals $D(\mathrm{P}_\rho^{\boldsymbol{M}}\|\mathrm{P}_\sigma^{\boldsymbol{M}})$. ■

Now, we consider the lower bounds of $D(\rho\|\sigma)$ as

$$D_{c,p}(\rho\|\sigma) := \max_{M:PVM} D(\mathrm{P}_\rho^M\|\mathrm{P}_\sigma^M), \quad D_c(\rho\|\sigma) := \max_{M:POVM} D(\mathrm{P}_\rho^M\|\mathrm{P}_\sigma^M). \tag{5.42}$$

When $D(\rho\|\sigma) < \infty$, since the function $M \mapsto D(\mathrm{P}_\rho^M\|\mathrm{P}_\sigma^M)$ is continuous and the set of PVMs is compact, the maximum for PVM exists. Also, Lemma A.11 guarantees the existence of the maximum for POVM.

In general, these quantities do not satisfy the additivity (3.106). By applying (2.29) to distributions $(\langle u_i|\rho|u_i\rangle)$ and $(\langle u_i|\sigma|u_i\rangle)$, the quantity $D_{c,p}(\rho\|\sigma)$ can be written as [21]

$$\begin{aligned} D_{c,p}(\rho\|\sigma) &= \max_{\{u_i\},\lambda=(\lambda_1,\ldots,\lambda_k)\in\mathbb{R}^k} \mathrm{Tr}\,\rho\sum_{i=1}^k \lambda_i|u_i\rangle\langle u_i| - \log \mathrm{Tr}\,\sigma\sum_{i=1}^k e^{\lambda_i}|u_i\rangle\langle u_i| \\ &= \max_X \mathrm{Tr}\,\rho X - \log \mathrm{Tr}\,\sigma e^X, \end{aligned} \tag{5.43}$$

where $\{u_i\}$ is a CONS and X is a Hermitian matrix.

Theorem 5.9

$$D_c(\rho\|\sigma) = D_{c,p}(\rho\|\sigma) = \max_X \mathrm{Tr}\,\rho X - \log \mathrm{Tr}\,\sigma e^X. \tag{5.44}$$

Proof Choose the optimal POVM M such that $D_c(\rho\|\sigma) = D(\mathrm{P}_\rho^M\|\mathrm{P}_\sigma^M)$. For the POVM M, we take the Naĭmark extension $(\mathcal{H}_B, \rho_0, E)$ given in Theorem 4.5. Then, we have

$$D_c(\rho\|\sigma) = D_{c,p}(\rho\otimes\rho_0\|\sigma\otimes\rho_0) = \max_{X'} \mathrm{Tr}\,\rho\otimes\rho_0 X' - \log \mathrm{Tr}\,\sigma\otimes\rho_0 e^{X'}. \tag{5.45}$$

Now, we choose the matrix X' attaining the maximum (5.45). Since $\log x$ is matrix concave, Corollary A.1 guarantees that $X := \log \mathrm{Tr}_B(I\otimes\rho_0)e^{X'} \geq \mathrm{Tr}_B(I\otimes\rho_0)\log e^{X'} = \mathrm{Tr}_B(I\otimes\rho_0)X'$. Then, we have $\mathrm{Tr}\,\rho X \geq \mathrm{Tr}(\rho\otimes\rho_0)X'$. Since $\mathrm{Tr}\,\sigma e^X = \mathrm{Tr}\,\sigma\,\mathrm{Tr}_B(I\otimes\rho_0)e^{X'} = \mathrm{Tr}(\sigma\otimes\rho_0)e^{X'}$, we have $D_{c,p}(\rho\|\sigma) \geq D_c(\rho\|\sigma)$. ■

Substituting $\log\sigma^{-\frac{1}{2}}\rho\sigma^{-\frac{1}{2}}$ into X in (5.44), we obtain [21]

$$D_c(\rho\|\sigma) \geq \mathrm{Tr}\,\rho\log\sigma^{-\frac{1}{2}}\rho\sigma^{-\frac{1}{2}}. \tag{5.46}$$

Similarly, substituting $2\log\sigma^{-\frac{1}{2}}(\sigma^{\frac{1}{2}}\rho\sigma^{\frac{1}{2}})^{\frac{1}{2}}\sigma^{-\frac{1}{2}}$ into X, we obtain

$$D_c(\rho\|\sigma) \geq 2\,\mathrm{Tr}\,\rho\log\sigma^{-\frac{1}{2}}(\sigma^{\frac{1}{2}}\rho\sigma^{\frac{1}{2}})^{\frac{1}{2}}\sigma^{-\frac{1}{2}}. \tag{5.47}$$

Now, using (5.46), we show the Golden-Thompson trace inequality [21].

Lemma 5.4 (Golden-Thompson trace inequality [22–24]) *Any two Hermitian matrices A and B satisfy*

$$\mathrm{Tr}\, e^A e^B \geq \mathrm{Tr}\, e^{A+B}. \tag{5.48}$$

Proof It is sufficient to show the case when $\mathrm{Tr}\, e^B = 1$. We choose $\rho = e^{A+B} / \mathrm{Tr}\, e^{A+B}$ and $\sigma = e^B$. Then, (5.43) implies that

$$\begin{aligned}
&\log \mathrm{Tr}\, e^B e^A = \mathrm{Tr}\, \rho A - \mathrm{Tr}\, \rho A + \log \mathrm{Tr}\, \sigma e^A \geq \mathrm{Tr}\, \rho A - D_{c,p}(\rho\|\sigma) \\
\geq &\mathrm{Tr}\, \rho A - D(\rho\|\sigma) = \mathrm{Tr}\, \rho(A - \log \rho + \log \sigma) \\
= &\mathrm{Tr}\, \rho \left(A - (A + B - \log \mathrm{Tr}\, e^{A+B}\right) + B) = \log \mathrm{Tr}\, e^{A+B}.
\end{aligned}$$

■

As will be shown in Corollary 8.4 of Sect. 8.2, the Bures distance $b(\rho, \sigma)$ also satisfies the **monotonicity** [25–27]

$$b(\rho, \sigma) \geq b(\kappa(\rho), \kappa(\sigma)) \tag{5.49}$$

with respect to an arbitrary TP-CP map κ. This inequality may be derived from Corollary 8.4 given later. From (5.49) we may also show its **joint convexity**

$$b^2\left(\sum_{i=1}^{k} p_i\rho_i, \sum_{i=1}^{k} p_i\sigma_i\right) \leq b^2(R, S) = \sum_{i=1}^{k} p_i b^2(\rho_i, \sigma_i) \tag{5.50}$$

in a similar way to (5.38). The variational distance $d_1(\rho, \sigma)$ also satisfies the **monotonicity**

$$d_1(\rho, \sigma) \geq d_1(\kappa(\rho), \kappa(\sigma)) \tag{5.51}$$

for an arbitrary TP-CP map κ [28] Exe. 3.29. Furthermore, as extensions of (3.19) and (3.20), the monotonicities

$$\phi(s|\rho\|\sigma) \leq \phi(s|\kappa(\rho)\|\kappa(\sigma)) \text{ for } 0 \leq s \leq 1 \tag{5.52}$$

$$\phi(s|\rho\|\sigma) \geq \phi(s|\kappa(\rho)\|\kappa(\sigma)) \text{ for } -1 \leq s \leq 0 \tag{5.53}$$

$$\widetilde{\phi}(s|\rho\|\sigma) \leq \widetilde{\phi}(s|\kappa(\rho)\|\kappa(\sigma)) \text{ for } 0 \leq s \leq \frac{1}{2} \tag{5.54}$$

$$\widetilde{\phi}(s|\rho\|\sigma) \geq \widetilde{\phi}(s|\kappa(\rho)\|\kappa(\sigma)) \text{ for } s \leq 0 \tag{5.55}$$

hold. The relations (5.52) and (5.53) will be proved in Appendix A.4 by using matrix convex or concave functions. For a proof of Relation (5.55), see Exercise 5.21. We omit the proof of (5.54), which is given in [29]. Notice that Inequality (5.53) does not hold in general with the parameter $s \in (-\infty, -1)$, as shown in Exercise A.16.

The inequalities (5.52), (5.53), (5.54), and (5.55) are rewritten as the monotonicity of the quantum relative Rényi entropies

$$D_{1-s}(\rho\|\sigma) \geq D_{1-s}(\kappa(\rho)\|\kappa(\sigma)) \text{ for } -1 \leq s \leq 1 \tag{5.56}$$

$$\underline{D}_{1-s}(\rho\|\sigma) \geq \underline{D}_{1-s}(\kappa(\rho)\|\kappa(\sigma)) \text{ for } s \leq \frac{1}{2}. \tag{5.57}$$

As the limit $s \to -\infty$, we have

$$D_{\max}(\rho\|\sigma) \geq D_{\max}(\kappa(\rho)\|\kappa(\sigma)). \tag{5.58}$$

Exercises

5.21 Show (5.55) by using Exercise 3.58.

5.22 Show that the equation $\phi(s|\rho\|\sigma) = \widetilde{\phi}(s|\rho\|\sigma)$ does not hold for $s \leq -1$ in general by following steps below.
(**a**) Derive inequality (5.53) for $s \leq -1$ by assuming $\phi(s|\rho\|\sigma) = \widetilde{\phi}(s|\rho\|\sigma)$ for $s \leq -1$.
(**b**) Show the above argument by using Exercise A.16.

5.23 Show the monotonicity of transmission information

$$I(p, W) \geq I(p, \kappa(W)) \tag{5.59}$$

for any TP-CP map κ and any c-q channel: $W = (W_x)$, where $\kappa(W) = (\kappa(W_x))$.

5.24 Let W, κ, and σ be a c-q channel, a TP-CP map, and a quantum state, respectively. Define the c-q channel $\kappa(W) : x \mapsto \kappa(W_x)$. Show the following inequalities for $s \in [-1, 1] \setminus \{0\}$ by using (4.74).

$$J(p, \kappa(\sigma), \kappa(W)) \leq J(p, \sigma, W) \tag{5.60}$$
$$J_{1+s}(p, \kappa(\sigma), \kappa(W)) \leq J_{1+s}(p, \sigma, W) \tag{5.61}$$
$$I(p, \kappa(W)) \leq I(p, W) \tag{5.62}$$
$$I_{1+s}(p, \kappa(\sigma), \kappa(W)) \leq I_{1+s}(p, \sigma, W) \tag{5.63}$$
$$I^{\downarrow}_{1+s}(p, \kappa(\sigma), \kappa(W)) \leq I^{\downarrow}_{1+s}(p, \sigma, W) \tag{5.64}$$
$$C_c(\kappa(W)) \leq C_c(W) \tag{5.65}$$
$$C^{\downarrow}_{1+s}(\kappa(W)) \leq C^{\downarrow}_{1+s}(W). \tag{5.66}$$

5.25 Extend the definition of $D_{1+s}(\rho\|\sigma)$ and $\underline{D}_{1+s}(\rho\|\sigma)$ by $\frac{1}{s} \log \operatorname{Tr} \rho^{1+s}\sigma^{-s}$ (3.9) and $\frac{1}{s} \log \operatorname{Tr}(\sigma^{-\frac{s}{2(1+s)}} \rho \sigma^{-\frac{s}{2(1+s)}})^{1+s}$ (3.13) to the case when σ satisfies only the condition $\sigma \geq 0$ although ρ satisfies the conditions $\rho \geq 0$ and $\operatorname{Tr} \rho = 1$. In this definition, the case with $s = 0$ is given with the limit $s \to 0$. Show the following items under this extension.

(**a**) Any TP-CP map κ satisfies

$$D_{1+s}(\kappa(\rho)\|\kappa(\sigma)) \leq D_{1+s}(\rho\|\sigma) \quad \text{for } s \in [-1, 1] \tag{5.67}$$

$$\underline{D}_{1+s}(\kappa(\rho)\|\kappa(\sigma)) \leq \underline{D}_{1+s}(\rho\|\sigma) \quad \text{for } s \in [-\frac{1}{2}, \infty). \tag{5.68}$$

(**b**) When a projection P satisfies $P\rho P = \rho$,

$$D_{1+s}(\rho\|P\sigma P) \leq D_{1+s}(\rho\|\sigma) \quad \text{for } s \in [-1, 1] \tag{5.69}$$

$$\underline{D}_{1+s}(\rho\|P\sigma P) \leq \underline{D}_{1+s}(\rho\|\sigma) \quad \text{for } s \in [-\frac{1}{2}, \infty). \tag{5.70}$$

In particular, the equality holds when $\sigma = P\sigma P + (I - P)\sigma(I - P)$.
(**c**) Any constant c satisfies that

$$D_{1+s}(\rho\|c\sigma) = D_{1+s}(\rho\|\sigma) - \log c \tag{5.71}$$

$$\underline{D}_{1+s}(\rho\|c\sigma) = \underline{D}_{1+s}(\rho\|\sigma) - \log c. \tag{5.72}$$

(**d**) Any isometry U satisfies

$$D_{1+s}(U\rho U^\dagger\|U\sigma U^\dagger) = D_{1+s}(\rho\|\sigma) \tag{5.73}$$

$$\underline{D}_{1+s}(U\rho U^\dagger\|U\sigma U^\dagger) = \underline{D}_{1+s}(\rho\|\sigma). \tag{5.74}$$

(**e**) When $\sigma \leq \sigma'$,

$$D_{1+s}(\rho\|\sigma) \geq D_{1+s}(\rho\|\sigma') \tag{5.75}$$

$$\underline{D}_{1+s}(\rho\|\sigma) \geq \underline{D}_{1+s}(\rho\|\sigma'). \tag{5.76}$$

5.5 Entropy Inequalities in Quantum Systems

In this section, we will derive various inequalities related to the von Neumann entropy from the properties of the quantum relative entropy.

Substituting $\sigma = \rho_{\text{mix}}$ into the joint convexity of the quantum relative entropy (5.38), we obtain the **concavity** of the von Neumann entropy as follows:

$$H\left(\sum_{i=1}^{k} p_i \rho_i\right) \geq \sum_{i=1}^{k} p_i H(\rho_i). \tag{5.77}$$

Further, as shown in Sect. 8.4, when a state $\rho^{A,B}$ on $\mathcal{H}_A \otimes \mathcal{W}_B$ is separable, the von Neumann entropy satisfies

$$H(\rho^{A,B}) \geq H(\rho^A), H(\rho^B) \tag{5.78}$$

for the reduced densities ρ^A and ρ^B of $\rho^{A,B}$. We apply the inequality (5.78) to the separable state R defined in (5.37). Since the von Neumann entropy of R is equal to $\sum_{i=1}^k p_i H(\rho_i) + H(p)$, we obtain the reverse inequality of (5.77):

$$H\left(\sum_{i=1}^k p_i \rho_i\right) \le \sum_{i=1}^k p_i H(\rho_i) + H(p) \le \sum_{i=1}^k p_i H(\rho_i) + \log k. \tag{5.79}$$

In particular, if the supports for the densities ρ_i are disjoint, the first inequality satisfies the equality.

Similar types of inequalities may also be obtained by examining the pinching κ_M of the PVM $\boldsymbol{M}$. The quantum relative entropy satisfies

$$H(\kappa_M(\rho)) - H(\rho) = D(\rho\|\kappa_M(\rho)) \ge 0. \tag{5.80}$$

Since the inequality

$$D(\rho\|\kappa_M(\rho)) \le \log|\boldsymbol{M}| \tag{5.81}$$

holds [30] Exe. 5.28, we obtain

$$H(\rho) \le H(\kappa_M(\rho)) \le H(\rho) + \log|\boldsymbol{M}|. \tag{5.82}$$

Let ρ^A, ρ^B, $\rho^{A,B}$, and $\rho^{A,C}$ be the reduced density matrices of the density matrix $\rho = \rho^{A,B,C}$ on $\mathcal{H}_A \otimes \mathcal{H}_B \otimes \mathcal{H}_C$. From the monotonicity of the quantum relative entropy, we obtain

$$D(\rho^{A,B,C}\|\rho^{A,C} \otimes \rho^B) \ge D(\rho^{A,B}\|\rho^A \otimes \rho^B).$$

Rewriting this inequality, we may derive the following theorem called the **strong subadditivity** of the von Neumann entropy.

Theorem 5.10 (Lieb and Ruskai [31, 32]) *The inequality*

$$H(\rho^{A,B,C}) + H(\rho^A) \le H(\rho^{A,B}) + H(\rho^{A,C}) \tag{5.83}$$

holds.

Further, the equality condition of (5.83) is given as follows.

Theorem 5.11 (Hayden et al. [33]) *The equality in (5.83) holds if and only if there is a decomposition of the system $\mathcal{H}_A$ as*

$$\mathcal{H}_A = \bigoplus_j \mathcal{H}_{A-B,j} \otimes \mathcal{H}_{A-C,j} \tag{5.84}$$

into a direct (orthogonal) sum of tensor products such that

$$\rho^{ABC} = \bigoplus_j q_j \rho_j^{AB} \otimes \rho_j^{AC} \tag{5.85}$$

with states ρ_j^{AB} on $\mathcal{H}_B \otimes \mathcal{H}_{A-B,j}$ and ρ_j^{AC} on $\mathcal{H}_C \otimes \mathcal{H}_{A-C,j}$, and probability distribution q_j.

In particular, when $\mathcal{H}_A$ is one-dimensional,

$$H(\rho^{B,C}) \leq H(\rho^B) + H(\rho^C), \tag{5.86}$$

which is called the **subadditivity**. Let us change the notation slightly and write $H(\rho^{A,B})$ as $H_\rho(A, B)$ in order to emphasize the quantum system rather than the quantum state. The **Strong subadditivity** is then written as

$$H_\rho(A, B, C) + H_\rho(A) \leq H_\rho(A, B) + H_\rho(A, C). \tag{5.87}$$

Now, using this notation, let us define the **conditional entropy** $H_\rho(A|B) \stackrel{\text{def}}{=} H_\rho(A, B) - H_\rho(B)$ using this notation. This quantity satisfies the following **concavity**:

$$H_\rho(A|B) \geq \sum_{i=1}^{k} p_i H_{\rho_i}(A|B), \tag{5.88}$$

where $\rho = \sum_i p_i \rho_i$.

Similarly to Sect. 2.1.1, we can define the quantum mutual information $I_\rho(A : B)$ and the quantum conditional mutual information $I_\rho(A : B|C)$ as

$$I_\rho(A : B) \stackrel{\text{def}}{=} H_\rho(A) + H_\rho(B) - H_\rho(AB) \tag{5.89}$$

$$I_\rho(A : B|C) \stackrel{\text{def}}{=} H_\rho(AC) + H_\rho(BC) - H_\rho(ABC) - H_\rho(C). \tag{5.90}$$

The positivity of quantum mutual information is equivalent to the subadditivity, and that of quantum conditional mutual information is equivalent to the strong subadditivity.

Theorem 2.10 shows that the entropy $H(p)$ satisfies the asymptotic continuity in the classical case. The same property holds even in the quantum case. To see the asymptotic continuity in the quantum case more precisely, we introduce the **Fannes inequality**, which is particularly useful.

Theorem 5.12 (Fannes [34]) *Define*

$$\eta_0(x) \stackrel{\text{def}}{=} \begin{cases} \eta(x) & 0 \leq x \leq 1/e \\ 1/e & 1/e < x, \end{cases} \tag{5.91}$$

where $\eta(x) \stackrel{\text{def}}{=} -x \log x$. *Then, for two states* ρ *and* σ *on* $\mathcal{H}$ $(\dim \mathcal{H} = d)$*, the inequality*

$$|H(\rho) - H(\sigma)| \le \epsilon \log d + \eta_0(\epsilon) \tag{5.92}$$

holds for $\epsilon \stackrel{\text{def}}{=} \|\rho - \sigma\|_1$.

Let us consider the following lemma before proving this theorem.

Lemma 5.5 *Write the eigenvalues of the Hermitian matrices A and B in decreasing order (largest first) including any degeneracies, i.e.,* $a_1, \ldots, a_d$, $b_1, \ldots, b_d$. *Then,* $\|A - B\|_1 \ge \sum_{i=1}^d |a_i - b_i|$.

Proof Let $P \stackrel{\text{def}}{=} \{A - B \ge 0\}$, $X \stackrel{\text{def}}{=} P(A - B)$, and $Y \stackrel{\text{def}}{=} -(I - P)(A - B)$. Then, $X \ge 0$, $Y \ge 0$, and $A - B = X - Y$. Let $C \stackrel{\text{def}}{=} A + Y = B + X$. Then, $C \ge A, B$. Now let c_i be the eigenvalues of C arranged in decreasing order. From Exercise A.12 we know that $c_i \ge a_i, b_i$. Therefore, if $a_i - b_i \ge 0$, then $2c_i - a_i - b_i - (a_i - b_i) = 2(c_i - a_i) \ge 0$, and we obtain $2c_i - a_i - b_i \ge |a_i - b_i|$. This also holds for $a_i - b_i \le 0$, and therefore

$$\sum_i |a_i - b_i| \le \sum_i (2c_i - a_i - b_i) = \mathrm{Tr}(2C - A - B) = \mathrm{Tr}(X + Y) = \mathrm{Tr}\,|A - B|.$$

■

Proof of Theorem 5.12 We only provide a proof for $\|\rho - \sigma\|_1 \le 1/e$. See Exercise 5.35 for the case when $\|\rho - \sigma\|_1 > 1/e$. Let a_i, b_i be the eigenvalues of ρ, σ placed in decreasing order. Define $\epsilon_i \stackrel{\text{def}}{=} |a_i - b_i|$. Then, according to Lemma 5.5 and the assumptions of the theorem, $\epsilon_i \le 1/e \le 1/2$. From Exercise 5.34 we obtain

$$|H(\rho) - H(\sigma)| \le \sum_{i=1}^d |\eta(a_i) - \eta(b_i)| \le \sum_{i=1}^d \eta(\epsilon_i).$$

Next, define $\epsilon' \stackrel{\text{def}}{=} \sum_{i=1}^d \epsilon_i$. We find that $\sum_{i=1}^d \eta(\epsilon_i) = \epsilon' \sum_{i=1}^d \eta\left(\frac{\epsilon_i}{\epsilon'}\right) + \eta(\epsilon')$. Since $\sum_{i=1}^d \eta\left(\frac{\epsilon_i}{\epsilon'}\right)$ represents the entropy of the probability distribution $(\frac{\epsilon_1}{\epsilon'}, \frac{\epsilon_2}{\epsilon}', \ldots, \frac{\epsilon_d}{\epsilon'})$, we see that this must be less than $\log d$. Exercise 5.34 (**b**) guarantees that η_0 is monotone increasing. Thus, the inequality $\epsilon = \|\rho - \sigma\|_1 \ge \epsilon'$ implies that $\eta(\epsilon') \le \eta_0(\|\rho - \sigma\|_1)$. Hence,

$$\sum_{i=1}^d \eta(\epsilon_i) \le \epsilon \log d + \eta_0(\epsilon) \tag{5.93}$$

Therefore, we obtain the inequality (5.92). ■

Finally, we address what axioms identify the von Neumann entropy $H(\rho)$. It is not difficult to generalize Axioms **K1-K3** and **A1-A2** and **A4** to the quantum case. **K4** can be regarded as the unitary invariance in the quantum case. However, it is not so easy to generalize Axioms **K5** and **A3** to the quantum case. Replacing **K5** by Subadditivity, we consider the following set of axioms.

Q1 (Normalization)

$$S(\rho_{\mathrm{mix},\mathbb{C}^k}) = \log k. \tag{5.94}$$

Q2 (Continuity) S is continuous on $\mathcal{S}(\mathcal{H})$.
Q3 (Nonnegativity) S is nonnegative.
Q4 (Invariance) For any unitary U, we have

$$S(\rho) = S(U\rho U^*). \tag{5.95}$$

Q5 (Additivity)

$$S(\rho \otimes \sigma) = S(\rho) + S(\sigma) \tag{5.96}$$

Q6 (Subadditivity)

$$S(\rho_{AB}) \le S(\rho_A) + S(\rho_B). \tag{5.97}$$

It is known that, when a quantity S satisfies all of the above axioms, it becomes the von Neumann entropy $H(\rho)$ [35].

Exercises

5.26 Show (5.83) using the monotonicity of the relative entropy.

5.27 Show (4.3) from the concavity of von Neumann entropy.

5.28 Show (5.81) following the steps below.
(**a**) Show (5.81) for a pure state.
(**b**) Show (5.81) for the general case using the joint convexity of the quantum relative entropy.

5.29 Show (5.88) using (5.87).

5.30 Show the Araki–Lieb inequality [36, 37] below using the subadditivity and the state purification introduced in Sect. 8.1

$$H(\rho^{A,B}) \ge |H(\rho^A) - H(\rho^B)|. \tag{5.98}$$

5.31 Show that the strong subadditivity (5.83) is equivalent to the following inequality:

$$H_\rho(AB|C) \le H_\rho(A|C) + H_\rho(B|C). \tag{5.99}$$

5.32 Show the following inequality using the strong subadditivity (5.87):

$$H_{|u\rangle\langle u|}(A, C) + H_{|u\rangle\langle u|}(A, D) \ge H_{|u\rangle\langle u|}(A) + H_{|u\rangle\langle u|}(B). \tag{5.100}$$

5.33 Using (5.98), show that

$$|H_\rho(A|B)| \le \log d_A. \tag{5.101}$$

5.34 Show that

$$|\eta(x) - \eta(y)| \le \eta(|x - y|) \tag{5.102}$$

if x and y satisfy $|x - y| \le 1/2$ following the steps below.
(**a**) Show that $\eta(x + \epsilon) - \eta(x) \le \eta(\epsilon)$ for $x \ge 0$ and $\epsilon \ge 0$.
(**b**) Show that $\eta(x)$ is strictly concave and has its maximum value when $x = 1/e$.
(**c**) Show that $\eta(\alpha - \epsilon) - \eta(\alpha) \le \eta(1 - \epsilon) - \eta(1)$ for $\epsilon < \alpha < 1$.
(**d**) Show that the function $\eta(x) - \eta(1 - x)$ is strictly concave and $\eta(x) - \eta(1 - x) > 0$ for $0 < x < 1/2$.
(**e**) Show that $\eta(x) - \eta(x + \epsilon) \le \eta(\epsilon)$ using (**c**) and (**d**), and hence show (5.102).

5.35 Prove Theorem 5.12 for $d_1(\rho, \sigma) > 1/e$ following the steps below with the notations given in the proof for $d_1(\rho, \sigma) \le 1/e$.
(**a**) Show (5.92) if $\epsilon_1 \le 1/e$, i.e., all the ϵ_i are less than $1/e$.
(**b**) Show that

$$|H(\rho) - H(\sigma)| \le 1/e + \epsilon' \log(d - 1) + \eta_0(\epsilon'),$$

where $\epsilon' \stackrel{\text{def}}{=} \sum_{i=2}^d \epsilon_i$ and if $\epsilon_1 > 1/e$.
(**c**) Show that $\epsilon \log d \ge \epsilon' \log(d - 1) + 1/e$ if $\epsilon_1 > 1/e$. Hence, show (5.92) in this case.

5.36 Show that $I(p, W) \le \delta \log d + \eta_0(\delta)$ using Theorem 5.12, where $\delta \stackrel{\text{def}}{=} \sum_x p(x) \| W_x - W_p \|_1$.

5.37 Let ρ and $\tilde{\rho}$ be two arbitrary states. For any real $0 \le \epsilon \le 1$, show that

$$|H_\rho(A|B) - H_\gamma(A|B)| \le 2\epsilon \log d_A + h(\epsilon), \tag{5.103}$$

following the steps, where $\gamma \stackrel{\text{def}}{=} (1 - \epsilon)\rho + \epsilon\tilde{\rho}$ [38].
(**a**) Using (5.88) and (5.101), show that $H_\rho(A|B) - H_\gamma(A|B) \le \epsilon(H_\rho(A|B) - H_{\tilde{\rho}}(A|B)) \le 2\epsilon \log d_A$.
(**b**) Show that $H_\gamma(B) \ge (1 - \epsilon)H_\rho(B) + \epsilon H_{\tilde{\rho}}(B)$.
(**c**) Show that $H_\gamma(AB) \le (1 - \epsilon)H_\rho(AB) + \epsilon H_{\tilde{\rho}}(AB) + h(\epsilon)$.
(**d** Using (5.101), show that $H_\rho(A|B) - H_\gamma(A|B) \ge \epsilon(H_\rho(A|B) - H_{\tilde{\rho}}(A|B)) - h(\epsilon) \ge -2\epsilon \log d_A - h(\epsilon)$.

5.38 Show that

$$|H_\rho(A|B) - H_\sigma(A|B)| \le 4\epsilon \log d_A + 2h(\epsilon) \tag{5.104}$$

for states ρ and σ on $\mathcal{H}_A \otimes \mathcal{H}_B$ and $\epsilon \stackrel{\text{def}}{=} \|\rho - \sigma\|_1$ following the steps below [38].
(**a**) Define the states $\tilde{\rho} \stackrel{\text{def}}{=} \frac{1}{\epsilon}|\rho - \sigma|$, $\tilde{\sigma} \stackrel{\text{def}}{=} \frac{1-\epsilon}{\epsilon}(\rho - \sigma) + \frac{1}{\epsilon}|\rho - \sigma|$, and $\gamma \stackrel{\text{def}}{=} (1 - \epsilon)\rho + \epsilon\tilde{\rho}$. Show that $\gamma = (1 - \epsilon)\sigma + \epsilon\tilde{\sigma}$.
(**b**) Using (5.103), show that $|H_\rho(A|B) - H_\sigma(A|B)| \le 4\epsilon \log d_A + 2h(\epsilon)$.

5.39 Using the above inequality, show that

$$|I_\rho(A : B) - I_\sigma(A : B)| \le 5\epsilon \log d_A + \eta_0(\epsilon) + 2h(\epsilon) \tag{5.105}$$

for states ρ and σ on $\mathcal{H}_A \otimes \mathcal{H}_B$ and $\epsilon \stackrel{\text{def}}{=} \|\rho - \sigma\|_1$.

5.40 Show that

$$|I_\rho(A : B|C) - I_\sigma(A : B|C)| \le 8\epsilon \log d_A d_B + 6h(\epsilon) \tag{5.106}$$

for states ρ and σ on $\mathcal{H}_A \otimes \mathcal{H}_B$ and $\epsilon \stackrel{\text{def}}{=} \|\rho - \sigma\|_1$ following the steps below [39].
(**a**) Show that $|I_\rho(A : B|C) - I_\sigma(A : B|C)| \le |H_\rho(A|C) - H_\sigma(A|C)| + |H_\rho(B|C) - H_\sigma(B|C)| + |H_\rho(AB|C) - H_\sigma(AB|C)|$.
(**b**) Show (5.106) using (5.104).

5.41 Show the **chain rules** of quantum mutual information and quantum conditional mutual information:

$$H_\rho(AB|C) = H_\rho(B|C) + H_\rho(A|BC), \tag{5.107}$$
$$I_\rho(A : BC) = I_\rho(A : C) + I_\rho(A : B|C), \tag{5.108}$$
$$I_\rho(A : BC|D) = I_\rho(A : C|D) + I_\rho(A : B|CD). \tag{5.109}$$

5.42 Show that the monotonicity $I_\rho(A : B) \ge I_{\kappa_A \otimes \kappa_B \rho}(A : B)$ for local TP-CP maps κ_A and κ_B.

5.43 Show that the monotonicity $I_\rho(A : B|C) \ge I_{\kappa_A \otimes \kappa_B \rho}(A : B|C)$ for local TP-CP maps κ_A and κ_B.

5.44 Using (5.82), show the Hiai–Petz theorem [30] for two arbitrary states ρ, σ

$$\lim_{n\to\infty} \frac{1}{n} D(\kappa_{\sigma^{\otimes n}}(\rho^{\otimes n}) \| \sigma^{\otimes n}) = D(\rho\|\sigma),$$

where $\kappa_{\sigma^{\otimes n}}$ represents the pinching of the measurement corresponding to the spectral decomposition of $\sigma^{\otimes n}$. Hence, the equality in (3.18) holds in an asymptotic sense when the POVM is the simultaneous spectral decomposition of $\kappa_{\sigma^{\otimes n}}(\rho^{\otimes n})$ and $\sigma^{\otimes n}$.

Combining this result with the classical Stein's lemma gives an alternate proof of Lemma 3.6.

5.45 Show Holevo's inequality $I(\boldsymbol{M}, p, W) \leq I(p, W)$.

5.46 For a classical-quantum channel $W = (W_i)$ and a TP-CP map κ, we define $\kappa(W) = (\kappa(W_i))$. Show the inequality $I(p, \kappa(W)) \leq I(p, W)$.

5.47 Given densities ρ_i^A and ρ_i^B on $\mathcal{H}_A$ and $\mathcal{H}_B$, show the strong concavity of von Neumann entropy:

$$H(\sum_i p_i \rho_i^A \otimes \rho_i^B) \geq H(\sum_i p_i \rho_i^A) + \sum_i p_i H(\rho_i^B) \tag{5.110}$$

from the joint convexity of quantum relative entropy (5.50) for states $\rho_i^A \otimes \rho_i^B$ and $\rho_{\mathrm{mix}}^A \otimes \rho_i^B$ [40].

5.48 Show that

$$H_{(\iota_A \otimes \kappa)(\rho)}(A|B) \geq H_\rho(A|B) \tag{5.111}$$

for any TP-CP map κ on $\mathcal{H}_B$.

5.6 Conditional Rényi Entropy and Duality

Finally, we consider the quantum version of the conditional extension of Rényi entropy. For generalization of the conditional entropy, we have four kinds of conditional Rényi entropies as

$$H_{1+s|\rho}(A|B) := -D_{1+s}(\rho \| I_A \otimes \rho_B) = -\frac{1}{s} \log \operatorname{Tr} \rho^{1+s} \left(I_A \otimes \rho_B^{-s} \right), \tag{5.112}$$

$$\begin{aligned} \widetilde{H}_{1+s|\rho}(A|B) &:= -\underline{D}_{1+s}(\rho \| I_A \otimes \rho_B) \\ &= -\frac{1}{s} \log \operatorname{Tr} \left\{ \left(\left(I_A \otimes \rho_B^{-\frac{s}{2(1+s)}} \right) \rho \left(I_A \otimes \rho_B^{-\frac{s}{2(1+s)}} \right) \right)^{\frac{1}{1+s}} \right\}, \end{aligned} \tag{5.113}$$

$$H^{\uparrow}_{1+s|\rho}(A|B) := \max_{\sigma_B} -D_{1+s}(\rho \| I_A \otimes \sigma_B), \tag{5.114}$$

$$\widetilde{H}^{\uparrow}_{1+s|\rho}(A|B) := \max_{\sigma_B} -\underline{D}_{1+s}(\rho \| I_A \otimes \sigma_B). \tag{5.115}$$

Due to the relations (5.56) and (5.57), any TP-CP map κ on $\mathcal{H}_B$ satisfies

$$H_{1+s|\rho}(A|B) \le H_{1+s|\kappa(\rho)}(A|B) \text{ for } s > -1, \tag{5.116}$$

$$\widetilde{H}_{1+s|\rho}(A|B) \le \widetilde{H}_{1+s|\kappa(\rho)}(A|B) \text{ for } s > -\frac{1}{2}, \tag{5.117}$$

$$H^{\uparrow}_{1+s|\rho}(A|B) \le H^{\uparrow}_{1+s|\kappa(\rho)}(A|B) \text{ for } s > -1, \tag{5.118}$$

$$\widetilde{H}^{\uparrow}_{1+s|\rho}(A|B) \le \widetilde{H}^{\uparrow}_{1+s|\kappa(\rho)}(A|B) \text{ for } s > -\frac{1}{2}. \tag{5.119}$$

Due to the properties of $D_{1+s}(\rho\|I_A \otimes \rho_B)$ and $\underline{D}_{1+s}(\rho\|I_A \otimes \rho_B)$, $H_{1+s|\rho}(A|B)$ and $\widetilde{H}_{1+s|\rho}(A|B)$ are monotone decreasing for s and $\lim_{s\to 0} H_{1+s|\rho}(A|B) = \lim_{s\to 0} \widetilde{H}_{1+s|\rho}(A|B) = H_\rho(A|B)$. In the case of $s = 0$, they are defined as $H_\rho(A|B)$ because[Exe. 2.29]

$$\begin{aligned}
&\lim_{s\to 0} H_{1+s|\rho}(A|B) = \lim_{s\to 0} H^{\uparrow}_{1+s|\rho}(A|B) \\
=&\lim_{s\to 0} \widetilde{H}_{1+s|\rho}(A|B) = \lim_{s\to 0} \widetilde{H}^{\uparrow}_{1+s|\rho}(A|B) = H(A|B).
\end{aligned} \tag{5.120}$$

From the definition, we find the relation

$$H_{1+s|\rho}(A|B) \le H^{\uparrow}_{1+s|\rho}(A|B), \quad \widetilde{H}_{1+s|\rho}(A|B) \le \widetilde{H}^{\uparrow}_{1+s|\rho}(A|B). \tag{5.121}$$

The relation (3.25) implies that

$$H_{1+s|\rho}(A|B) \le \widetilde{H}_{1+s|\rho}(A|B), \quad H^{\uparrow}_{1+s|\rho}(A|B) \le \widetilde{H}^{\uparrow}_{1+s|\rho}(A|B). \tag{5.122}$$

According to the relations (2.40), $H_{\min|\rho}(A|B)$, $H^{\uparrow}_{\min|\rho}(A|B)$, $\widetilde{H}_{\min|\rho}(A|B)$, $\widetilde{H}^{\uparrow}_{\min|\rho}(A|B)$, $H_{\max|\rho}(A|B)$, $H^{\uparrow}_{\max|\rho}(A|B)$ $\widetilde{H}_{\max|\rho}(A|B)$, and $\widetilde{H}^{\uparrow}_{\max|\rho}(A|B)$ are defined as

$$H_{\min|\rho}(A|B) \stackrel{\text{def}}{=} \lim_{s\to\infty} H_{1+s|\rho}(A|B), \quad H^{\uparrow}_{\min|\rho}(A|B) \stackrel{\text{def}}{=} \lim_{s\to\infty} H^{\uparrow}_{1+s|\rho}(A|B), \tag{5.123}$$

$$\widetilde{H}_{\min|\rho}(A|B) \stackrel{\text{def}}{=} \lim_{s\to\infty} \widetilde{H}_{1+s|\rho}(A|B), \quad \widetilde{H}^{\uparrow}_{\min|\rho}(A|B) \stackrel{\text{def}}{=} \lim_{s\to\infty} \widetilde{H}^{\uparrow}_{1+s|\rho}(A|B), \tag{5.124}$$

$$H_{\max|\rho}(A|B) \stackrel{\text{def}}{=} \lim_{s\to -1} H_{1+s|\rho}(A|B), \quad H^{\uparrow}_{\max|\rho}(A|B) \stackrel{\text{def}}{=} \lim_{s\to -1} H^{\uparrow}_{1+s|\rho}(A|B), \tag{5.125}$$

$$\widetilde{H}_{\max|\rho}(A|B) \stackrel{\text{def}}{=} \lim_{s\to -1} \widetilde{H}_{1+s|\rho}(A|B), \quad \widetilde{H}^{\uparrow}_{\max|\rho}(A|B) \stackrel{\text{def}}{=} \lim_{s\to -1} \widetilde{H}^{\uparrow}_{1+s|\rho}(A|B). \tag{5.126}$$

Unfortunately, these four conditional Rényi entropies are not the same in general. Thanks to the properties of the relative Rényi entropies $D(\rho\|\sigma)$ and $\underline{D}(\rho\|\sigma)$ given in Lemma 3.1, we have the following lemma[Exe. 2.31].

Lemma 5.6 *The functions* $s \mapsto sH_{1+s|\rho}(A|B)$, $s\widetilde{H}_{1+s|\rho}(A|B)$, $sH^{\uparrow}_{1+s|\rho}(A|B)$, *and* $s\widetilde{H}^{\uparrow}_{1+s|\rho}(A|B)$ *are concave for* $s \in (-1, \infty)$. *The functions* $s \mapsto H_{1+s|\rho}(A|B)$, $\widetilde{H}_{1+s|\rho}(A|B)$, $H^{\uparrow}_{1+s|\rho}(A|B)$, *and* $\widetilde{H}^{\uparrow}_{1+s|\rho}(A|B)$ *are monotonicallly decreasing.*

Lemma 5.7 *The quantity* $H^{\uparrow}_{1+s|\rho}(A|B)$ *has the following form.*

$$H^{\uparrow}_{1+s|\rho}(A|B) = -D_{1+s}(\rho\|I_A \otimes \sigma_B^{(1+s)}) = -\frac{1+s}{s}\log \mathrm{Tr}_B(\mathrm{Tr}_A\, \rho^{1+s})^{\frac{1}{1+s}}, \quad (5.127)$$

where $\sigma_B^{(1+s)} \stackrel{\mathrm{def}}{=} \frac{(\mathrm{Tr}_A\, \rho^{1+s})^{\frac{1}{1+s}}}{\mathrm{Tr}_B(\mathrm{Tr}_A\, \rho^{1+s})^{\frac{1}{1+s}}}$.

Proof Substituting $\mathrm{Tr}_A\, \rho^{1+s}$ and σ_B^{-s} into X and Y in the matrix reverse matrix Hölder inequality (A.28) with $p = \frac{1}{1+s}$ and $q = -\frac{1}{s}$, we obtain

$$\begin{aligned} e^{sD_{1+s}(\rho\|I\otimes\sigma_B)} &= \mathrm{Tr}_B\, \mathrm{Tr}_A\, \rho^{1+s}\sigma_B^{-s} \\ \geq &(\mathrm{Tr}_B(\mathrm{Tr}_A\, \rho^{1+s})^{1/(1+s)})^{1+s}(\mathrm{Tr}_B\, \sigma_B^{-s\cdot -1/s})^{-s} = (\mathrm{Tr}_B(\mathrm{Tr}_A\, \rho^{1+s})^{\frac{1}{1+s}})^{1+s} \end{aligned}$$

for $s \in (0, \infty]$. Since the equality holds when $\sigma_B = \sigma_B^{(1+s)}$, we obtain

$$e^{-sH^{\uparrow}_{1+s|\rho}(A|B)} = (\mathrm{Tr}_B(\mathrm{Tr}_A\, \rho^{1+s})^{\frac{1}{1+s}})^{1+s},$$

which implies (5.127) with $s \in (0, \infty]$.

The same substitution to the matrix Hölder inequality (A.26) yields

$$e^{sD_{1+s}(\rho\|I\otimes\sigma_B)} \leq (\mathrm{Tr}_B(\mathrm{Tr}_A\, \rho^{1+s})^{\frac{1}{1+s}})^{1+s}$$

for $s \in (-1, 0)$. Since the equality holds when $\sigma_B = \sigma_B^{(1+s)}$, we obtain (5.127) with $s \in (-1, 0)$.

Using (3.10), (3.11), and (3.16), we obtain the following lemma.

Lemma 5.8 *When* ρ *has the form* $\sum_a P_A(a)|a\rangle\langle a| \otimes \rho_{B|A=a}$, *the quantities* $H_{\min|\rho}(A|B)$, $H^{\uparrow}_{\min|\rho}(A|B)$, $H_{\max|\rho}(A|B)$, *and* $H^{\uparrow}_{\max|\rho}(A|B)$ *are characterized as*

$$\widetilde{H}_{\min|\rho}(A|B) = -\log \max_a P_A(a)\|\rho_B^{-\frac{1}{2}}\rho_{B|A=a}\rho_B^{-\frac{1}{2}}\|, \quad (5.128)$$

$$\widetilde{H}^{\uparrow}_{\min|\rho}(A|B) = -\log \min_{\sigma_B} \max_a P_A(a)\|\sigma_B^{-\frac{1}{2}}\rho_{B|A=a}\sigma_B^{-\frac{1}{2}}\|, \quad (5.129)$$

$$H_{\max|\rho}(A|B) = -\log \sum_{a:P_A(a)>0} \mathrm{Tr}\{\rho_{B|A=a} > 0\}\rho_B, \quad (5.130)$$

$$H^{\uparrow}_{\max|\rho}(A|B) = -\log \min_{\sigma_B} \sum_{a:P_A(a)>0} \mathrm{Tr}\{\rho_{B|A=a} > 0\}\sigma_B, \quad (5.131)$$

where σ_B is a density on $\mathcal{H}_B$.

Further, the quantity $\widetilde{H}^{\uparrow}_{\min|\rho}(A|B)$ has the following operational meaning with respect to state discrimination.

Lemma 5.9 (König et al. [41, Theorem 1]) *When $\rho = \sum_a p_a |a\rangle\langle a| \otimes \rho_a$, i.e., $P_A(a) = p_a$ and $\rho_{B|A=a} = \rho_a$, we have*

$$p_{\text{guess}} = e^{-\widetilde{H}^{\uparrow}_{\min|\rho}(A|B)}. \tag{5.132}$$

In this scenario, when we apply the POVM $\{M_a\}$ with $M_a := \rho_B^{-\frac{1}{2}} p_a \rho_a \rho_B^{-\frac{1}{2}}$, the correctly recovering probability is [41]

$$\sum_a p_a \operatorname{Tr} \rho_B^{-\frac{1}{2}} p_a \rho_a \rho_B^{-\frac{1}{2}} \rho_a = e^{-\widetilde{H}_{2|\rho}(A|B)}, \tag{5.133}$$

which gives a lower bound of p_{guess}. Hence, we have

$$\widetilde{H}_{2|\rho}(A|B) \geq \widetilde{H}^{\uparrow}_{\min|\rho}(A|B). \tag{5.134}$$

Proof Choosing $\sigma_B := \frac{F}{\operatorname{Tr} F}$ and $x := \operatorname{Tr} F$, we have

$$\begin{aligned}
&(\text{RHS of (3.84)}) = \min_{F \geq 0: I_A \otimes F \geq \rho_{AB}} \operatorname{Tr} F \\
= &\min_{\sigma_B \geq 0: \operatorname{Tr} \sigma_B = 1} \ \min_{x: I_A \otimes x\sigma_B \geq \rho_{AB}} x \\
\stackrel{(a)}{=} &\min_{\sigma_B \geq 0: \operatorname{Tr} \sigma_B = 1} \|(I_A \otimes \sigma_B)^{-\frac{1}{2}} \rho_{AB} (I_A \otimes \sigma_B)^{-\frac{1}{2}}\| \\
= &\min_{\sigma_B} \max_a P_A(a) \|\sigma_B^{-\frac{1}{2}} \rho_{B|A=a} \sigma_B^{-\frac{1}{2}}\| \stackrel{(b)}{=} e^{-\widetilde{H}^{\uparrow}_{\min|\rho}(A|B)},
\end{aligned}$$

where (a) and (b) follow from Exercise 3.13 and (5.129), respectively. Hence, 3.84 yields (5.132). ∎

Now, we give duality relations among four kinds of Rényi entropies. Consider tripartite system $\mathcal{H}_A \otimes \mathcal{H}_B \otimes \mathcal{H}_C$. When the state ρ of the composite system $\mathcal{H}_A \otimes \mathcal{H}_B \otimes \mathcal{H}_C$ is a pure state $|\psi\rangle\langle\psi|$, we can show that

$$H_\rho(A|B) + H_\rho(A|C) = 0, \tag{5.135}$$

which is a duality relation with respect to the conditional entropy. As a generalization of the duality relation, we have the following theorem.

Theorem 5.13 *[42–45] When the state ρ of the composite system $\mathcal{H}_A \otimes \mathcal{H}_B \otimes \mathcal{H}_C$ is a pure state $|\psi\rangle\langle\psi|$, the following holds.*

$$H_{\alpha|\rho}(A|B) + H_{\beta|\rho}(A|C) = 0 \qquad \textit{for} \quad \alpha, \beta \in [0, 2],\ \alpha + \beta = 2, \tag{5.136}$$

$$\widetilde{H}^{\uparrow}_{\alpha|\rho}(A|B) + \widetilde{H}^{\uparrow}_{\beta|\rho}(A|C) = 0 \qquad \textit{for} \quad \alpha, \beta \in \left[\frac{1}{2}, \infty\right],\ \frac{1}{\alpha} + \frac{1}{\beta} = 2, \tag{5.137}$$

$$H^{\uparrow}_{\alpha|\rho}(A|B) + \widetilde{H}_{\beta|\rho}(A|C) = 0 \qquad \textit{for} \quad \alpha, \beta \in [0, \infty],\ \alpha \cdot \beta = 1. \tag{5.138}$$

Proof Firstly, we can show (5.136) as follows:

$$\begin{aligned}
& - sH_{1+s|\rho}(A|B) = \log \operatorname{Tr} \rho_{AB}^{1+s}(I_A \otimes \rho_B^{-s}) = \log \operatorname{Tr} \rho_{AB}\rho_{AB}^{s}(I_A \otimes \rho_B^{-s}) \\
= & \log \langle \psi | (\rho_{AB}^{s} \otimes I_C)(I_{A,C} \otimes \rho_B^{-s}) | \psi \rangle \overset{(a)}{=} \log \langle \psi | (I_{AB} \otimes \rho_C^{s})(\rho_{A,C}^{-s} \otimes I_B) | \psi \rangle \\
= & \log \operatorname{Tr} \rho_{A,C}(I_A \otimes \rho_C^{s})\rho_{A,C}^{-s} = \log \operatorname{Tr} \rho_{A,C}^{1-s}(I_A \otimes \rho_C^{s}) = sH_{1-s|\rho}(A|C),
\end{aligned}$$

where (a) follows from Exercise 1.36.

Next, we show (5.138). Due to the expression of $H^{\uparrow}_{\alpha|\rho}(A|B)$ given in Lemma 5.7, it is sufficient to show that

$$\begin{aligned}
& \frac{\alpha}{1-\alpha} \log \operatorname{Tr} \left\{ \left(\operatorname{Tr}_A \{\rho_{AB}^{\alpha}\} \right)^{\frac{1}{\alpha}} \right\} \\
= & \frac{\alpha}{1-\alpha} \log \operatorname{Tr} \left\{ \left(\left(I_A \otimes \rho_C^{\frac{\alpha-1}{2}} \right) \rho_{AC} \left(I_A \otimes \rho_C^{\frac{\alpha-1}{2}} \right) \right)^{\frac{1}{\alpha}} \right\}
\end{aligned} \tag{5.139}$$

because $\beta = \frac{1}{\alpha}$. To prove (5.139), we show that the operators

$$\operatorname{Tr}_A \{\rho_{AB}^{\alpha}\} \quad \text{and} \quad \left(I_A \otimes \rho_C^{\frac{\alpha-1}{2}} \right) \rho_{AC} \left(I_A \otimes \rho_C^{\frac{\alpha-1}{2}} \right) \tag{5.140}$$

are unitarily equivalent, which is a stronger argument than (5.139). To see that this is indeed true, note the first operator in (5.140) can be rewritten as

$$\begin{aligned}
\operatorname{Tr}_A \{\rho_{AB}^{\alpha}\} = & \operatorname{Tr}_A \left\{ \rho_{AB}^{\frac{\alpha-1}{2}} \rho_{AB} \rho_{AB}^{\frac{\alpha-1}{2}} \right\} \\
= & \operatorname{Tr}_{AC} \left\{ \left(\rho_{AB}^{\frac{\alpha-1}{2}} \otimes I_C \right) \rho_{ABC} \left(\rho_{AB}^{\frac{\alpha-1}{2}} \otimes I_C \right) \right\} \\
\overset{(a)}{=} & \operatorname{Tr}_{AC} \left\{ \left(I_{AB} \otimes \rho_C^{\frac{\alpha-1}{2}} \right) \rho_{ABC} \left(I_{AB} \otimes \rho_C^{\frac{\alpha-1}{2}} \right) \right\},
\end{aligned} \tag{5.141}$$

where (a) follows from Exercise 1.36 because ρ is a pure state. Since

$$\left(I_{AB} \otimes \rho_C^{\frac{\alpha-1}{2}} \right) \rho_{ABC} \left(I_{AB} \otimes \rho_C^{\frac{\alpha-1}{2}} \right) \tag{5.142}$$

is a rank-1 matrix on the bipartite system B and AC, the RHS of (5.141) is unitarily equivalent with

$$\mathrm{Tr}_B\left(I_{AB}\otimes\rho_C^{\frac{\alpha-1}{2}}\right)\rho_{ABC}\left(I_{AB}\otimes\rho_C^{\frac{\alpha-1}{2}}\right)=\left(I_A\otimes\rho_C^{\frac{\alpha-1}{2}}\right)\rho_{AC}\left(I_A\otimes\rho_C^{\frac{\alpha-1}{2}}\right).$$

This concludes the proof of (5.138).

Finally, we show (5.137). For $\alpha < 1$, we have

$$\begin{aligned}\widetilde{H}^{\uparrow}_{\alpha|\rho}(A|B) &\stackrel{(a)}{=} \frac{\alpha}{1-\alpha}\log\max_{\sigma_B}\|\rho_{AB}^{\frac{1}{2}}(I_A\otimes\sigma_B^{\frac{1}{\alpha}-1})\rho_{AB}^{\frac{1}{2}}\|_\alpha\\ &\stackrel{(b)}{=}\frac{\alpha}{1-\alpha}\log\max_{\sigma_B}\min_{\tau_{AB}}\mathrm{Tr}\,\rho_{AB}^{\frac{1}{2}}(I_A\otimes\sigma_B^{\frac{1}{\alpha}-1})\rho_{AB}^{\frac{1}{2}}\tau_{AB}^{1-\frac{1}{\alpha}}\\ &\stackrel{(c)}{=}\frac{\alpha}{1-\alpha}\log\max_{\sigma_B}\min_{\tau_C}\langle\psi|(I_A\otimes\sigma_B^{\frac{1}{\alpha}-1})\otimes\tau_C^{1-\frac{1}{\alpha}}|\psi\rangle\\ &\stackrel{(d)}{=}\frac{\alpha}{1-\alpha}\log\min_{\tau_C}\max_{\sigma_B}\langle\psi|(I_A\otimes\sigma_B^{\frac{1}{\alpha}-1})\otimes\tau_C^{1-\frac{1}{\alpha}}|\psi\rangle,\end{aligned} \tag{5.143}$$

where (a), (b), (c), and (d) follow from Exercise 3.12, Exercise A.13, Exercise 1.37, and Lemma A.9, respectively. Similarly, for $\beta > 1$ with the relation $\frac{1}{\alpha}+\frac{1}{\beta}=2$, we have

$$\begin{aligned}\widetilde{H}^{\uparrow}_{\beta|\rho}(A|C) &= \frac{\beta}{1-\beta}\log\max_{\sigma_C}\min_{\tau_B}\langle\psi|(I_A\otimes\sigma_C^{\frac{1}{\beta}-1})\otimes\tau_B^{1-\frac{1}{\beta}}|\psi\rangle\\ &=-\frac{\alpha}{1-\alpha}\log\min_{\sigma_C}\max_{\tau_B}\langle\psi|(I_A\otimes\tau_B^{\frac{1}{\alpha}-1})\otimes\sigma_C^{1-\frac{1}{\alpha}}|\psi\rangle.\end{aligned} \tag{5.144}$$

The combination of (5.143) and (5.144) yields (5.137). ■

Considering inequalities opposite to (5.121) and (5.122), we obtain the following corollary, in which, the second inequality in (5.145) and the first inequality in (5.146) can be regarded as generalizations of Lemma 2.4.

Corollary 5.5 *[45] Let $\rho_{AB}\in\mathcal{S}(AB)$. Then, the following inequalities hold for $\alpha\in\left[\frac{1}{2},\infty\right]$:*

$$\widetilde{H}^{\uparrow}_{\alpha|\rho}(A|B)\le H^{\uparrow}_{2-\frac{1}{\alpha}|\rho}(A|B)\,,\quad \widetilde{H}_{\alpha|\rho}(A|B)\le H_{2-\frac{1}{\alpha}|\rho}(A|B)\,, \tag{5.145}$$

$$H^{\uparrow}_{\alpha|\rho}(A|B)\le H_{2-\frac{1}{\alpha}|\rho}(A|B)\,,\quad \widetilde{H}^{\uparrow}_{\alpha|\rho}(A|B)\le \widetilde{H}_{2-\frac{1}{\alpha}|\rho}(A|B)\,. \tag{5.146}$$

The preceding inequality (5.134) can be regarded as a special case of (5.146) with $\alpha=\infty$.

Proof Consider a purification ρ of ρ_{AB} with the reference system $\mathcal{H}_C$. The relations (5.145) follow from the combination of the relations (5.121) with the system $\mathcal{H}_A$ and $\mathcal{H}_C$ and the duality relations (5.136) and (5.138). The relations (5.146) follow from the combination of the relations (5.122) with the system $\mathcal{H}_A$ and $\mathcal{H}_C$ and the duality relations (5.137) and (5.138). ■

Exercise

5.49 Show the relations

$$H_{\alpha|\rho}(A|B_1) \geq H_{\alpha|\rho}(A|B_1B_2), \quad H^{\uparrow}_{\alpha|\rho}(A|B_1) \geq H^{\uparrow}_{\alpha|\rho}(A|B_1B_2) \tag{5.147}$$

for $\alpha \in [0, 2]$, and

$$\widetilde{H}_{\alpha|\rho}(A|B_1) \geq \widetilde{H}_{\alpha|\rho}(A|B_1B_2), \quad \widetilde{H}^{\uparrow}_{\alpha|\rho}(A|B_1) \geq \widetilde{H}^{\uparrow}_{\alpha|\rho}(A|B_1B_2) \tag{5.148}$$

for $\alpha \in [\frac{1}{2}, \infty)$.

5.7 Proof and Construction of Stinespring and Choi–Kraus Representations

In this section, we will prove Theorem 5.1 and construct the Stinespring and Choi–Kraus representations. First, let us consider the following theorem for completely positive maps, without the trace-preserving condition.

Theorem 5.14 *Given a linear map κ from the set of Hermitian matrices on the d-dimensional system $\mathcal{H}_A$ to that on the d'-dimensional system $\mathcal{H}_B$, the following conditions are equivalent.*

① *κ is a completely positive map.*
② *κ^* is a completely positive map.*
③ *κ is a $\min\{d, d'\}$-positive map.*
④ *The matrix $K(\kappa)$ on $\mathcal{H}_A \otimes \mathcal{H}_B$ is positive semidefinite.*
⑤ *(**Stinespring representation**) There exist a Hilbert space $\mathcal{H}_C$ with the same dimension as $\mathcal{H}_B$, a pure state $\rho_0 \in \mathcal{S}(\mathcal{H}_B \otimes \mathcal{H}_C)$, and a matrix W in $\mathcal{H}_A \otimes \mathcal{H}_B \otimes \mathcal{H}_C$ such that $\kappa(X) = \mathrm{Tr}_{A,C} W(X \otimes \rho_0)W^*$.*
⑥ *(**Choi–Kraus representation**) There exist dd' linear maps $F_1, \ldots, F_{dd'}$ from $\mathcal{H}_A$ to $\mathcal{H}_B$ such that $\kappa(X) = \sum_i F_i X F_i^*$.*

We also have Conditions ⑤′ and ⑥′ by deforming Conditions ⑤ and ⑥ as follows. These conditions are also equivalent to the above conditions.

⑤′ *There exist Hilbert spaces $\mathcal{H}_C$ and $\mathcal{H}'_C$, a positive semidefinite state $\rho'_0 \in \mathcal{S}(\mathcal{H}_C)$, and a linear map W from $\mathcal{H}_A\mathcal{H}_C$ to $\mathcal{H}_A\mathcal{H}'_C$ such that $\kappa(X) = \mathrm{Tr}_{C'} W(X \otimes \rho'_0)W^*$.*
⑥′ *There exist linear maps $F_1, \ldots, F_k$ from $\mathcal{H}_A$ to $\mathcal{H}_B$ such that $\kappa(X) = \sum_i F_i X F_i^*$.*

Proof Since ②⇔ ① has been shown in Sect. 5.1, we now show that ①⇒③⇒④⇒⑤⇒⑥⇒⑥′ ⇒① and ⑤⇒⑤′ ⇒①. Since ①⇒③, ⑤⇒⑤’, and ⑥⇒⑥’ by inspection, it suffices to prove the remaining relations.

First, we will derive ③⇒④ as follows. In the following, we will show ④ only when $d' \le d$ due to the following reason. The equivalence between ② and ① shows that Condition ③ implies that κ^* is a d-positive map for $d \le d'$. This fact derives Condition ④ for κ^* when ③⇒④ holds for $d' \le d$. Hence, we have $K(\kappa^*) \ge 0$, which is equivalent to $K(\kappa) \ge 0$. Thus, we obtain ④ for κ for $d \le d'$. from ③⇒④ $d' \le d$.

Since κ is a d-positive map, $\kappa \otimes \iota_B$ is a positive map (ι_B is the identity map in $\mathcal{T}(\mathcal{H}_B)$). Let X be a positive semidefinite Hermitian matrix on $\mathcal{H}_A \otimes \mathcal{H}_B$. Assume that $X = \sum_{i,k,j,l,} x^{(i,k),(j,l)} |e_i^A \otimes e_k^B\rangle\langle e_j^A \otimes e_l^B|$. Since $(\kappa \otimes \iota_B)(X) \ge 0$, we have

$$\begin{aligned} 0 \le & \langle I_B|(\kappa \otimes \iota_B)(X)|I_B\rangle \\ = & \sum_{i,j,k,l} x^{(i,k),(j,l)} \langle I_B| \left(\kappa(|e_i^A\rangle\langle e_j^A|) \otimes |e_k^B\rangle\langle e_l^B|\right) |I_B\rangle \\ = & \sum_{i,j,k,l} x^{(i,k),(j,l)} \langle e_k^B|\kappa(|e_i^A\rangle\langle e_j^A|)|e_l^B\rangle \\ = & \sum_{i,j,k,l} x^{(i,k),(j,l)} K(\kappa)^{(j,l),(i,k)} = \operatorname{Tr} X K(\kappa). \end{aligned} \tag{5.149}$$

Therefore, $K(\kappa) \ge 0$, and we obtain ④. In the above, we denote the vector $\sum_{k=1}^{d'} e_k^B \otimes e_k^B$ in the space $\mathcal{H}_B \otimes \mathcal{H}_B$ by I_B. In the derivation of (5.149), we used the fact that

$$\langle I_B| \left(|e_k^B\rangle\langle e_l^B| \otimes |e_s^B\rangle\langle e_t^B|\right) |I_B\rangle = \langle I_B| \left(|e_k^B \otimes e_s^B\rangle\langle e_l^B \otimes e_t^B|\right) |I_B\rangle = \delta_{k,s}\delta_{l,t}.$$

We now derive ④⇒⑤. Since $K(\kappa) \ge 0$, $\sqrt{K(\kappa)}$ exists. In what follows, we consider a space $\mathcal{H}_C$ with a basis $e_1^C, \ldots, e_{d'}^C$. Note that the space $\mathcal{H}_C$ is isometric to the space $\mathcal{H}_B$. Defining $U_{C,B} \stackrel{\text{def}}{=} \sum_{k=1}^{d'} e_k^C \otimes e_k^B$, we have

$$\begin{aligned} & \operatorname{Tr} |e_{i'}^A\rangle\langle e_{j'}^A| \otimes \left(|U_{C,B}\rangle\langle U_{C,B}|\right) |e_j^A \otimes e^C{}_k \otimes e_s^B\rangle\langle e_i^A \otimes e^C{}_l \otimes e_t^B| \\ = & \delta_{j',j}\delta_{i',i}\delta_{l,t}\delta_{k,s}, \end{aligned} \tag{5.150}$$

where the order of the tensor product is $\mathcal{H}_A \otimes \mathcal{H}_C \otimes \mathcal{H}_B$. Although $K(\kappa)$ is originally a Hermitian matrix on $\mathcal{H}_A \otimes \mathcal{H}_B$, we often regard it as a Hermitian matrix on $\mathcal{H}_A \otimes \mathcal{H}_C$ because $\mathcal{H}_C$ is isometric to $\mathcal{H}_B$. Using (5.150), we have

$$\begin{aligned} & \operatorname{Tr} \kappa(X)Y = \operatorname{Tr}(X \otimes Y)K(\kappa) = \operatorname{Tr} \left(X \otimes |U_{C,B}\rangle\langle U_{C,B}|\right) K(\kappa) \otimes Y \\ = & \operatorname{Tr} \left(X \otimes |U_{C,B}\rangle\langle U_{C,B}|\right) \left(\sqrt{K(\kappa)} \otimes I_B\right) \left(I_{A,C} \otimes Y\right) \left(\sqrt{K(\kappa)} \otimes I_B\right) \\ = & \operatorname{Tr}_B \operatorname{Tr}_{A,C} \left(\left(\sqrt{K(\kappa)} \otimes I_B\right) \left(X \otimes |U_{C,B}\rangle\langle U_{C,B}|\right) \left(\sqrt{K(\kappa)} \otimes I_B\right)\right) Y \end{aligned}$$

for $\forall X \in \mathcal{T}(\mathcal{H}_A)$, $\forall Y \in \mathcal{T}(\mathcal{H}_B)$. Therefore, we can show that

$$\kappa(X)=\mathrm{Tr}_{A,C}\left[\left(\sqrt{d'K(\kappa)}\otimes I_B\right)\left(X\otimes\frac{|U_{C,B}\rangle\langle U_{C,B}|}{d'}\right)\left(\sqrt{d'K(\kappa)}\otimes I_B\right)\right]. \tag{5.151}$$

Letting $\rho_0 = \frac{|U_{C,B}\rangle\langle U_{C,B}|}{d'}$ and $W = \sqrt{d'K(\kappa)} \otimes I_B$, we obtain ⑤.

Next, we show that ⑤⇒⑥. Let ρ_0 be $|x\rangle\langle x|$, P be a projection from $\mathcal{H}_A \otimes \mathcal{H}_B \otimes \mathcal{H}_C$ to $\mathcal{H}_A \otimes |x\rangle$, and $P_{i,k}$ be a projection from $\mathcal{H}_A \otimes \mathcal{H}_B \otimes \mathcal{H}_C$ to $\mathcal{H}_B \otimes |e_i^A \otimes e^C{}_k\rangle$. Using formula (1.29) of the partial trace, we have

$$\kappa(X) = \mathrm{Tr}_{A,C}\, W(X \otimes \rho_0)W^* = \sum_{i=1}^{d}\sum_{k=1}^{d'} P_{i,k} W P X P W^* P_{i,k}$$
$$= \sum_{i=1}^{d}\sum_{k=1}^{d'} (P_{i,k} W P) X (P_{i,k} W P)^*.$$

We thus obtain ⑥.

Finally, we show ⑤′ ⇒①. From Condition ⑤′ any positive semidefinite Hermitian matrix X on $\mathcal{H}_A \otimes \mathbb{C}^n$ satisfies

$$\kappa \otimes \iota_n(X) = \mathrm{Tr}_{\mathcal{H}_A \otimes \mathcal{H}_C} (W \otimes I_n)(X \otimes \rho_0')(W^* \otimes I_n) \geq 0,$$

where I_n is an identity matrix in $\mathbb{C}^n$. Therefore, κ is an n-positive map for arbitrary n. It follows that κ is a completely positive map from which we obtain ①.

Concerning a proof of ⑥′ ⇒①, we have

$$\kappa \otimes \iota_n(X) = \sum_i (F_i \otimes I_n) X (F_i^* \otimes I_n) \geq 0,$$

for a semipositive definite Hermitian matrix X on $\mathcal{H}_A \otimes \mathbb{C}^n$. Thus, we obtain ①. ■

Next, we prove Theorem 5.1. Thanks to Theorem 5.14, it is sufficient to show the equivalence of Conditions ① to ⑥′ in Theorem 5.1 when κ is a completely positive map. Indeed, ①⇒③, ⑤⇒⑤′, and ⑥⇒⑥′ by inspection. Concerning ⑤′ ⇒① and ⑥′ ⇒①, it is sufficient to show the trace-preserving property because of Theorem 5.14. Therefore, we only show ③⇒④⇒⑤⇒⑥ as follows.

We first show ③⇒④. From definition (1.26) of the partial trace we obtain

$$\mathrm{Tr}_A\, \rho = \mathrm{Tr}_B\, \kappa(\rho) = \mathrm{Tr}_{A,B} (\rho \otimes I_B) K(\kappa) = \mathrm{Tr}_A\, \rho\, (\mathrm{Tr}_B\, K(\kappa))$$

for arbitrary $\rho \in \mathcal{S}(\mathcal{H}_A)$. Hence, $\mathrm{Tr}_B\, K(\kappa) = I_A$, and thus we obtain ④.

Next, we show ④⇒⑤. Employing the notation used in the proof of ④⇒⑤ in Theorem 5.14, we let P be the projection from $\mathcal{H}_A \otimes \mathcal{H}_B \otimes \mathcal{H}_C$ to $\mathcal{H}_A \otimes |U_{C,B}\rangle$. Since any $\rho \in \mathcal{S}(\mathcal{H}_A)$ satisfies

$$\begin{aligned}\operatorname{Tr} \rho &= \operatorname{Tr}_B \operatorname{Tr}_{A,C} \left(\left(\sqrt{d' K(\kappa)} \otimes I_B \right) P \rho P \left(\sqrt{d' K(\kappa)} \otimes I_B \right) \right) \\ &= \operatorname{Tr}_{A,C,B} \left(\left(\sqrt{d' K(\kappa)} \otimes I_B \right) P \rho P \left(\sqrt{d' K(\kappa)} \otimes I_B \right) \right),\end{aligned}$$

we obtain

$$\operatorname{Tr}_{A,C,B} \left(\left(\sqrt{d' K(\kappa)} \otimes I_B \right) P \right)^* \left(\sqrt{d' K(\kappa)} \otimes I_B \right) P = P.$$

Let $\mathcal{H}_R$ be the range of $\left(\sqrt{d'K(\kappa)} \otimes I_B\right) P$ for $\mathcal{H}_A \otimes |U_{C,B}\rangle$. Then, the dimension of $\mathcal{H}_R$ is equal to that of $\mathcal{H}_A$. The matrix $\left(\sqrt{d'K(\kappa)} \otimes I_B\right) P$ can be regarded as a map from $\mathcal{H}_A \otimes |U_{C,B}\rangle$ to $\mathcal{H}_R$.

Let $\mathcal{H}_R{}^{\perp}$ be the orthogonal complementary space of $\mathcal{H}_R$ in $\mathcal{H}_A \otimes \mathcal{H}_B \otimes \mathcal{H}_C$, and $\mathcal{H}_A{}^{\perp}$ be the orthogonal complementary space of $\mathcal{H}_A \otimes |U_{C,B}\rangle$. Since the dimension of $\mathcal{H}_R{}^{\perp}$ is equal to that of $\mathcal{H}_A{}^{\perp}$, there exists a unitary (i.e., metric-preserving) linear mapping U' from $\mathcal{H}_R{}^{\perp}$ to $\mathcal{H}_A{}^{\perp}$. Then, $U_\kappa \stackrel{\text{def}}{=} \left(\sqrt{d'K(\kappa)} \otimes I_B\right) P \oplus U'$ is a unitary linear map from $\mathcal{H}_A \otimes \mathcal{H}_B \otimes \mathcal{H}_C = (\mathcal{H}_A \otimes |U_{C,B}\rangle) \oplus \mathcal{H}_A{}^{\perp}$ to $\mathcal{H}_A \otimes \mathcal{H}_B \otimes \mathcal{H}_C = \mathcal{H}_R \oplus \mathcal{H}_R{}^{\perp}$. Therefore, from (5.151) we have $\kappa(\rho) = \operatorname{Tr}_{A,C} U_\kappa \rho \otimes \frac{|U_{C,B}\rangle\langle U_{C,B}|}{d'} U_\kappa$, which gives Condition ⑤.

Next, we show ⑤⇒⑥ by employing the notation used in the proof of ⑤⇒⑥ in Theorem 5.14. Since

$$\begin{aligned}\operatorname{Tr} \rho &= \operatorname{Tr} \kappa(\rho) = \operatorname{Tr}_B \operatorname{Tr}_{A,C} U_\kappa (\rho \otimes \rho_0) U_\kappa^* = \sum_{i=1}^{d} \sum_{k=1}^{d'} \operatorname{Tr}_B P_{i,k} W P \rho P W^* P_{i,k} \\ &= \sum_{i=1}^{d} \sum_{k=1}^{d'} \operatorname{Tr}_B (P_{i,k} W P) \rho (P_{i,k} W P)^* = \operatorname{Tr}_A \sum_{i=1}^{d} \sum_{k=1}^{d'} (P_{i,k} W P)^* (P_{i,k} W P) \rho,\end{aligned}$$

we obtain $\sum_{i=1}^{d} \sum_{k=1}^{d'} (P_{i,k} W P)^* (P_{i,k} W P) = I_A$. Therefore, we obtain ⑥. Further, from the proof ⑤⇒⑥, we obtain Lemma 5.1.

Finally, we directly construct Stinespring representation ⑤′ from Choi–Kraus representation ⑥′. Define the map W from $\mathcal{H}_A$ to $\mathcal{H}_B \otimes \mathbb{C}^k$ as

$$W(x) \stackrel{\text{def}}{=} \sum_{i=1}^{k} F_i(x) \otimes e_i.$$

Then, W satisfies

$$\operatorname{Tr}_{\mathbb{C}^k} W \rho W^* = \sum_{i=1}^{k} F_i \rho F_i^*.$$

We obtain Condition ⑤′ from ⑥′ in Theorem 5.14. In Theorem 5.1, we have to check the unitarity. From the condition $\sum_{i=1}^{k} F_i^* F_i = I$, we obtain $W^* W = I$, i.e.,

W is an isometry map. Hence, it is possible to deform map W to a unitary map by extending the input space. In this case, the state in the environment $\kappa^E(\rho)$ equals $\mathrm{Tr}_B W\rho W^* = (\mathrm{Tr}\, F_j^* F_i \rho)_{i,j}$. Thus, we obtain Lemma 5.1.

5.8 Historical Note

5.8.1 Completely Positive Map and Quantum Relative Entropy

A completely positive map was initially introduced in the mathematical context; Stinespring [1] gave its representation theorem in the observable form, i.e., $\kappa^*(A) = PU_\kappa^*(A \otimes I)U_\kappa P$, where P is the projection from the extended space to the original space. Holevo [46] proposed that any state evolution in the quantum system could be described by a completely positive map based on the same reason considered in this text. After this, Lindblad [18] translated Stinespring's representation theorem to the state form. Then, he clarified that any state evolution by a completely positive map could be regarded as the interaction between the target system and the environment system.

Concerning other parts of Theorem 5.1, Jamiołkowski [5] showed the one-to-one correspondence between a CP map κ and a positive matrix $K(\kappa)$, firstly. After this study, Choi [3] obtained this correspondence. He also obtained the characterization ⑥ concerning CP maps. Kraus [2] also obtained this characterization. Choi [3] also characterize the extremal points as Lemma 5.2.

In this book, we proved the monotonicity of the quantum relative entropy based on the quantum Stein's Lemma. Using this property, we derived many inequalities in Sects. 5.4 and 5.5. However, historically, these were proved by completely different approaches. First, Lieb and Ruskai [31, 32] proved the strong subadditivity of the von Neumann entropy (5.83) based on Lieb's convex trace functions [47]. Using these functions, they derived the monotonicity of the quantum relative entropy only concerning the partial trace. During that period, Lindblad [20] proved the joint convexity of the quantum relative entropy (5.38). After this result, using the Stinespring's representation theorem in the state form, Lindblad [18] proved the monotonicity of the quantum relative entropy (5.36) from that concerning the partial trace. Later, Uhlmann [19] invented the interpolation theory, and proved the monotonicity of the quantum relative entropy based on this approach. As an extension of the quantum relative entropy, Petz [48] generalized this kind of monotonicity to quantum f-relative entropy for a matrix convex function f, as explained in Sect. 6.7.1. A more detailed history will be discussed in the end of the next chapter.

5.8.2 Quantum Relative Rényi Entropy

Now, as variants of quantum relative entropy, we discuss quantum relative Rényi entropy. The above mentioned Petz's [48] approach contains the case with $f(x) = x^{1+s}$. Hence, applying his method to the function x^{1+s}, we can derive the monotonicity (5.56) of the relative Rényi entropy $D_{1+s}(\rho\|\sigma)$. In this book, we prove the monotonicities of the relative Rényi entropy regarding measurements (3.19) and (3.20) using only elementary knowledge. Moreover, the monotonicity (3.20) holds with a larger parameter $s \le 0$ as compare with the monotonicity (5.53).

Recently, another kind of relative Rényi entropy $\underline{D}_{1+s}(\rho\|\sigma)$ was proposed by Wilde et al. [49] and Müller-Lennert et al. [44], independently. They showed the monotonicity (5.57) of $\underline{D}_{1+s}(\rho\|\sigma)$ for $2 \ge 1+s > 1$ by using Lieb's concavity theorem [47]. Then, the monotonicity (5.57) of $\underline{D}_{1+s}(\rho\|\sigma)$ was shown by Frank et al. [29] for $1+s \ge \frac{1}{2}$ and by Beigi [43] for $1+s \ge 1$, independently. Frank et al. [29] showed the case with $1 > 1+s \ge \frac{1}{2}$ by using Ando's convexity theorem [50] and the case with $1+s > 1$ by using Lieb's concavity theorem [47]. Beigi [43] showed the case with $1+s \ge 1$ by using Hölder inequalities and Riesz-Thorin theorem [51]. This book shows the case with $1+s \ge 1$ by using the equation $\tilde{\phi}(-s|\rho\|\sigma) = \lim_{n\to\infty} \frac{1}{n} \max_{\boldsymbol{M}} \phi(-s|\mathrm{P}^{\boldsymbol{M}}_{\rho^{\otimes n}}\|\mathrm{P}^{\boldsymbol{M}}_{\sigma^{\otimes n}})$, which is a simpler proof.

Indeed, the relative Rényi entropy $\underline{D}_{1+s}(\rho\|\sigma)$ produces the conditional Rényi entropies that is different from the conditional Rényi entropies by the relative Rényi entropy $D_{1+s}(\rho\|\sigma)$. Since each relative Rényi entropy produces two kinds of conditional Rényi entropies, we have four kinds of conditional Rényi entropies. These conditional Rényi entropies are linked to each other via the duality relation (Theorem 5.13). Firstly, Tomamichel et al. [42] showed Inequality (5.136). Then, Müller-Lennert [44] and Beigi [43] independently showed Inequality (5.137). Finally, Tomamichel et al. [45] linked the remaining two kinds of conditional Rényi entropies as (5.138). Indeed, as shown by König et al. [41, Theorem 1], the limit of one of conditional Rényi entropies $\widetilde{H}^{\uparrow}_{\min|\rho}(A|B)$ has an interesting operation meaning as Lemma 5.8.

5.9 Solutions of Exercises

Exercise 5.1 When κ is trace-preserving, (5.2) with $Y = I$ implies that $\mathrm{Tr}\, XI = \mathrm{Tr}\, \kappa(X)I = \mathrm{Tr}\, X\kappa^*(I)$ for $X \in \mathcal{T}(\mathcal{H}_A)$, which implies $\kappa^*(I) = I$. Conversely, when κ^* is identity-preserving, $\mathrm{Tr}\, \kappa(X)I = \mathrm{Tr}\, X\kappa^*(I) = \mathrm{Tr}\, XI$ for $X \in \mathcal{T}(\mathcal{H}_A)$.

Exercise 5.2 Considering the vector $\begin{pmatrix} x|a\rangle \\ \kappa^*(X)^*|a\rangle \end{pmatrix}$, we have

$$x^2\langle a|\kappa^*(X^*X)|a\rangle + 2x\langle a|\kappa^*(X)\kappa^*(X)^*|a\rangle + \langle a|\kappa^*(X)\kappa^*(X)^*|a\rangle \ge 0.$$

Since the discriminant is non-positive, we have $\langle a|\kappa^*(X^*X)|a\rangle \geq \langle a|\kappa^*(X)\kappa^*(X)^*|a\rangle$, which yields (5.3).

Exercise 5.3 Let $\mathcal{H}_D$ be $\mathcal{H}_C \otimes \mathcal{H}_B$, and consider the unitary matrix corresponding to the replacement $W : u \otimes v \mapsto v \otimes u$ in $\mathcal{H}_A \otimes \mathcal{H}_B$, and define $V \stackrel{\text{def}}{=} (W \otimes I_C)U$.

Exercise 5.4 Since $\sum_i u_{i,j} u^*_{j',i} = \delta_{j,j'}$, we have

$$\sum_j F_j \rho F_j^* = \sum_{j,j'} \delta_{j,j'} F_j \rho F_{j'}^* = \sum_{j,j'} \sum_i u_{i,j} u^*_{j',i} F_j \rho F_{j'}^*$$
$$= \sum_i \sum_j u_{i,j} F_j \rho \sum_{j'} u^*_{j',i} F_{j'}^* = \sum_i F'_i \rho F'^*_i.$$

Exercise 5.5

(**a**) Given a Choi–Kraus representation $\{F_i\}$ of the TP-CP map κ, there exists a unitary matrix $u_{j,i}$ such that all of non-zero matrices among $F'_j := \sum_i u_{j,i} F_i$ are linearly independent.
(**b**) Due to the condition for Choi–Kraus representation, the map $V : |x\rangle \mapsto (F_i|x\rangle)$ from $\mathcal{H}_A$ to $\mathcal{H}_B \otimes \mathbb{C}^d$ is an isometry. Similarly, we define the isometry V'. Similar to the proof of Theorem 5.2, we find that the states $V|\Phi\rangle\langle\Phi|V^*$ and $V'|\Phi\rangle\langle\Phi|V'^*$ are purifications of $\kappa(|\Phi\rangle\langle\Phi|)$. Hence, due to Lemma 8.1, there exists a partial isometry $\tilde{V} = (\tilde{v}_{j,i})$ from $\mathbb{C}^d$ to $\mathbb{C}^{d'}$ such that $\tilde{V} V|\Phi\rangle\langle\Phi|V^*\tilde{V}^* = V'|\Phi\rangle\langle\Phi|V'^*$. This relation shows that $F'_j = \sum_i \tilde{v}_{j,i} F_i$.

Exercise 5.6 Since $S_i S_j S_i^* = -S_j$ for $i \neq j$ and $S_i S_i S_i^* = S_i$, we have $\sum_{i=1}^3 S_i S_j S_i^* = -S_j$. When $\rho = \frac{1}{2} I + \frac{1}{2}\sum_{j=1}^3 x^j S_j$, we have

$$\frac{3\lambda+1}{4}\rho + \frac{1-\lambda}{4}\sum_{i=1}^3 S_i \rho S_i^*$$
$$= \frac{3\lambda+1}{4}\rho + \frac{1-\lambda}{4}\sum_{i=1}^3 S_i \left(\frac{1}{2}I + \frac{1}{2}\sum_{j=1}^3 x^j S_j\right) S_i^*$$
$$= \frac{3\lambda+1}{4}\rho + \frac{1-\lambda}{4}\left(\sum_{i=1}^3 S_i \frac{1}{2} I S_i^* - \frac{1}{2}\sum_{j=1}^3 x^j S_j\right)$$
$$= \frac{3\lambda+1}{4}\rho + \frac{1-\lambda}{4}\left(\frac{3}{2}I - \frac{1}{2}\sum_{j=1}^3 x^j S_j\right)$$
$$= \frac{3\lambda+1}{4}\rho + \frac{1-\lambda}{4}(2I - \rho) = \lambda\rho + (1-\lambda)(\mathrm{Tr}\,\rho)\rho_{\mathrm{mix}}$$

Exercise 5.7 Consider the Hilbert space $\mathcal{H}_B$ produced by $|\omega\rangle$. Apply Condition ⑤ of Theorem 5.1 to the entanglement breaking channel $\kappa_{M,W}$ given in Theorem 5.4 with $W_\omega = |\omega\rangle\langle\omega|$. Finally, consider the measurement $\{|\omega\rangle\langle\omega| \otimes I_C\}$.

Exercise 5.8 This relation follows from $\mathbf{X}_d\mathbf{Z}_d = \omega\mathbf{Z}_d\mathbf{X}_d$.

Exercise 5.9

(**a**) Since $(\mathbf{X}_d^j\mathbf{Z}_d^k)(\mathbf{X}_d^{j'}\mathbf{Z}_d^{k'}) = \omega^{-kj'}(\mathbf{X}_d^{j+j'}\mathbf{Z}_d^{k+k'})$, we have

$$\omega^{kj'}\left(\frac{1}{d^2}\sum_{j=0}^{d-1}\sum_{k=0}^{d-1}(\mathbf{X}_d^{j+j'}\mathbf{Z}_d^{k+k'})^* X\omega^{-kj'}(\mathbf{X}_d^{j+j'}\mathbf{Z}_d^{k+k'})\right)$$
$$=\left(\frac{1}{d^2}\sum_{j''=0}^{d-1}\sum_{k''=0}^{d-1}(\mathbf{X}_d^{j''}\mathbf{Z}_d^{k''})^* X(\mathbf{X}_d^{j''}\mathbf{Z}_d^{k''})\right),$$

where $j'' = j + j'$ and $k'' = k + k'$.
(**b**) $\mathbf{Z}_d A\mathbf{Z}_d^{-1} = \sum_{j,k} a^{j,k}\omega^{j-k}|u_j\rangle\langle u_k|$. The relation $\mathbf{Z}_d A = A\mathbf{Z}_d$ implies that $\sum_{j,k} a^{j,k}\omega^{j-k}|u_j\rangle\langle u_k| = \mathbf{Z}_d A\mathbf{Z}_d^{-1} = A = \sum_{j,k} a^{j,k}|u_j\rangle\langle u_k|$. Thus, $a^{j,k} = 0$ for $j \neq k$.
(**c**) The $j-1$-th diagonal element of $\mathbf{X}_d A\mathbf{X}_d^{-1}$ is the j-th diagonal element of A. The relation $\mathbf{X}_d A = A\mathbf{X}_d$ implies that $\mathbf{X}_d A\mathbf{X}_d^{-1} = A$. Thus, all of the diagonal elements of A are the same.
(**d**) Due to (**b**) and (**c**), (**a**) implies that $\left(\frac{1}{d^2}\sum_{j''=0}^{d-1}\sum_{k''=0}^{d-1}(\mathbf{X}_d^{j''}\mathbf{Z}_d^{k''})^* X(\mathbf{X}_d^{j''}\mathbf{Z}_d^{k''})\right)$ is a constant times of I. Comparing the traces of both sides of (5.23), we obtain (5.23).

Exercise 5.10 For any X and Y, we have

$$\mathrm{Tr}\,\frac{1}{d^2}\sum_{j=0}^{d-1}\sum_{k=0}^{d-1}(\mathbf{X}_A^j\mathbf{Z}_A^k \otimes I_B)^*\rho(\mathbf{X}_A^j\mathbf{Z}_A^k \otimes I_B)(X\otimes Y)$$
$$=\mathrm{Tr}\,\rho\frac{1}{d^2}\sum_{j=0}^{d-1}\sum_{k=0}^{d-1}(\mathbf{X}_A^j\mathbf{Z}_A^k \otimes I_B)(X\otimes Y)(\mathbf{X}_A^j\mathbf{Z}_A^k \otimes I_B)^*$$
$$=\mathrm{Tr}\,\rho\frac{1}{d^2}\sum_{j=0}^{d-1}\sum_{k=0}^{d-1}(\mathbf{X}_A^{-j}\mathbf{Z}_A^{-k} \otimes I_B)^*(X\otimes Y)(\mathbf{X}_A^{-j}\mathbf{Z}_A^{-k} \otimes I_B)$$
$$=\mathrm{Tr}\,\rho\left(\frac{1}{d^2}\sum_{j=0}^{d-1}\sum_{k=0}^{d-1}(\mathbf{X}_A^{-j}\mathbf{Z}_A^{-k})^* X(\mathbf{X}_A^{-j}\mathbf{Z}_A^{-k})\right)\otimes Y = \mathrm{Tr}\,\rho(\mathrm{Tr}\,X)\rho_{\mathrm{mix}}\otimes Y$$
$$=(\mathrm{Tr}\,X)\frac{1}{d}\mathrm{Tr}_B(\mathrm{Tr}_A\,\rho)Y = \mathrm{Tr}(\rho_{\mathrm{mix}}^A \otimes \mathrm{Tr}_A\,\rho)(X\otimes Y).$$

Exercise 5.11 Exercise 5.8 yields that

$$\langle u^{A,B}_{i,j}, u^{A,B}_{i',j'} \rangle = \frac{1}{d}\operatorname{Tr}(\mathbf{X}^i_A\mathbf{Z}^j_A)^\dagger \mathbf{X}^{i'}_A\mathbf{Z}^{j'}_A$$

$$= \frac{1}{d}\operatorname{Tr}\mathbf{Z}^{-j}_A\mathbf{X}^{-i}_A\mathbf{X}^{i'}_A\mathbf{Z}^{j'}_A = \frac{1}{d}\operatorname{Tr}\omega^{-ij'+i'j}\mathbf{X}^{i'-i}_A\mathbf{Z}^{j'-j}_A = \delta_{j,j'}\delta_{i,i'}.$$

Exercise 5.12 Since the largest eigenvalue of W_x is $\lambda + \frac{1-\lambda}{d} - \frac{1+(d-1)\lambda}{d}$ and remaining $d-1$ eigenvalues are $\frac{1-\lambda}{d}$, we have

$$\min_x H(W_x) = \frac{1+(d-1)\lambda}{d}\log\frac{d}{1+(d-1)\lambda)} + \frac{(d-1)(1-\lambda)}{d}\log\frac{d}{(1-\lambda)}.$$

This property holds for any input state x.

When we choose the input distribution p as the uniform distribution on a basis $\{|u_i\rangle\}_{i=1}^d$, the input mixture state $\sum_i p(i)|u_i\rangle\langle u_i| = \rho_{\mathrm{mix}}$. Thus, $H(\rho_{\mathrm{mix}}) = H(\rho_{\mathrm{mix}}) = \log d$, which attains $\max_p H(W_p)$. Therefore, the capacity is $\max_p H(W_p) - \sum_x p(x)H(W_x) = \log -(\min_x H(W_x)) = \frac{1+(d-1)\lambda}{d}\log(1+(d-1)\lambda) + \frac{(d-1)(1-\lambda)}{d}\log(1-\lambda)$.

Since all of output states $W_{|u_i\rangle\langle u_i|}$ commutative with each other, the depolarizing channel $\kappa_{d,\lambda}$ is pseudoclassical.

Exercise 5.13 The map $\tau^{\otimes n}$ is the transpose on the whole space $\mathcal{H}^{\otimes n}$. Thus, it keeps the positivity. Therefore, the transpose τ is tensor product positive.

Exercise 5.14

(**a**) Choosing $d_{k,l} := \sum_{j=0}^{d-1} p(0,j)\omega^{j(l-k)}$, we have

$$\kappa^{\mathrm{GP}}_p(\rho) = \sum_{j=0}^{d-1} p(0,j)(\mathbf{Z}^j_d)^*\rho\mathbf{Z}^j_d$$

$$= \sum_{j=0}^{d-1} p(0,j)\sum_{k,l}\omega^{j(l-k)}\rho_{k,l}|u_k\rangle\langle u_l| = \sum_{k,l} d_{k,l}\rho_{k,l}|u_k\rangle\langle u_l|.$$

(**b**) Assume that D satisfies (5.20). Then, $p(0,m) = \frac{1}{d}\operatorname{Tr} D\mathbf{X}^m_d = \frac{1}{d}\sum_{j=0}^{d-1} d_{j,0}\,\omega^{-jm}$. Since

$$\sum_{m=0}^{d-1}\frac{1}{d}\omega^{-jm}\omega^{m(l-k)} = \begin{cases} 1 & \text{if } l-k = j \\ 0 & \text{if } l-k \neq j \end{cases}$$

we have

$$\sum_{m=0}^{d-1} p(0,m)\omega^{m(l-k)} = \sum_{m=0}^{d-1}\frac{1}{d}\sum_{j=0}^{d-1} d_{j,0}\omega^{-jm}\omega^{m(l-k)} = d_{l-k,0},$$

which implies that $\kappa^{\mathrm{GP}}_p = \kappa^{\mathrm{PD}}_D$. This fact implies that κ^{PD}_D is a generalized Pauli channel.

(**c**) It is enough to show that the channel κ_D^{PD} is not a generalized Pauli channel if (5.20) does not hold. To show this fact, it is sufficient to show that the channel κ_p^{GP} is not a phase-damping channel when the condition given in (**a**) because there is one-to-one correspondence between generalized Pauli channels and phase-damping channels under the condition given in (**a**). When the condition given in (**a**) does not hold, the diagonal elements of $\kappa_p^{\mathrm{GP}}(\rho)$ are different from those of the state ρ. Hence, we obtain the desired argument.

Exercise 5.15 We focus on the input system $\mathcal{H}_A$ spanned by $\{|u_j\rangle\}_{j=0}^{d-1}$, the output system $\mathcal{H}_B$ spanned by $\{|u_j\rangle\}_{j=0}^{d}$, and the environment system $\mathcal{H}_E$ spanned by $\{|u_j^E\rangle\}_{j=0}^{d}$. Define the isometry U from $\mathcal{H}_A$ to $\mathcal{H}_B \otimes \mathcal{H}_E$ as $U(\sum_{j=0}^{d-1} v_j|u_j\rangle) := \sqrt{1-p}\sum_{j=0}^{d-1} v_j|u_j\rangle \otimes |u_d^E\rangle + \sqrt{p}\sum_{j=0}^{d-1} v_j|u_d\rangle \otimes |u_j^E\rangle$. Then, we have $\kappa_{d,p}^{\mathrm{era}}(\rho) = \mathrm{Tr}_E\, U\rho U^*$. Thus, $(\kappa_{d,p}^{\mathrm{era}})^E(\rho) = \mathrm{Tr}_B\, U\rho U^* = \kappa_{d,1-p}^{\mathrm{era}}$.

Exercise 5.16 Under the channel $\kappa_{d,n\to m}^{\mathrm{pns}}$, the environment system has the $n-m$-particle system. Hence, the channel to the environment system is $\kappa_{d,n\to n-m}^{\mathrm{pns}}$.

Exercise 5.17 Due to Condition (5.30), $1+\lambda_3$ and $1-\lambda_3$ are non-negative, (5.31) implies that $1+\lambda_3 \ge \pm(\lambda_1+\lambda_2)$ and $1-\lambda_3 \ge \pm(\lambda_1-\lambda_2)$, which implies (5.32).

Exercise 5.18 It is enough to consider the special case $p_0 \ge p_1 \ge p_2 \ge p_3$. We check whether the eigenvalues of (5.34) satisfy (5.33). Since the condition $p_0 \ge p_1 \ge p_2 \ge p_3$ implies that $p_0+p_1-p_2-p_3 \ge 0$ and $p_0-p_1+p_2-p_3 \ge 0$, the condition (5.33) is equivalent with $(p_0+p_1-p_2-p_3)+(p_0-p_1+p_2-p_3)+(p_0-p_1-p_2+p_3) \le 1$ and $(p_0+p_1-p_2-p_3)+(p_0-p_1+p_2-p_3)-(p_0-p_1-p_2+p_3) \le 1$. These two inequalities are equivalent with $p_0 \le \frac{1}{2}$ and $p_3 \ge 0$. Therefore, we obtain the desired argument.

Exercise 5.19 When we choose the coordinate $u_1 = u_1^A \otimes u_1^B$, $u_2 = u_1^A \otimes u_2^B$, $u_3 = u_2^A \otimes u_1^B$, $u_4 = u_2^A \otimes u_2^B$, we have
$$\mathrm{Inv}_\lambda \otimes \iota_{\mathbb{C}^2}(|\Phi_2\rangle\langle\Phi_2|) = \begin{pmatrix} 1-\lambda & 0 & 0 & 1-2\lambda \\ 0 & \lambda & 0 & 0 \\ 0 & 0 & \lambda & 0 \\ 1-2\lambda & 0 & 0 & 1-\lambda \end{pmatrix}.$$
This matrix is positive if and only if $(1-\lambda)^2-(1-2\lambda)^2 \ge 0$, i.e., $\frac{2}{3} \ge \lambda \ge 0$.

Exercise 5.20 It is enough to consider the case when $\boldsymbol{x} = (x,0,0)$ and $\boldsymbol{y} = (y,z,0)$.

$$\begin{aligned}
F(\rho_{\boldsymbol{x}},\rho_{\boldsymbol{y}}) &= \mathrm{Tr}\sqrt{\sqrt{\rho_{\boldsymbol{x}}}\rho_{\boldsymbol{y}}\sqrt{\rho_{\boldsymbol{x}}}} \\
&= \mathrm{Tr}\sqrt{\begin{pmatrix} \sqrt{\frac{1+x}{2}} & 0 \\ 0 & \sqrt{\frac{1-x}{2}} \end{pmatrix}\begin{pmatrix} \frac{1+y}{2} & \frac{z}{2} \\ \frac{z}{2} & \frac{1-y}{2} \end{pmatrix}\begin{pmatrix} \sqrt{\frac{1+x}{2}} & 0 \\ 0 & \sqrt{\frac{1-x}{2}} \end{pmatrix}} \\
&= \mathrm{Tr}\sqrt{\begin{pmatrix} \frac{(1+y)(1+x)}{4} & \frac{z\sqrt{1-x^2}}{4} \\ \frac{z\sqrt{1-x^2}}{4} & \frac{(1-y)(1-x)}{4} \end{pmatrix}}.
\end{aligned}$$

The eigenvalues of $\begin{pmatrix} \frac{(1+y)(1+x)}{4} & \frac{z\sqrt{1-x^2}}{4} \\ \frac{z\sqrt{1-x^2}}{4} & \frac{(1-y)(1-x)}{4} \end{pmatrix}$ are the solutions of $(\frac{(1+y)(1+x)}{4} - a)(\frac{(1-y)(1-x)}{4} - a) = (\frac{z\sqrt{1-x^2}}{4})^2$ for a. The equation is simplified to $a^2 - \frac{1+xy}{2}a + \frac{(1-y^2-z^2)(1-x^2)}{16} = 0$. The solutions $a_\pm$ satisfy $a_+ a_- = \frac{(1-y^2-z^2)(1-x^2)}{16}$ and $a_+ + a_- = \frac{1+xy}{2}$. Thus,

$$\begin{aligned} F(\rho_x, \rho_y)^2 &= (\sqrt{a_+} + \sqrt{a_-})^2 = a_+ + a_- + 2\sqrt{a_+ a_-} \\ &= \frac{1+xy}{2} + 2\frac{\sqrt{(1-y^2-z^2)(1-x^2)}}{4}, \end{aligned}$$

which equals the RHS of (5.35).

Exercise 5.21 For a POVM $\boldsymbol{M}) := \{M_i\}_i$, we define the POVM $\kappa^*(\boldsymbol{M}) := \{\kappa^*(M_i)\}_i$. Then, Exercise 3.58 implies that

$$\begin{aligned} \tilde{\phi}(s|\rho\|\sigma) &= \lim_{n\to\infty} \frac{1}{n} \max_{\boldsymbol{M}} \phi(s|\mathrm{P}^{\boldsymbol{M}}_{\rho^{\otimes n}} \| \mathrm{P}^{\boldsymbol{M}}_{\sigma^{\otimes n}}) \\ &\geq \lim_{n\to\infty} \frac{1}{n} \max_{\boldsymbol{M}} \phi(s|\mathrm{P}^{\kappa^*(\boldsymbol{M})}_{\rho^{\otimes n}} \| \mathrm{P}^{\kappa^*(\boldsymbol{M})}_{\sigma^{\otimes n}}) = \tilde{\phi}(s|\kappa(\rho)\|\kappa(\sigma)) \end{aligned}$$

for $s < 0$.

Exercise 5.22

(**a**) It follows from (5.55).
(**b**) Exercise A.16 guarantees that the inequality (5.53) does not necessarily holds for $s \leq -1$ in general. However, it contradicts the conclusion of (**a**). Thus, by contradiction, we can conclude that the equation $\phi(s|\rho\|\sigma) = \tilde{\phi}(s|\rho\|\sigma)$ does not hold for $s \leq -1$ in general.

Exercise 5.23

$$\begin{aligned} I(p, W) &= D\left(\sum_x p(x)|x\rangle\langle x| \otimes W_x \middle\| \left(\sum_x p(x)|x\rangle\langle x|\right) \otimes W_p\right) \\ &\geq D\left((\iota \otimes \kappa)\left(\sum_x p(x)|x\rangle\langle x| \otimes W_x\right) \middle\| (\iota \otimes \kappa)\left(\left(\sum_x p(x)|x\rangle\langle x|\right) \otimes W_p\right)\right) \\ &= D\left(\sum_x p(x)|x\rangle\langle x| \otimes \kappa(W_x) \middle\| \sum_x p(x)|x\rangle\langle x|) \otimes \kappa(W_p)\right) = I(p, \kappa(W)). \end{aligned}$$

Exercise 5.24 Equations (5.60) and (5.61) follow from (5.36) and (5.56), respectively. By substituting W_p into σ, (5.60) and (5.61) imply (5.62) and (5.63), respectively. By taking the infimum for σ, (5.61) implies (5.64). Finally, by taking the infimum for p, (5.62) and (5.64) imply (5.65) and (5.66), respectively.

Exercise 5.25

(**a**) These inequalities can be shown by replacing σ by $\frac{\sigma}{\operatorname{Tr}\sigma}$.

(**b**) Since the function $x \mapsto x^{-s}$ is matrix convex for $s \in [1, 0]$, we have $P\sigma^{-s}P \geq (P\sigma P)^{-s}$. Hence, we obtain (5.69) for $s \in [1, 0]$. Similarly, for $s \in [-1, 0]$, we have $P\sigma^{-s}P \leq (P\sigma P)^{-s}$, which implies (5.69).

Since $P\sigma^{-\frac{s}{1+s}}P \geq (P\sigma P)^{-\frac{s}{1+s}}$ for $s \in (0, \infty)$, we have $\rho^{\frac{1}{2}}\sigma^{-\frac{s}{1+s}}\rho^{\frac{1}{2}} \geq \rho^{\frac{1}{2}}(P\sigma P)^{-\frac{s}{1+s}}\rho^{\frac{1}{2}}$. Hence, using (3.9) and Lemma A.13, we obtain (5.70) for $s \in (0, \infty)$. Similarly, since $P\sigma^{-\frac{s}{1+s}}P \geq (P\sigma P)^{-\frac{s}{1+s}}$ for $s \in [-\frac{1}{2}, 0)$, using 3.39 and Lemma A.13, we obtain (5.70) for $s \in [-\frac{1}{2}, 0)$.

(**e**) Since the function $x \mapsto x^{-s}$ is matrix monotone for $s \in [1, 0]$, we have $\sigma^{-s} \geq (\sigma')^{-s}$. Hence, we obtain (5.75) for $s \in [1, 0]$. Similarly, $\sigma^{-s} \leq (\sigma')^{-s}$. for $s \in [-1, 0]$, we have $P\sigma^{-s}P \leq (P\sigma P)^{-s}$, which implies (5.75).

Since $\sigma^{-\frac{s}{1+s}} \geq (\sigma')^{-\frac{s}{1+s}}$ for $s \in (0, \infty)$, we have $\rho^{\frac{1}{2}}\sigma^{-\frac{s}{1+s}}\rho^{\frac{1}{2}} \geq \rho^{\frac{1}{2}}(\sigma')^{-\frac{s}{1+s}}\rho^{\frac{1}{2}}$. Hence, using (3.39) and Lemma A.13, we obtain (5.76) for $s \in (0, \infty)$. Similarly, since $\sigma^{-\frac{s}{1+s}} \geq (\sigma')^{-\frac{s}{1+s}}$ for $s \in [-\frac{1}{2}, 0)$, using (3.39) and Lemma A.13, we obtain (5.76) for $s \in [-\frac{1}{2}, 0)$.

Exercise 5.26 Let ρ_{mix} be a completely mixed state in $\mathcal{H}_B$. Consider the relative entropy $D(\rho_{A,B,C}\|\rho_{\mathrm{mix}} \otimes \rho_{A,C})$ and the partial trace of $\mathcal{H}_C$.

Exercise 5.27 The concavity of von Neumann entropy implies that

$$
\begin{aligned}
&I(\lambda p^1 + (1-\lambda)p^2, W) - (\lambda I(p^1, W) + (1-\lambda)I(p^2, W)) \\
=&H(W_{\lambda p^1+(1-\lambda)p^2} - \sum_x (\lambda p^1(x) + (1-\lambda)p^2(x))H(W_x), \\
&-\lambda(H(W_{p^1}) - \sum_x p^1(x)H(W_x)) - (1-\lambda)(H(W_{p^2}) - \sum_x p^2(x)H(W_x)) \\
=&H(W_{\lambda p^1+(1-\lambda)p^2} - \lambda H(W_{p^1}) - (1-\lambda)H(W_{p^2}) \geq 0.
\end{aligned}
$$

Exercise 5.28

(**a**) Use (5.80). Hence, $D(\rho\|\kappa_{\boldsymbol{M}}(\rho)) = H(\kappa_{\boldsymbol{M}}(\rho)) \leq \log|\boldsymbol{M}|$.

(**b**) Assume that $\rho = \sum_x p(x)|x\rangle\langle x|$. The joint convexity implies that $D(\rho\|\kappa_{\boldsymbol{M}}(\rho)) \leq \sum_x p(x)D(|x\rangle\langle x|\|\kappa_{\boldsymbol{M}}(|x\rangle\langle x|)) \leq \sum_x p(x)\log|\boldsymbol{M}| = \log|\boldsymbol{M}|$.

Exercise 5.29 Consider the state $\begin{pmatrix} p_1\rho_1 & & 0 \\ & \ddots & \\ 0 & & p_k\rho_k \end{pmatrix}$ in $\mathcal{H}_A \otimes \mathcal{H}_B \otimes \mathcal{H}_C$.

Exercise 5.30 Let the purification of $\rho_{A,B}$ be $\rho_{A,B,C}$ for a reference state $\mathcal{H}_R$. The subadditivity (5.86) implies that

$$
H(\rho_{A,B}) - H(\rho_A) + H(\rho_B) = H(\rho_C) - H(\rho_{B,C}) + H(\rho_B) \geq 0.
$$

Exercise 5.31

$$
\begin{aligned}
& H_\rho(AB|C) - H_\rho(A|C) - H_\rho(B|C) \\
= & H_\rho(ABC) - H_\rho(C) - H_\rho(AC) + H_\rho(C) - H_\rho(BC) + H_\rho(C) \\
= & H_\rho(ABC) + H_\rho(C) - H_\rho(AC) - H_\rho(BC).
\end{aligned}
$$

Exercise 5.32

$$
\begin{aligned}
& H_{|u\rangle\langle u|}(A, C) + H_{|u\rangle\langle u|}(A, D) - H_{|u\rangle\langle u|}(A) - H_{|u\rangle\langle u|}(B) \\
= & H_{|u\rangle\langle u|}(A, C) + H_{|u\rangle\langle u|}(A, D) - H_{|u\rangle\langle u|}(A) - H_{|u\rangle\langle u|}(A, C, D) \geq 0.
\end{aligned}
$$

Exercise 5.33 Inequality (5.98) implies that $H_\rho(A) \geq H_\rho(B) - H_\rho(A, B)$. Thus,

$$
-H_\rho(A|B) = H_\rho(B) - H_\rho(A, B) \leq H_\rho(A) \leq \log d_A.
$$

Since $H_\rho(A) + H_\rho(B) - H_\rho(A, B) = D(\rho^{A,B} \| \rho^A \otimes \rho^B) \geq 0$, we have

$$
H_\rho(A|B) = H_\rho(A, B) - H_\rho(B) \leq H_\rho(A) \leq \log d_A.
$$

Exercise 5.34

(**a**) $\frac{d\eta(x)}{dx} = -1 - \log x$. Since $\frac{d(\eta(x+\epsilon)-\eta(x)-\eta(\epsilon))}{d\epsilon} = -\log(x+\epsilon) + \log(\epsilon) \leq 0$, the function $\eta(x+\epsilon) - \eta(x) - \eta(\epsilon)$ with the variable $\epsilon \geq 0$ takes the minimum with $\epsilon = 0$. Since $\eta(x+0) - \eta(x) - \eta(0) = 0$, we obtain the desired argument.
(**b**) Since $\frac{d^2\eta(x)}{dx^2} = -\frac{1}{x} < 0$, $\eta(x)$ is strictly concave. $\frac{d\eta(x)}{dx} = -1 - \log x = 0$ if and only if $x = 1/e$. Hence, it takes the maximum value when $x = 1/e$.
(**c**) Since $\frac{d(\eta(\alpha-\epsilon)-\eta(\alpha)-\eta(1-\epsilon)+\eta(1))}{d\epsilon} = \log(\alpha - \epsilon) - \log(1 - \epsilon) \leq 0$, the function $\eta(\alpha - \epsilon) - \eta(\alpha) - \eta(1 - \epsilon) + \eta(1)$ with the variable $\epsilon \geq 0$ takes the minimum with $\epsilon = 0$. Since $\eta(\alpha - 0) - \eta(\alpha) - \eta(1 - 0) + \eta(1) = 0$, we obtain the desired argument.
(**d**) Since $\frac{d^2\eta(x)-\eta(1-x)}{dx^2} = -\frac{1}{x} + \frac{1}{1-x} < 0$ for $0 < x < 1/2$, the function $\eta(x) - \eta(1-x)$ is strictly concave for $0 < x < 1/2$. The function $\eta(x) - \eta(1-x)$ takes the value 0 at $x = 0, \frac{1}{2}$. Due to the strictly concavity, $\eta(x) - \eta(1-x) > 0$ for $0 < x < 1/2$.
(**e**): (**d**) implies that $\eta(x) > \eta(1-x)$. Thus, $\eta(\alpha - \epsilon) - \eta(\alpha) \leq \eta(1 - \epsilon) - \eta(1) \leq \eta(\epsilon) - \eta(1) = \eta(\epsilon)$. Combining this inequality and (**a**), we obtain (5.102).

Exercise 5.35

(**a**) Due to the condition $\epsilon_i \leq 1/2$, (**b**) of Exercise 5.34 implies $\eta(\epsilon_i) \geq \eta(1 - \epsilon_i)$. (**a**) and (**c**) of Exercise 5.34 implies $|\eta(a_i) - \eta(b_i)| \leq \max(\eta(\epsilon_i), \eta(1 - \epsilon_i))$. Therefore,

$$
|H(\rho) - H(\sigma)| \leq \sum_{i=1}^{d} |\eta(a_i) - \eta(b_i)| \leq \sum_{i=1}^{d} \eta(\epsilon_i).
$$

Combining (5.93), we obtain the desired argument.
(**b**) Note that $|\eta(a_1) - \eta(b_1)| \leq \max(\eta(\epsilon_1), \eta(1 - \epsilon_1)) \leq \eta(1/e) = 1/e$. Similar to (**a**), we have $\sum_{i=2}^{d} |\eta(a_i) - \eta(b_i)| \leq \sum_{i=2}^{d} \eta(\epsilon_i). \leq \epsilon' \log(d-1) + \eta_0(\epsilon')$.

(**c**) $1/e + \epsilon' \log(d-1) \le (1/e + \epsilon')\log d \le (\epsilon_1 + \epsilon')\log d = \epsilon \log d$. Since $\epsilon' \le \epsilon \le \|\rho - \sigma\|_1$ and η_0 is monotone increasing, we have $1/e + \epsilon' \log(d-1) + \eta_0(\epsilon') \le \|\rho - \sigma\|_1 \log d + \eta_0(\|\rho - \sigma\|_1)$.

Exercise 5.36 Since η_0 is concave, we have

$$
\begin{aligned}
I(p, W) =& \sum_x p(x)(H(W_p) - H(W_x)) \\
\le& \sum_x p(x)(\|W_x - W_p\|_1 \log d + \eta_0(\|W_x - W_p\|_1)) \\
\le& \sum_x p(x)\|W_x - W_p\|_1 \log d + \eta_0\left(\sum_x p(x)\|W_x - W_p\|_1\right) \\
=& \delta \log d + \eta_0(\delta).
\end{aligned}
$$

Exercise 5.37

(**a**) Since the concavity (5.88) implies that $H_\gamma(A|B) \ge (1-\epsilon)H_\rho(A|B) + \epsilon H_{\tilde{\rho}}(A|B)$. Hence, the inequality (5.101) implies that $H_\rho(A|B) - H_\gamma(A|B) \le \epsilon(H_\rho(A|B) - H_{\tilde{\rho}}(A|B)) \le 2\epsilon \log d_A$.
(**b**) It follows from the concavity (5.77).
(**c**) It follows from the first inequality of (5.79),
(**d**) (**b**) and (**c**) imply that

$$
\begin{aligned}
& H_\rho(A|B) - H_\gamma(A|B) = H_\rho(AB) - H_\gamma(AB) + H_\gamma(B) - H_\rho(B) \\
\ge& \epsilon(H_\rho(AB) - H_{\tilde{\rho}}(AB)) + \epsilon(H_{\tilde{\rho}}(B) - H_\rho(B)) - h(\epsilon) \\
=& \epsilon(H_\rho(A|B) - H_{\tilde{\rho}}(A|B)) - h(\epsilon) \ge -2\epsilon \log d_A - h(\epsilon).
\end{aligned}
$$

Exercise 5.38

(**a**) $(1-\epsilon)\sigma + \epsilon\tilde{\sigma} = (1-\epsilon)\sigma + \epsilon(\frac{1-\epsilon}{\epsilon}(\rho - \sigma) + \frac{1}{\epsilon}|\rho - \sigma|) = (1-\epsilon)\sigma + (1-\epsilon)(\rho - \sigma) + |\rho - \sigma| = (1-\epsilon)\rho + |\rho - \sigma| = \gamma$.
(**b**) Exercise 5.37 implies that

$$
\begin{aligned}
|H_\rho(A|B) - H_\sigma(A|B)| \le& |H_\rho(A|B) - H_\gamma(A|B)| + |H_\sigma(A|B) - H_\gamma(A|B)| \\
\le& 4\epsilon \log d_A + 2h(\epsilon).
\end{aligned}
$$

Exercise 5.39 Exercise 5.38 and (5.92) guarantee that

$$
\begin{aligned}
& |I_\rho(A:B) - I_\sigma(A:B)| = |H_\rho(A) - H_\sigma(A) - H_\rho(A|B) + H_\sigma(A|B)| \\
\le& |H_\rho(A) - H_\sigma(A) + |H_\rho(A|B) - H_\sigma(A|B)| \\
\le& \epsilon \log d_A + \eta_0(\epsilon) + 4\epsilon \log d_A + 2h(\epsilon) \\
=& 5\epsilon \log d_A + \eta_0(\epsilon) + 2h(\epsilon).
\end{aligned}
$$

Exercise 5.40

(a)

$$
\begin{aligned}
&|I_\rho(A:B|C) - I_\sigma(A:B|C)| \\
=&|H_\rho(A|C) + H_\rho(B|C) - H_\rho(AB|C) - H_\sigma(A|C) - H_\sigma(B|C) + H_\sigma(AB|C)| \\
\le&|H_\rho(A|C) - H_\sigma(A|C)| + |H_\rho(B|C) - H_\sigma(B|C)| + |H_\rho(AB|C) - H_\sigma(AB|C)|.
\end{aligned}
$$

(b) (5.104) implies that

$$
\begin{aligned}
&|H_\rho(A|C) - H_\sigma(A|C)| + |H_\rho(B|C) - H_\sigma(B|C)| + |H_\rho(AB|C) - H_\sigma(AB|C)| \\
\le&4\epsilon \log d_A + 2h(\epsilon). + 4\epsilon \log d_B + 2h(\epsilon). + 4\epsilon \log d_A d_B + 2h(\epsilon) \\
=&8\epsilon \log d_A d_B + 6h(\epsilon).
\end{aligned}
$$

Exercise 5.41

$$
\begin{aligned}
&H_\rho(AB|C) = H_\rho(ABC) - H_\rho(C) \\
=&H_\rho(ABC) - H_\rho(BC) + H_\rho(BC) - H_\rho(C) = H_\rho(B|C) + H_\rho(A|BC), \\
&I_\rho(A:BC) = H_\rho(A) + H_\rho(BC) - H_\rho(ABC) \\
=&H_\rho(A) + H_\rho(C) - H_\rho(AC) + H_\rho(BC) - H_\rho(ABC) - H_\rho(C) + H_\rho(AC) \\
=&I_\rho(A:C) + I_\rho(A:B|C), \\
&I_\rho(A:BC|D) = H_\rho(AD) + H_\rho(BCD) - H_\rho(ABCD) - H_\rho(D) \\
=&H_\rho(AD) + H_\rho(CD) - H_\rho(ACD) - H_\rho(D) \\
&+ H_\rho(BCD) - H_\rho(ABCD) - H_\rho(CD) + H_\rho(ACD) \\
=&I_\rho(A:C|D) + I_\rho(A:B|CD).
\end{aligned}
$$

Exercise 5.42

$$
\begin{aligned}
&I_\rho(A:B) = D(\rho_{AB}\|\rho_A \otimes \rho_B) \ge D(\kappa_A \otimes \kappa_B(\rho_{AB})\|\kappa_A \otimes \kappa_B(\rho_A \otimes \rho_B)) \\
=&D(\kappa_A \otimes \kappa_B(\rho_{AB})\|\kappa_A(\rho_A) \otimes \kappa_B(\rho_B)) = I_{\kappa_A \otimes \kappa_B \rho}(A:B).
\end{aligned}
$$

Exercise 5.43 It is sufficient to show $I_\rho(A:B|C) \ge I_{(\kappa_A \otimes \iota_{BC})(\rho)}(A:B|C)$. Since any TP-CP map can be regarded as the application of an isometry and the partial trace, it is enough to show $I_\rho(A:B_1B_2|C) \ge I_\rho(A:B_1|C)$. Equation (5.109) implies that $I_\rho(A:B_1B_2|C) = I_\rho(A:B_1|C) + I_\rho(A:B_2|CB_1) \ge I_\rho(A:B_1|C)$.

Exercise 5.44 Since $\mathrm{Tr}\,\rho^{\otimes n}\log\sigma^{\otimes n} = \mathrm{Tr}\,\kappa_{\sigma^{\otimes n}}(\rho^{\otimes n})\log\sigma^{\otimes n}$, we have

$$\begin{aligned}
&D(\rho\|\sigma) - \frac{1}{n}D(\kappa_{\sigma^{\otimes n}}(\rho^{\otimes n})\|\sigma^{\otimes n}) = \frac{1}{n}(D(\rho^{\otimes n}\|\sigma^{\otimes n}) - D(\kappa_{\sigma^{\otimes n}}(\rho^{\otimes n})\|\sigma^{\otimes n}))\\
=&\frac{1}{n}(-H(\rho^{\otimes n}) - \mathrm{Tr}\,\rho^{\otimes n}\log\sigma^{\otimes n} + H(\kappa_{\sigma^{\otimes n}}(\rho^{\otimes n})) + \mathrm{Tr}\,\kappa_{\sigma^{\otimes n}}(\rho^{\otimes n})\log\sigma^{\otimes n})\\
=&\frac{1}{n}(H(\kappa_{\sigma^{\otimes n}}(\rho^{\otimes n})) - H(\rho^{\otimes n})) \le \frac{1}{n}\log|\boldsymbol{E}_{\sigma^{\otimes n}}|,
\end{aligned}$$

where the final inequality follows from (5.82). Since Lemma 3.9 guarantees that $\frac{1}{n}\log|\boldsymbol{E}_{\sigma^{\otimes n}}| \to 0$, we obtain the desired argument.

Exercise 5.45 Denote the input classical system X and the output system $\mathcal{H}_A$. Define the state $\rho_{XA} := \sum_x p(x)|x\rangle\langle x| \otimes W_x$. Denoting the TP-CP map from the quantum system to the classical system due to the POVM $\boldsymbol{M}$ by κ', we have

$$I(p,W) = I_{\rho_{XA}}(X:A) \ge I_{\iota\otimes\kappa'\rho_{XA}}(X:A) = I(\boldsymbol{M},p,W).$$

Exercise 5.46

$$I(p,W) = I_{\rho_{XA}}(X:A) \ge I_{\iota\otimes\kappa\rho_{XA}}(X:A) = I(p,\kappa(W)).$$

Exercise 5.47

$$\begin{aligned}
&\log d_A - \sum_i p_i H(\rho_i^A) = \sum_i p_i(-H(\rho_i^A) - H(\rho_i^B) + \log d_A + H(\rho_i^B))\\
=&\sum_i p_i(-H(\rho_i^A\otimes\rho_i^B) + \log d_A + H(\rho_i^B))\\
=&\sum_i p_i D(\rho_i^A\otimes\rho_i^B\|\rho_{\mathrm{mix}}^A\otimes\rho_i^B) \ge D\left(\sum_i p_i\rho_i^A\otimes\rho_i^B\Big\|\rho_{\mathrm{mix}}^A\otimes\sum_i p_i\rho_i^B\right)\\
=&-H\left(\sum_i p_i\rho_i^A\otimes\rho_i^B\right) - \mathrm{Tr}\left(\sum_i p_i\rho_i^A\otimes\rho_i^B\right)\left(\log\rho_{\mathrm{mix}}^A + \log\sum_i p_i\rho_i^B\right)\\
=&-H\left(\sum_i p_i\rho_i^A\otimes\rho_i^B\right) + \log d_A + H\left(\sum_i p_i\rho_i^B\right),
\end{aligned}$$

which implies (5.110).

Exercise 5.48 Exercise 5.42 implies that

$$\begin{aligned}
&H_{(\iota_A\otimes\kappa)(\rho)}(A|B) = H_{(\iota_A\otimes\kappa)(\rho)}(A|B) - H_{(\iota_A\otimes\kappa)(\rho)}(A) + H_{(\iota_A\otimes\kappa)(\rho)}(A)\\
=&-I_{(\iota_A\otimes\kappa)(\rho)}(A:B) + H_\rho(A) \ge -I_\rho(A:B) + H_\rho(A)\\
=&H_\rho(A|B) - H_\rho(A) + H_\rho(A) = H_\rho(A|B).
\end{aligned}$$

Exercise 5.49 The monotonicity ((**a**) of Exercise 5.25) with respect to the partial trace yields

$$\begin{aligned} H_{\alpha|\rho}(A|B_1) &= -D_\alpha(\rho_{AB_1}\|I_A \otimes \rho_{B_1}) \\ &\geq - D_\alpha(\rho_{AB_1B_2}\|I_A \otimes \rho_{B_1B_2}) = H_{\alpha|\rho}(A|B_1B_2). \end{aligned}$$

We can show other relations.

References

1. W.F. Stinespring, Positive functions on C-algebras. Proc. Am. Math. Soc. **6**, 211 (1955)
2. K. Kraus, in *States, Effects, and Operations*, vol. 190, Lecture Notes in Physics (Springer, Berlin Heidelberg New York, 1983)
3. M.-D. Choi, Completely positive linear maps on complex matrices. Lin. Alg. Appl. **10**, 285–290 (1975)
4. A. Fujiwara, P. Algoet, One-to-one parametrization of quantum channels. Phys. Rev. A **59**, 3290–3294 (1999)
5. A. Jamiołkowski, Linear transformations which preserve trace and positive semidefiniteness of operators. Rep. Math. Phys. **3**, 275–278 (1972)
6. A. Fujiwara, Mathematics of quantum channels. Suurikagaku **474**, 28–35 (2002). (in Japanese)
7. G.M. D'Ariano, P.L. Presti, Imprinting complete information about a quantum channel on its output state. Phys. Rev. Lett. **91**, 047902 (2003)
8. D. Aharonov, A. Kitaev, N. Nisan, Quantum Circuits with Mixed States *Proceedings of the 30th Annual ACM Symposium on Theory of Computation (STOC)*, 20–30 (1997)
9. M. Horodecki, P. Shor, M.B. Ruskai, Entanglement breaking channels. Rev. Math. Phys. **15**, 1–13 (2003)
10. M. Horodecki, P. Horodecki, R. Horodecki, Separability of mixed states: necessary and sufficient conditions. Phys. Lett. **A 223**, 1–8 (1996)
11. P. Horodecki, Separability criterion and inseparable mixed states with positive partial transposition. Phys. Lett. A **232**, 333 (1997)
12. N. Datta, A.S. Holevo, Y. Suhov, Additivity for transpose depolarizing channels. Int. J. Quantum Inform. **4**, 85 (2006)
13. K. Matsumoto, F. Yura, Entanglement cost of antisymmetric states and additivity of capacity of some quantum channel. Jhys. A: Math. Gen. **37**, L167–L171 (2004)
14. R.F. Werner, A.S. Holevo, Counterexample to an additivity conjecture for output purity of quantum channels. J. Math. Phys. **43**, 4353 (2002)
15. M.B. Ruskai, S. Szarek, E. Werner, An analysis of completely-positive trace-preserving maps on 2×2 matrices. Lin. Alg. Appl. **347**, 159–187 (2002)
16. M.B. Ruskai, Qubit entanglement breaking channels. Rev. Math. Phys. **15**, 643–662 (2003)
17. A. Fujiwara, H. Nagaoka, Operational capacity and pseudoclassicality of a quantum channel. IEEE Trans. Inf. Theory **44**, 1071–1086 (1998)
18. G. Lindblad, Completely positive maps and entropy inequalities. Comm. Math. Phys. **40**, 147–151 (1975)
19. A. Uhlmann, Relative entropy and the Wigner-Yanase-Dyson-Lieb concavity in an interpolation theory. Comm. Math. Phys. **54**, 21–32 (1977)
20. G. Lindblad, Expectations and entropy inequalities for finite quantum systems. Comm. Math. Phys. **39**, 111–119 (1974)
21. F. Hiai, D. Petz, The golden-thompson trace inequality is complemented. Lin. Alg. Appl. **181**, 153–185 (1993)

22. S. Golden, Lower bounds for Helmholtz function. Phys. Rev. **137**, B1127–B1128 (1965)
23. K. Symanzik, Proof and refinements of an inequality of Feynman. J. Math. Phys. **6**, 1155–1156 (1965)
24. C.J. Thompson, Inequality with applications in statistical mechanics. J. Math. Phys. **6**, 1812–1813 (1965)
25. A. Uhlmann, The 'transition probability' in the state space of *-algebra. Rep. Math. Phys. **9**, 273–279 (1976)
26. R. Jozsa, Fidelity for mixed quantum states. J. Mod. Opt. **41**(12), 2315–2323 (1994)
27. H. Barnum, C.A. Fuchs, R. Jozsa, B. Schumacher, A general fidelity limit for quantum channels. Phys. Rev. A **54**, 4707–4711 (1996)
28. M.B. Ruskai, Beyond strong subadditivity? improved bounds on the contraction of generalized relative entropy. Rev. Math. Phys. **6**, 1147–1161 (1994)
29. R.L. Frank, E.H. Lieb, Monotonicity of a relative Renyi entropy. J. Math. Phys. **54**, 122201 (2013)
30. F. Hiai, D. Petz, The proper formula for relative entropy and its asymptotics in quantum probability. Comm. Math. Phys. **143**, 99–114 (1991)
31. E. Lieb, M.B. Ruskai, A fundamental property of quantum mechanical entropy. Phys. Rev. Lett. **30**, 434–436 (1973)
32. E. Lieb, M.B. Ruskai, Proof of the strong subadditivity of quantum mechanical entropy. J. Math. Phys. **14**, 1938–1941 (1973)
33. P. Hayden, R. Jozsa, D. Petz, A. Winter, Structure of states which satisfy strong subadditivity of quantum entropy with equality. Comm. Math. Phys. **246**, 359–374 (2004)
34. M. Fannes, A continuity property of the entropy density for spin lattice systems. Comm. Math. Phys. **31**, 291–294 (1973)
35. W. Ochs, A new axiomatic characterization of the von Neumann entropy. Rep. Math. Phys. **8**(1), 109–120 (1975)
36. H. Araki, E. Lieb, Entropy inequalities. Comm. Math. Phys. **18**, 160–170 (1970)
37. E. Lieb, Bull. Am. Math. Soc. **81**, 1–13 (1975)
38. R. Alicki, M. Fannes, Continuity of quantum mutual information, quant-ph/0312081 (2003)
39. M. Christandl, A. Winter, Squashed entanglement"-an additive entanglement measure. J. Math. Phys. **45**, 829–840 (2004)
40. H. Fan, A note on quantum entropy inequalities and channel capacities. J. Phys. A Math. Gen. **36**, 12081–12088 (2003)
41. R. König, R. Renner, C. Schaffner, The operational meaning of min- and max-entropy. IEEE Trans. Inf. Theory **55**(9), 4337–4347 (2009)
42. M. Tomamichel, R. Colbeck, R. Renner, A fully quantum asymptotic equipartition property. IEEE Trans. Inf. Theory **55**(12), 5840–5847 (2009)
43. S. Beigi, Sandwiched Rènyi divergence satisfies data processing inequality. J. Math. Phys. **54**(12), 122202 (2013)
44. M. Müller-Lennert, F. Dupuis, O. Szehr, S. Fehr, M. Tomamichel, On quantum Renyi entropies: a new generalization and some properties. J. Math. Phys. **54**, 122203 (2013)
45. M. Tomamichel, M. Berta, M. Hayashi, Relating different quantum generalizations of the conditional Rényi entropy. J. Math. Phys. **55**, 082206 (2014)
46. A.S. Holevo, Bounds for the quantity of information transmitted by a quantum communication channel. Problemy Peredachi Informatsii, **9**, 3–11 (1973) (in Russian). (English translation: Probl. Inf. Transm., **9**, 177–183 (1975))
47. E. Lieb, Convex trace functions and the Wigner-Yanase-Dyson conjecture. Adv. Math. **11**, 267–288 (1973)
48. D. Petz, Quasi-entropies for finite quantum systems. Rep. Math. Phys. **23**, 57–65 (1986)
49. M.M. Wilde, A. Winter, D. Yang, Strong converse for the classical capacity of entanglement-breaking and Hadamard channels via a sandwiched Renyi relative entropy. Comm. Math. Phys. **331**(2), 593 (2014)
50. T. Ando, Convexity of certain maps on positive definite matrices and applications to Hadamard products. Lin. Alg. and Appl. **26**, 203–241 (1979)
51. J. Bergh, J. Löfström, *Interpolation Spaces*. (Springer-Verlag, New York, 1976)

Chapter 6
Quantum Information Geometry and Quantum Estimation

Abstract In Chap. 3 we examined the discrimination of two unknown quantum states. This chapter will consider the estimation of a parameter θ, which labels an unknown state parameterized by a continuous variable θ. It is a remarkable property of quantum mechanics that a measurement inevitably leads to the state reduction. Therefore, when one performs a measurement for state estimation, it is necessary to choose the measurement that extracts as much information as possible. This problem is called quantum estimation, and the optimization of the measurement is an important topic in quantum information theory. In the classical theory of estimation (of probability distributions) discussed in Sect. 2.2, we saw that the estimation is intimately related to geometrical structures such as the inner product. We can expect that such geometrical structures will also play an important role in the quantum case. The study of geometrical structures in the space of quantum states is called quantum information geometry and is an important field in quantum information theory. This chapter will examine the geometrical structure of quantum systems and discuss its applications to estimation theory.

6.1 Inner Products in Quantum Systems

In any discussion about the geometry of quantum states, the metric plays a central role. To start talking about the metric, we must first discuss the quantum versions of the Fisher information and its associated inner product (2.95) examined in Sect. 2.2. Let A, B, p in (2.95) be the diagonal elements of the commuting Hermitian matrices Y, X, ρ, respectively. The inner product (2.95) is then equal to $\operatorname{Tr} Y(\rho X)$. Although the trace of a product of two matrices does not depend on the order of the multiplication, the trace of the product for three or more matrices is dependent on the order. If these matrices do not commute, then the inner product depends on the order of the product between ρ and X. At least, the product $E_\rho(X)$ should be defined by a linear map E_ρ satisfying the conditions

$$\operatorname{Tr} Y^* E_\rho(X) = \operatorname{Tr} E_\rho(Y)^* X, \tag{6.1}$$

$$\operatorname{Tr} X^* E_\rho(X) \geq 0, \tag{6.2}$$

M. Hayashi, *Quantum Information Theory*, Graduate Texts in Physics,
DOI 10.1007/978-3-662-49725-8_6

$$E_\rho(U^*XU) = U^* E_{U\rho U^*,x}(X) U, \tag{6.3}$$
$$E_\rho(I) = \rho, \tag{6.4}$$
$$E_{\rho\otimes\rho',x}(X \otimes X') = E_\rho(X) \otimes E_{\rho'}(X'), \tag{6.5}$$

which implies the following properties[Exe. 6.1]

$$\operatorname{Tr} E_\rho(X) = \operatorname{Tr} \rho X \tag{6.6}$$
$$E_{\rho\otimes\rho'}(X \otimes I) = E_\rho(X) \otimes \rho'. \tag{6.7}$$

There exist at least three possible ways of $E_\rho(X)$ to satisfy the above requirements.

$$E_{\rho,s}(X) \stackrel{\text{def}}{=} X \circ \rho \stackrel{\text{def}}{=} \frac{1}{2}(\rho X + X\rho), \tag{6.8}$$
$$E_{\rho,b}(X) \stackrel{\text{def}}{=} \int_0^1 \rho^\lambda X \rho^{1-\lambda} d\lambda, \tag{6.9}$$
$$E_{\rho,r}(X) \stackrel{\text{def}}{=} \rho X. \tag{6.10}$$

Here, $E_{\rho,s}$, $E_{\rho,b}$, and $E_{\rho,r}$ are defined as maps on $\mathcal{M}(\mathcal{H})$. Here, X is not necessarily Hermitian. These extensions are unified in the general form [1];

$$E_{\rho,p}(X) \stackrel{\text{def}}{=} \int_0^1 E_{\rho,\lambda}(X) p(d\lambda), \tag{6.11}$$
$$E_{\rho,\lambda}(X) \stackrel{\text{def}}{=} \rho^\lambda X \rho^{1-\lambda}, \tag{6.12}$$

where p is an arbitrary probability distribution on $[0, 1]$. When $\rho > 0$, these maps possess inverses. The case (6.8) is a special case of the case (6.11) with $p(1) = p(0) = 1/2$, and the case (6.10) is a special case of the case (6.11) with $p(1) = 1$. In particular, the map $E_{\rho,x}$ is called **symmetric** when $E_{\rho,x}(X)$ is Hermitian if and only if X is Hermitian. Hence, when the distribution p is symmetric, i.e., $p(\lambda) = p(1-\lambda)$, the map $E_{\rho,p}$ is symmetric. These maps $E_{\rho,x}$ satisfy Conditions (6.1)–(6.5). For example, when $x = s, b$, or $\frac{1}{2}$, the map $E_{\rho,x}$ is symmetric.

Now, we define the following types of inner products:

$$\langle Y, X\rangle^{(e)}_{\rho,x} \stackrel{\text{def}}{=} \operatorname{Tr} Y^* E_{\rho,x}(X) \quad x = s, b, r, \lambda, p. \tag{6.13}$$

If X, Y, ρ all commute, then these coincide with definition (2.95). These are called the SLD, Bogoljubov,[1] RLD, λ, and p inner products [1–8], respectively (reasons

[1]The Bogoljubov inner product is also called the canonical correlation in statistical mechanics. In linear response theory, it is often used to give an approximate correlation between two different physical quantities.

for this will be given in the next section). Due to Conditions (6.1) and (6.2), these inner products are positive semidefinite and Hermitian[Exe. 6.2], i.e.,

$$\left(\|X\|_{\rho,x}^{(e)}\right)^2 \stackrel{\text{def}}{=} \langle X, X\rangle_{\rho,x}^{(e)} \geq 0, \quad \langle Y, X\rangle_{\rho,x}^{(e)} = (\langle X, Y\rangle_{\rho,x}^{(e)})^*. \tag{6.14}$$

From property (6.3) we have

$$\langle X \otimes X', Y \otimes Y'\rangle_{\rho\otimes\rho',x}^{(e)} = \langle X, Y\rangle_{\rho,x}^{(e)} \langle X', Y'\rangle_{\rho',x}^{(e)}, \tag{6.15}$$
$$\langle U^* X U, U^* Y U\rangle_{\rho,x}^{(e)} = \langle X, Y\rangle_{U\rho U^*,x}^{(e)}, \quad \|I\|_{\rho,x}^{(e)} = 1.$$

In particular, the SLD inner product and the RLD inner product satisfy

$$\|X \otimes I_{\mathcal{H}'}\|_{\rho,x}^{(e)} = \|X\|_{\mathrm{Tr}_{\mathcal{H}'}\rho,x}^{(e)}, \quad x = s, r. \tag{6.16}$$

Generally, as is shown in Sect. 6.7.1, we have

$$\|X \otimes I_{\mathcal{H}'}\|_{\rho,x}^{(e)} \leq \|X\|_{\mathrm{Tr}_{\mathcal{H}'}\rho,x}^{(e)}, \quad x = b, \lambda, p. \tag{6.17}$$

From here, we assume that ρ is invertible. A dual inner product may be defined as

$$\langle A, B\rangle_{\rho,x}^{(m)} \stackrel{\text{def}}{=} \mathrm{Tr}(E_{\rho,x}^{-1}(A))^* B \tag{6.18}$$

with respect to the correspondence $A = E_{\rho,x}(X)$. Denote the norm of these inner products as

$$\left(\|A\|_{\rho,x}^{(m)}\right)^2 \stackrel{\text{def}}{=} \langle A, A\rangle_{\rho,x}^{(m)}. \tag{6.19}$$

Hence, the inner product $\langle A, B\rangle_{\rho,x}^{(m)}$ is positive semidefinite and Hermitian. In particular, the inner product is called **symmetric** when $\langle A, B\rangle_{\rho,x}^{(m)} = \langle B, A\rangle_{\rho,x}^{(m)}$ for two Hermitian matrices A and B. Similarly, the symmetricity is defined for the dual inner product $\langle X, Y\rangle_{\rho,x}^{(e)}$. That is, the inner product $\langle X, Y\rangle_{\rho,x}^{(e)}$ is called symmetric when $\langle X, Y\rangle_{\rho,x}^{(e)} = \langle Y, X\rangle_{\rho,x}^{(e)}$ for two Hermitian matrices X and Y. The symmetricity of the inner product $\langle X, Y\rangle_{\rho,x}^{(e)}$ is equivalent to not only the symmetricity of the dual inner product $\langle X, Y\rangle_{\rho,x}^{(e)}$, but also the symmetricity of map $E_{\rho,x}$[Exe. 6.3]. When the inner product $\langle A, B\rangle_{\rho,x}^{(m)}$ is symmetric, it can be symmetrized as $\langle A, B\rangle_{\rho,s(x)}^{(m)} \stackrel{\text{def}}{=} \frac{1}{2}\left(\langle A, B\rangle_{\rho,x}^{(m)} + \langle B, A\rangle_{\rho,x}^{(m)}\right)$, i.e., the symmetrized map $E_{\rho,s(x)}$ is defined as $E_{\rho,s(x)}^{-1}(A) = \frac{1}{2}\left(E_{\rho,x}^{-1}(A) + (E_{\rho,x}^{-1}(A))^*\right)$ for any Hermitian matrix A. Hence, we call the inner product $\langle A, B\rangle_{\rho,s(x)}^{(m)}$ the **symmetrized** inner product of $\langle A, B\rangle_{\rho,x}^{(m)}$. Note that the SLD inner product is not the symmetrized inner product of the RLD inner product.

Similarly to (6.3), we have

$$\langle X, Y\rangle_{\rho,x}^{(m)} = \langle UXU^*, UYU^*\rangle_{U\rho U^*,x}^{(m)} \tag{6.20}$$

for an arbitrary unitary matrix U. When a TP-CP map κ and a state ρ satisfy that $\kappa(\rho) > 0$, we also define the map $\kappa_{\rho,x}$ associated with κ and ρ by the relation

$$\kappa(E_{\rho,x}(X)) = E_{\kappa(\rho),x}(\kappa_{\rho,x}(X)), \quad x = s, b, r, \lambda, p, \tag{6.21}$$

where for a non-Hermitian matrix A, $\kappa(A)$ is defined as $\kappa(A) \stackrel{\text{def}}{=} \kappa((A + A^*)/2) - i\kappa(i(A - A^*)/2)$. This map satisfies the associativity

$$(\kappa_1 \circ \kappa_2)_{\rho,x}(X) = \kappa_{1\,\kappa_2(\rho),x} \circ \kappa_{2\,\rho,x}(X), \quad x = s, b, r, \lambda, p. \tag{6.22}$$

Also, the relation

$$\langle \kappa^*(Y), X\rangle = \mathrm{Tr}\, Y\kappa(E_{\rho,x}(X)) = \mathrm{Tr}\, Y E_{\kappa(\rho),x}(\kappa_{\rho,x}(X)) = \langle Y, \kappa_{\rho,x}(X)\rangle_{\kappa(\rho),x}^{(e)} \tag{6.23}$$

holds for any Y. Since (6.23) can be regarded as a quantum extension of (2.109), we call the map $\kappa_{\rho,x}$ as the **conditional expectation** with respect to the inner product x[Exe. 6.12]. Then, we have the following theorem.

Theorem 6.1 *The inequality*

$$\|A\|_{\rho,x}^{(m)} \ge \|\kappa(A)\|_{\kappa(\rho),x}^{(m)}, \quad x = s, b, r, \lambda, p \tag{6.24}$$

holds. This inequality (6.24) is also equivalent to

$$\|X\|_{\rho,x}^{(e)} \ge \|\kappa_{\rho,x}(X)\|_{\kappa(\rho),x}^{(e)}, \quad x = s, b, r, \lambda, p. \tag{6.25}$$

When an inner product satisfies property (6.24), it is called a **monotone metric**. Monotonicity implies that any operation does not increase the amount of information. That is, if the inner product is to be considered as a measure of information, this property should be satisfied because information processing does not cause any increase in the amount of information. It is also known that an arbitrary inner product $\|A\|_{\rho,x}^{(m)}$ satisfying property (6.24) and $\|\rho^{-1}\|_{\rho,x}^{(m)} = 1$ satisfies $\|A\|_{\rho,s}^{(m)} \le \|A\|_{\rho,x}^{(m)} \le \|A\|_{\rho,r}^{(m)}$, i.e., the SLD inner product is the minimum product and the RLD inner product is the maximum product [3].

Before proving Theorem 6.1, we need to discuss how to extend the above discussion to the case when ρ is non-invertible. Even though ρ is non-invertible, the map $E_{\rho,x}$ can defined. However, it has a non-trivial kernel $\mathcal{K}_{\rho,x}(\mathcal{H})$. So, the inner product $\langle\, ,\,\rangle_{\rho,x}^{(e)}$ is degenerate. Here, we introduce the quotient space $\mathcal{M}_{\rho,x}(\mathcal{H}) := \mathcal{M}(\mathcal{H})/\mathcal{K}_{\rho,x}(\mathcal{H})$. Then, the inner product $\langle\, ,\,\rangle_{\rho,x}^{(e)}$ is non-degenerate in $\mathcal{M}_{\rho,x}(\mathcal{H})$.

Next, to discuss the other inner product $\langle\ ,\ \rangle^{(m)}_{\rho,x}$, we focus on the image $\mathcal{M}^{(m)}_{\rho,x}(\mathcal{H})$ of the map $E_{\rho,x}$. Then, the inner product $\langle\ ,\ \rangle^{(m)}_{\rho,x}$, can be defined on the space $\mathcal{M}^{(m)}_{\rho,x}(\mathcal{H})$ as a non-degenerate inner product. For example, the space $\mathcal{M}^{(m)}_{\rho,r}(\mathcal{H})$ can be characterized by using the projection P_ρ to the range of ρ as $\{X \in \mathcal{M}(\mathcal{H}) | P_\rho X = X\}$. In this case, the space $\mathcal{M}^{(m)}_{\rho,r}(\mathcal{H})$ can be regarded as the set of representatives of the elements of the quotient space $\mathcal{M}_{\rho,r}(\mathcal{H})$. If there is no possibility for confusion, $\mathcal{M}_{\rho,x}(\mathcal{H})$ and $\mathcal{M}^{(m)}_{\rho,x}(\mathcal{H})$ are abbreviated to $\mathcal{M}_{\rho,x}$ and $\mathcal{M}^{(m)}_{\rho,x}$.

Proof of Theorem 6.1 Here, we prove (6.25) for $x = s, r$. The general case of (6.25) is shown assuming inequality (6.17), which will be proven in Sect. 6.7.1.

These inner products are invariant for the operations $\rho \mapsto \rho \otimes \rho_0$ and $\rho \mapsto U\rho U^*$. It is sufficient to show (6.25) in the case of partial trace because of the Stinespring representation and associativity (6.22). First, using property (6.16), we prove (6.25) for $x = s, r$. Letting κ be the partial trace from system $\mathcal{H} \otimes \mathcal{H}'$ to subsystem $\mathcal{H}'$, we have

$$\begin{aligned}&\langle Y \otimes I, \kappa_{\rho,x}(X) \otimes I\rangle^{(e)}_{\rho,x} = \langle Y, \kappa_{\rho,x}(X)\rangle_{\mathrm{Tr}_{\mathcal{H}'}\rho,x} = \mathrm{Tr}\, Y^* \kappa(E_{\rho,x}(X))\\ &= \mathrm{Tr}(Y \otimes I)^* E_{\rho,x}(X) = \langle Y \otimes I, X\rangle^{(e)}_{\rho,x}\end{aligned}$$

for any matrix X on $\mathcal{H} \otimes \mathcal{H}'$, any matrix Y on $\mathcal{H}$, and any state ρ on $\mathcal{H} \otimes \mathcal{H}'$. Hence, the map $\kappa_{\rho,x}$ is the projection from the space of all matrices on $\mathcal{H} \otimes \mathcal{H}'$ to the subspace of matrices $\{Y \otimes I\}$ with respect to the inner product $\langle\ ,\ \rangle^{(e)}_{\rho,x}$. Therefore, $\|X\|^{(e)}_{\rho,x} \geq \|\kappa_{\rho,x}(X)\|^{(e)}_{\kappa(\rho),x}$. Hence, we obtain (6.25) for $x = s, r$.

Next, we proceed to the general case, i.e., the case of $x = p, b$. Let $\mathcal{F}$ be the positive self adjoint map on the matrix space with respect to the inner product $\langle\ ,\ \rangle^{(e)}_{\mathrm{Tr}_{\mathcal{H}'}\rho,x}$ satisfying

$$\langle Y \otimes I, Y' \otimes I\rangle^{(e)}_{\rho,x} = \langle Y, \mathcal{F} Y'\rangle^{(e)}_{\mathrm{Tr}_{\mathcal{H}'}\rho,x}. \tag{6.26}$$

Since property (6.17) implies $\|Y \otimes I\|^{(e)}_{\rho,x} = \|\mathcal{F}^{1/2} Y\|^{(e)}_{\mathrm{Tr}_{\mathcal{H}'}\rho,x} \leq \|Y\|^{(e)}_{\mathrm{Tr}_{\mathcal{H}'}\rho,x}$, we have

$$\|(\mathcal{F}^{-1} Y) \otimes I\|^{(e)}_{\rho,x} = \|\mathcal{F}^{-1/2} Y\|^{(e)}_{\mathrm{Tr}_{\mathcal{H}'}\rho,x} \geq \|Y\|^{(e)}_{\mathrm{Tr}_{\mathcal{H}'}\rho,x}.$$

Hence,

$$\begin{aligned}&\langle Y \otimes I, (\mathcal{F}^{-1}\kappa_{\rho,x}(X)) \otimes I\rangle^{(e)}_{\rho,x} = \langle Y, \kappa_{\rho,x}(X)\rangle_{\mathrm{Tr}_{\mathcal{H}'}\rho,x} = \mathrm{Tr}\, Y^* \kappa(E_{\rho,x}(X))\\ &= \mathrm{Tr}(Y \otimes I)^* E_{\rho,x}(X) = \langle Y \otimes I, X\rangle^{(e)}_{\rho,x}.\end{aligned}$$

Similarly, we can show that $\|(\mathcal{F}^{-1}\kappa_{\rho,x}(X)) \otimes I\|^{(e)}_{\rho,x} \leq \|X\|^{(e)}_{\rho,x}$. Therefore, we obtain

$$\|\kappa_{\rho,x}(X)\|^{(e)}_{\mathrm{Tr}_{\mathcal{H}'}\rho,x} \leq \|X\|^{(e)}_{\rho,x}.$$

∎

Exercises

6.1 Show (6.6) and (6.7) by using Conditions (6.1)–(6.5).

6.2 Show (6.14) by using Conditions (6.1) and (6.2).

6.3 Show that the following conditions are equivalent. (Hint: Use Condition (6.1).)

① The inner product $\langle A, B\rangle^{(m)}_{\rho,x}$ is symmetric.
② The inner product $\langle X, Y\rangle^{(e)}_{\rho,x}$ is symmetric.
③ The map $E_{\rho,x}$ is symmetric.

6.4 Prove the following facts for a traceless Hermitian matrix A and a density matrix ρ of the form $\rho = \sum_{j=1}^{d} \lambda_j E_j$ and rank $E_j = 1$.
(**a**) Show that $(x + y)/2 \geq \mathrm{Lm}(x, y)$, where $\mathrm{Lm}(x, y)$ is the logarithmic average defined below.

$$\mathrm{Lm}(x, y) \stackrel{\text{def}}{=} \begin{cases} \frac{x-y}{\log x - \log y} & \text{if } x \neq y, \\ x & \text{if } x = y. \end{cases}$$

Also show that the equality holds if and only if $x = y$.
(**b**) Show the following [3]:

$$\|A\|^{(m)}_{\rho,s} = \sum_{j,k=1}^{d} \frac{2}{\lambda_j + \lambda_k} \operatorname{Tr} AE_jAE_k,$$

$$\|A\|^{(m)}_{\rho,b} = \sum_{j,k=1}^{d} \frac{1}{\mathrm{Lm}(\lambda_j, \lambda_k)} \operatorname{Tr} AE_jAE_k.$$

(**c**) Show that $\operatorname{Tr}(A\rho - \rho A)(A\rho - \rho A)^* = \sum_{j,k=1}^{d} (\lambda_j - \lambda_k)^2 \operatorname{Tr} AE_jAE_k$.
(**d**) Show the inequality $\|A\|^{(m)}_{\rho,b} \geq \|A\|^{(m)}_{\rho,s}$. Also, show the equivalence of the following.

① $\|A\|^{(m)}_{\rho,b} = \|A\|^{(m)}_{\rho,s}$.
② $[\rho, A] = 0$.

6.5 For the pinching $\kappa_{\boldsymbol{M}}$ of a PVM $\boldsymbol{M}$, we define $\kappa_{\boldsymbol{M},\rho,s} := (\kappa_{\boldsymbol{M}})_{\rho,s}$. Show the following facts.
(**a**) For any matrix X, show that $\kappa_{\boldsymbol{M},\rho,s}(X)$ commutes with every element M_i.
(**b**) Assume that $\kappa_{\boldsymbol{M}}(\rho) > 0$. Show that $\kappa_{\boldsymbol{M},\rho,s}(X) = X$ if and only if every M_i commutes with X.
(**c**) Show that $\kappa_{\boldsymbol{M},\rho,s} \circ \kappa_{\boldsymbol{M},\rho,s} = \kappa_{\boldsymbol{M},\rho,s}$, i.e., $\kappa_{\boldsymbol{M},\rho,s}$ can be regarded as a projection.
(**d**) Show that $\langle Y, X\rangle^{(e)}_{\rho,s} = \langle Y, \kappa_{\boldsymbol{M},\rho,s}(X)\rangle^{(e)}_{\kappa_{\boldsymbol{M}}(\rho),s}$ if every matrix M_i commutes with Y.
(**e**) Verify that the above is true for the RLD case.

6.6 Show that the following two conditions are equivalent for the Hermitian matrix A, the state $\rho > 0$, and the pinching $\kappa_{\boldsymbol{M}}$ corresponding to PVM $\boldsymbol{M} = \{M_i\}$.

① $\|A\|^{(m)}_{\rho,s} = \|\kappa_{M}(A)\|^{(m)}_{\kappa_{M}(\rho),s}$.
② $X := E^{-1}_{\rho,s}(A)$ is commutative with M_i for all i.

6.7 Show the inequality $\|A\|^{(m)}_{\rho,b} \geq \|\kappa_{M}(A)\|^{(m)}_{\kappa_{M}(\rho),s}$ with the same assumption as above. Also, show the equivalence of the following:

① $\|A\|^{(m)}_{\rho,b} = \|\kappa_{M}(A)\|^{(m)}_{\kappa_{M}(\rho),s}$.
② There exists a Hermitian matrix X such that it commutes with every M_i and satisfies $A = \rho X = X\rho$.

6.8 Show that $\|X\rho Y\|_1 \leq \sqrt{\operatorname{Tr}\rho Y Y^*}\sqrt{\operatorname{Tr}\rho X^* X}$ by the Schwarz inequality for the inner product $\langle X, Y\rangle^{(e)}_{\rho,r} \stackrel{\text{def}}{=} \operatorname{Tr}\rho Y X^*$, where X, Y are matrices and ρ is a density matrix. Note that $\|\cdot\|_1$ denotes the trace-norm (Sect. A.3).

6.9 Given a matrix X and a density matrix ρ, show that

$$\|X\|_1 \leq \sqrt{\operatorname{Tr}\rho^{-1}XX^*}\sqrt{\operatorname{Tr}\rho U^*U} = \sqrt{\operatorname{Tr}\rho^{-1}XX^*}, \tag{6.27}$$

where U is a unitary matrix satisfying $\|X\|_1 = \operatorname{Tr} XU$.

6.10 Given a matrix X and a density matrix ρ, show that

$$\|X\|_1 \leq \sqrt{\operatorname{Tr}\rho^{-1/2}X\rho^{-1/2}X^*}\sqrt{\operatorname{Tr}\rho^{1/2}U^*\rho^{1/2}U} \leq \sqrt{\operatorname{Tr}\rho^{-1/2}X\rho^{-1/2}X^*}, \tag{6.28}$$

where U is a unitary matrix satisfying $\|X\|_1 = \operatorname{Tr} XU$.

6.11 Assume that the distribution p has zero measure at $\lambda = 1, 0$. Let ρ be a pure state $|y\rangle\langle y|$. For the equality condition of the inequality (6.17), show the following. The equality of the inequality $\|X \otimes I_{\mathcal{H}'}\|^{(e)}_{\rho,p} \leq \|X\|^{(e)}_{\sigma,p}$ holds with $\operatorname{Tr}_{\mathcal{H}'}\rho = \sigma$ if and only if X is a constant times of I.

6.12 Let κ be the pinching κ_{M} of a PVM $M = \{M_i\}$. Define $\kappa_{M,\rho,x}$ as the same as Exercise 6.5 for $x = s, r$. Show that the map $\kappa_{M,\rho,x}$ can be regarded as the conditional expectation to the matrix subspace $\{X|[X, M_i] = 0\ \forall i\}$ for $x = s, r$. That is, show that the map $\kappa_{M,\rho,x}$ is the dual map of the inclusion of the matrix subspace $\{X|[X, M_i] = 0\ \forall i\}$ for $x = s, r$. (In general, the conditional expectation can be defined by (2.110) when the map κ is the dual map of the inclusion of a matrix subspace $\mathcal{U}$.)

6.13 Show that $E_{\rho,x}$ is a map from the set of Hermitian matrices to itself for $x = s, \frac{1}{2}, b$.

6.2 Metric-Induced Inner Products

In this section we treat the space of quantum states in a geometrical framework. In particular, we will discuss the properties of the metric, which will be defined in terms of the inner product discussed in the previous section. Consider a set of quantum states $\{\rho_\theta|\theta \in \mathbb{R}\}$ (a state family) parameterized by a single real number θ. We also assume that $\theta \mapsto \rho_\theta$ is continuous and differentiable up to the second order. The metric then represents the distance between two quantum states $\rho_{\theta_0}, \rho_{\theta_0+\epsilon}$ separated by a small $\epsilon > 0$. The difference in this case is approximately equal to $\frac{d\rho_\theta}{d\theta}(\theta_0)\epsilon$. When we focus on the norm $\| \ \|^{(m)}_{\rho_{\theta_0},s}$, the Fisher metric $J_{\theta_0,s}$ is defined to be $\left(\left\|\frac{d\rho_\theta}{d\theta}(\theta_0)\right\|^{(m)}_{\rho_{\theta_0},x}\right)^2$. In particular, the **SLD Fisher metric** $J_{\theta_0,s}$ is defined as the square of the size of $\frac{d\rho_\theta}{d\theta}(\theta_0)$ based on the SLD inner product at ρ_{θ_0}, i.e., $J_{\theta_0,s} \stackrel{\text{def}}{=} \left(\left\|\frac{d\rho_\theta}{d\theta}(\theta_0)\right\|^{(m)}_{\rho_{\theta_0},s}\right)^2$. The norm of the difference between two quantum states ρ_θ and $\rho_{\theta+\epsilon}$ is then approximately $\sqrt{J_{\theta_0,s}}\epsilon$. We can obviously define quantities such as the **Bogoljubov Fisher metric** $J_{\theta_0,b}$ [1, 3, 4, 8], the **RLD Fisher metric** $J_{\theta_0,r}$ [2, 7], and the p metric in a similar way for the Bogoljubov, RLD, and p inner products, respectively. Therefore, if $u_1, \ldots, u_k$ is an orthonormal basis in $\mathcal{H}$, the SLD, Bogoljubov, RLD, and p Fisher metrics of the state family $\{\rho_\theta \stackrel{\text{def}}{=} \sum_{i=1}^k p_\theta(i)|u_i\rangle\langle u_i||\theta \in \mathbb{R}\}$ are all equal to the Fisher metric for the probability family $\{p_\theta\}$.

Thus, we have a theorem equivalent to Theorem 6.1 as given below.

Theorem 6.2 *Let κ be a TP-CP map, and $J_{\theta_0,x,\kappa}$ be the $x = s, b, r, \lambda, p$ Fisher metric for the state family $\{\kappa(\rho_\theta)|\theta \in \mathbb{R}\}$. The following relation then holds:*

$$J_{\theta_0,x} \geq J_{\theta_0,x,\kappa}, \quad x = s, b, r, \lambda, p. \tag{6.29}$$

When a metric satisfies (6.29), it is called a monotone metric. Since the derivative $\frac{d\rho_\theta}{d\theta}(\theta_0)$ plays an important role in the definition of the metric, $\frac{d\rho_\theta}{d\theta}(\theta_0)$ will be called the m **representation** of the derivative. We shall also define an operator $L_{\theta_0,x}$ by the relation

$$E_{\rho_{\theta_0},x}(L_{\theta_0,x}) = \frac{d\rho_\theta}{d\theta}(\theta_0). \tag{6.30}$$

Such an operator is called the e **representation** of the derivative. If all the density matrices ρ_θ commute each other, the e representation is the same as a logarithmic derivative. On the other hand, if some of the density matrices ρ_θ do not commute each other, their logarithmic derivatives depend on the metric. The matrices $L_{\theta_0,s}$ and $L_{\theta_0,r}$ defined by

$$\frac{d\rho_\theta}{d\theta}(\theta_0) = \frac{1}{2}(\rho_{\theta_0}L_{\theta_0,s} + L_{\theta_0,s}\rho_{\theta_0}), \quad \frac{d\rho_\theta}{d\theta}(\theta_0) = \rho_{\theta_0}L_{\theta_0,r} \tag{6.31}$$

are called the symmetric logarithmic derivative (SLD) and the right logarithmic derivative (RLD), respectively. These matrices coincide with the e representations of the derivative concerning the SLD Fisher metric and the RLD Fisher metric, which are abbreviated to SLD e representation and RLD e representation, respectively.

Since the equation

$$\int_0^1 \rho_{\theta_0}^{\lambda} \left.\frac{d \log \rho_\theta}{d\theta}\right|_{\theta=\theta_0} \rho_{\theta_0}^{1-\lambda} d\lambda = \left.\frac{d\rho_\theta}{d\theta}\right|_{\theta=\theta_0} \tag{6.32}$$

holds [8][Exe. 6.18], the *e* representation of the derivative of the Bogoljubov Fisher metric $L_{\theta_0,b}$ is then equal to $\frac{d \log \rho_\theta}{d\theta}(\theta_0)$. Since $\operatorname{Tr} \frac{d\rho_\theta}{d\theta} = 0$, the *e* representation $L_{\theta,x}$ satisfies $\operatorname{Tr} \rho L_{\theta,x} = \operatorname{Tr} E_\rho(L_{\theta,x}) = 0$.

Theorem 6.3 *For a quantum state family* $\{\rho_\theta | \theta \in \mathbb{R}\}$, *the following relations hold [9–12]:*

$$\frac{1}{8} J_{\theta,s} = \lim_{\epsilon\to 0} \frac{b^2(\rho_\theta, \rho_{\theta+\epsilon})}{\epsilon^2}, \tag{6.33}$$

$$\frac{1}{2} J_{\theta,b} = \lim_{\epsilon\to 0} \frac{D(\rho_{\theta+\epsilon} \| \rho_\theta)}{\epsilon^2}. \tag{6.34}$$

Hence, we obtain another proof of Theorem 6.1 (Theorem 6.2) for the SLD (Bogoljubov) case by combining Theorem 6.3 and (5.49) (5.36).

Proof Define U_ϵ such that it satisfies

$$b^2(\rho_\theta, \rho_{\theta+\epsilon}) = \frac{1}{2} \operatorname{Tr}(\sqrt{\rho_\theta} - \sqrt{\rho_{\theta+\epsilon}} U_\epsilon)(\sqrt{\rho_\theta} - \sqrt{\rho_{\theta+\epsilon}} U_\epsilon)^*.$$

This can be rewritten as

$$\begin{aligned} 2b^2(\rho_\theta, \rho_{\theta+\epsilon}) &= \operatorname{Tr}(W(0) - W(\epsilon))(W(0) - W(\epsilon))^* \\ &\cong \operatorname{Tr}\left(-\frac{dW}{d\epsilon}(0)\epsilon\right)\left(-\frac{dW}{d\epsilon}(0)\epsilon\right)^* \cong \operatorname{Tr} \frac{dW}{d\epsilon}(0) \frac{dW}{d\epsilon}(0)^* \epsilon^2, \end{aligned}$$

where we defined $W(\epsilon) \stackrel{\text{def}}{=} \sqrt{\rho_{\theta+\epsilon}} U_\epsilon$. As will be shown later, the SLD $L_{\theta,s}$ satisfies

$$\frac{dW}{d\epsilon}(0) = \frac{1}{2} L W(0). \tag{6.35}$$

Therefore, $b^2(\rho_\theta, \rho_{\theta+\epsilon}) \cong \operatorname{Tr} \frac{1}{8} L W(0) W(0)^* L \epsilon^2 = \frac{1}{8} \operatorname{Tr} L^2 \rho_\theta \epsilon$, and we obtain (6.33). Thus, showing (6.35) will complete the proof.

From the definition of the Bures distance, we have

$$\begin{aligned} 2b^2(\rho_\theta, \rho_{\theta+\epsilon}) = &\min_{U:\text{unitary}} \operatorname{Tr}(\sqrt{\rho_\theta} - \sqrt{\rho_{\theta+\epsilon}} U)(\sqrt{\rho_\theta} - \sqrt{\rho_{\theta+\epsilon}} U)^* \\ = &2 - \operatorname{Tr}\left(\sqrt{\rho_\theta}\sqrt{\rho_{\theta+\epsilon}} U(\epsilon)^* + U(\epsilon)\sqrt{\rho_{\theta+\epsilon}}\sqrt{\rho_\theta}\right). \end{aligned}$$

Therefore, $\sqrt{\rho_\theta}\sqrt{\rho_{\theta+\epsilon}} U(\epsilon)^* = U(\epsilon)\sqrt{\rho_{\theta+\epsilon}}\sqrt{\rho_\theta}$. Hence, $W(0)W(\epsilon)^* = W(\epsilon)W(0)^*$. Taking the derivative, we obtain $W(0)\frac{dW}{d\epsilon}(0)^* = \frac{dW}{d\epsilon}(0) W(0)^*$. This shows that there is a Hermitian matrix L satisfying $\frac{dW}{d\epsilon}(0) = \frac{1}{2}LW(0)$. Since $\rho_{\theta+\epsilon} = W(\epsilon)W(\epsilon)^*$, we have $\frac{d\rho}{d\theta}(\theta) = \frac{1}{2}(LW(0)W(0)^* + W(0)W(0)^*L)$. We therefore see that L is an SLD.

We now prove (6.34). Since $L_{\theta,b}$ is equal to $\frac{d\log\rho_\theta}{d\theta}(\theta)$, we have

$$\begin{aligned} D(\rho_{\theta+\epsilon}\|\rho_\theta) &= \mathrm{Tr}\,(\rho_{\theta+\epsilon}(\log\rho_{\theta+\epsilon} - \log\rho_\theta)) \\ &\cong \mathrm{Tr}\left(\rho_\theta + \frac{d\rho_\theta}{d\theta}\epsilon\right)\left(\frac{d\log\rho_\theta}{d\theta}\epsilon + \frac{1}{2}\frac{d^2\log\rho_\theta}{d\theta^2}\epsilon^2\right) \\ &= \mathrm{Tr}(\rho_\theta L_{\theta,b})\epsilon + \left(\mathrm{Tr}\left(\frac{d\rho_\theta}{d\theta}L_{\theta,b}\right) + \frac{1}{2}\mathrm{Tr}\left(\rho_\theta\frac{d^2\log\rho_\theta}{d\theta^2}\right)\right)\epsilon^2. \end{aligned} \tag{6.36}$$

The first term on the right-hand side (RHS) may be evaluated as

$$\mathrm{Tr}\,(\rho_\theta L_{\theta,b}) = \int_0^1 \mathrm{Tr}\,(\rho_\theta^t L_{\theta,b}\rho_\theta^{1-t})\,dt = \mathrm{Tr}\left(\frac{d\rho_\theta}{d\theta}\right) = 0. \tag{6.37}$$

Using this equation, we obtain

$$\begin{aligned} \mathrm{Tr}\left(\rho_\theta\frac{d^2\log\rho_\theta}{d\theta^2}\right) &= \frac{d}{d\theta}\left(\mathrm{Tr}\left(\rho_\theta\frac{d\log\rho_\theta}{d\theta}\right)\right) - \mathrm{Tr}\left(\frac{d\rho_\theta}{d\theta}\frac{d\log\rho_\theta}{d\theta}\right) \\ &= -\mathrm{Tr}\left(\frac{d\rho_\theta}{d\theta}L_{\theta,b}\right) = -J_{\theta,b}. \end{aligned} \tag{6.38}$$

Combining (6.36)–(6.38), we obtain $D(\rho_{\theta+\epsilon}\|\rho_\theta) \cong \frac{1}{2}J_{\theta,b}\epsilon^2$. ∎

Next, let us consider a quantum state family $\{\rho_\theta|\theta \in \mathbb{R}^d\}$ with more than one parameter. The derivative at the point $\theta_0 = (\theta_0^1, \ldots, \theta_0^d)$ may be obtained by considering the partial derivative $\frac{\partial}{\partial\theta^1}|_{\theta=\theta_0}, \ldots, \frac{\partial}{\partial\theta^d}|_{\theta=\theta_0}$ with respect to each parameter. Since each partial derivative represents the size and direction of an infinitesimal transport, it may be regarded as a vector. We then call the vector space comprising these vectors the **tangent vector space** at θ_0, and its elements **tangent vectors**. The tangent vector $\frac{\partial}{\partial\theta^j}|_{\theta=\theta_0}$ can be represented as a matrix $\frac{\partial\rho_\theta}{\partial\theta^j}(\theta_0)$. This kind of representation of a tangent vector will be called an *m* **representation**. The matrix $L_{\theta_0,j,x}$ satisfying $E_{\rho_{\theta_0},x}(L_{\theta_0,j,x}) = \frac{\partial\rho_\theta}{\partial\theta^j}(\theta_0)$ will be called an *e* **representation** of the SLD (Bogoljubov, RLD) Fisher metric of $\frac{\partial}{\partial\theta^j}|_{\theta=\theta_0}$. The matrix $\boldsymbol{J}_{\theta_0,x} = [J_{\theta_0,x;i,j}]_{i,j}$

$$J_{\theta_0,x;i,j} \stackrel{\text{def}}{=} \left\langle \frac{\partial\rho_\theta}{\partial\theta^i}(\theta_0), \frac{\partial\rho_\theta}{\partial\theta^j}(\theta_0) \right\rangle_{\rho_0,x}^{(m)} \tag{6.39}$$

is called the SLD (Bogoljubov, RLD) Fisher information matrix [2, 7, 8], where $x = s, b, r$ corresponds to SLD, Bogoljubov, RLD, respectively. Note that the tangent vector refers to an infinitesimal change with respect to θ_0 and is different from the matrix represented by the *m* representation or *e* representation. The *m* representation and the e representation are nothing more than matrix representations of the infinitesimal change.

In summary, in this section we have defined the metric from the inner product given in Sect. 6.1 and investigated the relationship of this metric to the quantum relative entropy $D(\rho\|\sigma)$ and the Bures distance $b(\rho, \sigma)$. We also defined three types of Fisher information matrices for state families with more than one parameter.

Exercises

6.14 Define $\tilde{\phi}_\theta \stackrel{\text{def}}{=} \frac{d\phi_\theta}{d\theta} - \langle\phi_\theta|\frac{d\phi_\theta}{d\theta}\rangle\phi_\theta$ with respect to the pure state family $\{\rho_\theta \stackrel{\text{def}}{=} |\phi_\theta\rangle\langle\phi_\theta|\}$. Show that the SLD Fisher information $J_{\theta,s}$ is equal to $4\langle\tilde{\phi}_\theta|\tilde{\phi}_\theta\rangle$. Show that both the RLD Fisher information and the Bogoljubov Fisher information diverge.

6.15 Let $J_\theta^{\boldsymbol{M}}$ be the Fisher information of the probability family $\{\mathrm{P}_{\rho_\theta}^{\boldsymbol{M}}|\theta \in \mathbb{R}\}$ (Sect. 1.2) for a one-parameter state family $\{\rho_\theta|\theta \in \mathbb{R}\}$ and a POVM $\boldsymbol{M} = \{M_i\}$. Show that

$$J_{\theta,x} \geq J_\theta^{\boldsymbol{M}} \text{ for } x = s, r, b, \lambda, p. \tag{6.40}$$

6.16 Show that $J_\theta^{\boldsymbol{M}} = \sum_i \frac{\langle M_i, L_{\theta,s}\rangle_{\rho_\theta,s}^{(e)}\langle L_{\theta,s}, M_i\rangle_{\rho_\theta,s}^{(e)}}{\langle M_i, I\rangle_{\rho_\theta,s}^{(e)}}$, with respect to the POVM $\boldsymbol{M} = \{M_i\}$.

6.17 Show the following facts with respect to the PVM $\boldsymbol{M} = \{M_i\}$ of rank $M_i = 1$.
(a) Using Exercise 6.12, show that $\kappa_{\boldsymbol{M},\rho,s}(X) = \sum_i \langle M_i, X\rangle_{\rho,s}^{(e)} M_i$.
(b) Show that $J_\theta^{\boldsymbol{M}} = \left(\|\kappa_{\boldsymbol{M},\rho_\theta,s}(L_{\theta,s})\|_{\rho_\theta,s}^{(e)}\right)^2$.
(c) Assume that $\rho_\theta > 0$. Show that $J_{\theta,s} = J_\theta^{\boldsymbol{M}}$ if and only if every M_i commutes $L_{\theta,s}$.

6.18 Prove (6.32) following the steps below.
(a) Show that $\int_0^1 \lambda^n(1-\lambda)^m \, d\lambda = \frac{n!m!}{(n+m)!}$.
(b) For a matrix-valued function $X(\theta)$, show that

$$\int_0^1 \exp(\lambda X(\theta))\frac{dX(\theta)}{d\theta}\exp((1-\lambda)X(\theta))\, d\lambda = \frac{d\exp(X(\theta))}{d\theta}.$$

This is nothing other than (6.32).

6.19 Consider the state family $\{\rho_\theta^{\otimes n}|\theta \in \mathbb{R}\}$ consisting of the n-fold tensor product state of the state ρ_θ. Show that the metric $J_{\theta,x,n}$ of this state family $\{\rho_\theta^{\otimes n}|\theta \in \mathbb{R}\}$ is equal to n times the metric $J_{\theta,x}$ of the state family $\{\rho_\theta|\theta \in \mathbb{R}\}$, i.e., $J_{\theta,x,n} = nJ_{\theta,x}$ for $x = s, r, b, \lambda, p$.

6.20 Show that the Fisher information matrix $\boldsymbol{J}_{\theta,x}$ is Hermitian.

6.21 Show that the Fisher information matrix $\boldsymbol{J}_{\theta,x}$ is real symmetric for $x = s, \frac{1}{2}, b$.

6.22 Give an example of an RLD Fisher information matrix $J_{\theta,r}$ that is not real symmetric.

6.23 For a Hermitian matrix Y and a quantum state family $\{\rho_\theta = e^{-i\theta Y}\rho e^{i\theta Y}|\theta \in \mathbb{R}\}$, show that the derivative at $\theta = 0$ has the e representation $i[\log\rho, Y]$ with respect to the Bogoljubov metric.

6.24 Show that $i[\rho, Y] = E_{\rho,b}(i[\log\rho, Y])$ if Y is Hermitian.

6.25 Define the state family $S = \left\{\rho_\theta = \frac{1}{2}\left(I + \sum_{i=1}^3 \theta_i S^i\right)\middle| \|\theta\| \le 1\right\}$ on the two-dimensional system $\mathcal{H} = \mathbb{C}^2$. Show that the three Fisher information matrices $J_{\theta,s}$, $J_{\theta,b}$, $J_{\theta,r}$ can be written as

$$J_{\theta,s}^{-1} = I - |\theta\rangle\langle\theta|, \tag{6.41}$$

$$J_{\theta,b} = \frac{1}{1-\|\theta\|^2}|\theta\rangle\langle\theta| + \frac{1}{2\|\theta\|}\log\frac{1+\|\theta\|}{1-\|\theta\|}\left(I - \frac{1}{\|\theta\|^2}|\theta\rangle\langle\theta|\right), \tag{6.42}$$

$$J_{\theta,r}^{-1} = I - |\theta\rangle\langle\theta| + iR_\theta, \tag{6.43}$$

where R is defined in Sect. 5.3, following the steps below.
(**a**) Show the following for $\theta = (0, 0, \theta)$, and check (6.41)–(6.43) in this case using them.

$$L_{\theta,s,1} = S_1,\ L_{\theta,s,2} = S_2,\ L_{\theta,r,1} = \rho_\theta^{-1}S_1,\ L_{\theta,r,2} = \rho_\theta^{-1}S_2,$$

$$L_{\theta,b,1} = \frac{1}{2\theta}\log\frac{1+\theta}{1-\theta}S_1,\ L_{\theta,b,2} = \frac{1}{2\theta}\log\frac{1+\theta}{1-\theta}S_2,$$

$$L_{\theta,s,3} = L_{\theta,b,3} = L_{\theta,r,3} = \begin{pmatrix} \frac{1}{1+\theta} & 0 \\ 0 & -\frac{1}{1-\theta} \end{pmatrix},$$

$$J_{\theta,s} = \begin{pmatrix} 1 & 0 & 0 \\ 0 & 1 & 0 \\ 0 & 0 & \frac{1}{1-\theta^2} \end{pmatrix},\quad J_{\theta,r} = \begin{pmatrix} \frac{1}{1-\theta^2} & i\frac{\theta}{1-\theta^2} & 0 \\ -i\frac{\theta}{1-\theta^2} & \frac{1}{1-\theta^2} & 0 \\ 0 & 0 & \frac{1}{1-\theta^2} \end{pmatrix},$$

$$J_{\theta,b} = \begin{pmatrix} \frac{1}{2\theta}\log\frac{1+\theta}{1-\theta} & 0 & 0 \\ 0 & \frac{1}{2\theta}\log\frac{1+\theta}{1-\theta} & 0 \\ 0 & 0 & \frac{1}{1-\theta^2} \end{pmatrix}.$$

(**b**) Show that $O^T J_{O\theta} O = J_\theta$, where O is an orthogonal matrix.
(**c**) Show (6.41)–(6.43) for an arbitrary θ.

6.26 Let $J_{\theta,x}^1$ and $J_{\theta,x}^2$ be the Fisher metric of two state families $\{\rho_\theta^1|\theta \in \mathbb{R}\}$ and $\{\rho_\theta^2|\theta \in \mathbb{R}\}$ for $x = s, b, r, \lambda, p$, respectively. Show that the Fisher metric $J_{\theta,x}$ of the state family $\{\lambda\rho_\theta^1 + (1-\lambda)\rho_\theta^2|\theta \in \mathbb{R}\}$ satisfies $J_{\theta,x} \le \lambda J_{\theta,x}^1 + (1-\lambda)J_{\theta,x}^2$. Show that its equality holds when the space spanned by the supports of $\frac{d\rho_\theta^1}{d\theta}$ and ρ_θ^1 are orthogonal to those of $\frac{d\rho_\theta^2}{d\theta}$ and ρ_θ^2.

6.3 Geodesics and Divergences

In the previous section, we examined the inner product in the space of the quantum state. In this section, we will examine more advanced geometrical structures such as parallel transports, exponential family, and divergence. To introduce the concept of a parallel transport, consider an infinitesimal displacement in a one-parameter quantum state family $\{\rho_\theta|\theta \in \mathbb{R}\}$. The difference between $\rho_{\theta+\epsilon}$ and ρ_θ is approximately equal to $\frac{d\rho_\theta}{d\theta}(\theta)\epsilon$. Hence, the state $\rho_{\theta+\epsilon}$ can be regarded as the state transported from the state ρ_θ in the direction $\frac{d\rho_\theta}{d\theta}(\theta)$ by an amount ϵ. However, if the state $\rho_{\theta+\epsilon}$ coincides precisely with the state displaced from the state ρ_θ by ϵ in the direction of $\frac{d\rho_\theta}{d\theta}(\theta)$, the infinitesimal displacement at the intermediate states $\rho_{\theta+\epsilon'}$ $(0 < \epsilon' < \epsilon)$ must be equal to the infinitesimal displacement $\frac{d\rho_\theta}{d\theta}(\theta)\Delta$ at θ. In such a case, the problem is to ascertain which infinitesimal displacement at the point $\theta + \epsilon'$ corresponds to the given infinitesimal displacement $\frac{d\rho_\theta}{d\theta}(\theta)\Delta$ at the initial point θ. The rule for matching the infinitesimal displacement at one point to the infinitesimal displacement at another point is called parallel transport. The coefficient $\frac{d\rho_\theta}{d\theta}(\theta)$ of the infinitesimal displacement at θ is called the tangent vector, as it represents the slope of the tangent line of the state family $\{\rho_\theta|\theta \in \mathbb{R}\}$ at θ. We may therefore consider the parallel transport of a tangent vector instead of the parallel transport of an infinitesimal displacement.

Commonly used parallel transports can be classified into those based on the m representation (m **parallel translation**) and those based on the e representation (e **parallel translation**). The m parallel translation $\Pi^{(m)}_{\rho_\theta,\rho_{\theta'}}$ moves the tangent vector at one point ρ_θ to the tangent vector at another point $\rho_{\theta'}$ with the same m representation. On the other hand, the e parallel translation $\Pi^{(e)}_{x,\rho_\theta,\rho_{\theta'}}$ moves the tangent vector at one point ρ_θ with the e representation L to the tangent vector at another point $\rho_{\theta'}$ with the e representation $L - \operatorname{Tr}\rho_{\theta'} L$ [8]. Of course, this definition requires the agreement between the set of e representations at the point θ and that at another point θ'. Hence, this type of e parallel translation is defined only for the symmetric inner product $\langle X, Y\rangle^{(e)}_{\rho,x}$, and its definition depends on the choice of the metric. Indeed, the e parallel translation can be regarded as the dual parallel translation of the m parallel translation concerning the metric $\langle X, Y\rangle^{(e)}_{\rho,x}$ in the following sense:

$$\operatorname{Tr} X^* \Pi^{(m)}_{\rho_\theta,\rho_{\theta'}}(A) = \operatorname{Tr} \Pi^{(e)}_{x,\rho_{\theta'},\rho_\theta}(X)^* A,$$

where X is the e representation of a tangent vector at $\rho_{\theta'}$ and A is the m representation of another tangent vector at ρ_θ.

Further, a one-parameter quantum state family is called a **geodesic** or an **autoparallel curve** when the tangent vector (i.e., the derivative) at each point is given as a parallel transport of a tangent vector at a fixed point. In particular, the e geodesic is called a **one-parameter exponential family**. For example, in an e geodesic with respect to SLD $\{\rho_\theta|\theta \in \mathbb{R}\}$, any state ρ_θ coincides with the state transported from the state ρ_0 along the autoparallel curve in the direction L by an amount θ, where L denotes the SLD e representation of the derivative at ρ_0. We shall henceforth denote the state as $\Pi^{\theta}_{L,s}\rho_0$. Similarly, $\Pi^{\theta}_{L,b}\rho_0$ denotes the state transported autoparallely with respect to the Bogoljubov e representation from ρ_0 in the direction L by an amount θ.

When the given metric is not symmetric, the definition of the e parallel translation is more complicated. The e parallel translation moves the tangent vector with the e representation $\tilde{L}$ at one point θ to the tangent vector with the e representation $\tilde{L}' - \mathrm{Tr}\,\rho_{\theta'}\tilde{L}'$ at another point θ' so that the condition $\tilde{L}+\tilde{L}^* = \tilde{L}'+(\tilde{L}')^*$ holds. That is, we require the same Hermitian part for the e representation at the different points. Hence, the e parallel translation $\Pi^{(e)}_{x,\rho_\theta,\rho_{\theta'}}$ coincides with the e parallel translation $\Pi^{(e)}_{s(x),\rho_\theta,\rho_{\theta'}}$ with regard to its symmetrized inner product $s(x)$.

Therefore, we can define the state transported from the state ρ_0 along the autoparallel curve in the direction with the Hermitian part L by an amount θ with respect to RLD (λ, p), and we denote them by $\Pi^{\theta}_{L,r}\rho_0$ ($\Pi^{\theta}_{L,\lambda}\rho_0$, $\Pi^{\theta}_{L,p}\rho_0$). However, only the SLD one-parameter exponential family $\{\Pi^{\theta}_{L,s}\rho_0 | s \in \mathbb{R}\}$ plays an important role in quantum estimation examined in the next section. Notice that since the symmetrized inner product $s(r)$ of the RLD r is not the SLD s, $\Pi^{\theta}_{L,r}\rho_0$ is not the same as $\Pi^{\theta}_{L,s}\rho_0$.

Lemma 6.1 *$\Pi^{\theta}_{L,s}\sigma$, $\Pi^{\theta}_{L,b}\sigma$, $\Pi^{\theta}_{L,r}\sigma$, and $\Pi^{\theta}_{L,\frac{1}{2}}\sigma$ can be written in the following form [8, 13, 14]:*

$$\Pi^{\theta}_{L,s}\sigma = e^{-\mu_s(\theta)} e^{\frac{\theta}{2}L} \sigma e^{\frac{\theta}{2}L}, \tag{6.44}$$

$$\Pi^{\theta}_{L,b}\sigma = e^{-\mu_b(\theta)} e^{\log\sigma + \theta L}, \tag{6.45}$$

$$\Pi^{\theta}_{L,r}\sigma = e^{-\mu_r(\theta)} \sqrt{\sigma} e^{\theta L_r} \sqrt{\sigma}, \tag{6.46}$$

$$\Pi^{\theta}_{L,\frac{1}{2}}\sigma = e^{-\mu_{\frac{1}{2}}(\theta)} \sigma^{\frac{1}{4}} e^{\frac{\theta}{2}L_{\frac{1}{2}}} \sigma^{\frac{1}{2}} e^{\frac{\theta}{2}L_{\frac{1}{2}}} \sigma^{\frac{1}{4}}, \tag{6.47}$$

where we choose Hermitian matrices L_r and $L_{\frac{1}{2}}$ as $L = \frac{1}{2}(\sigma^{-\frac{1}{2}} L_r \sigma^{\frac{1}{2}} + \sigma^{\frac{1}{2}} L_r \sigma^{-\frac{1}{2}})$ and $L = \frac{1}{2}(\sigma^{-\frac{1}{4}} L_{\frac{1}{2}} \sigma^{\frac{1}{4}} + \sigma^{\frac{1}{4}} L_{\frac{1}{2}} \sigma^{-\frac{1}{4}})$, respectively, and

$$\mu_s(\theta) \stackrel{\text{def}}{=} \log \mathrm{Tr}\, e^{\frac{\theta}{2}L} \sigma e^{\frac{\theta}{2}L}, \qquad \mu_b(\theta) \stackrel{\text{def}}{=} \log \mathrm{Tr}\, e^{\log\sigma + \theta L}, \tag{6.48}$$

$$\mu_r(\theta) \stackrel{\text{def}}{=} \log \mathrm{Tr} \sqrt{\sigma} e^{\theta L_r} \sqrt{\sigma}, \qquad \mu_{1/2}(\theta) \stackrel{\text{def}}{=} \log \mathrm{Tr}\, \sigma^{\frac{1}{4}} e^{\frac{\theta}{2}L_{\frac{1}{2}}} \sigma^{\frac{1}{2}} e^{\frac{\theta}{2}L_{\frac{1}{2}}} \sigma^{\frac{1}{4}}.$$

Proof When $x = s, b$, or $\frac{1}{2}$, the map $E_{\rho,x}$ is symmetric. Hence, the definition of $\Pi^{\theta}_{L,x}\sigma$ implies that

$$\frac{d\Pi^{\theta}_{L,x}\sigma}{d\theta} = E_{\rho_\theta,x}(L - \mathrm{Tr}\, L\rho_\theta), \quad x = s, b, \frac{1}{2}. \tag{6.49}$$

Since the equation (6.49) is actually an ordinary differential equation, the uniqueness of the solution of the ordinary differential equation (6.49) guarantees that the only $\Pi^{\theta}_{L,x}\sigma$ satisfying $\Pi^{0}_{L,x}\sigma = \sigma$ is the solution of the above differential equation. Taking the derivative of the RHS of (6.44), (6.45), and (6.47), we see that the RHS satisfies (6.49) for $x = s, b, \frac{1}{2}$, respectively. So, we obtain (6.44), (6.45), and (6.47).

Since the RLD inner product is not symmetric, we need to more careful treatment for the case of $x = r$. In a one-parameter exponential family $\rho_\theta = \Pi^\theta_{L,r}\sigma$ for the RLD metric, any the RLD e representation at ρ_θ is written as $\hat{L}$, which is not necessarily Hermitian. Since $E_{\sigma,r}(\hat{L})$ is Hermitian, $\sigma\hat{L}$ is Hermitian. So, $\hat{L}$ is written as $\sqrt{\sigma}^{-1}\tilde{L}\sqrt{\sigma}$ with a Hermitian matrix $\tilde{L}$. As the RLD e representation at σ has the form $\sqrt{\sigma}^{-1}\tilde{L}\sqrt{\sigma}$ with a Hermitian matrix $\tilde{L}$, we only discuss $\rho_\theta = \Pi^\theta_{\sqrt{\sigma}^{-1}\tilde{L}\sqrt{\sigma},r}\sigma$. By taking its derivative, we have

$$\frac{\Pi^\theta_{\sqrt{\sigma}^{-1}\tilde{L}\sqrt{\sigma},r}\sigma}{d\theta} = \rho_\theta(\sqrt{\sigma}^{-1}\tilde{L}\sqrt{\sigma} - \operatorname{Tr}\rho_\theta\sqrt{\sigma}^{-1}\tilde{L}\sqrt{\sigma}). \tag{6.50}$$

On the RHS of (6.46), the RLD e representation of the derivative at each point is equal to the parallel transported e representation of the derivative $\sqrt{\sigma}^{-1}L_r\sqrt{\sigma}$ at σ. So, we find that only the state family (6.46) satisfies this condition. Similarly, the uniqueness of the solution of the ordinary differential equation (6.50) guarantees that only the state family (6.46) satisfies this condition. So, we obtain (6.46). ■

Now, using the concept of the exponential family, we extend the divergence based on the first equation in (2.129). For any two states ρ and σ, we choose the Hermitian matrix L such that the exponential family $\{\Pi^\theta_{L,x}\sigma\}_{\theta\in[0,1]}$ with regard to the inner product $J_{\theta,x}$ satisfies

$$\Pi^1_{L,x}\sigma = \rho. \tag{6.51}$$

Then, we define the x-e-divergences as follows:

$$D^{(e)}_x(\rho\|\sigma) = \int_0^1 J_{\theta,x}\theta d\theta, \tag{6.52}$$

where $J_{\theta,x}$ is the Fisher information for the exponential family $\Pi^\theta_{L,x}\sigma$. Since we can show that

$$\frac{d^2\mu_x(\theta)}{d\theta^2} = J_{\theta,x} \tag{6.53}$$

(For $x = b$, see (6.32).), $D^{(e)}_x(\rho\|\sigma)$ can be regarded as the Bregman divergence of $\mu_x(\theta)$, i.e.,

$$D^{(e)}_x(\rho\|\sigma) = D^{\mu_x}(0\|1). \tag{6.54}$$

Since $\Pi^\theta_{L^1\otimes I+I\otimes L^2,x}(\sigma_1\otimes\sigma_2)$ equals $(\Pi^\theta_{L^1,x}\sigma_1)\otimes(\Pi^\theta_{L^2,x}\sigma_2)$,

$$D^{(e)}_x(\rho_1\otimes\rho_2\|\sigma_1\otimes\sigma_2) = D^{(e)}_x(\rho_1\|\sigma_1) + D^{(e)}_x(\rho_2\|\sigma_2), \tag{6.55}$$

i.e., the e-divergence satisfies the **additivity** for any inner product.

Theorem 6.4 *When*

$$L = \begin{cases} 2\log \sigma^{-\frac{1}{2}}(\sigma^{\frac{1}{2}}\rho\sigma^{\frac{1}{2}})^{\frac{1}{2}}\sigma^{-\frac{1}{2}} & \text{for } x = s, \\ \log\rho - \log\sigma & \text{for } x = b, \\ \frac{1}{2}(\sigma^{-\frac{1}{2}}\log(\sigma^{-\frac{1}{2}}\rho\sigma^{-\frac{1}{2}})\sigma^{\frac{1}{2}} + \sigma^{\frac{1}{2}}\log(\sigma^{-\frac{1}{2}}\rho\sigma^{-\frac{1}{2}})\sigma^{-\frac{1}{2}}) & \text{for } x = r, \\ \sigma^{-\frac{1}{4}}\log(\sigma^{-\frac{1}{4}}\rho^{\frac{1}{2}}\sigma^{-\frac{1}{4}})\sigma^{\frac{1}{4}} + \sigma^{\frac{1}{4}}\log(\sigma^{-\frac{1}{4}}\rho^{\frac{1}{2}}\sigma^{-\frac{1}{4}})\sigma^{-\frac{1}{4}} & \text{for } x = \frac{1}{2}, \end{cases} \tag{6.56}$$

condition (6.51) holds. Hence, we obtain

$$D_s^{(e)}(\rho\|\sigma) = 2\operatorname{Tr}\rho\log\sigma^{-\frac{1}{2}}(\sigma^{\frac{1}{2}}\rho\sigma^{\frac{1}{2}})^{\frac{1}{2}}\sigma^{-\frac{1}{2}}, \tag{6.57}$$

$$D_b^{(e)}(\rho\|\sigma) = \operatorname{Tr}\rho(\log\rho - \log\sigma) = D(\rho\|\sigma), \tag{6.58}$$

$$D_r^{(e)}(\rho\|\sigma) = \operatorname{Tr}\rho\log(\rho^{\frac{1}{2}}\sigma^{-1}\rho^{\frac{1}{2}}), \tag{6.59}$$

$$D_{\frac{1}{2}}^{(e)}(\rho\|\sigma) = 2\operatorname{Tr}(\sigma^{\frac{1}{4}}\rho^{\frac{1}{2}}\sigma^{\frac{1}{4}})(\sigma^{-\frac{1}{4}}\rho^{\frac{1}{2}}\sigma^{-\frac{1}{4}})\log(\sigma^{-\frac{1}{4}}\rho^{\frac{1}{2}}\sigma^{-\frac{1}{4}}). \tag{6.60}$$

Proof When we substitute (6.56) into L, condition (6.51) can be checked by using Lemma 6.1. In this case, $L_r = \log(\sigma^{-\frac{1}{2}}\rho\sigma^{-\frac{1}{2}})$, $L_{\frac{1}{2}} = 2\log(\sigma^{-\frac{1}{4}}\rho^{\frac{1}{2}}\sigma^{-\frac{1}{4}})$. From (6.54), we can prove that

$$D_x^{(e)}(\rho\|\sigma) = \left.\frac{d\mu_x(\theta)}{d\theta}\right|_{\theta=1}(1-0) - \mu_x(1) + \mu_x(0) = \left.\frac{d\mu_x(\theta)}{d\theta}\right|_{\theta=1}, \tag{6.61}$$

where $\mu_x(\theta)$ is defined in Lemma 6.1. Using this relation, we can check (6.57), (6.58), and (6.60). Concerning (6.59), we obtain

$$D_r^{(e)}(\rho\|\sigma) = \operatorname{Tr}\sigma\sigma^{-\frac{1}{2}}\rho\sigma^{-\frac{1}{2}}\log(\sigma^{-\frac{1}{2}}\rho\sigma^{-\frac{1}{2}}) = \operatorname{Tr}\rho\log(\rho^{\frac{1}{2}}\sigma^{-1}\rho^{\frac{1}{2}}),$$

where the last equation follows from Exercise A.2. ■

Now we compare these quantum analogs of relative entropy given in (6.57)–(6.60). As is easily checked, these satisfy condition (3.104) for quantum analogs of relative entropy. Also, Inequality (5.47) shows that

$$D(\rho\|\sigma) \geq D_c(\rho\|\sigma) \geq 2\operatorname{Tr}\rho\log\sigma^{-\frac{1}{2}}(\sigma^{\frac{1}{2}}\rho\sigma^{\frac{1}{2}})^{\frac{1}{2}}\sigma^{-\frac{1}{2}} = D_s^{(e)}(\rho\|\sigma). \tag{6.62}$$

Alternative proof of the above relations is available in Exercise 6.30. Hence, from inequality (3.107) and additivity (6.55), $D_s^{(e)}(\rho\|\sigma)$ do not satisfy the monotonicity even for measurements because the equality of the first inequality in (6.62) does not always hold.

Further, we can extend the divergence based on equation (2.136). For any two states ρ and σ, the family $\{(1-t)\rho + t\sigma | 0 \leq t \leq 1\}$ is the m geodesic joining ρ and σ. Hence, as an extension of (2.136), we can define the x-m divergences as

$$D_x^{(m)}(\rho\|\sigma) \stackrel{\text{def}}{=} \int_0^1 J_{t,x} t dt\,. \tag{6.63}$$

Since the family $\{(1-t)\kappa(\rho)+t\kappa(\sigma)|0 \le t \le 1\}$ is the m geodesic joining $\kappa(\rho)$ and $\kappa(\sigma)$ for any TP-CP map κ, we have

$$D_x^{(m)}(\rho\|\sigma) \ge D_x^{(m)}(\kappa(\rho)\|\kappa(\sigma)), \tag{6.64}$$

i.e., the m divergence satisfies the **monotonicity**. Since the RLD is the largest inner product,

$$D_r^{(m)}(\rho\|\sigma) \ge D_x^{(m)}(\rho\|\sigma). \tag{6.65}$$

We can calculate the m divergences as[Exe. 6.29]

$$D_b^{(m)}(\rho\|\sigma) = \mathrm{Tr}\,\rho(\log\rho - \log\sigma) = D(\rho\|\sigma), \tag{6.66}$$

$$D_r^{(m)}(\rho\|\sigma) = \mathrm{Tr}\,\rho\log(\sqrt{\rho}\sigma^{-1}\sqrt{\rho}). \tag{6.67}$$

The Bogoljubov case follows from Theorem 6.5. Hence, from (6.64), $\mathrm{Tr}\,\rho\log(\sqrt{\rho}\sigma^{-1}\sqrt{\rho}) = D_r^{(m)}(\rho\|\sigma)$ satisfies the monotonicity for TP-CP maps. Further, from (6.65) we obtain $\mathrm{Tr}\,\rho\log(\sqrt{\rho}\sigma^{-1}\sqrt{\rho}) \ge D(\rho\|\sigma)$ [15].

Not all x-m divergences necessarily satisfy additivity (6.55). This fact can be shown as follows. Choose an inner product $J_{x,\theta}$ different from the Bogoljubov inner product $J_{b,\theta}$ such that $J_{\theta,x} \le J_{\theta,b}$. So, we have $D(\rho\|\sigma) \ge D_x^{(m)}(\rho\|\sigma)$. Further, since inner product $J_{x,\theta}$ different from the Bogoljubov inner product $J_{b,\theta}$, there exists a pair of states ρ and σ such that $D(\rho\|\sigma) > D_x^{(m)}(\rho\|\sigma)$. From (3.107) and monotonicity (6.64), $D_x^{(m)}(\rho\|\sigma)$ does not satisfy additivity (6.55). For example, since SLD Fisher information satisfies the above conditions, the SLD m divergence does not satisfy additivity (6.55).

Now, we consider whether it is possible in two-parameter-state families to have states that are e autoparallel transported in the direction of L_1 by θ^1 and in the direction L_2 by θ^2. In order to define such a state, we require that the following two states coincide with each other. (1) the state that be e autoparallel transported first in the L_1 direction by θ^1 from ρ_0, then further e autoparallel transported in the L_2 direction by θ^2, and (2) the state that is e autoparallel transported in the L_2 direction by θ^2 from ρ_0, then e autoparallel transported in the L_1 direction by θ^1. That is, if such a state were defined, the relation

$$\Pi_{L_2,x}^{\theta^2}\Pi_{L_1,x}^{\theta^1}\sigma = \Pi_{L_1,x}^{\theta^1}\Pi_{L_2,x}^{\theta^2}\sigma \tag{6.68}$$

should hold. Otherwise, the torsion $T(L_1, L_2)_x$ is defined as follows (Fig. 6.1):

$$T(L_1, L_2)_{\rho,x} \stackrel{\mathrm{def}}{=} \lim_{\epsilon\to 0}\frac{\Pi_{L_2,x}^{\epsilon}\Pi_{L_1,x}^{\epsilon}\rho - \Pi_{L_1,x}^{\epsilon}\Pi_{L_2,x}^{\epsilon}\rho}{\epsilon^2}.$$

Concerning condition (6.68), we have the following theorem.

Theorem 6.5 (Amari and Nagaoka [8]) *The following conditions for the inner product $J_{\theta,x}$ are equivalent.*

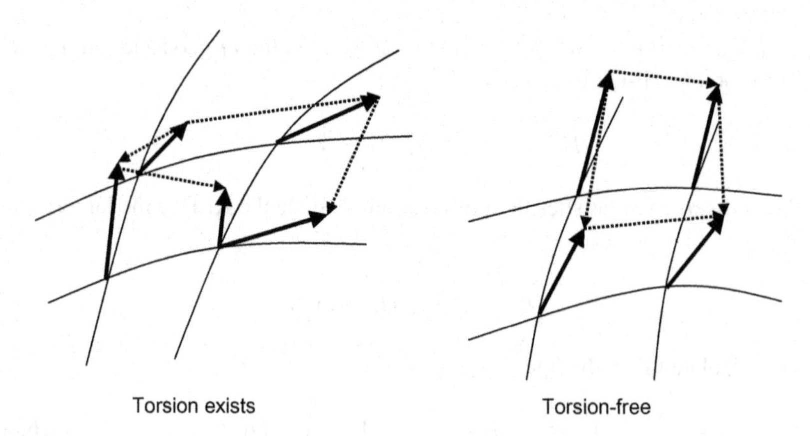

Fig. 6.1 Torsion

① *$J_{\theta,x}$ is the Bogoljubov inner product, i.e., $x = b$.*
② *Condition (6.68) holds for any two Hermitian matrices L_1 and L_2 and any state ρ_0.*
③ $D_x^{(e)}(\rho_{\bar{\theta}}\|\rho_\theta) = D^\mu(\bar{\theta}\|\theta)$.
④ $D_x^{(e)}(\rho\|\sigma) = D(\rho\|\sigma)$.
⑤ $D_x^{(m)}(\rho_{\bar{\eta}}\|\rho_\eta) = D^\nu(\eta\|\bar{\eta})$.
⑥ $D_x^{(m)}(\rho\|\sigma) = D(\rho\|\sigma)$.

Here, the convex functions $\mu(\theta)$ and $\nu(\eta)$ and the states ρ_θ and ρ_η are defined by

$$\rho_\theta \stackrel{\text{def}}{=} \exp\left(\sum_i \theta^i X_i - \mu(\theta)\right), \qquad \mu(\theta) \stackrel{\text{def}}{=} \log \operatorname{Tr} \exp\left(\sum_i \theta^i X_i\right), \tag{6.69}$$

$$\rho_\eta \stackrel{\text{def}}{=} \rho_{\text{mix}} + \sum_j \eta_j Y^j, \qquad \nu(\eta) \stackrel{\text{def}}{=} D_x^{(m)}(\rho_0\|\rho_\eta) = -H(\rho_\eta) + H(\rho_{\text{mix}}),$$

where $X_1, \ldots, X_k$ is a basis of the set of traceless Hermitian matrices, and $Y^1, \ldots, Y^k$ is its dual basis.

Proof First, we prove ①⇒②. Theorem 6.1 guarantees that the Bogoljubov e autoparallel transport satisfies

$$\Pi^{\theta^2}_{L_2,b}\Pi^{\theta^1}_{L_1,b}\rho = \Pi^{\theta^1}_{L_1,b}\Pi^{\theta^2}_{L_2,b}\rho = e^{-\mu_b(\theta^1,\theta^2)} e^{\log \rho + \theta^1 L_1 + \theta^2 L_2},$$

where $\mu_b(\theta) \stackrel{\text{def}}{=} \log \operatorname{Tr} e^{\log \rho + \theta^1 L_1 + \theta^2 L_2}$. Hence, we obtain ②.

Next, we prove that ②⇒③. We define $\tilde{\rho}_\theta \stackrel{\text{def}}{=} \Pi^{\theta^k}_{X_k,x}, \cdots, \Pi^{\theta^1}_{X_1,b}\rho_{\text{mix}}$ for $\theta = (\theta^1, \ldots, \theta^k)$. Then, condition ② guarantees that $\tilde{\rho}_{\bar{\theta}} = \Pi^1_{\sum_i (\bar{\theta}^i - \theta^i)X_i,x}\tilde{\rho}_\theta$. In particular, when $\theta = 0$, we obtain $\tilde{\rho}_{\bar{\theta}} = \Pi^1_{\sum_i \bar{\theta}^i X_i,x}\rho_{\text{mix}}$. Since $\sum_i \bar{\theta}^i X_i$ is commutative with

ρ_{mix}, we can apply the classical observation to this case. Hence, state $\tilde{\rho}_{\bar{\theta}}$ coincides with state $\rho_{\bar{\theta}}$ defined in (6.69).

Let $\tilde{X}_{j,\theta}$ be the x-e representation of the partial derivative concerning θ^j at ρ_θ. It can be expressed by using a skew-Hermitian matrix $\bar{X}_{j,\theta}$ as

$$\tilde{X}_{j,\theta} = X_j - \mathrm{Tr}\, \rho_\theta X_j + \bar{X}_{j,\theta}.$$

Thus,

$$\frac{\partial\, \mathrm{Tr}\, \rho_\theta X_j}{\partial \theta^i} = \mathrm{Tr} \left(\frac{\partial \rho_\theta}{\partial \theta^i} X_j \right) = \mathrm{Tr} \left(\frac{\partial \rho_\theta}{\partial \theta^i} (X_j - \mathrm{Tr}\, \rho_\theta X_j) \right)$$
$$= \mathbf{Re}\, \mathrm{Tr} \left(\frac{\partial \rho_\theta}{\partial \theta^i} (X_j - \mathrm{Tr}\, \rho_\theta X_j + \bar{X}_{j,\theta}) \right) = \mathbf{Re}\, J_{\theta,x;i,j}.$$

Note that the trace of the product of a Hermitian matrix and a skew-Hermitian matrix is an imaginary number. Since $\mathbf{Re}\, J_{\theta,x;i,j} = \mathbf{Re}\, J_{\theta,x;j,i}$, we have $\frac{\partial\, \mathrm{Tr}\, \rho_\theta X_j}{\partial \theta^i} = \frac{\partial\, \mathrm{Tr}\, \rho_\theta X_i}{\partial \theta^j}$. Thus, there exists a function $\bar{\mu}(\theta)$ such that $\bar{\mu}(0) = \mu(0)$ and

$$\frac{\partial \bar{\mu}(\theta)}{\partial \theta^i} = \mathrm{Tr}\, \rho_\theta X_i.$$

This function $\bar{\mu}$ satisfies the condition ③.

Moreover, since $\mathrm{Tr}\, \rho_{\mathrm{mix}} X_i = 0$, from definition (2.116), we have $\bar{\mu}(\theta) - \bar{\mu}(0) = D^{\bar{\mu}}(0\|\theta)$. Since the completely mixed state ρ_{mix} commutes with the state ρ_θ, the relation $D^{(e)}(\rho_{\mathrm{mix}}\|\rho_\theta) = \mu(\theta) - \mu(0)$ holds. Hence, we obtain $\bar{\mu}(\theta) = \mu(\theta)$.

Further, we have $D^\mu(\bar{\theta}\|\theta) = D(\rho\|\theta)$. Thus, the equivalence between ③ and ④ is trivial since the limit of $D(\rho_{\bar{\theta}}\|\rho_\theta)$ equals the Bogoljubov inner product $J_{b,\theta}$. Hence, we obtain ④⇒①.

Now we proceed to the proof of ①+②+③+④⇒⑤. In this case, the function $\nu(\eta)$ coincides with the Legendre transform of $\mu(\theta)$, and $\eta_i = \frac{\partial \mu}{\partial \theta^i}(\theta)$. Hence, $D^\nu(\eta\|\bar{\eta}) = D^\mu(\bar{\theta}\|\theta) = D(\rho_{\bar{\eta}}\|\rho_\eta)$. The second derivative matrix $\frac{\partial^2 \nu}{\partial \eta^i \partial \eta^j}$ coincides with the inverse of the second derivative matrix $\frac{\partial^2 \mu}{\partial \theta^i \partial \theta^j}$, which equals the Bogoljubov Fisher information matrix concerning the parameter θ. Since the Bogoljubov Fisher information matrix concerning the parameter η equals the inverse of the Bogoljubov Fisher information matrix concerning the parameter θ, the Bogoljubov Fisher information matrix concerning the parameter η coincides with the second derivative matrix $\frac{\partial^2 \nu}{\partial \eta^i \partial \eta^j}$. Hence, the relation (2.118) guarantees that $D^\nu(\eta\|\bar{\eta}) = D_b^{(m)}(\rho_{\bar{\theta}}\|\rho_\theta)$.

Next, we prove ⑤⇒⑥. Since $\rho_{\mathrm{mix}} = \rho_0$ commutes with ρ_η, the m divergence $D_x^{(m)}(\rho_0\|\rho_\eta)$ coincides with the Bogoljubov m divergence $D_b^{(m)}(\rho_0\|\rho_\eta)$, which equals the Legendre transform of $\mu(\theta)$ defined in (6.69). Thus, $D_x^{(m)}(\rho_{\bar{\eta}}\|\rho_\eta) = D^\nu(\eta\|\bar{\eta}) = D(\rho_{\bar{\eta}}\|\rho_\eta)$. Finally, taking the limit $\bar{\eta} \to \eta$, we obtain $J_{x,\eta} = J_{b,\eta}$, i.e., ⑥⇒①. ∎

This theorem shows that the Bogoljubov inner product is most natural from a geometrical point of view. However, from the viewpoint of estimation theory, the Bogoljubov metric is rather inconvenient, as will be shown in the next section.

In summary, this section has examined several geometrical structures that may be derived from the inner product. In the next section, we will discuss the connection between these structures and estimation theory.

Exercises

6.27 Define the SLD $L := S_1$ and the state $\rho_0 := 1/2(I + \sum_{i=1}^{3} x_i S^i)$ in the two-dimensional system $\mathcal{H} = \mathbb{C}^2$. Show that the SLD e geodesic $\Pi_{L,s}^t \rho_0$ is given by

$$\Pi_{L,s}^t \rho_0 = \frac{1}{2}\left(I + \sum_{i=1}^{3} x^i(t) S_i\right), \quad x_1(t) = \frac{e^t(1+x_1) - e^{-t}(1-x_1)}{e^t(1+x_1) + e^{-t}(1-x_1)},$$

$$x_2(t) = \frac{2x_2}{e^t(1+x_1) + e^{-t}(1-x_1)}, \quad x_3(t) = \frac{2x_3}{e^t(1+x_1) + e^{-t}(1-x_1)}.$$

6.28 Show that an arbitrary SLD e geodesic on the two-dimensional system $\mathcal{H} = \mathbb{C}^2$ is unitarily equivalent to $\mathcal{S}_\alpha$ if a suitable $\alpha \in [0, 1]$ is chosen, where $\mathcal{S}_\alpha \stackrel{\text{def}}{=} \left\{ \frac{1}{2} \begin{pmatrix} 1 + \alpha/\cosh t & \tanh t \\ \tanh t & 1 - \alpha/\cosh t \end{pmatrix} \middle| t \in \mathbb{R} \right\}$ [16].

6.29 Show equation (6.67) following the steps below.
(**a**) Show the equation $\int_0^1 (X^2 t)(I + tX)^{-1} dt = X - \log(I + X)$ for any Hermitian matrix X.
(**b**) Show the equation $\int_0^1 \operatorname{Tr}(\sigma - \rho)^2 (\rho + t(\sigma - \rho))^{-1} dt = \operatorname{Tr} \rho \log(\sqrt{\rho}\sigma^{-1}\sqrt{\rho})$.

6.30 Let $\boldsymbol{M}$ be a measurement corresponding to the spectral decomposition of $\sigma^{-1/2}(\sigma^{1/2}\rho\sigma^{1/2})^{1/2}\sigma^{-1/2}$. Show that $D_s^{(e)}(\rho\|\sigma) = D(\mathrm{P}_\rho^{\boldsymbol{M}} \| \mathrm{P}_\sigma^{\boldsymbol{M}})$ [14]. Show that $D_s^{(e)}(\rho\|\sigma) \geq -2 \log \operatorname{Tr} |\sqrt{\rho}\sqrt{\sigma}|$ from Exercise 3.21 and (2.26).

6.31 Show equation (6.66) following the steps below [17, 18].
(**a**) Show that $\int_0^1 \frac{d \log \rho_t}{dt} \frac{d\rho_t}{dt} t \, dt$
$= \left[(\log \rho_t) \frac{d\rho_t}{dt} t \right]_0^1 - \int_0^1 (\log \rho_t) \frac{d^2 \rho_t}{d^2 t} t \, dt - [(\log \rho_t)\rho_t]_0^1 + \int_0^1 \frac{d \log \rho_t}{dt} \rho_t \, dt.$
(**b**) Show that $J_{t,b} = \operatorname{Tr} \frac{d \log \rho_t}{dt} \frac{d\rho_t}{dt}$ for the Bogoljubov metric.
(**c**) Show that $\operatorname{Tr} \frac{d \log \rho_t}{dt} \rho_t = 0$.
(**d**) Show (6.66) for the m geodesic $\rho_t = t\sigma + (1-t)\rho$ connecting two states ρ and σ.

6.32 Show that the following three conditions are equivalent for two states ρ and σ and a PVM $\boldsymbol{M} = \{M_i\}$ of rank $M_i = 1$ [17, 18]. The equivalence of ① and ③ is nothing other than Theorem 3.6.

① $D(\rho\|\sigma) = D^{\boldsymbol{M}}(\rho\|\sigma)$.
② The m geodesic $\rho_\theta = \theta\sigma + (1-\theta)\rho$ satisfies $J_{\theta,b} = J_\theta^{\boldsymbol{M}}$ for $\theta \in [0, 1]$.

③ $[\sigma, \rho] = 0$, and there exists a set of real numbers $\{a_i\}_{i=1}^d$ satisfying

$$\rho = \sigma \left(\sum_{i=1}^{d} a_i M_i \right) = \left(\sum_{i=1}^{d} a_i M_i \right) \sigma. \tag{6.70}$$

6.33 Show that $\lim_{n\to\infty} \frac{1}{n} D_x^{(m)}(\rho^{\otimes n} \| \sigma^{\otimes n}) = D(\rho\|\sigma)$ when $J_{\theta,x} \le J_{\theta,b}$.

6.4 Quantum State Estimation

In Chap. 2, we only considered the case of two hypotheses existing for the quantum states. In this section, we will consider the problem of efficiently estimating an unknown quantum state that is included in a state family $\{\rho_\theta | \theta \in \mathbb{R}\}$ by performing a measurement. The goal is to find θ. We assume that a system has been prepared with n identical states in a similar manner to Chap. 2. In this case, the estimator is denoted by the pair $(\boldsymbol{M}^n, \hat{\theta}_n)$, where $\boldsymbol{M}^n$ is a POVM representing the measurement on the quantum system $\mathcal{H}^{\otimes n}$ (with the probability space Ω_n, which is the set of possible outcomes) and $\hat{\theta}_n$ is the map from Ω_n to the parameter space. In a similar way to the estimation of the probability distributions examined in Sect. 2.3, we assume the mean square error to be the measure of the error. If the parameter space is one-dimensional, an estimator with a smaller mean square error (MSE)

$$\hat{\mathrm{V}}_\theta(\boldsymbol{M}^n, \hat{\theta}_n) \stackrel{\text{def}}{=} \sum_{\omega} (\hat{\theta}_n(\omega) - \theta)^2 \operatorname{Tr} \rho_\theta^{\otimes n} \boldsymbol{M}^n(\omega) \tag{6.71}$$

results in a better estimation. We may then ask: what kind of estimator is then most appropriate for estimating the unknown state? One method is the following. First, we choose a POVM $\boldsymbol{M}$, and perform the corresponding measurement n times. Then, the problem is reduced to the estimation in the probability distribution family $\{\mathrm{P}_\theta^{\boldsymbol{M}} | \theta \in \mathbb{R}\}$. Second, we apply the maximum likelihood estimator $\hat{\theta}_{n,ML}$ of the probability distribution family to the obtained n outcomes. The mean square error is approximately $\frac{1}{n}(J_\theta^{\boldsymbol{M}})^{-1}$ in the asymptotic limit according to the discussion in Sect. 2.3, where $J_\theta^{\boldsymbol{M}}$ is the Fisher information at θ for the probability distribution $\{\mathrm{P}_\theta^{\boldsymbol{M}} | \theta \in \mathbb{R}\}$.

According to this argument there exists some arbitrariness in the choice of measurement $\boldsymbol{M}$. The essential point of quantum estimation is therefore to optimize this estimation procedure, including the choice of the POVM $\boldsymbol{M}$. It is evident that certain conditions must be imposed on the estimators. For example, consider an estimator $\hat{\theta}$ that always gives the value 0. If the true parameter is 0, the mean squared error is 0. On the other hand, if the true parameter is not 0, the mean squared error becomes large. Such an estimator is clearly not useful; this indicates that the problem of the formulation of the optimization problem for our estimator must be considered more carefully. A simple example of such a condition is the **unbiasedness condition**

$$\mathrm{E}_\theta(\boldsymbol{M}^n, \hat{\theta}_n) \stackrel{\text{def}}{=} \sum_\omega \hat{\theta}_n(\omega) \operatorname{Tr} \rho_\theta^{\otimes n} M^n(\omega) = \theta, \quad \forall \theta \in \Theta. \tag{6.72}$$

However, in general, the unbiasedness condition is too restrictive. In order to avoid this, we often consider the locally unbiased condition:

$$\mathrm{E}_\theta(\boldsymbol{M}^n, \hat{\theta}_n) = \theta, \quad \frac{d\mathrm{E}_\theta(\boldsymbol{M}^n, \hat{\theta}_n)}{d\theta} = 1 \tag{6.73}$$

at a fixed point $\theta \in \Theta$. However, since this condition depends on the true parameter, it is not so natural. As an intermediate condition, we often treat the asymptotic case, i.e., the asymptotic behavior when the number n of prepared spaces goes to infinity. In this case, the **asymptotic unbiasedness condition**:

$$\lim_{n\to\infty} \mathrm{E}_\theta(\boldsymbol{M}^n, \hat{\theta}_n) = \theta, \quad \lim_{n\to\infty} \frac{d}{d\theta}\mathrm{E}_\theta(\boldsymbol{M}^n, \hat{\theta}_n) = 1, \quad \forall \theta \in \Theta \tag{6.74}$$

is often imposed for a sequence of estimators $\{(\boldsymbol{M}^n, \hat{\theta}_n)\}_{n=1}^\infty$. The second condition guarantees a kind of uniformity of the convergence of $\mathrm{E}_\theta(\boldsymbol{M}^n, \hat{\theta}_n)$ to θ. We are now ready for the following theorem.

Theorem 6.6 (Helstrom [2], Nagaoka [13]) *If a sequence of estimators* $\{(\boldsymbol{M}^n, \hat{\theta}_n)\}$ *satisfies (6.74), then the following inequality holds:*

$$\underline{\lim}\, n\hat{\mathrm{V}}_\theta(\boldsymbol{M}^n, \hat{\theta}_n) \geq J_{\theta,s}^{-1}. \tag{6.75}$$

In the nonasymptotic case, when the locally unbiased condition (6.73) holds, inequality (6.75) also holds without limit [2]. Its proof is similar to the Proof of Theorem 6.6.

In the above discussion, we focus on the asymptotic unbiasedness condition. Using the van Trees inequality [19, 20], we can prove the same inequality almost everywhere without any assumption [21]. However, our method has the following advantage. Indeed, our method concerns one point. Hence, by choosing a coordinate suitable for one point, we can treat the general error function in the asymptotic setting. However, the van Trees method can be applied only to an error function with a quadratic form because the van Trees method concerns a Bayes prior distribution, i.e., all points.

Proof Define $O(\boldsymbol{M}^n, \hat{\theta}_n) \stackrel{\text{def}}{=} \sum_\omega \hat{\theta}_n(\omega) \boldsymbol{M}^n(\omega)$. Since $\sum_\omega \left(\hat{\theta}_n(\omega) - \theta\right) \boldsymbol{M}^n(\omega) = O(\boldsymbol{M}^n, \hat{\theta}_n) - \theta I$,

$$\begin{aligned} 0 \leq & \sum_\omega \left((\hat{\theta}_n(\omega) - \theta) - (O(\boldsymbol{M}^n, \hat{\theta}_n) - \theta I)\right) \boldsymbol{M}^n(\omega) \\ & \cdot \left((\hat{\theta}_n(\omega) - \theta) - (O(\boldsymbol{M}^n, \hat{\theta}_n) - \theta I)\right) \\ = & \sum_\omega \left(\hat{\theta}_n(\omega) - \theta\right) \boldsymbol{M}^n(\omega) \left(\hat{\theta}_n(\omega) - \theta\right) - (O(\boldsymbol{M}^n, \hat{\theta}_n) - \theta I)^2. \end{aligned} \tag{6.76}$$

The Schwarz inequality for the metric $\langle\ ,\ \rangle^{(e)}_{\rho_\theta^{\otimes n},s}$ yields

$$\hat{\mathrm{V}}_\theta(\boldsymbol{M}^n, \hat{\theta}_n) = \mathrm{Tr} \sum_\omega \left(\hat{\theta}_n(\omega) - \theta\right) \boldsymbol{M}^n(\omega) \left(\hat{\theta}_n(\omega) - \theta\right) \rho_\theta^{\otimes n}$$

$$\geq \mathrm{Tr}(O(\boldsymbol{M}^n, \hat{\theta}_n) - \theta I)^2 \rho_\theta^{\otimes n} = \left(\left\| O(\boldsymbol{M}^n, \hat{\theta}_n) - \theta I \right\|^{(e)}_{\rho_\theta^{\otimes n},s} \right)^2 \tag{6.77}$$

$$\geq \frac{\left(\langle L_{\theta,s,n}, O(\boldsymbol{M}^n, \hat{\theta}_n) - \theta I \rangle^{(e)}_{\rho_\theta^{\otimes n},s} \right)^2}{\left(\left\| L_{\theta,s,n} \right\|^{(e)}_{\rho_\theta^{\otimes n},s} \right)^2}, \tag{6.78}$$

where $L_{\theta,s,n}$ denotes the SLD e representation of the derivative of the state family $\{\rho_\theta^{\otimes n} | \theta \in \mathbb{R}\}$. Since $\mathrm{Tr}\, \theta_0 \frac{d\rho_\theta^{\otimes n}}{d\theta} = \theta_0 \,\mathrm{Tr}\, \frac{d\rho_\theta^{\otimes n}}{d\theta} = 0$, we have

$$\frac{d}{d\theta} \mathrm{E}_\theta(\boldsymbol{M}^n, \hat{\theta}_n)|_{\theta=\theta_0} = \mathrm{Tr}\, O(\boldsymbol{M}^n, \hat{\theta}_n) \frac{d\rho_\theta^{\otimes n}}{d\theta}|_{\theta=\theta_0}$$

$$= \mathrm{Tr} \left(O(\boldsymbol{M}^n, \hat{\theta}_n) - \theta_0 \right) \frac{d\rho_\theta^{\otimes n}}{d\theta}|_{\theta=\theta_0} = \langle L_{\theta_0,s,n}, O(\boldsymbol{M}^n, \hat{\theta}_n) - \theta_0 \rangle^{(e)}_{\rho_{\theta_0}^{\otimes n},s} \tag{6.79}$$

from the definition of $\mathrm{E}_\theta(\boldsymbol{M}^n, \hat{\theta}_n)$. Combining the above two formulas with Exercise 6.19, we obtain

$$n \hat{\mathrm{V}}_{\theta_0}(\boldsymbol{M}^n, \hat{\theta}_n) \geq n \frac{\left(\frac{d}{d\theta} \mathrm{E}_\theta(\boldsymbol{M}^n, \hat{\theta}_n)|_{\theta=\theta_0} \right)^2}{n J_{\theta,s}}.$$

Taking the limit, we obtain (6.75) with $\theta = \theta_0$. ■

According to the above proof, if equality (6.74) holds for a finite number n, then inequality (6.75) also holds for the same number n. Conversely, given a point $\theta_0 \in \mathbb{R}$, we choose the function $\hat{\theta}_{\theta_0,n}$ and projections $E_{\theta_0,n}(\omega)$ such that the following is the spectral decomposition of the matrix $\frac{L_{\theta_0,s,n}}{n J_{\theta_0,s}} + \theta_0$:

$$\frac{L_{\theta_0,s,n}}{n J_{\theta_0,s}} + \theta = \sum_\omega E_{\theta_0,n}(\omega) \hat{\theta}_{\theta_0,n}(\omega). \tag{6.80}$$

Then, $(\boldsymbol{E}_{\theta_0,n} = \{E_{\theta_0,n}(\omega)\}, \hat{\theta}_{\theta_0,n})$ gives an estimator satisfying

$$\hat{\mathrm{V}}_{\theta_0}(\boldsymbol{E}_{\theta_0,n}, \hat{\theta}_n) = \frac{1}{n J_{\theta_0,s}}, \quad \mathrm{E}_{\theta_0}(\boldsymbol{E}_{\theta_0,n}, \hat{\theta}_n) = \theta_0.$$

We may then show that

$$\left.\frac{d\mathrm{E}_\theta(\boldsymbol{E}_{\theta_0,n}, \hat{\theta}_{\theta_0,n})}{d\theta}\right|_{\theta=\theta_0} = \langle L_{\theta_0,s,n}, O(\boldsymbol{M}^n, \hat{\theta}_n) - \theta\rangle^{(e)}_{\rho_{\theta_0}^{\otimes n},s}$$
$$= \langle L_{\theta_0,s,n}, \frac{1}{nJ_{\theta_0,s}} L_{\theta_0,s,n}\rangle^{(e)}_{\rho_{\theta_0}^{\otimes n},s} = 1, \tag{6.81}$$

in a similar way to (6.79). This guarantees the existence of an estimator satisfying (6.75) under condition (6.74). However, it is crucial that the construction of $(E_{\theta_0,n}, \hat{\theta}_n)$ depends on θ_0. We may expect from (6.81) that if θ is in the neighborhood of θ_0, $\hat{\mathrm{V}}_\theta(E_{\theta_0,n}, \hat{\theta}_n)$ would not be very different from $\hat{\mathrm{V}}_{\theta_0}(E_{\theta_0,n}, \hat{\theta}_n)$. However, if θ is far away from θ_0, it is impossible to estimate $\hat{\mathrm{V}}_\theta(E_{\theta_0,n}, \hat{\theta}_n)$. The reason is that the SLD $L_{\theta_0,s,n}$ depends on θ_0. If the SLD $L_{\theta_0,s,n}$ did not depend on θ_0, one would expect that an appropriate estimator could be constructed independently of θ_0. This is the subject of the following theorem.

Theorem 6.7 (Nagaoka [13, 22]) *Assume that a distribution $p(\theta)$ satisfying*

$$\int \rho_\theta p(\theta)\, d\theta > 0 \tag{6.82}$$

exists. Then, the following two conditions for the quantum state family ρ_θ and the estimator $(\boldsymbol{M}, \hat{\theta})$ are equivalent.

① *The estimator $(\boldsymbol{M}, \hat{\theta})$ satisfies the unbiasedness condition (6.72), and the MSE $\hat{\mathrm{V}}_\theta(\boldsymbol{M}, \hat{\theta})$ satisfies*

$$\hat{\mathrm{V}}_\theta(\boldsymbol{M}, \hat{\theta}) = J_{\theta,s}^{-1}. \tag{6.83}$$

② *The state family is an SLD e geodesic $\rho_\theta = \Pi^\theta_{L,s}\rho_0$ given by (6.44); further, the parameter to be estimated equals the expectation parameter $\eta = \mathrm{Tr}\, L\rho_\theta$, and the estimator $(\boldsymbol{M}, \hat{\theta})$ equals the spectral decomposition of L.*

See Exercises 6.36 and 6.37 for the proof of the above theorem.

Therefore, the bound $\frac{1}{nJ_{\theta,s}}$ is attained in the nonasymptotic case only for the case (6.44), i.e., the SLD *e* geodesic curve. Another example is the case when a POVM $\boldsymbol{M}$ exists such that

$$J_\theta^{\boldsymbol{M}} = J_{\theta,s} \text{ for } \forall\theta, \tag{6.84}$$

where $J_\theta^{\boldsymbol{M}}$ is the Fisher information of the probability distribution family $\{\mathrm{P}^{\boldsymbol{M}}_{\rho_\theta} | \theta \in \mathbb{R}\}$. Then, if one performs the measurement $\boldsymbol{M}$ on n prepared systems and chooses the maximum likelihood estimator of the n outcomes of the probability distribution family $\{\mathrm{P}^{\boldsymbol{M}}_{\rho_\theta} | \theta \in \mathbb{R}\}$, the equality in inequality (6.75) is ensured according to the discussion in Sect. 2.3. Therefore, $J_{\theta,s}^{-1}$ is also attainable asymptotically in this case. In general, a POVM $\boldsymbol{M}$ satisfying (6.84) rarely exists. Such a state family is called a **quasiclassical**, and such a POVM is called a quasiclassical POVM [23].

Besides the above rare examples, the equality of (6.75) is satisfied in the limit $n \to \infty$ at all points, provided a sequence of estimators $\{(\boldsymbol{M}^n, \hat{\theta}_n)\}$ is constructed according to the following two-step estimation procedure [24, 25]. First, perform the measurement $\boldsymbol{M}$ satisfying $J_\theta^{\boldsymbol{M}} > 0$ for the first $\sqrt{n}$ systems. Next, perform the measurement $E_{\hat{\theta}_{ML,\sqrt{n}}, n-\sqrt{n}}$ (defined previously) for the remaining $n - \sqrt{n}$ systems, based on the maximum likelihood estimator $\hat{\theta}_{ML,\sqrt{n}}$ for the probability distribution $\{\mathrm{P}_{\rho_\theta}^{\boldsymbol{M}} | \theta \in \mathbb{R}\}$. Finally, choose the final estimate according to $\hat{\theta}_n \stackrel{\text{def}}{=} \hat{\theta}_{\hat{\theta}_{ML,\sqrt{n}}, n-\sqrt{n}}$, as given in (6.80). If n is sufficiently large, $\hat{\theta}_{ML,\sqrt{n}}$ will be in the neighborhood of the true parameter θ with a high probability. Hence, the expectation of $(\hat{\theta}_n - \theta)^2$ is approximately equal to $\frac{1}{(n-\sqrt{n})J_{\theta,s}}$. Since $\lim_{n\to\infty} n \frac{1}{(n-\sqrt{n})J_{\theta,s}} = \frac{1}{J_{\theta,s}}$, we can expect this estimator to satisfy the equality in (6.75). In fact, it is known that such an estimator does satisfy the equality in (6.75) [24, 25].

In summary, for the single-parameter case, it is the SLD Fisher metric and not the Bogoljubov Fisher metric that gives the tight bound in estimation theory. On the other hand, the Bogoljubov Fisher metric does play a role in large deviation evaluation, although it appears in a rather restricted way.

Exercises

6.34 Show that the measurement $\boldsymbol{E}_{\theta,n}$ defined in (6.80) satisfies $nJ_{\theta,s} = J_\theta^{\boldsymbol{E}_{\theta,n}}$.

6.35 Using the above result, show that an arbitrary inner product $\|A\|_{\rho,x}^{(m)}$ satisfies $\|A\|_{\rho,s}^{(m)} \le \|A\|_{\rho,x}^{(m)}$ when (6.24) and $\|\rho^{-1}\|_{\rho,x}^{(m)} = 1$ hold.

6.36 Prove Theorem 6.7 for $\rho_\theta > 0$ following the steps below [13, 22].
(**a**) Assume that the estimator $(\boldsymbol{M}, \hat{\theta})$ for the SLD e geodesic $\Pi_{L,s}^\theta \rho$ is given by the spectral decomposition of L. Show that the estimator $(\boldsymbol{M}, \hat{\theta})$ satisfies the unbiasedness condition with respect to the expectation parameter.
(**b**) For an SLD e geodesic, show that $\frac{d\eta}{d\theta} = J_{\theta,s}$.
(**c**) For an SLD e geodesic, show that the SLD Fisher information $J_{\eta,s}$ for an expectation parameter η is equal to the inverse $J_{\theta,s}^{-1}$ of the SLD Fisher information for the natural parameter θ.
(**d**) Show that ① follows from ②.
(**e**) Show that $\theta(\eta) = \int_0^\eta J_{\eta',s}\, d\eta'$ for an SLD e geodesic curve.
(**f**) Show that $\mu_s(\theta(\eta)) = \int_0^\eta \eta' J_{\eta',s}\, d\eta'$ for an SLD e geodesic curve.
(**g**) Show that if ① is true, $\frac{1}{J_{\theta,s}} L_{\theta,s} = O(\boldsymbol{M}, \hat{\theta}) - \theta$.
(**h**) Show that if ① is true, then $\frac{d\rho_\theta}{d\theta} = \frac{J_\eta}{2}((O(\boldsymbol{M}, \hat{\theta}) - \eta)\rho_\theta + \rho_\theta(O(\boldsymbol{M}, \hat{\theta}) - \eta))$, where η is the parameter to be estimated.
(**i**) Show that if $n = 1$ and $\rho_\theta > 0$, the equality in (6.77) is satisfied only if the estimator $(\boldsymbol{M}, \hat{\theta})$ is the spectral decomposition of $O(\boldsymbol{M}, \hat{\theta})$.
(**j**) Show that if ① holds, then ② holds.

6.37 Show that Theorem 6.7 holds even if $\rho_\theta > 0$ is not true, following the steps below. The fact that ②⇒① still follows from above.
(**a**) Show that (**h**) in Exercise 6.36 still holds for $\rho_\theta > 0$.
(**b**) Show that ① yields ② by using the condition (6.82).

6.38 Similarly to (6.77), show

$$\sum_{\omega} \mathrm{Tr}\left(\hat{\theta}_n(\omega) - \theta\right)^2 \boldsymbol{M}^n(\omega)\rho_\theta^{\otimes n} \geq \left(\left\| O(\boldsymbol{M}^n, \hat{\theta}_n) \right\|^{(e)}_{\rho_\theta^{\otimes n}, r}\right)^2.$$

6.5 Large Deviation Evaluation

In Sect. 2.4.2, we discussed the large deviation type estimation of a probability distribution for the case of a single parameter. In this section, we will examine the theory for large deviation in the case of quantum state estimation. As defined in (2.173) and (2.174), $\beta(\{\boldsymbol{M}^n, \hat{\theta}_n\})$ and $\alpha(\{\boldsymbol{M}^n, \hat{\theta}_n\})$ are defined as follows:

$$\beta(\{\boldsymbol{M}^n, \hat{\theta}_n\}, \theta, \epsilon) \stackrel{\text{def}}{=} \underline{\lim} -\frac{1}{n} \log \mathrm{Tr}\, \rho^{\otimes n} \boldsymbol{M}^n\{|\hat{\theta}_n - \theta| \geq \epsilon\}, \tag{6.85}$$

$$\alpha(\{\boldsymbol{M}^n, \hat{\theta}_n\}, \theta) \stackrel{\text{def}}{=} \underline{\lim}_{\epsilon \to 0} \frac{\beta(\{\boldsymbol{M}^n, \hat{\theta}_n\}, \theta, \epsilon)}{\epsilon^2}. \tag{6.86}$$

The notation $\boldsymbol{M}^n\{|\hat{\theta}_n - \theta| \geq \epsilon\}$ requires some explanation. For the general POVM $\boldsymbol{M} = \{M(\omega)\}$ and the set B, let us define $\boldsymbol{M}B$ according to

$$\boldsymbol{M}B \stackrel{\text{def}}{=} \sum_{\omega \in B} M(\omega). \tag{6.87}$$

Then, we have the following theorem in analogy to Theorem 2.9.

Theorem 6.8 (Nagaoka [12]) *Let the sequence of estimators* $\boldsymbol{M} = \{(\boldsymbol{M}^n, \hat{\theta}_n)\}$ *satisfy the weak consistency condition:*

$$\mathrm{Tr}\, \rho^{\otimes n} \boldsymbol{M}^n\{|\hat{\theta}_n - \theta| \geq \epsilon\} \to 0, \quad \forall \epsilon > 0, \forall \theta \in \mathbb{R}. \tag{6.88}$$

The following then holds:

$$\beta(\{\boldsymbol{M}^n, \hat{\theta}_n\}, \theta, \epsilon) \leq \inf_{\theta' : |\theta' - \theta| > \epsilon} D(\rho_{\theta'} \| \rho_\theta), \tag{6.89}$$

$$\alpha(\{\boldsymbol{M}^n, \hat{\theta}_n\}, \theta) \leq \frac{1}{2} J_{\theta, b}. \tag{6.90}$$

A different inequality that evaluates the performance of the estimator may be obtained by employing a slight reformulation. That is, a relation similar to (6.89) can be obtained as given by the following lemma.

Lemma 6.2 (Hayashi [26]) *Define* $\beta'(\boldsymbol{M}, \theta, \delta) \stackrel{\text{def}}{=} \lim_{\epsilon \to +0} \beta(\boldsymbol{M}, \theta, \delta - \epsilon)$ *for the sequence of estimators* $\boldsymbol{M} = \{(\boldsymbol{M}^n, \hat{\theta}_n)\}$. *The following inequality then holds:*

$$\inf_{\{s|1\ge s\ge 0\}} \left(\beta'(\boldsymbol{M}, \theta, s\delta) + \beta'(\boldsymbol{M}, \theta+\delta, (1-s)\delta)\right) \le -2 \log \operatorname{Tr} |\sqrt{\rho_\theta}\sqrt{\rho_{\theta+\delta}}|. \tag{6.91}$$

The essential part in the proof of this lemma is that the information $-\log \operatorname{Tr} |\sqrt{\rho}\sqrt{\sigma}|$ satisfies the information-processing inequality[Exe. 6.40].

The relation corresponding to (6.90) is then given by the following theorem.

Theorem 6.9 (Hayashi [26]) *Let the sequence of estimators* $\boldsymbol{M} = \{(\boldsymbol{M}^n, \hat{\theta}_n)\}$ *satisfy the weak consistency condition and the uniform convergence on the RHS of (6.86) with respect to* θ. *Define* $\alpha'(\boldsymbol{M}, \theta_0) \stackrel{\text{def}}{=} \underline{\lim}_{\theta \to \theta_0} \alpha(\boldsymbol{M}, \theta_0)$. *The following inequality then holds:*

$$\alpha'(\boldsymbol{M}, \theta) \le \frac{J_{\theta,s}}{2}. \tag{6.92}$$

Hence, the bound $\frac{J_{\theta,s}}{2}$ can be regarded as the bound under the following condition for a sequence of estimators:

$$\alpha(\boldsymbol{M}, \theta_0) = \underline{\lim}_{\theta \to \theta_0} \alpha(\boldsymbol{M}, \theta_0), \quad \beta(\boldsymbol{M}, \theta, \delta) = \lim_{\epsilon \to +0} \beta(\boldsymbol{M}, \theta, \delta - \epsilon). \tag{6.93}$$

So far, we have discussed the upper bound of $\alpha(\boldsymbol{M}, \theta)$ in two ways. The upper bound given here can be attained in both ways, as we now describe. Let us first focus on the upper bound $\frac{J_{\theta,s}}{2}$ given by (6.92), which is based on the SLD Fisher information. This upper bound can be attained by a sequence of estimators $\boldsymbol{M} = \{(\boldsymbol{M}^n, \hat{\theta}_n)\}$ such that $\alpha'(\boldsymbol{M}, \theta_0) = \alpha(\boldsymbol{M}, \theta_0)$ and the RHS of (6.86) converges uniformly concerning θ. This kind of estimator can be constructed according to the two-step estimator given in the previous section [26]. Let us now examine the upper bound given by (6.90) using the Bogoljubov Fisher information, which equals $\frac{J_{\theta,b}}{2}$ in this case. This bound can be attained by a sequence of estimators satisfying the weak coincidence condition but not the uniform convergence on the RHS of (6.86). However, this estimator can attain the bound $\frac{J_{\theta,b}}{2}$ only at a single point [26]. Although this method of construction is rather obtuse, the method is similar to the construction of the measurement that attains the bound $D(\rho\|\sigma)$ given in Stein's lemma for hypothesis testing. Of course, such an estimator is extremely unnatural and cannot be used in practice. Therefore, we see that the two bounds provide the respective answers for two completely separate problems. In a classical system, the bounds for these two problems are identical. This difference arises due to the quantum nature of the problem.

The above discussion indicates that geometrical characterization does not connect to quantum state estimation. However, there are two different approaches from the geometrical viewpoint. For example, Hayashi [27] focused on the scalar curvature of the Riemannian connection and clarified the relation between the scalar curvature and the second-order asymptotics of estimation error only for specific state families. These approaches treat the translation of the tangent bundle of state space. Matsumoto [28–30] focused on that of the line bundle and discovered the relation between the curvature and the bound of estimation error for the pure-state family. He pointed out that the difficulty rooted in two parameters is closely related to the curvature.

Exercises

6.39 Prove Theorem 6.8 referring to the Proof of Theorem 2.9.

6.40 Prove Lemma 6.2 following the steps below.
(a) Show that $\log \operatorname{Tr} |\sqrt{\rho^{\otimes n}}\sqrt{\sigma^{\otimes n}}| = n \log \operatorname{Tr} |\sqrt{\rho}\sqrt{\sigma}|$.
(b) Show that

$$
\begin{aligned}
&\log \operatorname{Tr} \left|\sqrt{\rho_\theta^{\otimes n}}\sqrt{\rho_{\theta+\delta}^{\otimes n}}\right| \\
\leq &\log\Bigg[\left(\operatorname{Tr} \rho_{\theta+\delta}^{\otimes n} \boldsymbol{M}^n \left\{|\hat{\theta}-(\theta+\delta)| \geq \frac{\delta(m-1)}{m}\right\}\right)^{\frac{1}{2}} + \left(\operatorname{Tr} \rho_\theta^{\otimes n} \boldsymbol{M}^n \left\{|\hat{\theta}-\theta| \geq \delta\right\}\right)^{\frac{1}{2}} \\
&+\sum_{i=1}^{m}\left(\operatorname{Tr}\rho_\theta^{\otimes n}\boldsymbol{M}^n\left\{|\hat{\theta}-\theta| > \frac{\delta(i-1)}{m}\right\}\right)^{\frac{1}{2}} \\
&\cdot \left(\operatorname{Tr}\rho_{\theta+\delta}^{\otimes n}\boldsymbol{M}^n\left\{|\hat{\theta}-(\theta+\delta)| \geq \frac{\delta(m-i-1)}{m}\right\}\right)^{\frac{1}{2}}\Bigg]
\end{aligned}
$$

for an arbitrary integer m from the fact that the amount of information $-\log \operatorname{Tr} |\sqrt{\rho}\sqrt{\sigma}|$ satisfies the information-processing inequality.
(c) Choosing a sufficiently large integer N for a real number $\epsilon > 0$ and an integer m, we have

$$
\frac{1}{n}\log \operatorname{Tr} \rho_\theta^{\otimes n} \boldsymbol{M}^n \left\{|\hat{\theta}-\theta| \geq \frac{\delta i}{m}\right\} \leq -\beta\left(\boldsymbol{M}, \theta, \frac{\delta i}{m}\right) + \epsilon
$$

$$
\frac{1}{n}\log \operatorname{Tr} \rho_{\theta+\delta}^{\otimes n} \boldsymbol{M}^n \left\{|\hat{\theta}-(\theta+\delta)| \geq \frac{\delta(m-i)}{m}\right\} \leq -\beta\left(\boldsymbol{M}, \theta+\delta, \frac{\delta(m-i)}{m}\right) + \epsilon,
$$

for $\forall n \geq N, 0 \leq \forall i \leq m$. Show that

$$
\begin{aligned}
&n \log \operatorname{Tr} |\sqrt{\rho_\theta}\sqrt{\rho_{\theta+\delta}}| \\
\leq &\log(m+2) - \frac{n}{2}\left(\min_{0\leq i\leq m}\beta\left(\boldsymbol{M}, \theta, \frac{\delta(i-1)}{m}\right) + \beta\left(\boldsymbol{M}, \theta+\delta, \frac{\delta(m-i-1)}{m}\right) - 2\epsilon\right).
\end{aligned}
$$

(d) Show the following for an arbitrary integer m:

$$
\begin{aligned}
&-\log \operatorname{Tr} |\sqrt{\rho_\theta}\sqrt{\rho_{\theta+\delta}}| \\
\geq &\frac{1}{2}\min_{0\leq i\leq m}\left(\beta\left(\boldsymbol{M}, \theta, \frac{\delta(i-1)}{m}\right) + \beta\left(\boldsymbol{M}, \theta+\delta, \frac{\delta(m-i-1)}{m}\right) - 2\epsilon\right)
\end{aligned}
$$

$$-\frac{1}{n}\log(m+2).$$

(**e**) Prove (6.91) using (**d**).

6.41 Prove Theorem 6.9 using Lemma 6.2.

6.6 Multiparameter Estimation

Let us now examine the case of a multidimensional parameter space Θ (dimension d). Assume that the unknown state lies in the multiparameter quantum state family $\{\rho_\theta|\theta \in \Theta \subset \mathbb{R}^d\}$. A typical estimation procedure is as follows. An appropriate POVM $\boldsymbol{M}$ is chosen in a manner similar to the previous one (excluding those for a Fisher matrix $\boldsymbol{J}_\theta^{\boldsymbol{M}}$ with zero eigenvalues). Then, a measurement corresponding to $\boldsymbol{M}$ is performed on each of n quantum systems, whose states are unknown but identical to the state of another system. The final estimate is then given by the maximum likelihood estimator for the probability distribution family $\{\mathrm{P}_\theta^{\boldsymbol{M}}|\theta \in \Theta \subset \mathbb{R}^d\}$. According to Sect. 2.3, the mean square error matrix asymptotically approaches $\frac{1}{n}(\boldsymbol{J}_\theta^{\boldsymbol{M}})^{-1}$ in this case. The maximum likelihood estimator then approaches the true parameter θ in probability. As mentioned in the previous section, our problem is the optimization of the quantum measurement $\boldsymbol{M}$ for our estimation. To this end, we need to find an estimator minimizing the mean square error $\hat{\mathrm{V}}_\theta^{i,j}(\boldsymbol{M}^n, \hat{\theta}_n)$ for the ith parameter θ^i or the mean square error matrix $\hat{\boldsymbol{V}}_\theta(\boldsymbol{M}^n, \hat{\theta}_n) = [\hat{\mathrm{V}}_\theta^{i,j}(\boldsymbol{M}^n, \hat{\theta}_n)]$ by taking into account the correlations between the θ^i, where

$$\hat{\mathrm{V}}_\theta^{i,j}(\boldsymbol{M}^n, \hat{\theta}_n) \stackrel{\text{def}}{=} \sum_\omega (\hat{\theta}_n^i(\omega) - \theta^i)(\hat{\theta}_n^j(\omega) - \theta^j) \operatorname{Tr} \rho_\theta^{\otimes n} M^n(\omega). \tag{6.94}$$

The unbiasedness condition is then given by

$$\mathrm{E}_\theta^i(\boldsymbol{M}^n, \hat{\theta}_n) \stackrel{\text{def}}{=} \sum_\omega \hat{\theta}_n^i(\omega) \operatorname{Tr} \rho_\theta^{\otimes n} M^n(\omega) = \theta^i, \quad \forall \theta \in \Theta.$$

In the asymptotic case, for a sequence of estimators $\{(\boldsymbol{M}^n, \hat{\theta}_n)\}$, we can also write down the asymptotic unbiasedness condition

$$\lim_{n\to\infty} \mathrm{E}_\theta^i(\boldsymbol{M}^n, \hat{\theta}_n) = \theta^i, \quad \lim_{n\to\infty} \frac{\partial}{\partial\theta^j}\mathrm{E}_\theta^i(\boldsymbol{M}^n, \hat{\theta}_n) = \delta_j^i, \quad \forall \theta \in \Theta. \tag{6.95}$$

Theorem 6.10 *Let the sequence of estimators* $\{(\boldsymbol{M}^n, \hat{\theta}_n)\}$ *satisfy the asymptotic unbiasedness condition (6.95) and have the limit* $\hat{\mathrm{V}}_\theta^{i,j}(\{\boldsymbol{M}^n, \hat{\theta}_n\}) \stackrel{\text{def}}{=} \lim_{n\to\infty} n\hat{\mathrm{V}}_\theta^{i,j}$

$(\boldsymbol{M}^n, \hat{\theta}_n)$. *The following matrix inequality then holds [2, 7]:*

$$\lim_{n\to\infty} n\hat{\boldsymbol{V}}_\theta(\boldsymbol{M}^n, \hat{\theta}_n) \geq (\boldsymbol{J}_{\theta,x})^{-1}, \quad x = s, r. \tag{6.96}$$

Proof First, assume that any two complex vectors $|b\rangle = (b_1, \ldots, b_d)^T \in \mathbb{C}^d$ and $|a\rangle \in \mathbb{C}^d$ satisfy

$$\langle b|\hat{\boldsymbol{V}}_\theta(\{\boldsymbol{M}^n, \hat{\theta}_n\})|b\rangle\langle a|\boldsymbol{J}_{\theta,x}|a\rangle \geq |\langle b|a\rangle|^2. \tag{6.97}$$

Substituting $a = (\boldsymbol{J}_{\theta,x})^{-1}b$ into (6.97), we have $\langle b|\hat{\boldsymbol{V}}_\theta(\{\boldsymbol{M}^n, \hat{\theta}_n\})|b\rangle \geq \langle b|(\boldsymbol{J}_{\theta,x})^{-1}|b\rangle$, since $(\boldsymbol{J}_{\theta,x})^{-1}$ is Hermitian. We therefore obtain (6.96).

We next show (6.97). Define $O_n \stackrel{\text{def}}{=} \sum_\omega \left(\sum_i (\hat{\theta}^i_n(\omega) - \theta^i)b_i\right) M^n(\omega)$ and $L_n \stackrel{\text{def}}{=} \sum_j L_{\theta,j,x,n}a_j$. Using (6.77) and Exercise 6.38, we can show that

$$\begin{aligned}\langle b|\hat{\boldsymbol{V}}_\theta(\{\boldsymbol{M}^n, \hat{\theta}_n\})|b\rangle &= \lim_{n\to\infty} n \sum_\omega \left|\sum_i (\hat{\theta}^i_n(\omega) - \theta^i)b_i\right|^2 \operatorname{Tr} \rho_\theta^{\otimes n} M^n(\omega) \\ &\geq \lim_{n\to\infty} n \left(\|O_n\|^{(e)}_{\rho_\theta^{\otimes n},x}\right)^2,\end{aligned}$$

in a manner similar to (6.76). Then

$$\langle b|a\rangle = \lim_{n\to\infty} \sum_{i,j} \overline{b}_i \frac{\partial}{\partial\theta^j} \mathrm{E}^i_\theta(\boldsymbol{M}^n, \hat{\theta}_n)a_j = \lim_{n\to\infty} \langle O_n, L_n\rangle^{(e)}_{\rho_\theta^{\otimes n},x},$$

in a manner similar to (6.81). Using the Schwarz inequality, we can show that

$$\left(\|O_n\|^{(e)}_{\rho_\theta^{\otimes n},x}\right)^2 \left(\|L_n\|^{(e)}_{\rho_\theta^{\otimes n},x}\right)^2 \geq \left|\langle O_n, L_n\rangle^{(e)}_{\rho_\theta^{\otimes n},x}\right|^2.$$

Inequality (6.97) can be obtained on taking the limit because $\langle a|\boldsymbol{J}_{\theta,x}|a\rangle = n \left(\|L_n\|^{(e)}_{\rho_\theta^{\otimes n},x}\right)^2$. ■

In general, there is no sequence of estimators that satisfies the equality in (6.96). Furthermore, as the matrix $\hat{\mathrm{V}}^{i,j}_\theta(\boldsymbol{M}^n, \hat{\theta}_n)$ is a real symmetric matrix and not a real number, there is no general minimum matrix $\hat{\mathrm{V}}^{i,j}_\theta(\boldsymbol{M}^n, \hat{\theta}_n)$ among the estimators satisfying (6.95). Instead, one can adopt the sum of MSE, i.e., the trace of $\hat{\mathrm{V}}^{i,j}_\theta(\boldsymbol{M}^n, \hat{\theta}_n)$ as our error criterion. It is therefore necessary to consider the minimum of $\operatorname{tr} \lim_{n\to\infty} n\hat{\boldsymbol{V}}_\theta(\boldsymbol{M}^n, \hat{\theta}_n)$ in the asymptotic case.

From (6.96) the lower bound of the minimum value of this quantity can be evaluated as

$$\operatorname{tr} \lim_{n\to\infty} n\hat{\boldsymbol{V}}_\theta(\boldsymbol{M}^n, \hat{\theta}_n) \geq \min\{\operatorname{tr} \boldsymbol{V} | \boldsymbol{V} : \text{real symmetric } \boldsymbol{V} \geq (\boldsymbol{J}_{\theta,x})^{-1}\} \tag{6.98}$$

because $\hat{V}_\theta(M^n, \hat{\theta}_n)$ is real symmetric. If $(J_{\theta,x})^{-1}$ is real symmetric, the RHS is equal to $\mathrm{tr}(J_{\theta,x})^{-1}$. If $(J_{\theta,x})^{-1}$ is a Hermitian matrix but contains imaginary elements, the RHS will be larger than $\mathrm{tr}(J_{\theta,x})^{-1}$. In this case, we may calculate [7]

$$\min\{\mathrm{tr}\, V | V : \text{real symmetric} V \geq J_{\theta,x}^{-1}\} = \mathrm{tr}\, \mathbf{Re}(J_{\theta,x}^{-1}) + \mathrm{tr}\, |\, \mathbf{Im}(J_{\theta,x}^{-1})|. \quad (6.99)$$

For example, $(J_{\theta,s})^{-1}$ and $(J_{\theta,b})^{-1}$ are real symmetric matrices, as discussed in Exercise 6.22. However, since the RLD Fisher information matrix $(J_{\theta,r})^{-1}$ possesses imaginary components, the RHS of (6.99) in the RLD case will be larger than $\mathrm{tr}(J_{\theta,r})^{-1}$. Moreover, in order to treat the set of the limits of MSE matrices, we often minimize $\mathrm{tr}\, G\hat{V}_\theta(\{M^n, \hat{\theta}_n\})$. From a discussion similar to (6.99), it can be shown that the minimum is greater than $\mathrm{tr}\, \sqrt{G}\, \mathbf{Re}(J_{\theta,r}^{-1})\sqrt{G} + \mathrm{tr}\, |\sqrt{G}\, \mathbf{Im}(J_{\theta,r}^{-1})\sqrt{G}|$. Its equality holds only when $\hat{V}_\theta(\{M^n, \hat{\theta}_n\}) = \mathbf{Re}(J_{\theta,r}^{-1}) + \sqrt{G}^{-1}|\sqrt{G}\, \mathbf{Im}(J_{\theta,r}^{-1})\sqrt{G}|\sqrt{G}^{-1}$. When the family in the two-dimensional space has the form $\{\rho_\theta | \|\theta\| = r\}$, the set of MSE matrices is restricted by the RLD Fisher information matrix, as shown in Fig. 6.2.

In Fig. 6.2, we use the parameterization $\begin{pmatrix} x_0 + x_1 & x_2 \\ x_2 & x_0 - x_1 \end{pmatrix}$ and assume that $J_{\theta,s}$ is a constant time of the identity matrix. In addition, it was shown that these limits of MSE matrices can be attained [31]. The above figure also illustrates that the set of MSE matrices can be realized by the adaptive estimators. See Exercises 6.25 and 6.50.

The following theorem gives the asymptotic lower bound of $\mathrm{tr}\, \hat{V}_\theta(M^n, \hat{\theta}_n)$.

Theorem 6.11 *Let the sequence of estimators $\{(M^n, \hat{\theta}_n)\}$ satisfy the same conditions as Theorem 6.10. The following inequality then holds:*

$$\mathrm{tr} \lim_{n\to\infty} n\hat{V}_\theta(M^n, \hat{\theta}_n) \geq \overline{\lim}\, n \inf_{M^n:\, POVM\ on\ \mathcal{H}^{\otimes n}} \mathrm{tr}(J_\theta^{M^n})^{-1}. \quad (6.100)$$

Conversely, we can construct the estimator attaining the bound $\min_M \mathrm{tr}(J_\theta^M)^{-1}$ by using the adaptive method in a manner similar to the one-parameter case. Moreover, applying this method to the n-fold tensor product system $\mathcal{H}^{\otimes n}$, we can construct

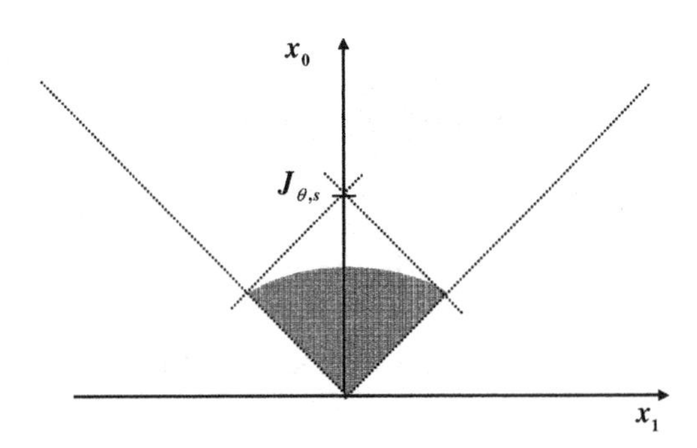

Fig. 6.2 Fisher information matrices

Fig. 6.3 MSE matrices

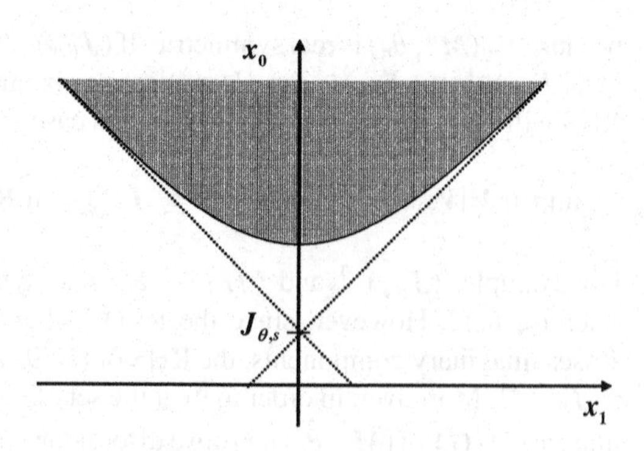

an estimator attaining the bound $n \min_{M^n} \operatorname{tr}(J_\theta^{M^n})^{-1}$. Hence, the set of realizable classical Fisher information J_θ^M and the set of $\frac{1}{n} J_\theta^{M^n}$ characterize the bound of estimation performance. When the family in the two-dimensional space has the form $\{\rho_\theta | \|\boldsymbol{\theta}\| = r\}$, they are as illustrated in Fig. 6.3. In Fig. 6.3, we assume that $J_{\theta,s}$ is a constant time of the identity matrix.

Proof Let us apply the same argument as in the Proof of Theorem 6.10 to the probability distribution family $\{\mathrm{P}_{\rho_\theta^{\otimes n}}^{M^n} | \theta\}$. Then

$$\langle b | \hat{V}_\theta(M^n, \hat{\theta}_n) | b \rangle \langle a | J_\theta^{M^n} | a \rangle \geq \left| \sum_{i,j} \overline{b_i} \frac{\partial}{\partial \theta^j} \mathrm{E}_\theta(M^n, \hat{\theta}_n)^i a_j \right|^2$$

for complex vectors $|b\rangle = (b_1, \ldots, b_d)^T \in \mathbb{C}^d$ and $|a\rangle \in \mathbb{C}^d$. Define $(A_n)^i_j \stackrel{\text{def}}{=} \frac{\partial}{\partial \theta^j} \mathrm{E}_\theta(M^n, \hat{\theta}_n)^i$ and substitute $a = (J_\theta^{M^n})^{-1} A_n b$. Then $\langle b | \hat{V}_\theta(M^n, \hat{\theta}_n) | b \rangle \geq \langle b | A_n^* (J_\theta^{M^n})^{-1} A_n | b \rangle$. Therefore,

$$\begin{aligned}
&\lim_{n\to\infty} \operatorname{tr} n \hat{V}_\theta(M^n, \hat{\theta}_n) \geq \overline{\lim} \operatorname{tr} A_n A_n^* n (J_\theta^{M^n})^{-1} \\
\geq &\overline{\lim} \inf_{M^n:\ \text{POVM on } \mathcal{H}^{\otimes n}} \operatorname{tr} A_n A_n^* n (J_\theta^{M^n})^{-1} \\
= &\overline{\lim}\, n \inf_{M^n:\ \text{POVM on } \mathcal{H}^{\otimes n}} \operatorname{tr} (J_\theta^{M^n})^{-1},
\end{aligned}$$

which completes the proof. ■

More generally, under the same conditions as in the previous theorem, we have [7]

$$\begin{aligned}
&\operatorname{tr} \lim_{n\to\infty} n \hat{V}_\theta(M^n, \hat{\theta}_n) \\
\geq &\min_{\mathbf{X}} \left\{ \operatorname{tr} \mathbf{Re}\, V_\theta(\mathbf{X}) + \operatorname{tr} | \mathbf{Im}\, V_\theta(\mathbf{X}) | \,\middle|\, \delta_i^j = \operatorname{Tr} \frac{\partial \rho_\theta}{\partial \theta^i} X^j \right\},
\end{aligned} \tag{6.101}$$

where X is a matrix, $\mathbf{Re}X$ is the matrix consisting of the real part of each component of X, $\mathbf{Im}X$ is the matrix consisting of the imaginary part of each component of X, and $\boldsymbol{V}_\theta(\mathbf{X}) \stackrel{\rm def}{=} (\mathrm{Tr}\, \rho_\theta X^i X^j)$ for a vector of matrices $\mathbf{X} = (X^1, \cdots, X^d)$. It is known that there exists a sequence of estimators satisfying the equality in (6.101) [32, 33]. In the proof of this argument, the quantum central limit theorem [34, 35] plays an essential role [32].

Such an argument can be given for infinite-dimensional systems. In particular, the quantum Gaussian state family is known as the quantum analog of the Gaussian distribution family and is a typical example in an infinite-dimensional system. In the classical case, the Gaussian distribution family has been extensively investigated. Similarly, the quantum Gaussian state family has been extensively investigated in the classical case [7, 16, 31, 32, 36–39].

Another related topic to state estimation is approximate state cloning. Of course, it is impossible to completely clone a given state. However, an approximate cloning is possible by first estimating the state to be cloned, then generating this estimated state twice. Although the initial state is changed in this case, it can be approximately recovered from the knowledge obtained via the estimation. An approximate cloning is therefore possible via state estimation. In fact, it is more convenient to treat the cloning process directly without performing the estimation. Then, the optimum cloning method is strictly better than the method via estimation [40]. In particular, the analysis for approximate state cloning is simplified for spaces having a group symmetry, e.g., sets of pure states [41, 42]. An investigation has also been done in an attempt to find the interaction that realizes the optimal cloning [43]. The analysis is more difficult for problems with less symmetry [44].

The probabilistic framework of mathematical statistics has been applied to many fields where statistical methods are necessary. In many cases, this probabilistic framework is merely a convenience for the applied field. That is, the probabilistic description is often used to supplement the lack of knowledge of the system of interest. In such a use of statistical methods, there is a possibility that statistical methods might be superseded by other methods due to further developments such as increasing computer speed and improvements in analysis. However, as discussed in Chap. 1, the probabilistic nature of quantum mechanics is intrinsic to the theory itself. Therefore, in fact, the framework of mathematical statistics can be naturally applied to quantum mechanics. Unfortunately, at present, it is not possible to operate a large number of quantum-mechanical particles as a collection of single quantum systems. Therefore, when we measure the order of 10^{23} particles, we often obtain only the average of the measured ensemble as the final outcome. The quantum-mechanical correlations cannot be controlled in this situation. Furthermore, quantum-mechanical effects such as those given in this text cannot be realized. Additionally, when an observable X is measured on a system in the state ρ, the measurement outcome coincides with $\mathrm{Tr}\, \rho X$ with a probability nearly equal to 1. Therefore, statistical methods are clearly not necessary in this case.

In proportion to experimental technology advances in microscopic systems, we can expect the growth in demand to individually operate a large number of quantum-mechanical particles. The measurement outcome will behave probabilistically in this situation, and therefore mathematical statistical methods will become more necessary. In fact, in several experiments, statistical methods have already been used to determine the generated quantum state [45]. Therefore, the theory presented here should become more important with future experimental progress.

Exercises

6.42 Show the following facts when a separable POVM $\boldsymbol{M}^n = \{M^n(\omega)\}_{\omega \in \Omega_n}$ in $\mathcal{H}^{\otimes n}$ is written as $M^n(\omega) = M_1^n(\omega) \otimes \cdots \otimes M_n^n(\omega)$.
(**a**) Show that a POVM $\boldsymbol{M}_{\theta:n,i}$ defined by (6.102) satisfies the conditions for a POVM and satisfies (6.103):

$$M_{\theta:n,i}(\omega) \stackrel{\text{def}}{=} M_i^n(\omega) \mathrm{Tr} \rho_\theta M_1^n(\omega) \cdots \mathrm{Tr} \rho_\theta M_{i-1}^n(\omega) \cdot \mathrm{Tr} \rho_\theta M_{i+1}^n(\omega) \cdots \mathrm{Tr} \rho_\theta M_n^n(\omega), \tag{6.102}$$

$$\sum_{i=1}^d J_\theta^{\boldsymbol{M}_{\theta:n,i}} = J_\theta^{\boldsymbol{M}_n}. \tag{6.103}$$

(**b**) Show that

$$\mathrm{tr} \lim_{n\to\infty} n \hat{V}_\theta(\boldsymbol{M}^n, \hat{\theta}_n) \geq \inf \left\{ \mathrm{tr}(J_\theta^{\boldsymbol{M}})^{-1} \middle| \boldsymbol{M} \text{ POVM on } \mathcal{H} \right\}. \tag{6.104}$$

6.43 Show the following given a POVM $\boldsymbol{M} = \{M_\omega\}$ in $\mathcal{H}$ of rank $M(\omega) = 1$ [24].
(**a**) Show that $\displaystyle\sum_\omega \frac{\langle M(\omega), M(\omega)\rangle_{\theta,s}^{(e)}}{\langle M(\omega), I\rangle_{\theta,s}^{(e)}} = \dim \mathcal{H}$.
(**b**) Show that $\mathrm{tr}\, J_{\theta,s}^{-1} J_\theta^{\boldsymbol{M}} = \displaystyle\sum_\omega \sum_{j=1}^d \frac{\langle M(\omega), L_{\theta,s}^j\rangle_{\theta,s}^{(e)} \langle L_{\theta,j,s}, M(\omega)\rangle_{\theta,s}^{(e)}}{\langle M(\omega), I\rangle_{\theta,s}^{(e)}}$, where $L_{\theta,s}^j \stackrel{\text{def}}{=} \sum_{i=1}^d (J_{\theta,s}^{-1})_{i,j} L_{\theta,j,s}$.
(**c**) Show that

$$\mathrm{tr}\, J_{\theta,s}^{-1} J_\theta^{\boldsymbol{M}} \leq \dim \mathcal{H} - 1. \tag{6.105}$$

When $\rho_\theta > 0$, show that the equality holds if and only if every element $M(\omega)$ can be written as a linear sum of $I, L_{\theta,1,s}, \ldots, L_{\theta,d,s}$.
(**d**) Give the condition for the equality in (6.105) for cases other than $\rho_\theta > 0$.
(**e**) Show that inequality (6.105) also holds if the POVM $\boldsymbol{M} = \{M_\omega\}$ is not of rank $M(\omega) = 1$.

6.44 When an estimator $(\boldsymbol{M}, \hat{\theta})$ for the state family $\{\rho_\theta | \theta \in \mathbb{R}^d\}$ in $\mathcal{H}$ satisfies

$$\left.\frac{\partial}{\partial \theta^j} \mathrm{E}_\theta^i(\boldsymbol{M}^n, \hat{\theta}_n)\right|_{\theta=\theta_0} = \delta_j^i, \quad \mathrm{E}_{\theta_0}^i(\boldsymbol{M}^n, \hat{\theta}_n) = \theta_0^i,$$

it is called a **locally unbiased estimator** at θ_0. Show that

$$\begin{aligned}&\inf\left\{\operatorname{tr}(J_\theta^{M})^{-1}\middle| M \text{ POVM on } \mathcal{H}\right\}\\&=\inf\left\{\operatorname{tr}\hat{V}_\theta(M,\hat{\theta})\middle|(M,\hat{\theta}):\text{ a locally unbiased estimator}\right\}.\end{aligned} \tag{6.106}$$

6.45 Show the following equation and that $J=\frac{(d-1)}{\operatorname{tr}J_{\theta,s}^{-\frac{1}{2}}}J_{\theta,s}^{\frac{1}{2}}$ gives the minimum value [24].

$$\min_{J:\text{ symmetric matrix}}\left\{\operatorname{Tr}(J^{-1})\middle|\operatorname{Tr}J_{\theta,s}^{-1}J=d-1\right\}=\frac{(\operatorname{tr}J_{\theta,s}^{-\frac{1}{2}})^2}{d-1}.$$

6.46 Fix a normalized vector u on $\in\mathbb{R}^d$, i.e., assume that $\|u=(u^1,\ldots,u^d)\|=1$. Let M^u be a measurement corresponding to the spectral decomposition of $L(u)\stackrel{\text{def}}{=}\sum_{j=1}^d u^j L_{\theta,j,s}$. Show that the Fisher information satisfies

$$J_\theta^{M^u}\geq\frac{1}{\langle u|J_{\theta,s}|u\rangle}J_{\theta,s}|u\rangle\langle u|J_{\theta,s}. \tag{6.107}$$

6.47 Let M and M' be POVMs $\{M_\omega\}$ and $\{M'_{\omega'}\}$ with probability spaces Ω and Ω', respectively. Let M'' be a POVM that performs the measurements M, M' with probability λ, $(1-\lambda)$. Show that $\lambda J_\theta^{M}+(1-\lambda)J_\theta^{M'}=J_\theta^{M''}$.

6.48 Consider the set of vectors $u_1,\ldots,u_k\in\mathbb{R}^d$ with norm 1 in parameter space. Let M^p be the POVM corresponding to the probabilistic mixture of spectral decomposition $L(u^i)$ with probability p_i. Show that the Fisher information matrix satisfies

$$J_\theta^{M^p}\geq\sum_{i=1}^k p_i\frac{1}{\langle u_i|J_{\theta,s}|u_i\rangle}J_{\theta,s}|u_i\rangle\langle u_i|J_{\theta,s}. \tag{6.108}$$

6.49 Using the result of the preceding exercise, show the following equation regardless of the number of parameters when $\dim\mathcal{H}=2$ [24, 46–48]:

$$\inf\left\{\operatorname{tr}(J_\theta^{M})^{-1}\middle| M\text{ POVM on }\mathcal{H}\right\}=\left(\operatorname{tr}J_{\theta,s}^{-\frac{1}{2}}\right)^2.$$

6.50 Using the result of the preceding exercise, show the following equation under the above assumption [24, 46–48].

$$\inf\left\{\operatorname{tr}G(J_\theta^{M})^{-1}\middle| M\text{ POVM on }\mathcal{H}\right\}=\left(\operatorname{tr}\left(\sqrt{G}^{-1}J_{\theta,s}\sqrt{G}^{-1}\right)^{-\frac{1}{2}}\right)^2. \tag{6.109}$$

6.51 Let $\{p_\theta(i)|\theta=(\theta^1,\ldots,\theta^d)\in\mathbb{R}^d\}$ be a probability family and $U_i (i=1,\ldots,k)$ be unitary matrices in $\mathcal{H}$. Consider the TP-CP map $\kappa_\theta:\rho\mapsto\sum_{i=1}^k p_\theta(i)$

$U_i \rho U_i^*$. Then, let $\boldsymbol{J}_{\theta,\rho,x}$ be the Fisher information matrix of the quantum state family $\{\kappa_\theta(\rho)|\theta \in \mathbb{R}^d\}$ for $x = s, r, b, \lambda, p$, and $\boldsymbol{J}_\theta$ be the Fisher information matrix of the probability distribution family $\{p_\theta(i)|\theta = (\theta^1, \ldots, \theta^d) \in \mathbb{R}^d\}$. Show that

$$\boldsymbol{J}_\theta \geq \boldsymbol{J}_{\theta,\rho,x}. \tag{6.110}$$

6.52 Show that the equality in (6.110) holds if

$$\operatorname{Tr} U_i \rho U_i^* U_j \rho U_j^* = 0 \tag{6.111}$$

holds for $i \neq j$. In addition, let P_i be the projection to the range of $U_i \rho U_i^*$. Show that the output distribution of the PVM $\{P_i\}$ is equal to $p_\theta(i)$, i.e., the problem reduces to estimating this probability distribution.

6.53 Consider the problem of estimating the probability distribution θ with the generalized Pauli channel κ_{p_θ} given in Example 5.8 as the estimation of the channel $\kappa_{p_\theta} \otimes \iota_A$. Show that (6.111) holds if $\rho = |\Phi_d\rangle\langle\Phi_d|$, where $|\Phi_d\rangle = \frac{1}{d}\sum_{i=0}^d |u_i\rangle \otimes |u_i\rangle$ and $d = \dim \mathcal{H}_A$.

6.54 As in the preceding problem, show that no estimator can improve the estimation accuracy of the estimator with the input state $|\Phi_d\rangle\langle\Phi_d|^{\otimes n}$, even though any entangled state is input to a channel $\left(\kappa_{p_\theta} \otimes \iota_A\right)$ defined with respect to the generalized Pauli channel κ_{p_θ}.

6.55 Prove (6.99) following the steps below [7].
(**a**) Show that an arbitrary antisymmetric matrix Y may be rewritten as

$$VYV^t = \begin{pmatrix} 0 & \alpha_1 & & & \\ -\alpha_1 & 0 & & & \\ & & 0 & \alpha_2 & \\ & & -\alpha_2 & 0 & \\ & & & & \ddots \end{pmatrix}$$

for a suitable real orthogonal matrix V.
(**b**) Show that

$$\max_{X:\text{real antisymmetric matrix}} \{\operatorname{Tr} X | X - iY \geq 0\} = \operatorname{Tr} |iY| \tag{6.112}$$

for a real antisymmetric matrix Y.
(**c**) Show (6.99) using (**b**).

6.56 Define the $\mathcal{D}_\theta$ operator as

$$E_{\rho_\theta,s}\mathcal{D}_\theta(X) = i[X, \rho_\theta]. \tag{6.113}$$

Let T_θ be the space spanned by $\{L_{\theta,1,s}, \ldots, L_{\theta,d,s}\}$. Let $\bar{T}_\theta$ be the orbit of T_θ with respect to the action of $\mathcal{D}_\theta$.

Show the following items.

(**a**) Show the following:

$$\min_{\mathbf{X}} \left\{ \operatorname{tr} \mathbf{Re}\, \boldsymbol{V}_\theta(\mathbf{X}) + \operatorname{tr} |\mathbf{Im}\, \boldsymbol{V}_\theta(\mathbf{X})| \,\middle|\, \delta_i^j = \operatorname{Tr} \frac{\partial \rho_\theta}{\partial \theta^i} X^j \right\}$$
$$= \min_{\mathbf{X}: X^j \in \bar{T}_\theta} \left\{ \operatorname{tr} \mathbf{Re}\, \boldsymbol{V}_\theta(\mathbf{X}) + \operatorname{tr} |\mathbf{Im}\, \boldsymbol{V}_\theta(\mathbf{X})| \,\middle|\, \delta_i^j = \operatorname{Tr} \frac{\partial \rho_\theta}{\partial \theta^i} X^j \right\}. \quad (6.114)$$

(**b**) Show the following:

$$\min_{\mathbf{X}: X^j \in \bar{T}_\theta} \left\{ \operatorname{tr} \mathbf{Re}\, \boldsymbol{V}_\theta(\mathbf{X}) + \operatorname{tr} |\mathbf{Im}\, \boldsymbol{V}_\theta(\mathbf{X})| \,\middle|\, \delta_i^j = \operatorname{Tr} \frac{\partial \rho_\theta}{\partial \theta^i} X^j \right\}$$
$$= \frac{1}{n} \min_{\mathbf{X}: X^j \in \bar{T}_\theta} \left\{ \operatorname{tr} \mathbf{Re}\, \boldsymbol{V}_\theta(\mathbf{X}) + \operatorname{tr} |\mathbf{Im}\, \boldsymbol{V}_\theta(\mathbf{X})| \,\middle|\, \delta_i^j = \operatorname{Tr} \frac{\partial \rho_\theta^{\otimes n}}{\partial \theta^i} X^j \right\}. \quad (6.115)$$

6.57 Show inequality (6.101) following the steps below [7, 16].

(**a**) Show that the RHS of (6.101) has the same value as the original even if we added the constraint condition $\operatorname{Tr} \rho_\theta X^i = 0$ to the minimization.

(**b**) Assume that an estimator $(\boldsymbol{M}, \hat{\theta})$ and a vector $\mathbf{X} = (X^i)$ of Hermitian matrices satisfy $X^i = \sum_i \hat{\theta}^i(\omega) M_\omega$. Show that $\boldsymbol{V}_\theta(\boldsymbol{M}, \hat{\theta}) \geq \boldsymbol{V}_\theta(\mathbf{X})$.

(**c**) Show that $\operatorname{tr} \boldsymbol{V}_\theta(\boldsymbol{M}, \hat{\theta}) \geq \operatorname{tr}(\mathbf{Re}\, \boldsymbol{V}_\theta(\mathbf{X}) + |\mathbf{Im}\, \boldsymbol{V}_\theta(\mathbf{X})|)$.

(**d**) Show (6.101).

6.58 Show that the equality in (6.101) holds for a pure state following the steps below [49–52].

(**a**) Let $\rho_\theta = |u\rangle\langle u|$ and let $\mathbf{X} = (X^i)$ be a vector of Hermitian matrices satisfying $\operatorname{Tr} \rho_\theta X^i = 0$. Show that the vectors $x^i \stackrel{\text{def}}{=} X^i u$ are orthogonal to $|u\rangle$ and satisfy $\boldsymbol{V}_\theta(\mathbf{X}) = (\langle x^i, x^j \rangle)$.

(**b**) Choose u_i such that $\frac{\partial \rho_\theta}{\partial \theta^i} = (|u_i\rangle\langle u| + |u\rangle\langle u_i|)/2$ with the condition $\langle u|u_i\rangle = 0$. Define the matrix $\boldsymbol{V}(\boldsymbol{x}) := (\langle x^i, x^j \rangle)$. Show that

$$\min_{\mathbf{X}} \left\{ \operatorname{tr} \mathbf{Re}\, \boldsymbol{V}_\theta(\mathbf{X}) + \operatorname{tr} |\mathbf{Im}\, \boldsymbol{V}_\theta(\mathbf{X})| \,\middle|\, \delta_i^j = \operatorname{Tr} \frac{\partial \rho_\theta}{\partial \theta^i} X^j \right\}$$
$$= \min_{\boldsymbol{x} := (x^1, \ldots, x^d)} \left\{ \operatorname{tr} \mathbf{Re}\, \boldsymbol{V}(\boldsymbol{x}) + \operatorname{tr} |\mathbf{Im}\, \boldsymbol{V}(\boldsymbol{x})| \,\middle|\, \delta_i^j = \langle x^i | u_j \rangle \right\}.$$

(**c**) Consider the case where all $\langle x^i | x^j \rangle$ are real. Suppose that orthogonal vectors v_k are real linear sums of the vectors $u, x^1, \ldots, x^d$, and each $\langle v_i | u \rangle$ is nonzero. Then, we make the POVM $\{|v_k\rangle\langle v_k|\}$. Show that the estimator $(\boldsymbol{M}, \hat{\theta}) \stackrel{\text{def}}{=} (\{|v_k\rangle\langle v_k|\}, \frac{\langle v_k | x^j \rangle}{\langle v_k | u \rangle})$ satisfies $\mathrm{E}_\theta^i(\boldsymbol{M}, \hat{\theta}) u = x^j$. Also, show that $\hat{\boldsymbol{V}}_\theta(\boldsymbol{M}, \hat{\theta}) = \boldsymbol{V}(\boldsymbol{x})$.

(**d**) Let $x^1, \ldots, x^d$ be a set of vectors such that $\langle x^i | x^j \rangle$ are not necessarily real. Show that there exists a set of vectors $w^1, \ldots, w^d$ in another d-dimensional space such that $|\mathbf{Im}\, \boldsymbol{V}(\boldsymbol{x})| - \mathbf{Im}\, \boldsymbol{V}(\boldsymbol{x}) = (\langle w^i | w^j \rangle)$.

(e) Under the same assumption as **(d)**, show that $\langle y^i | y^j \rangle$ are all real, where $y^i \stackrel{\text{def}}{=} x^i \oplus w^i$ and $\oplus$ denotes the direct sum product.
(f) For a given set of vectors $x^1, \ldots, x^d$, show that there exists an estimator $(\boldsymbol{M}, \hat{\theta})$ such that $\mathrm{E}^i_\theta(\boldsymbol{M}, \hat{\theta})u = x^j$ and $\hat{V}_\theta(\boldsymbol{M}, \hat{\theta}) = \mathbf{Re}(V(\boldsymbol{x})) + |\,\mathbf{Im}(V(\boldsymbol{x}))|$.
(g) Show that the equality in (6.101) holds for a pure state.
(h) Examine the state family consisting of pure states with $2l$ parameters, where l is the dimension of the Hilbert space. Show that the RHS of (6.101) is equal to $\mathrm{tr}(\mathbf{Re}\, J)^{-1} + \mathrm{tr}\,|(\mathbf{Re}\, J)^{-1}\, \mathbf{Im}\, J (\mathbf{Re}\, J)^{-1}|$, where $J = (J_{i,j} \stackrel{\text{def}}{=} \langle u_i | u_j \rangle)$.

6.7 Relative Modular Operator and Quantum f-Relative Entropy

6.7.1 Monotonicity Under Completely Positivity

In this section, we introduce the relative modular operator and quantum f-relative entropy, and investigate their properties. The content require only that of Sect. 6.1 among this chapter, but has so different taste from the main topic of this chapter that the content cannot be put in the next of Sect. 6.1. Since the topic is not related to other sections in this chapter, and is related only to Sect. 5.4, we discuss this topic in the end of this chapter.

For this purpose, we focus on two density matrices $\rho = \sum_i a_i |u_i\rangle\langle u_i|$ and $\sigma = \sum_j b_j |v_j\rangle\langle v_j|$. Given a matrix convex function f defined on $[0, \infty)$, we define the **quantum f-relative entropy** $D_f(\rho\|\sigma) \stackrel{\text{def}}{=} \sum_{i,j} f(\frac{a_i}{b_j}) Q_{(\rho\|\sigma)}(i, j)$, where $Q_{(\rho\|\sigma)}(i, j) = b_j |\langle v_j | u_i \rangle|$ [53] (See Sect. 3.2 for the detail of the notation $Q_{(\rho\|\sigma)}$.). When the matrix convex function f is defined only on $(0, \infty)$, i.e., it diverges at 0, the quantum f-relative entropy $D_f(\rho\|\sigma)$ can be defined only when $P_\rho \geq P_\sigma$, P_ρ is the image of ρ.

To analyze quantum f-relative entropy, we define the two super operators L_σ^{-1} and R_ρ as a linear map on the matrix space $\mathcal{M}_{\sigma,r}$:

$$L_\sigma(X) \stackrel{\text{def}}{=} \sigma X, \quad R_\rho(X) \stackrel{\text{def}}{=} X\rho.$$

Using these super operators, we define the relative modular operator $\Delta_{\rho,\sigma} \stackrel{\text{def}}{=} R_\rho L_\sigma^{-1}$. By using the **relative modular operator**, we define another super operator $f(\Delta_{\rho,\sigma})$. So, the quantum f-relative entropy $D_f(\rho\|\sigma)$ can be rewritten as

$$D_f(\rho\|\sigma) = \mathrm{Tr}\, f(\Delta_{\rho,\sigma})\sigma. \tag{6.116}$$

Similar to Sect. 2.1.2, when we choose the matrix convex function $f(x) = x \log x$ defined on $[0, \infty)$, the quantum f-relative entropy $D_f(\rho\|\sigma)$ is the quantum relative entropy $D(\rho\|\sigma)$.

Then, we have the following **monotonicity** for quantum f-relative entropy.

Theorem 6.12 (Petz [53]) *For a TP-CP map κ, the monotonicity relation*

$$D_f(\kappa(\rho)\|\kappa(\sigma)) \le D_f(\rho\|\sigma) \tag{6.117}$$

holds for a matrix convex function f defined on $[0,\infty)$ *when $P_\rho \le P_\sigma$. When the matrix convex function f defined only on* $(0,\infty)$, (6.117) *holds under the assumption when $P_\rho = P_\sigma$.*

As another choice, the function $f(x) = x^\alpha$ with $\alpha \in [1,2]$ is a matrix convex function defined on $[0,\infty)$. Then, the quantum f-relative entropy $D_f(\rho\|\sigma)$ is $e^{(\alpha-1)D_\alpha(\rho\|\sigma)} = e^{\phi(\alpha-1|\rho\|\sigma)}$. The function $f(x) = -x^\alpha$ with $\alpha \in [0,1]$ is also a matrix convex function defined on $[0,\infty)$. The quantum f-relative entropy $D_f(\rho\|\sigma)$ is $-e^{(\alpha-1)D_\alpha(\rho\|\sigma)} = -e^{\phi(\alpha-1|\rho\|\sigma)}$. Then, we obtain (5.52) and (5.53). The functions $f(x) = x^\alpha$ with $\alpha \in [-1,0)$ and $f(x) = -\log x$ are matrix convex functions defined only on $(0,\infty)$.

To show Theorem 6.12, we focus on the inner product $\langle\ ,\ \rangle^{(e)}_{\sigma,r}$ and the space $\mathcal{M}^{(m)}_{\sigma,r}(\mathcal{H})$, which can be identified with the quotient space $\mathcal{M}_{\sigma,r}(\mathcal{H})$. The map $\Delta_{\rho,\sigma}$ is positive Hermitian under the inner product $\langle Y, X\rangle^{(e)}_{\sigma,r} = \operatorname{Tr} Y^*\sigma X$ because

$$\langle X, \Delta_{\rho,\sigma} X\rangle^{(e)}_{\sigma,r} = \operatorname{Tr} Y^*\sigma\sigma^{-1}X\rho = \operatorname{Tr} Y^* P_\sigma X\rho \ge 0$$

Then, the quantum f-relative entropy $D_f(\rho\|\sigma)$ can be rewritten as

$$D_f(\rho\|\sigma) = \langle P_\sigma, \Delta_{\rho,\sigma} P_\sigma\rangle^{(e)}_{\sigma,r} \tag{6.118}$$

when $P_\rho \le P_\sigma$.

Now, we prepare the following lemmas.

Lemma 6.3 *When κ is the partial trace from $\mathcal{H}_{A,B} = \mathcal{H}_A \otimes \mathcal{H}_B$ to $\mathcal{H}_A$, a matrix convex function f defined on* $[0,\infty)$ *satisfies that*

$$\operatorname{Tr} X^*\kappa(\sigma) f(\Delta_{\kappa(\rho),\kappa(\sigma)})(X) \le \operatorname{Tr}(X^*\otimes I)\sigma f(\Delta_{\rho,\sigma})(X\otimes I) \tag{6.119}$$

for $X \in \mathcal{M}^{(m)}_{\sigma,r}(\mathcal{H}_{A,B})$ when $P_\rho \le P_\sigma$.

Lemma 6.4 *When κ is the partial trace from $\mathcal{H}_{A,B} = \mathcal{H}_A \otimes \mathcal{H}_B$ to $\mathcal{H}_A$, a matrix convex function f defined on* $[0,\infty)$ *satisfies that*

$$\sigma^s f(\Delta_{\rho,\sigma})(\sigma^t) = f(\Delta_{\rho,\sigma})(\sigma^{s+t}) = \sigma^{s+t} f(\Delta_{\rho,\sigma})(P_\sigma) = \sigma^{s+t} f(\Delta_{\rho,\sigma})(P) \tag{6.120}$$

for $s,t > 0$ when P is a projection satisfying $P \ge P_\sigma$.

Proof of Theorem 6.12 Now, we show Theorem 6.12 by using Lemma 6.3. Choose the Stinespring representation of the TP-CP map κ as $\kappa(\rho) = \operatorname{Tr}_B U(\rho\otimes\rho_0)U^*$. Substituting $P_{\kappa(\sigma)}$ into X in Lemma 6.3, we have

$$\begin{aligned}
&D_f(\kappa(\rho)\|\kappa(\sigma)) = \operatorname{Tr} f(\Delta_{\kappa(\rho),\kappa(\sigma)})(\kappa(\sigma)) \\
&\overset{(a)}{=} \operatorname{Tr} P_{\kappa(\sigma)}\kappa(\sigma) f(\Delta_{\kappa(\rho),\kappa(\sigma)})(P_{\kappa(\sigma)}) \\
&\overset{(b)}{\le} \operatorname{Tr}(P_{\kappa(\sigma)} \otimes I)U(\sigma \otimes \rho_0)U^* f(\Delta_{U(\rho\otimes\rho_0)U^*,U(\sigma\otimes\rho_0)U^*})(P_{\kappa(\sigma)} \otimes I) \\
&\overset{(c)}{=} \operatorname{Tr} f(\Delta_{U(\rho\otimes\rho_0)U^*,U(\sigma\otimes\rho_0)U^*})(U(\sigma \otimes \rho_0)U^*) = \operatorname{Tr} f(\Delta_{\rho\otimes\rho_0,\sigma\otimes\rho_0})(\sigma \otimes \rho_0) \\
&= \operatorname{Tr}(f(\Delta_{\rho,\sigma})(\sigma)) \otimes \rho_0 = D_f(\rho\|\sigma).
\end{aligned} \tag{6.121}$$

Here, (a) and (c) follow from Lemma 6.4, and (b) follows from Lemma 6.3. ■

Now, we show (6.17) by using Lemma 6.3.

Proof of (6.17) Substituting $\rho^{A,B}$ $(-x^\lambda)$ into σ $(f(x))$, we have

$$\begin{aligned}
&- \operatorname{Tr}_A X^*(\operatorname{Tr}_B \rho^{A,B})^{1-\lambda} X(\operatorname{Tr}_B \rho^{A,B})^\lambda \\
&\le - \operatorname{Tr}_{A,B}(X \otimes I_B)^*(\rho^{A,B})^{1-\lambda}(X \otimes I_B)(\rho^{A,B})^\lambda
\end{aligned} \tag{6.122}$$

because $f(L^{-1}_{\rho^{A,B}} R_{\sigma^{A,B}})X = (\rho^{A,B})^{-\lambda} X(\sigma^{A,B})^\lambda$. Hence, we obtain (6.17). ■

Proof of Lemma 6.3 Although the map $\kappa^*(X) = X \otimes I$ is defined as the dual of κ with respect to the Hilbert Schmidt inner product, the dual $\kappa^*_{\sigma,r}$ of $\kappa_{\sigma,r}$ with respect to the inner products $\langle\ ,\ \rangle^{(e)}_{\sigma,r}$ and $\langle\ ,\ \rangle^{(e)}_{\kappa(\sigma),r}$ is also κ^*. Note that $\kappa_{\sigma,r}$ is a map from $\mathcal{M}^{(m)}_{\sigma,r}(\mathcal{H}_{A,B})$ to $\mathcal{M}^{(m)}_{\kappa(\sigma),r}(\mathcal{H}_A)$. This is because

$$\begin{aligned}
&\langle \kappa^*_{\sigma,r}(Y), X\rangle^{(e)}_{\sigma,r} = \langle Y, \kappa_{\sigma,r}(X)\rangle^{(e)}_{\kappa(\sigma),r} = \operatorname{Tr}_A Y^*\kappa(\sigma)\kappa_{\sigma,r}(X) \\
&= \operatorname{Tr}_A Y^*\kappa(\sigma X) = \operatorname{Tr}_A Y^* \operatorname{Tr}_B(\sigma X) = \operatorname{Tr}(Y^* \otimes I)(\sigma X) = \langle Y \otimes I, X\rangle^{(e)}_{\sigma,r}
\end{aligned} \tag{6.123}$$

for $X, Y \in \mathcal{M}^{(m)}_{\sigma,r}(\mathcal{H}_A)$. Since $P_\rho \le P_\sigma$, we have

$$\begin{aligned}
&\kappa_{\sigma,r} \circ \Delta_{\rho,\sigma} \circ \kappa^*_{\sigma,r}(Y) = \kappa_{\sigma,r} \circ \Delta_{\rho,\sigma}(Y \otimes I) = \kappa_{\sigma,r}(\sigma^{-1}(Y \otimes I)\rho) \\
&=\kappa(\sigma)^{-1}\kappa(\sigma\sigma^{-1}(Y \otimes I)\rho) = \kappa(\sigma)^{-1}\kappa(P_\sigma(Y \otimes I)\rho) \\
&=(\operatorname{Tr}_B \sigma)^{-1} Y(\operatorname{Tr}_B \rho) = \Delta_{\kappa(\rho),\kappa(\sigma)} Y
\end{aligned} \tag{6.124}$$

for $Y \in \mathcal{M}^{(m)}_{\sigma,r}(\mathcal{H}_A)$.

Since $\kappa_{\sigma,r} \circ \kappa^*_{\sigma,r}(Y) = \kappa_{\sigma,r}(Y \otimes I) = (\operatorname{Tr}_B(\sigma))^{-1} \operatorname{Tr}_B \sigma(Y \otimes I) = (\operatorname{Tr}_B(\sigma))^{-1}$ $(\operatorname{Tr}_B \sigma)Y = Y$ for $Y \in \mathcal{M}^{(m)}_{\sigma,r}(\mathcal{H}_A)$, $\kappa_{\sigma,r} \circ \kappa^*_{\sigma,r}$ is the identity operator on the space $\mathcal{M}^{(m)}_{\sigma,r}(\mathcal{H}_A)$.

Applying Condition ② in Theorem A.1 to operators $\Delta_{\rho,\sigma}$ and $\kappa^*_{\sigma,r}$ on the space $\mathcal{M}^{(m)}_{\sigma,r}(\mathcal{H}_{A,B})$, we obtain

$$\begin{aligned}
&\operatorname{Tr} X^*\kappa(\sigma)\Delta_{\kappa(\rho),\kappa(\sigma)}(X) = \operatorname{Tr} X^*\kappa(\sigma) f(\kappa_{\sigma,r}\Delta_{\rho,\sigma}\kappa^*_{\sigma,r})(X) \\
&=\langle X, f(\kappa_{\sigma,r}\Delta_{\rho,\sigma}\kappa^*_{\sigma,r})(X)\rangle^{(e)}_{\kappa(\sigma),r} \le \langle X, \kappa_{\sigma,r} f(\Delta_{\rho,\sigma}) \circ \kappa^*_{\sigma,r}(X)\rangle^{(e)}_{\kappa(\sigma),r}
\end{aligned}$$

$$=\langle \kappa_{\sigma,r}^*(X), f(\Delta_{\rho,\sigma}) \circ \kappa_{\sigma,r}^*(X)\rangle_{\sigma,r}^{(e)} = \mathrm{Tr}\, \kappa^*(X^*)\sigma f(\Delta_{\rho,\sigma}) \circ \kappa^*(X)$$
$$= \mathrm{Tr}(X^* \otimes I)\sigma f(\Delta_{\rho,\sigma})(X \otimes I).$$

■

Proof of Lemma 6.4 We have

$$\sigma^s f(\Delta_{\rho,\sigma})(\sigma^t) = \sigma^s f(L_\sigma^{-1}) \circ f(R_\rho)(\sigma^t) = \sigma^s f(\sigma^{-1})\sigma^t f(\rho)$$
$$= f(\sigma^{-1})\sigma^{s+t} f(\rho) = f(\Delta_{\rho,\sigma})(\sigma^{s+t}).$$

Since

$$\sigma^s f(\sigma^{-1})\sigma^t f(\rho) = \sigma^{s+t} f(\sigma^{-1}) P f(\rho) = \sigma^{s+t} f(\sigma^{-1}) P_\sigma f(\rho),$$

we can show the remaining relations. ■

6.7.2 Monotonicity Under 2-Positivity

Next, we relax the condition for the map κ to 2-positivity. This relaxation for our analysis seems to have no physical meaning because it is too mathematical. However, this analysis is very useful for deriving the equality condition discussed in Theorem 5.8 even under the completely positivity. For this purpose, a function f defined on $(0, \infty)$ is called sub-linear when $\lim_{x\to\infty} f(x)/x = 0$.

Theorem 6.13 ([54, 55]) *For any TP-2-positive map κ, the monotonicity relation*

$$D_f(\kappa(\rho)\|\kappa(\sigma)) \le D_f(\rho\|\sigma) \tag{6.125}$$

holds when one of the following conditions holds.

① *f is a sub-linear matrix convex function defined on $[0, \infty)$, e.g., $f(x) = -x^\alpha$ with $\alpha \in [0, 1)$.*
② *f is a sub-linear matrix convex function defined on $(0, \infty)$, (e.g., $f(x) = x^\alpha$ with $\alpha \in [-1, 0)$, $f(x) = -\log x$), and $P_\rho \ge P_\sigma$.*
③ *f is a matrix convex function defined on $[0, \infty)$, (e.g., $f(x) = x^\alpha$ with $\alpha \in (1, 2]$, $f(x) = x \log x$), and $P_\rho \le P_\sigma$.*
④ *f is a matrix convex function defined on $(0, \infty)$, and $P_\rho = P_\sigma$.*

Indeed the advantage of Theorem 6.13 over Theorem 6.12 is not limited to the condition for the map κ. Theorem 6.13 also relaxes the condition for the projections P_ρ and P_σ when the matrix convex function f is sub-linear. The detail treatment of Theorem 6.13 enables such a subtle analysis. Further, we have the following equality condition. In the following discussion, the extremal decomposition given in Theorem A.2 plays an essential role.

Theorem 6.14 ([54, 55]) *For a TP-2-positive map κ, the following conditions are equivalent*

① *Equality in* (6.125) *holds for any* $f_t(x) = \frac{1}{x+t}$ *with an arbitrary* $t > 0$.
② *There exists a real number* $\alpha \in (0, 1)$ *such that the equality in* (6.125) *holds for* $f(x) = -x^{\alpha}$.
③ *Equality in* (6.125) *holds when the matrix convex function* f *is a sub-linear matrix convex function defined on* $[0, \infty)$.
④ *The relation* $P_\sigma \kappa^*(\kappa(\sigma)^{-t} P_{\kappa(\sigma)} \kappa(\rho)^t)) = \sigma^{-t} P_\sigma \rho^t$ *holds for any* $t > 0$.

When the relation $P_\rho \le P_\sigma$ *holds, Condition* ④ *be simplified as follows.*

④' *The relation* $P_\sigma \kappa^*(\kappa(\sigma)^{-t} \kappa(\rho)^t)) = \sigma^{-t} \rho^t$ *holds for any* $t > 0$.

Under this assumption, Conditions ①–④ *are equivalent to the following conditions.*

⑤ *Equality in* (6.125) *holds for* $f(x) = x \log x$.
⑥ *There exists a real number* $\alpha \in (1, 2)$ *such that the equality in* (6.125) *holds for* $f(x) = x^{\alpha}$.
⑦ *Equality in* (6.125) *holds when the matrix convex function* f *is a convex function defined on* $[0, \infty)$.
⑧ $P_\sigma \kappa^*((\log \kappa(\rho) - \log \kappa(\sigma)) P_{\kappa(\rho)}) = (\log \rho - \log \sigma) P_\rho$.

When the relation $P_\rho \ge P_\sigma$ *holds, Conditions* ①–④ *are equivalent to the following conditions.*

❶ *Equality in* (6.125) *holds for any* $f_t(x) = \frac{1}{x+t}$ *with an arbitrary* $t \ge 0$.
❷ *Equality in* (6.125) *holds for* $f(x) = -\log x$.
❸ *There exists a real number* $\alpha \in [-1, 0)$ *such that the equality in* (6.125) *holds for* $f(x) = x^{\alpha}$.
❹ *Equality in* (6.125) *holds when the matrix convex function* f *is a sub-linear matrix convex function defined on* $(0, \infty)$.
❺ *The relation* $P_\sigma \kappa^*(\kappa(\sigma)^{-t} P_{\kappa(\sigma)} \kappa(\rho)^t)) = \sigma^{-t} P_\sigma \rho^t$ *holds for any real number* t.
❻ $P_\sigma \kappa^*(P_{\kappa(\sigma)}(\log \kappa(\rho) - \log \kappa(\sigma))) = P_\sigma(\log \rho - \log \sigma)$.

When the relation $P_\rho = P_\sigma$ *holds, Conditions* ①–⑧, ❶–❻ *are equivalent to the following condition.*

❼ *Equality in* (6.125) *holds when the matrix convex function* f *is a convex function defined on* $(0, \infty)$.

As the special case of TP-CP maps, we obtain the following corollary.

Corollary 6.1 ([54]) *For any TP-CP map κ, the following conditions are equivalent when* $P_\rho \le P_\sigma$.

① *Conditions given in Theorem 6.14 with* $P_\rho \le P_\sigma$ *hold.*
② *The relation* $\sigma^{1/2} \kappa^*(\kappa(\sigma)^{-1/2} \kappa(\rho) \kappa(\sigma)^{-1/2}) \sigma^{1/2} = \rho$ *holds.*

When σ is invertible, Condition ② of Corollary 6.1 is rewritten as $\kappa^*_{\sigma,\frac{1}{2}}(\rho) = \rho$. That is, this can be interpreted via the conditional expectation with respect to the inner product $x = \frac{1}{2}$. However, when σ is not invertible, we cannot define the dual map $\kappa^*_{\sigma,\frac{1}{2}}$.

Proof of Corollary 6.1 We show ②⇒①. Assume ② for a TP-CP map κ from $\mathcal{H}_A$ to $\mathcal{H}_B$ and states ρ and σ on $\mathcal{H}_A$. Then, we define the TP-CP map τ_1 from $\mathcal{H}_B$ to $\mathcal{H}_A$ as $\tau_1(\rho') := P_{\kappa(\sigma)}\rho' P_{\kappa(\sigma)} + (\operatorname{Tr}\rho'(I - P_{\kappa(\sigma)}))\kappa(\sigma)$. We denote the image of $P_{\kappa(\sigma)}$ by $\mathcal{H}'_B$. So, we define the TP-CP map τ_2 from $\mathcal{H}'_B$ to $\mathcal{H}_A$ as $\tau_2(\rho'') := \sigma^{1/2}\kappa^*(\kappa(\sigma)^{-1/2}\rho''\kappa(\sigma)^{-1/2})\sigma^{1/2}$. Since $\tau_2^*(I) = \kappa(\sigma)^{-1/2}\kappa(\sigma^{1/2}I\sigma^{1/2})\kappa(\sigma)^{-1/2} = I$, τ_2 is trace-preserving. So, we have $\tau_2\circ\tau_1(\kappa(\rho)) = \rho$ and $\tau_2\circ\tau_1(\kappa(\sigma)) = \sigma$. Since

$$D(\rho\|\sigma) \ge D(\kappa(\rho)\|\kappa(\sigma)) \ge D(\tau_2\circ\tau_1(\kappa(\rho))\|\tau_2\circ\tau_1(\kappa(\sigma))) = D(\rho\|\sigma),$$

we obtain ①.

Next, we show ①⇒②. Firstly, we show it only when κ is the partial trace from $\mathcal{H}_{A,B} = \mathcal{H}_A\otimes\mathcal{H}_B$ to $\mathcal{H}_A$. In this case, the map κ satisfies Condition ② of Theorem 5.3. So, ⑤ with $t = 1/2$ implies that

$$\begin{aligned}
&P_\sigma\kappa^*(\kappa(\sigma)^{-1/2}P_{\kappa(\sigma)}\kappa(\rho)P_{\kappa(\sigma)}\kappa(\sigma)^{-1/2})P_\sigma\\
=&P_\sigma\kappa^*(\kappa(\sigma)^{-1/2}P_{\kappa(\sigma)}\kappa(\rho)^{1/2}))\kappa^*(\kappa(\rho)^{1/2})P_{\kappa(\sigma)}\kappa(\sigma)^{-1/2})P_\sigma\\
=&\sigma^{-1/2}P_\sigma\rho^{1/2}\rho^{1/2}P_\sigma\sigma^{-1/2} = \sigma^{-1/2}P_\sigma\rho P_\sigma\sigma^{-1/2}.
\end{aligned} \tag{6.126}$$

Multiplying $\sigma^{1/2}$ from both sides, we obtain ②.

Next, we consider the general case. We choose the Stinespring representation of the TP-CP map κ as $\kappa(\rho) = \operatorname{Tr}_B U(\rho\otimes\rho_0)U^*$. So, we have

$$\begin{aligned}
&P_{U(\sigma\otimes\rho_0)U^*}(\kappa(\sigma)^{-1/2}P_{\kappa(\sigma)}\kappa(\rho)P_{\kappa(\sigma)}\kappa(\sigma)^{-1/2})\otimes I_B P_{U(\sigma\otimes\rho_0)U^*}\\
=&(U(\sigma\otimes\rho_0)U^*)^{-1/2}P_{U(\sigma\otimes\rho_0)U^*}U(\rho\otimes\rho_0)U^*P_{U(\sigma\otimes\rho_0)U^*}(U(\sigma\otimes\rho_0)U^*)^{-1/2}.
\end{aligned} \tag{6.127}$$

Since $P_{\sigma\otimes\rho_0} = P_\sigma\otimes P_{\rho_0}$, applying the unitary U and U^*, we have

$$\begin{aligned}
&(P_\sigma\otimes I)(I\otimes P_{\rho_0})U^*((\kappa(\sigma)^{-1/2}P_{\kappa(\sigma)}\kappa(\rho)P_{\kappa(\sigma)}\kappa(\sigma)^{-1/2})\otimes I_B)\\
&\cdot U(I\otimes P_{\rho_0})(P_\sigma\otimes I)\\
=&P_{\sigma\otimes\rho_0}U^*((\kappa(\sigma)^{-1/2}P_{\kappa(\sigma)}\kappa(\rho)P_{\kappa(\sigma)}\kappa(\sigma)^{-1/2})\otimes I_B)UP_{\sigma\otimes\rho_0}\\
=&(\sigma\otimes\rho_0)^{-1/2}P_{\sigma\otimes\rho_0}(\rho\otimes\rho_0)P_{\sigma\otimes\rho_0}(\sigma\otimes\rho_0)^{-1/2}\\
=&(\sigma^{-1/2}P_\sigma\rho P_\sigma\sigma^{-1/2})\otimes P_{\rho_0}.
\end{aligned} \tag{6.128}$$

Since $\kappa^*(X)$ is given as $\operatorname{Tr}_B(I\otimes P_{\rho_0})U^*(X\otimes I_B)U(I\otimes P_{\rho_0})$, taking the partial trace, we obtain

$$P_\sigma\kappa^*(\kappa(\sigma)^{-1/2}P_{\kappa(\sigma)}\kappa(\rho)P_{\kappa(\sigma)}\kappa(\sigma)^{-1/2})P_\sigma = \sigma^{-1/2}P_\sigma\rho P_\sigma\sigma^{-1/2}. \tag{6.129}$$

Since $P_\rho \le P_\sigma$, we have $P_{\kappa(\rho)} \le P_{\kappa(\sigma)}$, which implies that

$$P_\sigma \kappa^*(\kappa(\sigma)^{-1/2}\kappa(\rho)\kappa(\sigma)^{-1/2})P_\sigma = \sigma^{-1/2}\rho\sigma^{-1/2}. \tag{6.130}$$

Multiplying $\sigma^{1/2}$ from both sides, we obtain ②. ■

To prove the above arguments. we prepare the map $\hat{\kappa}$ from the matrix space $\mathcal{M}_{\kappa(\sigma),r}$ to the matrix space $\mathcal{M}_{\sigma,r}$ as

$$\hat{\kappa}(\kappa(\sigma)^{1/2}X) = \sigma^{1/2}\kappa^*(X), \tag{6.131}$$

for $X \in \mathcal{M}_{\kappa(\sigma),r}$. That is, $\hat{\kappa}(X)$ is defined to be $\sigma^{1/2}\kappa^*(\kappa(\sigma)^{-1/2}X)$. From here, we focus on the Hilbert Schmidt inner product $\langle\, ,\,\rangle$ instead of $\langle\, ,\,\rangle^{(e)}_{\sigma,r}$ and $\langle\, ,\,\rangle^{(e)}_{\kappa(\sigma),r}$. Then, we prepare two lemmas.

Lemma 6.5 *Given a TP-2-positive map κ, we have the following items.*

(1) The monotonicity relation (6.125) *with $f = f_t$ holds for $t > 0$.*
(2) When $P_\rho \geq P_\sigma$, the monotonicity relation (6.125) *with $f = f_t$ holds for $t = 0$.*
(3) When $P_\rho \leq P_\sigma$, the equality $D_f(\rho\|\sigma) = D_f(\kappa(\rho)\|\kappa(\sigma))$ holds with a function $f = ax + b$.
(4) When $P_\rho \leq P_\sigma$, the monotonicity relation (6.125) *holds with a quadratic function $f(x) = x^2$.*

Lemma 6.6 *For any TP-2-positive map κ, we have the following matrix inequalities on the matrix space $\mathcal{M}_{\sigma,r}$ with respect to the Hilbert Schmidt inner product $\langle\, ,\,\rangle$;*

$$\hat{\kappa}^*\hat{\kappa} \leq I_{\mathcal{M}_{\sigma,r}} \tag{6.132}$$

$$\hat{\kappa}^*\Delta_{\rho,\sigma}\hat{\kappa} \leq \Delta_{\kappa(\rho),\kappa(\sigma)} \tag{6.133}$$

$$(\Delta_{\kappa(\rho),\kappa(\sigma)} + t)^{-1} \leq \hat{\kappa}^*(\Delta_{\rho,\sigma} + t)^{-1}\hat{\kappa} \tag{6.134}$$

for $t > 0$.

Proof of Theorem 6.13 Now, using Lemma 6.5, we show Theorem 6.13. (A.47) of Theorem A.2 guarantees that any sub-linear matrix convex function defined on $[0, \infty)$ can be written as a positive sum of functions f_t with $t > 0$ and a constant. So, (1) of Lemma 6.5 yields the desired argument under Condition ①.

Also, (A.45) of Theorem A.2 guarantees that any sub-linear matrix convex function defined on $(0, \infty)$ can be written as a positive sum of functions f_t with $t \geq 0$ and a constant. So, (2) of Lemma 6.5 yields the desired argument under Condition ②.

Similarly, the combination of (A.46) of Theorem A.2 and (1), (3), (4) of Lemma 6.5 yields the desired argument under Condition ③.

Finally, the combination of (A.44) of Theorem A.2 and (2), (3), (4) of Lemma 6.5 yields the desired argument under Condition ④. ■

Proof of Theorem 6.14 **Step 1:** Firstly, we discuss the case without the assumption $P_\rho \leq P_\sigma$ nor $P_\rho \geq P_\sigma$. The decomposition (A.37) guarantees the equivalence between ① and ②. Due to the same reason, the decomposition (A.47) in Theorem

A.2 guarantees the equivalence between ① and ③. Since ④ with $t \in (0, 1)$ yields that

$$\begin{aligned}&\operatorname{Tr} \kappa(\sigma)^{1-t}\kappa(\rho)^t = \operatorname{Tr} \kappa(\sigma P_\sigma)\kappa(\sigma)^{-t} P_{\kappa(\sigma)}\kappa(\rho)^t))\\ =&\operatorname{Tr} \sigma P_\sigma \kappa^*(\kappa(\sigma)^{-t} P_{\kappa(\sigma)}\kappa(\rho)^t)) = \operatorname{Tr} \sigma\sigma^{-t} P_\sigma \rho^t = \operatorname{Tr} \sigma^{1-t}\rho^t,\end{aligned} \tag{6.135}$$

which implies ②. Hence, it is sufficient to show that ①⇒④.

Assume that Condition ① holds. So, (6.149) implies that

$$\begin{aligned}&\langle \kappa(\sigma)^{1/2}, (\Delta_{\kappa(\rho),\kappa(\sigma)} + t)^{-1}(\kappa(\sigma)^{1/2})\rangle\\ =&\langle \kappa(\sigma)^{1/2}, \hat{\kappa}^* \circ (\Delta_{\rho,\sigma} + t)^{-1} \circ \hat{\kappa}(\kappa(\sigma)^{1/2})\rangle\end{aligned} \tag{6.136}$$

for $t > 0$. (6.134) and (6.136) imply that

$$\begin{aligned}&(\Delta_{\kappa(\rho),\kappa(\sigma)} + t)^{-1}(\kappa(\sigma)^{1/2}) = \hat{\kappa}^* \circ (\Delta_{\rho,\sigma} + t)^{-1} \circ \hat{\kappa}(\kappa(\sigma)^{1/2})\\ =&\hat{\kappa}^* \circ (\Delta_{\rho,\sigma} + t)^{-1}(\sigma^{1/2}).\end{aligned} \tag{6.137}$$

Taking the derivative for t in this equation, we have

$$(\Delta_{\kappa(\rho),\kappa(\sigma)} + t)^{-2}(\kappa(\sigma)^{1/2}) = \hat{\kappa}^* \circ (\Delta_{\rho,\sigma} + t)^{-2}(\sigma^{1/2}). \tag{6.138}$$

Thus,

$$\begin{aligned}&\langle \hat{\kappa}^* \circ (\Delta_{\rho,\sigma} + t)^{-1}(\sigma^{1/2}), \hat{\kappa}^* \circ (\Delta_{\rho,\sigma} + t)^{-1}(\sigma^{1/2})\rangle\\ \stackrel{(a)}{=}&\langle (\Delta_{\kappa(\rho),\kappa(\sigma)} + t)^{-1}(\kappa(\sigma)^{1/2}), (\Delta_{\kappa(\rho),\kappa(\sigma)} + t)^{-1}(\kappa(\sigma)^{1/2})\rangle\\ =&\langle \kappa(\sigma)^{1/2}, (\Delta_{\kappa(\rho),\kappa(\sigma)} + t)^{-2}(\kappa(\sigma)^{1/2})\rangle \stackrel{(b)}{=} \langle \kappa(\sigma)^{1/2}, \hat{\kappa}^* \circ (\Delta_{\rho,\sigma} + t)^{-2}(\sigma^{1/2})\rangle\\ =&\langle \hat{\kappa}(\kappa(\sigma)^{1/2}), (\Delta_{\rho,\sigma} + t)^{-2}\sigma^{1/2}\rangle = \langle \sigma^{1/2}, (\Delta_{\rho,\sigma} + t)^{-2}(\sigma^{1/2})\rangle\\ =&\langle (\Delta_{\rho,\sigma} + t)^{-1}(\sigma^{1/2}), (\Delta_{\rho,\sigma} + t)^{-1}(\sigma^{1/2})\rangle,\end{aligned} \tag{6.139}$$

where (a) and (b) follow from (6.137) and (6.138), respectively. Thus, the combination of (6.132) and (6.139) implies that

$$\hat{\kappa} \circ \hat{\kappa}^* \circ (\Delta_{\rho,\sigma} + t)^{-1}(\sigma^{1/2}) = (\Delta_{\rho,\sigma} + t)^{-1}(\sigma^{1/2}) \stackrel{(a)}{=} \sigma^{1/2}(\Delta_{\rho,\sigma} + t)^{-1}(P_\sigma), \tag{6.140}$$

where (a) follows from Lemma 6.4. Since (6.137) and (6.131) imply that $\hat{\kappa} \circ \hat{\kappa}^* \circ (\Delta_{\rho,\sigma} + t)^{-1}(\sigma^{1/2}) = \hat{\kappa} \circ (\Delta_{\kappa(\rho),\kappa(\sigma)} + t)^{-1}(\kappa(\sigma)^{1/2}) = \sigma^{1/2}\kappa^* \circ ((\Delta_{\kappa(\rho),\kappa(\sigma)} + t)^{-1}(P_{\kappa(\sigma)}))$, we have

$$\sigma^{1/2}\kappa^*((\Delta_{\kappa(\rho),\kappa(\sigma)} + t)^{-1}(P_{\kappa(\sigma)})) = \sigma^{1/2}(\Delta_{\rho,\sigma} + t)^{-1}(P_\sigma). \tag{6.141}$$

That is,

$$P_\sigma \kappa^*((\Delta_{\kappa(\rho),\kappa(\sigma)} + t)^{-1}(P_{\kappa(\sigma)})) = P_\sigma(\Delta_{\rho,\sigma} + t)^{-1}(P_\sigma). \tag{6.142}$$

Due to Stone-Weierstrass theorem, any continuous function can be approximated by a sum of f_t. So, any continuous function f defined on $[0, \infty)$ satisfies

$$P_\sigma \kappa^*(f(\Delta_{\kappa(\rho),\kappa(\sigma)})(P_{\kappa(\sigma)})) = P_\sigma f(\Delta_{\rho,\sigma})(P_\sigma). \tag{6.143}$$

Applying $f(x) = x^t$, we obtain ④. So, we obtain the required equivalence relations in this case.
Step 2: Next, we discuss the case with the assumption $P_\rho \leq P_\sigma$. We have already shown the equivalence from ① to ④. ④ is trivially simplified to ④' due to the condition $P_\rho \leq P_\sigma$. In this case, a linear function f satisfies the equality $D_f(\rho\|\sigma) = D_f(\kappa(\rho)\|\kappa(\sigma))$ due to (3) of Lemma 6.5. So, the decomposition (A.41) guarantees the equivalence between ① and ⑤. Due to the same reason, the decomposition (A.43) guarantees the equivalence between ① and ⑥. When ④'is assumed, the relation (6.135) with $t = 2$ shows the equality $D_f(\rho\|\sigma) = D_f(\kappa(\rho)\|\kappa(\sigma))$ for $f(x) = x^2$. So, the decomposition (A.46) of Theorem A.2 guarantees ④'+①⇒⑦. Also, the relation ⑦⇒① is trivial. Taking the derivative in ④' at $t = 0$, we obtain ④⇒⑧.

Assume ⑧. Multiplying σ and taking the trace, we have ⑤. So, we obtain the required equivalence relations in this case.
Step 3: Next, we discuss the case with the assumption $P_\rho \leq P_\sigma$. Notice that the equivalence from ① to ④ has been already shown.

Assume ①. Due to the assumption $P_\rho \leq P_\sigma$, we can apply (6.143) to any continuous function f defined on $(0, \infty)$. So, we choose $f(x) = x^t$ for any real number t. Hence, we obtain ❺, which implies ① ⇒ ❺.

Assume ❺. the relation (6.135) with $t = -1$ shows the equality $D_{f_0}(\rho\|\sigma) = D_{f_0}(\kappa(\rho)\|\kappa(\sigma))$ for $f(x) = x^2$. Since ❺ ⇒ ④ ⇒ ①, we have ❶, which implies ❺ ⇒ ❶. Trivially, ❶ ⇒ ①.

The decomposition (A.42) guarantees the equivalence between ❶ and ❷. Also, the decomposition (A.41) guarantees the equivalence between ❶ and ❸. Similarly, the decomposition (A.45) of Theorem A.2 guarantees ❶ ⇒ ❹.

Taking the derivative at $t = 0$ in ❺, we obtain ❺ ⇒ ❻. Assume ❺. Multiplying σ and taking the trace, we obtain ❷. So, we obtain the required equivalence relations in this case.
Step 4: Finally, we discuss the case with the assumption $P_\rho = P_\sigma$. Notice that the equivalence from ① to ⑧ and from ❶ to ❻ has been already shown. The decomposition (A.47) of Theorem A.2 guarantees ❶+⑦ with $f(x) = x^2$ ⇒ ❼. Since ❼ is stronger requirement than ③, we obtain the required equivalence relations. ■

Proof of Lemma 6.6 For $X \in \mathcal{M}_{\kappa(\sigma),r}$, we have

$$
\begin{aligned}
&\langle \hat{\kappa}(\kappa(\sigma)^{1/2}X), \hat{\kappa}(\kappa(\sigma)^{1/2}X)\rangle = \langle \sigma^{1/2}\kappa^*(X), \sigma^{1/2}\kappa^*(X)\rangle \\
=&\operatorname{Tr} \kappa^*(X)^*\sigma^{1/2}\sigma^{1/2}\kappa^*(X) = \operatorname{Tr} \sigma\kappa^*(X)\kappa^*(X)^* \\
\le&\operatorname{Tr} \sigma\kappa^*(XX^*) = \operatorname{Tr} \kappa(\sigma)XX^* = \langle \kappa(\sigma)^{1/2}\hat{\kappa}(X), \kappa(\sigma)^{1/2}\hat{\kappa}(X)\rangle. \qquad (6.144)
\end{aligned}
$$

Since any element of $\mathcal{M}_{\kappa(\sigma),r}$ can be written with the form $\kappa(\sigma)^{1/2}X$, (6.144) implies (6.132).

For $X \in \mathcal{M}_{\kappa(\sigma),r}$, we also have

$$
\begin{aligned}
&\langle \kappa(\sigma)^{1/2}X, \hat{\kappa}^* \circ \Delta_{\rho,\sigma} \circ \hat{\kappa}(\kappa(\sigma)^{1/2}X)\rangle = \langle \hat{\kappa}(\kappa(\sigma)^{1/2}X), \Delta_{\rho,\sigma} \circ \hat{\kappa}(\kappa(\sigma)^{1/2}X)\rangle \\
=&\langle \sigma^{1/2}\kappa^*(X), \Delta_{\rho,\sigma}(\sigma^{1/2}\kappa^*(X))\rangle = \langle \sigma^{1/2}\kappa^*(X), \sigma^{-1/2}P_\sigma\kappa^*(X)\rho\rangle \\
=&\operatorname{Tr} \kappa^*(X)^*\sigma^{1/2}\sigma^{-1/2}P_\sigma\kappa^*(X)\rho = \operatorname{Tr} \kappa^*(X)^*P_\sigma\kappa^*(X)\rho \\
=&\operatorname{Tr} \kappa^*(X)^*P_\sigma\kappa^*(X)\rho \le \operatorname{Tr} \kappa^*(X)^*\kappa^*(X)\rho = \operatorname{Tr} \kappa^*(X^*)\kappa^*(X)\rho, \qquad (6.145) \\
&\langle \kappa(\sigma)^{1/2}X, \Delta_{\kappa(\rho),\kappa(\sigma)}(\kappa(\sigma)^{1/2}X)\rangle = \langle \kappa(\sigma)^{1/2}X, \kappa(\sigma)^{-1/2}X\kappa(\rho)\rangle \\
=&\operatorname{Tr} X^*\kappa(\sigma)^{1/2}\kappa(\sigma)^{-1/2}X\kappa(\rho) = \operatorname{Tr} X^*X\kappa(\rho) = \operatorname{Tr} \rho\kappa^*(X^*X). \qquad (6.146)
\end{aligned}
$$

So, the inequalities (5.3), (6.146), and (6.145) imply (6.133).

(6.133) implies that $\hat{\kappa}^*\Delta_{\rho,\sigma}\hat{\kappa} + t \ge \Delta_{\kappa(\rho),\kappa(\sigma)} + t$. Since $x \mapsto -x^{-1}$ is matrix monotone, we have

$$
(\Delta_{\kappa(\rho),\kappa(\sigma)} + t)^{-1} \le (\hat{\kappa}^*\Delta_{\rho,\sigma}\hat{\kappa} + t)^{-1}. \qquad (6.147)
$$

Since the function $x \mapsto x^{-1}$ satisfies the condition of Corollary A.2, (6.132) implies that

$$
(\hat{\kappa}^*\Delta_{\rho,\sigma}\hat{\kappa} + t)^{-1} \le \hat{\kappa}^*(\Delta_{\rho,\sigma} + t)^{-1}\hat{\kappa}. \qquad (6.148)
$$

Thus, (6.147) and (6.148) yields (6.134). ∎

Proof of Lemma 6.5 Now, we show Lemma 6.5 by using Lemma 6.6. When $t > 0$, we have

$$
\begin{aligned}
&D_{f_t}(\kappa(\rho)\|\kappa(\sigma)) = \operatorname{Tr} f_t(\Delta_{\kappa(\rho),\kappa(\sigma)})(\kappa(\sigma)) \\
\overset{(a)}{=}&\operatorname{Tr} \kappa(\sigma)^{1/2} f_t(\Delta_{\kappa(\rho),\kappa(\sigma)})(\kappa(\sigma)^{1/2}) = \langle \kappa(\sigma)^{1/2}, (\Delta_{\kappa(\rho),\kappa(\sigma)} + t)^{-1}(\kappa(\sigma)^{1/2})\rangle \\
\overset{(b)}{\le}&\langle \kappa(\sigma)^{1/2}, \hat{\kappa}^* \circ (\Delta_{\rho,\sigma} + t)^{-1} \circ \hat{\kappa}(\kappa(\sigma)^{1/2})\rangle \\
=&\langle \hat{\kappa}(\kappa(\sigma)^{1/2}), (\Delta_{\rho,\sigma} + t)^{-1} \circ \hat{\kappa}(\kappa(\sigma)^{1/2})\rangle \\
=&\langle \sigma^{1/2}, (\Delta_{\rho,\sigma} + t)^{-1}(\sigma^{1/2})\rangle \overset{(c)}{=} D_{f_t}(\rho\|\sigma). \qquad (6.149)
\end{aligned}
$$

Here, (a) and (c) follow from Lemma 6.4, and (b) follows from (6.134) in Lemma 6.6. Thus, we obtain the first argument.

When $P_\rho \ge P_\sigma$, we have $P_{\kappa(\rho)} \ge P_{\kappa(\sigma)}$. So, the matrices σ and $\kappa(\sigma)$ belong to the spaces spanned by eigen spaces corresponding to non-zero eigenvalue of the super operators $\Delta_{\rho,\sigma}$ and $\Delta_{\kappa(\rho),\kappa(\sigma)}$, respectively. So, we have the relation (6.149) with $t = 0$. Thus, we obtain the second argument.

When $P_\rho \leq P_\sigma$ and $f(x) = ax + b$, we have the equality in (6.149). Choosing $f(x) = x^2$, we have $D_f(\rho\|\sigma) = D_{f_0}(\sigma\|\rho) \leq D_{f_0}(\kappa(\sigma)\|\kappa(\rho)) = D_f(\kappa(\sigma)\|\kappa(\rho))$. Hence, we obtain the third argument. ■

6.8 Historical Note

6.8.1 Quantum State Estimation

Research on quantum state estimation was initiated by Helstrom [2] in 1967. He derived the one-parameter Cramér–Rao inequality (6.75) for the nonasymptotic version. He also proved the multiparameter SLD Cramér–Rao inequality (6.96) for the nonasymptotic version [6]. Yuen and Lax [37] developed the RLD version of the Cramér–Rao inequality for estimation with a complex multiparameter. They applied it to the estimation of the complex amplitude of the Gaussian state. Belavkin [56] derived a necessary and sufficient condition for the achievement of this bound. Further, Holevo [7] derived the RLD Cramér–Rao inequality (6.96) with a real multiparameter and obtained the lower bound (6.101) with the locally unbiased condition in the nonasymptotic case [7].

Young introduced the concept of quasiclassical POVM concerning the state family [23]. Nagaoka [13] focused on (6.106) and derived the SLD one-parameter Cramér–Rao inequality (6.75) with an asymptotic framework. He derived its lower bound based on inequality (7.33) [57]. This bound is called the Nagaoka bound. Applying it to the quantum two-level system, he obtained (6.109) for the two-parameter case [58]. Hayashi [46, 47] applied the duality theorem in infinite-dimensional linear programming to quantum state estimation and obtained (6.109) in the three-parameter case as well as in the two-parameter case. After these developments, Gill and Massar [24] derived the same equation by a simpler method, which is explained in Exercise 6.50. Fujiwara and Nagaoka [51] defined the coherent model as a special case of pure-state families and showed that bound (6.101) can be attained with the locally unbiased and nonasymptotic framework in this case. Following this result, Matsumoto [29, 52] extended it to the general pure-state case. Further, Hayashi and Matsumoto [31] showed that bound (6.101) can be attained with the asymptotic framework in the quantum two-level system using the Cramér–Rao approach. The achievability of bound (6.101) is discussed in Matsumoto [33] in a general framework using irreducible decompositions of group representation. It has also been examined in Hayashi [32] using the quantum central limit theorem.

As a nonasymptotic extension of the quantum Crámer–Rao inequality, Tsuda and Matsumoto [59] treated its nondifferentiable extension (Hammersley–Chapman–Robbins–Kshiragar bound). They also derived the lower bound of mean square errors of unbiased estimators based on higher-order derivatives (quantum Bhattacharyya bound). The quantum Bhattacharyya bound has also been obtained by Brody and

Hughston [60] in the pure-state case. Using this bound, Tsuda [61] derived an interesting bound for the estimation of polynomials of complex amplitude of quantum Gaussian states. Further, nonparametric estimation has been researched by D'Ariano [62] and Artiles et al. [63].

The group covariant approach was initiated by Helstrom [64]. He treated the estimation problem of one-parameter covariant pure-state families. Holevo has established the general framework of this approach [65] and applied it to several problems. Ozawa [66] and Bogomolov [67] extended it to the case of the noncompact parameter space. Holevo applied it to the estimation of the shifted one-parameter pure-state family [68]. Holevo [7] and Massar and Popescu [69] treated the estimation of a pure qubit state with n-i.i.d. samples using the Fidelity risk function. Hayashi [70] extended it to an arbitrary dimensional case with the general invariant risk function. Bruß et al. [40] discussed its relation with approximate cloning. Further, Hayashi [71] applied this method to the estimation of the squeezed parameter with vacuum-squeezed-state families. Hayashi and Matsumoto [31] also treated the estimation of the full-parameter model in quantum two-level systems using this approach. Bagan et al. [72] treated the same problem by the covariant and Bayesian approach.

Nagaoka [12] extended Bahadur's large deviation approach to the quantum estimation and found that the estimation accuracy with condition (6.88) is bounded by the Bogoljubov Fisher information in this approach. Hayashi [26] introduced a more strict condition (6.93) and showed that the estimation accuracy with condition (6.93) is bounded by the SLD Fisher information.

6.8.2 Quantum Channel Estimation

Fujiwara [73] started to treat the estimation of a quantum channel within the framework of quantum state estimation. Sasaki et al. [74] discussed a similar estimation problem with the Bayesian approach in a nonasymptotic setting. Fischer et al. [75] focused on the use of the maximally entangled input state for the estimation of the Pauli channel. Fujiwara and Imai [76] showed that in the estimation of the Pauli channel κ_θ, the best estimation performance is obtained if and only if the input state is the n-fold tensor product of the maximally entangled state $|\Phi_d\rangle\langle\Phi_d|^{\otimes n}$. Exercise 6.54 treats the same problem using a different approach. After this result, Fujiwara [77] and Tanaka [78] treated, independently, the estimation problem of the amplitude damping channel. Especially, Fujiwara [77] proceeded to the estimation problem of the generalized amplitude damping channel, which is the more general and difficult part. De Martini et al. [79] implemented an experiment for the estimation of an unknown unitary.

Concerning the estimation of unitary operations, Bužek et al. [80] focused on estimating an unknown one-parameter unitary action first time. They showed that the error goes to 0 with the order $\frac{1}{n^2}$, where n is the number of applications of the unknown operation. Acín et al. [81] characterized the optimal input state for the SU(d) estimation where the input state is entangled with the reference system.

On the other hand, Fujiwara [82] treats this problem using the Cramér–Rao approach in the SU(2) case. This result was extended by Ballester [83]. Bagan et al. [84] treated the estimation of the unknown n-identical SU(2) operations using entanglement with the reference system. They also showed that the optimal error goes to 0 at a rate of $\frac{\pi^2}{n^2}$ and effectively applied the Clebsch–Gordan coefficient method to this problem. Hayashi [85, 86] treated the same problem using a different method. He derived a relation between this problem and that of Bužek et al. [80] and applied the obtained relation to this problem. He also pointed out that the multiplicity of the same irreducible representations can be regarded as the reference system, i.e., the effect of "self-entanglement." Indeed, independently of Hayashi, Chiribella et al. [87] and Bagan et al. [88] also pointed out this effect of the multiplicity based on the idea of Chiribella et al. [89]. That is, these three groups proved that the error of the estimation of SU(2) goes to 0 at a rate of $\frac{\pi^2}{n^2}$. The role of this "self-entanglement" is widely discussed in Chiribella et al. [90]. Note that, as was mentioned by Hayashi [85], the Cramér–Rao approach does not necessarily provide the optimal coefficient in the estimation of unitary operations by the use of entanglement. In particular, as was shown in [91], under the phase estimation with energy constraint, the Cramér–Rao approach does not work because the maximum Fisher information is infinity while the true minimum error can be characterized by using group covariant approach. Chiribella et al. [92] derived the optimal estimator in the Bayesian setup. Recently, Hayashi [93] discussed the Cramer-Rao approach more deeply. He showed the additivity of the maximum of the RLD Fisher information in the case of channel estimation. This fact shows that when the maximum of the RLD Fisher information exists, the maximum SLD Fisher information increase only linearly, i.e., the minimum error behaves as $O(\frac{1}{n})$.

6.8.3 Geometry of Quantum States

The study of monotone metric in quantum state family was initiated by Morozowa and Chentsov [94]. Following this research, Petz [3] showed that every monotone metric is constructed from the matrix monotone function or the matrix average. Nagaoka introduced an SLD one-parameter exponential family [13] and a Bogoljubov one-parameter exponential family [14], characterized them as (6.44) and (6.45), respectively, and calculated the corresponding divergences (6.57) and (6.58) [14]. He also calculated the Bogoljubov m divergence as (6.66) [18]. Other formulas (6.60), and (6.67) for divergences were first obtained by Hayashi [95]. Further, Matsumoto [96] obtained an interesting characterization of RLD (m)-divergence. Moreover, he showed that an efficient estimator exists only in the SLD one-parameter exponential family (Theorem 6.7) [13, 22]. However, before this study, Belavkin [56] introduced a complex-parameterized exponential family and showed that the RLD version of the Cramér–Rao bound with the complex multiparameter could be attained only in special cases. Theorem 6.7 coincides with its real-one-parameter case. Following this result, Fujiwara [97] showed that any unitary SLD one-parameter exponential family is generated by an observable satisfying the canonical commutation relation.

In addition, Amari and Nagaoka [8] introduced the torsion concerning the e parallel translation as the limit of the RHS–LHS in (6.68) and showed that the torsion-free inner product is only a Bogoljubov metric. They proved that the torsions of e-connection vanish only for a Bogoljubov inner product. They also showed that the divergence can be defined by a convex function if and only if the torsions of e-connection and m-connection vanish. Combining these facts, we can derive Theorem 6.5. However, their proof is based on the calculation of Christoffel symbols. In this textbook, Theorem 6.5 is proved without any use of Christoffel symbols.

Further, Nagaoka [11, 12] showed that the Bogoljubov metric is characterized by the limit of the quantum relative entropy as (6.34). Concerning the SLD inner product, Uhlmann [9] showed that the SLD metric is the limit of the Bures distance in the mixed-state case as (6.33). Matsumoto [10] extended it to the general case.

6.8.4 *Equality Condition for Monotonicity of Relative Entropy*

Although Petz [54] derived Corollary 6.1 in terms of operator algebra, he assumed that ρ and σ are invertible. Also, he assume that the dual map κ^* is given as the inclusion of a subalgebra, which corresponds to the case when the original map κ is the partial trace. For a general TP-CP map κ, we have a Stinespring representation as $\kappa(\rho) = \mathrm{Tr}_{AC} U_\kappa(\rho \otimes \rho_0)U_\kappa^*$. Then, the dual map κ^* is given as the combination of the inclusion of a subalgebra and the multiplication of an isometry. To reduce it to the case of the partial trace, we need to treat two states $\rho \otimes \rho_0$ and $\sigma \otimes \rho_0$, which are not invertible. So, Petz's proof for Corollary 6.1 does not work even for invertible states ρ and σ when the TP-CP map κ is not a partial trace. To avoid to assume the invertible property for ρ and σ, we introduce the matrix space $\mathcal{M}_{\sigma,r}(\mathcal{H})$ although Petz's original proof employed only the full matrix space $\mathcal{M}(\mathcal{H})$.

When ρ and σ are invertible, Petz [54] also derived an equivalent Condition ❺ of Theorem 6.14 by replacing t by it (which is called the modified Condition ❺). Since Petz [54] treated the infinite-dimensional case, σ^{-1} might be unbounded. To avoid the difficulty for unboundedness, he employed σ^{-it} instead of σ^{-t}. However, when ρ and σ are not invertible, we need to treat 0^{-it}, which cannot be defined. So, in this book, we employ σ^{-t} instead of σ^{-it} with careful treatment of the projections P_ρ and P_σ.

His derivation ❺ requires only the inequality (5.3), which can be derived from the trace-preserving property and 2-positivity for κ. So, in another paper [55], he rewrote the derivation with the modified Condition ❺ in terms of linear algebra by assuming this weaker condition when ρ and σ are invertible. Theorems 6.13 and 6.14 are extensions of this part in the following sense. Petz [54, 55] assumed that ρ and σ are invertible and treated only the quantum relative entropy $D(\rho\|\sigma)$.

However, Theorems 6.13 and 6.14 can treat non-invertible ρ and σ and general quantum f-relative entropies, where the possible matrix convex function f depends on the relation of images of ρ and σ. Then, Corollary 6.1 gives the same equivalence condition under the TP-CP condition for general quantum f-relative entropies.

In addition, Ohya and Petz [98] applied the result of Petz [54] to the case with measurement. Then, they characterized the existence of a measurement attaining equality in the monotonicity of the relative entropy (3.18) when ρ and σ are invertible. Indeed, once we obtain Theorem 5.8, it is not difficult to derive Theorem 3.6, as shown in Sect. 5.4. However, it is not easy to prove this argument without use of Theorem 5.8. Nagaoka [18] showed the same fact without assuming the invertible condition by using information geometrical method (Exercise 6.32 and Theorem 3.6). Fujiwara [17] improved these discussions further.

6.9 Solutions of Exercises

Exercise 6.1 Equation (6.1) and (6.4) yield (6.6). (6.4) and (6.5) yield (6.7).

Exercise 6.2 Condition (6.1) implies that

$$\begin{aligned}\langle Y, X\rangle^{(e)}_{\rho,x} &= \operatorname{Tr} Y^* E_{\rho,x}(X) = \operatorname{Tr} E_{\rho,x}(Y)^* X\\ =&(\operatorname{Tr} X^* E_{\rho,x}(Y))^* = (\langle X, Y\rangle^{(e)}_{\rho,x})^*.\end{aligned}$$

Also, Condition (6.2) yields that

$$\langle X, X\rangle^{(e)}_{\rho,x} = \operatorname{Tr} X^* E_{\rho,x}(X) \ge 0.$$

Exercise 6.3 Assume ③. Then, for two Hermitian matrices X and Y, $E_{\rho,x}(X)$ and $E_{\rho,x}(Y)$ are also Hermitian. Then, we have

$$\langle X, Y\rangle^{(e)}_{\rho,x} = \operatorname{Tr} X^* E_{\rho,x}(Y) \overset{(a)}{=} \operatorname{Tr} E_{\rho,x}(X)^* Y = \operatorname{Tr} Y^* E_{\rho,x}(X)\langle Y, X\rangle^{(e)}_{\rho,x},$$

where (a) follows from Condition (6.1). Hence, we obtain ②.

Assume ②. Then, two Hermitian matrices X and Y satisfy that

$$\begin{aligned}\operatorname{Tr} Y E_{\rho,x}(X)^* &\overset{(a)}{=} \operatorname{Tr} X^* E_{\rho,x}(Y) = \langle X, Y\rangle^{(e)}_{\rho,x}\\ =&\langle Y, X\rangle^{(e)}_{\rho,x} = \operatorname{Tr} Y^* E_{\rho,x}(X) = \operatorname{Tr} Y E_{\rho,x}(X),\end{aligned}$$

where (a) follows from Condition (6.1). Hence, we obtain ③.

Assume ③. We choose two matrices X and Y such that $E_{\rho,x}(X)$ and $E_{\rho,x}(Y)$ are Hermitian. So, the two matrices X and Y also are Hermitian. Denoting $E_{\rho,x}(X)$ and $E_{\rho,x}(Y)$ by A and B, we have

$$\begin{aligned}\langle A, B\rangle^{(m)}_{\rho,x} &= \langle X, Y\rangle^{(e)}_{\rho,x} = \operatorname{Tr} X^* E_{\rho,x}(Y) \overset{(a)}{=} \operatorname{Tr} E_{\rho,x}(X)^* Y\\ =& \operatorname{Tr} Y^* E_{\rho,x}(X) = \langle Y, X\rangle^{(e)}_{\rho,x} = \langle B, A\rangle^{(m)}_{\rho,x},\end{aligned}$$

where (a) follows from Condition (6.1). Hence, we obtain ①.

Assume ①. We choose two matrices X and Y such that $E_{\rho,x}(X)$ and $E_{\rho,x}(Y)$ are Hermitian. Denoting $E_{\rho,x}(X)$ and $E_{\rho,x}(Y)$ by A and B, we have

$$\begin{aligned}\operatorname{Tr} Y^* E_{\rho,x}(X) &= \langle Y, X\rangle^{(e)}_{\rho,x} = \langle B, A\rangle^{(m)}_{\rho,x} = \langle A, B\rangle^{(m)}_{\rho,x} = \langle X, Y\rangle^{(e)}_{\rho,x}\\ &= \operatorname{Tr} X^* E_{\rho,x}(Y) \stackrel{(a)}{=} \operatorname{Tr} E_{\rho,x}(X)^* Y = \operatorname{Tr} Y E_{\rho,x}(X),\end{aligned}$$

where (a) follows from Condition (6.1). Hence, we obtain ③.

Exercise 6.4

(**a**) We consider only the case when $x \ge y$ because the opposite case can be treated by swapping x and y.

$$\begin{aligned}&\frac{1}{\operatorname{Lm}(x,y)} - \frac{2}{x+y} = \frac{1}{x-y}\left(\log x - \log\frac{x+y}{2} \log\frac{x+y}{2} - \log y\right) - \frac{2}{x+y}\\ &= \frac{1}{x-y}\int_0^{\frac{x-y}{2}} \frac{1}{\frac{x+y}{2}+t} + \frac{1}{\frac{x+y}{2}-t} dt - \frac{2}{x+y}\\ &= \frac{1}{x-y}\int_0^{\frac{x-y}{2}} \frac{x+y}{(\frac{x+y}{2})^2 - t^2} - \frac{4}{x+y} dt\\ &= \frac{1}{x-y}\int_0^{\frac{x-y}{2}} \frac{4}{x+y}\left(\frac{(\frac{x+y}{2})^2}{(\frac{x+y}{2})^2 - t^2} - 1\right) dt\\ &= \frac{4}{x^2-y^2}\int_0^{\frac{x-y}{2}} \frac{t^2}{(\frac{x+y}{2})^2 - t^2} dt \ge 0.\end{aligned}$$

The equality in the final inequality holds if and only if $x = y$.
(**b**) We have $E^{-1}_{\rho,s}(A) = \sum_{j,k=1}^d \frac{2}{\lambda_j+\lambda_k} E_j A E_k$ because

$$\begin{aligned}&\frac{1}{2}\left(\sum_{j,k=1}^d \frac{2}{\lambda_j+\lambda_k} E_j A E_k \rho + \rho \sum_{j,k=1}^d \frac{2}{\lambda_j+\lambda_k} E_j A E_k\right)\\ &= \left(\sum_{j,k=1}^d \frac{\lambda_k}{\lambda_j+\lambda_k} E_j A E_k + \sum_{j,k=1}^d \frac{\lambda_j}{\lambda_j+\lambda_k} E_j A E_k\right)\\ &= \sum_{j,k=1}^d \left(\frac{\lambda_k}{\lambda_j+\lambda_k} + \frac{\lambda_j}{\lambda_j+\lambda_k}\right) E_j A E_k = \sum_{j,k=1}^d E_j A E_k = A.\end{aligned}$$

Thus,

$$\|A\|_{\rho,s}^{(m)} = \operatorname{Tr} A E_{\rho,s}^{-1}(A) = \operatorname{Tr} A \sum_{j,k=1}^{d} \frac{2}{\lambda_j + \lambda_k} E_j A E_k$$

$$= \sum_{j,k=1}^{d} \frac{2}{\lambda_j + \lambda_k} \operatorname{Tr} A E_j A E_k.$$

Since

$$\int_0^1 x^\lambda y^{1-\lambda} d\lambda = \int_0^1 y e^{(\log x - \log y)\lambda} d\lambda = y \left[\frac{e^{(\log x - \log y)\lambda}}{\log x - \log y} \right]_0^1$$

$$= y \frac{\frac{x}{y} - 1}{\log x - \log y} = \frac{x - y}{\log x - \log y} = \operatorname{Lm}(x, y),$$

we have

$$\int_0^1 \rho^\lambda \sum_{j,k=1}^{d} \frac{1}{\operatorname{Lm}(\lambda_j, \lambda_k)} E_j A E_k \rho^{1-\lambda} d\lambda$$

$$= \int_0^1 \sum_{j,k=1}^{d} \frac{1}{\operatorname{Lm}(\lambda_j, \lambda_k)} \lambda_j^\lambda \lambda_k^{1-\lambda} E_j A E_k d\lambda$$

$$= \sum_{j,k=1}^{d} \frac{1}{\operatorname{Lm}(\lambda_j, \lambda_k)} \operatorname{Lm}(\lambda_j, \lambda_k) E_j A E_k = \sum_{j,k=1}^{d} E_j A E_k = A.$$

Thus, $E_{\rho,b}^{-1}(A) = \sum_{j,k=1}^{d} \frac{1}{\operatorname{Lm}(\lambda_j,\lambda_k)} E_j A E_k$. Hence,

$$\|A\|_{\rho,b}^{(m)} = \operatorname{Tr} A E_{\rho,b}^{-1}(A) = \sum_{j,k=1}^{d} \frac{1}{\operatorname{Lm}(\lambda_j, \lambda_k)} \operatorname{Tr} A E_j A E_k.$$

(**c**) Since $A\rho - \rho A = \sum_{j,k=1}^{d} E_j A E_k (\lambda_k - \lambda_j)$, we have

$$\operatorname{Tr} (A\rho - \rho A)(A\rho - \rho A)^* = -\operatorname{Tr} (A\rho - \rho A)(A\rho - \rho A)$$

$$= -\operatorname{Tr} \left(\sum_{j,k=1}^{d} E_j A E_k (\lambda_k - \lambda_j) \right) \left(\sum_{j',k'=1}^{d} E_{j'} A E_{k'} (\lambda_{k'} - \lambda_{j'}) \right)$$

$$= -\operatorname{Tr} \sum_{j,k=1}^{d} E_j A E_k A (\lambda_k - \lambda_j)(\lambda_j - \lambda_k) = \operatorname{Tr} \sum_{j,k=1}^{d} (\lambda_k - \lambda_j)^2 E_j A E_k A.$$

(**d**) The statements (**a**) and (**b**) guarantee that $\|A\|_{\rho,b}^{(m)} \geq \|A\|_{\rho,s}^{(m)}$. The equality $\frac{2}{\lambda_j+\lambda_k} = \frac{1}{\operatorname{Lm}(\lambda_j,\lambda_k)}$ holds only when $\lambda_j = \lambda_k$. Therefore, ① holds if and only if $E_j A E_k A = 0$ or $\lambda_j = \lambda_k$ holds for any $k \neq j$. Due to (**c**), the latter condition is equivalent with ②.

Exercise 6.5

(**a**) $\kappa_{\boldsymbol{M}}(E_{\rho,s}(X))$ is commutative with M_i. Since $\kappa_{\boldsymbol{M}}(E_{\rho,s}(X)) = E_{\kappa_{\boldsymbol{M}}(\rho),s}(\kappa_{\boldsymbol{M},\rho,s}(X))$, $E_{\kappa_{\boldsymbol{M}}(\rho),s}(\kappa_{\boldsymbol{M},\rho,s}(X))$ is commutative with M_i. Since $[\kappa_{\boldsymbol{M}}(\rho), M_i] = 0$,

$$
\begin{aligned}
0 =& [E_{\kappa_{\boldsymbol{M}}(\rho),s}(\kappa_{\boldsymbol{M},\rho,s}(X)), M_i] \\
=&[\frac{1}{2}(\kappa_{\boldsymbol{M}}(\rho)\kappa_{\boldsymbol{M},\rho,s}(X) + \kappa_{\boldsymbol{M},\rho,s}(X)\kappa_{\boldsymbol{M}}(\rho)), M_i] \\
=&\frac{1}{2}(\kappa_{\boldsymbol{M}}(\rho)[\kappa_{\boldsymbol{M},\rho,s}(X), M_i] + [\kappa_{\boldsymbol{M},\rho,s}(X), M_i]\kappa_{\boldsymbol{M}}(\rho)) \\
=&E_{\kappa_{\boldsymbol{M}}(\rho),s}([\kappa_{\boldsymbol{M},\rho,s}(X), M_i]).
\end{aligned} \tag{6.150}
$$

Since the map $E_{\kappa_{\boldsymbol{M}}(\rho),s}$ is injective, $[\kappa_{\boldsymbol{M},\rho,s}(X), M_i] = 0$.
(**b**) Assume that every M_i commutes with X. Thus,

$$
\begin{aligned}
&E_{\kappa_{\boldsymbol{M}}(\rho),s}(X) = \frac{1}{2}(\kappa_{\boldsymbol{M}}(\rho)X + X\kappa_{\boldsymbol{M}}(\rho))) = \kappa_{\boldsymbol{M}}(\frac{1}{2}(\rho X + X\rho)) \\
=&\kappa_{\boldsymbol{M}}(E_{\rho,s}(X)) = E_{\kappa_{\boldsymbol{M}}(\rho),s}(\kappa_{\boldsymbol{M},\rho,s}(X)).
\end{aligned} \tag{6.151}
$$

Since the map $E_{\kappa_{\boldsymbol{M}}(\rho),s}$ is injective, $X = \kappa_{\boldsymbol{M},\rho,s}(X)$.

Conversely, when $X = \kappa_{\boldsymbol{M},\rho,s}(X)$, we have $\kappa_{\boldsymbol{M}}(E_{\rho,s}(X)) = E_{\kappa_{\boldsymbol{M}}(\rho),s}(X)$. Thus,

$$
\begin{aligned}
&0 = [M_i, \kappa_{\boldsymbol{M}}(E_{\rho,s}(X))] = [M_i, E_{\kappa_{\boldsymbol{M}}(\rho),s}(X)] \\
=&[M_i, \frac{1}{2}(\kappa_{\boldsymbol{M}}(\rho)X + X\kappa_{\boldsymbol{M}}(\rho))] \\
=&\frac{1}{2}(\kappa_{\boldsymbol{M}}(\rho)[M_i, X] + [M_i, X]\kappa_{\boldsymbol{M}}(\rho)) = E_{\kappa_{\boldsymbol{M}}(\rho),s}([M_i, X]).
\end{aligned} \tag{6.152}
$$

That is, $[M_i, X] = 0$.
(**c**) Since $\kappa_{\boldsymbol{M},\rho,s}$ is commutative with M_i, the statement (**b**) implies that $\kappa_{\boldsymbol{M},\rho,s} \circ \kappa_{\boldsymbol{M},\rho,s}(X) = \kappa_{\boldsymbol{M},\rho,s}(\kappa_{\boldsymbol{M},\rho,s}(X)) = \kappa_{\boldsymbol{M},\rho,s}(X)$.
(**d**) Assume that every matrix M_i commutes with Y.

$$
\begin{aligned}
\langle Y, X\rangle^{(e)}_{\rho,s} =& \operatorname{Tr} Y E_{\rho,s}(X) = \operatorname{Tr} Y \kappa_{\boldsymbol{M}}(E_{\rho,s}(X)) \\
=& \operatorname{Tr} Y E_{\kappa(\rho),s}(\kappa_{\boldsymbol{M},\rho,s}(X)) = \langle Y, \kappa_{\boldsymbol{M},\rho,s}(X)\rangle^{(e)}_{\kappa_{\boldsymbol{M}}(\rho),s}.
\end{aligned} \tag{6.153}
$$

(**e**) Similar to (6.150), we have

$$
[E_{\kappa_{\boldsymbol{M}}(\rho),r}(\kappa_{\boldsymbol{M},\rho,s}(X)), M_i] = E_{\kappa_{\boldsymbol{M}}(\rho),r}([\kappa_{\boldsymbol{M},\rho,s}(X), M_i]).
$$

Hence, the statement (**a**) holds for the RLD.

Similar to (6.150), we have $E_{\kappa_{\boldsymbol{M}}(\rho),r}(X) = E_{\kappa_{\boldsymbol{M}}(\rho),r}(\kappa_{\boldsymbol{M},\rho,r}(X))$. Hence, when $[M_i, X] = 0$, $X = \kappa_{\boldsymbol{M},\rho,r}(X)$. When $X = \kappa_{\boldsymbol{M},\rho,r}(X)$, similar to (6.152), we have $0 = [M_i, \kappa_{\boldsymbol{M}}(E_{\rho,r}(X))] = [M_i, E_{\kappa_{\boldsymbol{M}}(\rho),r}(X)] = E_{\kappa_{\boldsymbol{M}}(\rho),r}([M_i, X])$. That is, $[M_i, X] = 0$. Hence, the statement (**b**) holds for the RLD.

The statement (**c**) for the RLD follows from the statements (**a**) and (**b**) for the RLD. Similar to (6.153), we have $\langle Y, X\rangle^{(e)}_{\rho,r} = \langle Y, \kappa_{M,\rho,r}(X)\rangle^{(e)}_{\kappa(\rho),r}$, i.e., the statement (**d**) holds for the RLD.

Exercise 6.6 We have

$$\begin{aligned}&(\|\kappa_M(A)\|^{(m)}_{\kappa_M(\rho),s})^2\\ =&\operatorname{Tr}\frac{1}{2}(\kappa_M(\rho)\kappa_{M,\rho,s}(X)+\kappa_{M,\rho,s}(X)\kappa_M(\rho))\kappa_{M,\rho,s}(X)\\ =&\operatorname{Tr}\frac{1}{2}(\rho\kappa_{M,\rho,s}(X)+\kappa_{M,\rho,s}(X)\rho)\kappa_{M,\rho,s}(X)=(\|\kappa_{M,\rho,s}(X)\|^{(e)}_{\rho,s})^2.\end{aligned}$$

As shown in the statement (**c**) of Exercise 6.5, $\kappa_{M,\rho,s}$ is a projection. Since $\|A\|^{(m)}_{\rho,s} = \|X\|^{(e)}_{\rho,s}$, ① is equivalent with $\kappa_{M,\rho,s}(X) = X$. Due to (**b**) of Exercise 6.5, the latter condition is equivalent with ②.

Exercise 6.7 The statement (**d**) of Exercise 6.4 and Theorem 6.1 show that $\|A\|^{(m)}_{\rho,b} \geq \|A\|^{(m)}_{\rho,s}$ and $\|A\|^{(m)}_{\rho,s} \geq \|\kappa_M(A)\|^{(m)}_{\kappa_M(\rho),s}$, respectively. Hence, $\|A\|^{(m)}_{\rho,b} \geq \|\kappa_M(A)\|^{(m)}_{\kappa_M(\rho),s}$.

Therefore, ① is equivalent with $\|A\|^{(m)}_{\rho,b} = \|A\|^{(m)}_{\rho,s}$ and $\|A\|^{(m)}_{\rho,s} = \|\kappa_M(A)\|^{(m)}_{\kappa_M(\rho),s}$. Due to (**d**) of Exercise 6.4, the former is equivalent with $[\rho, A] = 0$. Due to Exercise 6.6, the latter is equivalent with ② of Exercise 6.6. Hence, ① is equivalent with ②.

Exercise 6.8 Choose unitary U such that $\|X\rho Y\|_1 = \operatorname{Tr} UX\rho Y$. Thus, the Schwarz inequality implies that

$$\begin{aligned}&\|X\rho Y\|_1 = |\langle UX, Y\rangle^{(e)}_{\rho,r}| \leq \sqrt{\operatorname{Tr}\rho YY^*}\sqrt{\operatorname{Tr}\rho(UX)^*UX}\\ =&\sqrt{\operatorname{Tr}\rho YY^*}\sqrt{\operatorname{Tr}\rho X^*X}.\end{aligned}$$

Exercise 6.9 The Schwarz inequality implies that

$$\begin{aligned}&|\|X\|_1| = |\operatorname{Tr} XU| = |\operatorname{Tr}\rho\rho^{-1}XU| = |\langle\rho^{-1}X, U^*\rangle^{(e)}_{\rho,r}|\\ \leq&\sqrt{\operatorname{Tr}\rho\rho^{-1}X(\rho^{-1}X)^*}\sqrt{\operatorname{Tr}\rho U^*U} = \sqrt{\operatorname{Tr}\rho^{-1}XX^*}\sqrt{\operatorname{Tr}\rho U^*U}\\ =&\sqrt{\operatorname{Tr}\rho^{-1}XX^*}.\end{aligned}$$

Exercise 6.10 The Schwarz inequality for the inner product $\operatorname{Tr} X\rho^{1/2}Y^*\rho^{1/2}$ implies that

$$\begin{aligned}&\|X\|_1 = |\operatorname{Tr} XU| = \operatorname{Tr}(\rho^{-1/2}X\rho^{-1/2})\rho^{1/2}U\rho^{1/2}\\ \leq&\sqrt{\operatorname{Tr}\rho^{-1/2}X\rho^{-1/2}\rho^{1/2}(\rho^{-1/2}X\rho^{-1/2})^*\rho^{1/2}}\sqrt{\operatorname{Tr}\rho^{1/2}U^*\rho^{1/2}U}\\ =&\sqrt{\operatorname{Tr}\rho^{-1/2}X\rho^{-1/2}X^*}\sqrt{\operatorname{Tr}\rho^{1/2}U^*\rho^{1/2}U} \leq \sqrt{\operatorname{Tr}\rho^{-1/2}X\rho^{-1/2}X^*}.\end{aligned}$$

Exercise 6.11

$$
\begin{aligned}
&(\|X \otimes I_{\mathcal{H}'}\|^{(e)}_{|y\rangle\langle y|,p})^2 = \int_0^1 \langle y|X \otimes I_{\mathcal{H}'}|y\rangle^2 p(\lambda)d\lambda \\
=&|\operatorname{Tr} X\sigma|^2 = |\langle I, X\rangle^{(e)}_{\sigma,p}|^2 \le (\|I\|^{(e)}_{\sigma,p})^2(\|X\|^{(e)}_{\sigma,p})^2 = (\|X\|^{(e)}_{\sigma,p})^2.
\end{aligned}
$$

Due to the equality condition of the Schwartz inequality, the equality holds if and only if X is a constant times of I.

Exercise 6.12 For $X \in \{X|[X, M_i] = 0\ \forall i\}$ and a matrix Y, the statements (**d**) and (**e**) of Exercise 6.5 imply that

$$
\langle \kappa^*_{M,\rho,x}(X), Y\rangle^{(e)}_{\rho,x} = \langle X, \kappa_{M,\rho,x}(Y)\rangle^{(e)}_{\rho,x} = \langle X, Y\rangle^{(e)}_{\rho,x}
$$

for $x = s, r$. The above relations guarantee that $\kappa^*_{M,\rho,x}(X) = X$. That is, $\kappa_{M,\rho,x}$ is is the dual map of the inclusion of the matrix subspace $\{X|[X, M_i] = 0\ \forall i\}$ for $x = s, r$.

Exercise 6.13 When X is Hermitian, $E_{\rho,s}(X) = \frac{1}{2}(X\rho+\rho X)$, $E_{\rho,\frac{1}{2}}(X) = \sqrt{\rho}X\sqrt{\rho}$, and $E_{\rho,b}(X) = \int_0^1 \rho^\lambda X\rho^{1-\lambda}d\lambda$ are Hermitian.

Exercise 6.14 Since $0 = \frac{d}{d\theta}\langle\phi_\theta|\phi_\theta\rangle = \langle\phi_\theta|\frac{d\phi_\theta}{d\theta}\rangle + \langle\frac{d\phi_\theta}{d\theta}|\phi_\theta\rangle$, $\langle\phi_\theta|\frac{d\phi_\theta}{d\theta}\rangle$ is a pure imaginary number, which is denoted by ia. Thus,

$$
\begin{aligned}
&\frac{d}{d\theta}|\phi_\theta\rangle\langle\phi_\theta| = |\frac{d\phi_\theta}{d\theta}\rangle\langle\phi_\theta| + |\phi_\theta\rangle\langle\frac{d\phi_\theta}{d\theta}| \\
=&|\tilde{\phi}_\theta\rangle\langle\phi_\theta| + ia|\phi_\theta\rangle\langle\phi_\theta| + |\phi_\theta\rangle\langle\tilde{\phi}_\theta| - ia|\phi_\theta\rangle\langle\phi_\theta| \\
=&|\tilde{\phi}_\theta\rangle\langle\phi_\theta| + |\phi_\theta\rangle\langle\tilde{\phi}_\theta| = E_{|\phi_\theta\rangle\langle\phi_\theta|,s}(2(|\tilde{\phi}_\theta\rangle\langle\phi_\theta| + |\phi_\theta\rangle\langle\tilde{\phi}_\theta|)).
\end{aligned}
$$

Hence,

$$
J_{\theta,s} = \operatorname{Tr} |\tilde{\phi}_\theta\rangle\langle\phi_\theta| + |\phi_\theta\rangle\langle\tilde{\phi}_\theta|2(|\tilde{\phi}_\theta\rangle\langle\phi_\theta| + |\phi_\theta\rangle\langle\tilde{\phi}_\theta|) = 4\langle\tilde{\phi}_\theta|\tilde{\phi}_\theta\rangle.
$$

However, there is no matrix X such that $E_{|\phi_\theta\rangle\langle\phi_\theta|,x}(X) = |\tilde{\phi}_\theta\rangle\langle\phi_\theta| + |\phi_\theta\rangle\langle\tilde{\phi}_\theta|$ for $x = r, b$. Hence, both the RLD Fisher information and the Bogoljubov Fisher information diverge.

Exercise 6.15 Apply Theorem 6.2 to the entanglement-breaking channel $\rho \mapsto \sum_i(\operatorname{Tr} M_i\rho)|u_i\rangle\langle u_i|$ with the CONS $\{u_i\}$.

Exercise 6.16 Since $\frac{d}{d\theta}\operatorname{Tr} M_i\rho_\theta = \operatorname{Tr} M_i\frac{d}{d\theta}\rho_\theta = \langle M_i, L_{\theta,s}\rangle^{(e)}_{\rho_\theta,s}$ is real number, $\langle M_i, L_{\theta,s}\rangle^{(e)}_{\rho_\theta,s} = \langle L_{\theta,s}, M_i\rangle^{(e)}_{\rho_\theta,s}$. Thus,

$$
J^M_\theta = \sum_i \frac{(\frac{d}{d\theta}\operatorname{Tr} M_i\rho_\theta)^2}{\operatorname{Tr} M_i\rho_\theta} = \sum_i \frac{\langle M_i, L_{\theta,s}\rangle^{(e)}_{\rho_\theta,s}\langle L_{\theta,s}, M_i\rangle^{(e)}_{\rho_\theta,s}}{\langle M_i, I\rangle^{(e)}_{\rho_\theta,s}}.
$$

Exercise 6.17

(**a**) Exercise 6.12 shows that $\kappa_{\boldsymbol{M},\rho,s}(X)$ is the projection to the space spanned by $\{M_i\}_i$ with respect to the inner product $\langle\, ,\, \rangle^{(e)}_{\rho,s}$. Thus, $\kappa_{\boldsymbol{M},\rho,s}(X) = \sum_i \langle M_i, X\rangle^{(e)}_{\rho,s} M_i$.
(**b**) Since $\langle M_i, M_j\rangle^{(e)}_{\rho_\theta,s} = \frac{\delta_{i,j}}{\langle M_i, I\rangle^{(e)}_{\rho_\theta,s}}$, Exercise 6.16 yields that

$$\left(\|\kappa_{\boldsymbol{M},\rho_\theta,s}(L_{\theta,s})\|^{(e)}_{\rho_\theta,s}\right)^2 = \left(\|\sum_i \langle M_i, L_{\theta,s}\rangle^{(e)}_{\rho,s} M_i\|^{(e)}_{\rho_\theta,s}\right)^2$$

$$= \sum_{i,j} \langle M_i, M_j\rangle^{(e)}_{\rho_\theta,s} \langle M_i, L_{\theta,s}\rangle^{(e)}_{\rho,s} \langle L_{\theta,s}, M_j\rangle^{(e)}_{\rho,s}$$

$$= \sum_i \frac{\langle M_i, L_{\theta,s}\rangle^{(e)}_{\rho_\theta,s} \langle L_{\theta,s}, M_i\rangle^{(e)}_{\rho_\theta,s}}{\langle M_i, I\rangle^{(e)}_{\rho_\theta,s}} = J^{\boldsymbol{M}}_\theta.$$

(**c**) The statement (**b**) guarantees that the equation $J_{\theta,s} = J^{\boldsymbol{M}}_\theta$ is equivalent with the equation $\|\kappa_{\boldsymbol{M},\rho_\theta,s}(L_{\theta,s})\|^{(e)}_{\rho_\theta,s} = \|L_{\theta,s}\|^{(e)}_{\rho_\theta,s}$. Since $\kappa_{\boldsymbol{M},\rho_\theta,s}$ is a projection (See Exercise 6.12), the latter condition is equivalent with $\kappa_{\boldsymbol{M},\rho_\theta,s}(L_{\theta,s}) = L_{\theta,s}$. Due to (**b**) of Exercise 6.5, the final condition holds if and only if every M_i commutes with $L_{\theta,s}$.

Exercise 6.18

(**a**) Use the formula of the Beta function.
(**b**) Use $\exp(X(\theta)) = \sum_{n=0}^\infty \frac{X(\theta)^n}{n!}$ and (**a**).

$$\int_0^1 \exp(\lambda X(\theta)) \frac{dX(\theta)}{d\theta} \exp((1-\lambda)X(\theta))\, d\lambda$$

$$= \int_0^1 \sum_{n=0}^\infty \frac{\lambda^n X(\theta)^n}{n!} \frac{dX(\theta)}{d\theta} \sum_{m=0}^\infty \frac{(1-\lambda)^m X(\theta)^m}{m!} d\lambda$$

$$= \sum_{n=0}^\infty \sum_{m=0}^\infty \frac{n!m!}{(n+m)!} \frac{X(\theta)^n}{n!} \frac{dX(\theta)}{d\theta} \frac{X(\theta)^m}{m!}$$

$$= \sum_{n=0}^\infty \sum_{m=0}^\infty \frac{1}{(n+m)!} X(\theta)^n \frac{dX(\theta)}{d\theta} X(\theta)^m$$

$$= \sum_{k=0}^\infty \frac{1}{k!} \sum_{m=0}^k X(\theta)^{k-m} \frac{dX(\theta)}{d\theta} X(\theta)^m = \sum_{k=0}^\infty \frac{1}{k!} \frac{dX(\theta)^k}{d\theta} = \frac{d\exp(X(\theta))}{d\theta}.$$

Exercise 6.19 First, show that $\frac{d\rho_\theta^{\otimes n}}{d\theta} = \sum_{i=1}^n \underbrace{\rho_\theta \otimes \rho_\theta}_{i-1} \otimes \frac{d\rho_\theta}{d\theta} \otimes \underbrace{\rho_\theta \otimes \rho_\theta}_{n-i}$. Hence, we have $\frac{d\rho_\theta^{\otimes n}}{d\theta} = E_{\rho_\theta^{\otimes},x}\left(\sum_{i=1}^n \underbrace{I \otimes I}_{i-1} \otimes L_{\theta,x} \otimes \underbrace{I \otimes I}_{n-i}\right)$. Since $\langle I, L_{\theta,x}\rangle^{(e)}_{\rho_\theta,x} = 0$ and $\langle I, I\rangle^{(e)}_{\rho_\theta,x} = 1$, the relation (6.15) guarantees that

$$J_{\theta,x,n} = \left(\| \sum_{i=1}^{n} \underbrace{I \otimes I}_{i-1} \otimes L_{\theta,x} \otimes \underbrace{I \otimes I}_{n-i} \|^{(e)}_{\rho_\theta^{\otimes n},x} \right)^2 = n \langle L_{\theta,x}, L_{\theta,x} \rangle^{(e)}_{\rho_\theta,x} = n J_{\theta,x}.$$

Exercise 6.20 Since $\langle L^*_{\theta,i,x}, L_{\theta,j,x} \rangle^{(e)}_{\rho_\theta,x} = (\langle L^*_{\theta,j,x}, L_{\theta,i,x} \rangle^{(e)}_{\rho_\theta,x})^*$, $\boldsymbol{J}_{\theta,x}$ is Hermitian.

Exercise 6.21 Due to Exercise 6.13, $L_{\rho_\theta,i,x}$ is Hermitian for $x = s, \frac{1}{2}, b$. Hence, $\boldsymbol{J}_{\theta,s;i,j} = \langle L^*_{\rho_\theta,i,x}, L_{\rho_\theta,j,x} \rangle^{(e)}_{\rho_\theta,x} = \mathrm{Tr}\, \frac{\partial \rho_\theta}{\partial \theta^i} L_{\rho_\theta,j,x}$ is a real number. Thus, Exercise 6.20 guarantees that the Fisher information matrix $\boldsymbol{J}_{\theta,x}$ is real symmetric.

Exercise 6.22 Consider the following example: $\rho_\theta = \begin{pmatrix} \frac{1+r}{2} & 0 \\ 0 & \frac{1-r}{2} \end{pmatrix}$, $\frac{\partial \rho_\theta}{\partial \theta^1} = \begin{pmatrix} 0 & 1 \\ 1 & 0 \end{pmatrix}$, $\frac{\partial \rho_\theta}{\partial \theta^2} = \begin{pmatrix} 0 & -i \\ i & 0 \end{pmatrix}$.

Exercise 6.23 The relation (6.32) guarantees that $L_{\theta,b} = \frac{d \log \rho_\theta}{d\theta}$. Since $\log \rho_\theta = e^{-i\theta Y} \log \rho e^{i\theta Y}$, $L_{\theta,b} = \frac{d e^{-i\theta Y} \log \rho e^{i\theta Y}}{d\theta} = i[\log \rho, Y]$.

Exercise 6.24 Consider the quantum state family $\{\rho_\theta = e^{-i\theta Y} \rho e^{i\theta Y}\}$ in Exercise 6.23. The e and m representations of the derivative are $i[\log \rho, Y]$ and $i[\rho, Y]$. Then, we have $i[\rho, Y] = E_{\rho,b}(i[\log \rho, Y])$.

Exercise 6.25

(**a**) Simple calculations.
(**b**) Use (6.20) and (5.26).
(**c**) Use the relation given in (**b**).

Exercise 6.26 Consider the TP-CP map $\tilde{\rho}_\theta := \lambda \rho^1_\theta \otimes |1\rangle\langle 1| + (1-\lambda)\rho^2_\theta \otimes |2\rangle\langle 2| \to \lambda \rho^1_\theta + (1-\lambda)\rho^2_\theta$. Then, we denote the Fisher information of the family $\{\tilde{\rho}_\theta\}$ by $\tilde{J}_{\theta,x}$. Theorem 6.2 implies that $\tilde{J}_{\theta,x} \geq J_{\theta,x}$.

Now, we choose $L_{\theta,i,x}$ as $\frac{d\rho^i_\theta}{d\theta} = E_{\rho^i_\theta,x}(L_{\theta,i,x})$. Then, $\tilde{L}_{\theta,x} := \sum_{i=1}^2 L_{\theta,i,x} \otimes |i\rangle\langle i|$ satisfies $\frac{d\tilde{\rho}_\theta}{d\theta} = E_{\tilde{\rho}_\theta,x}(\tilde{L}_{\theta,x})$. Thus, $\tilde{J}_{\theta,x} = \lambda J^1_{\theta,x} + (1-\lambda) J^2_{\theta,x}$, which implies $J_{\theta,x} \leq \lambda J^1_{\theta,x} + (1-\lambda) J^2_{\theta,x}$.

Next, we assume that the space spanned by the supports of $\frac{d\rho^1_\theta}{d\theta}$ and ρ^1_θ are orthogonal to those of $\frac{d\rho^2_\theta}{d\theta}$ and ρ^2_θ. Then, $E_{\lambda\rho^1_\theta+(1-\lambda)\rho^2_\theta,x}(\sum_{i=1}^2 L_{\theta,i,x}) = \lambda \frac{d\rho^1_\theta}{d\theta} + (1-\lambda)\frac{d\rho^2_\theta}{d\theta}$. Thus, we have $J_{\theta,x} = \lambda J^1_{\theta,x} + (1-\lambda) J^2_{\theta,x}$.

Exercise 6.27 First, notice that

$$e^{\frac{t}{2}S_1} S_i e^{\frac{t}{2}S_1} = S_i \tag{6.154}$$

for $i = 2, 3$ and

$$\begin{aligned} & e^{\frac{t}{2}S_1}(I + x_1 S_1) e^{\frac{t}{2}S_1} \\ = & \frac{e^t(1+x_1) + e^{-t}(1-x_1)}{2} I + \frac{e^t(1+x_1) - e^{-t}(1-x_1)}{2} S_1. \end{aligned}$$

Thus,

$$\frac{1}{2}e^{\frac{t}{2}S_1}(I+\sum_{i=1}^{3}x_iS^i)e^{\frac{t}{2}S_1}$$
$$=\frac{e^t(1+x_1)+e^{-t}(1-x_1)}{2}I+\frac{e^t(1+x_1)-e^{-t}(1-x_1)}{2}S_1+x_2+x_3S_3.$$

Since $\mathrm{Tr}\,\frac{1}{2}e^{\frac{t}{2}S_1}(I+\sum_{i=1}^{3}x_iS^i)e^{\frac{t}{2}S_1}=\frac{e^t(1+x_1)+e^{-t}(1-x_1)}{2}$, we have $\mu_s(t)=\log\frac{e^t(1+x_1)+e^{-t}(1-x_1)}{2}$. Therefore, we obtain the desired argument.

Exercise 6.28 First, for a given SLD geodesic $\Pi^\theta_{L,s}\sigma$, choose a unitary matrix U_1 such that ULU^* is equal to the constant times of S_1. Then, the SLD geodesic $U\Pi^\theta_{L,s}\sigma U^*$ has the form given in Exercise 6.27. Next, choose another unitary matrix U_2 such that

$$U_2S_1U_2^*=S_1,\ U_2(x_2S_2+x_3S_3)U_2^*=\sqrt{x_2^2+x_3^2}S_3. \tag{6.155}$$

Then, $\{U_2U\Pi^\theta_{L,s}\sigma U^*U_2^*\}$ is $\mathcal{S}_\alpha$ when $\alpha=\sqrt{x_2^2+x_3^2}$.

Exercise 6.29

(a) It follows from $\int_0^1(x^2t)(1+tx)^{-1}dt=x-\log(1+x)$.
(b) $\int_0^1\mathrm{Tr}(\sigma-\rho)^2(\rho+t(\sigma-\rho))^{-1}dt$
$=\int_0^1\mathrm{Tr}\,\rho(\sqrt{\rho}^{-1}\sigma\sqrt{\rho}^{-1}-I)(I+t(\sqrt{\rho}^{-1}\sigma\sqrt{\rho}^{-1}-I))^{-1}(\sqrt{\rho}^{-1}\sigma\sqrt{\rho}^{-1}-I)dt$
$=\mathrm{Tr}\,\rho\left((\sqrt{\rho}^{-1}\sigma\sqrt{\rho}^{-1}-I)-\log(I+(\sqrt{\rho}^{-1}\sigma\sqrt{\rho}^{-1}-I))\right)$
$=-\mathrm{Tr}\,\rho\log(\sqrt{\rho}^{-1}\sigma\sqrt{\rho}^{-1})=\mathrm{Tr}\,\rho\log(\sqrt{\rho}\sigma^{-1}\sqrt{\rho})$.

Exercise 6.30 We denote the pinching corresponding to the spectral decomposition of $\sigma^{-1/2}(\sigma^{1/2}\rho\sigma^{1/2})^{1/2}\sigma^{-1/2}$ by κ. Then, it is enough to show $D_s^{(e)}(\rho\|\sigma)=D_s^{(e)}(\kappa(\rho)\|\kappa(\sigma))$ and $D_s^{(e)}(\kappa(\rho)\|\kappa(\sigma))\geq-2\log\mathrm{Tr}\,|\sqrt{\rho}\sqrt{\sigma}|$. Since $\kappa(L)=L$, $\kappa(\Pi^\theta_{L,s}\rho)=\Pi^\theta_{\kappa(L),s}\kappa(\rho)$ and $J_{\theta,s}$ equals the Fisher information of $\Pi^\theta_{\kappa(L),s}\kappa(\rho)$, which is calculated to J_θ^M. Hence, we obtain $D_s^{(e)}(\rho\|\sigma)=D_s^{(e)}(\kappa(\rho)\|\kappa(\sigma))$.

Since Exercise 3.21 shows $\mathrm{Tr}\,|\sqrt{\rho}\sqrt{\sigma}|=\mathrm{Tr}\,\kappa(\rho)^{1/2}\kappa(\sigma)^{1/2}$, the relation (2.26) yields that $D_s^{(e)}(\kappa(\rho)\|\kappa(\sigma))\geq-2\log\mathrm{Tr}\,|\sqrt{\rho}\sqrt{\sigma}|$.

Exercise 6.31

(a) Use the partial integration formula twice.
(c) Use (6.32).
(d) Use **(a)**, **(b)**, **(c)** and the fact that $\frac{d^2\rho_\theta}{d\theta^2}=0$.

Exercise 6.32 The equivalence of ① and ② follows from (6.63). The equivalence of ② and ③ follows from Exercise 6.7.

Exercise 6.33 Apply (6.64) to the pinching $\kappa_{\sigma^{\otimes n}}$. Then, combining Exercise 5.44, we have $\lim_{n\to\infty}\frac{1}{n}D_x^{(m)}(\rho^{\otimes n}\|\sigma^{\otimes n})\geq\lim_{n\to\infty}\frac{1}{n}D(\kappa_{\sigma^{\otimes n}}(\rho^{\otimes n})\|\sigma^{\otimes n})=D(\rho\|\sigma)$. Due to the relation $J_{\theta,x} \le J_{\theta,b}$, (6.63) implies that $D_x^{(m)}(\rho^{\otimes n}\|\sigma^{\otimes n}) \le D_b^{(m)}(\rho^{\otimes n}\|\sigma^{\otimes n}) = nD(\rho\|\sigma)$. Thus, we obtain the desired argument.

Exercise 6.34 Similar to Exercise 6.30, since $\kappa_{\boldsymbol{E}_{\theta,n}}(L_{\theta,s,n}) = L_{\theta,s,n}$, we have $nJ_{\theta,s} = J_\theta^{\boldsymbol{E}_{\theta,n}}$.

Exercise 6.35 Consider a state family where $A = \frac{d\rho_\theta}{d\theta}$ and $\rho_{\theta_0} = \rho$, and let κ be given by a POVM M. From property (6.24) and $\|\rho^{-1}\|_{\rho,x}^{(m)} = 1$, we have $J_{\theta_0}^M = \|\kappa(A)\|_{\kappa(\rho),x}^{(m)} \le \|A\|_{\rho,x}^{(m)}$. Using Exercise 6.34 with $n = 1$, we have $\|\kappa(A)\|_{\kappa(\rho),x}^{(m)} = \|A\|_{\rho,s}^{(m)}$. Thus, we obtain the desired argument.

Exercise 6.36 This exercise can be shown as the same way as Exercise 2.40.

Exercise 6.37

(**a**) Let K be the difference K between $O(\boldsymbol{M},\hat{\theta}) - \theta$ and $\frac{1}{J_{\theta,s}}L_{\theta,s}$. Then, $K\rho + \rho K = 0$ when Condition ① holds.
(**b**) In the proof given in Exercise 6.36, the bottleneck is showing that the POVM $\boldsymbol{M}^n$ is the spectral decomposition of $O(\boldsymbol{M}^n,\hat{\theta}_n)$ from ① because this step uses the condition $\rho_\theta > 0$ in Exercise 6.36. Hence, it is sufficient to show this step. Assume that ① holds. Then, the equality in (6.77) holds. Thus, we obtain

$$\begin{aligned} 0 &= \operatorname{Tr} \rho_\theta \left(\sum_\omega \left(\hat{\theta}_n(\omega) - \theta\right) \boldsymbol{M}^n(\omega) \left(\hat{\theta}_n(\omega) - \theta\right) - (O(\boldsymbol{M}^n,\hat{\theta}_n) - \theta I)^2 \right) \\ &= \operatorname{Tr} \rho_\theta \left(\sum_\omega \hat{\theta}_n(\omega) \boldsymbol{M}^n(\omega) \hat{\theta}_n(\omega) - O(\boldsymbol{M}^n,\hat{\theta}_n)^2 \right). \end{aligned}$$

Since

$$\begin{aligned} 0 &= \int p\left(\theta) \operatorname{Tr} \rho_\theta (\sum_\omega \hat{\theta}_n(\omega) \boldsymbol{M}^n(\omega) \hat{\theta}_n(\omega) - O(\boldsymbol{M}^n,\hat{\theta}_n)^2 \right) d\theta \\ &= \operatorname{Tr} \left(\int p(\theta)\rho_\theta d\theta \right) \left(\sum_\omega \hat{\theta}_n(\omega) \boldsymbol{M}^n(\omega) \hat{\theta}_n(\omega) - O(\boldsymbol{M}^n,\hat{\theta}_n)^2 \right), \end{aligned}$$

the condition (6.82) guarantees that $\sum_\omega \hat{\theta}_n(\omega)\boldsymbol{M}^n(\omega)\hat{\theta}_n(\omega) - O(\boldsymbol{M}^n,\hat{\theta}_n)^2 = 0$. Hence, the POVM $\boldsymbol{M}^n$ is the spectral decomposition of $O(\boldsymbol{M}^n,\hat{\theta}_n)$.

Exercise 6.38 Apply the same discussion as (6.77) to the state family $\{\rho_\theta^{\otimes n}\}$.

Exercise 6.39 Due to the relation $\lim_{\epsilon\to 0} \frac{1}{\epsilon^2} \inf_{|\theta'-\theta|>\epsilon} D(\rho_{\theta'}\|\rho_\theta) = \frac{1}{2}J_{\theta,b}$, Theorem 6.8 can be shown as the same way the Proof of Theorem 2.9.

Exercise 6.40

(**b**) Define $B_0 \stackrel{\text{def}}{=} \left\{\hat{\theta} \le \theta\right\}$, $B_i \stackrel{\text{def}}{=} \left\{\theta + \frac{\delta(i-1)}{m} \le \hat{\theta} \le \theta + \frac{\delta i}{m}\right\}$, $(i = 1, \ldots, m)$, and $B_{m+1} \stackrel{\text{def}}{=} \left\{\theta + \delta \le \hat{\theta}\right\}$, and consider a POVM $M_i \stackrel{\text{def}}{=} M(B_i)$ composed of $m + 1$

outcomes. Then, applying the monotonicity for fidelity (3.55), we obtain the desired inequality.
(**c**) The desired inequality follows from the combination of the preceding inequalities.
(**d**) This statement follows from (**c**).
(**e**) In the inequality given in (**d**), take the limit $n \to \infty$. Next, take the limit $m \to \infty$. Finally, taking the limit $\epsilon \to 0$, we obtain (6.91).

Exercise 6.41 Taking the limit $\delta \to 0$, we have

$$\frac{1}{2}\alpha'(\boldsymbol{M},\theta) = \lim_{\delta\to 0}\frac{1}{\delta^2}\inf_{\{s|1\ge s\ge 0\}}\left(\beta'(\boldsymbol{M},\theta,s\delta)+\beta'(\boldsymbol{M},\theta+\delta,(1-s)\delta)\right).$$

(6.33) implies that

$$\frac{1}{4}J_\theta = \lim_{\delta\to 0}\frac{1}{\delta^2} - 2\log \operatorname{Tr}|\sqrt{\rho_\theta}\sqrt{\rho_{\theta+\delta}}|.$$

Hence, combining (6.91), we obtain (6.92).

Exercise 6.42

(**b**) Since the set $\inf\left\{J_\theta^{\boldsymbol{M}}\middle|\boldsymbol{M}\text{ POVM on }\mathcal{H}\right\}$ is a convex set, $\operatorname{tr} n(J_\theta^{\boldsymbol{M}_n})^{-1} = \operatorname{tr}(\sum_{i=1}^d \frac{1}{n} J_\theta^{M_{\theta:n,i}})^{-1} \ge \inf\left\{\operatorname{tr}(J_\theta^{\boldsymbol{M}})^{-1}\middle|\boldsymbol{M}\text{ POVM on }\mathcal{H}\right\}$. Combining (6.100), we obtain (6.104).

Exercise 6.43

(**a**) First, notice that $\frac{\langle M(\omega),M(\omega)\rangle_{\theta,s}^{(e)}}{\langle M(\omega),I\rangle_{\theta,s}^{(e)}} = \operatorname{Tr} M(\omega)$. Then, taking the sum for ω, we obtain the desired argument.

(**b**) We notice that $J_{i,j}^{\boldsymbol{M}} = \sum_\omega \frac{\langle M(\omega), L_{\theta,i,s}\rangle_{\theta,s}^{(e)}\langle L_{\theta,j,s}, M(\omega)\rangle_{\theta,s}^{(e)}}{\langle M(\omega), I\rangle_{\theta,s}^{(e)}}$. Thus,

$$\begin{aligned}\operatorname{tr} \boldsymbol{J}_{\theta,s}^{-1}\boldsymbol{J}_\theta^{\boldsymbol{M}} &= \sum_{i,j}(\boldsymbol{J}_{\theta,s}^{-1})_{i,j}\sum_\omega \frac{\langle M(\omega), L_{\theta,j,s}\rangle_{\theta,s}^{(e)}\langle L_{\theta,j,s}, M(\omega)\rangle_{\theta,s}^{(e)}}{\langle M(\omega), I\rangle_{\theta,s}^{(e)}}\\ &= \sum_\omega\sum_{j=1}^d \frac{\langle M(\omega), L_{\theta,s}^j\rangle_{\theta,s}^{(e)}\langle L_{\theta,j,s}, M(\omega)\rangle_{\theta,s}^{(e)}}{\langle M(\omega), I\rangle_{\theta,s}^{(e)}}.\end{aligned}$$

(**c**) We have $1 = \sum_\omega \frac{\langle M(\omega), I\rangle_{\theta,s}^{(e)}\langle I, M(\omega)\rangle_{\theta,s}^{(e)}}{\langle M(\omega), I\rangle_{\theta,s}^{(e)}}$. Hence, as shown in Exercise A.1, $\sum_{j=1}^d |L_{\theta,i,s}\rangle_{\theta,s}^{(e)}\langle L_{\theta,j,s}| + |I\rangle_{\theta,s}^{(e)}\langle I|$ can be regarded as the projection to the subspace spanned by $I, L_{\theta,1,s},\ldots,L_{\theta,d,s}$. Thus,

$$\begin{aligned}&\sum_\omega\sum_{j=1}^d \frac{\langle M(\omega), L_{\theta,s}^j\rangle_{\theta,s}^{(e)}\langle L_{\theta,j,s}, M(\omega)\rangle_{\theta,s}^{(e)}}{\langle M(\omega), I\rangle_{\theta,s}^{(e)}} + \sum_\omega \frac{\langle M(\omega), I\rangle_{\theta,s}^{(e)}\langle I, M(\omega)\rangle_{\theta,s}^{(e)}}{\langle M(\omega), I\rangle_{\theta,s}^{(e)}}\\ \le &\sum_\omega \frac{\langle M(\omega), M(\omega)\rangle_{\theta,s}^{(e)}}{\langle M(\omega), I\rangle_{\theta,s}^{(e)}} = \dim\mathcal{H},\end{aligned}$$

which implies (6.105). Due to the above discussion, the equality holds only when every element $M(\omega)$ belongs to the subspace spanned by $I, L_{\theta,1,s}, \ldots, L_{\theta,d,s}$.
(**d**) The equality in (6.105) holds only when every element $M(\omega)$ satisfying $\operatorname{Tr} M(\omega) \rho_\theta = \langle M(\omega), I\rangle^{(e)}_{\theta,s} > 0$ belongs to the subspace spanned by $I, L_{\theta,1,s}, \ldots, L_{\theta,d,s}$.
(**e**) Consider a POVM $\boldsymbol{M}' = \{M'_i\}$ of rank $Mi' = 1$ and a stochastic transition matrix $Q = (Q^i_\omega)$ such that $M_\omega = \sum_i Q^i_\omega M'_i$. Due to Exercise 2.42, we have $\boldsymbol{J}^{\boldsymbol{M}'} \geq \boldsymbol{J}^{\boldsymbol{M}}$. Hence, we obtain (6.105).

Exercise 6.44 If $(\boldsymbol{M}, \hat{\theta})$ is a locally unbiased estimator, we can show that $\boldsymbol{V}_\theta(\boldsymbol{M}, \hat{\theta}) \geq (\boldsymbol{J}^{\boldsymbol{M}}_\theta)^{-1}$ in the same way as (2.139).

Then, we can show that for each POVM $\boldsymbol{M}$ there exists a function $\hat{\theta}$ such that $(\boldsymbol{M}, \hat{\theta})$ is a locally unbiased estimator and $\boldsymbol{V}_\theta(\boldsymbol{M}, \hat{\theta}) = (\boldsymbol{J}^{\boldsymbol{M}}_\theta)^{-1}$.

Exercise 6.45 Use the method of Lagrange multipliers.

Exercise 6.46 First, notice that $\left(\|L(u)\|^{(e)}_{\rho_\theta,s}\right)^2 = \langle u|\boldsymbol{J}_{\theta,s}|u\rangle$. Then, we can show that

$$\begin{aligned}
\langle x|\boldsymbol{J}^{\boldsymbol{M}^u}_\theta|x\rangle &= \operatorname{Tr} L(x)\kappa_{\boldsymbol{M}^u}(L(x)) = \langle L(x)|\kappa_{\boldsymbol{M}^u,\rho_\theta,s}(L(x))\rangle^{(e)}_{\rho_\theta,s} \\
=&\langle L(x)|\kappa_{\boldsymbol{M}^u,\rho_\theta,s}|L(x)\rangle^{(e)}_{\rho_\theta,s} \geq \langle L(x)|\frac{|L(u)\rangle^{(e)}_{\rho_\theta,s}\langle L(u)|}{\langle u|\boldsymbol{J}_{\theta,s}|u\rangle}|L(x)\rangle^{(e)}_{\rho_\theta,s}
\end{aligned}$$

for $x \in \mathbb{R}^d$. Hence, we obtain (6.107).

Exercise 6.47 Since $J^{\boldsymbol{M}}_\theta = \sum_{\omega\in\Omega} \frac{(\operatorname{Tr} \frac{d\rho_\theta}{d\theta} M_\omega)^2}{\operatorname{Tr} \rho_\theta M_\omega}$, we have

$$\begin{aligned}
J^{\boldsymbol{M}''}_\theta &= \sum_{\omega\in\Omega} \frac{(\lambda \operatorname{Tr} \frac{d\rho_\theta}{d\theta} M_\omega)^2}{\lambda \operatorname{Tr} \rho_\theta M_\omega} + \sum_{\omega'\in\Omega'} \frac{((1-\lambda) \operatorname{Tr} \frac{d\rho_\theta}{d\theta} M'_{\omega'})^2}{(1-\lambda) \operatorname{Tr} \rho_\theta M'_{\omega'}} \\
=&\lambda \sum_{\omega\in\Omega} \frac{(\operatorname{Tr} \frac{d\rho_\theta}{d\theta} M_\omega)^2}{\operatorname{Tr} \rho_\theta M_\omega} + (1-\lambda) \sum_{\omega'\in\Omega'} \frac{(\operatorname{Tr} \frac{d\rho_\theta}{d\theta} M'_{\omega'})^2}{\operatorname{Tr} \rho_\theta M'_{\omega'}} = \lambda J^{\boldsymbol{M}}_\theta + (1-\lambda) J^{\boldsymbol{M}'}_\theta.
\end{aligned}$$

Exercise 6.48 (6.108) follows from (6.107) and Exercise 6.47.

Exercise 6.49 The inequality $\geq$ follows from Exercise 6.45 and (6.105). Hence, it is sufficient to show the existence of a POVM $\boldsymbol{M}$ such that $\operatorname{tr}(\boldsymbol{J}^{\boldsymbol{M}}_\theta)^{-1} = \left(\operatorname{tr} \boldsymbol{J}^{-\frac{1}{2}}_{\theta,s}\right)^2$. Let $u_1, \ldots, u_d$ be the eigenvectors of $\boldsymbol{J}_{\theta,s}$, and let p_i be the eigenvalues of $\frac{1}{\operatorname{tr} \boldsymbol{J}^{-\frac{1}{2}}_{\theta,s}} \boldsymbol{J}^{-\frac{1}{2}}_{\theta,s}$. Then, the RHS of (6.108) is equal to $\frac{1}{\operatorname{tr} \boldsymbol{J}^{-\frac{1}{2}}_{\theta,s}} \boldsymbol{J}^{\frac{1}{2}}_{\theta,s}$.

Exercise 6.50 Choose the new coordinate θ' such that the SLD Fisher information matrix is $\sqrt{G}^{-1} \boldsymbol{J}_{\theta,s} \sqrt{G}^{-1}$. Hence, we have

$$\inf\left\{\operatorname{tr}(\boldsymbol{J}^{\boldsymbol{M}}_{\theta'})^{-1} \middle| \boldsymbol{M} \text{ POVM on } \mathcal{H}\right\} = \left(\operatorname{tr}\left(\sqrt{G}^{-1} \boldsymbol{J}_{\theta,s} \sqrt{G}^{-1}\right)^{-\frac{1}{2}}\right)^2.$$

Since $\sqrt{G}^{-1} J_\theta^M \sqrt{G}^{-1} = J_{\theta'}^M$, we have $\operatorname{tr} G(J_\theta^M)^{-1} = \operatorname{tr}(J_{\theta'}^M)^{-1}$. Thus, we obtain (6.109)

Exercise 6.51 Apply Theorem 6.2 to the TP-CP map $p_\theta \mapsto \kappa_\theta(\rho)$.

Exercise 6.52 In this case, the outcome i' by the PVM $\{P_i\}$ coincides with the initial i. Hence, the output distribution of the PVM $\{P_i\}$ is equal to $p_\theta(i)$. Therefore, applying Theorem 6.2 to the measurement by the PVM $\{P_i\}$, we obtain the inequality opposite to (6.110), which implies the equality of (6.110).

Exercise 6.53 It follows from the fact that $\langle \Phi_d | \mathbf{X}^i \mathbf{Z}^j | \Phi_d \rangle = 0$ unless i and j are 0.

Exercise 6.54 As shown in Exercise 6.51, the Fisher information is bounded by that of the distribution family $\{p_\theta\}$. This bound can be attained if and only if the input is the maximally entangled state because the condition holds only in this case.

Exercise 6.55

(**b**) First, we consider the case when Y is given as the RHS of the equation in (**a**). If X satisfies $X \geq iY$, the diagonal elements of X is greater than the following:

$$\begin{pmatrix} \alpha_1 & & & & \\ & \alpha_1 & & & \\ & & \alpha_1 & & \\ & & & \alpha_2 & \\ & & & & \ddots \end{pmatrix}.$$

This fact can be shown by applying the projection to the 2-dimensional space spanned by the $2i$th and $2j-1$th components.

Hence, the LHS of (6.112) is greater than $\operatorname{Tr} |iY|$. When $X = |iY|$, the condition $X \geq iY$ holds. So, we obtain (6.112). The general case can be reduced to this special case by applying the orthogonal matrix V given in (**a**).

(**c**)

$$\begin{aligned} &\min\{\operatorname{tr} \boldsymbol{V} | \boldsymbol{V} : \text{real symmetric} \boldsymbol{V} \geq \boldsymbol{J}_{\theta,x}^{-1}\} \\ =& \operatorname{tr} \mathbf{Re}(\boldsymbol{J}_{\theta,x}^{-1}) + \min\{\operatorname{tr} X | X : \text{real symmetric} X \geq i\, \mathbf{Im}(\boldsymbol{J}_{\theta,x}^{-1})\} \\ =& \operatorname{tr} \mathbf{Re}(\boldsymbol{J}_{\theta,x}^{-1}) + \operatorname{tr} |\mathbf{Im}(\boldsymbol{J}_{\theta,x}^{-1})|. \end{aligned}$$

Exercise 6.56

(**a**) Let P be the projection to $\bar{\mathcal{T}}_\theta$. Then, define $P(\mathbf{X}) := (PX^1, \ldots, PX^d)$. When $\delta_i^j = \operatorname{Tr} \frac{\partial \rho_\theta}{\partial \theta^i} X^j$, we have $\operatorname{Tr} \frac{\partial \rho_\theta}{\partial \theta^i} PX^j = \langle L_{\theta,i,s}, PX^j \rangle_{\rho_\theta,s}^{(e)} = \langle PL_{\theta,i,s}, X^j \rangle_{\rho_\theta,s}^{(e)} = \langle L_{\theta,i,s}, X^j \rangle_{\rho_\theta,s}^{(e)} = \delta_i^j$. Since $\langle PX, (I-P)X \rangle_{\rho_\theta,s}^{(e)} = 0$, we have $\boldsymbol{V}_\theta(\mathbf{X}) = \boldsymbol{V}_\theta(P(\mathbf{X})) + \boldsymbol{V}_\theta((I-P)(\mathbf{X}))$, which implies $\operatorname{tr} \mathbf{Re}\, \boldsymbol{V}_\theta(\mathbf{X}) + \operatorname{tr} |\mathbf{Im}\, \boldsymbol{V}_\theta(\mathbf{X})| \geq \operatorname{tr} \mathbf{Re}\, \boldsymbol{V}_\theta(P(\mathbf{X})) + \operatorname{tr} |\mathbf{Im}\, \boldsymbol{V}_\theta(P(\mathbf{X}))|$. Then, we can show the desired argument.

(**b**) For any matrix X on $\mathcal{H}$, we define $X^{(n)} := \frac{1}{n}\sum_{i=1}^{n} \underbrace{I \otimes I}_{i-1} \otimes X \otimes \underbrace{I \otimes I}_{n-i}$. The space spanned by the orbits of SLD of the state $\rho_\theta^{\otimes n}$ is given as $\{X^{(n)} | X \in \bar{\mathcal{T}}_\theta\}$. When $\mathbf{X}$ satisfies $\mathrm{Tr}\, \frac{\partial \rho_\theta}{\partial \theta^i} X^j = \delta_i^j$, we have $\mathrm{Tr}\, \frac{\partial \rho_\theta^{\otimes n}}{\partial \theta^i} X^{j(n)} = \delta_i^j$. Then, $\mathrm{tr}\, \mathbf{Re}\, \boldsymbol{V}_\theta((X^{j(n)})) + \mathrm{tr}\, |\, \mathbf{Im}\, \boldsymbol{V}_\theta((X^{j(n)}))| = \frac{1}{n}(\mathrm{tr}\, \mathbf{Re}\, \boldsymbol{V}_\theta(\mathbf{X}) + \mathrm{tr}\, |\, \mathbf{Im}\, \boldsymbol{V}_\theta(\mathbf{X})|)$. Thus, we obtain the desired argument.

Exercise 6.57

(**a**) Define the vector $E_\theta(\mathbf{X}) := ((\mathrm{Tr}\, X^i \rho_\theta) I)$ and the matrix $v_\theta(\mathbf{X}) := (\mathrm{Tr}\, X^i \rho_\theta\, \mathrm{Tr}\, X^j \rho_\theta)_{i,j}$. Then, we have

$$\boldsymbol{V}_\theta(\mathbf{X}) = \boldsymbol{V}_\theta(\mathbf{X} - E_\theta(\mathbf{X})) + v_\theta(\mathbf{X}), \tag{6.156}$$

which implies $\mathrm{tr}\, \mathbf{Re}\, \boldsymbol{V}_\theta(\mathbf{X}) + \mathrm{tr}\, |\, \mathbf{Im}\, \boldsymbol{V}_\theta(\mathbf{X})| = \mathrm{tr}\, \mathbf{Re}\, \boldsymbol{V}_\theta(\mathbf{X} - E_\theta(\mathbf{X})) + \mathrm{tr}\, |\, \mathbf{Im}\, \boldsymbol{V}_\theta(\mathbf{X} - E_\theta(\mathbf{X}))| + \mathrm{tr}\, v_\theta(\mathbf{X})$. When the matrix-valued vector $\mathbf{X}$ satisfies the condition $\delta_i^j = \mathrm{Tr}\, \frac{\partial \rho_\theta}{\partial \theta^i} X^j$, the matrix-valued vector $\mathbf{X}' = \mathbf{X} - E_\theta(\mathbf{X})$ also satisfies it. The matrix-valued vector $\mathbf{X}'$ satisfies the additional condition $\mathrm{Tr}\, \rho_\theta X'^i = 0$ So, the minimum value is realized when the additional condition $\mathrm{Tr}\, \rho_\theta X^i = 0$ holds.
(**b**) Apply (6.76) to the case when $O(\boldsymbol{M}^n, \hat{\theta}_n)$ is replaced by $\sum_i a_i X^i$, where (a_i) is an arbitrary complex-valued vector. Hence, we obtain $\langle a | \boldsymbol{V}_\theta(\boldsymbol{M}, \hat{\theta}) | a \rangle \geq \langle a | \boldsymbol{V}_\theta(\mathbf{X}) | a \rangle$. Since (a_i) is an arbitrary complex-valued vector, we have $\boldsymbol{V}_\theta(\boldsymbol{M}, \hat{\theta}) \geq \boldsymbol{V}_\theta(\mathbf{X})$.
(**c**) Since $\boldsymbol{V}_\theta(\boldsymbol{M}, \hat{\theta}) \geq \boldsymbol{V}_\theta(\mathbf{X})$, we can show that $\mathrm{tr}\, \boldsymbol{V}_\theta(\boldsymbol{M}, \hat{\theta}) \geq \mathrm{tr}(\mathbf{Re}\, \boldsymbol{V}_\theta(\mathbf{X}) + |\, \mathbf{Im}\, \boldsymbol{V}_\theta(\mathbf{X})|)$ in the same way as Exercise 6.55.
(**d**) Combining (6.100) and (6.106), we have

$$\begin{aligned} &\mathrm{tr} \lim_{n\to\infty} n \hat{\boldsymbol{V}}_\theta(\boldsymbol{M}^n, \hat{\theta}_n) \\ \geq &\overline{\lim}\, n \inf \left\{ \mathrm{tr}\, \hat{\boldsymbol{V}}_\theta(\boldsymbol{M}^n, \hat{\theta}) \middle| (\boldsymbol{M}, \hat{\theta}) : \text{ a locally unbiased estimator} \right\}. \end{aligned} \tag{6.157}$$

When $(\boldsymbol{M}, \hat{\theta})$ is a locally unbiased estimator, the matrix-valued vector $\mathbf{X}' := (O(\boldsymbol{M}^n, \hat{\theta}_n^i))$ satisfies the condition $\delta_i^j = \mathrm{Tr}\, \frac{\partial \rho_\theta^{\otimes n}}{\partial \theta^i} X'^j$. Due to (**c**), we have $\mathrm{tr}\, \boldsymbol{V}_\theta(\boldsymbol{M}, \hat{\theta}) \geq \mathrm{tr}(\mathbf{Re}\, \boldsymbol{V}_\theta(\mathbf{X}') + |\, \mathbf{Im}\, \boldsymbol{V}_\theta(\mathbf{X}')|)$. Thus, we obtain

$$\begin{aligned} &\inf \left\{ \mathrm{tr}\, \hat{\boldsymbol{V}}_\theta(\boldsymbol{M}^n, \hat{\theta}) \middle| (\boldsymbol{M}, \hat{\theta}) : \text{ a locally unbiased estimator} \right\} \\ \geq &\min_{\mathbf{X}} \left\{ \mathrm{tr}\, \mathbf{Re}\, \boldsymbol{V}_\theta(\mathbf{X}) + \mathrm{tr}\, |\, \mathbf{Im}\, \boldsymbol{V}_\theta(\mathbf{X})| \middle| \delta_i^j = \mathrm{Tr}\, \frac{\partial \rho_\theta^{\otimes n}}{\partial \theta^i} X^j \right\}. \end{aligned} \tag{6.158}$$

Combining (6.114), (6.115), (6.157), and (6.158), we obtain (6.101)

Exercise 6.58

(**d**) Due to (**b**) of Exercise 6.55, we can see that $|\, \mathbf{Im}\, \boldsymbol{V}(\boldsymbol{x})| - \mathbf{Im}\, \boldsymbol{V}(\boldsymbol{x}) \geq 0$.

(e) $V(y) = V(x) + V(w) = \mathbf{Re}\, V(x) + \mathbf{Im}\, V(x) + |\mathbf{Im}\, V(x)| - \mathbf{Im}\, V(x) = \mathbf{Re}\, V(x) + |\mathbf{Im}\, V(x)|$.

(f) Let $(M = \{M_k\}, \hat{\theta})$ be the estimator given in **(c)** for the set of vectors (y^i) in the extended system given. Let P be the projection to the original space. Define another POVM $M' = \{PM_kP\}$. Then, $\mathrm{E}^i_\theta(M', \hat{\theta})u = P\mathrm{E}^i_\theta(M, \hat{\theta})u = x^j$. Also, $\hat{V}_\theta(M', \hat{\theta}) = \hat{V}_\theta(M, \hat{\theta}) = V(y) = \mathbf{Re}\, V(x) + |\mathbf{Im}\, V(x)|$.

(f) For any x satisfying $\delta_i^j = \langle x^i|u_j\rangle$, the above argument shows the existence of locally unbiased estimator M such that $V(M) = \mathbf{Re}\, V(x) + |\mathbf{Im}\, V(x)|$.

(h) A vector (x^i) satisfying $\langle x^i|u_j\rangle = \delta^i_j$ is limited to the vector $x^i = ((\mathbf{Re}\, J)^{-1})^{i,j} u_j$. In this case, we have $\mathbf{Re}\, V(x) + |\mathbf{Im}\, V(x)| = \mathrm{tr}(\mathbf{Re}\, J)^{-1} + \mathrm{tr}\, |(\mathbf{Re}\, J)^{-1}\, \mathbf{Im}\, J (\mathbf{Re}\, J)^{-1}|$.

References

1. D. Petz, G. Toth, Lett. Math. Phys. **27**, 205 (1993)
2. C.W. Helstrom, Minimum mean-square error estimation in quantum statistics. Phys. Lett. **25A**, 101–102 (1976)
3. D. Petz, Monotone metrics on matrix spaces. Lin. Alg. Appl. **224**, 81–96 (1996)
4. D. Petz, C. Sudár, Extending the Fisher metric to density matrices, eds. by O.E. Barndorff-Nielsen, E.B.V. Jensenin. *Geometry in Present Day Science*, vol. 21 (World Scientific, Singapore, 1998)
5. H. Nagaoka, The world of quantum information geometry. IEICE Trans. **J88-A**(8), 874–885 (2005) (in Japanese)
6. C.W. Helstrom, *Quantum Detection and Estimation Theory* (Academic, New York, 1976)
7. A.S. Holevo, *Probabilistic and Statistical Aspects of Quantum Theory* (North-Holland, Amsterdam, 1982); originally published in Russian (1980)
8. S. Amari, H. Nagaoka, *Methods of Information Geometry* (AMS & Oxford University Press, Oxford, 2000)
9. A. Uhlmann, Density operators as an arena for differential geometry. Rep. Math. Phys. **33**, 253–263 (1993)
10. K. Matsumoto, *Geometry of a Quantum State*, Master's Thesis (Graduate School of Engineering, University of Tokyo, Japan, Department of Mathematical Engineering and Information Physics, 1995). (in Japanese)
11. H. Nagaoka, An asymptotically efficient estimator for a one-dimensional parametric model of quantum statistical operators, *Proceedings 1988 IEEE International Symposium on Information Theory*, vol. 198 (1988)
12. H. Nagaoka, On the relation between Kullback divergence and Fisher information—from classical systems to quantum systems, in *Proceedings Joint Mini-Workshop for Data Compression Theory and Fundamental Open Problems in Information Theory* (1992), pp. 63–72. (Originally in Japanese; also appeared as Chap. 27 of *Asymptotic Theory of Quantum Statistical Inference*, M. Hayashi eds.)
13. H. Nagaoka, On the parameter estimation problem for quantum statistical models, in *Proceedings 12th Symposium on Information Theory and Its Applications (SITA)* (1989), pp. 577–582. (Also appeared as Chap. 10 of *Asymptotic Theory of Quantum Statistical Inference*, M. Hayashi eds.)
14. H. Nagaoka, Differential geometrical aspects of quantum state estimation and relative entropy, eds. by V.P. Belavkin, O. Hirota, R.L. Hudson. *Quantum Communications and Measurement* (Plenum, New York, 1995), pp. 449–452
15. F. Hiai, D. Petz, The proper formula for relative entropy and its asymptotics in quantum probability. Commun. Math. Phys. **143**, 99–114 (1991)

16. A. Fujiwara, *Statistical Estimation Theory for Quantum States*, Master's Thesis (Graduate School of Engineering, University of Tokyo, Japan, Department of Mathematical Engineering and Information Physics, 1993). (in Japanese)
17. A. Fujiwara, Private communication to H. Nagaoka (1996)
18. H. Nagaoka, Private communication to A. Fujiwara (1991)
19. H.L. van Trees, *Detection, Estimation and Modulation Theory, Part 1* (Wiley, New York, 1968)
20. R.D. Gill, B.Y. Levit, Applications of the van Tree inequality: a Bayesian Cramér-Rao bound. Bernoulli **1**, 59–79 (1995)
21. R.D. Gill, Conciliation of Bayes and pointwise quantum state estimation: asymptotic information bounds in quantum statistics, eds. by V.P. Belavkin, M. Guta. *Quantum Stochastics & Information: Statistics, Filtering & Control* (World Scientific, 2008), pp. 239–261
22. H. Nagaoka, On fisher information of quantum statistical models, in *Proceedings 10th Symposium on Information Theory and Its Applications (SITA)*, Enoshima, Kanagawa, Japan, 19–21 November 1987, pp. 241–246. (Originally in Japanese; also appeared as Chap. 9 of *Asymptotic Theory of Quantum Statistical Inference,* M. Hayashi eds.)
23. T.Y. Young, Asymptotically efficient approaches to quantum-mechanical parameter estimation. Inf. Sci. **9**, 25–42 (1975)
24. R. Gill, S. Massar, State estimation for large ensembles. Phys. Rev. A **61**, 042312 (2000)
25. M. Hayashi, K. Matsumoto, Statistical model with measurement degree of freedom and quantum physics, in *RIMS koukyuroku Kyoto University*, vol. 1055 (1998), pp. 96–110 (in Japanese). (Also appeared as Chap. 13 of *Asymptotic Theory of Quantum Statistical Inference,* M. Hayashi eds.)
26. M. Hayashi, Two quantum analogues of Fisher information from a large deviation viewpoint of quantum estimation. J. Phys. A Math. Gen. **35**, 7689–7727 (2002); quant-ph/0202003 (2002). (Also appeared as Chap. 28 of *Asymptotic Theory of Quantum Statistical Inference,* M. Hayashi eds.)
27. M. Hayashi, Second-order asymptotics in fixed-length source coding and intrinsic randomness. IEEE Trans. Inf. Theory **54**, 4619–4637 (2008)
28. K. Matsumoto, Uhlmann's parallelism in quantum estimation theory. quant-ph/9711027 (1997)
29. K. Matsumoto, *A Geometrical Approach to Quantum Estimation Theory*, Ph.D. Thesis (Graduate School of Mathematical Sciences, University of Tokyo, 1997)
30. K. Matsumoto, The asymptotic efficiency of the consistent estimator, Berry-Uhlmann' curvature and quantum information geometry, eds. by P. Kumar, G.M. D'ariano, O. Hirotain. *Quantum Communication, Computing, and Measurement 2* (Plenum, New York, 2000), pp. 105–110
31. M. Hayashi, K. Matsumoto, Asymptotic performance of optimal state estimation in qubit system. J. Math. Phys. **49**, 102101 (2008); quant-ph/0411073 (2004)
32. M. Hayashi, Quantum estimation and quantum central limit theorem. Sugaku **55**, 4, 368–391 (2003) (in Japanese); English translation is in *Selected Papers on Probability and Statistics* (American Mathematical Society Translations Series 2) vol. 277 (2009), pp 95–123
33. K. Matsumoto, Seminar notes (1999)
34. N. Giri, W. von Waldenfels, An algebraic version of the central limit theorem. Z. Wahrscheinlichkeitstheorie Verw. Gebiete **42**, 129–134 (1978)
35. D. Petz, *An Invitation to the Algebra of Canonical Commutation Relations*. Leuven Notes in Mathematical and Theoretical Physics, vol. 2 (1990)
36. M. Hayashi, Asymptotic quantum estimation theory for the thermal states family, eds. by P. Kumar, G.M. D'ariano, O. Hirota. *Quantum Communication, Computing, and Measurement 2* (Plenum, New York, 2000) pp. 99–104; quant-ph/9809002 (1998). (Also appeared as Chap. 14 of *Asymptotic Theory of Quantum Statistical Inference*, M. Hayashi eds.)
37. H.P. Yuen, M. Lax, Multiple-parameter quantum estimation and measurement of nonselfadjoint observables. IEEE Trans. Inf. Theory **19**, 740 (1973)
38. M. Hayashi, Asymptotic large deviation evaluation in quantum estimation theory, in *Proceedings Symposium on Statistical Inference and Its Information-Theoretical Aspect* (1998), pp. 53–82 (in Japanese)

39. A.S. Holevo, Some statistical problems for quantum Gaussian states. IEEE Trans. Inf. Theory **21**, 533–543 (1975)
40. D. Bruß, A. Ekert, C. Machiavello, Optimal universal quantum cloning and state estimation. Phys. Rev. Lett. **81**, 2598–2601 (1998). (also appeared as Chap. 24 of *Asymptotic Theory of Quantum Statistical Inference*, M. Hayashi eds.)
41. R.F. Werner, Optimal cloning of pure states. Phys. Rev. A **58**, 1827 (1998)
42. M. Keyl, R.F. Werner, Optimal cloning of pure states, judging single clones. J. Math. Phys. **40**, 3283–3299 (1999)
43. H. Fan, K. Matsumoto, M. Wadati, Quantum cloning machines of a d-level system. Phys. Rev. A **64**, 064301 (2001)
44. H. Fan, K. Matsumoto, X. Wang, M. Wadati, Quantum cloning machines for equatorial qubits. Phys. Rev. A **65**, 012304 (2002)
45. K. Usami, Y. Nambu, Y. Tsuda, K. Matsumoto, K. Nakamura, Accuracy of quantum-state estimation utilizing Akaike's information criterion. Phys. Rev. A **68**, 022314 (2003)
46. M. Hayashi, *Minimization of deviation under quantum local unbiased measurements*, Master's Thesis (Graduate School of Science, Kyoto University, Japan, Department of Mathematics, 1996)
47. M. Hayashi, A linear programming approach to attainable Cramer–Rao type bound, eds. by O. Hirota, A.S. Holevo, C.M. Cavesin. *Quantum Communication, Computing, and Measurement* (Plenum, New York, 1997), pp. 99–108. (Also appeared as Chap. 12 of *Asymptotic Theory of Quantum Statistical Inference,* M. Hayashi eds.)
48. M. Hayashi, A linear programming approach to attainable cramer-rao type bound and randomness conditions. Kyoto-Math 97–08; quant-ph/9704044 (1997)
49. A. Fujiwara, H. Nagaoka, Coherency in view of quantum estimation theory, eds. by K. Fujikawa, Y.A. Ono. *Quantum Coherence and Decoherence* (Elsevier, Amsterdam, 1996), pp. 303–306
50. A. Fujiwara, H. Nagaoka, Quantum Fisher metric and estimation for pure state models. Phys. Lett. **201A**, 119–124 (1995)
51. A. Fujiwara, H. Nagaoka, An estimation theoretical characterization of coherent states. J. Math. Phys. **40**, 4227–4239 (1999)
52. K. Matsumoto, A new approach to the Cramér-Rao type bound of the pure state model. J. Phys. A Math. Gen. **35**, 3111–3123 (2002)
53. D. Petz, Quasi-entropies for finite quantum systems. Rep. Math. Phys. **23**, 57–65 (1986)
54. D. Petz, Sufficient subalgebras and the relative entropy of states of a von Neumann algebra. Commun. Math. Phys. **105**, 123–131 (1986)
55. D. Petz, Monotonicity of quantum relative entropy revisited. Rev. Math. Phys. **15**, 29–91 (2003)
56. V.P. Belavkin, Generalized uncertainty relations and efficient measurements in quantum systems. Teor. Mat. Fiz. **26**(3), 316–329 (1976); quant-ph/0412030 (2004)
57. H. Nagaoka, A new approach to Cramér–Rao bound for quantum state estimation. IEICE Tech. Rep. **IT 89-42**(228), 9–14 (1989)
58. H. Nagaoka, A generalization of the simultaneous diagonalization of Hermitian matrices and its relation to quantum estimation theory. Trans. Jpn. Soc. Ind. Appl. Math. **1**, 43–56 (1991) (Originally in Japanese; also appeared as Chap. 11 of *Asymptotic Theory of Quantum Statistical Inference*, M. Hayashi eds.)
59. Y. Tsuda, K. Matsumoto, Quantum estimation for non-differentiable models. J. Phys. A Math. Gen. **38**(7), 1593–1613 (2005)
60. D.C. Brody, L.P. Hughston, R. Soc, Lond. Proc. A **454**, 2445–2475 (1998)
61. Y. Tsuda, Estimation of Polynomial of Complex Amplitude of Quantum Gaussian States, in *Annual Meeting of the Japan Statistical Society, Proceedings*, 2005. (in Japanese)
62. G.M. D'Ariano, Homodyning as universal detection, eds. by O. Hirota, A.S. Holevo, C.A. Caves. *Quantum Communication, Computing, and Measurement* (Plenum, New York, 1997), pp. 253–264; quant-ph/9701011 (1997). (also appeared as Chap. 33 of *Asymptotic Theory of Quantum Statistical Inference,* M. Hayashi eds.)
63. L.M. Artiles, R.D. Gill, M.I. Guţă, An invitation to quantum tomography (II). J. R. Stat. Soc. Ser. B (Stat. Methodol.) **67**, 109–134 (2005)

64. C.W. Helstrom, Int. J. Theor. Phys. **11**, 357 (1974)
65. A.S. Holevo, Covariant measurements and uncertainty relations. Rep. Math. Phys. **16**, 385–400 (1979)
66. M. Ozawa, On the noncommutative theory of statistical decisions. Res. Reports Inform. Sci. **A-74** (1980)
67. N.A. Bogomolov, Minimax measurements in a general statistical decision theory. Teor. Veroyatnost. Primenen. **26**, 798–807 (1981); (English translation: *Theory Probab. Appl.*, **26**, 4, 787–795 (1981))
68. A.S. Holevo, Covariant Measurements and Optimality, in Chap. IV of *Probabilistic and Statistical Aspects of Quantum Theory* (North-Holland, Amsterdam, 1982). Originally published in Russian (1980): "Bounds for generalized uncertainty of the shift parameter. Lecture Notes in Mathematics **1021**, 243–251 (1983)
69. S. Massar, S. Popescu, Optimal extraction of information from finite quantum ensembles. Phys. Rev. Lett. **74**, 1259 (1995)
70. M. Hayashi, Asymptotic estimation theory for a finite dimensional pure state model. J. Phys. A Math. Gen. **31**, 4633–4655 (1998). (Also appeared as Chap. 23 of *Asymptotic Theory of Quantum Statistical Inference*, M. Hayashi eds.)
71. M. Hayashi, On the second order asymptotics for pure states family. IEICE Trans. **E88-JA**, 903–916 (2005) (in Japanese)
72. E. Bagan, M.A. Ballester, R.D. Gill, A. Monras, R. Muñoz-Tapia, O. Romero-Isart, Optimal full estimation of qubit mixed states. Phys. Rev. A **73**, 032301 (2006)
73. A. Fujiwara, Quantum channel identification problem. Phys. Rev. A **63**, 042304 (2001)
74. M. Sasaki, M. Ban, S.M. Barnett, Optimal parameter estimation of a depolarizing channel. Phys. Rev. A **66**, 022308 (2002)
75. D.G. Fischer, H. Mack, M.A. Cirone, M. Freyberger, Enhanced estimation of a noisy quantum channel using entanglement. Phys. Rev. A **64**, 022309 (2001)
76. A. Fujiwara, H. Imai, Quantum parameter estimation of a generalized Pauli channel. J. Phys. A Math. Gen. **36**, 8093–8103 (2003)
77. A. Fujiwara, Estimation of a generalized amplitude-damping channel. Phys. Rev. A **70**, 012317 (2004)
78. F. Tanaka, *Investigation on fisher metric on the classical statistical models and the quantum ones*, Master's thesis (Graduate School of Information Science and Technology, University of Tokyo, Japan, Department of Mathematical Informatics, 2004). (in Japanese)
79. F. De Martini, A. Mazzei, M. Ricci, G.M. D'Ariano, Pauli tomography: complete characterization of a single qubit device. Fortschr. Phys. **51**, 342–348 (2003)
80. V. Bužek, R. Derka, S. Massar, Optimal quantum clocks. Phys. Rev. Lett. **82**, 2207 (1999)
81. A. Acín, E. Jané, G. Vidal, Optimal estimation of quantum dynamics. Phys. Rev. A **64**, 050302(R) (2001)
82. A. Fujiwara, Estimation of SU(2) operation and dense coding: an information geometric approach. Phys. Rev. A **65**, 012316 (2002)
83. M.A. Ballester, Estimation of unitary quantum operations. Phys. Rev. A **69**, 022303 (2004)
84. E. Bagan, M. Baig, R. Muñoz-Tapia, Entanglement-assisted alignment of reference frames using a dense covariant coding. Phys. Rev. A **69**, 050303(R) (2004)
85. M. Hayashi, Parallel treatment of estimation of SU(2) and phase estimation. Phys. Lett. A **354**, 183–189 (2006)
86. M. Hayashi, Estimation of SU(2) action by using entanglement, in *Proceedings 9th Quantum Information Technology Symposium (QIT9)*, NTT Basic Research Laboratories, Atsugi, Kangawa, Japan, 11–12 December 2003, pp. 9–13 (in Japanese)
87. G. Chiribella, G.M. D'Ariano, P. Perinotti, M.F. Sacchi, Efficient use of quantum resources for the transmission of a reference frame. Phys. Rev. Lett. **93**, 180503 (2004)
88. E. Bagan, M. Baig, R. Muñoz-Tapia, Quantum reverse-engineering and reference-frame alignment without nonlocal correlations. Phys. Rev. A **70**, 030301(R) (2004)
89. G. Chiribella, G.M. D'Ariano, P. Perinotti, M.F. Sacchi, Covariant quantum measurements which maximize the likelihood. Phys. Rev. A **70**, 061205 (2004)

90. G. Chiribella, G.M. D'Ariano, P. Perinotti, M.F. Sacchi, Maximum likelihood estimation for a group of physical transformations. Int. J. Quantum Inform. **4**, 453 (2006)
91. M. Hayashi, S. Vinjanampathy, L.C. Kwek, Unattainable & attainable bounds for quantum sensors (2016). arXiv:1602.07131
92. G. Chiribella, G.M. D'Ariano, M.F. Sacchi, Optimal estimation of group transformations using entanglement. Phys. Rev. A **72**, 042338 (2005)
93. M. Hayashi, Comparison between the Cramer-Rao and the mini-max approaches in quantum channel estimation. Commun. Math. Phys. **304**(3), 689–709 (2011)
94. M.A. Morozowa, N.N. Chentsov, Itogi Nauki i Tekhniki **36**, 289–304 (1990). (in Russian)
95. M. Hayashi, Characterization of several kinds of quantum analogues of relative entropy. Quant. Inf. Comput. **6**, 583–596 (2006)
96. K. Matsumoto, Reverse estimation theory, complementarity between RLD and SLD, and monotone distances. quant-ph/0511170 (2005); *Proceedings 9th Quantum Information Technology Symposium (QIT13)*, Touhoku University, Sendai, Miyagi, Japan, 24–25 November 2005, pp. 81–86
97. A. Fujiwara, *A Geometrical Study in Quantum Information Systems*, Ph.D. Thesis (Department of Mathematical Engineering and Information Physics, Graduate School of Engineering, University of Tokyo, Japan, 1995)
98. M. Ohya, D. Petz, *Quantum Entropy and Its Use* (Springer, Berlin, 1993)

Chapter 7
Quantum Measurements and State Reduction

Abstract In quantum mechanics, the state reduction due to a measurement is called the collapse of a wavefunction. Its study is often perceived as a somewhat mystical phenomenon because of the lack of proper understanding. As a result, the formalism for the state reduction is often somewhat inadequately presented. However, as will be explained in Sect. 7.1, the state reduction due to a measurement follows automatically from the formulation of quantum mechanics, as described in Sect. 1.2. Starting with the formulation of quantum mechanics given in Sects. 1.2 and 1.4, we give a detailed formulation of the state reduction due to a measurement. In Sect. 7.2, we discuss the relation with the uncertainty relation using these concepts. Finally, in Sect. 7.4, we propose a measurement with negligible state reduction.

7.1 State Reduction Due to Quantum Measurement

In previous chapters, we examined several issues related to quantum measurement; these issues were concerned only with the probability distribution of the measurement outcomes. However, when we examine an application of a measurement after another application of a measurement to the same quantum system, we need to describe the state reduction due to the first measurement. First, we discuss the state reduction due to a typical measurement corresponding to a POVM $\boldsymbol{M} = \{M_\omega\}$. Then, we give the general conditions for state reduction from the axiomatic framework given in Sect. 1.2.

Assume that we perform a measurement corresponding to the POVM $\boldsymbol{M} = \{M_\omega\}$ and obtain a measurement outcome ω. When the state reduction has the typical form due to the POVM $\boldsymbol{M} = \{M_\omega\}$, the resultant state is

$$\frac{1}{\operatorname{Tr} \rho M_\omega} \sqrt{M_\omega} \rho \sqrt{M_\omega}, \tag{7.1}$$

where $\frac{1}{\operatorname{Tr} \rho M_\omega}$ is the normalization factor.[1] In particular, if $\boldsymbol{M}$ is a PVM, then M_ω is a projection and therefore the above state is [1]

[1]Normalization here implies the division of the matrix by its trace such that its trace is equal to 1.

M. Hayashi, *Quantum Information Theory*, Graduate Texts in Physics,
DOI 10.1007/978-3-662-49725-8_7

$$\frac{1}{\operatorname{Tr} \rho M_\omega} M_\omega \rho M_\omega. \tag{7.2}$$

Since (7.2) is sandwiched by projection operators, the above-mentioned state reduction is called the **projection hypothesis**. In many books on quantum mechanics, the state reduction due to a measurement is restricted only to that satisfying the projection hypothesis. However, this is in fact incorrect, and such a state reduction is merely typical. That is, it is not necessarily true that any state reduction corresponding to the POVM $\boldsymbol{M}$ satisfies the above (7.2). In fact, other types of state reductions can occur due to a single POVM $\boldsymbol{M}$, as will be described later.

Now, we assume that we are given an initial state ρ on a composite system $\mathcal{H}_A \otimes \mathcal{H}_B$. When we perform a measurement corresponding to the POVM $\boldsymbol{M}^B = \{M_\omega^B\}_\omega$ on the system $\mathcal{H}_B$ and obtain the measurement outcome ω, the resultant state of $\mathcal{H}_A$ is then

$$\frac{1}{\operatorname{Tr} \rho (I_A \otimes M_\omega^B)} \operatorname{Tr}_B (I_A \otimes \sqrt{M_\omega^B}) \rho (I_A \otimes \sqrt{M_\omega^B}), \tag{7.3}$$

regardless of the type of state reduction on $\mathcal{H}_B$, as long as the measurement outcome ω obeys the distribution $\operatorname{Tr}_B (\operatorname{Tr}_A \rho) M_\omega^B$. To prove this fact, we consider an arbitrary POVM $\boldsymbol{M}^A = \{M_x^A\}_{x \in \mathcal{X}}$ on $\mathcal{H}_A$. Since $(M_x^A \otimes M_\omega^B) = (I_A \otimes \sqrt{M_\omega^B})(M_x^A \otimes I_B)(I_A \otimes \sqrt{M_\omega^B})$, the joint distribution of (x, ω) is given by

$$\begin{aligned} &\operatorname{Tr} \rho (M_x^A \otimes M_\omega^B) = \operatorname{Tr} (I_A \otimes \sqrt{M_\omega^B}) \rho (I_A \otimes \sqrt{M_\omega^B})(M_x^A \otimes I_B) \\ = &\operatorname{Tr} [\operatorname{Tr}_B (I_A \otimes \sqrt{M_\omega^B}) \rho (I_A \otimes \sqrt{M_\omega^B})] M_x^A, \end{aligned}$$

according to the discussion of Sect. 1.4, e.g., (1.26). When the measurement outcome ω is observed, the probability distribution of the other outcome x is

$$\frac{1}{\operatorname{Tr} \rho (I_A \otimes M_\omega^B)} \operatorname{Tr} [\operatorname{Tr}_B (I_A \otimes \sqrt{M_\omega^B}) \rho (I_A \otimes \sqrt{M_\omega^B})] M_x^A,$$

which is the conditional distribution of x when the the measurement outcome on $\mathcal{H}_A$ is ω. Since this condition holds for an arbitrary POVM $\boldsymbol{M}^A = \{M_x^A\}_{x \in \mathcal{X}}$ on $\mathcal{H}_A$, (7.3) gives the resultant state of $\mathcal{H}_A$ when ω is the outcome of the measurement $\boldsymbol{M}^B$ on $\mathcal{H}_B$.

However, since (7.3) only describes the state reduction of a system that is not directly measured, the above discussion does not directly deal with the state reduction on the system $\mathcal{H}_B$, e.g., (7.2). As shown from the Naĭmark–Ozawa extension [2, 3] given below, it is theoretically possible to perform a measurement such that the state reduction follows (7.1) or (7.2).

Theorem 7.1 (Naĭmark [2], Ozawa [3]) *Consider an arbitrary POVM* $\boldsymbol{M} = \{M_\omega\}_{\omega \in \Omega}$ *on* $\mathcal{H}$ *and arbitrary outcome* $\omega_0 \in \Omega$. *Let* $\mathcal{H}_0$ *be the additional space* *with the orthonormal basis* $\{u_\omega\}_\omega$, *and let us define the PVM* $E_\omega = |u_\omega\rangle\langle u_\omega|$ *on* $\mathcal{H}_0$. *There exists a unitary matrix* U *such that*

$$\operatorname{Tr} M_\omega \rho = \operatorname{Tr}(I \otimes E_\omega) U(\rho \otimes \rho_0) U^*, \tag{7.4}$$

$$\sqrt{M_\omega}\rho\sqrt{M_\omega} = \operatorname{Tr}_{\mathcal{H}_0}(I \otimes E_\omega) U(\rho \otimes \rho_0) U^*(I \otimes E_\omega), \tag{7.5}$$

where $\rho_0 = |u_{\omega_0}\rangle\langle u_{\omega_0}|$.

Theorem 4.5 can be obtained from this theorem by considering a PVM $\{U^*(I \otimes E_\omega)U\}_\omega$.

In the following, we make several observations based on this theorem. As described by (7.4), a measurement corresponding to an arbitrary POVM $\boldsymbol{M}$ can be realized by a PVM $\boldsymbol{E} = \{E_\omega\}$ with an appropriate time evolution U between $\mathcal{H}$ and $\mathcal{H}_0$. Furthermore, according to the above arguments, when a measurement corresponding to the PVM $\boldsymbol{E}$ on the system $\mathcal{H}_0$ is performed, the resultant state of $\mathcal{H}$ with the measurement outcome ω is given by [3]

$$\frac{1}{\operatorname{Tr} U(\rho \otimes \rho_0)U^*(I \otimes E_\omega)} \operatorname{Tr}_{\mathcal{H}_0}(I \otimes E_\omega) U(\rho \otimes \rho_0) U^*(I \otimes E_\omega).$$

Theorem 7.1 therefore shows that the above procedure produces a measurement corresponding to the resultant state (7.1). This model of measurement is called an **indirect measurement** model. The additional space $\mathcal{H}_0$ is called an **ancilla**, which interacts directly with the macroscopic system. In this way, this model describes the resultant state of system $\mathcal{H}$ of our interest when the measurement outcome ω is obtained in the ancilla. However, it does not reveal anything about the process whereby the measurement outcome is obtained in the ancilla. Hence, there remains an undiscussed part in the process whereby the measurement outcome is obtained via an ancilla. This is called the measurement problem.

In almost all real experiments, the target system $\mathcal{H}$ is not directly but indirectly measured via the measurement on an ancilla. For example, consider the Stern–Gerlach experiment, which involves the measurement of the spin of silver atoms. In this case, the spin is measured indirectly by measuring the momentum of the atom after the interaction between the spin system and the angular momentum system of the atom. Therefore, in such experiments, it is natural to apply the indirect measurement model to the measurement process.

The above theorem can be regarded as the refinement of the Naĭmark extension for the measurement of real quantum systems [3]. As this construction was firstly given by Ozawa, the triple $(\mathcal{H}_0, \rho_0, U)$ given in Theorem 7.1 is called the **Naĭmark–Ozawa extension**.

Proof of Theorem 7.1 For simplicity, we consider the case when the probability space Ω has a finite cardinality. Without loss of generality, we can assume that the probability space Ω is $\{1, \dots, n\}$, the orthonormal basis of $\mathcal{H}_0$ is given by $\{u_i\}_{i=1}^n$, and ρ_0 is given by $|u_1\rangle\langle u_1|$, for simplicity. First, let us check that an arbitrary matrix

U on $\mathcal{H} \otimes \mathcal{H}_0$ can be written as $(U^{i,j})_{i,j}$, using a matrix $U^{i,j}$ on $\mathcal{H}$. This implies that $(I \otimes |u_i\rangle\langle u_i|)U(I \otimes |u_j\rangle\langle u_j|) = U^{i,j} \otimes |u_i\rangle\langle u_j|$. Accordingly, (7.5) is equivalent to $\sqrt{M_i}\rho\sqrt{M_i} = U^{i,1}\rho(U^{i,1})^*$ with $\omega = i$. Choosing $U^{i,1} = \sqrt{M_i}$, we have $\sum_{i=1}^n U^{i,1}(U^{i,1})^* = I$, and therefore, it is possible to choose the remaining elements such that $U = (U^{i,j})_{i,j}$ is a unitary matrix, according to Exercise 1.2. This confirms the existence of a unitary matrix U that satisfies (7.5). Taking the trace in (7.5) gives us (7.4). ∎

We next consider the possible state reductions according to the framework for a quantum measurement given in Sect. 1.2. Perform the measurement corresponding to a POVM $\boldsymbol{M}$ on a quantum system in a state ρ. Then, using the map κ_ω, we describe the resultant state with a measurement outcome ω by

$$\frac{1}{\operatorname{Tr} \rho M_\omega} \kappa_\omega(\rho). \tag{7.6}$$

The map κ_ω can be restricted to a completely positive map as shown below. In order to show this fact, we prove that

$$\kappa_\omega(\lambda\rho_1 + (1-\lambda)\rho_2) = \lambda\kappa_\omega(\rho_1) + (1-\lambda)\kappa_\omega(\rho_2) \tag{7.7}$$

for two arbitrary states ρ_1 and ρ_2 and an arbitrary real number λ satisfying $0 \le \lambda \le 1$. Consider an application of a measurement corresponding to another arbitrary POVM $\{M'_{\omega'}\}_{\omega' \in \Omega'}$ after the first measurement. This is equivalent to performing a measurement with the probability space $\Omega \times \Omega'$ on the system in the initial state. The joint probability distribution of ω and ω' is then given by $\operatorname{Tr} \kappa_\omega(\rho)M'_{\omega'}$ (Exercise 7.3). Consider the convex combination of the density matrix. Then, similarly to (1.11), the equation

$$\operatorname{Tr} \kappa_\omega(\lambda\rho_1 + (1-\lambda)\rho_2)M'_{\omega'} = \lambda \operatorname{Tr} \kappa_\omega(\rho_1)M'_{\omega'} + (1-\lambda) \operatorname{Tr} \kappa_\omega(\rho_2)M'_{\omega'}$$

should hold. Since $\{M'_{\omega'}\}_{\omega' \in \Omega'}$ is an arbitrary POVM, it is also possible to choose $M'_{\omega'}$ to be an arbitrary one-dimensional projection. Therefore, we obtain (7.7). Taking entangled input states into account, we then require κ_ω to be a completely positive map. This statement can be shown by using arguments similar to that of Sect. 5.1. That is, it can be verified by adding a reference system. Since (7.6) is a density matrix, we obtain [4]

$$\operatorname{Tr} \kappa_\omega(\rho) = \operatorname{Tr} \rho M_\omega, \quad \forall \rho, \tag{7.8}$$

which is equivalent to $M_\omega = \kappa_\omega^*(I)$. Thus $\operatorname{Tr} \sum_\omega \kappa_\omega(\rho) = 1$. The measurement with the state reduction is represented by the set of completely positive maps $\kappa = \{\kappa_\omega\}_{\omega \in \Omega}$, where the map $\sum_\omega \kappa_\omega$ preserves the trace [3]. Henceforth, we shall call $\kappa = \{\kappa_\omega\}_{\omega \in \Omega}$ an **instrument**. In this framework, if ρ is the initial state, the probability of obtaining a measurement outcome ω is given by $\operatorname{Tr} \kappa_\omega(\rho)$. Once a measurement outcome ω is obtained, the resultant state is given by $\frac{1}{\mathrm{Tr}\, \kappa_\omega(\rho)} \kappa_\omega(\rho)$. We can also regard the Choi–Kraus representation $\{A_i\}$ of a TP-CP map as an instrument $\{\kappa_i\}$ with the correspondence $\kappa_i(\rho) = A_i \rho A_i^*$. The notation "an instrument $\{A_i\}$" then actually implies an instrument $\{\kappa_i\}$ with the above correspondence. Therefore, the state evolution with a Choi–Kraus representation $\{A_i\}$ can be regarded as a state reduction given by the instrument $\{A_i\}$ when the measurement outcome is not recorded. (The measurement is performed, but the experimenter does not read the outcome.)

When the instrument is in the form of the square roots $\{\sqrt{M_\omega}\}$ of a POVM $\boldsymbol{M} = \{M_\omega\}_{\omega \in \Omega}$, the resultant state is given by (7.1) and will be denoted by $\boldsymbol{\kappa}_{\boldsymbol{M}}$. If the instrument $\boldsymbol{\kappa} = \{\kappa_\omega\}_{\omega \in \Omega}$ and the POVM $\boldsymbol{M} = \{M_\omega\}_{\omega \in \Omega}$ satisfy condition (7.8), we shall call the instrument $\boldsymbol{\kappa}$ an **instrument corresponding to the POVM $\boldsymbol{M}$**. We can characterize an instrument corresponding to a POVM $\boldsymbol{M}$ as follows.

Theorem 7.2 *Let $\boldsymbol{\kappa} = \{\kappa_\omega\}_{\omega \in \Omega}$ be an instrument corresponding to a POVM $\boldsymbol{M} = \{M_\omega\}$ in a quantum system $\mathcal{H}$. There exists a TP-CP map κ'_ω for each measurement outcome ω such that*

$$\kappa_\omega(\rho) = \kappa'_\omega \left(\sqrt{M_\omega} \rho \sqrt{M_\omega} \right). \tag{7.9}$$

According to this theorem, it is possible to represent any state reduction $\boldsymbol{\kappa}$ as a combination of the state reduction given by the joint of the typical state reduction (7.1) and the state evolution κ'_ω that depends on the outcome ω of the POVM $\boldsymbol{M}$.[2]

Proof From Condition ⑥ (the Choi–Kraus representation) of Theorem 5.14, there exists a set of matrices $E_1, \ldots, E_k$ such that $\kappa_\omega(\rho) = \sum_{i=1}^{k} E_i \rho E_i^*$. Since $\mathrm{Tr} \kappa_\omega(\rho) = \mathrm{Tr}\, \rho \sum_{i=1}^{k} E_i^* E_i = \mathrm{Tr}\, \rho M_\omega$ for an arbitrary state ρ, then $\sum_{i=1}^{k} E_i^* E_i = M_\omega$. Using the generalized inverse matrix $\sqrt{M_\omega}^{-1}$ defined in Sect. 1.5, and letting P be the projection to the range of M_ω (or $\sqrt{M_\omega}$), we have

$$\sum_{i=1}^{k} \sqrt{M_\omega}^{-1} E_i^* E_i \sqrt{M_\omega}^{-1} = P.$$

Hence, the matrices $E_1 \sqrt{M_\omega}^{-1}, \ldots, E_k \sqrt{M_\omega}^{-1}, I - P$ are the Choi–Kraus representations of the TP-CP map. Denoting this TP-CP map as κ'_ω, we have

---

[2] It can also be understood as follows: the state reduction due to any measurement by PVM can be characterized as the state reduction satisfying the projection hypothesis, followed by the state evolution κ_ω. Indeed, many texts state that the state reduction due to any measurement is given by the projection hypothesis. Theorem 7.2 guarantees their correctness in a sense.

$$\begin{aligned}&\kappa'_\omega\left(\sqrt{M_\omega}\rho\sqrt{M_\omega}\right)\\ =&(I-P)\sqrt{M_\omega}\rho\sqrt{M_\omega}(I-P)+\sum_{i=1}^{k}E_i\sqrt{M_\omega}^{-1}\sqrt{M_\omega}\rho\sqrt{M_\omega}\sqrt{M_\omega}^{-1}E_i^*\\ =&\sum_{i=1}^{k}E_i\rho E_i^*.\end{aligned}$$

Therefore, we see that (7.9) holds. ∎

Combining Theorems 7.1 and 7.2, we can construct a model of indirect measurement in a manner similar to Theorem 7.1 for an arbitrary instrument $\boldsymbol{\kappa}=\{\kappa_\omega\}$.

Theorem 7.3 *Let $\mathcal{H}_A$ and $\mathcal{H}_B$ be two quantum systems. The following two conditions are equivalent for the set of linear maps $\boldsymbol{\kappa}=\{\kappa_\omega\}_{\omega\in\Omega}$ from $\mathcal{T}(\mathcal{H}_A)$ to $\mathcal{T}(\mathcal{H}_B)$.*

① *$\boldsymbol{\kappa}$ is an instrument.*
② *$\boldsymbol{\kappa}$ can be expressed as*

$$\kappa_\omega(\rho)=\operatorname*{Tr}_{A,C}\left(I_{A,B}\otimes E_\omega\right)U\left(\rho\otimes\rho_0\right)U^*\left(I_{A,B}\otimes E_\omega\right), \tag{7.10}$$

where $\mathcal{H}_C$ is a quantum system with the dimension $\dim\mathcal{H}_B\times$ *(Number of elements in Ω), ρ_0 is a pure state on $\mathcal{H}_B\otimes\mathcal{H}_C$, $\boldsymbol{E}=\{E_\omega\}_\omega$ is a PVM on $\mathcal{H}_C$, and U is a unitary matrix on $\mathcal{H}_A\otimes\mathcal{H}_B\otimes\mathcal{H}_C$.*

The above is also equivalent to Condition ② with arbitrary-dimensional space $\mathcal{H}_C$, which is called Condition ②'.

If $\mathcal{H}_A=\mathcal{H}_B$, the following corollary holds.

Corollary 7.1 (Ozawa [3]) *The following two conditions for the set of linear maps $\boldsymbol{\kappa}=\{\kappa_\omega\}_{\omega\in\Omega}$ from $\mathcal{T}(\mathcal{H}_A)$ to $\mathcal{T}(\mathcal{H}_A)$ are equivalent [3].*

① *$\boldsymbol{\kappa}$ is an instrument.*
② *$\boldsymbol{\kappa}$ can be expressed as*

$$\kappa_\omega(\rho)=\operatorname*{Tr}_{D}\left(I_A\otimes E_\omega\right)V\left(\rho\otimes\rho_0\right)V^*\left(I_A\otimes E_\omega\right), \tag{7.11}$$

where $\mathcal{H}_D$ is a $(\dim\mathcal{H}_A)^2\times$ *(Number of elements in Ω)-dimensional quantum system, ρ_0 is a pure state on $\mathcal{H}_D$, $\boldsymbol{E}=\{E_\omega\}_\omega$ is a PVM on $\mathcal{H}_D$, and V is a unitary matrix on $\mathcal{H}_A\otimes\mathcal{H}_D$.*

Therefore, a model of indirect measurement exists for an arbitrary instrument $\boldsymbol{\kappa}=\{\kappa_\omega\}$. Henceforth, we shall call $(\mathcal{H}_D,V,\rho_0,\boldsymbol{E})$ and $(\mathcal{H}_C,U,\rho_0,\boldsymbol{E})$ **indirect measurements** and denote them by $\mathcal{I}$. Let us now rewrite the above relation among the three different notations for measurements $\boldsymbol{M}=\{M_\omega\}$, $\boldsymbol{\kappa}=\{\kappa_\omega\}$, and $\mathcal{I}=(\mathcal{H}_D,V,\rho_0,\boldsymbol{E})$. The POVM $\boldsymbol{M}$ only describes the probability distribution of the measurement outcomes and contains the least amount of information among the three notations. The instrument $\boldsymbol{\kappa}$ refers not only to the measurement outcome itself, but also to the resultant state of the measurement. Hence, a POVM $\boldsymbol{M}$ corresponds uniquely to an instrument $\boldsymbol{\kappa}$; however, the converse is not unique. Furthermore, the indirect measurement $\mathcal{I}$ denotes the unitary evolution required to realize the measurement device as well as the resultant state and the probability distribution of the observed outcome. This is the most detailed of the three notations (i.e., it contains the most information). Hence, a POVM $\boldsymbol{M}$ and an instrument $\boldsymbol{\kappa} = \{\kappa_\omega\}$ correspond uniquely to an indirect measurement $\mathcal{I}$, although the converse is not unique.

The proof of Theorem 7.3 is as follows: ②′⇒① and ②⇒②′ follows from inspection. See Exercise 7.1 for ①⇒②.

Exercises

7.1 Show that Condition ② in Theorem 7.3 may be derived from the Naĭmark–Ozawa extension $(\mathcal{H}_0, \rho_1', U')$ and Condition ① by using the Stinespring representation $(\mathcal{H}_C, \rho_0, U_{\kappa_{\omega}'})$ for the TP-CP map κ_ω' given in Theorem 7.2.

7.2 Prove Corollary 7.1 using Theorem 7.3.

7.3 Consider the situation when we apply a measurement $\boldsymbol{M} = \{M_{\omega'}\}$ after application of an instrument $\boldsymbol{\kappa} = \{\kappa_\omega\}$ to a quantum system $\mathcal{H}$. Define the POVM $\boldsymbol{M}' = \{M'_{\omega,\omega'}\}$ by $M'_{\omega,\omega'} \stackrel{\text{def}}{=} \kappa_\omega^*(M_{\omega'})$, where κ_ω^* is the dual map of κ_ω. Show that $\operatorname{Tr} \kappa_\omega(\rho) M_{\omega'} = \operatorname{Tr} \rho M'_{\omega,\omega'}$ for an arbitrary input state ρ.

7.4 Given an initial state given by a pure state $x = (x^{k,i}) \in \mathcal{H}_A \otimes \mathcal{H}_B$, perform a measurement given by the PVM $\{|u_i\rangle\langle u_i|\}$ (where $u_i = (u_i^j) \in \mathcal{H}_B$) on $\mathcal{H}_B$. Show that the resultant state on $\mathcal{H}_A$ with the measurement outcome i is given by $\frac{v_i}{\|v_i\|}$, assuming that $v_i \stackrel{\text{def}}{=} (\sum_j u_i^j x^{k,j}) \in \mathcal{H}_A$.

7.5 Prove Theorem 5.4 following the steps below.
(**a**) Using formula (7.3) for the state reduction due to a measurement, show that ②⇒①.
(**b**) Consider a quantum system $\mathcal{H}_C$. When a maximally entangled state is input to an entanglement-breaking channel κ, show that (i) the output is a separable state and (ii) ①⇒② using relationship (5.5).

7.2 Uncertainty and Measurement

7.2.1 Uncertainties for Observable and Measurement

The concept of uncertainty is often discussed in quantum mechanics in various contexts. In fact, there are no less than four distinct implications of the word *uncertainty*. Despite this, the differences between these implications are rarely discussed, and consequently, “uncertainty” is often used in a somewhat confused manner. In particular, there appears to be some confusion regarding its implication in the context of the Heisenberg uncertainty principle. We define the four meanings of uncertainty in the following and discuss the Heisenberg uncertainty principle and related topics in some detail [see (7.28), (7.31), and (7.33)].

First, let us define the **uncertainty of an observable** $\Delta_1(X, \rho)$ for a Hermitian matrix X (this can be considered an observable) and the state ρ by

$$\Delta_1^2(X, \rho) \stackrel{\text{def}}{=} \operatorname{Tr} \rho X^2 - (\operatorname{Tr} \rho X)^2 = \operatorname{Tr} \rho (X - \operatorname{Tr} \rho X)^2. \tag{7.12}$$

Next, let us define the **uncertainty of a measurement** $\Delta_2(\boldsymbol{M}, \rho)$ for a POVM $\boldsymbol{M} = \{(M_i, x_i)\}$ with real-valued measurement outcomes and a state ρ by

$$\Delta_2^2(\boldsymbol{M}, \rho) \stackrel{\text{def}}{=} \sum_i (x_i - \mathrm{E}_\rho(\boldsymbol{M}))^2 \operatorname{Tr} \rho M_i, \quad \mathrm{E}_\rho(\boldsymbol{M}) \stackrel{\text{def}}{=} \sum_i x_i \operatorname{Tr} \rho M_i. \tag{7.13}$$

By defining the **average matrix** $O(\boldsymbol{M})$ for the POVM $\boldsymbol{M}$ as below, the inequality

$$\Delta_2^2(\boldsymbol{M}, \rho) \geq \Delta_1^2(O(\boldsymbol{M}), \rho), \quad O(\boldsymbol{M}) \stackrel{\text{def}}{=} \sum_i x_i M_i \tag{7.14}$$

holds, and the equality holds when $\boldsymbol{M}$ is a PVM[Exe. 7.6]. Inequality (7.14) can be shown as

$$\operatorname{Tr} \sum_i (x_i - \mathrm{E}_\rho(\boldsymbol{M}))^2 M_i \rho \geq \operatorname{Tr}(O(\boldsymbol{M}) - \mathrm{E}_\rho(\boldsymbol{M}))^2 \rho$$

because $\sum_i (x_i - \mathrm{E}_\rho(\boldsymbol{M}))^2 M_i - (O(\boldsymbol{M}) - \mathrm{E}_\rho(\boldsymbol{M}))^2 = \sum_i (x_i - O(\boldsymbol{M})) M_i (x_i - O(\boldsymbol{M})) \geq 0$. In particular, an indirect measurement $\mathcal{I} = (\mathcal{H}_D, V, \rho_0, \boldsymbol{E})$ corresponding to $\boldsymbol{M}$ satisfies

$$\Delta_2(\boldsymbol{M}, \rho) = \Delta_2(\boldsymbol{E}, V(\rho \otimes \rho_0) V^\dagger) = \Delta_1(O(\boldsymbol{E}), V(\rho \otimes \rho_0) V^\dagger) \tag{7.15}$$

because $\boldsymbol{E}$ is a PVM. Similarly, the Naĭmark extension $(\mathcal{H}_B, \boldsymbol{E}, \rho_0)$ of the POVM $\boldsymbol{M}$ satisfies

$$\Delta_2(\boldsymbol{M}, \rho) = \Delta_2(\boldsymbol{E}, \rho \otimes \rho_0) = \Delta_1(O(\boldsymbol{E}), \rho \otimes \rho_0). \tag{7.16}$$

Let us define the deviation $\Delta_3(\boldsymbol{M}, X, \rho)$ of the POVM $\boldsymbol{M}$ from the observable X for the state ρ by

$$\Delta_3^2(\boldsymbol{M}, X, \rho) \stackrel{\text{def}}{=} \sum_i \operatorname{Tr}(x_i - X) M_i (x_i - X) \rho. \tag{7.17}$$

It then follows that the deviation of $\boldsymbol{M}$ from $O(\boldsymbol{M})$ becomes zero if $\boldsymbol{M}$ is a PVM. The square of the uncertainty $\Delta_2^2(\boldsymbol{M}, \rho)$ of the measurement $\boldsymbol{M}$ can be decomposed into the sum of the square of the uncertainty of the average matrix $O(\boldsymbol{M})$ and the square of the deviation of $\boldsymbol{M}$ from $O(\boldsymbol{M})$ as follows[Exe. 7.7]:

$$\Delta_2^2(\boldsymbol{M}, \rho) = \Delta_3^2(\boldsymbol{M}, O(\boldsymbol{M}), \rho) + \Delta_1^2(O(\boldsymbol{M}), \rho). \tag{7.18}$$

When the POVM $\boldsymbol{M}$ and the observable X do not necessarily satisfy $O(\boldsymbol{M}) = X$, the square of their deviation can be written as the sum of the square of the uncertainty of the observable $X - O(\boldsymbol{M})$ and the square of the deviation of the POVM $\boldsymbol{M}$ from $O(\boldsymbol{M})$ as follows[Exe. 7.8]:

$$\Delta_3^2(\boldsymbol{M}, X, \rho) = \Delta_3^2(\boldsymbol{M}, O(\boldsymbol{M}), \rho) + \Delta_1^2(O(\boldsymbol{M}) - X, \rho). \tag{7.19}$$

7.2.2 Disturbance

Now, consider the disturbance caused by the state evolution κ from quantum system $\mathcal{H}_A$ to quantum system $\mathcal{H}_B$. For this purpose, we examine how well the POVM $\boldsymbol{M} = \{(M_i, x_i)\}_i$ on the final system $\mathcal{H}_B$ recovers the observable X on $\mathcal{H}_A$. Its quality can be measured by the quantity $\Delta_3(\kappa^*(\boldsymbol{M}), X, \rho)$, where $\kappa^*(\boldsymbol{M})$ denotes the POVM $\{(\kappa^*(M_i), x_i)\}_i$ on the initial system $\mathcal{H}_A$. Since κ^* is the dual map of the map κ, the minimum value $\Delta_4(\kappa, X, \rho) \stackrel{\text{def}}{=} \min_{\boldsymbol{M}} \Delta_3(\kappa^*(\boldsymbol{M}), X, \rho)$ is thought to present the **disturbance** with respect to the observable X caused by the state evolution κ. Using the Stinespring representation, (7.18) yields[Exe. 7.9]

$$\Delta_3^2(\kappa^*(\boldsymbol{M}), X, \rho) = \Delta_3^2(\boldsymbol{M}, O(\boldsymbol{M}), \kappa(\rho)) + \Delta_3^2(\kappa^*(\boldsymbol{E}_{O(\boldsymbol{M})}), X, \rho). \tag{7.20}$$

Thus, our minimization can be reduced to $\min_Y \Delta_3(\kappa^*(\boldsymbol{E}_Y), X, \rho)$. Interestingly, using the Stinespring representation $(\mathcal{H}_C, \rho_0, U_\kappa)$ of κ, we can express the quantity $\Delta_3(\kappa^*(\boldsymbol{E}_Y), X, \rho)$ as[Exe. 7.10]

$$\Delta_3^2(\kappa^*(\boldsymbol{E}_Y), X, \rho) = \operatorname{Tr}\left(U_\kappa(X \otimes I_{B,C})U_\kappa^* - (I_{A,C} \otimes Y)\right)^2 U_\kappa(\rho \otimes \rho_0)U_\kappa^*. \tag{7.21}$$

As discussed in Sect. 6.1, the matrix $\kappa_{\rho,s}(X)$ can be regarded as the image of $U_\kappa(X \otimes I_{B,C})U_\kappa^*$ by the projection to the space $\{Y \otimes I\}$. Hence, by using property (6.16), the above can be calculated as

$$\begin{aligned}
&\operatorname{Tr}\left(U_\kappa(X \otimes I_{B,C})U_\kappa^* - (I_{A,C} \otimes \kappa_{\rho,s}(X))\right)^2 U_\kappa(\rho \otimes \rho_0)U_\kappa^* \\
&+ \operatorname{Tr}\left((I_{A,C} \otimes \kappa_{\rho,s}(X)) - (I_{A,C} \otimes Y)\right)^2 U_\kappa(\rho \otimes \rho_0)U_\kappa^* \\
&= \operatorname{Tr} X^2\rho - \operatorname{Tr}\left(\left(\kappa_{\rho,s}(X)\right)^2 \kappa(\rho) + \left(Y - \kappa_{\rho,s}(X)\right)^2 \kappa(\rho)\right).
\end{aligned} \tag{7.22}$$

Thus, this quantity gives the minimum when $Y = \kappa_{\rho,s}(X)$. That is, the matrix $\kappa_{\rho,s}(X)$ on the output system gives the best approximation of the matrix X on the input system. In particular, when κ is the partial trace, $\kappa_{\rho,s}$ can be regarded as the conditional expectation, i.e., the quantum version of conditional expectation. Therefore, the disturbance of X caused by κ has the form

$$\begin{aligned} \Delta_4(\kappa, X, \rho) &= \min_{\boldsymbol{M}} \Delta_3(\kappa^*(\boldsymbol{M}), X, \rho) = \min_{Y} \Delta_3(\kappa^*(\boldsymbol{E}_Y), X, \rho) \\ &= \Delta_3(\kappa^*(\boldsymbol{E}_{\kappa_{\rho,s}(X)}), X, \rho) = \sqrt{\left(\|X\|^{(e)}_{\rho,s}\right)^2 - \left(\|\kappa_{\rho,s}(X)\|^{(e)}_{\kappa(\rho),s}\right)^2}. \end{aligned} \quad (7.23)$$

Hence, if X is the SLD e representation $L_{\theta,s}$ of the derivative, this can be regarded as the loss of the SLD Fisher metric.

Remember $\kappa_{\rho,s}(X)$ is a kind of conditional expectation. So, $\left(\|\kappa_{\rho,s}(X)\|^{(e)}_{\kappa(\rho),s}\right)^2$ is lower bounded by $(\mathrm{Tr}\,\rho X)^2$. That is,

$$\Delta_4(\kappa, X, \rho) \le \Delta_1(X, \rho). \quad (7.24)$$

Furthermore, when an instrument $\boldsymbol{\kappa} = \{\kappa_\omega\}$ is used, the disturbance of the observable X is defined as

$$\Delta_4(\boldsymbol{\kappa}, X, \rho) \stackrel{\text{def}}{=} \Delta_4\left(\overline{\boldsymbol{\kappa}}, X, \rho\right), \quad \overline{\boldsymbol{\kappa}} \stackrel{\text{def}}{=} \sum_\omega \kappa_\omega. \quad (7.25)$$

Letting $\mathcal{I} = (\mathcal{H}_C, U, \rho_0, \boldsymbol{E})$ be an indirect measurement corresponding to the instrument $\boldsymbol{\kappa}$, we can describe the disturbance of the observable X by

$$\begin{aligned} \Delta_4^2(\boldsymbol{\kappa}, X, \rho) &= \mathrm{Tr}((X \otimes I_{B,C}) - U^*(I_{A,C} \otimes \overline{\boldsymbol{\kappa}}_{\rho,s}(X))U)^2(\rho \otimes \rho_0) \\ &= \left(\|X\|^{(e)}_{\rho,s}\right)^2 - \left(\|\overline{\boldsymbol{\kappa}}_{\rho,x}(X)\|^{(e)}_{\overline{\boldsymbol{\kappa}}(\rho),s}\right)^2, \end{aligned} \quad (7.26)$$

which may be found in a manner similar to (7.22). The four uncertainties given here are often confused and are often denoted by $\Delta^2(X)$. Some care is therefore necessary to ensure these quantities.

7.2.3 Uncertainty Relations

The most famous uncertainty relation by Robertson [5] is

$$\Delta_1(X, |u\rangle\langle u|)\Delta_1(Y, |u\rangle\langle u|) \ge \frac{|\langle u|[X, Y]|u\rangle|}{2}. \quad (7.27)$$

This may be generalized to

$$\Delta_1(X, \rho)\Delta_1(Y, \rho) \ge \frac{\mathrm{Tr}\,|\sqrt{\rho}[X, Y]\sqrt{\rho}|}{2}. \quad (7.28)$$

Indeed, the above inequality still holds if the right-hand side (RHS) is replaced by $\frac{|\mathrm{Tr}\,\rho[X,Y]|}{2}$. However, if the state is not a pure state, inequality (7.28) is a stronger requirement. For the rest of this section, we assume that ρ is a density matrix, although the essential point is that $\rho \geq 0$ and not that its trace is equal to 1.

Now, let us prove (7.28). The problem is reduced to the case of $\mathrm{Tr}\,\rho X = 0$ by replacing X by $X - \mathrm{Tr}\,\rho X$. Since

$$0 \leq (X \pm iY)(X \pm iY)^* = (X \pm iY)(X \mp iY) = X^2 + Y^2 \mp i[X, Y],$$

we have $\sqrt{\rho}(X^2 + Y^2)\sqrt{\rho} \geq \pm i\sqrt{\rho}[X, Y]\sqrt{\rho}$. From Exercise 1.34 we thus obtain

$$\Delta_1^2(X, \rho) + \Delta_1^2(Y, \rho) \geq \mathrm{Tr}\,|i\sqrt{\rho}[X, Y]\sqrt{\rho}| = \mathrm{Tr}\,|\sqrt{\rho}[X, Y]\sqrt{\rho}|. \tag{7.29}$$

Replacing X by tX, we see that

$$\Delta_1^2(X, \rho)t^2 - \mathrm{Tr}\,|\sqrt{\rho}[X, Y]\sqrt{\rho}|t + \Delta_2^2(Y, \rho) \geq 0.$$

Equation (7.28) can then be obtained from the discriminant equation for t.

In addition, when the equality of (7.28) holds and $\mathrm{Tr}\,\rho X = \mathrm{Tr}\,\rho Y = 0$, the relation $(X + iY)\sqrt{\rho} = 0$ or $(X - iY)\sqrt{\rho} = 0$ holds.

The original uncertainty relation proposed by Heisenberg [6] was obtained through a gedanken experiment, and it relates the accuracy of measurements to the disturbance of measurements. The implications of the accuracy and the disturbance due to the measurement are not necessarily clear in (7.28). At least, it is incorrect to call (7.28) the Heisenberg uncertainty relation because it does not involve quantities related to measurement [7, 8].

One may think that Heisenberg's argument would be formulated as

$$\Delta_3(\boldsymbol{M}, X, \rho)\Delta_4(\boldsymbol{\kappa}, Y, \rho) \geq \frac{\mathrm{Tr}\,|\sqrt{\rho}[X, Y]\sqrt{\rho}|}{2} \tag{7.30}$$

for a POVM $\boldsymbol{M}$ and an instrument $\boldsymbol{\kappa}$ satisfying (7.8). However, this is in fact incorrect, for the following reason [7, 9, 10]. Consider the POVM $\boldsymbol{M}$ that always gives the measurement outcome 0 without making any measurement. Then, $\Delta_3(\boldsymbol{M}, X, \rho)$ is finite while $\Delta_4(\boldsymbol{\kappa}, Y, \rho)$ is 0. Therefore, this inequality does not hold in general. The primary reason for this is that the RHS has two quantities having no connection to the POVM $\boldsymbol{M}$. Hence, we need to seek a more appropriate formulation.

Indeed, Heisenberg's gedanken experiment does not treat the above unnatural case. He considered the case when the measurement has the proper relation with the observable. That is, it seems that he considered the observable measured by the measurement $\boldsymbol{M}$. In this case, it is more appropriate to address $\Delta_2(\boldsymbol{M}, \rho)$ rather than $\Delta_3(\boldsymbol{M}, X, \rho)$. Since the quantity $\Delta_2(\boldsymbol{M}, \rho)$ is defined only with a POVM $\boldsymbol{M}$ and a state ρ, it is better to lower bound the product $\Delta_2(\boldsymbol{M}, \rho)\Delta_4(\boldsymbol{\kappa}, Y, \rho)$ by a amount determined by the POVM $\boldsymbol{M}$, the observable Y, and the state ρ. Then, we can reformulate Heisenberg's argument as follows.

Theorem 7.4 *Let κ be an instrument corresponding to the POVM M. A state ρ on $\mathcal{H}_A$ then satisfies*[3]

$$\Delta_2(M, \rho)\Delta_4(\kappa, Y, \rho) \geq \frac{\operatorname{Tr} |\sqrt{\rho}[O(M), Y]\sqrt{\rho}|}{2}. \tag{7.31}$$

A proof of this relation will be given later. This inequality means that the product of the error $\Delta_2(M, \rho)$ and the disturbance $\Delta_4(\kappa, Y, \rho)$ is lower bounded by a quantity involving M and Y.

Remember $\Delta_2(M, \rho)$ is lower bounded by $\Delta_1(O(M), \rho)$ as (7.18) and $\Delta_4(\kappa, Y, \rho)$ is upper bounded by $\Delta_1(Y, \rho)$ as (7.24). When the POVM M corresponding to κ is the spectral decomposition of $O(M)$, we have $\Delta_2(M, \rho) = \Delta_1(O(M), \rho)$. In this case, (7.31) implies that

$$\Delta_1(Y, \rho) \geq \Delta_4(\kappa, Y, \rho) \geq \frac{\operatorname{Tr} |\sqrt{\rho}[O(M), Y]\sqrt{\rho}|}{2\Delta_1(O(M), \rho)}. \tag{7.32}$$

When the equality in (7.28) hold for $O(M)$ and Y, we have $\Delta_1(Y, \rho) = \Delta_4(\kappa, Y, \rho)$.

In particular, when Y is the SLD e representation $L_{\theta,s}$ of the derivative, due to (7.31), the information loss $\Delta_4(\kappa, L_{\theta,s}, \rho)$ satisfies

$$\Delta_2(M, \rho)\Delta_4(\kappa, L_{\theta,s}, \rho) \geq \frac{\operatorname{Tr} |\sqrt{\rho}[O(M), L_{\theta,s}]\sqrt{\rho}|}{2}.$$

Next, we consider a simultaneous measurement of two observables. For this purpose, we denote a POVM with two outcomes by $M = (\{M_\omega\}, \{x^1_\omega\}, \{x^2_\omega\})$. Then, the two average matrices are given by

$$O^1(M) \stackrel{\text{def}}{=} \sum_\omega x^1_\omega M_\omega, \quad O^2(M) \stackrel{\text{def}}{=} \sum_\omega x^2_\omega M_\omega.$$

Theorem 7.5 *The POVM $M = (\{M_\omega\}, \{x^1_\omega\}, \{y^2_\omega\})$ satisfies*[4]

$$\begin{aligned} &\Delta_3(M, O^1(M), \rho)\Delta_3(M, O^2(M), \rho) \\ \geq &\frac{\operatorname{Tr} |\sqrt{\rho}[O^1(M), O^2(M)]\sqrt{\rho}|}{2}. \end{aligned} \tag{7.33}$$

[3]As discussed in Sect. 6.2, $\Delta_4(\kappa, Y, \rho)$ also has the meaning of the amount of the loss of the SLD Fisher information. Therefore, this inequality is interesting from the point of view of estimation theory. It indicates the naturalness of the SLD inner product. This is in contrast to the naturalness of the Bogoljubov inner product from a geometrical viewpoint.

[4]The equality holds when an appropriate POVM M is performed in a quantum two-level system [11]. For its more general equality condition, see Exercise 7.17.

There have also been numerous discussions relating to uncertainties, including its relation to quantum computation [9, 10].

Proof of Theorem 7.4 The theorem is proven by considering an indirect measurement $\mathcal{I} = (\mathcal{H}_C, U, \rho_0, \boldsymbol{E})$ corresponding to the instrument $\boldsymbol{\kappa}$. Let $Z = \overline{\boldsymbol{\kappa}}_{\rho,s}(Y)$. Then, from (7.26),

$$\Delta_4(\boldsymbol{\kappa}, Y, \rho) = \Delta_1(Y \otimes I_{B,C} - U^*(Z \otimes I_{A,C})U, \rho \otimes \rho_0). \tag{7.34}$$

Since the indirect measurement $\mathcal{I} = (\mathcal{H}_C, U, \rho_0, \boldsymbol{E})$ corresponds to the POVM $\boldsymbol{M}$, we have $\mathrm{Tr}_{B,C}(I \otimes \sqrt{\rho_0})U^*(I_{A,B} \otimes O(\boldsymbol{E}))U(I \otimes \sqrt{\rho_0}) = O(\boldsymbol{M})^{\text{Exe. 7.19}}$. Referring to Exercise 1.25, we have

$$\begin{aligned}
&\mathop{\mathrm{Tr}}_{B,C} \sqrt{\rho\otimes\rho_0}[U^*(I_{A,B}\otimes O(\boldsymbol{E}))U, Y\otimes I_{B,C} - U^*(Z\otimes I_{A,C})U]\sqrt{\rho\otimes\rho_0} \\
=&\mathop{\mathrm{Tr}}_{B,C} \sqrt{\rho\otimes\rho_0}[U^*(I_{A,B}\otimes O(\boldsymbol{E}))U, Y\otimes I_{B,C}]\sqrt{\rho\otimes\rho_0} \\
=&\sqrt{\rho}[O(\boldsymbol{M}), Y]\sqrt{\rho}.
\end{aligned}$$

Thus,

$$\begin{aligned}
&\mathop{\mathrm{Tr}}_{A} |\sqrt{\rho}[O(\boldsymbol{M}), Y]\sqrt{\rho}| \\
=&\mathop{\mathrm{Tr}}_{A} | \mathop{\mathrm{Tr}}_{B,C} \sqrt{\rho\otimes\rho_0}[U^*(I_{A,B}\otimes O(\boldsymbol{E}))U, Y\otimes I_{B,C} - U^*(Z\otimes I_{A,C})U]\sqrt{\rho\otimes\rho_0}| \\
\le&\mathrm{Tr} |\sqrt{\rho\otimes\rho_0}[U^*(I_{A,B}\otimes O(\boldsymbol{E}))U, Y\otimes I_{B,C} - U^*(Z\otimes I_{A,C})U]\sqrt{\rho\otimes\rho_0}| \\
\le&\Delta_1(U^*(I_{A,B}\otimes O(\boldsymbol{E}))U, \rho\otimes\rho_0)\Delta_1(Y\otimes I_{B,C} - U^*(Z\otimes I_{A,C})U, \rho\otimes\rho_0).
\end{aligned}$$

Finally, (7.15) implies the equation $\Delta_2(\boldsymbol{M}, \rho) = \Delta_1(U^*(I_{A,B} \otimes O(\boldsymbol{E}))U, \rho \otimes \rho_0)$. Combining these relations with (7.34), we obtain (7.31). ■

Proof of Theorem 7.5 We apply Exercise 5.7. Let us choose $\mathcal{H}_B$, a PVM on $\mathcal{H} \otimes \mathcal{H}_B$, and a state ρ_0 on $\mathcal{H}_B$ such that

$$\mathrm{Tr}\, \rho M_\omega = \mathrm{Tr}(\rho \otimes \rho_0)E_\omega,$$

with respect to an arbitrary state ρ on $\mathcal{H}$. Let $(\mathcal{H}_B, \boldsymbol{E}, \rho_0)$ be the Naĭmark extension of $\boldsymbol{M}$. Then,

$$\Delta_3(\boldsymbol{M}, O^i(\boldsymbol{M}), \rho) = \Delta_1(O^i(\boldsymbol{E}) - O^i(\boldsymbol{M}) \otimes I_B, \rho \otimes \rho_0).$$

Since $[O^1(\boldsymbol{E}), O^2(\boldsymbol{E})] = 0$, we have

$$\begin{aligned}
&[O^1(\boldsymbol{E}) - O^1(\boldsymbol{M}) \otimes I_B, O^2(\boldsymbol{E}) - O^2(\boldsymbol{M}) \otimes I_B] \\
=& - [O^1(\boldsymbol{E}), O^2(\boldsymbol{M}) \otimes I_B] - [O^1(\boldsymbol{M}) \otimes I_B, O^2(\boldsymbol{E}) - O^2(\boldsymbol{M}) \otimes I_B].
\end{aligned}$$

Accordingly,

$$\begin{aligned}
&\Delta_1(O^1(\boldsymbol{E}) - O^1(\boldsymbol{M}) \otimes I_B, \rho \otimes \rho_0)\Delta_1(O^2(\boldsymbol{E}) - O^2(\boldsymbol{M}) \otimes I_B, \rho \otimes \rho_0) \\
\geq & \operatorname{Tr} |\sqrt{\rho \otimes \rho_0}[O^1(\boldsymbol{E}) - O^1(\boldsymbol{M}) \otimes I_B, O^2(\boldsymbol{E}) - O^2(\boldsymbol{M}) \otimes I_B]\sqrt{\rho \otimes \rho_0}| \\
\geq & \operatorname{Tr}_A | \operatorname{Tr}_B \sqrt{\rho \otimes \rho_0}[O^1(\boldsymbol{E}) - O^1(\boldsymbol{M}) \otimes I_B, O^2(\boldsymbol{E}) - O^2(\boldsymbol{M}) \otimes I_B]\sqrt{\rho \otimes \rho_0}| \\
= & \operatorname{Tr}_A | \operatorname{Tr}_B \sqrt{\rho \otimes \rho_0}(-[O^1(\boldsymbol{E}), O^2(\boldsymbol{M}) \otimes I_B] \\
& \quad - [O^1(\boldsymbol{M}) \otimes I_B, O^2(\boldsymbol{E}) - O^2(\boldsymbol{M}) \otimes I_B])\sqrt{\rho \otimes \rho_0}| \\
= & \operatorname{Tr}_A | \operatorname{Tr}_B \sqrt{\rho \otimes \rho_0}[O^1(\boldsymbol{E}), O^2(\boldsymbol{M}) \otimes I_B]\sqrt{\rho \otimes \rho_0}|.
\end{aligned}$$

This completes the proof. In the last equality, we used the fact that $\operatorname{Tr}_B(I \otimes \sqrt{\rho_0})(O^2(\boldsymbol{E}) - O^2(\boldsymbol{M}) \otimes I_B)(I \otimes \sqrt{\rho_0}) = 0$ and Exercise 1.25. ∎

Exercises

7.6 Show that the equality in (7.14) holds for the PVM $\boldsymbol{M}$.

7.7 Show (7.18).

7.8 Show (7.19).

7.9 Show (7.20).

7.10 Show (7.21) following the steps below.
(a) Let $(\mathcal{H}_C, \rho_0, U)$ be a Stinespring representation of κ. Show that any projection E satisfies

$$\begin{aligned}
&\operatorname{Tr}(X \otimes I - x)U^*(I \otimes E)U(X \otimes I - x)\rho \otimes \rho_0 \\
= &\operatorname{Tr}(X - x)\kappa^*(E)(X - x)\rho.
\end{aligned}$$

(b) Show (7.21).

7.11 Let $(\mathcal{H}_B, \boldsymbol{E}, \rho_0)$ be a Naĭmark extension[Exe. 5.7] of $\boldsymbol{M} = (\{M_\omega\}, \{x_\omega\})$. Show that $\Delta_3^2(\boldsymbol{M}, X, \rho) = \Delta_1^2(O(\boldsymbol{E}) - X \otimes I_B, \rho \otimes \rho_0)$.

7.12 Show that

$$\Delta_2(\boldsymbol{M}, \rho) \leq \Delta_3(\boldsymbol{M}, X, \rho) + \Delta_1(O(\boldsymbol{M}), \rho) \tag{7.35}$$

by following steps below.
(a) Show that $\sqrt{a^2 + b^2 - c^2} \leq a + d$ when $a, b, c, d \geq, a \geq c$, and $c + d \geq b$.
(b) Show (7.35) using the above.

7.13 Show the following using (7.31) [7].

$$\Delta_2(\boldsymbol{M}, \rho)\Delta_4(\boldsymbol{\kappa}, Y, \rho) + \Delta_1(O(\boldsymbol{M}) - X, \rho)\Delta_1(Y, \rho) \geq \frac{\operatorname{Tr} |\sqrt{\rho}[X, Y]\sqrt{\rho}|}{2}. \tag{7.36}$$

7.14 Show the following using (7.36) [7].

$$\begin{aligned} &\Delta_3(\boldsymbol{M}, X, \rho)\Delta_4(\boldsymbol{\kappa}, Y, \rho) + \Delta_1(X, \rho)\Delta_4(\boldsymbol{\kappa}, Y, \rho) + \Delta_3(\boldsymbol{M}, X, \rho)\Delta_1(Y, \rho) \\ \geq &\frac{\operatorname{Tr} |\sqrt{\rho}[X, Y]\sqrt{\rho}|}{2}. \end{aligned} \tag{7.37}$$

7.15 Define the correlation between two Hermitian matrices X and Y under the state ρ as

$$\operatorname{Cov}_\rho(X, Y) \stackrel{\text{def}}{=} \operatorname{Tr}(X - \operatorname{Tr} X\rho) \circ (Y - \operatorname{Tr} Y\rho)\rho, \tag{7.38}$$

which is a quantum analogue of the covariance $\operatorname{Cov}_p(X, Y)$ defined in (2.93). Show that

$$\det \begin{pmatrix} \operatorname{Cov}_\rho(X, X) & \operatorname{Cov}_\rho(X, Y) \\ \operatorname{Cov}_\rho(X, Y) & \operatorname{Cov}_\rho(Y, Y) \end{pmatrix} \geq \left(\frac{\operatorname{Tr} |\sqrt{\rho}[X, Y]\sqrt{\rho}|}{2} \right)^2, \tag{7.39}$$

which is a stronger inequality than (7.28), using (7.28).

7.16 Show that the equality in (7.39) always holds if $\mathcal{H} = \mathbb{C}^2$ [11] by following the steps below. This fact shows the equality in (7.28) when $\operatorname{Cov}_\rho(X, Y) = 0$ for $\mathcal{H} = \mathbb{C}^2$. In the following proof, we fist treat the case where $\operatorname{Tr} X = \operatorname{Tr} Y = 0$, which implies that X and Y are written as $X = \sum_{i=1}^3 x_i S_i$, $Y = \sum_{i=1}^3 y_i S_i$, $\rho = \frac{1}{2}(I + \sum_{i=1}^3 a_i S_i)$. After this special case, we consider the general case.

(**a**) Show that $\operatorname{Cov}_\rho(X, Y) = \langle \boldsymbol{x}, \boldsymbol{y} \rangle - \langle \boldsymbol{x}, \boldsymbol{a} \rangle \langle \boldsymbol{a}, \boldsymbol{y} \rangle$.

(**b**) Let z be the vector product (outer product) of $\boldsymbol{x}$ and $\boldsymbol{y}$, i.e., $z = \boldsymbol{x} \times \boldsymbol{y} \stackrel{\text{def}}{=} (x_2 y_3 - x_3 y_2, x_3 y_1 - x_1 y_3, x_1 y_2 - x_2 y_1)$. Show that $\frac{-i}{2}[X, Y] = Z \stackrel{\text{def}}{=} \sum_{i=1}^3 z_i S_i$.

(**c**) Show that $\operatorname{Tr} |\sqrt{\rho} Z \sqrt{\rho}| = \sqrt{\|z\|^2 - \|z \times \boldsymbol{a}\|^2}$.

(**d**) Show that (7.39) is equivalent to

$$\begin{aligned} &(\|\boldsymbol{x}\|^2 - \langle \boldsymbol{x}, \boldsymbol{a} \rangle^2)(\|\boldsymbol{y}\|^2 - \langle \boldsymbol{y}, \boldsymbol{a} \rangle^2) - (\langle \boldsymbol{x}, \boldsymbol{y} \rangle - \langle \boldsymbol{x}, \boldsymbol{a} \rangle \langle \boldsymbol{a}, \boldsymbol{y} \rangle)^2 \\ \geq &\|\boldsymbol{x} \times \boldsymbol{y}\|^2 - \|(\boldsymbol{x} \times \boldsymbol{y}) \times \boldsymbol{a}\|^2 \end{aligned} \tag{7.40}$$

when $\operatorname{Tr} X = \operatorname{Tr} Y = 0$ in a quantum two-level system.

(**e**) Show (7.40) if $\langle \boldsymbol{x}, \boldsymbol{y} \rangle = \langle \boldsymbol{x}, \boldsymbol{a} \rangle = 0$.

(**f**) Show that there exists a 2×2 matrix $(b_{i,j})$ with determinant 1 such that the vectors $\tilde{\boldsymbol{x}} \stackrel{\text{def}}{=} b_{1,1}\boldsymbol{x} + b_{1,2}\boldsymbol{y}$ and $\tilde{\boldsymbol{y}} \stackrel{\text{def}}{=} b_{2,1}\boldsymbol{x} + b_{2,2}\boldsymbol{y}$ satisfy $\langle \tilde{\boldsymbol{x}}, \boldsymbol{y} \rangle = \langle \tilde{\boldsymbol{x}}, \boldsymbol{a} \rangle = 0$.

(**g**) Show (7.40) for arbitrary vectors $\boldsymbol{x}, \boldsymbol{y}, \boldsymbol{a}$.

(**h**) Show that (7.39) still holds even if $\operatorname{Tr} X = \operatorname{Tr} Y = 0$ does not hold.

7.17 Show that the POVM $\boldsymbol{M}_{X,Y,\rho}$ below satisfies $O^1(\boldsymbol{M}) = X$, $O^2(\boldsymbol{M}) = Y$. Further, show the equality in (7.33) when X, Y, ρ satisfy the equality in (7.28).
Construction of the POVM $\boldsymbol{M}_{X,Y,\rho}$: Let the spectral decomposition of X and Y be $X = \sum_i x_i E_{X,i}$ and $Y = \sum_j y_j E_{Y,j}$, respectively. Define the POVM $\boldsymbol{M}_{X,Y,\rho}$ with the probability space $\Omega = \{i\} \cup \{j\}$ for $p \in (0, 1)$ as follows. Let $M_{X,Y,\rho,i} = pE_{X,i}$ and $M_{X,Y,\rho,j} = (1-p)E_{Y,j}$. Define $x_i^1 \stackrel{\text{def}}{=} \frac{1}{p}(x_i - \operatorname{Tr} \rho X) + \operatorname{Tr} \rho X$, $x_j^1 \stackrel{\text{def}}{=} \operatorname{Tr} \rho X$, $x_i^2 \stackrel{\text{def}}{=} \operatorname{Tr} \rho Y$, $x_j^2 \stackrel{\text{def}}{=} \frac{1}{1-p}(y_i - \operatorname{Tr} \rho Y) + \operatorname{Tr} \rho Y$. The POVM is then defined as $\boldsymbol{M}_{X,Y,\rho} \stackrel{\text{def}}{=} \{(M_{X,Y,\rho,i}, x_i^1, x_i^2)\} \cup \{(M_{X,Y,\rho,j}, x_j^1, x_j^2)\}$.

7.18 Using (7.31), show that the following two conditions are equivalent for two Hermitian matrices X and Y.

① $[X, Y] = 0$.
② There exist an instrument $\boldsymbol{\kappa} = \{\kappa_\omega\}$ and a set $\{x_\omega\}$ such that the following two conditions

$$\operatorname{Tr} \rho X = \sum_\omega x_\omega \operatorname{Tr} \kappa_\omega(\rho), \quad \Delta_4(\boldsymbol{\kappa}, Y, \rho) = 0$$

hold for an arbitrary state ρ. The first equation implies that the instrument $\boldsymbol{\kappa}$ corresponds to the observable X. The second equation implies that the instrument $\boldsymbol{\kappa}$ does not disturb the observable Y.

7.19 Show that

$$\operatorname*{Tr}_{B,C} \left((I_A \otimes \sqrt{\rho_0})\, U^* \left(I_{A,B} \otimes O(\boldsymbol{E}) \right) U \left(I_A \otimes \sqrt{\rho_0} \right) \right) = O(\boldsymbol{M}) \tag{7.41}$$

for an indirect measurement $\mathcal{I} = (\mathcal{H}_C, U, \rho_0, \boldsymbol{E})$ corresponding to $\boldsymbol{M}$.

7.20 Given two state evolutions κ_1 and κ_2, show the following items.
(a) Show that $\Delta_4^2(\lambda\kappa_1 + (1-\lambda)\kappa_2, X, \rho) \geq \lambda\Delta_4^2(\kappa_1, X, \rho) + (1-\lambda)\Delta_4^2(\kappa_2, X, \rho)$.
(b) Show that the equality holds when the space spanned by the supports of $\kappa_1(X \circ \rho)$ and $\kappa_1(\rho)$ is orthogonal to the space spanned by the supports of $\kappa_2(X \circ \rho)$ and $\kappa_2(\rho)$.

7.21 Let two Hermitian matrices X and Y on $\mathcal{H}$ satisfy the equality in (7.28). Let $X = \sum_{i=1}^k x_i E_i$, and define the POVM $\boldsymbol{M} = \{M_i, \hat{x}_i\}_{i=0}^k$ according to the conditions $\hat{x}_0 = \operatorname{Tr} X\rho$, $\hat{x}_i = \operatorname{Tr} X\rho + \frac{1}{p}(x_i - \operatorname{Tr} X\rho)$, $M_0 = (1-p)I$, and $M_i = pE_i$. Now consider another equivalent space $\mathcal{H}'$ to $\mathcal{H}$, and the unitary map U from $\mathcal{H}$ to $\mathcal{H}'$. Define $\boldsymbol{\kappa} = \{\kappa_i\}$ according to $\kappa_0(\rho) = (1-p)U\rho U^*$ and $\kappa_i(\rho) = p\sqrt{M_i}\rho\sqrt{M_i}$. That is, the output system of $\boldsymbol{\kappa}$ is $\mathcal{H} \oplus \mathcal{H}'$. Show that $O(\boldsymbol{M}) = X$ and that the equality of (7.31) holds.

7.3 Entropic Uncertainty Relation

Even though the initial state ρ is pure, the outcome of a measurement $\boldsymbol{M}$ is not deterministic. In this case, the uncertainty of the outcome can be evaluated by use of the entropy $H(\mathrm{P}_\rho^{\boldsymbol{M}})$. Of course, if the the element of the PVM $\boldsymbol{M}$ is commutative with the state $|x\rangle\langle x|$, the outcome is deterministic, i.e., the entropy $H(\mathrm{P}_\rho^{\boldsymbol{M}})$ is zero. However, when two non-commutative PVMs $\boldsymbol{M}$ and $\boldsymbol{M}'$ are given, there is no pure state $|x\rangle\langle x|$ satisfying that $H(\mathrm{P}_{|x\rangle\langle x|}^{\boldsymbol{M}}) = H(\mathrm{P}_{|x\rangle\langle x|}^{\boldsymbol{M}'}) = 0$. Similar to the uncertainty relation concerning the square errors, we have the following relation between two quantities $H(\mathrm{P}_{|x\rangle\langle x|}^{\boldsymbol{M}})$ and $H(\mathrm{P}_{|x\rangle\langle x|}^{\boldsymbol{M}'})$.

Theorem 7.6 (Entropic Uncertainty Relation, Maassen and Uffink [12]) *Let $\boldsymbol{M}$ and $\boldsymbol{M}'$ be the PVMs composing of bases $\{u_j\}$ and $\{v_l\}$ of $\mathcal{H}_A$. Any state ρ satisfies*

$$H(\kappa_{\boldsymbol{M}}(\rho)) + H(\kappa_{\boldsymbol{M}'}(\rho)) = H(\mathrm{P}_\rho^{\boldsymbol{M}}) + H(\mathrm{P}_\rho^{\boldsymbol{M}'}) \geq -\log c + H(\rho), \tag{7.42}$$

where $c := \max_{l,j} |\langle u_j|v_l\rangle|^2$. The equality holds when $|\langle u_l|v_j\rangle|$ does not depend on l, j and the matrix $A + B$ is commutative with ρ, where

$$A := \sum_j \log\langle u_j|\rho|u_j\rangle |u_j\rangle\langle u_j|, \quad B := \sum_l \log\langle v_l|\rho|v_l\rangle |v_l\rangle\langle v_l|. \tag{7.43}$$

Proof Golden-Thompson trace inequality (Lemma 5.4) yields that $\mathrm{Tr}\, e^A e^B \geq \mathrm{Tr}\, e^{A+B}$. We also have

$$\begin{aligned}\mathrm{Tr}\, e^A e^B &= \sum_{l,j} \langle u_j|\rho|u_j\rangle\langle v_l|\rho|v_l\rangle \,\mathrm{Tr}\, |v_l\rangle\langle v_l|u_j\rangle\langle u_j| \\ &= \sum_{l,j} \langle u_j|\rho|u_j\rangle\langle v_l|\rho|v_l\rangle |\langle v_l|u_j\rangle|^2 \leq \sum_{l,j} \langle u_j|\rho|u_j\rangle\langle v_l|\rho|v_l\rangle c = c.\end{aligned} \tag{7.44}$$

Choosing the other state $\sigma := e^{A+B}/\mathrm{Tr}\, e^{A+B}$, we have $\mathrm{Tr}\,\rho\log\rho - \mathrm{Tr}\,\rho\sigma = D(\rho\|\sigma) \geq 0$. Hence, combining the above relations, we obtain

$$\begin{aligned}&H(\mathrm{P}_\rho^{\boldsymbol{M}}) + H(\mathrm{P}_\rho^{\boldsymbol{M}'}) = -\mathrm{Tr}\,\rho(A+B) = -\mathrm{Tr}\,\rho\log\sigma - \log\mathrm{Tr}\, e^{A+B} \\ \geq& -\mathrm{Tr}\,\rho\log\rho - \log\mathrm{Tr}\, e^A e^B = H(\rho) - \log\mathrm{Tr}\, e^A e^B \geq H(\rho) - \log c,\end{aligned}$$

which implies (7.42). The equality of the first inequality holds when the matrix $A+B$ is commutative with ρ. The equality of the second inequality, i.e., that of (7.44) holds when $|\langle u_j|v_l\rangle|$ does not depend on l and j. Hence, we obtain the required equality condition. ∎

As a generalization of Theorem 7.6, we can show the following theorem by replacing the entropies by the conditional Rényi entropies.

Theorem 7.7 ([13–15]) *Given a state ρ on the system $\mathcal{H}_A \otimes \mathcal{H}_B \otimes \mathcal{H}_C$, let $\{M_x\}_x$ and $\{N_y\}_y$ be two POVMs on $\mathcal{H}_A$. We define* the overlap $c := \max_{x,y} \left\| \sqrt{M_x}\sqrt{N_y} \right\|^2$ *and consider the post-measurement states*[5]

$$\rho_{XB} := \sum_x |x\rangle_X\, {}_X\langle x| \otimes \operatorname{Tr}_{AC}(M_x \otimes I_{BC})\rho, \tag{7.45}$$

$$\rho_{YC} := \sum_y |y\rangle_Y\, {}_Y\langle y| \otimes \operatorname{Tr}_{AB}(N_y \otimes I_{BC})\rho. \tag{7.46}$$

Then, the following relations hold:

$$H^{\downarrow}_{\alpha|\rho}(X|B) + H^{\downarrow}_{\beta|\rho}(Y|C) \geq \log \frac{1}{c}, \quad \textit{for} \quad \alpha, \beta \in [0, 2],\ \alpha + \beta = 2, \tag{7.47}$$

$$\widetilde{H}^{\uparrow}_{\alpha|\rho}(X|B) + \widetilde{H}^{\uparrow}_{\beta|\rho}(Y|C) \geq \log \frac{1}{c}, \quad \textit{for} \quad \alpha, \beta \in \left[\frac{1}{2}, \infty\right],\ \frac{1}{\alpha} + \frac{1}{\beta} = 2, \tag{7.48}$$

$$H^{\uparrow}_{\alpha|\rho}(X|B) + \widetilde{H}^{\downarrow}_{\beta|\rho}(Y|C) \geq \log \frac{1}{c}, \quad \textit{for} \quad \alpha \in [0, 2],\ \beta \in \left[\frac{1}{2}, \infty\right],\ \alpha \cdot \beta = 1. \tag{7.49}$$

In particular, when $\alpha = \beta = 1$, we have [16, 17]

$$H_\rho(X|B) + H_\rho(Y|C) \geq \log \frac{1}{c}. \tag{7.50}$$

In Theorem 7.7, we introduce additional classical systems $\mathcal{H}_X$ and $\mathcal{H}_Y$ to give our arguments (7.47)–(7.50). When the POVMs $\boldsymbol{M} = \{M_x\}_x$ and $\boldsymbol{N} = \{N_y\}_y$ are rank-one PVMs, these relations can be stated without additional classical systems $\mathcal{H}_X$ and $\mathcal{H}_Y$. Then, (7.50) can be written as

$$H_{\kappa_{\boldsymbol{M}}(\rho)}(A|B) + H_{\kappa_{\boldsymbol{N}}(\rho)}(A|C) \geq \log \frac{1}{c}. \tag{7.51}$$

Now, we focus on mutually unbiased bases. Two bases $\{u_j\}$ and $\{v_l\}$ on a d-dimensional system $\mathcal{H}_A$ are called **mutually unbiased** when $|\langle u_i | v_j \rangle|^2 = 1/d$ for i and j. Now, we apply (7.51) to the POVMs $\boldsymbol{M}$ and $\boldsymbol{N}$ given by the measurement based on the mutually unbiased bases $\{u_j\}$ and $\{v_l\}$. Then, we have $c = 1/d$. When the relation $H_{\kappa_{\boldsymbol{M}}(\rho)}(A|B) = 0$ holds, i.e., the outcome of $\boldsymbol{M}$ of the system $\mathcal{H}_A$ is completely determined by the optimal measurement of $\mathcal{H}_B$, we obtain $H_{\kappa_{\boldsymbol{N}}(\rho)}(A|C) = \log d$, which implies that $D(\kappa_{\boldsymbol{M}}(\rho) \| \rho_{A,\mathrm{mix}} \otimes \rho_E) = 0$, i.e.,

$$\kappa_{\boldsymbol{N}}(\rho_{A,C}) = \rho_{A,\mathrm{mix}} \otimes \rho_C. \tag{7.52}$$

Proof Since any conditional Rényi entropy satisfies the relations given in Exercise 5.49, it is sufficient to show the case when ρ is a pure state. We define

[5] This definition of c is the generalization of that in Theorem 7.6. See Exercise 7.22.

the isometries V_X and V_Y from $\mathcal{H}_A$ to either $\mathcal{H}_A \otimes \mathcal{H}_X \otimes \mathcal{H}_{X'}$ or $\mathcal{H}_A \otimes \mathcal{H}_Y \otimes \mathcal{H}_{Y'}$, respectively, as

$$V_X|a\rangle := \sum_x |x\rangle_X \otimes |x\rangle_{X'} \otimes \sqrt{M_x}|a\rangle, \tag{7.53}$$

$$V_Y|a\rangle := \sum_x |y\rangle_Y \otimes |y\rangle_{Y'} \otimes \sqrt{N_y}|a\rangle. \tag{7.54}$$

We apply the duality relation given in Theorem 5.13 to the pure state $V_Y \rho V_Y^\dagger$ on the system $\mathcal{H}_Y \otimes \mathcal{H}_{Y'} \otimes \mathcal{H}_A \otimes \mathcal{H}_B \otimes \mathcal{H}_C$. Then, we have

$$H_{\beta|\rho}(Y|C) = H_{\beta|V_Y\rho V_Y^\dagger}(Y|C) = -H_{\alpha|V_Y\rho V_Y^\dagger}(Y|Y'AB),$$
$$\widetilde{H}^\uparrow_{\beta|\rho}(Y|C) = -\widetilde{H}^\uparrow_{\alpha|V_Y\rho V_Y^\dagger}(Y|Y'AB), \quad \widetilde{H}_{\beta|\rho}(Y|C) = -H^\uparrow_{\alpha|V_Y\rho V_Y^\dagger}(Y|Y'AB)$$

when α and β satisfy the respective condition. Since

$$H_{\alpha|\rho}(X|B) = H_{\alpha|V_X\rho V_X^\dagger}(X|B), \quad \widetilde{H}^\uparrow_{\alpha|\rho}(X|B) = \widetilde{H}^\uparrow_{\alpha|V_X\rho V_X^\dagger}(X|B),$$
$$H^\uparrow_{\alpha|\rho}(X|B) = H^\uparrow_{\alpha|V_X\rho V_X^\dagger}(X|B),$$

it is sufficient to show that

$$H_{\alpha|V_X\rho V_X^\dagger}(X|B) + \log c \ge H_{\alpha|V_Y\rho V_Y^\dagger}(Y|Y'AB), \tag{7.55}$$

$$\widetilde{H}^\uparrow_{\alpha|V_X\rho V_X^\dagger}(X|B) + \log c \ge \widetilde{H}^\uparrow_{\alpha|V_Y\rho V_Y^\dagger}(Y|Y'AB), \tag{7.56}$$

$$H^\uparrow_{\alpha|V_X\rho V_X^\dagger}(X|B) + \log c \ge H^\uparrow_{\alpha|V_Y\rho V_Y^\dagger}(Y|Y'AB). \tag{7.57}$$

Here, we will show (7.57). For this purpose, using the relation $\sqrt{N_y}M_x\sqrt{N_y} \le cI_A$, we evaluate $\mathrm{Tr}_{X'A} V_X V_Y^\dagger (I_Y \otimes \sigma_{Y'AB}) V_Y V_X^\dagger$ as follows:

$$\begin{aligned}
&\mathrm{Tr}_{X'A}\, V_X V_Y^\dagger (I_Y \otimes \sigma_{Y'AB}) V_Y V_X^\dagger \\
=&\mathrm{Tr}_{X'A}\, V_X \sum_{y,y'} {}_Y\langle y| \otimes {}_{Y'}\langle y| \otimes \sqrt{N_y}(I_Y \otimes \sigma_{Y'AB})|y'\rangle_Y \otimes |y'\rangle_{Y'} \otimes \sqrt{N_y} V_X^\dagger \\
=&\sum_x |x\rangle_X\, {}_X\langle x| \otimes \mathrm{Tr}_A \sqrt{M_x}\left(\sum_y {}_{Y'}\langle y| \otimes \sqrt{N_y}\sigma_{Y'AB}|y\rangle_{Y'} \otimes \sqrt{N_y}\right)\sqrt{M_x} \\
=&\sum_x |x\rangle_X\, {}_X\langle x| \otimes \sum_y \mathrm{Tr}_{Y'A}[(|y\rangle_{Y'}\, {}_{Y'}\langle y| \otimes \sqrt{N_y}M_x\sqrt{N_y})\sigma_{Y'AB}] \\
\le&\sum_x |x\rangle_X\, {}_X\langle x| \otimes \sum_y \mathrm{Tr}_{Y'A}[(|y\rangle_{Y'}\, {}_{Y'}\langle y| \otimes cI_A)\sigma_{Y'AB}] \\
=&I_X \otimes \mathrm{Tr}_{Y'A}[(I_Y \otimes cI_A)\sigma_{Y'AB}] = cI_X \otimes \sigma_B.
\end{aligned} \tag{7.58}$$

Since V_Y is an isometry, $V_Y V_Y^\dagger$ is a projection to the image of V_Y, which can be regarded as a subspace of $\mathcal{H}_Y \otimes \mathcal{H}_{Y'} \otimes \mathcal{H}_A \otimes \mathcal{H}_B$. Since $V_Y \rho_{AB} V_Y^\dagger = (V_Y \rho V_Y^\dagger)_{YY'AB}$, we have $(V_Y V_Y^\dagger)(V_Y \rho V_Y^\dagger)_{YY'AB}(V_Y V_Y^\dagger) = (V_Y V_Y^\dagger) V_Y \rho_{AB} V_Y^\dagger (V_Y V_Y^\dagger) = V_Y \rho_{AB} V_Y^\dagger$. Hence, (**b**) of Exercise 5.25 yields the following relation (a):

$$
\begin{aligned}
& D_\alpha((V_Y \rho V_Y^\dagger)_{YY'AB} \| I_Y \otimes \sigma_{Y'AB}) \\
\overset{(a)}{\ge} & D_\alpha(V_Y \rho_{AB} V_Y^\dagger \| ((V_Y V_Y^\dagger)(I_Y \otimes \sigma_{Y'AB})(V_Y V_Y^\dagger)) \\
\overset{(b)}{=} & D_\alpha(\rho_{AB} \| V_Y^\dagger (I_Y \otimes \sigma_{Y'AB}) V_Y) \\
\overset{(c)}{=} & D_\alpha(V_X \rho_{AB} V_X^\dagger \| V_X V_Y^\dagger (I_Y \otimes \sigma_{Y'AB}) V_Y V_X^\dagger) \\
\overset{(d)}{\ge} & D_\alpha(\operatorname{Tr}_{X'A} V_X \rho_{AB} V_X^\dagger \| \operatorname{Tr}_{X'A} V_X V_Y^\dagger (I_Y \otimes \sigma_{Y'AB}) V_Y V_X^\dagger) \\
\overset{(e)}{\ge} & D_\alpha((V_X \rho V_X^\dagger)_{XB} \| c I_X \otimes \sigma_B) \\
\overset{(f)}{=} & -\log c + D_\alpha(\operatorname{Tr}_{X'A} V_X \rho_{AB} V_X^\dagger \| I_X \otimes \sigma_B),
\end{aligned}
$$

where (b), (c), (d), (e), and (f), follow from (**d**), (**d**), (**a**), (**e**), and (**c**) of Exercise 5.25, respectively. In particular, to derive (e), we employ (7.58) as well as (**e**) of Exercise 5.25. Thus,

$$
\begin{aligned}
& -H^{\uparrow}_{\alpha|V_Y \rho V_Y^\dagger}(Y|Y'AB) = \min_{\sigma_{Y'AB}} D_\alpha((V_Y \rho V_Y^\dagger)_{YY'AB} \| I_Y \otimes \sigma_{Y'AB}) \\
\ge & -\log c + \min_{\sigma_{Y'AB}} D_\alpha(\operatorname{Tr}_{X'A} V_X \rho_{AB} V_X^\dagger \| I_X \otimes \sigma_B) \\
= & -H^{\uparrow}_{\alpha|V_X \rho V_X^\dagger}(X|B) - \log c,
\end{aligned}
$$

which implies (7.57). (7.55) can be shown by replacing $\sigma_{Y'AB}$ by $\rho_{Y'AB}$. (7.56) can be shown by replacing D_α by $\underline{D}_\alpha$. ■

Exercise

7.22 Show that $\left\| \sqrt{M_x} \sqrt{N_y} \right\| = |\langle u_x | v_y \rangle|$ when $M_x = |u_x\rangle\langle u_x|$ and $N_y = |v_y\rangle\langle v_y|$.

7.4 Measurements with Negligible State Reduction

As discussed previously, any measurement inevitably changes the state of the measured system. In this section, we propose a method constructing a measurement with negligible state reduction. When a measurement described by an instrument κ is applied on a system in a state ρ, the amount of the state reduction is characterized by

$$
\varepsilon(\rho, \kappa) \overset{\text{def}}{=} \sum_\omega \operatorname{Tr} \kappa_\omega(\rho) b^2 \left(\rho, \frac{1}{\operatorname{Tr} \kappa_\omega(\rho)} \kappa_\omega(\rho) \right), \tag{7.59}
$$

where b is the Bures distance. In the following discussion, we consider the typical state reduction κ_M of a POVM $M = \{M_i\}$. Then, the amount of the state reduction can be found to be [18]

$$\begin{aligned}\varepsilon(\rho, \kappa_M) = &\sum_i \operatorname{Tr} \rho M_i \left(1 - \operatorname{Tr} \sqrt{\rho^{1/2} \frac{\sqrt{M_i}\rho\sqrt{M_i}}{\operatorname{Tr} \rho M_i} \rho^{1/2}}\right)\\ =&1 - \sum_i \sqrt{\operatorname{Tr} \rho M_i} \operatorname{Tr} \sqrt{\rho^{1/2}\sqrt{M_i}\rho\sqrt{M_i}\rho^{1/2}}\\ =&1 - \sum_i \sqrt{\operatorname{Tr} \rho M_i} \operatorname{Tr} \rho^{1/2}\sqrt{M_i}\rho^{1/2}\\ =&1 - \sum_i \sqrt{\operatorname{Tr} \rho M_i} \operatorname{Tr} \rho\sqrt{M_i} \le 1 - \sum_i \left(\operatorname{Tr} \rho\sqrt{M_i}\right)^2, \end{aligned} \tag{7.60}$$

where we used the quantum version of Jensen's inequality[Exe. 3.3] in the last formula. Conversely, since $M_i \le I$, we have $M_i \le \sqrt{M_i}$, and therefore $\varepsilon(\rho, \kappa_M) \ge 1 - \sum_i \left(\operatorname{Tr} \rho\sqrt{M_i}\right)^{3/2}$. When there is no possibility of ambiguity, we will abbreviate $\varepsilon(\rho, \kappa_M)$ to $\varepsilon(\rho, M)$. In particular, if ρ_j is generated with a probability distribution $p = \{p_j\}$, the average of $\varepsilon(\rho_j, \kappa_M)$ can be evaluated as

$$\begin{aligned}\varepsilon(p, \kappa_M) \stackrel{\text{def}}{=} &\sum_j p_j \varepsilon(\rho_j, \kappa_M) \le 1 - \sum_j p_j \sum_i \left(\operatorname{Tr} \rho_j\sqrt{M_i}\right)^2\\ \le&1 - \sum_i \Big(\sum_j p_j \operatorname{Tr} \rho_j\sqrt{M_i}\Big)^2 = 1 - \sum_i \left(\operatorname{Tr} \bar{\rho}_p\sqrt{M_i}\right)^2, \end{aligned} \tag{7.61}$$

where $\bar{\rho}_p \stackrel{\text{def}}{=} \sum_j p_j \rho_j$. Hence, the analysis of $\varepsilon(p, \kappa_M)$ is reduced to that of $1 - \sum_i \left(\operatorname{Tr} \bar{\rho}_p\sqrt{M_i}\right)^2$.

Let us now consider which POVM M has a negligible state reduction. For this analysis, we focus on the number $i_m \stackrel{\text{def}}{=} \operatorname{argmax}_i \operatorname{Tr} M_i \rho$ and the probability $\mathrm{P}_\rho^{M,\max} \stackrel{\text{def}}{=} \operatorname{Tr} M_{i_m} \rho$. Then, we obtain

$$\begin{aligned}&(1 - \mathrm{P}_\rho^{M,\max})(1 + \mathrm{P}_\rho^{M,\max}) = 1 - (\mathrm{P}_\rho^{M,\max})^2 \ge 1 - \left(\operatorname{Tr} \rho\sqrt{M_{i_m}}\right)^2\\ \ge&1 - \sum_i \left(\operatorname{Tr} \rho\sqrt{M_i}\right)^2 \ge \varepsilon(\rho, M).\end{aligned}$$

Therefore, we see that $\varepsilon(\rho, M)$ approaches 0 when $\mathrm{P}_\rho^{M,\max}$ approaches 1.

However, a meaningful POVM does not necessarily have the above property, but usually has the following property in the asymptotic case, i.e., in the case of n-fold tensor product state $\rho^{\otimes n}$ on $\mathcal{H}^{\otimes n}$. Let $(M^{(n)}, x^n) = \{(M^{(n)}, x^n)\}$ be a sequence of pairs of POVMs and functions to $\mathbb{R}^d$. Hence, the vector $x^n(i) = (x^{n,k}(i))$ is the

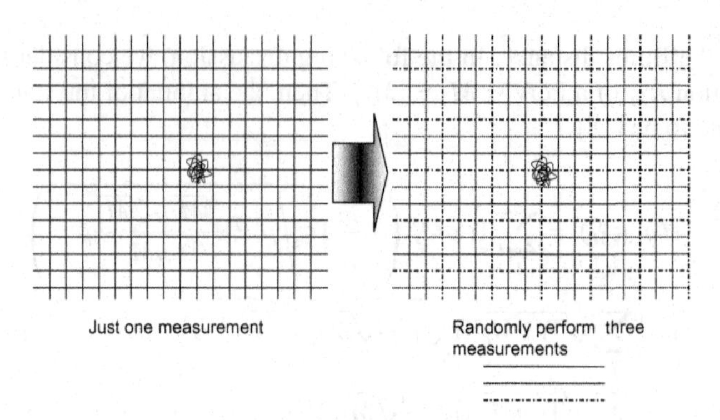

Fig. 7.1 Measurement with negligible state reduction

measurement outcome subject to the probability distribution $\mathrm{P}^{\boldsymbol{M}^{(n)}}_{\rho^{\otimes n}}$. Suppose that the measurement outcome $\boldsymbol{x}^n(i) = (x^{n,k}(i))$ satisfies the weak law of large numbers as a random variable. That is, for a given density ρ, there exists a vector $\boldsymbol{a} \in \mathbb{R}^d$ such that

$$\operatorname{Tr} \rho^{\otimes n} \boldsymbol{M}^{(n)}\{\|\boldsymbol{x}^n(i) - \boldsymbol{a}\| \geq \epsilon'\} \to 0, \quad \forall \epsilon' > 0. \tag{7.62}$$

For the definition of the notation $\boldsymbol{M}^{(n)}\{\|\boldsymbol{x}^n(i) - \boldsymbol{a}\| \geq \epsilon'\}$, see (6.87). Therefore, we propose a method to perform a measurement with negligible state reduction from a POVM satisfying (7.62) as follows.

Theorem 7.8 *For a given positive real number δ and a given positive integer l, we define the modified POVM $\boldsymbol{M}^{(n),\delta,l}$ taking values in $\mathbb{Z}^d$ in the following way. We also define the function $\boldsymbol{x}^n_\delta$ from the set $\mathbb{Z}^d$ to $\mathbb{R}^d$ as $\boldsymbol{x}^n_\delta(\boldsymbol{j}) = \delta \boldsymbol{j}$. If a sequence $\{\delta_n\}$ of real numbers and another sequence $\{l_n\}$ of integers satisfy $\delta_n \to 0$, $l_n \to \infty$, and*

$$\operatorname{Tr} \rho^{\otimes n} \boldsymbol{M}^{(n)}\{\|\boldsymbol{x}^n(i) - \boldsymbol{a}\| \geq \delta_n\} \to 0, \tag{7.63}$$

$$l_n \delta_n \to 0, \tag{7.64}$$

we then have

$$\varepsilon(\rho^{\otimes n}, \boldsymbol{M}^{(n),\delta_n,l_n}) \to 0, \tag{7.65}$$

$$\operatorname{Tr} \rho^{\otimes n} \boldsymbol{M}^{(n),\delta_n,l_n}\{\|\boldsymbol{x}^n_{\delta_n}(\boldsymbol{j}) - \boldsymbol{a}\| \geq \epsilon'\} \to 0, \quad \forall \epsilon' > 0. \tag{7.66}$$

Construction of $\boldsymbol{M}^{(n),\delta,l}$ *Define*

$$U_{\boldsymbol{y},\epsilon} \stackrel{\text{def}}{=} \left\{ \boldsymbol{x} \in \mathbb{R}^d \,\middle|\, \forall k,\ y^k - \frac{1}{2}\epsilon \leq x^k < y^k + \frac{1}{2}\epsilon \right\},$$

$$\tilde{U}_{\boldsymbol{y},\epsilon} \stackrel{\text{def}}{=} \left\{ \boldsymbol{x} \in \mathbb{R}^d \,\middle|\, \forall k,\ y^k - \frac{1}{2}\epsilon < x^k \leq y^k + \frac{1}{2}\epsilon \right\}$$

for $\boldsymbol{y} = (y^k) \in \mathbb{R}^d$. *Define* $M^{(n)}_{\boldsymbol{y},\delta} \stackrel{\text{def}}{=} \sum_{\boldsymbol{x}^n_i \in U_{\boldsymbol{y},\delta}} M^{(n)}_i$. *Then,* $\{M^{(n)}_{\delta(\boldsymbol{j}+\boldsymbol{j}'),l\delta}\}_{\boldsymbol{j}\in(l\mathbb{Z})^d}$ *is a POVM since* $\sum_{\boldsymbol{j}\in(l\mathbb{Z})^d} M^{(n)}_{\delta(\boldsymbol{j}+\boldsymbol{j}'),l\delta} = I$ *for arbitrary* $\delta > 0$ *and* $\boldsymbol{j}' \in \{0,\ldots,l-1\}^d$. *Moreover, we define* $M^{(n),\delta,l}_{\boldsymbol{j}} \stackrel{\text{def}}{=} \frac{1}{l^d} M^{(n)}_{\delta\boldsymbol{j},l\delta}$. *So,* $\boldsymbol{M}^{(n),\delta,l} = \{M^{(n),\delta,l}_{\boldsymbol{j}}\}_{\boldsymbol{j}\in\mathbb{Z}^d}$ *is a POVM with measurement outcomes in* $\mathbb{Z}^d$ *because it is the randomized mixture of the POVMs* $\{M^{(n)}_{\delta(\boldsymbol{j}+\boldsymbol{j}'),l\delta}\}_{\boldsymbol{j}\in(l\mathbb{Z})^d}$ *as Fig. 7.1.*

The existence of a sequence $\{\delta_n\}$ that satisfies condition (7.63) and $\delta_n \to 0$ can be verified from Lemma A.3 in Appendix. Note that the choice of the POVM $\boldsymbol{M}^{(n),\delta,l}$ depends only on the choice of δ and not $\rho^{\otimes n}$. If the convergence of (7.62) is uniform for every $\epsilon > 0$, then the convergences of (7.65) and (7.66) also does not depend on ρ.

Proof of (7.65) Let $\delta\boldsymbol{j} \in \tilde{U}_{\boldsymbol{a},(l-1)\delta} \cap \delta\mathbb{Z}^d$. Since $\{\boldsymbol{x}^n_i \mid \|\boldsymbol{x}^n(i)-\boldsymbol{a}\| < \delta\} \subset U_{\delta\boldsymbol{j},l\delta}$, we obtain

$$\frac{1}{l^d}\boldsymbol{M}^{(n)}\{\|\boldsymbol{x}^n(i)-\boldsymbol{a}\| < \delta\} \le M^{(n),\delta,l}_{\boldsymbol{j}}. \tag{7.67}$$

From the matrix monotonicity of $x \mapsto \sqrt{x}$ and the fact that $0 \le \boldsymbol{M}^{(n)}\{\|\boldsymbol{x}^n(i)-\boldsymbol{a}\| < \delta\} \le I$, we obtain

$$\frac{1}{l^{d/2}}\boldsymbol{M}^{(n)}\{\|\boldsymbol{x}^n(i)-\boldsymbol{a}\| < \delta\} \le \frac{1}{l^{d/2}}\sqrt{\boldsymbol{M}^{(n)}\{\|\boldsymbol{x}^n(i)-\boldsymbol{a}\| < \delta\}} \le \sqrt{M^{(n),\delta,l}_{\boldsymbol{j}}}.$$

Meanwhile, since $\#(\tilde{U}_{\boldsymbol{a},(l-1)\delta} \cap \delta\mathbb{Z}^d) = (l-1)^d$, we have

$$\begin{aligned}
\varepsilon(\rho^{\otimes n}, \boldsymbol{M}^{(n),\delta,l}) &\le 1 - \sum_{\boldsymbol{j}\in\mathbb{Z}^d}\left(\operatorname{Tr}\rho^{\otimes n}\sqrt{M^{(n),\delta,l}_{\boldsymbol{j}}}\right)^2 \\
&\le 1 - \sum_{\delta\boldsymbol{j}\in\tilde{U}_{\boldsymbol{a},(l-1)\delta}}\left(\operatorname{Tr}\rho^{\otimes n}\sqrt{M^{(n),\delta,l}_{\boldsymbol{j}}}\right)^2 \\
&\le 1 - \sum_{\delta\boldsymbol{j}\in\tilde{U}_{\boldsymbol{a},(l-1)\delta}}\left(\frac{1}{l^{d/2}}\operatorname{Tr}\rho^{\otimes n}\boldsymbol{M}^{(n)}\{\|\boldsymbol{x}^n(i)-\boldsymbol{a}\| < \delta\}\right)^2 \\
&= 1 - \frac{(l-1)^d}{l^d}\left(\operatorname{Tr}\rho^{\otimes n}\boldsymbol{M}^{(n)}\{\|\boldsymbol{x}^n(i)-\boldsymbol{a}\| < \delta\}\right)^2.
\end{aligned}$$

From (7.63) and the fact that $l_n \to \infty$, substituting δ_n and l_n in δ and l, respectively, we obtain $\varepsilon(\rho^{\otimes n}, \boldsymbol{M}^{(n),\delta_n,l_n}) \to 0$. ∎

Proof of (7.66) If $\|\delta\boldsymbol{j}-\boldsymbol{a}\| \ge \epsilon'$ and $\boldsymbol{x}^n(i) \in U_{\delta\boldsymbol{j},l\delta}$, then $\|b\boldsymbol{x}^n(i)-\boldsymbol{a}\| \ge \epsilon' - \sqrt{d}l\delta$. It then follows that

$$M^{(n),\delta,l}\{\|x^n_\delta(j) - a\| \geq \epsilon'\} \leq M^{(n)}\{\|x^n(i) - a\| \geq \epsilon' - \sqrt{d}l\delta\}.$$

Therefore, if δ_n and l_n are chosen such that condition (7.64) is satisfied, then (7.66) is satisfied. ∎

Note that similarly we can show that $\varepsilon(p^n, \kappa_{M^{(n),\delta_n,l_n}}) \to 0$, which will be used in Chap. 10.

As discussed in Chap. 6, the asymptotic performance of an estimator can be treated with at least two criteria. One is large deviation, wherein we focus on the decreasing exponential rate of the probability that the estimate does not belong to the neighborhood of the true value with a fixed radius. The other is small deviation, wherein we focus on the asymptotic behavior of mean square error. In mathematical statistics, it is known that the latter discussion is essentially equivalent to that of the probability that the estimate does not belong to the the neighborhood of the true value with a radius proportional to $\frac{1}{\sqrt{n}}$. That is, the difference between two criteria is essentially expressed by the difference of the asymptotic behavior of the radius of the neighborhood of interest.

As mentioned in Exercise 7.23, if the original POVM is optimal in the sense of a large deviation, the deformed one is also optimal in the same sense. However, even if the original estimator is optimal in the sense of a small deviation, the estimator deformed by the presented method is not necessarily optimal in the same sense. That is, when $\lim_{n\to\infty} \varepsilon(\rho^{\otimes n}, M^{(n),\delta_n,l_n})$ is different from the original quantity, the modification affects the accuracy of the deformed estimator in the sense of the small deviation, but not in the sense of the large deviation. Therefore, it is expected that there exists a tradeoff relation between the limit of the mean square error of the estimator. and the difference between $\lim_{n\to\infty} \varepsilon(\rho^{\otimes n}, M^{(n)})$ and the original quantity.

Moreover, since the measurement with negligible state reduction has not been realized in the experiment, its realization is strongly desired.

Exercise

7.23 Consider the sequence $M = \{(M^n, \hat{\theta}^n)\}$ of estimators for the state family with one parameter $\{\rho_\theta | \theta \in \mathbb{R}\}$. Let $\beta(M, \theta, \epsilon)$ be continuous with respect to ϵ. Show that $\beta(M, \theta, \epsilon) = \beta(\{(M^{(n),\delta_n,l_n}, x^n_{\delta_n})\}, \theta, \epsilon)$ when $l_n\delta_n \to 0$, where $M^{(n),\delta,l}$ is defined in the above discussion.

7.5 Historical Note

The mathematical description of a measurement process was initiated by von Neumann [1]. In his formulation, the measurement is described by a projection-valued measure. From the mathematical viewpoint, Naĭmark [2] showed that any POVM can be characterized as the restriction of the projection-valued measure. This projection-valued measure is called a Naĭmark extension. Holevo applied this argument to quantum measurements [19].

Further, Davis and Lewis [20] formulated the state reduction as a positive-map-valued measure. Following this research, Ozawa [3] proved that any measurement reduction should be described by a CP-map-valued measure, i.e., an instrument. He also proposed the indirect measurement model as a description of the interaction with the macroscopic system [21, 22]. Indeed, a positive-map-valued measure $\{\kappa_\omega\}$ is a CP-map-valued measure if and only if it can be described by an indirect measurement model [3]. For any POVM, its indirect measurement model gives a Naĭmark extension of this POVM (Theorem 7.1). For example, an indirect measurement model of the joint measurement of the position Q and the momentum P is known (see Holevo [19]). Further, Hayashi et al. [23] constructed indirect measurements for a meaningful POVM for squeezed states. For a more precise description of state reduction, see Holevo [24], who discusses state reductions due to continuous-variable measurements using semigroup theory. Busch et al. [25] discuss the connection between this formulation and experiments. In addition, Ozawa characterized the instrument given by (7.1) as a minimal-disturbance measurement [22, 26]. Furthermore, this book treats state reductions where the input and output systems are different systems because such a reduction is common in quantum information processing. Hence, this book focuses on Theorem 7.3 as a generalization of Corollary 7.1 obtained by Ozawa [3].

The uncertainty relation between conjugate observables was discussed in the context of gedanken experiments by Heisenberg [6]. It was first treated mathematically by Robertson [5], who was not, however, concerned with the effect of measurement. Recently, Ozawa [7–10, 22] formulated the disturbance by measurement, and treated the uncertainty relation concerning measurement, mathematically. These are undoubtably the first attempts at a mathematically rigorous treatment of Heisenberg uncertainty. In this book, we mathematically formulate the same problem, but in a different way. In particular, the definition of disturbance in this text is different from that by Ozawa. Hence, the inequality given in this text is a sightly stronger requirement than that of Ozawa. However, the methods of Ozawa's and our proofs are almost identical. For further discussion of the historical perspective of this topic, see Ozawa [9]. Indeed, Ozawa considered inequality (7.30) to be the mathematical formulation of Heisenberg's uncertainty relation, and he gave its counterexample. He also proposed another type of uncertainty relation—(7.36) and (7.37)—due to measurement. However, in this book, inequality (7.31) is treated as the mathematical formulation of Heisenberg's uncertainty relation. Therefore, the discussion in this book is different from that of Ozawa.

Concerning the mixed-state case, Nagaoka [11] generalized inequality (7.27) to inequality (7.28). [Indeed, the RHS of Nagaoka's original inequality has a different expression; however, it is equal to the RHS of (7.28).] This is a stronger inequality than the trivial generalization $\Delta_1(X,\rho)\Delta_1(Y,\rho) \geq \frac{|\operatorname{Tr}\rho[X,Y]|}{2}$. All inequalities in Sect. 7.2 are based on the former,

Further, using inequality (7.28), Nagaoka [11] derived inequality (7.33) in the mixed-state case. The same inequality with the RHS $\frac{|\operatorname{Tr}\rho[X,Y]|}{2}$ has been discussed by many researchers [27–30]. Nagaoka applied this inequality to the Cramér–Rao-type bound and obtained the bound (6.109) in the two-parameter case, first. Hayashi [31] extended this inequality to the case with more than two parameters.

The extension of the uncertainty relation to the entropic uncertainty relation was firstly done by Maassen and Uffink [12] as Theorem 7.6. The proof presented in this book is based on Golden-Thompson trace inequality (Lemma 5.4) and was given by Frank and Lieb [32]. This inequality was extended to the case with conditional entropies as Theorem 7.7. Renes and Boileau [16] showed Inequality (7.50) in the mutually unbiased case, and conjectured it in the general form. Then, Berta et al. [17] showed it in the general form by representing the theorem in a different form. Then, Coles et al. [13] showed (7.47), and Müller-Lennert et al. [14] did (7.48). Then, following the framework of [13], Tomamichel et al. [15] showed (7.49).

The study of measurements with a negligible state reduction has been motivated by quantum universal variable-length source coding because quantum universal variable-length source coding requires determination of the compression rate with small state reduction (Sect. 10.5). Hayashi and Matsumoto [18] treated this problem and obtained the main idea of Sect. 7.4. This method is useful for estimating the state of the system without considerable state reduction. This method is often called gentle tomography. Bennett et al. [33] considered the complexity of this kind of measurement.

7.6 Solutions of Exercises

Exercise 7.1 Choose the unitary matrix $\left(\sum_\omega |u_\omega\rangle\langle u_\omega| \otimes U_{\kappa'_\omega}\right)\left(I_{B,C} \otimes U'\right)$ and the state $\rho'_1 \otimes \rho_0$ as the unitary U and the state ρ_0 in Condition ①. Then, the unitary U and the state ρ_0 satisfy (7.10).

Exercise 7.2 Let $\mathcal{H}_D$ be $\mathcal{H}_C \otimes \mathcal{H}_B$, consider the unitary matrix corresponding to the replacement $W : u \otimes v \mapsto v \otimes u$ in $\mathcal{H}_A \otimes \mathcal{H}_B$, and define $V \stackrel{\rm def}{=} (W \otimes I_C)U$. Then, V satisfies (7.11).

Exercise 7.3 Equation (5.2) yields that $\operatorname{Tr} \rho M'_{\omega,\omega} = \operatorname{Tr} \rho \kappa^*_\omega(M_{\omega'}) = \operatorname{Tr} \kappa^*_\omega(\rho) M_{\omega'}$.

Exercise 7.4 Since

$$\operatorname{Tr}_B I_A \otimes |u_i\rangle\langle u_i||x\rangle\langle x| = |v_i\rangle\langle v_i|,$$

the resultant state is $\frac{1}{\|v_i\|^2}|v_i\rangle\langle v_i|$.

Exercise 7.5

(**a**) Assume that κ is given in ②. When the input state is an entangled state $|\Psi\rangle := \sum_i a_i |u_i^A, u_i^B\rangle$, we have

$$\kappa \otimes \iota_C(|\Psi\rangle\langle\Psi|) = \sum_\omega \operatorname{Tr}_A M_\omega |\Psi\rangle\langle\Psi| \otimes W_\omega,$$

which is a separable form between $\mathcal{H}_B$ and $\mathcal{H}_C$.
(b) Let $\sum_i p_i \rho_i^A \otimes \rho_i^B = (\kappa \otimes \iota_{A'})(|x_M\rangle\langle x_M|)$. Then, due to (5.5), the Choi-Jamiołkowski matrix has the separable form, i.e., $K(\kappa) = d\sum_i p_i \rho_i^A \otimes \rho_i^B$. From the definition of $K(\kappa)$, we have $\mathrm{Tr}\,\kappa(\rho)\sigma = \mathrm{Tr}\,K(\kappa)\rho\otimes\sigma = d\sum_i p_i(\mathrm{Tr}\,\rho\rho_i^A)(\mathrm{Tr}\,\sigma\rho_i^B)$. Thus, we obtain $\kappa(\rho) = d\sum_i p_i(\mathrm{Tr}\,\rho\rho_i^A)\rho_i^B$.

Exercise 7.6

$$\Delta_2^2(\boldsymbol{M},\rho) = \sum_i (x_i - \mathrm{E}_\rho(\boldsymbol{M}))^2\,\mathrm{Tr}\,\rho M_i = \mathrm{Tr}\sum_i (x_i - \mathrm{Tr}\,\rho O(\boldsymbol{M}))^2 M_i\rho$$
$$= \mathrm{Tr}\left(\sum_i (x_i - \mathrm{Tr}\,\rho O(\boldsymbol{M}))M_i\right)\left(\sum_j (x_j - \mathrm{Tr}\,\rho O(\boldsymbol{M}))M_j\right)\rho$$
$$= \mathrm{Tr}(O(\boldsymbol{M}) - \mathrm{Tr}\,\rho O(\boldsymbol{M}))^2\rho = \Delta_1^2(O(\boldsymbol{M}),\rho).$$

Exercise 7.7

$$\Delta_3^2(\boldsymbol{M},O(\boldsymbol{M}),\rho) = \sum_i \mathrm{Tr}(x_i - O(\boldsymbol{M}))M_i(x_i - O(\boldsymbol{M}))\rho$$
$$= \sum_i \mathrm{Tr}\,x_i^2 M_i\rho - \mathrm{Tr}\,O(\boldsymbol{M})^2\rho$$
$$= \sum_i \mathrm{Tr}\,x_i)^2 M_i\rho - (\mathrm{Tr}\,\rho O(\boldsymbol{M}))^2 - \mathrm{Tr}\,O(\boldsymbol{M})^2\rho + (\mathrm{Tr}\,\rho O(\boldsymbol{M}))^2$$
$$= \Delta_3^2(\boldsymbol{M},O(\boldsymbol{M}),\rho) + \Delta_1^2(O(\boldsymbol{M}) - X,\rho).$$

Exercise 7.8

$$\Delta_3^2(\boldsymbol{M},X,\rho) = \sum_i \mathrm{Tr}(x_i - X)M_i(x_i - X)\rho$$
$$= \sum_i \mathrm{Tr}\,x_i^2 M_i\rho - \mathrm{Tr}\,X\rho O(\boldsymbol{M}) - \mathrm{Tr}\,O(\boldsymbol{M})\rho X + \mathrm{Tr}\,X^2\rho$$
$$= \sum_i \mathrm{Tr}\,x_i^2 M_i\rho - \mathrm{Tr}\,O(\boldsymbol{M})^2\rho + \mathrm{Tr}\,O(\boldsymbol{M})^2\rho$$
$$- \mathrm{Tr}\,X\rho O(\boldsymbol{M}) - \mathrm{Tr}\,O(\boldsymbol{M})\rho X + \mathrm{Tr}\,X^2\rho$$
$$= \sum_i \mathrm{Tr}(x_i - O(\boldsymbol{M}))M_i(x_i - O(\boldsymbol{M}))\rho + \mathrm{Tr}(X - O(\boldsymbol{M}))^2\rho$$
$$= \Delta_3^2(\boldsymbol{M},O(\boldsymbol{M}),\rho) + \Delta_1^2(O(\boldsymbol{M}) - X,\rho).$$

Exercise 7.9 We denote the spectral decomposition of $O(\boldsymbol{M})$ by $\{(E_{O(\boldsymbol{M}),j}, y_j)\}_j$. Then, we have $\mathrm{Tr}\,O(\boldsymbol{M})^2\kappa(\rho) = \sum_j y_j^2\,\mathrm{Tr}\,E_{O(\boldsymbol{M}),j}\kappa(\rho) = \sum_j y_j^2\,\mathrm{Tr}\,\kappa^*(E_{O(\boldsymbol{M}),j})\rho$ and $\sum_i x_i\kappa^*(M_i) = \kappa^*(\sum_i x_i M_i) = \kappa^*(O(\boldsymbol{M})) = \kappa^*(\sum_j y_j E_{O(\boldsymbol{M}),j}) = \sum_j y_j\kappa^*(E_{O(\boldsymbol{M}),j})$. Using these relations, we have

$$\begin{aligned}
&\Delta_3^2(\kappa^*(\boldsymbol{M}), X, \rho) = \sum_i \operatorname{Tr}(x_i - X)\kappa^*(M_i)(x_i - X)\rho \\
=&\sum_i \operatorname{Tr} x_i^2 \kappa^*(M_i)\rho - \operatorname{Tr}\left(\sum_i x_i \kappa^*(M_i)\right)\rho X - \operatorname{Tr} X\rho\left(\sum_i x_i \kappa^*(M_i)\right) + \operatorname{Tr} X^2\rho \\
=&\sum_i \operatorname{Tr} x_i^2 M_i \kappa(\rho) - \operatorname{Tr} O(\boldsymbol{M})^2\kappa(\rho) + \sum_j y_j^2 \operatorname{Tr} \kappa^*(E_{O(\boldsymbol{M}),j})\rho \\
&- \operatorname{Tr}\left(\sum_j y_j \kappa^*(E_{O(\boldsymbol{M}),j})\right)\rho X - \operatorname{Tr} X\rho\left(\sum_j y_j \kappa^*(E_{O(\boldsymbol{M}),j})\right) + \operatorname{Tr} X^2\rho \\
=&\sum_i \operatorname{Tr}(x_i - O(\boldsymbol{M}))M_i(x_i - O(\boldsymbol{M}))\kappa(\rho) \\
&+ \sum_j \operatorname{Tr}(y_j - X)\kappa^*(E_{O(\boldsymbol{M}),j})(y_j - X)\rho \\
=&\Delta_3^2(\boldsymbol{M}, O(\boldsymbol{M}), \kappa(\rho)) + \Delta_3^2(\kappa^*(\boldsymbol{E}_{O(\boldsymbol{M})}), X, \rho).
\end{aligned}$$

Exercise 7.10

(**a**) Since $\operatorname{Tr}_A \rho \operatorname{Tr}_C[U^*(I \otimes E)UI \otimes \rho_0] = \operatorname{Tr} I \otimes EU\rho \otimes \rho_0 U^* = \operatorname{Tr}_C E \operatorname{Tr}_A U\rho \otimes \rho_0 U^* = \operatorname{Tr}_C E\kappa(\rho) = \operatorname{Tr}_C \kappa^*(E)\rho$ for any state ρ on $\mathcal{H}_A$, we have $\operatorname{Tr}_C U^*(I \otimes E)UI \otimes \rho_0 = \kappa^*(E)$. Thus,

$$\begin{aligned}
&\operatorname{Tr}(X \otimes I - x)U^*(I \otimes E)U(X \otimes I - x)\rho \otimes \rho_0 \\
=&\operatorname{Tr} X \otimes IU^*(I \otimes E)UX \otimes I\rho \otimes \rho_0 - \operatorname{Tr}(X \otimes I)xU^*(I \otimes E)U\rho \otimes \rho_0 \\
&- \operatorname{Tr} xU^*(I \otimes E)U(X \otimes I)\rho \otimes \rho_0 + \operatorname{Tr} x^2 U^*(I \otimes E)UU^*(I \otimes E)U\rho \otimes \rho_0 \\
=&\operatorname{Tr} X \otimes IU^*(I \otimes E)UX \otimes I\rho \otimes \rho_0 - \operatorname{Tr}(X \otimes I)xU^*(I \otimes E)U\rho \otimes \rho_0 \\
&- \operatorname{Tr} xU^*(I \otimes E)U(X \otimes I)\rho \otimes \rho_0 + \operatorname{Tr} x^2 U^*(I \otimes E)U\rho \otimes \rho_0 \\
=&\operatorname{Tr}_A X\rho X \operatorname{Tr}_C[U^*(I \otimes E)UI \otimes \rho_0] - \operatorname{Tr}_A \rho X x \operatorname{Tr}_C[U^*(I \otimes E)UI \otimes \rho_0] \\
&- \operatorname{Tr}_A \operatorname{Tr}_C[U^*(I \otimes E)UI \otimes \rho_0]xX\rho + \operatorname{Tr}_A x^2\rho \operatorname{Tr}_C[U^*(I \otimes E)UI \otimes \rho_0] \\
=&\operatorname{Tr}_A X\kappa^*(E)X\rho - \operatorname{Tr}_A \rho X x\kappa^*(E) - \operatorname{Tr}_A \kappa^*(E)xX\rho + \operatorname{Tr}_A x^2\rho\kappa^*(E) \\
=&\operatorname{Tr}(X - x)\kappa^*(E)(X - x)\rho.
\end{aligned}$$

(**b**) We denote the spectral decomposition of Y by $\{(E_{Y,j}, y_j)\}_j$. Then,

$$\begin{aligned}
&\Delta_3^2(\kappa^*(\boldsymbol{E}_Y), X, \rho) \\
=&\operatorname{Tr} \sum_j (y_j - X)\kappa^*(E_{Y,j})(y_j - X)\rho
\end{aligned}$$

$$= \mathrm{Tr} \sum_j (X \otimes I - y_j) U^*(I \otimes E_{Y,j}) U (X \otimes I - y_j) \rho \otimes \rho_0$$

$$= \mathrm{Tr} \sum_j [(X \otimes I) U^*(I \otimes E_{Y,j}) U (X \otimes I) - X^{\otimes} I U^*(I \otimes y_j E_{Y,j}) U$$

$$- U^*(I \otimes y_j E_{Y,j}) U X^{\otimes} I + U^*(I \otimes y_j^2 E_{Y,j}) U] \rho \otimes \rho_0$$

$$= \mathrm{Tr}(X^2 \otimes I - X^{\otimes} I U^*(I \otimes Y) U - U^*(I \otimes Y) U X^{\otimes} I + U^*(I \otimes Y^2) U) \rho \otimes \rho_0$$

$$= \mathrm{Tr}\left(U(X^2 \otimes I) U^* - U(X^{\otimes} I) U^*(I \otimes Y) - (I \otimes Y) U (X^{\otimes} I) U^*\right.$$

$$\left. + (I \otimes Y^2)\right) U \rho \otimes \rho_0 U^*$$

$$= \mathrm{Tr}\left(U(X \otimes I_{B,C}) U^* - (I_{A,C} \otimes Y)\right)^2 U(\rho \otimes \rho_0) U^*.$$

Exercise 7.11

$$\Delta_3^2(\boldsymbol{M}, X, \rho) = \sum_i \mathrm{Tr}(x_i - X) M_i (x_i - X) \rho = \sum_i \mathrm{Tr}\, M_i (x_i - X) \rho (x_i - X)$$

$$= \sum_i \mathrm{Tr}\, E_i ((x_i - X \otimes I) \rho \otimes \rho_0 (x_i - X \otimes I))$$

$$= \sum_i \mathrm{Tr}(x_i - X \otimes I) E_i (x_i - X \otimes I) \rho \otimes \rho_0$$

$$= \mathrm{Tr} \sum_i [(x_i^2 E_i - x_i E_i (X \otimes I) - (X \otimes I) x_i E_i + (X \otimes I) E_i (X \otimes I)] \rho \otimes \rho_0$$

$$= \mathrm{Tr}[(O(\boldsymbol{E})^2 - O(\boldsymbol{E})(X \otimes I) - (X \otimes I) O(\boldsymbol{E}) + (X^2 \otimes I)] \rho \otimes \rho_0$$

$$= \mathrm{Tr}(O(\boldsymbol{E}) - (X \otimes I))^2 \rho \otimes \rho_0$$

$$= \Delta_1^2(O(\boldsymbol{E}) - X \otimes I_B, \rho \otimes \rho_0).$$

Exercise 7.12

(**a**) Since $c + d \geq b$, we have $b^2 \leq d^2 + c^2 + 2cd \leq d^2 + c^2 + 2ad$. Thus, $a^2 + b^2 - c^2 \leq a^2 + d^2 + 2ad$, which implies $\sqrt{a^2 + b^2 - c^2} \leq a + d$.
(**b**) Using (7.19) and (7.18), we have

$$\Delta_2^2(\boldsymbol{M}, \rho) = \Delta_3^2(\boldsymbol{M}, O(\boldsymbol{M}), \rho) + \Delta_1^2(O(\boldsymbol{M}), \rho)$$
$$= \Delta_3^2(\boldsymbol{M}, X, \rho) - \Delta_1^2(O(\boldsymbol{M}) - X, \rho) + \Delta_1^2(O(\boldsymbol{M}), \rho),$$

and $\Delta_3(\boldsymbol{M}, X, \rho) \geq \Delta_1^2(O(\boldsymbol{M}) - X, \rho)$. Since $\Delta_1(X, \rho)$ can be regarded as a norm for X, we have $\Delta_1(O(\boldsymbol{M}) - X, \rho) + \Delta_1(X, \rho) \geq \Delta_1(O(\boldsymbol{M}), \rho)$.

Applying (**a**) to the case when $a = \Delta_3(\boldsymbol{M}, X, \rho)$, $b = \Delta_1(O(\boldsymbol{M}), \rho)$ $c = \Delta_1(O(\boldsymbol{M}) - X, \rho)$, and $d = \Delta_1(X, \rho)$, we have

$$\begin{aligned}\Delta_2(\boldsymbol{M},\rho)=&\sqrt{\Delta_3^2(\boldsymbol{M},X,\rho)-\Delta_1^2(O(\boldsymbol{M})-X,\rho)+\Delta_1^2(O(\boldsymbol{M}),\rho)}\\ \le&\Delta_3(\boldsymbol{M},X,\rho)+\Delta_1(X,\rho).\end{aligned}$$

Exercise 7.13 Since $\frac{\mathrm{Tr}\,|\sqrt{\rho}[O(\boldsymbol{M})-X,Y]\sqrt{\rho}|}{2} \le \Delta_1(O(\boldsymbol{M})-X,\rho)\Delta_1(Y,\rho)$, (7.31) implies that

$$\begin{aligned}\Delta_2(\boldsymbol{M},\rho)\Delta_4(\boldsymbol{\kappa},Y,\rho)\ge&\frac{\mathrm{Tr}\,|\sqrt{\rho}[O(\boldsymbol{M}),Y]\sqrt{\rho}|}{2}\\ =&\frac{\mathrm{Tr}\,|\sqrt{\rho}[X,Y]\sqrt{\rho}+\sqrt{\rho}[O(\boldsymbol{M})-X,Y]\sqrt{\rho}|}{2}\\ \ge&\frac{\mathrm{Tr}\,|\sqrt{\rho}[X,Y]\sqrt{\rho}|}{2}-\frac{\mathrm{Tr}\,|\sqrt{\rho}[O(\boldsymbol{M})-X,Y]\sqrt{\rho}|}{2}\\ \ge&\frac{\mathrm{Tr}\,|\sqrt{\rho}[X,Y]\sqrt{\rho}|}{2}-\Delta_1(O(\boldsymbol{M})-X,\rho)\Delta_1(Y,\rho).\end{aligned}$$

Exercise 7.14 Notice that (7.19) implies $\Delta_3(\boldsymbol{M},X,\rho)\ge\Delta_1(O(\boldsymbol{M})-X,\rho)$. Using (7.35) and this inequality, we have

$$\begin{aligned}&(\Delta_3(\boldsymbol{M},X,\rho)+\Delta_1(O(\boldsymbol{M}),\rho))\Delta_4(\boldsymbol{\kappa},Y,\rho)+\Delta_3(\boldsymbol{M},X,\rho)\Delta_1(Y,\rho)\\ \ge&\Delta_2(\boldsymbol{M},\rho)\Delta_4(\boldsymbol{\kappa},Y,\rho)+\Delta_3(\boldsymbol{M},X,\rho)\Delta_1(Y,\rho)\\ \ge&\Delta_2(\boldsymbol{M},\rho)\Delta_4(\boldsymbol{\kappa},Y,\rho)+\Delta_1(O(\boldsymbol{M})-X,\rho)\Delta_1(Y,\rho)\\ \ge&\frac{\mathrm{Tr}\,|\sqrt{\rho}[X,Y]\sqrt{\rho}|}{2}.\end{aligned}$$

Exercise 7.15 Choose a 2×2 orthogonal matrix $(a_{i,j})$ such that the two matrices $\tilde{X}\stackrel{\mathrm{def}}{=}a_{1,1}X+a_{1,2}Y$ and $\tilde{Y}\stackrel{\mathrm{def}}{=}a_{2,1}X+a_{2,2}Y$ satisfy $\mathrm{Cov}_\rho(\tilde{X},\tilde{Y})=0$. Then, (7.39) for $\tilde{X},\tilde{Y}$ is equivalent with (7.28) for $\tilde{X},\tilde{Y}$. Since the fact that both sides of (7.39) are invariant under the orthogonal matrix transformation $(X,Y)\mapsto(\tilde{X},\tilde{Y})$, (7.39) holds for X,Y.

Exercise 7.16 Use the fact $S_i\circ S_j=\delta_{i,j}S_i$ and $\mathrm{Tr}\,S_iS_j=2\delta_{i,j}$.

(e) Since the relation $\|\boldsymbol{x}\|^2\langle\boldsymbol{y},\boldsymbol{a}\rangle^2=\|(\boldsymbol{x}\times\boldsymbol{y})\times\boldsymbol{a}\|^2$ holds in this special case, (7.40) holds.
(g) Both sides of (7.40) are invariant under the transformation $(\boldsymbol{x},\boldsymbol{y})\mapsto(\tilde{\boldsymbol{x}},\tilde{\boldsymbol{y}})$ because the determinant of $(b_{i,j})$ is 1. Then, the statement **(e)** implies (7.40) for arbitrary vectors $\boldsymbol{x},\boldsymbol{y},\boldsymbol{a}$.
(h) Both sides of (7.39) are invariant under the transformation $(X,Y)\mapsto(X-xI,Y-yI)$.

Exercise 7.17 We have

$$O^1(\boldsymbol{M}) = \sum_i \left(\frac{1}{p}(x_i - \operatorname{Tr}\rho X) + \operatorname{Tr}\rho X\right) pE_{X,i} + \sum_j (\operatorname{Tr}\rho X)(1-p)E_{Y,j}$$
$$= X - (\operatorname{Tr}\rho X)I + p(\operatorname{Tr}\rho)I + (1-p)(\operatorname{Tr}\rho)I = X.$$

Similarly, we have $O^2(\boldsymbol{M}) = Y$. We also have

$$\begin{aligned}
&\Delta_3(\boldsymbol{M}_{X,Y,\rho}, X, \rho) \\
=&\sum_i \left(\frac{1}{p}(x_i - \operatorname{Tr}\rho X) + \operatorname{Tr}\rho X\right)^2 \operatorname{Tr} pE_{X,i}\rho \\
&+ \sum_j (\operatorname{Tr}\rho X)^2 \operatorname{Tr}(1-p)E_{Y,j}\rho - (\operatorname{Tr}\rho X)^2 - \operatorname{Tr}\rho(X - (\operatorname{Tr}\rho X))^2 \\
=&\sum_i \left(\frac{1}{p^2}(x_i - \operatorname{Tr}\rho X)^2 + (\operatorname{Tr}\rho X)^2 + \frac{2}{p}(x_i - \operatorname{Tr}\rho X)(\operatorname{Tr}\rho X)\right) \operatorname{Tr} pE_{X,i}\rho \\
&+ (1-p)(\operatorname{Tr}\rho X)^2 - (\operatorname{Tr}\rho X)^2 - \operatorname{Tr}\rho(X - (\operatorname{Tr}\rho X))^2 \\
=&\sum_i \left(\frac{1}{p}(x_i - \operatorname{Tr}\rho X)^2 + p(\operatorname{Tr}\rho X)^2 + 2(x_i - \operatorname{Tr}\rho X)(\operatorname{Tr}\rho X)\right) \operatorname{Tr} E_{X,i}\rho \\
&+ (1-p)(\operatorname{Tr}\rho X)^2 - (\operatorname{Tr}\rho X)^2 - \operatorname{Tr}\rho(X - (\operatorname{Tr}\rho X))^2 \\
=&\frac{1}{p} \operatorname{Tr}\rho(X - \operatorname{Tr}\rho X)^2 + p(\operatorname{Tr}\rho X)^2 + 2\operatorname{Tr}\rho(X - \operatorname{Tr}\rho X)(\operatorname{Tr}\rho X) \\
&+ (1-p)(\operatorname{Tr}\rho X)^2 - (\operatorname{Tr}\rho X)^2 - \operatorname{Tr}\rho(X - (\operatorname{Tr}\rho X))^2 \\
=&\frac{1}{p} \operatorname{Tr}\rho(X - \operatorname{Tr}\rho X)^2 - \operatorname{Tr}\rho(X - (\operatorname{Tr}\rho X))^2 = \frac{1-p}{p}\Delta_1(X, \rho).
\end{aligned}$$

Similarly, we have

$$\Delta_3(\boldsymbol{M}_{X,Y,\rho}, Y, \rho) = \frac{p}{1-p}\Delta_1(Y, \rho).$$

Hence, when X, Y, ρ satisfy the equality in (7.28), the equality in (7.33) holds.

Exercise 7.18 When ① does not hold, ② does not hold due to (7.31). Hence, ② implies ①.

Assume ①. Choose the spectral decomposition $\{(E_{X,\omega}, x_\omega)\}_\omega$ of X. Then, choosing κ_ω as $\kappa_\omega(\rho) := E_{X,\omega}\rho E_{X,\omega}$, we obtain the conditions for ②.

Exercise 7.19 Since

$$\operatorname*{Tr}_A \rho M_i = \operatorname*{Tr}_A \operatorname*{Tr}_{B,C} \left(\left(I_A \otimes \sqrt{\rho_0}\right) U^* \left(I_{A,B} \otimes E_i\right) U \left(I_A \otimes \sqrt{\rho_0}\right)\right),$$

we have

$$
\begin{aligned}
O(\boldsymbol{M}) &= \sum_i x_i M_i \\
&= \sum_i x_i \operatorname*{Tr}_{B,C} \left(\left(I_A \otimes \sqrt{\rho_0}\right) U^* \left(I_{A,B} \otimes E_i\right) U \left(I_A \otimes \sqrt{\rho_0}\right)\right) \\
&= \operatorname*{Tr}_{B,C} \left(\left(I_A \otimes \sqrt{\rho_0}\right) U^* \left(I_{A,B} \otimes O(\boldsymbol{E})\right) U \left(I_A \otimes \sqrt{\rho_0}\right)\right).
\end{aligned}
$$

Exercise 7.20

(**a**) Consider the exponential family ρ_θ generated by state ρ with the SLD X, where $\rho_0 = \rho$. Then, the SLD Fisher informations of the families $\lambda\kappa_1(\rho_\theta) + (1-\lambda)\kappa_2(\rho_\theta)$, $\kappa_1(\rho_\theta)$ and $\kappa_2(\rho_\theta)$ at $\theta = 0$ are $\left(\|\lambda\kappa_1(X) + (1-\lambda)\kappa_2(X)\|^{(e)}_{\lambda\kappa_1(\rho)+(1-\lambda)\kappa_2(\rho),s}\right)^2$, $\left(\|\kappa_1(X)\|^{(e)}_{\kappa_1(\rho),s}\right)^2$, and $\left(\|\kappa_2(X)\|^{(e)}_{\kappa_2(\rho),s}\right)^2$. Thus, Exercise 6.26 implies that

$$
\left(\|\lambda\kappa_1(X) + (1-\lambda)\kappa_2(X)\|^{(e)}_{\lambda\kappa_1(\rho)+(1-\lambda)\kappa_2(\rho),s}\right)^2 \tag{7.68}
$$

$$
\leq \left(\|\kappa_1(X)\|^{(e)}_{\kappa_1(\rho),s}\right)^2 + \left(\|\kappa_2(X)\|^{(e)}_{\kappa_2(\rho),s}\right)^2. \tag{7.69}
$$

Hence, (7.25) implies that

$$
\begin{aligned}
&\Delta_4^2(\lambda\kappa_1 + (1-\lambda)\kappa_2, X, \rho) \\
=& \left(\|X\|^{(e)}_{\rho,s}\right)^2 - \left(\|\lambda\kappa_1(X) + (1-\lambda)\kappa_2(X)\|^{(e)}_{\lambda\kappa_1(\rho)+(1-\lambda)\kappa_2(\rho),s}\right)^2 \\
\geq& \left(\|X\|^{(e)}_{\rho,s}\right)^2 - \lambda\left(\|\kappa_1(X)\|^{(e)}_{\kappa_1(\rho),s}\right)^2 - (1-\lambda)\left(\|\kappa_2(X)\|^{(e)}_{\kappa_2(\rho),s}\right)^2 \\
=& \lambda\Delta_4^2(\kappa_1, X, \rho) + (1-\lambda)\Delta_4^2(\kappa_2, X, \rho).
\end{aligned}
$$

(**b**) Due to Exercise 6.26, the assumption of (**b**) satisfies the equality condition for (7.69). So, the desired equality holds.

Exercise 7.21 We can show $O(\boldsymbol{M}) = X$ by the same way as $O^1(\boldsymbol{M}) = X$ in Exercise 7.17. We can also show that $\Delta_2(\boldsymbol{M}, \rho) = \Delta_1(X, \rho)/p$ by the same way as $O^1(\boldsymbol{M}) = X$ in Exercise 7.17.

Define two TP-CP maps $\kappa_a := \kappa_0/(1-p)$ and $\kappa_b := \sum_{i=1}^k \kappa_i/p$. Since κ_a is an operation making nothing, we have

$$
\Delta_4(\kappa_a, Y, \rho) = 0.
$$

Then, due to (7.32), the equality in (7.28) implies that

$$
\Delta_4(\kappa_b, Y, \rho) = \Delta_1(Y, \rho).
$$

Since $\mathcal{H}$ and $\mathcal{H}'$ are orthogonal to each other, Exercise 7.20 guarantees that

$$\Delta_4(\boldsymbol{\kappa}, Y, \rho) = (1-p)\Delta_4(\kappa_a, Y, \rho) + p\Delta_4(\kappa_b, Y, \rho) = p\Delta_1(Y, \rho).$$

Therefore, we have

$$\Delta_2(\boldsymbol{M}, \rho)\Delta_4(\boldsymbol{\kappa}, Y, \rho) = \Delta_1(X, \rho)\Delta_1(Y, \rho).$$

Hence, the equality in (7.28) guarantees the equality in (7.31).

Exercise 7.22 The relation (A.17) implies that $\|\sqrt{M_x}\sqrt{N_y}\| = \|\sqrt{\sqrt{N_y}M_x\sqrt{N_y}}\| = \|\sqrt{|\langle u_x|v_y\rangle|^2|v_y\rangle\langle v_y|}\| = \||\langle u_x|v_y\rangle||v_y\rangle\langle v_y|\| = |\langle u_x|v_y\rangle|$.

Exercise 7.23 From the definition of $\boldsymbol{M}^{(n),\delta,l}$, we have

$$\begin{aligned}\boldsymbol{M}^n\{|\theta - \hat{\theta}^n| \ge \epsilon + \frac{\delta_n(l_n+1)}{2}\} \subset & \boldsymbol{M}^{(n),\delta_n,l_n}\{|\theta - x^n_{\delta_n}| \ge \epsilon\} \\ \subset & \boldsymbol{M}^n\{|\theta - \hat{\theta}^n| \ge \epsilon - \frac{\delta_n(l_n+1)}{2}\}.\end{aligned}$$

Since $\frac{\delta_n(l_n+1)}{2} \to 0$, we obtain

$$\lim_{\delta\to 0}\beta(\boldsymbol{M}, \theta, \epsilon+\delta) \ge \beta(\{(\boldsymbol{M}^{(n),\delta_n,l_n}, x^n_{\delta_n})\}, \theta, \epsilon) \ge \lim_{\delta\to 0}\beta(\boldsymbol{M}, \theta, \epsilon-\delta).$$

As $\beta(\boldsymbol{M}, \theta, \epsilon)$ is continuous with respect to ϵ, we obtain the desired argument.

References

1. J. von Neumann, *Mathematical Foundations of Quantum Mechanics* (Princeton University Press, Princeton, 1955). (Originally appeared in German in 1932)
2. M.A. Naĭmark, *Comptes Rendus (Doklady) de l'Acadenie des Sience de l'URSS*, **41**, 9, 359 (1943)
3. M. Ozawa, Quantum measuring processes of continuous observables. J. Math. Phys. **25**, 79 (1984)
4. M. Ozawa, An operational approach to quantum state reduction. Ann. Phys. **259**, 121–137 (1997)
5. H.P. Robertson, The uncertainty principle. Phys. Rev. **34**, 163 (1929)
6. W. Heisenberg, Über den anschaulichen Inhalt der quantentheoretischen Kinematik und Mechanik. Z. Phys. **43**, 172–198 (1927)
7. M. Ozawa, Universally valid reformulation of the Heisenberg uncertainty principle on noise and disturbance in measurement. Phys. Rev. A **67**, 042105 (2003)
8. M. Ozawa, Physical content of Heisenberg's uncertainty relation: limitation and reformulation. Phys. Lett. A **318**, 21–29 (2003)
9. M. Ozawa, Uncertainty relations for noise and disturbance in generalized quantum measurements. Ann. Phys. **311**, 350–416 (2004)
10. M. Ozawa, Uncertainty principle for quantum instruments and computing. Int. J. Quant. Inf. **1**, 569–588 (2003)

11. H. Nagaoka, A generalization of the simultaneous diagonalization of Hermitian matrices and its relation to quantum estimation theory. Trans. Jpn. Soc. Ind. Appl. Math. **1**, 43–56 (1991) (Originally in Japanese; also appeared as Chap. 11 of *Asymptotic Theory of Quantum Statistical Inference*, ed. by M. Hayashi)
12. H. Maassen, J. Uffink, Generalized entropic uncertainty relations. Phys. Rev. Lett. **60**, 1103 (1988)
13. P.J. Coles, R. Colbeck, L. Yu, M. Zwolak, Uncertainty relations from simple entropic properties. Phys. Rev. Lett. **108**, 210405 (2012)
14. M. Müller-Lennert, F. Dupuis, O. Szehr, S. Fehr, M. Tomamichel, On quantum Renyi entropies: a new generalization and some properties. J. Math. Phys. **54**, 122203 (2013)
15. M. Tomamichel, M. Berta, M. Hayashi, Relating different quantum generalizations of the conditional Rényi entropy. J. Math. Phys. **55**, 082206 (2014)
16. J.M. Renes, J.-C. Boileau, Conjectured strong complementary information tradeoff. Phys. Rev. Lett. **103**, 020402 (2009)
17. M. Berta, M. Christandl, R. Colbeck, J.M. Renes, R. Renner, The uncertainty principle in the presence of quantum memory. Nat. Phys. **6**, 659–662 (2010)
18. M. Hayashi, K. Matsumoto, Simple construction of quantum universal variable-length source coding. Quant. Inf. Comput. **2**, Special Issue, 519–529 (2002)
19. A.S. Holevo, *Probabilistic and Statistical Aspects of Quantum Theory* (North-Holland, Amsterdam, 1982); originally published in Russian (1980)
20. E.B. Davies, J.T. Lewis, An operational approach to quantum probability. Commun. Math. Phys. **17**, 239 (1970)
21. M. Ozawa, Measurements of nondegenerate discrete observables. Phys. Rev. A **63**, 062101 (2000)
22. M. Ozawa, Operations, disturbance, and simultaneous measurability. Phys. Rev. A **62**, 032109 (2001)
23. M. Hayashi, F. Sakaguchi, Subnormal operators regarded as generalized observables and compound-system-type normal extension related to su(1,1). J. Phys. A Math. Gen. **33**, 7793–7820 (2000)
24. A.S. Holevo, *Statistical Structure of Quantum Theory*, vol. 67, Lecture Notes in Physics (Springer, Berlin, 2001)
25. P. Busch, M. Grabowski, P.J. Lahti, *Operational Quantum Physics*, vol. 31, Lecture Notes in Physics (Springer, Berlin, 1997)
26. M. Ozawa, Operational characterization of simultaneous measurements in quantum mechanics. Phys. Lett. A **275**, 5–11 (2000)
27. E. Arthurs, J.L. Kelly Jr., On the simultaneous measurement of a pair of conjugate observables. Bell Syst. Tech. **44**, 725–729 (1965)
28. E. Arthurs, M.S. Goodman, Quantum correlations: a generalized Heisenberg uncertainty relation. Phys. Rev. Lett. **60**, 2447–2449 (1988)
29. S. Ishikawa, Uncertainty relations in simultaneous measurements for arbitrary observables. Rep. Math. Phys. **29**, 257–273 (1991)
30. M. Ozawa, Quantum limits of measurements and uncertainty principle, in *Quantum Aspects of Optical Communications*, Lecture Notes in Physics, vol. 378, eds. by C. Bendjaballah, O. Hirota, S. Reynaud. (Springer, Berlin, 1991), pp. 3–17
31. M. Hayashi, Simultaneous measurements of non-commuting physical quantities. RIMS koukyuroku Kyoto University, No. **1099**, 96–118 (1999). (in Japanese)
32. R.L. Frank, E.H. Lieb, Monotonicity of a relative Renyi entropy. J. Math. Phys. **54**, 122201 (2013)
33. C.H. Bennett, A.W. Harrow, S. Lloyd, Universal quantum data compression via nondestructive tomography. Phys. Rev. A **73**, 032336 (2006)

Chapter 8
Entanglement and Locality Restrictions

Abstract Quantum mechanics violates daily intuition not only because the measured outcome can only be predicted probabilistically but also because of a quantum-specific correlation called entanglement. It is believed that this type of correlation does not exist in macroscopic objects. Entanglement can be used to produce nonlocal phenomena. States possessing such correlations are called entangled states (or states that possess entanglement). A state on a bipartite system is called called a maximally entangled state or an EPR state when it has the highest degree of entanglement among these states. Historically, the idea of a nonlocal effect due to entanglement was pointed out by Einstein, Podolsky, and Rosen; hence, the name EPR state. In order to transport a quantum state over a long distance, we have to retain its coherence during its transmission. However, it is often very difficult because the transmitted system can be easily correlated with the environment system. If the sender and receiver share an entangled state, the sender can transport his/her quantum state to the receiver without transmitting it, as explained in Chap. 9. This protocol is called quantum teleportation and clearly explains the effect of entanglement in quantum systems. Many other effects of entanglement have also been examined, some of which are given in Chap. 9. However, it is difficult to take advantage of entanglement if the shared state is insufficiently entangled. Therefore, we investigate how much of a maximally entangled state can be extracted from a state with a partially entangled state. Of course, if we allow quantum operations between two systems, we can always produce maximally entangled states. Therefore, we examine cases where locality conditions are imposed to our possible operations.

8.1 Entanglement and Local Quantum Operations

As explained in Chap. 9, when there are two players, they can perform several magical protocols using a maximally entangled state defined in Sect. 1.4. However, all entangled states are not necessarily a maximally entangled state. To perform such magical protocols based on a partially entangled state, we need to convert the partially entangled state. If we are allowed to any quantum operation, we can generate a maximally entangled state. Hence, it is usual to impose a locality condition for our

M. Hayashi, *Quantum Information Theory*, Graduate Texts in Physics,
DOI 10.1007/978-3-662-49725-8_8

operations. The most usual condition is the condition for local quantum operations and classical communications (LOCC). That is, we are allowed to perform local quantum operation on each player and classical communication between two players. Now, we consider what operations are possible on an entangled state under this condition.

As stated in Sect. 1.2, a pure state on $\mathcal{H}_A$ can be represented by an element u on $\mathcal{H}_A$ with the norm $\|u\|$ equal to 1. A pure entangled state in the composite system is represented by an element x on $\mathcal{H}_A \otimes \mathcal{H}_B$. Using the basis $u_1, \ldots, u_d$ for $\mathcal{H}_A$ and the basis $v_1, \ldots, v_{d'}$ for $\mathcal{H}_B$, x can be written as $x = \sum_{j,i} x^{i,j} |u_i\rangle \otimes |v_j\rangle$. Let us define the matrix form of x as the linear map X_x from $\mathcal{H}_B$ to $\mathcal{H}_A$ with respect to x by $|x\rangle = |X_x\rangle$. Then,

$$X_x = \sum_{j,i} x^{i,j} |u_i\rangle\langle v_j|. \tag{8.1}$$

Therefore, the correspondence $x \mapsto X_x$ gives a one-to-one relationship between these matrices and the elements of $\mathcal{H}_A \otimes \mathcal{H}_B$ under fixed bases on $\mathcal{H}_A$ and $\mathcal{H}_B$. From (1.23), any tensor product $u \otimes v$ satisfies

$$\langle u \otimes v | x \rangle = \langle u \otimes v | X_x \rangle = \mathrm{Tr}\, |\bar{v}\rangle\langle u | X_x = \langle u | X_x | \bar{v} \rangle.$$

Now, let X'_x be the same as X_x but defined in a different basis v'_j for B, and define $U \stackrel{\text{def}}{=} \sum_{j,k,l} \langle v'_k | v_j \rangle \langle v'_k | v_l \rangle |v_j\rangle\langle v_l|$. Since $|v_j\rangle = \sum_k \langle v'_k | v_j \rangle |v'_k\rangle$ and $\langle v'_k| = \sum_l \langle v'_k | v_l \rangle \langle v_l|$, we have

$$X'_x = \sum_{i,j,k} x^{i,j} \langle v'_k | v_j \rangle |u_i\rangle\langle v'_k| = \sum_{i,j,k,l} x^{i,j} \langle v'_k | v_j \rangle \langle v'_k | v_l \rangle |u_i\rangle\langle v_l| = X_x U.$$

That is, the definition of X_x depends on the orthonormal basis of $\mathcal{H}_B$.

Further, we have

$$\begin{aligned} \rho_x \stackrel{\text{def}}{=} \mathrm{Tr}_B\, |x\rangle\langle x| &= \sum_j \left(\sum_i x^{i,j} |u_i\rangle \right) \left(\sum_{i'} \overline{x^{i',j}} \langle u_{i'}| \right) \\ &= \sum_{i',i} \left(\sum_j \overline{x^{i',j}} x^{i,j} \right) |u_i\rangle\langle u_{i'}| \qquad (8.2) \\ &= X_x X_x^*. \qquad (8.3) \end{aligned}$$

Now, let us denote the nonzero eigenvalues of ρ_x in (8.3) by $\lambda_1, \ldots, \lambda_l$. Then, we can apply the arguments given in Sect. A.2 to the matrix $x^{i,j}$. Choosing sets of orthogonal vectors of length 1 as $u'_1, \ldots, u'_l$ and $v'_1, \ldots, v'_l$, we obtain $x = \sum_{i=1}^{l} \sqrt{\lambda_i} |u'_i\rangle \otimes |v'_i\rangle$. The right-hand side (RHS) of this equation is often called the **Schmidt decomposition**, and $\sqrt{\lambda_i}$ is called the **Schmidt coefficient**. Of course,

$$\operatorname{Tr}_B |x\rangle\langle x| = \sum_i \lambda_i |u_i'\rangle\langle u_i'|, \quad \operatorname{Tr}_A |x\rangle\langle x| = \sum_i \lambda_i |v_i'\rangle\langle v_i'|.$$

Hence, both have the same eigenvalues and entropies. However, the above is true only if the state on the composite system is a pure state. The number of nonzero Schmidt coefficients $\sqrt{\lambda_i}$ is called the **Schmidt rank** and is equal to the ranks of both $\operatorname{Tr}_B |x\rangle\langle x|$ and X_x.

Conversely, for a general state ρ^A on $\mathcal{H}_A$, a pure state $|u\rangle\langle u|$ on $\mathcal{H}_A \otimes \mathcal{H}_R$ satisfying $\rho^A = \operatorname{Tr}_R |u\rangle\langle u|$ is called the **purification** of ρ^A. The quantum system $\mathcal{H}_R$ used for the purification is called the **reference** and is denoted R.

Lemma 8.1 *For two pure states $|x\rangle\langle x|$ and $|y\rangle\langle y|$ on the composite system $\mathcal{H}_A \otimes \mathcal{H}_R$, the following two conditions are equivalent.*

① *The Schmidt coefficients of $|x\rangle\langle x|$ and $|y\rangle\langle y|$ coincide, i.e.,*

$$x = \sum_{i=1}^{l} \sqrt{\lambda_i} |u_i\rangle \otimes |v_i\rangle, \quad y = \sum_{i=1}^{l} \sqrt{\lambda_i} |u_i'\rangle \otimes |v_i'\rangle.$$

② *There exist unitary matrices U^A, U^R in A and R such that*

$$x = \left(U^A \otimes U^R\right) y. \tag{8.4}$$

Furthermore, if

$$\operatorname{Tr}_R |x\rangle\langle x| = \operatorname{Tr}_R |y\rangle\langle y|, \tag{8.5}$$

then

$$x = \left(I \otimes U^R\right) y \tag{8.6}$$

for a suitable unitary matrix U^R on R. Therefore, the purification of the mixed state ρ^A on $\mathcal{H}_A$ can be transferred by the operation of the unitary matrix on R. When the states $\operatorname{Tr}_A |x\rangle\langle x|$ and $\operatorname{Tr}_A |y\rangle\langle y|$ are not full rank, U^R is need to be chosen as an partial isometry.

Proof ②⇒① by inspection. If unitary matrices U^A and U^R on $\mathcal{H}_A$ and $\mathcal{H}_R$, respectively, are chosen such that $U^A(u_i) = u_i'$ and $U^R(v_i) = v_i'$, then (8.4) is satisfied, and hence ①⇒②. From (8.5), $X_x X_x^* = X_y X_y^*$. From (A.9), choosing appropriate unitary matrices U_x and U_y, we have $X_x = \sqrt{X_x X_x^*} U_x$ and $X_y = \sqrt{X_x X_x^*} U_y$. Therefore, $X_x = X_y U_y^* U_x$. Then, (8.6) can be obtained from (1.22). ■

Therefore, pure entangled states can be classified according to their Schmidt coefficients. In particular, when all the Schmidt coefficients are equal to $\sqrt{\frac{1}{L}}$, the state is called a **maximally entangled state** of size L. Any maximally entangled state of size

L may be transformed from a maximally entangled state $|\Phi_L\rangle\langle\Phi_L|$ by local operations. Hence, we can examine the properties of maximally entangled states of size L by treating a typical maximally entangled state $|\Phi_L\rangle\langle\Phi_L|$. A maximally entangled state is separated from separable states as follows.

$$\max_{\sigma \in S} \langle \Phi_L|\sigma|\Phi_L\rangle = \frac{1}{L}, \tag{8.7}$$

where S is the set of separable states.

Since $\langle\Phi_L|u \otimes v\rangle = \langle X_{\Phi_L}\bar{v}|v\rangle$ and $\|X_{\Phi_L}\bar{v}\|^2 = \langle v X^*_{\Phi_L} X_{\Phi_L}\bar{v}\rangle = \langle v \frac{1}{L}\bar{v}\rangle = \frac{1}{L}$, we have

$$\langle\Phi_L|u \otimes v\rangle\langle u \otimes v|\Phi_L\rangle \le \|X_{\Phi_L}\bar{v}\|^2\|u\|^2 \le \frac{1}{L}.$$

Since any separable state σ is written as a mixture of separable pure states, we have $\langle\Phi_L|\sigma|\Phi_L\rangle \le \frac{1}{L}$. When u is equal to $\sqrt{L}X_{\Phi_L}\bar{v}$, $\langle\Phi_L|u \otimes v\rangle = \frac{1}{\sqrt{L}}$, which implies (8.7).

Next, we discuss state operations consisting of local operations (LO) and classical communications (CC). This can be classified into three classes as Fig. 8.1: (i) only classical communications from A to B or from B to A are allowed (This class is called **one-way LOCC**. It is denoted by $\rightarrow$ when classical communications from A to B is allowed, and is denoted by $\leftarrow$ when classical communications from B to A is allowed.); (ii) classical communications from A to B and B to A are allowed (This class is called **two-way LOCC** and is denoted by $\leftrightarrow$); and (iii) no classical communications are allowed (only local quantum operations are allowed) (This class is denoted by $\emptyset$).[1] In terms of the Choi–Kraus representation of the TP-CP map given in ⑥ of Theorem 5.1, the state evolutions may be written

$$\kappa(\rho) = \sum_i \left(E_{A,i} \otimes E_{B,i}\right) \rho \left(E^*_{A,i} \otimes E^*_{B,i}\right). \tag{8.8}$$

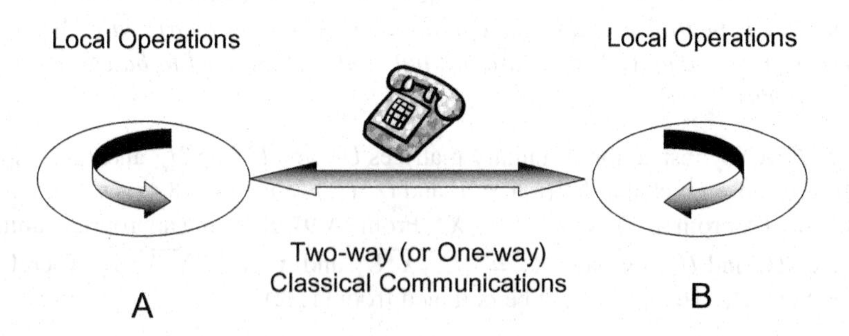

Fig. 8.1 Two-way LOCC (or one-way LOCC)

[1] When the measurement is employed in the class $\emptyset$, it is required that Alice and Bob obtain the same outcome.

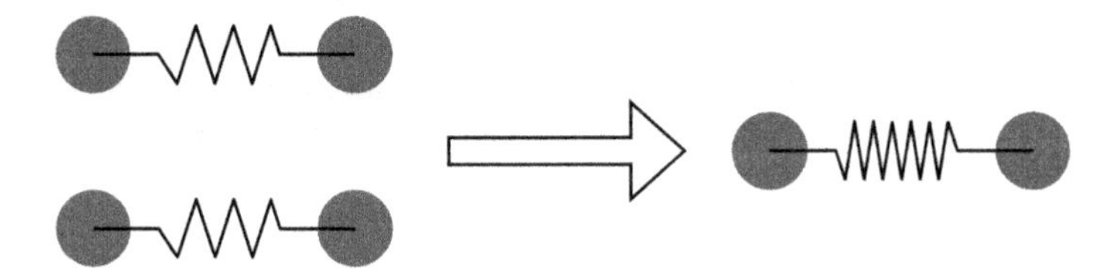

Fig. 8.2 Two partially entangled states (*left*) and one completely entangled state (*right*)

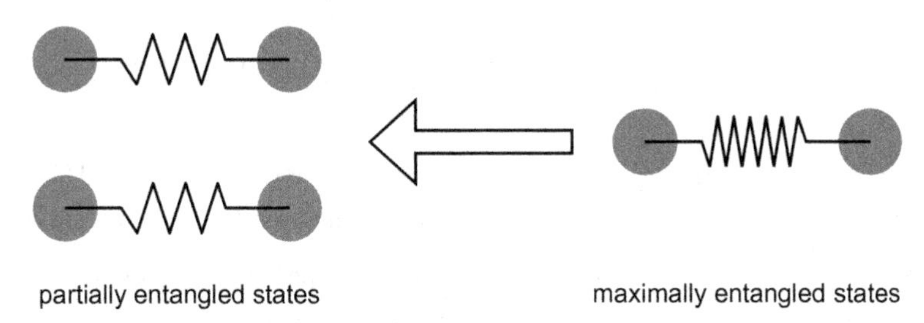

Fig. 8.3 Entanglement dilution

If a TP-CP map can be written in the form (8.8), it is called a **separable TP-CP map** (**S-TP-CP map**) [1]. A TP-CP map κ is an S-TP-CP map if and only if the matrix $K(\kappa)$ defined in (5.4) can be regarded as a separable state in the composite system $(\mathcal{H}_A \otimes \mathcal{H}_{A'}) \otimes (\mathcal{H}_B \otimes \mathcal{H}_{B'})$, where we assume that the map $E_{A,i}$ $(E_{B,i})$ is a map from $\mathcal{H}_A$ $(\mathcal{H}_B)$ to $\mathcal{H}_{A'}$ $(\mathcal{H}_{B'})$. Since the set of separable TP-CP maps forms a class of localized operations, we denote it by S. There are two typical types of LOCC operation. One is distillation, which converts a partially entangled state to a maximally entangled state (Fig. 8.2). The other is entanglement dilution, which converts a maximally entangled state to a given partially entangled state (Fig. 8.3). The following theorem discusses the possibility of entanglement dilution.

Theorem 8.1 (Lo and Popescu [2]) *Let the initial state of a composite system $\mathcal{H}_A \otimes \mathcal{H}_B$ be a known pure state $|x\rangle\langle x|$. LOCC state operations consisting of two-way classical communications can be realized by state operations consisting of one-way classical communications from A to B.*

Proof For the proof of this theorem, it is sufficient to show that any final state realized by operation (1) can be realized by operation (2), where operations (1) and (2) are given as follows. In operation (1), we (i) perform a measurement in system B, (ii) transmit B's measured outcome to system A, and (iii) finally apply state evolutions to each system. In operation (2), we (i) perform a measurement in system A, (ii) transmit A's measured outcome to system B, and (iii) finally apply state evolutions to each system.

From Theorem 7.2, any operation with class (1) can be described by the the state reduction

$$I_A \otimes \sqrt{M_i^B}|x\rangle\langle x|I_A \otimes \sqrt{M_i^B}, \tag{8.9}$$

and local TP-CP maps on A and B depend on the measurement datum i. Hence, it is sufficient to prove that the state reduction (8.9) can be realized by a state reduction by a measurement on A and local TP-CP maps on A and B depending on the measurement datum i.

Using (1.22) and (A.8), we have

$$\begin{aligned} I_A \otimes \sqrt{M_i^B}|X_x\rangle &= |X_x \sqrt{M_i^B}^T\rangle = |U_i \overline{\sqrt{M_i^B}}(X_x)^* U\rangle \\ &= |U_i \overline{\sqrt{M_i^B}} V^* X_x V^* U_i\rangle = U_i V_i^* (V_i \overline{\sqrt{M_i^B}} V_i^*) \otimes (V_i^* U_i)^T |X_x\rangle, \end{aligned}$$

where U_i and V_i are unitary matrices satisfying $X_x \sqrt{M_i^B}^T = U_i |X_x \sqrt{M_i^B}^T|$ and $X_x = V|X_x|$. This equation implies that the state reduction (8.9) is realized by the state reduction on A by the instrument $\{V_i \overline{\sqrt{M_i^B}} V_i^*\}_i$ and the local unitaries $U_i V_i^*$ and $(V_i^* U_i)^T$ depending on the datum i on A and B, respectively. ∎

Exercise

8.1 Let x be the purification of state ρ on $\mathcal{H}_A$. Show that $H(\rho) = H(\mathrm{Tr}_A |x\rangle\langle x|)$.

8.2 Fidelity and Entanglement

We can characterize the fidelity of two states on $\mathcal{H}_A$ using the purification of mixed states in the following way.

Lemma 8.2 (Uhlmann [3]) *Consider two mixed states ρ_1 and ρ_2 on $\mathcal{H}_A$. Let $|u_1\rangle\langle u_1|$ and $|u_2\rangle\langle u_2|$ be their purifications, respectively. Then,*

$$F(\rho_1, \rho_2) \left(\stackrel{\text{def}}{=} \mathrm{Tr}\, |\sqrt{\rho_1}\sqrt{\rho_2}| \right) = \max_{u_1, u_2} |\langle u_1 | u_2 \rangle|, \tag{8.10}$$

where the max *on the RHS is with respect to the purifications of ρ_1 and ρ_2.*

Proof First, we choose the matrix X_{u_i} according to (8.1) in the previous section as a matrix from the reference system $\mathcal{H}_R$ to the system $\mathcal{H}_A$. (Note that the map $u \mapsto X_u$ depends upon the basis of $\mathcal{H}_R$). Since $\rho_i = X_{u_i} X_{u_i}^*$, from (A.9) we obtain $X_{u_i} = \sqrt{\rho_i} U_i$ choosing an appropriate unitary matrix U_i on $\mathcal{H}_R$. From (1.23) and (A.18) we have

$$\begin{aligned} &|\langle u_1 | u_2 \rangle| = |\mathrm{Tr}\, X_{u_2} X_{u_1}^*| = |\mathrm{Tr}\, \sqrt{\rho_2} U_2 U_1^* \sqrt{\rho_1}| \\ &= |\mathrm{Tr}\, \sqrt{\rho_1}\sqrt{\rho_2} U_2 U_1^*| \le \mathrm{Tr}\, |\sqrt{\rho_1}\sqrt{\rho_2}|, \end{aligned} \tag{8.11}$$

which proves the $\geq$ part of (8.10). The equality follows from the existence of $U_2 U_1^*$ satisfying the equality of (8.11). ∎

From (8.6), for an arbitrary purification x of ρ_1, there exists a purification y of ρ_2 such that

$$F(\rho_1, \rho_2) = |\langle x|y\rangle| = \langle x|y\rangle, \tag{8.12}$$

where the second equation follows from choosing suitable phase factor $e^{i\theta}$ in y. Vectors $v_1, \ldots v_n$ satisfying $\sum_{i=1}^{n} |v_i\rangle\langle v_i| = \rho$ are called a **decomposition** of ρ. Using this fact, we obtain the following corollary regarding decompositions.

Corollary 8.1 *Let ρ_1 and ρ_2 be two mixed states on $\mathcal{H}_A$. For an arbitrary decomposition $u_1, \ldots, u_l$ of ρ_1, there exists a decomposition $v_1, \ldots v_l$ of ρ_2 such that $F(\rho_1, \rho_2) = \sum_{i=1}^{l} \langle u_i|v_i\rangle$.*

Proof Let $w_1, \ldots, w_l$ be an orthonormal basis for the space $\mathcal{H}_R$. Let $x = \sum_{i=1}^{l} u_i \otimes w_i$. Choose a purification $y \in \mathcal{H}_A \otimes \mathcal{H}_R$ of ρ_2 satisfying (8.12). Since $w_1, \ldots, w_l$ is an orthonormal basis, there exist appropriate elements $v_1, \ldots, v_l$ of $\mathcal{H}_A$ such that $y = \sum_{i=1}^{l} v_i \otimes w_i$. Therefore, $|\langle x|y\rangle| = \sum_{i=1}^{l} \langle u_i|v_i\rangle$. ∎

Corollary 8.2 (Uhlmann [3]) *Let $\rho = \sum_i p_i \rho_i$ for the states ρ_i and σ, and the probability p_i. The following concavity holds:*

$$F^2(\rho, \sigma) \geq \sum_i p_i F^2(\rho_i, \sigma). \tag{8.13}$$

If σ is a pure state, then

$$F^2(\rho, |u\rangle\langle u|) = \langle u|\rho|u\rangle, \tag{8.14}$$

and the equality in (8.13) holds.

Proof The validity of (8.14) follows from the fact that $F(\rho, |u\rangle\langle u|) = \mathrm{Tr} \sqrt{|u\rangle\langle u|\rho|u\rangle\langle u|}$. Let y be the purification of σ, and x_i be the purification of ρ_i satisfying $\langle x_i|y\rangle = F(\rho_i, \sigma)$. Then,

$$\sum_i p_i F^2(\rho_i, \sigma) = \sum_i p_i \langle y|x_i\rangle\langle x_i|y\rangle = F^2\left(\sum_i p_i |x_i\rangle\langle x_i|, |y\rangle\langle y|\right) \leq F^2(\rho, \sigma)$$

completes the proof. The last inequality can be proved by the relation for the partial trace as follows. Two densities ρ_1 and ρ_2 on the composite system $\mathcal{H}_A \otimes \mathcal{H}_B$ satisfy

$$F(\rho_1, \rho_2) = \max_{u_1, u_2} |\langle u_1|u_2\rangle| \leq \max_{u_1', u_2'} |\langle u_1'|u_2'\rangle| = F(\mathrm{Tr}_B\, \rho_1, \mathrm{Tr}_B\, \rho_2),$$

where u_1, u_2 are purifications of ρ_1, ρ_2 and u_1', u_2' are purifications of $\mathrm{Tr}_B\, \rho_1$, $\mathrm{Tr}_B\, \rho_2$. ∎

By applying the Jensen's inequality to the function $x \mapsto -\sqrt{x}$, Corollary 8.2 yields that

$$F(\rho, \sigma) \geq \sum_i p_i F(\rho_i, \sigma). \tag{8.15}$$

A stronger statement (**strong concavity of the fidelity**) than (8.15) holds regarding the concavity of $F(\rho, \sigma)$.

Corollary 8.3 (Nielsen and Chuang [4]) *For states ρ_i and σ_i and probabilities $\{p_i\}$ and $\{q_i\}$, the following concavity property holds:*

$$F\left(\sum_i p_i \rho_i, \sum_i q_i \sigma_i\right) \geq \sum_i \sqrt{p_i q_i} F(\rho_i, \sigma_i). \tag{8.16}$$

Proof Let x_i and y_i be the purifications of ρ_i and σ_i, respectively, satisfying $F(\rho_i, \sigma_i) = \langle x_i | y_i \rangle$. Consider the space spanned by the orthonormal basis $\{u_i\}$. The purifications of $\sum_i p_i \rho_i$ and $\sum_i q_i \sigma_i$ are then $x \stackrel{\text{def}}{=} \sum_i \sqrt{p_i} x_i \otimes u_i$ and $y \stackrel{\text{def}}{=} \sum_i \sqrt{q_i} y_i \otimes u_i$. Therefore,

$$F\left(\sum_i p_i \rho_i, \sum_i q_i \sigma_i\right) \geq |\langle x | y \rangle| = \sum_i \sqrt{p_i q_i} \langle x_i | y_i \rangle,$$

completing the proof. ∎

Monotonicity is the subject of the following corollary.

Corollary 8.4 *For an arbitrary TP-CP map κ from $\mathcal{H}_A$ to $\mathcal{H}_{A'}$,*

$$F(\rho_1, \rho_2) \leq F(\kappa(\rho_1), \kappa(\rho_2)). \tag{8.17}$$

This corollary is called the **monotonicity**. Further, the **monotonicity** (5.49), i.e., $b(\rho, \sigma) \geq b(\kappa(\rho), \kappa(\sigma))$, can be derived from this.

Proof Choose the Stinespring representation $(\mathcal{H}_C, |u\rangle\langle u|, U)$ of κ, i.e., choose $(\mathcal{H}_C, |u\rangle\langle u|, U)$, such that it satisfies $\kappa(\rho) = \mathrm{Tr}_{A,C} U(\rho \otimes |u\rangle\langle u|)U^*$. Let two pure states u_1 and u_2 be purifications of ρ_1 and ρ_2 on $\mathcal{H}_A \otimes \mathcal{H}_R$ maximizing the RHS of (8.10). Since

$$\kappa(\rho_i) = \mathrm{Tr}_{A,C,R}(U \otimes I_R)(|u_i\rangle\langle u_i| \otimes |u\rangle\langle u|)(U \otimes I_R)^*,$$

$(U \otimes I_R)(u_i \otimes u)$ is the purification of $\kappa(\rho_i)$; therefore, it satisfies $|\langle u_1 \otimes u | u_2 \otimes u \rangle| = |\langle u_1 | u_2 \rangle| = F(\rho_1, \rho_2)$. Then, (8.17) can be obtained from (8.10). ∎

Let us next examine a quantity called the **entanglement fidelity**, which expresses how much entanglement is preserved in a TP-CP map κ from $\mathcal{H}_A$ to $\mathcal{H}_A$ [5]. Let R

be the reference system with respect to the CP map κ and the mixed state ρ on $\mathcal{H}_A$. The entanglement fidelity is then defined as

$$F_e(\rho, \kappa) \stackrel{\text{def}}{=} \sqrt{\langle x|\kappa \otimes \iota_R(|x\rangle\langle x|)|x\rangle}, \tag{8.18}$$

where x is the purification of ρ. At first glance, this definition seems to depend on the choice of the purification x. Using the Choi–Kraus representation $\{E_j\}_j$ of κ, we can show that[Exe. 8.4] [6]

$$F_e^2(\rho, \kappa) = \langle x|\kappa \otimes \iota_R(|x\rangle\langle x|)|x\rangle = \sum_j |\operatorname{Tr} E_j\rho|^2. \tag{8.19}$$

Hence, $F_e(\rho, \kappa)$ is independent of the purification x and of the Choi–Kraus representation $\{E_i\}_i$. From the monotonicity of the fidelity, we have

$$F_e(\rho, \kappa) \leq F(\rho, \kappa(\rho)). \tag{8.20}$$

The equality holds if ρ is a pure state. The entanglement fidelity satisfies the following properties,[2] which will be applied in later sections.

① Let κ' be a TP-CP map from $\mathcal{H}_A$ to $\mathcal{H}_B$, and κ be a TP-CP map from $\mathcal{H}_B$ to $\mathcal{H}_A$. When $\dim \mathcal{H}_A \leq \dim \mathcal{H}_B$, given a state ρ on $\mathcal{H}_A$, there exists an isometry U from $\mathcal{H}_A$ to $\mathcal{H}_B$ such that[Exe. 8.5] [6]

$$F_e^2(\rho, \kappa \circ \kappa') \leq F_e(\rho, \kappa \circ \kappa_U). \tag{8.21}$$

When $\dim \mathcal{H}_A > \dim \mathcal{H}_B$, given a state ρ on $\mathcal{H}_A$, there exist a subspace $\mathcal{H}_C \subset \mathcal{H}_A$ with the dimension $\dim \mathcal{H}_B$ and a unitary U from $\mathcal{H}_C$ to $\mathcal{H}_B$ such that

$$F_e^2(\rho, \kappa \circ \kappa') \leq (\operatorname{Tr} P_C\rho) F_e\left(\frac{P_C\rho P_C}{\operatorname{Tr} P_C\rho}, \kappa \circ \kappa_U\right), \tag{8.22}$$

where P_C is the projection to the subspace $\mathcal{H}_C$ and the minimum is taken with respect to the projection with the rank $\dim \mathcal{H}_B$.

② If $\rho = \sum_i p_i\rho_i$, we have [6][Exe. 8.6]

$$F_e^2(\rho, \kappa) \leq \sum_i p_i F_e^2(\rho_i, \kappa). \tag{8.23}$$

In particular, when all the ρ_i are pure states, the following holds [5]:

$$F_e^2(\rho, \kappa) \leq \sum_i p_i F^2(\rho_i, \kappa(\rho_i)). \tag{8.24}$$

③ Let $\mathcal{H}_B$ be a subspace of $\mathcal{H}_A$. Given a real number a such that $1 > a > 0$, there exists a subspace $\mathcal{H}_C$ of $\mathcal{H}_B$ with a dimension $\lfloor (1-a) \dim \mathcal{H}_B \rfloor$ such that [6, 7]

[2] A large part of the discussion relating to entanglement fidelity and information quantities relating to entanglement (to be discussed in later sections) was first done by Schumacher [5].

$$\max_{x\in\mathcal{H}_C^1}\left\{1-F^2(x,\kappa(x))\right\}\le\frac{1-F_e^2(\rho_{\rm mix}^B,\kappa)}{a}. \tag{8.25}$$

④ Let the support of ρ be included in the subspace $\mathcal{H}_B$ of $\mathcal{H}_A$. The following then holds [6]:

$$\frac{2}{3}\left(1-F_e^2(\rho,\kappa)\right)\le\max_{x\in\mathcal{H}_B}\left\{1-F^2(x,\kappa(x))\right\}. \tag{8.26}$$

The completely mixed-state $\rho_{\rm mix}$ on $\mathcal{H}$ satisfies

$$\frac{d}{d+1}\left(1-F_e^2(\rho_{\rm mix},\kappa)\right)={\rm E}_{\mu,x}\left[1-F^2(x,\kappa(x))\right], \tag{8.27}$$

where ${\rm E}_{\mu,x}$ denotes the expectation with respect to the pure state x under the invariant distribution μ on $\mathcal{H}^1$ and d is the dimension of $\mathcal{H}$.

The property ① evaluates the entanglement fidelity when we replace the recovery CP map by a suitable isometry map. Other properties of the entanglement fidelity can be used for evaluating the fidelities between the input and output states for a given channel κ. When we focus on the worst fidelity, (8.25) and (8.26) are useful. When we focus on the average of the fidelity, (8.24) and (8.27) are useful. In fact, the average $\sum_i p_i F^2(\rho_i,\kappa(\rho_i))$ depends on the choice of the decomposition $\rho=\sum_i p_i\rho_i$, however, the entanglement fidelity does not depend on it because the entanglement fidelity reflects how the map κ preserves the coherence of the input states.

From the definition, for a general CP map κ_i and a positive real number f_i, we have

$$\sum_i f_i F_e^2(\rho,\kappa_i)=F_e^2\left(\rho,\sum_i f_i\kappa_i\right). \tag{8.28}$$

Therefore, we can define the entanglement fidelity $F_e(\rho,\boldsymbol{\kappa})$ as

$$F_e^2(\rho,\boldsymbol{\kappa})\stackrel{\rm def}{=}\sum_\omega F_e^2(\rho,\kappa_\omega)\left(=F_e^2\left(\rho,\sum_\omega\kappa_\omega\right)\right) \tag{8.29}$$

for an instrument $\boldsymbol{\kappa}=\{\kappa_\omega\}$ with an input and output $\mathcal{H}_A$ and a state ρ on $\mathcal{H}_A$. Since $\varepsilon(\rho,\boldsymbol{\kappa})\le 1-F_e^2(\rho,\boldsymbol{\kappa})$ from (8.24), combining these properties gives (7.60) and (7.61).

In fact, the purification is useful only for treating a single state. In order to analyze a mixed-state ρ on $\mathcal{H}_A$, we often focus on the **probabilistic decomposition** of ρ; this is defined as the set $\{(p_i,\rho_i)\}$ satisfying

$$\rho=\sum_i p_i\rho_i,$$

where p_i is a probability distribution and ρ_i is a state on $\mathcal{H}_A$. In a quantum system, the probabilistic decomposition is not unique for a given mixed state ρ. Now, we let $|X\rangle$ be a purification of ρ with the reference system $\mathcal{H}_R$. (Here, we choose the reference $\mathcal{H}_R$ whose dimension is equal to the rank of ρ.) We choose a suitable coordinate of $\mathcal{H}_R$ so that the reduced density $\operatorname{Tr}_A |X\rangle\langle X|$ is ρ. When we perform a POVM $\boldsymbol{M} = \{M_i\}$ on the reference $\mathcal{H}_R$, the outcome i is obtained with the probability:

$$p_i \stackrel{\text{def}}{=} \langle X|(I_A \otimes M_i)|X\rangle = \operatorname{Tr} M_i^T \rho = \operatorname{Tr} X M_i^T X^*. \tag{8.30}$$

The final state on $\mathcal{H}_A$ is given as

$$\rho_i \stackrel{\text{def}}{=} \frac{1}{p_i} \operatorname{Tr}_R (I_A \otimes \sqrt{M_i})|X\rangle\langle X|(I_A \otimes \sqrt{M_i}) = \frac{1}{p_i} X M_i^T X^*. \tag{8.31}$$

Since

$$\sum_i p_i \rho_i = \sum_i X M_i^T X^* = X\left(\sum_i M_i^T\right) X^* = X X^* = \rho,$$

any POVM $\boldsymbol{M}$ on $\mathcal{H}_R$ gives a probabilistic decomposition. Conversely, for any probabilistic decomposition $\{(p_i, \rho_i)\}$ of ρ, the matrix $M_i = X^{-1} p_i \rho_i (X^*)^{-1}$ on $\mathcal{H}_R$ forms a POVM as

$$\sum_i M_i = \sum_i X^{-1} p_i \rho_i (X^*)^{-1} = X^{-1} \rho (X^*)^{-1} = I.$$

Moreover, this POVM $\{M_i\}$ satisfies (8.30) and (8.31). Hence, we obtain the following lemma.

Lemma 8.3 *Any probabilistic decomposition $\{(p_i, \rho_i)\}$ of ρ is given by a POVM $\boldsymbol{M}$ on the reference system as (8.30) and (8.31).*

Indeed, using this discussion, we can characterize the TP-CP map to the environment based on the output state $(\kappa \otimes \iota_R)(|\Phi_d\rangle\langle\Phi_d|)$ of the given channel κ as follows. In this case, since the initial state of the total system of the reference system, the output system, and the environment system is pure, its final state is also pure. That is, the final state of the total system is given as the purification $|u\rangle\langle u|$ of $(\kappa \otimes \iota_R)(|\Phi_d\rangle\langle\Phi_d|)$. Since any state ρ can be described as $d_A \operatorname{Tr}_R I_A \otimes \rho^T |\Phi_d\rangle\langle\Phi_d|$, the output state with the input state ρ on $\mathcal{H}_A$ is given as

$$d_A \operatorname{Tr}_{A,R} (I_{A,E} \otimes \rho^T)|u\rangle\langle u|. \tag{8.32}$$

Exercises

8.2 Show that $1 - \left(\operatorname{Tr} |\sqrt{\rho}\sqrt{\sigma}|\right)^2 \geq d_1^2(\rho, \sigma)$ using (8.10) and Exercise 3.18 for two mixed states ρ and σ.

8.3 Show that

$$F^2(\rho,\sigma) \le \operatorname{Tr} \sqrt{\rho}\sqrt{\sigma} \le F(\rho,\sigma) \tag{8.33}$$

using the purifications and the monotonicity of $\phi(1/2|\rho,\sigma)$.

8.4 Prove (8.19) noting that $\operatorname{Tr}(E_i \otimes I)|x\rangle\langle x| = \operatorname{Tr}_E E_i\rho$.

8.5 Prove property ① of the entanglement fidelity by following the steps below.
(**a**) Show that there exist Choi–Kraus representations $\{E_i\}_i$ and $\{A_i\}_j$ of κ and κ', respectively, such that the matrix $\{\operatorname{Tr} E_i A_j \rho\}_{i,j}$ can be written in diagonal form with positive and real diagonal elements.
(**b**) Using (**a**) and (8.19), show that there exist a matrix A and a Choi–Kraus representation $\{E_i\}_i$ of κ such that $\operatorname{Tr} A\rho A^* = 1$ and $F_e^2(\rho, \kappa \circ \kappa') \le |\operatorname{Tr} E_1 A\rho|^2$.
(**c**) Let E be a matrix from $\mathcal{H}_B$ to $\mathcal{H}_A$. Assume that $E^*E \le I$ and $\operatorname{Tr} A\rho A^* = \operatorname{Tr} \rho = 1$. Take U^* to be partially isometric under the polar decomposition $E = U|E|$. Show that $|\operatorname{Tr} EA\rho|^2 \le \operatorname{Tr} U|E|U^*\rho = \operatorname{Tr} EU^*\rho$.
(**d**) Assume that $\dim \mathcal{H}_A \le \dim \mathcal{H}_B$. Take the polar decomposition $E_1 = U|E_1|$ such that U^* is an isometry from $\mathcal{H}_A$ to $\mathcal{H}_B$. Show that $F_e^2(\rho, \kappa \circ \kappa') \le F_e(\rho, \kappa \circ \kappa_{U^*})$.
(**e**) Assume that $\dim \mathcal{H}_A > \dim \mathcal{H}_B$. Take the polar decomposition $E_1 = U|E_1|$ such that U is an isometry from $\mathcal{H}_B$ to $\mathcal{H}_A$. Then, choose the subspace $\mathcal{H}_C \subset \mathcal{H}_A$ as the range of U. So, U can be regarded as a unitary from $\mathcal{H}_B$ to $\mathcal{H}_C$. Show that $F_e^2(\rho, \kappa \circ \kappa') \le (\operatorname{Tr} P_C\rho) F_e(\frac{P_C\rho P_C}{\operatorname{Tr} P_C\rho}, \kappa \circ \kappa_{U^*})$.

8.6 Show ②, using (8.19) and showing the fact that the function $\rho \mapsto |\operatorname{Tr} A\rho|^2$ is a convex function.

8.7 Prove ③ by following the steps below. As the first step, determine the orthogonal basis $x_1, \ldots, x_d$ of $\mathcal{H}_B$ inductively. Let x_1 be the vector $\operatorname{argmax}_{x \in \mathcal{H}_B^1} \{1 - F^2(x, \kappa(x))\}$. Given $x_1, \ldots, x_j$, let $\mathcal{H}_j$ be the orthogonal complement space to the space spanned by $x_1, \ldots, x_j$. Let x_{j+1} be $\operatorname{argmax}_{x \in \mathcal{H}_j^1} \{1 - F^2(x, \kappa(x))\}$. Then, let $\mathcal{H}_C$ be the space spanned by $x_{d_B}, \ldots, x_{d_B - d_C + 1}$, where $d_C = \lfloor (1-a) \dim \mathcal{H}_B \rfloor$. Show that the space $\mathcal{H}_C$ satisfies (8.25) using Markov's inequality and ②.

8.8 Show (8.26) in ④ by following the steps below.
(**a**) Show that $F_e^2(\rho, \kappa) = \sum_{i,j} p_i p_j \langle u_i|\kappa(|u_i\rangle\langle u_j|)|u_j\rangle$ for $\rho = \sum_i p_i |u_i\rangle\langle u_i|$, where $p_1 \ge p_2 \ge \ldots \ge p_d$.
(**b**) Let $\phi = (\phi_1, \ldots, \phi_d)$. Define $u(\phi) \stackrel{\text{def}}{=} \sum_j \sqrt{p_j} e^{i\phi_j} u_j$. Show that $F_e^2(\rho, \kappa) + \sum_{j \ne k} p_j p_k \langle u_k|\kappa(|u_j\rangle\langle u_j|)|u_k\rangle$ is equal to the expectation of $F^2(u(\phi), \kappa(u(\phi)))$ under the uniform distribution with respect to $\phi = (\phi_1, \ldots, \phi_d)$.
(**c**) Let δ be the RHS of (8.26). Show that

$$\sum_{k=2}^{d} p_k \langle u_k|\kappa(|u_1\rangle\langle u_1|)|u_k\rangle \le p_2\delta,$$

$$\sum_{j=2}^{d} \sum_{k \ne j} p_j p_k \langle u_k|\kappa(|u_j\rangle\langle u_j|)|u_k\rangle \le \sum_{j=2}^{d} p_k p_1 \delta.$$

(**d**) Show (8.26) using (**a**) to (**c**).

8.9 Show that the equality of (8.26) in ④ holds when κ is a depolarizing channel for a quantum two-level system and ρ is the completely mixed-state ρ_{mix}.

8.10 Prove (8.27) following the steps below.
(**a**) Prove (8.27) when κ is a depolarizing channel.
(**b**) Given a channel κ, we choose the depolarizing channel $\kappa_{d,\lambda}$ as

$$\kappa_{d,\lambda}(\rho) = \int_{\mathrm{SU}(d_A)} U^* \kappa(U \rho U^*) U \nu(dU),$$

where $\nu(dU)$ is the invariant distribution. Show that $\mathrm{E}_{\mu,x} F^2(x, \kappa(x)) = \mathrm{E}_x F^2(y, \kappa_{d,\lambda}(y))$ for any element $y \in \mathcal{H}_A$, where $\mathrm{E}_{\mu,x}$ is the expectation with respect to the pure state x under the invariant distribution μ.
(**c**) Show that $F_e(\rho_{\mathrm{mix}}, \kappa) = F_e(\rho_{\mathrm{mix}}, \kappa_{d,\lambda})$.
(**d**) Prove (8.27) for any channel κ.

8.11 Verify (8.28).

8.12 Show the following for the states ρ_i in (8.31) when the state ρ is full rank and the reference system $\mathcal{H}_R$ has the same dimension as $\mathcal{H}$.
(**a**) Show that the states ρ_i in (8.31) are pure if and only if rank $M_i = 1$.
(**b**) Show that the states ρ_i in (8.31) are orthogonal to each other if and only if the POVM $\boldsymbol{M} = \{M_i\}$ is a PVM and commutative with $(X^* X)^T$.

8.13 Let κ be a TP-CP map from $\mathbb{C}^d$ to $\mathbb{C}^{d'}$ and κ' be a TP-CP map from $\mathbb{C}^{d'}$ to $\mathbb{C}^d$. Show that $F_e(\rho_{\mathrm{mix}}, \kappa' \circ \kappa) \le \sqrt{\frac{d'}{d}}$.

8.14 Let ρ be a bipartite state on $\mathcal{H}_A \otimes \mathcal{H}_B$. Show that the state ρ is separable if and only if ρ has a purification $\sum_i \sqrt{p_i} |x_i^A\rangle \otimes |x_i^B\rangle \otimes |u_i^R\rangle$ with the reference system $\mathcal{H}_R$ such that $\{|u_i^R\rangle\}$ is a CONS of $\mathcal{H}_R$.

8.3 Entanglement and Information Quantities

So far, we have examined the transmission information for a classical-quantum channel, but not the quantum version of the mutual information $I(X : Y)$, defined by (2.30) in Sect. 2.1.1. In Sect. 5.5, we defined the **quantum mutual information** $I_\rho(A : B)$ as

$$I_\rho(A : B) = H_\rho(A) + H_\rho(B) - H_\rho(AB) = D(\rho \| \rho^A \otimes \rho^B) \tag{8.34}$$

with respect to a state ρ on $\mathcal{H}_{A,B}$ for quantum systems $\mathcal{H}_A$ and $\mathcal{H}_B$. We used the notation introduced in Sect. 5.5 for the second expression above. Confusingly, the transmission information $I(p, W)$ for classical-quantum channels is also occasionally called the quantum mutual information. However, since $I_\rho(A : B)$ is a more natural generalization of the mutual information defined in (2.30), we shall call the quantity $I_\rho(A : B)$ the quantum mutual information in this text.

As discussed in Sect. 2.1.1, there is a precise relationship between the mutual information and the transmission information for classical systems. Similarly, there is a relationship between the classical-quantum transmission information and the quantum mutual information. To see this relation, let us consider a classical-quantum channel W with an input system $\mathcal{X}$ and an output system $\mathcal{H}_A$. Let $\{u_x\}$ be the orthonormal basis states of the Hilbert space $\mathcal{H}_X$. Let us consider a state on the composite system $\mathcal{H}_X \otimes \mathcal{H}_A$ given by $\rho = \sum_x p_x |u_x\rangle\langle u_x| \otimes W_x$, where p is a probability distribution in $\mathcal{X}$. The quantum mutual information is then given by $I_\rho(X : A) = I(p, W)$. Therefore, this is equal to the transmission information of a classical-quantum channel.

It is possible to find a connection between the transmission information and the quantum mutual information of a classical-quantum channel by appropriately defining the composite system. Let us now define the transmission information of the quantum-quantum channel κ (which is a TP-CP map) from the quantum mutual information using a similar method. Here it is necessary to find the quantum-mechanical correlation between the input and output systems. For this purpose, similar to the entanglement fidelity, we consider the purification x of the state ρ on the input system $\mathcal{H}_A$ because the final state of the purification x reflects how the map κ preserves the coherence of the input states. The **transmission information** $I(\rho, \kappa)$ of the quantum-quantum channel κ can then be defined using the quantum mutual information as [8]

$$I(\rho, \kappa) \stackrel{\text{def}}{=} I_{(\kappa \otimes \iota_R)(|x\rangle\langle x|)}(R : B), \tag{8.35}$$

where R is the reference system and B is the output system. Since $H(\rho)$ is equal to the entropy of the reference system, this can also be written as

$$I(\rho, \kappa) = H(\kappa(\rho)) + H(\rho) - H(\kappa \otimes \iota_R(|x\rangle\langle x|)). \tag{8.36}$$

This quantity will play an important role in Sect. 9.3.

Let us now consider the following quantity called the **coherent information**, which expresses how much coherence is preserved through a quantum-quantum channel κ [9].

$$I_c(\rho, \kappa) \stackrel{\text{def}}{=} H(\kappa(\rho)) - H(\kappa \otimes \iota_R(|x\rangle\langle x|)) = -H_{\kappa \otimes \iota_R(|x\rangle\langle x|)}(R|B) \tag{8.37}$$

for a TP-CP map κ from $\mathcal{H}_A$ to $\mathcal{H}_B$, a state ρ on $\mathcal{H}_A$, and a purification x of ρ. Therefore, the coherent information is equal to the negative conditional entropy. Of course, in the classical case, the conditional entropy can only take either positive values or 0. Therefore, a negative conditional entropy indicates the existence of some quantum features in the system. For example, in an entanglement-breaking channel, the conditional entropy is nonnegative, as can be seen in (8.62).

The coherent information can be related to the entanglement fidelity if $\sqrt{2(1-F_e(\rho,\kappa))} \le 1/e$ as follows[Exe. 8.16,8.7] [6]:

$$0 \le H(\rho) - I_c(\rho,\kappa) \le \sqrt{2(1-F_e(\rho,\kappa))}\left(3\log d - 2\log\sqrt{2(1-F_e(\rho,\kappa))}\right). \tag{8.38}$$

The first inequality holds without any assumption. Therefore, we can expect that the difference between $H(\rho)$ and the coherent information $I_c(\rho,\kappa)$ will express how the TP-CP map κ preserves the coherence. This will be justified in Sect. 9.6.

The above information quantities also satisfy the **monotonicity** [8, 9]

$$I_c(\rho,\kappa'\circ\kappa) \le I_c(\rho,\kappa), \tag{8.39}$$

$$I(\rho,\kappa'\circ\kappa) \le I(\rho,\kappa), \tag{8.40}$$

$$I(\rho,\kappa\circ\kappa') \le I(\kappa'(\rho),\kappa). \tag{8.41}$$

If U is an isometric matrix, then the coherent information satisfies[Exe. 8.22] [9, 10]

$$I_c(\rho,\kappa\circ\kappa_U) = I_c(U\rho U^*,\kappa). \tag{8.42}$$

If $\kappa = \sum_i p_i\kappa_i$, these quantities satisfy the **convexity** for channels [8, 10]

$$I_c(\rho,\kappa) \le \sum_i p_i I_c(\rho,\kappa_i), \tag{8.43}$$

$$I(\rho,\kappa) \le \sum_i p_i I(\rho,\kappa_i). \tag{8.44}$$

The transmission information satisfies the **concavity** for states [8]

$$I\Big(\sum_{i=1}^k p_i\rho_i,\kappa\Big) \ge \sum_{i=1}^k p_i I(\rho_i,\kappa). \tag{8.45}$$

Conversely, the following reverse inequality also holds:

$$I\Big(\sum_{i=1}^k p_i\rho_i,\kappa\Big) \le \sum_{i=1}^k p_i I(\rho_i,\kappa) + 2\log k. \tag{8.46}$$

Let κ^A (κ^B) be a TP-CP map from $\mathcal{H}_A$ ($\mathcal{H}_B$) to $\mathcal{H}_{A'}$ ($\mathcal{H}_{B'}$). Let $\rho^{A,B}$ be a state on $\mathcal{H}_A\otimes\mathcal{H}_B$. Let ρ^A and ρ^B be the partially traced state of $\rho^{A,B}$. The transmission information of a quantum-quantum channel then satisfies

$$I(\rho^{A,B},\kappa^A\otimes\kappa^B) \le I(\rho^A,\kappa^A) + I(\rho^B,\kappa^B) \tag{8.47}$$

in a similar way to (4.5) for the transmission information of a classical-quantum channel [8].

In addition to the types of information defined up until now, we may also define the **pseudocoherent information**

$$\tilde{I}_c(\rho, \kappa) \stackrel{\text{def}}{=} H(\rho) - H(\kappa \otimes \iota_R(|x\rangle\langle x|)). \tag{8.48}$$

Although it is difficult to interpret the above quantity as information, it does possess the following useful properties [11], which will be used in Sect. 9.3.

$$\tilde{I}_c(\rho, \kappa \circ \kappa') \le \tilde{I}_c(\kappa'(\rho), \kappa), \tag{8.49}$$

$$\tilde{I}_c\Big(\sum_j p_j \rho_j, \kappa\Big) \ge \sum_j p_j \tilde{I}_c(\rho_j, \kappa). \tag{8.50}$$

The first property (8.49) is the **monotonicity**, and can be derived immediately from property (8.41) and definitions. The second inequality (8.50) is the **concavity** with respect to a state. The following reverse inequality also holds, i.e.,

$$\tilde{I}_c\Big(\sum_{j=1}^k p_j \rho_j, \kappa\Big) \le \sum_{j=1}^k p_j \tilde{I}_c(\rho_j, \kappa) + \log k. \tag{8.51}$$

The derivations for (8.50) and (8.51) are rather difficult (Exercises 8.24 and 8.25). We can also obtain the following relationship by combining (8.49) and (8.50):

$$\tilde{I}_c\Big(\sum_j p_j \kappa_j(\rho), \kappa\Big) \ge \sum_j p_j \tilde{I}_c(\rho, \kappa \circ \kappa_j). \tag{8.52}$$

Finally, we focus on the entropy $H((\kappa \otimes \iota_R)(|x\rangle\langle x|))$, which is called the **entropy exchange** [5] and is denoted by $H_e(\kappa, \rho)$. This is equal to the entropy of the environment system $\mathcal{H}_E$ after the state ρ is transmitted. Its relationship to the entanglement fidelity $F_e(\rho, \kappa)$ is given by the **quantum Fano inequality** as [5][3]

$$H_e(\rho, \kappa) \le h(F_e^2(\rho, \kappa)) + (1 - F_e^2(\rho, \kappa)) \log(d^2 - 1), \tag{8.53}$$

where d is the dimension of $\mathcal{H}$.

Exercises

8.15 Let $\mathcal{H}_{E'}$ be the environment system after performing a state evolution given by the TP-CP map κ from $\mathcal{H}_A$ to $\mathcal{H}_{A'}$. Let x be the purification of the state ρ on $\mathcal{H}_A$. Let the reference system be $\mathcal{H}_R$. Show that

[3]Since the form of this inequality is similar to the Fano inequality, it is called quantum Fano inequality. However, it cannot be regarded as a quantum extension of the Fano inequality (2.35). The relationship between the two formulas is still unclear.

$$I_c(\rho, \kappa) = H_{x'}(A') - H_{x'}(E'), \tag{8.54}$$
$$I(\rho, \kappa) = H_{x'}(A') + H_{x'}(A'E') - H_{x'}(E'),$$

where x' is the final state of x.

8.16 Show the first inequality in (8.38) by considering the Stinespring representation of κ and (5.86) with respect to the composite system of the environment system E and the reference system R.

8.17 Show the second inequality of (8.38) by considering the purification of ρ and Fannes inequality (Theorem 5.12).

8.18 Prove (8.39) based on the Stinespring representations of κ and κ' and the strong subadditivity (5.83) of the von Neumann entropy.

8.19 Prove (8.40) by following steps below.
(**a**) Let $|x\rangle$ be a purification of ρ with the reference system $\mathcal{H}_R$. Show that

$$I(\rho, \kappa) = D(\kappa \otimes \iota_R(|x\rangle\langle x|)\|\kappa(\rho) \otimes \mathrm{Tr}_A |x\rangle\langle x|). \tag{8.55}$$

(**b**) Show (8.40).

8.20 Prove (8.43) and (8.44) using the concavity (5.88) of the conditional entropy.

8.21 Prove (8.41) based on (8.55) and the monotonicity of the quantum relative entropy by considering the Stinespring representation of κ'.

8.22 Prove (8.42).

8.23 Let x be the purification of ρ with respect to the reference system $\mathcal{H}_R$. Let $\mathcal{H}_{E'_A}$ and $\mathcal{H}_{E'_B}$ be the environment systems after the state evolutions κ^A and κ^B. Let x' be the final state of x. Show (8.47) by following the steps below.
(**a**) Show the following, using Exercise 8.15.

$$I(\rho^A, \kappa^A) = H_{x'}(A') + H_{x'}(A'E'_A) - H_{x'}(E'_A)$$
$$I(\rho, \kappa^A \otimes \kappa^B) = H_{x'}(A'B') + H_{x'}(A'B'E'_AE'_B) - H_{x'}(E'_AE'_B).$$

(**b**) Show that

$$\begin{aligned}&I(\rho^A, \kappa^A) + I(\rho^B, \kappa^B) - I(\rho, \kappa^A \otimes \kappa^B)\\ =&H_{x'}(A') + H_{x'}(B') - H_{x'}(A'B') - \left(H_{x'}(E'_A) + H_{x'}(E'_B) - H_{x'}(E'_AE'_B)\right)\\ &+ \left(H_{x'}(A'E'_A) + H_{x'}(B'E'_B) - H_{x'}(A'E'_AB'E'_B)\right).\end{aligned}$$

(**c**) Prove (8.47) by combining (8.34) with (**b**).

8.24 Let κ be the state evolution from $\mathcal{H}_A$ and from $\mathcal{H}_{A'}$. Show (8.50) following the steps below.

(**a**) Let x_j be the purification of ρ_j with respect to the reference system $\mathcal{H}_R$. Let $\{u_j\}$ be an orthonormal basis of another system $\mathcal{H}_{R'}$. Show that the pure state $x \stackrel{\text{def}}{=} \sum_j \sqrt{p_j} x_j \otimes u_j$ on $\mathcal{H}_A \otimes \mathcal{H}_R \otimes \mathcal{H}_{R'}$ is the purification of $\rho \stackrel{\text{def}}{=} \sum_j p_j \rho_j$.
(**b**) Show that the pinching $\kappa_{\boldsymbol{E}}$ of the measurement $\boldsymbol{E} = \{|u_j\rangle\langle u_j|\}$ on $\mathcal{H}_{R'}$ satisfies

$$\begin{aligned}&D((\kappa_{\boldsymbol{E}} \otimes \iota_{A,R})(\kappa \otimes \iota_{R,R'})(|x\rangle\langle x|)\|(\kappa_{\boldsymbol{E}} \otimes \iota_{A,R})(\kappa(\rho) \otimes \mathrm{Tr}_A(|x\rangle\langle x|)))\\ =&H(\kappa(\rho)) + \sum_j p_j H(\rho_j) - \sum_j p_j H((\kappa \otimes \iota_R)(|x_j\rangle\langle x_j|)).\end{aligned}$$

(**c**) Prove (8.50) by considering the monotonicity of the quantum relative entropy for the pinching $\kappa_{\boldsymbol{E}}$.

8.25 Prove (8.51) using the same symbols as Exercise 8.24 by following the steps below.
(**a**) Show that

$$\begin{aligned}&\sum_{j=1}^{k} p_j \tilde{I}_c(\rho_j, \kappa)\\ =&H(\kappa_{\boldsymbol{E}}(\mathrm{Tr}_A(|x\rangle\langle x|))) - H(\kappa_{\boldsymbol{E}} \otimes \iota_{A,R})(\kappa \otimes \iota_{R,R'})(|x\rangle\langle x|)).\end{aligned}$$

(**b**) Verify that

$$\begin{aligned}&\tilde{I}_c(\rho, \kappa) - \sum_{j=1}^{k} p_j \tilde{I}_c(\rho_j, \kappa)\\ =&H(\mathrm{Tr}_A |x\rangle\langle x|) - H(\kappa_{\boldsymbol{E}}(\mathrm{Tr}_A |x\rangle\langle x|)) - H(\kappa \otimes \iota_{R,R'})(|x\rangle\langle x|))\\ &+ H(\kappa_{\boldsymbol{E}} \otimes \iota_{A,R})(\kappa \otimes \iota_{R,R'})(|x\rangle\langle x|))\\ \le&H(\kappa_{\boldsymbol{E}} \otimes \iota_{A,R})(\kappa \otimes \iota_{R,R'})(|x\rangle\langle x|)) - H(\kappa \otimes \iota_{R,R'})(|x\rangle\langle x|)).\end{aligned} \tag{8.56}$$

(**c**) Prove (8.51) using (5.81) and the above results.

8.26 Prove (8.45) using (8.50) and (5.77).

8.27 Prove (8.46) using (8.51) and (5.79).

8.28 Show that

$$\max\{H(\rho) | \langle u|\rho|u\rangle = f\} = h(f) + (1 - f)\log(d - 1) \tag{8.57}$$

for a pure state $|u\rangle\langle u|$ on $\mathcal{H}$ ($\dim \mathcal{H} = d$). Then, prove (8.53) using this result.

8.29 Show that

$$H_e(\kappa_{\boldsymbol{p}}, \rho_{\mathrm{mix}}) = H(\boldsymbol{p}), \quad H_e(\kappa_{\boldsymbol{p}}, |e_0\rangle\langle e_0|) = H_e(\kappa_{\boldsymbol{p}}, |e_1\rangle\langle e_1|) = h(p_0 + p_3)$$

for the entropy exchange of a Pauli channel $\kappa_{\boldsymbol{p}}$.

8.4 Entanglement and Majorization

In this section we consider what kind of state evolutions are possible using only local quantum operations and classical communications given an entangled state between two systems. Before tackling this problem, let us first consider a partial ordering called **majorization** defined between two d-dimensional vectors $a = (a_i), b = (b_i)$ with positive real-number components. This will be useful in the discussion that follows. If a and b satisfy

$$\sum_{j=1}^{k} a_j^{\downarrow} \leq \sum_{j=1}^{k} b_j^{\downarrow}, \quad (1 \leq \forall k \leq n), \quad \sum_{j=1}^{n} a_j^{\downarrow} = \sum_{j=1}^{n} b_j^{\downarrow},$$

we say that b majorizes a, which we denote as $a \preceq b$. In the above, $(a_j^{\downarrow})$ and $(b_j^{\downarrow})$ are the reordered versions of the elements of a and b, respectively, largest first. If $x \preceq y$ and $y \preceq x$, we represent it as $x \cong y$. If $\frac{1}{\sum_i x_i} x \cong \frac{1}{\sum_i y_i} y$, we write $x \approx y$. If $\frac{1}{\sum_i x_i} x = \frac{1}{\sum_i y_i} y$, we represent it as $x \propto y$. The following theorem discusses the properties of this partial ordering. The relation with entanglement will be discussed after this theorem.

Theorem 8.2 *The following conditions for two d-dimensional vectors $x = (x_i)$ and $y = (y_i)$ with positive real components are equivalent [12].*

① $x \preceq y$.
② *There exists a finite number of T-transforms $T_1, \ldots, T_n$ such that $x = T_n \cdots T_1 y$. A T-transform is defined according to a matrix $A = (a^{i,j})$ satisfying $a^{i_1,i_1} = a^{i_2,i_2} = 1 - t$ and $a^{i_1,i_2} = a^{i_2,i_1} = t$ for some pair i_1 and i_2, and $a^{i,j} = \delta_{i,j}$ otherwise, where t is a real number between $0 \leq t \leq 1$.*
③ *There exists a double stochastic matrix A such that $x = Ay$.*
④ *There exists a stochastic matrix $B = (b^{i,j})$ such that $(B^j)^T \circ x \approx y$ for all integers j. $(B^j)^T$ is the column vector obtained by transposing B^j. The product of the two vectors x and y is defined as $(y \circ x)_i \stackrel{\text{def}}{=} y_i x_i$.*

The product $\circ$ satisfies the associative law. A vector e with each of its components equal to 1 satisfies $e \circ x = x$ and $\sum_j (B^j)^T = e$.

From the concavity of the entropy, we can show that a T-transform T and a probability distribution satisfy $H(T(p)) \geq H(p)$. Therefore, if $q \preceq p$, then

$$H(q) \geq H(p). \tag{8.58}$$

Since a double stochastic matrix Q and a probability distribution p satisfy $Q(p) \preceq p$, we have

$$H(Q(p)) \geq H(p), \tag{8.59}$$

from which we obtain (2.27).

Further, any double stochastic matrix A can be written by a distribution p on the permutations S_k as $(Ax)_i = \sum_{s \in S_k} p_s x_{s^{-1}(i)}$. Thus, when two positive-valued vectors x and y have decreasing ordered elements, we can show that

$$\langle x, y \rangle \geq \langle x, Ay \rangle. \tag{8.60}$$

Let us now consider how majorization can be defined for two density matrices ρ and σ. The eigenvalues of ρ and σ form the respective vectors with real-number components. Therefore, majorization can be defined with respect to these vectors. Letting $\rho = \sum_i a_i |u_i\rangle\langle u_i|$ and $\sigma = \sum_i b_i |v_i\rangle\langle v_i|$, we can write $\rho \preceq \sigma$ if $a \preceq b$. If ρ and σ come from different Hilbert spaces, let us define $\rho \preceq \sigma$ by adding zero eigenvalues to the smaller Hilbert space until the size of the spaces are identical. The relations $\rho \cong \sigma$ and $\rho \approx \sigma$ can be defined in a similar way.

As this is a partial ordering, if $\rho \preceq \rho'$ and $\rho' \preceq \sigma$, then $\rho \preceq \sigma$. Since the entropy $H(\rho)$ of a density matrix ρ depends only on its eigenvalues, if $\rho \preceq \sigma$, then $H(\rho) \geq H(\sigma)$ due to (8.58). Further, we can also show that for a unital channel κ (e.g., pinching),

$$\kappa(\rho) \preceq \rho. \tag{8.61}$$

Hence, we find that $H(\kappa(\rho)) \geq H(\rho)$, and therefore the first inequality in (5.82) is satisfied even if $\boldsymbol{M}$ is a general POVM. Thus, the following theorem can be shown from Theorem 8.2.

Theorem 8.3 (Nielsen and Kempe [13]) *Let $\rho^{A,B}$ be a separable state on $\mathcal{H}_A \otimes \mathcal{H}_B$. Then, $\rho^{A,B} \preceq \rho^A \stackrel{\text{def}}{=} \mathrm{Tr}_B\, \rho^{A,B}$.*

Combining (8.58) with this theorem, we find that $H(\rho^{A,B}) \geq H(\rho^A)$ [i.e., (5.78)] if $\rho^{A,B}$ is separable. This shows that any separable state ρ satisfies [14]

$$H_\rho(B|A) \geq 0. \tag{8.62}$$

The following theorem shows how two entangled states can be transformed between each other.

Theorem 8.4 (Nielsen [15], Vidal [16]) *Let $|u\rangle\langle u|$ and $|v_j\rangle\langle v_j|$ be pure states on $\mathcal{H}_A \otimes \mathcal{H}_B$. It is possible to transform the state $|u\rangle\langle u|$ into $|v_j\rangle\langle v_j|$ using a two-way LOCC with probability p_j if and only if the condition*

$$\sum_{i=1}^{k} \lambda_i^{\downarrow} \leq \sum_{i=1}^{k} \sum_j p_j \lambda_i^{j,\downarrow}, \quad \forall k \tag{8.63}$$

holds, where $\sqrt{\lambda_i^j}$ is the Schmidt coefficient of $|v_j\rangle$ and $\sqrt{\lambda_i}$ is the Schmidt coefficient of $|u\rangle$. This operation can be realized by performing a measurement at A and then performing a unitary state evolution at B dependently of the measurement outcome j at A. Of course,

$$H(\mathrm{Tr}_B\, |u\rangle\langle u|) \geq \sum_j p_j H(\mathrm{Tr}_B\, |v_j\rangle\langle v_j|). \tag{8.64}$$

In particular, it is possible to transform $|u\rangle\langle u|$ into $|v\rangle\langle v|$ using a two-way LOCC with probability 1 *if and only if the condition*

$$\mathrm{Tr}_B |u\rangle\langle u| \preceq \mathrm{Tr}_B |v\rangle\langle v| \tag{8.65}$$

holds. These conditions still hold even if the two-way LOCC is restricted to a one-way LOCC.

Proof **Step 1: Proof of the part "only if"** First, we show that (8.63) holds if it is possible to transform the pure state $|u\rangle\langle u|$ into $|v_j\rangle\langle v_j|$ with probability p_j. According to the discussion concerning instruments in Sect. 7.1, an arbitrary state evolution κ can be regarded as an instrument given by the Choi–Kraus representation $\{A_j\}_j$. Therefore, we see that if the initial state is a pure state, the final state for each measurement outcome j must also be a pure state.

Now, consider local operations and two-way communications from A to B and from B to A. This operation consists of repetitions of the following procedure. First, A performs a measurement $\{A'_j\}_j$ and then sends this measurement outcome j to B. Then, B performs a measurement $\{B_i^j\}_i$ at B corresponding to A's measurement outcome j. Finally, B sends his or her measurement outcome i to A. Since the final state after the measurement is also a pure state, the measurement at B may be written as A's measurement and a unitary operation at B corresponding to A's measurement outcome, according to Theorem 8.1.

Therefore, we see that the whole operation is equivalent to performing a measurement $\{A_j\}_j$ at A and then performing a unitary state operation at B dependently of the measurement outcome j at A. By defining $\rho_u \stackrel{\mathrm{def}}{=} \mathrm{Tr}_B |u\rangle\langle u|$, the probability of obtaining the measurement outcome j is then $p_j \stackrel{\mathrm{def}}{=} \mathrm{Tr}\, A_j \rho_u A_j^*$. The final state is a pure state, and the partially traced state is equal to $\frac{1}{\mathrm{Tr}\, A_j \rho_u A_j^*} A_j \rho_u A_j^*$. Taking the unitary matrix U_j giving the polar decomposition $\sqrt{\rho_u} A_j^* = U_j \sqrt{A_j \rho_u A_j^*}$, we obtain

$$U_j A_j \rho_u A_j^* U_j^* = U_j \sqrt{A_j \rho_u A_j^*} \sqrt{A_j \rho_u A_j^*} U_j^* = \sqrt{\rho_u} A_j^* A_j \sqrt{\rho_u}.$$

If P is a projection with rank k and satisfies the equation $\mathrm{Tr}\, \rho_u P = \sum_{i=1}^k \lambda_i^{\downarrow}$, then

$$\sum_{i=1}^k \sum_j p_j \lambda_i^{j,\downarrow} = \sum_j \max\{\mathrm{Tr}\, A_j \rho_u A_j^* P_j | P_j \text{ is a projection of rank } k\}$$

$$\geq \sum_j \mathrm{Tr}\, U_j A_j \rho_u A_j^* U_j^* P = \sum_j \sqrt{\rho_u} A_j^* A_j \sqrt{\rho_u} P = \mathrm{Tr}\, \rho_u P = \sum_{i=1}^k \lambda_i^{\downarrow}.$$

Therefore, we obtain (8.63).

Step 2: Proof of the part "if" with the deterministic case Next, let us construct the operation that evolves $|u\rangle\langle u|$ into $|v\rangle\langle v|$ with probability 1 when (8.65) is satisfied. Let the Schmidt coefficients of $|u\rangle$ and $|v\rangle$ be $\sqrt{\lambda_i}$ and $\sqrt{\lambda'_i}$, respectively. Let a stochastic matrix (b^{ij}) satisfy Condition ④ of Theorem 8.2 when $x = \lambda = (\lambda_i)$ and $y = \lambda' = (\lambda'_i)$. Now, let us define an orthonormal basis $\{u_i\}$ and E_j by

$$\rho_u = \sum_i \lambda_i |u_i\rangle\langle u_i|, \quad E_j \stackrel{\text{def}}{=} \sum_i b^{i,j} |u_i\rangle\langle u_i|. \tag{8.66}$$

Then, we have $\sum_j E_j = I$ because $B = (b^{i,j})$ is a stochastic matrix. The probability of obtaining the measurement outcome j for the measurement $\{E_j\}$ is $\operatorname{Tr} \rho_u E_j$. The final state for this measurement outcome is a pure state, and the partially traced state is $\frac{1}{\operatorname{Tr} \rho_u E_j} \sqrt{E_j} \rho_u \sqrt{E_j}$. Since $(B^j)^T \circ \lambda \approx \lambda'$, we have $\frac{1}{\operatorname{Tr} \rho_u E_j} \sqrt{E_j} \rho_u \sqrt{E_j} \cong \operatorname{Tr}_B |v\rangle\langle v|$. Therefore, when an appropriate unitary state evolution is applied dependently of the measurement outcome j, the final state will be $|v\rangle\langle v|$ with probability 1.

Step 3: Proof of the part "if" with the stochastic case Finally, we construct the operation that evolves the pure state $|u\rangle\langle u|$ on $\mathcal{H}_A \otimes \mathcal{H}_B$ into $|v_j\rangle\langle v_j|$ with probability p_j when the inequality (8.63) holds. Let $\lambda' = (\lambda'_i)$ be a probability distribution such that

$$\sum_{i=1}^{k} \lambda'_i = \sum_{i=1}^{k} \sum_j p_j \lambda_i^{j,\downarrow}, \quad \forall k.$$

Then, the pure state $|v\rangle\langle v|$ is defined as the pure entangled state with the Schmidt coefficient $\lambda' = (\lambda'_i)$. The discussion of the deterministic case guarantees that there exists an LOCC operation transforming $|u\rangle\langle u|$ into $|v\rangle\langle v|$ with the Schmidt coefficient $\sqrt{x_i}$. Therefore, it is sufficient to construct the required operation when the equality of (8.65) holds, i.e., $\lambda'_i = \lambda_i^{\downarrow}$. Let us define $b^{i,j} \stackrel{\text{def}}{=} p_j \lambda_i^{j,\downarrow} / \lambda_i^{\downarrow}$. Then, $B = (b^{i,j})$ is a stochastic matrix. Defining E_j using (8.66), we have

$$p_j = \operatorname{Tr} \rho_u E_j, \quad \operatorname{Tr}_B |v_j\rangle\langle v_j| \cong \sum_i \lambda_i^{j,\downarrow} |u_i\rangle\langle u_i| \cong \frac{1}{\operatorname{Tr} \rho_u E_j} \sqrt{E_j} \rho_u \sqrt{E_j}.$$

This completes the proof. ■

Using this theorem, we obtain the following characterization.

Lemma 8.4 (Vidal et al. [17]) *Let v and u be entangled pure states with Schmidt coefficient $\sqrt{p_j}$ and $\sqrt{q_j}$ in decreasing order. Then, we have*

$$|\langle u|v\rangle|^2 \le \Big(\sum_j \sqrt{p_j}\sqrt{q_j}\Big)^2. \tag{8.67}$$

The equality holds when vectors v *and* u *have the Schmidt decompositions* $v = \sum_i \sqrt{p_j}|e_i^A\rangle \otimes |e_i^B\rangle$ *and* $u = \sum_i \sqrt{q_j}|e_i^A\rangle \otimes |e_i^B\rangle$ *by the same Schmidt basis. Further, we have*

$$\max_{\kappa \in \leftrightarrow} \langle u|\kappa(|v\rangle\langle v|)|u\rangle = \max_{q':q \preceq q'} \Big(\sum_j \sqrt{p_j}\sqrt{q'_j}\Big)^2. \tag{8.68}$$

Proof Let ρ and σ be the reduced density matrix on $\mathcal{H}_A$ of v and u. Then, there exisis a unitary matrix U such that

$$\langle u|v\rangle = \operatorname{Tr} \sqrt{\rho}\sqrt{\sigma}U.$$

Assume that σ is diagonalized as $\sigma = \sum_j q_j |e_j\rangle\langle e_j|$. Thus,

$$\begin{aligned}
|\operatorname{Tr} \sqrt{\rho}\sqrt{\sigma}U|^2 &= \Big|\sum_j \sqrt{q_j}\langle e_j|\sqrt{\rho}U|e_j\rangle\Big|^2 \le \Big(\sum_j \sqrt{q_j}\Big|\langle e_j|\sqrt{\rho}U|e_j\rangle\Big|\Big)^2 \\
&\le \Big(\sum_j \sqrt{\sqrt{q_j}\langle e_j|\sqrt{\rho}|e_j\rangle}\sqrt{\sqrt{q_j}\langle e_j|U^*\sqrt{\rho}U|e_j\rangle}\Big)^2 \\
&\le \Big(\sum_j \sqrt{q_j}\langle e_j|\sqrt{\rho}|e_j\rangle\Big)\Big(\sum_j \sqrt{q_j}\langle e_j|U^*\sqrt{\rho}U|e_j\rangle\Big).
\end{aligned}$$

Now, we diagonalize ρ as $\rho = \sum_i p_i |f_i\rangle\langle f_i|$. Hence,

$$\sum_j \sqrt{q_j}\langle e_j|\sqrt{\rho}|e_j\rangle = \sum_{i,j} \sqrt{q_j}\sqrt{p_i}|\langle e_j|f_i\rangle|^2.$$

Since $|\langle e_j|f_i\rangle|^2$ is a double stochastic matrix, (8.60) implies $\sum_{i,j} \sqrt{q_j}\sqrt{p_i}|\langle e_j|f_i\rangle|^2 \le \sum_i \sqrt{q_i}\sqrt{p_i}$. Thus,

$$\sum_j \sqrt{q_j}\langle e_j|\sqrt{\rho}|e_j\rangle \le \sum_i \sqrt{q_i}\sqrt{p_i}.$$

Similarly, we have

$$\sum_j \sqrt{q_j}\langle e_j|U^*\sqrt{\rho}U|e_j\rangle \le \sum_i \sqrt{q_i}\sqrt{p_i}.$$

Therefore, we obtain (8.67).

Next, we prove (8.68). From the equality condition of (8.67) and Theorem 8.4, we can easily verify the $\ge$ part of (8.68). Assume that the LOCC operation κ generates the state v_j with probability r_j from the initial pure state v. When the Schmidt coefficient of v_j is $(\sqrt{p_i^j})_i$, Corollary 8.2 and (8.67) imply

$$\langle u|\kappa(|v\rangle\langle v|)|u\rangle = \sum_j r_j |\langle u|v_j\rangle|^2 \le \sum_j r_j \left(\sum_i \sqrt{p_i^j}\sqrt{q_i} \right)^2$$
$$\le \left(\sum_i \sqrt{\sum_j r_j p_i^j}\sqrt{q_i} \right)^2.$$

Since Theorem 8.4 guarantees that $(p_i)_i \preceq (\sum_j r_j p_i^j)_i$, we obtain the $\le$ part of (8.68). ■

Exercises

8.30 Choose the CONSSs $\{|u_i\rangle\}$ and $\{|v_j\rangle\}$ and the distributions $p = (p_i)$ and $q = (q_i)$ such that $\rho = \sum_i p_i |u_i\rangle\langle u_i|$ and $\kappa(\rho) = \sum_j q_j |v_j\rangle\langle v_j|$. Then, prove (8.61) by using the map $p = (p_i) \mapsto (\langle v_i|\kappa(\sum_j p_j |u_j\rangle\langle u_j|)|v_i\rangle)$.

8.31 Given a pure entangled state with Schmidt coefficients $\sqrt{\lambda_i}$, show that a maximally entangled state can be produced with error probability 0 if and only if the size of the maximally entangled state is less than $1/\lambda_1^{\downarrow}$.

8.5 Distillation of Maximally Entangled States

In order to use the merit of entanglement, we often require maximally entangled states, not partially entangled states. Then, one encounters the distillation problem of maximally entangled states from partially entangled states. Such an operation is called **entanglement distillation** and is one of the established fields in quantum information theory. If the initial state is pure, it is called entanglement concentration. It has also been verified experimentally [18, 19]. Other experimental models have also been proposed by combining other protocols [20].

Consider the problem of creating a maximally entangled state $|\Phi_L\rangle\langle\Phi_L|$ on $\mathbb{C}^L \otimes \mathbb{C}^L$ from a pure state $|u\rangle\langle u|$ on the composite system $\mathcal{H}_A \otimes \mathcal{H}_B$. If the relation

$$\mathrm{Tr}_B\, |u\rangle\langle u| \preceq \mathrm{Tr}_B\, |\Phi_L\rangle\langle\Phi_L|$$

does not hold, it is impossible to create $|\Phi_L\rangle\langle\Phi_L|$ with probability 1. Therefore, we must allow some failure probability in our scheme for creating $|\Phi_L\rangle\langle\Phi_L|$.

Theorem 8.5 ([21]) *Consider the two-way LOCC operation κ converting the initial state $|u\rangle\langle u|$ to a maximally entangled state $|\Phi_L\rangle\langle\Phi_L|$. The optimal failure probability $\varepsilon_1(\kappa, |u\rangle\langle u|)$ is less than $f(x) \stackrel{\mathrm{def}}{=} \mathrm{Tr}(\rho_u - xI)\{\rho_u - xI \ge 0\}$ if and only if*

$$L \le \frac{1 - f(x)}{x}. \tag{8.69}$$

Proof Since our operation has two outcomes "success" and "failure," the distribution of the outcome is described by the two-valued POVM $\{T, I - T\}$. Hence, from

Theorem 7.2, our operation is given by the combination of the state evolution (7.1) due to a measurement $\{T, I-T\}$ and the TP-CP map dependently of its outcome of "success" or "failure". The final state $|v\rangle\langle v|$ corresponding to "success" should satisfy $\rho_v (\stackrel{\text{def}}{=} \mathrm{Tr}_B |v\rangle\langle v|) \preceq \mathrm{Tr}_B |\Phi_L\rangle\langle\Phi_L|$ because of Theorem 8.4. Thus, Theorem 8.1 characterizes the minimum probability that the creation of $|\Phi_L\rangle\langle\Phi_L|$ fails as

$$\begin{aligned}&\min_{T\geq 0 \text{ on } \mathcal{H}_A} \left\{ \mathrm{Tr}\, \rho_u (I-T) \,\middle|\, \frac{1}{\mathrm{Tr}\, \rho_u T} \sqrt{T}\rho_u\sqrt{T} \preceq \mathrm{Tr}_B |\Phi_L\rangle\langle\Phi_L| \right\} \\ =&\min_{0\leq T\leq I:\text{ on } \mathcal{H}_A} \left\{ \mathrm{Tr}\, \rho_u (I-T) \,\middle|\, \sqrt{\rho_u} T \sqrt{\rho_u} \leq x \right\},\end{aligned}$$

where we used (A.7) to rewrite the above equation: henceforth, we abbreviate xI to x. Now, let L be the size of the maximally entangled state to be created and the ratio $\frac{\mathrm{Tr}\, \rho_u T}{L}$ be fixed to x. Since $\mathrm{Tr}\, \rho_u T$ is the success probability, the minimum failure probability can be calculated from the following equation:

$$\begin{aligned}&\min_{0\leq T\leq I:\text{ on } \mathcal{H}_A} \left\{ \mathrm{Tr}\, \rho_u (I-T) \,\middle|\, \sqrt{\rho_u} T \sqrt{\rho_u} \leq x \right\} \\ =&\min_{0\leq S\leq \rho:\text{ on } \mathcal{H}_A} \{ 1 - \mathrm{Tr}\, S \,|\, S \leq x \} \\ =&\min_{S \text{ on } \mathcal{H}} \left\{ 1 - \sum_i \langle u_i|S|u_i\rangle \,\middle|\, \langle u_i|S|u_i\rangle \leq \lambda_i, x \right\} \\ =&1 - \sum_{i:\lambda_i\leq x} \lambda_i - \sum_{i:\lambda_i > x} x = \mathrm{Tr}(\rho_u - x)\{\rho_u - x \geq 0\} = f(x),\end{aligned} \tag{8.70}$$

where $S = \sqrt{\rho_u} T \sqrt{\rho_u}$. Therefore, if the failure probability is less than $f(x)$, the size L of the maximally entangled state satisfies

$$L \leq \max_{x'} \left\{ \frac{1}{x'}(1 - f(x')) \,\middle|\, f(x') \leq f(x) \right\} = \frac{1}{x}(1 - f(x)).$$

In the last equality, we used the fact that $f(x)$ is strictly monotonically increasing and continuous.

Conversely, if (8.69) is true, then by choosing a projection T that attains the minimum value in (8.70) and performing a two-valued projective measurement $\{T, I-T\}$ on the system $\mathcal{H}_A$, the outcome corresponding to T will be obtained with a probability $1 - f(x)$. Since the final state u satisfies

$$\frac{1}{1-f(x)} T\rho_u T \leq \frac{x}{1-f(x)} \leq \frac{I}{L},$$

we may construct a maximally entangled state of size L according to Theorem 8.4. ■

On the other hand, Lo and Popescu [2] characterized the optimal success probability $P^{opt}(u \to |\Phi_L\rangle)$ for obtaining a maximally entangled state $|\Phi_L\rangle$ as follows:

$$P^{opt}(u \to |\Phi_L\rangle) = \max_{r:1\le r\le L} \frac{L}{L-r-1}\sum_{i=r}^{L}\lambda_i^{\downarrow}. \tag{8.71}$$

Next, we consider the problem of determining how large a maximally entangled state we can distill from a tensor product state $\rho^{\otimes n}$ of a partially entangled state ρ on $\mathcal{H}_A \otimes \mathcal{H}_B$, in the asymptotic case. Here, we formulate this problem in the mixed-state case as well as in the pure-state case. In such problems, we require that our operation κ_n be optimized for a given partially entangled state $\rho^{\otimes n}$ and hence treat the first type of **entanglement of distillation**:

$$E_{d,1}^{C}(\rho) \stackrel{\text{def}}{=} \sup_{\{\kappa_n\}\subset C}\left\{ \underline{\lim}\frac{1}{n}\log L(\kappa_n) \,\middle|\, \lim_{n\to\infty}\varepsilon_1(\kappa_n,\rho)=0\right\} \tag{8.72}$$

$$E_{d,1}^{C,\dagger}(\rho) \stackrel{\text{def}}{=} \sup_{\{\kappa_n\}\subset C}\left\{ \underline{\lim}\frac{1}{n}\log L(\kappa_n) \,\middle|\, \lim_{n\to\infty}\varepsilon_1(\kappa_n,\rho)<1\right\}, \tag{8.73}$$

where C denotes the set of local operations, i.e., the notations $C = \to$, $C = \emptyset$, $C = \leftarrow$, $C = \leftrightarrow$, and $C = S$ imply the set of one-way ($\mathcal{H}_A \to \mathcal{H}_B$) LOCC operations, only local operations, one-way ($\mathcal{H}_A \leftarrow \mathcal{H}_B$) LOCC operations, two-way LOCC operations, and S-TP-CP maps, respectively. Here, we denote the size of the maximally entangled state produced by the operation κ by $L(\kappa)$. If ρ is a mixed state, it is extremely difficult to produce a maximally entangled state perfectly, even allowing some failure probability. Therefore, let us relax our conditions and aim to produce a state close to the desired maximally entangled state. Hence, for our operation κ', we will evaluate the error $\varepsilon_2(\kappa',\rho) \stackrel{\text{def}}{=} 1-\langle\Phi_L|\kappa'(\rho)|\Phi_L\rangle$ between the final state $\kappa'(\rho)$ and the maximally entangled state $|\Phi_L\rangle\langle\Phi_L|$ of size L. When the initial state is a pure state v with Schmidt coefficient $\sqrt{\lambda_i^{\downarrow}}$, Lemma 8.4 gives the optimum fidelity:

$$\max_{\kappa\in\leftrightarrow}\langle\Phi_L|\kappa(|v\rangle\langle v|)|\Phi_L\rangle = \left(\max_{p\preceq p'}\sum_{i=1}^{L}\sqrt{\frac{p_i'}{L}}\right)^2. \tag{8.74}$$

In the asymptotic case, we optimize the operation κ_n' for a given $\rho^{\otimes n}$; thus, we focus on the second type of **entanglement of distillation**:

$$E_{d,2}^{C}(\rho) \stackrel{\text{def}}{=} \sup_{\{\kappa_n\}\subset C}\left\{ \underline{\lim}\frac{1}{n}\log L(\kappa_n) \,\middle|\, \lim_{n\to\infty}\varepsilon_2(\kappa_n,\rho)=0\right\}, \tag{8.75}$$

$$E_{d,2}^{C,\dagger}(\rho) \stackrel{\text{def}}{=} \sup_{\{\kappa_n\}\subset C}\left\{ \underline{\lim}\frac{1}{n}\log L(\kappa_n) \,\middle|\, \lim_{n\to\infty}\varepsilon_2(\kappa_n,\rho)<1\right\}. \tag{8.76}$$

The following trivial relations follow from their definitions:

$$E_{d,2}^{C}(\rho) \ge E_{d,1}^{C}(\rho),\ E_{d,2}^{C,\dagger}(\rho) \ge E_{d,1}^{C,\dagger}(\rho),\ E_{d,i}^{C,\dagger}(\rho) \ge E_{d,i}^{C}(\rho),$$

for $i = 1, 2$. The following theorem holds under these definitions.

Theorem 8.6 (Bennett et al. [22]) *The two kinds of entanglement of distillation of any pure state $|u\rangle\langle u|$ in the composite system $\mathcal{H}_A \otimes \mathcal{H}_B$ can be expressed by the reduced density $\rho_u = \mathrm{Tr}_B |u\rangle\langle u|$ as*

$$E_{d,i}^{C}(|u\rangle\langle u|) = E_{d,i}^{C,\dagger}(|u\rangle\langle u|) = H(\rho_u),$$

for $i = 1, 2$ and $C = \emptyset, \rightarrow, \leftarrow, \leftrightarrow, S$.

The proof of this theorem will be given later, except for the case of $C = \emptyset$. This case is proved in Exercise 8.33. This theorem states that the entropy of the reduced density matrix $\rho_u = \mathrm{Tr}_B |u\rangle\langle u|$ gives the degree of entanglement when the state of the total system is a pure state. Further, as shown by Hayashi and Matsumoto [23], there exists an LO protocol that attains this bound without any knowledge about the pure state u, as long as the given state is its tensor product state. That is, there exists a local operation protocol (without any communication) that produces a maximally entangled state of size $e^{nH(\rho_n)}$ and is independent of u. This protocol is often called a **universal concentration protocol**.

For a general mixed state ρ on the composite system $\mathcal{H}_A \otimes \mathcal{H}_B$, the entropy of the reduced density does not have the same implication. Consider

$$E_{r,S}(\rho) \stackrel{\text{def}}{=} \min_{\sigma \in S} D(\rho\|\sigma) \tag{8.77}$$

as its generalization for a mixed state ρ. This is called the **entanglement of relative entropy**. Any pure state $|u\rangle\langle u|$ satisfies

$$E_{r,S}(|u\rangle\langle u|) = H(\mathrm{Tr}_B |u\rangle\langle u|). \tag{8.78}$$

Lemma 8.5 (Vedral and Plenio [1]) *The entanglement of the relative entropy satisfies the monotonicity property*

$$E_{r,S}(\kappa(\rho)) \le E_{r,S}(\rho) \tag{8.79}$$

for any S-TP-CP map κ. Hence, any LOCC operation satisfies the above monotonicity because it is an S-TP-CP map.

Proof Let σ be a separable state such that $D(\rho\|\sigma) = E_r(\rho)$, then $\kappa(\sigma)$ is separable. From the monotonicity of the relative entropy (5.36),

$$E_{r,S}(\kappa(\rho)) \le D(\kappa(\rho)\|\kappa(\sigma)) \le D(\rho\|\sigma) = E_{r,S}(\rho),$$

which gives (8.79). ∎

The following theorem may be proved by using the method in the proof of Lemma 3.7.

Theorem 8.7 (Vedral and Plenio [1]) *Any mixed state ρ on the composite system $\mathcal{H}_A \otimes \mathcal{H}_B$ and any separable state σ satisfy*

$$E_{d,2}^{S,\dagger}(\rho) \le D(\rho\|\sigma). \tag{8.80}$$

Hence, we obtain

$$E_{d,2}^{S,\dagger}(\rho) \le E_{r,S}(\rho), \tag{8.81}$$

$$E_{d,2}^{S,\dagger}(\rho) \le E_{r,S}^{\infty}(\rho) \stackrel{\text{def}}{=} \lim_{n\to\infty} \frac{E_{r,S}(\rho^{\otimes n})}{n}. \tag{8.82}$$

Proof Consider an S-TP-CP map κ_n' on $\mathcal{H}_A \otimes \mathcal{H}_B$ and a real number $r > D(\rho\|\sigma)$. Since $\kappa_n'(\sigma^{\otimes n})$ is also separable, (8.7) implies that $\langle \Phi_{e^{nr}}|\kappa_n'(\sigma^{\otimes n})|\Phi_{e^{nr}}\rangle \le e^{-nr}$. From $I - |\Phi_{e^{nr}}\rangle\langle\Phi_{e^{nr}}| \ge 0$ we have

$$\begin{aligned} &I - (\kappa_n')^*(|\Phi_{e^{nr}}\rangle\langle\Phi_{e^{nr}}|) = (\kappa_n')^*(I) - (\kappa_n')^*(|\Phi_{e^{nr}}\rangle\langle\Phi_{e^{nr}}|) \\ =&(\kappa_n')^*(I - |\Phi_{e^{nr}}\rangle\langle\Phi_{e^{nr}}||) \ge 0, \end{aligned}$$

where $(\kappa_n')^*$ is the dual map of κ_n' (see ④ of Theorem 5.1). Moreover,

$$\begin{aligned} &(\kappa_n')^*(|\Phi_{e^{nr}}\rangle\langle\Phi_{e^{nr}}|) \ge 0, \\ \operatorname{Tr} \sigma^{\otimes n} &(\kappa_n')^*(|\Phi_{e^{nr}}\rangle\langle\Phi_{e^{nr}}|) = \langle \Phi_{e^{nr}}|\kappa_n'(\sigma^{\otimes n})|\Phi_{e^{nr}}\rangle \le e^{-nr}. \end{aligned} \tag{8.83}$$

Since the matrix $(\kappa_n')^*(|\Phi_{e^{nr}}\rangle\langle\Phi_{e^{nr}}|)$ satisfies the condition for the test $0 \le (\kappa_n')^*(|\Phi_{e^{nr}}\rangle\langle\Phi_{e^{nr}}|) \le I$, the inequality (3.138) in Sect. 3.8 yields

$$\langle \Phi_{e^{nr}}|\kappa_n'(\rho^{\otimes n})|\Phi_{e^{nr}}\rangle = \operatorname{Tr} \rho^{\otimes n}(\kappa_n')^*(|\Phi_{e^{nr}}\rangle\langle\Phi_{e^{nr}}|) \le e^{-n\frac{-\phi(s)-sr}{1-s}}, \tag{8.84}$$

for $s \le 0$, where $\phi(s) \stackrel{\text{def}}{=} \phi(s|\rho\|\sigma) = \log \operatorname{Tr} \rho^{1-s}\sigma^s$. Using arguments similar to those used for the Proof of Lemma 3.7, we have $\langle \Phi_{e^{nr}}|\kappa_n'(\rho^{\otimes n})|\Phi_{e^{nr}}\rangle \to 0$. We thus obtain (8.80). Applying the same arguments to $\rho^{\otimes k}$, we have

$$E_{d,2}^{S,\dagger}(\rho) \le \frac{E_{r,S}(\rho^{\otimes k})}{k}.$$

Combining this relation with Lemma A.1 in Appendix, we obtain (8.82). ■

Conversely, the following lemma holds for a pure state.

Lemma 8.6 *Any pure state $|u\rangle\langle u|$ on the composite system $\mathcal{H}_A \otimes \mathcal{H}_B$ satisfies*

$$E_{d,1}^{\rightarrow}(|u\rangle\langle u|) \ge H(\rho_u). \tag{8.85}$$

Proof When $R < H(\rho_u)$, according to Theorem 8.5, there exists an operation κ_n satisfying

$$L(\kappa_n) = \lfloor \frac{1 - \mathrm{Tr}(\rho_u^{\otimes n} - e^{-nR}) \{\rho_u^{\otimes n} - e^{-nR} \geq 0\}}{e^{-nR}} \rfloor \tag{8.86}$$

$$\varepsilon_1(\kappa_n, |u\rangle\langle u|^{\otimes n}) = \mathrm{Tr}(\rho_u^{\otimes n} - e^{-nR}) \{\rho_u^{\otimes n} - e^{-nR} \geq 0\}. \tag{8.87}$$

Define $\psi(s) \stackrel{\text{def}}{=} \log \mathrm{Tr}\, \rho_u^{1-s}$. The failure probability $\varepsilon_1(\kappa_n, |u\rangle\langle u|^{\otimes n})$ can then be calculated as

$$\begin{aligned}\varepsilon_1(\kappa_n, |u\rangle\langle u|^{\otimes n}) &\leq \mathrm{Tr}\, \rho_u^{\otimes n} \{\rho_u^{\otimes n} - e^{-nR} \geq 0\} \leq \mathrm{Tr}\left(\rho_u^{\otimes n}\right)^{1+s} e^{snR} \\ &= e^{n(\psi(-s)+sR)},\end{aligned}$$

for $s \geq 0$. Since $R < H(\rho_u)$, we can show that $\varepsilon(\kappa_n) \to 0$ using arguments similar to those given in Sect. 2.1. Based on this relation, we can show that

$$1 - \mathrm{Tr}(\rho_u^{\otimes n} - e^{-nR}) \{\rho_u^{\otimes n} - e^{-nR} \geq 0\} \to 1,$$

which proves that $\lim_{n\to\infty} \frac{\log L(\kappa_n)}{n} = R$ for $L(\kappa_n)$. Hence, we obtain (8.85). ∎

Proofs of Theorem 8.6 **and** (8.78) Let $|u\rangle = \sum_i \sqrt{p_i} |u_i \otimes u_i'\rangle$ and $\sigma = \sum_i p_i |u_i \otimes u_i'\rangle\langle u_i \otimes u_i'|$. Since u_i and u_i' are orthogonal,

$$E_{r,S}(|u\rangle\langle u|) \leq D(|u\rangle\langle u| \| \sigma) = H(\rho_u).$$

Combining (8.81) and (8.85), we obtain

$$E_{r,S}(|u\rangle\langle u|) \leq H(\mathrm{Tr}_B\, |u\rangle\langle u|) \leq E_{d,1}^{\rightarrow}(|u\rangle\langle u|) \leq E_{d,2}^{S,\dagger}(|u\rangle\langle u|) \leq E_{r,S}(|u\rangle\langle u|).$$

This proves Theorem 8.6 and (8.78). ∎

When we treat the optimum outcome case, the following value is important:

$$E_{d,L}^C(\rho) \stackrel{\text{def}}{=} \max_{\kappa=\{\kappa_\omega\}\in C} \max_{\omega} \frac{\langle \Phi_L | \kappa_\omega(\rho) | \Phi_L \rangle}{\mathrm{Tr}\, \kappa_\omega(\rho)}.$$

It can easily be checked that

$$E_{d,L}^C(\rho) = \max_{A,B} \frac{\langle \Phi_L | (A \otimes B)\rho(A \otimes B)^* | \Phi_L \rangle}{\mathrm{Tr}\, \rho(A^*A \otimes B^*B)} \tag{8.88}$$

for $C = \rightarrow \leftrightarrow, S$. This value is called conclusive teleportation fidelity and was introduced by Horodecki et al. [24]; it describes the relation between this value and the conclusive teleportation.

Exercises

8.32 Define the **entanglement of exact distillation** $E_{d,e}^C(\rho)$ and the **asymptotic entanglement of exact distillation** $E_{d,e}^{C,\infty}(\rho)$

$$E_{d,e}^{C,\infty}(\rho) \stackrel{\text{def}}{=} \lim_{n\to\infty} \frac{E_{d,e}^{C}(\rho^{\otimes n})}{n}, \quad E_{d,e}^{C}(\rho) \stackrel{\text{def}}{=} \max_{\kappa\in C} \{\log L(\kappa) | \varepsilon_2(\kappa,\rho)=0\}, \tag{8.89}$$

and show [21, 25]

$$E_{d,e}^{\leftrightarrow,\infty}(|u\rangle\langle u|) = -\log \lambda_1^{\downarrow}.$$

(This bound can be attained by a tournamentlike method for $d = 2$, but such a method is known to be impossible for $d > 2$ [26, 27].)

8.33 Let $u = \sum_i \sqrt{\lambda_i} u_i^A \otimes u_i^B$ be a Schmidt decomposition, and define the POVM $\boldsymbol{M}^{X,n} = \{M_q^{X,n}\}_{q\in T_n}$ as $M_q^{X,n} \stackrel{\text{def}}{=} \sum_{i\in T_q^n} |u_i^X\rangle\langle u_i^X|$ for $X = A, B$. Apply the measurements $\boldsymbol{M}^{A,n}$ and $\boldsymbol{M}^{B,n}$ to the both sides when the state is $|u\rangle^{\otimes n}$. Show that the resultant state with the measurement outcome q is a maximally entangled state with the size $|T_q^n|$. Using this protocol, show $E_{d,1}^{\emptyset}(|u\rangle\langle u|) \geq H(\rho_u)$. This protocol is called a Procrustean method [22].

8.34 Define the generalized Bell states $u_{i,j}^{A,B} \stackrel{\text{def}}{=} (I_A \otimes \mathbf{X}_B^i \mathbf{Z}_B^j) u_{0,0}^{A,B}$, and the **generalized Bell diagonal state** $\rho_{\text{Bell},p} \stackrel{\text{def}}{=} \sum_{i,j} p_{i,j} |u_{i,j}^{A,B}\rangle\langle u_{i,j}^{A,B}|$, where $u_{0,0}^{A,B} \stackrel{\text{def}}{=} \sum_i u_i^A \otimes u_i^B$. Show that $E_{r,S}(\rho_{\text{Bell},p}) \leq \log d - H(p)$. For the definition of $\mathbf{X}_B$ and $\mathbf{Z}_B$, see Example 5.8.

8.35 Define the quantity

$$E_{d,i}^{C}(r|\rho) \stackrel{\text{def}}{=} \sup_{\{\kappa_n\}\subset C} \left\{ \underline{\lim} -\frac{1}{n} \log \varepsilon_i(\kappa_n,\rho) \middle| \underline{\lim} -\frac{1}{n} \log L(\kappa_n) \geq r \right\} \tag{8.90}$$

for $i = 1, 2$ and any class C. Then, show [21, 25]

$$E_{d,1}^{C}(r||u\rangle\langle u|) = \max_{s\geq 0} s(H_{1+s}(\rho_u) - r) \text{ for } C = \to, \leftarrow, \leftrightarrow .$$

8.36 Define the quantity

$$E_{d,i}^{C,*}(r|\rho) \stackrel{\text{def}}{=} \inf_{\{\kappa_n\}\subset C} \left\{ \overline{\lim} -\frac{1}{n} \log(1-\varepsilon_i(\kappa_n,\rho)) \middle| \underline{\lim} -\frac{1}{n} \log L(\kappa_n) \geq r \right\} \tag{8.91}$$

for $i = 1, 2$ and any class C. Then, show [21, 25]

$$E_{d,2}^{S,*}(r||u\rangle\langle u|) \geq \max_{t\geq 0} \frac{t(r - H_{1-t}(\rho_u))}{1+t} \tag{8.92}$$

8.37 Show the following equation:

$$E_{d,e}^{S,\infty}(|u\rangle\langle u|) = -\log \lambda_1^{\downarrow}. \tag{8.93}$$

(**a**) Show

$$\max_{\sigma \in S_s} \mathrm{Tr}\, |\Phi_d\rangle\langle\Phi_d| \kappa(\sigma) \le \frac{1}{d} \tag{8.94}$$

(**b**) Show

$$\max_{\sigma \in S_s} \mathrm{Tr}\, |u\rangle\langle u| \sigma = \lambda_1^{\downarrow}. \tag{8.95}$$

(**c**) Check the following relation:

$$\begin{aligned}
&\max_{\kappa \in S} \{d |\, \mathrm{Tr}\, \kappa(|u\rangle\langle u|) |\Phi_d\rangle\langle\Phi_d| = 1\} \\
\overset{(a)}{\le}\, &\max_{\kappa \in S} \{\min_{\sigma \in S_s} (\mathrm{Tr}\, |\Phi_d\rangle\langle\Phi_d| \kappa(\sigma))^{-1} |\, \mathrm{Tr}\, \kappa(|u\rangle\langle u|) |\Phi_d\rangle\langle\Phi_d| = 1\} \\
\overset{(b)}{=}\, &\max_{\kappa \in S} \{\min_{\sigma \in S_s} (\mathrm{Tr}\, \kappa^*(|\Phi_d\rangle\langle\Phi_d|) \sigma)^{-1} |\, \mathrm{Tr}\, |u\rangle\langle u| \kappa^*(|\Phi_d\rangle\langle\Phi_d|) = 1\} \\
\overset{(c)}{\le}\, &\max_{0 \le T \le I} \{\min_{\sigma \in S_s} (\mathrm{Tr}\, T\sigma)^{-1} |\, \mathrm{Tr}\, |u\rangle\langle u| T = 1\} \\
\overset{(d)}{=}\, &\min_{\sigma \in S_s} (\mathrm{Tr}\, |u\rangle\langle u| \sigma)^{-1} \overset{(e)}{=} (\lambda_1^{\downarrow})^{-1},
\end{aligned} \tag{8.96}$$

where S_s is the set of separable states.
(**d**) Prove (8.93).

8.6 Dilution of Maximally Entangled States

In the previous section, we considered the problem of producing a maximally entangled state from the tensor product of a particular entangled state. In this section, we examine the converse problem, i.e., to produce a tensor product state of a particular entangled state from a maximally entangled state in the composite system $\mathcal{H}_A \otimes \mathcal{H}_B$. In this book, we call this problem **entanglement dilution** even if the required state is mixed while historically it has been called this only for the pure-state case.

For an analysis of the mixed state ρ on the composite system $\mathcal{H}_A \otimes \mathcal{H}_B$, we define the **entanglement of formation** $E_f(\rho)$ for a state ρ in the composite system $\mathcal{H}_A \otimes \mathcal{H}_B$ based on the probabilistic decomposition $\{(p_i, \rho_i)\}$ of ρ [28]:

$$E_f(\rho) \overset{\mathrm{def}}{=} \min_{\{(p_i, \rho_i)\}} \sum_i p_i H(\mathrm{Tr}_B\, \rho_i). \tag{8.97}$$

Since this minimum value is attained when all ρ_i are pure, this minimization can be replaced by the minimization for probabilistic decompositions by pure states. From the above definition, a state ρ_1 on $\mathcal{H}_{A_1} \otimes \mathcal{H}_{B_1}$ and a state ρ_2 on $\mathcal{H}_{A_2} \otimes \mathcal{H}_{B_2}$ satisfy

$$E_f(\rho_1) + E_f(\rho_2) \geq E_f(\rho_1 \otimes \rho_2). \tag{8.98}$$

Theorem 8.8 (Vidal [16]) *Perform an operation corresponding to the S-TP-CP map κ with respect to a maximally entangled state $|\Phi_L\rangle\langle\Phi_L|$ of size L (the initial state). The fidelity between the final state and the target pure state $|x\rangle\langle x|$ on $\mathcal{H}_A \otimes \mathcal{H}_B$ then satisfies*

$$\begin{aligned}&\max_{\kappa\in S} F(\kappa(|\Phi_L\rangle\langle\Phi_L|), |x\rangle\langle x|)\\ =&\max_{\kappa\in\rightarrow} F(\kappa(|\Phi_L\rangle\langle\Phi_L|), |x\rangle\langle x|) = \sqrt{P(x, L)},\end{aligned} \tag{8.99}$$

where $P(u, L)$ is defined using the Schmidt coefficients $\sqrt{\lambda_i}$ of $|u\rangle$ as follows:

$$P(u, L) \stackrel{\text{def}}{=} \sum_{i=1}^{L} \lambda_i^{\downarrow}. \tag{8.100}$$

Note the similarity between $P(u, L)$ and $P(p, L)$ given in Sect. 2.1.4. Furthermore, the fidelity between the final state and a general mixed state ρ on $\mathcal{H}_A \otimes \mathcal{H}_B$ satisfies

$$\max_{\kappa\in S} F(\kappa(|\Phi_L\rangle\langle\Phi_L|), \rho) = \max_{\{(p_i,x_i)\}} \sqrt{\sum_i p_i P(x_i, L)}, \tag{8.101}$$

where $\{(p_i, x_i)\}$ is the probabilistic decomposition of ρ.

Using (2.50), we obtain

$$1 - P(x, [e^R]) \leq e^{\frac{\log \operatorname{Tr} \rho_x^{1-s} - sR}{1-s}}, \tag{8.102}$$

where $0 \leq s \leq 1$ and $\rho_x \stackrel{\text{def}}{=} \operatorname{Tr}_B |x\rangle\langle x|$.

Proof The proof only considers the case of a pure state $|x\rangle\langle x|$. Let $\{E_{A,i} \otimes E_{B,i}\}_i$ be the Choi–Kraus representation of the S-TP-CP map κ. Then

$$\kappa(|\Phi_L\rangle\langle\Phi_L|) = \sum_i \left(E_{A,i} \otimes E_{B,i}\right) |\Phi_L\rangle\langle\Phi_L| \left(E_{A,i} \otimes E_{B,i}\right)^*.$$

Taking the partial trace inside of the summation $\sum_i$ on the RHS, we have

$$\begin{aligned}&\operatorname{Tr}_B \left(E_{A,i} \otimes E_{B,i}\right) |\Phi_L\rangle\langle\Phi_L| \left(E_{A,i} \otimes E_{B,i}\right)^*\\ =&(E_{A,i} X_{\Phi_L} E_{B,i}^T)(E_{A,i} X_{\Phi_L} E_{B,i}^T)^*\end{aligned}$$

from (1.22). Its rank is less than L. Let y be a pure state on the composite system such that the rank of the reduced density of y is equal to L. Thus, by proving that

$$|\langle y|x\rangle| \le \sqrt{P(x,L)}, \tag{8.103}$$

the proof can be completed. To this end, we define the pure state $|y_i\rangle\langle y_i|$ as

$$q_i|y_i\rangle\langle y_i| = \left(E_{A,i}\otimes E_{B,i}\right)|\Phi_L\rangle\langle\Phi_L|\left(E_{A,i}\otimes E_{B,i}\right)^*,$$

where q_i is a normalized constant. Then,

$$F^2(\kappa(|\Phi_L\rangle\langle\Phi_L|), |x\rangle\langle x|) = \sum_i q_i F^2(|y_i\rangle\langle y_i|, |x\rangle\langle x|) \le \sum_i q_i P(x,L).$$

We can use this relation to show that the fidelity does not exceed $\sqrt{P(x,L)}$. In a proof of (8.103), it is sufficient to show that $F(\rho_x,\sigma) \le \sqrt{P(x,L)}$ for $\rho_x \stackrel{\text{def}}{=} \mathrm{Tr}_B\,|x\rangle\langle x|$ and a density matrix σ of rank L. First, let $\sigma \stackrel{\text{def}}{=} \sum_{i=1}^L p_i|v_i\rangle\langle v_i|$ and let P be the projection to the range of σ. Since the rank of P is L, choosing an appropriate unitary matrix U, we obtain the following relation:

$$\mathrm{Tr}\,|\sqrt{\rho_x}\sqrt{\sigma}| = \mathrm{Tr}\,\sqrt{\rho_x}\sqrt{\sigma}U = \mathrm{Tr}\,\sqrt{\rho_x}\left(\sum_{i=1}^L \sqrt{p_i}|v_i\rangle\langle v_i|U\right)$$

$$= \sum_{i=1}^L \sqrt{p_i}\langle v_i|U\sqrt{\rho_x}|v_i\rangle \le \sqrt{\sum_{i=1}^L p_i}\sqrt{\sum_{i=1}^L |\langle v_i|U\sqrt{\rho_x}|v_i\rangle|^2} \tag{8.104}$$

$$= \sqrt{\sum_{i=1}^L \langle v_i|\sqrt{\rho_x}U^*|v_i\rangle\langle v_i|U\sqrt{\rho_x}|v_i\rangle} \le \sqrt{\sum_{i=1}^L \langle v_i|\sqrt{\rho_x}\sqrt{\rho_x}|v_i\rangle} \tag{8.105}$$

$$= \sqrt{\mathrm{Tr}\,P\rho_x} \le \sqrt{P(x,L)}. \tag{8.106}$$

This evaluation can be checked as follows. Inequality (8.104) follows from the Schwarz inequality. Inequality (8.105) follows from $U^*|v_i\rangle\langle v_i|U \le I$. The final inequality (8.106) can be derived from the fact that P is a projection of rank L. Thus, we obtain

$$\max_{\kappa\in S} F(\kappa(|\Phi_L\rangle\langle\Phi_L|), |x\rangle\langle x|) \le \sqrt{P(x,L)}.$$

Conversely, we can verify the existence of an S-TP-CP map with a fidelity of $\sqrt{P(x,L)}$ by the following argument. There exists a pure state y satisfying the equality in (8.103); this can be confirmed by considering the conditions for equality in the above inequalities. Since the pure state $|y\rangle\langle y|$ satisfies $\mathrm{Tr}_B\,|\Phi_L\rangle\langle\Phi_L| \preceq \mathrm{Tr}_B\,|y\rangle\langle y|$, according to Theorem 8.4, there exists a one-way LOCC that produces the pure state $|y\rangle\langle y|$ from the maximally entangled state $|\Phi_L\rangle\langle\Phi_L|$, i.e., it attains the RHS of (8.99). This proves the existence of an S-TP-CP map with a fidelity of $\sqrt{P(x,L)}$. ■

Next, let us consider how large a maximally entangled state is required for producing n tensor products of the entangled state ρ. In order to examine its asymptotic case, we focus on the S-TP-CP map κ_n to produce the state $\rho^{\otimes n}$. The following **entanglement of cost** $E_c^C(\rho)$ expresses the asymptotic conversion rate

$$E_c^C(\rho) \stackrel{\text{def}}{=} \inf_{\{\kappa_n\}\subset C} \left\{ \overline{\lim} \frac{1}{n} \log L_n \middle| \lim_{n\to\infty} F(\rho^{\otimes n}, \kappa_n(|\Phi_{L_n}\rangle\langle\Phi_{L_n}|)) = 1 \right\}, \quad (8.107)$$

which is the subject of the following theorem.

Theorem 8.9 (Bennett et al. [22], Hayden et al. [29]) *For any state ρ on the composite system $\mathcal{H}_A \otimes \mathcal{H}_B$,*

$$E_c^C(\rho) = \lim_{n\to\infty} \frac{E_f(\rho^{\otimes n})}{n} = \inf_n \frac{E_f(\rho^{\otimes n})}{n}, \quad (8.108)$$

for $C = \rightarrow, \leftarrow, \leftrightarrow, S$.

above theorem implies that the entanglement cost $E_c^C(\rho)$ has the same value for $C = \rightarrow, \leftarrow, \leftrightarrow, S$. Hence, we denote it by $E_c(\rho)$.

Proof of Pure-State Case. We prove Theorem 8.9 by analyzing the pure state $|x\rangle\langle x|$ in greater detail and by noting that $\psi(s) = \psi(s|\operatorname{Tr}_B |x\rangle\langle x|)$. For any $R > H(x)$, we can calculate how fast the quantity (1 – Optimal fidelity) approaches 0 according to

$$\begin{aligned}\lim_{n\to\infty} -\frac{1}{n}\log\left(1-\sqrt{P(p^n, e^{nR})}\right) &= \lim_{n\to\infty} -\frac{1}{n}\log(1-P(p^n, e^{nR})) \\ &= -\min_{0\le s\le 1} \frac{\psi(s)-sR}{1-s},\end{aligned}$$

where we used (2.188) for $P(x^{\otimes n}, e^{nR})$ and $1-\sqrt{1-\epsilon} \cong \frac{1}{2}\epsilon$. If $R < H(x)$, the fidelity approaches zero for any LOCC (or separable) operation. The speed of this approach is

$$\lim_{n\to\infty} -\frac{1}{n}\log\sqrt{P(p^n, e^{nR})} = -\frac{1}{2}\min_{s\le 0} \frac{\psi(s)-sR}{1-s}, \quad (8.109)$$

where we used (2.190). From these inequalities, we have $E_c^C(\rho) = H(\rho_u)$ for $C = \rightarrow, \leftarrow, \leftrightarrow, S$ using the relationship between $H(p)$ and $\psi(s)$ given in Sect. 2.1.4. That is, we may also derive Theorem 8.9 in the pure-state case. ■

Since the additivity relation

$$\frac{E_f(\rho^{\otimes n})}{n} = E_f(\rho) \quad (8.110)$$

holds in the pure-state case, the entanglement of cost has a simple expression:

$$E_c^C(\rho) = E_f(\rho), \quad (8.111)$$

for $C = \rightarrow, \leftarrow, \leftrightarrow, S$. However, it is not known whether this formula holds for mixed states, except in a few cases, which will be treated later. Certainly, this problem is closely connected with other open problems, as will be discussed in Sect. 9.2.

Similar to (8.89), we define the **entanglement of exact cost** $E^C_{c,e}(\rho)$ and the **asymptotic entanglement of exact cost** $E^{C,\infty}_{c,e}(\rho)$

$$E^{C,\infty}_{c,e}(\rho) \stackrel{\text{def}}{=} \lim_{n\to\infty} \frac{E^C_{c,e}(\rho^{\otimes n})}{n}, \quad E^C_{c,e}(\rho) \stackrel{\text{def}}{=} \min_{\kappa \in C} \{\log L | \, F(\rho, \kappa(|\Phi_L\rangle\langle\Phi_L|)) = 1\} \tag{8.112}$$

and the logarithm of the **Schmidt rank** for a mixed state ρ:

$$E_{sr}(\rho) \stackrel{\text{def}}{=} \min_{\{(p_i,\rho_i)\}} \max_{i: p_i > 0} \log \operatorname{rank} \rho_i, \tag{8.113}$$

where $\sum_i p_i \rho_i = \rho$ and ρ_i is a pure state. Due to Theorem 8.4, any operation κ satisfying $F(\rho, \kappa(|\Phi_L\rangle\langle\Phi_L|)) = 1$ makes a decomposition $\sum_i p_i \rho_i = \rho$ such that ρ_i is a pure state and the rank of ρ_i is less than L. So, we have $E^C_{c,e}(\rho) = E_{sr}(\rho)$. Hence,

$$E^{C,\infty}_{c,e}(\rho) = \lim_{n\to\infty} \frac{E_{sr}(\rho^{\otimes n})}{n} \quad \text{for } C = \rightarrow, \leftarrow, \leftrightarrow, S. \tag{8.114}$$

Any pure state $|u\rangle\langle u|$ satisfies the additivity $E_{sr}(|u\rangle\langle u|^{\otimes n}) = n E_{sr}(|u\rangle\langle u|)$. However, the quantity $E_{sr}(\rho)$ with a mixed state ρ does not necessarily satisfy the additivity. Moreover, such that $E_{sr}(\rho) = E_{sr}(\rho^{\otimes 2})$ [30].

Exercises

8.38 Let $\mathcal{H}_A = \mathcal{H}_B = \mathbb{C}^3$. Choose a state ρ on $\mathcal{H}_A \otimes \mathcal{H}_B$ such that the support belongs to $\{v \otimes u - u \otimes v | u, v \in \mathbb{C}^3\}$. Show that $E_f(\rho) = \log 2$ [31].

8.39 Show that E_f satisfies the **monotonicity** for a two-way LOCC κ.

$$E_f(\rho) \geq E_f(\kappa(\rho)). \tag{8.115}$$

8.40 Show that $E^{C,\dagger}_c(|u\rangle\langle u|) \geq E(\operatorname{Tr}_B |u\rangle\langle u|)$ for any pure state $|u\rangle\langle u|$ by defining $E^{C,\dagger}_c(\rho)$ in a similar way to Theorem 8.7. This argument can be regarded as the strong converse version of Theorem 8.9.

8.7 Unified Approach to Distillation and Dilution

In this section, we derive the converse parts of distillation and dilution based on the following unified method. In this method, for a class of local operations C, we focus on the entanglement measure $E^C(\rho)$ of a state $\rho \in \mathcal{S}(\mathcal{H}_A \otimes \mathcal{H}_B)$ that satisfies the following axioms.

E1 (Normalization) $E^C(\rho) = \log d$ when ρ is a maximally entangled state of size d.

E2*C* (Monotonicity) $E^C(\kappa(\rho)) \le E^C(\rho)$ holds for any local operation κ in class C.

E3 (Continuity) When any two states ρ_n and σ_n of system $\mathcal{H}_n$ satisfy $\|\rho_n - \sigma_n\|_1 \to 0$, the convergence $\frac{|E^C(\rho_n)-E^C(\sigma_n)|}{\log \dim \mathcal{H}_n} \to 0$ holds.

E4 (Convergence) The quantity $\frac{E^C(\rho^{\otimes n})}{n}$ converges as $n \to \infty$.

Based on the above conditions only, we can prove the following theorem.

Theorem 8.10 (Donald et al. [32]) *When the quantity $E^C(\rho)$ satisfies the above conditions,*

$$E_{d,2}^C(\rho) \le E^{C,\infty}(\rho) \left(\stackrel{\text{def}}{=} \lim_{n\to\infty} \frac{E^C(\rho^{\otimes n})}{n} \right) \le E_c^C(\rho). \tag{8.116}$$

Proof Let κ_n be a local operation κ_n in class C from $(\mathcal{H}_A)^{\otimes n} \otimes (\mathcal{H}_B)^{\otimes n}$ to $\mathbb{C}^{d_n} \otimes \mathbb{C}^{d_n}$ such that[4]

$$\| \, |\Phi_{d_n}\rangle\langle\Phi_{d_n}| - \kappa_n(\rho^{\otimes n})\|_1 \to 0, \tag{8.117}$$

where $\frac{\log d_n}{n} \to E_{d,2}^C(\rho)$. From Conditions **E1** and **E3** and (8.117), we have

$$\begin{aligned} &\left| \frac{E^C(\kappa_n(\rho^{\otimes n}))}{n} - E_{d,2}^C(\rho) \right| \\ \le &\frac{|E^C(\kappa_n(\rho^{\otimes n})) - E^C(|\Phi_{d_n}\rangle\langle\Phi_{d_n}|)|}{n} + \left| \frac{\log d_n}{n} - E_{d,2}^C(\rho) \right| \to 0. \end{aligned}$$

Hence, Condition **E2*C*** guarantees that

$$\lim_{n\to\infty} \frac{E^C(\rho^{\otimes n})}{n} \ge \lim_{n\to\infty} \frac{E^C(\kappa_n(\rho^{\otimes n}))}{n} = E_{d,2}^C(\rho). \tag{8.118}$$

We obtain the first inequality.

Next, we choose a local operation κ_n in class C from $\mathbb{C}^{d_n} \otimes \mathbb{C}^{d_n}$ to $(\mathcal{H}_A)^{\otimes n} \otimes (\mathcal{H}_B)^{\otimes n}$ such that

$$\|\kappa_n(|\Phi_{d_n}\rangle\langle\Phi_{d_n}|) - \rho^{\otimes n}\|_1 \to 0,$$

where $\frac{\log d_n}{n} \to E_c^C(\rho)$. Similarly, we can show $|\frac{E^C(\kappa_n(|\Phi_{d_n}\rangle\langle\Phi_{d_n}|))}{n} - \frac{E^C(\rho^{\otimes n})}{n}| \to 0$. Since $\frac{E^C(\kappa_n(|\Phi_{d_n}\rangle\langle\Phi_{d_n}|))}{n} \le \frac{\log d_n}{n}$, we obtain

[4] If the operation κ_n in C has a larger output system than $\mathbb{C}^{d_n} \otimes \mathbb{C}^{d_n}$, there exists an operation κ_n' in C with the output system $\mathbb{C}^{d_n} \otimes \mathbb{C}^{d_n}$ such that $\| \, |\Phi_{d_n}\rangle\langle\Phi_{d_n}| - \kappa_n(\rho^{\otimes n})\|_1 \ge \| \, |\Phi_{d_n}\rangle\langle\Phi_{d_n}| - \kappa_n'(\rho^{\otimes n})\|_1$.

$$\lim_{n\to\infty} \frac{E^C(\rho^{\otimes n})}{n} \le E_c^C(\rho).$$

■

For example, the entanglement of formation $E_f(\rho)$ satisfies Conditions **E1**, **E2**$\leftrightarrow$ (Exercise 8.39), **E3** (Exercise 8.42), and **E4** (Lemma A.1). Using this fact, Theorem 8.10 yields an alternative proof of the converse part of Theorem 8.9. The entanglement of relative entropy $E_{r,S}(\rho)$ also satisfies Conditions **E1**, **E2**s (Lemma 8.5), and **E4** (Lemma A.1). Further, Donald and Horodecki [33] showed Condition **E3** for entanglement of relative entropy $E_{r,S}(\rho)$. Similarly, by using this fact, Theorem 8.10 yields an alternative proof of (8.82) of Theorem 8.7.

In addition, the **maximum of the negative conditional entropy**

$$E_m^C(\rho) \stackrel{\text{def}}{=} \max_{\kappa\in C} -H_{\kappa(\rho)}(A|B) \tag{8.119}$$

satisfies Conditions **E1**, **E2**C, **E3** (Exercise 8.44), and **E4** (Lemma A.1) for $C = \to, \leftrightarrow, S$. Thus,

$$E_{d,2}^C(\rho) \le \lim_{n\to\infty} \frac{E_m^C(\rho^{\otimes n})}{n} \le E_c^C(\rho) \tag{8.120}$$

for $C =\to, \leftrightarrow, S$. Conversely, as will be proved in Sect. 9.6, the opposite inequality (**Hashing inequality**)

$$E_{d,2}^{\to}(\rho) \ge -H_\rho(A|B) \tag{8.121}$$

holds, i.e., there exists a sequence of one-way LOCC operations producing an approximate maximally entangled state of an approximate size of $e^{-nH_\rho(A|B)}$. Performing the local operation κ_n in class C, we can prepare the state $\kappa_n(\rho^{\otimes n})$. Applying this sequence of one-way LOCC operations to the state $\kappa_n(\rho^{\otimes n})$, we can show that $E_{d,2}^C(\rho) \ge E_{d,2}^C(\rho^{\otimes n}) \ge -H_{\kappa_n(\rho^{\otimes n})}(A|B)$, which implies $E_{d,2}^C(\rho) \ge \frac{E_m^C(\rho^{\otimes n})}{n}$. Thus, we obtain

$$E_{d,2}^C(\rho) = \lim_{n\to\infty} \frac{E_m^C(\rho^{\otimes n})}{n}. \tag{8.122}$$

Therefore, since the relation $E_{r,S}(\rho) \le E_f(\rho)$ holds[Exe. 8.41], we obtain

$$\begin{aligned} E_{d,2}^C(\rho) &= \lim_{n\to\infty} \frac{E_m^C(\rho^{\otimes n})}{n} \le E_{d,2}^{C,\dagger}(\rho) \le \lim_{n\to\infty} \frac{E_{r,S}(\rho^{\otimes n})}{n} \\ &\le \lim_{n\to\infty} \frac{E_f(\rho^{\otimes n})}{n} = E_c^C(\rho). \end{aligned} \tag{8.123}$$

We also have the following relations without the limiting forms:

$$-H_\rho(A|B) \le E_m^C(\rho) \le E_{d,2}^C(\rho) \le E_{d,2}^{C,\dagger}(\rho) \le E_{r,S}(\rho) \le E_f(\rho). \tag{8.124}$$

The above quantities are the same in the pure-state case. However, the equalities do not necessarily hold in the mixed-state case.

Indeed, the expression of $E_m^C(\rho)$ can be slightly simplified as follows. Consider a TP-CP κ with the Choi–Kraus representation $\{F_i\}$. This operation is realized by the following process. First, we perform the measurement $\boldsymbol{M} = \{M_i\}_{i=1}^k$ and obtain the outcome i with probability $p_i = \mathrm{P}_\rho^{\boldsymbol{M}}(i)$, where $F_i = U_i\sqrt{M_i}$. Next, we perform the isometry matrix U_i dependently of i. All outcomes are sent to system B. Finally, we take the partial trace with respect to the measurement outcome on $\mathcal{H}_B$. Hence, Inequality (5.88) yields

$$\begin{aligned} &- H_{\kappa(\rho)}(A|B) \le -H_{\bigoplus_{i=1}^k U_i\sqrt{M_i}\rho\sqrt{M_i}U_i^*}(A|B) \\ &= -\sum_i p_i H_{\frac{U_i\sqrt{M_i}\rho\sqrt{M_i}U_i^*}{p_i}}(A|B). \end{aligned}$$

Since operation κ is separable at least, the unitary U_i has the form $U_i^A \otimes U_i^B$. Hence,

$$H_{\frac{U_i\sqrt{M_i}\rho\sqrt{M_i}U_i^*}{p_i}}(A|B) = H_{\frac{\sqrt{M_i}\rho\sqrt{M_i}}{p_i}}(A|B).$$

Therefore,

$$E_m^C(\rho) = \max_{M\in C} -\sum_i p_i H_{\frac{\sqrt{M_i}\rho\sqrt{M_i}}{p_i}}(A|B) = \max_{M\in C} -H_{\hat{\kappa}_M(\rho)}(A|BE), \tag{8.125}$$

where $p_i \stackrel{\text{def}}{=} \mathrm{P}_\rho^{\boldsymbol{M}}(i)$, and $\mathcal{H}_E$ is the space spanned by $\{e_i^E\}$ because

$$\hat{\kappa}_{\boldsymbol{M}}(\rho) = \sum_i p_i \frac{\sqrt{M_i}\rho\sqrt{M_i}}{p_i} \otimes |e_i^E\rangle\langle e_i^E|. \tag{8.126}$$

As another measure of entanglement, Christandl and Winter [34] introduced **squashed entanglement**:

$$E_{sq}(\rho) \stackrel{\text{def}}{=} \inf\left\{ \frac{1}{2} I_{\rho_{A,B,E}}(A:B|E) \middle| \rho_{A,B,E} : \mathrm{Tr}_E\, \rho_{A,B,E} = \rho \right\}. \tag{8.127}$$

It satisfies Conditions **E1**, **E2**$\leftrightarrow$ (see [34]), **E3** (Exercise 8.43), and **E4** and the additivity (Exercise 8.45)

$$E_{sq}(\rho) + E_{sq}(\sigma) = E_{sq}(\rho\otimes\sigma). \tag{8.128}$$

Hence, Theorem 8.10 implies that

$$E_{d,2}^{\leftrightarrow}(\rho) = \lim_{n\to\infty} \frac{E_m^{\leftrightarrow}(\rho^{\otimes n})}{n} \le E_{sq}(\rho) \le \lim_{n\to\infty} \frac{E_f(\rho^{\otimes n})}{n} = E_c^C(\rho).$$

Now, we give a theorem to calculate $E_{d,2}(\rho)$.

Theorem 8.11 *For a given state ρ on $\mathcal{H}_A \otimes \mathcal{H}_B$, there exists a TP-CP map κ from system $\mathcal{H}_B$ to system the reference $\mathcal{H}_R$ such that the equation*

$$\begin{aligned}&\kappa(\mathrm{Tr}_{A,R}(M_i \otimes I_{B,R})|x\rangle\langle x|)\\ =&\mathrm{Tr}_{A,B}(M_i \otimes I_{B,R})|x\rangle\langle x|)\end{aligned} \tag{8.129}$$

holds for any POVM $\boldsymbol{M} = \{M_i\}$ on $\mathcal{H}_A$, where $|x\rangle$ is a purification of ρ with the reference system $\mathcal{H}_R$. Then, the quantity $E_m^{\rightarrow}(\rho)$ is calculated as

$$-H_\rho(A|B) = E_m^{\rightarrow}(\rho) = \frac{E_m^{\rightarrow}(\rho^{\otimes n})}{n} = E_{d,2}^{\rightarrow}(\rho). \tag{8.130}$$

Further, the condition for Theorem 8.11 holds in the following case. The system $\mathcal{H}_B$ can be decomposed as a composite system $\mathcal{H}_{B,1} \otimes \mathcal{H}_{B,2}$ such that the system $\mathcal{H}_{B,1}$ is unitarily equivalent to $\mathcal{H}_R$ by a unitary U. Moreover, the state $\mathrm{Tr}_{\mathcal{H}_A,\mathcal{H}_{B,2}} |x\rangle\langle x|$ commutes the projection to the symmetric subspace of $\mathcal{H}_R \otimes \mathcal{H}_{B,1}$, which is spanned by the set $\{U(x) \otimes y + U(y) \otimes x | x, y \in \mathcal{H}_{B,1}\}$. In this case, any state ρ_s on the symmetric subspace and any state ρ_a on the antisymmetric subspace satisfy $U \,\mathrm{Tr}_R\, \rho_i U^* = \mathrm{Tr}_{B,1}\, \rho_i$ for $i = s, a$. Note that the antisymmetric subspace is spanned by the set $\{U(x) \otimes y - U(y) \otimes x | x, y \in \mathcal{H}_{B,1}\}$. Hence, the map κ satisfying (8.129) is given as the map $\rho \mapsto U \,\mathrm{Tr}_{B,2}\, \rho U^*$ with the state ρ on the system $\mathcal{H}_B = \mathcal{H}_{B,1} \otimes \mathcal{H}_{B,2}$.

Proof of Theorem 8.11 For any one-way ($\mathcal{H}_A \rightarrow \mathcal{H}_B$) LOCC operation κ', the local operation on $\mathcal{H}_A$ can be described by the Choi–Kraus representation $\{F_i\}$ on $\mathcal{H}_A$, and the operation on $\mathcal{H}_B$ can be described by a set of TP-CP maps $\{\kappa_i\}$ on $\mathcal{H}_B$. Let $|x\rangle$ be the purification of ρ with the reference $\mathcal{H}_R$. Then, the measurement outcome i is obtained with the probability $p_i = \mathrm{Tr}\, F_i(\mathrm{Tr}_B\, \rho) F_i^*$, and the resultant states with the measurement outcome i on $\mathcal{H}_B$ and $\mathcal{H}_R$ are the states $\frac{1}{p_i} \mathrm{Tr}_{A,R}(M_i \otimes I_{B,R})|x\rangle\langle x|$ and $\frac{1}{p_i} \mathrm{Tr}_{A,B}(M_i \otimes I_{B,R})|x\rangle\langle x|$, respectively, where $M_i \stackrel{\text{def}}{=} F_i^* F_i$. Since the monotonicity of transmission information (Exercise 5.23) for the TP-CP map κ given in (8.129) implies that

$$\begin{aligned}&H\left(\sum_i p_i \,\mathrm{Tr}_{A,R}(M_i \otimes I_{B,R})|x\rangle\langle x|\right) - \sum_i p_i H(\mathrm{Tr}_{A,R}(M_i \otimes I_{B,R})|x\rangle\langle x|)\\ \geq &H\left(\sum_i p_i \,\mathrm{Tr}_{A,B}(M_i \otimes I_{B,R})|x\rangle\langle x|\right) - \sum_i p_i H(\mathrm{Tr}_{A,B}(M_i \otimes I_{B,R})|x\rangle\langle x|).\end{aligned}$$

Since $\sum_i p_i \,\mathrm{Tr}_{A,R}(M_i \otimes I_{B,R})|x\rangle\langle x|) = \mathrm{Tr}_{A,R}\, |x\rangle\langle x|$ and $\sum_i p_i \,\mathrm{Tr}_{A,B}(M_i \otimes I_{B,R})|x\rangle\langle x| = \mathrm{Tr}_{A,B}\, |x\rangle\langle x|$, we have

$$
\begin{aligned}
&- H_\rho(A|B) \\
=&H\left(\sum_i p_i \operatorname{Tr}_{A,R}(M_i \otimes I_{B,R})|x\rangle\langle x|\right) - H\left(\sum_i p_i \operatorname{Tr}_{A,B}(M_i \otimes I_{B,R})|x\rangle\langle x|\right) \\
\geq &\sum_i p_i H(\operatorname{Tr}_{A,R}(M_i \otimes I_{B,R})|x\rangle\langle x|) - \sum_i p_i H(\operatorname{Tr}_{A,B}(M_i \otimes I_{B,R})|x\rangle\langle x|) \\
= &-\sum_i p_i H_{\rho_i}(A|B).
\end{aligned}
$$

Further, from inequality (5.111)

$$
-H_{\kappa'(\rho)}(A|B) = -\sum_i p_i H_{(\iota_A \otimes \kappa_i)(\rho_i)}(A|B) \leq -\sum_i p_i H_{\rho_i}(A|B).
$$

Hence, we obtain $-H_{\kappa'(\rho)}(A|B) \leq -H_\rho(A|B)$.

Further, in this case, the tensor product state $\rho^{\otimes n}$ also satisfies this condition. Hence, using (8.122), we obtain $-H_\rho(A|B) = \frac{E_m^{\rightarrow}(\rho^{\otimes n})}{n} = E_{d,2}^{\rightarrow}(\rho)$. ∎

As generalizations of $E_m^C(\rho)$ and $E_{r,S}(\rho)$, we define

$$
E_{1+s|m}^C(\rho) \stackrel{\text{def}}{=} \max_{\kappa \in C} -H_{1+s|\kappa(\rho)}^{\uparrow}(A|B) \tag{8.131}
$$

$$
\tilde{E}_{1+s|m}^C(\rho) \stackrel{\text{def}}{=} \max_{\kappa \in C} -\tilde{H}_{1+s|\kappa(\rho)}^{\uparrow}(A|B) \tag{8.132}
$$

$$
E_{1+s|S}(\rho) \stackrel{\text{def}}{=} \min_{\sigma \in S} D_{1+s}(\rho\|\sigma) \tag{8.133}
$$

$$
\tilde{E}_{1+s|S}(\rho) \stackrel{\text{def}}{=} \min_{\sigma \in S} \underline{D}_{1+s}(\rho\|\sigma). \tag{8.134}
$$

Then, we can show the following lemma.

Lemma 8.7 *When C is $\rightarrow$, $\rightleftarrows$, or S, we have*

$$
E_{1+s|m}^C(\rho) \leq E_{1+s|S}(\rho) \text{ for } s \in [-1, 1] \tag{8.135}
$$

$$
\tilde{E}_{1+s|m}^C(\rho) \leq \tilde{E}_{1+s|S}(\rho) \text{ for } s \in [-\frac{1}{2}, \infty). \tag{8.136}
$$

Proof Any separable state $\sigma = \sum_i p_i |u_i\rangle\langle u_i| \otimes |v_i\rangle\langle v_i|$ with $\|u_i\| = \|v_i\| = 1$ satisfies

$$
\sigma \leq \sum_i p_i I_A \otimes |v_i\rangle\langle v_i| = I_A \otimes \sigma_B. \tag{8.137}
$$

Since $\kappa(\sigma)$ is separable for $\sigma \in S$ and $\kappa \in S$, we have

$$
\begin{aligned}
& - H^{\uparrow}_{1+s|\kappa(\rho)}(A|B) = \min_{\sigma_B} D_{1+s}(\kappa(\rho)\|I_A \otimes \sigma_B) \\
& \le \min_{\sigma \in S} D_{1+s}(\kappa(\rho)\|I_A \otimes \kappa(\sigma)_B) \\
& \overset{(a)}{\le} \min_{\sigma \in S} D_{1+s}(\kappa(\rho)\|\kappa(\sigma)) \overset{(b)}{\le} \min_{\sigma \in S} D_{1+s}(\rho\|\sigma) = E_{1+s|S}(\rho),
\end{aligned}
$$

where (a) follows from (8.137) and (**e**) of Exercise 5.25, and (b) follows from (**a**) of Exercise 5.25. Thus, we obtain (8.135). Similarly, we can show (8.136). ■

Exercises

8.41 Show that $E_{r,S}(\rho) \le E_f(\rho)$ using the joint convexity of the quantum relative entropy.

8.42 Show that the entanglement of formation $E_f(\rho)$ satisfies Condition **E3** (continuity) (Nielsen [35]) following the steps below.
(**a**) Let states ρ_n and σ_n on the bipartite system $\mathcal{H}_{A,n} \otimes \mathcal{H}_{B,n}$ satisfy $\|\rho_n - \sigma_n\|_1 \to 0$. Here, we assume that $\dim \mathcal{H}_{A,n} = \dim \mathcal{H}_{B,n}$. Show that there exists a decomposition $\rho_n = \sum_i p_{n,i}|x_{n,i}\rangle\langle x_{n,i}|$ such that $\frac{1}{\log \dim \mathcal{H}_{A,n}}$ $|\sum_i p_{n,i} H(\mathrm{Tr}_B |x_{n,i}\rangle\langle x_{n,i}|) - E_f(\sigma_n)| \to 0$. Here, choose the purifications x_n and y_n of ρ_n and σ_n based on Lemma 8.2.
(**b**) Prove Condition **E3** (continuity).

8.43 Show that the squashed entanglement $E_{sq}(\rho)$ satisfies Condition **E3** following the steps below.
(**a**) Let states ρ_n and σ_n on the bipartite system $\mathcal{H}_{A,n} \otimes \mathcal{H}_{B,n}$ satisfy $\|\rho_n - \sigma_n\|_1 \to 0$. Here, we assume that $\dim \mathcal{H}_{A,n} = \dim \mathcal{H}_{B,n}$. Show that there exists an extension ρ_n^{ABE} of ρ_n such that $\frac{1}{n}|\frac{1}{2} I_{\rho_n^{ABE}}(A:B|E) - E_{sq}(\sigma_n)| \to 0$ using (5.106).
(**b**) Show Condition **E3** (continuity).

8.44 Show that $E_m^C(\rho)$ satisfies Condition **E3** (continuity) for $C = \to, \leftrightarrow, S$ using (8.125), (5.104), and the monotonicity of d_1.

8.45 Show the additivity of squashed entanglement (8.128) using chain rule (5.109) for quantum conditional mutual information.

8.46 Let $|x\rangle\langle x|$ be a purification of ρ with the reference system $\mathcal{H}_R$. Assume that the state $\mathrm{Tr}_B |x\rangle\langle x|$ is separable between $\mathcal{H}_A$ and $\mathcal{H}_R$. Prove the equation

$$
E_f(\rho) + E_f(\sigma) = E_f(\rho \otimes \sigma) \tag{8.138}
$$

by using a similar discussion to the proof of (8.144) [31].

8.47 Show that the inequality $E_{d,2}^C(\rho) \le \lim_{n\to\infty} \frac{E^C(\rho^{\otimes n})}{n}$ holds even though we replace Condition **E3** in Theorem 8.10 by the following condition:

E3′ (Weak lower continuity) When the sequence of states ρ_n on $\mathbb{C}^{d_n} \otimes \mathbb{C}^{d_n}$ satisfies $\|\rho_n - |\Phi_{d_n}\rangle\langle\Phi_{d_n}|\|_1 \to 0$, then $\lim_{n\to\infty} \frac{E^C(\rho^{\otimes n})}{n} \ge \lim_{n\to\infty} \frac{\log d_n}{n}$.

8.48 Show the following relations

$$E_{d,2}^{S,*}(r|\rho) \geq \max_{t\geq 0} \frac{t(r - E_{1+t|S}(\rho))}{1+t} \tag{8.139}$$

$$E_{d,2}^{S,*}(r|\rho) \geq \max_{t\geq 0} \frac{t(r - \tilde{E}_{1+t|S}(\rho))}{1+t}. \tag{8.140}$$

8.8 Maximally Correlated State

Next, we introduce an important class of entangled states. When a state on the composite system $\mathcal{H}_A \otimes \mathcal{H}_B$ has the form

$$\rho_\alpha \stackrel{\text{def}}{=} \sum_{i,j} \alpha_{i,j} |u_i^A \otimes u_i^B\rangle\langle u_j^A \otimes u_j^B|, \tag{8.141}$$

where $(\alpha_{i,j})$ is a matrix and $\{u_i^A\}(\{u_i^B\})$ is an orthonormal basis of $\mathcal{H}_A(\mathcal{H}_B)$, it is called a **maximally correlated state**. A state ρ is maximally correlated if and only if there exist CONSs $\{u_i^A\}$ and $\{u_i^B\}$ of $\mathcal{H}_A$ and $\mathcal{H}_B$ such that the outcome of the measurement $\{|u_i^A\rangle\langle u_i^A|\}$ coincide with those of $\{|u_i^B\rangle\langle u_i^B|\}$. Evidently, any pure entangled state belongs to this class. Under this class, many entanglement measures introduced above can be calculated as follows.

To calculate these quantities, we consider the separable state

$$\sigma_\alpha \stackrel{\text{def}}{=} \sum_{i} \alpha_{i,i} |u_i^A \otimes u_i^B\rangle\langle u_i^A \otimes u_i^B|, \tag{8.142}$$

which satisfies

$$E_{r,S}(\rho_\alpha) \leq D(\rho_\alpha\|\sigma_\alpha) = H(\sigma_\alpha) - H(\rho_\alpha) = -H_{\rho_\alpha}(A|B).$$

Hence, we obtain

$$-H_{\rho_\alpha}(A|B) = E_m^C(\rho_\alpha) = E_{d,2}^C(\rho_\alpha) = E_{d,2}^{C,\dagger}(\rho_\alpha) = E_{r,S}(\rho_\alpha), \tag{8.143}$$

for $C = \rightarrow, \leftarrow, \leftrightarrow, S$. Regarding the entanglement formation, as is shown latter, the equation

$$E_f(\rho_\alpha) + E_f(\sigma) = E_f(\rho_\alpha \otimes \sigma) \tag{8.144}$$

holds for any maximally correlated state ρ_α on $\mathcal{H}_{A,1} \otimes \mathcal{H}_{B,1}$ and any state σ on $\mathcal{H}_{A,2} \otimes \mathcal{H}_{B,2}$. Hence, any maximally correlated state ρ_α satisfies $E_f(\rho_\alpha) = E_c(\rho_\alpha)$. Indeed, many researchers [31, 36] conjectured that the equation (8.144) holds for arbitrary two states: The conjecture is called the *additivity of entanglement formation*.

This relation can be generalized to the *superadditivity of entanglement formation* [37] as follows.

$$E_f(\mathrm{Tr}_2\,\rho) + E_f(\mathrm{Tr}_1\,\rho) \le E_f(\rho). \tag{8.145}$$

While the superadditivity of entanglement formation trivially derives the additivity of entanglement formation, as shown in Sect. 9.2, the converse relation holds [36]. However, as shown in Sect. 8.13, there is a counterexample for superadditivity of entanglement formation. Hence, the additivity of entanglement formation does not hold for general two states. However, the additivity of entanglement formation for the tensor product case remains unsolved, i.e., it is still open whether the equation $E_f(\rho^{\otimes n}) = nE_f(\rho)$ holds in general.

One might consider that $E_f(\rho)$ equals $E_{r,S}(\rho)$ for a maximally correlated state ρ because this relation holds for pure states. However, this relation does not hold in general, as disproved in $\mathbb{C}^2 \otimes \mathbb{C}^2$ by (8.321) and (8.322) in Sect. 8.16.1.

A state ρ is maximally correlated if and only if it has a probabilistic decomposition of pure states $(p_i, |x_i\rangle)$ such that all $|x_i\rangle$ have the common Schmidt bases on $\mathcal{H}_A$ and $\mathcal{H}_B$. Its necessary and sufficient condition was obtained by Hiroshima and Hayashi [38]. For example, any mixture of two maximally entangled states is maximally correlated.

We also have another characterization of maximally correlated states.

Lemma 8.8 *Let $|x\rangle$ be a pure state on the composite system $\mathcal{H}_A \otimes \mathcal{H}_B \otimes \mathcal{H}_R$. Then, the following conditions are equivalent.*

① $\rho^{AB} \stackrel{\text{def}}{=} \mathrm{Tr}_R\,|x\rangle\langle x|$ *is maximally correlated.*

② $\rho^{BR} \stackrel{\text{def}}{=} \mathrm{Tr}_A\,|x\rangle\langle x|$ *has the following form*

$$\rho^{BR} = \sum_i p_i |u_i^B \otimes x_i^R\rangle\langle u_i^B \otimes x_i^R|, \tag{8.146}$$

where $\{u_i^B\}$ is a CONS of $\mathcal{H}_B$, but $\{x_i^R\}$ is not necessarily a CONS of $\mathcal{H}_R$.

Using this property of maximally correlated states, we can show that any maximally correlated state satisfies the condition for Theorem 8.11[Exe. 8.49]. Hence, we obtain another proof for a part of (8.143) with $C = \rightarrow$.

Proof of (8.144) Let ρ_α be a maximally correlated state on $\mathcal{H}_{A_1} \otimes \mathcal{H}_{B_1}$ and σ be a state on $\mathcal{H}_{A_2} \otimes \mathcal{H}_{B_2}$. Then, Let y_1 and y_2 be the purifications of ρ_α and σ with the reference systems $\mathcal{H}_{R,1}$ and $\mathcal{H}_{R,2}$, respectively. Then, any probabilistic decomposition of $\rho_\alpha \otimes \sigma$ is given by a POVM $\boldsymbol{M} = \{M_i\}$ on $\mathcal{H}_{R,1} \otimes \mathcal{H}_{R,2}$ (Lemma 8.3). Due to (8.146) in Lemma 8.8, the matrix $\mathrm{Tr}_{A_1}\,|u_j^{A_1}\rangle\langle u_j^{A_1}| \otimes I_{B_1,R_1}|y_1\rangle\langle y_1|$ can be written as $p_i|u_i^{B_1}, x_i^{R_1}\rangle\langle u_i^{B_1}, x_i^{R_1}|$. For a POVM $\{M_i\}$ on $\mathcal{H}_R$ and each j, we define the distribution $Q_i^j := \mathrm{Tr}_R\, M_i(|x_j^{R_1}\rangle\langle x_j^{R_1}| \otimes \mathrm{Tr}_{A_2,B_2}\,|y_2\rangle\langle y_2|)$ and the state σ_i^j on $\mathcal{H}_{A_2} \otimes \mathcal{H}_{B_2}$ by $\mathrm{Tr}_R(M_i \otimes I_{A_2,B_2})|x_j^{R_1}, y_2\rangle\langle x_j^{R_1}, y_2| = Q_i^j\sigma_i^j$. From the definition of maximally correlated states, we have the expression of the conditional state on system $\mathcal{H}_B$ as follows: Using the notation $\sigma_{B,i}^j := \mathrm{Tr}_{A_2}\,\sigma_i^j$, we have

$$
\begin{aligned}
&\operatorname{Tr}_{R,A}(M_i \otimes I_{A,B})|y_1 \otimes y_2\rangle\langle y_1 \otimes y_2| \\
=&\operatorname{Tr}_{R,A_2} \sum_j (M_i \otimes |u_j^{A_1}\rangle\langle u_j^{A_1} \otimes I_{A_2,B})|y_1 \otimes y_2\rangle\langle y_1 \otimes y_2| \\
=&\operatorname{Tr}_{R,A_2} \sum_j (M_i \otimes I_{A_2,B_1,B_2})(p_i|u_i^{B_1}, x_i^{R_1}\rangle\langle u_i^{B_1}, x_i^{R_1}| \otimes |y_2,\rangle\langle y_2|) \\
=&\sum_j p_j|u_j^{B_1}\rangle\langle u_j^{B_1}| \otimes (\operatorname{Tr}_{R,A_2}(M_i \otimes I_{A_2,B_2})|x_j^{R_1}, y_2\rangle\langle x_j^{R_1}, y_2|) \\
=&\sum_j p_j|u_j^{B_1}\rangle\langle u_j^{B_1}| \otimes (\operatorname{Tr}_{R,A_2}(M_i \otimes I_{A_2,B_2})|x_j^{R_1}, y_2\rangle\langle x_j^{R_1}, y_2|) \\
=&\sum_j p_j Q_i^j |u_j^B\rangle\langle u_j^B| \otimes \sigma_{B,i}^j.
\end{aligned}
\tag{8.147}
$$

Now, we define the distribution P_I and the conditional distribution $P_{J|I}$ as $P_I(i)P_{J|I}(j|i) = p_j Q_i^j$.

The strong concavity of von Neumann entropy (5.110) yields that

$$
\begin{aligned}
&H\left(\sum_j P_{J|I}(j|i)|u_j^B\rangle\langle u_j^B| \otimes \sigma_{B,i}^j\right) \\
\ge &H\left(\sum_j P_{J|I}(j|i)|u_j^B\rangle\langle u_j^B|\right) + \sum_j P_{J|I}(j|i)H(\sigma_{B,i}^j).
\end{aligned}
\tag{8.148}
$$

Hence, the probabilistic decomposition by the POVM $\boldsymbol{M}$ yields the following average entropy on $\mathcal{H}_{B,1} \otimes \mathcal{H}_{B,2}$:

$$
\begin{aligned}
&\sum_i P_I(i)H\left(\sum_j P_{J|I}(j|i)|u_j^B\rangle\langle u_j^B| \otimes \sigma_{B,i}^j\right) \\
\ge &\sum_i P_I(i)H\left(\sum_j P_{J|I}(j|i)|u_j^B\rangle\langle u_j^B|) + \sum_{i,j} P_I(i)P_{J|I}(j|i)H(\sigma_{B,i}^j\right).
\end{aligned}
$$

Using the state ρ_i defined by $p_I(i)\rho_i = \operatorname{Tr}_R M_i \otimes I_{A_1,B_2}|y_1\rangle\langle y_1| \otimes \operatorname{Tr}_{A_2,B_2}|y_2\rangle\langle y_2|$, we have the decompositions

$$
\rho_\alpha = \sum_i P_I(i)\rho_i, \quad \sigma = \sum_{i,j} P_I(i)P_{J|I}(j|i)\sigma_i^j,
\tag{8.149}
$$

and the inequalities

$$\sum_i P_I(i)H\left(\sum_j P_{J|I}(j|i)|u_j^B\rangle\langle u_j^B|\right) \geq \sum_i P_I(i)E_f(\rho_i) \geq E_f(\rho_\alpha) \quad (8.150)$$

$$\sum_{i,j} P_I(i)P_{J|I}(j|i)H(\sigma_{B,i}^j) \geq \sum_{i,j} P_I(i)P_{J|I}(j|i)E_f(\sigma_i^j) \geq E_f(\sigma). \quad (8.151)$$

Therefore, we obtain (8.144). ∎

Evidently, (8.147) is essential to the proof of (8.144) in the following sense [31]. In fact, if the vectors $u_1^A, \ldots, u_{d_A}^A$ in (8.142) are not orthogonal, the state ρ_α does not satisfy (8.147) in general. As shown in Sect. 9.2, such a state is essentially related to entanglement-breaking channels.

Proof of Lemma 8.8 Assume Condition ①. Perform the POVM $\{|u_i^A\rangle\langle u_i^A|\}$. The final state on $\mathcal{H}_B \otimes \mathcal{H}_R$ is a pure state $|u_i^B \otimes x_i^R\rangle\langle u_i^B \otimes x_i^R|$. In this case, $\{u_i^B\}$ is a CONS of $\mathcal{H}_B$. Since any measurement on $\mathcal{H}_A$ gives a probabilistic decomposition on $\mathcal{H}_B \otimes \mathcal{H}_R$ (Lemma 8.3), we have (8.146).

Next, assume Condition ②. There exists a CONS $\{u_i^A\}$ of $\mathcal{H}_A$ such that

$$|x\rangle = \sum_i \sqrt{p_i} u_i^A \otimes u_i^B \otimes x_i^R. \quad (8.152)$$

Thus, when we perform the measurements $\{|u_i^A\rangle\langle u_i^A|\}$ and $\{|u_i^B\rangle\langle u_i^B|\}$, we obtain the same outcome. That is, $\rho^{A,B}$ is maximally correlated. ∎

Further, as a generalization of a part of (8.143), we have the following lemma.

Lemma 8.9 *The maximally correlated state ρ_α given in* (8.141) *satisfies the equality in* (8.135) *and* (8.136), *i.e.,*

$$\begin{aligned} E^C_{1+s|m}(\rho_\alpha) =& -H^\uparrow_{1+s|\rho_\alpha}(A|B) = D_{1+s}(\rho_\alpha\|\sigma_\alpha) \\ =& D_{1+s}(\alpha\|D(\alpha)) = E_{1+s|S}(\rho_\alpha) \end{aligned} \quad (8.153)$$

$$\begin{aligned} \tilde{E}^C_{1+s|m}(\rho_\alpha) =& -\tilde{H}^\uparrow_{1+s|\rho_\alpha}(A|B) = \underline{D}_{1+s}(\rho_\alpha\|\sigma_\alpha) \\ =& \underline{D}_{1+s}(\alpha\|D(\alpha)) = \tilde{E}_{1+s|S}(\rho_\alpha), \end{aligned} \quad (8.154)$$

where $D(\alpha)$ is the diagonal matrix with the diagonal elements $\alpha_{i,i}$ and σ_α is given in (8.142). *In particular, when ρ_α is pure, i.e., α is pure, we have*

$$E_{1+s|S}(\rho_\alpha) = H_{1-s|\rho_\alpha}(A), \quad \tilde{E}_{1+s|S}(\rho_\alpha) = H_{\frac{1}{1+s}|\rho_\alpha}(A). \quad (8.155)$$

Proof Due to Lemma 8.7, it is sufficient to show

$$-H^\uparrow_{1+s|\rho_\alpha}(A|B) = D_{1+s}(\rho_\alpha\|\sigma_\alpha) = D_{1+s}(\alpha\|D(\alpha)) \quad (8.156)$$

$$-\tilde{H}^\uparrow_{1+s|\rho_\alpha}(A|B) = \underline{D}_{1+s}(\rho_\alpha\|\sigma_\alpha) = \underline{D}_{1+s}(\alpha\|D(\alpha)). \quad (8.157)$$

The second equations of (8.156) and (8.157) follow from the definitions of ρ_α and σ_α. Now, we consider the projection P to the subspace spanned by the CONS $\{|u_i^A \otimes u_i^B\rangle\}_i$. Since $PI_A \otimes (\sigma_\alpha)_B P + (I-P)I_A \otimes (\sigma_\alpha)_B(I-P) = I_A \otimes (\sigma_\alpha)_B$, $P\rho_\alpha P = \rho_\alpha$, and $PI_A \otimes (\sigma_\alpha)_B P = \sigma_\alpha$, (**b**) of Exercise 5.25 implies that

$$D_{1+s}(\rho_\alpha \| I_A \otimes (\sigma_\alpha)_B) = D_{1+s}(\rho_\alpha \| \sigma_\alpha). \tag{8.158}$$

Since $(\rho_\alpha)_B = (\sigma_\alpha)_B$

$$D_{1+s}(\rho_\alpha \| I_A \otimes (\sigma_\alpha)_B) = -H^{\uparrow}_{1+s|\rho_\alpha}(A|B). \tag{8.159}$$

Hence, we obtain (8.156). Similarly, we can show (8.157).

Finally, we shown (8.155) when α is pure. In this case, we have

$$\begin{aligned}
&D_{1+s}(\alpha \| D(\alpha)) = \frac{1}{s} \log \operatorname{Tr} \alpha^{1+s} D(\alpha)^{-s} = \frac{1}{s} \log \operatorname{Tr} \alpha D(\alpha)^{-s} \\
=&\frac{1}{s} \log \operatorname{Tr} D(\alpha)^{1-s} = H_{1-s}(\alpha) = H_{1-s|\rho_\alpha}(A).
\end{aligned}$$

and

$$\begin{aligned}
&\underline{D}_{1+s}(\alpha \| D(\alpha)) \stackrel{(a)}{=} \frac{1}{s} \log \operatorname{Tr}(\alpha D(\alpha)^{-\frac{s}{1+s}} \alpha)^{1+s} \\
=&\frac{1}{s} \log \operatorname{Tr}((\operatorname{Tr} D(\alpha)^{1-\frac{s}{1+s}})\alpha)^{1+s} = \frac{1}{s} \log \operatorname{Tr}((\operatorname{Tr} D(\alpha)^{1-\frac{s}{1+s}})^{1+s}\alpha) \\
=&\frac{1}{s} \log(\operatorname{Tr} D(\alpha)^{1-\frac{s}{1+s}})^{1+s} = H_{\frac{1}{1+s}}(\alpha) = H_{\frac{1}{1+s}|\rho_\alpha}(A),
\end{aligned}$$

where (a) follows from Exercise 3.12. Hence, we obtain (8.155). ■

Exercises

8.49 Consider the maximally correlated state ρ_α given in (8.141). Employing the notation in (8.146) of Lemma 8.8, we define the TP-CP map κ from the state on $\mathcal{H}_B$ to the state $\mathcal{H}_R$ by $\kappa(\rho) \stackrel{\text{def}}{=} \sum_i \langle u_i^B | \rho | u_i^B \rangle |x_i^R\rangle\langle x_i^R|$. Show that the TP-CP map κ satisfies the condition (8.129).

8.50 Show that the maximally correlated state ρ_α satisfies

$$E^{S,*}_{d,2}(r|\rho_\alpha) \ge \max_{t \ge 0} \frac{t(r - D_{1+t}(\rho_\alpha \| \sigma_\alpha)}{1+t} \tag{8.160}$$

where $E^{S,*}_{d,2}(r|\rho_\alpha)$ is defined in (8.91) and σ_α is given in (8.142). Inequality (8.160) is a generalization of (8.92).

8.9 Dilution with Zero-Rate Communication

In this section, we treat entanglement dilution with small communication costs. When d is the Schmidt number of the initial pure entangled state $|u\rangle\langle u|$, from the proof of the part "if" of Theorem 8.4, we can convert the pure entangled state $|u\rangle\langle u|$ to the other pure entangled state $|v\rangle\langle v|$ satisfying the condition (8.63) by using the measurement whose outcomes are at most d elements. That is, the required amount of classical communication is at most $\log_2 d$ bits. In this case, we call the number of measurement outcomes the size of classical communication. Now, we consider the size of classical communication in the asymptotic situation. For this analysis, we focus on the **entanglement of cost with zero-rate communication**:

$$E_c^{\dashrightarrow}(\rho) \stackrel{\text{def}}{=} \inf_{\{\kappa_n\}\subset C} \left\{ \overline{\lim_{n\to\infty}} \frac{\log L_n}{n} \middle| \begin{array}{l} \lim_{n\to\infty} F(\rho^{\otimes n}, \kappa_n(|\Phi_{L_n}\rangle\langle\Phi_{L_n}|)) = 1 \\ \frac{\log \mathrm{CC}(\kappa_n)}{n} \to 0 \end{array} \right\}, \tag{8.161}$$

where $\mathrm{CC}(\kappa)$ is the size of classical communication. $\dashrightarrow$ denotes the set of LOCCs with zero-rate classical communications. This value is calculated in the pure-state case as follows.

Lemma 8.10 (Lo and Popescu [39])

$$E_c^{\dashrightarrow}(|x\rangle\langle x|) = H(\mathrm{Tr}_B |x\rangle\langle x|).$$

Proof To prove this, we first assume that there exist two probability spaces Ω_n and Ω_n' and a distribution p_n' on Ω_n' for a given distribution p on Ω and $\epsilon > 0$ such that

$$d_1((p^n)^{\downarrow}, (p_{\mathrm{mix},\Omega_n} \times p_n')^{\downarrow}) \to 0, \ \underline{\lim_{n\to\infty}} \frac{\log |\Omega_n'|}{n} < \epsilon, \ \underline{\lim_{n\to\infty}} \frac{\log |\Omega_n|}{n} \le H(p). \tag{8.162}$$

Indeed, the pure state with the Schmidt coefficients $\sqrt{(p_{\mathrm{mix},\Omega_n} \times p_n')_i^{\downarrow}}$ can be realized from the maximally entangled state with the size $|\Omega_n| \times |\Omega_n'|$ by classical communication with a size of at most $|\Omega_n'|$. Therefore, if the state $|u\rangle\langle u|$ has the Schmidt coefficients $\sqrt{p_i}$, its n-fold tensor product $|u\rangle\langle u|^{\otimes n}$ can be asymptotically realized from the maximally entangled state with asymptotically zero-rate classical communication.

It is sufficient to prove (8.162) by replacing a distribution p_n' on Ω_n' by a positive measure p_n' on Ω_n'. Letting $l_n \le l_n'$ be integers, we construct a measure $\tilde{p}_n$ on Ω^n as follows. For a type $q \in T_n$ satisfying $l_n \le |T_q^n| \le l_n'$, we choose a subset $T_q^{n\prime} \subset T_q^n$ such that $|T_q^n \setminus T_q^{n\prime}| < l_n$. We define a measure $\tilde{p}_n \stackrel{\text{def}}{=} p^n 1_{\Omega_n'}$, where $\Omega_n' \stackrel{\text{def}}{=} \cup_{q\in T_n: l_n \le |T_q^n| \le l_n'} T_q^{n\prime}$. Then,

$$d(\tilde{p}_n, p^n) \leq \sum_{q \in T_n : l_n \leq |T_q^n| \leq l'_n} l_n e^{n \sum_\omega q_\omega \log p_\omega} + \sum_{q \in T_n : |T_q^n| < l_n} p^n(T_q^n) + \sum_{q \in T_n : |T_q^n| > l'_n} p^n(T_q^n). \tag{8.163}$$

In this case, the measure $\tilde{p}_n$ has the form $p_{\mathrm{mix}, \Omega_n} \times p'_n$ with $|\Omega_n| = l_n$ and $|\Omega'_n| = \frac{l'_n}{l_n} |T^n|$. When we choose $l_n = e^{n(H(p)-\epsilon)}$ and $l'_n = e^{n(H(p)+\epsilon)}$, $\lim_{n\to\infty} \frac{\log |\Omega'_n|}{n} = 2\epsilon$ and $\lim_{n\to\infty} \frac{\log |\Omega_n|}{n} = H(p) - \epsilon$. From the discussion in Sect. 2.4.1, the right-hand side (RHS) of (8.163) goes to 0. ∎

Next, we focus on the mixed state $\mathrm{Tr}_{A_2, B_2} |x\rangle\langle x|$. Using this theorem, we can check that

$$E_c^{\dashrightarrow}(\mathrm{Tr}_{A_2, B_2} |x\rangle\langle x|) \leq H(\mathrm{Tr}_B |x\rangle\langle x|).$$

Hence, defining the **entanglement of purification** for a state ρ on $\mathcal{H}_{A_1} \otimes \mathcal{H}_{B_1}$:

$$E_p(\rho) \stackrel{\mathrm{def}}{=} \min_{x : \mathrm{Tr}_{A_2, B_2} |x\rangle\langle x| = \rho} H(\mathrm{Tr}_B |x\rangle\langle x|), \tag{8.164}$$

we obtain

$$E_c^{\dashrightarrow}(\rho) \leq \lim_{n\to\infty} \frac{E_p(\rho^{\otimes n})}{n}.$$

Conversely, the opposite inequality follows from generalizing Theorem 8.10. Hence, we have the following theorem.

Theorem 8.12 (Terhal et al. [40])

$$E_c^{\dashrightarrow}(\rho) = \lim_{n\to\infty} \frac{E_p(\rho^{\otimes n})}{n} \tag{8.165}$$

To generalize Theorem 8.10, we prepare the following condition.

E2′ (Weak monotonicity) Let κ be an operation containing quantum communication with size d. Then,

$$E(\kappa(\rho)) \leq E(\rho) + \log d.$$

Lemma 8.11 *When the quantity $E(\rho)$ satisfies Conditions* **E1**, **E2′**, **E3**, *and* **E4**,

$$E^\infty(\rho) \left(\stackrel{\mathrm{def}}{=} \lim_{n\to\infty} \frac{E(\rho^{\otimes n})}{n} \right) \leq E_c^{\dashrightarrow}(\rho). \tag{8.166}$$

This inequality holds even if we replace the one-way classical communication in the definition of $E_c^{\dashrightarrow}(\rho)$ with quantum communication.

Proof of Theorem 8.12 In fact, the entanglement of purification $E_p(\rho)$ satisfies Conditions **E1**, **E2∅** (Exercise 8.51), **E2′** (Exercise 8.51), **E3** (Exercise 8.52), and **E4**. Hence, (8.165) holds. ∎

Further, using relation (8.165), we can slightly modify Lemma 8.11. For this purpose, we introduce the following condition.

E1′ (Strong normalization)

$$E(|u\rangle\langle u|) = H(\mathrm{Tr}_B |u\rangle\langle u|).$$

Lemma 8.12 *When the quantity $E(\rho)$ satisfies Conditions* **E1′** *and* **E2∅**,

$$E(\rho) \leq E_p(\rho). \tag{8.167}$$

Hence,

$$E^{\infty}(\rho) \left(\stackrel{\mathrm{def}}{=} \lim_{n\to\infty} \frac{E(\rho^{\otimes n})}{n} \right) \leq \lim_{n\to\infty} \frac{E_p(\rho^{\otimes n})}{n} = E_c^{--\to}(\rho). \tag{8.168}$$

Proof of Lemma 8.11 We choose a local operation κ_n with one-way classical communication with a size l_n such that

$$\|\kappa_n(|\Phi_{d_n}\rangle\langle\Phi_{d_n}|) - \rho^{\otimes n}\|_1 \to 0, \quad \frac{\log l_n}{n} \to 0,$$

where $\frac{\log d_n}{n} \to E_c^{--\to}(\rho)$. Condition **E3** guarantees that $|\frac{E_c^{--\to}(\kappa_n(|\Phi_{d_n}\rangle\langle\Phi_{d_n}|))}{n} - \frac{E_c^{--\to}(\rho^{\otimes n})}{n}|$ $\to 0$. Combining Conditions **E2∅** and **E2′**, we have

$$E(\kappa_n(|\Phi_{d_n}\rangle\langle\Phi_{d_n}|)) \leq \log d_n + \log l_n.$$

Therefore, we obtain (8.166). ∎

Proof of Lemma 8.12 Let $|x\rangle$ be a purification of ρ attaining the minimum of $H(\mathrm{Tr}_B |x\rangle\langle x|)$. Hence, from Conditions **E1′** and **E2∅**,

$$E(\rho) \leq E(|x\rangle\langle x|) = H(\mathrm{Tr}_B |x\rangle\langle x|) = E_p(\rho).$$

∎

The entanglement of purification $E_p(\rho)$ is bounded as follows.

Lemma 8.13 (Terhal et al. [40])

$$\begin{aligned} E_p(\rho) &\left(= \min_x \left\{ H(\mathrm{Tr}_B |x\rangle\langle x|) | \mathrm{Tr}_{A_2,B_2} |x\rangle\langle x| = \rho \right\} \right) \\ &\leq \min\{H(\rho^{A_1}), H(\rho^{B_1})\}, \end{aligned} \tag{8.169}$$

where $\dim \mathcal{H}_{A_2} \leq d_{A_1} d_{B_1}$, $\dim \mathcal{H}_{B_2} \leq (d_{A_1} d_{B_1})^2$, *and* $\rho^{A_1} = \mathrm{Tr}_{B_1} \rho$, $\rho^{B_1} = \mathrm{Tr}_{A_1} \rho$.

Proof Let $|x\rangle$ be a purification of ρ and $\mathcal{H}'_{B_2}$ be its reference space. Then, any purification is given by an isometry U from $\mathcal{H}'_{B_2} \otimes \mathcal{H}_{A_2}$ to $\mathcal{H}_{B_2} \otimes \mathcal{H}_{A_2}$ as

$$U \otimes I_{A_1,B_1}(|x\rangle\langle x| \otimes \rho_0)(U \otimes I_{A_1,B_1})^*,$$

where ρ_0 is a pure state on $\mathcal{H}_{A_2}$. Hence,

$$E_p(\rho) = \min_{\kappa} H(\kappa(\mathrm{Tr}_{A_1} |x\rangle\langle x|)),$$

where κ is a TP-CP map from $\mathcal{H}'_{B_2}$ to $\mathcal{H}_{B_2}$. Since the minimum value is attained with an extremal point, from Corollary 5.2 we can restrict $\mathcal{H}_{B_2}$ to a $(d_{A_1} d_{B_1})^2$-dimensional space. Further, we can restrict $\mathcal{H}_{A_2}$ to a $d_{A_1} d_{B_1}$-dimensional space.

In addition, substituting $\kappa = \iota$, we obtain $E_p(\rho) \le H(\rho^{A_1})$. Similarly, the inequality $E_p(\rho) \le H(\rho^{B_1})$ holds. ∎

Now, we apply the above discussion to the evaluation of the quantum mutual information $I_\rho(A : B)$. The quantity $\frac{I_\rho(A:B)}{2}$ satisfies Conditions **E1**$'$ and **E2**$'$ (Exercise 5.42) and the additivity $\frac{I_{\rho_1}(A:B)}{2} + \frac{I_{\rho_2}(A:B)}{2} = \frac{I_{\rho_1 \otimes \rho_2}(A:B)}{2}$. Hence,

$$\frac{I_\rho(A : B)}{2} \le E_c^{--\to}(\rho).$$

Exercises

8.51 Show that the entanglement of purification $E_p(\rho)$ satisfies Conditions **E2∅** and **E2**$'$.

8.52 Show that the entanglement of purification $E_p(\rho)$ satisfies Condition **E3** based on the discussions in the proof of Lemma 8.13 and Exercise 8.42.

8.10 Discord

Next, we consider non-classical correlation. For this purpose, we prepare the quantity $C_d^{A\to B}(\rho)$ as a measure of the classical correlation as Fig. 8.4.

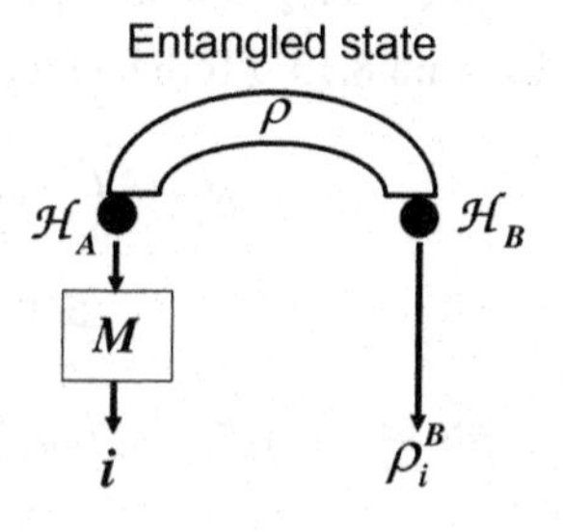

Fig. 8.4 Discord

$$C_d^{A\to B}(\rho) \stackrel{\text{def}}{=} \max_{M} H(\rho^B) - \sum_i \mathrm{P}^{M}_{\rho^A}(i) H(\rho_i^B)$$
$$= \max_{M} H(\rho^B) - H_{\mathrm{Tr}_A \hat{\kappa}_M \otimes \iota_B(\rho)}(B|E), \tag{8.170}$$
$$\rho_i^B \stackrel{\text{def}}{=} \frac{1}{\mathrm{P}^{M}_{\rho^A}(i)} \mathrm{Tr}_A(M_i \otimes I_B)\rho,$$

where $M = \{M_i\}$ is a POVM on the system $\mathcal{H}_A$ [41] and ρ_i^B is the resultant state on $\mathcal{H}_B$ with the measurement outcome i.

The quantity $C_d^{A\to B}(\rho)$ has another form as (Exercise 8.56)

$$C_d^{A\to B}(\rho) = \max_{M=\{M_i\}:\mathrm{rank}\, M_i=1} H(\rho^B) - \sum_i \mathrm{P}^{M}_{\rho^A}(i) H(\rho_i^B). \tag{8.171}$$

For the derivation of this equation, see (8.126). It is easily checked that it satisfies Conditions **E1**′ (Exercise 8.54) and **E2∅** (Exercise 8.54). Thus,

$$\lim_{n\to\infty} \frac{C_d^{A\to B}(\rho^{\otimes n})}{n} \le E_c^{--\to}(\rho). \tag{8.172}$$

In fact, $C_d^{A\to B}(\rho)$ satisfies Condition **E3** (continuity) (Exercise 8.55).

For example, when state ρ is a maximally correlated state, $C_d^{A\to B}(\rho)$ is calculated as

$$C_d^{A\to B}(\rho) = H(\rho^B)$$

because there exists a POVM M such that $H(\rho_i^B) = 0$. Moreover,

$$C_d^{A\to B}(\rho^{\otimes n}) = nH(\rho^B).$$

Hence, from (8.169),

$$E_c^{--\to}(\rho) = H(\rho^B) = C_d^{A\to B}(\rho).$$

Further, we have an interesting characterization of $C_d^{A\to B}(\rho)$ when $|x\rangle$ is a purification of ρ with the reference system $\mathcal{H}_R$. Since any probabilistic decomposition of the state on $\mathcal{H}_B \otimes \mathcal{H}_R$ can be given by a POVM on $\mathcal{H}_A$ (Lemma 8.3 and (8.31)), the relation

$$C_d^{A\to B}(\rho) = H(\rho^B) - E_f(\rho^{B,R}) \tag{8.173}$$

holds, where $\rho^{B,R} \stackrel{\text{def}}{=} \mathrm{Tr}_A |x\rangle\langle x|$ [42]. This quantity is different from the usual entanglement measure in representing the amount of classical correlation because the separable state $\sum_{i=1}^d \frac{1}{d} |u_i^A, u_i^B\rangle\langle u_i^A, u_i^B|$ with the CONSs $\{u_i^A\}$ and $\{u_i^B\}$ has the same amount of $C_d^{A\to B}(\rho)$ as the maximally entangled state. This issue will be revisited in Sect. 9.5.

Indeed, when $\mathcal{H}_A = \mathcal{H}_B$, we can define the **flip operator(swapping operator)** F as $F(u \otimes v) \stackrel{\text{def}}{=} v \otimes u$. Operator F has the form $F = P_s - P_a$, where $P_s(P_a)$ is the projection to the **symmetric space** $\mathcal{H}_s$ (the **antisymmetric space** $\mathcal{H}_a$), which is spanned by $\{u \otimes v + v \otimes u\}$ ($\{u \otimes v - v \otimes u\}$). The flip operator F satisfies the following property

$$\begin{aligned}\operatorname{Tr} A \otimes BF &= \sum_{i,j} \langle u_i, u_j | A \otimes BF | u_i, u_j \rangle = \sum_{i,j} \langle u_i, u_j | A \otimes B | u_j, u_i \rangle \\ &= \sum_{i} \langle u_i | A | u_j \rangle \langle u_j | B | u_i \rangle = \sum_{i,j} \langle u_i | AB | u_i \rangle = \operatorname{Tr} AB. \end{aligned} \tag{8.174}$$

As is shown latter, when the support $\operatorname{supp}(\rho)$ is contained by $\mathcal{H}_s$ or $\mathcal{H}_a$, the equation

$$E_p(\rho) = H(\rho^B) = H(\rho^A) \tag{8.175}$$

holds as follows [43]. Since $\rho^{\otimes n}$ also satisfies this condition, we have

$$E_c^{-\to}(\rho) = E_p(\rho) = H(\rho^B) = H(\rho^A). \tag{8.176}$$

Proof of (8.175) Let $|u\rangle$ be a purification of ρ with the reference systems A_2 and B_2. Then, $F|u\rangle\langle u|F^* = |u\rangle\langle u|$. $H_{|u\rangle\langle u|}(B_1B_2) = H_{F|u\rangle\langle u|F}(B_1B_2) = H_{|u\rangle\langle u|}(A_1B_2) = H_{|u\rangle\langle u|}(B_1A_2)$. Hence, from inequality (5.100) we have

$$\begin{aligned}&H_{|u\rangle\langle u|}(A_1A_2) + H_{|u\rangle\langle u|}(B_1B_2) = H_{|u\rangle\langle u|}(A_1A_2) + H_{|u\rangle\langle u|}(B_1A_2) \\ \geq &H_{|u\rangle\langle u|}(A_1) + H_{|u\rangle\langle u|}(B_1).\end{aligned}$$

Since $H_{|u\rangle\langle u|}(A_1A_2) = H_{|u\rangle\langle u|}(B_1B_2)$ and $H_{|u\rangle\langle u|}(A_1) = H_{|u\rangle\langle u|}(B_1)$, we obtain

$$H_{|u\rangle\langle u|}(A_1A_2) \geq H_{|u\rangle\langle u|}(A_1),$$

which implies (8.175). ∎

Now, in order to measure non-classical correlation, we introduce the **discord** as the discrepancy between the two measures $C_d^{A\to B}(\rho)$ and $I_\rho(A:B)$

$$D(B|A)_\rho := I_\rho(A:B) - C_d^{A\to B}(\rho) \tag{8.177}$$

because $I_\rho(A:B)$ expresses the whole correlation and $C_d^{A\to B}(\rho)$ expresses only the classical correlation. Using the monotonicity of quantum relative entropy for the measurement, we can show the non-negativity of the discord $D(B|A)_\rho$:

$$D(B|A)_\rho \geq 0. \tag{8.178}$$

The discord $D(B|A)_\rho$ has another form (Exercise 8.58)

$$D(B|A)_\rho = H_{\rho^{BR}}(B|R) + E_f(\rho^{BR}). \tag{8.179}$$

Hence, (8.178) can be also checked from (8.124).

Consider the case where ρ has the following specific separable form:

$$\rho = \sum_i p_i |u_i^A\rangle\langle u_i^A| \otimes \rho_i^B, \tag{8.180}$$

where $\{u_i^A\}$ is a CONS on $\mathcal{H}_A$. In this case, the optimal POVM $\boldsymbol{M}$ on $\mathcal{H}_A$ is $|u_i^A\rangle\langle u_i^A|$ because $H(\rho^B) - \sum_i p_i H(\rho_i^B) = I_\rho(A:B)$. Thus,

$$C_d^{A\to B}(\rho) = H(\rho^B) - \sum_i p_i H(\rho_i^B), \tag{8.181}$$

which implies the equality of (8.178). In fact, the converse argument holds without assuming the form (8.180).

Lemma 8.14 *The equality of (8.178) holds if and only if ρ has the specific separable form of (8.180).*

Due to this lemma, even though the state ρ^{AB} is separable, if it does not have the form (8.180), the separable state ρ^{AB} has non-zero non-classical correlation $D(B|A)_\rho > 0$.

Proof As is above shown, a state with the form of (8.180) satisfies the equality in (8.178). Hence, we will prove (8.180) from the equality in (8.178). Due to (8.171), from the equality in (8.178), there exists a POVM $\boldsymbol{M} = \{M_i\}$ such that rank $M_i = 1$

$$H(\rho^B) - \sum_i \mathrm{P}_{\rho^A}^{\boldsymbol{M}}(i) H(\rho_i^B) = I_\rho(A:B). \tag{8.182}$$

We denote M_i as $\alpha_i|v_i\rangle\langle v_i|$, where $\|v_i\| = 1$. We can assume that $\rho^A > 0$ without loss of generality. Now, we focus on the entanglement-breaking channel $\tilde{\kappa}_{\boldsymbol{M}}$ from system $\mathcal{H}_A$ to system $\mathbb{C}^k$ for a POVM $\boldsymbol{M} = \{M_i\}_{i=1}^k$ on $\mathcal{H}_A$:

$$\tilde{\kappa}_{\boldsymbol{M}}(\rho) \stackrel{\text{def}}{=} \sum_i (\mathrm{Tr}\, \rho M_i)|u_i\rangle\langle u_i|,$$

where $\{u_i\}$ is a CONS of $\mathbb{C}^k$. Then, the left-hand side (LHS) of (8.182) is equal to $I_{(\tilde{\kappa}_{\boldsymbol{M}}\otimes\iota_B)(\rho)}(A:B)$, i.e.,

$$D((\tilde{\kappa}_{\boldsymbol{M}} \otimes \iota_B)(\rho)\|(\tilde{\kappa}_{\boldsymbol{M}} \otimes \iota_B)(\rho^A \otimes \rho^B)) = D(\rho\|\rho^A \otimes \rho^B),$$

where $\rho^A = \mathrm{Tr}_B\, \rho$, $\rho^B = \mathrm{Tr}_A\, \rho$. Applying Theorem 5.8, we have

$$\rho = \sum_i \frac{\sqrt{M_i}\rho^A\sqrt{M_i}}{\operatorname{Tr} M_i\rho^A} \otimes (\operatorname{Tr}_A \rho(M_i \otimes I_B))$$
$$= \sum_i \frac{\alpha|u_i\rangle\langle u_i|\rho^A|u_i\rangle\langle u_i|}{\alpha\langle u_i|\rho^A|u_i\rangle} \otimes (\operatorname{Tr}_A \rho(M_i \otimes I_B))$$
$$= \sum_i |u_i\rangle\langle u_i| \otimes (\operatorname{Tr}_A \rho(M_i \otimes I_B)). \tag{8.183}$$

Now, we denote the resulting state on the system $\mathcal{H}_B$ with the measurement outcome j by ρ_j^B. Then,

$$(\operatorname{Tr} \rho(M_j \otimes I_B))\rho_j^B = \operatorname{Tr}_A(M_j \otimes I_B)\rho$$
$$= \operatorname{Tr}_A(M_j \otimes I_B) \sum_i |u_i\rangle\langle u_i| \otimes (\operatorname{Tr}_A \rho(M_i \otimes I_B))$$
$$= \sum_i \langle u_i|M_j|u_i\rangle(\operatorname{Tr}_A \rho(M_i \otimes I_B)), \tag{8.184}$$

which implies $\sum_i \langle u_i|M_j|u_i\rangle \operatorname{Tr}_\rho(M_i \otimes I_B) = \operatorname{Tr} \rho(M_j \otimes I_B)$. Thus, $P(i|j) := \frac{\operatorname{Tr} \rho(M_i \otimes I_B)}{\operatorname{Tr} \rho(M_j \otimes I_B)}\langle u_i|M_j|u_i\rangle$ gives a conditional distribution. Then, we have

$$I_\rho(A : B) = H(\rho^B) - \sum_i \operatorname{Tr} \rho(M_i \otimes I_B) H\left(\frac{\operatorname{Tr}_A \rho(M_i \otimes I_B))}{\operatorname{Tr} \rho(M_i \otimes I_B)}\right)$$
$$= H(\rho^B) - \sum_i \sum_j \langle u_i|M_j|u_i\rangle \operatorname{Tr} \rho(M_i \otimes I_B) H\left(\frac{(\operatorname{Tr}_A \rho(M_i \otimes I_B))}{\operatorname{Tr} \rho(M_i \otimes I_B)}\right)$$
$$= H(\rho^B) - \sum_j \operatorname{Tr} \rho(M_j \otimes I_B) \sum_i P(i|j) H\left(\frac{(\operatorname{Tr}_A \rho(M_i \otimes I_B))}{\operatorname{Tr} \rho(M_i \otimes I_B)}\right). \tag{8.185}$$

The relation (8.184) yields

$$C_d^{A\to B}(\rho) = H(\rho^B) - \sum_j \operatorname{Tr} \rho(M_j \otimes I_B) H\left(\frac{\sum_i \langle u_i|M_j|u_i\rangle(\operatorname{Tr}_A \rho(M_i \otimes I_B))}{\operatorname{Tr} \rho(M_j \otimes I_B)}\right)$$
$$= H(\rho^B) - \sum_j \operatorname{Tr} \rho(M_j \otimes I_B) H\left(\sum_i P(i|j) \frac{(\operatorname{Tr}_A \rho(M_i \otimes I_B))}{\operatorname{Tr} \rho(M_i \otimes I_B)}\right). \tag{8.186}$$

Combining (8.185) and (8.186), we have

$$D(B|A)_\rho$$
$$= \sum_j \operatorname{Tr} \rho(M_j \otimes I_B) H\left(\sum_i P(i|j) \frac{(\operatorname{Tr}_A \rho(M_i \otimes I_B))}{\operatorname{Tr} \rho(M_i \otimes I_B)}\right)$$
$$- \sum_j \operatorname{Tr} \rho(M_j \otimes I_B) \sum_i P(i|j) H\left(\frac{(\operatorname{Tr}_A \rho(M_i \otimes I_B))}{\operatorname{Tr} \rho(M_i \otimes I_B)}\right)$$

$$= \sum_j \operatorname{Tr} \rho(M_j \otimes I_B)\Big(H\Big(\sum_i P(i|j)\frac{(\operatorname{Tr}_A \rho(M_i \otimes I_B))}{\operatorname{Tr} \rho(M_i \otimes I_B)}\Big) - \sum_i P(i|j) H\Big(\frac{(\operatorname{Tr}_A \rho(M_i \otimes I_B))}{\operatorname{Tr} \rho(M_i \otimes I_B)}\Big)\Big). \tag{8.187}$$

Since

$$H\Big(\sum_i P(i|j)\frac{(\operatorname{Tr}_A \rho(M_i \otimes I_B))}{\operatorname{Tr} \rho(M_i \otimes I_B)}\Big) \geq \sum_i P(i|j) H\Big(\frac{(\operatorname{Tr}_A \rho(M_i \otimes I_B))}{\operatorname{Tr} \rho(M_i \otimes I_B)}\Big) \tag{8.188}$$

and $D(B|A)_\rho = 0$, the equality in (8.188) holds for all j. Now, we decompose the set of indexes i to the collection of disjoint subsets S_a such that the relation

$$\frac{(\operatorname{Tr}_A \rho(M_i \otimes I_B))}{\operatorname{Tr} \rho(M_i \otimes I_B)} = \frac{(\operatorname{Tr}_A \rho(M_j \otimes I_B))}{\operatorname{Tr} \rho(M_j \otimes I_B)} \tag{8.189}$$

holds for $i \neq j \in S_a$ and the relation (8.189) holds for $i \in S_a$ and $j \notin S_a$. By denoting the state $\frac{(\operatorname{Tr}_A \rho(M_j \otimes I_B))}{\operatorname{Tr} \rho(M_j \otimes I_B)}$ for $j \in S_a$ by ρ_a^B, the state ρ is written as

$$\rho = \sum_a \sum_{i \in S_a} \operatorname{Tr}_A \rho(M_i \otimes I_B)|u_i\rangle\langle u_i| \otimes \rho_a^B = \sum_a P_A(a)\rho_a^A \otimes \rho_a^B,$$

where $P_A(a) := \sum_{i \in S_a} \operatorname{Tr} \rho(M_i \otimes I_B)$ and $\rho_a^A := \frac{\operatorname{Tr}_A \rho(M_i \otimes I_B)}{P_A(a)}|u_i\rangle\langle u_i|$. Since the equality in (8.188) implies (8.189) for $P(i|j) \neq 0$, the support of ρ_a^A is orthogonal to that of $\rho_{a'}^A$ for $a \neq a'$. Considering the spectral decomposition of each ρ_a^A, we obtain the form (8.180). ∎

Now, we calculate $E_c^{-\to}(\rho)$ and $E_p(\rho)$ in another case by treating $C_d^{A\to B}(\rho)$. We assume that ρ has the form (8.180) and ρ_i^B is pure. Then, we have[Exe. 8.57]

$$I_\rho(A:B) = H(\rho^B) = E_p(\rho). \tag{8.190}$$

Further, for its purification $|x\rangle$ with the reference system $\mathcal{H}_R$, Exercise 8.14 guarantees that the state $\rho^{B,R} \stackrel{\text{def}}{=} \operatorname{Tr}_A |x\rangle\langle x|$ is separable. So, (8.173) yields that $C_d^{A\to B}(\rho) = H(\rho^B)$. Therefore,

$$C_d^{A\to B}(\rho) = H(\rho^B) = I_\rho(A:B) = E_p(\rho) \leq H(\rho^A). \tag{8.191}$$

In this case, we also have $I_{\rho^{\otimes n}}(A:B) = nH(\rho^B)$. Hence, (8.190), (8.165), and (8.169) imply

$$H(\rho^B) = \lim_{n\to\infty} \frac{I_{\rho^{\otimes n}}(A:B)}{n} \le E_c^{-\to}(\rho) = \lim_{n\to\infty} \frac{E_p(\rho^{\otimes n})}{n} \le H(\rho^B). \quad (8.192)$$

Hence,

$$E_c^{-\to}(\rho) = C_d^{A\to B}(\rho) = H(\rho^B). \quad (8.193)$$

Exercises

8.53 Show that

$$C_d^{A\to B}(\rho) + C_d^{A\to B}(\sigma) = C_d^{A\to B}(\rho\otimes\sigma) \quad (8.194)$$

for a separable state ρ using Exercise 8.46.

8.54 Show that the quantity $C_d^{A\to B}(\rho)$ satisfies Conditions **E1**$'$ and **E2∅**.

8.55 Show that the quantity $C_d^{A\to B}(\rho)$ satisfies Condition **E3** using (8.173).

8.56 Show the equation (8.171).

8.57 Prove (8.190) for a separable state ρ of the form (8.180) with rank $\rho_i^B = 1$ following the steps below.
(**a**) Let $|X\rangle\rangle\langle\langle X|$ be a pure entangled state on $\mathcal{H}_A\otimes\mathcal{H}_B$ and $\boldsymbol{M} = \{M_i\}$ be a rank-one PVM on $\mathcal{H}_A$. Show that the state $\rho \stackrel{\text{def}}{=} \sum M_i \otimes XM_iX^*$ satisfies $H(\mathrm{Tr}_A |X\rangle\rangle\langle\langle X|) = I_\rho(A:B)$.
(**b**) Show $H(\mathrm{Tr}_A \rho) = I_\rho(A:B)$.
(**c**) Show $E_p(\rho) = H(\mathrm{Tr}_A \rho)$.

8.58 Show the equation (8.179) by using (8.173).

8.11 State Generation from Shared Randomness

In this section, we address the state generation from minimum shared random numbers in an asymptotic formulation. If the desired state ρ is nonseparable between $\mathcal{H}_A$ and $\mathcal{H}_B$, it is impossible to generate state ρ only from shared random numbers. Hence, we treat a separable state:

$$\rho = \sum_i p_i \rho_i^A \otimes \rho_i^B. \quad (8.195)$$

In particular, when the conditions

$$[\rho_i^A, \rho_j^A] = 0 \quad \forall i, j, \quad (8.196)$$
$$[\rho_i^B, \rho_j^B] = 0 \quad \forall i, j \quad (8.197)$$

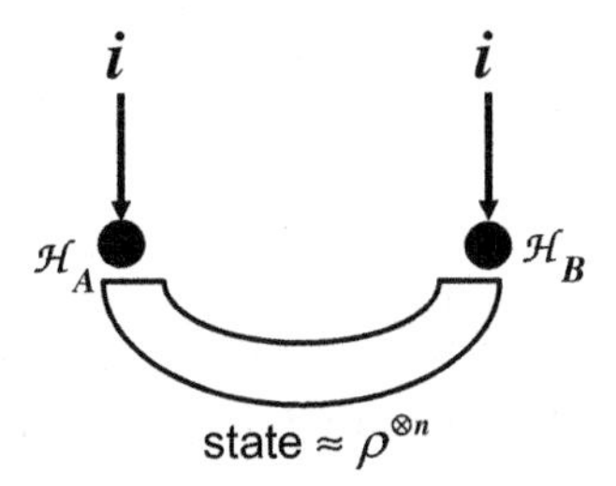

Fig. 8.5 State generation from shared randomness

hold, the problem is essentially classical. In this problem, our operation is described by the size of shared random numbers M and the local states σ_i^A and σ_i^B dependently of the shared random number $i = 1, \ldots, M$, i.e., we focus on our operation $\Phi \stackrel{\text{def}}{=} \{\sigma_i^A \otimes \sigma_i^B\}_{i=1}^M$ for approximating state ρ. Its performance is characterized by the size $|\Phi| \stackrel{\text{def}}{=} M$ and the quality of the approximation $\left\| \frac{1}{|\Phi|} \sum_{i=1}^{|\Phi|} \sigma_i^A \otimes \sigma_i^B - \rho \right\|_1$. Hence, the minimum size of shared random numbers is asymptotically characterized by $C_c(\rho)$ (Fig. 8.5)[5]

$$C_c(\rho) \stackrel{\text{def}}{=} \inf_{\{\Phi_n\}} \left\{ \overline{\lim_{n\to\infty}} \frac{1}{n} \log |\Phi_n| \,\middle|\, \lim_{n\to\infty} \left\| \frac{1}{|\Phi_n|} \sum_{i=1}^{|\Phi_n|} \rho_{n,i}^A \otimes \rho_{n,i}^B - \rho^{\otimes n} \right\|_1 = 0 \right\}, \tag{8.198}$$

where $\Phi_n = \{\sigma_{n,i}^A \otimes \sigma_{n,i}^B\}$. Since a shared random number with size M can be simulated by a maximally entangled state with size M, we have

$$C_c(\rho) \geq E_c^{-\to}(\rho). \tag{8.199}$$

For this analysis, we define the quantities

$$C(\rho, \delta) \stackrel{\text{def}}{=} \inf \left\{ I_{\rho_{ABE}}(AB : E) \,\middle|\, \begin{array}{l} \mathrm{Tr}_E\, \rho_{ABE} = \rho,\, I_{\rho_{ABE}}(A : B|E) \leq \delta \\ \rho_{ABE} = \sum_x p_x \rho_x^{A,B} \otimes |u_x^E\rangle\langle u_x^E| \end{array} \right\}, \tag{8.200}$$

$$\tilde{C}(\rho, \delta) \stackrel{\text{def}}{=} \inf \left\{ I_{\rho_{ABE}}(AB : E) \,\middle|\, \mathrm{Tr}_E\, \rho_{ABE} = \rho,\, I_{\rho_{ABE}}(A : B|E) \leq \delta \right\}, \tag{8.201}$$

where $\{u_x^E\}$ is a CONS on $\mathcal{H}_E$. From the definitions, the inequality

$$C(\rho, \delta) \geq \tilde{C}(\rho, \delta) \tag{8.202}$$

holds. In particular, we can prove[Exe. 8.61]

$$C(\rho) := C(\rho, 0) = \tilde{C}(\rho, 0). \tag{8.203}$$

---

[5] The subscript c denotes "common randomness."

Further, this quantity satisfies the following properties.

CM1 (Monotonicity) Operations κ_A and κ_B on $\mathcal{H}_A$ and $\mathcal{H}_B$ satisfy the monotonicity[Exe. 8.59]

$$C(\rho, \delta) \geq C((\kappa_A \otimes \kappa_B)(\rho), \delta), \quad \tilde{C}(\rho, \delta) \geq \tilde{C}((\kappa_A \otimes \kappa_B)(\rho), \delta). \tag{8.204}$$

CM2 (Additivity) The quantity $C(\rho)$ satisfies the **additivity**[Exe. 8.60]:

$$C(\rho \otimes \sigma) = C(\rho) + C(\sigma). \tag{8.205}$$

CM3 (Continuity) The former quantity $C(\rho, \delta)$ satisfies two kinds of continuity, i.e., if ρ_n is separable and $\rho_n \to \rho$, then

$$\lim_{n\to\infty} C(\rho_n) = C(\rho), \tag{8.206}$$

$$\lim_{\delta\to 0} C(\rho, \delta) = C(\rho, 0). \tag{8.207}$$

In particular, the convergence in (8.207) is locally uniform concerning ρ.

CM4 (Asymptotic weak-lower-continuity) When $\|\rho_n - \rho^{\otimes n}\|_1 \to 0$, the inequality

$$\underline{\lim}_{n\to\infty} \frac{C(\rho_n)}{n} \geq C(\rho) \tag{8.208}$$

holds.

CM5 $C(\rho)$ satisfies

$$C(\rho) \geq I_\rho(A : B) \tag{8.209}$$

because

$$\begin{aligned} &I_{\rho^{ABE}}(AB : E) \geq I_{\rho^{ABE}}(A : E) = I_{\rho^{ABE}}(A : E) + I_{\rho^{ABE}}(A : B|E) \\ =&I_{\rho^{ABE}}(A : BE) \geq I_{\rho^{ABE}}(A : B) \end{aligned}$$

for any extension ρ^{ABE} of ρ satisfying $I_{\rho^{ABE}}(A : B|E) = 0$.

CM6 When condition (8.196) holds, $C(\rho)$ is upper bounded as

$$C(\rho) \leq H_\rho(A). \tag{8.210}$$

This can be checked by substituting $\mathcal{H}_A$ into $\mathcal{H}_E$ in the definition of $\tilde{C}(\rho, 0)$.

Using the quantity $C(\rho)$, we can characterize $C_c(\rho)$ as follows.

Theorem 8.13 *When ρ is separable, then*

$$C_c(\rho) = C(\rho). \tag{8.211}$$

Hence, from (8.199),

$$E_c^{--\to}(\rho) \le C(\rho). \tag{8.212}$$

Further, there exists an example of separable states ρ such that conditions (8.196) and (8.197) hold and $C_c(\rho) > E_c^{--\to}(\rho)$ [44].

Proof Since the direct part follows from the discussion in Sect. 9.4, its proof will be given in Sect. 9.4. Hence, we only prove the converse part here. Now, we choose the state $\rho_n \stackrel{\text{def}}{=} \frac{1}{|\Phi_n|}\sum_{i=1}^{|\Phi_n|} \sigma_{n,i}^A \otimes \sigma_{n,i}^B \otimes |u_i^E\rangle\langle u_i^E|$ such that $\| \operatorname{Tr}_E \rho_n - \rho^{\otimes n} \|_1 \to 0$. Then, we have

$$\log |\Phi_n| \ge I_{\rho_n}(AB : E) \ge C(\operatorname{Tr}_E \rho_n). \tag{8.213}$$

Hence, combining (8.208), we obtain

$$\varliminf_{n\to\infty} \frac{1}{n} \log |\Phi_n| \ge C(\rho).$$

■

Proof of (8.207) We first characterize the quantity $C(\rho)$ as follows. Since the state $\rho^{(AB)E}$ is restricted to a separable state between AB and E, the state $\rho^{(AB)E}$ is given by a probabilistic decomposition $(p_i \rho_i)$ of ρ. Now, recall that any probabilistic decomposition of ρ on $\mathcal{H}_A \otimes \mathcal{H}_B$ is given by POVM $\boldsymbol{M}$ on the reference system as (8.30) and (8.31). In order to satisfy the condition $I_{\rho^{ABE}}(A : B|E) = 0$, any component ρ_i has a tensor product form. Hence,

$$C(\rho) = \inf_{\boldsymbol{M}} \left\{ I_{\rho_{\boldsymbol{M}}^{ABE}}(AB : E) \middle| I_{\rho_{\boldsymbol{M}}^{ABE}}(A : B|E) = 0 \right\},$$

where

$$\rho_{\boldsymbol{M}}^{ABE} \stackrel{\text{def}}{=} \sum_i \operatorname{Tr}_R(\sqrt{M_i} \otimes I)|x\rangle\langle x|(\sqrt{M_i} \otimes I) \otimes |u_i^E\rangle\langle u_i^E|.$$

Therefore, from Lemma A.12 we can restrict the range of the above infimum to the POVM $\boldsymbol{M}$ with at most $2(\dim \mathcal{H}_{A,B})^2$ elements. Since the set of POVMs with at most $2(\dim \mathcal{H}_{A,B})^2$ elements is compact, the above infimum can be replaced by the maximum. Further, we define

$$C(\rho, \delta) = \inf_{\boldsymbol{M}} \left\{ I_{\rho_{\boldsymbol{M}}^{ABE}}(AB : E) \middle| I_{\rho_{\boldsymbol{M}}^{ABE}}(A : B|E) = \delta \right\}. \tag{8.214}$$

Since $I_{\rho_{\boldsymbol{M}}^{ABE}}(A : B|E)$ is written as

$$I_{\rho_{\boldsymbol{M}}^{ABE}}(A : B|E) = \sum_i p_i I_\rho(M_i), \quad p_i \stackrel{\text{def}}{=} \operatorname{Tr}(M_i \otimes I)|x\rangle\langle x|,$$

$$I_\rho(M_i) \stackrel{\text{def}}{=} I_{\frac{\operatorname{Tr}_R(\sqrt{M_i}\otimes I)|x\rangle\langle x|(\sqrt{M_i}\otimes I)}{p_i}}(A : B),$$

from Lemma A.12, we can restrict the range of the infimum in (8.214) to the POVMs $\boldsymbol{M}$ satisfying that $|\boldsymbol{M}| \leq 2(\dim \mathcal{H}_{A,B})^2$. Since the set of POVMs with at most $2(\dim \mathcal{H}_{A,B})^2$ elements is compact, from Lemma A.4, we have

$$\lim_{\delta \to 0} C(\rho, \delta) = C(\rho). \tag{8.215}$$

Indeed, the above convergence is locally uniform for ρ. From (5.106), the functions $I_{\rho_{\boldsymbol{M}}^{ABE}}(AB:E)$ and $I_{\rho_{\boldsymbol{M}}^{ABE}}(A:B|E)$ satisfy

$$|I_{\rho_{\boldsymbol{M}}^{ABE}}(AB:E) - I_{\sigma_{\boldsymbol{M}}^{ABE}}(AB:E)| \leq 5\epsilon \log \dim \mathcal{H}_{A,B} + \eta_0(\epsilon) + 2h(\epsilon),$$
$$|I_{\rho_{\boldsymbol{M}}^{ABE}}(A:B|E) - I_{\sigma_{\boldsymbol{M}}^{ABE}}(A:B|E)| \leq 8\epsilon \log \dim \mathcal{H}_{A,B} + 6h(\epsilon),$$

where $\epsilon = \|\sigma - \rho\|_1$. Hence, the local uniformality follows by checking the discussion in the proof of Lemma A.4. ∎

Proof of (8.206) Now, we prove (8.206). Let $|x_n\rangle$ ($|x\rangle$) be a purification of ρ_n (ρ) such that $|\langle x|x_n\rangle| = F(\rho, \rho_n)$. From (3.48),

$$\| \, |x\rangle\langle x| - |x_n\rangle\langle x_n| \, \|_1 \to 0. \tag{8.216}$$

We choose a POVM $\boldsymbol{M}_n$ with at most $2(\dim \mathcal{H}_{A,B})^2$ elements such that

$$I_{\rho_{n,\boldsymbol{M}_n}^{ABE}}(AB:E) = C(\rho_n), \quad I_{\rho_{n,\boldsymbol{M}_n}^{ABE}}(A:B|E) = 0.$$

From (8.216), (5.105), and (5.106),

$$\delta_n \stackrel{\text{def}}{=} I_{\rho_{\boldsymbol{M}_n}^{ABE}}(A:B|E) \to 0,$$
$$\delta_n' \stackrel{\text{def}}{=} |I_{\rho_{n,\boldsymbol{M}_n}^{ABE}}(AB:E) - I_{\rho_{\boldsymbol{M}_n}^{ABE}}(AB:E)| \to 0.$$

Hence,

$$C(\rho_n) + \delta_n' \geq C(\rho, \delta_n).$$

From (8.215) we obtain the inequality $\lim_{n\to\infty} C(\rho_n) \geq C(\rho)$.

Conversely, we choose a POVM $\boldsymbol{M}$ with at most $2(\dim \mathcal{H}_{A,B})^2$ elements such that

$$I_{\rho_{\boldsymbol{M}_n}^{ABE}}(AB:E) = C(\rho), \quad I_{\rho_{\boldsymbol{M}_n}^{ABE}}(A:B|E) = 0.$$

From (8.216), (5.105), and (5.106),

$$\epsilon_n \stackrel{\text{def}}{=} I_{\rho^{ABE}_{n,M}}(A:B|E) \to 0,$$
$$\epsilon'_n \stackrel{\text{def}}{=} |I_{\rho^{ABE}_{n,M}}(AB:E) - I_{\rho^{ABE}_{M}}(AB:E)| \to 0.$$

Hence,

$$C(\rho) + \epsilon'_n \geq C(\rho_n, \epsilon_n).$$

Since the convergence of (8.215) is locally uniform, we obtain the opposite inequality $\lim_{n\to\infty} C(\rho_n) \leq C(\rho)$. ∎

Proof of (8.208) Let ρ_n^{ABE} be a state satisfying $\mathrm{Tr}_E\, \rho_n^{ABE} = \rho_n$, $I_{\rho_n^{ABE}}(A:B|E) = 0$, and $I_{\rho_n^{ABE}}(AB:E) = C(\rho_n)$. From (5.99), the state ρ_n^{ABE} satisfies $H_{\rho_n^{ABE}}(AB|E) \leq \sum_i H_{\rho_n^{ABE}}(A_i B_i|E)$. Hence,

$$\begin{aligned}
C(\rho_n) =& H_{\rho_n^{ABE}}(AB) - H_{\rho_n^{ABE}}(AB|E) \\
\geq& H_{\rho_n^{ABE}}(AB) - \sum_i H_{\rho_n^{ABE}}(A_i B_i|E) \\
=& H_{\rho_n}(AB) - \sum_i H_{\rho_n^{ABE}}(A_i B_i) + \sum_i (H_{\rho_n}(A_i B_i) - H_{\rho_n^{ABE}}(A_i B_i|E)) \\
\geq& H_{\rho_n}(AB) - \sum_i H_{\rho_n}(A_i B_i) + \sum_i C(\rho_{n,i}),
\end{aligned}$$

where $\rho_{n,i}$ is the reduced density matrix on $A_i B_i$. The final inequality follows from the definition of $C(\rho_{n,i})$. Since $\rho_{n,i}$ approaches ρ, properties (8.206) and (5.92) yield

$$\underline{\lim}_{n\to\infty} \frac{C(\rho_n)}{n} \geq C(\rho).$$

∎

Exercises

8.59 Prove inequality (8.204).

8.60 Prove (8.205) following the steps below.
(**a**) Assume that an extension ρ^{ABE} of $\rho^{A_1B_1} \otimes \rho^{A_2B_2}$ satisfies $I_{\rho^{ABE}}(A_1A_2 : B_1B_2|E) = 0$. Show that $I_{\rho^{ABE}}(A_1 : B_1|E) = I_{\rho^{ABE}}(A_2 : B_2|A_1B_1E) = 0$ using (5.109).
(**b**) Prove (8.205) using (**a**).

8.61 Prove (8.203) following the steps below.
(**a**) Assume that an extension ρ^{ABE} of ρ satisfies $I_{\rho^{ABE}}(A : B|E) = 0$. Show that $I_{(\kappa_M \otimes \iota_{AB})(\rho^{ABE})}(A : B|E) = 0$ for any PVM M on $\mathcal{H}_E$.
(**b**) Prove (8.203) using (**a**).

8.12 Positive Partial Transpose (PPT) Operations

In this section, we treat the class of positive partial transpose (PPT) maps (operations) as a wider class of local operations than the class of S-TP-CP maps. Remember that τ^A is defined as a transpose concerning the system $\mathcal{H}_A$ with the basis $\{u_1, \ldots, u_d\}$, as defined in Example 5.7 in Sect. 5.2. As was mentioned in Sect. 5.2, any separable state ρ satisfies $\tau^A(\rho) = (\tau^A \otimes \iota_B)(\rho) \geq 0$. These states are called **positive partial transpose (PPT) states** positive partial transpose (PPT) state. Note that the PPT condition $\tau^A(\rho) \geq 0$ does not depend on the choice of the basis of $\mathcal{H}_A$ Exe. 8.62. A TP-CP map κ from a system $\mathcal{H}_A \otimes \mathcal{H}_B$ to another system $\mathcal{H}_{A'} \otimes \mathcal{H}_{B'}$ is called a **positive partial transpose (PPT) map (operation)** if the map $\tau^{A'} \circ \kappa \circ \tau^A$ is a TP-CP map. As is easily checked, any PPT map κ can transform a PPT state into another PPT state. This condition is equivalent to the condition that the matrix $K(\kappa)$ defined in (5.4) has a PPT state form similar to a state on the composite system $(\mathcal{H}_A \otimes \mathcal{H}_{A'}) \otimes (\mathcal{H}_B \otimes \mathcal{H}_{B'})$. Hence, any PPT state can be produced by a PPT operation without any entangled state. Note that S-TP-CP maps also have a similar characterization. Since any separable state on the composite system $(\mathcal{H}_A \otimes \mathcal{H}_{A'}) \otimes (\mathcal{H}_B \otimes \mathcal{H}_{B'})$ is a PPT state on the composite system $(\mathcal{H}_A \otimes \mathcal{H}_{A'}) \otimes (\mathcal{H}_B \otimes \mathcal{H}_{B'})$, all S-TP-CP maps are PPT maps [45]. Hence, the class of PPT maps $C =$ PPT is the largest class of local operations among $C = \emptyset, \rightarrow, \leftarrow, \leftrightarrow, S,$ PPT. Further, the definition of PPT maps does not depend on the choice of the basis Exe. 8.65. In addition, Cirac et al. [46] showed that any PPT operation could be realized by a bound entangled state and an LOCC operation.

As an entanglement measure related to PPT maps, we often focus on the **log negativity** $\log \|\tau^A(\rho)\|_1$, which does not depend on the choice of the basis Exe. 8.66. For a pure state $u = \sum_i \sqrt{\lambda_i} u_i^A \otimes u_i^B$,

$$\tau^A(|u\rangle\langle u|) = \sum_{i,j} \sqrt{\lambda_i}\sqrt{\lambda_j} |u_j^A \otimes u_i^B\rangle\langle u_i^A \otimes u_j^B|.$$

Then,

$$\begin{aligned}|\tau^A(|u\rangle\langle u|)| &= \sqrt{\tau^A(|u\rangle\langle u|)^* \tau^A(|u\rangle\langle u|)} \\ &= \sum_{i,j} \sqrt{\lambda_i}\sqrt{\lambda_j} |u_i^A \otimes u_j^B\rangle\langle u_i^A \otimes u_j^B| = \left(\sum_i \sqrt{\lambda_i} |u_i\rangle\langle u_i|\right)^{\otimes 2}.\end{aligned} \tag{8.217}$$

Therefore, $\|\tau^A(|u\rangle\langle u|)\|_1 = \left(\sum_i \sqrt{\lambda_i}\right)^2$, i.e.,

$$-2\log \|\tau^A(|u\rangle\langle u|)\|_1 = H_{\frac{1}{2}}(\mathrm{Tr}_B\, |u\rangle\langle u|). \tag{8.218}$$

In particular,

$$\tau^A(|\Phi_L\rangle\langle\Phi_L|) = \frac{1}{L}F, \quad |\tau^A(|\Phi_L\rangle\langle\Phi_L|)| = \frac{1}{L}, \tag{8.219}$$

where F is the flip operator $P_s - P_a$. Moreover, the log negativity satisfies the additivity $\log \|\tau^A(\rho \otimes \sigma)\|_1 = \log \|\tau^A(\rho)\|_1 + \log \|\tau^A(\sigma)\|_1$ and the **monotonicity** regarding the PPT operations κ

$$\|\tau^A(\kappa(\rho))\|_1 \leq \|\tau^A(\rho)\|_1, \tag{8.220}$$

i.e.,

$$\log \|\tau^A(\kappa(\rho))\|_1 \leq \log \|\tau^A(\rho)\|_1. \tag{8.221}$$

Using (8.219), we can generalize relation (8.7) as

$$\begin{aligned}\langle \Phi_L|\rho|\Phi_L\rangle &= \mathrm{Tr}\, \rho |\Phi_L\rangle\langle \Phi_L| = \mathrm{Tr}\, \tau^A(\rho)\tau^A(|\Phi_L\rangle\langle\Phi_L|) \\ \leq & \|\tau^A(\rho)\|_1 \|\tau^A(|\Phi_L\rangle\langle\Phi_L|)\| = \frac{\|\tau^A(\rho)\|_1}{L}.\end{aligned} \tag{8.222}$$

This relation implies that

$$E_{d,2}^{\mathrm{PPT},\dagger}(\rho) \leq D(\rho\|\sigma) + \log \|\tau^A(\sigma)\|_1. \tag{8.223}$$

The RHS is called an SDP (semidefinite programming) bound [45] and satisfies the **monotonicity**, i.e.,

$$D(\rho\|\sigma) + \log \|\tau^A(\sigma)\|_1 \geq D(\kappa(\rho)\|\kappa(\sigma)) + \log \|\tau^A(\kappa(\sigma))\|_1$$

for a PPT operation κ. It implies inequality (8.220). As a consequence, we have

$$E_{d,2}^{\mathrm{PPT},\dagger}(\rho) \leq \log \|\tau^A(\rho)\|_1, \tag{8.224}$$

$$E_{d,2}^{\mathrm{PPT},\dagger}(\rho) \leq D(\rho\|\sigma), \text{ for a PPT state } \sigma. \tag{8.225}$$

Hence, the entanglement of relative entropy with PPT states $E_{r,\mathrm{PPT}}(\rho) \stackrel{\mathrm{def}}{=} \min_{\sigma:\mathrm{PPT}} D(\rho\|\sigma)$ and the **semi-definite programming (SDP) bound** $E_{\mathrm{SDP}}(\rho) \stackrel{\mathrm{def}}{=} \min_\sigma D(\rho\| \sigma) + \log \|\tau^A(\sigma)\|_1$ do not increase for a PPT operation, i.e., **SDP bound** satisfies the **monotonicity**. Further, from (8.223) we obtain

$$E_{d,2}^{\mathrm{PPT},\dagger}(\rho) \leq \lim_{n\to\infty} \frac{E_{\mathrm{SDP}}(\rho^{\otimes n})}{n} \leq \lim_{n\to\infty} \frac{E_{r,\mathrm{PPT}}(\rho^{\otimes n})}{n}. \tag{8.226}$$

This relation implies that

$$E_{r,S}(\rho) = E_{r,\mathrm{PPT}}(\rho) = E_{\mathrm{SDP}}(\rho) = \lim_{n\to\infty} \frac{E_{\mathrm{SDP}}(\rho^{\otimes n})}{n} = \lim_{n\to\infty} \frac{E_{r,\mathrm{PPT}}(\rho^{\otimes n})}{n} \tag{8.227}$$

when $-H_\rho(A|B) = E_{r,S}(\rho)$ because $E_{r,S}(\rho)$ is not smaller than $E_{r,\mathrm{PPT}}(\rho)$, $E_{\mathrm{SDP}}(\rho)$, $\lim_{n\to\infty} \frac{E_{\mathrm{SDP}}(\rho^{\otimes n})}{n}$, and $\lim_{n\to\infty} \frac{E_{r,\mathrm{PPT}}(\rho^{\otimes n})}{n}$.

Regarding the direct part, since the quantity $E_{d,2}^{\rm PPT}(\rho)$ satisfies Condition **E3'** (weak lower continuity) because of Fannes' inequality (5.92), from Exercise 8.47 and the Hashing inequality (8.121), we can show

$$E_{d,2}^{\rm PPT}(\rho) = \lim_{n\to\infty} \frac{E_m^{\rm PPT}(\rho^{\otimes n})}{n}. \tag{8.228}$$

For any state σ, we choose the positive semidefinite matrix

$$\sigma' := \frac{1}{\|\tau^A(\sigma)\|_1}\sigma. \tag{8.229}$$

Then,

$$D(\rho\|\sigma) + \log \|\tau^A(\sigma)\|_1 = D(\rho\|\sigma'). \tag{8.230}$$

Since we have one-to-one correspondence between the state σ and the positive semidefinite matrix σ' satisfying the condition $\|\tau^A(\sigma')\|_1 = 1$, we have another formula for SDP bound as

$$E_{\rm SDP}(\rho) = \min_{\sigma'\ge 0:\|\tau^A(\sigma')\|_1=1} D(\rho\|\sigma') = \min_{\sigma'\ge 0:\|\tau^A(\sigma')\|_1\le 1} D(\rho\|\sigma'). \tag{8.231}$$

Using this notation, we can show the convexity.[6] That is, for two states ρ_1 and ρ_2 and a real number $\lambda \in (0,1)$, we have

$$E_{\rm SDP}(\lambda\rho_1 + (1-\lambda)\rho_2) \le \lambda E_{\rm SDP}(\rho_1) + (1-\lambda)E_{\rm SDP}(\rho_2). \tag{8.232}$$

To show this inequality, we consider the state $\lambda\rho_1|0,0\rangle\langle 0,0| + (1-\lambda)\rho_2|1,1\rangle\langle 1,1|$. Choose $\sigma_i' := \mathrm{argmin}_{\sigma'\ge 0:\|\tau^A(\sigma')\|_1=1} D(\rho\|\sigma')$. Applying the monotonicity (**a**) of Exercise 5.25 to the partial trace, we have

$$\begin{aligned}
&E_{\rm SDP}(\lambda\rho_1 + (1-\lambda)\rho_2) \le D(\lambda\rho_1 + (1-\lambda)\rho_2\|\lambda\sigma_1' + (1-\lambda)\sigma_2')\\
\le& D(\lambda\rho_1|0,0\rangle\langle 0,0| + (1-\lambda)\rho_2|1,1\rangle\langle 1,1|\|\lambda\sigma_1'|0,0\rangle\langle 0,0| + (1-\lambda)\sigma_2'|1,1\rangle\langle 1,1|)\\
=&\lambda D(\rho_1\|\sigma_1') + (1-\lambda)D(\rho_2\|\sigma_2') = \lambda E_{\rm SDP}(\rho_1) + (1-\lambda)E_{\rm SDP}(\rho_2).
\end{aligned}$$

Proof of (8.223) Consider a PPT operation κ_n' on $\mathcal{H}_A \otimes \mathcal{H}_B$ and a real number $r > D(\rho\|\sigma) + \log\|\tau^A(\sigma)\|_1$. Inequalities (8.222) and (8.220) imply that $\langle\Phi_{e^{nr}}|\kappa_n'(\sigma^{\otimes n})|\Phi_{e^{nr}}\rangle \le e^{-nr}\|\tau^A(\kappa_n'(\sigma^{\otimes n}))\|_1 \le e^{-nr}\|\tau^A(\sigma)\|_1^n$. From $I - |\Phi_{e^{nr}}\rangle\langle\Phi_{e^{nr}}| \ge 0$ we have

$$\begin{aligned}
&I - (\kappa_n')^*(|\Phi_{e^{nr}}\rangle\langle\Phi_{e^{nr}}|) = (\kappa_n')^*(I) - (\kappa_n')^*(|\Phi_{e^{nr}}\rangle\langle\Phi_{e^{nr}}|)\\
=&(\kappa_n')^*(I - |\Phi_{e^{nr}}\rangle\langle\Phi_{e^{nr}}|) \ge 0,
\end{aligned}$$

[6] If we employ the original definition, the inequality (8.232) cannot shown by the concavity of log.

where $(\kappa'_n)^*$ is the dual map of κ'_n (see ④ of Theorem 5.1). Moreover,

$$(\kappa'_n)^*(|\Phi_{e^{nr}}\rangle\langle\Phi_{e^{nr}}|) \geq 0,$$
$$\operatorname{Tr}\sigma^{\otimes n}(\kappa'_n)^*(|\Phi_{e^{nr}}\rangle\langle\Phi_{e^{nr}}|) = \langle\Phi_{e^{nr}}|\kappa'_n(\sigma^{\otimes n})|\Phi_{e^{nr}}\rangle \leq e^{-n(r-\log\|\tau^A(\sigma)\|_1)}.$$

Since the matrix $(\kappa'_n)^*(|\Phi_{e^{nr}}\rangle\langle\Phi_{e^{nr}}|)$ satisfies the condition of test $0 \leq (\kappa'_n)^*(|\Phi_{e^{nr}}\rangle\langle\Phi_{e^{nr}}|) \leq I$, inequality (3.138) in Sect. 3.8 yields

$$\langle\Phi_{e^{nr}}|\kappa'_n(\rho^{\otimes n})|\Phi_{e^{nr}}\rangle = \operatorname{Tr}\rho^{\otimes n}(\kappa'_n)^*(|\Phi_{e^{nr}}\rangle\langle\Phi_{e^{nr}}|) \leq e^{n\frac{-\phi(s)-s(r-\log\|\tau^A(\sigma)\|_1)}{1-s}} \tag{8.233}$$

for $s \leq 0$, where $\phi(s) \stackrel{\text{def}}{=} \phi(s|\rho\|\sigma)$. Using arguments similar to those used for the Proof of Lemma 3.7, the condition $r - \log\|\tau^A(\sigma)\|_1 > D(\rho\|\sigma)$ implies $\langle\Phi_{e^{nr}}|\kappa'_n(\rho^{\otimes n})|\Phi_{e^{nr}}\rangle \to 0$. We thus obtain (8.223). ∎

Further, using the log negativity, Rains [45] showed that

$$E_{d,2}^{\mathrm{PPT}}(\rho_1) + E_{d,2}^{\mathrm{PPT}}(\rho_2) \leq E_{d,2}^{\mathrm{PPT}}(\rho_1\otimes\rho_2) \leq E_{d,2}^{\mathrm{PPT}}(\rho_1) + \log\|\tau^A(\rho_2)\|_1. \tag{8.234}$$

Indeed, Donald and Horodecki [33] proved Condition **E3** for $E_{r,\mathrm{PPT}}(\rho)$. Therefore, since $E_{r,\mathrm{PPT}}(\rho)$ satisfies Conditions **E1**, **E2**PPT, and **E4** in a manner similar to $E_{r,S}(\rho)$, Theorem 8.10 guarantees the inequality

$$\lim_{n\to\infty}\frac{E_{r,\mathrm{PPT}}(\rho^{\otimes n})}{n} \leq E_c^{\mathrm{PPT}}(\rho). \tag{8.235}$$

In inequality (8.224), the log negativity gives the upper bound of the entanglement of distillation; however, it does not give the lower bound of the entanglement of cost because $\log\|\tau^A(|u\rangle\langle u|)\|_1 = 2\log(\sum_i\sqrt{\lambda_i}) > -\sum_i\lambda_i\log\lambda_i = E_c^S(|u\rangle\langle u|)$. Thus, it does not satisfy Condition **E3** (continuity) because Theorem 8.10 leads to a contradiction if it holds (Exercise 8.69). In this case, we can show the following lemma as its alternative.

Lemma 8.15 *When the quantity $\tilde{E}_C(\rho)$ satisfies Conditions* **E1** *and* **E2***C, the entanglement of exact distillation and the entanglement of exact cost are evaluated as*

$$E_{d,e}^C(\rho) \leq \tilde{E}^C(\rho) \leq E_{c,e}^C(\rho).$$

Further, if it satisfies Condition **E4** *also, their asymptotic version are evaluated as*

$$E_{d,e}^{C,\infty}(\rho) \leq \lim_{n\to\infty}\frac{\tilde{E}^C(\rho^{\otimes n})}{n} \leq E_{c,e}^{C,\infty}(\rho).$$

Hence, from (8.221) we have the following formula for the exact cost with PPT operations [47]:

$$\log \|\tau^A(\rho)\|_1 \le E_{c,e}^{\mathrm{PPT},\infty}(\rho). \tag{8.236}$$

Further, Audenaert et al. [47] showed the opposite inequality

$$E_{c,e}^{\mathrm{PPT},\infty}(\rho) \le \log(\|\tau^A(\rho)\|_1 + d_A d_B \max(0, -\lambda_{\min}(\tau^A(|\tau^A(\rho)|))),$$

where $\lambda_{\min}(X)$ denotes the minimum eigenvalue of X. Hence, when

$$\tau^A(|\tau^A(\rho)|) \ge 0, \tag{8.237}$$

we obtain

$$\log \|\tau^A(\rho)\|_1 = E_{c,e}^{\mathrm{PPT},\infty}(\rho).$$

For example, from (8.217) any pure state satisfies condition (8.237). Further, Ishizaka [48] proved that all states on the system $\mathbb{C}^2 \otimes \mathbb{C}^2$ satisfy this condition. Therefore, the entanglement measures for a pure state $\rho = |u\rangle\langle u|$ are summarized as follows. Let λ be a probability distribution of the eigenvalues of the reduced density $\mathrm{Tr}_B\, \rho$. Then, each entanglement measure is described by the Rényi entropy $H_{1-s}(\lambda) = \frac{1}{s}\psi(s|\lambda) = \frac{1}{s}\log \sum_i \lambda^{1-s}$ as[Exe. 8.32,8.68]

$$\begin{array}{ccccccc} E_{d,e}^{C_1,\infty}(\rho) \le & E_{d,i}^{C_2}(\rho) = & E_c^{C_3}(\rho) \le & E_{c,e}^{\mathrm{PPT},\infty}(\rho) \le & E_{c,e}^{C_4,\infty}(\rho) \\ \| & \| & & \| & \| \\ H_{\min}(\lambda) \le & H(\lambda) & \le & H_{\frac{1}{2}}(\lambda) & \le H_{\max}(\lambda), \end{array} \tag{8.238}$$

where $i=1,2, C_1 = \ \rightarrow, \leftrightarrow, S, \mathrm{PPT}, C_2 = \emptyset, \rightarrow, \leftrightarrow, S, \mathrm{PPT},\ C_3 = \dashrightarrow, \rightarrow, \leftrightarrow, S, \mathrm{PPT}$, and $C_4 = \rightarrow, \leftrightarrow, S$. Remember that the quantity $H_{1-s}(\lambda)$ is monotone increasing for s (Sect. 2.1.4).

To conclude this section, we briefly discuss the relationship between $E_{d,2}^C(\rho^{A,B})$ and Theorem 8.3 [49]. In Theorem 8.3, we derived $\rho^{A,B} \preceq \rho^A$ from the fact that $\rho^{A,B}$ is separable. In fact, there exist several other conditions regarding less entanglement:

LE1 (Separability) $\rho^{A,B}$ is separable.
LE2 (PPT) $\tau^A \otimes \iota_B(\rho^{A,B}) \ge 0$.
LE3 (Nondistillability) $E_{d,2}^{\leftrightarrow}(\rho^{A,B}) = 0$.
LE4 (Reduction) $\rho^A \otimes I_B \ge \rho^{A,B}$ and $I_A \otimes \rho^B \ge \rho^{A,B}$.
LE5 (Majorization) $\rho^{A,B} \preceq \rho^A$ and $\rho^{A,B} \preceq \rho^B$.

The relations between these conditions can be summarized as follows:

$$\mathbf{LE1} \Rightarrow \mathbf{LE2} \overset{\text{Horodecki [50]}}{\Rightarrow} \mathbf{LE3} \overset{\text{Horodecki [51]}}{\Rightarrow} \mathbf{LE4} \overset{\text{Hiroshima [52]}}{\Rightarrow} \mathbf{LE5}$$

In particular, a nondistillable state is called a bound entangled state when it is not separable. The relation **LE2⇒LE1** (Theorem 5.5) has been shown only for $\mathbb{C}^2 \otimes \mathbb{C}^2$ and $\mathbb{C}^2 \otimes \mathbb{C}^3$. Hence, there is no bound entangled state on the $\mathbb{C}^2 \otimes \mathbb{C}^2$ system.

However, a counterexample, i.e., a bound entangled state, exists for **LE1⇐LE2** on $\mathbb{C}^2 \otimes \mathbb{C}^4$ and $\mathbb{C}^3 \otimes \mathbb{C}^3$ [53]. Since any PPT state can be produced by a PPT operation without any entangled state, this counterexample provides an interesting insight. That is, there exists a separable state ρ' and a PPT operation κ' such that $\kappa'(\rho')$ is not separable. Further, it is known that the relation **LE2⇐LE4** holds for $\mathbb{C}^2 \otimes \mathbb{C}^n$ Exe. 8.70 [51, 54, 55].

As easily checked, Condition **LE1** is equivalent to the conditions $E^s_{c,e}(\rho) = 0$ and $E_f(\rho) = 0$. Since $E_f(\kappa'(\rho'))$ is not 0, E_f is not monotone for PPT operations. Further, examining the quantity $C_d^{A\to B}(\rho)$, Yang et al. [56] showed that if the entanglement of cost $E_c(\rho)$ is zero, ρ is separable. That is, a nonseparable state has nonzero entanglement of cost. Hence, E_c is not monotone for PPT operations.

Further, for any nonseparable state σ, there exist a state ρ and an integer L such that [57]

$$E^C_{d,L}(\rho) < E^C_{d,L}(\rho \otimes \sigma),$$

which implies that $E_c(\sigma) > 0$.

In addition, a counterexample also exists for **LE4⇐LE5** when $\mathbb{C}^2 \otimes \mathbb{C}^2$ [58]. However, it is an open problem whether the opposite relation **LE2⇐LE3** holds. To discuss it in greater detail, we focus on the following relation:

$$\begin{aligned} &E^{\leftrightarrow}_{d,2}(\rho) \le E^S_{d,2}(\rho) \le E^{\mathrm{PPT}}_{d,2}(\rho) \le \lim_{n\to\infty} \frac{E_{\mathrm{SDP}}(\rho^{\otimes n})}{n} \le \lim_{n\to\infty} \frac{E_{r,\mathrm{PPT}}(\rho^{\otimes n})}{n} \\ \le &E^{\mathrm{PPT}}_c(\rho) \le E^{\mathrm{PPT},\infty}_{c,e}(\rho) \le E^{\mathrm{PPT}}_{c,e}(\rho). \end{aligned}$$

Since any PPT state can be produced by a PPT operation without any entangled state, Condition **LE2** is equivalent to the condition $E^{\mathrm{PPT}}_{c,e}(\rho) = 0$. From (8.236) it is also equivalent to the condition $E^{\mathrm{PPT},\infty}_{c,e}(\rho) = 0$. Therefore, if Condition **LE2** is equivalent to Condition **LE3**, these conditions hold if and only if one of the above values is equal to zero.

Exercises

8.62 Let $\tilde{\tau}^A$ be a partial transpose concerning another (the second) basis. Show that there exists a unitary U such that $\tilde{\tau}^A(\rho) = U(\tau^A(U^*\rho U))U^*$.

8.63 Let κ be a map from the set of Hermitian matrices on $\mathcal{H}$ to that on $\mathcal{H}'$. Show that $\tau' \circ \kappa \circ \tau$ is a CP map if and only if κ is a CP map, where τ and τ' are the transposes on $\mathcal{H}$ and $\mathcal{H}'$, respectively. Show that $\|\tau(X)\|_1 = \|X\|_1$ for a Hermitian matrix X on $\mathcal{H}$.

8.64 Show that $\tau^{A'} \circ \kappa \circ \tau^A$ is TP-CP if and only if $\tau^{B'} \circ \kappa \circ \tau^B$ is TP-CP.

8.65 Show that $\tau^{A'} \circ \kappa \circ \tau^A$ is TP-CP if and only if $\tilde{\tau}^{A'} \circ \kappa \circ \tilde{\tau}^A$ is TP-CP when $\tilde{\tau}^A$ and $\tilde{\tau}^{A'}$ are the partial transposes for other bases.

8.66 Show that $\|\tilde{\tau}^A(\rho)\|_1 = \|\tau^A(\rho)\|_1$ when $\tilde{\tau}^A$ is the partial transposes for other bases.

8.67 Show that the maximally correlated state ρ_α satisfies

$$E_{d,2}^{\mathrm{PPT},*}(r|\rho_\alpha) \geq \max_{t \geq 0} \frac{t(r - D_{1+t}(\rho_\alpha\|\sigma_\alpha))}{1+t} \tag{8.239}$$

where $E_{d,2}^{\mathrm{PPT},*}(r|\rho_\alpha)$ is defined in (8.91) and σ_α is given in (8.142). (8.239) is a generalization of (8.160).

8.68 Prove the following equation:for an entangled pure state $|u\rangle$:

$$E_{d,e}^{\mathrm{PPT},\infty}(|x\rangle\langle x|) = -\log \lambda_1^{\downarrow}. \tag{8.240}$$

(**a**) Prove the following inequality as a generalization of (8.222):

$$\mathrm{Tr}\, \sigma\rho \leq \|\tau^A(\sigma)\| \|\tau^A(\rho)\|_1. \tag{8.241}$$

(**b**) Prove

$$\max_{\sigma \in S_{\mathrm{PPT}}} \mathrm{Tr}\, |\Phi_d\rangle\langle\Phi_d|\kappa(\sigma) \leq \frac{1}{d} \tag{8.242}$$

for $\kappa \in \mathrm{PPT}$, where S_{PPT} is the set of positive partial transpose states.
(**c**) Show the following by using (8.217):

$$\max_{\sigma \in S_{\mathrm{PPT}}} \mathrm{Tr}\, |x\rangle\langle x|\sigma = \lambda_1^{\downarrow}. \tag{8.243}$$

(**d**) Prove (8.240) by combining (8.242) and (8.243) in a way similar to (8.96).

8.69 Check the following counterexample of the continuity of $2\log\|\tau^A(|x\rangle\langle x|)\|_1$ as follows [59].
(**a**) Show that $\|\rho^{\otimes n} - \rho_n\|_1 \to 0$, i.e., $F(\rho^{\otimes n}, \rho_n) \to 1$, where $\rho_n \stackrel{\mathrm{def}}{=} \frac{\{e^{-n(H(\rho)+\epsilon)} \leq \rho^{\otimes n} \leq e^{-n(H(\rho)-\epsilon)}\}\rho^{\otimes n}}{\mathrm{Tr}\{e^{-n(H(\rho)+\epsilon)} \leq \rho^{\otimes n} \leq e^{-n(H(\rho)-\epsilon)}\}\rho^{\otimes n}}$.
(**b**) Show that $H(\rho) - \epsilon \leq \frac{1}{n} H_\alpha(\rho_n) \leq H(\rho) + 2\epsilon$ for $\alpha \geq 0$ and sufficiently large n.
(**c**) Check that the purifications x_n, y_n of $\rho^{\otimes n}, \rho_n$ give a counterexample of the continuity of $2\log\|\tau^A(|x\rangle\langle x|)\|_1$.

8.70 Let A, B, and C be $n \times n$ matrices. Show that $\begin{pmatrix} A & B^* \\ B & C \end{pmatrix} \geq 0$ when $\begin{pmatrix} A+C & 0 \\ 0 & A+C \end{pmatrix} \geq \begin{pmatrix} A & B \\ B^* & C \end{pmatrix}$ by using the unitary $\begin{pmatrix} 0 & I \\ -I & 0 \end{pmatrix}$. This argument means that **LE4** Reduction criterion implies **LE1** Separability on the system $\mathbb{C}^2 \otimes \mathbb{C}^n$ [55].

8.71 Define the SDP bounds with relative Rényi entropy.

$$E_{1+s|\mathrm{SDP}}(\rho) \stackrel{\mathrm{def}}{=} \min_{\sigma} D_{1+s}(\rho\|\sigma) + \log \|\tau^A(\sigma)\|_1, \tag{8.244}$$

$$\tilde{E}_{1+s|\mathrm{SDP}}(\rho) \stackrel{\mathrm{def}}{=} \min_{\sigma} \underline{D}_{1+s}(\rho\|\sigma) + \log \|\tau^A(\sigma)\|_1 \tag{8.245}$$

Show the following relations similar to (8.231)

$$\begin{aligned} E_{1+s|\mathrm{SDP}}(\rho) =& \min_{\sigma'\ge 0:\|\tau^A(\sigma')\|_1=1} D_{1+s|\mathrm{SDP}}(\rho\|\sigma') \\ =& \min_{\sigma'\ge 0:\|\tau^A(\sigma')\|_1\le 1} D_{1+s|\mathrm{SDP}}(\rho\|\sigma'), \end{aligned} \tag{8.246}$$

$$\begin{aligned} \tilde{E}_{1+s|\mathrm{SDP}}(\rho) =& \min_{\sigma'\ge 0:\|\tau^A(\sigma')\|_1=1} \underline{D}_{1+s|\mathrm{SDP}}(\rho\|\sigma') \\ =& \min_{\sigma'\ge 0:\|\tau^A(\sigma')\|_1\le 1} \underline{D}_{1+s|\mathrm{SDP}}(\rho\|\sigma'). \end{aligned} \tag{8.247}$$

8.72 Show the following relations similar to (8.231)

$$\begin{aligned} & e^{sE_{1+s|\mathrm{SDP}}(\lambda\rho_1+(1-\lambda)\rho_2)} \\ \ge & \lambda e^{sE_{1+s|\mathrm{SDP}}(\rho_1)} + (1-\lambda)e^{sE_{1+s|\mathrm{SDP}}(\rho_2)} \text{ for } s \in [-1,0], \end{aligned} \tag{8.248}$$

$$\begin{aligned} & e^{sE_{1+s|\mathrm{SDP}}(\lambda\rho_1+(1-\lambda)\rho_2)} \\ \le & \lambda e^{sE_{1+s|\mathrm{SDP}}(\rho_1)} + (1-\lambda)e^{sE_{1+s|\mathrm{SDP}}(\rho_2)} \text{ for } s \in [0,1], \end{aligned} \tag{8.249}$$

$$\begin{aligned} & e^{s\tilde{E}_{1+s|\mathrm{SDP}}(\lambda\rho_1+(1-\lambda)\rho_2)} \\ \ge & \lambda e^{s\tilde{E}_{1+s|\mathrm{SDP}}(\rho_1)} + (1-\lambda)e^{s\tilde{E}_{1+s|\mathrm{SDP}}(\rho_2)} \text{ for } s \in [-\frac{1}{2},0], \end{aligned} \tag{8.250}$$

$$\begin{aligned} & e^{s\tilde{E}_{1+s|\mathrm{SDP}}(\lambda\rho_1+(1-\lambda)\rho_2)} \\ \le & \lambda e^{s\tilde{E}_{1+s|\mathrm{SDP}}(\rho_1)} + (1-\lambda)e^{s\tilde{E}_{1+s|\mathrm{SDP}}(\rho_2)} \text{ for } s \in [0,\infty). \end{aligned} \tag{8.251}$$

8.73 We define $E_{d,2}^{\mathrm{PPT},*}(r|\rho)$ similar to $E_{d,2}^{S,*}(r|\rho)$ defined in (8.239), Show the following relations

$$E_{d,2}^{S,*}(r|\rho) \ge \max_{t\ge 0} \frac{t(r - E_{1+t|\mathrm{SDP}}(\rho))}{1+t}, \tag{8.252}$$

$$E_{d,2}^{S,*}(r|\rho) \ge \max_{t\ge 0} \frac{t(r - \tilde{E}_{1+t|\mathrm{SDP}}(\rho))}{1+t}. \tag{8.253}$$

8.13 Violation of Superadditivity of Entanglement Formation

8.13.1 Counter Example for Superadditivity of Entanglement Formation

In this section, we give a counter example for superadditivity of entanglement formation by Fukuda [60] while the first counter example was given in Hastings [61]. In order to give a counter example for superadditivity of entanglement formation (8.145), we consider a large bipartite system $\mathbb{C}^k \otimes \mathbb{C}^n = \mathbb{C}^{nk}$, in which, the system $\mathcal{H}_{A,1}$ is given as $\mathbb{C}^k$ and the other system $\mathcal{H}_{B,1}$ is given as $\mathbb{C}^n$. Then, we focus on a $\lceil cn \rceil$-dimensional subspace $\mathcal{K}$ and its complex conjugate subspace $\overline{\mathcal{K}}$ defined as

$$\overline{\mathcal{K}} := \{x \in \mathbb{C}^{nk} | \overline{x} \in \mathcal{K}\}. \tag{8.254}$$

In the following, we consider that the space $\overline{\mathcal{K}}$ as the subspace of $\mathcal{H}_{A,2} \otimes \mathcal{H}_{B,2}$, in which, the system $\mathcal{H}_{A,2}$ is given as $\mathbb{C}^k$ and the other system $\mathcal{H}_{B,2}$ is given as $\mathbb{C}^n$. Then, we obtain the following lemma for the bipartite system $\mathcal{H}_A \otimes \mathcal{H}_B$, where $\mathcal{H}_A := \mathcal{H}_{A,1} \otimes \mathcal{H}_{A,2}$ and $\mathcal{H}_B := \mathcal{H}_{B,1} \otimes \mathcal{H}_{B,2}$.

Lemma 8.16 *Any $\lceil cn \rceil$-dimensional subspace $\mathcal{K}$ satisfies*

$$\min_{|x\rangle \in \mathcal{K} \otimes \overline{\mathcal{K}}} E(|x\rangle\langle x|) \le 2\left(1 - \frac{c}{k}\right) \log k + h\left(\frac{c}{k}\right). \tag{8.255}$$

Proof First, let V be the isometry from $\mathcal{H}_{C,1} := \mathbb{C}^{\lceil cn \rceil}$ to $\mathcal{H}_{A,1} \otimes \mathcal{H}_{B,1}$, whose image is the subspace $\mathcal{K}$. Then, the complex conjugate $\overline{V}$ is the isometry from $\mathcal{H}_{C,2} := \mathbb{C}^{\lceil cn \rceil}$ to $\mathcal{H}_{A,2} \otimes \mathcal{H}_{B,2}$, whose image is the subspace $\overline{\mathcal{K}}$. Then, we have

$$\mathrm{Tr}\, V V^* = \lceil cn \rceil. \tag{8.256}$$

In this proof, we denote the maximally entangle states $\frac{1}{\sqrt{k}} \sum_{i=1}^{k} |u_i, u_i\rangle$, $\frac{1}{\sqrt{n}} \sum_{i=1}^{n} |u_i, u_i\rangle$, and $\frac{1}{\sqrt{\lceil cn \rceil}} \sum_{i=1}^{\lceil cn \rceil} |u_i, u_i\rangle$ on $\mathcal{H}_{A,1} \otimes \mathcal{H}_{A,2}$, $\mathcal{H}_{B,1} \otimes \mathcal{H}_{B,2}$, and $\mathcal{H}_{C,1} \otimes \mathcal{H}_{C,2}$ by $|\Phi_A\rangle$, $|\Phi_B\rangle$, and $|\Phi_C\rangle$, respectively, where u_i is the canonical basis.

Now, we focus on the state $(V \otimes \overline{V})|\Phi_C\rangle\langle\Phi_C|(V^* \otimes V^T)$ in $\mathcal{H}_A \otimes \mathcal{H}_B$. Due to (8.256), the maximal eigenvalue of $\mathrm{Tr}_B (V \otimes \overline{V})|\Phi_C\rangle\langle\Phi_C|(V^* \otimes V^T)$ is bounded by the following quantity.

$$\begin{aligned}
&\langle\Phi_A|(\mathrm{Tr}_B (V \otimes \overline{V})|\Phi_C\rangle\langle\Phi_C|(V^* \otimes V^T))|\Phi_A\rangle \\
\ge&\langle\Phi_A, \Phi_B|(V \otimes \overline{V})|\Phi_C\rangle\langle\Phi_C|(V^* \otimes V^T)|\Phi_A, \Phi_B\rangle \\
=&|\langle\Phi_A, \Phi_B|(V \otimes \overline{V})|\Phi_C\rangle|^2 = |\frac{1}{\sqrt{nk\lceil cn \rceil}} \mathrm{Tr}\, V V^*|^2 = |\sqrt{\frac{\lceil cn \rceil}{nk}}|^2 = \frac{\lceil cn \rceil}{nk}.
\end{aligned}$$

Recall Exercise 2.3. Then, the above constraint for maximal eigenvalue yields that

$$\begin{aligned} & H((\mathrm{Tr}_B(V \otimes \overline{V})|\Phi_C\rangle\langle\Phi_C|(V^* \otimes V^T))) \\ \le & h\left(\frac{\lceil cn \rceil}{nk}\right) + \left(1 - \frac{\lceil cn \rceil}{nk}\right) \log(k^2 - 1) \\ \le & h\left(\frac{c}{k}\right) + \left(1 - \frac{c}{k}\right) \log(k^2 - 1) \le 2\left(1 - \frac{c}{k}\right) \log k + h\left(\frac{c}{k}\right). \end{aligned}$$

■

On the other hand, we have the following theorem with respect to the Rényi entropy of order 2, $H_2(\mathrm{Tr}_B |x\rangle\langle x|) \le H(\mathrm{Tr}_B |x\rangle\langle x|)$.

Theorem 8.14 *For given constants $c \in (0, 1)$, $\epsilon > 0$, $\epsilon' > 0$, and a positive integer k, there exist a sufficiently large n and a $\lceil cn \rceil$-dimensional subspace $\mathcal{K} \subset \mathbb{C}^{nk} = \mathcal{H}_A \otimes \mathcal{H}_B$ such that*

$$\begin{aligned} & \min_{x \in \mathcal{K} \cap S^{2\lceil cn \rceil - 1}} H_2(\mathrm{Tr}_B |x\rangle\langle x|) \\ \ge & \log k - \frac{\epsilon}{k} - \log\left(1 + \left(\frac{(-4c \log \frac{\epsilon'}{4}) + 2\sqrt{-2c \log \frac{\epsilon'}{4}}\sqrt{1 - 2c \log \frac{\epsilon'}{4}}}{1 - \epsilon'}\right)^2 \frac{1}{k}\right) \\ \ge & \log k - \frac{\epsilon}{k} - \left(\frac{(-4c \log \frac{\epsilon'}{4}) + 2\sqrt{-2c \log \frac{\epsilon'}{4}}\sqrt{1 - 2c \log \frac{\epsilon'}{4}}}{1 - \epsilon'}\right)^2 \frac{1}{k}. \end{aligned} \tag{8.257}$$

Any state ρ on $\mathcal{K}$ satisfies that

$$\begin{aligned} E_f(\rho) \ge & \min_{x \in \mathcal{K} \cap S^{2\lceil cn \rceil - 1}} E(|x\rangle\langle x|) = \min_{x \in \mathcal{K} \cap S^{2\lceil cn \rceil - 1}} H(\mathrm{Tr}_B |x\rangle\langle x|) \\ \ge & \min_{x \in \mathcal{K} \cap S^{2\lceil cn \rceil - 1}} H_2(\mathrm{Tr}_B |x\rangle\langle x|). \end{aligned}$$

When the subspace $\mathcal{K}$ is chosen by the above theorem, any state ρ on $\mathcal{K}$ satisfies that

$$E_f(\rho) \ge \log k - \frac{\epsilon}{k} - \left(\frac{(-4c \log \frac{\epsilon'}{4}) + 2\sqrt{-2c \log \frac{\epsilon'}{4}}\sqrt{1 - 2c \log \frac{\epsilon'}{4}}}{1 - \epsilon'}\right)^2 \frac{1}{k}.$$

The complex conjugate subspace $\overline{\mathcal{K}}$ also has this property.

Now, we choose a $\lceil cn \rceil$-dimensional subspace $\mathcal{K} \subset \mathbb{C}^{nk} = \mathcal{H}_A \otimes \mathcal{H}_B$ given in Theorem 8.14. Then, we choose a pure state ρ in $\mathcal{K} \otimes \overline{\mathcal{K}}$ that realizes the minimum entropy evaluated in Lemma 8.16. Therefore, fixing a real number c and taking k to be large, we obtain

$$\begin{aligned} & E_f(\rho) = E(\rho) \le 2\left(1 - \frac{c}{k}\right) \log k + h\left(\frac{c}{k}\right) = 2 \log k - \frac{c}{k} \log k + o\left(\frac{1}{k} \log k\right) \\ & < 2 \log k + o\left(\frac{1}{k} \log k\right) \end{aligned}$$

$$
\begin{aligned}
=&2\log k - \frac{2\epsilon}{k} - 2\left(\frac{(-4c\log\frac{\epsilon'}{4}) + 2\sqrt{-2c\log\frac{\epsilon'}{4}}\sqrt{1-2c\log\frac{\epsilon'}{4}}}{1-\epsilon'}\right)^2\frac{1}{k}\\
\le& E_f(\mathrm{Tr}_2\,\rho) + E_f(\mathrm{Tr}_1\,\rho),
\end{aligned} \tag{8.258}
$$

which contradicts the superadditivity of entanglement formation (8.145).

8.13.2 Proof of Theorem 8.14

Firstly, we notice that

$$
\begin{aligned}
e^{-H_2(\mathrm{Tr}_B |x\rangle\langle x|)} =& \mathrm{Tr}(\mathrm{Tr}_B |x\rangle\langle x|)^2 = \|\mathrm{Tr}_B |x\rangle\langle x|\|_2^2\\
=&\frac{1}{k} + \|\mathrm{Tr}_B |x\rangle\langle x| - \rho_{\mathrm{mix},A}\|_2^2
\end{aligned} \tag{8.259}
$$

for any unit vector $x \in \mathcal{H}_A \otimes \mathcal{H}_B$. Hence, we obtain

$$
H_2(\mathrm{Tr}_B |x\rangle\langle x|) = \log k - \log(1 + k\|\mathrm{Tr}_B |x\rangle\langle x| - \rho_{\mathrm{mix},A}\|_2^2). \tag{8.260}
$$

Next, we prepare the following two lemmas.

Lemma 8.17 *For a given $|x\rangle, |y\rangle \in \mathcal{H}_A \otimes \mathcal{H}_B$, we have*

$$
\frac{|\|\mathrm{Tr}_B |x\rangle\langle x| - \rho_{\mathrm{mix},A}\|_2 - \|\mathrm{Tr}_B |y\rangle\langle y| - \rho_{\mathrm{mix},A}\|_2|}{d(x,y)} \le 2, \tag{8.261}
$$

$$
\begin{aligned}
&\frac{|\|\mathrm{Tr}_B |x\rangle\langle x| - \rho_{\mathrm{mix},A}\|_2 - \|\mathrm{Tr}_B |y\rangle\langle y| - \rho_{\mathrm{mix},A}\|_2|}{d(x,y)}\\
\le&2\sqrt{\max(\|\mathrm{Tr}_B |x\rangle\langle x| - \rho_{\mathrm{mix},A}\|_2, \|\mathrm{Tr}_B |y\rangle\langle y| - \rho_{\mathrm{mix},A}\|_2) + \frac{1}{k}}.
\end{aligned} \tag{8.262}
$$

Proof Due to (8.259), (8.261) is a quantum extension of (2.209) in Exercise 2.51 with a modification. However, we show it in a different way. We chose $k \times n$ matrices X and Y as $|x\rangle = |X\rangle$ and $|y\rangle = |Y\rangle$. Since $\|X\|, \|Y\| \le 1$ and $\|X^* - Y^*\|_2 = \|X - Y\|_2 = \||x\rangle - |y\rangle\|$, using Exercise 2.54 and (A.22) with $i = 2$, we obtain

$$
\begin{aligned}
&|\|\mathrm{Tr}_B |x\rangle\langle x| - \rho_{\mathrm{mix},A}\|_2 - \|\mathrm{Tr}_B |y\rangle\langle y| - \rho_{\mathrm{mix},A}\|_2|\\
=&|\|XX^* - \rho_{\mathrm{mix},A}\|_2 - \|YY^* - \rho_{\mathrm{mix},A}\|_2|\\
\le&\|XX^* - YY^*\|_2 = \|X(X^* - Y^*) + (X - Y)Y^*\|_2\\
\le&\|X(X^* - Y^*)\|_2 + \|(X - Y)Y^*\|_2 = \|X(X^* - Y^*)\|_2 + \|Y^*(X - Y)\|_2\\
\le&\|X\|\|X^* - Y^*\|_2 + \|Y^*\|\|X - Y\|_2 \le 2\max(\|X\|, \|Y\|)\||x\rangle - |y\rangle\|\\
\overset{(a)}{\le}&2\max(\|X\|, \|Y\|)d(x,y),
\end{aligned} \tag{8.263}
$$

where (a) follows from Exercise 2.54. Due to the relation $\max(\|X\|, \|Y\|) \le 1$, we have (8.261).

Since $\|X\|^2 = \|X\| \cdot \|X^*\| \le \|XX^*\|$ and

$$\|XX^*\| \le \|XX^* - \rho_{\mathrm{mix},A}\| + \|\rho_{\mathrm{mix},A}\| \le \|XX^* - \rho_{\mathrm{mix},A}\|_2 + \frac{1}{k}$$
$$= \| \mathrm{Tr}_B |x\rangle\langle x| - \rho_{\mathrm{mix},A}\|_2 + \frac{1}{k},$$

we have

$$\|X\| \le \sqrt{\| \mathrm{Tr}_B |x\rangle\langle x| - \rho_{\mathrm{mix},A}\|_2 + \frac{1}{k}}. \tag{8.264}$$

Combining (8.263) and (8.264), we obtain (8.262) ∎

Now we prepare the following lemma.

Lemma 8.18 *We assume that the pure state $|x\rangle \in \mathcal{H}_A \otimes \mathcal{H}_B$ is generated subject to the invariant measure $\mu_{\mathcal{H}}$ given in Sect. 2.6. Then, we have*

$$\mathrm{E}_{\mathcal{H}} \mathrm{Tr}(\mathrm{Tr}_B |x\rangle\langle x|)^2 = \frac{k+n}{nk+1} = \frac{1+k/n}{k+1/n}, \tag{8.265}$$

$$\mathrm{E}_{\mathcal{H}} \| \mathrm{Tr}_B |x\rangle\langle x| - \rho_{\mathrm{mix},A}\|_2 \le \sqrt{\frac{1+1/k^2}{n+1/k}}, \tag{8.266}$$

where $\mathrm{E}_{\mathcal{H}}$ is the simplification of the expectation $\mathrm{E}_{\mu_{\mathcal{H}}}$.

Using Theorem 2.11, and Lemmas 8.17 and 8.18 we can show the following lemma.

Lemma 8.19 *For a given $\delta > 0$, we choose $C_\delta := 2\delta + 2\sqrt{\delta^2 + \frac{1}{k} + \sqrt{\frac{1-1/k^2}{n+1/k}} + \sqrt{\frac{\pi}{nk-1}}}$. When the pure state $|x\rangle \in \mathcal{H}_A \otimes \mathcal{H}_B$ generated subject to the invariant distribution, the relation*

$$\| \mathrm{Tr}_B |x\rangle\langle x| - \rho_{\mathrm{mix},A}\|_2 > \sqrt{\frac{1-1/k^2}{n+1/k}} + \sqrt{\frac{\pi}{nk-1}} + C_\delta \delta \tag{8.267}$$

holds at most with the probability $e^{-\delta^2(nk-1)}/2$.

Proof Due to Lemma 8.18, it is sufficient to show that the relation

$$\| \mathrm{Tr}_B |x\rangle\langle x| - \rho_{\mathrm{mix},A}\|_2 > \mathrm{E}_{\mathcal{H}} \| \mathrm{Tr}_B |x\rangle\langle x| - \rho_{\mathrm{mix},A}\|_2 + \sqrt{\frac{\pi}{nk-1}} + C_\delta \delta$$

holds at most with the probability $e^{-\delta^2(nk-1)}/2$. The relation (8.261) guarantees that the function $f : |x\rangle \mapsto \| \mathrm{Tr}_B |x\rangle\langle x| - \rho_{\mathrm{mix},A}\|_2$ has the Lipschitz constant $C_0 = 2$ (See (2.233).). If the Lipschitz constant of the function f is bounded by C_δ on the subset $\{|x\rangle \in S^{2nk-1} \cap \mathcal{H}_A \otimes \mathcal{H}_B | \| \mathrm{Tr}_B |x\rangle\langle x| - \rho_{\mathrm{mix},A}\|_2 \le \sqrt{\frac{1-1/k^2}{n+1/k}} + \sqrt{\frac{\pi}{nk-1}} + C_\delta \delta\}$, Theorem 2.11 yields (8.267). Hence, it is sufficient to show the above relation for the Lipschitz constant of the function f. For this purpose, due to (8.262), it is enough to show that

$$2\sqrt{\max(\| \mathrm{Tr}_B |x\rangle\langle x| - \rho_{\mathrm{mix},A}\|_2, \| \mathrm{Tr}_B |y\rangle\langle y| - \rho_{\mathrm{mix},A}\|_2) + \frac{1}{k}} \le C_\delta \tag{8.268}$$

for elements x, y of the subset, which is equivalent with

$$2\sqrt{\sqrt{\frac{1-1/k^2}{n+1/k}} + \sqrt{\frac{\pi}{nk-1}} + C_\delta \delta + \frac{1}{k}} \le C_\delta. \tag{8.269}$$

Solving the quadratic equation for C_δ, we can check that the above inequality holds under the our choice of C_δ. ■

Lemma 8.20 *Given an l-dimensional subspace $\mathcal{K}$ of $\mathcal{H}_A \otimes \mathcal{H}_B$, any ϵ-net Ω of $\mathcal{K} \cap S^{2l-1}$ satisfies that*

$$\max_{x \in \mathcal{K} \cap S^{2l-1}} \| \mathrm{Tr}_B |x\rangle\langle x| - \rho_{\mathrm{mix},A}\|_2 \le \max_{x \in \Omega \cap S^{2l-1}} \frac{\| \mathrm{Tr}_B |x\rangle\langle x| - \rho_{\mathrm{mix},A}\|_2}{1 - 2\sin\epsilon}. \tag{8.270}$$

Proof For any $x \in \mathcal{K} \cap S^{2l-1}$, we choose $y \in \Omega \cap S^{2l-1}$ such that $d(x, y) \le \epsilon$. Thus, Exercise 2.52 implies that $\| |x\rangle\langle x| - |y\rangle\langle y| \|_1 \le 2\sin\epsilon$. Since the rank of $|x\rangle\langle x| - |y\rangle\langle y|$ is two, there exist two unit vectors $w_1, w_2 \in \mathcal{K}$ and a positive real number $c \le \sin\epsilon$ such that

$$|x\rangle\langle x| - |y\rangle\langle y| = c|w_1\rangle\langle w_1| - c|w_2\rangle\langle w_2|. \tag{8.271}$$

Thus, using (A.21), we have

$$\begin{aligned}
&\| \mathrm{Tr}_B |x\rangle\langle x| - \rho_{\mathrm{mix},A}\|_2 \\
=&\| \mathrm{Tr}_B |y\rangle\langle y| - \rho_{\mathrm{mix},A} + c(\mathrm{Tr}_B |w_1\rangle\langle w_1| - \rho_{\mathrm{mix},A}) - c(\mathrm{Tr}_B |w_2\rangle\langle w_2| - \rho_{\mathrm{mix},A})\|_2 \\
\le&\| \mathrm{Tr}_B |y\rangle\langle y| - \rho_{\mathrm{mix},A}\|_2 + c\| \mathrm{Tr}_B |w_1\rangle\langle w_1| - \rho_{\mathrm{mix},A}\|_2 \\
&+ c\| \mathrm{Tr}_B |w_2\rangle\langle w_2| - \rho_{\mathrm{mix},A}\|_2 \\
\le&\| \mathrm{Tr}_B |y\rangle\langle y| - \rho_{\mathrm{mix},A}\|_2 + 2c \max_{w \in \mathcal{K} \cap S^{2l-1}} \| \mathrm{Tr}_B |w\rangle\langle w| - \rho_{\mathrm{mix},A}\|_2 \\
\le& \max_{y \in \Omega \cap S^{2l-1}} \| \mathrm{Tr}_B |y\rangle\langle y| - \rho_{\mathrm{mix},A}\|_2 \\
&+ 2\sin\epsilon \max_{w \in \mathcal{K} \cap S^{2l-1}} \| \mathrm{Tr}_B |w\rangle\langle w| - \rho_{\mathrm{mix},A}\|_2.
\end{aligned}$$

Taking the maximum with respect to $x \in \mathcal{K} \cap S^{2l-1}$, we obtain (8.270). ■

We choose an ϵ-net Ω of $\mathcal{K} \cap S^{2l-1}$ by using Lemma 2.11. That is, $|\Omega| < \pi\sqrt{2l-1}\,(\frac{2}{\sin\epsilon})^{2l-1}$. Now, we choose the unitary matrix $U \in U(\mathbb{C}^{nk})$ subject to the invariant distribution. Then, for any element $y \in \Omega$, the unit vector Uy obeys the uniform distribution μ on S^{nk-1}. Now, we apply Theorem 2.11 with $f(x) := \|\operatorname{Tr}_B |x\rangle\langle x| - \rho_{\mathrm{mix},A}\|_2$. We choose $\delta > 0$ such that $e^{-\delta^2(nk-1)}/2 = \pi\sqrt{2l-1}(\frac{2}{\sin\epsilon})^{2l-1} < |\Omega|^{-1}$, i.e., $\delta^2 = -\frac{2l-1}{nk-1}\log\frac{\sin\epsilon}{2} - \frac{1}{nk-1}\log(2\pi\sqrt{2l-1})$. As $\frac{2^5}{3^2\pi} \ge 1$, we have

$$\delta^2 \le \delta_n^2 := -\frac{2l-1}{nk-1}\log\frac{\sin\epsilon}{2} - \frac{1}{nk-1}\log(2\pi\sqrt{2l-1}). \tag{8.272}$$

Due to Lemma 8.19, the probability of the following event is greater than $1 - |\Omega|e^{-\delta^2(l-1)}/2$, which is strictly greater than zero: The relation

$$\|\operatorname{Tr}_B |Uy\rangle\langle Uy| - \rho_{\mathrm{mix},A}\|_2 \le \sqrt{\frac{1-1/k^2}{n+1/k}} + \sqrt{\frac{\pi}{nk-1}} + C_{\delta_n}\delta_n, \quad \forall y \in \Omega. \tag{8.273}$$

Thus, we can choose a unitary U satisfying (8.273) for any $y \in \Omega$, and define the subspace $\mathcal{K}' := U\mathcal{K}$. Therefore, due to Lemma 8.20, any unit vector y of $\mathcal{K}' \cap S^{2l-1}$ satisfies

$$\max_{y \in \mathcal{K}' \cap S^{2l-1}} \|\operatorname{Tr}_B |y\rangle\langle y| - \rho_{\mathrm{mix},A}\|_2 \le \frac{\sqrt{\frac{1-1/k^2}{n+1/k}} + \sqrt{\frac{\pi}{nk-1}} + C_{\delta_n}\delta_n}{1 - 2\sin\epsilon}. \tag{8.274}$$

When we choose ϵ to be $2\sin\epsilon = \epsilon'$ and l to be $\lceil cn \rceil$, we have $\lim_{n\to\infty}\delta_n = \sqrt{-\frac{2c}{k}\log\frac{\epsilon'}{4}}$, $\lim_{n\to\infty} C_{\delta_n} = 2\sqrt{-\frac{2c}{k}\log\frac{\epsilon'}{4}} + 2\sqrt{\frac{1}{k} - \frac{2c}{k}\log\frac{\epsilon'}{4}}$ because of (8.272). Thus, the RHS of (8.274) goes to

$$\begin{aligned}
&\frac{2\left(\sqrt{-\frac{2c}{k}\log\frac{\epsilon'}{4}}\right)^2 + 2\sqrt{-\frac{2c}{k}\log\frac{\epsilon'}{4}}\sqrt{\frac{1}{k} - \frac{2c}{k}\log\frac{\epsilon'}{4}}}{1-\epsilon'} \\
=&\frac{\left(-\frac{4c}{k}\log\frac{\epsilon'}{4}\right) + 2\sqrt{-\frac{2c}{k}\log\frac{\epsilon'}{4}}\sqrt{\frac{1}{k} - \frac{2c}{k}\log\frac{\epsilon'}{4}}}{1-\epsilon'}.
\end{aligned} \tag{8.275}$$

Therefore, since

$$\begin{aligned}
&k\left(\frac{(-4\frac{c}{k}\log\frac{\epsilon'}{4}) + 2\sqrt{-\frac{2c}{k}\log\frac{\epsilon'}{4}}\sqrt{\frac{1}{k} - \frac{2c}{k}\log\frac{\epsilon'}{4}}}{1-\epsilon'}\right)^2 \\
=&\left(\frac{(-4c\log\frac{\epsilon'}{4}) + 2\sqrt{-2c\log\frac{\epsilon'}{4}}\sqrt{1 - 2c\log\frac{\epsilon'}{4}}}{1-\epsilon'}\right)^2 \frac{1}{k},
\end{aligned}$$

combining (8.260), we obtain (8.257) because $\frac{\epsilon}{k}$ is an arbitrary constant independent of n. More precisely, as a lower bound dependent of n, we obtain

$$\min_{x\in\mathcal{K}\cap S^{2\lceil cn\rceil-1}} H_2(\mathrm{Tr}_B |x\rangle\langle x|) \geq \log k - \log\left(1+\left(\frac{\sqrt{\frac{1-1/k^2}{n+1/k}}+\sqrt{\frac{\pi}{nk-1}}+C_{\delta_n}\delta_n}{1-2\sin\epsilon}\right)^2\frac{1}{k}\right). \tag{8.276}$$

Proof of Lemma 8.18 In order to show Lemma 8.18, we assume that the pure state $|x\rangle\in\mathcal{H}_A\otimes\mathcal{H}_B$ is generated subject to the invariant measure μ given in Sect. 2.6. We denote the flip operator on $\mathcal{H}_{A,1}\otimes\mathcal{H}_{A,2}$ by F_A, and the projection to the symmetric (anti-symmetric) space on $\mathcal{H}_{A,1}\otimes\mathcal{H}_{A,2}$ by $P_{s,A}$ ($P_{a,A}$). Similarly, we define F_B, $P_{s,B}$, $P_{a,B}$, F_{AB}, $P_{s,AB}$, and $P_{a,AB}$. Then, we have

$$P_{s,AB}=P_{s,A}\otimes P_{s,B}+P_{a,A}\otimes P_{a,B}. \tag{8.277}$$

By using (8.174), the value (8.259) is calculated as

$$\begin{aligned}\mathrm{Tr}(\mathrm{Tr}_B |x\rangle\langle x|)^2 &= \mathrm{Tr}\,\mathrm{Tr}_{B,1} |x\rangle\langle x|\otimes \mathrm{Tr}_{B,2}|x\rangle\langle x|F_A\\ &=\mathrm{Tr}\,|x\rangle\langle x|\otimes|x\rangle\langle x|F_A\otimes I_{\mathcal{H}_B}.\end{aligned} \tag{8.278}$$

Using (8.277) and (8.278), we can calculate the expectation of $\mathrm{Tr}(\mathrm{Tr}_B |x\rangle\langle x|)^2$ as

$$\begin{aligned}&\mathrm{E}_{\mathcal{H}}\,\mathrm{Tr}(\mathrm{Tr}_B |x\rangle\langle x|)^2=\mathrm{E}_{\mathcal{H}}\,\mathrm{Tr}\,|x\rangle\langle x|\otimes|x\rangle\langle x|F_A\otimes I_{n^2}\\ =&\mathrm{Tr}\,\frac{1}{nk(nk+1)/2}P_{s,AB}F_k\otimes I_{n^2}\\ =&\mathrm{Tr}\,\frac{1}{nk(nk+1)/2}P_{s,AB}(P_{s,A}-P_{a,A})\otimes(P_{s,B}+P_{a,B})\\ =&\mathrm{E}_{\mathcal{H}}\,\mathrm{Tr}\,\frac{1}{nk(nk+1)/2}P_{s,A}\otimes P_{s,B}-P_{a,A}\otimes P_{a,B}\\ =&\frac{1}{nk(nk+1)/2}\left(\frac{k(k+1)}{2}\cdot\frac{n(n+1)}{2}-\frac{k(k-1)}{2}\cdot\frac{n(n-1)}{2}\right)\\ =&\frac{1}{nk(nk+1)/2}\frac{kn(k+n)}{2}=\frac{k+n}{nk+1}=\frac{1+k/n}{k+1/n},\end{aligned}$$

which implies (8.265). Using (8.259), we have

$$\mathrm{E}_{\mathcal{H}}\|\,\mathrm{Tr}_B |x\rangle\langle x|-\rho_{\mathrm{mix},A}\|_2^2=\frac{1+1/k^2}{n+1/k}. \tag{8.279}$$

Using Jensen inequality with respect to $x\mapsto x^2$, we obtain (8.266). ∎

Exercises

8.74 Show the following inequality (8.280) instead of Lemma 8.17 by following the steps below.

$$\frac{|\|\operatorname{Tr}_{\mathbb{C}^n}|x\rangle\langle x|-\rho_{\mathrm{mix},A}\|_2^2-\|\operatorname{Tr}_{\mathbb{C}^n}|y\rangle\langle y|-\rho_{\mathrm{mix},A}\|_2^2|}{\tilde{d}(x,y)}\le 2\sqrt{2}, \tag{8.280}$$

where $\tilde{d}(x,y)$ is defined as $\cos\tilde{d}(x,y)=|\langle x|y\rangle|$. Compare that $\cos d(x,y)=\mathbf{Re}\langle x|y\rangle$.
(**a**) Show $\frac{\||x\rangle\langle x|-|y\rangle\langle y|\|_1}{\tilde{d}(x,y)}\le 2$.
(**b**) Show the inequality (8.280) by using (8.278) and (**a**). (Use a similar discussion to Exercise 2.51.)

8.14 Secure Random Number Generation

8.14.1 Security Criteria and Their Evaluation

When a given classical random number A is partially leaked to the third party, the random number is not secure. In this case, it is possible to increase the secrecy by applying a hash function. Now, we assume that the third party, Eve, has the quantum system $\mathcal{H}_E$ correlated to the classical random number A, which is described by the d-dimensional system $\mathcal{H}_A$ spanned by the CONS $\{u_j\}_{j=1}^d$. Then, the state of the composite system $\mathcal{H}_A\otimes\mathcal{H}_E$ is written as

$$\rho=\sum_{j=1}^{d}P_A(j)|u_j\rangle\langle u_j|\otimes\rho_{E|j}. \tag{8.281}$$

The leaked information can be evaluated by the mutual information.

$$I_\rho(A:E)=D(\rho\|\rho_A\otimes\rho_E). \tag{8.282}$$

When we employ the trace norm or the fidelity instead of the relative entropy, the criterion is given as

$$d_1(A:E|\rho):=\|\rho-\rho_A\otimes\rho_E\|_1 \tag{8.283}$$
$$F(A:E|\rho):=F(\rho,\rho_A\otimes\rho_E). \tag{8.284}$$

When we take the uniformity of A into account as well as the independence, we employ the quantities

$$I'_\rho(A:E) := D(\rho\|\rho_{\mathrm{mix},A}\otimes\rho_E) = \log d - H_\rho(A|E) \tag{8.285}$$
$$= I(A:E|\rho) + D(\rho_A|\rho_{\mathrm{mix},A}), \tag{8.286}$$
$$d'_1(A:E|\rho) := \|\rho - \rho_{\mathrm{mix},A}\otimes\rho_E\|_1, \tag{8.287}$$
$$F'(A:E|\rho) := F(\rho, \rho_{\mathrm{mix},A}\otimes\rho_E). \tag{8.288}$$

The quantity $I'_\rho(A:E)$ is called the modified mutual information, and satisfies the uniqueness under a suitable collection of axioms [62].

Now, we focus on an ensemble of the hash functions f_X from $\{1,\ldots,d\}$ to $\{1,\ldots,M\}$, where X is a random variable describing the stochastic behavior of the hash function and subject to the distribution P_X because a randomized choice of the hash function makes the evaluation of the above values easy. In this case, the random variable X is independent of the state ρ in the composite system $\mathcal{H}_A\otimes\mathcal{H}_E$, and we denote the system describing the random variable X by $\mathcal{H}_X$. The state of the total system $\mathcal{H}_A\otimes\mathcal{H}_E\otimes\mathcal{H}_X$ is written as

$$\sum_x P_X(x)|x\rangle\langle x|\otimes\rho_{A,E}, \tag{8.289}$$

which is denoted by $\rho\otimes P_X$, Then, the total system is composed of the quantum system $\mathcal{H}_E$ and the classical systems $f_X(A)$ and X, the state is given as $\sum_x P_X(x)|x\rangle\langle x|\otimes\rho_{f_X(A),E}$, where $\rho_{f(A)E} := \sum_b |b\rangle\langle b|\otimes(\sum_{a:f(a)=b} P_A(a)\rho_{E|a})$. The security is evaluated by $I'_{\rho,P_X}(f_X(A):E,X)$, which can be expressed as

$$I'_{\rho\otimes P_X}(f_X(A):E,X) = \mathrm{E}_X I'_\rho(f_X(A):E). \tag{8.290}$$

That is, when we employ the random choice of the hash function, it is sufficient to focus on the expectation $\mathrm{E}_X I'(f_X(A):E|\rho)$.

An ensemble of the functions f_X is called universal$_2$ when it satisfies the following condition [63]:

Condition 8.1 *For arbitrary two distinct elements $a_1\neq a_2\in\{1,\ldots,d\}$, the probability that $f_X(a_1)=f_X(a_2)$ is at most $\frac{1}{M}$.*

Indeed, when the cardinality d is a power of a prime p and M is another power of the same prime p, an ensemble $\{f_X\}$ satisfying the both conditions is given by the the concatenation of Toeplitz matrix and the identity (X,I) [64] only with $\log_p(d-1)$ random variables taking values in the finite filed $\boldsymbol{F}_p = \mathbf{Z}/p\mathbf{Z}$. That is, the matrix (X,I) has a small calculation complexity.

Theorem 8.15 ([65]) *When the ensemble of the functions $\{f_X\}$ is universal$_2$, it satisfies*

$$I_{\rho\otimes P_X}(f_X(A):E,X) \le I'_{\rho\otimes P_X}(f_X(A):E,X) = \mathrm{E}_X I'_\rho(f_X(A):E)$$
$$\le \frac{v^s M^s}{s} e^{-s\tilde{H}_{1+s|\rho}(A|E)} = v^s \frac{e^{s(\log M - \tilde{H}_{1+s|\rho}(A|E))}}{s}, \tag{8.291}$$

where v is the number of eigenvalues of ρ_E.

That is, there exists a function $f : \mathcal{A} \to \{1, \ldots, M\}$ such that

$$I'_\rho(f(A) : E) \le v^s \frac{e^{s(\log M - \tilde{H}_{1+s|\rho}(A|E))}}{s}. \tag{8.292}$$

Next, we consider the case when our state is given by the n-fold independent and identical state ρ, i.e., $\rho^{\otimes n}$. We define the optimal generation rate

$$\begin{aligned} G(\rho) :=& \sup_{\{(f_n, M_n)\}} \left\{ \lim_{n\to\infty} \frac{\log M_n}{n} \middle| \begin{array}{l} \lim_{n\to\infty} \frac{I_{\rho^{\otimes n}}(f_n(A) : E)}{n} = 0 \\ \lim_{n\to\infty} \frac{H_{\rho^{\otimes n}}(f_n(A))}{\log M_n} = 1 \end{array} \right\} \\ =& \sup_{\{(f_n, M_n)\}} \left\{ \lim_{n\to\infty} \frac{\log M_n}{n} \middle| \lim_{n\to\infty} \frac{I'_{\rho^{\otimes n}}(f_n(A) : E)}{n} = 0 \right\}, \end{aligned}$$

whose classical version is treated by [66]. The second equation holds as follows. the condition $\lim_{n\to\infty} \frac{H_{\rho^{\otimes n}}(f_n(A))}{\log M_n} = 1$ is equivalent with $\lim_{n\to\infty} \frac{D(\rho^{f_n(A)} \| \rho_{\text{mix}, f_n(A)})}{n} = 0$. Hence, $\lim_{n\to\infty} \frac{I_{\rho^{\otimes n}}(f_n(A):E)}{n} = 0$ and $\lim_{n\to\infty} \frac{H_{\rho^{\otimes n}}(f_n(A))}{\log M_n} = 1$ if and only if $\lim_{n\to\infty} \frac{I'_{\rho^{\otimes n}}(f_n(A):E)}{n} = 0$.

When the generation rate $R = \lim_{n\to\infty} \frac{\log M_n}{n}$ is smaller than $H(A|E)$, there exists a sequence of functions $f_n : \mathcal{A} \to \{1, \ldots, e^{nR}\}$ such that

$$I'_{\rho^{\otimes n}}(f_n(A) : E) \le v_n^s \frac{e^{s(R - \tilde{H}_{1+s|\rho^{\otimes n}}(A|E))}}{s}, \tag{8.293}$$

where v_n is the number of eigenvalues of $\rho_E^{\otimes n}$, which is a polynomial increasing for n because of (3.9). Since $\lim_{s\to 0} \tilde{H}_{1+s|\rho}(A|E)) = H_\rho(A|E))$, there exists a number $s \in (0, 1]$ such that $s(R - \tilde{H}_{1+s|\rho}(A|E)) > 0$. Thus, the right hand side of (8.293) goes to zero exponentially. Conversely, due to (8.12), any sequence of functions $f_n : \mathcal{A}^n \mapsto \{1, \ldots, e^{nR}\}$ satisfies that

$$\lim_{n\to\infty} \frac{H_{\rho^{\otimes n}}(f_n(A)|E)}{n} \le \frac{H_{\rho^{\otimes n}}(A|E)}{n} = H_\rho(A|E). \tag{8.294}$$

When $\lim_{n\to\infty} \frac{H_{\rho^{\otimes n}}(f_n(A))}{nR} = 1$,

$$\begin{aligned} \lim_{n\to\infty} \frac{I_{\rho^{\otimes n}}(f_n(A) : E)}{n} =& R - \lim_{n\to\infty} \frac{H_{\rho^{\otimes n}}(f_n(A)|E)}{n} \\ \ge& R - H_\rho(A|E). \end{aligned} \tag{8.295}$$

That is, when $R > H_\rho(A|E)$, $\frac{I_{\rho^{\otimes n}}(f_n(A):E)}{n}$ does not go to zero. Hence, we derive the formula by [67, 68]:

$$G(\rho) = H_\rho(A|E). \tag{8.296}$$

In order to treat the speed of this convergence, we focus on the supremum of the *exponentially decreasing rate (exponent)* of $I'_{\rho^{\otimes n}}(f_n(A) : E)$ for a given R

$$e_I(\rho|R) := \sup_{\{(f_n, M_n)\}} \left\{ \lim_{n\to\infty} \frac{-\log I'_{\rho^{\otimes n}}(f_n(A) : E)}{n} \middle| \lim_{n\to\infty} \frac{-\log M_n}{n} \leq R \right\}.$$

Since the relation $s\tilde{H}_{1+s|\rho^{\otimes n}}(A|E) = ns\tilde{H}_{1+s|\rho}(A|E)$ holds, the inequality (8.293) implies that

$$\begin{aligned} e_I(\rho|R) \geq e_H(\rho|R) &:= \max_{0\leq s\leq 1} s\tilde{H}_{1+s|\rho}(A|E) - sR \\ &= \max_{0\leq s\leq 1} s(\tilde{H}_{1+s|\rho}(A|E) - R), \end{aligned} \tag{8.297}$$

whose commutative version coincides with the bound given in [69].

Next, we apply our evaluation to the criterion $d'_1(A : E|\rho)$. When $\{f_X\}$ satisfies Condition 8.1, combining (3.53), (3.50) and (8.291), we obtain

$$\mathrm{E}_X d'_1(f_X(A) : E|\rho) \leq \sqrt{\mathrm{E}_X d'_1(f_X(A) : E|\rho)^2} \leq \frac{\sqrt{2}v^{s/2}M^{s/2}}{\sqrt{s}} e^{-\frac{s}{2}\tilde{H}_{1+s|\rho}(A|E)}. \tag{8.298}$$

$$\mathrm{E}_X F'(f_X(A) : E|\rho) \geq 1 - \frac{v^s M^s}{2s} e^{-s\tilde{H}_{1+s|\rho}(A|E)}. \tag{8.299}$$

Then, similarly we can derive their exponentially decreasing rate (exponent) in the n-fold asymptotic setting.

8.14.2 Proof of Theorem 8.15

In order to show Theorem 8.15, we prepare the following two lemmas.

Lemma 8.21 *The matrix inequality* $(I + X)^s \leq I + X^s$ *holds with a non-negative matrix* X *and* $s \in (0, 1]$.

Proof Since I is commutative with X, it is sufficient to show that $(1+x)^s \leq 1+x^s$ for $x \geq 0$. Sp, we obtain the matrix inequality. ■

Lemma 8.22 *The matrix inequality* $\log(I + X) \leq \frac{1}{s}X^s$ *holds with a non-negative matrix* X *and* $s \in (0, 1]$.

Proof Since I is commutative with X, it is sufficient to show that $\log(1 + x) \leq \frac{x^s}{s}$ for $x \geq 0$. Since the inequalities $(1 + x)^s \leq 1 + x^s$ and $\log(1 + x) \leq x$ hold for $x \geq 0$ and $0 < s \leq 1$, the inequalities

$$\log(1+x) = \frac{\log(1+x)^s}{s} \le \frac{\log(1+x^s)}{s} \le \frac{x^s}{s} \tag{8.300}$$

hold. ■

Now, we prove Theorem 8.15.

$$\begin{aligned}
&\mathrm{E}_{\mathbf{X}} I'(f_{\mathbf{X}}(A) : E|\rho) \\
=&\mathrm{E}_{\mathbf{X}} D\left(\sum_{i=1}^{M} |i\rangle\langle i| \otimes \sum_{a: f_{\mathbf{X}}(a)=i} P_A(a)\rho_E^a \| \frac{1}{M} I \otimes \rho_E\right) \\
=&\mathrm{E}_{\mathbf{X}} \sum_a \mathrm{Tr}\, P_A(a)\rho_E^a (\log\left(\sum_{a': f_{\mathbf{X}}(a')=f_{\mathbf{X}}(a)} P_A(a')\rho_{a'}^E) - \log \frac{1}{M}\rho_E\right) \\
\le&\sum_a P_A(a)\, \mathrm{Tr}\, \rho_E^a (\log\left(\mathrm{E}_{\mathbf{X}} \sum_{a': f_{\mathbf{X}}(a')=f_{\mathbf{X}}(a)} P_A(a')\rho_{a'}^E) - \log \frac{1}{M}\rho_E\right) &(8.301)\\
=&\sum_a P_A(a)\, \mathrm{Tr}\, \rho_E^a \left(\log\left(P_A(a)\rho_E^a + \mathrm{E}_{\mathbf{X}} \sum_{a': f_{\mathbf{X}}(a')=f_{\mathbf{X}}(a), a'\neq a} P_A(a')\rho_{a'}^E\right) - \log \frac{1}{M}\rho_E\right) \\
\le&\sum_a P_A(a)\, \mathrm{Tr}\, \rho_E^a \left(\log\left(P_A(a)\rho_E^a + \frac{1}{M} \sum_{a': a'\neq a} P_A(a')\rho_{a'}^E\right) - \log \frac{1}{M}\rho_E\right) &(8.302)\\
\le&\sum_a P_A(a)\, \mathrm{Tr}\, \rho_E^a \left(\log\left(P_A(a)\rho_E^a + \frac{1}{M}\rho_E\right) - \log \frac{1}{M}\rho_E\right) \\
\le&\sum_a P_A(a)\, \mathrm{Tr}\, \rho_E^a \left(\log\left(v P_A(a)\kappa_{\rho_E}(\rho_E^a) + \frac{1}{M}\rho_E\right) - \log \frac{1}{M}\rho_E\right) &(8.303)\\
=&\sum_a P_A(a)\, \mathrm{Tr}\, \rho_E^a \log(vMP_A(a)\kappa_{\rho_E}(\rho_E^a)\rho_E^{-1} + I),
\end{aligned}$$

where (8.301) follows from the matrix convexity of $x \mapsto \log x$, (8.302) follows from Condition 8.1 and the matrix monotonicity of $x \mapsto \log x$, and (8.303) follows from (3.146) and the matrix monotonicity of $x \mapsto \log x$.

Using Lemma 8.22, we obtain

$$\begin{aligned}
&\sum_a P_A(a)\, \mathrm{Tr}\, \rho_E^a \log(vMP_A(a)\kappa_{\rho_E}(\rho_E^a)\rho_E^{-1} + I) \\
\le&\frac{1}{s} \sum_a P_A(a)\, \mathrm{Tr}\, \rho_E^a (vMP_A(a)\kappa_{\rho_E}(\rho_E^a)\rho_E^{-1})^s \\
=&\frac{v^s M^s}{s} \sum_a P_A(a)^{1+s}\, \mathrm{Tr}\, \kappa_{\rho_E}(\rho_E^a)^{1+s} (\rho_E)^{-s} \\
=&\frac{v^s M^s}{s} e^{-s\tilde{H}_{1+s|\kappa_{I\otimes\rho_E}(\rho)}(A|E)} \le \frac{v^s M^s}{s} e^{-s\tilde{H}_{1+s|\rho}(A|E)}, &(8.304)
\end{aligned}$$

where (8.304) follows from (5.57).

8.15 Duality Between Two Conditional Entropies

8.15.1 Recovery of Maximally Entangled State from Evaluation of Classical Information

Firstly, for a given state ρ on the composite system $\mathcal{H}_A \otimes \mathcal{H}_B$, we consider a sufficient condition to approximately generate the maximally entangled state $|\Phi\rangle := \sum_{j=1}^{d} \frac{1}{\sqrt{d}} |u_j\rangle \otimes |u'_j\rangle \in \mathcal{H}_A \otimes \mathcal{H}_{A'}$, where $\{u_j\}$ and $\{u'_j\}$ are the CONSs of $\mathcal{H}_A$ and $\mathcal{H}_{A'}$, respectively. For this purpose, we focus on the following two conditions for a pure state $\rho = |\Psi\rangle\langle\Psi|$ on the composite system $\mathcal{H}_A \otimes \mathcal{H}_B \otimes \mathcal{H}_R$.

ϵ_1-bit security: The PVM $\boldsymbol{E} := \{|u_j\rangle\langle u_j|\}_{j=1}^{d}$ satisfies $F(\kappa_{\boldsymbol{E}} \otimes \iota_R(\rho_{AR}), \rho_{\mathrm{mix},A} \otimes \rho_R) \geq 1 - \epsilon_1$.

ϵ_2-bit recoverability: There exists a POVM $\boldsymbol{M} = \{M_j\}_{j=1}^{d}$ on $\mathcal{H}_B$ such that $\sum_{j=1}^{d} \mathrm{Tr}\, \rho_{AB} |u_j\rangle\langle u_j| \otimes M_j \geq 1 - \epsilon_2$.

Then, we obtain the following theorem.

Theorem 8.16 (Renes [70]) *Assume that a pure state $\rho = |\Psi\rangle\langle\Psi|$ on the system $\mathcal{H}_A \otimes \mathcal{H}_B \otimes \mathcal{H}_R$ satisfies the above both conditions. Then, there is a TP-CP map $\kappa : \mathcal{S}(\mathcal{H}_B) \to \mathcal{S}(\mathcal{H}_{A'})$ such that*

$$F(\iota_A \otimes \kappa(\rho_{AB}), |\Phi\rangle\langle\Phi|) \geq 1 - (\sqrt{\epsilon_2} + \sqrt{\epsilon_1})^2. \tag{8.305}$$

Theorem 8.16 guarantees that we can approximately generate the maximally entangled state between two systems $\mathcal{H}_A$ and $\mathcal{H}_B$ only by the operation on the system $\mathcal{H}_B$ if the classical information on the specific basis on the system $\mathcal{H}_A$ can be recovered by the system $\mathcal{H}_B$, is close to the uniform random number, and is almost independent of the environment system $\mathcal{H}_E$. These conditions can be easily checked because only the classical information of the specific basis concerns all of these conditions. That is, we do not have to care about other informations for constructing the maximally entangled state.

Proof **Step 1: Case when** $\epsilon_1 = 0$: First, we choose an isometry $U_B : \mathcal{H}_B \to \mathcal{H}_{A'} \otimes \mathcal{H}_{B'}$ such that $M_j = U^* |u_j\rangle\langle u_j| \otimes I_{B'} U$. Then, the state $U_B|\Psi\rangle$ can be written as

$$U_B|\Psi\rangle = \sum_{j=1}^{d} \sqrt{q_j} |u_j, u'_j, x_{B'|j}, x_{R|j}\rangle.$$

Next, we choose the purification $|\psi_{B'R|j}\rangle$ of ρ_R such that $F(|x_{R|j}\rangle\langle x_{R|j}|, \rho_R) = F(|x_{B'|j}, x_{R|j}\rangle\langle x_{B'|j}, x_{R|j}|, |\psi_{B'R|j}\rangle\langle\psi_{B'R|j}|)$. Now, we denote $|\psi_{B'R|1}\rangle$ by $|\psi_{B'R}\rangle$. Thus, there is a unitary $U_{B'|j}$ on $\mathcal{H}_{B'}$ such that $U_{B'|j}|\psi_{B'R|j}\rangle = |\psi_{B'R}\rangle$. Then, we define the pure state $|\xi\rangle := \sqrt{\frac{1}{d}} \sum_{j=1}^{d} |u_j, u'_j, \psi_{B'R|j}\rangle$ and the unitary U'_B on

$\mathcal{H}_{A'} \otimes \mathcal{H}_{B'}$ such that $U'_B := \oplus_{j=1}^d |u'_j\rangle\langle u'_j| \otimes U_{B'|j}$. Thus, $U'_B|\xi\rangle = |\Phi\rangle \otimes |\psi_{B'R}\rangle$. Since $\kappa_E \otimes \iota_R(\rho_{AR}) = \sum_{j=1}^d q_j |u_j\rangle\langle u_j| \otimes |x_{R|j}\rangle\langle x_{R|j}|$,

$$
\begin{aligned}
1-\epsilon_1 \le & F(\kappa_E \otimes \iota_R(\rho_{AR}), \rho_{\mathrm{mix},A} \otimes \rho_R) = \sum_{j=1}^d \sqrt{\frac{q_j}{d}} F(|x_{R|j}\rangle\langle x_{R|j}|, \rho_R) \\
= & \sum_{j=1}^d \sqrt{\frac{q_j}{d}} F(|x_{B'|j}, x_{R|j}\rangle\langle x_{B'|j}, x_{R|j}|, |\psi_{B'R|j}\rangle\langle\psi_{B'R|j}|) \\
= & F(U_B|\Psi\rangle\langle\Psi|U_B^*, |\xi\rangle\langle\xi|) = F(U'_B U_B|\Psi\rangle\langle\Psi|U_B^* U_B'^*, |\Phi\rangle\langle\Phi| \otimes |\psi_{B'R}\rangle\langle\psi_{B'R}|) \\
\le & F(\mathrm{Tr}_{B'R}\, U'_B U_B \rho U_B^* U_B'^*, |\Phi\rangle\langle\Phi|).
\end{aligned} \tag{8.306}
$$

Hence, defining the TP-CP map $\kappa : \mathcal{S}(\mathcal{H}_B) \to \mathcal{S}(\mathcal{H}_{A'} \otimes \mathcal{H}_{B'})$ by

$$
\kappa(\sigma) := \mathrm{Tr}_{B'}\, U'_B U_B \sigma (U'_B U_B)^*, \tag{8.307}
$$

we obtain (8.305).

Step 2: General case: In the general case, the state $U_B|\Psi\rangle$ can be written as

$$
U_B|\Psi\rangle = \sum_{j=1}^d \sqrt{q_j} |u_j, x_{A'|j}, x_{B'|j}, x_{R|j}\rangle
$$

satisfying that $\langle u'_j | x_{A'|j}\rangle \ge 0$. We define another state by

$$
|\Psi'\rangle := U_B^* \sum_{j=1}^d \sqrt{q_j} |u_j, u'_j, x_{B'|j}, x_{R|j}\rangle
$$

and $\rho' := |\Psi'\rangle\langle\Psi'|$, which satisfies the assumption of Step 1. Since $\sum_{j=1}^d q_j |\langle u'_j|x_{A'|j}\rangle|^2 = \sum_{j=1}^d \mathrm{Tr}\, \rho_{AB} |u_j\rangle\langle u_j| \otimes M_j$, the ϵ_2-bit recoverability guarantees that

$$
\begin{aligned}
& F(|\Psi\rangle\langle\Psi|, |\Psi'\rangle\langle\Psi'|) = F(U_B|\Psi\rangle\langle\Psi|U_B{}^*, U_B|\Psi'\rangle\langle\Psi'|U_B^*) = \sum_{j=1}^d q_j \langle u'_j|x_{A'|j}\rangle \\
\ge & \sum_{j=1}^d q_j |\langle u'_j|x_{A'|j}\rangle|^2 \ge 1-\epsilon_2.
\end{aligned} \tag{8.308}
$$

Since $\kappa_E \otimes \iota_R(\rho'_{AR}) = \sum_{j=1}^d q_j |u_j\rangle\langle u_j| \otimes |x_{R|j}\rangle\langle x_{R|j}| = \kappa_E \otimes \iota_R(\rho_{AR})$, by applying Step 1 to the state ρ', the TP-CP map $\kappa : \mathcal{S}(\mathcal{H}_B) \to \mathcal{S}(\mathcal{H}_{A'} \otimes \mathcal{H}_{B'})$ defined in (8.307) satisfies

$$
1-\epsilon_1 \le F(\iota_A \otimes \kappa(\rho'_{AB}), |\Phi\rangle\langle\Phi|). \tag{8.309}
$$

Then, (8.308) and (8.309) yield that

$$\begin{aligned}
&b(\iota_A \otimes \kappa(\rho_{AB}), |\Phi\rangle\langle\Phi|) \\
=&b(\iota_A \otimes \kappa(\rho_{AB}), \iota_A \otimes \kappa(\rho'_{AB})) + b(\iota_A \otimes \kappa(\rho'_{AB}), |\Phi\rangle\langle\Phi|) \\
\le&b(\rho, \rho') + b(\iota_A \otimes \kappa(\rho'_{AB}), |\Phi\rangle\langle\Phi|) \le \sqrt{\epsilon_1} + \sqrt{\epsilon_2},
\end{aligned}$$

which implies (8.305) ■

Theorem 8.16 requires ϵ_1 bit security. However, in a case, the ϵ_1 bit security holds only with the partial trace in a part of the system $\mathcal{H}_A$. In order to address such a case, we generalize Theorem 8.16 as follows. Now, we consider the following conditions for a pure state $\rho = |\Psi\rangle\langle\Psi|$ on the system $\mathcal{H}_{A_1} \otimes \mathcal{H}_{A_2} \otimes \mathcal{H}_B \otimes \mathcal{H}_R$.

ϵ_1-bit security for $\mathcal{H}_{A_1}$: The PVM $\boldsymbol{E}^1 := \{|u^1_{j_1}\rangle\langle u^1_{j_1}|\}_{j_1=1}^d$ on $\mathcal{H}_{A_1}$ satisfies $F(\kappa_{\boldsymbol{E}^1} \otimes \iota_R(\rho_{A_1R}), \rho_{\mathrm{mix},A_1} \otimes \rho_R) \ge 1 - \epsilon_1$.

ϵ_2-bit recoverability for $\mathcal{H}_{A_1} \otimes \mathcal{H}_{A_2}$: There exists a POVM $\boldsymbol{M} = \{M_{j_1,j_1}\}$ on $\mathcal{H}_B$ such that $\sum_{j_1,j_2} \operatorname{Tr} \rho_{AB} |u^1_{j_1}, u^2_{j_2}\rangle\langle u^1_{j_1}, u^2_{j_2}| \otimes M_{j_1,j_2} \ge 1 - \epsilon_2$.

Theorem 8.17 *Assume that a pure state $\rho = |\Psi\rangle\langle\Psi|$ on the system $\mathcal{H}_{A_1} \otimes \mathcal{H}_{A_2} \otimes \mathcal{H}_B \otimes \mathcal{H}_R$ satisfies above both conditions. Let $\{v_l\}$ be a basis mutually unbiased to $\{u_{j_2}\}$ of $\mathcal{H}_{A_2}$. We can choose a TP-CP map $\kappa_l : \mathcal{S}(\mathcal{H}_B) \to \mathcal{S}(\mathcal{H}_{A_1})$ dependently of l such that*

$$\begin{aligned}
&F\left(\sum_{l=1}^{d_2} \iota_{A_1} \otimes \kappa_l(\langle v_l|\rho_{A_1A_2B}|v_l\rangle) \otimes |v_l\rangle\langle v_l|, |\Phi\rangle\langle\Phi| \otimes \rho_{\mathrm{mix},A_2}\right) \\
\ge&1 - (\sqrt{\epsilon_2} + \sqrt{\epsilon_1})^2 \ge 1 - 2(\epsilon_2 + \epsilon_1).
\end{aligned} \tag{8.310}$$

In particular, when the PVM $\boldsymbol{F}^2 := \{|v_l\rangle\langle v_l|\}_{l=1}^{d_2}$ on $\mathcal{H}_{A_2}$ satisfies $\kappa_{\boldsymbol{F}^2}(\rho_{A_2}) = \rho_{\mathrm{mix},A_2}$,

$$\begin{aligned}
&\sum_{l=1}^{d_2} \frac{1}{d_2} F(\iota_{A_1} \otimes \kappa_l(d_2\langle v_l|\rho_{A_1A_2B}|v_l\rangle), |\Phi\rangle\langle\Phi|) \\
\ge&1 - \left(\sqrt{\epsilon_2} + \sqrt{\epsilon_1}\right)^2 \ge 1 - 2(\epsilon_2 + \epsilon_1).
\end{aligned} \tag{8.311}$$

Theorem 8.17 relaxes the conditions of Theorem 8.16. That is, Theorem 8.17 has a wider applicability than Theorem 8.16. In fact, Theorem 8.17 plays an important role in Sect. 9.6. In Sect. 9.6, Lemma 9.7 will be shown by Theorem 8.17. The Hashing inequality (8.121) for entanglement distillation will be also shown in Sect. 9.6, Lemma 9.7 plays an essential role in this proof.

Proof **Step 1: Case when** $\epsilon_2 = 0$: Due to the relation (7.52), since $\epsilon_2 = 0$, the PVM $\boldsymbol{F}^2$ satisfies

$$\iota_{A_1R} \otimes \kappa_{\boldsymbol{F}^2}(\rho_{A_1A_2R}) = \rho_{\mathrm{mix},A_2} \otimes \rho_{A_1R}, \tag{8.312}$$

which implies that $\langle v_l|\rho_{A_2}|v_l\rangle = \frac{1}{d_2}$. Thus, we have

$$\sum_{l=1}^{d_2} \frac{1}{d_2} F(\kappa_{E^1} \otimes \iota_R(d_2 \langle v_l | \rho_{A_1 A_2 R} | v_l \rangle), \rho_{\mathrm{mix},A_1} \otimes \rho_R)$$
$$= F(\kappa_{E^1} \otimes \kappa_{F^2} \otimes \iota_R(\rho_{A_1 A_2 R}), \rho_{\mathrm{mix},A_1} \otimes \rho_{\mathrm{mix},A_2} \otimes \rho_R)$$
$$= F(\kappa_{E^1} \otimes \iota_R(\rho_{A_1 R}), \rho_{\mathrm{mix},A_1} \otimes \rho_R) \geq 1 - \epsilon_1.$$

Due to Theorem 8.16 with $\epsilon_1 = 0$, there exists a TP-CP map $\kappa_l : \mathcal{S}(\mathcal{H}_B) \to \mathcal{S}(\mathcal{H}_{A_1})$ dependently of l such that

$$F\left(\sum_l \iota_A \otimes \kappa_l(d_2 \langle v_l | \rho_{A_1 A_2 R} | v_l \rangle \right), |\Phi\rangle\langle\Phi|)$$
$$\geq F(\kappa_{E^1} \otimes \iota_R(d_2 \langle v_l | \rho_{A_1 A_2 R} | v_l \rangle), \rho_{\mathrm{mix},A_1} \otimes \rho_R). \tag{8.313}$$

Thus,

$$F\left(\sum_{l=1}^{d_2} \iota_{A_1} \otimes \kappa_l(\langle v_l | \rho_{A_1 A_2 B} | v_l \rangle) \otimes |v_l\rangle\langle v_l|, \rho_{\mathrm{mix},A_2} \otimes |\Phi\rangle\langle\Phi|\right)$$
$$= \sum_{l=1}^{d_2} \frac{1}{d_2} F\left(\sum_l \iota_A \otimes \kappa_l(d_2 \langle v_l | \rho_{A_1 A_2 R} | v_l \rangle), |\Phi\rangle\langle\Phi|\right)$$
$$\geq \sum_{j_2=1}^{d_2} \frac{1}{d_2} F(\kappa_{E^1} \otimes \iota_R(d_2 \langle u^2_{j_2} | \rho_{A_1 A_2 R} | u^2_{j_2} \rangle), \rho_{\mathrm{mix},A_1} \otimes \rho_R) \geq 1 - \epsilon_1.$$

Step 2: General case: In the general case, we choose an isometry $U_B : \mathcal{H}_B \to \mathcal{H}_{A'_1} \otimes \mathcal{H}_{A'_2} \otimes \mathcal{H}_{B'}$ such that $M_{j_1,j_2} = U^* |u^1_{j_1}, u^2_{j_2}\rangle\langle u^1_{j_1}, u^2_{j_2}| \otimes I_{B'} U$. Then, the state $U_B|\Psi\rangle$ can be written as

$$U_B|\Psi\rangle = \sum_{j_1,j_2} \sqrt{q_{j_1,j_2}} |u^1_{j_1}, u^2_{j_2}, x^1_{A'_1|j_1}, x^2_{A'_2|j_2}, x_{B'|j_1,j_2}, x_{R|j_1,j_2}\rangle$$

satisfying that $\langle {u^1_{j_1}}' | x_{A'_1|j_1}\rangle, \langle {u^2_{j_2}}' | x_{A'_2|j_2}\rangle \geq 0$. We define another state

$$|\Psi'\rangle = U_B^* \sum_{j_1,j_2}^{d} \sqrt{q_{j_1,j_2}} |u^1_{j_1}, u^2_{j_2}, {u^1_{j_1}}', {u^2_{j_2}}', x_{B'|j_1,j_2}, x_{R|j_1,j_2}\rangle$$

and $\rho' := |\Psi'\rangle\langle\Psi'|$. Since

$$\sum_{j_1,j_2}^{d} q_{j_1,j_2} |\langle {u^1_{j_1}}', {u^2_{j_2}}' | x^1_{A'_1|j_1}, x^2_{A'_2|j_2}\rangle|^2 = \sum_{j_1,j_2}^{d} \mathrm{Tr}\, \rho_{A_1 A_2 B} |u^1_{j_1}, u^2_{j_2}\rangle\langle u^1_{j_1}, u^2_{j_2}| \otimes M_{j_1,j_2},$$

similar to (8.308), the ϵ_2-bit recoverability guarantees that

$$F(|\Psi\rangle\langle\Psi|, |\Psi'\rangle\langle\Psi'|) \geq 1 - \epsilon_2. \tag{8.314}$$

Since $\kappa_{E^1,E^2} \otimes \iota_R(\rho'_{A_1A_2R}) = \sum_{j_1,j_2} q_{j_1,j_2} |u^1_{j_1}, u^2_{j_2}\rangle\langle u^1_{j_1}, u^2_{j_2}| \otimes |x_{R|j_1,j_2}\rangle\langle x_{R|j_1,j_2}| = \kappa_{E^1,E^2} \otimes \iota_R(\rho_{A_1A_2R})$, applying Step 1 to the state ρ', we can choose the TP-CP map $\kappa_l : \mathcal{S}(\mathcal{H}_B) \to \mathcal{S}(\mathcal{H}_{A_1} \otimes \mathcal{H}_{B'})$ such that

$$F\left(\sum_{l=1}^{d_2} \iota_{A_1} \otimes \kappa_l(\langle v_l|\rho'_{A_1A_2B}|v_l\rangle) \otimes |v_l\rangle\langle v_l|, \rho_{\mathrm{mix},A_2} \otimes |\Phi\rangle\langle\Phi|\right) \geq 1 - \epsilon_1. \tag{8.315}$$

We denote the TP-CP map $\sigma_{A_1A_2B} \mapsto \sum_{l=1}^{d_2} \iota_{A_1} \otimes \kappa_l(\langle v_l|\sigma_{A_1A_2B}|v_l\rangle) \otimes |v_l\rangle\langle v_l|$ by κ. Then, (8.314) and (8.315) yield that

$$\begin{aligned}
&b\left(\sum_{l=1}^{d_2} \iota_{A_1} \otimes \kappa_l(\langle v_l|\rho_{A_1A_2B}|v_l\rangle) \otimes |v_l\rangle\langle v_l|, \rho_{\mathrm{mix},A_2} \otimes |\Phi\rangle\langle\Phi|\right)\\
\leq &b(\kappa(\rho_{A_1A_2B}), \kappa(\rho'_{A_1A_2B})) + b(\kappa(\rho_{A_1A_2B})', \rho_{\mathrm{mix},A_2} \otimes |\Phi\rangle\langle\Phi|)\\
\leq &b(\rho, \rho') + \sqrt{\epsilon_2} \leq \sqrt{\epsilon_1} + \sqrt{\epsilon_2},
\end{aligned}$$

which implies (8.310). ■

8.15.2 *Duality Between Two Conditional Entropies of Mutually Unbiased Basis*

Now, we revisit (7.51) with the mutually unbiased bases $\{u_j\}_{j=1}^d$ and $\{v_l\}_{l=1}^d$. In this case, we have $c = 1/\sqrt{d}$. So, if the outcome of $\boldsymbol{E}' = \{|v_l\rangle\langle v_l|\}$ is almost determined by the information in $\mathcal{H}_B$, i.e., the conditional entropy $H_{\kappa_{E'} \otimes \iota_B(\rho_{A,B})}(A|B)$ is small, the other conditional entropy $H_{\kappa_E \otimes \iota_E(\rho_{A,E})}(A|E)$ with $\boldsymbol{E} = \{|u_j\rangle\langle u_j|\}$ is almost equal to the maximum value $\log d$. That is, the information in the basis $\{u_j\}$ is almost independent of the information in $\mathcal{H}_E$.

Now, we arise the reverse question: whether the outcome of $\boldsymbol{E}'$ is almost determined by the information in $\mathcal{H}_B$ when the outcome of $\boldsymbol{E}$ is almost independent of the information in $\mathcal{H}_E$. Theorem 8.16 gives the solution when the outcome of $\boldsymbol{E}$ is almost determined by the information in $\mathcal{H}_B$. That is, we can show the following theorem by using Theorem 8.16.

Theorem 8.18 *Assume that* $\mathbf{E}$ *and* $\mathbf{E}'$ *are the PVMs given by arbitrary two bases* $\{u_j\}_{j=1}^d$ *and* $\{v_l\}_{l=1}^d$*, respectively. When* $H_{\kappa_E \otimes \iota_E(\rho_{A,E})}(A|E) \geq \log d - \epsilon_1$ *and the* ϵ_2*-bit recoverability holds for the state* ρ*, we have*

$$H_{\kappa_{E'} \otimes \iota_B(\rho_{A,B})}(A|B) \leq \log d (\sqrt{\epsilon_2} + \sqrt{\sqrt{\epsilon_1}/2})^2 + h((\sqrt{\epsilon_2} + \sqrt{\sqrt{\epsilon_1}/2})^2). \tag{8.316}$$

Then, under the ϵ_2-bit recoverability, the two conditional entropies $H_{\kappa_E \otimes \iota_E(\rho_{A,E})}(A|E)$ and $H_{\kappa_{E'} \otimes \iota_B(\rho_{A,B})}(A|B)$ satisfy the following equivalent conditions; $H_{\kappa_E \otimes \iota_E(\rho_{A,E})}(A|E)$

is close to the maximal value, $\log d$, if and only if $H_{\kappa_{E'}\otimes\iota_B(\rho_{A,B})}(A|B)$ is close to zero. This relation can be regarded as a kind of duality relation. That is, we obtain the argument reverse to (7.51) under the ϵ_2-bit recoverability for the state ρ when the two bases $\{u_j\}_{j=1}^d$ and $\{v_l\}_{l=1}^d$ are mutually unbiased. Although Theorem 8.18 holds with arbitrary two bases, the relation (7.51) gives weaker evaluation in the general case. Hence, the above equivalence relation cannot be shown from (7.51) and Theorem 8.18 in the general case.

Proof Since

$$
\begin{aligned}
&\epsilon_1 \geq \log d - H_{\kappa_E\otimes\iota_E(\rho_{A,E})}(A|E) = D(\kappa_{\boldsymbol{E}} \otimes \iota_E(\rho_{A,E})\|\rho_{\mathrm{mix},A}\otimes\rho_E) \\
\geq &\|\kappa_{\boldsymbol{E}} \otimes \iota_E(\rho_{A,E}) - \rho_{\mathrm{mix},A}\otimes\rho_E)\|^2 \\
\geq &4(1 - F(\kappa_{\boldsymbol{E}} \otimes \iota_E(\rho_{A,E}), \rho_{\mathrm{mix},A}\otimes\rho_E)))^2,
\end{aligned}
$$

we have

$$
1 - \sqrt{\epsilon_1}/2 \leq F(\kappa_{\boldsymbol{E}} \otimes \iota_E(\rho_{A,E}), \rho_{\mathrm{mix},A}\otimes\rho_E).
$$

Thus,

$$
\begin{aligned}
&\sum_{j=1}^{d} \langle u_j, u'_j|\kappa_{\boldsymbol{E}'} \otimes \kappa(\rho_{AB})|u_j, u'_j\rangle \\
\geq &\sum_{j=1}^{d} \frac{1}{\sqrt{d}} \sqrt{\langle u_j, u'_j|\kappa_{\boldsymbol{E}'} \otimes \kappa(\rho_{AB})|u_j, u'_j\rangle} \\
= &F(\kappa_{\boldsymbol{E}'} \otimes \iota_B(\iota_A \otimes \kappa(\rho_{AB})), \kappa_{\boldsymbol{E}'} \otimes \iota_B(|\Phi\rangle\langle\Phi|)) \\
\geq &F(\kappa(\rho_{AB}), |\Phi\rangle\langle\Phi|) \geq 1 - (\sqrt{\epsilon_2} + \sqrt{\sqrt{\epsilon_1}/2})^2.
\end{aligned}
$$

When X and Y are the random variables subject to the distribution $P(X = x, Y = x) := \langle u_x, u'_y|\kappa_{\boldsymbol{E}'} \otimes \kappa(\rho_{AB})|u_x, u'_y\rangle = \operatorname{Tr} \kappa_{\boldsymbol{E}'} \otimes \iota_B(\rho_{A,B})|u_x\rangle\langle|u_x| \otimes \kappa^*(|u'_y\rangle\langle u'_y|)$, Fano inequality guarantees that

$$
\begin{aligned}
&H_{\kappa_{E'}\otimes\iota_B(\rho_{A,B})}(A|B) \leq H(X|Y) \\
\leq &\log d(\sqrt{\epsilon_2} + \sqrt{\sqrt{\epsilon_1}/2})^2 + h((\sqrt{\epsilon_2} + \sqrt{\sqrt{\epsilon_1}/2})^2).
\end{aligned}
$$

∎

8.16 Examples

In this section, we summarize the preceding calculation of entanglement measures using several examples in the mixed-state case.

8.16.1 2 × 2 System

In the case of $\mathbb{C}^2 \otimes \mathbb{C}^2$, Wootters [71] calculated the entanglement of formation as

$$E_f(\rho) = h\left(\frac{1+\sqrt{1-C_o(\rho)^2}}{2}\right), \quad C_o(\rho) \stackrel{\text{def}}{=} \max\{0, \lambda_1 - \lambda_2 - \lambda_3 - \lambda_4\}, \tag{8.317}$$

where λ_i is the square root of the eigenvalue of $\rho(S_2 \otimes S_2)\bar{\rho}(S_2 \otimes S_2)$ in decreasing order. The function $C_o(\rho)$ is called **concurrence**. When we perform an instrument $\{\kappa_\omega\}_\omega$ with the separable form $\kappa_\omega(\rho) = (A_\omega \otimes B_\omega)\rho(A_\omega \otimes B_\omega)^*$, the final state $\frac{(A_\omega \otimes B_\omega)\rho(A_\omega \otimes B_\omega)^*}{\mathrm{Tr}(A_\omega \otimes B_\omega)\rho(A_\omega \otimes B_\omega)^*}$ has the following concurrence [72, 73][Exe. 8.75]:

$$C_o\left(\frac{(A_\omega \otimes B_\omega)\rho(A_\omega \otimes B_\omega)^*}{\mathrm{Tr}(A_\omega \otimes B_\omega)\rho(A_\omega \otimes B_\omega)^*}\right) = C_o(\rho)\frac{|\det A_\omega||\det B_\omega|}{\mathrm{Tr}(A_\omega \otimes B_\omega)\rho(A_\omega \otimes B_\omega)^*}. \tag{8.318}$$

For example, the concurrence of the Bell diagonal state $\rho_{\mathrm{Bell},p} \stackrel{\text{def}}{=} \sum_{i=0}^3 p_i |e_i^{A,B}\rangle\langle e_i^{A,B}|$ is calculated as[Exe. 8.76]

$$C_o(\rho_{\mathrm{Bell},p}) = 2\max_i p_i - 1, \tag{8.319}$$

and it does not increase by any stochastic operation [74][Exe. 8.77]:

$$C_o(\rho_{\mathrm{Bell},p}) \geq C_o\left(\frac{(A_\omega \otimes B_\omega)\rho_{\mathrm{Bell},p}(A_\omega \otimes B_\omega)^*}{\mathrm{Tr}(A_\omega \otimes B_\omega)\rho_{\mathrm{Bell},p}(A_\omega \otimes B_\omega)^*}\right). \tag{8.320}$$

This state satisfies

$$-H_{\rho_{\mathrm{Bell},p}}(A|B) = \log 2 - H(p), \quad I_{\rho_{\mathrm{Bell},p}}(A:B) = 2\log 2 - H(p).$$

Further, the maximally correlated state $\rho_{a,b} \stackrel{\text{def}}{=} a|00\rangle\langle 00| + b|00\rangle\langle 11| + b|11\rangle\langle 00| + (1-a)|11\rangle\langle 11|$[Exe. 8.78] has the concurrence $2b$[Exe. 8.79]. Hence,

$$E_c(\rho_{a,b}) = E_f(\rho_{a,b}) = h\left(\frac{1+\sqrt{1-4b^2}}{2}\right). \tag{8.321}$$

Regarding distillation, from (8.143) and (8.223) we have[Exe. 8.80]

$$\begin{aligned} E_{d,2}^{C,\dagger}(\rho_{a,b}) &= E_{d,2}^{C}(\rho_{a,b}) = E_{r,S}(\rho_{a,b}) = E_{r,\mathrm{PPT}}(\rho_{a,b}) = -H_{\rho_{a,b}}(A|B) \\ &= h(a) - h\left(\frac{1+\sqrt{(2a-1)^2+4b^2}}{2}\right), \end{aligned} \tag{8.322}$$

for $C = \to, \leftarrow, \leftrightarrow, S$, and PPT. Further,

$$I_{\rho_{a,b}}(A:B) = 2h(a) - h\left(\frac{1+\sqrt{(2a-1)^2+4b^2}}{2}\right)$$
$$C_d^{A\to B}(\rho_{a,b}) = E_c^{-\,-\to}(\rho_{a,b}) = h(a).$$

Since Ishizaka [48] proved $\tau^A(|\tau^A(\rho)|) \geq 0$ for the 2×2 case, the relation

$$E_{c,e}^{\mathrm{PPT},\infty}(\rho) = \log \|\tau^A(\rho_{a,b})\|_1 = \log(1+2b)$$

holds. Hence, comparing these values, we obtain the inequality

$$\log(1+2b) \geq h\left(\frac{1+\sqrt{1-4b^2}}{2}\right) \geq h(a) - h\left(\frac{1+\sqrt{(2a-1)^2+4b^2}}{2}\right)$$

for $\sqrt{a(1-a)} \geq b$. In particular, the second equality holds only when $\sqrt{a(1-a)} = b$, i.e., the state $\rho_{a,b}$ is pure.

8.16.2 Werner State

Next, we consider the **Werner state**:

$$\rho_{W,p} \stackrel{\mathrm{def}}{=} (1-p)\rho_{\mathrm{mix}}^s + p\rho_{\mathrm{mix}}^a = \frac{p}{d(d-1)}(I-F) + \frac{1-p}{d(d+1)}(I+F), \quad (8.323)$$

where ρ_{mix}^s (ρ_{mix}^a) is the completely mixed state on the symmetric space (antisymmetric space). We can easily check that

$$-H_{\rho_{W,p}}(A|B) = \log d + p\log\frac{2p}{d(d-1)} + (1-p)\log\frac{2(1-p)}{d(d+1)}$$
$$I_{\rho_{W,p}}(A:B) = 2\log d + p\log\frac{2p}{d(d-1)} + (1-p)\log\frac{2(1-p)}{d(d+1)}. \quad (8.324)$$

Further, any pure state $|u\rangle\langle u|$ on $\mathcal{H}_A$ satisfies

$$\mathrm{Tr}_A(|u\rangle\langle u| \otimes I_B)\rho_{\mathrm{mix}}^a = \frac{I_B - |u\rangle\langle u|}{d-1}, \quad \mathrm{Tr}_A(|u\rangle\langle u| \otimes I_B)\rho_{\mathrm{mix}}^s = \frac{I_B + |u\rangle\langle u|}{d+1}.$$

Thus,

$$\mathrm{Tr}_A(|u\rangle\langle u| \otimes I_B)\rho_{W,p} = p\frac{I_B - |u\rangle\langle u|}{d-1} + (1-p)\frac{I_B + |u\rangle\langle u|}{d+1},$$

which has entropy $-\frac{(d+1)p+(d-1)(1-p)}{d+1}\log\frac{(d+1)p+(d-1)(1-p)}{d^2-1}-\frac{2(1-p)}{d+1}\log\frac{2(1-p)}{d+1}$. Since this entropy is independent of $|u\rangle$,

$$\begin{aligned}C_d^{A\to B}(\rho_{W,p}) =& \log d + \frac{2(1-p)}{d+1}\log\frac{2(1-p)}{d+1}\\&+\frac{(d+1)p+(d-1)(1-p)}{d+1}\log\frac{(d+1)p+(d-1)(1-p)}{d^2-1}.\end{aligned}$$

Using the symmetry of this state, Vollbrecht and Werner [37] showed that

$$E_f(\rho_{W,p}) = \begin{cases} h\left(\frac{1+2\sqrt{p(1-p)}}{2}\right) & \text{if } p \ge \frac{1}{2} \\ 0 & \text{if } p < \frac{1}{2}. \end{cases}$$

Rains [75] showed

$$E_{r,S}(\rho_{W,p}) = E_{r,\mathrm{PPT}}(\rho_{W,p}) = \log 2 - h(p).$$

Rains [45] and Audenaert et al. [76] proved

$$\begin{aligned}&\lim_{n\to\infty}\frac{1}{n}E_{r,S}((\rho_{W,p})^{\otimes n}) = E_{\mathrm{SDP}}(\rho_{W,p})\\ =&\begin{cases} 0 & \text{if } p \le \frac{1}{2} \\ \log 2 - h(p) & \text{if } \frac{1}{2} < p \le \frac{1}{2}+\frac{1}{d} \\ \log\frac{d-2}{d} + p\log\frac{d+2}{d-2} & \text{if } \frac{1}{2}+\frac{1}{d} < p \le 1, \end{cases}\end{aligned}$$

where d is the dimension of the local system. Note that $\frac{1}{2}+\frac{1}{d}=1$ when $d=2$. Hence, $E_{r,S}(\rho)$ does not satisfy the additivity. This also implies $\lim_{n\to\infty}\frac{1}{n}E_{r,\mathrm{PPT}}((\rho_{W,p})^{\otimes n}) = E_{\mathrm{SDP}}(\rho_{W,p})$.

Further, Rains [45] also showed that

$$E_{d,2}^{\mathrm{PPT}}(\rho_{W,1}) = \log\frac{d+2}{d}.$$

Since $\tau^A(|\tau^A(\rho_{W,p})|) \ge 0$ Exe. 8.82 [47], we obtain

$$E_{c,e}^{\mathrm{PPT},\infty}(\rho_{W,p}) = \log\|\tau^A(\rho_{W,p})\|_1 = \log\left(\frac{2(2p-1)}{d}+1\right).$$

In particular,

$$\begin{aligned}&E_{d,2}^{\mathrm{PPT}}(\rho_{W,1}) = E_{d,2}^{\mathrm{PPT},\dagger}(\rho_{W,1}) = E_{\mathrm{SDP}}(\rho_{W,1}) = E_c^{\mathrm{PPT}}(\rho_{W,1})\\ =&E_{c,e}^{\mathrm{PPT},\infty}(\rho_{W,1}) = \log\frac{d+2}{d} \le \log 2 = E_f(\rho_{W,1}).\end{aligned}$$

The equality of the inequality $\log \frac{d+2}{d} \leq \log 2$ holds only when $d = 2$. From (8.234) the entanglement of distillation of the state $\rho_{W,1}$ satisfies the additivity

$$E_{d,2}^{\mathrm{PPT}}(\rho_{W,1}) + E_{d,2}^{\mathrm{PPT}}(\rho) = E_{d,2}^{\mathrm{PPT}}(\rho_{W,1} \otimes \rho)$$

for any state ρ.

On the other hand, Yura [77] calculated the entanglement of cost of any state ρ in the antisymmetric space of the system $\mathbb{C}^3 \otimes \mathbb{C}^3$ as

$$E_c(\rho) = \log 2, \tag{8.325}$$

which is equal to its entanglement of formation $E_f(\rho)^{\text{Exe. 8.38}}$. Hence, in this case,

$$E_c^{\mathrm{PPT}}(\rho_{W,1}) = E_{c,e}^{\mathrm{PPT},\infty}(\rho_{W,1}) = \log \frac{5}{3} < \log 2 = E_c(\rho_{W,1}).$$

Further, Matsumoto and Yura [78] focused on $\rho_{W,1}^{B,R} \stackrel{\text{def}}{=} \mathrm{Tr}_A |x\rangle\langle x|$, where $|x\rangle$ is a purification of $\rho_{W,1}$ with the reference system $\mathcal{H}_R$, and showed that

$$E_c(\rho_{W,1}^{B,R}) = \frac{E_f((\rho_{W,1}^{B,R})^{\otimes n})}{n} = E_f(\rho_{W,1}^{B,R}) = \log(d-1). \tag{8.326}$$

Hence, using (8.173), we have

$$C_d^{A\to B}(\rho_{W,1}^{\otimes n}) = \log \frac{d}{d-1}.$$

Since $\rho_{W,1}$ and $\rho_{W,0}$ satisfy the condition for (8.175), the relation (8.176) holds, i.e.,

$$E_c^{-\to}(\rho_{W,1}) = E_c^{-\to}(\rho_{W,0}) = \log d. \tag{8.327}$$

The entanglement purification $E_p(\rho_{W,p})$ of the other cases has been numerically calculated by Terhal et al. [40].

8.16.3 Isotropic State

Next, we consider the **isotropic state**

$$\rho_{I,p} \stackrel{\text{def}}{=} (1-p)\frac{I - |\Phi_d\rangle\langle\Phi_d|}{d^2-1} + p|\Phi_d\rangle\langle\Phi_d| \tag{8.328}$$

$$= \frac{(1-p)d^2}{d^2-1}\rho_{\mathrm{mix}} + \frac{d^2 p - 1}{d^2-1}|\Phi_d\rangle\langle\Phi_d|, \tag{8.329}$$

where $|\Phi_d\rangle = \frac{1}{\sqrt{d}}\sum_i u_i \otimes u_i$. We can easily check that

$$-H_{\rho_{I,p}}(A|B) = \log d + p \log p + (1-p) \log \frac{1-p}{d^2-1}$$
$$I_{\rho_{I,p}}(A:B) = 2\log d + p \log p + (1-p) \log \frac{1-p}{d^2-1}. \tag{8.330}$$

Further, any pure state $|u\rangle\langle u|$ on $\mathcal{H}_A$ satisfies

$$\begin{aligned}\mathrm{Tr}_A\, |u\rangle\langle u| \otimes I_B \rho_{I,p} =& \frac{(1-p)d^2}{d^2-1}\rho_{\mathrm{mix}}^B + \frac{d^2p-1}{d^2-1}|u\rangle\langle u| \\ =& \frac{(1-p)d}{d^2-1}(I - |u\rangle\langle u|) + \frac{dp+1}{d+1}|u\rangle\langle u|,\end{aligned}$$

which has entropy $-\frac{(1-p)d}{d^2-1}\log\frac{(1-p)d}{d+1} - \frac{dp+1}{d+1}\log\frac{dp+1}{d+1}$. Since this entropy is independent of $|u\rangle$,

$$C_d^{A\to B}(\rho_{W,p}) = \log d + \frac{(1-p)d}{d^2-1}\log\frac{(1-p)d}{d+1} + \frac{dp+1}{d+1}\log\frac{dp+1}{d+1}.$$

Further, King [79] showed that

$$C_d^{A\to B}((\rho_{W,p})^{\otimes n}) = nC_d^{A\to B}(\rho_{W,p}).$$

Define $\rho_{I,p}^{B,R} \stackrel{\mathrm{def}}{=} \mathrm{Tr}_A\, |x\rangle\langle x|$, where $|x\rangle$ is a purification of $\rho_{I,p}$ with the reference system $\mathcal{H}_R$. Then, using (8.173), we have

$$\begin{aligned}E_c(\rho_{I,p}^{B,R}) =& \frac{E_f((\rho_{I,p}^{B,R})^{\otimes n})}{n} = E_f(\rho_{I,p}^{B,R}) \\ =& -\frac{(1-p)d}{d^2-1}\log\frac{(1-p)d}{d+1} - \frac{dp+1}{d+1}\log\frac{dp+1}{d+1}.\end{aligned}$$

Using the symmetry of this state, Terhal and Vollbrecht [80] showed that

$$\begin{aligned}E_f(\rho_{I,p}) = \min_{x,y,p\ge 0}\, & p(h(\gamma(x)) + (1-\gamma(x))\log(d-1)) \\ & + (1-p)(h(\gamma(y)) + (1-\gamma(y))\log(d-1)),\end{aligned}$$

where we take the minimum with the condition $p = px + (1-p)y$, and

$$\gamma(p) = \frac{1}{d}(\sqrt{p} + \sqrt{(d-1)(1-p)})^2.$$

They also showed the following relation for the $d = 3$ case and conjectured it in the $d > 3$ case as follows:

$$E_f(\rho_{I,p}) = \begin{cases} 0 & \text{if } p \le \frac{1}{d} \\ h(\gamma(p)) + (1-\gamma(p))\log(d-1)) & \text{if } \frac{1}{d} < p \le \frac{4(d-1)}{d^2} \\ \frac{(p-1)d}{d-2}\log(d-1) + \log d & \text{if } \frac{4(d-1)}{d^2} < p \le 1. \end{cases}$$

Note that the isotropic state is locally unitarily equivalent to the Werner state when $d = 2$.

Further, Rains [75] showed that

$$E_{r,S}(\rho_{I,p}) = E_{r,\mathrm{PPT}}(\rho_{I,p}) = \begin{cases} \log d - (1-p)\log(d-1) - h(p) & \text{if } p \ge \frac{1}{d} \\ 0 & \text{if } p < \frac{1}{d}. \end{cases}$$

Rains [45] also proved

$$E_{d,2}^{\mathrm{PPT}}(\rho_{I,p}) \ge \log d - (1-p)\log(d+1) - h(p).$$

Since $\tau^A(|\tau^A(\rho_{I,p})|) \ge 0^{\text{Exe. 8.83}}$, we obtain

$$E_{c,e}^{\mathrm{PPT},\infty}(\rho_{I,p}) = \log \|\tau^A(\rho_{I,p})\|_1 = \begin{cases} \log dp & \text{if } p \ge \frac{1}{d} \\ 0 & \text{if } p < \frac{1}{d}. \end{cases} \tag{8.331}$$

In the system $\mathbb{C}^2 \otimes \mathbb{C}^2$, Terhal and Horodecki [30] proved

$$E_{sr}(\rho_{I,\frac{1}{\sqrt{2}}}) = \log 2, \quad E_{sr}(\rho_{I,\frac{1}{\sqrt{2}}}^{\otimes 2}) = \log 2.$$

Since $E_{c,e}^{\mathrm{PPT},\infty}(\rho_{I,p}) \le E_{c,e}^{S,\infty}(\rho_{I,p})$, from (8.331) and (8.114) we obtain

$$E_{c,e}^{C,\infty}(\rho_{I,\frac{1}{\sqrt{2}}}) = \log\sqrt{2} \text{ for } C = \rightarrow, \leftarrow, \leftrightarrow, S, \mathrm{PPT}.$$

Exercises

8.75 Prove (8.318) following the steps below.
(**a**) Show that $A^T S_2 A = S_2 \det A$ for a 2×2 matrix A.
(**b**) Show that $(A\otimes B)\rho(A\otimes B)^*(S_2\otimes S_2)\overline{(A\otimes B)\rho(A\otimes B)^*(S_2\otimes S_2)} = |\det A|^2|\det B|^2(A\otimes B)\rho(S_2\otimes S_2)\bar{\rho}(S_2\otimes S_2)(A\otimes B)^{-1}$.
(**c**) Show that $C_o(\frac{(A\otimes B)\rho(A\otimes B)^*}{\mathrm{Tr}(A\otimes B)\rho(A\otimes B)^*}) = \frac{|\det A||\det B|C_o(\rho)}{\mathrm{Tr}(A\otimes B)\rho(A\otimes B)^*}$.
(**d**) Prove (8.318).

8.76 Prove (8.319) following the steps below.
(**a** Show that $\overline{\rho_{\mathrm{Bell},p}} = \rho_{\mathrm{Bell},p}$.
(**b**) Show that $(S_2\otimes S_2)\rho_{\mathrm{Bell},p}(S_2\otimes S_2) = \rho_{\mathrm{Bell},p}$.
(**c**) Prove (8.319).

8.77 Prove (8.320) following the steps below.
(**a**) Show that $\mathrm{Tr}(A\otimes B)\rho_{\mathrm{Bell},p}(A\otimes B)^* = \sum_{i=0}^3 \frac{p_i}{2}\mathrm{Tr}\, A^*AS_iB^T\bar{B}S_i$.
(**b**) Show that $\frac{1}{2}\mathrm{Tr}\, A^*AS_iB^T\bar{B}S_i \ge |\det A||\det B|$.

(**c**) Prove (8.320).

8.78 Show that any maximally correlated state can be written as $\rho_{a,b} \stackrel{\text{def}}{=} a|00\rangle\langle 00| + b|00\rangle\langle 11| + b|11\rangle\langle 00| + (1-a)|11\rangle\langle 11|$ with two non-negative numbers a and b in a 2×2 system by choosing suitable bases.

8.79 Show that $C_o(\rho_{a,b}) = 2b$ following the steps below.
(**a**) Show that $(S_2 \otimes S_2)\overline{\rho_{a,b}}(S_2 \otimes S_2) = \rho_{1-a,b}$.
(**b**) Show that $\rho_{a,b}(S_2 \otimes S_2)\overline{\rho_{a,b}}(S_2 \otimes S_2) = (a(1-a)+b^2)|00\rangle\langle 00| + 2ab|00\rangle\langle 11| + 2(1-a)b|11\rangle\langle 00| + (a(1-a)+b^2)|11\rangle\langle 11|$.
(**c**) Show that $C_o(\rho_{a,b}) = 2b$.

8.80 Show that $H(\rho_{a,b}) = h\left(\frac{1+\sqrt{(2a-1)^2+4b^2}}{2}\right)$.

8.81 Assume that the input state is the maximally entangled state $|\Phi_d\rangle\langle\Phi_d|$ between the channel input system $\mathcal{H}_A$ and the reference system $\mathcal{H}_R$. Show that the output state of depolarizing channel $\kappa_{d,\lambda}$ (Example 5.3) (transpose depolarizing channel $\kappa^T_{d,\lambda}$ (Example 5.9)) is equal to the isotropic state (Werner state) as

$$(\kappa_{d,\lambda} \otimes \iota_R)(|\Phi_d\rangle\langle\Phi_d|) = \rho_{I,\frac{1-\lambda(d^2-1)}{d^2}} \tag{8.332}$$

$$(\kappa^T_{d,\lambda} \otimes \iota_R)(|\Phi_d\rangle\langle\Phi_d|) = \rho_{W,\frac{(1-(d+1)\lambda)(d-1)}{2d}}. \tag{8.333}$$

8.82 Show that $\tau^A(|\tau^A(\rho_{W,p})|) \geq 0$ following the steps below.
(**a**) Show that $\rho_{W,p} = qI + r\tau^A(|\Phi_d\rangle\langle\Phi_d|)$, where $q = \frac{1-p}{d(d+1)} + \frac{p}{d(d-1)}$, $r = \frac{1-p}{d+1} - \frac{p}{d-1}$.
(**b**) Show that $\tau^A(\rho_{W,p}) = q(I - |\Phi_d\rangle\langle\Phi_d|) + (q+r)|\Phi_d\rangle\langle\Phi_d|$.
(**c**) Show that $\tau^A(|\tau^A(\rho_{W,p})|) = q(I - \frac{1}{d}F) + \frac{|q+r|}{d}F \geq 0$.

8.83 Show that $\tau^A(|\tau^A(\rho_{I,p})|) \geq 0$ for $p > \frac{1}{d}$ following the steps below. (This inequality is trivial when $p \leq \frac{1}{d}$ because $\tau^A(\rho_{I,p}) \geq 0$.)
(**a**) Show that $\tau^A(\rho_{I,p}) = \frac{1-p}{d^2-1}I + \frac{d^2p-1}{d(d^2-1)}F = \frac{1-p}{d^2-1}(I+F) + \frac{dp-1}{d(d-1)}F$.
(**b**) Show that $|\tau^A(\rho_{I,p})| = \frac{1-p}{d^2-1}(I+F) + \frac{dp-1}{d(d-1)}I$.
(**c**) Show that $\tau^A(|\tau^A(\rho_{I,p})|) = \frac{1-p}{d^2-1}(I + d|\Phi_d\rangle\langle\Phi_d|) + \frac{dp-1}{d(d-1)}I \geq 0$.

8.84 Show that $(1-\lambda)\rho_{\text{mix}} + \lambda\tau^A(|\Phi_d\rangle\langle\Phi_d|) \geq 0$ if and only if $-\frac{1}{d-1} \leq \lambda \leq \frac{1}{d+1}$, where $\mathcal{H}_A = \mathcal{H}_B = \mathbb{C}^d$.

8.17 Proof of Theorem 8.2

We prove this theorem in the following steps: ①⇒②⇒③⇒①, ②⇒④⇒①. The proof given here follows from Bhatia [12].

We first show ①⇒② for dimension d by induction. Let $t \stackrel{\text{def}}{=} (y_1 - x_1)/(y_1 - y_2) = (x_2 - y_2)/(y_1 - y_2)$ for $d = 2$. Since $x \preceq y$, we have $0 \le t \le 1$. Further, the relation

$$\begin{pmatrix} x_1 \\ x_2 \end{pmatrix} = \begin{pmatrix} 1-t & t \\ t & 1-t \end{pmatrix} \begin{pmatrix} y_1 \\ y_2 \end{pmatrix} \tag{8.334}$$

proves the case for $d = 2$. In the following proof, assuming that the result holds for $d \le n - 1$, we prove the case for $d = n$. Any permutation is expressed by a product of T transforms. Hence, it is sufficient to show ② when $x_1 \ge x_2 \ge \ldots \ge x_n$ and $y_1 \ge y_2 \ge \ldots \ge y_n$. Since $x \preceq y$, we have $y_n \le x_1 \le y_1$. Choosing an appropriate k, we have $y_k \le x_1 \le y_{k-1}$. When t satisfies $x_1 = t y_1 + (1-t) y_k$, the relation $0 \le t \le 1$ holds. Let T_1 be the T transform among the first and kth elements defined by t. Define

$$x' \stackrel{\text{def}}{=} (x_2, \ldots, x_n)^T, \tag{8.335}$$

$$y' \stackrel{\text{def}}{=} (y_2, \ldots, y_{k-1}, (1-t)y_1 + t y_k, y_{k+1}, \ldots, y_n)^T. \tag{8.336}$$

Then, $T_1 y = (x_1, y')$. Since $x' \preceq y'$ (as shown below), from the assumptions of the induction there exist T transforms $T_f, \ldots, T_2$ such that $T_f \cdots T_2 y' = x'$. Therefore, $T_f \cdots T_2 T_1 y = T_f \cdots T_2 (x_1, y') = (x_1, x') = x$, which completes the proof for this part. We now show that $x' \preceq y'$. For an integer m satisfying $2 \le m \le k - 1$, we have

$$\sum_{j=2}^{m} x_j \le \sum_{j=2}^{m} y_j. \tag{8.337}$$

If $k \le m \le n$, then

$$\begin{aligned}
\sum_{j=2}^{m} y'_j &= \left(\sum_{j=2}^{k-1} y_j \right) + (1-t)y_1 + t y_k + \left(\sum_{j=k+1}^{m} y_j \right) \\
&= \left(\sum_{j=1}^{m} y_j \right) - t y_1 + (t-1) y_k = \sum_{j=1}^{m} y_j - x_1 \ge \sum_{j=1}^{m} x_j - x_1 = \sum_{j=2}^{m} x_j,
\end{aligned}$$

which shows that $x' \preceq y'$.

Next, we show ②⇒③. The product of two double stochastic transition matrices A_1 and A_2 is also a double stochastic transition matrix $A_1 A_2$. Since a T transform is a double stochastic transition matrix, we obtain ③.

For proving ③⇒①, it is sufficient to show that

$$\sum_{t=1}^{k} \sum_{j=1}^{d} x^{i_t, j} a_j \le \sum_{j=1}^{k} a_j^{\downarrow}$$

for an arbitrary integer k and a set of k arbitrary integers $i_1, \ldots, i_k$ from 1 to d. This can be shown from the fact that $\sum_{j=1}^{d} \sum_{t=1}^{k} x^{i_t,j} = k$ and $\sum_{t=1}^{k} x^{i_t,j} \le 1$ for each j.

We now show ②⇒④. For simplicity, we consider $d = 2$ and let

$$B = \begin{pmatrix} \frac{(y_1/y_2)^2-(x_2/x_1)(y_1/y_2)}{(y_1/y_2)^2-1} & \frac{(y_1/y_2)(x_1/x_2)-1}{(y_1/y_2)^2-1} \\ \frac{(x_2/x_1)(y_1/y_2)-1}{(y_1/y_2)^2-1} & \frac{(y_1/y_2)^2-(y_1/y_2)(x_1/x_2)}{(y_1/y_2)^2-1} \end{pmatrix}. \tag{8.338}$$

It can be verified that this is a stochastic transition matrix. Since

$$\begin{pmatrix} \frac{(y_1/y_2)^2-(x_2/x_1)(y_1/y_2)}{(y_1/y_2)^2-1} x_1 \\ \frac{(y_1/y_2)(x_1/x_2)-1}{(y_1/y_2)^2-1} x_2 \end{pmatrix} = \frac{(y_1/y_2)x_1 - x_2}{(y_1/y_2)^2-1} \begin{pmatrix} (y_1/y_2) \\ 1 \end{pmatrix}$$

$$\begin{pmatrix} \frac{(x_2/x_1)(y_1/y_2)-1}{(y_1/y_2)^2-1} x_1 \\ \frac{(y_1/y_2)^2-(y_1/y_2)(x_1/x_2)}{(y_1/y_2)^2-1} x_2 \end{pmatrix} = \frac{(y_1/y_2)x_2 - x_1}{(y_1/y_2)^2-1} \begin{pmatrix} 1 \\ (y_1/y_2) \end{pmatrix},$$

we observe that $B^1 \circ x \approx B^2 \circ x \approx y$.

Let T_0 be a T transform defined with respect to t between kth and lth elements $(k < l)$, and define B^1 and B^2 as

$$\begin{pmatrix} b^{1,k} & b^{1,l} \\ b^{2,k} & b^{2,l} \end{pmatrix} = \begin{pmatrix} \frac{(y_k/y_l)^2-(x_l/x_k)(y_k/y_l)}{(y_k/y_l)^2-1} & \frac{(y_k/y_l)(x_k/x_l)-1}{(y_k/y_l)^2-1} \\ \frac{(x_l/x_k)(y_k/y_l)-1}{(y_k/y_l)^2-1} & \frac{(y_k/y_l)^2-(y_k/y_l)(x_k/x_l)}{(y_k/y_l)^2-1} \end{pmatrix}$$

$$b^{1,i} = \frac{(y_k/y_l)x_k - x_l}{(y_k/y_l)-1}, \quad b^{2,i} = \frac{(y_k/y_l)x_l - x_k}{(y_k/y_l)-1} \quad \text{if } i \neq k, l.$$

Then, $B^1 \circ x \approx B^2 \circ x \approx y$, if $x = T_0 y$.

Further, if two stochastic transition matrices B, C satisfy $y \approx (B^j)^* \circ x$ and $z \approx (C^i)^* \circ y$ for arbitrary integers i and j, then there exists an appropriate substitution $s(j)$ such that

$$y \propto s(j)((B^j)^* \circ x),$$

where we identify the permutation $s(j)$ and the matrix that represents it. Since $s(j)^* = (s(j))^{-1}$,

$$\begin{aligned} z \approx & (C^i)^* \circ y = s(j)\left(((s(j))^{-1}(C^i)^*) \circ (s(j))^{-1} y\right) \\ \propto & s(j)\left(((s(j))^{-1}(C^i)^*) \circ (B^j)^* \circ x\right) = s(j)\left((C^i s(j))^* \circ (B^j)^* \circ x\right) \\ = & s(j)\left((C^i s(j))^* \circ (B^j)^* \circ x\right) \approx (C^i s(j))^* \circ (B^j)^* \circ x. \end{aligned}$$

Therefore,

$$\sum_{i,j}(C^i s(j))^* \circ (B^j)^* = \sum_j \left(\sum_i C^i s(j)\right)^* \circ (B^j)^*$$
$$= \sum_j (e^* s(j))^* \circ (B^j)^* = \sum_j e \circ (B^j)^* = \sum_j (B^j)^* = e.$$

When we define the matrix D by $(D^{i,j})^* \stackrel{\text{def}}{=} (C^i s(j))^* \circ (B^j)^*$ (note that the pair i, j refers to one column), this matrix is a stochastic transition matrix and satisfies

$$(D^{i,j})^* \circ x = z. \tag{8.339}$$

Using this and the previous facts, we obtain ②⇒④.

Finally, we show ④⇒①. It is sufficient to show the existence of a d-dimensional vector $c = (c_i)$ with positive real elements such that

$$c_i \le 1, \quad \sum_{i=1}^{d} c_i = k, \quad \sum_{j=1}^{k} y_{i_j} \ge \sum_{j=1}^{k} x_j^{\downarrow} \tag{8.340}$$

for arbitrary k. For this purpose, we choose k different integers $i_1, \ldots, i_k$ such that

$$\sum_{j=1}^{k} x_j^{\downarrow} = \sum_{j=1}^{k} x_{i_j}.$$

For each j, we choose the permutation $s(j)$ and the positive real number d_j such that $(B^j)^* \circ x = d_j s(j) y$. Note that $\sum_{j=1}^{d} d_j = 1$. Since

$$\sum_{j=1}^{d} b^{j,i} x_i = x_i,$$

we have

$$\sum_{j=1}^{k} x_{i_j} = \sum_{t=1}^{d}\sum_{j=1}^{k} b^{t,i_j} x_{i_j} = \sum_{t=1}^{d}\sum_{j=1}^{k} d_t (s(t)y)_{i_j} = \sum_{t=1}^{d}\sum_{j=1}^{k}\sum_{l=1}^{d} d_t s(t)_{i_j,l} y_l.$$

Since

$$\sum_{j=1}^{k} s(t)_{i_j,l} \le 1, \quad \sum_{l=1}^{d}\sum_{j=1}^{k} s(t)_{i_j,l} = k,$$

we obtain

$$\sum_t \sum_{j=1}^{k} d_t s(t)_{i_j,l} \leq 1, \quad \sum_{l=1}^{d} \sum_t \sum_{j=1}^{k} s(t)_{i_j,l} = k$$

where we used $\sum_t d_t = 1$. These relations show the existence of a vector $c = (c_i)$ satisfying (8.340).

8.18 Proof of Theorem 8.3

Let ρ be a separable state of $\mathcal{H}_A \otimes \mathcal{H}_B$. We can choose an appropriate set of vectors $\{u_i\}_i$ in $\mathcal{H}_A$ and $\{v_i\}_i$ in $\mathcal{H}_B$ such that $\rho = \sum_i |u_i \otimes v_i\rangle\langle u_i \otimes v_i| = \sum_j \lambda_j |e_j\rangle\langle e_j|$, where the RHS is the diagonalized form of ρ. From Lemma A.5 we can take an isometric matrix $W = (w_{i,j})$ such that $u_i \otimes v_i = \sum_j w_{i,j}\sqrt{\lambda_j} e_j$. Since $W^* W = I$, we have

$$\sum_i w_{i,j}^* u_i \otimes v_i = \sqrt{\lambda_j} e_j. \tag{8.341}$$

Similarly, we diagonalize $\mathrm{Tr}_B\, \rho$ such that $\mathrm{Tr}_B\, \rho = \sum_k \lambda'_k |f_k\rangle\langle f_k|$. Then, we can take an isometric matrix $W' = (w'_{i,k})$ such that $u_i = \sum_k w'_{i,k}\sqrt{\lambda'_k} f_k$.

Substituting this into (8.341), we obtain

$$\sqrt{\lambda_j} e_j = \sum_i \sum_k w'_{i,k} w_{i,j}^* \sqrt{\lambda'_k} f_k \otimes v_i.$$

Taking the norm on both sides, we have

$$\lambda_j = \sum_k D_{j,k} \lambda'_k, \quad D_{j,k} \stackrel{\text{def}}{=} \left(\sum_{i,i'} w'_{i,k} w_{i,j}^* (w'_{i',k})^* w_{i',j} \langle v_{i'} | v_i \rangle \right).$$

If we can show that $D_{j,k}$ is a double stochastic transition matrix, Condition ③ in Theorem 8.2 implies $(\lambda'_k) \preceq (\lambda_j)$. Since

$$\left(\sum_{i,i'} w'_{i,k} w_{i,j}^* (w'_{i',k})^* w_{i',j} \langle v_{i'} | v_i \rangle \right) = \left\langle \sum_{i'} w'_{i',k} w_{i',j}^* v_{i'} \middle| \sum_i w'_{i,k} w_{i,j}^* v_i \right\rangle \geq 0$$

and $W'^* W' = I$, $W^* W = I$, we obtain

$$\sum_k \left(\sum_{i,i'} w'_{i,k} w^*_{i,j} (w'_{i',k})^* w_{i',j} \langle v_{i'} | v_i \rangle \right) = \sum_{i,i'} \delta_{i,i'} w^*_{i,j} w_{i',j} \langle v_{i'} | v_i \rangle$$
$$= \sum_i w^*_{i,j} w_{i,j} = 1.$$

We may similarly show that $\sum_j D_{j,k} = 1$. Hence, $D_{j,k}$ is a double stochastic transition matrix.

8.19 Proof of Theorem 8.8 for Mixed States

We show the $\leq$ part of (8.101) for a general state ρ. Let $\{E_{A,i} \otimes E_{B,i}\}_i$ be the Choi–Kraus representation of an S-TP-CP map κ. Then,

$$\kappa(|\Phi_L\rangle\langle\Phi_L|) = \sum_i \left(E_{A,i} \otimes E_{B,i}\right) |\Phi_L\rangle\langle\Phi_L| \left(E_{A,i} \otimes E_{B,i}\right)^*.$$

Now, choose y_i such that

$$p'_i \stackrel{\text{def}}{=} \operatorname{Tr} \left(E_{A,i} \otimes E_{B,i}\right) |\Phi_L\rangle\langle\Phi_L| \left(E_{A,i} \otimes E_{B,i}\right)^*$$
$$p'_i |y_i\rangle\langle y_i| = \left(E_{A,i} \otimes E_{B,i}\right) |\Phi_L\rangle\langle\Phi_L| \left(E_{A,i} \otimes E_{B,i}\right)^*.$$

From Corollary 8.1 there exists a probabilistic decomposition $\{(p_i, x_i)\}$ of ρ such that

$$F(\kappa(|\Phi_L\rangle\langle\Phi_L|), \rho) = \sum_i \sqrt{p_i p'_i} |\langle x_i | y_i \rangle|.$$

Since the Schmidt rank of y_i is at most L,

$$|\langle x_i | y_i \rangle| \leq \sqrt{P(x_i, L)}. \tag{8.342}$$

From the Schwarz inequality,

$$F(\kappa(|\Phi_L\rangle\langle\Phi_L|), \rho) \leq \sum_i \sqrt{p_i p'_i} \sqrt{P(x_i, L)}$$
$$\leq \sqrt{\sum_i p'_i} \sqrt{\sum_i p_i P(x_i, L)} = \sqrt{\sum_i p_i P(x_i, L)}. \tag{8.343}$$

Thus, we obtain the $\leq$ part of (8.101).

Conversely, if there exists a vector y_i with a Schmidt rank of at most L that satisfies the equality in (8.342) and

$$p'_i = \frac{p_i P(x_i, L)}{\sum_j p_j P(x_j, L)},$$

the equality in (8.343) holds. Therefore, according to Theorem 8.4, there exists a one-way LOCC satisfying the RHS of (8.99).

8.20 Proof of Theorem 8.9 for Mixed States

8.20.1 Proof of Direct Part

The second equality in (8.108) holds according to (8.98) and Lemma A.1. We therefore show the $\leq$ part of the first equality. Let us first show that

$$\min_{(p_i, x_i)} \left\{ \sum_i p_i (1 - P(x_i, [e^{nR}])) \middle| \sum_i p_i |x_i\rangle\langle x_i| = \rho^{\otimes n} \right\}$$

converges to zero exponentially for $R > E_f(\rho)$. The convergence of this expression to zero is equivalent to that of the value inside of $\sqrt{\ }$ on the RHS of (8.101) to one. Hence, we consider the latter quantity, i.e., the value inside of $\sqrt{\ }$. Choose a decomposition $\{(p_i, x_i)\}$ such that $R > \sum_i p_i E(|x_i\rangle\langle x_i|)$. Let $\rho_i \stackrel{\text{def}}{=} \operatorname{Tr}_B |x_i\rangle\langle x_i|$. From (8.102),

$$\sum_i p_i P(x_i, [e^R]) \leq \sum_i p_i e^{\frac{\log \operatorname{Tr} \rho_i^{1-s} - sR}{1-s}}.$$

In particular, since $\frac{\log \operatorname{Tr}(\rho_i \otimes \rho_j)^{1-s} - 2sR}{1-s} = \frac{\log \operatorname{Tr} \rho_i^{1-s} - sR}{1-s} + \frac{\log \operatorname{Tr} \rho_j^{1-s} - sR}{1-s}$, we obtain

$$\begin{aligned} \sum_{i^n} p^n_{i^n} P(x^n_{i^n}, [e^{nR}]) &\leq \sum_{i^n} p^n_{i^n} e^{\frac{\log \operatorname{Tr}(\rho^n_{i^n})^{1-s} - snR}{1-s}} \\ &= \left(\sum_i p_i e^{\frac{\log \operatorname{Tr} \rho_i^{1-s} - sR}{1-s}} \right)^n, \end{aligned} \tag{8.344}$$

where we define $x^n_{i^n} \stackrel{\text{def}}{=} x_{i_1} \otimes \cdots \otimes x_{i_n}$, $\rho^n_{i^n} \stackrel{\text{def}}{=} \rho_{i_1} \otimes \cdots \otimes \rho_{i_n}$ with respect to $i^n \stackrel{\text{def}}{=} (i_1, \ldots, i_n)$, and p^n is the independent and identical distribution of p. Further, we obtain

$$\lim_{s\to 0}\frac{1}{s}\log\Bigl(\sum_i p_i e^{\frac{\log \operatorname{Tr}\rho_i^{1-s}-sR}{1-s}}\Bigr) = \frac{d}{ds}\log\Bigl(\sum_i p_i e^{\frac{\log \operatorname{Tr}\rho_i^{1-s}-sR}{1-s}}\Bigr)\Bigr|_{s=0}$$
$$=\sum_i p_i(H(\rho_i)-R) < 0.$$

Note that the inside of the logarithmic on the left-hand side (LHS) of the above equation is equal to 1 when $s = 0$. Taking an appropriate $1 > s_0 > 0$, we have $\log\Bigl(\sum_i p_i e^{\frac{\log \operatorname{Tr}\rho_i^{1-s_0}-s_0 R}{1-s_0}}\Bigr) < 0$. Thus, the RHS of (8.344) exponentially converges to zero. Therefore, we obtain $E_c^{\to}(\rho) \le E_f(\rho)$. Similarly, $E_c^{\to}(\rho^{\otimes k}) \le E_f(\rho^{\otimes k})$.

Next, we choose a sequence $\{m_n\}$ such that $(m_n-1)k \le n \le m_n k$ with respect to n. Denote the partial trace of $(\mathcal{H}_A\otimes\mathcal{H}_B)^{\otimes m_n k-n}$ by C_n. Then,

$$F(\rho^{\otimes m_n k}, \kappa_{m_n}(|\Phi_{L_{m_n}}\rangle\langle\Phi_{L_{m_n}}|)) \ge F(\rho^{\otimes n}, C_n\circ\kappa_{m_n}(|\Phi_{L_{m_n}}\rangle\langle\Phi_{L_{m_n}}|)) \tag{8.345}$$

for κ_m, L_m. Therefore, if the LHS of (8.345) converges to zero, then the RHS also converges to zero. Since

$$\overline{\lim}\frac{1}{n}\log L_{m_n} = \frac{1}{k}\overline{\lim_{m\to\infty}}\frac{1}{m}\log L_m,$$

C_n is a local quantum operation, and $E_c^{\to}(\rho^{\otimes k}) \le E_f(\rho^{\otimes k})$, and we have

$$E_c^{\to}(\rho) \le \frac{E_c^{\to}(\rho^{\otimes k})}{k} \le \frac{E_f(\rho^{\otimes k})}{k}.$$

Considering $\inf_k$, we obtain the $\le$ part of (8.108).

8.20.2 Proof of Converse Part

Let us first consider the following lemma as a preparation.

Lemma 8.23 *Let p be a probability distribution $p = \{p_i\}_{i=1}^d$. Then,*

$$\sum_{i=L+1}^{d} p_i^{\downarrow} \ge \frac{H(p)-\log L-\log 2}{\log(d-L)-\log L}. \tag{8.346}$$

Proof By defining the double stochastic transition matrix $A=(a_{i,j})$

$$a_{i,j} \stackrel{\text{def}}{=} \begin{cases} \frac{1}{L} & \text{if } i,j\le L \\ \frac{1}{d-L} & \text{if } i,j > L \\ 0 & \text{otherwise,}\end{cases}$$

the image Ap satisfies $(Ap)_i = \begin{cases} \frac{P(p,L)}{L} & \text{if } i \le L \\ \frac{1-P(p,L)}{d-L} & \text{if } i > L \end{cases}$. From Condition ③ in Theorem 8.2 we have $Ap \preceq p$. Therefore,

$$H(p) \ge H(Ap) = -P(p, L) \log \frac{P(p, L)}{L} - (1 - P(p, L)) \log \frac{1 - P(p, L)}{d - L}.$$

Since the binary entropy $h(x)$ is less than $\log 2$, we have

$$\begin{aligned} & P^c(p, L)(\log(d - L) - \log L) + \log 2 \\ \ge & P^c(p, L)(\log(d - L) - \log L) + h(P(p, L)) \ge H(p) - \log L. \end{aligned}$$

We thus obtain (8.346). ■

We now show the $\ge$ part of Theorem 8.9 by using Lemma 8.23. Consider the sequence of S-TP-CP maps $\{\kappa_n\}$ and the sequence of maximally entangled states $\{|\Phi_{L_n}\rangle\}$ satisfying

$$F(\kappa_n(|\Phi_{L_n}\rangle\langle\Phi_{L_n}|), \rho^{\otimes n}) \to 1. \tag{8.347}$$

Combining (8.100) and (8.101) in Theorem 8.8 and Lemma 8.23, we have

$$\begin{aligned} & 1 - F^2(\kappa_n(|\Phi_{L_n}\rangle\langle\Phi_{L_n}\|), \rho^{\otimes n}) \\ \ge & \min_{(p_i, x_i)} \left\{ \sum_i p_i P^c(x_i, L_n) \middle| \sum_i p_i |x_i\rangle\langle x_i| = \rho^{\otimes n} \right\} \\ \ge & \min_{(p_i, x_i)} \left\{ \sum_i p_i \frac{E(|x_i\rangle\langle x_i|) - \log L_n - \log 2}{\log(d^n - L_n) - \log L_n} \middle| \sum_i p_i |x_i\rangle\langle x_i| = \rho^{\otimes n} \right\} \\ = & \frac{E_f(\rho^{\otimes n}) - \log L_n - \log 2}{\log(d^n - L_n) - \log L_n} = \frac{\frac{E_f(\rho^{\otimes n})}{n} - \frac{\log L_n}{n} - \frac{\log 2}{n}}{\frac{\log(d^n - L_n)}{n} - \frac{\log L_n}{n}}. \end{aligned}$$

Using (8.347) and Lemma A.1, we obtain

$$\begin{aligned} 0 = & \lim_{n\to\infty} \left(\frac{\log(d^n - L_n)}{n} - \frac{\log L_n}{n} \right) \left(1 - \left(F(\kappa_n(|\Phi_{L_n}\rangle\langle\Phi_{L_n}|), \rho^{\otimes n}) \right)^2 \right) \\ \ge & \overline{\lim} \left(\frac{E_f(\rho^{\otimes n})}{n} - \frac{\log L_n}{n} - \frac{\log 2}{n} \right) = \lim_{n\to\infty} \frac{E_f(\rho^{\otimes n})}{n} - \underline{\lim} \frac{\log L_n}{n}. \end{aligned}$$

Thus, we obtain

$$\lim_{n\to\infty} \frac{E_f(\rho^{\otimes n})}{n} \le \underline{\lim} \frac{\log L_n}{n} \le \overline{\lim} \frac{\log L_n}{n} \le E_c(\rho),$$

which completes the Proof of Theorem 8.9.

8.21 Historical Note

8.21.1 Entanglement Distillation

The study of conversion among entangled states in an asymptotic setting was initiated by Bennett et al. [22]. These researchers derived the direct and converse parts of Theorem 8.6 in the pure-state case (Exercise 8.33). After this research, Lo and Popescu [2] considered convertibility among two pure states with LOCC. They found that the two-way LOCC could be simulated by the one-way LOCC (Theorem 8.1) in the finite regime when the initial state was pure. They also obtained the optimal value of the probability that we will succeed in converting a given pure partially entangled state into a desired maximally entangled state by LOCC (8.71). Following this research, Nielsen [15] completely characterized the LOCC convertibility between two pure states by use of majorization (pure-state case of Theorem 8.4). Vidal [16] extended this result to the mixed-state case, i.e., he showed the mixed-state case of Theorem 8.4. Using Nielsen's condition, Morikoshi and Koashi [26] proved that the optimal deterministic distillation with an initial pure state can be realized only by two-pair collective manipulations in each step. Applying the method of type to the optimal failure probability (the optimal successful probability) for distillation with an initial pure state, Hayashi et al. [25] derived the optimal generation rate with an exponential constraint for the failure probability (for the successful probability). They also treated this problem with the fidelity criterion. Further, Hayashi [21] extended this result to the non-i.i.d. case. Regarding the mixed-state case, Bennett et al. [28] discussed the relation between distillation and quantum error correction, which will be mentioned in Sect. 9.6. They derived several characterizations of the two-way LOCC distillation as well as of the one-way LOCC distillation. They also conjectured the Hashing inequality (8.121). Rains [75] showed this inequality in the maximally correlated case and the relation (8.143). Horodecki et al. [81] showed that (8.122) holds if this inequality holds. They also initiated a unified approach, which has been established by Donald et al. [32] as Theorem 8.10. Modifying the discussion by Devetak [82], Devetak and Winter [67] proved the inequality for any mixed state.

For the converse part, Bennett et al. [22] proved the converse part of the pure-state case by constructing the dilution protocol attaining the entropy rate. Then, proposing the entanglement of relative entropy $E_r(\rho)$, Vedral and Plenio [1] proved the inequality $E^S_{d,2}(\rho) \leq E_r(\rho)$. In this book, its improved version (Theorem 8.7) is derived by combining their idea and the strong converse of quantum Stein's lemma. Then, we obtain the strong converse inequality $E^{S,\dagger}_{d,2}(\rho) \leq E_r(\rho)$ even for the mixed case. Horodecki et al. [81] obtained the first inequality in (8.120). Further, establishing a unified approach, Donald et al. [32] simplified its proof.

Christandl and Winter [34] introduced squashed entanglement and proved the inequality $E^{\leftrightarrow}_{d,2}(\rho) \leq E_{sq}(\rho)$. Concerning PPT operations, Rains [45] proved inequality (8.223). This book extended this result to the strong converse inequality (8.226).

8.21.2 Entanglement Dilution and Related Topics

Regarding the dilution, as mentioned above, Bennett et al. [22] proved Theorem 8.9 in the pure-state case. Bennett et al. [28] introduced the entanglement formation. Following these results, Hayden et al. [29] proved Theorem 8.9. In this book, we proved Theorem 8.9 in a little different way to that given in [29]. In Sect. 8.20.2, we rigorously optimized the fidelity with the finite regime and proved Theorem 8.9 by taking its limit.

Further, Lo and Popescu [39] showed that the bound can be attained by classical communication with the square root of n bits in the pure-state case. Further, Hayden and Winter [83] and Harrow and Lo [84] proved the optimality of Lo and Popescu's protocol. Using their results, Terhal et al. [40] showed that the optimal rate of dilution with zero-rate communication can be characterized by the entanglement of purification. They also showed that it is lower bounded by the quantity $C_d^{A\to B}(\rho)$, which was introduced by Henderson and Vedral [41]. As a problem related to dilution with zero-rate communication, we may consider the problem generating a given separable state from common randomness. This problem with the classical setting has been solved by Wyner [85]. Theorem 8.13 is its quantum extension.

For entanglement of exact cost for PPT operations, Audenaert et al. [47] derived its lower bound. Concerning entanglement of exact cost for LOCC operations, Terhal and Horodecki [30] focused on the Schmidt rank and calculated it for the two-tensor product of the two-dimensional isotropic state. Joining these, we derived the entanglement of exact cost for these settings in this example.

As a related problem, we often consider how to characterize a pure entangled state producing a given state with nonzero probability by LOCC. This problem is called stochastic convertibility. Owari et al. [86] treated this problem in infinite-dimensional systems using the partial order. Miyake [87] treated this problem in tripartite systems using a hyperdeterminant. Ishizaka [88] focused on PPT operations and showed that any pure entangled state can be stochastically converted from another pure entangled state by PPT operations.

For the discord $D(B|A)_\rho$, we can easily show Inequality (8.178). The equality condition (Lemma 8.14) was given in the first edition of this book. However, at that time, the proof was not perfect. Then, Datta [89] and Dakic et al. [90] showed this argument latter. The Proof of Lemma 8.14 has been given by filling in the gap in the first edition of this book.

8.21.3 Additivity

Many researchers [31, 36] conjectured *additivity of entanglement formation*, i.e., the equation (8.144) holds for arbitrary two bipartite states. This relation can be generalized to the *superadditivity of entanglement formation* [37] as (8.145). Shimono [91] showed this conjecture when the states are in the antisymmetric space of the system

$\mathbb{C}^3 \otimes \mathbb{C}^3$, and Yura [77] extended it as (8.325). Matsumoto and Yura [78] extended this result to a more general case as (8.326). Then, Shimono et al. [92] numerically checked that there is no counter example for superadditivity of entanglement formation. However, Hastings [61] showed the existence of a counter example for superadditivity of entanglement formation. So, we find that the dimension of Hilbert spaces discussed in the numerical demonstration by [92] is not enough high. That is, the counter example requires higher dimensions. In this book, we discuss a counter example based on Fukuda [60]. Fukuda [60] employed the large deviation on the sphere with the Haar measure, which is summarized in Sect. 2.6. Using Theorem 2.11 in Sect. 2.6, we show Lemma 8.19, which plays an important role in this counter example. In fact, a lemma similar to Lemma 8.19 is often employed in quantum information theory. However, they sometimes drop an important factor in such a lemma. We need to be careful to use this type lemma.

8.21.4 Security and Related Topics

Theorem 8.15 plays an important role in the security evaluation. When $s = 1$, we can replace v by 1 in (8.291) of Theorem 8.15. This argument is called Left over hashing lemma. Its classical version has been shown by Bennett et al. [93] and Håstad et al. [94]. Renner [68] extended it to the quantum case when the security criterion is given by $d_1'(A : E|\rho)$. However, to derive an exponential upper bound for the security criterion like (8.293), we need Theorem 8.15. Its classical version was shown by Hayashi [69] and the quantum version was shown by Hayashi [65]. Recently, the tightness of the exponential evaluation was shown in the classical case by Hayashi et al [95].

In fact, there is a duality relation between the security and the coherence. To clarify this relation, Renes [70] showed Theorem 8.16. This kind of relation can be used for showing the performance of the code for the quantum-state transmission and the Hashing inequality (8.121) for entanglement distillation. For this purpose, we need to relax the condition of Theorem 8.16. So, we derive Theorem 8.17 as a generalization of Theorem 8.16. That is, Theorem 8.17 has a wider applicability than Theorem 8.16. Then, Theorem 8.17 will be employed in Sect. 9.6. Further, using this idea, we can derive the opposite inequality to the entropic uncertainty relation (7.51) as Theorem 8.18.

8.22 Solutions of Exercises

Exercise 8.1 Diagonalize ρ as $\sum_i p_i |u_i^A\rangle\langle u_i^A|$. Then, the purification is given as $|x\rangle = \sum_i \sqrt{p_i}|u_i^A, u_i^B\rangle$. So, we have $\mathrm{Tr}_A |x\rangle\langle x| = \sum_i p_i |u_i^B\rangle\langle u_i^B|$, which implies that $H(\rho) = H(\mathrm{Tr}_A |x\rangle\langle x|)$.

Exercise 8.2 Choose the purifications $|x\rangle$ and $|y\rangle$ of ρ and σ satisfying (8.10), i.e., $1-\left(\mathrm{Tr}\,|\sqrt{\rho}\sqrt{\sigma}|\right)^2 = 1-|\langle x|y\rangle|^2$. Exercise 3.18 implies that $1-|\langle x|y\rangle|^2 = d_1^2(|x\rangle\langle x|, |y\rangle\langle y|)$. The monotonicity of d_1 for the partial trace implies that $d_1^2(|x\rangle\langle x|, |y\rangle\langle y|) \ge d_1^2(\rho,\sigma)$. The combination of these relations yields the desired inequality.

Exercise 8.3 Let u and v be purifications of ρ and σ such that $F(\rho,\sigma) = F(|u\rangle\langle u|, |v\rangle\langle v|) = \mathrm{Tr}\sqrt{|u\rangle\langle u|}\sqrt{|v\rangle\langle v|} = |\langle u|v\rangle|^2 = F^2(|u\rangle\langle u|, |v\rangle\langle v|)$. Using the monotonicity of $\phi(1/2,\rho,\sigma)$, we have $F^2(\rho,\sigma) \le \mathrm{Tr}\sqrt{\rho}\sqrt{\sigma}$.

In addition, $F^2(\rho,\sigma) = \mathrm{Tr}\,|\sqrt{\rho}\sqrt{\sigma}| \ge \mathrm{Tr}\sqrt{\rho}\sqrt{\sigma}$.

Exercise 8.4 $F_e^2(\rho,\kappa) = \langle x|\kappa\otimes\iota_R(|x\rangle\langle x|)|x\rangle = \sum_i \langle x|E_i\otimes I|x\rangle\langle x|E_i\otimes I|x\rangle = \sum_i |\mathrm{Tr}_E\, E_i\rho|^2$.

Exercise 8.5

(**a**) Consider the singular value decomposition of the matrix $\{\mathrm{Tr}\, E_i A_j \rho\}_{i,j}$. That is, choose unitaries $U=(u_{i,j})$ and $V=(v_{i,j})$ such that $\sum_{i',j'} u_{i,i'}\,\mathrm{Tr}\, E_{i'} A_{j'}\rho v_{j',j}$ is a diagonal matrix with positive and real diagonal elements. Then, we define $E_i' := \sum_{i'} u_{i,i'} E_{i'}$ and $A_j' := \sum_{j'} E_{i'} v_{j',j} A_{j'}$. Due to Exercise 5.4, $\{E_i'\}_i$ and $\{A_j'\}_j$ are Choi–Kraus representations of κ and κ', respectively.
(**b**) We retake Choi–Kraus representations $\{E_i\}_i$ and $\{A_j\}_j$ of κ and κ' based on (**a**). Hence, $\{E_i A_j\}_{i,j}$ is a Choi–Kraus representation of $\kappa\circ\kappa'$. Define $p_i \stackrel{\mathrm{def}}{=} \mathrm{Tr}\, A_i\rho A_i^*$ and $A_i' \stackrel{\mathrm{def}}{=} A_i/\sqrt{p_i}$. (8.19) yields that $F_e^2(\rho,\kappa\circ\kappa') = \sum_{i,j}|\mathrm{Tr}\, E_i A_j\rho|^2 = \sum_i |\mathrm{Tr}\, E_i A_i\rho|^2 = \sum_i p_i |\mathrm{Tr}\, E_i A_i'\rho|^2$. Now, we choose the largest $|\mathrm{Tr}\, E_{i_0} A_{i_0}'\rho|^2$ one among $\{|\mathrm{Tr}\, E_i A_i'\rho|^2\}_i$. Hence, we obtain $F_e^2(\rho,\kappa\circ\kappa') = \sum_i p_i|\mathrm{Tr}\, E_i A_i'\rho|^2 \le |\mathrm{Tr}\, E_{i_0} A_{i_0}'\rho|^2$.
(**c**) Note that $EA = (U|E|^{1/2})(|E|^{1/2}A)$. Apply the Schwarz inequality for the inner product $\mathrm{Tr}\, X^*Y\rho$. Then, $|\mathrm{Tr}\, EA\rho|^2 = |\mathrm{Tr}(U|E|^{1/2})(|E|^{1/2}A)\rho|^2 \le \mathrm{Tr}\, U|E|U^*\rho\,\mathrm{Tr}\, A^*|E|A\rho$. Since $\mathrm{Tr}\, A^*|E|A\rho = \mathrm{Tr}\,|E|A\rho A^* \le \mathrm{Tr}\, A\rho A^* = 1$, we have $|\mathrm{Tr}\, EA\rho|^2 \le \mathrm{Tr}\, U|E|U^*\rho = \mathrm{Tr}\, EU^*\rho$.
(**d**) First, note that $F_e^2(\rho,\kappa\circ\kappa_U) \ge |\mathrm{Tr}\, E_1 U^*\rho|^2$. We choose a Choi–Kraus representation $\{E_i\}_i$ of κ and a matrix A according to (**b**). Take U^* to be an isometry from $\mathcal{H}_A$ to $\mathcal{H}_B$ under the polar decomposition $E_1 = U|E_1|$. (**c**) implies that $F_e^2(\rho,\kappa\circ\kappa') \le |\mathrm{Tr}\, E_1 A\rho|^2 \le \mathrm{Tr}\, U|E_1|U^*\rho = \mathrm{Tr}\, E_1 U^*\rho \le F_e(\rho,\kappa\circ\kappa_{U^*})$.
(**e**) In this case, we can take the polar decomposition $E_1 = U|E_1|$ such that U is an isometry from $\mathcal{H}_B$ to $\mathcal{H}_A$. Hence, (**c**) implies that

$$\begin{aligned}
&F_e^2(\rho,\kappa\circ\kappa') \le |\mathrm{Tr}\, E_1 A\rho|^2 \le \mathrm{Tr}\, U|E_1|U^*\rho \\
=&\mathrm{Tr}\, P_C U|E_1|U^* P_C\rho = \mathrm{Tr}\, E_1 U^* P_C\rho P_C \\
=&(\mathrm{Tr}\, P_C\rho)\,\mathrm{Tr}\, E_1 U^*\frac{P_C\rho P_C}{\mathrm{Tr}\, P_C\rho} \le (\mathrm{Tr}\, P_C\rho) F_e\Big(\frac{P_C\rho P_C}{\mathrm{Tr}\, P_C\rho}, \kappa\circ\kappa_{U^*}\Big).
\end{aligned}$$

Exercise 8.6 We have $|\mathrm{Tr}(A_1 + A_2 i)\rho|^2 = |\mathrm{Tr}\, A_1\rho + \mathrm{Tr}\, A_2\rho i|^2 = (\mathrm{Tr}\, A_1\rho)^2 + (\mathrm{Tr}\, A_2\rho)^2$, where A_1 and A_2 are Hermitian matrices. Hence, the function $\rho \mapsto |\mathrm{Tr}\, A\rho|^2$ is a convex function. Using (8.19) and a Choi–Kraus representation $\{A_j\}_j$

of κ, we have

$$F_e^2(\rho, \kappa) = \sum_j |\operatorname{Tr} A_j \rho|^2 \le \sum_i p_i \sum_j |\operatorname{Tr} A_j \rho_i|^2 = \sum_i p_i F_e^2(\rho_i, \kappa).$$

Exercise 8.7 Markov's inequality implies that there are d_C elements x_j among $\{x_i\}_{j=1}^{d_B}$ such that $(1 - F^2(x_j, \kappa(x_j))) \le \frac{1}{a} \sum_{i=1}^{d_B} \frac{1}{d_B}(1 - F^2(x_i, \kappa(x_i)))$. Since $(1 - F^2(x_i, \kappa(x_i))) \ge (1 - F^2(x_{i+1}, \kappa(x_{i+1})))$, we have

$$\begin{aligned}&(1 - F^2(x_{d_B - d_C + 1}, \kappa(x_{d_B - d_C + 1})))\\ \le& \frac{1}{a} \sum_{i=1}^{d_B} \frac{1}{d_B} \le (1 - F^2(x_i, \kappa(x_i))) = \max_{x \in \mathcal{H}_C} \left\{1 - F^2(x, \kappa(x))\right\},\end{aligned}$$

where the final equation follows from the construction of x_i.

Exercise 8.8

(**a**) Choose a purification $|x\rangle = \sum_i \sqrt{p_i} |u_i, u_i^R\rangle$, where u_i^R is a CONS of the reference system $\mathcal{H}_R$. Hence, (8.19) implies that

$$\begin{aligned}F_e^2(\rho, \kappa) =& \langle x | \kappa \otimes \iota_R(|x\rangle\langle x|) |x\rangle = \sum_{i,j} \sqrt{p_i} \langle u_i, u_i^R | \kappa \otimes \iota_R(|x\rangle\langle x|) \sqrt{p_j} |u_i, u_i^R\rangle\\ =& \sum_{i,j} p_i p_j \langle u_i | \kappa(|u_i\rangle\langle u_j|) |u_j\rangle.\end{aligned}$$

(**b**) We denote the expectation under the uniform distribution with respect to $\phi = (\phi_1, \ldots, \phi_d)$ by E. The average $\mathrm{E} \sum_{j,j',j'',j'''} e^{i(-\phi_j + \phi_{j'} - \phi_{j''} + \phi_{j'''})}$ is nonzero only when $j = j'$ and $j'' = j'''$ or $j = j''$ and $j' = j'''$. Hence,

$$\begin{aligned}&\mathrm{E} F^2(u(\phi), \kappa(u(\phi)))\\ =&\mathrm{E} \sum_{j,j',j'',j'''} p_i p_j e^{i(-\phi_j + \phi_{j'} - \phi_{j''} + \phi_{j'''})} \langle u_j | \kappa(|u_{j'}\rangle\langle u_{j''}|) |u_{j'''}\rangle\\ =&\mathrm{E} \sum_{j,j''} p_i p_j \langle u_j | \kappa(|u_j\rangle\langle u_{j''}|) |u_{j''}\rangle + \mathrm{E} \sum_{j,j'} p_i p_j \langle u_j | \kappa(|u_{j'}\rangle\langle u_j|) |u_{j'}\rangle\\ =&F_e^2(\rho, \kappa) + \sum_{j \ne k} p_j p_k \langle u_k | \kappa(|u_j\rangle\langle u_j|) |u_k\rangle.\end{aligned}$$

(**c**) We have

$$\begin{aligned}&\sum_{k=2}^{d} p_k \langle u_k | \kappa(|u_1\rangle\langle u_1|) |u_k\rangle \le p_2 \operatorname{Tr} \left(\sum_{k=2}^{d} |u_k\rangle\langle u_k| \right) \kappa(|u_1\rangle\langle u_1|)\\ =&p_2 \operatorname{Tr}(I - |u_1\rangle\langle u_1|) \kappa(|u_1\rangle\langle u_1|) \le p_2 \delta,\end{aligned}$$

and

$$\sum_{j=2}^{d}\sum_{k\neq j} p_j p_k \langle u_k|\kappa(|u_j\rangle\langle u_j|)|u_k\rangle = \sum_{j=2}^{d} p_j \sum_{k\neq j} p_k \langle u_k|\kappa(|u_j\rangle\langle u_j|)|u_k\rangle$$

$$\leq \sum_{j=2}^{d} p_j p_1 \operatorname{Tr}\left(\sum_{k\neq j} |u_k\rangle\langle u_k|)\kappa(|u_j\rangle\langle u_j|\right) \leq \sum_{j=2}^{d} p_j p_1 \operatorname{Tr}(I - |u_j\rangle\langle u_j|)\kappa(|u_j\rangle\langle u_j|)$$

$$\leq \sum_{j=2}^{d} p_j p_1 \delta.$$

(**d**) We show that $(1+p_2-p_1)p_1 \leq \frac{1}{2}$ as follows. When $p_1 \leq 1/2$, $\max_{p_2}(1+p_2-p_1)p_1 = p_1 \leq 1/2$. When $p_1 \geq 1/2$, $\max_{p_2}(1+p_2-p_1)p_1 = 2(1-p_1)p_1 \leq 1/2$.
Using (**b**) and (**c**), we have

$$\begin{aligned}
&1 - F_e^2(\rho,\kappa) + \sum_{j\neq k} p_j p_k \\
=&\langle u_k|\kappa(|u_j\rangle\langle u_j|)|u_k\rangle + \mathrm{E}(1 - F^2(u(\phi), \kappa(u(\phi)))) \\
\leq& p_1 \sum_{k=2}^{d} p_k \langle u_k|\kappa(|u_1\rangle\langle u_1|)|u_k\rangle + \sum_{j=2}^{d}\sum_{k\neq j} p_j p_k \langle u_k|\kappa(|u_j\rangle\langle u_j|)|u_k\rangle \\
&+ \mathrm{E}(1 - F^2(u(\phi), \kappa(u(\phi)))) \\
\leq& p_1 p_2 \delta + \sum_{j=2}^{d} p_j p_1 \delta + \mathrm{E}(1 - F^2(u(\phi), \kappa(u(\phi)))) \\
\leq& \left(p_1 p_2 + \sum_{j=2}^{d} p_j p_1 + 1\right)\delta \\
\leq& ((1+p_2-p_1)p_1 + 1)\delta \leq \frac{3}{2}\delta
\end{aligned}$$

Exercise 8.9 Consider the depolarizing channel $\kappa_{2,\lambda}$. (5.12) implies that

$$\max_{x\in\mathcal{H}_B^1}\left\{1 - F^2(x, \kappa_{2,\lambda}(x))\right\} = 1 - \left(\lambda + \frac{1}{2}(1-\lambda)\right) = \frac{1}{2}(1-\lambda).$$

We also have

$$\left(1 - F_e^2(\rho, \kappa_{2,\lambda})\right) = 1 - \frac{3\lambda+1}{4} = \frac{3(1-\lambda)}{4}.$$

Hence, we obtain the equality in (8.26).

Exercise 8.10

(**a**) (5.12) implies that

$$\left(1 - F_e^2(\rho_{\text{mix}}, \kappa_{d,\lambda})\right) = 1 - \left(\lambda + \frac{1-\lambda}{d^2}\right) = \frac{(1-\lambda)(d^2-1)}{d^2}$$
$$= \frac{(1-\lambda)(d+1)(d-1)}{d^2}$$

and

$$\mathrm{E}_{\mu,x}\left[1 - F^2(x, \kappa_{d,\lambda}(x))\right], = 1 - \left(\lambda + \frac{1-\lambda}{d}\right) = \frac{(1-\lambda)(d-1)}{d}.$$

Thus,

$$\frac{d}{d+1}\left(1 - F_e^2(\rho_{\text{mix}}, \kappa_{d,\lambda})\right) = \mathrm{E}_{\mu,x}\left[1 - F^2(x, \kappa_{d,\lambda}(x))\right],$$

(**b**) For any element $y \in \mathcal{H}_A$, we have

$$\mathrm{E}_{\mu,x} F^2(x, \kappa(x)) = \int_{\mathrm{SU}(d_A)} \langle y|U^* \kappa(U|y\rangle\langle y|U^*)U|y\rangle \nu(dU)$$
$$= F^2(y, \kappa_{d,\lambda}(y)).$$

(**c**) Let $|z\rangle$ be a purification of ρ_{mix}. Since $\overline{U}U^T = I$, (8.18) implies that

$$F_e^2(\rho_{\text{mix}}, \kappa_{d,\lambda}) = \langle z|\kappa_{d,\lambda} \otimes \iota_R(|z\rangle\langle z|)|z\rangle$$
$$= \int_{\mathrm{SU}(d_A)} \langle z|U^* \otimes I_R \kappa \otimes \iota_R(U \otimes I_R|y\rangle\langle y|U^* \otimes I_R)U \otimes I_R|y\rangle \nu(dU)$$
$$= \int_{\mathrm{SU}(d_A)} \langle z|I_A \otimes \overline{U} \kappa \otimes \iota_R(I_A \otimes U^T|y\rangle\langle y|I_A \otimes \overline{U})I_A \otimes U^T|y\rangle \nu(dU)$$
$$= \int_{\mathrm{SU}(d_A)} \langle z|\kappa \otimes \iota_R(|y\rangle\langle y|)|y\rangle \nu(dU)$$
$$= \langle z|\kappa \otimes \iota_R(|z\rangle\langle z|)|z\rangle = F_e^2(\rho_{\text{mix}}, \kappa).$$

(**d**) Using (**a**), (**b**), and (**c**), we have

$$\frac{d}{d+1}\left(1 - F_e^2(\rho_{\text{mix}}, \kappa)\right) = \frac{d}{d+1}\left(1 - F_e^2(\rho_{\text{mix}}, \kappa_{d,\lambda})\right)$$
$$= \mathrm{E}_{\mu,x}\left[1 - F^2(x, \kappa_{d,\lambda}(x))\right] = \mathrm{E}_{\mu,x}\left[1 - F^2(x, \kappa(x))\right].$$

Exercise 8.11 Let $|x\rangle$ is a purification of ρ. Then,

$$F_e^2\left(\rho, \sum_i f_i \kappa_i\right) = \langle x| \sum_i f_i \kappa_i(|x\rangle\langle x|)|x\rangle$$
$$= \sum_i f_i \langle x|\kappa_i(|x\rangle\langle x|)|x\rangle = \sum_i f_i F_e^2(\rho, \kappa_i).$$

Exercise 8.12

(**a**) Since X is a full rank matrix, $\rho_i = \frac{1}{p_i} X M_i^T X^*$ is pure if and only if M_i^T is rank-one.

(**b**) Assume that the states ρ_i in (8.31) are orthogonal to each other. Then, $X M_i^T X^* X M_j^T X^* = 0$, which is equivalent with $M_i^T X^* X M_j^T = 0$, i.e., $M_j (X^* X)^T M_i = 0$. This condition holds if and only if POVM $\boldsymbol{M} = \{M_i\}$ is a PVM and commutative with $(X^* X)^T$. We can also show the converse argument.

Exercise 8.13 Denote the input and output systems of κ by $\mathcal{H}_A$ and $\mathcal{H}_B$, respectively. Let $|x\rangle\langle x|$ be a purification of ρ_{mix}. Choose the probalistic decomposition as $\kappa \otimes \iota_R(|x\rangle\langle x|) = \sum_i p_i |y_i\rangle\langle y_i|$. Then, the Schmidt rank of $|y_i\rangle$ is less than d'. That is, the rank of $\mathrm{Tr}_B |y_i\rangle\langle y_i|$ is less than d' Therefore, $\mathrm{Tr}_R \sqrt{\mathrm{Tr}_A |y_i\rangle\langle y_i|} \le \sqrt{d'}$. Thus,

$$\langle x|(\kappa' \circ \kappa) \otimes \iota_R(|x\rangle\langle x|)|x\rangle = \langle x| \sum_i p_i (\kappa' \otimes \iota_R)(|y_i\rangle\langle y_i|)|x\rangle$$
$$\le \sum_i p_i F^2(\mathrm{Tr}_A(\kappa' \otimes \iota_R)(|y_i\rangle\langle y_i|), \mathrm{Tr}_A |x\rangle\langle x|)$$
$$= \sum_i p_i F^2(\mathrm{Tr}_A |y_i\rangle\langle y_i|, \rho_{\text{mix},R}) = \sum_i p_i (\mathrm{Tr}_R |\sqrt{\mathrm{Tr}_A |y_i\rangle\langle y_i|}\sqrt{\rho_{\text{mix},R}}|)^2$$
$$= \sum_i p_i \frac{(\mathrm{Tr}_R \sqrt{\mathrm{Tr}_A |y_i\rangle\langle y_i|})^2}{d} \le \frac{d'}{d}.$$

Exercise 8.14 When ρ has a purification $|x\rangle$ with the given form, $\mathrm{Tr}_R |x\rangle\langle x| = \sum_i p_i |x_i^A\rangle\langle x_i^A| \otimes |x_i^B\rangle\langle x_i^B|$. Conversely, we assume that ρ is a separable state with the form $\sum_i p_i |x_i^A\rangle\langle x_i^A| \otimes |x_i^B\rangle\langle x_i^B|$. The above given state $|x\rangle$ is a purification of ρ.

Exercise 8.15 Since x' is a pure state on $\mathcal{H}_{A'} \otimes \mathcal{H}_{E'} \otimes \mathcal{H}_R$, $H_{x'}(A'R) = H_{x'}(E')$ and $H_{x'}(R) = H_{x'}(A'E')$.

Exercise 8.16 Since the final state on $\mathcal{H}_R \otimes \mathcal{H}_E \otimes \mathcal{H}_B$ is a pure state, $H(\rho)$ is equal to the entropy of the final state on the reference system $\mathcal{H}_R$. $H(\kappa(\rho))$ is equal to the entropy of the final state on $\mathcal{H}_R \otimes \mathcal{H}_E$. $H_e(\rho, \kappa)$ is therefore equal to the entropy of the final state on the environment $\mathcal{H}_E$, respectively. Then, we denote the final state on $\mathcal{H}_R \otimes \mathcal{H}_E$ by $\sigma_{R,E}$. Thus,

$$H(\rho) - I_c(\rho, \kappa) = H(\rho) - (H(\kappa(\rho)) - H_e(\rho, \kappa))$$
$$= H(\sigma_R) - H(\sigma_{R,E}) + H(\sigma_E) = I_\sigma(R : E) \ge 0.$$

Exercise 8.17 Let $|x\rangle$ be the purification of ρ. Use the second inequality in (3.48) and the monotonicity of the trace norm concerning the partial trace on the reference system. Then,

$$\|\rho - \kappa(\rho)\|_1 \le \|\kappa \otimes \iota(|x\rangle\langle x|) - |x\rangle\langle x|\|_1 \le \sqrt{2(1 - F_e(\rho, \kappa))} \le 1/e.$$

Thus, Fannes inequality (5.92) yields that

$$\begin{aligned}
&H(\rho) - I_c(\rho, \kappa) = H(\rho) - H(\kappa(\rho)) + H(\kappa \otimes \iota(|x\rangle\langle x|)) \\
\le& |H(\rho) - H(\kappa(\rho))| + |H(\kappa \otimes \iota(|x\rangle\langle x|)) - 0| \\
\le& \|\rho - \kappa(\rho)\|_1(\log d - \log \|\rho - \kappa(\rho)\|_1) \\
&+ \|\kappa \otimes \iota(|x\rangle\langle x|) - |x\rangle\langle x|\|_1(\log d^2 - \log \|\kappa \otimes \iota(|x\rangle\langle x|) - |x\rangle\langle x|\|_1) \\
\le& \sqrt{2(1 - F_e(\rho, \kappa))}\left(3\log d - 2\log \sqrt{2(1 - F_e(\rho, \kappa))}\right).
\end{aligned}$$

Exercise 8.18 Let κ and κ' be the TP-CP maps from $\mathcal{H}_A$ to $\mathcal{H}_B$ and from $\mathcal{H}_B$ to $\mathcal{H}_C$. Let $\mathcal{H}_R$, $\mathcal{H}_E$, and $\mathcal{H}_{E'}$ be the reference system of $\mathcal{H}_A$ and the environment systems of κ and κ'. Then, we denote the output state of κ on the whole system $\mathcal{H}_R \otimes \mathcal{H}_E \otimes \mathcal{H}_B$ by ρ', and denote the output state of $\kappa' \circ \kappa$ on the whole system $\mathcal{H}_R \otimes \mathcal{H}_E \otimes \mathcal{H}_{E'} \otimes \mathcal{H}_C$ by ρ''. Since the strong subadditivity (5.83) of the von Neumann entropy implies that $H_{\rho''}(C) - H_{\rho''}(RC) \le H_{\rho''}(E'C) - H_{\rho''}(RE'C)$, we have

$$\begin{aligned}
&I_c(\rho, \kappa) = H_{\rho'}(B) - H_{\rho'}(E) = H_{\rho''}(E'C) - H_{\rho''}(E) \\
=& H_{\rho''}(E'C) - H_{\rho''}(RE'C) \ge H_{\rho''}(C) - H_{\rho''}(RC) \\
=& H_{\rho''}(C) - H_{\rho''}(EE') = I_c(\rho, \kappa' \circ \kappa).
\end{aligned}$$

Exercise 8.19

(a)

$$\begin{aligned}
&I(\rho, \kappa) = H(\kappa(\rho)) + H(\rho) - H(\kappa \otimes \iota_R(|x\rangle\langle x|)) \\
=& D(\kappa \otimes \iota_R(|x\rangle\langle x|)\|\kappa(\rho) \otimes \mathrm{Tr}_A |x\rangle\langle x|).
\end{aligned}$$

(b)

$$\begin{aligned}
&I(\rho, \kappa' \circ \kappa) = D((\kappa' \circ \kappa) \otimes \iota_R(|x\rangle\langle x|)\|(\kappa' \circ \kappa)(\rho) \otimes \mathrm{Tr}_A |x\rangle\langle x|) \\
\le& D(\kappa \otimes \iota_R(|x\rangle\langle x|)\|\kappa(\rho) \otimes \mathrm{Tr}_A |x\rangle\langle x|) = I(\rho, \kappa).
\end{aligned}$$

Exercise 8.20 Let κ be a TP-CP map from $\mathcal{H}_A$ to $\mathcal{H}_B$. Let $\mathcal{H}_R$ and $\mathcal{H}_E$ be the reference system of $\mathcal{H}_A$ and the environment systems of κ, respectively. Let $|x\rangle$ be a purification of ρ with the reference system $\mathcal{H}_R$. We also denote the unitary from $\mathcal{H}_A$ to $\mathcal{H}_B \otimes \mathcal{H}_E$ as the Stinespring representation of κ by U. Since $\sum_i p_i \kappa_i = \kappa$, we have

$$
\begin{aligned}
I_c(\rho,\kappa) =& H(\mathrm{Tr}_R \kappa \otimes \iota_R(|x\rangle\langle x|)) - H(\kappa \otimes \iota_R(|x\rangle\langle x|))\\
=& - H_{\kappa\otimes\iota_R(|x\rangle\langle x|)}(R|B) \le - \sum_i p_i H_{\kappa_i\otimes\iota_R(|x\rangle\langle x|)}(B|R) = \sum_i p_i I_c(\rho,\kappa_i).
\end{aligned}
$$

Hence,

$$
I(\rho,\kappa) = H(\rho) + I_c(\rho,\kappa) \le H(\rho) + \sum_i p_i I_c(\rho,\kappa_i) = \sum_i p_i I(\rho,\kappa_i).
$$

Exercise 8.21 Let κ' and κ be the TP-CP maps from $\mathcal{H}_A$ to $\mathcal{H}_B$ and from $\mathcal{H}_B$ to $\mathcal{H}_C$, respectively. Let $\mathcal{H}_R$, $\mathcal{H}_E$, and $\mathcal{H}_{E'}$ be the reference system of $\mathcal{H}_A$ and the environment systems of κ and κ', respectively. Let $|x\rangle$ be a purification of ρ with the reference system $\mathcal{H}_R$. We also denote the unitary from $\mathcal{H}_A$ to $\mathcal{H}_B \otimes \mathcal{H}_{E'}$ as the Stinespring representation of κ' by U.

The monotonicity with respect to the partial trace with $\mathcal{H}_{E'}$ implies that

$$
\begin{aligned}
&I(\kappa'(\rho),\kappa). = D(\kappa \otimes \iota_{RE'}(U|x\rangle\langle x|U^*)\|\kappa(\rho) \otimes \mathrm{Tr}_B U^*|x\rangle\langle x|U^*)\\
\ge& D(\kappa \otimes \iota_R(U|x\rangle\langle x|U^*)\|\kappa(\rho) \otimes \mathrm{Tr}_{BE'} U^*|x\rangle\langle x|U^*)\\
=& D((\kappa\circ\kappa') \otimes \iota_R(|x\rangle\langle x|)\|(\kappa\circ\kappa')(\rho) \otimes \mathrm{Tr}_A |x\rangle\langle x|)\\
=& I(\rho,\kappa\circ\kappa').
\end{aligned}
$$

Exercise 8.22 Let $|x\rangle$ be a purification of ρ. Then, $U|x\rangle$ is a purification of $U\rho U^*$.

$$
\begin{aligned}
&I_c(\rho,\kappa\circ\kappa_U) = H(\kappa\circ\kappa_U(\rho)) - H(\kappa\circ\kappa_U \otimes \iota_R(|x\rangle\langle x|))\\
=& H(\kappa(U\rho U^*)) - H(\kappa \otimes \iota_R(U|x\rangle\langle x|U^*)) = I_c(U\rho U^*,\kappa).
\end{aligned}
$$

Exercise 8.23

(**c**) We denote the density $|x'\rangle\langle x'|$ by ρ'. Then, $I_{x'}(A'E'_A : B'E'_B) = D(\rho'_{A'E'_AB'E'_B}\| \rho'_{A'E'_A} \otimes \rho'_{B'E'_B}) \ge D(\rho'_{E'_AE'_B}\|\rho'_{E'_A} \otimes \rho'_{E'_B}) = I_{x'}(E'_A : E'_B)$. Hence, (8.34) implies that

$$
\begin{aligned}
&I(\rho^A,\kappa^A) + I(\rho^B,\kappa^B) - I(\rho,\kappa^A\otimes\kappa^B)\\
=& H_{x'}(A') + H_{x'}(B') - H_{x'}(A'B') - \left(H_{x'}(E'_A) + H_{x'}(E'_B) - H_{x'}(E'_AE'_B)\right)\\
&+ \left(H_{x'}(A'E'_A) + H_{x'}(B'E'_B) - H_{x'}(A'E'_AB'E'_B)\right)\\
=& I_{x'}(A' : B') - I_{x'}(E'_A : E'_B) + I_{x'}(A'E'_A : B'E'_B)\\
\ge& I_{x'}(A' : B') \ge 0.
\end{aligned}
$$

Exercise 8.24

(**a**)

$$
\mathrm{Tr}_{R,R'} |x\rangle\langle x| = \sum_j p_j \mathrm{Tr}_R |x_j\rangle\langle x_j| = \sum_j p_j \rho_j.
$$

(b) Since $\mathrm{Tr}_A(\kappa \otimes \kappa_E \otimes \iota_R)(|x\rangle\langle x|) = (\kappa_E \otimes \iota_R)\mathrm{Tr}_A(|x\rangle\langle x|)$ and $\mathrm{Tr}_{R,R'}(\kappa \otimes \kappa_E \otimes \iota_R)(|x\rangle\langle x|) = \kappa(\rho)$, we have

$$
\begin{aligned}
& D((\kappa_E \otimes \iota_{A,R})(\kappa \otimes \iota_{R,R'})(|x\rangle\langle x|)\|(\kappa_E \otimes \iota_{A,R})(\kappa(\rho) \otimes \mathrm{Tr}_A(|x\rangle\langle x|))) \\
=& D((\kappa \otimes \kappa_E \otimes \iota_R)(|x\rangle\langle x|)\|\kappa(\rho) \otimes (\kappa_E \otimes \iota_R)\mathrm{Tr}_A(|x\rangle\langle x|)) \\
=& H(\kappa(\rho)) + H((\kappa_E \otimes \iota_R)\mathrm{Tr}_A(|x\rangle\langle x|)) - H((\kappa \otimes \kappa_E \otimes \iota_R)(|x\rangle\langle x|)) \\
=& H(\kappa(\rho)) + \sum_j p_j(H(\rho_j) - \log p_j) - \sum_j p_j(H((\kappa \otimes \iota_R)(|x_j\rangle\langle x_j|)) - \log p_j) \\
=& H(\kappa(\rho)) + \sum_j p_j H(\rho_j) - \sum_j p_j H((\kappa \otimes \iota_R)(|x_j\rangle\langle x_j|)).
\end{aligned}
$$

(c)

$$
\begin{aligned}
& \tilde{I}_c\Big(\sum_j p_j\rho_j, \kappa\Big) + H(\kappa(\rho)) \\
=& D((\kappa \otimes \iota_{R,R'})(|x\rangle\langle x|)\|(\kappa(\rho) \otimes \mathrm{Tr}_A(|x\rangle\langle x|))) \\
\ge& D((\kappa_E \otimes \iota_{A,R})(\kappa \otimes \iota_{R,R'})(|x\rangle\langle x|)\|(\kappa_E \otimes \iota_{A,R})(\kappa(\rho) \otimes \mathrm{Tr}_A(|x\rangle\langle x|))) \\
=& H(\kappa(\rho)) + \sum_j p_j H(\rho_j) - \sum_j p_j H((\kappa \otimes \iota_R)(|x_j\rangle\langle x_j|)) \\
=& \sum_j p_j \tilde{I}_c(\rho_j, \kappa) + H(\kappa(\rho)),
\end{aligned}
$$

which implies (8.50).

Exercise 8.25

(a)

$$
\begin{aligned}
& \sum_{j=1}^k p_j \tilde{I}_c(\rho_j, \kappa) = \sum_j p_j H(\rho_j) - \sum_j p_j H((\kappa \otimes \iota_R)(|x_j\rangle\langle x_j|)) \\
=& \sum_j p_j(H(\rho_j) - \log p_j) - \sum_j p_j(H((\kappa \otimes \iota_R)(|x_j\rangle\langle x_j|)) - \log p_j) \\
=& H(\kappa_E(\mathrm{Tr}_A(|x\rangle\langle x|))) - H(\kappa_E \otimes \iota_{A,R})(\kappa \otimes \iota_{R,R'})(|x\rangle\langle x|)).
\end{aligned}
$$

(b) Inequality (5.80) implies that $H(\mathrm{Tr}_A |x\rangle\langle x|) \le H(\kappa_E(\mathrm{Tr}_A |x\rangle\langle x|))$, which yields the last inequality in (8.56).
(c) Inequality (5.81) implies that $H(\kappa_E(\mathrm{Tr}_A(|x\rangle\langle x|))) - H(\kappa_E \otimes \iota_{A,R})(\kappa \otimes \iota_{R,R'})(|x\rangle\langle x|)) \le \log k$. Hence, (8.56) implies (8.51).

Exercise 8.26 Since (5.77) implies that $H(\kappa(\sum_{i=1}^k p_i\rho_i)) \ge \sum_{i=1}^k p_i H(\kappa(\rho_i))$, (8.50) guarantees that

$$
\begin{aligned}
I\Big(\sum_{i=1}^k p_i\rho_i, \kappa\Big) &= \tilde{I}_c\Big(\sum_{i=1}^k p_i\rho_i, \kappa\Big) + H\left(\kappa\left(\sum_{i=1}^k p_i\rho_i\right)\right) \\
&\ge \sum_{i=1}^k p_i \tilde{I}_c(\rho_i, \kappa) + \sum_{i=1}^k p_i H(\kappa(\rho_i)) = \sum_{i=1}^k p_i I(\rho_i, \kappa).
\end{aligned}
$$

Exercise 8.27 Since (5.79) guarantees that $H(\kappa(\sum_{i=1}^k p_i\rho_i)) \le \sum_{i=1}^k p_i H(\kappa(\rho_i)) + \log k$, (8.51) implies that

$$
\begin{aligned}
I\Big(\sum_{i=1}^k p_i\rho_i, \kappa\Big) &= \tilde{I}_c\Big(\sum_{i=1}^k p_i\rho_i, \kappa\Big) + H\left(\kappa\left(\sum_{i=1}^k p_i\rho_i\right)\right) \\
&\le \sum_{i=1}^k p_i \tilde{I}_c(\rho_i, \kappa) + \sum_{i=1}^k p_i H(\kappa(\rho_i)) + 2\log k \\
&= \sum_{i=1}^k p_i I(\rho_i, \kappa) + 2\log k.
\end{aligned}
$$

Exercise 8.28 Let κ_E be a pinching of a PVM $\{|u_i\rangle\langle u_i|\}_{i=1}^d$ satisfying that $|u_1\rangle\langle u_1| = |u\rangle\langle u|$. For any permutation g on $\{2,\ldots,d\}$, we define the unitary $U_g := \sum_{i=1}^d |u_{g(i)}\rangle \langle u_i|$. Assume that $\langle u|\rho|u\rangle = f$. Then, we have

$$
\begin{aligned}
H(\rho) &\overset{(a)}{\le} H(\kappa_E(\rho)) \overset{(a)}{\le} H\left(\sum_g \frac{1}{(d-1)!} U_g \kappa_E(\rho) U_g^*\right) \\
&= h(f) + (1-f)\log(d-1),
\end{aligned}
$$

where (a) and (b) follow from (5.80) and the concavity of the entropy, respectively. Hence, we have (8.57).

Applying (8.57) to the case when $|u\rangle$ is the purification $|x\rangle\langle x|$ of ρ and ρ is $\kappa \otimes \iota_R(|x\rangle\langle x|)$, we obtain (8.53).

Exercise 8.29 Consider the unitary matrix

$$
\begin{pmatrix}
S_0 & 0 & 0 & 0 \\
0 & S_1 & 0 & 0 \\
0 & 0 & S_2 & 0 \\
0 & 0 & 0 & S_3
\end{pmatrix}
\begin{pmatrix}
\sqrt{p_0} I & * & * & * \\
\sqrt{p_1} I & * & * & * \\
\sqrt{p_2} I & * & * & * \\
\sqrt{p_3} I & * & * & *
\end{pmatrix}
$$

as a Stinespring representation in $\mathbb{C}^2 \otimes \mathbb{C}^4$, where the elements $*$ of the second matrix are chosen appropriately to preserve the unitarity. Then, the channel κ^E given in (5.7) is given as

$$\kappa^E(\rho)_{i,j} = \sqrt{p_i p_j}\,\mathrm{Tr}\, S_i \rho S_j.$$

Since $\kappa^E(\rho_{\mathrm{mix}})_{i,j} = \delta_{i,j} p_i$, we have $H_e(\kappa, \rho_{\mathrm{mix}}) = H(\kappa^E(\rho_{\mathrm{mix}})) = H(p)$.

Since $\kappa^E(|e_0\rangle\langle e_0|) = \begin{pmatrix} p_0 + p_3 & 0 \\ 0 & p_1 + p_2 \end{pmatrix}$ and $\kappa^E(|e_1\rangle\langle e_1|) = \begin{pmatrix} p_0 + p_3 & 0 \\ 0 & p_1 + p_2 \end{pmatrix}$, we have $H_e(\kappa, |e_0\rangle\langle e_0|) = H(\kappa^E(|e_0\rangle\langle e_0|)) = h(p_0 + p_3)$ and $H_e(\kappa, |e_1\rangle\langle e_1|) = \kappa^E(|e_1\rangle\langle e_1|) = h(p_0 + p_3)$.

Exercise 8.30 The map $p = (p_i) \mapsto (\langle v_i|\kappa(\sum_j p_j |u_j\rangle\langle u_j|)|v_i\rangle)$ is double stochastic. Further, we have $q_i = \langle v_i|\kappa(\sum_j p_j |u_j\rangle\langle u_j|)|v_i\rangle$. Hence, we have $q \preceq p$.

Exercise 8.31 Let p_{mix} be the uniform distribution. $(\lambda_i) \preceq p_{\mathrm{mix}}$ if and only if the size of the uniform distribution p_{mix} is less than $1/\lambda_1^{\downarrow}$. Then, Theorem 8.4 with $p_j = \delta_{1,j}$ guarantees the desired argument.

Exercise 8.32 This follows immediately from Exercise 8.31.

Exercise 8.33 The state $\sqrt{M_q^{A,n}}\,\mathrm{Tr}_B\, |u\rangle\langle u|^{\otimes n}\sqrt{M_q^{A,n}}$ is a completely mixed state whose support is that of $M_q^{A,n}$. The measurement outcome on A is the same as that on B. So, the resultant state with the measurement outcome q is a maximally entangled state with the size $|T_q^n|$.

For an arbitrary real number $\epsilon > 0$, due to Theorem 2.6, the probability of the case when $H(q) > H(\rho_u) - \epsilon$ goes to zero exponentially. So, we have $E_{d,1}^{\emptyset}(|u\rangle\langle u|) \geq H(\rho_u) - \epsilon$. Since ϵ is arbitrary, we obtain the desired argument.

Exercise 8.34 For any separable state σ, we have

$$\begin{aligned}
& D(\rho_{\mathrm{Bell},p}\|\sigma) \\
=& \mathrm{Tr} \sum_{i,j} p_{i,j} |u_{i,j}^{A,B}\rangle\langle u_{i,j}^{A,B}| \left(\log \left(\sum_{i',j'} p_{i',j'} |u_{i',j'}^{A,B}\rangle\langle u_{i',j'}^{A,B}| \right) - \log \sigma \right) \\
=& \sum_{i,j} p_{i,j}\, \mathrm{Tr}\, |u_{i,j}^{A,B}\rangle\langle u_{i,j}^{A,B}| (\log(|u_{i,j}^{A,B}\rangle\langle u_{i,j}^{A,B}|) - \log \sigma) + \sum_{i,j} p_{i,j} \log p_{i,j} \\
=& \sum_{i,j} p_{i,j} D(|u_{i,j}^{A,B}\rangle\langle u_{i,j}^{A,B}| \|\sigma) - H(p) \\
=& \sum_{i,j} p_{i,j} E_{r,S}(|u_{i,j}^{A,B}\rangle\langle u_{i,j}^{A,B}|) - H(p) \\
=& \log d - H(p).
\end{aligned}$$

Exercise 8.35 The optimal protocol satisfying $-\frac{1}{n} \log L(\kappa_n) = r$ is given as the operation κ_n given in (8.86) and (8.87) with $r = r_n'$ satisfying

$$e^{nr} = \frac{1 - \mathrm{Tr}(\rho_u^{\otimes n} - e^{-nr_n'}) \{\rho_u^{\otimes n} - e^{-nr_n'} \geq 0\}}{e^{-nr_n'}}. \tag{8.348}$$

In this case, we have

$$\varepsilon_1(\kappa_n, |u\rangle\langle u|^{\otimes n}) = \mathrm{Tr}(\rho_u^{\otimes n} - e^{-nr_n'}) \left\{\rho_u^{\otimes n} - e^{-nr_n'} \geq 0\right\}. \tag{8.349}$$

Since $r_n \to r$, (2.187) implies that

$$\begin{aligned} -\frac{1}{n}\log \varepsilon_1(\kappa_n, |u\rangle\langle u|^{\otimes n}) &\to \max_{s\leq 0} -\psi(s|\operatorname{Tr}_B |u\rangle\langle u|) + sr \\ &= \max_{s\geq 0} s(H_{1+s}(\rho_u) - r) \end{aligned}$$

Exercise 8.36 Consider the case when $L(\kappa_n) = e^{nr}$. Then, we apply (8.84) at the Proof of Theorem 8.7 to the case with $\sigma = \sum_i \lambda_i |u_i^A \otimes u_i^B\rangle\langle u_i^A \otimes u_i^B|$. We choose $s = -t$. Then, we have $\phi(s) = \phi(-t) = tH_{1-t}(\rho_u)$. Hence, we obtain (8.92).

Exercise 8.37

(**a**) Since $\kappa(\sigma)$ is separable, (8.94) follows from (8.7).
(**b**) Assume that $|u\rangle$ has the Schmidt decomposition $\sum_i \sqrt{\lambda_i}|u_i^A\rangle|u_i^B\rangle$ and that $\lambda_i = \lambda_i^{\downarrow}$. When $\sigma = |u_1^A\rangle\langle u_1^A| \otimes |u_1^B\rangle\langle u_1^B|$, $\operatorname{Tr}|u\rangle\langle u|\sigma = \lambda_1^{\downarrow}$. Hence, it is enough to show $\operatorname{Tr}|u\rangle\langle u|\sigma \leq \lambda_1^{\downarrow}$ when σ is a pure state $|x\rangle|y\rangle$.
This argument can be shown by

$$\begin{aligned} &|\langle u|x\rangle|y\rangle|^2 = |\sum_i \sqrt{\lambda_i}\langle u_i^A|x\rangle\langle u_i^B|y\rangle|^2 \\ =&\left|\langle\overline{y}|\sum_i \sqrt{\lambda_i}|\overline{u_i^B}\rangle\langle u_i^A|x\rangle\right|^2 \leq \|\sum_i \sqrt{\lambda_i}|\overline{u_i^B}\rangle\langle u_i^A|\|^2 = \lambda_1, \end{aligned}$$

where $|\overline{y}\rangle$ is the complex conjugate of $|y\rangle$.
(**c**) (a) follows from the relation $\min_{\sigma\in S_s}(\operatorname{Tr}|\Phi_d\rangle\langle\Phi_d|\kappa(\sigma))^{-1} \geq \frac{1}{d}$ shown by (**a**). (b) follows from the definition of the dual map κ^*. (c) follows from the fact that $0 \leq \kappa^*(|\Phi_d\rangle\langle\Phi_d|) \leq I$. (e) follows from (**b**). (d) can be shown as follows. The maximum $\max_{0\leq T\leq I:\operatorname{Tr}|u\rangle\langle u|T=1}(\operatorname{Tr} T\sigma)^{-1}$ is realized when $T = |u\rangle\langle u|$, which does not depend on σ. So, we have

$$\begin{aligned} &\max_{0\leq T\leq I}\{\min_{\sigma\in S_s}(\operatorname{Tr} T\sigma)^{-1}|\operatorname{Tr}|u\rangle\langle u|T = 1\} \\ =&\max_{0\leq T\leq I:\operatorname{Tr}|u\rangle\langle u|T=1}\ \min_{\sigma\in S_s}(\operatorname{Tr} T\sigma)^{-1} = \min_{\sigma\in S_s}(\operatorname{Tr}|u\rangle\langle u|\sigma)^{-1}. \end{aligned}$$

(**d**) The separable TP-CP map achieving the rate $-\log\lambda_1^{\downarrow}$ is given in Exercise 8.31. The inequality $\leq$ in (8.93) can be shown the combination of (8.96) and the relation

$$\max_{\kappa\in S}\{\log L(\kappa)|\,\varepsilon_2(\kappa,\rho) = 0\} = \max_{\kappa\in S}\{\log d|\operatorname{Tr}\kappa(|u\rangle\langle u|)|\Phi_d\rangle\langle\Phi_d| = 1\}.$$

Exercise 8.38 First, notice that any pure state on the subspace $\{v\otimes u - u\otimes v|u, v\in\mathbb{C}^3\}$ has the form $\frac{1}{\sqrt{2}}(v_1\otimes v_2 - v_2\otimes v_1)$ with orthogonal normalized vectors v_1 and v_2. Thus, when ρ has a decomposition $\sum_i p_i|x_i\rangle\langle x_i|$, we have $H(\operatorname{Tr}_B|x_i\rangle\langle x_i|) = \log 2$, which implies that $E_f(\rho) = \log 2$.

Exercise 8.39 When ρ is a pure state case, (8.115) follows from (8.64) in Theorem 8.4. In the mixed state case, we choose the decomposition $\rho = \sum_i p_i |x_i\rangle\langle x_i|$ such that $E_f(\rho) = \sum_i p_i H(\mathrm{Tr}_B |x_i\rangle\langle x_i|)$. Then,

$$\sum_i p_i H(\mathrm{Tr}_B |x_i\rangle\langle x_i|) \geq \sum_i p_i E_f(\kappa(|x_i\rangle\langle x_i|))$$
$$\geq E_f\left(\sum_i p_i \kappa(|x_i\rangle\langle x_i|)\right) = E_f(\kappa(\rho)),$$

which implies (8.115).

Exercise 8.40 Due to (8.109), the RHS of (8.99) in Theorem 8.8 approaches 0 exponentially when $R > E(\rho)$ and $L = [e^{nR}]$. Hence, we obtain $E_c^{C,\dagger}(|u\rangle\langle u|) \geq E(\mathrm{Tr}_B |u\rangle\langle u|)$.

Exercise 8.41 We make a decomposition $\rho = \sum_i p_i |x_i\rangle\langle x_i|$ such that $\sum_i p_i H(\mathrm{Tr}_B | x_i\rangle\langle x_i|) = E_f(\rho)$. Then, we choose a separable state σ_i such that $H(\mathrm{Tr}_B |x_i\rangle\langle x_i|) = E_{r,S}(|x_i\rangle\langle x_i|) = D(|x_i\rangle\langle x_i| \| \sigma_i)$. The joint convexity of the relative entropy guarantees that

$$E_{r,S}(\rho) \leq D\left(\rho \| \sum_i p_i \sigma_i\right) = D\left(\sum_i p_i |x_i\rangle\langle x_i| \| \sum_i p_i \sigma_i\right)$$
$$\leq \sum_i p_i D(|x_i\rangle\langle x_i| \| \sigma_i) = \sum_i p_i H(\mathrm{Tr}_B |x_i\rangle\langle x_i|) = E_f(\rho).$$

Exercise 8.42

(**a**) Based on Lemma 8.2, we choose purifications x_n and y_n with the reference system $\mathcal{H}_{R,n}$ such that $b(|x_n\rangle\langle x_n|, |y_n\rangle\langle y_n|) = b(\rho_n, \sigma_n)$. Hence, due to Exercises 3.24 and 3.25, we have

$$d_1(|x_n\rangle\langle x_n|, |y_n\rangle\langle y_n|) \to 0 \tag{8.350}$$

because $d_1(\rho_n, \sigma_n) \to 0$. Then, we choose a decomposition $\sigma_n = \sum_i p_{n,i} |y_{n,i}\rangle\langle y_{n,i}|$ such that $\sum_i p_{n,i} H(\mathrm{Tr}_B |y_{n,i}\rangle\langle y_{n,i}|) = E_f(\sigma_n)$. Using Lemma 8.3, we find a POVM $M^n = \{M_i^n\}$ on the reference system $\mathcal{H}_{R,n}$ such that $\mathrm{Tr}_R(I \otimes M_i^n)|y_n\rangle\langle y_n| = p_{n,i}|y_{n,i}\rangle\langle y_{n,i}|$. Then, we make the decomposition $\rho_n = \sum_i p'_{n,i}|x_{n,i}\rangle\langle x_{n,i}|$ by $\mathrm{Tr}_R(I \otimes M_i^n)|x_n\rangle\langle x_n| = p'_{n,i}|x_{n,i}\rangle\langle x_{n,i}|$. Thus, we have

$$E_f(\sigma_n) = \sum_i p_{n,i} H(\mathrm{Tr}_B |y_{n,i}\rangle\langle y_{n,i}|) = H_{\hat{\kappa}_{M^n}(\mathrm{Tr}_B |y_n\rangle\langle y_n|)}(A|R) \tag{8.351}$$

$$\sum_i p'_{n,i} H(\mathrm{Tr}_B |x_{n,i}\rangle\langle x_{n,i}|) = H_{\hat{\kappa}_{M^n}(\mathrm{Tr}_B |x_n\rangle\langle x_n|)}(A|R). \tag{8.352}$$

Since the monotonicity for d_1 implies

$$\begin{aligned} & d_1(\hat{\kappa}_{M^n} \otimes \iota_A(\mathrm{Tr}_B |x_n\rangle\langle x_n|), \hat{\kappa}_{M^n} \otimes \iota_A(\mathrm{Tr}_B |y_n\rangle\langle y_n|)) \\ \le & d_1(|x_n\rangle\langle x_n|, |y_n\rangle\langle y_n|), \end{aligned}$$

(5.104) and (8.350) imply that

$$\begin{aligned} & \frac{1}{\log \dim \mathcal{H}_{A,n}} |H_{\hat{\kappa}_{M^n} \otimes \iota_A(\mathrm{Tr}_B |y_n\rangle\langle y_n|)}(A|R) - H_{\hat{\kappa}_{M^n} \otimes \iota_A(\mathrm{Tr}_B |x_n\rangle\langle x_n|)}(A|R)| \\ & \to 0. \end{aligned}$$

Thus, (8.351) and (8.352) yield the desired argument.
(**b**) From (**a**), we find that $E_f(\rho_n) \le E_f(\sigma_n) + o(\log \dim \mathcal{H}_{A,n})$ when $d_1(\rho_n, \sigma_n) \to 0$. Replacing ρ_n and σ_n, we have $E_f(\rho_n) \ge E_f(\sigma_n) + o(\log \dim \mathcal{H}_{A,n})$. Hence, we have $E_f(\rho_n) = E_f(\sigma_n) + o(\log \dim \mathcal{H}_{A,n})$.

Exercise 8.43

(**a**) Choose an extension σ_n^{ABE} of σ_n such that $I_{\sigma_n^{ABE}}(A : B|E) = E_{sq}(\sigma_n)$. We choose a purification $|y_n\rangle$ of σ_n^{ABE} with the reference system $\mathcal{H}_{R,n}$. Based on on Lemma 8.2, we choose a purification x_n of ρ_n with the reference system $\mathcal{H}_E \otimes \mathcal{H}_{R,n}$ such that $b(|x_n\rangle\langle x_n|, |y_n\rangle\langle y_n|) = b(\rho_n, \sigma_n)$. Hence, we have (8.350), which implies that

$$d_1(\rho_n^{ABE}, \sigma_n^{ABE}) \to 0, \tag{8.353}$$

where $\rho_n^{ABE} = \mathrm{Tr}_R |x_n\rangle\langle x_n|$. Thus, (5.106) in Exercise 5.40 guarantees that

$$\frac{1}{\log \dim \mathcal{H}_{A,n}} |I_{\rho_n^{ABE}}(A : B|E) - I_{\sigma_n^{ABE}}(A : B|E)| \to 0, \tag{8.354}$$

which implies the desired argument.
(**b**) The desired argument can be shown by the same way as (**b**) of Exercise 8.42.

Exercise 8.44 Let states ρ_n and σ_n on the bipartite system $\mathcal{H}_{A,n} \otimes \mathcal{H}_{B,n}$ satisfy $\|\rho_n - \sigma_n\|_1 \to 0$. Based on (8.125), we choose $\boldsymbol{M}_n \in C$ such that $E_m^C(\sigma_n) = -H_{\hat{\kappa}_{\boldsymbol{M}_n}(\sigma_n)}(A|BE)$. Since the monotonicity of d_1 yields that $d_1(\hat{\kappa}_{\boldsymbol{M}_n}(\sigma_n), \hat{\kappa}_{\boldsymbol{M}_n}(\rho_n)) \le d_1(\sigma_n, \rho_n)$, (5.104) implies that

$$\frac{1}{\log \dim \mathcal{H}_{A,n}} |H_{\hat{\kappa}_{\boldsymbol{M}_n}(\rho_n)}(A|BE) - H_{\hat{\kappa}_{\boldsymbol{M}_n}(\sigma_n)}(A|BE)| \to 0. \tag{8.355}$$

Thus, $E_m^C(\rho_n) \ge E_m^C(\sigma_n) + o(\log \dim \mathcal{H}_{A,n})$. Therefore, we can show Condition **E3** (continuity) by the same way as (**b**) of Exercise 8.42.

Exercise 8.45 Since the $\ge$ part is obvious, we show the $\le$ part. Assume that ρ and σ are states on $\mathcal{H}_{A_1} \otimes \mathcal{H}_{B_1}$ and $\mathcal{H}_{A_2} \otimes \mathcal{H}_{B_2}$, respectively. We choose an arbitrary extension $\rho_{A_1,A_2,B_1,B_2,E}$ of $\rho \otimes \sigma$. Chain rule (5.109) implies that

$$
\begin{aligned}
& I_{\rho_{A_1,A_2,B_1,B_2,E}}(A_1A_2 : B_1B_2|E) \\
=& I_{\rho_{A_1,A_2,B_1,B_2,E}}(A_1A_2 : B_1|B_2E) + I_{\rho_{A_1,A_2,B_1,B_2,E}}(A_1A_2 : B_2|E) \\
=& I_{\rho_{A_1,A_2,B_1,B_2,E}}(A_1 : B_1|A_2B_2E) + I_{\rho_{A_1,A_2,B_1,B_2,E}}(A_2 : B_1|B_2E) \\
& + I_{\rho_{A_1,A_2,B_1,B_2,E}}(A_2 : B_2|A_1E) + I_{\rho_{A_1,A_2,B_1,B_2,E}}(A_1 : B_2|E) \\
\stackrel{(a)}{\ge} & I_{\rho_{A_1,A_2,B_1,B_2,E}}(A_1 : B_1|A_2B_2E) + I_{\rho_{A_1,A_2,B_1,B_2,E}}(A_2 : B_2|A_1E),
\end{aligned}
$$

where (a) follows from the non-negativity of the conditional mutual information (See (5.90).). Thus, we obtain the $\le$ part of (8.128).

Exercise 8.46 The difference from (8.144) is only the point that $|u_j^{B_1}\rangle$ is not necessarily orthogonal. However, the proof of (8.144) does not require the orthogonality of $|u_j^{B_1}\rangle$. So, (8.138) can be shown by the same way.

Exercise 8.47 Choose a local operation κ_n in class C from $(\mathcal{H}_A)^{\otimes n} \otimes (\mathcal{H}_B)^{\otimes n}$ to $\mathbb{C}^{d_n} \otimes \mathbb{C}^{d_n}$ satisfying that $\frac{\log d_n}{n} \to E_{d,2}^C(\rho)$ and (8.117). Then, the monotonicity (**E2**C) and **E3**$'$ yield that $\lim_{n\to\infty} \frac{E^C(\rho^{\otimes n})}{n} \ge \lim_{n\to\infty} \frac{E^C(\kappa_n(\rho^{\otimes n}))}{n} \ge \lim_{n\to\infty} \frac{\log d_n}{n} = E_{d,2}^C(\rho)$.

Exercise 8.48 Due to (8.84), when a separable operation κ_n satisfies that $L(\kappa_n) = e^{nr}$, any separable state σ satisfies that

$$
\langle \Phi_{e^{nr}} | \kappa_n'(\rho^{\otimes n}) | \Phi_{e^{nr}} \rangle \le e^{-ns \frac{-D_{1+s}(\rho\|\sigma)+r}{1+s}} \tag{8.356}
$$

for $s > 0$. Taking the minimum for σ, we have

$$
\langle \Phi_{e^{nr}} | \kappa_n'(\rho^{\otimes n}) | \Phi_{e^{nr}} \rangle \le e^{-n \frac{-E_{1+s|S}(\rho)+sr}{1+s}}, \tag{8.357}
$$

which implies (8.139). Since the same discussion holds with $\underline{D}_{1+s}(\rho\|\sigma)$, we obtain (8.140).

Exercise 8.49 We choose the purification $|x\rangle = \sum_i \sqrt{p_i} |u_i^A, u_i^B, x_i^R\rangle$. Then,

$$
\mathrm{Tr}_A (M_{i'} \otimes I_{B,R}) |x\rangle\langle x| = \sum_{i,j} \sqrt{p_i}\sqrt{p_j} \langle u_i^A | M_{i'} | u_j^A \rangle |u_i^B, x_i^R\rangle\langle u_j^B, x_j^R|.
$$

Thus,

$$
\begin{aligned}
\mathrm{Tr}_{A,R} (M_{i'} \otimes I_{B,R}) |x\rangle\langle x|) =& \sum_{i,j} \sqrt{p_i}\sqrt{p_j} \langle u_i^A | M_{i'} | u_j^A \rangle \langle x_j^R | x_i^R \rangle |u_i^B\rangle\langle u_j^B|, \\
\mathrm{Tr}_{A,B} (M_{i'} \otimes I_{B,R}) |x\rangle\langle x| =& \sum_i p_i \langle u_i^A | M_{i'} | u_i^A \rangle |x_i^R\rangle\langle x_i^R|.
\end{aligned}
$$

Therefore,

$$\begin{aligned}&\kappa(\mathrm{Tr}_{A,R}(M_{i'} \otimes I_{B,R})|x\rangle\langle x|)\\
=&\sum_{i''}\sum_{i,j}\sqrt{p_i}\sqrt{p_j}\langle u_i^A|M_{i'}|u_j^A\rangle\langle x_j^R|x_i^R\rangle\langle u_{i''}^B|u_i^B\rangle\langle u_j^B|u_{i''}^B\rangle|x_{i''}^R\rangle\langle x_{i''}^R|\\
=&\sum_i p_i\langle u_i^A|M_{i'}|u_i^A\rangle|x_i^R\rangle\langle x_i^R| = \mathrm{Tr}_{A,B}(M_{i'} \otimes I_{B,R})|x\rangle\langle x|.\end{aligned}$$

Exercise 8.50 Consider the case when $L(\kappa_n) = e^{nr}$. Then, we apply (8.84) at the proof of Theorem 8.7 to the case with $\sigma = \sigma_\alpha$. We choose $s = -t$. Then, we have $\phi(s) = \phi(-t) = tD_{1+t}(\rho_\alpha\|\sigma_\alpha)$. Hence, we obtain (8.160).

Exercise 8.51 Let ρ be a state on $\mathcal{H}_A \otimes \mathcal{H}_B$ and κ_X be a TP-CP map on $\mathcal{H}_X$ for $X = A, B$. Choose the Stinespring representation U_{X,X_3} with the environment $\mathcal{H}_{X_3}$ of κ_X for $X = A, B$. We choose a purification $|x\rangle$ of ρ with the reference system $\mathcal{H}_{A_2} \otimes \mathcal{H}_{B_2}$ such that $E_p(\rho) = H(\mathrm{Tr}_B |x\rangle\langle x|)$. Then, $U_A \otimes U_B|x\rangle$ is a purification of $\kappa_A \otimes \kappa_B(\rho)$ and satisfies $H(U_A \otimes U_B|x\rangle\langle x|U_A^* \otimes U_B^*) = H(\mathrm{Tr}_B |x\rangle\langle x|)$. Hence, $E_p(\kappa_A \otimes \kappa_B(\rho)) \le E_p(\rho)$, which implies **E2∅**.

Let κ be an operation containing quantum communication with size d. Alice's operation is given as instrument $\{\kappa_i\}_{i=1}^d$ with Alice's resultant system $\mathcal{H}_{A_3}$. Take its indirect measurement $(\mathcal{H}_E, U, |z\rangle\langle z|, \{E_i\}_{i=1}^d)$ given in Theorem 7.3. Then, considering the Stinespring representation $U_{B,i}$ of Bob's operation for i, we have a purification $(\sum_i E_i \otimes U_{B,i})(U \otimes I_B|x\rangle \otimes |v_i^B\rangle$ of $\kappa(\rho)$. Hence, (5.82) yields

$$\begin{aligned}H(\mathrm{Tr}_B\,\kappa(\rho)) =& H\left(\mathrm{Tr}_B \sum_{i=1}^d E_i U(\mathrm{Tr}_B\,|x\rangle\langle x|)U^* E_i\right)\\
\le& H(U(\mathrm{Tr}_B\,|x\rangle\langle x|)U^*) + \log d = H((\mathrm{Tr}_B\,|x\rangle\langle x|)) + \log d = E_p(\rho) + \log d.\end{aligned}$$

Exercise 8.52 Let states ρ_n and σ_n on the bipartite system $\mathcal{H}_{A,n} \otimes \mathcal{H}_{B,n}$ satisfy $d_1(\rho_n, \sigma_n) \to 0$. We choose a purification $|y_n\rangle$ of σ_n with the reference system $\mathcal{H}_{A_2,n} \otimes \mathcal{H}_{B_2,n}$ such that $E_p(\sigma_n) = H(\mathrm{Tr}_B\,|y_n\rangle\langle y_n|)$. Due to the discussion in the proof of Lemma 8.13, we can assume that

$$\dim \mathcal{H}_{A_2,n} \le (\dim \mathcal{H}_{A,n} \dim \mathcal{H}_{B,n})^2. \tag{8.358}$$

Based on Lemma 8.2, we choose a purification $|x_n\rangle$ of ρ_n such that $b(|x_n\rangle\langle x_n|, |y_n\rangle\langle y_n|) = b(\rho_n, \sigma_n)$. Hence, due to Exercises 3.24 and 3.25, we have

$$d_1(|x_n\rangle\langle x_n|, |y_n\rangle\langle y_n|) \to 0 \tag{8.359}$$

because $d_1(\rho_n, \sigma_n) \to 0$. Thus, (5.92) in Theorem 5.12, (8.358), and (8.359) imply that

$$\frac{1}{\log \max(\dim \mathcal{H}_{A,n}, \dim \mathcal{H}_{B,n})} |H(\mathrm{Tr}_B |x_n\rangle\langle y_n|) - E_p(\sigma_n)|$$
$$= \frac{1}{\log \max(\dim \mathcal{H}_{A,n}, \dim \mathcal{H}_{B,n})} |H(\mathrm{Tr}_B |x_n\rangle\langle y_n|) - H(\mathrm{Tr}_B |y_n\rangle\langle y_n|)| \to 0.$$

Hence, combining the discussion in (**b**) of Exercise 8.42, we obtain the condition **E3**.

Exercise 8.53 Due to (8.173), it is sufficient to show

$$E_f(\rho^{B,R}) + E_f(\sigma^{E,R}) = E_f(\rho^{B,R} \otimes \sigma^{B,R}). \tag{8.360}$$

Since ρ is separable, $\rho^{B,R}$ satisfies the condition for Exercise 8.46. Hence, we obtain (8.360)

Exercise 8.54 For any bipartite pure state $|u\rangle\langle u|$, the entanglement formation E_f between the system $\mathcal{H}_B$ and the reference is zero. So, (8.173) guarantees that $C_d^{A\to B}(|u\rangle\langle u|) = H(\mathrm{Tr}_A |u\rangle\langle u|)$, which implies Condition **E1**$'$.

Consider a bipartite state ρ on $\mathcal{H}_A \otimes \mathcal{H}_B$, and TP-CP maps κ_A and κ_B on $\mathcal{H}_A$ and $\mathcal{H}_B$. Hence, there exists a POVM $\{M_i\}$ on $\mathcal{H}_A$ satisfying the following equation (a). Then, the following equation (b) follows from the definition of the dual map κ_A^* and the following inequality (c) follows from Exercise 5.46.

$$\begin{aligned}
&C_d^{A\to B}(\kappa_A \otimes \kappa_B(\rho)) \\
\overset{(a)}{=}& H(\kappa_B(\rho^B)) - \sum_i \mathrm{Tr}\, M_i \kappa_A(\rho^A) H\left(\frac{\mathrm{Tr}_A (M_i \otimes I_B)(\kappa_A \otimes \kappa_B(\rho))}{\mathrm{Tr}\, M_i \kappa_A(\rho^A)}\right) \\
\overset{(b)}{=}& H(\kappa_B(\rho^B)) - \sum_i \mathrm{Tr}\, \kappa_A^*(M_i)\rho^A H\left(\frac{\kappa_B(\mathrm{Tr}_A(\kappa_A^*(M_i) \otimes I_B)\rho)}{\mathrm{Tr}\, \kappa_A^*(M_i)\rho^A}\right) \\
\overset{(c)}{\le}& H(\rho^B) - \sum_i \mathrm{Tr}\, \kappa_A^*(M_i)\rho^A H\left(\frac{1}{\mathrm{Tr}\, \kappa_A^*(M_i)\rho^A} \mathrm{Tr}_A(\kappa_A^*(M_i) \otimes I_B)\rho\right) \\
\le& C_d^{A\to B}(\rho).
\end{aligned}$$

Exercise 8.55 The dimension of the reference system is less than $\dim \mathcal{H}_A \dim \mathcal{H}_B$. $H(\rho^B)$ satisfies Condition **E3**. Thus, due to Relation (8.173), Condition **E3** for $C_d^{A\to B}(\rho)$ follows from Condition **E3** for $E_f(\rho^{B,R})$.

Exercise 8.56 For a POVM $\{M_i\}$ on $\mathcal{H}_A$, we choose another POVM $\{M_{i,j}\}$ on $\mathcal{H}_A$ such that rank $M_{i,j} = 1$ and $\sum_j M_{i,j} = M_i$. Since $\mathrm{Tr}_A(M_i \otimes I_B)\rho) = \sum_j \mathrm{Tr}_A(M_{i,j} \otimes I_B)\rho)$, the concavity of von Neumann entropy yields

$$\begin{aligned}
&H(\rho^B) - \sum_i \mathrm{Tr}\, M_i \rho^A H\left(\frac{1}{\mathrm{Tr}\, M_i \rho^A} \mathrm{Tr}_A(M_i \otimes I_B)\rho)\right) \\
\le& H\left(\rho^B) - \sum_{i,j} \mathrm{Tr}\, M_{i,j} \rho^A H(\frac{1}{\mathrm{Tr}\, M_{i,j}\rho^A} \mathrm{Tr}_A(M_{i,j} \otimes I_B)\rho)\right).
\end{aligned}$$

Exercise 8.57

(**a**) $\{M_i\}$ is given by a CONS $\{|u_i\rangle\}$. Hence, $I_\rho(A:B) = H_\rho(B) - H_\rho(B|A) = H(\mathrm{Tr}_A |X\rangle\rangle\langle\langle X|) - \sum_i \mathrm{Tr}\, XM_iX^* H(\frac{1}{XM_iX^*} XM_iX^*) = H(\mathrm{Tr}_A |X\rangle\rangle\langle\langle X|)$.
(**b**) Let ρ be a separable state of the form (8.180) with rank $\rho_i^B = 1$. Then, we denote ρ_i^B by $|x_i^B\rangle\langle x_i^B|$. We define the map $X \stackrel{\text{def}}{=} \sum_i \sqrt{p_i}|x_i^B\rangle\langle u_i^A$. Then, the state $\sum M_i \otimes XM_iX^*$ is ρ. Due to (**a**), we have $H(\mathrm{Tr}_A \rho) = I_\rho(A:B)$.
(**c**) Now, we denote the original systems $\mathcal{H}_A$ and $\mathcal{H}_B$ by $\mathcal{H}_{A_1}$ and $\mathcal{H}_{B_1}$. We choose a purification $|x\rangle$ of ρ with the reference system $\mathcal{H}_{A_2} \otimes \mathcal{H}_{B_2}$ such that $E_p(\rho) = H(\mathrm{Tr}_A |x\rangle\langle x|)$. Using the CONS $\{|u_i^{A_1}\rangle\}$ on $\mathcal{H}_{A_1}$, we define PVM $\boldsymbol{M} \stackrel{\text{def}}{=} \{M_i\}$ with $M_i \stackrel{\text{def}}{=} |u_i^{A_1}\rangle\langle u_i^{A_1}|$. Then, the pinching map $\kappa_{\boldsymbol{M}}$ given in (1.13) satisfies $\mathrm{Tr}_{A_1} \kappa_{\boldsymbol{M}}(|x\rangle\langle x|) = \mathrm{Tr}_{A_1} |x\rangle\langle x|$. Since $\kappa_{\boldsymbol{M}}(|x\rangle\langle x|)$ can be written as the form $\sum_i |u_i^{A_1}\rangle\langle u_i^{A_1}| \otimes \rho_i^{A_2,B_1,B_2}$, we have

$$\mathrm{Tr}_{A_1,A_2} |x\rangle\langle x| = \sum_i \mathrm{Tr}_{A_2} \rho_i^{A_2,B_1,B_2}. \tag{8.361}$$

Since rank $\mathrm{Tr}_{A_2B_2} \rho_i^{A_2,B_1,B_2} = 1$, the state $\sum_i \mathrm{Tr}_{A_2} \rho_i^{A_2,B_1,B_2}$ is separable as a bipartite state on $\mathcal{H}_{B_1} \otimes \mathcal{H}_{B_2}$. Thus, Theorem 8.3 guarantees that $H(\sum_i \mathrm{Tr}_{A_2} \rho_i^{A_2,B_1,B_2}) \geq H(\sum_i \mathrm{Tr}_{A_2,B_2} \rho_i^{A_2,B_1,B_2}) = H(\mathrm{Tr}_{A_1} \rho)$. Hence, (8.361) implies that $E_p(\rho) = H(\mathrm{Tr}_{A_1} \rho)$.

Exercise 8.58 Using (8.173), we have

$$\begin{aligned}
D(B|A)_\rho =& H(\rho^A) + H(\rho^B) - H(\rho^{AB}) - (H(\rho^B) - E_f(\rho^{BR})) \\
=& H(\rho^A) - H(\rho^{AB}) + E_f(\rho^{BR}) = H(\rho^{RB}) - H(\rho^R) + E_f(\rho^{BR}) \\
=& H_{\rho^{BR}}(B|R) + E_f(\rho^{BR}).
\end{aligned}$$

Exercise 8.59 Given a state ρ on $\mathcal{H}_A \otimes \mathcal{H}_B$, we choose a state ρ_{ABE} with the form $\sum_x p_x \rho_x^{A,B} \otimes |u_x^E\rangle\langle u_x^E|$ satisfying the condition given in (8.200). We also choose operations κ_A and κ_B on $\mathcal{H}_A$ and $\mathcal{H}_B$, respectively. Then, we have $I_{(\kappa_A\otimes\kappa_B)(\rho_{ABE})}(A:B|E) \leq I_{\rho_{ABE}}(A:B|E) \leq \delta$. We can check that the state $(\kappa_A \otimes \kappa_B)(\rho_{ABE})$ satisfies other conditions in (8.200) for $(\kappa_A \otimes \kappa_B)(\rho_{AB})$. Since $I_{\rho_{ABE}}(AB:E) \geq I_{(\kappa_A\otimes\kappa_B)(\rho_{ABE})}(AB:E)$, we have the first inequality of (8.204). Similarly, we can show the second inequality of (8.204).

Exercise 8.60

(**a**) Chain rule for the conditional mutual information (5.109) implies that

$$\begin{aligned}
0 =& I_{\rho^{ABE}}(A_1A_2 : B_1B_2|E) \\
=& I_{\rho^{ABE}}(A_1 : B_1B_2|E) + I_{\rho^{ABE}}(A_2 : B_1B_2|A_1E) \\
=& I_{\rho^{ABE}}(A_1 : B_1|E) + I_{\rho^{ABE}}(A_1 : B_2|B_1E) \\
&+ I_{\rho^{ABE}}(A_2 : B_1|A_1E) + I_{\rho^{ABE}}(A_2 : B_2|A_1B_1E).
\end{aligned}$$

Since the conditional mutual information is non-negative, we obtain the desired argument.
(**b**) We show only $C(\rho^{A_1B_1} \otimes \rho^{A_2B_2}) \geq C(\rho^{A_1B_1}) + C(\rho^{A_2B_2})$ because the opposite inequality is oblivious. For this purpose, we show

$$\tilde{C}(\rho^{A_1B_1} \otimes \rho^{A_2B_2}, 0) \geq \tilde{C}(\rho^{A_1B_1}, 0) + \tilde{C}(\rho^{A_2B_2}, 0). \tag{8.362}$$

As shown in (**a**), when an extension ρ^{ABE} of $\rho^{A_1B_1} \otimes \rho^{A_2B_2}$ satisfies the condition in (8.201) with $\delta = 0$, the state $\mathrm{Tr}_{A_2,B_2} \rho^{ABE}$ satisfies the condition for ρ^{A_1,B_1} and the state ρ^{ABE} satisfies the condition for ρ^{A_2,B_2} by regarding the system $\mathcal{H}_{A_1} \otimes \mathcal{H}_{B_1} \otimes \mathcal{H}_E$ as the environment.

Since $H_{\rho^{ABE}}(A_2B_2A_1B_1) = H_{\rho^{ABE}}(A_2B_2) + H_{\rho^{ABE}}(A_1B_1)$, we have

$$\begin{aligned}
&I_{\rho^{ABE}}(A_2B_2 : E|A_1B_1) \\
=&H_{\rho^{ABE}}(A_2B_2A_1B_1) + H_{\rho^{ABE}}(EA_1B_1) \\
&- H_{\rho^{ABE}}(EA_2B_2A_1B_1) - H_{\rho^{ABE}}(A_1B_1) \\
=&H_{\rho^{ABE}}(A_2B_2) + H_{\rho^{ABE}}(EA_1B_1) - H_{\rho^{ABE}}(EA_2B_2A_1B_1) \\
=&I_{\rho^{ABE}}(A_2B_2 : EA_1B_1).
\end{aligned}$$

Thus, chain rule for the mutual information (5.108) implies that

$$\begin{aligned}
&I_{\rho^{ABE}}(A_1A_2B_1B_2 : E) = I_{\rho^{ABE}}(A_1B_1 : E) + I_{\rho^{ABE}}(A_2B_2 : E|A_1B_1) \\
=&I_{\rho^{ABE}}(A_1B_1 : E) + I_{\rho^{ABE}}(A_2B_2 : EA_1B_1).
\end{aligned}$$

Hence, we obtain (8.362).

Exercise 8.61

(**a**) Use Exercise 5.43.
(**b**) It is sufficient to show $C(\rho, 0) \leq \tilde{C}(\rho, 0)$ because the opposite inequality is oblivious. Choose an extension ρ^{ABE} of ρ satisfies $I_{\rho^{ABE}}(A : B|E) = 0$. Then, $(\kappa_M \otimes \iota_{AB})(\rho^{ABE})$ is an extension ρ^{ABE} of ρ satisfying the conditions for $C(\rho, 0)$. Exercise 5.42 implies that $I_{(\kappa_M \otimes \iota_{AB})(\rho^{ABE})}(AB : E) \leq I_{\rho^{ABE}}(AB : E)$. Hence, $C(\rho, 0) \leq \tilde{C}(\rho, 0)$.

Exercise 8.62 Choose the unitary U as the unitary matrix transforming every base of the first basis to every base of the second basis.

Exercise 8.63 $\tau' \circ \kappa \circ \tau$ is a CP map if and only if the following holds for any integer n: The inequality $\mathrm{Tr}\, \sigma \tau' \circ \kappa \circ \tau(\rho) \geq 0$ holds for any states ρ and σ on $\mathcal{H} \otimes \mathbb{C}^n$ and $\mathcal{H}' \otimes \mathbb{C}^n$. Let τ^n denote the transpose on $\mathbb{C}^n$. Then, since τ^n commutes κ,

$$\mathrm{Tr}(\tau' \otimes \tau^n)(\sigma)\tau^n \circ \kappa \circ \tau^n((\tau \otimes \tau^n)(\rho)) = \mathrm{Tr}(\tau' \otimes \tau^n)(\sigma)\kappa(\tau \otimes \tau^n)(\rho)).$$

Since $(\tau \otimes \tau^n)(\rho)$ and $(\tau' \otimes \tau^n)(\sigma)$ are states on on $\mathcal{H} \otimes \mathbb{C}^n$ and $\mathcal{H}' \otimes \mathbb{C}^n$, we obtain the desired equivalence.

The second argument can be shown by (A.18) as follows.

$$\|X\|_1 = \mathrm{Tr}_{U:\text{unitary}} \mathrm{Tr}\, UX = \mathrm{Tr}_{U:\text{unitary}} \mathrm{Tr}\, \tau(U)\tau(X) = \|\tau(X)\|_1.$$

Exercise 8.64 Due to Exercise 8.63, $\tau^{A'} \circ \kappa \circ \tau^A$ is TP-CP if and only if $(\tau^{A'} \otimes \tau^{B'}) \circ \tau^{B'} \circ \kappa \circ \tau^B \circ (\tau^{A'} \otimes \tau^{B'})$ is TP-CP. Since $(\tau^{A'} \otimes \tau^{B'}) \circ \tau^{B'} \circ \kappa \circ \tau^B \circ (\tau^{A'} \otimes \tau^{B'}) = \tau^{B'} \circ \kappa \circ \tau^B$, we obtain the desired equivalence.

Exercise 8.65 Due to Exercise 8.64, the completely positivity of $\tau^{A'} \circ \kappa \circ \tau^A$ is equivalent with that of $\tau^{B'} \circ \kappa \circ \tau^B$. Since this equivalence does not depend on the choice of the bases on $\mathcal{H}_A$ and $\mathcal{H}_{A'}$, we obtain the desired equivalence.

Exercise 8.66 The second argument of Exercise 8.63 implies that

$$\|\tau^A(\rho)\|_1 = \|(\tau^A \otimes \tau^B)(\tau^A(\rho))\|_1 = \|\tau^B(\rho)\|_1.$$

Since the above equation holds for any basis on $\mathcal{H}_A$, we obtain the desired equation.

Exercise 8.67 Equation (8.239) can be shown by the same way as Exercise 8.50 by replacing the role of (8.84) at the proof of Theorem 8.7 by that of (8.233) at the proof of (8.223).

Exercise 8.68

(a)

$$\mathrm{Tr}\, \sigma\rho = \mathrm{Tr}\, \tau^A(\sigma)\tau^A(\rho) \le \|\tau^A(\sigma)\| \|\tau^A(\rho)\|_1.$$

(b) Since $\|\tau^A(\kappa(\sigma))\|_1 = \mathrm{Tr}\, \tau^A(\kappa(\sigma)) = 1$, using (8.241) and the second equation of (8.219), we have

$$\max_{\sigma \in S_{\mathrm{PPT}}} \mathrm{Tr}\, |\Phi_d\rangle\langle\Phi_d|\kappa(\sigma) \le \max_{\sigma \in S_{\mathrm{PPT}}} \|\tau^A(|\Phi_d\rangle\langle\Phi_d|)\| \|\tau^A(\kappa(\sigma))\|_1 \le \frac{1}{d}.$$

(c) Since $\tau^A(\sigma) \ge 0$ and $\tau^A(|u\rangle\langle u|) \le |\tau^A(|u\rangle\langle u|)|$, we have

$$\begin{aligned}
\max_{\sigma \in S_{\mathrm{PPT}}} \mathrm{Tr}\, |x\rangle\langle x|\sigma &= \max_{\sigma \in S_{\mathrm{PPT}}} \mathrm{Tr}\, \tau^A(|x\rangle\langle x|)\tau^A(\sigma) \\
&\le \max_{\sigma \in S_{\mathrm{PPT}}} \mathrm{Tr}\, |\tau^A(|x\rangle\langle x|)|\tau^A(\sigma) \\
&\overset{(a)}{=} \max_{\sigma \in S_{\mathrm{PPT}}} \mathrm{Tr} \left(\sum_i \sqrt{\lambda_i} |u_i\rangle\langle u_i| \right)^{\otimes 2} \tau^A(\sigma) = \lambda_1^{\downarrow},
\end{aligned}$$

where (a) follows from (8.217). The equality follows from the choice of σ given in **(b)** of Exercise 8.37.

(d) We have

$$\max_{\kappa\in \mathrm{PPT}} \{\log L(\kappa) | \varepsilon_2(\kappa,\rho)=0\} = \max_{\kappa\in \mathrm{PPT}} \{\log d | \operatorname{Tr} \kappa(|x\rangle\langle x|)|\Phi_d\rangle\langle\Phi_d| = 1\}$$

$$\stackrel{(a)}{\le} \max_{\kappa\in S_{\mathrm{PPT}}} \{ \min_{\sigma\in S_{\mathrm{PPT}}} (\operatorname{Tr} |\Phi_d\rangle\langle\Phi_d|\kappa(\sigma))^{-1} | \operatorname{Tr} \kappa(|x\rangle\langle x|)|\Phi_d\rangle\langle\Phi_d| = 1\}$$

$$\stackrel{(b)}{\le} \min_{\sigma\in S_{\mathrm{PPT}}} (\operatorname{Tr} |x\rangle\langle x|\sigma)^{-1} \stackrel{(c)}{=} (\lambda_1^{\downarrow})^{-1},$$

where (a), (b) and (c) follow from (8.242), the same discussion as (8.96), and (8.243), respectively. The PPT operation achieving the rate $-\log \lambda_1^{\downarrow}$ is given in Exercise 8.31.

Exercise 8.69

(**a**) Since $\operatorname{Tr}\{e^{-n(H(\rho)+\epsilon)} \le \rho^{\otimes n} \le e^{-n(H(\rho)-\epsilon)}\}\rho^{\otimes n} \to 1$, we have

$$\|\rho^{\otimes n} - \rho_n\|_1 = \operatorname{Tr} \rho^{\otimes n} \left(I - \frac{\{e^{-n(H(\rho)+\epsilon)} \le \rho^{\otimes n} \le e^{-n(H(\rho)-\epsilon)}\}}{\operatorname{Tr}\{e^{-n(H(\rho)+\epsilon)} \le \rho^{\otimes n} \le e^{-n(H(\rho)-\epsilon)}\}\rho^{\otimes n}} \right)$$

$$= \operatorname{Tr} \rho^{\otimes n} \Bigg((I - \{e^{-n(H(\rho)+\epsilon)} \le \rho^{\otimes n} \le e^{-n(H(\rho)-\epsilon)}\})$$

$$+ \left(1 - \frac{1}{\operatorname{Tr}\{e^{-n(H(\rho)+\epsilon)} \le \rho^{\otimes n} \le e^{-n(H(\rho)-\epsilon)}\}\rho^{\otimes n}}\right)$$

$$\cdot \{e^{-n(H(\rho)+\epsilon)} \le \rho^{\otimes n} \le e^{-n(H(\rho)-\epsilon)}\} \Bigg)$$

$$\to 0.$$

(**b**) All of eigenvalues of $\operatorname{Tr}\{e^{-n(H(\rho)+\epsilon)} \le \rho^{\otimes n} \le e^{-n(H(\rho)-\epsilon)}\}\rho^{\otimes n}\rho_n$ belong to the interval $[e^{-n(H(\rho)+\epsilon)}, e^{-n(H(\rho)-\epsilon)}]$. So, for sufficiently large n, all of eigenvalues of ρ_n belong to the interval $[e^{-n(H(\rho)+\epsilon)}, e^{-n(H(\rho)-2\epsilon)}]$. Hence, Exercise 2.27 guarantees the desired inequality.
(**c**) Choose purifications x_n, y_n of $\rho^{\otimes n}$, ρ_n such that $F(|x_n\rangle\langle x_n|, |y_n\rangle\langle y_n|) = F(\rho^{\otimes n}, \rho_n)$. Hence, (**a**) and (3.52) guarantee that $d_1(|x_n\rangle\langle x_n|, |y_n\rangle\langle y_n|) \to 0$. Thus, we find that the purifications x_n, y_n of $\rho^{\otimes n}$, ρ_n give a counterexample of the continuity of $2\log \|\tau^A(|x\rangle\langle x|)\|_1$. However, (8.218) and (**a**) imply that

$$\frac{-2}{n} \log \|\tau^A(|x_n\rangle\langle x_n|)\|_1 = H_{\frac{1}{2}}(\rho)$$
$$\frac{-2}{n} \log \|\tau^A(|y_n\rangle\langle y_n|)\|_1 = \frac{1}{n} H_{\frac{1}{2}}(\rho_n) \le H(\rho) + \epsilon.$$

Exercise 8.70 Since $\begin{pmatrix} C & B \\ -B^* & A \end{pmatrix} = \begin{pmatrix} A+C & 0 \\ 0 & A+C \end{pmatrix} - \begin{pmatrix} A & B \\ B^* & C \end{pmatrix} \ge 0$, we have $\begin{pmatrix} A & B^* \\ B & C \end{pmatrix} = \begin{pmatrix} 0 & I \\ -I & 0 \end{pmatrix} \begin{pmatrix} C & B \\ -B^* & A \end{pmatrix} \begin{pmatrix} 0 & I \\ -I & 0 \end{pmatrix} \ge 0.$

Exercise 8.71 Under the correspondence (8.229), we have

$$D_{1+s}(\rho\|\sigma) + \log \|\tau^A(\sigma)\|_1 = D_{1+s}(\rho\|\sigma').$$

Hence, similar to (8.231), we have (8.246). Similarly, we can show (8.247).

Exercise 8.72 To show the inequality (8.249), we consider the state $\lambda\rho_1|0,0\rangle\langle 0,0| + (1-\lambda)\rho_2|1,1\rangle\langle 1,1|$. Choose $\sigma_i' := \mathrm{argmin}_{\sigma'\geq 0:\|\tau^A(\sigma')\|_1=1} D_{1+s}(\rho\|\sigma')$. Applying the monotonicity (**a**) of Exercise 5.25 to the partial trace, we have

$$\begin{aligned}
& e^{sE_{1+s|\mathrm{SDP}}(\lambda\rho_1+(1-\lambda)\rho_2)} \leq e^{sD_{1+s}(\lambda\rho_1+(1-\lambda)\rho_2\|\lambda\sigma_1'+(1-\lambda)\sigma_2')} \\
\leq & e^{sD_{1+s}(\lambda\rho_1\otimes|0,0\rangle\langle 0,0|+(1-\lambda)\rho_2\otimes|1,1\rangle\langle 1,1|\|\lambda\sigma_1'\otimes|0,0\rangle\langle 0,0|+(1-\lambda)\sigma_2'\otimes|1,1\rangle\langle 1,1|)} \\
= & \lambda e^{sD_{1+s}(\rho_1\|\sigma_1')} + (1-\lambda)e^{sD_{1+s}(\rho_2\|\sigma_2')} \\
= & \lambda e^{sE_{1+s|\mathrm{SDP}}(\rho_1)} + (1-\lambda)e^{sE_{1+s|\mathrm{SDP}}(\rho_2)}.
\end{aligned}$$

Other inequalities can be shown in the same way.

Exercise 8.73 Due to (8.233), when a separable operation κ_n satisfies that $L(\kappa_n) = e^{nr}$, any state σ satisfies that

$$\langle \Phi_{e^{nr}}|\kappa_n'(\rho^{\otimes n})|\Phi_{e^{nr}}\rangle \leq e^{-ns\frac{-D_{1+s}(\rho\|\sigma)-\log\|\tau^A(\sigma)\|_1+r}{1+s}} \tag{8.363}$$

for $s > 0$. Taking the minimum for σ, we have

$$\langle \Phi_{e^{nr}}|\kappa_n'(\rho^{\otimes n})|\Phi_{e^{nr}}\rangle \leq e^{-n\frac{-E_{1+s|\mathrm{SDP}}(\rho)+sr}{1+s}}, \tag{8.364}$$

which implies (8.252). Since the same discussion holds with $\underline{D}_{1+s}(\rho\|\sigma)$, we obtain (8.253).

Exercise 8.74

(**a**) When $\theta = \tilde{d}(x, y)$, Exercise 3.18 guarantees that

$$\frac{\||x\rangle\langle x| - |y\rangle\langle y|\|_1}{\tilde{d}(x,y)} = \frac{2\sqrt{1-|\langle x|y\rangle|^2}}{\tilde{d}(x,y)} = \frac{2\sqrt{1-\cos^2\theta}}{\theta} = \frac{2\sin\theta}{\theta} \overset{(a)}{\leq} 2,$$

where (a) follows from $\frac{\sin\theta}{\theta} \leq 1$.
(**b**) (8.278) and triangle inequality imply that

$$\begin{aligned}
& \frac{|\|\operatorname{Tr}_{\mathbb{C}^n}|x\rangle\langle x| - \rho_{\mathrm{mix},A}\|_2^2 - \|\operatorname{Tr}_{\mathbb{C}^n}|y\rangle\langle y| - \rho_{\mathrm{mix},A}\|_2^2|}{\tilde{d}(x,y)} \\
= & \frac{\operatorname{Tr}_A(\operatorname{Tr}_B|x\rangle\langle x|)^2 - 2\operatorname{Tr}_A(\operatorname{Tr}_B|x\rangle\langle x|)\rho_{\mathrm{mix},A}}{\tilde{d}(x,y)} \\
+ & \frac{-\operatorname{Tr}_A(\operatorname{Tr}_B|y\rangle\langle y|)^2 + 2\operatorname{Tr}_A(\operatorname{Tr}_B|y\rangle\langle y|)\rho_{\mathrm{mix},A}}{\tilde{d}(x,y)}
\end{aligned}$$

$$
\begin{aligned}
&= \frac{\mathrm{Tr}_A[(\mathrm{Tr}_B |x\rangle\langle x|) - (\mathrm{Tr}_B |y\rangle\langle y|)][(\mathrm{Tr}_B |x\rangle\langle x|) + (\mathrm{Tr}_B |y\rangle\langle y|) - 2\rho_{\mathrm{mix},A}]}{\tilde{d}(x,y)} \\
&\le \frac{\| \mathrm{Tr}_B |x\rangle\langle x| - \mathrm{Tr}_B |y\rangle\langle y| \|_1 \| \mathrm{Tr}_B |x\rangle\langle x|) + (\mathrm{Tr}_B |y\rangle\langle y|) - 2\rho_{\mathrm{mix},A} \|}{\tilde{d}(x,y)} \\
&\le \frac{2\| |x\rangle\langle x| - |y\rangle\langle y| \|_1}{\tilde{d}(x,y)} \overset{(a)}{\le} 4,
\end{aligned}
$$

where (a) follows from (**a**).

Exercise 8.75

(**a**) Consider the matrix $A = \begin{pmatrix} a & b \\ c & d \end{pmatrix}$. Then, we have

$$
\begin{aligned}
A^T S_2 A &= \begin{pmatrix} a & c \\ b & d \end{pmatrix} \begin{pmatrix} 0 & -i \\ i & 0 \end{pmatrix} \begin{pmatrix} a & b \\ c & d \end{pmatrix} \\
&= \begin{pmatrix} 0 & -i(ad-bc) \\ i(ad-bc) & 0 \end{pmatrix} = S_2 \det A.
\end{aligned}
$$

(**b**) We have

$$
\begin{aligned}
& (A \otimes B)\rho(A \otimes B)^*(S_2 \otimes S_2)\overline{(A \otimes B)\rho(A \otimes B)^*}(S_2 \otimes S_2) \\
=& (A \otimes B)\rho(\bar{A}^T \otimes \bar{B}^T)(S_2 \otimes S_2)(\bar{A} \otimes \bar{B})\bar{\rho}(A^T \otimes B^T)(S_2 \otimes S_2) \\
=& (\det \bar{A})(\det \bar{B})(A \otimes B)\rho(S_2 \otimes S_2)\bar{\rho}(A^T \otimes B^T)(S_2 \otimes S_2)(A \otimes B)(A \otimes B)^{-1} \\
=& (\det \bar{A})(\det \bar{B})(\det A)(\det B)(A \otimes B)\rho(S_2 \otimes S_2)(S_2 \otimes S_2)(A \otimes B)^{-1} \\
=& |\det A|^2 |\det B|^2 (A \otimes B)\rho(S_2 \otimes S_2)\bar{\rho}(S_2 \otimes S_2)(A \otimes B)^{-1}.
\end{aligned}
$$

(**c**) The eigenvalues of $(A \otimes B)\rho(S_2 \otimes S_2)\bar{\rho}(S_2 \otimes S_2)(A \otimes B)^{-1}$ is the same as those of $\rho(S_2 \otimes S_2)\bar{\rho}(S_2 \otimes S_2)$. Hence, due to (**b**), the eigenvalues of $(A \otimes B)\rho(A \otimes B)^*(S_2 \otimes S_2)\overline{(A \otimes B)\rho(A \otimes B)^*}(S_2 \otimes S_2)$ is the same as those of $(S_2 \otimes S_2)\overline{(|\det A||\det B|\rho)}(S_2 \otimes S_2)$. Thus, $C_o((A \otimes B)\rho(A \otimes B)^*) = C_o(|\det A||\det B|\rho)$. Since $C_o(c\rho) = cC_o(\rho)$ for any constant $c > 0$, we obtain the desired argument.
(**d**) Substituting A_ω and B_ω into A and B, we obtain the desired argument.

Exercise 8.76

(**a**) The definitions of $|e_i^{AB}\rangle$ and $|u_{i,j}^{A,B}\rangle$ in (1.20) and Exercise 8.34 imply that

$$
\begin{aligned}
|e_0^{AB}\rangle\langle e_0^{AB}| &= |u_{0,0}^{A,B}\rangle\langle u_{0,0}^{A,B}|, \quad |e_1^{AB}\rangle\langle e_1^{AB}| = |u_{1,0}^{A,B}\rangle\langle u_{1,0}^{A,B}| \\
|e_2^{AB}\rangle\langle e_2^{AB}| &= |u_{1,1}^{A,B}\rangle\langle u_{1,1}^{A,B}|, \quad |e_3^{AB}\rangle\langle e_3^{AB}| = |u_{0,1}^{A,B}\rangle\langle u_{0,1}^{A,B}|.
\end{aligned}
$$

Since all of entries of $|u_{i,j}^{A,B}\rangle$ are real numbers, $\overline{|e_i^{AB}\rangle\langle e_i^{AB}|} = |e_i^{AB}\rangle\langle e_i^{AB}|$, which implies that $\overline{\rho_{\mathrm{Bell},p}} = \rho_{\mathrm{Bell},p}$.

(**b**) It is sufficient to show $(S_2 \otimes S_2)|e_i^{AB}\rangle = |e_i^{AB}\rangle$. Due to the definition (1.20) of $|e_i^{AB}\rangle$, the relation follows from the relation $S_2 S_i S_2^T = S_i$.
(**c**) The statement (**b**) implies that

$$\rho_{\mathrm{Bell},p}(S_2 \otimes S_2)\overline{\rho_{\mathrm{Bell},p}}(S_2 \otimes S_2) = \rho_{\mathrm{Bell},p}^2.$$

So, the eigenvalues of $\rho_{\mathrm{Bell},p}(S_2 \otimes S_2)\overline{\rho_{\mathrm{Bell},p}}(S_2 \otimes S_2)$ are p_0^2, p_1^2, p_2^2, and p_3^2. That is, the square roots are p_0, p_1, p_2, and p_3. Hence, $C_o(\rho_{\mathrm{Bell},p})$ is $(2\max_i p_i) - 1$.

Exercise 8.77

(**a**) It is sufficient to show that $\mathrm{Tr}(A \otimes B)|e_i^{AB}\rangle\langle e_i^{AB}|(A \otimes B)^* = \frac{1}{2}\mathrm{Tr}\, A^* A S_i^T B^T \bar{B} S_i^T$. We have

$$\begin{aligned}
&\mathrm{Tr}(A \otimes B)|e_i^{AB}\rangle\langle e_i^{AB}|(A \otimes B)^* = \langle e_i^{AB}|(A \otimes B)^*(A \otimes B)|e_i^{AB}\rangle \\
=&\langle e_0^{AB}|(S_i \otimes I)^*(A \otimes B)^*(A \otimes B)(S_i \otimes I)|e_0^{AB}\rangle \\
=&\langle e_0^{AB}|(AS_i \otimes B)^*(AS_i \otimes B)|e_0^{AB}\rangle = \langle e_0^{AB}|(AS_iB^T \otimes I)^*(AS_iB^T \otimes I)|e_0^{AB}\rangle \\
=&\frac{1}{2}\mathrm{Tr}(AS_iB^T)^* AS_iB^T = \frac{1}{2}\mathrm{Tr}\, A^*AS_iB^T\bar{B}S_i.
\end{aligned}$$

(**b**) For any positive semidefinite matrix C, we have $\frac{1}{2}\mathrm{Tr}\, C \geq \sqrt{\det C}$. Hence, $\frac{1}{2}\mathrm{Tr}\, A^*AS_iB^T\bar{B}S_i = \frac{1}{2}\mathrm{Tr}\, AS_iB^T\bar{B}S_i^T A^* \geq \sqrt{\det AS_iB^T\bar{B}S_i^T A^*} = |\det A||\det B|$.
(**c**) The statements (**a**) and (**b**) imply that $\frac{|\det A_\omega||\det B_\omega|}{\mathrm{Tr}(A_\omega \otimes B_\omega)\rho_{\mathrm{Bell},p}(A_\omega \otimes B_\omega)^*} \leq 1$. Hence, (8.318) yields (8.320).

Exercise 8.78 From the definition (8.141), any maximally correlated state ρ can be written as

$$\begin{aligned}
\rho =& a|u_0, u_0\rangle\langle u_0, u_0| + t|u_0, u_0\rangle\langle u_1, u_1| \\
&+ \bar{t}|u_1, u_1\rangle\langle u_0, u_0| + (1-a)|u_1, u_1\rangle\langle u_1, u_1|
\end{aligned}$$

with $a \geq 0$ and $t \in \mathbb{C}$. Assume that $t = be^{i\theta}$ with $b \geq 0$. We choose the new basis $|0\rangle := |u_0\rangle$ and $|1\rangle = e^{-i\theta}|u_1\rangle$. Then, $\rho = \rho_{a,b}$.

Exercise 8.79

(**a**) Since $(S_2 \otimes S_2)|00\rangle = -|11\rangle$ and $(S_2 \otimes S_2)|11\rangle = -|00\rangle$, we have $(S_2 \otimes S_2)\overline{\rho_{a,b}}(S_2 \otimes S_2) = (S_2 \otimes S_2)\rho_{a,b}(S_2 \otimes S_2) = \rho_{1-a,b}$.
(**b**) Since $\begin{pmatrix} a & b \\ b & 1-a \end{pmatrix}\begin{pmatrix} 1-a & b \\ b & a \end{pmatrix} = \begin{pmatrix} a(1-a)+b^2 & 2ab \\ 2(1-a)b & a(1-a)+b^2 \end{pmatrix}$, the statement (**a**) implies that

$$\begin{aligned}
&\rho_{a,b}(S_2 \otimes S_2)\overline{\rho_{a,b}}(S_2 \otimes S_2) = \rho_{a,b}\rho_{1-a,b} \\
=&(a(1-a)+b^2)|00\rangle\langle 00| + 2ab|00\rangle\langle 11| \\
&+ 2(1-a)b|11\rangle\langle 00| + (a(1-a)+b^2)|11\rangle\langle 11|.
\end{aligned}$$

(**c**) The eigenvalues of $\begin{pmatrix} a(1-a)+b^2 & 2ab \\ 2(1-a)b & a(1-a)+b^2 \end{pmatrix}$ are $a(1-a)+b^2 \pm 2b\sqrt{a(1-a)} = (\sqrt{a(1-a)} \pm b)^2$ because $\sqrt{a(1-a)} \geq b$. Thus,

$$C_o(\rho_{a,b}) = \sqrt{a(1-a)} + b - (\sqrt{a(1-a)} - b) = 2b.$$

Exercise 8.80 The eigenvalues of $\rho_{a,b}$ are $\frac{1\pm\sqrt{(2a-1)^2+4b^2}}{2}$. Hence, we have $H(\rho_{a,b}) = h\left(\frac{1+\sqrt{(2a-1)^2+4b^2}}{2}\right)$.

Exercise 8.81 The relation (5.12) implies that

$$\begin{aligned}
&(\kappa_{d,\lambda} \otimes \iota_R)(|\Phi_d\rangle\langle\Phi_d|) = \lambda|\Phi_d\rangle\langle\Phi_d| + (1-\lambda)\rho_{\text{mix},A} \otimes \operatorname{Tr}_A |\Phi_d\rangle\langle\Phi_d| \\
=&\lambda|\Phi_d\rangle\langle\Phi_d| + (1-\lambda)\rho_{\text{mix},A} \otimes \rho_{\text{mix},R} = \rho_{I,\frac{1-\lambda(d^2-1)}{d^2}}.
\end{aligned}$$

Similarly, (5.18) and the first equation in (8.219) imply that

$$\begin{aligned}
&(\kappa^T_{d,\lambda} \otimes \iota_R)(|\Phi_d\rangle\langle\Phi_d|) = \lambda\tau_A(|\Phi_d\rangle\langle\Phi_d|) + (1-\lambda)\rho_{\text{mix},A} \otimes \rho_{\text{mix},R} \\
=&\frac{\lambda}{d}F + \frac{1-\lambda}{d^2}I = \rho_{W,\frac{(1-(d+1)\lambda)(d-1)}{2d}}.
\end{aligned}$$

Exercise 8.82

(**a**) The first equation in (8.219) implies that

$$qI + r\tau^A(|\Phi_d\rangle\langle\Phi_d|) = qI + \frac{r}{d}F = \rho_{W,p}.$$

(**b**) We have

$$\begin{aligned}
&\tau^A(\rho_{W,p}) = qI + r\tau^A(\tau^A(|\Phi_d\rangle\langle\Phi_d|)) = qI + r|\Phi_d\rangle\langle\Phi_d| \\
=&q(I - |\Phi_d\rangle\langle\Phi_d|) + (q+r)|\Phi_d\rangle\langle\Phi_d|.
\end{aligned}$$

(**c**) Since $q \geq 0$, the statement (**b**) implies that $|\tau^A(\rho_{W,p})| = q(I - |\Phi_d\rangle\langle\Phi_d|) + |q+r||\Phi_d\rangle\langle\Phi_d|$. Thus,

$$\begin{aligned}
\tau^A(|\tau^A(\rho_{W,p})|) &= q(I - \tau^A(|\Phi_d\rangle\langle\Phi_d|)) + |q+r|\tau^A(|\Phi_d\rangle\langle\Phi_d|) \\
&= q(I - \frac{1}{d}F) + \frac{|q+r|}{d}F = qI + \frac{|q+r|-q}{d}F.
\end{aligned}$$

Since $\frac{q-|q+r|}{d} \le \frac{q-(q+r)}{d} = \frac{-r}{d} \le q$, we have $\tau^A(|\tau^A(\rho_{W,p})|) = qI + \frac{|q+r|-q}{d} F \ge 0$.

Exercise 8.83

(**a**) Equation (8.329) and the first equation in (8.219) imply that

$$\tau^A(\rho_{I,p}) = \frac{1-p}{d^2-1} I + \frac{d^2 p - 1}{(d^2-1)} \tau^A(|\Phi_d\rangle\langle\Phi_d|) = \frac{1-p}{d^2-1} I + \frac{d^2 p - 1}{d(d^2-1)} F$$
$$= \frac{1-p}{d^2-1}(I+F) + \frac{dp-1}{d(d-1)} F.$$

(**b**) Since $\frac{1-p}{d^2-1} \ge 0$ and $\frac{dp-1}{d(d-1)} \ge 0$, we have

$$\left|\frac{1-p}{d^2-1}(I+F) + \frac{dp-1}{d(d-1)} F\right| = \frac{1-p}{d^2-1}(I+F) + \frac{dp-1}{d(d-1)} |F|$$
$$= \frac{1-p}{d^2-1}(I+F) + \frac{dp-1}{d(d-1)} I.$$

Hence, combining (**a**), we obtain the desired argument.
(**c**) The statement (**b**) and the first equation in (8.219) imply that $\tau^A(|\tau^A(\rho_{I,p})|) = \frac{1-p}{d^2-1}(I+\tau^A(F)) + \frac{dp-1}{d(d-1)} I = \frac{1-p}{d^2-1}(I+d|\Phi_d\rangle\langle\Phi_d|) + \frac{dp-1}{d(d-1)} I$. The second inequality is trivial.

Exercise 8.84 The first equation in (8.219) implies that $(1-\lambda)\rho_{\mathrm{mix}} + \lambda\tau^A(|\Phi_d\rangle\langle\Phi_d|) = \frac{1-\lambda}{d^2} I + \frac{\lambda}{d} F$. $\frac{1-\lambda}{d^2} I + \frac{\lambda}{d} F \ge 0$ if and only if $\frac{1-\lambda}{d^2} \ge |\frac{\lambda}{d}|$, which is equivalent to $-\frac{1}{d-1} \le \lambda \le \frac{1}{d+1}$.

References

1. V. Vedral, M.B. Plenio, Entanglement measures and purification procedures. Phys. Rev. A **57**, 822 (1998)
2. H.-K. Lo, S. Popescu, Concentrating entanglement by local actions: beyond mean values. Phys. Rev. A **63**, 022301 (2001)
3. A. Uhlmann, The 'transition probability' in the state space of *-algebra. Rep. Math. Phys. **9**, 273–279 (1976)
4. M.A. Nielsen, I.L. Chuang, *Quantum Computation and Quantum Information* (Cambridge University Press, Cambridge, 2000)
5. B. Schumacher, Sending quantum entanglement through noisy channels. Phys. Rev. A **54**, 2614–2628 (1996)
6. H. Barnum, E. Knill, M.A. Nielsen, On quantum fidelities and channel capacities. IEEE Trans. Inf. Theory **46**, 1317–1329 (2000)
7. M. Hamada, Lower bounds on the quantum capacity and highest error exponent of general memoryless channels. IEEE Trans. Inf. Theory **48**, 2547–2557 (2002)
8. C. Adami, N.J. Cerf, On the von Neumann capacity of noisy quantum channels. Phys. Rev. A **56**, 3470 (1997)
9. B. Schumacher, M.A. Nielsen, Quantum data processing and error correction. Phys. Rev. A **54**, 2629 (1996)

10. H. Barnum, M.A. Nielsen, B. Schumacher, Information transmission through a noisy quantum channel. Phys. Rev. A **57**, 4153–4175 (1997)
11. A.S. Holevo, On entanglement-assisted classical capacity. J. Math. Phys. **43**, 4326–4333 (2002)
12. R. Bhatia, *Matrix Analysis* (Springer, Berlin, 1997)
13. M.A. Nielsen, J. Kempe, Separable states are more disordered globally than locally. Phys. Rev. Lett. **86**, 5184–5187 (2001)
14. N.J. Cerf, C. Adami, Negative entropy and information in quantum mechanics. Phys. Rev. Lett. **79**, 5194 (1997)
15. M.A. Nielsen, Conditions for a class of entanglement transformations. Phys. Rev. Lett. **83**, 436 (1999)
16. G. Vidal, Entanglement of pure states for a single copy. Phys. Rev. Lett. **83**, 1046–1049 (1999)
17. G. Vidal, D. Jonathan, M.A. Nielsen, Approximate transformations and robust manipulation of bipartite pure state entanglement. Phys. Rev. A **62**, 012304 (2000)
18. T. Yamamoto, M. Koashi, Ş. Özdemir, N. Imoto, Experimental extraction of an entangled photon pair from two identically decohered pairs. Nature **421**, 343–346 (2003)
19. J.-W. Pan, S. Gasparoni, R. Ursin, G. Weihs, A. Zeilinger, Experimental entanglement purification of arbitrary unknown states. Nature **423**, 417–422 (2003)
20. X. Wang, H. Fan, Non-post-selection entanglement concentration by ordinary linear optical devices. Phys. Rev. A **68**, 060302(R) (2003)
21. M. Hayashi, General formulas for fixed-length quantum entanglement concentration. IEEE Trans. Inf. Theory **52**, 1904–1921 (2006)
22. C.H. Bennett, H.J. Bernstein, S. Popescu, B. Schumacher, Concentrating partial entanglement by local operations. Phys. Rev. A **53**, 2046 (1996)
23. M. Hayashi, K. Matsumoto, Variable length universal entanglement concentration by local operations and its application to teleportation and dense coding. quant-ph/0109028 (2001); K. Matsumoto, M. Hayashi, Universal entanglement concentration. Phys. Rev. A **75**, 062338 (2007)
24. M. Horodecki, P. Horodecki, R. Horodecki, General teleportation channel, singlet fraction and quasi-distillation. Phys. Rev. A **60**, 1888 (1999)
25. M. Hayashi, M. Koashi, K. Matsumoto, F. Morikoshi A. Winter, Error exponents for entangle concentration. J. Phys. A Math. Gen. **36**, 527–553 (2003)
26. F. Morikoshi, M. Koashi, Deterministic entanglement concentration. Phys. Rev. A **64**, 022316 (2001)
27. F. Morikoshi, Recovery of entanglement lost in entanglement manipulation. Phys. Rev. Lett. **84**, 3189 (2000)
28. C.H. Bennett, D.P. DiVincenzo, J.A. Smolin, W.K. Wootters, Mixed state entanglement and quantum error correction. Phys. Rev. A **54**, 3824–3851 (1996)
29. P.M. Hayden, M. Horodecki, B.M. Terhal, The asymptotic entanglement cost of preparing a quantum state. J. Phys. A Math. Gen. **34**, 6891–6898 (2001)
30. B.M. Terhal, P. Horodecki, A Schmidt number for density matrices. Phys. Rev. A **61**, 040301(R) (2000)
31. G. Vidal, W. Dür, J.I. Cirac, Entanglement cost of antisymmetric states. quant-ph/0112131v1 (2001)
32. M. Donald, M. Horodecki, O. Rudolph, The uniqueness theorem for entanglement measures. J. Math. Phys. **43**, 4252–4272 (2002)
33. M.J. Donald, M. Horodecki, Continuity of relative entropy of entanglement. Phys. Lett. A **264**, 257–260 (1999)
34. M. Christandl, A. Winter, Squashed entanglement-an additive entanglement measure. J. Math. Phys. **45**, 829–840 (2004)
35. M.A. Nielsen, Continuity bounds for entanglement. Phys. Rev. A **61**, 064301 (2000)
36. K. Matsumoto, T. Shimono, A. Winter, Remarks on additivity of the Holevo channel capacity and of the entanglement of formation. Commun. Math. Phys. **246**(3), 427–442 (2004)
37. K.G.H. Vollbrecht, R.F. Werner, Entanglement measures under symmetry. Phys. Rev. A **64**, 062307 (2001)

38. T. Hiroshima, M. Hayashi, Finding a maximally correlated state-simultaneous Schmidt decomposition of bipartite pure states. Phys. Rev. A **70**, 030302(R) (2004)
39. H.-K. Lo, S. Popescu, Classical communication cost of entanglement manipulation: is entanglement an interconvertible resource? Phys. Rev. Lett. **83**, 1459 (1999)
40. B.M. Terhal, M. Horodecki, D.W. Leung, D.P. DiVincenzo, The entanglement of purification. J. Math. Phys. **43**, 4286 (2002)
41. L. Henderson, V. Vedral, Classical, quantum and total correlations. J. Phys. A Math. Gen. **34**, 6899 (2001)
42. M. Koashi, A. Winter, Monogamy of quantum entanglement and other correlations. Phys. Rev. A **69**, 022309 (2004)
43. M. Christandl, A. Winter, Uncertainty, monogamy, and locking of quantum correlations. IEEE Trans. Inf. Theory **51**, 3159–3165 (2005)
44. A. Winter, Secret, public and quantum correlation cost of triples of random variables, in *Proceedings 2005 IEEE International Symposium on Information Theory* (2005), p. 2270
45. E.M. Rains, A semidefinite program for distillable entanglement. IEEE Trans. Inf. Theory **47**, 2921–2933 (2001)
46. J.I. Cirac, W. Dür, B. Kraus, M. Lewenstein, Entangling operations and their implementation using a small amount of entanglement. Phys. Rev. Lett. **86**, 544 (2001)
47. K. Audenaert, M.B. Plenio, J. Eisert, Entanglement cost under positive-partial-transpose-preserving operations. Phys. Rev. Lett. **90**, 027901 (2003)
48. S. Ishizaka, Binegativity and geometry of entangled states in two states. Phys. Rev. A **69**, 020301(R) (2004)
49. M. Horodecki, P. Horodecki, R. Horodecki, Mixed-state entanglement and quantum communication, in *Quantum Information: An Introduction to Basic Theoretical Concepts and Experiments* (Springer Tracts in Modern Physics, 173), G. Alber, T. Beth, M. Horodecki, P. Horodecki, R. Horodecki, M. Rotteler, H. Weinfurter, R. Werner, A. Zeilinger (eds.), (Springer, Berlin Heidelberg New York, 2001)
50. M. Horodecki, P. Horodecki, R. Horodecki, Mixed-state entanglement and distillation: is there a “bound” entanglement in nature? Phys. Rev. Lett. **80**, 5239 (1998)
51. M. Horodecki, P. Horodecki, Reduction criterion of separability and limits for a class of distillation protocols. Phys. Rev. A **59**, 4206 (1999)
52. T. Hiroshima, Majorization criterion for distillability of a bipartite quantum state. Phys. Rev. Lett. **91**, 057902 (2003)
53. P. Horodecki, Separability criterion and inseparable mixed states with positive partial transposition. Phys. Lett. A **232**, 333 (1997)
54. W. Dür, J.I. Cirac, M. Lewenstein, D. Bruß, Distillability and partial transposition in bipartite systems. Phys. Rev. A **61**, 062313 (2000)
55. N.J. Cerf, C. Adami, R.M. Gingrich, Reduction criterion for separability. Phys. Rev. A **60**, 898 (1999)
56. D. Yang, M. Horodecki, R. Horodecki, B. Synak-Radtke, Irreversibility for all bound entangled state. Phys. Rev. Lett. **95**, 190501 (2005)
57. L. Masanes, All entangled states are useful for information processing. Phys. Rev. Lett. **96**, 150501 (2006)
58. S. Ishizaka, T. Hiroshima, Maximally entangled mixed states under nonlocal unitary operations in two qubits. Phys. Rev. A **62**, 022310 (2000)
59. S. Lloyd, The capacity of the noisy quantum channel. Phys. Rev. A **56**, 1613 (1997)
60. M. Fukuda, Revisiting additivity violation of quantum channels. Commun. Math. Phys. **332**, 713–728 (2014)
61. M.B. Hastings, Superadditivity of communication capacity using entangled inputs. Nat. Phys. **5**, 255 (2009)
62. M. Hayashi, Security analysis of ε-almost dual universal$_2$ hash functions: smoothing of min entropy vs. smoothing of Rényi entropy of order 2 (2013). IEEE Trans. Inf. Theory **62**, 3451–3476 (2016)

63. L. Carter, M. Wegman, Universal classes of hash functions. J. Comput. Sys. Sci. **18**, 143–154 (1979)
64. H. Krawczyk, LFSR-based hashing and authentication, in *Advances in Cryptology — CRYPTO '94*, Lecture Notes in Computer Science, vol. 839 (Springer-Verlag, 1994), pp. 129–139
65. M. Hayashi, Precise evaluation of leaked information with secure randomness extraction in the presence of quantum attacker. Commun. Math. Phys. **333**(1), 335–350 (2015)
66. R. Ahlswede, I. Csiszár, Common randomness in information theory and cryptography part 1: Secret sharing. IEEE Trans. Inform. Theory **39**, 1121–1132 (1993)
67. I. Devetak, A. Winter, Distillation of secret key and entanglement from quantum states. Proc. R. Soc. Lond. A **461**, 207–235 (2005)
68. R. Renner, Security of quantum key distribution, PhD thesis, Dipl. Phys. ETH, Switzerland, 2005; arXiv:quantph/0512258; Int. J. Quant. Inf. **6**, 1–127 (2008)
69. M. Hayashi, Exponential decreasing rate of leaked information in universal random privacy amplification. IEEE Trans. Inf. Theory **57**, 3989–4001 (2011)
70. J.M. Renes, Duality of privacy amplification against quantum adversaries and data compression with quantum side information. Proc. Roy. Soc. A **467**(2130), 1604–1623 (2011)
71. W.K. Wootters, Entanglement of formation of an arbitrary state of two qubits. Phys. Rev. Lett. **80**, 2245 (1998)
72. F. Verstraete, J. Dehaene, B. DeMorr, Local filtering operations on two qubits. Phys. Rev. A **64**, 010101(R) (2001)
73. N. Linden, S. Massar, S. Popescu, Purifying noisy entanglement requires collective measurements. Phys. Rev. Lett. **81**, 3279 (1998)
74. A. Kent, N. Linden, S. Massar, Optimal entanglement enhancement for mixed states. Phys. Rev. Lett. **83**, 2656 (1999)
75. E.M. Rains, Bound on distillable entanglement. Phys. Rev. A **60**, 179–184 (1999)
76. K. Audenaert, J. Eisert, E. Jané, M.B. Plenio, S. Virmani, B. De Moor, The asymptotic relative entropy of entanglement. Phys. Rev. Lett. **87**, 217902 (2001)
77. F. Yura, Entanglement cost of three-level antisymmetric states. J. Phys. A Math. Gen. **36**, L237–L242 (2003)
78. K. Matsumoto, F. Yura, Entanglement cost of antisymmetric states and additivity of capacity of some quantum channel. J. Phys. A: Math. Gen. **37**, L167–L171 (2004)
79. C. King, The capacity of the quantum depolarizing channel. IEEE Trans. Inf. Theory **49**, 221–229 (2003)
80. B.M. Terhal, K.G.H. Vollbrecht, Entanglement of formation for isotropic states. Phys. Rev. Lett. **85**, 2625 (2000)
81. M. Horodecki, P. Horodecki, R. Horodecki, Unified approach to quantum capacities: towards quantum noisy coding theorem. Phys. Rev. Lett. **85**, 433–436 (2000)
82. I. Devetak, The private classical capacity and quantum capacity of a quantum channel. IEEE Trans. Inf. Theory **51**, 44–55 (2005)
83. P. Hayden, A. Winter, On the communication cost of entanglement transformations. Phys. Rev. A **67**, 012326 (2003)
84. A. Harrow, H.K. Lo, A tight lower bound on the classical communication cost of entanglement dilution. IEEE Trans. Inf. Theory **50**, 319–327 (2004)
85. A.D. Wyner, The common information of two dependent random variables. IEEE Trans. Inf. Theory **21**, 163–179 (1975)
86. M. Owari, K. Matsumoto, M. Murao, Entanglement convertibility for infinite dimensional pure bipartite states. Phys. Rev. A **70**, 050301 (2004); quant-ph/0406141; Existence of incomparable pure bipartite states in infinite dimensional systems. quant-ph/0312091 (2003)
87. A. Miyake, Classification of multipartite entangled states by multidimensional determinants. Phys. Rev. A **67**, 012108 (2003)
88. S. Ishizaka, Bound entanglement provides convertibility of pure entangled states. Phys. Rev. Lett. **93**, 190501 (2004)
89. A. Datta, A condition for the nullity of quantum discord (2010). arXiv:1003.5256

90. B. Dakic, V. Vedral, C. Brukner, Necessary and sufficient condition for non-zero quantum discord. Phys. Rev. Lett. **105**, 190502 (2010)
91. T. Shimono, Additivity of entanglement of formation of two three-level-antisymmetric states. Int. J. Quant. Inf. **1**, 259–268 (2003)
92. T. Shimono, H. Fan, Numerical test of the superadditivity of entanglement of formation for four-partite qubits, in *Proceedings ERATO Conference on Quantum Information Science (EQIS)***2003**, 119–120 (2003)
93. C.H. Bennett, G. Brassard, C. Crepeau, U.M. Maurer, Generalized privacy amplification. IEEE Trans. Inform. Theory **41**, 1915–1923 (1995)
94. J. Håstad, R. Impagliazzo, L.A. Levin, M. Luby, A pseudorandom generator from any one-way function. SIAM J. Comput. **28**, 1364 (1999)
95. M. Hayashi, V.Y.F. Tan, Equivocations, exponents and second-order coding rates under various renyi information measures (2015). arXiv:1504.02536

Chapter 9
Analysis of Quantum Communication Protocols

Abstract The problems of transmitting a classical message via a quantum channel (Chap. 4) and estimating a quantum state (Chaps. 3 and 6) have a classical analog. These are not intrinsically quantum-specific problems but quantum extensions of classical problems. The difficulties of these quantum extensions are mainly caused by the non-commutativity of quantum mechanics. However, quantum information processing is not merely a non-commuting version of classical information processing. There exist many quantum protocols without any classical analog. In this context, quantum information theory covers a greater field than a noncommutative analog of classical information theory. The key to these additional effects is the advantage of using entanglement treated in Chap. 8, where we examined mainly the quantification of entanglement. In this chapter, we will introduce several quantum communication protocols that are possible only by using entanglement and are therefore classically impossible. (Some of protocols introduced in this section have classical analogs.) We also examine the transmission of quantum states (quantum error correction), communication in the presence of eavesdroppers, and several other types of communication that we could not handle in Chap. 4. As seen in this chapter, the transmission of a quantum state is closely related to communication with no information leakage to eavesdroppers. The noise in the transmission of a quantum state clearly corresponds to the eavesdropper in a quantum communication.

9.1 Quantum Teleportation

The curious properties of entangled states were first examined by Einstein et al. [1] in an attempt to show that quantum mechanics was incomplete. Recently, the entangled states have been treated in a manner rather different than when it was first introduced. For example, by regarding these states as the source of a quantum advantage, Bennett et al. [2] proposed quantum teleportation. Since this topic can be understood without any complicated mathematics, we introduce it in this section.

In quantum teleportation, an entangled state is first shared between two parties. Then, by sending a classical message from one party to the other, it is possible to transmit a quantum state without directly sending it. Let us look at this protocol in

M. Hayashi, *Quantum Information Theory*, Graduate Texts in Physics,
DOI 10.1007/978-3-662-49725-8_9

more detail. First, we prepare an entangled state $e_0^{A,B} = \frac{1}{\sqrt{2}}(u_0^A \otimes u_0^B + u_1^A \otimes u_1^B)$ on the composite system $\mathcal{H}_A \otimes \mathcal{H}_B$ composed of two qubits $\mathcal{H}_A$ and $\mathcal{H}_B$ spanned by u_0^A, u_1^A and u_0^B, u_1^B, respectively. The sender possesses a qubit $\mathcal{H}_C$ spanned by u_0^C, u_1^C as well as the quantum system $\mathcal{H}_A$. The sender sends qubit $\mathcal{H}_C$ to the receiver. The receiver possesses the other qubit $\mathcal{H}_B$. Then, we have the following theorem.

Theorem 9.1 (BBCJPW [2]) *Let the sender perform a measurement corresponding to the CONS* $e_i^{A,C} \stackrel{\text{def}}{=} (I_A \otimes S_i^C)e_0^{A,C}$ $(i = 0, 1, 2, 3)$ *on the composite system* $\mathcal{H}_A \otimes \mathcal{H}_C$ *and its result be sent to the receiver. [From (1.21), it satisfies the conditions for a PVM.] Let the receiver perform a unitary time evolution corresponding to* $\overline{S_i^B}$ *on the quantum system* $\mathcal{H}_B$. *Then, the final state on* $\mathcal{H}_B$ *is the same state as the initial state on* $\mathcal{H}_C$.

This argument holds irrespective of the initial state on $\mathcal{H}_C$ and the measurement outcome i, as proved below.

Proof Let us first consider the case where the measurement outcome is 0. Let the initial state on $\mathcal{H}_C$ be the pure state $x = \sum_i x^i u_i^C$. Then, the state on the composite system $\mathcal{H}_A \otimes \mathcal{H}_B \otimes \mathcal{H}_C$ is $\frac{1}{\sqrt{2}} \sum_{i,j,k} x^i \delta^{j,k} u_j^A \otimes u_k^B \otimes u_i^C$. Therefore, the final state on $\mathcal{H}_B$ is $\sum_k \sum_{i,j} \frac{1}{\sqrt{2}} x^i \delta^{j,k} \frac{1}{\sqrt{2}} \delta^{i,j} u_k^B = \sum_k \frac{1}{2} x^k u_k^B$, following Exercise 7.4. Normalizing this vector, we can prove that the final state on $\mathcal{H}_B$ equals $\sum_k x^k u_k^B$, which is the same state as the initial state on $\mathcal{H}_C$.

Now, consider the case in which the measurement outcome i is obtained. Since $(S_i^C)^* = S_i^C$,

$$
\begin{aligned}
&\mathrm{Tr}_{A,C}\left(|e_i^{A,C}\rangle\langle e_i^{A,C}| \otimes I_B\right) |x \otimes e_0^{A,B}\rangle\langle x \otimes e_0^{A,B}| \\
=&\mathrm{Tr}_{A,C}(\overline{S_i^C} \otimes I_{A,B}) \left(|e_0^{A,C}\rangle\langle e_0^{A,C}| \otimes I_B\right) (\overline{S_i^C} \otimes I_{A,B})|x \otimes e_0^{A,B}\rangle\langle x \otimes e_0^{A,B}| \\
=&\mathrm{Tr}_{A,C}\left(|e_0^{A,C}\rangle\langle e_0^{A,C}| \otimes I_B\right) |(\overline{S_i}x) \otimes e_0^{A,B}\rangle\langle(\overline{S_i}x) \otimes e_0^{A,B}| = \frac{1}{4}|\overline{S_i}x\rangle\langle\overline{S_i}x|.
\end{aligned}
$$

Operating $\overline{S_i}$ on $\mathcal{H}_B$ ($\overline{S_i}$ is its own inverse), the final state on $\mathcal{H}_B$ is x. ∎

It is noteworthy that this protocol has been experimentally demonstrated [3–5]. Other protocols that combine quantum teleportation with cloning have also been proposed [6].

Exercise

9.1 Show that quantum teleportation in any dimension d is possible by following the steps below. Let $\mathcal{H}_A, \mathcal{H}_B, \mathcal{H}_C$ be the spaces spanned by $u_1^A, \ldots, u_d^A, u_1^B, \ldots, u_d^B, u_1^C, \ldots, u_d^C$, respectively. Prepare an entangled state $u_{0,0}^{A,B} \stackrel{\text{def}}{=} \frac{1}{\sqrt{d}} \sum_{i=1}^d u_i^A \otimes u_i^B$ in $\mathcal{H}_A \otimes \mathcal{H}_B$. Now perform a measurement corresponding to $\{u_{i,j}^{A,C} \stackrel{\text{def}}{=} (I_A \otimes \mathbf{X}_C^i \mathbf{Z}_C^j) u_{0,0}^{A,C}\}_{i,j}$, and then an operation $\overline{\mathbf{X}_B^i \mathbf{Z}_B^j}$ depending on the measurement outcome (i, j). Show that the final state on $\mathcal{H}_B$ is the same as the initial state on $\mathcal{H}_C$. (For the definitions of $\mathbf{X}$ and $\mathbf{Z}$, see Example 5.8.)

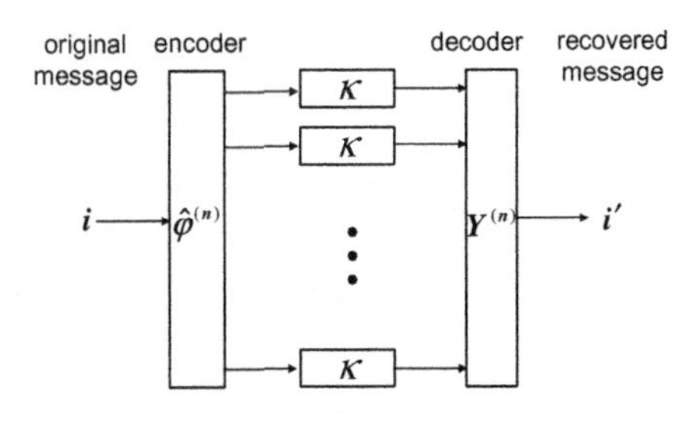

Fig. 9.1 C-Q channel coding with entangled inputs

9.2 C-Q Channel Coding with Entangled Inputs

In this section, we treat classical message transmission via a quantum channel κ from $\mathcal{H}_A$ to $\mathcal{H}_B$. When we use only tensor product states in $\mathcal{S}(\mathcal{H}_A^{\otimes n})$, the problem becomes that of classical-quantum (c-q) channel coding discussed in Chap. 4 by setting $\mathcal{X}$ to $\mathcal{S}(\mathcal{H}_A)$ and $W_\rho \stackrel{\text{def}}{=} \kappa(\rho)$. However, when we are allowed to use any state in $\mathcal{S}(\mathcal{H}_A^{\otimes n})$ as the input state, our problem cannot be regarded as a special case of Chap. 4. In this case, the optimal rate of the sending classical information is called the **classical capacity**, which is given by

$$C_c(\kappa) = \max_{p \in \mathcal{P}(\mathcal{S}(\mathcal{H}_A))} I(p, \kappa). \tag{9.1}$$

When we are allowed to send any state entangled between n systems of the input, the classical capacity is $C_c(\kappa^{\otimes n})/n$. When any entangled state is available as an input state as Fig. 9.1, the code $\hat{\Phi}^{(n)} = (N_n, \hat{\varphi}^{(n)}, Y^{(n)})$ is expressed by the triplet of the size N_n, the encoder $\hat{\varphi}^{(n)}$ mapping from $\{1, \dots, N_n\}$ to $\mathcal{S}(\mathcal{H}_A^{\otimes n})$, and the POVM $Y^{(n)} = \{Y_i^{(n)}\}_{i=1}^{N_n}$ taking values in $\{1, \dots, N_n\}$ on the output space $\mathcal{H}_B^{\otimes n}$. The error probability is given by

$$\varepsilon[\hat{\Phi}^{(n)}] \stackrel{\text{def}}{=} \frac{1}{N_n} \sum_{i=1}^{N_n} \left(1 - \operatorname{Tr} Y_i^{(n)} \kappa^{\otimes n}(\hat{\varphi}^{(n)}(i))\right).$$

Then, we can show the following theorem by using Theorem 4.1 and the discussion in the proof of Theorem 4.2[Exe. 9.3].

Theorem 9.2 *Define the* **entanglement-assisted classical capacity** $C_c^e(\kappa)$[1]*:*

$$C_c^e(\kappa) \stackrel{\text{def}}{=} \sup_{\{\hat{\Phi}^{(n)}\}} \left\{ \underline{\lim} \frac{1}{n} \log |\hat{\Phi}^{(n)}| \,\middle|\, \lim_{n\to\infty} \varepsilon[\hat{\Phi}^{(n)}] = 0 \right\} \tag{9.2}$$

for a quantum channel κ from $\mathcal{H}_A$ to $\mathcal{H}_B$. Then, we have

[1]The superscript e of C_c^e indicates that "entangled" input is allowed.

$$C_c^e(\kappa) = \sup_n \frac{C_c(\kappa^{\otimes n})}{n} = \lim_{n\to\infty} \frac{C_c(\kappa^{\otimes n})}{n}.$$

Since the inequality

$$C_c(\kappa^{\otimes n}) \geq nC_c(\kappa) \tag{9.3}$$

holds, the inequality

$$C_c^e(\kappa) \geq C_c(\kappa) \tag{9.4}$$

holds. For a past period, many people [7–12] conjectured the **additivity** of the classical capacity for two arbitrary TP-CP maps κ^1 and κ^2:

$$C_c(\kappa^1) + C_c(\kappa^2) = C_c(\kappa^1 \otimes \kappa^2), \tag{9.5}$$

which implies the equality in (9.3). Here, remember the relation (8.173). This relation indicates the relation between the classical capacity and the entanglement formation. Matsumoto et al. [13], Shor [14], and Pomeransky [15] showed the equivalence of the additivity of the classical capacity (9.5) and the additivity of the entanglement formation. However, as stated in Sect. 8.13, the additivity of the entanglement formation does not hold in general. Hence, the additivity of the classical capacity (9.5) also does not hold in general. Also, the equality in (9.3) does not hold in general [16].

Here, we see the equivalence of the additivities of the classical capacity and the entanglement formation in the detail. For this purpose, we prepare some notations. The classical capacity $C_c(\kappa)$ is described by

$$C_c(\kappa) = \max_\rho \chi_\kappa(\rho),$$

where **Holevo information** $\chi_\kappa(\rho)$ and **minimum average output entropy** $H_\kappa(\rho)$ are defined by

$$\chi_\kappa(\rho) \stackrel{\text{def}}{=} H(\kappa(\rho)) - H_\kappa(\rho), \tag{9.6}$$

$$H_\kappa(\rho) \stackrel{\text{def}}{=} \min_{(p_x,\rho_x):\sum_x p_x\rho_x=\rho} \sum_x p_x H(\kappa(\rho_x)). \tag{9.7}$$

When κ is the partial trace from the system $\mathcal{H}_A \otimes \mathcal{H}_B$ to $\mathcal{H}_B$, the relation (**MSW correspondence** [13])

$$H_\kappa(\rho) = E_f(\rho) \tag{9.8}$$

holds, i.e.,

$$\chi_{\mathrm{Tr}_A}(\rho) = H(\mathrm{Tr}_A\,\rho) - E_f(\rho). \tag{9.9}$$

Now, we state the equivalence relations for several kinds of additivity conditions.

Theorem 9.3 (Matsumoto et al. [13], Shor [14], Pomeransky [15]) *The following 14 conditions are equivalent.*

HM *Additivity of classical capacity of q-q channel (additivity of maximum Holevo information):*

$$\max_{\rho^1} \chi_{\kappa^1}(\rho^1) + \max_{\rho^2} \chi_{\kappa^2}(\rho^2) = \max_{\rho^{1,2}} \chi_{\kappa^1 \otimes \kappa^2}(\rho^{1,2}) \tag{9.10}$$

holds for arbitrary channels κ^1 *and* κ^2.

HA *Additivity of Holevo information:*

$$\chi_{\kappa^1}(\rho^1) + \chi_{\kappa^2}(\rho^2) = \chi_{\kappa^1 \otimes \kappa^2}(\rho^1 \otimes \rho^2) \tag{9.11}$$

holds for arbitrary channels κ^1 *and* κ^2 *and arbitrary states* ρ^1 *and* ρ^2.

HL *Additivity of classical capacity of q-q channel with linear cost constraint (additivity of maximum Holevo information with linear cost constraint):*

$$\max_{\lambda} C_{X^1 \le \lambda K}(\kappa^1) + C_{X^2 \le (1-\lambda)K}(\kappa^2) = C_{X^1 + X^1 \le K}(\kappa^1 \otimes \kappa^2), \tag{9.12}$$

i.e.,

$$\begin{aligned} &\max_{\lambda} \max_{\rho^1 : \mathrm{Tr}\, \rho^1 X^1 \le \lambda K} \chi_{\kappa^1}(\rho^1) + \max_{\rho^2 : \mathrm{Tr}\, \rho^2 X^2 \le (1-\lambda)K} \chi_{\kappa^2}(\rho^2) \\ = &\max_{\rho^{1,2} : \mathrm{Tr}\, \rho^{1,2}(X^1 + X^2) \le K} \chi_{\kappa^1 \otimes \kappa^2}(\rho^{1,2}) \end{aligned} \tag{9.13}$$

holds for arbitrary channels κ^1 *and* κ^2*, arbitrary Hermitian matrices* X^1 *and* X^2 *on the respective input system, and an arbitrary constant* K*. Here we identify* X^1 (X^2) *with* $X^1 \otimes I^2$ ($I^1 \otimes X^2$)*. Note that the classical capacity of a q-q channel with linear cost constraint has the form* $C_{X \le \lambda K}(\kappa) = \max_{\rho : \mathrm{Tr}\, \rho X \le K} \chi_{\kappa}(\rho)$.

HC *Additivity of conjugate Holevo information:*

$$\chi^*_{\kappa^1}(X^1) + \chi^*_{\kappa^2}(X^2) = \chi^*_{\kappa^1 \otimes \kappa^2}(X^1 + X^2) \tag{9.14}$$

holds for Hermitian matrices X^1 *and* X^2 *on systems* $\mathcal{H}_1$ *and* $\mathcal{H}_2$*, where conjugate Holevo information* $\chi^*_{\kappa}(X)$ *is defined as the Legendre transform of* $\chi_{\kappa}(\rho)$ *as*

$$\chi^*_{\kappa}(X) \stackrel{\text{def}}{=} \max_{\rho} \mathrm{Tr}\, X\rho + \chi_{\kappa}(\rho).$$

HS *Subadditivity of Holevo information:*

$$\chi_{\kappa^1}(\rho^1) + \chi_{\kappa^2}(\rho^2) \le \chi_{\kappa^1 \otimes \kappa^2}(\rho^{1,2}) \tag{9.15}$$

holds for arbitrary channels κ^1 *and* κ^2 *and arbitrary states* $\rho^{1,2}$*, where* $\rho^1 = \mathrm{Tr}_2\, \rho^{1,2}$ *and* $\rho^2 = \mathrm{Tr}_1\, \rho^{1,2}$.

EM *Additivity of minimum output entropy:*

$$\min_{\rho^1} H(\kappa^1(\rho^1)) + \min_{\rho^2} H(\kappa^2(\rho^2)) = \min_{\rho^{1,2}} H(\kappa^1 \otimes \kappa^2(\rho^{1,2})), \tag{9.16}$$

i.e.,

$$\min_{\rho^1} H_{\kappa^1}(\rho^1) + \min_{\rho^2} H_{\kappa^2}(\rho^2) = \min_{\rho^{1,2}} H_{\kappa^1 \otimes \kappa^2}(\rho^{1,2}) \tag{9.17}$$

holds for arbitrary channels κ^1 and κ^2. Note that the minimum output entropy has the form $\min_\rho H(\kappa(\rho)) = \min_\rho H_\kappa(\rho)$.

EA *Additivity of minimum average output entropy:*

$$H_{\kappa^1}(\rho^1) + H_{\kappa^2}(\rho^2) = H_{\kappa^1 \otimes \kappa^2}(\rho^1 \otimes \rho^2) \tag{9.18}$$

holds for arbitrary channels κ^1 and κ^2 and arbitrary states ρ^1 and ρ^2.

EL *Additivity of minimum average output entropy with linear cost constraint:*

$$\begin{aligned} &\min_{\lambda} \min_{\rho^1 : \mathrm{Tr}\, \rho^1 X^1 \le \lambda K} H_{\kappa^1}(\rho^1) + \min_{\rho^2 : \mathrm{Tr}\, \rho^2 X^2 \le (1-\lambda) K} H_{\kappa^2}(\rho^2) \\ = &\min_{\rho^{1,2} : \mathrm{Tr}\, \rho^{1,2}(X^1 + X^2) \le K} H_{\kappa^1 \otimes \kappa^2}(\rho^{1,2}) \end{aligned} \tag{9.19}$$

holds for arbitrary channels κ^1 and κ^2, arbitrary Hermitian matrices X^1 and X^2 on the respective input system, and an arbitrary constant K.

EC *Additivity of conjugate minimum average output entropy:*

$$H^*_{\kappa^1}(X^1) + H^*_{\kappa^2}(X^2) = H^*_{\kappa^1 \otimes \kappa^2}(X^1 + X^2) \tag{9.20}$$

holds for Hermitian matrices X^1 and X^2 on systems $\mathcal{H}_1$ and $\mathcal{H}_2$, where the conjugate minimum average output entropy $H^*_\kappa(X)$ is defined as the Legendre transform of $H_\kappa(\rho)$ as

$$H^*_\kappa(X) \stackrel{\mathrm{def}}{=} \max_\rho \mathrm{Tr}\, X\rho - H_\kappa(\rho).$$

ES *Superadditivity of minimum average output entropy:*

$$H_{\kappa^1}(\rho^1) + H_{\kappa^2}(\rho^2) \ge H_{\kappa^1 \otimes \kappa^2}(\rho^{1,2}) \tag{9.21}$$

holds for arbitrary channels κ^1, κ^2 and arbitrary states $\rho^{1,2}$.

FA *Additivity of entanglement of formation:*

$$E_f(\rho^1) + E_f(\rho^2) = E_f(\rho^1 \otimes \rho^2) \tag{9.22}$$

holds for a state ρ^1 on $\mathcal{H}_1 \stackrel{\mathrm{def}}{=} \mathcal{H}_{A_1} \otimes \mathcal{H}_{B_1}$ and a state ρ^2 on $\mathcal{H}_2 \stackrel{\mathrm{def}}{=} \mathcal{H}_{A_2} \otimes \mathcal{H}_{B_2}$.

FL *Additivity of minimum entanglement of formation with linear cost constraint:*

$$\min_{\lambda} \min_{\rho^1:\mathrm{Tr}\,\rho^1 X^1 \le \lambda K} E_f(\rho^1) + \min_{\rho^2:\mathrm{Tr}\,\rho^2 X^2 \le (1-\lambda)K} E_f(\rho^2)$$
$$= \min_{\rho^{1,2}:\mathrm{Tr}\,\rho^{1,2}(X^1+X^2)\le K} E_f(\rho^{1,2}) \tag{9.23}$$

holds for arbitrary Hermitian matrices X^1 and X^2 and an arbitrary constant K.

FC *Additivity of conjugate entanglement of formation:*

$$E_f^*(X^1) + E_f^*(X^2) = E_f^*(X^1 + X^2) \tag{9.24}$$

holds for Hermitian matrices X^1 and X^2 on systems $\mathcal{H}_1$ and $\mathcal{H}_2$, where the conjugate entanglement of formation $E_f^*(X)$ is defined as the Legendre transform of $E_f(\rho)$ as

$$E_f^*(X) \stackrel{\mathrm{def}}{=} \max_{\rho} \mathrm{Tr}\, X\rho - E_f(\rho).$$

FS *Superadditivity of entanglement of formation:*

$$E_f(\rho^1) + E_f(\rho^2) \le E_f(\rho^{1,2}) \tag{9.25}$$

holds for a state $\rho^{1,2}$ on $(\mathcal{H}_{A_1} \otimes \mathcal{H}_{A_2}) \otimes (\mathcal{H}_{B_1} \otimes \mathcal{H}_{B_2})$.

However, as shown in Sect. 8.13, **FS** does not hold. Hence, all of the above conditions are invalid. However, the papers [8, 17] numerically verified **HM** in the qubit case. **HM** has been shown in the following cases.

(**a**) When $C_c(\kappa)$ is equal to the dimension of the output system, additivity (9.3) holds (trivial case).

(**b**) Any entanglement-breaking channel κ_1 (Example 5.4) satisfies additivity (9.5) with an arbitrary channel κ^2 [12].

(**c**) Any depolarizing channel $\kappa_{d,\lambda}$ (Example 5.3) satisfies additivity (9.5) with $\kappa^1 = \kappa_{d,\lambda}$ and an arbitrary channel κ^2 [11].

(**d**) Any unital qubit channel κ^1 satisfies additivity (9.5) with an arbitrary channel κ^2 [10].

(**e**) Any antisymmetric channel $\kappa^T_{d,-\frac{1}{d-1}}$ (Werner–Holevo channels, Example 5.9) satisfies additivity (9.3) with $\kappa = \kappa^T_{d,-\frac{1}{d-1}}$ [18].

(**f**) All transpose depolarizing channels $\kappa^T_{d,\lambda}$ and $\kappa^T_{d',\lambda}$ satisfy additivity (9.5) with $\kappa^1 = \kappa^T_{d,\lambda}$ and $\kappa^2 = \kappa^T_{d',\lambda}$ [19, 20].

(**g1**) Channels $\kappa_{d,\lambda} \circ \kappa_D^{\mathrm{PD}}$ satisfy additivity (9.5) with $\kappa^1 = \kappa_{d,\lambda} \circ \kappa_D^{\mathrm{PD}}$ and an arbitrary channel κ^2 [9, 21].

(**g2**) Channels $\kappa^T_{d,-\frac{1}{d-1}} \circ \kappa_D^{\mathrm{PD}}$ and $\kappa_D^{\mathrm{PD}} \circ \kappa^T_{d,-\frac{1}{d-1}}$ satisfy additivity (9.3) with $\kappa = \kappa^T_{d,-\frac{1}{d-1}} \circ \kappa_D^{\mathrm{PD}}$ or $\kappa_D^{\mathrm{PD}} \circ \kappa^T_{d,-\frac{1}{d-1}}$ [9, 22].

(g3) Channels $\kappa_{d,\lambda}^T \circ \kappa_D^{\text{PD}}$ and $\kappa_D^{\text{PD}} \circ \kappa_{d,\lambda}^T$ satisfy additivity (9.5) with $\kappa_1 = \kappa_{d,\lambda}^T \circ \kappa_D^{\text{PD}}$ or $\kappa_D^{\text{PD}} \circ \kappa_{d,\lambda}^T$ and $\kappa_2 = \kappa_{d',\lambda}^T \circ \kappa_{D'}^{\text{PD}}$ or $\kappa_{D'}^{\text{PD}} \circ \kappa_{d',\lambda}^T$ [9, 22].

Therefore, we obtain $C_c^e(\kappa) = C_c(\kappa)$ in the cases **(a)**, **(b)**, **(c)**, **(d)**, **(e)**, **(g1)**, and **(g2)**. Indeed, since $C_c(\kappa_{d,\lambda} \circ \kappa_D^{\text{PD}}) = C_c(\kappa_{d,\lambda})$, **(c)** yields that $C_c(\kappa_{d,\lambda} \circ \kappa_D^{\text{PD}} \otimes \kappa_2) \leq C_c(\kappa_{d,\lambda} \otimes \kappa_2)$
$= C_c(\kappa_{d,\lambda}) + C_c(\kappa_2) = C_c(\kappa_{d,\lambda} \circ \kappa_D^{\text{PD}}) + C_c(\kappa_2)$, which implies **(g1)**. Similarly, we can show **(g2)** and **(g3)** from **(e)** and **(f)**.

Moreover, the additivity of minimum output entropy **EM** holds not only in the above cases but also in the more extended cases of **(c)**, **(e)**, and **(f)** as opposed to **(g1)**, **(g2)**, and **(g3)** [9, 22]. Since the condition **EM** is simple, it has been mainly discussed for verifying these conjectures.

Before proceeding to the proof of the equivalence, we give a counter example of **EM** by modifying the discussion in Sect. 8.13 [23]. First, for a given $\lceil cn \rceil$-dimensional subspace $\mathcal{K}$, we choose the isometry V and the spaces $\mathcal{H}_{C,1}$ and $\mathcal{H}_{C,2}$ as in the proof of Lemma 8.16. We also use the notations $\mathcal{H}_{A,1}$, $\mathcal{H}_{A,2}$, $\mathcal{H}_{B,1}$, and $\mathcal{H}_{B,2}$ given in Sect. 8.13. We define the TP-CP map κ_i from the system $\mathcal{H}_{C,i}$ to $\mathcal{H}_{A,i}$ for $i = 1, 2$ as

$$\kappa_1(\rho) := \operatorname{Tr}_{B,1} V\rho V^*, \quad \kappa_2(\rho) := \operatorname{Tr}_{B,2} \overline{V}\rho V^T. \tag{9.26}$$

Then, Lemma 8.16 implies that

$$\min_\rho H((\kappa_1 \otimes \kappa_2)(\rho)) \leq 2(1 - \frac{c}{k}) \log k + h(\frac{c}{k}). \tag{9.27}$$

Next, for given $\epsilon, \epsilon' > 0$ and $c > 0$, we choose a sufficiently large n. Then, we choose a $\lceil cn \rceil$-dimensional subspace $\mathcal{K}$ given in Theorem 8.14 such that

$$\min_\rho H(\kappa_i(\rho)) \geq \log k - \frac{\epsilon}{k} - \left(\frac{(-4c \log \frac{\epsilon'}{4}) + 2\sqrt{-2c \log \frac{\epsilon'}{4}}\sqrt{1 - 2c \log \frac{\epsilon'}{4}}}{1 - \epsilon'} \right)^2 \frac{1}{k}$$

for $i = 1, 2$. Hence, with a sufficiently large k, the relation (8.258) guarantees that

$$\min_\rho H(\kappa_1(\rho)) + \min_\rho H(\kappa_2(\rho)) > \min_\rho H((\kappa_1 \otimes \kappa_2)(\rho)), \tag{9.28}$$

which contradicts **EM**.

Now, we start to show the equivalence for these conditions. Among the above conditions, the relations **HC⇒HM** and **EC⇒EM** are trivial. From MSW correspondence (9.8) we obtain **HA⇒FA** and **EA⇒FA**. Next, we focus on the Stinespring representation $(\mathcal{H}_C, \rho_0, U_\kappa)$ of κ mapping from a system $\mathcal{H}_A$ to another system $\mathcal{H}_B$. In this case, the MSW correspondence (9.8) can be generalized as

$$H_\kappa(\rho) = \min_{(p_i,\rho_i):\sum_i p_i\rho_i = \rho} \sum_i p_i H(\operatorname{Tr}_{A,C} U_\kappa(\rho_i \otimes \rho_0)U_\kappa^*) = E_f(\bar{\kappa}(\rho)),$$
$$\bar{\kappa}(\rho) \stackrel{\text{def}}{=} U_\kappa(\rho \otimes \rho_0)U_\kappa^*,$$

i.e.,

$$\chi_{\kappa}(\rho) = H(\kappa(\rho)) - E_f(\bar{\kappa}(\rho)), \tag{9.29}$$

where we use the notation E_f as the entanglement of formation between the output system $\mathcal{H}_B$ and the environment $\mathcal{H}_A \otimes \mathcal{H}_C$. Hence, if Condition **FS** holds, for $\rho^{1,2}$, we have

$$\begin{aligned}
&\chi_{\kappa^1 \otimes \kappa^2}(\rho^{1,2}) = H(\kappa^1 \otimes \kappa^2(\rho^{1,2})) - E_f(\overline{\kappa^1 \otimes \kappa^2}(\rho^{1,2})) \\
\le & H(\kappa^1(\rho^1)) + H(\kappa^2(\rho^2)) - E_f(\overline{\kappa^1 \otimes \kappa^2}(\rho^{1,2})) \\
\le & H(\kappa^1(\rho^1)) + H(\kappa^2(\rho^2)) - (E_f(\overline{\kappa^1}(\rho^1)) + E_f(\overline{\kappa^2}(\rho^2))) \\
= & \chi_{\kappa^1}(\rho^1) + \chi_{\kappa^2}(\rho^2).
\end{aligned}$$

Hence, we have **FS**⇒**HS**. Similarly, the relation **FS**⇒**MS** holds.

The following lemma is useful for proofs of the remaining relations.

Lemma 9.1 *Let f^i be a convex function defined on $\mathcal{S}(\mathcal{H}_i)$ $(i = 1, 2)$ and $f^{1,2}$ be a convex function defined on $\mathcal{S}(\mathcal{H}_1 \otimes \mathcal{H}_2)$ satisfying*

$$f^1(\rho^1) + f^2(\rho^2) \ge f^{1,2}(\rho^1 \otimes \rho^2). \tag{9.30}$$

The relations **L**⇔**C**⇔**S**⇒**A** *hold among the following conditions.*

S *Superadditivity:*

$$f^1(\rho^1) + f^2(\rho^2) \le f^{1,2}(\rho^{1,2}) \tag{9.31}$$

holds for a state $\rho^{1,2}$ on $(\mathcal{H}_{A_1} \otimes \mathcal{H}_{A_2}) \otimes (\mathcal{H}_{B_1} \otimes \mathcal{H}_{B_2})$.

C *Additivity of conjugate function:*

$$f^{1*}(X^1) + f^{2*}(X^2) = f^{1,2*}(X^1 + X^2) \tag{9.32}$$

holds for Hermitian matrices X^1 and X^2 on the systems $\mathcal{H}_1$ and $\mathcal{H}_2$, where conjugate entanglement of formation $f^*(X)$ is defined as the Legendre transform of $f(\rho)$ as

$$f^*(X) \stackrel{\text{def}}{=} \max_{\rho} \operatorname{Tr} X\rho - f(\rho).$$

L *Additivity of minimum value with linear cost constraint:*

$$\begin{aligned}
&\min_{\lambda} \min_{\rho^1 : \operatorname{Tr} \rho^1 X^1 \le \lambda K} f^1(\rho^1) + \min_{\rho^2 : \operatorname{Tr} \rho^2 X^2 \le (1-\lambda) K} f^2(\rho^2) \\
= & \min_{\rho^{1,2} : \operatorname{Tr} \rho^{1,2}(X^1 + X^2) \le K} f^{1,2}(\rho^{1,2})
\end{aligned} \tag{9.33}$$

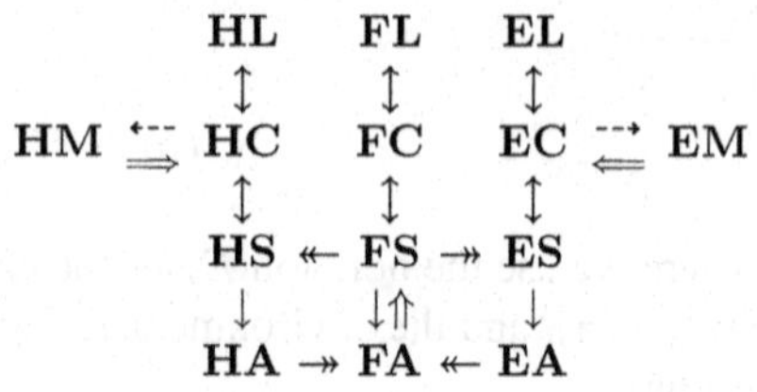

Fig. 9.2 $\rightarrow$: Lemma 9.1, $\dashrightarrow$: easy, $\twoheadrightarrow$: MSW correspondence, $\Longrightarrow$: hard

$$
\begin{array}{ccccccc}
 & & \mathbf{HL} & & \mathbf{FL} & & \mathbf{EL} & \\
 & & \updownarrow & & \updownarrow & & \updownarrow & \\
\mathbf{HM} & \overset{\dashleftarrow}{\Longrightarrow} & \mathbf{HC} & & \mathbf{FC} & & \mathbf{EC} & \overset{\dashrightarrow}{\Longleftarrow} & \mathbf{EM} \\
 & & \updownarrow & & \updownarrow & & \updownarrow & \\
 & & \mathbf{HS} & \twoheadleftarrow & \mathbf{FS} & \twoheadrightarrow & \mathbf{ES} & \\
 & & \downarrow & & \downarrow\Uparrow & & \downarrow & \\
 & & \mathbf{HA} & \twoheadrightarrow & \mathbf{FA} & \twoheadleftarrow & \mathbf{EA} &
\end{array}
$$

holds for arbitrary Hermitian matrices X^1 and X^2 and an arbitrary constant K.

A *Additivity:*

$$f^1(\rho^1) + f^2(\rho^2) = f^{1,2}(\rho^1 \otimes \rho^2) \tag{9.34}$$

holds for a state ρ^1 on $\mathcal{H}_1 \stackrel{\text{def}}{=} \mathcal{H}_{A_1} \otimes \mathcal{H}_{B_1}$ and a state ρ^2 on $\mathcal{H}_2 \stackrel{\text{def}}{=} \mathcal{H}_{A_2} \otimes \mathcal{H}_{B_2}$.

Lemma 9.1 yields the relations **HL$\Leftrightarrow$HC$\Leftrightarrow$HS$\Rightarrow$HA**, **EL$\Leftrightarrow$EC$\Leftrightarrow$ES$\Rightarrow$EA**, and **FL$\Leftrightarrow$FC$\Leftrightarrow$FS$\Rightarrow$FA**.

Hence, if we prove the relations **HM$\Rightarrow$HC**, **EM$\Rightarrow$EC**, and **FA$\Rightarrow$FS**, we obtain the equivalence among the above 14 conditions, as explained in Fig. 9.2. These proofs will be given in Sect. 9.8. The relations are summarized as follows.

Finally, we prove the additivity for the classical capacity for entanglement-breaking channels by using inequality (5.110). From the definition, any entanglement-breaking channel κ^1 has the form of the output state for any input state ρ_x as

$$(\kappa^1 \otimes \iota)(\rho_x) = \sum_y Q^x_y \rho^1_{x,y} \otimes \rho^2_{x,y},$$

which implies that

$$(\kappa^1 \otimes \kappa^2)(\rho_x) = (\iota \otimes \kappa^2) \sum_y Q^x_y \rho^1_{x,y} \otimes \rho^2_{x,y} = \sum_y Q^x_y \rho^1_{x,y} \otimes \kappa^2(\rho^2_{x,y}). \tag{9.35}$$

Hence, using (5.110) and (5.86), we have

$$
\begin{aligned}
& C_c(\kappa^1 \otimes \kappa^2) \\
= & \max_\rho H((\kappa^1 \otimes \kappa^2)(\rho)) - \min_{(p_x,\rho_x):\sum_x p_x \rho_x = \rho} \sum_x p_x H((\kappa^1 \otimes \kappa^2)(\rho_x)) \\
\stackrel{(a)}{=} & \max_\rho H((\kappa^1 \otimes \kappa^2)(\rho)) \\
& - \min_{(p_x,\rho_x):\sum_x p_x \rho_x = \rho} \sum_x p_x H\Big(\sum_y Q^x_y \rho^1_{x,y} \otimes \kappa^2(\rho^1_{x,y})\Big) \\
\stackrel{(b)}{\le} & H(\mathrm{Tr}_2(\kappa^1 \otimes \kappa^2)(\rho)) + H(\mathrm{Tr}_1(\kappa^1 \otimes \kappa^2)(\rho))
\end{aligned}
$$

$$
\begin{aligned}
&- \min_{(p_x,\rho_x):\sum_x p_x\rho_x=\rho} \sum_x p_x \Big(\sum_y Q_y^x H(\kappa^2(\rho_{x,y}^2)) + H\Big(\sum_y Q_y^x \rho_{x,y}^1\Big)\Big) \\
\le & H(\kappa^1(\mathrm{Tr}_2\,\rho)) - \min_{(p_x,\rho_x):\sum_x p_x\rho_x=\rho} \sum_x p_x H\Big(\sum_y Q_y^x \rho_{x,y}^1\Big) \\
& + H(\kappa^2(\mathrm{Tr}_1\,\rho)) - \min_{(p_x,\rho_x):\sum_x p_x\rho_x=\rho} \sum_x p_x \sum_y Q_y^x H(\kappa^2(\rho_{x,y}^2)) \\
\overset{(c)}{\le} & C_c(\kappa^1) + C_c(\kappa^2),
\end{aligned} \tag{9.36}
$$

where (a) follows from (9.35), (b) does from (5.110) and (5.86), and (c) does from the relations $\mathrm{Tr}_2\,\rho = \sum_x p_x \mathrm{Tr}_2\,\rho_x$, $\kappa^1(\mathrm{Tr}_2\,\rho_x) = \mathrm{Tr}_2\,\iota \otimes \kappa^1(\rho_x) = \sum_y Q_y^x \rho_{x,y}^1$, and $\mathrm{Tr}_1\,\rho = \sum_x p_x \sum_y Q_y^x \rho_{x,y}^2$. Then, (9.36) implies (9.5) for entanglement breaking channel κ_1 and arbitrary channel κ_2.

Exercises

9.2 Using a discussion similar to (9.36), show that the additivity of minimum output entropy when κ^1 is entanglement breaking.

9.3 Prove Theorem 9.2 by referring to Theorem 4.1 and the proof of Theorem 4.2.

9.3 C-Q Channel Coding with Shared Entanglement

In the preceding section, we considered the effectiveness of using the input state entangled between systems that are to be sent. In this section, we will consider the usefulness of entangled states $\rho^{A,B}$ on a composite system $\mathcal{H}_A \otimes \mathcal{H}_B$ that is a priori shared between the sender and the receiver. If the sender wishes to send some information corresponding to an element i of $\{1, \dots, N\}$, he or she must perform an operation $\varphi_e(i)$ on the system $\mathcal{H}_A$ according to the element i, then send the system $\mathcal{H}_A$ to the receiver using the quantum channel κ. Then, the receiver performs a measurement (POVM) $Y = \{Y_i\}_{i=1}^N$ on the composite system $\mathcal{H}_{A'} \otimes \mathcal{H}_B$. Note that this measurement is performed not only on the output system $\mathcal{H}_{A'}$ of the quantum channel κ but also on the composite system $\mathcal{H}_{A'} \otimes \mathcal{H}_B$.

Consider the simple case in which the systems $\mathcal{H}_A$, $\mathcal{H}_{A'}$, and $\mathcal{H}_B$ are all quantum two-level systems. Let the initial state $\rho^{A,B}$ be a pure state $\frac{1}{\sqrt{2}}\left(|u_0^A \otimes u_0^B\rangle + |u_1^A \otimes u_1^B\rangle\right)$. Assume that there is no noise in the quantum channel, which enables the perfect transmission of the quantum state. In this case, we send the message $i \in \{0, \dots, 3\}$ by applying the unitary transformation S_i^A on system $\mathcal{H}_A$. Then, the receiver possesses the transmitted system as well as the initially shared system. The state of the composite system $(\mathbb{C}^2)^{\otimes 2}$ of the receiver is given by $(S_i^A \otimes I_B)\frac{1}{\sqrt{2}}\left(|u_0^A \otimes u_0^B\rangle + |u_1^A \otimes u_1^B\rangle\right)$. Since the vectors form an orthogonal basis with $i = 0, 1, 2, 3$, we can perform a measurement Y comprising this basis. Hence, this measurement provides error-free

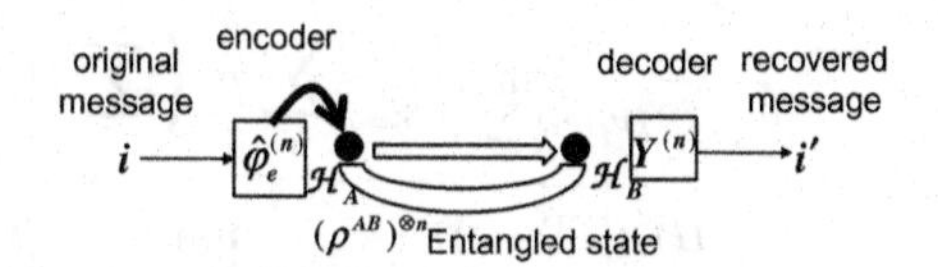

Fig. 9.3 C-q channel coding with shared entanglement with noiseless channel

decoding. According to this protocol, two bits of information may be sent through only one qubit channel. We observe that by sharing an entangled state between two parties a priori, more information can be sent than simply by sending a quantum state [24]. This protocol is often called superdense coding.

However, the initially shared entangled state is not necessarily a maximally entangled state such as $\frac{1}{\sqrt{2}}(|u_0^A \otimes u_0^B\rangle + |u_1^A \otimes u_1^B\rangle)$ in general. Hence, it is an important question to determine how much a partially entangled state shared between the sender and the receiver improves the classical capacity. This will give a quantitative measure of the utilizable entanglement of a partially entangled state.

Assume that the sender and the receiver share the partially entangled state $(\rho^{A,B})^{\otimes n}$ on $\mathcal{H}_A^{\otimes n} \otimes \mathcal{H}_B^{\otimes n}$. The **code** is then given by the set $\Phi_e^{(n)} = (N_n, \mathcal{H}_{A_n'}, \varphi_e^{(n)}, Y^{(n)})$ consisting of its size N_n, the quantum system $\mathcal{H}_{A_n'}$ transmitted by the sender to the receiver, the operation $\varphi_e^{(n)}(i)$ from the quantum system $\mathcal{H}_A^{\otimes n}$ to $\mathcal{H}_{A_n'}$ dependently of each message i, and the measurement $Y^{(n)}$ on the composite system $\mathcal{H}_{A_n'} \otimes \mathcal{H}_B^{\otimes n}$ as Fig. 9.3.

Further, the effectiveness of an entangled state $\rho^{A,B}$, i.e., the increase of the transmitted message, is given by

$$|\Phi_e^{(n)}| \stackrel{\text{def}}{=} \frac{N_n}{\dim \mathcal{H}_{A_n'}},$$

and the error probability is given by

$$\varepsilon[\Phi_e^{(n)}] \stackrel{\text{def}}{=} \frac{1}{N_n} \sum_{i=1}^{N_n} \left(1 - \operatorname{Tr}\left[\varphi_e^{(n)}(i) \otimes \iota_B^{\otimes n}((\rho^{A,B})^{\otimes n}) Y_i^{(n)}\right]\right).$$

Hence, the amount of assistance for sending information by the state $\rho^{A,B}$ can be quantified as[2]

$$C_a(\rho^{A,B}) \stackrel{\text{def}}{=} \sup\left\{ \overline{\lim} \frac{1}{n} \log |\Phi_e^{(n)}| \,\middle|\, \lim_{n\to\infty} \varepsilon[\Phi_e^{(n)}] = 0 \right\}. \tag{9.37}$$

Then, we obtain the following theorem.

Theorem 9.4 *The quantity* $\frac{1}{n} \min_\kappa H_{\kappa\otimes\iota_B((\rho^{A,B})^{\otimes n})}(A|B)$ *converges as* $n \to \infty$ *and*

---

[2] The subscript a expresses "assistance."

$$C_a(\rho^{A,B}) = - \lim_{n\to\infty} \frac{1}{n} \min_\kappa H_{\kappa\otimes\iota_B((\rho^{A,B})^{\otimes n})}(A|B), \tag{9.38}$$

where κ is a TP-CP map from $\mathcal{H}_A^{\otimes n}$ to $\mathcal{H}_{A'_n}$ [25–29]. We assume that the output system $\mathcal{H}_{A'_n}$ can be chosen dependently on κ.

When the initial state $\rho^{A,B}$ is a maximally correlated state, $\min_\kappa H_{\kappa\otimes\iota_B((\rho^{A,B})^{\otimes n})}(A|B) = nH_{\rho^{A,B}}(A|B)$, i.e., $C_a(\rho^{A,B}) = -H_{\rho^{A,B}}(A|B)$. Certainly, this equation holds when condition (8.129) is satisfied. In particular, if $\rho^{A,B}$ is a pure state, we have $C_a(\rho^{A,B}) = H(\rho^B)$.

Proof We first show that

$$C_a(\rho^{A,B}) \geq H(\rho^B) - \min_\kappa H((\kappa\otimes\iota_B)(\rho^{A,B})) \tag{9.39}$$

in order to obtain the $\geq$ part of (9.38). Let κ_m be the channel $\operatorname{argmin}_\kappa H((\kappa\otimes\iota_B)(\rho^{A,B}))$. We denote the output system of κ_m and its dimension by $\mathcal{H}_{A'}$ and d, respectively. Now, we focus on the c-q channel $(i,j)\mapsto W_{(i,j)} \stackrel{\text{def}}{=} (\mathbf{X}^i_{A'}\mathbf{Z}^j_{A'}\otimes I_B)^*(\kappa\otimes\iota_B)(\rho^{A,B})(\mathbf{X}^i_{A'}\mathbf{Z}^j_{A'}\otimes I_B)$ with the set of input signals $\mathcal{X} \stackrel{\text{def}}{=} \{(i,j)\}_{1\leq i,j\leq d}$. Using Theorem 4.1 and Exercise 5.10, we see that the capacity of this channel is larger than

$$\begin{aligned}
&H\left(\sum_{(i,j)} \frac{1}{d^2}(\mathbf{X}^i_{A'}\mathbf{Z}^j_{A'}\otimes I_B)^*(\kappa\otimes\iota_B)(\rho^{A,B})(\mathbf{X}^i_{A'}\mathbf{Z}^j_{A'}\otimes I_B)\right)\\
&\quad -\sum_{(i,j)} \frac{1}{d^2} H\left((\mathbf{X}^i_{A'}\mathbf{Z}^j_{A'}\otimes I_B)^*(\kappa\otimes\iota_B)(\rho^{A,B})(\mathbf{X}^i_{A'}\mathbf{Z}^j_{A'}\otimes I_B)\right)\\
=&H\left(\rho^{A'}_{\text{mix}}\otimes(\operatorname{Tr}_{A'}(\kappa\otimes\iota_B)(\rho^{A,B}))\right) - \sum_{(i,j)}\frac{1}{d^2} H\left((\kappa\otimes\iota_B)(\rho^{A,B})\right)\\
=&H\left(\rho^{A'}_{\text{mix}}\otimes \operatorname{Tr}_A \rho^{A,B}\right) - H\left((\kappa\otimes\iota_B)(\rho^{A,B})\right)\\
=&\log d + H\left(\operatorname{Tr}_A\rho^{A,B}\right) - H\left((\kappa\otimes\iota_B)(\rho^{A,B})\right).
\end{aligned}$$

From the definition of $|\Phi_e^{(n)}|$, we immediately obtain (9.39). Fixing n and applying the same argument to $\kappa_n \stackrel{\text{def}}{=} \operatorname{argmin}_\kappa H((\kappa\otimes\iota_B^{\otimes n})((\rho^{A,B})^{\otimes n}))$, we obtain $C_a(\rho^{A,B}) \geq H(\operatorname{Tr}_A\rho^{A,B}) - \frac{1}{n}\min_\kappa H((\kappa\otimes\iota_B^{\otimes n})((\rho^{A,B})^{\otimes n}))$. Therefore, we have $C_a(\rho^{A,B}) \geq H(\operatorname{Tr}_A\rho^{A,B}) - \inf_n \frac{1}{n}\min_\kappa H((\kappa\otimes\iota_B^{\otimes n})((\rho^{A,B})^{\otimes n}))$. Since $nH(\operatorname{Tr}_A\rho^{A,B}) - \min_\kappa H((\kappa\otimes\iota_B^{\otimes n})((\rho^{A,B})^{\otimes n}))$ satisfies the assumptions of Lemma A.1, this converges with $n\to\infty$. We therefore obtain (9.38) with the $\geq$ sign.

Next, we prove the $\leq$ part of (9.38). Let X be a random variable taking values in $\{1,\ldots,N_n\}$ and following the uniform distribution. Let Y be the decoded message at the receiver as the random variable taking values in $\{1,\ldots,N_n\}$. Since $H(X) = \log N_n$, the Fano inequality yields that

$$I(X:Y) \geq H(X) - \log 2 - \varepsilon[\Phi^{(n)}] \log N_n$$
$$= -\log 2 + \log N_n (1 - \varepsilon[\Phi_e^{(n)}]). \tag{9.40}$$

Using the monotonicity of the quantum relative entropy and (5.86), it can be shown that [Exe. 9.4]

$$I(X:Y) \leq nH\left(\mathrm{Tr}_A \rho^{A,B}\right) + \log \dim \mathcal{H}_{A'_n} - \min_{\kappa} H((\kappa \otimes \iota_B^{\otimes n})((\rho^{A,B})^{\otimes n})). \tag{9.41}$$

Combining this inequality with (9.40), we obtain

$$H\left(\mathrm{Tr}_A \rho^{A,B}\right) - \frac{1}{n} \min_{\kappa} H((\kappa \otimes \iota_B^{\otimes n})((\rho^{A,B})^{\otimes n})) + \frac{\log 2}{n}$$
$$\geq \frac{\log N_n}{n}(1 - \varepsilon[\Phi_e^{(n)}]) - \frac{\log \dim \mathcal{H}'_{A,n}}{n}.$$

Taking the limit $n \to \infty$, we have

$$H\left(\mathrm{Tr}_A \rho^{A,B}\right) - \overline{\lim} \frac{1}{n} \min_{\kappa} H((\kappa \otimes \iota_B^{\otimes n})((\rho^{A,B})^{\otimes n}))$$
$$\geq \overline{\lim} \frac{\log N_n - \log \dim \mathcal{H}_{A'_n}}{n},$$

which gives the $\leq$ part of (9.38). ■

We assumed above that there was no noise in the quantum channel. Since real quantum channels always contain some noise, we often restrict our channel to a given TP-CP map κ. Now, consider the case in which the quantum channel κ has some noise, but the sender and the receiver are allowed access to any entangled state. Let us also say that the quantum channel κ can be used n times (i.e., $\kappa^{\otimes n}$), as considered previously.

First, we prepare an entangled pure state $x^{(n)}$ on the composite system $\mathcal{H}_{A'_n} \otimes \mathcal{H}_{R_n}$, comprising quantum system $\mathcal{H}_{A'_n}$ at the sender and quantum system $\mathcal{H}_{R_n}$ at the receiver. Let the size of the code be N_n, and let an element $i \in \{1, \ldots, N_n\}$ be transmitted. Next, the sender performs the operation $\varphi_e^{(n)}(i)$ from the system $\mathcal{H}_{A'_n}$ to the other system $\mathcal{H}_A^{\otimes n}$ dependently on $i = 1, \ldots, N_n$. Then, the state on $\mathcal{H}_A^{\otimes n}$ is transmitted to the receiver via the given quantum channel $\kappa^{\otimes n}$. The receiver performs a measurement $Y^{(n)}$ on the composite system $\mathcal{H}_B^{\otimes n} \otimes \mathcal{H}_{Rn}$, thereby recovering the original signal i. In this case, our code can be described by the set $(\mathcal{H}_{A'_n}, \mathcal{H}_{R_n}, x^{(n)}, N_n, \varphi_e^{(n)}, Y^{(n)})$, and is denoted by $\Phi_e^{(n),2}$. Hence, the size of the code and its error probability are given by

$$|\Phi_e^{(n),2}| \stackrel{\text{def}}{=} N_n,$$
$$\varepsilon[\Phi_e^{(n),2}] \stackrel{\text{def}}{=} \frac{1}{N_n} \sum_{i=1}^{N_n} \left(1 - \mathrm{Tr}\left[\varphi_e^{(n)}(i) \otimes \iota_{R,n}((\rho^{A,B})^{\otimes n}) Y_i^{(n)}\right]\right).$$

The entanglement-assisted classical capacity $C^e_{c,e}(\kappa)$[3] is given by

$$C^e_{c,e}(\kappa) \stackrel{\text{def}}{=} \sup \left\{ \underline{\lim} \frac{1}{n} \log |\Phi_e^{(n),2}| \,\middle|\, \lim_{n\to\infty} \varepsilon[\Phi_e^{(n),2}] = 0 \right\}. \tag{9.42}$$

Theorem 9.5 (Bennett et al. [30, 31], Holevo [32]) *The entanglement-assisted classical capacity $C^e_{c,e}(\kappa)$ of a quantum-quantum channel κ from $\mathcal{H}_A$ to $\mathcal{H}_B$ is*

$$C^e_{c,e}(\kappa) = \max_\rho I(\rho, \kappa), \tag{9.43}$$

where $I(\rho, \kappa)$ is the transmission information of a quantum-quantum channel defined in (8.36).

In a manner similar to $J(p, \sigma, W)$, we define $J(\rho, \sigma, \kappa)$ as

$$\begin{aligned} J(\rho, \sigma, \kappa) &\stackrel{\text{def}}{=} \operatorname{Tr}(\kappa \otimes \iota_R)(|x\rangle\langle x|)(\log(\kappa \otimes \iota_R)(|x\rangle\langle x|)) - \log \rho \otimes \sigma) \\ &= H(\rho) - \operatorname{Tr} \kappa(\rho) \log \sigma - H_e(\rho, \kappa) = \tilde{I}_c(\rho, \kappa) - \operatorname{Tr} \kappa(\rho) \log \sigma, \end{aligned} \tag{9.44}$$

where x is a purification of ρ. Then, $J(\rho, \sigma, \kappa)$ is concave for ρ because of (8.50), and is convex for σ. Since

$$J(\rho, \sigma, \kappa) = I(\rho, \kappa) + D(\kappa(\rho)\|\sigma), \tag{9.45}$$

in a manner similar to (4.71), Lemma A.9 guarantees that

$$C^e_{c,e}(\kappa) = \max_\rho I(\rho, \kappa) = \max_\rho \min_\sigma J(\rho, \sigma, \kappa) = \min_\sigma \max_\rho J(\rho, \sigma, \kappa). \tag{9.46}$$

Proof We first construct a code attaining the right-hand side (RHS) of (9.43), i.e., we prove the $\geq$ part in (9.43) for $\operatorname{argmax}_\rho I(\rho, \kappa) = \rho^A_{\text{mix}}$. Let $\rho^{A,R}$ be the purification of ρ^A_{mix}. Perform the encoding operation using the operation $\rho^{A,R} \mapsto \rho^{A,R}_{(i,j)} \stackrel{\text{def}}{=} (\mathbf{X}^i_A \mathbf{Z}^j_A \otimes I)(\rho^{A,R})(\mathbf{X}^i_A \mathbf{Z}^j_A \otimes I)^*$ at A, as in the case of a noise-free channel. Since

$$\begin{aligned} &\sum_{i,j} \frac{1}{d^2} (\kappa \otimes \iota_R)(\rho^{A,R}_{(i,j)}) = (\kappa \otimes \iota_R)\Big(\sum_{i,j} \frac{1}{d^2} (\rho^{A,R}_{(i,j)})\Big) \\ =&(\kappa \otimes \iota_R)(\rho^A_{\text{mix}} \otimes \rho^R_{\text{mix}}) = \kappa(\rho^A_{\text{mix}}) \otimes \rho^R_{\text{mix}}, \end{aligned}$$

we obtain

$$\sum_{i,j} \frac{1}{d^2} D\Big((\kappa \otimes \iota_R)(\rho^{A,R}_{(i,j)}) \Big\| \sum_{i,j} \frac{1}{d^2} (\kappa \otimes \iota_R)(\rho^{A,R}_{(i,j)})\Big)$$

[3]The second subscript, e, of $C^e_{c,e}$ indicates the shared "entanglement." The superscript e indicates "entangled" operations between sending systems.

$$= \sum_{i,j} \frac{1}{d^2} D((\kappa \otimes \iota_R)(\rho^{A,R}_{(i,j)}) \| \kappa(\rho^A_{\text{mix}}) \otimes \rho^R_{\text{mix}}) = I(\rho^A_{\text{mix}}, \kappa).$$

Combining this equation with the argument given in Theorem 4.1, we find a code attaining $I(\rho^A_{\text{mix}}, \kappa)$.

Now, consider the case when $I(\rho^A_{\text{mix}}, \kappa) = \max_\rho I(\rho, \kappa)$ does not hold. Let $\rho^{\mathcal{K}_n}_{\text{mix}}$ be the completely mixed state on a subspace $\mathcal{K}_n$ of $\mathcal{H}^{\otimes n}$. If we can take the state $\rho^{\mathcal{K}_n}_{\text{mix}}$ such that

$$\lim_{n\to\infty} \frac{1}{n} I(\rho^{\mathcal{K}_n}_{\text{mix}}, \kappa^{\otimes n}) = \max_\rho I(\rho, \kappa), \tag{9.47}$$

we can construct a code satisfying $\max_\rho I(\rho, \kappa)$. To choose such a subspace $\mathcal{K}_n$, let $\rho_M \stackrel{\text{def}}{=} \operatorname{argmax}_\rho I(\rho, \kappa)$, and we take the spectral decomposition $\rho_M^{\otimes n} = \sum_{j=1}^{v_n} \lambda_{j,n} E_{j,n}$, where v_n represents the number of eigenvalues of $\rho_M^{\otimes n}$. Let $\rho^{j,n}_{\text{mix}}$ be the completely mixed state in the range of $E_{j,n}$, and let $p_{j,n} \stackrel{\text{def}}{=} \lambda_{j,n} \operatorname{rank} E_{j,n}$. Then, we have $\rho_M^{\otimes n} = \sum_{j=1}^{v_n} p_{j,n} \rho^{j,n}_{\text{mix}}$; therefore, $p_{j,n}$ is a probability distribution. Applying (8.46), we have

$$\sum_{j=1}^{v_n} p_{j,n} I(\rho^{j,n}_{\text{mix}}, \kappa^{\otimes n}) + 2 \log v_n \geq I(\rho_M^{\otimes n}, \kappa^{\otimes n}).$$

Thus, there exists an integer $j_n \in [1, v_n]$ such that

$$I(\rho^{j_n,n}_{\text{mix}}, \kappa^{\otimes n}) + 2 \log v_n \geq I(\rho_M^{\otimes n}, \kappa^{\otimes n}).$$

From Lemma 3.9, since $\frac{2}{n} \log v_n \to 0$, we obtain (9.47). This shows the existence of a code attaining the bound.

Next, we show that there is no code that exceeds the RHS of (9.43). Given any pure state $\rho^{A',R}$ on $\mathcal{H}_{A'} \otimes \mathcal{H}_R$, a set of TP-CP maps $\{\varphi_e(j)\}$ from $\mathcal{H}_{A'}$ to $\mathcal{H}_A$, and a probability distribution p_j, we have

$$\begin{aligned} &\sum_j p_j D((\kappa \circ \varphi_e(j) \otimes \iota_R)(\rho^{A',R}) \| \sum_j p_j (\kappa \circ \varphi_e(j) \otimes \iota_R)(\rho^{A',R})) \\ &\leq \max_\rho I(\rho, \kappa), \end{aligned} \tag{9.48}$$

where we used (4.7) and (8.52). From (8.47) we obtain

$$\max_\rho I(\rho, \kappa^{\otimes n}) = n \max_\rho I(\rho, \kappa). \tag{9.49}$$

Using the Fano inequality appropriately as in (9.40), we show the $\leq$ part in (9.43)[Exe. 9.6]. ∎

Next, we examine the relation between the entanglement-assisted classical capacity $C^e_{c,e}(\kappa)$ and the classical capacity $C_c(\kappa)$. Let $\mathcal{H}_B$ be the output system of κ, $\mathcal{H}_R$ be the input system of κ, and $\mathcal{H}_R$ be a reference system of $\mathcal{H}_A$. Due to the relation (8.173), the quantity $C_d^{R\to B}(\rho)$ characterizes the classical capacity $C_c(\kappa)$ as

$$C_c(\kappa) = \sup_{|x\rangle\langle x|} C_d^{R\to B}((\kappa\otimes\iota_R)(|x\rangle\langle x|)).$$

Hence, from (8.178)

$$C_d^{R\to B}((\kappa\otimes\iota_R)(|x\rangle\langle x|)) \le I_{(\kappa\otimes\iota_R)(|x\rangle\langle x|)}(R:B), \tag{9.50}$$

i.e.,

$$C_c(\kappa) \le C^e_{c,e}(\kappa). \tag{9.51}$$

For the equality condition, the following lemma holds.

Theorem 9.6 *When channel κ is entanglement breaking and is written by a CONS $\{u_i^A\}$ on $\mathcal{H}_A$ as*

$$\kappa(\rho) = \sum_i \langle u_i^A|\rho|u_i^A\rangle \rho_i^B, \tag{9.52}$$

the equality of (9.51) holds. Conversely, when the equality of (9.51) holds, the channel essentially has the form of (9.52), i.e., there exists a state $\rho_{\max}$ such that $I(\rho_{\max},\kappa) = C^e_{c,e}(\kappa)$ and $\kappa|_{\mathrm{supp}(\rho_{\max})}$ has the form of (9.52), where $\mathrm{supp}(\rho_{\max})$ is the support of $\rho_{\max}$. Further, in the case of (9.52), the classical capacity is calculated as

$$C^e_{c,e}(\kappa) = \max_p H\left(\sum_i p_i\rho_i\right) - \sum_i p_i H(\rho_i). \tag{9.53}$$

Notice that the channel $\kappa|_{\mathrm{supp}(\rho_{\max})}$ is not necessarily the same as the channel κ. Indeed, there exists a counterexample κ such that the channel κ does not have the form of (9.52) while the equality of (9.51) holds, hence, the channel $\kappa|_{\mathrm{supp}(\rho_{\max})}$ has the form of (9.52)[Exe. 9.8].

From Theorem 9.6, we see that even if the channel κ is entanglement breaking, the equality does not necessarily hold, i.e., it is advantageous to use shared entanglement. This is because an entanglement breaking channel does not necessarily have the form (9.52). For example, we consider an entanglement breaking channel κ with the form

$$\kappa(\rho) = \sum_i (\mathrm{Tr}\, M_i\rho)|u_i^B\rangle\langle u_i^B|, \tag{9.54}$$

where $\{u_i^B\}$ is a CONS on $\mathcal{H}_B$ and $\boldsymbol{M} = \{M_i\}$ is a POVM one rank on $\mathcal{H}_A$. Then, the classical capacity $C^e_{c,e}(\kappa)$ is calculated as

$$
\begin{aligned}
C^e_{c,e}(\kappa) &= \sup_\rho H(\rho) + H\left(\sum_i (\operatorname{Tr} M_i \rho)|u^B_i\rangle\langle u^B_i|\right) \\
&\quad - H\left(\sum_i \operatorname{Tr}_A (M_i \otimes I_R |x\rangle\langle x|) \otimes |u^B_i\rangle\langle u^B_i|\right) \\
&= \sup_\rho H(\rho) = \log d_A,
\end{aligned}
$$

where $|x\rangle\langle x|$ is a purification of ρ. However, when the POVM $\boldsymbol{M}$ is given as

$$
M_0 = \frac{1}{2}|0\rangle\langle 0|,\ M_1 = \frac{1}{2}|1\rangle\langle 1|,\ M_2 = \frac{1}{2}|+\rangle\langle +|,\ M_3 = \frac{1}{2}|-\rangle\langle -|, \tag{9.55}
$$

the classical capacity without shared entanglement is calculated as Exe. 9.7

$$
C_c(\kappa) = C^e_c(\kappa) = \frac{1}{2}\log 2. \tag{9.56}
$$

Proof of Theorem 9.6 Assume that condition (9.52) holds. Let $U_{\theta_1,\ldots,\theta_{d_A}}$ be defined by $U_\theta \stackrel{\text{def}}{=} \sum_j e^{i\theta_j}|u^A_i\rangle\langle u^A_i|,\quad \theta = (\theta_1,\ldots,\theta_{d_A})$. Then, the channel κ has the invariance $\kappa(\rho) = \kappa(U_\theta \rho U^*_\theta)$. Hence, $I(\rho,\kappa) = I(U_\theta\rho U^*_\theta,\kappa)$. From (8.45)

$$
I(\rho,\kappa) \le I\left(\int U_\theta \rho U^*_\theta d\theta, \kappa\right).
$$

Since $\int U_\theta\rho U^*_\theta d\theta$ has eigenvectors $\{u^A_j\}$, we have

$$
\begin{aligned}
C^e_{c,e}(\kappa) &= \sup_p H\left(\sum_j p_j |u^A_j\rangle\langle u^A_j|\right) + H\left(\sum_i \langle u^A_i| \left(\sum_j p_j |u^A_j\rangle\langle u^A_j|\right)|u^A_i\rangle \rho_i\right) \\
&\quad - H\left(\sum_i (\operatorname{Tr}_A |u^A_i\rangle\langle u^A_i| \otimes I_R |x\rangle\langle x|) \otimes \rho_i\right) \\
&= \sup_p H(p) + H\left(\sum_i p_i\rho_i\right) - H\left(\sum_i p_i |u^R_i\rangle\langle u^R_i| \otimes \rho_i\right) \\
&= \sup_p H(p) + H\left(\sum_i p_i\rho_i\right) - H(p) - \sum_i p_i H(\rho_i) \\
&= \sup_p H\left(\sum_i p_i\rho_i\right) - \sum_i p_i H(\rho_i),
\end{aligned}
$$

where $|x\rangle$ is a purification of $\sum_j p_j|u^A_j\rangle\langle u^A_j|$. Hence, we obtain (9.53). In particular, the classical capacity is equal to $C_c(\kappa)$. That is, the equality of inequality (9.51) holds.

Next, we assume that the equality of (9.51) holds. Then, there exists a state $\rho_{\max}$ such that $I(\rho_{\max}, \kappa) = C^e_{c,e}(\kappa)$ and its purification $|x\rangle$ satisfies the equality in (9.50). Lemma 8.14 guarantees that there exist a CONS $\{u_i^R\}$ on $\mathcal{H}_R$, states ρ_i on $\mathcal{H}_B$, and a probability distribution p such that

$$\sum_i p_i |u_i^R\rangle\langle u_i^R| \otimes \rho_i^B = (\kappa \otimes \iota_R)(|x\rangle\langle x|).$$

Now, we let ρ^R be the reduced density of $|x\rangle\langle x|$. Using relation (5.9),

$$(\kappa|_{\mathrm{supp}(\rho_{\max})} \otimes \iota_R)(|\Phi_d\rangle\langle\Phi_d|) = \sum_i d p_i \left(\sqrt{\rho^R}^{-1} |u_i^R\rangle\langle u_i^R| \sqrt{\rho^R}^{-1}\right) \otimes \rho_i^B,$$

where d is the dimension of $\mathrm{supp}(\rho_{\max})$. Since $\sum_i d p_i \sqrt{\rho^R}^{-1} |u_i^R\rangle\langle u_i^R| \sqrt{\rho^R}^{-1}$ is the completely mixed state on $\mathrm{supp}(\rho_{\max})$, each u_i^R is an eigenvector of ρ^R with the eigenvalue q_i. Hence,

$$(\kappa|_{\mathrm{supp}(\rho_{\max})} \otimes \iota_R)(|\Phi_d\rangle\langle\Phi_d|) = \sum_i \frac{d p_i}{q_i} |u_i^R\rangle\langle u_i^R| \otimes \rho_i^B.$$

The discussion in Theorem 5.1 guarantees that the channel κ has the form (9.52). ■

Exercises

9.4 Show (9.41) using (5.86) and the monotonicity of the quantum relative entropy.

9.5 Show (9.48) using (8.52) and the inequality $D(\sum_j p_j (\kappa \circ \varphi_e(j) \otimes \iota_R)(\rho^{A',R}) \| \sum_j p_j (\kappa \circ \varphi_e(j)(\rho^{A'}) \otimes \rho^R) \geq 0$.

9.6 Show that the $\leq$ part of (9.43) by combining (9.48) and (9.49) with the Fano inequality.

9.7 Show that the channel κ defined by (9.53) and (9.55) satisfies the equation (9.56) following the steps below.

(**a**) Show that $C_c(\kappa) = \log 4 - \min_\theta f(\theta)$, where $f(\theta) \stackrel{\text{def}}{=} -\frac{1+\cos\theta}{4} \log \frac{1+\cos\theta}{4} - \frac{1-\cos\theta}{4} \log \frac{1-\cos\theta}{4} - \frac{1+\sin\theta}{4} \log \frac{1+\sin\theta}{4} - \frac{1-\sin\theta}{4} \log \frac{1-\sin\theta}{4}$.

(**b**) Show that $\frac{df}{d\theta}(\theta) = \frac{\sin\theta}{4} \log \frac{1+\cos\theta}{1-\cos\theta} + \frac{\cos\theta}{4} \log \frac{1-\sin\theta}{1+\sin\theta}$, and $\frac{d^2 f}{d\theta^2}(\theta) = \frac{\cos\theta}{4} \log \frac{1+\cos\theta}{1-\cos\theta} + \frac{\sin\theta}{4} \log \frac{1+\sin\theta}{1-\sin\theta} + 1$.

(**c**) Show the following table and the equation (9.56).

θ	0		$\frac{\pi}{4}$		$\frac{\pi}{2}$
$f(\theta)$	$\frac{3}{2}\log 2$	$\nearrow$	$\frac{2+\sqrt{2}}{4} \log \frac{8}{2+\sqrt{2}} + \frac{2-\sqrt{2}}{4} \log \frac{8}{2-\sqrt{2}}$	$\searrow$	$\frac{3}{2}\log 2$
$\frac{df}{d\theta}(\theta)$	0	$+$	0	$-$	0
$\frac{d^2 f}{d\theta^2}(\theta)$	∞	$\searrow$	-1	$\nearrow$	∞

9.8 Let κ_1 and κ_2 be channels from systems $\mathcal{H}_{A,1}$ and $\mathcal{H}_{A,2}$ to system $\mathcal{H}_B$, respectively. Assume that the states $\rho_{\max,1} \stackrel{\text{def}}{=} \operatorname{argmax}_\rho I(\rho, \kappa_1)$ and $\rho_{\max,2} \stackrel{\text{def}}{=} \operatorname{argmax}_\rho I(\rho, \kappa_2)$ satisfy that $\kappa_1(\rho_{\max,1}) = \kappa_2(\rho_{\max,2})$. Define the channel κ from the system $\mathcal{H}_{A,1} \oplus \mathcal{H}_{A,2}$ to the system $\mathcal{H}_B$ as $\kappa(\rho) = \kappa_1(P_1 \rho P_1) + \kappa_2(P_2 \rho P_2)$, where P_i is the projection to $\mathcal{H}_{A,i}$. Show that $C^e_{c,e}(\kappa) = \max(C^e_{c,e}(\kappa_1), C^e_{c,e}(\kappa_2))$ by using (9.46).

Further, show the equality of (9.51), i.e., $C_c(\kappa) = C^e_{c,e}(\kappa)$, even though κ_2 does not satisfy (9.52) if κ_1 satisfies (9.52) and $C^e_{c,e}(\kappa_1) \geq C^e_{c,e}(\kappa_2)$.

9.4 Quantum Channel Resolvability

In this section, we examine the problem of approximating a given quantum state on the output of a c-q channel. In this problem, we choose a finite number of input signals and approximate a desired quantum state by the average output state with the uniform distribution on the chosen input signals. Then, the task of this problem is to choose the support of the uniform distribution at the input system as small as possible while approximating the desired state by the average output state as accurately as possible.

The classical version of this problem is called **channel resolvability**. It was proposed by Han and Verdú [33, 34] in order to examine another problem called the **identification code** proposed by Ahlswede and Dueck [35]. The problem of approximating a quantum state at the output system of a c-q channel is analogously called **quantum-channel resolvability**. Hence, quantum-channel resolvability is expected to be useful for examining identification codes [36] for (classical-) quantum channels. Indeed, this problem essentially has been treated by Wyner [37] in order to evaluate the information of the eavesdropper. Hence, it is also a fundamental tool for the discussion of communications in the presence of an eavesdropper for the following reason. Regarding the channel connecting the sender to the eavesdropper, approximating two states on the output system is almost equivalent to making these two states indistinguishable for the eavesdropper. Its detail will be discussed in the next section.

Quantum-channel resolvability may be formulated as follows (Fig. 9.4) [33, 34]. Consider a c-q channel $W : \mathcal{X} \to \mathcal{S}(\mathcal{H})$ and a quantum state $\sigma \in W(\mathcal{P}(\mathcal{X}))$, and prepare a map φ from $\{1, \ldots, M\}$ to the alphabet set $\mathcal{X}$. Now, the sender chooses an element i of $\{1, \ldots, M\}$ according to the uniform distribution and sends the state $W_{\varphi(i)}$. The problem is then to determine how many (M) elements are required for

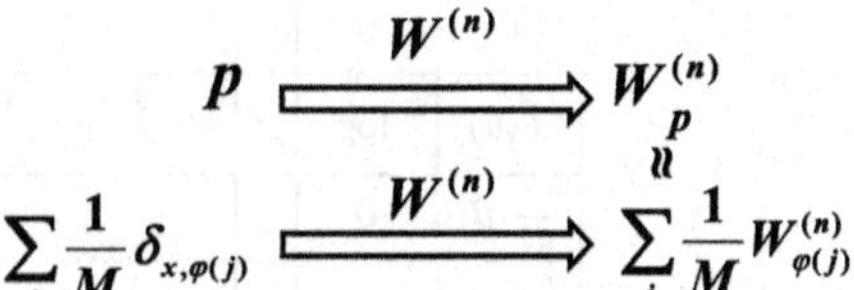

Fig. 9.4 Channel resolvability

sufficiently approximating the quantum state σ by the output average state $W_\varphi \stackrel{\rm def}{=} \frac{1}{M}\sum_{j=1}^{M} W_{\varphi(j)}$ of the c-q channel W. (Here, we are allowed to use input elements duplicately.) The quality of the approximation is evaluated by the trace norm $\| W_\varphi - \sigma\|_1$. Here, we choose the trace norm as the criterion of the approximation because it represents how well two states can be discriminated, as seen in Lemma 3.2. If the number M is sufficiently large, we can easily approximate the state $W_p = \sigma$ by the output average state W_φ. However, our aim is to approximate the state $W_p = \sigma$ with a small number M. One of the features of this problem is the following; Even when the distribution $p_\varphi(x) = \#\{\varphi^{-1}\{x\}\}/M$ at the input system is not close to p, the state $\sigma = W_p$ can be approximated by $W_{p_\varphi} = W_\varphi$ using the noise of channel W. In this case, our protocol is represented by $\Phi \stackrel{\rm def}{=} (M, \varphi)$, and its performance is by $M = |\Phi|$ and $\varepsilon[\sigma, \Phi] \stackrel{\rm def}{=} \left\| \left(\frac{1}{M}\sum_{j=1}^{M} W_{\varphi(j)}\right) - \sigma\right\|_1$. Here, we consider the performance of the approximation in the worst case as $\max_{p\in\mathcal{P}(\mathcal{X}^n)} \min_{\Phi:|\Phi|=M} \varepsilon[W_p^{(n)}, \Phi]$. Then, the asymptotic rate of its performance is given as the quantum-channel resolvability capacity[4];

$$C_r(W) \stackrel{\rm def}{=} \left\{ R \,\middle|\, \lim_{n\to\infty} \sup_{p\in\mathcal{P}(\mathcal{X}^n)} \inf_{\Phi:|\Phi|=e^{nR}} \varepsilon[W_p^{(n)}, \Phi] = 0 \right\}. \tag{9.57}$$

Theorem 9.7 *The quantum-channel resolvability capacity $C_r(W)$ satisfies*

$$C_r(W) \le C_c(W) = \sup_p I(p, W). \tag{9.58}$$

To show Theorem 9.7, we prepare two lemmas as follows.

Lemma 9.2 *For a given state σ, a distribution p on the set $\mathcal{X}$, and a real number $s \in [0, 1]$, there exists a map φ from $\{1, \ldots, M\}$ to $\mathcal{X}$ satisfying*

$$\begin{aligned} &\left\| \left(\frac{1}{M}\sum_{i=1}^{M} W_{\varphi(i)}\right) - W_p \right\|_1 \\ \le & 4\sqrt{\sum_x p(x) \operatorname{Tr} W_x\{\kappa_\sigma(W_x) \ge \mathsf{M}\sigma\}} \\ &+ \sqrt{\frac{v}{\mathsf{M}} \mathrm{E}_x \operatorname{Tr} \sigma^{-1}\kappa_\sigma(W_x)^2\{\kappa_\sigma(W_x) < \mathsf{M}\sigma\}} \qquad (9.59)\\ \le & \max(4\sqrt{2}, \sqrt{2v})\mathsf{M}^{-\frac{s}{2}} e^{\frac{s}{2}J_{1+s}(p,\sigma,W)}, \qquad (9.60) \end{aligned}$$

where v is the number of eigenvalues of σ, E_x denotes the expectation under the distribution p, and κ_σ is the pinching map concerning the matrix σ, which is defined in (1.14).

[4]The subscript r indicates "resolvability".

Lemma 9.2 will be shown after the proof of Theorem 9.7. Using Lemma 9.2, we can show the following lemma.

Lemma 9.3 *For a given state σ, a distribution p on the set $\mathcal{X}$, and a real number $s \in [0, 1]$, there exists a map φ from $\{1, \dots, M\}$ to $\mathcal{X}$ satisfying*

$$\sup_{p \in \mathcal{P}(\mathcal{X})} \inf_{\Phi:|\Phi|=M} \varepsilon[W_p, \Phi] \le \max\left(4\sqrt{2}, \sqrt{2v}\right) \mathsf{M}^{-\frac{s}{2}} e^{\frac{s}{2} C^{\downarrow}_{1+s}(W)}, \tag{9.61}$$

where v is the number of eigenvalues of $\sigma_{1+s|p_{1+s}}$. Remember that p_{1+s} and $\sigma_{1+s|p}$ are defined in (4.62) and (4.23), respectively.

Proof of Lemma 9.3 We apply Lemma 9.2 to the case with $\sigma = \sigma_{1+s|p_{1+s}}$. Hence, we obtain

$$\inf_{\Phi:|\Phi|=M} \varepsilon[W_p, \Phi] \le \max\left(4\sqrt{2}, \sqrt{2v}\right) \mathsf{M}^{-s/2} e^{\frac{s}{2} J_{1+s}(p, \sigma_{1+s|p_{1+s}}, W)}.$$

Taking the supremum for p, we have

$$\begin{aligned}
&\sup_{p \in \mathcal{P}(\mathcal{X})} \inf_{\Phi:|\Phi|=M} \varepsilon[W_p, \Phi] \\
\le &\max\left(4\sqrt{2}, \sqrt{2v}\right) \mathsf{M}^{-s/2} e^{\frac{s}{2} \sup_{p \in \mathcal{P}(\mathcal{X})} J_{1+s}(p, \sigma_{1+s|p_{1+s}}, W)} \\
\overset{(a)}{=} &\max\left(4\sqrt{2}, \sqrt{2v}\right) \mathsf{M}^{-s/2} e^{\frac{s}{2} C^{\downarrow}_{1+s}(W)},
\end{aligned}$$

where (a) follows from (4.74). ■

Proof of Theorem 9.7 Assume $R > C_c(W)$, and choose $M = e^{nR}$. Now, we denote the state $\sigma_{1+s|p_{1+s}}$ for the channel $W^{(n)}$ by $\sigma^{(n)}$. Since the additivity of $C^{\downarrow}_{1+s}(W)$ (4.76) implies $C^{\downarrow}_{1+s}(W^{(n)}) = nC^{\downarrow}_{1+s}(W)$, applying Lemma 9.3 to the channel $W^{(n)}$, we have

$$\sup_{p \in \mathcal{P}(\mathcal{X}^n)} \inf_{\Phi:|\Phi|=e^{nR}} \varepsilon[W^{(n)}_p, \Phi] \le \max\left(4\sqrt{2}, \sqrt{2v_n}\right) e^{\frac{s}{2} n (C^{\downarrow}_{1+s}(W) - R)} \tag{9.62}$$

for $s \in [0, 1]$. Then, due to the discussion in the solution of Exercise 4.74, we find that $\sigma^{(n)} = (\sigma^{(1)})^{\otimes n}$. Hence, the number v_n increases only polynomially. Therefore, the RHS of (9.62) goes to zero exponentially. ■

Proof of Lemma 9.2 We prove (9.59) and (9.60) by employing the random coding method. Let $X \overset{\text{def}}{=} (x_1, \dots, x_M)$ be M independent random variables subject to a probability distribution p in $\mathcal{X}$. Consider a protocol (M, φ) such that $\varphi(i) = x_i$. Denoting the expectation by E_X, we will show that

$$\mathrm{E}_X \left\| \left(\frac{1}{M} \sum_{i=1}^{M} W_{x_i} \right) - W_p \right\|_1$$

$$\leq 4\sqrt{\sum_x p(x) \operatorname{Tr} W_x\{\kappa_\sigma(W_x) \geq \mathsf{M}\sigma\}}$$

$$+ \sqrt{\frac{v}{\mathsf{M}} \mathrm{E}_x \operatorname{Tr} \sigma^{-1} \kappa_\sigma(W_x)^2 \{\kappa_\sigma(W_x) < \mathsf{M}\sigma\}}, \tag{9.63}$$

$$\leq \max\left(4\sqrt{2}, \sqrt{2v}\right) \mathsf{M}^{-\frac{s}{2}} e^{\frac{s}{2} J_{1+s}(p,\sigma,W)}. \tag{9.64}$$

Now, define $P_x \stackrel{\text{def}}{=} \{\kappa_{W_p}(W_x) \geq C W_p\}$, $P_x^c \stackrel{\text{def}}{=} I - P_x$, and $W_p' \stackrel{\text{def}}{=} \sum_x p(x) P_x^c W_x P_x^c$. Exercise 6.8 implies

$$\|W_x P_x\|_1 \leq \sqrt{\operatorname{Tr} W_x P_x} \tag{9.65}$$

$$\|P_{x_i} W_{x_i} P_x^c\|_1 \leq \sqrt{(\operatorname{Tr} P_{x_i} W_{x_i})(\operatorname{Tr} W_{x_i} P_x^c)} \leq \sqrt{\operatorname{Tr} W_x P_x}. \tag{9.66}$$

Since $W_p' - W_p = \sum_x p(x)(W_{x_i} P_x + P_{x_i} W_{x_i} P_x^c)$, we have $\|W_p' - W_p\|_1 \leq \sum_x p(x)(\|W_{x_i} P_x\|_1 + \|P_{x_i} W_{x_i} P_x^c\|_1) \leq 2\sum_x p(x)\sqrt{\operatorname{Tr} W_x P_x} \leq 2\sqrt{\sum_x p(x) \operatorname{Tr} W_x P_x} = 2\sqrt{\mathrm{E}_x \operatorname{Tr} W_x P_x}$. Thus,

$$\begin{aligned}
&\mathrm{E}_X \left\| \left(\frac{1}{\mathsf{M}} \sum_{i=1}^{\mathsf{M}} W_{x_i}\right) - W_p \right\|_1 \\
=&\mathrm{E}_X \left\| \left(\frac{1}{\mathsf{M}} \sum_{i=1}^{\mathsf{M}} W_{x_i} P_x + P_{x_i} W_{x_i} P_x^c\right) + (W_p' - W_p) + \left(\frac{1}{\mathsf{M}} \sum_{i=1}^{\mathsf{M}} P_{x_i}^c W_{x_i} P_{x_i}^c\right) - W_p' \right\|_1 \\
\leq&\mathrm{E}_X \left\| \frac{1}{\mathsf{M}} \sum_{i=1}^{\mathsf{M}} (P_{x_i}^c W_{x_i} P_{x_i}^c - W_p') \right\|_1 + 2\sqrt{\mathrm{E}_x \operatorname{Tr} W_x P_x} \\
&+ \mathrm{E}_X \left[\frac{1}{\mathsf{M}} \sum_{i=1}^{\mathsf{M}} \|W_{x_i} P_{x_i}\|_1 + \|P_{x_i} W_{x_i} P_{x_i}^c\|_1\right] \\
\leq&\mathrm{E}_X \left\| \frac{1}{\mathsf{M}} \sum_{i=1}^{\mathsf{M}} (P_{x_i}^c W_{x_i} P_{x_i}^c - W_p') \right\|_1 + 4\sqrt{\mathrm{E}_x \operatorname{Tr} W_x P_x}.
\end{aligned} \tag{9.67}$$

Thus, Exercise 6.10 yields

$$\begin{aligned}
&\left\| \frac{1}{M} \sum_{i=1}^{M} \left(P_{x_i}^c W_{x_i} P_{x_i}^c - W_p'\right) \right\|_1 \\
\leq &\sqrt{\operatorname{Tr} \sigma^{-1/2} \left(\frac{1}{M} \sum_i P_{x_i}^c W_{x_i} P_{x_i}^c - W_p'\right) \sigma^{-1/2} \left(\frac{1}{M} \sum_j P_{x_j}^c W_{x_j} P_{x_j}^c - W_p'\right)}.
\end{aligned} \tag{9.68}$$

Since the random variables x_i are independent of each other,

$$\begin{aligned}
&\mathrm{E}_X \operatorname{Tr} \sigma^{-1/2}\left(\frac{1}{\mathsf{M}}\sum_{i=1}^{\mathsf{M}} P^c_{x_i} W_{x_i} P^c_{x_i} - W'_p\right)\sigma^{-1/2}\left(\frac{1}{\mathsf{M}}\sum_{j=1}^{\mathsf{M}} P^c_{x_j} W_{x_j} P^c_{x_j} - W'_p\right) \\
=&\mathrm{E}_X \frac{1}{\mathsf{M}^2}\sum_{i=1}^{\mathsf{M}} \operatorname{Tr} \sigma^{-1/2} P^c_{x_i} W_{x_i} P^c_{x_i} \sigma^{-1/2} P^c_{x_i} W_{x_i} P^c_{x_i} - \operatorname{Tr} \sigma^{-1/2} W'_p \sigma^{-1/2} W'_p \\
\le&\mathrm{E}_x \frac{1}{\mathsf{M}} \operatorname{Tr} \sigma^{-1/2} P^c_x W_x P^c_x \sigma^{-1/2} P^c_x W_x P^c_x
\end{aligned} \tag{9.69}$$

$$\begin{aligned}
\le&\frac{1}{\mathsf{M}}\mathrm{E}_x \operatorname{Tr} \sigma^{-1/2} P^c_x v \kappa_\sigma(W_x) P^c_x \sigma^{-1/2} P^c_x W_x P^c_x \\
=&\frac{1}{\mathsf{M}}\mathrm{E}_x v \operatorname{Tr} \sigma^{-1/2} P^c_x \kappa_\sigma(W_x) P^c_x \sigma^{-1/2} P^c_x \kappa_\sigma(W_x) P^c_x \\
=&\frac{v}{\mathsf{M}}\mathrm{E}_x \operatorname{Tr} \sigma^{-1} \kappa_\sigma(W_x)^2 \{\kappa_\sigma(W_x) < \mathsf{M}\sigma\}.
\end{aligned} \tag{9.70}$$

Thus, using these relations as well as Jensen's inequality for $x \mapsto -\sqrt{x}$, we obtain

$$\begin{aligned}
&\mathrm{E}_X \left\| \frac{1}{\mathsf{M}^2}\sum_{i=1}^{\mathsf{M}} (P^c_{x_i} W_{x_i} P^c_{x_i} - W'_p) \right\|_1 \\
\le&\mathrm{E}_X \sqrt{\frac{v}{\mathsf{M}^2}\sum_{i=1}^{\mathsf{M}} \operatorname{Tr} \sigma^{-1} \kappa_\sigma(W_{x_i})^2 \{\kappa_\sigma(W_{x_i}) < \mathsf{M}\sigma\}} \\
\le&\sqrt{\mathrm{E}_X \frac{v}{\mathsf{M}^2}\sum_{i=1}^{\mathsf{M}} \operatorname{Tr} \sigma^{-1} \kappa_\sigma(W_{x_i})^2 \{\kappa_\sigma(W_{x_i}) < \mathsf{M}\sigma\}} \\
\le&\sqrt{\frac{v}{\mathsf{M}}\mathrm{E}_x \operatorname{Tr} \sigma^{-1} \kappa_\sigma(W_x)^2 \{\kappa_\sigma(W_x) < \mathsf{M}\sigma\}}.
\end{aligned}$$

Since $\sqrt{a} + \sqrt{b} = 2(\sqrt{a}/2 + \sqrt{b}/2) \le 2\sqrt{a/2 + b/2} = \sqrt{2(a+b)}$ for $a, b \ge 0$,

$$\begin{aligned}
&\mathrm{E}_X \left\| \left(\frac{1}{M}\sum_{i=1}^{M} W_{x_i}\right) - W_p \right\|_1 \\
\le&4\sqrt{\mathrm{E}_x \operatorname{Tr} W_x \{\kappa_\sigma(W_x) \ge \mathsf{M}\sigma\}} + \sqrt{\frac{v}{\mathsf{M}}\mathrm{E}_x \operatorname{Tr} \sigma^{-1} \kappa_\sigma(W_x)^2 \{\kappa_\sigma(W_x) < \mathsf{M}\sigma\}} \\
\le&\sqrt{32\mathrm{E}_x \operatorname{Tr} W_x \{\kappa_\sigma(W_x) \ge \mathsf{M}\sigma\} + 2\frac{v}{\mathsf{M}}\mathrm{E}_x \operatorname{Tr} \sigma^{-1} \kappa_\sigma(W_x)^2 \{\kappa_\sigma(W_x) < \mathsf{M}\sigma\}}.
\end{aligned}$$

Since

$$\begin{aligned}
&32\mathrm{E}_x \operatorname{Tr} W_x \{\kappa_\sigma(W_x) \ge \mathsf{M}\sigma\} + 2\frac{v}{\mathsf{M}}\mathrm{E}_x \operatorname{Tr} \sigma^{-1} \kappa_\sigma(W_x)^2 \{\kappa_\sigma(W_x) < \mathsf{M}\sigma\} \\
\le&32\mathsf{M}^{-s}\mathrm{E}_x \operatorname{Tr} \sigma^{-s} \kappa_\sigma(W_x)^{1+s} \{\kappa_\sigma(W_x) \ge \mathsf{M}\sigma\}
\end{aligned}$$

$$
\begin{aligned}
&+ 2v\mathsf{M}^{-s}\mathrm{E}_x \operatorname{Tr} \sigma^{-s} \kappa_\sigma(W_x)^{1+s}\{\kappa_\sigma(W_x) < C\sigma\} \\
&\le \max(32, 2v)\mathsf{M}^{-s}\mathrm{E}_x \operatorname{Tr} \sigma^{-s} \kappa_\sigma(W_x)^{1+s} \\
&= \max(32, 2v)\mathsf{M}^{-s} e^{s J_{1+s}(p,\sigma,\kappa_\sigma(W))} = \max(32, 2v)\mathsf{M}^{-s} e^{s J_{1+s}(p,\kappa_\sigma(\sigma),\kappa_\sigma(W))} \\
&\overset{(a)}{\le} \max(32, 2v)\mathsf{M}^{-s} e^{s J_{1+s}(p,\sigma,W)},
\end{aligned}
$$

where $\kappa_\sigma(W)$ is the c-q channel $x \mapsto \kappa_\sigma(W_x)$, and (a) follows from (5.60). Hence, we obtain (9.64). ■

When channel W is a classical channel and the map $p \mapsto W_p$ is one-to-one, $C_r(W)$ is equal to $C_c(W)$. To see the detail of this fact, we discuss the relation between the identification codes and the channel resolvability. Ahlswede-Dueck [35] introduced the identification capacity $C_i(W)$ as the upper limit of the rate of identification codes, and showed that $C_i(W) = C_c(W)$ for the classical case. Han-Verdú [33] tackled the strong converse capacity $C_i(W)^\dagger$ for identification codes. For this purpose, they introduced the channel resolvability, they showed that $C_i^\dagger(W) \le C_r(W)$ and $C_r(W) \le C_c(W)$ for the classical case. The combination of these relations yields that $C_i(W) = C_i^\dagger(W) = C_r(W) = C_c(W)$. The same relation can be expected. Now, we consider the c-q channel W when the map $p \mapsto W_p$ is one-to-one. The proof for $C_i^\dagger(W) \le C_r(W)$ by Han-Verdú [33] is still valid even for the c-q channel. Using the general method by Han-Verdú [33], we can show that $C_i^\dagger(W) \ge C_i(W) \ge C_c(W)$ for the c-q channel. Hence, combining Theorem 9.7, we can show $C_i(W) = C_i^\dagger(W) = C_r(W) = C_c(W)$ even for the c-q channel.[5] Finally, we prove the direct part of Theorem 8.13 by using quantum-channel resolvability.

Proof of Direct Part of Theorem 8.13 Choose a probabilistic decomposition $(p_x, \rho_x^A \otimes \rho_x^B)_{x\in\mathcal{X}}$ of the separable state ρ as

$$
C(\rho) = I_{\rho^{ABE}}(AB : E), \quad \rho^{ABE} \stackrel{\text{def}}{=} \sum_i p_i \rho_i^A \otimes \rho_i^B \otimes |u_i^E\rangle\langle u_i^E|. \tag{9.71}
$$

For any $\epsilon > 0$, we let $M_n = e^{n(C(\rho)+\epsilon)}$. Due to Lemma 9.2, we can choose M_n indexes $\varphi(1), \dots, \varphi(M_n)$ in $\mathcal{X}^n$ such that

$$
\left\| \frac{1}{M_n} \sum_{i=1}^{M_n} \rho_{\varphi(i)}^{A,(n)} \otimes \rho_{\varphi(i)}^{B,(n)} - \rho^{\otimes n} \right\|_0 \to 0,
$$

which implies the direct part of Theorem 8.13. ■

Exercises

9.9 Show that

$$
\sup_{\{\Phi^{(n)}\}} \left\{ \lim_{n\to\infty} \frac{-\log \varepsilon[W_p^{\otimes n}, \Phi^{(n)}]}{n} \middle| \lim_{n\to\infty} \frac{\log |\Phi^{(n)}|}{n} \le R \right\}
$$

[5] Ahlswede-Winter [36] also showed that $C_i(W) = C_i^\dagger(W) = C_c(W)$ in a different way.

$$\geq \max_{s \leq 0} \frac{s}{2}(R - I^{\downarrow}_{1+s}(p, W)) \tag{9.72}$$

by using Lemma 9.2.

9.10 Assume that all the output states W_x commute. Show that there exists a map $\varphi : \{1, \ldots, M\} \to \mathcal{X}$ such that

$$\begin{aligned} &\left\| \left(\frac{1}{M} \sum_{i=1}^{M} W_{\varphi(i)} \right) - W_p \right\|_1 \\ \leq & 2 \sum_x p(x) \operatorname{Tr} W_x \{W_x - C W_p \geq 0\} + \sqrt{\frac{C}{M}} \end{aligned}$$

as a modification of (9.59) of Lemma 9.2.

9.5 Quantum-Channel Communications with an Eavesdropper

9.5.1 C-Q Wiretap Channel

The **BB84 protocol** [38] enables us to securely distribute a secret key using a quantum system. Experiments realizing this protocol have been performed [39, 40], with successful transmissions over 150 km [41, 42] via optical fibers. Therefore, the protocol is almost at a practically usable stage. In the original proposal of the BB84 protocol, it was assumed that there was no noise in the channel. However, a real channel always has some noise. In the presence of noise, the noise can be used by an eavesdropper to mask his/her presence while obtaining information from the channel. Therefore, it is necessary to communicate on the assumption that an eavesdropper may obtain a certain amount of information.

This type of communication is called a **wiretap channel** and was first considered by Wyner [37] for the classical case. Its quantum-mechanical extension, i.e., a **classical-quantum wiretap channel** (**c-q wiretap channel**) was examined by Devetak [43]. In this communication, we require a code such that the authorized receiver can accurately recover the original message and the eavesdropper cannot obtain any information concerning the original message. Hence, one of the main problems in this communication is to find the bound of the communication rate of the code. Although this problem is not the same problem as the BB84 protocol itself, it will lead us to a proof of its security even in the presence of noise.

Let $\mathcal{H}_B$ be the system received by the authorized receiver, $\mathcal{H}_E$ be the system received by the eavesdropper, and W_x be an output state on the composite system $\mathcal{H}_B \otimes \mathcal{H}_E$ when the sender sends an alphabet $x \in \mathcal{X}$. Hence, the authorized

Fig. 9.5 Wiretap channel

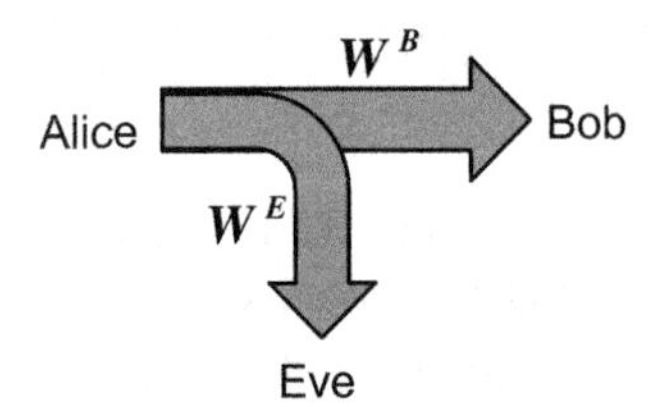

receiver receives the state $W_x^B \stackrel{\text{def}}{=} \operatorname{Tr}_E W_x$, and the eavesdropper receives the state $W_x^E \stackrel{\text{def}}{=} \operatorname{Tr}_B W_x$ as Fig. 9.5. In this case, we use a probabilistic code as follows. When the sender wishes to send a message $m \in \{1, \ldots, M\}$, he or she transmits the alphabet $x \in \mathcal{X}$ according to the probability distribution Q^m in $\mathcal{X}$ dependently on the message m. That is, the encoding process is described by a stochastic transition matrix Q from $\{1, \ldots, M\}$ to $\mathcal{X}$. Then, the authorized receiver performs the M-valued POVM $Y = \{Y_{m'}\}_{m'=1}^M$ and receives the signal m'. Therefore, our protocol is described by $\Phi = (M, Q, Y)$ and evaluated by the following three quantities. The first quantity is the size of the protocol $|\Phi| \stackrel{\text{def}}{=} M$, and the second is the error probability of the authorized receiver $\varepsilon[\Phi] \stackrel{\text{def}}{=} \frac{1}{M}\sum_{m=1}^M \left(1 - \operatorname{Tr}(W^B Q)_m Y_m\right)$, where $(W^B Q)_m \stackrel{\text{def}}{=} \sum_{x\in\mathcal{X}} W_x^B Q_x^m$. The third quantity is the upper bound of the eavesdropper's information $I_E(\Phi) \stackrel{\text{def}}{=} I(p_{\text{mix}}^M, W^E Q)$, where $(W^E Q)_m \stackrel{\text{def}}{=} \sum_{x\in\mathcal{X}} W_x^E Q_x^m$. Instead of $I_E(\Phi)$, we often employ $d_1(\Phi) := \min_\sigma \sum_{m=1}^{\mathrm{M}} \frac{1}{2\mathrm{M}} \|(W^E Q)_m - \sigma\|_1$.

Let us now examine the **wiretap channel capacity**, i.e., the bound of the communication rate $\frac{1}{n}\log|\Phi^{(n)}|$, for the asymptotically reliable protocol $\{\Phi^{(n)}\}$ with the stationary memoryless channel $W^{(n)}$, i.e., n times use of W. This is given by

$$C_c^{B,E}(W) \stackrel{\text{def}}{=} \sup_{\{\Phi^{(n)}\}} \left\{ \underline{\lim} \frac{1}{n}\log|\Phi^{(n)}| \,\middle|\, \varepsilon[\Phi^{(n)}] \to 0,\, I_E(\Phi^{(n)}) \to 0 \right\}. \tag{9.73}$$

Theorem 9.8 (Devetak [43]) *The wiretap channel capacity $C_c^{B,E}(W)$ satisfies*

$$C_c^{B,E}(W) = \overline{\lim} \frac{1}{n} \sup_Q \sup_p \left(I(p, W^{B,(n)}Q) - I(p, W^{E,(n)}Q) \right). \tag{9.74}$$

If every W_x^E can be written as $W_x^E = \kappa(W_x^B)$, using a completely positive map κ from $\mathcal{H}_B$ to $\mathcal{H}_E$, it is called a **quantum degraded channel**, and it satisfies

$$C_c^{B,E}(W) = \sup_p \left(I(p, W^B) - I(p, W^E) \right). \tag{9.75}$$

It is also proved in Sect. 9.5.6. Further, a quantum degraded channel (W^B, W^E) satisfies [Exe. 9.19]

$$I(Qp, W^B) - I(Qp, W^E) \geq \sum_i p_i (I(Q^i, W^B) - I(Q^i, W^E)). \tag{9.76}$$

That is, $I(p, W^B) - I(p, W^E)$ satisfies the concavity in this case.

Let us suppose that W^B is given by a TP-CP map κ from $\mathcal{H}_A$ to $\mathcal{H}_B$, and the channel to the eavesdropper W^E is given by a channel κ_E to the environment of κ. Under this assumption, the eavesdropper's state is always a state reduced from the state on the environment. That is, he/she has less information than the environment system. Hence, the eavesdropper's information can be sufficiently estimated by treating the case where the eavesdropper's state is equal to the state on the environment. Now, we consider the set of input signals $\mathcal{X}$ given as the set of pure states on the input system. Then, for any input pure state $|x\rangle$ in $\mathcal{H}_A^{\otimes n}$, the states $W_x^{B,(n)}$ and $W_x^{E,(n)}$ are given by $\kappa^{\otimes n}(|x\rangle\langle x|)$ and $\kappa_E^{\otimes n}(|x\rangle\langle x|)$, respectively. In this scheme, any entangled state is allowed as the input state. From $H(W_x^B) = H(W_x^E)$ and (8.54), any state $\rho = \sum_i p_i |u_i\rangle\langle u_i|$ satisfies

$$\begin{aligned} & I(p, W^B) - I(p, W^E) \\ =& H(\kappa(\rho)) - \sum_x p_x H(W_x^B) - \left(H(\mathrm{Tr}_B(U_\kappa \rho U_\kappa^*)) - \sum_x p_x H(W_x^E) \right) \\ =& H(\kappa(\rho)) - H(\mathrm{Tr}_B(U_\kappa \rho U_\kappa^*)) = I_c(\rho, \kappa). \end{aligned} \tag{9.77}$$

Hence, letting $C_c^{e,B,E}(\kappa)$ be the asymptotic bound of the secure communication rate when any state on $\mathcal{H}_A^{\otimes n}$ is allowed as an input state, we can show that[Exe. 9.13]

$$C_c^{e,B,E}(\kappa) \geq \lim_{n \to \infty} \frac{1}{n} \max_{\rho \in \mathcal{S}(\mathcal{H}_A^{\otimes n})} I_c(\rho, \kappa^{\otimes n}). \tag{9.78}$$

In addition, the following **monotonicity** also holds with respect to the eavesdropper's information:

$$I(\{p_i\}, \{\kappa'_E \circ \kappa(\rho_i)\}) \leq I(\{p_i\}, \{(\kappa' \circ \kappa)_E(\rho_i)\}), \tag{9.79}$$

$$I(\{p_i\}, \{\kappa_E(\rho_i)\}) \leq I(\{p_i\}, \{(\kappa' \circ \kappa)_E(\rho_i)\}). \tag{9.80}$$

9.5.2 Relation to BB84 Protocol

Let us now relate these arguments to the BB84 protocol discussed earlier. In the BB84 protocol, the sender A transmits a state chosen from $e_0, e_1, e_+ \stackrel{\mathrm{def}}{=} \frac{1}{\sqrt{2}}(e_0 + e_1)$, and $e_- \stackrel{\mathrm{def}}{=} \frac{1}{\sqrt{2}}(e_0 - e_1)$ with an equal probability. The receiver B then chooses one of the two measurement bases $\{|e_0\rangle\langle e_0|, |e_1\rangle\langle e_1|\}$ and $\{|e_+\rangle\langle e_+|, |e_-\rangle\langle e_-|\}$ with an equal probability and performs this measurement on the received quantum system. Then, the authorized receiver B sends his/her measurement outcome to the sender A via a public channel. The sender A tells the authorized receiver B whether the original state belongs to the set $\{e_0, e_1\}$ or $\{e_+, e_-\}$ via a public channel. This determines whether the basis used by the sender A coincides with the basis by the authorized receiver B. The bases should agree for approximately half number of the transmitted states, which is numbered by n. They choose ϵn bits randomly among these obtained n bits and announce the information of these ϵn bits using the public channel in order to verify whether these bits coincide with each other (ϵ is a suitably chosen positive real number). When they find a bit with disagreement, the sender A and the authorized receiver B conclude that an eavesdropper was present. Otherwise, both parties can conclude that they succeeded in sharing a secret key X without divulging information to any third party. Finally, the sender encrypts the information Y_A to be sent according to the conversion $Z = X + Y_A$ (mod 2). The encrypt message Y may be decrypted according to the conversion $Y_B = Z + X$, thereby obtaining secure communication.

In reality, the bits held by A and B may not agree due to noise even if an eavesdropper is not present. In this case, we must estimate the quantum channel κ connecting the sender to the receiver is partially leaked to the third party. Consider a case in which the sender sends bits based on the basis $\{e_0, e_1\}$, and the receiver detects the bits through the measurement $E = \{|e_i\rangle\langle e_i|\}_{i=0}^1$. Now, let X_A and X_B be the random bits sent by the sender and the random bits detected by the authorized receiver through the measurement, respectively. When the sender transmits bit i, the authorized receiver obtains his/her outcome subject to the distribution $\mathrm{P}^E_{\kappa(e_i)}$. By performing the communication steps as described above, the stochastic transition matrix Q joining Y_A and Y_B is given by

$$Q_0^0 = Q_1^1 = \frac{1}{2}\mathrm{P}^E_{\kappa(e_0)}(0) + \frac{1}{2}\mathrm{P}^E_{\kappa(e_1)}(1), \; Q_1^0 = Q_0^1 = \frac{1}{2}\mathrm{P}^E_{\kappa(e_0)}(1) + \frac{1}{2}\mathrm{P}^E_{\kappa(e_1)}(0),$$

which is the same as that for a noisy classical channel. Using a suitable coding protocol, the sender and the authorized receiver can communicate with almost no error and almost no information leakage.

Let us now estimate the amount of information leaked to the eavesdropper. In this case, it is impossible to distinguish the eavesdropping from the noise in the channel. For this reason, we assume that any information lost has been caused by the interception by the eavesdropper. Consider the case in which each bit is independently eavesdropped, i.e., the quantum channel from the state inputted by the sender to the state intercepted by the eavesdropper is assumed to be stationary memoryless. Therefore, if the sender transmits the state e_i, the eavesdropper obtains the state $\kappa_E(|e_i\rangle\langle e_i|)$, where κ_E was defined in (5.7). Since the eavesdropper knows Z, he/she possesses the state on the composite system $\mathcal{H}_E \otimes \mathbb{C}^2$ consisting of the quantum system $\mathcal{H}_E$ and the classical system $\mathbb{C}^2$ corresponding to Z. For example, if $Y_A = i$, the state W_i^E obtained by the eavesdropper is

$$W_0^E = \begin{pmatrix} \frac{1}{2}\kappa_E(|e_0\rangle\langle e_0|) & 0 \\ 0 & \frac{1}{2}\kappa_E(|e_1\rangle\langle e_1|) \end{pmatrix},$$
$$W_1^E = \begin{pmatrix} \frac{1}{2}\kappa_E(|e_1\rangle\langle e_1|) & 0 \\ 0 & \frac{1}{2}\kappa_E(|e_0\rangle\langle e_0|) \end{pmatrix}.$$

We may therefore reduce this problem to the c-q wiretap channel problem discussed previously [44]. In particular, if κ is a Pauli channel κ_p,

$$I(p_{\mathrm{mix}}, Q) - I(p_{\mathrm{mix}}, W^E) = \log 2 - H(p), \tag{9.81}$$

which is a known quantity in quantum key distribution [45].

In practice, it is not possible to estimate κ completely using communications that use only e_0, e_1, e_+, e_-. However, it is possible to estimate $I(p, W^E)$.[6] Since the encoding constructed in the proof of Theorem 9.8 depends on the form of W^E, it is desirable to construct a protocol that depends only on the value of $I(p, W^E)$.

9.5.3 Secret Sharing

Let us consider an application of the above discussion to a protocol called **secret sharing**. In secret sharing, there are m receivers, and the encoded information sent by the sender can be recovered only by combining the information of m receivers. Therefore, a single receiver cannot obtain the encoded information [49, 50].

Denote the channel from the sender to each receiver by W [51, 52]. The transmission information possessed by one receiver is equal to $I(p, W)$. The transmission information possessed by m receivers is therefore $mI(p, W)$. Theorem 9.8 guarantees that performing the communication n times, the sender can transmit almost $n(m-1)I(p, W)$ bits of information with no leakage to each receiver. That is, the problem is to ensure that the information possessed by an individual receiver approaches zero asymptotically. The random coding method used in the proof of Lemma 9.4 may be used to show the existence of such a code. Let $I_i(\Phi_X)$ be the information possessed by the ith receiver for the code Φ_X. Let $\varepsilon[\Phi_X]$ be the average decoding error probability of combining the m receivers. Then, $\mathrm{E}_X[\varepsilon[\Phi_X]]$ satisfies (9.93), and it can be shown that $\sum_{i=1}^m \mathrm{E}_X[I_i(\Phi_X)] \le m(\epsilon_2 \log d + \eta_0(\epsilon_2))$. Therefore, $\mathrm{E}_X[\varepsilon[\Phi_X]]$ satisfies (9.93), and we can show that there exists a code Φ such that $\varepsilon[\Phi] + \sum_{i=1}^m I_i(\Phi_X) \le \epsilon_1 + m(\epsilon_2 \log d + \eta_0(\epsilon_2))$. Therefore, it is possible to securely transmit $n(m-1)I(p, W)$ bits of information asymptotically. Further, we can consider the capacity with the following requirement: There are m receivers, and the information can be recovered from composite quantum states by any n_1 receivers. However, it can be recovered not only by n_2 receivers. In this case, the capacity is

[6] By adding the states $e_+^* \stackrel{\mathrm{def}}{=} \frac{1}{\sqrt{2}}(e_0 + ie_1)$ and $e_-^* \stackrel{\mathrm{def}}{=} \frac{1}{\sqrt{2}}(e_0 - ie_1)$ in the transmission, and by adding the measurement $\{|e_+^*\rangle\langle e_+^*|, |e_-^*\rangle\langle e_-^*|\}$, it is possible to estimate κ. This is called the six-state method [46–48].

$(n_1 - n_2)C(W)$. It can be shown by the combination of the proofs of Corollary 4.1 and Theorem 9.8.

9.5.4 Distillation of Classical Secret Key

In addition, this approach can be applied to the distillation of a classical secret key from shared state ρ on the the composite system $\mathcal{H}_A \otimes \mathcal{H}_B \otimes \mathcal{H}_E$ as Fig. 9.6. Although we discussed a related topic in Sect. 8.14, the discussion in Sect. 8.14 considers only the information leakage, i.e., assumes that the information on the system $\mathcal{H}_A$ is the same as that on the system $\mathcal{H}_B$. In this subsection, the information on the system $\mathcal{H}_A$ is not necessarily the same as that on the system $\mathcal{H}_B$, i.e., there might exist a noise between the two systems $\mathcal{H}_A$ and $\mathcal{H}_B$.

In the distillation of a classical secret key, it is our task to generate a secret uniform random number shared by the two systems $\mathcal{H}_A$ and $\mathcal{H}_B$. That is, it is required that the eavesdropper's system $\mathcal{H}_E$ cannot hold any information concerning the distilled random number. Then, the optimal key rate with one-way ($A \to B$) communication is defined by

$$C_k^{A\to B-E}(\rho) \stackrel{\text{def}}{=} \sup_{\kappa_n} \left\{ \overline{\lim} \frac{\log L_n}{n} \left| \begin{array}{l} \| \operatorname{Tr}_E \kappa_n(\rho_n) - \rho_{\text{mix},L_n} \|_1 \to 0 \\ I_{\kappa_n(\rho_n)}(AB : E) \to 0 \end{array} \right. \right\}, \tag{9.82}$$

where $\rho_{\text{mix},L} = \frac{1}{L}\sum_{i=1}^{L} |e_i^A\rangle\langle e_i^A|$. For this analysis, we define the quantity $C^{A\to}(\rho)$:

$$\begin{aligned} & C_d^{A\to B-E}(\rho) \\ \stackrel{\text{def}}{=} & \max_M (H(\rho^B) - \sum_i \mathrm{P}_{\rho^A}^M(i) H(\rho_i^B) - H(\rho^E) + \sum_i \mathrm{P}_{\rho^A}^M(i) H(\rho_i^E)). \end{aligned} \tag{9.83}$$

From this definition, the quantity $C_d^{A\to B-E}(\rho)$ satisfies the monotonicity concerning the one-way $A \to B$ operation. Further, we can show Condition **E2** (continuity) similarly to $C_d^{A\to B}(\rho)$.

Using Theorem 9.8 and a discussion similar to Theorem 8.10, we obtain the following theorem.

Theorem 9.9 (Devetak and Winter [53])

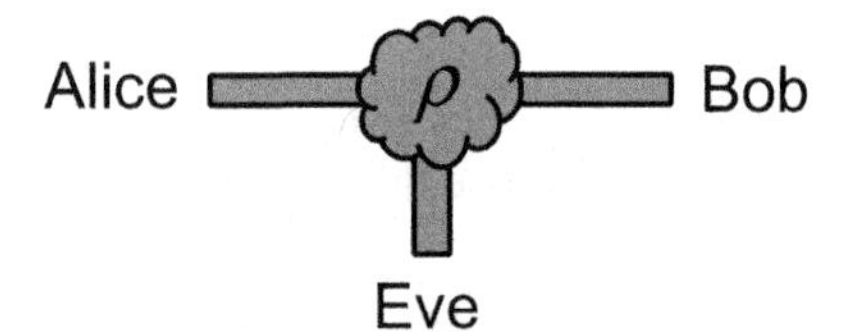

Fig. 9.6 Tripartite state

$$C_k^{A\to B-E}(\rho) = \lim_{n\to\infty} \frac{C_d^{A\to B-E}(\rho^{\otimes n})}{n}. \tag{9.84}$$

Further, if there exists a TP-CP map κ from $\mathcal{H}_B$ to $\mathcal{H}_E$ such that

$$\mathrm{Tr}_{AB}\,\rho(M\otimes I_{BE}) = \kappa(\mathrm{Tr}_{AE}\,\rho(M\otimes I_{BE})), \quad \forall M \geq 0, \tag{9.85}$$

we have

$$C_k^{A\to B-E}(\rho) = C_d^{A\to B-E}(\rho). \tag{9.86}$$

In particular, when ρ has the form $\rho^{AB}\otimes\rho^E$,

$$C_k^{A\to B-E}(\rho) = C_d^{A\to B}(\rho^{AB}). \tag{9.87}$$

Proof First, we prove the direct part:

$$C_k^{A\to B-E}(\rho) \geq \lim_{n\to\infty} \frac{C_d^{A\to B-E}(\rho^{\otimes n})}{n}. \tag{9.88}$$

For this purpose, we consider the following operation. Let M be a POVM on $\mathcal{H}_A$ attaining its maximum on (9.83), and $\{1,\ldots,l\}$ be its probability space. First, we define the channel W^B, W^E as the sender prepares the classical information $j \in \{1,\ldots,l\}$ and perform the measurement $\boldsymbol{M}$ on $\mathcal{H}_A$. Hence, the sender obtains the datum i as its outcome. He sends the classical information $k = i + j \bmod l$. Then, systems B and E receive this information k. Since the channel W^B, W^E is described as

$$W_j^B = \sum_i \mathrm{P}_{\rho^A}^{\boldsymbol{M}}(i)\rho_i^B \otimes |e_{i+j}\rangle\langle e_{i+j}|, \quad W_j^E = \sum_i \mathrm{P}_{\rho^A}^{\boldsymbol{M}}(i)\rho_i^E \otimes |e_{i+j}\rangle\langle e_{i+j}|,$$

we obtain

$$I(p_{\mathrm{mix}}, W^B) = I(\mathrm{P}_{\rho^A}^{\boldsymbol{M}}, \rho_\cdot^B) + H(p_{\mathrm{mix}}) - H(\mathrm{P}_{\rho^A}^{\boldsymbol{M}}),$$
$$I(p_{\mathrm{mix}}, W^E) = I(\mathrm{P}_{\rho^A}^{\boldsymbol{M}}, \rho_\cdot^E) + H(p_{\mathrm{mix}}) - H(\mathrm{P}_{\rho^A}^{\boldsymbol{M}}).$$

Hence, Theorem 9.8 yields

$$C_k^{A\to B-E}(\rho) \geq C_d^{A\to B-E}(\rho).$$

Thus, we obtain (9.88).

Next, we prove the converse part:

$$C_k^{A\to B-E}(\rho) \leq \lim_{n\to\infty} \frac{C_d^{A\to B-E}(\rho^{\otimes n})}{n}. \tag{9.89}$$

As was mentioned above, the quantity $C_d^{A\to B-E}(\rho)$ satisfies the monotonicity and the continuity. Hence, from a discussion similar to that concerning Theorem 8.10, we can show inequality (9.89). Further, we can prove (9.86) based on a similar derivation as for (9.75). ∎

9.5.5 Proof of Direct Part of C-Q Wiretap Channel Coding Theorem

We consider the attainability of the RHS of (9.74). Given a map φ from $\{1,\ldots,M\}\times\{1,\ldots,L\}$ to $\mathcal{X}$, we define the distribution Q^m by $\sum_{l=1}^{L}\frac{1}{L}\delta_{\varphi(m,l)}$, where $\delta_{\varphi(m,l)}$ is the deterministic distribution taking values only in $\{\varphi(m,l)\}$. Then, for a POVM $Y=\{Y_{(m,l)}\}_{(m,l)}$, we denote the code (M,Q,Y) by $\Phi(\varphi,Y)$. Now, let us examine the following lemma.

Lemma 9.4 *Given a distribution p, let v be the number of eigenvalues of W_p^E. Define*

$$\epsilon_1 \stackrel{\rm def}{=} \min_{s\in[0,1]} 2^{1+s}(\mathsf{ML})^s e^{-sI_{1-s}(p,W^B)}$$

$$\epsilon_2 \stackrel{\rm def}{=} \min_{s\in[0,1]} \max\left(4\sqrt{2},\sqrt{2v}\right)\mathsf{L}^{-\frac{s}{2}}e^{\frac{s}{2}I_{1+s}^{\downarrow}(p,W^E)}$$

for integers M and L, where v is the number of eigenvalues of $\sigma_{1+s|p_{1+s}}$. There exist a map φ from $\{1,\ldots,M\}\times\{1,\ldots,L\}$ to $\mathcal{X}$ and a POVM $Y=\{Y_{(m,l)}\}_{(m,l)}$ such that

$$\epsilon[\Phi(\varphi,Y)]\le 3\epsilon_1,\quad d_1(\Phi(\varphi,Y))\le 3\epsilon_2, \tag{9.90}$$

$$I_E(\Phi(\varphi,Y))\le 3(\epsilon_2\log d+\eta_0(\epsilon_2)). \tag{9.91}$$

When we only focus on $\epsilon[\Phi(\varphi,Y)]$ and $d_1(\Phi(\varphi,Y))$, there exist a one-to-one map φ from $\{1,\ldots,M\}\times\{1,\ldots,L\}$ to $\mathcal{X}$ and a POVM $Y=\{Y_{(m,l)}\}_{(m,l)}$ such that

$$\epsilon[\Phi(\varphi,Y)]\le 2\epsilon_1,\quad d_1(\Phi(\varphi,Y))\le 2\epsilon_2. \tag{9.92}$$

Proof Apply the random coding method. That is, for each pair (m,l), let $\varphi(m,l)$ be given by the independent and identical random variables $x_{m,l}$ subject to the probability distribution p. Using the random variable $X=(x_{m,l})$, we denote $\varphi=(\varphi(m,l))$ by φ_X. Hence, this protocol is determined by X and is denoted by $\Phi_X=\Phi(\varphi_X,Y_X)$. Denoting the expectation by E_X, (4.53) and (4.49) yield

$$\mathrm{E}_X\left[\varepsilon[\Phi_X]\right]\le\epsilon_1. \tag{9.93}$$

This is because the error probability can be reduced further than the case when ML messages are transmitted since only M messages $\{\sum_{l=1}^{L}W_{\varphi(m,l)}\}_m$ are transmitted and decoded. Also (9.64) with $\sigma=\sigma_{1+s|p_{1+s}}$ yields

$$
\mathrm{E}_X d_1(\Phi_X) \leq \mathrm{E}_X \sum_{m=1}^{\mathsf{M}} \frac{1}{2\mathsf{M}} \| \frac{1}{L} \sum_{l=1}^{L} W^E_{\varphi_X(m,l)} - W^E_p \|_1 \leq \epsilon_2. \tag{9.94}
$$

Applying Exercise 5.36 to (9.60), we immediately obtain

$$
\mathrm{E}_X \left[\frac{1}{M} D((W^E Q_X)_m \| W^E_p) \right] \leq \epsilon_2 \log d + \eta_0(\epsilon_2).
$$

Thus,

$$
\begin{aligned}
\mathrm{E}_X [I_E(\Phi_X)] =& \mathrm{E}_X \left[H\left(\sum_{m'=1}^{M} \frac{1}{M} (W^E Q_X)_{m'} \right) - \sum_{m=1}^{M} \frac{1}{M} H((W^E Q_X)_m) \right] \\
=& \mathrm{E}_X \left[\sum_{m=1}^{M} \frac{1}{M} D\left((W^E Q_X)_m \| W^E_p\right) \right] - \mathrm{E}_X \left[D\left(\sum_{m=1}^{M} \frac{1}{M} (W^E Q_X)_m \middle\| W^E_p \right) \right] \\
\leq & \epsilon_2 \log d + \eta_0(\epsilon_2).
\end{aligned} \tag{9.95}
$$

Since

$$
\begin{aligned}
& \Pr\{3\mathrm{E}_X [\varepsilon[\Phi_X]] < \varepsilon[\Phi_X]\} \cup \{3\mathrm{E}_X d_1(\Phi_X) < d_1(\Phi_X)\} \\
& \quad \cup \{\mathrm{E}_X [I_E(\Phi_X)] < I_E(\Phi_X)\} \\
& < 1,
\end{aligned} \tag{9.96}
$$

the above evaluations prove the existence of a protocol Φ satisfying (9.90) and (9.91).

Replacing the role of (9.96) by the inequality

$$
\Pr\{2\mathrm{E}_X [\varepsilon[\Phi_X]] < \varepsilon[\Phi_X]\} \cup \{2\mathrm{E}_X d_1(\Phi_X) < d_1(\Phi_X)\} < 1, \tag{9.97}
$$

we can show the existence of a map φ from $\{1, \ldots, M\} \times \{1, \ldots, L\}$ to $\mathcal{X}$ and a POVM $Y = \{Y_{(m,l)}\}_{(m,l)}$ satisfying (9.92). ∎

We now prove the direct part by applying Lemma 9.4 to $W^{B,(n)}$, $W^{E,(n)}$, and p^n. First, we define $R \stackrel{\text{def}}{=} I(p, W^B) - I(p, W^E)$ and $R_1 \stackrel{\text{def}}{=} I(p, W^E)$. Let $M = M_n \stackrel{\text{def}}{=} e^{n(R-3\delta)}$, $C = C_n \stackrel{\text{def}}{=} e^{n(R_1+\delta)}$, and $L = L_n \stackrel{\text{def}}{=} e^{n(R_1+2\delta)}$ for arbitrary $\delta > 0$. In this case, we denote the $\epsilon_1, \epsilon_2, d$ by $\epsilon_1^{(n)}, \epsilon_2^{(n)}, d_n$, respectively.

We show that the RHSs of (9.90) and (9.91) converge to zero under the above conditions. Since $M_n L_n = e^{n(I(p,W^B)-\delta)}$, the discussion in Sect. 4.5 guarantees that $\epsilon_1^{(n)}$ converges to zero. Using an argument similar to that in Sect. 9.4, we observe that $\epsilon_2^{(n)}$ approaches zero exponentially. Since d_n is the dimension, the equations $\log d_n = \log \dim \mathcal{H}^{\otimes n} = n \log \dim \mathcal{H}$ hold. Hence, the quantity $\epsilon_2^{(n)} \log \dim \mathcal{H}^{\otimes n} + \eta_0(\epsilon_2^{(n)})$ converges to zero. We can therefore show that $C_c^{B,E}(W) \geq I(p, W^B) - I(p, W^E)$. Finally, replacing W^B and W^E with $W^{B,(n)} Q$ and $W^{E,(n)} Q$, respectively, we can show that $C_c^{B,E}(W) \geq \frac{1}{n} \left(I(p, W^{B,(n)} Q) - I(p, W^{E,(n)} Q) \right)$, which implies that the RHS of (9.74) $\geq$ the LHS of (9.74).

9.5.6 *Proof of Converse Part of C-Q Wiretap Channel Coding Theorem*

We prove the converse part of the theorem following Devetak [43]. Consider the sequence of protocols $\Phi^{(n)} = (M_n, Q_n, Y_n)$. Let X_n be the random variables taking values in $\{1, \ldots, M_n\}$ subject to the uniform distributions p^n_{mix}. Let Z_n be the random variable corresponding to the message decoded by the receiver.

Then, the equations $\log M_n = H(X_n) = I(X_n : Z_n) + H(X_n : Z_n)$ hold. The Fano inequality yields $H(X_n : Z_n) \le \varepsilon[\Phi^{(n)}] \log M_n + \log 2$. We can evaluate $I(X_n : Z_n)$ to be

$$
\begin{aligned}
& I(X_n : Z_n) = I(Y_n, p^n_{\mathrm{mix}}, (W^{B,(n)} Q_n)) \le I(p^n_{\mathrm{mix}}, (W^{B,(n)} Q_n)) \\
=& I(p^n_{\mathrm{mix}}, (W^{B,(n)} Q_n)) - I(p^n_{\mathrm{mix}}, (W^{E,(n)} Q_n)) + I_E(\Phi^{(n)}) \\
\le & \sup_Q \sup_p I(p, (W^{B,(n)} Q)) - I(p, (W^{E,(n)} Q)) + I_E(\Phi^{(n)}).
\end{aligned}
$$

Therefore,

$$
\begin{aligned}
\frac{1}{n} \log M_n \le & \frac{1}{n} \sup_Q \sup_p I(p, (W^{B,(n)} Q)) - I(p, (W^{E,(n)} Q)) + \frac{1}{n} I_E(\Phi^{(n)}) \\
& + \varepsilon[\Phi^{(n)}] \frac{1}{n} \log M_n + \frac{1}{n} \log 2. \qquad (9.98)
\end{aligned}
$$

Since $I_E(\Phi^{(n)}) \to 0$ and $\varepsilon[\Phi^{(n)}] \to 0$, the $\le$ part of (9.74) can be shown.

Proof of (9.75) In what follows, we prove (9.75). If we can write $W^E_x = \kappa(W^B_x)$ using a completely positive map κ, then $I(Q^m, W^{B,(n)}) \ge I(Q^m, W^{E,(n)})$. Defining $(Qp)_x \stackrel{\mathrm{def}}{=} \sum_m p(m) Q^m_x$, we have

$$
\begin{aligned}
& I(p, (W^{B,(n)} Q)) - I(p, (W^{E,(n)} Q)) \\
=& H\left(\sum_x (Qp)_x W^{B,(n)}_x \right) - \sum_m p(m) H((W^{B,(n)} Q)^m) \\
& - H\left(\sum_x (Qp)_x W^{E,(n)}_x) + \sum_m p(m) H((W^{E,(n)} Q)^m \right) \\
=& H\left(\sum_x (Qp)_x W^{B,(n)}_x) - \sum_m p(m) Q^m_x H(W^{B,(n)}_x \right) \\
& - \sum_m p(m) I(Q^m, W^{B,(n)}) - H\left(\sum_x (Qp)_x W^{E,(n)}_x \right)
\end{aligned}
$$

$$+\sum_m p(m) Q_x^m H(W_x^{E,(n)}) + \sum_m p(m) I(Q^m, W^{E,(n)})$$
$$\leq I(Qp, W^{B,(n)}) - I(Qp, W^{E,(n)}). \tag{9.99}$$

Using Exercise 9.12, we can obtain $\frac{1}{n}\sup_p I(p, W^{B,(n)}) - I(p, W^{E,(n)}) \leq \sup_p I(p, W^B) - I(p, W^E)$, from which we obtain (9.75). ■

Exercises

9.11 Show (9.81) using Exercise 8.29.

9.12 Consider two c-q channels W^B and $W^{B'}$ defined in $\mathcal{X}$ and $\mathcal{X}'$ and two TP-CP maps κ and κ'. Define two c-q channels $W_x^E = \kappa(W_x^B)$ and $W_x^{E'} = \kappa'(W_{x'}^{B'})$. Consider an arbitrary probability distribution q in $\mathcal{X} \times \mathcal{X}'$, and let p and p' be its marginal distributions in $\mathcal{X}$ and $\mathcal{X}'$, respectively. Show that

$$I(q, W^B \otimes W^{B'}) - I(q, W^E \otimes W^{E'})$$
$$\leq I(p, W^B) - I(p, W^E) + I(p', W^{B'}) - I(p', W^{E'}). \tag{9.100}$$

9.13 Prove (9.78) referring to the discussions in Sects. 9.5.5 and 9.5.6.

9.14 Replace the condition $I_E(\Phi^{(n)}) \to 0$ by another condition $\frac{I_E(\Phi^{(n)})}{n} \to 0$ in the definitions of $C_c^{B,E}(W)$ and $C_c^{e,B,E}(\kappa)$. Show that the capacity is the same as the original one.

9.15 Consider the secure communication via a quantum channel κ when unlimited shared entanglement between the sender and the receiver is available. We denote the asymptotic bound of the secure communication rate by $C_{c,e}^{e,B,E}(\kappa)$. Show that $C_{c,e}^{e,B,E}(\kappa) = C_{c,e}^{e}(\kappa) = \max_\rho I(\rho, \kappa)$.

9.16 Prove (9.79) and (9.80) by expressing the environment of the composite map $\kappa' \circ \kappa$ in terms of the environment systems of the maps κ' and κ.

9.17 Show that the capacity is equal to the original one even though the condition $I_E(\Phi^{(n)}) \to 0$ in the definition is replaced by another condition $\varepsilon_{E,a}[\Phi^{(n)}] \stackrel{\text{def}}{=} \sum_i \sum_{j \neq i} \frac{d_1((W^E Q)_i, (W^E Q)_j)}{M(M-1)} \to 0$. Here, use the Fannes inequality (5.92).

9.18 Show that the capacity is equal to the original one even though the above condition is replaced with another condition $\varepsilon_{E,w}[\Phi] \stackrel{\text{def}}{=} \sup_i \sup_j d_1((W^E Q)_i, (W^E Q)_j)$, which converges to 0.

9.19 Show that $I(Qp, W^B) - I(Qp, W^E) - \sum_i p_i (I(Q^i, W^B) - I(Q^i, W^E)) = I(p, W^B Q) - I(p, W^E Q)$ for a quantum degraded channel (W^B, W^E). Also show (9.76).

9.6 Channel Capacity for Quantum-State Transmission

9.6.1 Conventional Formulation

Let us consider the problem of finding how a large quantum system can be transmitted with negligible error via a given noisy quantum channel κ using encoding and decoding quantum operations. This problem is important for preventing noise from affecting a quantum state for a reliable quantum computation. Hence, this problem is a crucial one for realizing quantum computers and is called **quantum error correction**.

It is a standard approach in this problem to algebraically construct particular codes [54–60] [61, Chapter 9]. However, the achievability of the optimal rate is shown only by employing the random coding method [43, 62, 63]. Although this method is not directly applicable in a practical sense, it is still nevertheless an important theoretical result. In the discussion below, we will not discuss the former algebraic approach and concentrate only on the theoretical bounds.

Let us now formally state the problem of transmitting quantum systems accurately via a quantum channel κ from an input quantum system $\mathcal{H}_A$ to an output quantum system $\mathcal{H}_B$. When the quantum system $\mathcal{H}$ is to be sent, the encoding and decoding operations are given as TP-CP maps τ and ν from $\mathcal{H}$ to $\mathcal{H}_A$ and from $\mathcal{H}_B$ to $\mathcal{H}$, respectively. By combining these operations, it is possible to protect the quantum state from noise during transmission. We may therefore express our protocol by $\Phi = (\mathcal{H}, \tau, \nu)$. The quality of our protocol may be measured by the size $|\Phi| \stackrel{\text{def}}{=} \dim \mathcal{H}$ of the system to be sent. The accuracy of transmission is measured by $\varepsilon_1[\Phi] \stackrel{\text{def}}{=} \max_{u \in \mathcal{H}^1} \left[1 - F^2(u, \nu \circ \kappa \circ \tau(u))\right]$ ($\mathcal{H}^1 \stackrel{\text{def}}{=} \{u \in \mathcal{H} | \|u\| = 1\}$). We often focus on $\varepsilon_2[\Phi] \stackrel{\text{def}}{=} \left[1 - F_e^2(\rho_{\text{mix}}, \nu \circ \kappa \circ \tau)\right]$ as another criterion of accuracy. Let us now examine how a large communication rate $\frac{1}{n} \log |\Phi^{(n)}|$ of our code $\Phi^{(n)}$ is possible for a given channel $\kappa^{\otimes n}$ under the condition that $\varepsilon_1[\Phi^{(n)}]$ or $\varepsilon_2[\Phi^{(n)}]$ approaches zero asymptotically. Then, two kinds of **quantum capacities** $C_{q,1}$ and $C_{q,2}$[7] are defined as

$$C_{q,i}(\kappa) \stackrel{\text{def}}{=} \sup_{\{\Phi^{(n)}\}} \left\{ \underline{\lim} \frac{1}{n} \log |\Phi^{(n)}| \,\middle|\, \lim_{n \to \infty} \varepsilon_i[\Phi^{(n)}] = 0 \right\}, \quad i = 1, 2. \tag{9.101}$$

Theorem 9.10 *Two channel capacities $C_{q,1}$ and $C_{q,2}$ are calculated as*

$$C_{q,1}(\kappa) = C_{q,2}(\kappa) = \lim_{n \to \infty} \frac{1}{n} \max_{\rho \in \mathcal{S}(\mathcal{H}_A^{\otimes n})} I_c(\rho, \kappa^{\otimes n}). \tag{9.102}$$

[7] Since "quantum" states are to be sent, the capacities are called quantum capacities. The subscript q indicates that "quantum" states are to be sent.

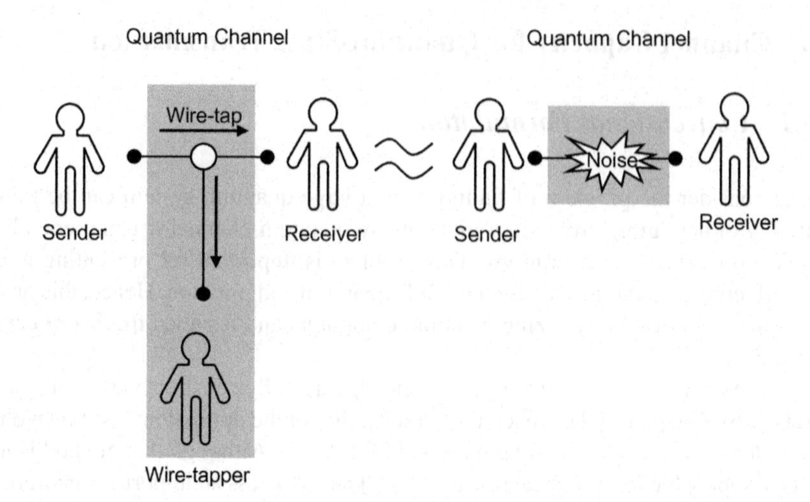

Fig. 9.7 A quantum channel with an eavesdropper and a quantum channel with noise

We now give a proof of the above theorem. Our strategy will be to relate this theorem to the problem of transmitting classical information in the presence of an eavesdropper, as examined in Sect. 9.5. In this approach, as shown in the proof of Theorem, protecting a quantum state from noise is equivalent in a sense to sending classical information without wiretapping when regarding the noise as the wiretapper (Fig. 9.7). To this end, consider the Stinespring representation $(\mathcal{H}_C, \rho_0 = |u_0\rangle\langle u_0|, U_\kappa)$ of κ. We fix a state ρ_A on $\mathcal{H}_A$ and diagonalize ρ_A as $\rho_{A,\mathrm{fix}} = \sum_{x\in\mathcal{X}} p_x |\tilde{u}_x\rangle\langle\tilde{u}_x|$. For the given basis $\tilde{u}_x$ of $\mathcal{H}_A$, we focus on c-q channels $W_x^B := \kappa(|\tilde{u}_x\rangle\langle\tilde{u}_x|)$ and $W_x^E := \kappa_E(|\tilde{u}_x\rangle\langle\tilde{u}_x|)$ on $\mathcal{H}_B$ and $\mathcal{H}_E$, respectively. Then, the following lemma holds.

Lemma 9.5 *Now, we regard as p as a distribution on $\mathcal{X}$. Let M and L be arbitrary integers and v be the number of eigenvalues of W_p^E. There exists an isometric map V from $\mathbb{C}^M$ to $\mathcal{H}_A$ and a TP-CP map ν from $\mathcal{H}_B$ to $\mathbb{C}^M$ such that*

$$F_e(\rho_{\mathrm{mix}}, \nu \circ \kappa \circ \kappa_V) \geq 1 - 4\left(\sqrt{2\epsilon_2} + \sqrt{2\epsilon_1}\right)^2, \tag{9.103}$$

where

$$\epsilon_1 \stackrel{\mathrm{def}}{=} \min_{s\in[0,1]} 2^{1+s}(\mathsf{ML})^s e^{-sI_{1-s}(p,W^B)}$$
$$\epsilon_2 \stackrel{\mathrm{def}}{=} \min_{s\in[0,1]} \max\left(4\sqrt{2}, \sqrt{2v}\right) \mathsf{L}^{-s/2} e^{\frac{s}{2} I_{1+s}^{\downarrow}(p,W^E)}.$$

To show Lemma 9.5, we prepare the following lemmas.

Lemma 9.6 *Let κ be a TP-CP map from a system $\mathcal{H}_A$ to $\mathcal{H}_B := \mathbb{C}^M$ and $|\Phi\rangle$ be the maximally entangled state on $\mathcal{H}_C \otimes \mathcal{H}_B$, where $\mathcal{H}_B = \mathbb{C}^M$. When a pure entangled* *state $|\Psi\rangle$ between $\mathcal{H}_C$ and $\mathcal{H}_A$ satisfies*

$$F(\kappa(|\Psi\rangle\langle\Psi|), |\Phi\rangle\langle\Phi|) \geq 1 - \delta^2, \tag{9.104}$$

there exists an isometric map V from $\mathcal{H}_C$ to $\mathcal{H}_A$ such that

$$F_e(\rho_{\mathrm{mix},C}, \kappa \circ \kappa_V) \geq 1 - 4\delta^2. \tag{9.105}$$

Proof We choose the isometric map V from $\mathbb{C}^M$ to $\mathcal{H}_A$ such that

$$F(\rho_{\mathrm{mix},C}, \mathrm{Tr}_A |\Psi\rangle\langle\Psi|) = F(V|\Phi\rangle\langle\Phi|V^*, |\Psi\rangle\langle\Psi|).$$

Thus, we have

$$\begin{aligned}
&F(\kappa(|\Psi\rangle\langle\Psi|), \kappa \circ \kappa_V(|\Phi\rangle\langle\Phi|)) \geq F(|\Psi\rangle\langle\Psi|, \kappa_V(|\Phi\rangle\langle\Phi|)) \\
=&F(\rho_{\mathrm{mix},C}, \mathrm{Tr}_A |\Psi\rangle\langle\Psi|) = F(\mathrm{Tr}_B |\Phi\rangle\langle\Phi|, \mathrm{Tr}_B \kappa(|\Psi\rangle\langle\Psi|)) \\
\geq&F(|\Phi\rangle\langle\Phi|, \kappa(|\Psi\rangle\langle\Psi|)).
\end{aligned}$$

Applying the triangle inequality of Bures' distance $b(\rho, \sigma)^2 = 1 - F(\rho, \sigma)$, we have

$$\begin{aligned}
&\sqrt{1 - F_e(\rho_{\mathrm{mix},C}, \kappa \circ \kappa_V)} = b(|\Phi\rangle\langle\Phi|, \kappa \circ \kappa_V(|\Phi\rangle\langle\Phi|)) \\
\leq&b(|\Phi\rangle\langle\Phi|, \kappa(|\Psi\rangle\langle\Psi|)) + b(\kappa(|\Psi\rangle\langle\Psi|), \kappa \circ \kappa_V(|\Phi\rangle\langle\Phi|)) \\
\leq&2b(\kappa(|\Psi\rangle\langle\Psi|), \kappa \circ \kappa_V(|\Phi\rangle\langle\Phi|)) = 2\sqrt{1 - F(\kappa(|\Psi\rangle\langle\Psi|), |\Phi\rangle\langle\Phi|)} \leq 2\delta.
\end{aligned}$$

Hence, we obtain (9.105). ∎

Lemma 9.7 *Under the same assumption as Lemma 9.5, there exist an entangled state $|\Psi\rangle$ between $\mathcal{H}_C := \mathbb{C}^M$ and $\mathcal{H}_A$ and a TP-CP map ν from $\mathcal{H}_B$ to $\mathcal{H}_D := \mathbb{C}^M$ such that*

$$F_e(\rho_{\mathrm{mix},C}, \nu \circ \kappa(|\Psi\rangle\langle\Psi|)) \geq 1 - \left(\sqrt{2\epsilon_2} + \sqrt{2\epsilon_1}\right)^2. \tag{9.106}$$

Proof Lemma 9.7 is shown by the combination of Theorem 8.17 and Lemma 9.4. First, for the given basis $\tilde{u}_x$ of $\mathcal{H}_A$, we focus on c-q channels $\tilde{W}_x^B := \kappa(|\tilde{u}_x\rangle\langle\tilde{u}_x|)$ and $\tilde{W}_x^E := \kappa_E(|\tilde{u}_x\rangle\langle\tilde{u}_x|)$ on $\mathcal{H}_B$ and $\mathcal{H}_E$, respectively. That is, κ and κ_E can be written by the partial trace and the isometry U from $\mathcal{H}_A$ to $\mathcal{H}_B \otimes \mathcal{H}_E$, respectively.

Due to Lemma 9.4, there exist a map φ from $\{1, \dots, M\} \times \{1, \dots, L\}$ to $\mathcal{X}$ and a POVM $Y = \{Y_{(m,l)}\}_{(m,l)}$ satisfying

$$\epsilon[\Phi(\varphi, Y)] \leq 2\epsilon_1, \quad d_1(\Phi(\varphi, Y)) \leq 2\epsilon_2. \tag{9.107}$$

Now, we apply Theorem 8.17 in the following way. $\mathcal{H}_{A_1}$ and $\mathcal{H}_{A_2}$ are defined as the Hilbert space spanned by $\{|m\rangle_{A_1}\}_{m=1}^M$ and $\{|l\rangle_{A_2}\}_{l=1}^L$, respectively. The reference system $\mathcal{H}_R$ is chosen to be $\mathcal{H}_E$. Define the state $|\Psi'\rangle := \frac{1}{\sqrt{ML}} \sum_{m,l} |m\rangle_{A_1} |l\rangle_{A_2} |\tilde{u}_{\varphi(m,l)}\rangle_A$.

Then, we define the state $\rho := U_\kappa|\Psi', u_0\rangle\langle\Psi', u_0|U_\kappa^*$ on the system $\mathcal{H}_{A_1} \otimes \mathcal{H}_{A_2} \otimes \mathcal{H}_B \otimes \mathcal{H}_R$. Then, the state ρ satisfies $2\epsilon_1$-bit security for $\mathcal{H}_{A_1}$ and $2\epsilon_2$-bit recoverability for $\mathcal{H}_{A_1} \otimes \mathcal{H}_{A_2}$. Let $\{v_l\}$ be a basis mutually unbiased to $\{|l\rangle_{A_2}\}_{l=1}^L$ of $\mathcal{H}_{A_2}$. Therefore, Theorem 8.17 guarantees the existence of a TP-CP map $\kappa_l : \mathcal{S}(\mathcal{H}_B) \to \mathcal{S}(\mathcal{H}_{A_1})$ dependently of l such that

$$\sum_{l=1}^{L} \frac{1}{L} F(L\kappa_l(\langle v_l|\rho_{A_1A_2B}|v_l\rangle) \otimes |v_l\rangle\langle v_l|, |\Phi\rangle\langle\Phi|)$$
$$\geq 1 - \left(\sqrt{2\epsilon_2} + \sqrt{2\epsilon_1}\right)^2. \tag{9.108}$$

Then, we choose an integer l' such that

$$F(\kappa_{l'}(L\langle v_{l'}|\rho_{A_1A_2B}|v_{l'}\rangle), |\Phi\rangle\langle\Phi|) \geq 1 - \left(\sqrt{2\epsilon_2} + \sqrt{2\epsilon_1}\right)^2. \tag{9.109}$$

Define the entangled state $|\Psi\rangle$ between $\mathcal{H}_C := \mathbb{C}^M$ and $\mathcal{H}_{A,1}$ as

$$|\Psi\rangle := \sum_{l=1}^{L} \langle v_{l'}|l\rangle_{A_2} \frac{1}{\sqrt{M}} \sum_{m=1}^{M} |m\rangle_{A_1} |\tilde{u}_{\varphi(m,l)}\rangle_A = \sqrt{L}\langle v_{l'}|\Phi'\rangle. \tag{9.110}$$

Then, we have

$$F(\kappa_{l'}(|\Psi\rangle\langle\Psi|), |\Phi\rangle\langle\Phi|) = F(L\kappa_{l'}(\langle v_{l'}|\rho_{A_1A_2B}|v_{l'}\rangle), |\Phi\rangle\langle\Phi|), \tag{9.111}$$

which implies (9.106). ∎

Proof of Lemma 9.5 Lemma 9.5 can be shown by the direct combination of Lemmas 9.6 and 9.7. ∎

Proof of the direct part of Theorem 9.10 We are now ready to prove the direct part of Theorem 9.10, i.e., $C_{q,2}(\kappa) \geq$ (RHS of (9.102)). Now, we fix the distribution p on $\mathcal{X}$. Assume $R < I(p, W^B) - I(p, W^E)$ and choose $\delta := I(p, W^B) - I(p, W^E) - R$. We choose M and L to be $e^{n(R-\delta)}$ and $e^{n(I(p,W^E)+\delta/2)}$. Then, we have $ML = e^{n(I(p,W^B)-\delta/2)}$. Then, we choose a sufficiently large n such that

$$s(I_{1+s}^{\downarrow}(p, W^E) - I(p, W^E) - \frac{\delta}{2}) < 0,$$
$$s(-I_{1-s}(p, W^B) + I(p, W^B) - \frac{\delta}{2}) < 0 \tag{9.112}$$

with a suitable $s \in (0, 1)$. Now, we apply Lemma 9.5 with $\sigma = (W_p^E)^{\otimes n}$. Then, there exists a code satisfying (9.103) when the distribution is given as the independent and identical distribution of p. Due to (9.112), ϵ_1 and ϵ_2 go to zero exponentially. Therefore, the rate $I(p, W^B) - I(p, W^E)$ is achievable. That is, we obtain $C_{q,2}(\kappa) \geq I(p, W^B) - I(p, W^E)$.

Since $I(p, W^B) - I(p, W^E) = I_c(\rho, \kappa)$ from (9.99), we may apply the same argument to $\kappa^{\otimes n}$ to obtain $C_{q,2}(\kappa) \geq$ (RHS of (9.102)), which completes the proof. The proof of $C_{q,2}(\kappa) = C_{q,1}(\kappa)$ is left as Exercises 9.20 and 9.21. ■

Next, we give a proof of the converse part by a method similar to that in Theorem 8.10, i.e., we show that

$$C_{q,2}(\kappa) \leq \lim_{n\to\infty} \frac{1}{n} \max_{\rho} I_c(\rho, \kappa^{\otimes n}). \tag{9.113}$$

In this method, for any class of local operations C, we focus on the function $C(\kappa)$ of a channel κ from $\mathcal{H}_A$ to $\mathcal{H}_B$ that satisfies the following conditions.

C1 (Normalization) $C(\iota_d) = \log d$ for an identical map ι_d with a size d.

C2 (Monotonicity) $C(\kappa' \circ \kappa \circ \kappa_U) \leq C(\kappa)$ holds for any TP-CP map κ' and any isometry U.

C3 (Continuity) When any two channels $\kappa_{1,n}$ and $\kappa_{2,n}$ from a system $\mathcal{H}_n$ to another system $\mathcal{H}'_n$ satisfy $\max_x 1 - F^2((\kappa_{1,n} \otimes \iota_R)(|x\rangle\langle x|), (\kappa_{2,n} \otimes \iota_R)(|x\rangle\langle x|)) \to 0$, $\frac{|C(\kappa_{1,n}) - C(\kappa_{2,n})|}{\log(\dim \mathcal{H}_n \dim \mathcal{H}'_n)} \to 0$, where the system $\mathcal{H}_R$ is the reference system of $\mathcal{H}_n$.

C4 (Convergence) The quantity $\frac{C(\kappa^{\otimes n})}{n}$ converges as $n \to \infty$.

Based on only the above conditions, we can prove the following theorem.

Lemma 9.8 *When C satisfies all of the above conditions, we have*

$$C_{q,2}(\kappa) \leq C^{\infty}(\kappa) \left(\stackrel{\text{def}}{=} \lim_{n\to\infty} \frac{C(\kappa^{\otimes n})}{n} \right). \tag{9.114}$$

Since $\max_\rho I_c(\rho, \kappa)$ satisfies Conditions **C1**, **C2** (8.39), **C3** (Exercise 9.22), and **C4** (Lemma A.1), we obtain (9.113).

Proof According to Condition ④ in Sect. 8.2, we can choose an encoding τ_n and a decoding ν_n with d_n-dimensional space $\mathcal{K}_n$ such that

$$1 - F_e^2(\rho_{\text{mix}}, \nu_n \circ \kappa^{\otimes n} \circ \tau_n) \to 0$$

and $\lim_{n\to\infty} \frac{\log d_n}{n} = C_{q,2}(\kappa)$. Condition ① in Sect. 8.2 guarantees that there exists isometry U_n such that $F_e^2(\rho_{\text{mix}}, \nu_n \circ \kappa^{\otimes n} \circ \tau_n) \leq F_e(\rho_{\text{mix}}, \nu_n \circ \kappa^{\otimes n} \circ \kappa_{U_n})$. From Condition ③ in Sect. 8.2 there exists a subspace $\mathcal{K}'_n \subset \mathcal{K}_n$ with the dimension $\frac{d_n}{2}$ such that

$$\max_{x\in\mathcal{K}'_n} 1 - F^2(x, \nu_n \circ \kappa^{\otimes n} \circ \kappa_{U_n}(x)) \leq \frac{1 - F_e^2(\rho_{\text{mix}}, \nu_n \circ \kappa^{\otimes n} \circ \kappa_{U_n})}{2}.$$

Therefore, from Condition ④ in Sect. 8.2 we have

$$\frac{2}{3} \max_{\rho\in\mathcal{S}(\mathcal{K}'_n)} \left(1 - F^2(\rho, \nu_n \circ \kappa^{\otimes n} \circ \kappa_{U_n})\right) \to 0.$$

Letting $\kappa_{2,n}$ be a noiseless channel, we have $\max_x 1 - F^2((\nu_n \circ \kappa^{\otimes n} \circ \kappa_{U_n} \otimes \iota_R)(|x\rangle\langle x|), (\kappa_{2,n} \otimes \iota_R)(|x\rangle\langle x|)) \to 0$. Thus, Condition **C3** implies

$$\frac{|C(\nu_n \circ \kappa^{\otimes n} \circ \kappa_{U_n}) - C(\iota_{d_n})|}{n} \to 0.$$

From Condition **C1** we have

$$\left| \frac{C(\nu_n \circ \kappa^{\otimes n} \circ \kappa_{U_n})}{n} - C_{q,2}(\kappa) \right| \\ \le \frac{|C(\nu_n \circ \kappa^{\otimes n} \circ \kappa_{U_n}) - C(\iota_{\frac{d_n}{2}})|}{n} + \left| \frac{\log d_n - \log 2}{n} - C(\kappa) \right| \to 0.$$

Hence, Condition **C2** guarantees that

$$\lim_{n\to\infty} \frac{C(\kappa^{\otimes n})}{n} \ge \lim_{n\to\infty} \frac{C(\nu_n \circ \kappa^{\otimes n} \circ \kappa_{U_n})}{n} = C_{q,2}(\kappa). \tag{9.115}$$

We obtain the desired inequality. ■

Further, Theorem 9.10 brings us the following corollary.

Corollary 9.1 *We have the following relations*

$$\lim_{n\to\infty} \frac{1}{n} \max_{\rho\in\mathcal{S}(\mathcal{H}_A^{\otimes n})} I_c(\rho, \kappa^{\otimes n}) = \lim_{n\to\infty} \frac{1}{n} \max_{\rho\in\mathcal{S}(\mathcal{H}_A^{\otimes n}),\kappa} I_c(\rho, \kappa^{\otimes n} \circ \kappa) \\ = \lim_{n\to\infty} \frac{1}{n} \max_{\rho\in\mathcal{S}(\mathcal{H}_A^{\otimes n}\otimes\mathcal{H}_R^{(n)})} -H_{\kappa^{\otimes n}(\rho)}(R|B), \tag{9.116}$$

where κ is a TP-CP map on $\mathcal{H}_A^{\otimes n}$.

Proof Since the discussion in the direct part of Theorem 9.10 yields that $C_{q,2}(\kappa) \ge \frac{1}{n} \max_{\rho\in\mathcal{S}(\mathcal{H}_A^{\otimes n}),\kappa} I_c(\rho, \kappa^{\otimes n} \circ \kappa)$, we obtain the first equation in (9.116). For a state $\rho \in \mathcal{S}(\mathcal{H}_A^{\otimes n} \otimes \mathcal{H}_R^{(n)})$, we choose a pure state $|x\rangle$ on $\mathcal{H}_A^{\otimes n} \otimes \mathcal{H}_R^{(n)}$ such that $\mathrm{Tr}_A |x\rangle\langle x| = \mathrm{Tr}_A \rho$. Hence, there exists a TP-CP map κ on $\mathcal{H}_A^{\otimes n}$ such that $\rho = \kappa(|x\rangle\langle x|)$. So, we have

$$-H_{\kappa^{\otimes n}(\rho)}(R|B) = -H_{\kappa^{\otimes n}\circ\kappa(|x\rangle\langle x|)}(R|B) = I_c(\mathrm{Tr}_R |x\rangle\langle x|, \kappa^{\otimes n} \circ \kappa),$$

which implies the second equation in (9.116) ■

Moreover, using Exercise 9.12, we can simplify the RHS of (9.102) in the following special case.

Lemma 9.9 *When there exists another channel κ' from the output system $\mathcal{H}_B$ of channel κ to its environment system $\mathcal{H}_E$ such that*

$$\kappa'(\kappa(\rho)) = \kappa_E(\rho) \text{ for } \forall \rho \in \mathcal{S}(\mathcal{H}_A), \tag{9.117}$$

then

$$\max_{\rho \in \mathcal{S}(\mathcal{H}_A)} I_c(\rho, \kappa) = \lim_{n \to \infty} \frac{1}{n} \max_{\rho \in \mathcal{S}(\mathcal{H}_A^{\otimes n})} I_c(\rho, \kappa^{\otimes n}). \tag{9.118}$$

Further, when (9.117) holds, (9.76) implies the concavity

$$\sum_i p_i I_c(\rho_i, \kappa) \le I_c\left(\sum_i p_i \rho_i, \kappa\right). \tag{9.119}$$

For example, the following case satisfies the above condition. Suppose that there exist a basis $\{u_1, \ldots, u_d\}$ of the input system $\mathcal{H}_A$ of channel κ and the POVM $\boldsymbol{M} = \{M_i\}_{i=1}^d$ on the output system $\mathcal{H}_B$ such that

$$\operatorname{Tr} \kappa(|u_i\rangle\langle u_i|) M_j = \delta_{i,j}. \tag{9.120}$$

As checked as follows, this channel satisfies the condition (9.117). For example, the phase-damping channel κ_D^{PD} satisfies condition (9.120).

Now, consider the Naĭmark–Ozawa extension $(\mathcal{H}_0, \rho_0, U)$ with the ancilla $\mathcal{H}_D$ given in Theorem 7.1. Note that we measure the system $\mathcal{H}_0$ with the measurement basis $\{v_1, \ldots, v_d\}$. We also use the Stinespring representation $(\mathcal{H}_C, \rho_0', U_\kappa)$ of κ. Since for any input pure state ρ, the state $(U \otimes I_E)((U_\kappa(\rho \otimes \rho_0')U_\kappa^*) \otimes \rho_0)(U \otimes I_E)^*$ is pure, there exists a basis $\{v_1', \ldots, v_d'\}$ on the space $\mathcal{H}_D \otimes \mathcal{H}_E$, where $\mathcal{H}_E = \mathcal{H}_A \otimes \mathcal{H}_C$. This is ensured by (9.120). Hence, the unitary $U' \stackrel{\text{def}}{=} \sum_{i=1}^d |v_i'\rangle\langle v_i|$ satisfies

$$\begin{aligned}
&\operatorname{Tr}_{\mathcal{H}_0}(U \otimes I_E)((U_\kappa(\rho \otimes \rho_0')U_\kappa^*) \otimes \rho_0)(U \otimes I_E)^* \\
=&U' \operatorname{Tr}_{D,E}(U \otimes I_E)((U_\kappa(\rho \otimes \rho_0')U_\kappa^*) \otimes \rho_0)(U \otimes I_E)^* U'^*.
\end{aligned}$$

Therefore,

$$\begin{aligned}
\kappa_E(\rho) =& \operatorname{Tr}_D \operatorname{Tr}_{\mathcal{H}_0}(U \otimes I_E)((U_\kappa(\rho \otimes \rho_0')U_\kappa^*) \otimes \rho_0)(U \otimes I_E)^* \\
=& \operatorname{Tr}_D U' \operatorname{Tr}_{D,E}(U \otimes I_E)((U_\kappa(\rho \otimes \rho_0')U_\kappa^*) \otimes \rho_0)(U \otimes I_E)^* U'^*,
\end{aligned}$$

which implies condition (9.117).

9.6.2 Proof of Hashing Inequality (8.121)

Finally, we prove the Hashing inequality (8.121) based on a remarkable relation between the transmission of the quantum state and the distillation of a partially entangled state.

Given a channel κ, we can define the partially entangled state $\kappa \otimes \iota_{A'}(|u_{0,0}^{A,A'}\rangle\langle u_{0,0}^{A,A'}|)$, where $\mathcal{H}_{A'}$ is the reference system of the input system $\mathcal{H}_A$ and $u_{0,0}^{A,A'}$ is a maximally entangled state between $\mathcal{H}_A$ and $\mathcal{H}_{A'}$. Conversely, given a partially entangled state ρ on the composite system $\mathcal{H}_A \otimes \mathcal{H}_B$, we can define channel κ_ρ with the same dimensional input system $\mathcal{H}_{A'}$ as the system $\mathcal{H}_A$ (with the dimension d) via quantum teleportation as follows [64]. Perform the generalized Bell measurement $\{|u_{i,j}^{A,A'}\rangle\langle u_{i,j}^{A,A'}|\}$ on the composite system $\mathcal{H}_A \otimes \mathcal{H}_{A'}$ and transmit the outcome (i,j), where $u_{i,j}^{A,A'} \stackrel{\text{def}}{=} (I_{A'} \otimes \mathbf{X}_A^i \mathbf{Z}_A^j) u_{0,0}^{A,A'}$. In this channel, the output system is given by $\mathcal{H}_B \otimes \mathbb{C}^{d^2}$. Now, we consider the purification of ρ with the reference system $\mathcal{H}_R$. The environment $\kappa_{\rho,E}$ of the channel κ_ρ is $\mathcal{H}_R \otimes \mathbb{C}^{d^2}$. Now, we consider the wiretap channels in the way as Lemma 9.7. Then, we choose the distribution p to be the uniform distribution on a CONS of $\mathcal{H}_{A'}$. Then, $I(p, \kappa_\rho) - I(p, \kappa_{\rho,E}) = I_c(\rho_{\text{mix},A'}, \kappa_\rho) = H(\rho_B) - H(\rho_R) = -H_\rho(A|B)$. Applying Lemma 9.7 to the channel $\kappa_\rho^{\otimes n}$, similar to the proof of the direct part of Theorem 9.10, we find that $E_{d,2}^{\to}(\rho) \geq -H_\rho(A|B)$.

9.6.3 Decoder with Assistance by Local Operations

Next, we consider the quantum capacity when the decoder is allowed to employ several classes of local operations, etc, two-way LOCC $\leftrightarrow$, separable operation S, and PPT operation PPT.

In this case, we need to describe sender's system $\mathcal{H}_{A'}$ after the encoding. When the quantum system $\mathcal{H}$ is to be sent, the encoding operation are given as TP-CP maps τ from $\mathcal{H}$ to $\mathcal{H}_A \otimes \mathcal{H}_{A'}$. Then, a decoder $\nu_\leftrightarrow$ assisted by two-way LOCC operations is given as a two-way LOCC operation from $\mathcal{H}_B \otimes \mathcal{H}_{A'}$ to $\mathcal{H} \otimes \mathcal{H}_{A''}$, where $\mathcal{H}_{A''}$ is the one-dimensional system. Decoders ν_S and ν_{PPT} assisted by separable operation and PPT operation are given as a separable operation and a PPT operation from $\mathcal{H}_B \otimes \mathcal{H}_{A'}$ to $\mathcal{H} \otimes \mathcal{H}_{A''}$, respectively. For one-way LOCC assistance, there are two cases. One is the one-way from the sender to the receiver. The other is the one-way from the receiver to the sender. However, the latter is meaningless for decoding because to improve the quantum capacity, transmitted information needs to be used by the receiver. So, we consider only the former case, and denote it by $\to$.

Then, for $C = \to, \leftrightarrow, S$, PPT, we express our protocol by $\Phi_C = (\mathcal{H}, \tau, \nu_C)$. The quality of our protocol may be measured by the size $|\Phi_C| \stackrel{\text{def}}{=} \dim \mathcal{H}$ of the system to be sent. The accuracy of transmission is measured by $\varepsilon_1[\Phi_C] \stackrel{\text{def}}{=} \max_{u \in \mathcal{H}^1} \left[1 - F^2(u, \nu_C \circ \kappa \circ \tau(u))\right]$ $(\mathcal{H}^1 \stackrel{\text{def}}{=} \{u \in \mathcal{H} | \|u\| = 1\})$. We also employ $\varepsilon_2[\Phi_C] \stackrel{\text{def}}{=} \left[1 - F_e^2(\rho_{\text{mix}}, \nu_C\right.$

$\circ \kappa \circ \tau)]$ as another criterion of accuracy. Similar to (9.101), the capacities are defined as

$$C_{q,C,i}(\kappa) \stackrel{\text{def}}{=} \sup_{\{\Phi_C^{(n)}\}} \left\{ \underline{\lim} \frac{1}{n} \log |\Phi_C^{(n)}| \,\middle|\, \lim_{n\to\infty} \varepsilon_i[\Phi_C^{(n)}] = 0 \right\} \tag{9.121}$$

for $i = 1, 2$ and $C =\rightarrow, \leftrightarrow, S$, PPT, where the supremum is taken among codes $\Phi_C^{(n)}$ with decoding assistance by the class C. Similar to $C_c^\dagger$, we also define the **strong converse quantum capacity** by using the second criterion $\varepsilon_2[\Phi_C^{(n)}]$ as

$$C_{q,C}^\dagger(\kappa) \stackrel{\text{def}}{=} \sup_{\{\Phi_C^{(n)}\}} \left\{ \underline{\lim} \frac{1}{n} \log |\Phi_C^{(n)}| \,\middle|\, \underline{\lim}\, \varepsilon_2[\Phi_C^{(n)}] < 1 \right\} \tag{9.122}$$

for $C = \emptyset, \rightarrow, \leftrightarrow, S$, PPT, where $C = \emptyset$ means the non-assistance case. Then, we have the following theorem.

Theorem 9.11 *The capacity $C_{q,C,i}(\kappa)$ is characterized as*

$$C_{q,C,1}(\kappa) = C_{q,C,2}(\kappa) = \lim_{n\to\infty} \frac{1}{n} \max_{\rho\in\mathcal{S}(\mathcal{H}_A^{\otimes n}\otimes\mathcal{H}_R)} E_m^C(\kappa^{\otimes n} \otimes \iota_R(\rho)) \tag{9.123}$$

for $C =\rightarrow, \leftrightarrow, S$, PPT.

Although we explicitly describe ι_R in (9.123), we will omit it latter.

Proof We can show that $C_{q,C,1}(\kappa) = C_{q,C,2}(\kappa)$ by the same way as $C_{q,1}(\kappa) = C_{q,2}(\kappa)$. Hence, we discuss only $C_{q,2}(\kappa)$.

Direct part: First, we show the relation

$$C_{q,C,2}(\kappa) \geq \max_{\rho\in\mathcal{S}(\mathcal{H}_A\otimes\mathcal{H}_R)} E_m^C(\kappa(\rho)).$$

Consider the following protocol. The sender prepares the state $\rho \in \mathcal{S}(\mathcal{H}_A \otimes \mathcal{H}_R)$. Next, the sender sends the part $\mathcal{H}_A$ of ρ via the channel κ. After the receiver receives the state in $\mathcal{H}_B$, the sender and the receiver apply distillation protocol to achieve the rate $E_m^C(\kappa \otimes \iota_R(\rho))$. Finally, the sender and the receiver perform the teleportation protocol. This method achieves the rate $E_m^C(\kappa(\rho))$. Applying the same method to the state $\rho \in \mathcal{S}(\mathcal{H}_A^{\otimes n} \otimes \mathcal{H}_R)$, we can achieve the rate $\lim_{n\to\infty} \frac{1}{n} \max_{\rho\in\mathcal{S}(\mathcal{H}_A^{\otimes n}\otimes\mathcal{H}_R)} E_m^C(\kappa^{\otimes n}(\rho))$.

Converse part: Consider a sequence of codes $\Phi_C^{(n)} = (\mathcal{H}^{\otimes n}, \tau_n, \nu_{C,n})$ such that $\lim_{n\to\infty} \varepsilon_2[\Phi_C^{(n)}] = 0$. Let $|\Phi_n\rangle\langle\Phi_n|$ be the maximally entangled state on $\mathcal{H}^{\otimes n} \otimes \mathcal{H}_R^{\otimes n}$. We define the bipartite state $\rho_n := \tau_n(|\Phi_n\rangle\langle\Phi_n|)$ on $\mathcal{H}_A^{\otimes n} \otimes (\mathcal{H}_{A',n} \otimes \mathcal{H}_R^{\otimes n})$, where τ_n is a TP-CP map from $\mathcal{H}^{\otimes n}$ to $\mathcal{H}_A^{\otimes n} \otimes \mathcal{H}_{A',n}$. Then, we have

$$\langle\Phi_n|\nu_{C,n}(\kappa^{\otimes n}(\rho_n))|\Phi_n\rangle = 1 - \varepsilon_2[\Phi_C^{(n)}] \to 1.$$

So, due to (5.103), we have

$$\lim_{n\to\infty}\frac{1}{n}\max_{\rho\in\mathcal{S}(\mathcal{H}_A^{\otimes n}\otimes\mathcal{H}_R)} E_m^C(\kappa^{\otimes n}\otimes\iota_R(\rho)) \ge \lim_{n\to\infty} -\frac{1}{n} H_{\nu_{C,n}(\kappa^{\otimes n}(\rho_n))}(RA'|B)$$
$$= \lim_{n\to\infty}\frac{1}{n}\log\dim\mathcal{H}^{\otimes n} = \lim_{n\to\infty}\frac{1}{n}\log|\Phi_C^{(n)}|. \tag{9.124}$$

■

As shown latter, the relation

$$\lim_{n\to\infty}\frac{1}{n}\max_{\rho\in\mathcal{S}(\mathcal{H}_A^{\otimes n})} I_c(\rho,\kappa^{\otimes n}) = \lim_{n\to\infty}\frac{1}{n}\max_{\rho\in\mathcal{S}(\mathcal{H}_A^{\otimes n}\otimes\mathcal{H}_R)} E_m^{\rightarrow}(\kappa^{\otimes n}\otimes\iota_R(\rho)) \tag{9.125}$$

holds, where $\rightarrow$ is the one-way LOCC from R to B. So, combining (9.102) and (9.123), we have

$$C_{q,2}(\kappa) = C_{q,\rightarrow,2}(\kappa), \tag{9.126}$$

which implies that the post one-way communication does not improve the capacity. *Proof of* (9.125) Since the inequality $\le$ in (9.125) is trivial, we show only the part $\ge$. Choose a state $\rho\in\mathcal{S}(\mathcal{H}_A^{\otimes n}\otimes\mathcal{H}_R)$ and an instrument $\boldsymbol{\kappa}=\{\kappa_i\}_i$ on the system $\mathcal{H}_R$ such that $E_m^{\rightarrow}(\kappa^{\otimes n}\otimes\iota_R(\rho)) = -H_{\sum_i \iota_B\otimes\kappa_i(\kappa^{\otimes n}\otimes\iota_R(\rho)))\otimes|i,i\rangle\langle i,i|}(R|B)$. Define the probability $p_i := \mathrm{Tr}\,\kappa_i(\rho)$. Then,

$$-H_{\sum_i \iota_B\otimes\kappa_i(\kappa^{\otimes n}\otimes\iota_R(\rho)))\otimes|i,i\rangle\langle i,i|}(R|B)$$
$$= -H_{\sum_i \kappa^{\otimes n}\otimes\iota_R(\iota_A\otimes\kappa_i(\rho)))\otimes|i,i\rangle\langle i,i|}(R|B)$$
$$= -\sum_i p_i H_{\kappa^{\otimes n}\otimes\iota_R(\frac{1}{p_i}\iota_A\otimes\kappa_i(\rho)))}(R|B)$$
$$\le \max_{\rho\in\mathcal{S}(\mathcal{H}_A^{\otimes n}\otimes\mathcal{H}_R^{(n)})} -H_{\kappa^{\otimes n}(\rho)}(R|B) = \max_{\rho\in\mathcal{S}(\mathcal{H}_A^{\otimes n}\otimes\mathcal{H}_R^{(n)})} I_c(\rho,\kappa^{\otimes n}).$$

Hence, (9.116) of Corollary 9.1 yields (9.125). ■

Next, we address the strong converse quantum capacity $C_{q,C}^{\dagger}(\kappa)$. For this purpose, we define the **SDP bound**:

$$C_{\mathrm{SDP}}(\kappa) := \max_{\rho\in\mathcal{S}(\mathcal{H}_A\otimes\mathcal{H}_R)} E_{\mathrm{SDP}}(\kappa(\rho))$$
$$\overset{(a)}{=} \max_{|y\rangle\langle y|\in\mathcal{S}(\mathcal{H}_A\otimes\mathcal{H}_R)} E_{\mathrm{SDP}}(\kappa(|y\rangle\langle y|)) = \max_{\rho\in\mathcal{S}(\mathcal{H}_A)} E_{\mathrm{SDP}}(\kappa(|x\rangle\langle x|)), \tag{9.127}$$

where $|x\rangle$ is the purification of ρ. The equation (a) holds the convexity (8.232). We also have another expression for the SDP bound

$$C_{\mathrm{SDP}}(\kappa) = \max_{p} \min_{\sigma' \geq 0: \|\tau^A(\sigma')\|_1 \leq 1} \sum_y p(y) D(\kappa(|y\rangle\langle y|)\|\sigma'), \tag{9.128}$$

where p is a distribution taking values in the set of pure states on $\mathcal{H}_A \otimes \mathcal{H}_R$. Here, $D(\rho\|\sigma)$ is defined as $\mathrm{Tr}\, \rho(\log \rho - \log \sigma)$ even though $\mathrm{Tr}\, \sigma \neq 1$. (See Exercise 5.25.) Then, we can apply the minimax theorem (Theorem A.9), which implies that

$$\begin{aligned} C_{\mathrm{SDP}}(\kappa) &= \min_{\sigma' \geq 0: \|\tau^A(\sigma')\|_1 \leq 1} \max_p \sum_y p(y) D(\kappa(|y\rangle\langle y|)\|\sigma') \\ &= \min_{\sigma' \in \mathcal{S}(\mathcal{H}_A \otimes \mathcal{H}_R)} \max_{|y\rangle\langle y| \in \mathcal{S}(\mathcal{H}_A \otimes \mathcal{H}_R)} D(\kappa(|y\rangle\langle y|)\|\sigma') \\ &\overset{(a)}{=} \min_{\sigma' \in \mathcal{S}(\mathcal{H}_A \otimes \mathcal{H}_R)} \max_{\rho \in \mathcal{S}(\mathcal{H}_A \otimes \mathcal{H}_R)} D(\kappa(\rho)\|\sigma'), \end{aligned} \tag{9.129}$$

where (a) follows from the convexity of the function $\rho \to D(\kappa(\rho)\|\sigma')$.

Now, we introduce another expression for the SDP bound. Choosing a state ρ on the input system, we consider its purification $|y\rangle\langle y|$. Then, we have the concavity of the quantity $I_{\mathrm{SDP}}(\rho, \kappa) := E_{\mathrm{SDP}}(\kappa(|\Phi\rangle\langle\Phi|))$ with respect to ρ (Lemma 9.11). This property can be also shown by the monotonicity for local TP-CP maps because $I_{\mathrm{SDP}}(\rho, \kappa)$ is independent of the choice of the purification[Exe. 9.24].

Lemma 9.10 *The SDP bound $C_{\mathrm{SDP}}(\kappa)$ satisfies the subadditivity.*

$$C_{\mathrm{SDP}}(\kappa_1 \otimes \kappa_2) \leq C_{\mathrm{SDP}}(\kappa_1) + C_{\mathrm{SDP}}(\kappa_2). \tag{9.130}$$

Proof Choose the pure state $|x^i\rangle\langle x^i|$ and the positive semidefinite matrix $\sigma_i' \geq 0$ as $\|\tau^A(\sigma_i')\|_1 = 1$ and

$$C_{\mathrm{SDP}}(\kappa_i) = D(\kappa_i(|x^i\rangle\langle x^i|)\|\sigma_i). \tag{9.131}$$

Then, for a pure state $|x\rangle\langle x| \in \mathcal{S}(\mathcal{H}_{A_1} \otimes \mathcal{H}_{A_2} \otimes \mathcal{H}_{R_1} \otimes \mathcal{H}_{R_2})$, using (5.86) in the following step (a), we have

$$\begin{aligned} & D(\kappa_1 \otimes \kappa_2(|x\rangle\langle x|)\|\sigma_1' \otimes \sigma_2') \\ =& - \mathrm{Tr}\, \kappa_1 \otimes \kappa_2(|x\rangle\langle x|)(\log(\sigma_1') \otimes I_2 + I_1 \otimes \log(\sigma_2')) - H(\kappa_1 \otimes \kappa_2(|x\rangle\langle x|) \\ \overset{(a)}{\leq}& - \mathrm{Tr}_1(\mathrm{Tr}_2\, \kappa_1 \otimes \kappa_2(|x\rangle\langle x|)) \log(\sigma_1') - \mathrm{Tr}_2(\mathrm{Tr}_1\, \kappa_1 \otimes \kappa_2(|x\rangle\langle x|)) \log(\sigma_2') \\ & - H(\mathrm{Tr}_1\, \kappa_1 \otimes \kappa_2(|x\rangle\langle x|)) - H(\mathrm{Tr}_2\, \kappa_1 \otimes \kappa_2(|x\rangle\langle x|)) \\ =& D(\kappa_1(\mathrm{Tr}_2\, |x\rangle\langle x|)\|\sigma_1') + D(\kappa_2(\mathrm{Tr}_1\, |x\rangle\langle x|)\|\sigma_2') \\ \overset{(b)}{\leq}& C_{\mathrm{SDP}}(\kappa_1) + C_{\mathrm{SDP}}(\kappa_2), \end{aligned}$$

where (b) follows from (9.129). ■

We have the following theorem.

Theorem 9.12 ([65]) *The strong converse quantum capacity $C^{\dagger}_{q,C}(\kappa)$ is characterized as follows.*

$$C^{\dagger}_{q,C}(\kappa) \le C_{\mathrm{SDP}}(\kappa). \tag{9.132}$$

This theorem means that the SDP bound $C_{\mathrm{SDP}}(\kappa)$ upper bounds the strong converse quantum capacity $C^{\dagger}_{q,C}(\kappa)$. Since the SDP bound $C_{\mathrm{SDP}}(\kappa)$ has single-letterized forms (9.127), (9.128), and (9.129), this formula is helpful for evaluating $C^{\dagger}_{q,C}(\kappa)$.

To show Theorem 9.12, we prepare the following quantities.

$$I_{\alpha|\,\mathrm{SDP}}(\rho,\kappa) := E_{\alpha}(\kappa(|y\rangle\langle y|)) \tag{9.133}$$

$$C_{\alpha|\,\mathrm{SDP}}(\kappa) := \max_{\rho\in\mathcal{S}(\mathcal{H}_A)} I_{\alpha|\,\mathrm{SDP}}(\rho,\kappa) = \max_{|\Psi\rangle\langle\Psi|\in\mathcal{S}(\mathcal{H}_A\otimes\mathcal{H}_R)} E_{\alpha}(\kappa(|\Psi\rangle\langle\Psi|)), \tag{9.134}$$

$$\tilde{I}_{\alpha|\,\mathrm{SDP}}(\rho,\kappa) := \tilde{E}_{\alpha}(\kappa(|y\rangle\langle y|)), \tag{9.135}$$

$$\tilde{C}_{\alpha|\,\mathrm{SDP}}(\kappa) := \max_{\rho\in\mathcal{S}(\mathcal{H}_A)} \tilde{I}_{\alpha|\,\mathrm{SDP}}(\rho,\kappa) = \max_{|\Psi\rangle\langle\Psi|\in\mathcal{S}(\mathcal{H}_A\otimes\mathcal{H}_R)} \tilde{E}_{\alpha}(\kappa(|\Psi\rangle\langle\Psi|)), \tag{9.136}$$

where $|y\rangle$ is a purification of $\rho\in\mathcal{S}(\mathcal{H}_A)$.

Lemma 9.11 *For states ρ_i on $\mathcal{H}_A$ and a distribution p_i, we have*

$$I_{1+s|\,\mathrm{SDP}}\left(\sum_i \lambda_i\rho_i,\kappa\right) \ge \frac{1+s}{s}\log\left(\sum_i \lambda_i e^{\frac{s}{1+s}R_{1+s}(\rho_i,\kappa)}\right) \text{ for } s\in[-1,1]\setminus\{0\}, \tag{9.137}$$

$$\tilde{I}_{1+s|\,\mathrm{SDP}}\left(\sum_i \lambda_i\rho_i,\kappa\right) \ge \frac{1+s}{s}\log\left(\sum_i \lambda_i e^{\frac{s}{1+s}\tilde{R}_{1+s}(\rho_i,\kappa)}\right) \text{ for } s\in[-\frac{1}{2},\infty)\setminus\{0\}. \tag{9.138}$$

As the limit $s\to 0$, we have the concavity

$$I_{\mathrm{SDP}}\left(\sum_i \lambda_i\rho_i,\kappa\right) = I_{1|\,\mathrm{SDP}}\left(\sum_i \lambda_i\rho_i,\kappa\right) \ge \sum_i \lambda_i I_{1|\,\mathrm{SDP}}(\rho_i,\kappa). \tag{9.139}$$

Proof We show (9.137) with $s\in(0,1]$. We choose the purification $|x_i\rangle$ of ρ_i on $\mathcal{H}_A\otimes\mathcal{H}_R$ and the positive semidefinite matrix σ_i' such that $\|\tau^A(\sigma_i')\|_1 = 1$ and $E_{1+s}(\kappa(|x_i\rangle\langle x_i|)) = D_{1+s}(\kappa(|x_i\rangle\langle x_i|)\|\sigma_i')$.

Now, we consider the bipartite system $\mathcal{H}_A\otimes(\mathcal{H}_R\otimes\mathbb{C}^d)$ and focus on the state $|y\rangle := \sum_{i=1}^d \sqrt{\lambda_i}|x_i,i\rangle$ and an arbitrary positive semidefinite matrix σ' satisfying that $\|\tau^A(\sigma')\|_1 = 1$. Using $P_i = I\otimes|i\rangle\langle i|$, we define $p_i := \|\tau^A(P_i\sigma' P_i)\|_1$ and $\sigma_i' := \frac{1}{p_i}\langle i|P_i\sigma' P_i|i\rangle$. Applying (**a**) of Exercise 5.25 to the TP-CP map $\sigma'\mapsto\sum_i P_i\sigma' P_i$, we have

$$
\begin{aligned}
& e^{sI_{1+s|\text{SDP}}(\sum_i \lambda_i \rho_i, \kappa)} \ge e^{sD_{1+s}(\kappa(|y\rangle\langle y|)\|\sigma')} \\
\ge & e^{sD_{1+s}(\kappa(\sum_i P_i|y\rangle\langle y|P_i)\|\sum_i P_i\sigma' P_i)} = \sum_i \lambda_i^{1+s} p_i^{-s} e^{sD_{1+s}(\kappa(|x_i\rangle\langle x_i|)\|\sigma_i')}.
\end{aligned}
$$

By taking the maximum for σ_i', the reverse Hölder inequality (A.27) yields that

$$
\begin{aligned}
e^{sI_{1+s|\text{SDP}}(\sum_i \lambda_i \rho_i, \kappa)} \ge & \sum_i \lambda_i^{1+s} p_i^{-s} e^{sI_{1+s|\text{SDP}}(\rho_i, \kappa)} \\
\ge & \left(\sum_i \lambda_i e^{\frac{s}{1+s} I_{1+s|\text{SDP}}(\rho_i, \kappa)} \right)^{1+s},
\end{aligned}
$$

which implies (9.137) with $s \in [0, 1]$. Replacing the role of the reverse Hölder inequality (A.27) by the Hölder inequality (A.25), we obtain (9.137) with $s \in [-1, 0]$. Similarly, we can show (9.138). ■

Lemma 9.12 *[65] We have*

$$
I_{\alpha|\text{SDP}}(\kappa^{\otimes n}) \le n I_{\alpha|\text{SDP}}(\kappa) + \frac{\alpha d_A}{\alpha - 1} \log n \text{ for } \alpha \in [1, 2] \tag{9.140}
$$

$$
\tilde{I}_{\alpha|\text{SDP}}(\kappa^{\otimes n}) \le n \tilde{I}_{\alpha|\text{SDP}}(\kappa) + \frac{\alpha d_A}{\alpha - 1} \log n \text{ for } \alpha \in [1, \infty), \tag{9.141}
$$

where d_A is the dimension of $\mathcal{H}_A$.

Proof For an input state ρ on $\mathcal{H}_A^{\otimes}$, we define $\overline{\rho} := \frac{1}{n!} \sum_\pi U_\pi \rho U_\pi^\dagger$, where π is a permutation among n letters and U_π is defined in (2.213). Since $I_{\alpha|\text{SDP}}(\rho, \kappa^{\otimes n}) = I_{\alpha|\text{SDP}}(U_\pi \rho U_\pi^\dagger, \kappa^{\otimes n})$, Lemma 9.11 implies that

$$
I_{\alpha|\text{SDP}}(\rho, \kappa^{\otimes n}) \le I_{\alpha|\text{SDP}}(\overline{\rho}, \kappa^{\otimes n}). \tag{9.142}
$$

We choose the purification $|y\rangle$ of $\overline{\rho}$. (2.215) implies that

$$
|y\rangle\langle y| \le (n+1)^{|d_A|^2 - 1} \int |\Psi\rangle\langle\Psi|^{\otimes n} \mu(d\Psi), \tag{9.143}
$$

where μ is the Haar measure on $\mathcal{H}_A \otimes \mathcal{H}_R$ and $\mathcal{H}_R$ is the reference system of $\mathcal{H}_A$. For a positive semidefinite matrix $\sigma' \ge 0$, we have

$$
\begin{aligned}
& D_\alpha(\kappa^{\otimes n}(|\Psi'\rangle\langle\Psi'|)\|\sigma') \\
\le & \frac{\alpha(|d_A|^2 - 1)}{\alpha - 1} \log(n+1) + D_\alpha(\kappa^{\otimes n} \left(\int |\Psi\rangle\langle\Psi|^{\otimes n}) \mu(d\Psi) \|\sigma' \right).
\end{aligned} \tag{9.144}
$$

Taking the minimum for σ' with the condition $\|\tau^A(\sigma')\|_1 = 1$, we have

$$
\begin{aligned}
& E_\alpha(\kappa^{\otimes n}(|\Psi'\rangle\langle\Psi'|)) \\
\le & \frac{\alpha(|d_A|^2-1)}{\alpha-1}\log(n+1) + E_\alpha\Big(\kappa^{\otimes n}\Big(\int |\Psi\rangle\langle\Psi|^{\otimes n})\mu(d\Psi)\Big) \\
\overset{(a)}{\le} & \frac{\alpha(|d_A|^2-1)}{\alpha-1}\log(n+1) + \max_{|\Psi\rangle} E_\alpha(\kappa^{\otimes n}(|\Psi\rangle\langle\Psi|^{\otimes n})) \\
\le & \frac{\alpha(|d_A|^2-1)}{\alpha-1}\log(n+1) + n\max_{|\Psi\rangle} E_\alpha(\kappa(|\Psi\rangle\langle\Psi|)),
\end{aligned}
\tag{9.145}
$$

where (a) follows from Exercise 8.72. Taking the maximum for $\langle\Psi|$, the second equation of (9.134) implies (9.140). We can show (9.141) in the same way. ■

Proof of Theorem 9.12 Consider a sequence of codes $\Phi_C^{(n)} = (\mathcal{H}^{\otimes n}, \tau_n, \nu_{C,n})$ such that $\dim \mathcal{H}^{\otimes n} = e^{nr'}$. Let $|\Phi_n\rangle\langle\Phi_n|$ be the maximally entangled state on $\mathcal{H}^{\otimes n} \otimes \mathcal{H}_R^{\otimes n}$. We define the bipartite state $\rho_n := \tau_n(|\Phi_n\rangle\langle\Phi_n|)$ on $\mathcal{H}_A^{\otimes n} \otimes (\mathcal{H}_{A',n} \otimes \mathcal{H}_R^{\otimes n})$, where τ_n is a TP-CP map from $\mathcal{H}^{\otimes n}$ to $\mathcal{H}_A^{\otimes n} \otimes \mathcal{H}_{A',n}$. Then, since $|\Phi_n\rangle$ is the maximally entangled state with the size e^{nr}, the relation (8.364) with $n = 1, r = nr', \rho^{\otimes n} = \kappa^{\otimes n}(\rho_n)$, and $\kappa'_n = \nu_{C,n}$ implies that

$$
\begin{aligned}
\langle\Phi_n|\nu_{C,n}(\kappa^{\otimes n}(\rho_n))|\Phi_n\rangle \le & e^{-\frac{-E_{1+s|\mathrm{SDP}}(\kappa^{\otimes n}(\rho_n))+snr'}{1+s}} \\
\le & e^{-\frac{-C_{1+s|\mathrm{SDP}}(\kappa^{\otimes n})+snr'}{1+s}} \le n^{\frac{(1+s)d_A}{s}} e^{-n\frac{-C_{1+s|\mathrm{SDP}}(\kappa)+sr'}{1+s}}
\end{aligned}
$$

for $s \in [0, 1]$, where (a) follows from (9.140) of Lemma 9.12. Since

$$
\max_{s\in\in[0,1]} \frac{-C_{1+s|\mathrm{SDP}}(\kappa)+sr'}{1+s} > 0,
\tag{9.146}
$$

when $r' > C_{\mathrm{SDP}}(\kappa)$, we obtain

$$
C_{q,C}^{\dagger}(\kappa) \le C_{1+s|\mathrm{SDP}}(\kappa) \text{ for } s \in [0,1],
\tag{9.147}
$$

which implies (9.132).

Similarly, we can show

$$
\langle\Phi_n|\nu_{C,n}(\kappa^{\otimes n}(\rho_n))|\Phi_n\rangle \le n^{\frac{(1+s)d_A}{s}} e^{-n\frac{-\tilde{C}_{1+s|\mathrm{SDP}}(\kappa)+sr'}{1+s}}
$$

for $s \in [0, \infty)$. So, we can show

$$
C_{q,C}^{\dagger}(\kappa) \le \tilde{C}_{1+s|\mathrm{SDP}}(\kappa) \text{ for } s \in [0,\infty),
\tag{9.148}
$$

which gives another proof of (9.132). ■

Exercises

9.20 Show that $C_{q,2}(\kappa) \le C_{q,1}(\kappa)$ using (8.25).

9.21 Show that $C_{q,1}(\kappa) \geq C_{q,2}(\kappa)$ using (8.26).

9.22 Show that $\max_\rho I_c(\rho, \kappa)$ satisfies Condition **C3** similarly to Exercise 8.44. Here, use Fannes inequality (Theorem 5.12) for two states $(\kappa_{1,n} \otimes \iota_R)(|x\rangle\langle x|)$ and $(\kappa_{2,n} \otimes \iota_R)(|x\rangle\langle x|)$.

9.23 Give an alternative proof of $C_{q,2}(\kappa) \leq \lim_{n\to\infty} \frac{1}{n} \max_{\rho\in\mathcal{S}(\mathcal{H}_A^{\otimes n})} I_c(\rho, \kappa^{\otimes n})$ by following the steps below [66].
(**a**) Let κ be a quantum channel with the input system $\mathcal{H}_A$, and Φ be a code for the channel κ. Show the existence of a code Φ' such that $|\Phi'| = \min(\dim \mathcal{H}_A, |\Phi|)$ and $\varepsilon_2[\Phi'] \leq 2\varepsilon_2[\Phi]$, by using property ① of Sect. 8.2.
(**b**) Let $\Phi = (\mathcal{H}_A, \kappa_U, \nu)$ be a code with an isometric encoder κ_U for a channel κ. Let ρ_{mix} be a completely mixed state in $\mathcal{H}_A$. Define $\delta \stackrel{\mathrm{def}}{=} \sqrt{2(1 - F_e(\rho_{\mathrm{mix}}, \nu \circ \kappa \circ \kappa_U))}$. Show that

$$\max_{\rho\in\mathcal{S}(\mathcal{H}_A)} I_c(\rho, \kappa) \geq I_c(U\rho_{\mathrm{mix}}U^*, \kappa) \geq I_c(\rho_{\mathrm{mix}}, \nu \circ \kappa \circ \kappa_U)$$
$$\geq \log|\Phi| - 2\delta\left(\log|\Phi| - \log\delta\right),$$

by using (8.39), (8.42), and (8.38).
(**c**) Given that a sequence of codes $\Phi^{(n)} = (\mathcal{H}^{\otimes n}, \tau^{(n)}, \nu^{(n)})$ satisfying $\varepsilon_2[\Phi^{(n)}] \to 0$, show that

$$\lim_{n\to\infty} \frac{1}{n} \max_{\rho\in\mathcal{S}(\mathcal{H}_A^{\otimes n})} I_c(\rho, \kappa^{\otimes n}) \geq \limsup_{n\to\infty} \frac{1}{n} \log \min\{|\Phi^{(n)}|, d_A^n\}. \tag{9.149}$$

(**d**) Complete the alternate proof of $C_{q,2}(\kappa) \leq \lim_{n\to\infty} \frac{1}{n} \max_{\rho\in\mathcal{S}(\mathcal{H}_A^{\otimes n})} I_c(\rho, \kappa^{\otimes n})$.

9.24 Show the concavity of the map $\rho \mapsto I_{\mathrm{SDP}}(\rho, \kappa)$ (Hint: use the monotonicity of E_{SDP} for local TP-CP maps.)

9.7 Examples

In this section, we calculate the capacities $\tilde{C}_c(\kappa)$, $C_c(\kappa)$, $C_c^e(\kappa)$, $C_{c,e}^e(\kappa)$, and $C_{q,1}(\kappa)$ in several cases.

9.7.1 Group Covariance Formulas

For this purpose, we derive forlumas for $C_c(\kappa)$, $C_c^e(\kappa)$, $C_{c,e}^e(\kappa)$, and $C_{q,1}(\kappa)$ with the group covariance. Let κ be a TP-CP map from system $\mathcal{H}_A$ to $\mathcal{H}_B$. Assume that there exists an irreducible (projective) representation U_A of a group G on the space $\mathcal{H}_A$ satisfying the following. There exist unitary matrices (not necessarily a representation) $\{U_B(g)\}_{g\in G}$ on $\mathcal{H}_B$ such that

$$\kappa(U_A(g)\rho U_A(g)^*) = U_B(g)\kappa(\rho)U_B(g)^* \tag{9.150}$$

for any density ρ and element $g \in G$. In what follows, we derive useful formulas in the above assumption. Then,

$$I(p, \kappa) = I(p_{U_A(g)}, \kappa),$$

where $p_U(\rho) \stackrel{\text{def}}{=} p(U\rho U^*)$. Hence,

$$\begin{aligned}
I(p, \kappa) = &\int_G I(p_{U_A(g)}, \kappa)\nu(dg) \le I\Big(\int_G p_{U_A(g)}\nu(dg), \kappa\Big) \\
=&H(\kappa(\rho_{\text{mix}})) - \int_G \sum_x p_x H(\kappa(U_A(g)^*\rho_x U_A(g)))\nu(dg) \\
\le&H(\kappa(\rho_{\text{mix}})) - \min_\rho H(\kappa(\rho)).
\end{aligned}$$

This upper bound is attained by the distribution $(\nu(dg), U_A(g)^*\rho_{\min}U_A(g))$. Thus,

$$C_c(\kappa) = H(\kappa(\rho_{\text{mix}})) - \min_\rho H(\kappa(\rho)). \tag{9.151}$$

Next, we define the representation $U_A^{(n)}$ of the group $G^n \stackrel{\text{def}}{=} \underbrace{G \times G \times \cdots \times G}_{n}$ on the n-fold tensor product system $\mathcal{H}_A^{\otimes n}$ as

$$U_A^{(n)}(g_1, \ldots, g_n) \stackrel{\text{def}}{=} U_A(g_1) \otimes \cdots \otimes U_A(g_n).$$

Then, the set of unitary matrices

$$U_B^{(n)}(g_1, \ldots, g_n) \stackrel{\text{def}}{=} U_B(g_1) \otimes \cdots \otimes U_B(g_n)$$

satisfies

$$\begin{aligned}
&\kappa^{\otimes n}(U_A^{(n)}(g_1, \ldots, g_n)\rho(U_A^{(n)}(g_1, \ldots, g_n))^*) \\
=&U_B^{(n)}(g_1, \ldots, g_n)\kappa^{\otimes n}(\rho)(U_B^{(n)}(g_1, \ldots, g_n))^*
\end{aligned}$$

for any density ρ on the n-fold tensor product system $\mathcal{H}_A^{\otimes n}$. If U_A is irreducible, then $U_A^{(n)}$ is also irreducible. Hence, we have the formula

$$C_c(\kappa^{\otimes n}) = nH(\kappa(\rho_{\text{mix}})) - \min_\rho H(\kappa^{\otimes n}(\rho)).$$

Thus,

$$C_c^e(\kappa) = H(\kappa(\rho_{\mathrm{mix}})) - \lim_{n\to\infty} \frac{\min_\rho H(\kappa^{\otimes n}(\rho))}{n}. \tag{9.152}$$

Further, relation (9.150) yields that

$$\begin{aligned}&(\kappa \otimes \iota_R)((U_A(g) \otimes I_R)|u\rangle\langle u|(U_A(g)^* \otimes I_R))\\ =&(U_B(g) \otimes I_R)(\kappa \otimes \iota_R)(|u\rangle\langle u|)(U_B(g)^* \otimes I_R)).\end{aligned}$$

Hence, we have

$$I(\rho, \kappa) = I(U_A(g)\rho U_A(g)^*, \kappa), \quad I_c(\rho, \kappa) = I_c(U_A(g)\rho U_A(g)^*, \kappa). \tag{9.153}$$

Concavity (8.45) of the transmission information guarantees that

$$\begin{aligned}I(\rho, \kappa) =& \int_G I(U_A(g)\rho U_A(g)^*, \kappa)\nu(dg)\\ \le& I\Big(\int_G U_A(g)\rho U_A(g)^*\nu(dg), \kappa\Big) = I(\rho_{\mathrm{mix}}, \kappa),\end{aligned} \tag{9.154}$$

which implies

$$C_{c,e}^e(\kappa) = I(\rho_{\mathrm{mix}}, \kappa) = \log d_A + \log d_B - H(\kappa \otimes \iota(|\Phi_d\rangle\langle\Phi_d|)),$$

where $|\Phi_d\rangle\langle\Phi_d|$ is the maximally entangled state.

Next, we consider the quantum capacity $C_{1,q}(\kappa)$ when the wiretap channel (κ, κ^E) is a degraded channel. In this case, concavity (9.119) holds. Hence, using the second relation in (9.153), we have

$$C_{1,q}(\kappa) = \max_\rho I_c(\rho, \kappa) = I_c(\rho_{\mathrm{mix}}, \kappa) = \log d_B - H((\kappa \otimes \iota_R)(|\Phi_d\rangle\langle\Phi_d|)). \tag{9.155}$$

Since the SDP bound $I_{\mathrm{SDP}}(\rho, \kappa)$ is concave for ρ[Exe. 9.24] (Lemma 9.11), the SDP bound $C_{\mathrm{SDP}}(\kappa_D^{\mathrm{PD}})$ is calculated to $I_{\mathrm{SDP}}(\rho_{\mathrm{mix}}, \kappa)$.

9.7.2 *d-Dimensional Depolarizing Channel*

When κ is the d-dimensional depolarizing channel $\kappa_{d,\lambda}$, the natural representation of $SU(d)$ satisfies the above condition. Hence, using (9.151), we have

$$\begin{aligned}&C_c(\kappa_{d,\lambda}) = H(\rho_{\mathrm{mix}}) - H(\lambda|u\rangle\langle u| + (1-\lambda)\rho_{\mathrm{mix}})\\ =&\frac{\lambda d + (1-\lambda)}{d}\log(\lambda d + (1-\lambda)) + \frac{(1-\lambda)(d-1)}{d}\log(1-\lambda).\end{aligned}$$

Indeed, we can easily check that this bound is attained by commutative input states. Thus,

$$\tilde{C}_c(\kappa_{d,\lambda}) = \frac{\lambda d + (1-\lambda)}{d} \log(\lambda d + (1-\lambda)) + \frac{(1-\lambda)(d-1)}{d} \log(1-\lambda).$$

For entangled input states, King [11] showed that

$$\min_{\rho} H(\kappa_{d,\lambda}^{\otimes n}(\rho)) = nH(\lambda|u\rangle\langle u| + (1-\lambda)\rho_{\mathrm{mix}}). \tag{9.156}$$

Thus, formula (9.152) yields [11]

$$C_c^e(\kappa_{d,\lambda}) = C_c(\kappa_{d,\lambda}).$$

Further, from (8.332) and (8.330),

$$\begin{aligned} C_{c,e}^e(\kappa_{d,\lambda}) &= I(\rho_{\mathrm{mix}}, \kappa_{d,\lambda}) = 2\log d - H(\rho_{I,\frac{1-\lambda(d^2-1)}{d^2}}) \\ &= 2\log d + \frac{1-\lambda(d^2-1)}{d^2} \log \frac{1-\lambda(d^2-1)}{d^2} + \frac{(1-\lambda)(d^2-1)}{d^2} \log \frac{1-\lambda}{d^2}. \end{aligned}$$

9.7.3 Transpose Depolarizing Channel

In the transpose depolarizing channel $\kappa_{d,\lambda}^T$, the natural representation of $SU(d)$ satisfies the above condition. However, U_B is not a representation. As with a depolarizing channel, using (9.151), we have

$$\begin{aligned} & C_c(\kappa_{d,\lambda}^T) = \tilde{C}_c(\kappa_{d,\lambda}) \\ = & \frac{\lambda d + (1-\lambda)}{d} \log(\lambda d + (1-\lambda)) + \frac{(1-\lambda)(d-1)}{d} \log(1-\lambda). \end{aligned}$$

Further, relation (9.156) yields

$$\begin{aligned} & \min_{\rho} H((\kappa_{d,\lambda}^T)^{\otimes n}(\rho)) = \min_{\rho} H((\kappa_{d,\lambda})^{\otimes n}(\rho^T)) = \min_{\rho^T} H((\kappa_{d,\lambda})^{\otimes n}(\rho)) \\ = & nH(\lambda|u\rangle\langle u| + (1-\lambda)\rho_{\mathrm{mix}}) \end{aligned}$$

for $\lambda \geq 0$. Hence,

$$C_c^e(\kappa_{d,\lambda}^T) = C_c(\kappa_{d,\lambda}^T)$$

for $\lambda \geq 0$. Matsumoto and Yura [18] proved this relation for $\lambda = -\frac{1}{1-d}$.

Further, from (8.333) and (8.324), its entanglement-assisted capacity is

$$C^e_{c,e}(\kappa^T_{d,\lambda}) = I(\rho_{\text{mix}}, \kappa^T_{d,\lambda}) = 2\log d - H(\rho_{W,\frac{(1-(d+1)\lambda)(d-1)}{2d}})$$
$$=2\log d + \frac{(1-(d+1)\lambda)(d-1)}{2d}\log\frac{1-(d+1)\lambda}{d^2}$$
$$+\frac{(1+(d-1)\lambda)(d+1)}{2d}\log\frac{1+(d-1)\lambda}{d^2}.$$

9.7.4 Generalized Pauli Channel

In the generalized Pauli channel κ^{GP}_p, the representation of the group $(i,j) \in \mathbb{Z}_d \times \mathbb{Z}_d \mapsto \mathbf{X}^i_d\mathbf{Z}^j_d$ satisfies condition (9.150). Its entanglement-assisted capacity can be calculated as

$$C^e_{c,e}(\kappa^{\text{GP}}_p) = I(\rho_{\text{mix}}, \kappa^{\text{GP}}_p) = 2\log d - H(p).$$

When the dimension d is equal to 2, using (5.34), we can check

$$\tilde{C}_c(\kappa^{\text{GP}}_p) = C_c(\kappa^{\text{GP}}_p) = \log 2 - \min_{i\neq j} h(p_i + p_j).$$

In this case, as mentioned in (**d**) in Sect. 9.2, King [10] showed that

$$C_c(\kappa^{\text{GP}}_p) = \tilde{C}_c(\kappa^{\text{GP}}_p) = C^e_c(\kappa^{\text{GP}}_p).$$

When the distribution $p = (p_{i,j})$ satisfies $p_{i,j} = 0$ for $j \neq 0$ in the d-dimensional system, we have

$$C_c(\kappa^{\text{GP}}_p) = \tilde{C}_c(\kappa^{\text{GP}}_p) = C^e_c(\kappa^{\text{GP}}_p) = \log d.$$

In this case, the channel κ^{GP}_p is a phase-damping channel. As proved in Sect. 9.7.7, it satisfies condition (9.117). Hence, (9.155) yields

$$C_{q,1}(\kappa^{\text{GP}}_p) = I_c(\rho_{\text{mix}}, \kappa^{\text{GP}}_p) = \log d - H(p).$$

9.7.5 PNS Channel

The PNS channel satisfies condition (9.150). Hence, using (9.151), we have

$$C_c(\kappa^{\text{pns}}_{d,n\to m}) = H(\kappa^{\text{pns}}_{d,n\to m}(\rho_{\text{mix}})) - \min_{\rho} H(\kappa^{\text{pns}}_{d,n\to m}(\rho)) = \log\binom{m+d-1}{d-1}.$$

Since $C^e_c(\kappa^{\text{pns}}_{d,n\to m})$ is less than the dimension of the input system, $C^e_c(\kappa^{\text{pns}}_{d,n\to m}) = C_c(\kappa^{\text{pns}}_{d,n\to m})$. Its entanglement-assisted capacity is calculated as

$$\begin{aligned}&C^e_{c,e}(\kappa^{\text{pns}}_{d,n\to m}) = I(\rho_{\text{mix}}, \kappa^{\text{pns}}_{d,n\to m})\\
=&\log\binom{m+d-1}{d-1} + \log\binom{n+d-1}{d-1} - \log\binom{(n-m)+d-1}{d-1}.\end{aligned}$$

From Exercise 5.16, the wiretap channel $(\kappa^{\text{pns}}_{d,n\to m}, (\kappa^{\text{pns}}_{d,n\to m})^E)$ is a degraded channel. Hence, from (9.155), its quantum capacity is calculated as

$$\begin{aligned}&C_{q,1}(\kappa^{\text{pns}}_{d,n\to m}) = I_c(\rho_{\text{mix}}, \kappa^{\text{pns}}_{d,n\to m})\\
=&-\log\binom{(n-m)+d-1}{d-1} + \log\binom{m+d-1}{d-1}.\end{aligned} \tag{9.157}$$

9.7.6 Erasure Channel

The erasure channel also satisfies condition (9.150). Hence, using (9.151), we have

$$\begin{aligned}&C_c(\kappa^{\text{era}}_{d,p}) = I_c(\rho_{\text{mix}}, \kappa^{\text{era}}_{d,p}) = H(\kappa^{\text{era}}_{d,p}(\rho_{\text{mix}})) - \min_\rho H(\kappa^{\text{era}}_{d,p}(\rho))\\
=&-(1-p)\log\frac{1-p}{d} - p\log p - h(p) = (1-p)\log d.\end{aligned}$$

Since it is attained by commutative input states, $C_c(\kappa^{\text{era}}_{d,p}) = \tilde{C}_c(\kappa^{\text{era}}_{d,p})$.

Next, we consider the capacity $C^e_c(\kappa^{\text{era}}_{d,p})$. Because

$$(\kappa^{\text{era}}_{d,p})^{\otimes n}(\rho) = \bigoplus_{\{i_1,\ldots,i_k\}\subset\{1,\ldots,n\}} (1-p)^k p^{n-k}\,\mathrm{Tr}_{i_1,\ldots,i_k}\rho\otimes|u_d\rangle\langle u_d|^{\otimes(n-k)},$$

the minimum entropy $\min_\rho H((\kappa^{\text{era}}_{d,p})^{\otimes n}(\rho))$ is calculated as

$$\begin{aligned}&\min_\rho H((\kappa^{\text{era}}_{d,p})^{\otimes n}(\rho))\\
=&\min_\rho \sum_{\{i_1,\ldots,i_k\}\subset\{1,\ldots,n\}} (1-p)^k p^{n-k}(-\log(1-p)^k p^{n-k} + H(\mathrm{Tr}_{i_1,\ldots,i_k}\rho))\\
=&nh(p) + \min_\rho \sum_{\{i_1,\ldots,i_k\}\subset\{1,\ldots,n\}} (1-p)^k p^{n-k} H(\mathrm{Tr}_{i_1,\ldots,i_k}\rho) = nh(p).\end{aligned}$$

Hence, from (9.152),

$$C^e_c(\kappa^{\text{era}}_{d,p}) = C_c(\kappa^{\text{era}}_{d,p}) = \tilde{C}_c(\kappa^{\text{era}}_{d,p}) = (1-p)\log d.$$

The entanglement-assisted capacity is calculated as

$$
\begin{aligned}
&C^e_{c,e}(\kappa^{\rm era}_{d,p}) = I(\rho_{\rm mix}, \kappa^{\rm era}_{d,p}) \\
=&\log d + (1-p)\log\frac{d}{1-p} - p\log p - p\log\frac{d}{p} + (1-p)\log(1-p) \quad (9.158)\\
=&2(1-p)\log d,
\end{aligned}
$$

where we used Exercise 5.15 in (9.158).

From Exercise 5.15, the wiretap channel $(\kappa^{\rm era}_{d,p}, (\kappa^{\rm era}_{d,p})^E)$ is a degraded channel. Hence, from (9.155), its quantum capacity is calculated as [67]

$$
C_{q,1}(\kappa^{\rm era}_{d,p}) = I_c(\rho_{\rm mix}, \kappa^{\rm era}_{d,p}) = (1-2p)\log d \text{ for } p \le 1/2.
$$

9.7.7 *Phase-Damping Channel*

Any phase-damping channel $\kappa^{\rm PD}_D$ clearly satisfies

$$
C_c(\kappa^{\rm PD}_D) = \tilde{C}_c(\kappa^{\rm PD}_D) = C^e_c(\kappa^{\rm PD}_D) = \log d.
$$

Indeed, we can show that the wiretap channel $(\kappa^{\rm PD}_D, (\kappa^{\rm PD}_D)^E)$ is a degraded channel as follows. When the input state is the maximally entangled state $\frac{1}{d}\sum_{k,l}|e_k, e^R_k\rangle\langle e_l, e^R_l|$, the purification of the output state $\sum_{k,l} d_{k,l}|e_k, e^R_k\rangle\langle e_l, e^R_l|$ is given as

$$
\frac{1}{d_A}\sum_{k,k,',l,l'} y_{k,k'}\overline{y_{l,l'}}|e_k, e^R_k, e^E_{k'}\rangle\langle e_l, e^R_l, e^E_{l'}|,
$$

where $Y = (y_{k,k'})$ satisfies $Y^*Y = D$. From the condition $X_{k,k} = 1$, the positive semidefinite matrix $\rho^E_k \stackrel{\rm def}{=} \sum_{k'l'} y_{k,k'}\overline{y_{k,l'}}|e^E_{k'}\rangle\langle e^E_{l'}|$ satisfies the condition of states $\operatorname{Tr}\rho^E_k = 1$. Then, by applying (8.32), the channel $(\kappa^{\rm PD}_D)^E$ to the environment is described as

$$
\begin{aligned}
&(\kappa^{\rm PD}_D)^E(\rho) = \operatorname{Tr}_{R,A}(I_{A,E}\otimes\rho^T)\sum_{k,k,',l,l'} y_{k,k'}\overline{y_{l,l'}}|e_k, e^R_k, e^E_{k'}\rangle\langle e_l, e^R_l, e^E_{l'}| \\
=&\sum_k \rho_{k,k} y_{k,k'}\overline{y_{k,l'}}|e^E_{k'}\rangle\langle e^E_{l'}| = \sum_k \langle e_k|\kappa^{\rm PD}_D(\rho)|e_k\rangle y_{k,k'}\overline{y_{k,l'}}|e^E_{k'}\rangle\langle e^E_{l'}| \\
=&\sum_k \langle e_k|\kappa^{\rm PD}_D(\rho)|e_k\rangle \rho^E_k.
\end{aligned}
$$

Hence, the wiretap channel $(\kappa^{\rm PD}_D, (\kappa^{\rm PD}_D)^E)$ is a degraded channel. Further, the phase-damping channel $\kappa^{\rm PD}_D$ satisfies the invariance

$$I_c(U_\theta \rho U_\theta^*, \kappa_D^{\mathrm{PD}}) = I_c(\rho, \kappa_D^{\mathrm{PD}}), \quad U_\theta \stackrel{\mathrm{def}}{=} \sum_k e^{i\theta_k} |e_k\rangle\langle e_k|.$$

Hence, using concavity (9.119), we have

$$C_{q,1}(\kappa_D^{\mathrm{PD}}) = \max_\rho I_c(\rho, \kappa_D^{\mathrm{PD}}) = \max_p I_c\left(\sum_k p_k |e_k\rangle\langle e_k|, \kappa_D^{\mathrm{PD}}\right)$$

$$= \max_p H(p) - H\left(\sum_k p_k \rho_k^E\right) \geq \log d - H\left(\frac{1}{d} D\right).$$

Further, since

$$C_{\mathrm{SDP}}(\kappa_D^{\mathrm{PD}}) = \max_p I_c\left(\sum_k p_k |e_k\rangle\langle e_k|, \kappa_D^{\mathrm{PD}}\right) \tag{9.159}$$

is shown below, Theorem 9.12 implies that

$$C_{q,1}(\kappa_D^{\mathrm{PD}}) = C_{q,C}^{\dagger}(\kappa_D^{\mathrm{PD}}).$$

That is, the channel κ_D^{PD} satisfies the strong converse property for quantum state transmission [65].

(9.159) can be shown as follows. Notice that the channel κ_D^{PD} is the covariant with respect to the unitary $\sum_k e^{i\theta_k} |e_k\rangle\langle e_k|$. Since the SDP bound $I_{\mathrm{SDP}}(\rho, \kappa_D^{\mathrm{PD}})$ is concave for ρ[Exe. 9.24] (Lemma 9.11), the maximum $C_{\mathrm{SDP}}(\kappa_D^{\mathrm{PD}}) = \max_\rho I_{\mathrm{SDP}}(\rho, \kappa_D^{\mathrm{PD}})$ is realized when ρ is invariant with respect to the unitary $\sum_k e^{i\theta_k} |e_k\rangle\langle e_k|$. $C_{\mathrm{SDP}}(\kappa_D^{\mathrm{PD}})$ is calculated to $\max_p I_{\mathrm{SDP}}(\sum_k p_k |e_k\rangle\langle e_k|, \kappa_D^{\mathrm{PD}})$. Since the state $\rho_p := \kappa_D^{\mathrm{PD}}(\sum_{k,l} \sqrt{p_k}\sqrt{p_l} |e_k, e_k^R\rangle\langle e_l, e_l^R|)$ is maximally correlated, (8.143) and (8.227) imply that

$$I_c\left(\sum_k p_k |e_k\rangle\langle e_k|, \kappa_D^{\mathrm{PD}}\right) = H\left(\kappa_D^{\mathrm{PD}}\left(\sum_k p_k |e_k\rangle\langle e_k|\right)\right) - H(\rho_p) = E_{r,S}(\rho_p)$$

$$= E_{\mathrm{SDP}}(\rho_p) = I_{\mathrm{SDP}}\left(\sum_k p_k |e_k\rangle\langle e_k|, \kappa_D^{\mathrm{PD}}\right).$$

So, we obtain (9.159).

9.8 Proof of Theorem 9.3

First we prove Lemma 9.1.

Proof of Lemma 9.1

S⇒**A:** From (9.30), **S** implies (9.34), i.e., **A**.
S⇒**L:** From (9.31),

$$
\begin{aligned}
& \min_{\rho^{1,2}:\operatorname{Tr}\rho^{1,2}(X^1+X^2)\le K} f^{1,2}(\rho^{1,2}) \\
\ge & \min_{\rho^{1,2}:\operatorname{Tr}\rho^{1,2}(X^1+X^2)\le K} f^1(\rho^1)+f^2(\rho^2) \\
= & \min_{\rho^1,\rho^2:\operatorname{Tr}\rho^1X^1+\operatorname{Tr}\rho^2X^2\le K} f^1(\rho^1)+f^2(\rho^2) \\
= & \min_{0\le\lambda\le 1}\min_{\rho^1:\operatorname{Tr}\rho^1X^1\le\lambda K} f^1(\rho^1)+\min_{\rho^2:\operatorname{Tr}\rho^2X^2\le(1-\lambda)K} f^2(\rho^2).
\end{aligned}
$$

On the other hand, since $f^1(\rho^1)+f^2(\rho^2)\ge f^{1,2}(\rho^1\otimes\rho^2)$, we have

$$
\begin{aligned}
& \min_{\rho^1,\rho^2:\operatorname{Tr}\rho^1X^1+\operatorname{Tr}\rho^2X^2\le K} f^1(\rho^1)+f^2(\rho^2) \\
\ge & \min_{\rho^1,\rho^2:\operatorname{Tr}\rho^1X^1+\operatorname{Tr}\rho^2X^2\le K} f^{1,2}(\rho^1\otimes\rho^2) \\
\ge & \min_{\rho^{1,2}:\operatorname{Tr}\rho^{1,2}(X^1+X^2)\le K} f^{1,2}(\rho^{1,2}).
\end{aligned}
$$

Hence, we obtain (9.33).
L⇒C: Choose $\rho_0^{1,2}$ such that $\operatorname{Tr}\rho_0^{1,2}(X^1+X^2)-f^{1,2}(\rho_0^{1,2})=\max_{\rho^{1,2}}\operatorname{Tr}\rho^{1,2}(X^1+X^2)-f^{1,2}(\rho^{1,2})$. Then, the real number $K\stackrel{\mathrm{def}}{=}\operatorname{Tr}\rho_0^{1,2}(X^1+X^2)$ satisfies

$$
\begin{aligned}
& \max_{\rho^{1,2}}\operatorname{Tr}\rho^{1,2}(X^1+X^2)-f^{1,2}(\rho^{1,2}) \\
= & \max_{\rho^{1,2}:\operatorname{Tr}\rho^{1,2}(X^1+X^2)\ge K}\operatorname{Tr}\rho^{1,2}(X^1+X^2)-f^{1,2}(\rho^{1,2}) \\
=& K+\max_{\rho^{1,2}:\operatorname{Tr}\rho^{1,2}(X^1+X^2)\ge K} -f^{1,2}(\rho^{1,2}) \\
=& K+\max_{\rho^1,\rho^2:\operatorname{Tr}\rho^1X^1+\operatorname{Tr}\rho^2X^2\ge K} -f^1(\rho^1)-f^2(\rho^2) \\
= & \max_{\rho^1,\rho^2:\operatorname{Tr}\rho^1X^1+\operatorname{Tr}\rho^2X^2\ge K}\operatorname{Tr}\rho^1X^1-f^1(\rho^1)+\operatorname{Tr}\rho^2X^2-f^2(\rho^2) \\
\le & \max_{\rho^1,\rho^2}\operatorname{Tr}\rho^1X^1-f^1(\rho^1)+\operatorname{Tr}\rho^2X^2-f^2(\rho^2).
\end{aligned}
$$

Conversely, from (9.30),

$$
\begin{aligned}
& \max_{\rho^{1,2}}\operatorname{Tr}\rho^{1,2}(X^1+X^2)-f^{1,2}(\rho^{1,2}) \\
\ge & \max_{\rho^1,\rho^2}\operatorname{Tr}\rho^{1,2}(X^1+X^2)-f^{1,2}(\rho^1\otimes\rho^2) \\
\ge & \max_{\rho^1,\rho^2}\operatorname{Tr}\rho^1X^1-f^1(\rho^1)+\operatorname{Tr}\rho^2X^2-f^2(\rho^2).
\end{aligned}
$$

Hence, we obtain (9.32).
C⇒S: For any $\rho_0^{1,2}$, from Lemma A.8, we choose Hermitian matrices X^1 and X^2 such that $\operatorname{Tr}\rho_0^iX^i-f^i(\rho_0^i)=\max_{\rho^i}\operatorname{Tr}\rho^iX^i-f^i(\rho^i)$. Hence,

$$\sum_{i=1}^{2} \operatorname{Tr} \rho_0^i X^i - f^i(\rho_0^i) = \sum_{i=1}^{2} \max_{\rho^i} \operatorname{Tr} \rho^i X^i - f^i(\rho^i)$$
$$= \max_{\rho^{1,2}} \operatorname{Tr} \rho^{1,2}(X^1 + X^2) - f^{1,2}(\rho^{1,2}) \geq \operatorname{Tr} \rho_0^{1,2}(X^1 + X^2) - f^{1,2}(\rho_0^{1,2}).$$

Since $\operatorname{Tr} \rho_0^{1,2}(X^1 + X^2) = \operatorname{Tr} \rho_0^1 X^1 + \operatorname{Tr} \rho_0^2 X^2$, we have (9.31). ■

Proof of HM⇒HC First, we assume that there exists a channel $\kappa_{X,p}$ for any channel κ, any positive semidefinite Hermitian matrix X on the input system $\mathcal{H}_A$, and any probability p, such that

$$C_c(\kappa_{X,p}) = \max_{\rho}(1-p)(\chi_\kappa(\rho)) + p \operatorname{Tr} H\rho, \tag{9.160}$$

and

$$\begin{aligned} &C_c(\kappa^1_{X^1,p} \otimes \kappa^2_{X^2,p}) \\ =&\max_{\rho}\Big((1-p)^2\chi_{\kappa^1\otimes\kappa^2}(\rho) + (1-p)p(\chi_{\kappa^1}(\rho^1) + \operatorname{Tr} X^2\rho^2) \\ &+ (1-p)p(\chi_{\kappa^2}(\rho^2) + \operatorname{Tr} X^1\rho^1) + p^2(\operatorname{Tr} X^1\rho^1 + \operatorname{Tr} X^2\rho^2)\Big). \end{aligned} \tag{9.161}$$

The channel $\kappa_{X,p}$ is called **Shor extension** [14] of κ. Apply Condition **HM** to the channel $\kappa^1_{\frac{1}{p}X^1,p} \otimes \kappa^2_{\frac{1}{p}X^2,p}$, then we have

$$\begin{aligned} &\max_{\rho^{1,2}}\left((1-p)^2\chi_{\kappa^1\otimes\kappa^2}(\rho^{1,2}) + (1-p)p(\chi_{\kappa^1}(\rho^1) + \operatorname{Tr} \frac{1}{p}X^2\rho^2\right) \\ &+ (1-p)p\left(\chi_{\kappa^2}(\rho^2) + \operatorname{Tr}\frac{1}{p}X^1\rho^1) + p^2(\operatorname{Tr}\frac{1}{p}X^1\rho^1 + \operatorname{Tr}\frac{1}{p}X^2\rho^2)\right) \\ \leq&\max_{\rho^1}(1-p)\left(\chi_{\kappa^1}(\rho^1) + \operatorname{Tr}\frac{1}{p}X^1\rho^1\right) + \max_{\rho^2}(1-p)\left(\chi_{\kappa^2}(\rho^2) + \operatorname{Tr}\frac{1}{p}X^2\rho^2\right). \end{aligned}$$

Taking the limit $p \to 0$, we obtain

$$\begin{aligned} &\max_{\rho^{1,2}} \chi_{\kappa^1\otimes\kappa^2}(\rho^{1,2}) + \operatorname{Tr}(X^1 + X^2)\rho^{1,2} \\ \leq&\max_{\rho^1}(\chi_{\kappa^1}(\rho^1) + \operatorname{Tr} X^1\rho^1) + \max_{\rho^2}(\chi_{\kappa^2}(\rho^2) + \operatorname{Tr} X^2\rho^2), \end{aligned}$$

which implies Condition **HC**.

Next, we define the channel $\kappa_{X,p}$ with the input system $\mathcal{H}_A \otimes \mathbb{C}^k$, where $k \geq \|X\|$, and check (9.161). First, we generate one-bit random number X with probability $P_0 = 1 - p$ and $P_1 = p$. When $X = 0$, the output state is $\kappa(\operatorname{Tr}_{\mathbb{C}^k} \rho)$ for the input state ρ. Otherwise, we perform the measurement of the spectral decomposition of X, and send the receiver the state $\tilde{\kappa}^y(\mathrm{Tr}_A \rho)$ dependently of its measurement outcome y, which is eigenvalue of X. Here, we defined the channel $\tilde{\kappa}^y$ and the stochastic transition matrix Q_l^j such that

$$\tilde{\kappa}^y(\sigma) = \sum_l \sum_j Q_l^j |u_l\rangle\langle u_l|\langle u_j|\sigma|u_j\rangle, \quad y = C_c(Q) = I(p_{\mathrm{mix}}, Q).$$

In this case, we assume that the receiver received the information X and y. Then, the relation (9.160) holds. From these discussions, we can check the equation (9.161). ■

Proof of EM⇒EC We define the channel $\tilde{\kappa}_{H,p}$ with the input system $\mathcal{H}_A$ as follows First, we generate one-bit random number X with probabilities $P_0 = 1 - p$ and $P_1 = p$. When $X = 0$, the output state is $\kappa(\mathrm{Tr}_{\mathbb{C}^k} \rho)$ for the input state ρ. When $X = 1$, we perform the measurement of the spectral decomposition of H, and obtain the eigenvalue y of H. Then, the output state is ρ_y, where ρ_y satisfies $H(\rho_y) = y$. In this case, the receiver is assumed to receive the information X and y. Then, the output entropy of the channel $\tilde{\kappa}_{H,p}$ can be calculated as

$$H(\tilde{\kappa}_{X,p}(\rho)) = (1-p)H(\kappa(\rho)) + p \,\mathrm{Tr}\, X\rho + h(p) - pH(\mathrm{P}_\rho^{E_X}).$$

Further,

$$\begin{aligned} & H(\tilde{\kappa}^1_{X^1,p} \otimes \tilde{\kappa}^2_{X^2,p}(\rho)) \\ =&(1-p)^2(H(\kappa^1 \otimes \kappa^2(\rho))) + p(1-p)(\mathrm{Tr}\, X^1\rho^1 + H(\kappa^2(\rho^2))) \\ &+ p(1-p)(\mathrm{Tr}\, X^2\rho^2 + H(\kappa^1(\rho^1))) + p^2(\mathrm{Tr}\, X^1\rho^1 + \mathrm{Tr}\, X^2\rho^2) + 2h(p) \\ &- pH(\mathrm{P}_{\rho^1}^{E_{X^1}}) - pH(\mathrm{P}_{\rho^2}^{E_{X^2}}). \end{aligned}$$

Condition **EM** implies

$$\min_{\rho^{1,2}} H_{\tilde{\kappa}^1_{\frac{1}{p}X^1,p} \otimes \tilde{\kappa}^2_{\frac{1}{p}X^2,p}}(\rho^{1,2}) = \min_{\rho^1} H_{\tilde{\kappa}^1_{\frac{1}{p}X^1,p}}(\rho^1) + \min_{\rho^2} H_{\tilde{\kappa}^2_{\frac{1}{p}X^2,p}}(\rho^2).$$

Since $H(\mathrm{P}_{\rho^2}^{E_{X^2}}) \le \log d_A d_B$, taking the limit $p \to 0$, we have

$$\begin{aligned} H(\tilde{\kappa}_{\frac{1}{p}X,p}, \rho) &\to H_\kappa(\rho) + \mathrm{Tr}\, X\rho \\ H_{\tilde{\kappa}^1_{\frac{\lambda}{p}X^1,p} \otimes \tilde{\kappa}^2_{\frac{1}{p}X^2,p}}(\rho) &\to H_{\kappa^1 \otimes \kappa^2}(\rho) + (\mathrm{Tr}\, X^1\rho^1 + \mathrm{Tr}\, X^2\rho^2). \end{aligned}$$

Since the set of density matrices is compact, we obtain

$$\min_{\rho^{1,2}} H_{\kappa^1 \otimes \kappa^2}(\rho^{1,2}) + \mathrm{Tr}(X^1 + X^2)\rho^{1,2}$$
$$= \min_{\rho^1,\rho^2} (H_{\kappa^1}(\rho^1) + H_{\kappa^2}(\rho^2) + \mathrm{Tr}\, X^1\rho^1 + \mathrm{Tr}\, X^2\rho^2,$$

which implies **EC**. ∎

Proof of FA⇒FS See Pomeransky [15].

9.9 Historical Note

9.9.1 Additivity Conjecture

Bennett et al. [68] consider the transmission of classical information by using entangled states as input states. After this research, in order to consider the additivity of the classical channel capacity, Nagaoka [69] proposed quantum analogs of the Arimoto–Blahut algorithms [70, 71], and Nagaoka and Osawa [7] numerically analyzed two-tensor product channels in the qubit case with quantum analogs based on this algorithms. In this numerical analysis, all the examined channels κ satisfy $C(\kappa^{\otimes 2}) = 2C(\kappa)$. This numerical analysis strongly suggests Conjecture **HM**. This research was published by Osawa and Nagaoka [17]. Independently, King proved Conditions **HM**, **EM**, and **RM** with κ^1 as a unital channel in the qubit system and κ^2 as an arbitrary channel [10]. Following this result, Fujiwara and Hashizume [72] showed **HM** and **EM** with κ^1 and κ^2 as depolarizing channels. Further, King [11] proved **HM**, **EM**, and **RM** with only κ^1 as a depolarizing channel. Shor [12] also proved **HM** with only κ_1 as an entanglement-breaking channel.

On the other hand, Vidal et al. [73] pointed out that the entanglement of formation is $\log 2$ when the support of the state is contained by the antisymmetric space of $\mathbb{C}^3$. Following this research, Shimono [74] proved **FA** when the supports of ρ_1 and ρ_2 are contained by the antisymmetric space of $\mathbb{C}^3$; Yura [75] proved that $E_f(\rho) = E_c(\rho)$ for this case. Further, using the idea in Vidal et al. [73], Matsumoto et al. [13] introduced the MSW correspondence (9.8) or (9.29). Using this correspondence, they proved **FS⇒HM** and **FS⇒HL**.

Following this result, Shor [14] proved **HL⇒FA** and **HM⇒HL**. Audenaert and Braunstein [76] pointed out the importance of the conjugate function in this problem. Further, Pomeransky [15] proved the equivalence among **FA**, **FC**, and **FS** by employing the idea by Audenaert and Braunstein [76]. Shor also showed **FA⇒FS** independently. He also proved **EM⇒FA** and (**HM** or **FA**)⇒**EM**. Further, applying this idea, Koashi and Winter [77] obtained relation (8.173). Recently, Matsumoto [78] found short proofs of **EM⇒EL** and **EL⇔ML**. In this textbook, based on his idea, we analyze the structure of equivalence among these conditions and derive 14 conditions (Theorem 9.3). Matsumoto [79] also introduced another measure of entanglement and showed that its additivity is equivalent to the additivity of entanglement of formation.

Further, Matsumoto and Yura [18] showed $E_f(\rho) = E_c(\rho)$ for antisymmetric states. Applying the concept of channel states to antisymmetric states, they proved that $C(\kappa^{\otimes n}) = C(\kappa)$ for antisymmetric channels. Indeed, this channel has been proposed by Werner and Holevo [80] as a candidate for a counterexample of Additivity **HM** or **EM** because they showed that it does not satisfy Condition **RM** for sufficiently large s. Vidal et al. implicitly applied the same concept to entanglement-breaking channels and proved **FA** when only ρ_1 satisfies condition (8.147). Following discovery of this equivalence, Datta et al. [20] and Fannes et al. [19] showed **HM** and **EM** when κ_1 and κ_2 are transpose depolarizing channels. Wolf and Eisert [22], Fukuda [9], and Datta and Ruskai [21] extended the above results to larger classes of channels.

However, besides of so many equivalent conditions, Hastings [81] showed the existence of a counter example for **FA** superadditivity of entanglement formation. Hence, it was shown that all of these equivalent conditions do not hold.

9.9.2 Channel Coding with Shared Entanglement

Concerning the channel coding with shared entanglement, Bennett and Wiesner [24] found the effectiveness of shared entanglement. Assuming Theorem 4.1 in the nonorthogonal two-pure-state case,[8] Barenco and Ekert [83] proved the direct part of Theorem 9.4 in the two-dimensional pure-state case. Hausladen et al. [84] independently proved the unitary coding version of Theorem 9.4 in the two-dimensional pure-state case. Bose et al. [25] showed the direct part of Theorem 9.4 in the two-dimensional mixed-state case. Hiroshima [26] showed the unitary coding version of Theorem 9.4 in the general mixed-state case. Bowen [27] independently showed the same fact in the two-dimensional case. Finally, Horodecki et al. [28] and Winter [29] independently proved Theorem 9.4 in the form presented in this book. When the channel has noise, Bennett et al. [30] showed the direct part of Theorem 9.5 in the general case and its converse part in the generalized Pauli case. In this converse part, they introduced the reverse Shannon theorem. Following this result, Bennett et al. [31] and Holevo [32] completed the proof of Theorem 9.5. In this book, we proved this theorem in a way similar to Holevo [32].

[8] In their paper, it is mentioned that Levitin [82] showed the direct part of Theorem 4.1 in this special case.

9.9.3 *Quantum-State Transmission*

Many researchers have treated the capacity of quantum-state transmission via a noisy quantum channel by algebraic methods first [54–60]. This approach is called quantum error correction. Using these results, Bennett et al. [64] discussed the relation between quantum error correction and entanglement of distillation. Following these studies, Schumacher [85] introduced many information quantities for noisy channels (Sect. 8.2). Barnum et al. [86] showed that a capacity with the error $\varepsilon_2[\Phi^{(n)}]$ is less than $\lim_{n\to\infty}\frac{1}{n}\max_{\rho\in\mathcal{S}(\mathcal{H}_A^{\otimes n})} I_c(\rho,\kappa^{\otimes n})$ if the encoding is restricted to being isometry. Barnum et al. [66] proved the coincidence with two capacities $C_1(\kappa)$ and $C_2(\kappa)$. They also showed that these capacities are less than $\lim_{n\to\infty}\frac{1}{n}\max_{\rho\in\mathcal{S}(\mathcal{H}_A^{\otimes n})} I_c(\rho,\kappa^{\otimes n})$. On the other hand, Lloyd [63] predicted that the bound $I_c(\rho,\kappa)$ could be achieved without a detailed proof, and Shor [62] showed its achievability. Then, the capacity theorem for quantum-state transmission (Theorem 9.10) was obtained. Further, Devetak [43] formulated a capacity theorem for quantum wiretap channels (Theorem 9.8). Applying this discussion, he gave an alternative proof of Theorem 9.10. Here, the bit error of state transmission corresponds to the error of normal receiver in wiretap channels, and the phase error of state transmission corresponds to information obtained by the eavesdropper in a wiretap channel. Indeed, the analysis of information obtained by the eavesdropper is closely related to the channel resolvability. Hence, in this book, we analyze quantum-channel resolvability first. Then we proceed to quantum wiretap channels and quantum-state transmission. Indeed, Devetak [43] also essentially showed the direct part of quantum-channel resolvability (Theorem 9.7) in the tensor product case; however, our proof of it is slightly different from the proof by Devetak.

Indeed, to obtain the capacity theorem for quantum-state transmission from the capacity theorem for quantum wiretap channels, we need an additional discussion. That is, we need to show that the entanglement fidelity is close to 1 when the bit error and the phase error of state transmission are close to 0. To clarify this point, we employ a duality relation between the security and the coherence. That is, combining Theorem 8.17 and Lemma 9.4, we show the capacity theorem for quantum-state transmission. Further, Devetak and Shor [87] studied the asymptotic tradeoff between the transmission rates of transmissions of quantum-state and classical information.

When a channel is degraded, the wiretap capacity can be single-letterized as (9.75). This formula was obtained independently in the original Japanese version of this book in 2004 and by Devetak and Shor [87]. By applying this relation to quantum capacity, the single-letterized capacity is derived independently in several examples in this English version and Yard [88].

The strong converse of quantum-state transmission is more difficult. Morgan et al. [89] demonstrated that a "pretty strong converse" holds for degradable quantum channels, i.e., they showed that there is (at least) a jump in the quantum error from zero to 1/2 once the communication rate exceeds the quantum capacity. Then, Tomamichel et al. [65] showed Theorem 9.12. The proof essentially employed the strong converse of entanglement distillation (8.226), which was shown in the first edition of this book. To obtain Theorem 9.12, we additionally need to show Lemma 9.12. Note that the bound given in Theorem 9.12 is not necessarily attained in general. They showed that the bound is attained for the phase-damping channel defined in Example 5.10.

9.10 Solutions of Exercises

Exercise 9.1 Assume that the initial state on $\mathcal{H}_C$ is $\sum_{k=1}^d x_k|u_k^C\rangle$. Then, the initial state in the total system is $\sum_{k,l}\frac{1}{\sqrt{d}}x_k|u_l^A,u_l^B,u_k^C\rangle$. When the measurement outcome is (i,j), the resultant state of this measurement is

$$\begin{aligned}
&d\langle u_{i,j}^{A,C}|\sum_{k,l}\frac{1}{\sqrt{d}}x_k|u_l^A,u_l^B,u_k^C\rangle=\sum_t\langle u_t^A,u_t^C|(\mathbf{X}_C^i\mathbf{Z}_C^j)^T\sum_{k,l}x_k|u_l^A,u_l^B,u_k^C\rangle\\
=&\sum_t\sum_{k,l}\langle u_t^A,u_t^C|((\mathbf{X}^i\mathbf{Z}^j)^T x)_k|u_l^A,u_l^B,u_k^C\rangle=\sum_{k,l}((\mathbf{X}^i\mathbf{Z}^j)^T x)_k\langle u_k^A||u_l^A,u_l^B\rangle\\
=&\sum_k((\mathbf{X}^i\mathbf{Z}^j)^T x)_k|u_k^B\rangle=(\mathbf{X}_B^i\mathbf{Z}_B^j)^T\sum_k x_k|u_k^B\rangle
\end{aligned}$$

because all of outcomes occur with the equal probability $\frac{1}{d^2}$. Thus, the final state is $\overline{\mathbf{X}_B^i\mathbf{Z}_B^j}(\mathbf{X}_B^i\mathbf{Z}_B^j)^T\sum_k x_k|u_k^B\rangle=\sum_k x_k|u_k^B\rangle$.

Exercise 9.2 For any input state ρ_x on the composite system, we have

$$\begin{aligned}
&H((\kappa^1\otimes\kappa^2)(\rho_x))\overset{(a)}{=}H\Big(\sum_y Q_y^x\rho_{x,y}^1\otimes\kappa^2(\rho_{x,y}^1)\Big)\\
\overset{(b)}{\geq}&\Big(\sum_y Q_y^x H(\kappa^2(\rho_{x,y}^2))+H\Big(\sum_y Q_y^x\rho_{x,y}^1\Big)\Big)\\
=&H\Big(\sum_y Q_y^x\rho_{x,y}^1\Big)+\sum_y Q_y^x H(\kappa^2(\rho_{x,y}^2))\\
\geq&\min_{\rho^1}H(\kappa^1(\rho^1))+\min_{\rho^2}H(\kappa^2(\rho^2)),
\end{aligned}$$

(a) and (b) follow from (9.35) and (5.110), respectively.

Exercise 9.3 Applying Theorem 4.1 to the channel $\kappa^{\otimes n}$, we find that $C_c^e(\kappa)\geq\frac{C_c(\kappa^{\otimes n})}{n}$, which implies that $C_c^e(\kappa)\geq\sup_n\frac{C_c(\kappa^{\otimes n})}{n}$. Since $\sup_n\frac{C_c(\kappa^{\otimes n})}{n}=\lim_{n\to\infty}\frac{C_c(\kappa^{\otimes n})}{n}$ follows from Lemma A.1, it is sufficient to show

$$C_c^e(\kappa)\leq\lim_{n\to\infty}\frac{C_c(\kappa^{\otimes n})}{n}.\tag{9.162}$$

Consider the code $\hat{\Phi}^{(n)}=(N_n,\hat{\varphi}^{(n)},Y^{(n)})$ satisfying that $\varepsilon[\hat{\Phi}^{(n)}]\to 0$. Since $C_c(\kappa^{\otimes n})=\sup_p I(p,\kappa^{\otimes n})$, similar to (4.32) in the proof of Theorem 4.2, the Fano inequality (2.35) yields that

$$\frac{1}{n}\log N_n \leq \frac{(\log 2)/n + C_c(\kappa^{\otimes n})}{1-\varepsilon[\hat{\Phi}^{(n)}]}. \tag{9.163}$$

Since $\varepsilon[\hat{\Phi}^{(n)}] \to 0$, we have, $\overline{\lim}_{n\to\infty}\frac{1}{n}\log N_n \leq \lim_{n\to\infty}\frac{C_c(\kappa^{\otimes n})}{n}$, which implies (9.162).

Exercise 9.4 Using the monotonicity of the quantum relative entropy and (5.86), we obtain

$$\begin{aligned}
& I(X:Y)\\
\leq & \frac{1}{N_n}\sum_{i=1}^{N_n} D\left(\left(\varphi_e^{(n)}(i)\otimes \iota_B^{\otimes n}\right)(\rho_{A,B}^{\otimes n})\middle\| \frac{1}{N_n}\sum_{i=1}^{N_n}\left(\varphi_e^{(n)}(i)\otimes \iota_B^{\otimes n}\right)(\rho_{A,B}^{\otimes n})\right)\\
= & H\left(\frac{1}{N_n}\sum_{i=1}^{N_n}\left(\varphi_e^{(n)}(i)\otimes \iota_B^{\otimes n}\right)(\rho_{A,B}^{\otimes n})\right) - \frac{1}{N_n}\sum_{i=1}^{N_n} H(\left(\varphi_e^{(n)}(i)\otimes \iota_B^{\otimes n}\right)(\rho_{A,B}^{\otimes n}))\\
\leq & H\left(\mathrm{Tr}_{A_n'}\frac{1}{N_n}\sum_{i=1}^{N_n}\left(\varphi_e^{(n)}(i)\otimes \iota_B^{\otimes n}\right)(\rho_{A,B}^{\otimes n})\right)\\
& + H\left(\mathrm{Tr}_B\frac{1}{N_n}\sum_{i=1}^{N_n}\left(\varphi_e^{(n)}(i)\otimes \iota_B^{\otimes n}\right)(\rho_{A,B}^{\otimes n})\right) - \min_\kappa H((\kappa\otimes \iota_B^{\otimes n})(\rho_{A,B}^{\otimes n})).
\end{aligned}$$

Using

$$\mathrm{Tr}_{A_n'}\left(\varphi_e^{(n)}(i)\otimes \iota_B^{\otimes n}\right)(\rho_{A,B}^{\otimes n}) = \mathrm{Tr}_A \rho_{A,B}^{\otimes n}$$

$$H\left(\mathrm{Tr}_B\frac{1}{N_n}\sum_{i=1}^{N_n}\left(\varphi_e^{(n)}(i)\otimes \iota_B^{\otimes n}\right)(\rho_{A,B}^{\otimes n})\right) \leq \log\dim\mathcal{H}_{A_n'},$$

we obtain

$$\begin{aligned}
I(X:Y) \leq & H\left(\mathrm{Tr}_A \rho_{A,B}^{\otimes n}\right) + \log\dim\mathcal{H}_{A_n'} - \min_\kappa H((\kappa\otimes \iota_B^{\otimes n})(\rho_{A,B}^{\otimes n}))\\
= & nH\left(\mathrm{Tr}_A \rho_{A,B}\right) + \log\dim\mathcal{H}_{A_n'} - \min_\kappa H((\kappa\otimes \iota_B^{\otimes n})(\rho_{A,B}^{\otimes n})).
\end{aligned}$$

Exercise 9.5 The LHS of (9.48) can be rewritten as

$$\begin{aligned}
& \sum_j p_j D\left((\kappa\circ\varphi_e(j)\otimes\iota_R)(\rho_{A',R})\| \sum_j p_j(\kappa\circ\varphi_e(j)\otimes\iota_R)(\rho_{A',R})\right)\\
\leq & \sum_j p_j D((\kappa\circ\varphi_e(j)\otimes\iota_R)(\rho_{A',R})\| \left(\sum_j p_j(\kappa\circ\varphi_e(j))\rho_{A'}\right)\otimes\rho_R\right)
\end{aligned}$$

$$
\begin{aligned}
=&H\left(\sum_j p_j(\kappa\circ\varphi_e(j))\rho_{A'}\right)+H(\rho_R)-\sum_j p_j H((\kappa\circ\varphi_e(j)\otimes\iota_R)\rho_{A',R})\\
=&H(\kappa\left(\sum_j p_j\varphi_e(j)(\rho_{A'}))+\sum_j p_j\tilde{I}_c(\rho_{A'},\kappa\circ\varphi_e(j)\right)\\
\le&H(\kappa\left(\sum_j p_j\varphi_e(j)(\rho_{A'})))+\tilde{I}_c(\sum_j p_j\varphi_e(j)(\rho_{A'}),\kappa\right)\\
=&I\left(\sum_j p_j\varphi_e(j)(\rho_{A'}),\kappa\right),
\end{aligned}
$$

from (4.7) and (8.52). Since $\rho_{A',R}$ is a pure state, we may write $H(\rho_{A'})=H(\rho_R)$.

Exercise 9.6 Consider the code $\Phi_e^{(n),2}:=(\mathcal{H}_{A'_n},\mathcal{H}_{R_n},x^{(n)},N_n,\varphi_e^{(n)},Y^{(n)})$ satisfying that $\varepsilon[\Phi_e^{(n),2}]\to 0$. Due to (9.48) and (9.49) similar to (4.32) in the proof of Theorem 4.2, the Fano inequality (2.35) yields that

$$
\frac{1}{n}\log N_n\le\frac{(\log 2)/n+\max_\rho I(\rho,\kappa)}{1-\varepsilon[\Phi_e^{(n),2}]}. \tag{9.164}
$$

Since $\varepsilon[\Phi_e^{(n),2}]\to 0$, we have, $\overline{\lim}\,\frac{1}{n}\log N_n\le\max_\rho I(\rho,\kappa)$, which implies the $\le$ part of (9.43).

Exercise 9.7

(**a**) Since all of POVM elements of $\boldsymbol{M}$ consist of real entries, $H(\kappa(\rho))=H(\kappa(\mathbf{Re}\,\rho))$. An arbitrary real pure state is given as the form $|\theta\rangle\langle\theta|$, where $|\theta\rangle=\cos\frac{\theta}{2}|0\rangle+\sin\frac{\theta}{2}|1\rangle$. Then, $H(\kappa(|\theta\rangle\langle\theta|))=f(\theta)$. Hence, we have

$$
\min_\rho H(\kappa(\rho))=\min_\theta H(\kappa(|\theta\rangle\langle\theta|))=\min_\theta f(\theta). \tag{9.165}
$$

Thus,

$$
\begin{aligned}
C_c(\kappa)\le&\max_{\rho'}H(\kappa(\rho'))-\min_\rho H(\kappa(\rho))\\
=&H(\kappa(\rho_{\mathrm{mix}}))-\min_\theta f(\theta)=\log 4-\min_\theta f(\theta).
\end{aligned} \tag{9.166}
$$

Further, when we generate the state $|\theta_0\rangle\langle\theta_0|$ and the state $|\theta_0+\pi\rangle\langle\theta_0+\pi|$ with the same probability $\frac{1}{2}$ for $\theta_0:=\mathrm{argmax}_\theta f(\theta)$, the transmission information $I(p,\kappa)=\log 4-\min_\theta f(\theta)$ holds.

(**b**) Since the calculation of $\frac{df}{d\theta}(\theta)$ is easy, we calculate only $\frac{d^2 f}{d\theta^2}(\theta)$ as follows.

$$\begin{aligned}
&\frac{d^2 f}{d\theta^2}(\theta) \\
=&\frac{\cos\theta}{4}\log\frac{1+\cos\theta}{1-\cos\theta}+\frac{\sin\theta}{4}\log\frac{1+\sin\theta}{1-\sin\theta}-\frac{\sin\theta}{4}\left(\frac{\sin\theta}{1-\cos\theta}+\frac{\sin\theta}{1+\cos\theta}\right) \\
&-\frac{\cos\theta}{4}\left(\frac{\cos\theta}{1+\sin\theta}+\frac{\cos\theta}{1-\sin\theta}\right) \\
=&\frac{\cos\theta}{4}\log\frac{1+\cos\theta}{1-\cos\theta}+\frac{\sin\theta}{4}\log\frac{1+\sin\theta}{1-\sin\theta}-\frac{1-\cos^2\theta}{4}\left(\frac{1}{1-\cos\theta}+\frac{1}{1+\cos\theta}\right) \\
&-\frac{1-\sin^2\theta}{4}\left(\frac{1}{1+\sin\theta}+\frac{1}{1-\sin\theta}\right) \\
=&\frac{\cos\theta}{4}\log\frac{1+\cos\theta}{1-\cos\theta}+\frac{\sin\theta}{4}\log\frac{1+\sin\theta}{1-\sin\theta}-\frac{1}{4}(1+\cos\theta+1-\cos\theta) \\
&-\frac{1}{4}(1-\sin\theta+1+\sin\theta) \\
=&\frac{\cos\theta}{4}\log\frac{1+\cos\theta}{1-\cos\theta}+\frac{\sin\theta}{4}\log\frac{1+\sin\theta}{1-\sin\theta}-1.
\end{aligned}$$

(**c**) Since $\frac{d^2 f}{d\theta^2}(\theta)=\frac{\cos\theta}{4}\log\frac{1+\cos\theta}{1-\cos\theta}+\frac{\sin\theta}{4}\log\frac{1+\sin\theta}{1-\sin\theta}-1$, we find that $\frac{d^2 f}{d\theta^2}(\theta)$ is monotonically decreasing from ∞ ($\theta=0$) to -1 ($\theta=\frac{\pi}{4}$) and that it is monotonically increasing from -1 ($\theta=\frac{\pi}{4}$) to ∞ ($\theta=\frac{\pi}{2}$). Thus, we find that $\frac{df}{d\theta}(\theta)$ is positive for $\theta\in(0,\frac{\pi}{4})$ because $\frac{df}{d\theta}(0)=\frac{df}{d\theta}(\frac{\pi}{4})=0$. Similarly, we find that $\frac{df}{d\theta}(\theta)$ is negative for $\theta\in(\frac{\pi}{4},\frac{\pi}{2})$ because $\frac{df}{d\theta}(\frac{\pi}{4})=\frac{df}{d\theta}(\frac{\pi}{2})=0$. Since $f(\frac{\pi}{4})=\frac{2+\sqrt{2}}{4}\log\frac{8}{2+\sqrt{2}}+\frac{2-\sqrt{2}}{4}\log\frac{8}{2-\sqrt{2}}$ and $f(0)=f(\frac{\pi}{2})=\frac{3}{2}\log 2$, we obtain the behavior of $f(\theta)$ given in the table, which implies the equation (9.56).

Exercise 9.8 Since $J(\rho,\sigma,\kappa)$ is concave for ρ and $\kappa(U_\theta\rho U_\theta^*)=\kappa(\rho)$, we have

$$J(\rho,\sigma,\kappa)\le J\left(\int_\theta U_\theta\rho U_\theta^* d\theta,\sigma,\kappa\right)=J(P_1\rho P_1+P_2\rho P_2,\sigma,\kappa), \tag{9.167}$$

where $U_\theta=P_1+e^{i\theta}P_2$. The definition of $J(\rho,\sigma,\kappa)$ given in (9.44) implies that

$$J(\lambda\rho_1\oplus(1-\lambda)\rho_2,\sigma,\kappa)=\lambda J(\rho_1,\sigma,\kappa)+(1-\lambda)J(\rho_2,\sigma,\kappa). \tag{9.168}$$

Hence, we obtain

$$\max_{\rho} I(\rho, \kappa) \overset{(a)}{=} \min_{\sigma} \max_{\rho} J(\rho, \sigma, \kappa)$$
$$\overset{(b)}{=} \min_{\sigma} \max_{\lambda, \rho_1, \rho_2} J(\lambda \rho_1 \oplus (1-\lambda)\rho_2, \sigma, \kappa)$$
$$\overset{(c)}{=} \min_{\sigma} \max_{\lambda, \rho_1, \rho_2} \lambda J(\rho_1, \sigma, \kappa) + (1-\lambda) J(\rho_2, \sigma, \kappa)$$
$$\le \max_{\lambda, \rho_1, \rho_2} \lambda J(\rho_1, \kappa_1(\rho_{\max,1}), \kappa) + (1-\lambda) J(\rho_2, \kappa_1(\rho_{\max,1}), \kappa)$$
$$= \max_{\lambda} \lambda C^e_{c,e}(\kappa_1) + (1-\lambda) C^e_{c,e}(\kappa_2) = \max(C^e_{c,e}(\kappa_1), C^e_{c,e}(\kappa_2)),$$

where (a), (b), and (c) follow from (9.46), (9.167), and (9.168), respectively.

Now, we assume that κ_1 satisfies (9.52) and $C^e_{c,e}(\kappa_1) \ge C^e_{c,e}(\kappa_2)$. Then, $C^e_{c,e}(\kappa) = C^e_{c,e}(\kappa_1)$ and $C_c(\kappa_1) = C^e_{c,e}(\kappa_1)$. Since $C_c(\kappa) \le C^e_{c,e}(\kappa)$ and $C_c(\kappa_1) \le C_c(\kappa)$, we have $C_c(\kappa) = C^e_{c,e}(\kappa)$.

Exercise 9.9 Apply Lemma 9.2 to the case with $M = e^{nR}$ and $\sigma = \sigma_{1+s|p_{1+s}}^{\otimes n}$ for the channel $W^{(n)}$. Since $I^{\downarrow}_{1+s}(p^n, W^{(n)}) = n I^{\downarrow}_{1+s}(p, W)$, there exists a code Φ^n for the channel $W^{(n)}$ such that $|\Phi^n| = e^{nR}$ and

$$\varepsilon[W_p^{\otimes n}, \Phi^n] \le \max\left(4\sqrt{2}, \sqrt{2v_n}\right) e^{\frac{s}{2} n (I^{\downarrow}_{1+s}(p,W) - R)}, \tag{9.169}$$

where v_n is the number of eigenvalues of $\sigma_{1+s|p_{1+s}}^{\otimes n}$. Since v_n is polynomial, taking the limit, we obtain (9.72).

Exercise 9.10 In this case, we can replace (9.65) by $\|P_x W_x\|_1 \le \operatorname{Tr} W_x P_x$.

Exercise 9.11 Since $\mathrm{P}^E_{\kappa(e_0)}(0) = \mathrm{P}^E_{\kappa(e_1)}(1) = p_0 + p_3$ and $\mathrm{P}^E_{\kappa(e_0)}(1) = \mathrm{P}^E_{\kappa(e_1)}(0) = p_1 + p_2$, we have $I(p_{\mathrm{mix}}, Q) = \log 2 - h(p_0 + p_3)$. Since Exercise 8.29 yields that

$$\begin{aligned} I(p_{\mathrm{mix}}, W^E) =& H_e(\kappa_p, \rho_{\mathrm{mix}}) - \frac{1}{2} H_e(\kappa_p, |e_0\rangle\langle e_0|) - \frac{1}{2} H_e(\kappa_p, |e_1\rangle\langle e_1|) \\ =& H(p) - h(p_0 + p_3), \end{aligned}$$

we have

$$I(p_{\mathrm{mix}}, Q) - I(p_{\mathrm{mix}}, W^E) = \log 2 - H(p). \tag{9.170}$$

Exercise 9.12 Since

$$\begin{aligned} & I(p, W^B) + I(p', W^{B'}) - I(q, W^B \otimes W^{B'}) \\ =& D\left(\sum_{x,x'} q(x,x') W^B_x \otimes W^{B'}_{x'} \middle\| \left(\sum_x p(x) W^B_x\right) \otimes \left(\sum_{x'} p'(x') W^{B'}_{x'}\right)\right), \end{aligned}$$

we have

$$
\begin{aligned}
& I(p, W^E) + I(p', W^{E'}) - I(q, W^E \otimes W^{E'}) \\
=& D\Bigg((\kappa \otimes \kappa') \Big(\sum_{x,x'} q(x,x') W_x^B \otimes W_{x'}^{B'} \Big) \Bigg\| \\
& \qquad (\kappa \otimes \kappa') \Big(\sum_x p(x) W_x^B \Big) \otimes \Big(\sum_{x'} p'(x') W_{x'}^{B'} \Big) \Bigg) \\
\le & D\Bigg(\sum_{x,x'} q(x,x') W_x^B \otimes W_{x'}^{B'} \Bigg\| \Big(\sum_x p(x) W_x^B \Big) \otimes \Big(\sum_{x'} p'(x') W_{x'}^{B'} \Big) \Bigg).
\end{aligned}
$$

Thus,

$$
\begin{aligned}
& I(p, W^E) + I(p', W^{E'}) - I(q, W^E \otimes W^{E'}) \\
\le & I(p, W^B) + I(p', W^{B'}) - I(q, W^B \otimes W^{B'}),
\end{aligned}
$$

which implies (9.100).

Exercise 9.13 Applying Lemma 9.4 to the channel $\kappa^{\otimes n}$, we have $C_c^{e,B,E}(\kappa) \ge \frac{1}{n} \max_{\rho \in \mathcal{S}(\mathcal{H}_A^{\otimes n})} I_c(\rho, \kappa^{\otimes n})$, which implies (9.78).

Exercise 9.14 Using the same discussion as the proof of Theorem 9.8, we find that $C_c^{e,B,E}(\kappa) = \overline{\lim} \frac{1}{n} \sup_Q \sup_p \left(I(p, \kappa^{\otimes n} Q) - I(p, {\kappa_E}^{\otimes n} Q) \right)$.

Denote the new capacities by $\bar{C}_c^{B,E}(W)$ and $\bar{C}_c^{e,B,E}(\kappa)$, respectively. Since the condition $\frac{I_E(\Phi^{(n)})}{n} \to 0$ is weaker than the condition $I_E(\Phi^{(n)}) \to 0$, we have $\bar{C}_c^{B,E}(W) \ge C_c^{B,E}(W)$ and $\bar{C}_c^{e,B,E}(\kappa) \ge C_c^{e,B,E}(\kappa)$. Due to (9.98), the condition $\frac{I_E(\Phi^{(n)})}{n} \to 0$ implies

$$
\bar{C}_c^{B,E}(W) \le \overline{\lim} \frac{1}{n} \sup_Q \sup_p \left(I(p, W^{B,(n)} Q) - I(p, W^{E,(n)} Q) \right).
$$

We find that $\bar{C}_c^{B,E}(W) = C_c^{B,E}(W)$.

Similarly, we have

$$
\bar{C}_c^{e,B,E}(\kappa) \le \overline{\lim} \frac{1}{n} \sup_Q \sup_p \left(I(p, \kappa^{\otimes n} Q) - I(p, {\kappa_E}^{\otimes n} Q) \right),
$$

which implies that $\bar{C}_c^{e,B,E}(\kappa) = C_c^{e,B,E}(\kappa)$.

Exercise 9.15 From the definition, we find that $C_{c,e}^{e,B,E}(\kappa) \le C_{c,e}^{e}(\kappa)$. It is sufficient to show $C_{c,e}^{e,B,E}(\kappa) \ge \max_\rho I(\rho, \kappa)$. Consider the code given in the proof of Theorem 9.5. In this code, the eavesdropper's state does not depend on the message to be sent. So, we obtain $C_{c,e}^{e,B,E}(\kappa) \ge \max_\rho I(\rho, \kappa)$.

Exercise 9.16 Let $\mathcal{H}_E$ and $\mathcal{H}_{E'}$ be the environment system of κ and κ', respectively. Then, the environment system of $\kappa' \circ \kappa$ is $\mathcal{H}_E \otimes \mathcal{H}_{E'}$. Thus, we have $\kappa'_E \circ \kappa(\rho) = \mathrm{Tr}_{\mathcal{H}_E}(\kappa' \circ \kappa)_E(\rho)$. Hence, (5.59) implies (9.79). Similarly, we have $\kappa_E(\rho) = \mathrm{Tr}_{\mathcal{H}_{E'}}(\kappa' \circ \kappa)_E(\rho)$. Thus, (5.59) implies (9.80).

Exercise 9.17 Denote the modified capacity by $\hat{C}_c^{B,E}(W)$. First, note that

$$\begin{aligned}\varepsilon_{E,a}[\Phi] =& \sum_i \sum_{j \neq i} \frac{1}{M(M-1)} d_1((W^E Q)_i, (W^E Q)_j) \\ \geq & \frac{1}{M} \sum_i d_1((W^E Q)_i, \frac{1}{M} \sum_{j=1} (W^E Q)_j).\end{aligned}$$

The concavity and monotonicity (Exercise 5.34) of η_0 imply that

$$\begin{aligned} I_E(\Phi) =& \frac{1}{M} \sum_{i=1}^M H(\frac{1}{M} \sum_{j=1} (W^E Q)_j)) - H(W^E Q)_i) \\ \leq & \frac{1}{M} \sum_{i=1}^M |H\left(\frac{1}{M} \sum_{j=1} (W^E Q)_j\right) - H(W^E Q)_i)| \\ \leq & \frac{1}{M} \sum_{i=1}^M d_1\left((W^E Q)_i, \frac{1}{M} \sum_{j=1} (W^E Q)_j\right) \log d \\ & + \eta_0\left(d_1\left((W^E Q)_i, \frac{1}{M} \sum_{j=1} (W^E Q)_j\right)\right) \\ \leq & \varepsilon_{E,a}[\Phi] \log d + \eta_0(\varepsilon_{E,a}[\Phi]), \end{aligned} \tag{9.171}$$

where the final inequality follows from Fannes' inequality (5.92). Due to (9.171), when $\varepsilon_{E,a}[\Phi^{(n)}] \to 0$, we have $\frac{I_E(\Phi^{(n)})}{n} \to 0$. Exercise 9.14 yields that $C_c^{B,E}(W) \leq \hat{C}_c^{B,E}(W)$. The opposite inequality also holds from the following relation:

$$\begin{aligned} \varepsilon_{E,a}[\Phi^{(n)}] =& \sum_i \sum_{j \neq i} \frac{d_1((W^E Q)_i, (W^E Q)_j)}{M(M-1)} \\ \overset{(a)}{\leq} & \sum_i \sum_{j \neq i} \frac{d_1((W^E Q)_i, \frac{1}{M}\sum_k (W^E Q)_k) + d_1((W^E Q)_j, \frac{1}{M}\sum_k (W^E Q)_k)}{M(M-1)} \\ \overset{(b)}{\leq} & \sum_i \sum_{j \neq i} \frac{D((W^E Q)_i \| \frac{1}{M}\sum_k (W^E Q)_k)^{1/2} + D((W^E Q)_j \| \frac{1}{M}\sum_k (W^E Q)_k)^{1/2}}{2M(M-1)} \\ =& \sum_i \frac{1}{M} D((W^E Q)_i \| \frac{1}{M} \sum_k (W^E Q)_k)^{1/2} \end{aligned}$$

$$\overset{(c)}{\le} \left(\sum_i \frac{1}{M} D((W^E Q)_i \| \frac{1}{M} \sum_k (W^E Q)_k) \right)^{1/2} = I_E(\Phi^{(n)})^{1/2}, \tag{9.172}$$

where (a), (b), and (c) follow from the triangle inequality, quantum Pinsker inequality (3.53), and the Jensen inequality for the function $x \mapsto x^{1/2}$, respectively.

Exercise 9.18 Denote the modified capacity by $\tilde{C}_c^{B,E}(W)$. Since $\varepsilon_{E,a}[\Phi^{(n)}] \ge \varepsilon_{E,w}[\Phi]$, we have $\tilde{C}_c^{B,E}(W) \le \hat{C}_c^{B,E}(W) = C_c^{B,E}(W)$. So, we will show the opposite inequality. Now, we consider i be a random variable subject to the uniform distribution on $\{1, \ldots, M\}$. Then, Markov inequality (2.158) implies that

$$\frac{1}{M} \# \left\{ i \middle| D\left((W^E Q)_i \middle\| \frac{1}{M} \sum_k (W^E Q)_k \right) \le 2 I_E(\Phi) \right\} \ge \frac{1}{2}. \tag{9.173}$$

Then, we number all elements of $\{i | D((W^E Q)_i \| \frac{1}{M} \sum_k (W^E Q)_k) \le 2 I_E(\Phi)\}$ as $i_1, \ldots, i_K$, and denote the code whose message set consists of $i_1, \ldots, i_K$, by $\hat{\Phi}$. Then, $K \ge \frac{M}{2}$. Similar to (9.172), we can show that

$$\begin{aligned}
&\varepsilon_{E,w}[\hat{\Phi}] = \sup_j \sup_{j'} d_1((W^E Q)_{i_j}, (W^E Q)_{i_{j'}}) \\
\le& \sup_j \sup_{j'} \left(d_1((W^E Q)_{i_j}, \frac{1}{M} \sum_k (W^E Q)_k) + d_1((W^E Q)_{i_{j'}}, \frac{1}{M} \sum_k (W^E Q)_k) \right) \\
=&2 \sup_j d_1\left((W^E Q)_{j'}, \frac{1}{M} \sum_k (W^E Q)_k \right) \\
\le& \sup_j D((W^E Q)_{i_j} \| \frac{1}{M} \sum_k (W^E Q)_k))^{1/2} \le \sqrt{2 I_E(\Phi)}.
\end{aligned}$$

Hence, we have $\tilde{C}_c^{B,E}(W) \ge C_c^{B,E}(W)$.

Exercise 9.19 We have

$$\begin{aligned}
&I(Qp, W^B) - I(Qp, W^E) - \sum_i p_i (I(Q^i, W^B) - I(Q^i, W^E)) \\
=& \sum_j (Qp)_j D(W_j^B \| W_{Qp}^B) - D(W_j^E \| W_{Qp}^E) \\
&- \sum_i \sum_j p_i Q_j^i (D(W_j^B \| W_{Q^i}^B) - D(W_j^E \| W_{Q^i}^E)) \\
=& \sum_i \sum_j p_i Q_j^i (D(W_{Q^i}^B \| W_{Qp}^B) - D(W_{Q^i}^E \| W_{Qp}^E)) = I(p, W^B Q) - I(p, W^E Q).
\end{aligned}$$

Since (5.59) implies $I(p, W^B Q) - I(p, W^E Q) = I(p, W^B Q) - I(p, \kappa(W^B Q)) \ge 0$, we have (9.76).

Exercise 9.20 Consider (8.25) with $a = 1/2$. When $\varepsilon_2[\Phi] \to 0$, we have $\varepsilon_1[\Phi] \to 0$. Hence, $C_{q,2}(\kappa) \le C_{q,1}(\kappa)$.

Exercise 9.21 Due to (8.26), when $\varepsilon_1[\Phi] \to 0$, we have $\varepsilon_2[\Phi] \to 0$. Hence, $C_{q,2}(\kappa) \ge C_{q,1}(\kappa)$.

Exercise 9.22 Assume that any two channels $\kappa_{1,n}$ and $\kappa_{2,n}$ from a system $\mathcal{H}_n$ to another system $\mathcal{H}'_n$ satisfy $\max_x 1 - F^2((\kappa_{1,n} \otimes \iota_R)(|x\rangle\langle x|), (\kappa_{2,n} \otimes \iota_R)(|x\rangle\langle x|)) \to 0$. Then, (3.48) implies that

$$\max_x \|\kappa_{1,n} \otimes \iota_R(|x\rangle\langle x|) - \kappa_{2,n} \otimes \iota_R(|x\rangle\langle x|)\|_1 \to 0. \tag{9.174}$$

Hence,

$$\max_x \|\kappa_{1,n}(\mathrm{Tr}_R |x\rangle\langle x|) - \kappa_{2,n}(\mathrm{Tr}_R |x\rangle\langle x|)\|_1 \to 0. \tag{9.175}$$

By applying Fannes inequality (Theorem 5.12) to two states $(\kappa_{1,n} \otimes \iota_R)(|x\rangle\langle x|)$ and $(\kappa_{2,n} \otimes \iota_R)(|x\rangle\langle x|)$, (9.174) yields

$$\frac{|H(\kappa_{1,n} \otimes \iota_R(|x\rangle\langle x|)) - H(\kappa_{2,n} \otimes \iota_R(|x\rangle\langle x|))|}{\log(\dim \mathcal{H}_n \dim \mathcal{H}'_n)} \to 0.$$

Similarly, (9.175) yields

$$\frac{|H(\kappa_{1,n}(\mathrm{Tr}_R |x\rangle\langle x|)) - H(\kappa_{2,n}(\mathrm{Tr}_R |x\rangle\langle x|))|}{\log(\dim \mathcal{H}_n \dim \mathcal{H}'_n)} \to 0.$$

Since $I_c(\mathrm{Tr}_R |x\rangle\langle x|, \kappa) = H(\kappa(\mathrm{Tr}_R |x\rangle\langle x|)) - H(\kappa \otimes \iota_R(|x\rangle\langle x|))$, we obtain

$$\frac{|\max_\rho I_c(\rho, \kappa_{1,n}) - \max_\rho I_c(\rho, \kappa_{2,n})|}{\log(\dim \mathcal{H}_n \dim \mathcal{H}'_n)} \le \frac{\max_\rho |I_c(\rho, \kappa_{1,n}) - I_c(\rho, \kappa_{2,n})|}{\log(\dim \mathcal{H}_n \dim \mathcal{H}'_n)} \to 0.$$

Exercise 9.23

(**a**) Let $(\mathcal{H}, \tau, \nu)$ be Φ. Apply the property ② of Sect. 8.2. Let $\mathcal{H}_C$ be the subspace of $\mathcal{H}$ with the dimension $\min(\mathcal{H}, \mathcal{H}_A)$. Then, we can choose an isometry U from $\mathcal{H}_C$ to $\mathcal{H}_A$ such that

$$F_e^2(\rho_{\mathrm{mix}}, \nu \circ \kappa \circ \tau) \le F_e(\rho_{\mathrm{mix}}, \nu \circ \kappa \circ \kappa_U),$$

which implies that

$$2(1 - F_e(\rho_{\mathrm{mix}}, \nu \circ \kappa \circ \tau)) \leq 1 - F_e(\rho_{\mathrm{mix}}, \nu \circ \kappa \circ \kappa_U). \tag{9.176}$$

Then, the code $(\mathcal{H}_A, \kappa_U, \nu)$ satisfies the desired properties for Φ'.
(b) Since the first inequality is trivial, we will show the remaining inequalities. (8.39) and (8.42) imply

$$I_c(U\rho_{\mathrm{mix}}U^*, \kappa) = I_c(\rho_{\mathrm{mix}}, \kappa \circ \kappa_U) \geq I_c(\rho_{\mathrm{mix}}, \nu \circ \kappa \circ \kappa_U).$$

Applying (8.38) to the case with ρ_{mix} and $\nu \circ \kappa \circ \kappa_U$, we have

$$\log|\Phi| - I_c(\rho_{\mathrm{mix}}, \nu \circ \kappa \circ \kappa_U) \leq 2\delta\left(\log|\Phi| - \log\delta\right).$$

(c) Combining **(a)** and **(b)**, we have

$$\begin{aligned}&\lim_{n\to\infty}\frac{1}{n}\max_{\rho\in\mathcal{S}(\mathcal{H}_A^{\otimes n})} I_c(\rho, \kappa^{\otimes n})\\ \geq&\frac{1}{n}\log\min(|\Phi^{(n)}|, d_A^n) - \frac{2}{n}\delta_n\left(\log\min(|\Phi^{(n)}|, d_A^n) - \log\delta_n\right),\end{aligned}$$

where $\delta_n \stackrel{\mathrm{def}}{=} \sqrt{4\varepsilon_2[\Phi^{(n)}]}$. Since Lemma A.1 implies

$$\lim_{n\to\infty}\frac{1}{n}\max_{\rho\in\mathcal{S}(\mathcal{H}_A^{\otimes n})} I_c(\rho, \kappa^{\otimes n}) = \limsup_{n\to\infty}\frac{1}{n} = \limsup_{n\to\infty}\frac{1}{n}\max_{\rho\in\mathcal{S}(\mathcal{H}_A^{\otimes n})} I_c(\rho, \kappa^{\otimes n}),$$

taking the limit, we obtain (9.149).
(d) The first inequality of (8.38) implies that $\frac{1}{n}I_c(\rho, \kappa^{\otimes n}) \leq \log d_A$. When $\lim_{n\to\infty}\frac{1}{n}\max_{\rho\in\mathcal{S}(\mathcal{H}_A^{\otimes n})} I_c(\rho, \kappa^{\otimes n}) < \log d_A$, (9.149) implies that

$$\limsup_{n\to\infty}\frac{1}{n}\log\min\{|\Phi^{(n)}|, d_A^n\} = \limsup_{n\to\infty}\frac{1}{n}\log|\Phi^{(n)}|.$$

Hence, we have $\lim_{n\to\infty}\frac{1}{n}\max_{\rho\in\mathcal{S}(\mathcal{H}_A^{\otimes n})} I_c(\rho, \kappa^{\otimes n}) \geq C_{q,2}(\kappa)$.

When $\lim_{n\to\infty}\frac{1}{n}\max_{\rho\in\mathcal{S}(\mathcal{H}_A^{\otimes n})} I_c(\rho, \kappa^{\otimes n}) = \log d_A$, we can show that $\log d_A \geq C_{q,2}(\kappa)$ as follows. Consider a sequence of codes $\Phi^{(n)} = (\mathcal{H}^{\otimes n}, \tau^{(n)}, \nu^{(n)})$. Then, (8.22) implies that $1 - \varepsilon_2[\Phi^{(n)}] = F_e(\rho_{\mathrm{mix}}, \nu^{(n)} \circ \kappa^{\otimes n} \circ \tau^{(n)}) \leq \sqrt{\frac{d_A^n}{|\Phi^{(n)}|}}$. When $\varepsilon_2[\Phi^{(n)}] \to 0$, we have $\limsup_{n\to\infty}\frac{1}{n}\log|\Phi^{(n)}| \leq d_A$, which implies $\log d_A \geq C_{q,2}(\kappa)$.

Exercise 9.24 Consider two states ρ_1 and ρ_2. Choose their purifications $|\Phi_1\rangle\langle\Phi_1|$ and $|\Phi_2\rangle\langle\Phi_2|$ so that their reference systems $\mathcal{H}_{R1}$ and $\mathcal{H}_{R2}$ are disjoint to each other. So, we choose a purification $|\Phi\rangle\langle\Phi|$ of $\lambda\rho_1 + (1-\lambda)\rho_2$ such that the reference system is $\mathcal{H}_{R1} \oplus \mathcal{H}_{R2}$, $P_1|\Phi\rangle\langle\Phi|P_1 = \lambda|\Phi_1\rangle\langle\Phi_1|$, and $P_2|\Phi\rangle\langle\Phi|P_2 = (1-\lambda)|\Phi_2\rangle\langle\Phi_2|$, where P_i is the projection to $\mathcal{H}_{Ri}$. Hence, we have

$$I_{\mathrm{SDP}}(\rho, \kappa) = E_{\mathrm{SDP}}(\kappa(|\Phi\rangle\langle\Phi|)) \geq E_{\mathrm{SDP}}(\kappa(\lambda|\Phi_1\rangle\langle\Phi_1| + (1-\lambda)|\Phi_2\rangle\langle\Phi_2|))$$
$$= \lambda E_{\mathrm{SDP}}(\kappa(|\Phi_1\rangle\langle\Phi_1|)) + (1-\lambda) E_{\mathrm{SDP}}(\kappa(|\Phi_2\rangle\langle\Phi_2|))$$
$$= \lambda I_{\mathrm{SDP}}(\rho_1, \kappa) + (1-\lambda) I_{\mathrm{SDP}}(\rho_2, \kappa).$$

References

1. A. Einstein, R. Podolsky, N. Rosen, Can quantum-mechanical descriptions of physical reality be considered complete? Phys. Rev. **47**, 777–780 (1935)
2. C.H. Bennett, G. Brassard, C. Crepeau, R. Jozsa, A. Peres, W.K. Wootters, Teleporting an unknown quantum state via dual classical and Einstein-Podolsky-Rosen channels. Phys. Rev. Lett. **70**, 1895 (1993)
3. D. Bouwmeester, J.-W. Pan, K. Mattle, M. Eibl, H. Weinfurter, A. Zeilinger, Experimental quantum teleportation. Nature **390**, 575–579 (1997)
4. A. Furusawa, J.L. Sørensen, S.L. Braunstein, C.A. Fuchs, H.J. Kimble, E.J. Polzik, Unconditional quantum teleportation. Science **282**, 706 (1998)
5. J.-W. Pan, S. Gasparoni, M. Aspelmeyer, T. Jennewein, A. Zeilinger, Experimental realization of freely propagating teleported qubits. Nature **421**, 721–725 (2003)
6. M. Murao, D. Jonathan, M.B. Plenio, V. Vedral, Quantum telecloning and multiparticle entanglement. Phys. Rev. A **59**, 156–161 (1999)
7. H. Nagaoka, S. Osawa, Theoretical basis and applications of the quantum Arimoto-Blahut algorithms, in *Proceedings 2nd Quantum Information Technology Symposium (QIT2)* (1999), pp. 107–112
8. M. Hayashi, H. Imai, K. Matsumoto, M.B. Ruskai, T. Shimono, Qubit channels which require four inputs to achieve capacity: implications for additivity conjectures. Quant. Inf. Comput. **5**, 13–31 (2005)
9. M. Fukuda, Extending additivity from symmetric to asymmetric channels. J. Phys. A Math. Gen. **38**, L753–L758 (2005)
10. C. King, Additivity for a class of unital qubit channels. J. Math. Phys. **43**, 4641–4653 (2002)
11. C. King, The capacity of the quantum depolarizing channel. IEEE Trans. Inf. Theory **49**, 221–229 (2003)
12. P.W. Shor, Additivity of the classical capacity of entanglement-breaking quantum channels. J. Math. Phys. **43**, 4334–4340 (2002)
13. K. Matsumoto, T. Shimono, A. Winter, Remarks on additivity of the Holevo channel capacity and of the entanglement of formation. Commun. Math. Phys. **246**(3), 427–442 (2004)
14. P.W. Shor, Equivalence of additivity questions in quantum information theory. Commun. Math. Phys. **246**(3), 453–473 (2004)
15. A.A. Pomeransky, Strong superadditivity of the entanglement of formation follows from its additivity. Phys. Rev. A **68**, 032317 (2003)
16. M. Fukuda, M.M. Wolf, Simplifying additivity problems using direct sum constructions. J. Math. Phys. **48**(7), 072101 (2007)
17. S. Osawa, H. Nagaoka, Numerical experiments on the capacity of quantum channel with entangled input states. IEICE Trans. **E84-A**, 2583–2590 (2001)
18. K. Matsumoto, F. Yura, Entanglement cost of antisymmetric states and additivity of capacity of some quantum channel. J. Phys. A: Math. Gen. **37**, L167–L171 (2004)
19. M. Fannes, B. Haegeman, M. Mosonyi, D. Vanpeteghem, Additivity of minimal entropy output for a class of covariant channels. quant-ph/0410195 (2004)
20. N. Datta, A.S. Holevo, Y. Suhov, Additivity for transpose depolarizing channels. Int. J. Quantum Inform. **4**, 85 (2006)
21. N. Datta, M.B. Ruskai, Maximal output purity and capacity for asymmetric unital qudit channels. J. Phys. A: Math. Gen. **38**, 9785 (2005)

22. M.M. Wolf, J. Eisert, Classical information capacity of a class of quantum channels. New J. Phys. **7**, 93 (2005)
23. M. Fukuda, Revisiting additivity violation of quantum channels. Commun. Math. Phys. **332**, 713–728 (2014)
24. C.H. Bennett, S.J. Wiesner, Communication via one- and two-particle operators on Einstein-Podolsky-Rosen states. Phys. Rev. Lett. **69**, 2881 (1992)
25. S. Bose, M.B. Plenio, B. Vedral, Mixed state dense coding and its relation to entanglement measures. J. Mod. Opt. **47**, 291 (2000)
26. T. Hiroshima, Optimal dense coding with mixed state entanglement. J. Phys. A Math. Gen. **34**, 6907–6912 (2001)
27. G. Bowen, Classical information capacity of superdense coding. Phys. Rev. A **63**, 022302 (2001)
28. M. Horodecki, P. Horodecki, R. Horodecki, D.W. Leung, B.M. Terhal, Classical capacity of a noiseless quantum channel assisted by noisy entanglement. Quant. Inf. Comput. **1**, 70–78 (2001)
29. A. Winter, Scalable programmable quantum gates and a new aspect of the additivity problem for the classical capacity of quantum channels. J. Math. Phys. **43**, 4341–4352 (2002)
30. C.H. Bennett, P.W. Shor, J.A. Smolin, A.V. Thapliyal, Entanglement-assisted classical capacity of noisy quantum channels. Phys. Rev. Lett. **83**, 3081 (1999)
31. C.H. Bennett, P.W. Shor, J.A. Smolin, A.V. Thapliyal, Entanglement-assisted capacity of a quantum channel and the reverse Shannon theorem. IEEE Trans. Inf. Theory **48**(10), 2637–2655 (2002)
32. A.S. Holevo, On entanglement-assisted classical capacity. J. Math. Phys. **43**, 4326–4333 (2002)
33. T.S. Han, S. Verdú, Approximation theory of output statistics. IEEE Trans. Inf. Theory **39**, 752–772 (1993)
34. T.S. Han, S. Verdú, Spectrum invariancy under output approximation for full-rank discrete memoryless channels. Problemy Peredachi Informatsii **29**(2), 9–27 (1993)
35. R. Ahlswede, G. Dueck, Identification via channels. IEEE Trans. Inf. Theory **35**, 15–29 (1989)
36. R. Ahlswede, A. Winter, Strong converse for identification via quantum channels. IEEE Trans. Inf. Theory **48**, 569–579 (2002)
37. A.D. Wyner, The wire-tap channel. Bell. Syst. Tech. J. **54**, 1355–1387 (1975)
38. C.H. Bennett, G. Brassard, Quantum cryptography: public key distribution and coin tossing, in *Proceedings IEEE International Conference on Computers, Systems and Signal Processing* (Bangalore, India, 1984), pp. 175–179
39. D. Stucki, N. Gisin, O. Guinnard, G. Ribordy, H. Zbinden, Quantum key distribution over 67 km with a plug & play system. New J. Phys. **4**, 41.1–41.8 (2002)
40. E. Klarreich, Quantum cryptography: can you keep a secret? Nature **418**, 270–272 (2002)
41. H. Kosaka, A. Tomita, Y. Nambu, N. Kimura, K. Nakamura, Single-photon interference experiment over 100 km for quantum cryptography system using a balanced gated-mode photon detector. Electron. Lett. **39**(16), 1199–1201 (2003)
42. C. Gobby, Z.L. Yuan, A.J. Shields, Quantum key distribution over 122 km of standard telecom fiber. Appl. Phys. Lett. **84**, 3762–3764 (2004)
43. I. Devetak, The private classical capacity and quantum capacity of a quantum channel. IEEE Trans. Inf. Theory **51**, 44–55 (2005)
44. I. Devetak, A. Winter, Distillation of secret key and entanglement from quantum states. Proc. R. Soc. Lond. A **461**, 207–235 (2005)
45. H.-K. Lo, Proof of unconditional security of six-state quantum key distribution scheme. Quant. Inf. Comput. **1**, 81–94 (2001)
46. N. Gisin, contribution to the Torino Workshop, 1997
47. D. Bruß, Optimal eavesdropping in quantum cryptography with six states. Phys. Rev. Lett. **81**, 3018–3021 (1998)
48. H. Bechmann-Pasquinucci, N. Gisin, Incoherent and coherent eavesdropping in the six-state protocol of quantum cryptography. Phys. Rev. A **59**, 4238–4248 (1999)
49. G. Blakely, Safeguarding cryptographic keys. Proc. AFIPS **48**, 313 (1979)

50. A. Shamir, How to share a secret. Commun. ACM **22**, 612 (1979)
51. R. Cleve, D. Gottesman, H.-K. Lo, How to share a quantum secret. Phys. Rev. Lett. **82**, 648 (1999)
52. D. Gottesman, On the theory of quantum secret sharing. Phys. Rev. A **61**, 042311 (2000)
53. I. Devetak, A. Winter, Classical data compression with quantum side information. Phys. Rev. A **68**, 042301 (2003)
54. P.W. Shor, Scheme for reducing decoherence in quantum computer memory. Phys. Rev. A **52**, 2493 (1995)
55. A.R. Calderbank, P.W. Shor, Good quantum error-correcting codes exist. Phys. Rev. A **54**, 1098 (1996)
56. A.M. Steane, Multiple particle interference and quantum error correction. Proc. R. Soc. Lond. A **452**, 2551 (1996)
57. E. Knill, R. Laflamme, Theory of quantum error-correcting codes. Phys. Rev. A **55**, 900 (1997)
58. D. Gottesman, Class of quantum error-correcting codes saturating the quantum Hamming bound. Phys. Rev. A **54**, 1862 (1996)
59. A.R. Calderbank, E.M. Rains, P.W. Shor, N.J.A. Sloane, Quantum error correction and orthogonal geometry. Phys. Rev. Lett. **78**, 405 (1996)
60. M. Hamada, Notes on the fidelity of symplectic quantum error-correcting codes. Int. J. Quant. Inf. **1**, 443–463 (2003)
61. M. Hayashi, S. Ishizaka, A. Kawachi, G. Kimura, T. Ogawa, *Introduction to Quantum Information Science*, Graduate Texts in Physics (2014)
62. P. W. Shor, The quantum channel capacity and coherent information, in *Lecture Notes, MSRI Workshop on Quantum Computation* (2002). http://www.msri.org/publications/ln/msri/2002/quantumcrypto/shor/1/
63. S. Lloyd, The capacity of the noisy quantum channel. Phys. Rev. A **56**, 1613 (1997)
64. C.H. Bennett, D.P. DiVincenzo, J.A. Smolin, W.K. Wootters, Mixed state entanglement and quantum error correction. Phys. Rev. A **54**, 3824–3851 (1996)
65. M. Tomamichel, M. M. Wilde, A. Winter, Strong converse rates for quantum communication (2014). arXiv:1406.2946
66. H. Barnum, E. Knill, M.A. Nielsen, On quantum fidelities and channel capacities. IEEE Trans. Inf. Theory **46**, 1317–1329 (2000)
67. C.H. Bennett, D.P. DiVincenzo, J.A. Smolin, Capacities of quantum erasure channels. Phys. Rev. Lett. **78**, 3217–3220 (1997)
68. C.H. Bennett, C.A. Fuchs, J.A. Smolin, Entanglement-enhanced classical communication on a noisy quantum channel, eds. by O. Hirota, A. S. Holevo, C. M. Cavesby. *Quantum Communication, Computing, and Measurement* (Plenum, New York, 1997), pp. 79–88
69. H. Nagaoka, Algorithms of Arimoto-Blahut type for computing quantum channel capacity, in *Proceedings 1998 IEEE International Symposium on Information Theory* (1998), p. 354
70. S. Arimoto, An algorithm for computing the capacity of arbitrary discrete memoryless channels. IEEE Trans. Inf. Theory **18**, 14–20 (1972)
71. R. Blahut, Computation of channel capacity and rate-distortion functions. IEEE Trans. Inf. Theory **18**, 460–473 (1972)
72. A. Fujiwara, T. Hashizume, Additivity of the capacity of depolarizing channels. Phys. Lett A **299**, 469–475 (2002)
73. G. Vidal, W. Dür, J.I. Cirac, Entanglement cost of antisymmetric states. quant-ph/0112131v1 (2001)
74. T. Shimono, Additivity of entanglement of formation of two three-level-antisymmetric states. Int. J. Quant. Inf. **1**, 259–268 (2003)
75. F. Yura, Entanglement cost of three-level antisymmetric states. J. Phys. A Math. Gen. **36**, L237–L242 (2003)
76. K.M.R. Audenaert, S.L. Braunstein, On strong superadditivity of the entanglement of formation. Commun. Math. Phys. **246**(3), 443–452 (2004)
77. M. Koashi, A. Winter, Monogamy of quantum entanglement and other correlations. Phys. Rev. A **69**, 022309 (2004)

78. K. Matsumoto, private communication (2005)
79. K. Matsumoto, Yet another additivity conjecture. Phys. Lett. A **350**, 179–181 (2006)
80. R.F. Werner, A.S. Holevo, Counterexample to an additivity conjecture for output purity of quantum channels. J. Math. Phys. **43**, 4353 (2002)
81. M.B. Hastings, Superadditivity of communication capacity using entangled inputs. Nat. Phys. **5**, 255 (2009)
82. L.B. Levitin, *Information, Complexity and Control in Quantum Physics*, eds. by A. Blaquière, S. Diner, G. Lochak. (Springer, Vienna), pp. 15–47
83. A. Barenco, A.K. Ekert, Dense coding based on quantum entanglement. J. Mod. Opt. **42**, 1253 (1995)
84. P. Hausladen, R. Jozsa, B. Schumacher, M. Westmoreland, W. Wooters, Classical information capacity of a quantum channel. Phys. Rev. A **54**, 1869–1876 (1996)
85. B. Schumacher, Sending quantum entanglement through noisy channels. Phys. Rev. A **54**, 2614–2628 (1996)
86. H. Barnum, M.A. Nielsen, B. Schumacher, Information transmission through a noisy quantum channel. Phys. Rev. A **57**, 4153–4175 (1997)
87. I. Devetak, P.W. Shor, The capacity of a quantum channel for simultaneous transmission of classical and quantum information. Commun. Math. Phys. **256**, 287–303 (2005)
88. J. Yard, in preparation
89. C. Morgan, A. Winter, “Pretty strong" converse for the quantum capacity of degradable channels. IEEE Trans. Inf. Theory **60**, 317–333 (2014)

Chapter 10
Source Coding in Quantum Systems

Abstract Nowadays, data compression software has become an indispensable tool for current network system. Why is such a compression possible? Information commonly possesses redundancies. In other words, information possesses some regularity. If one randomly types letters of the alphabet, it is highly unlikely letters that form a meaningful sentence or program. Imagine that we are assigned a task of communicating a sequence of 1000 binary digits via telephone. Assume that the binary digits satisfies the following rule: The $2n$th and $(2n + 1)$th digits of this sequence are the same. Naturally, we would not read out all 1000 digits of the sequence; we would first explain that the $2n$th and $(2n + 1)$th digits are the same, and then read out the even-numbered (or odd-numbered) digits. We may even check whether there is any further structure in the sequence. In this way, compression software works by changing the input sequence of letters (or numbers) into another sequence of letters that can reproduce the original sequence, thereby reducing the necessary storage. The compression process may therefore be regarded as an encoding. This procedure is called source coding in order to distinguish it from the channel coding examined in Chap. 4. Applying this idea to the quantum scenario, the presence of any redundant information in a quantum system may be similarly compressed to a smaller quantum memory for storage or communication. However, in contrast to the classical case, we have at least two distinct scenarios. The task of the first scenario is saving memory in a quantum computer. This will be relevant when quantum computers are used in practice. In this case, a given quantum state is converted into a state on a system of lower size (dimension). The original state must then be recoverable from the compressed state. Note that the encoder does not know what state is to be compressed. The task in the second scenario is to save the quantum system to be sent for quantum cryptography. In this case, the sender knows what state to be sent. This provides the encoder with more options for compression. In the decompression stage, there is no difference between the first and second scenarios, since their tasks are conversions from one quantum system to another. In this chapter, the two scenarios of compression outlined above are discussed in detail.

M. Hayashi, *Quantum Information Theory*, Graduate Texts in Physics,
DOI 10.1007/978-3-662-49725-8_10

10.1 Four Kinds of Source Coding Schemes in Quantum Systems

As discussed above, source coding can be formulated in two ways. In the encoding process of the first scheme, we perform a state evolution from an original quantum system to a system of lower dimension. In that of the second, the encoder prepares a state in a system of lower dimension depending on the input signal. In the first scenario, the state is unknown since only the quantum system is given. Hence, the first scheme is called **blind**. In the second scenario, the state is known, and this scheme is called **visible**. The quality of the compression is evaluated by its compression rate. Of course, a lower dimension of the compressed quantum system produces a better encoding in terms of its compression rate. We may choose the compression rate to be either fixed or dependent on the input state. Coding with a fixed compression rate is called fixed-length coding, while it is called variable-length coding when the compression rate that depends on the input state. Therefore, there exist four schemes for the problem, i.e., fixed-/variable-length and visible/blind coding.

Let us summarize the known results on fixed- and variable-length compression in classical systems. In fixed-length compression, it is not possible to completely recover all input signals. Decoders may erroneously recover some input signals. However, when the state on the input system is subject to a certain probability distribution and the compression rate is larger than a threshold, an application of a proper code reduces the probability of erroneously recovering the state so that the error probability is sufficiently close to zero [1, 2]. This threshold is called the **minimum admissible rate**. In order to treat this problem precisely, we often assume that the input data is subject to the n-fold independent and identical distribution of a given probability distribution with sufficiently large n.

In variable-length compression, it is possible to construct a code recovering all input signals perfectly. This is an advantage of variable-length encoding over fixed-length encoding. In this case, since there is no occurrence of erroneously decoding, we measure the quality of the variable-length encoding by the coding length. The worst-case scenario in this type of coding occurs when the coding length is greater than the input information. However, when the input is subject to a certain probability distribution, the average coding length can be shorter than the number of bits in the input. For an independent and identical distribution, it has been shown that the average coding length is equal to its entropy in the optimal case [1, 2].

Let us now turn to quantum systems. As for the classical case, for fixed-length coding, it is possible to construct a coding protocol with an error of sufficiently small size for both visible and blind cases, provided the compression rate is larger than a certain value [3, 4]. This construction will be examined in more detail later. In fact such an encoding has already been realized experimentally [5]. For variable-length coding, in many cases there does not exist a code with zero error of a smaller coding length than the size of the input information [6]. However, when we replace the condition "zero error" by "almost zero error," it is possible to construct codes with the admissible compression rate. Therefore, if the information source is a quantum state that is generated by an n-fold independent and identical distribution of a "known" distribution, variable-length encoding does not offer any advantage.

On the other hand, if we do not know the probability distribution to generate the quantum state, the situation is entirely different. In fixed-length coding, since the compression rate is fixed a priori, it is impossible to recover the input state with a small error when the compression rate is less than the minimum admissible rate. In this case, it is preferable to use variable-length encoding wherein the compression rate depends on the input state [7, 8]. However, as a measurement is necessary to determine the compression rate, the determination of the compression rate causes the state reduction due to the quantum mechanical nature. Consider a method to determine the compression rate based on the approximate estimate of the input state. This approximate estimation requires a measurement. If the initial state is changed considerably due to the measurement, clearly we cannot expect that the decoding error is close to zero. It is therefore necessary to examine the trade-off between the degree of state reduction and the estimation error of the distribution of the input state, which is required for determining the encoding method.

As will be discussed later, both this estimation error and the degree of the state reduction can be made to approach zero simultaneously and asymptotically. Therefore, even when the probability distribution for the quantum state is unknown, we can asymptotically construct a variable-length code such that the minimum admissible rate is achieved with a probability close to 1 and the decoding error is almost 0 [9, 10]. In particular, when a given coding protocol is effective for all probability distributions, it is called **universality**; this is an important topic in information theory. Various other types of source compression problems have also been studied [11, 12].

10.2 Quantum Fixed-Length Source Coding

The source of quantum system $\mathcal{H}$ is denoted by

$$W : \mathcal{X} \to \mathcal{S}(\mathcal{H}) \ (x \mapsto W_x) \tag{10.1}$$

(which is the same notation as that in a quantum channel) and a probability distribution p in $\mathcal{X}$. That is, the quantum information source is described by the ensemble $(p_x, W_x)_{x\in\mathcal{X}}$. Let $\mathcal{K}$ be the compressed quantum system. For the blind case, the **encoder** is represented by the TP-CP map τ from $\mathcal{S}(\mathcal{H})$ to $\mathcal{S}(\mathcal{K})$. The **decoder** is represented by a TP-CP map ν from $\mathcal{S}(\mathcal{K})$ to $\mathcal{S}(\mathcal{H})$. The triplet $\psi \stackrel{\mathrm{def}}{=} (\mathcal{K}, \tau, \nu)$ is then called a **blind code**. In the visible case, the encoder is not as restricted as in the blind case. In this case, the **encoder** is given by a map T from $\mathcal{X}$ to $\mathcal{S}(\mathcal{K})$. Any blind encoder τ can be converted into a visible encoder according to $\tau \circ W$. The triplet $\Psi \stackrel{\mathrm{def}}{=} (\mathcal{K}, T, \nu)$ is then called a **visible code**. That is, the information is stored by a quantum memory. The errors $\varepsilon_{p,W}(\psi)$ and $\varepsilon_{p,W}(\Psi)$ and **sizes** $|\psi|$ and $|\Psi|$ of the codes ψ and Ψ, respectively, are defined as follows:

$$\varepsilon_{p,W}(\psi) \stackrel{\text{def}}{=} \sum_{x \in \mathcal{X}} p_x \left(1 - F^2(W_x, \nu \circ \tau(W_x))\right), \quad |\psi| \stackrel{\text{def}}{=} \dim \mathcal{K} \tag{10.2}$$

$$\varepsilon_{p,W}(\Psi) \stackrel{\text{def}}{=} \sum_{x \in \mathcal{X}} p_x \left(1 - F^2(W_x, \nu \circ T(x))\right), \quad |\Psi| \stackrel{\text{def}}{=} \dim \mathcal{K}. \tag{10.3}$$

We used $1 - F^2(\cdot, \cdot)$ in our definition of the decoding error.

Now, let the source be given by the quantum system $\mathcal{H}^{\otimes n}$ and its candidate states be given by $W^{(n)} : \mathcal{X}^{(n)} \to \mathcal{S}(\mathcal{H}^{\otimes n})$ $(x^n = (x_1, \ldots, x_n) \mapsto W^{(n)}_{x^n} \stackrel{\text{def}}{=} W_{x_1} \otimes \cdots \otimes W_{x_n})$. Further, let the probability distribution for these states be given by the nth-order independent and identical distribution of the probability distribution p in $\mathcal{X}$. Denote the blind and visible codes by $\psi^{(n)}$ and $\Psi^{(n)}$, respectively. Define[1]

$$R_{B,q}(p, W) \stackrel{\text{def}}{=} \inf_{\{\psi^{(n)}\}} \left\{ \overline{\lim} \frac{1}{n} \log |\psi^{(n)}| \,\middle|\, \varepsilon_{p^n, W^{(n)}}(\psi^{(n)}) \to 0 \right\}, \tag{10.4}$$

$$R_{V,q}(p, W) \stackrel{\text{def}}{=} \inf_{\{\Psi^{(n)}\}} \left\{ \overline{\lim} \frac{1}{n} \log |\Psi^{(n)}| \,\middle|\, \varepsilon_{p^n, W^{(n)}}(\Psi^{(n)}) \to 0 \right\}, \tag{10.5}$$

$$R^{\dagger}_{B,q}(p, W) \stackrel{\text{def}}{=} \inf_{\{\psi^{(n)}\}} \left\{ \overline{\lim} \frac{1}{n} \log |\psi^{(n)}| \,\middle|\, \overline{\lim}\, \varepsilon_{p^n, W^{(n)}}(\psi^{(n)}) < 1 \right\}, \tag{10.6}$$

$$R^{\dagger}_{V,q}(p, W) \stackrel{\text{def}}{=} \inf_{\{\Psi^{(n)}\}} \left\{ \overline{\lim} \frac{1}{n} \log |\Psi^{(n)}| \,\middle|\, \overline{\lim}\, \varepsilon_{p^n, W^{(n)}}(\Psi^{(n)}) < 1 \right\}. \tag{10.7}$$

Since a blind code $\psi^{(n)}$ can be regarded as a visible code, we have

$$R_{B,q}(p, W) \geq R_{V,q}(p, W), \quad R^{\dagger}_{B,q}(p, W) \geq R^{\dagger}_{V,q}(p, W). \tag{10.8}$$

From the definitions it is also clear that

$$R_{B,q}(p, W) \geq R^{\dagger}_{B,q}(p, W), \quad R_{V,q}(p, W) \geq R^{\dagger}_{V,q}(p, W). \tag{10.9}$$

The following theorem holds with respect to the above.

Theorem 10.1 *If all of states W_x are pure states, then the quantities defined above are equal. We have*

$$R_{B,q}(p, W) = R^{\dagger}_{B,q}(p, W) = R_{V,q}(p, W) = R^{\dagger}_{V,q}(p, W) = H(W_p). \tag{10.10}$$

This theorem can be proved by combining the following two lemmas.

Lemma 10.1 (Direct Part) *There exists a sequence of blind codes $\{\psi^{(n)}\}$ satisfying*

$$\frac{1}{n} \log |\psi^{(n)}| \leq H(W_p) - \delta \tag{10.11}$$

[1]The subscript q indicates the "quantum" memory.

$$\varepsilon_{p^n,W^{(n)}}(\psi^{(n)}) \to 0 \tag{10.12}$$

for arbitrary real number $\delta > 0$.

Lemma 10.2 (Converse Part) *If all of states* W_x *are pure states and the sequence of visible codes* $\{\Psi^{(n)}\}$ *satisfies*

$$\overline{\lim} \frac{1}{n} \log |\Psi^{(n)}| < H(W_p), \tag{10.13}$$

then

$$\varepsilon_{p^n,W^{(n)}}(\Psi^{(n)}) \to 1. \tag{10.14}$$

Lemma 10.1 tells us that $R_{B,q}(p, W) \le H(W_p)$, and Lemma 10.2 tells us that $R^{\dagger}_{V,q}(p, W) \ge H(W_p)$. Using (10.8) and (10.9), we thus obtain (10.10).

Further, we have another fixed-length coding scheme. In this scheme, the state is given as a pure state $|x\rangle\langle x|$ on the composite system $\mathcal{H}_A \otimes \mathcal{H}_R$, and encoder and decoder can treat only the local system $\mathcal{H}_A$. Then, our task is recovering the state $|x\rangle\langle x|$ on the composite system $\mathcal{H}_A \otimes \mathcal{H}_R$. Hence, the code of this scheme is the triplet $\psi \stackrel{\text{def}}{=} (\mathcal{K}, \tau, \nu)$, which is the same as that of the blind scheme. The error is given as

$$\varepsilon'_{\rho}(\psi) \stackrel{\text{def}}{=} 1 - \langle x|(\nu \otimes \iota) \circ (\tau \otimes \iota)(|x\rangle\langle x|)|x\rangle = 1 - F_e^2(\rho, \nu \circ \tau),$$

where $\rho = \mathrm{Tr}_R|x\rangle\langle x|$. Recall the definition of the entanglement fidelity (8.19). Hence, the quality depends only on the reduced density ρ. This scheme is called the **purification scheme**, while the former scheme with the visible case or the blind case is called the **ensemble scheme**. Hence, we define the minimum compression rate as

$$R_{P,q}(\rho) \stackrel{\text{def}}{=} \inf_{\{\psi^{(n)}\}} \left\{ \overline{\lim} \frac{1}{n} \log |\psi^{(n)}| \,\middle|\, \varepsilon'_{\rho^{\otimes n}}(\psi^{(n)}) \to 0 \right\}, \tag{10.15}$$

$$R^{\dagger}_{P,q}(\rho) \stackrel{\text{def}}{=} \inf_{\{\psi^{(n)}\}} \left\{ \overline{\lim} \frac{1}{n} \log |\psi^{(n)}| \,\middle|\, \overline{\lim}\, \varepsilon'_{\rho^{\otimes n}}(\psi^{(n)}) < 1 \right\}. \tag{10.16}$$

When all of states W_x are pure, Inequality (8.24) implies that $1 - \varepsilon_{W_p}(\Psi) \le 1 - \varepsilon_{p,W}(\Psi)$. Hence, we have

$$R_{P,q}(W_p) \le R_{B,q}(p, W), \quad R^{\dagger}_{P,q}(W_p) \le R^{\dagger}_{B,q}(p, W). \tag{10.17}$$

Using this relation, we can show the following theorem[Exe. 9.36].

Theorem 10.2

$$R_{P,q}(\rho) = R^{\dagger}_{P,q}(\rho) = H(\rho). \tag{10.18}$$

Exercises

10.1 Show that the condition $\sum_{x\in\mathcal{X}^n} p^n_x F(W^{(n)}_x, \nu_n(T(x))) \to 1$ is equivalent to the condition $\sum_{x\in\mathcal{X}^n} p^n_x F(W^{(n)}_x, \nu_n(T(x)))^2 \to 1$.

10.2 Define other error functions

$$\bar{\varepsilon}_{p,W}(\Psi) \stackrel{\text{def}}{=} \sum_{x\in\mathcal{X}} p_x d_1(W_x, \nu\circ\tau(W_x)), \tag{10.19}$$

$$\bar{\varepsilon}_{p,W}(\Psi) \stackrel{\text{def}}{=} \sum_{x\in\mathcal{X}} p_x d_1(W_x, \nu\circ T(x)). \tag{10.20}$$

Show that the optimal rates $R_{B,q}(p, W)$ and $R_{V,q}(p, W)$ given in (10.4) and (10.5) are not changed even when the conditions $\varepsilon_{p^n,W^{(n)}}(\psi^{(n)}) \to 0$ and $\varepsilon_{p^n,W^{(n)}}(\Psi^{(n)}) \to 0$ are replaced by $\bar{\varepsilon}_{p^n,W^{(n)}}(\psi^{(n)}) \to 0$ and $\bar{\varepsilon}_{p^n,W^{(n)}}(\Psi^{(n)}) \to 0$.

10.3 Construction of a Quantum Fixed-Length Source Code

Let us construct a blind fixed-length code that attains the minimum compression rate $H(W_p)$ when the quantum state is generated subject to the independent and identical distribution of the probability distribution p. (Since any blind code can be regarded as a visible code, it is sufficient to construct a blind code.) Since formula (8.24) guarantees that

$$\begin{aligned}\varepsilon_{p,W}(\Psi) =& \sum_{x\in\mathcal{X}} p(x)(1 - F^2(W_x, \nu\circ\tau(W_x)))\\ \le& 1 - F_e^2(W_p, \nu\circ\tau) = \varepsilon'_{W_p}(\psi),\end{aligned} \tag{10.21}$$

it is sufficient to treat the purification scheme.

Now define $\rho^P_{\text{mix}} \stackrel{\text{def}}{=} \frac{P}{\operatorname{Tr} P}$. Let the encoder $\tau_P : \mathcal{S}(\mathcal{H}) \to \mathcal{S}(\operatorname{Ran} P)$, using the projection P in $\mathcal{H}$, be given by

$$\tau_P(\rho) \stackrel{\text{def}}{=} P\rho P + \operatorname{Tr}[(I-P)\rho]\rho^P_{\text{mix}}. \tag{10.22}$$

Define the decoder ν_P as the natural embedding from $\mathcal{S}(\operatorname{Ran} P)$ to $\mathcal{S}(\mathcal{H})$, where $\operatorname{Ran} A$ is the range of A.

Let x be the purification of ρ. Then,

$$\begin{aligned}&F_e^2(\rho, \nu_P\circ\tau_P)\\ =&\langle x|(I\otimes P)|x\rangle\langle x|(I\otimes P)|x\rangle\\ &+ \langle x|\rho^P_{\text{mix}} \otimes \operatorname{Tr}_{\mathcal{H}}\left[(I\otimes(I-P))|x\rangle\langle x|(I\otimes(I-P))\right]|x\rangle\end{aligned}$$

$$
\begin{aligned}
&\geq \langle x|(I \otimes P)|x\rangle \langle x|(I \otimes P)|x\rangle = (\operatorname{Tr} P\rho)^2 \\
&= (1 - (1 - \operatorname{Tr} P\rho))^2 \geq 1 - 2(1 - \operatorname{Tr} P\rho).
\end{aligned} \tag{10.23}
$$

We now define $b(s, R) \stackrel{\text{def}}{=} \frac{R-\psi(s)}{1-s}$, $\psi(s) = \psi(s|\rho)$, $s_0 \stackrel{\text{def}}{=} \operatorname{argmax}_{0<s<1} \frac{sR-\psi(s)}{1-s}$ for $R > H(\rho)$ and $0 < s < 1$. We choose P such that

$$
P_n \stackrel{\text{def}}{=} \left\{ W_p^{\otimes n} - e^{-nb(s_0,R)} > 0 \right\}. \tag{10.24}
$$

Then, from (3.2), (3.4), and (10.23), the code $\Phi^{(n)} = (\mathcal{K}_n, \tau_{P_n}, \nu_{P_n})$ satisfies

$$
\begin{aligned}
&\dim \mathcal{K}_n \stackrel{\text{def}}{=} \operatorname{Ran} P_n = \operatorname{Tr} \left\{ \rho^{\otimes n} - e^{-nb(s,R)} > 0 \right\} \leq e^{nR} \\
&\varepsilon'_{\rho^{\otimes n}}(\Phi^{(n)}) \leq 2 \operatorname{Tr} \rho^{\otimes n} \left\{ \rho^{\otimes n} - e^{-nb(s_0,R)} \leq 0 \right\} \leq 2e^{-n\frac{s_0 R - \psi(s_0)}{1-s_0}}.
\end{aligned} \tag{10.25}
$$

The combination of (10.21) and (10.25) proves the existence of a code that attains the compression rate $H(\rho) + \delta$ for arbitrary $\delta > 0$. Hence, we have proven Lemma 10.1. Note that the code constructed here depends only on the state ρ and the rate R. In order to emphasize this dependence, we denote this encoding and decoding as $\tau_{n,\rho,R}$ and $\nu_{n,\rho,R}$, respectively.

We next show that the code given above still works even when the true density ρ' is slightly different from the predicted density ρ. This property is called **robustness** and is important for practical applications.

Let us consider the case where the true density ρ' is close to the predicted one ρ. Choosing a real number $\alpha > 0$, we have

$$
\rho' \leq \rho e^{\alpha}. \tag{10.26}
$$

Hence,

$$
\operatorname{Tr} \rho'^{\otimes n} \leq e^{n\alpha} \operatorname{Tr} \rho^{\otimes n}.
$$

Using the same argument as that in the derivation of (10.25), we obtain

$$
\begin{aligned}
&\varepsilon'_{\rho'^{\otimes n}}(\Phi^{(n)}) \leq 2 \operatorname{Tr} \rho'^{\otimes n} \{\rho^{\otimes n} - e^{-na} < 0\} \\
&\leq 2e^{n\alpha} \operatorname{Tr} \rho^{\otimes n} \{\rho^{\otimes n} - e^{-na} < 0\} \leq 2e^{n(\alpha + \frac{\psi(s_0) - s_0 R}{1-s_0})}.
\end{aligned} \tag{10.27}
$$

Therefore, if $\alpha < \max_{0 \leq s \leq 1} \frac{sR-\psi(s)}{1-s}$, then $\varepsilon'_{\rho'^{\otimes n}}(\Phi^{(n)}) \to 0$.

Let us now prove the converse part of the theorem, i.e., Lemma 10.2. For a proof of Lemma 10.2, we prepare the following lemma, which is proved in Sect. 10.10.

Lemma 10.3 (Hayashi [13]) *Any visible code* $\Psi = (\mathcal{K}, T, \nu)$ *satisfies*

$$1 - \varepsilon_{p,W}(\Psi) \leq a|\Psi| + \mathrm{Tr}\, W_p\{W_p - a \geq 0\} \tag{10.28}$$

for $\forall a > 0$.

Proof of Lemma 10.2 The above inequality (10.28) can be regarded as the "dual" inequality of inequality (10.25) given in the proof of the direct part of the theorem. Inequality (10.28) shows that the quality of any code is evaluated by use of $\mathrm{Tr}\, W_p\{W_p - e^{\lambda} \geq 0\}$. Inequality (10.28) plays the same role as (2.5.2) in Sect. 2.1.4, and thus any sequence of codes $\{\Psi^{(n)}\}$ satisfies

$$1 - \varepsilon_{p^n, W^{(n)}}(\Psi^{(n)}) \leq 2e^{n\frac{\psi(s)-sR}{1-s}}. \tag{10.29}$$

Choosing an appropriate $s_0 < 0$, we have $\frac{\psi(s_0)-s_0 R}{1-s_0} < 0$. Therefore, we obtain (10.14), which gives us Lemma 10.2. ■

In order to construct a code with the compression rate $H(W_p)$, we replace R by $H(W_p) - \frac{1}{n^{1/4}}$ in (10.29). Approximating $\psi(s)$ as $H(W_p)s + \frac{1}{2}\psi''(0)s^2$, we obtain

$$\min_{s<0} \frac{\psi(s) - s(H(W_p) - \frac{1}{n^{1/4}})}{1-s} \cong -\frac{C^2}{\psi''(0)\sqrt{n}}. \tag{10.30}$$

Hence,

$$\varepsilon_{p^n, W^{(n)}}(\Psi^{(n)}) \leq 2e^{-\sqrt{n}\frac{C^2}{\psi''(0)}} \to 0. \tag{10.31}$$

Finally, let us focus on the property of the state on the compressed system when the asymptotic compression rate is $H(W_p)$.

Theorem 10.3 (Han [14]) *When a sequence of codes* $\{\Psi_n = (\mathcal{K}_n, T_n, \nu_n)\}$ *satisfies*

$$\varepsilon_{p^n, W^{(n)}}(\Psi_n) \to 0, \quad \frac{1}{n}\log|\Psi_n| \to H(W_p), \tag{10.32}$$

we obtain

$$\frac{1}{n}D\left(\sum_x p_x^n T_n(x) \| \rho_{\mathrm{mix}}^{\mathcal{K}_n}\right) \to 0. \tag{10.33}$$

That is, the compressed state is almost completely mixed in the sense of the normalized quantum relative entropy $\frac{1}{n}D(\rho_n\|\sigma_n)$. However, the compressed state is different from the completely mixed state if we focus on the Bures distance, trace norm, or quantum relative entropy. This fact has been shown in the classical case [15].

Proof From the monotonicity of the transmission information (5.59) we have

$$\log|\Phi_n| \geq H\Big(\sum_x p_x^n T_n(x)\Big) \geq H\Big(\sum_x p_x^n T_n(x)\Big) - \sum_x p_x^n H(T_n(x))$$
$$\geq H\Big(\sum_x p_x^n \nu(T_n(x))\Big) - \sum_x p_x^n H(\nu(T_n(x))).$$

From condition (10.32) the two conditions in (5.105) yield

$$\lim_{n\to\infty} \frac{1}{n}\left(H\Big(\sum_x p_x^n \nu(T_n(x))\Big) - \sum_x p_x^n H(\nu(T_n(x))\right) = H(W_p).$$

Hence, we obtain

$$\lim_{n\to\infty} \frac{1}{n} H\Big(\sum_x p_x^n T_n(x)\Big) = H(W_p).$$

Since $\mathrm{Tr} \sum_x p_x^n T_n(x) \log \rho_{\mathrm{mix}}^{\mathcal{K}_n} = -\log|\Phi_n|$, relation (10.33) holds. ■

Exercise

10.3 Prove (10.18) by using (10.25) and (10.17).

10.4 Universal Quantum Fixed-Length Source Codes

The code given in Sect. 10.3 depends on the quantum state W_p. For the classical case, there exists a code that depends on the entropy $H(p)$ and works when the data are generated subject to the independent and identical information source of p. Such codes are called **universal** codes. To construct such universal codes, we often use the method of types, which is discussed in Sect. 2.4.1 [16]. Similarly, for the quantum case, there exists a code that depends only on the entropy $H(W_p)$ of the average state W_p and works well provided the states are generated according to an independent and identical distribution of p [17]. In this subsection, we propose such a kind of universal code.

For this purpose, the projection P_n given by (10.24) should depend only on the compression rate. That is, we have to construct a subspace $\Upsilon_n(R)$ of $\mathcal{H}^{\otimes n}$ depending only on the compression rate R. As the first step, we construct a code depending only on the compression rate and the basis $B = (u_1, \ldots, u_d)$ comprising the eigenvectors of W_p. Let us consider a set of types T_n with the probability space $\mathbb{N}_d = \{1, \ldots, d\}$. Define the subspace $\Upsilon_n(R, B)$ of the n-fold tensor product space $\mathcal{H}^{\otimes n}$ to be the space spanned by $\cup_{q\in T^n : H(q)\leq R}\{u(\mathbf{i}_n)\}_{\mathbf{i}_n \in T_q^n}$, where $u(\mathbf{i}_n) \stackrel{\mathrm{def}}{=} u_{i_1} \otimes \cdots \otimes u_{i_n} \in \mathcal{H}^{\otimes n}$ and $\boldsymbol{i}_n = (i_1, \ldots, i_n)$. Let $P_{n,R,B}$ be a projection to $\Upsilon_n(R, B)$. Then, according to the discussion in Sect. 2.4.1, we can show that[Exe. 10.4]

$$\dim \Upsilon_n(R, B) \le (n+1)^d e^{nR}, \tag{10.34}$$
$$\mathrm{Tr}(I - P_{n,R,B}) W_p^{\otimes n} \le (n+1)^d \exp(-n \inf_{q:H(q)>R} D(q\|r)), \tag{10.35}$$

where r is the distribution that consists of the diagonal elements of W_p. Hence, the code $\{(\Upsilon_n(R, B), \nu_{P_{n,R,B}}, \tau_{P_{n,R,B}})\}$ almost has the compression rate R. Since $\min_{0\le s\le 1} \frac{\psi(s)-sR}{1-s} = \min_{q:H(q)\ge R} D(q\|r)$, its entanglement fidelity $F_e(W_p^{\otimes n}, \nu_{P_{n,R,B}} \circ \tau_{P_{n,R,B}})$ asymptotically approaches 1 when $R > H(W_p)$. This code is effective when the basis $B = \{u_1, \dots, u_d\}$ is known.

However, when the basis B is unknown, we need a subspace depending only on the compression rate R. For this purpose, we define the subspace $\Upsilon_n(R)$ as the subspace spanned by $\cup_B \Upsilon_n(R, B)$. That is, we consider the union for all of bases B in $\mathcal{H}$. Then, the projection $P_{n,R}$ is defined as the projection to $\Upsilon_n(R)$. Thus, we can show that the space $\Upsilon_n(R)$ and the projection $P_{n,R}$ satisfy

$$\dim \Upsilon_n(R) \le (n+1)^{d+d^2} e^{nR}, \tag{10.36}$$
$$\mathrm{Tr}(I - P_{n,R}) W_p^{\otimes n} \le (n+1)^d \exp(-n \inf_{q:H(q)>R} D(q\|r)). \tag{10.37}$$

Hence, the entanglement fidelity $F_e(W_p^{\otimes n}, \nu_{P_{n,R}} \circ \tau_{P_{n,R}})$ asymptotically approaches 1 when $R > H(W_p)$. Then, we can conclude that the blind code $(\Upsilon_n(R), \tau_{P_{n,R}}, \nu_{P_{n,R}})$ works when $R > H(W_p)$. Since the blind code $(\Upsilon_n(R), \tau_{P_{n,R}}, \nu_{P_{n,R}})$ does not depend on the basis of W_p, it can be regarded as a universal code.

Proofs of (10.36) and (10.37) Since (10.37) follows immediately from $P_{n,R} \ge P_{n,R,B}$, we prove inequality (10.36) as follows. For simplicity, we consider the case of $d = 2$, but this discussion may be easily extended to the general case. First, we fix the basis $B = \{u_1, u_2\}$. Then, an arbitrary basis $B' = \{u_1', u_2'\}$ may be written as $u_1' = au_1 + bu_2$, $u_2' = cu_1 + du_2$ using $d^2 = 4$ complex numbers a, b, c, d. Thus,

$$u_1' \otimes u_2' \otimes \cdots \otimes u_1' = (au_1 + bu_2) \otimes (cu_1 + du_2) \otimes \cdots \otimes (au_1 + bu_2).$$

Choosing an appropriate vector $v_{n_1,n_2,n_3,n_4} \in \mathcal{H}^{\otimes n}$, we have

$$u_1' \otimes u_2' \otimes \cdots \otimes u_1' = \sum_{n_1,n_2,n_3,n_4} a^{n_1} b^{n_2} c^{n_3} d^{n_4} v_{n_1,n_2,n_3,n_4}.$$

The vector v_{n_1,n_2,n_3,n_4} does not depend on a, b, c, and d. Hence, the vector $u_1' \otimes u_2' \otimes \cdots \otimes u_1'$ belongs to the subspace spanned by the vectors v_{n_1,n_2,n_3,n_4} with the condition $n_1 + n_2 + n_3 + n_4 = n$. The dimension of this subspace is at most $(n+1)^{2^2-1} = (n+1)^{d^2-1}$. Since the dimension of the space $\Upsilon_n(R)$ is at most this number multiplied by the dimension of the space $\Upsilon_n(R, B)$ with a fixed basis, we obtain (10.36). ∎

Finally, we show an inequality complementary to (10.37), which will be used in the next section. Choosing $s_0 \stackrel{\mathrm{def}}{=} \mathrm{argmin}_{s\le 0} \frac{\psi(s)-sR}{1-s}$ with $\psi(s) = sH_{1-s}(W_p)$, we obtain

$$
\begin{aligned}
\operatorname{Tr} P_{n,R} W_p^{\otimes n} &\overset{(a)}{\leq} 2 \exp\left(\min_{s \leq 0} \frac{n\psi(s) - s \log \dim \Upsilon_n(R)}{1-s} \right) \\
&\overset{(b)}{\leq} 2 \exp\left(n \frac{\psi(s_0) - s_0 R}{1 - s_0} + \frac{-s_0}{1-s_0}(d + d^2) \log(n+1) \right) \\
&\overset{(c)}{=} 2(n+1)^{\frac{-s_0}{1-s_0}(d+d^2)} \exp(-n \inf_{q:H(q) \leq R} D(q\|r)),
\end{aligned} \tag{10.38}
$$

where (a), (b), and (c) follow from (2.54) and (10.36), and (2.65), respectively. Inequality (10.38) implies that the blind code $(\Upsilon_n(R), \tau_{P_{n,R}}, \nu_{P_{n,R}})$ does not work when $R < H(W_p)$. However, this conclusion can be trivially shown from Theorem 10.1.

Exercise

10.4 Show (10.34) and (10.35) by using (2.154), (2.155) and (2.156).

10.5 Universal Quantum Variable-Length Source Codes

Let us construct a code that has a sufficiently small error and achieves the entropy rate $H(W_p)$, even though the entropy rate $H(W_p)$ is unknown, provided the source follows an independent and identical distribution of p. Such codes are called universal quantum variable-length source codes. For these codes, it is essential to determine the compression rate dependently of the input state.[2] If nonorthogonal states are included in the source, then the state reduction inevitably occurs due to the determination of the compression rate. Hence, the main problem is to reduce the amount of the state reduction as much as possible [9, 10].

Let us first construct a measurement to determine the compression rate by using the projection $P_{n,R}$ given in the previous section. Consider the projection $E_{n,R} \overset{\text{def}}{=} \lim_{\epsilon \to +0}(P_{n,R} - P_{n,R-\epsilon})$. Let $\Omega_n = \{H(p)\}_{p \in T^n}$ be a set of R such that $E_{n,R}$ is nonzero. Then, $\sum_{R \in \Omega_n} E_{n,R} = I$. Due to (10.37) and (10.38), the probability distribution for the outcome of the measurement $E_n = \{E_{n,R_i}\}_i$ satisfies

$$
\begin{aligned}
& \mathrm{P}^{E_n}_{W_p^{\otimes n}} \{|H(W_p) - R_i| \geq \epsilon\} \\
\leq & 2 \max \Big\{ (n+1)^{d+d^2} \exp(-n \inf_{q:H(q) \leq R - \epsilon} D(q\|r)), \\
& (n+1)^d \exp(-n \inf_{q:H(q) \geq R+\epsilon} D(q\|r)) \Big\}.
\end{aligned}
$$

[2] Even though the error of universal quantum variable-length source code converges to 0, its convergence rate is not exponential [9].

We may therefore apply the arguments of Sect. 7.4 as follows. We choose l_n, δ_n so that they satisfy (7.63) and (7.64). Then, the POVM $\boldsymbol{M}^{(n),\delta_n,l_n}$ given from E_n in Theorem 7.8 satisfies

$$F_e^2(W_p^{\otimes n}, \kappa_{\boldsymbol{M}^{(n),\delta_n,l_n}}) \to 1.$$

Since the measurement $\boldsymbol{M}^{(n),\delta_n,l_n}$ takes values in $[0, \log d]$ with spacing δ_n, the number of its possible outcomes is $(\frac{\log d}{\delta_n}+1)$. Hence, we choose δ_n such that $\frac{1}{n}\log\delta_n \to 0$.
Construction of universal quantum variable-length source code We now construct a universal variable-length code based on this measurement. In the encoding step, we perform a measurement corresponding to the instrument $\kappa_{\boldsymbol{M}^{(n),\delta_n,l_n}}$. When the measurement outcome is R_i, the resulting state is a state in Ran $M_i^{(n),\delta_n,l_n}$. The state in the space Ran $M_i^{(n),\delta_n,l_n}$ is sent with the outcome R_i. Then, the coding length is $\log\dim \mathrm{Ran}\, M_i^{(n),\delta_n,l_n} + \log(\frac{\log d}{\delta_n}+1)$. The compression rate is this value divided by n.

Analysis of our code Since the second term converges to zero after the division by n, we only consider the first term. Since

$$\begin{aligned}\dim M_R^{(n),\delta_n,l_n} &= \sum_{R-\delta_n<R'<R+\delta_n} \mathrm{rank}\, E_{n,R'} \\ &\le \dim \Upsilon_n(R+\delta_n) \le (n+1)^{d+d^2} e^{n(R+\delta_n)},\end{aligned}$$

we obtain

$$\frac{1}{n}\left\{\log\dim\mathrm{Ran}\, M_i^{(n),\delta_n,l_n} + \log\left(\frac{\log d}{\delta_n}+1\right)\right\} \le R.$$

Therefore, in this protocol, the compression rate is asymptotically less than the entropy $H(W_p)$ with a probability of approximately 1 [more precisely the compression rate converges to the entropy $H(W_p)$ in probability]; the error also approaches zero asymptotically. That is, we can conclude that the above protocol is a universal quantum variable-length source code.

10.6 Mixed-State Case and Bipartite State Generation

So far, we have treated quantum data compression when W_x is pure. That is, in this case, the optimal compression rate in the blind case coincides with that in the visible case. However, when W_x is not pure, these are different. In this problem, one may think that the quantity $H(W_p)$ or $I(p, W)$ is a good candidate for the optimal rate. This intuition is not entirely inaccurate. In the blind scheme, if the ensemble (p_x, W_x) has no trivial redundancy, the optimal compression rate is given as follows [18]:

$$R_{B,q}(p, W) = H(W_p). \tag{10.39}$$

On the other hand, in the visible scheme, the inequality

$$R_{V,q}(p, W) \geq I(p, W) \tag{10.40}$$

holds[Exe. 10.7] [19]. However, it does not give the optimal rate in general:

Theorem 10.4

$$R_{V,q}(p, W) = E_c^{-\to}(\tilde{W}_p) = \lim_{n\to\infty} \frac{1}{n} E_p(\tilde{W}_p^{\otimes n}), \quad \tilde{W}_p \stackrel{\text{def}}{=} \sum_x p_x |e_x^A\rangle\langle e_x^A| \otimes W_x. \tag{10.41}$$

In fact, Horodecki [20] focused on the quantity

$$H^{ext}(p, W) \stackrel{\text{def}}{=} \inf_{W_x^{ext}:\text{purification of } W_x} H\left(\sum_x p_x W_x^{ext}\right)$$

and proved

$$R_{V,q}(p, W) = \lim_{n\to\infty} \frac{H^{ext}(W^{(n)}, p^n)}{n}. \tag{10.42}$$

From the definition of $E_p(\tilde{W}_p)$, we can easily check that $E_p(\tilde{W}_p) \leq H^{ext}(p, W)$. When all of states W_x are pure, $R_{V,q}(p, W) = H(W_p)$. This fact matches (8.193).

Before the proof of Theorem 10.4, we address this problem in a more general framework. Suppose that, given a bipartite state ρ on the bipartite system $\mathcal{H}_A \otimes \mathcal{H}_B$, Alice and Bob intend to share the state ρ by using limited amount of noiseless quantum communication. This task is called bipartite state generation. The following operations are allowed for this task. First, Alice generates a bipartite state ρ' on the bipartite system $\mathcal{H}_A \otimes \mathcal{K}$. Second, Alice sends the system $\mathcal{K}$ to Bob. Finally, Bob applies a TP-CP map ν from the system $\mathcal{K}$ to $\mathcal{H}_B$. Then, Alice and Bob share the state $\nu \otimes \iota_A(\rho')$ on the bipartite system $\mathcal{H}_A \otimes \mathcal{H}_B$. In this case, our operation $\tilde{\Psi}$ is given as the triple $(\mathcal{K}, \rho', \nu)$, which is called a code. The performance of the code $\tilde{\Psi}$ is characterized by the following quantities. One is the dimension of $\mathcal{K}$, which is denoted by $|\tilde{\Psi}|$. The other is the error $1 - F(\rho, \nu \otimes \iota_A(\rho'))$, which is denoted by $\tilde{\varepsilon}_\rho(\tilde{\Psi})$.

Now, we slightly modify the formulation of visible compression. For a code $\Psi = (\mathcal{K}, T, \nu)$, we consider another error:

$$\tilde{\varepsilon}_{p,W}(\Psi) := \sum_{x\in\mathcal{X}} p_x \left(1 - F(W_x, \nu \circ \tau(W_x))\right), \tag{10.43}$$

which has the relation:

$$\frac{1}{2}\varepsilon_{p,W}(\Psi) \leq \tilde{\varepsilon}_{p,W}(\Psi) \leq \varepsilon_{p,W}(\Psi). \tag{10.44}$$

So, even if we replace $\varepsilon_{p,W}(\Psi)$ by $\tilde{\varepsilon}_{p,W}(\Psi)$, our definition of (10.5) is not changed.

When ρ is given as $\tilde{W}_p$ defined in (10.41), the code $\Psi = (\mathcal{K}, T, \nu)$ for visible compression is converted to the code $\tilde{\Psi} = (\mathcal{K}, \rho', \nu)$ for bipartite state generation, where ρ' is given as follows.

$$\rho' := \sum_x p_x |e_x^A\rangle\langle e_x^A| \otimes T(x). \tag{10.45}$$

In this correspondence,

$$\tilde{\varepsilon}_{p,W}(\Psi) = \tilde{\varepsilon}_{\tilde{W}_p}(\tilde{\Psi}). \tag{10.46}$$

Hence, we have

$$\min_{\Psi}\{\tilde{\varepsilon}_{p,W}(\Psi) | |\Psi| = M\} \geq \min_{\tilde{\Psi}}\{\varepsilon_{\tilde{W}_p}(\tilde{\Psi}) | |\tilde{\Psi}| = M\}. \tag{10.47}$$

For the bipartite state generation, we define the following quantity:

$$R_{g,q}(\rho) \stackrel{\text{def}}{=} \inf_{\{\tilde{\Psi}^{(n)}\}} \left\{ \overline{\lim} \frac{1}{n} \log |\tilde{\Psi}^{(n)}| \,\middle|\, \tilde{\varepsilon}_{\rho^{\otimes n}}(\tilde{\Psi}^{(n)}) \to 0 \right\}. \tag{10.48}$$

Then, the following theorem holds.

Theorem 10.5

$$R_{g,q}(\rho) = E_c^{-\to}(\rho) = \lim_{n\to\infty} \frac{1}{n} E_p(\rho^{\otimes n}). \tag{10.49}$$

Before proving Theorem 10.4, we show Theorem 10.5. Due to (10.47), Theorem 10.5 implies the converse type inequality $R_{V,q}(p, W) \geq E_c^{-\to}(\tilde{W}_p)$. So, after the proof of Theorem 10.5, we show the direct part of Theorem 10.4.

The direct part of Theorem 10.5 essentially is obtained by the following lemma.

Lemma 10.4 *Let κ be a one-way LOCC operation from Alice to Bob. There exists a code Ψ such that*

$$\tilde{\varepsilon}_\rho(\tilde{\Psi}) \leq 1 - F(\rho, \kappa(|\Phi_L\rangle\langle\Phi_L|)), \tag{10.50}$$

$$|\tilde{\Psi}| = L \cdot \mathrm{CC}(\kappa), \tag{10.51}$$

where $\mathrm{CC}(\kappa)$ is the size of the classical communication of κ.

This lemma can be shown as follows. Firstly, Alice prepares the maximally entangled state $|\Phi_L\rangle\langle\Phi_L|$, and sends a part of the maximally entangled state to Bob via noiseless quantum channel. Then, Alice and Bob apply the one-way LOCC operation κ from Alice to Bob. This operation satisfies the condition for the above code $\tilde{\Phi}$.

Proof of Theorem 10.5 First, we prove the direct part. Using Lemma 10.4, we obtain the direct part as follows. Let κ_n be a one-way LOCC operation satisfying

$$\lim_{n\to\infty} F(\rho^{\otimes n}, \kappa_n(|\Phi_{L_n}\rangle\langle\Phi_{L_n}|)) = 1, \quad \frac{\log \mathrm{CC}(\kappa_n)}{n} \to 0,$$
$$\lim_{n\to\infty} \frac{\log L_n}{n} \le E_c^{-\to}(\rho) + \epsilon$$

for any $\epsilon > 0$. Thus, the application of Lemma 10.4 indicates that there exists a sequence of codes $\{\tilde{\Psi}_n\}$ such that

$$\tilde{\varepsilon}_{\rho^{\otimes n}}(\tilde{\Psi}_n) \to 0, \quad \lim_{n\to\infty} \frac{\log|\tilde{\Psi}_n|}{n} \le E_c^{-\to}(\rho) + \epsilon.$$

Therefore, we obtain

$$R_{g,q}(\rho) \le E_c^{-\to}(\rho).$$

Next, we prove the converse part. For any $\epsilon > 0$, we choose a sequence of codes $\tilde{\Psi}_n = (\mathcal{K}_n, \rho_n', \nu_n)$ such that

$$R \stackrel{\text{def}}{=} \overline{\lim} \frac{1}{n} \log|\tilde{\Psi}_n| \le R_{g,q}(\rho) + \epsilon, \quad \tilde{\varepsilon}_{\rho^{\otimes n}}(\tilde{\Psi}_n) \to 0. \tag{10.52}$$

Then, we have

$$\log|\tilde{\Psi}_n| = \log\dim\mathcal{K}_n \ge H(\mathrm{Tr}_A \rho_n') \stackrel{(a)}{\ge} E_p(\rho_n') \stackrel{(b)}{\ge} E_p(\iota_A \otimes \nu_n(\rho_n')), \tag{10.53}$$

where (a) and (b) follow from Lemma 8.13 and Condition **E2'**, respectively. Since (10.52) yields that

$$F(\rho^{\otimes n}, \iota_A \otimes \nu_n(\rho_n')) \to 1$$

and E_p satisfies Condition **E3** (Exercise 8.52), (10.53) implies that

$$\underline{\lim} \frac{1}{n} \log|\tilde{\Psi}_n| \ge \lim_{n\to\infty} \frac{1}{n} E_p(\rho^{\otimes n}).$$

Hence, using Theorem 8.12, we obtain

$$R_{g,q}(\rho) \geq E_c^{\dashrightarrow}(\rho) = \lim_{n\to\infty} \frac{1}{n} E_p(\rho^{\otimes n}).$$

■

Next, to show the direct part of Theorem 10.4, we prepare the following lemma.

Lemma 10.5 *Let κ be a one-way LOCC operation. There exists a code Ψ such that*

$$\varepsilon_{p,W}(\Psi) \leq (1 - F^2(\tilde{W}_p, \kappa(|\Phi_L\rangle\langle\Phi_L|))) + \frac{1}{2}\|\tilde{W}_p - \kappa(|\Phi_L\rangle\langle\Phi_L|)\|_1, \quad (10.54)$$

$$|\Psi| = L \cdot \mathrm{CC}(\kappa), \quad (10.55)$$

where $\mathrm{CC}(\kappa)$ is the size of the classical communication of κ.

(Note that any two-way LOCC operation can be simulated by one-way LOCC when the initial state is pure [21].) Lemma 10.5 will be shown latter.

Proof of Theorem 10.4 Using Lemma 10.5, we obtain the direct part as follows. Let κ_n be a one-way LOCC operation satisfying

$$\lim_{n\to\infty} F(\tilde{W}_p^{\otimes n}, \kappa_n(|\Phi_{L_n}\rangle\langle\Phi_{L_n}|)) = 1, \quad \frac{\log \mathrm{CC}(\kappa_n)}{n} \to 0,$$

$$\lim_{n\to\infty} \frac{\log L_n}{n} \leq E_c^{\dashrightarrow}(\tilde{W}_p) + \epsilon$$

for any $\epsilon > 0$. Thus, the application of this lemma indicates that there exists a sequence of codes $\{\Psi_n\}$ such that

$$\varepsilon_{p^n,W^{(n)}}(\Psi_n) \to 0, \quad \lim_{n\to\infty} \frac{\log |\Psi_n|}{n} \leq E_c^{\dashrightarrow}(\tilde{W}_p) + \epsilon.$$

Therefore, we obtain

$$R_{V,q}(p,W) \leq E_c^{\dashrightarrow}(\tilde{W}_p).$$

■

Proof of Lemma 10.5 *Construction of code Ψ satisfying* (10.54) *and* (10.55) Assume that the operation κ has the form $\kappa = \sum_i \kappa_{A,i} \otimes \kappa_{B,i}$, where $\{\kappa_{A,i}\}_{i=1}^{l_n}$ is an instrument (a TP-CP-map-valued measure) on $\mathcal{H}_A$ and $\kappa_{B,i}$ is a TP-CP map on $\mathcal{H}_B$ for each i. Define the probability q_x

$$\begin{aligned} q_x &\stackrel{\text{def}}{=} \mathrm{Tr}\, |e_x^A\rangle\langle e_x^A| \otimes I_B \sum_i \kappa_{A,i} \otimes \kappa_{B,i}(|\Phi_L\rangle\langle\Phi_L|) \quad (10.56)\\ &= \sum_i \mathrm{Tr}\, \kappa_{A,i}^*(|e_x^A\rangle\langle e_x^A|) \otimes I_B(|\Phi_L\rangle\langle\Phi_L|), \end{aligned}$$

the probability $p_{i,x}$, and the state $\rho_{i,x}$ as

$$p_{i,x} \stackrel{\text{def}}{=} \frac{\operatorname{Tr} \kappa_{A,i}^*(|e_x^A\rangle\langle e_x^A|) \otimes I_B(|\Phi_L\rangle\langle\Phi_L|)}{q_x}$$

$$\rho_{i,x} \stackrel{\text{def}}{=} \frac{\operatorname{Tr}_A \kappa_{A,i}^*(|e_x^A\rangle\langle e_x^A|) \otimes I_B(|\Phi_L\rangle\langle\Phi_L|)}{q_x p_{i,x}}.$$

Now we construct the coding protocol Ψ. When the encoder receives the input signal x, he sends the state $\rho_{i,x}$ with the probability $p_{i,x}$ and sends the classical information i. The decoder performs the TP-CP map $\kappa_{B,i}$ depending on the classical signal i. Then, Inequality (10.55) follows from this construction. Also, Inequality (10.54) holds under this construction of Ψ, as shown below.

Proof of (10.54) First, we have the following inequality:

$$\begin{aligned}
& F^2(\tilde{W}_p, \kappa(|\Phi_L\rangle\langle\Phi_L|)) \\
=& F^2\left(\sum_x p_x |e_x^A\rangle\langle e_x^A| \otimes W_x, \sum_i \kappa_{A,i} \otimes \kappa_{B,i}(|\Phi_L\rangle\langle\Phi_L|)\right) \\
\overset{(a)}{\le}& \operatorname{Tr} \sqrt{\sum_x p_x |e_x^A\rangle\langle e_x^A| \otimes W_x} \sqrt{\sum_i \kappa_{A,i} \otimes \kappa_{B,i}(|\Phi_L\rangle\langle\Phi_L|))} \\
=& \operatorname{Tr} \sum_x \sqrt{p_x} |e_x^A\rangle\langle e_x^A| \otimes \sqrt{W_x} \sqrt{\sum_i \kappa_{A,i} \otimes \kappa_{B,i}(|\Phi_L\rangle\langle\Phi_L|)} \\
=& \sum_x \sqrt{p_x} \operatorname{Tr}_B \sqrt{W_x} (\operatorname{Tr}_A |e_x^A\rangle\langle e_x^A| \otimes I_B \sqrt{\sum_i \kappa_{A,i} \otimes \kappa_{B,i}(|\Phi_L\rangle\langle\Phi_L|))} \\
\overset{(b)}{\le}& \sum_x \sqrt{p_x} \operatorname{Tr}_B \sqrt{W_x} \sqrt{(\operatorname{Tr}_A |e_x^A\rangle\langle e_x^A| \otimes I_B \sum_i \kappa_{A,i} \otimes \kappa_{B,i}(|\Phi_L\rangle\langle\Phi_L|))} \\
\overset{(c)}{=}& \sum_x \sqrt{p_x q_x} \operatorname{Tr}_B \sqrt{W_x} \sqrt{\sum_i p_{i,x} \kappa_{B,i}(\rho_{i,x})} \\
=& \sum_x p_x \operatorname{Tr}_B \sqrt{W_x} \sqrt{\sum_i p_{i,x} \kappa_{B,i}(\rho_{i,x})} \\
& + \sum_x \left(\sqrt{p_x q_x} - p_x\right) \operatorname{Tr}_B \sqrt{W_x} \sqrt{\sum_i p_{i,x} \kappa_{B,i}(\rho_{i,x})},
\end{aligned} \tag{10.57}$$

where (a) follows from a basic inequality $F^2(\rho, \sigma) \le \operatorname{Tr} \sqrt{\rho}\sqrt{\sigma}$, and (b) follows from Exercise 1.26 and Condition ② of Theorem A.1 because $\sqrt{t}$ is matrix concave. Equation (c) follows from

$$
\begin{aligned}
&q_x \sum_i p_{i,x} \kappa_{B,i}(\rho_{i,x}) \\
=&\sum_i \kappa_{B,i}(\mathrm{Tr}_A(\kappa_{A,i}^*(|e_x^A\rangle\langle e_x^A|) \otimes I_B)|\Phi_L\rangle\langle\Phi_L|) \\
=&\sum_i \kappa_{B,i}(\mathrm{Tr}_A(|e_x^A\rangle\langle e_x^A| \otimes I_B)(\kappa_{A,i} \otimes \iota_B)(|\Phi_L\rangle\langle\Phi_L|)) \\
=&\mathrm{Tr}_A\, |e_x^A\rangle\langle e_x^A| \otimes I_B \sum_i \kappa_{A,i} \otimes \kappa_{B,i}(|\Phi_L\rangle\langle\Phi_L|).
\end{aligned}
$$

Then, we have

$$
\begin{aligned}
\tilde{\varepsilon}_{p,W}(\Psi) =& 1 - \sum_x p_x F(\sqrt{W_x}, \sqrt{\sum_i p_{i,x} \kappa_{B,i}(\rho_{i,x})}) \\
\le& 1 - \sum_x p_x \mathrm{Tr}_B \sqrt{W_x} \sqrt{\sum_i p_{i,x} \kappa_{B,i}(\rho_{i,x})} \\
\overset{(a)}{\le}& (1 - F^2(\tilde{W}_p, \kappa|\Phi_L\rangle\langle\Phi_L|)) \\
&+ \sum_x (\sqrt{p_x q_x} - p_x) \mathrm{Tr}_B \sqrt{W_x} \sqrt{\sum_i p_{i,x} \kappa_{B,i}(\rho_{i,x})}.
\end{aligned}
\tag{10.58}
$$

where (a) follows from (10.57).

Further, the RHS of the second term of (10.58) is evaluated by

$$
\begin{aligned}
&\sum_x (\sqrt{p_x q_x} - p_x) \mathrm{Tr}_B \sqrt{W_x} \sqrt{\sum_i p_{i,x} \kappa_{B,i}(\rho_{i,x})} \\
\le& \sum_x (\sqrt{p_x q_x} - p_x)_+ = \sum_x \Big(\sqrt{\frac{q_x}{p_x}} - 1\Big)_+ p_x \le \sum_x \Big(\frac{q_x}{p_x} - 1\Big)_+ p_x \\
=& \sum_x (q_x - p_x)_+ = \frac{1}{2}\|q - p\|_1 \le \frac{1}{2}\|\tilde{W}_p - \kappa(|\Phi_L\rangle\langle\Phi_L|)\|_1,
\end{aligned}
\tag{10.59}
$$

where $(t)_+$ is t when t is positive and 0 otherwise. The final inequality follows from the definition of distribution q (10.56). Hence, (10.54) follows from (10.58) and (10.59). ∎

10.7 Compression with Classical Memory

In the previous section, we treated visible compression with quantum memory. In this section, we consider the compression rate with classical memory. This problem was first discussed by Hayden et al. [12]. In this problem, when the state W_x is to be sent to the decoder, the **encoder** is given by a stochastic transition matrix Q with the input system $\mathcal{X}$ and the output system $\{1, \ldots, M\}$. The **decoder** is represented by a c-q channel $\{W_i'\}_{i=1}^M$ with the output system $\mathcal{H}$. Hence, our code in this problem is given by the triplet $\Psi_c \stackrel{\text{def}}{=} (M, Q, W')$, which can be regarded as a code in the visible scheme. Then, the optimal compression rate is defined as[3]

$$R_{V,c}(p, W) \stackrel{\text{def}}{=} \inf_{\{\Psi_c^{(n)}\}} \left\{ \overline{\lim} \frac{1}{n} \log |\Psi_c^{(n)}| \, \middle| \, \varepsilon_{p^n, W^{(n)}}(\Psi_c^{(n)}) \to 0 \right\}. \tag{10.60}$$

Clearly, the inequality

$$R_{V,c}(p, W) \geq R_{V,q}(p, W)$$

holds.

Theorem 10.6 *[45]*

$$R_{V,c}(p, W) = C(\tilde{W}_p) = C_c(\tilde{W}_p). \tag{10.61}$$

Note that the quantities $C(\tilde{W}_p)$ and $C_c(\tilde{W}_p)$ are defined in Sect. 8.11.

Similar to the previous section, we can consider the bipartite state generation via classical channel. This task can be formulated by restricting the channel to the classical channel in the bipartite state generation. That is, an operation $\tilde{\Psi} = (\mathcal{K}, rho', \nu)$ for the bipartite state generation can be regarded as an operation for the bipartite state generation via classical channel when $\rho' = \sum_i P_i \rho' P_i$, where $P_i := I_A \otimes |i\rangle\langle i|$ and the CONS $\{|i\rangle\}$ spans the space $\mathcal{K}$. In this case, the state ρ' is written as $\sum_{i=1}^M \rho_i^{A'} \otimes |i\rangle\langle i|$, where $M = \dim \mathcal{K}$. So, we write the operation by the triple $\tilde{\Psi}_c = (M, \rho', \nu)$.

Similar to the previous section, when ρ is given as $\tilde{W}_p$, the code $\Psi_c = (M, Q, \nu)$ for visible compression with classical memory is converted to the code $\tilde{\Psi}_c = (M, \rho', \nu)$ for bipartite state generation, where ρ' is given as follows.

$$\rho' := \sum_{i=1}^{M} Q_i^x \sum_x p_x |e_x^A\rangle\langle e_x^A| \otimes |i\rangle\langle i|. \tag{10.62}$$

In this correspondence,

$$\tilde{\varepsilon}_{p,W}(\Psi_c) = \tilde{\varepsilon}_{\tilde{W}_p}(\tilde{\Psi}_c). \tag{10.63}$$

Hence, we have

$$\min_{\Psi_c}\{\tilde{\varepsilon}_{p,W}(\Psi_c) || \Psi_c| = M\} \geq \min_{\tilde{\Psi}_c}\{\varepsilon_{\tilde{W}_p}(\tilde{\Psi}_c) || \tilde{\Psi}_c| = M\}. \tag{10.64}$$

---

[3]The subscript c denotes classical memory.

For the bipartite state generation via classical channel, we define the following quantity:

$$R_{g,c}(\rho) \stackrel{\text{def}}{=} \inf_{\{\tilde{\Psi}_c^{(n)}\}} \left\{ \overline{\lim} \frac{1}{n} \log |\tilde{\Psi}_c^{(n)}| \,\middle|\, \tilde{\varepsilon}_{\rho^{\otimes n}}(\tilde{\Psi}_c^{(n)}) \to 0 \right\}. \tag{10.65}$$

Then, the following theorem holds.

Theorem 10.7

$$R_{g,c}(\rho) = C(\rho) = C_c(\rho). \tag{10.66}$$

Before proving Theorem 10.6, we show Theorem 10.7. Due to (10.64), Theorem 10.7 yields the converse type inequality $R_{V,c}(p, W) \geq C(\tilde{W}_p)$. So, after the proof of Theorem 10.7, we show the direct part of Theorem 10.6.

The direct part of Theorem 10.7 essentially is obtained by the following lemma.

Lemma 10.6 *Given M states ρ_i^A on $\mathcal{H}_A$ and M states ρ_i^B on $\mathcal{H}_B$ for $i = 1, \ldots, M$, There exists a code $\tilde{\Psi}_c$ such that*

$$\tilde{\varepsilon}_\rho(\tilde{\Psi}_c) \leq \frac{1}{2} \left\| \frac{1}{M} \sum_{i=1}^{M} \rho_i^A \otimes \rho_i^B - \rho \right\|_1, \tag{10.67}$$

$$|\tilde{\Psi}_c| = M. \tag{10.68}$$

This lemma can be shown as follows. Firstly, Alice prepares the random variable X subject to the uniform distribution on $\{1, \ldots, M\}$, and sends it to Bob via noiseless classical channel. Then, when the random variable is i, Alice and Bob generate the state ρ_i^A and ρ_i^B, respectively. Due to (3.48), this operation satisfies the condition for the above operation $\tilde{\Phi}_c$.

Proof of Theorem 10.7 First, we prove the direct part. Lemma 10.6 and the definition (8.198) of $C_c(\rho)$ yield that

$$R_{g,c}(\rho) \leq C_c(\rho)$$

Next, we prove the converse part. Now, we prove the converse inequality. For any $\epsilon > 0$, we choose a sequence of codes $\tilde{\Psi}_c^{(n)} = (M_n, \rho_n', W'^{(n)})$ such that

$$\overline{\lim} \frac{1}{n} \log |\tilde{\Psi}_c^{(n)}| \leq R_{g,c}(\rho) + \epsilon,$$

$$F\Big(\sum_{i=1}^{M_n} \rho_{i,n}^{A\,\prime} \otimes W_i'^{(n)}, \rho^{\otimes n}\Big) \to 1,$$

where $\rho_n' = \sum_{i=1}^{M_n} \rho_{i,n}^{A\,\prime} \otimes |i\rangle\langle i|$. Hence,

$$\left\| \sum_{i=1}^{M_n} \rho_{i,n}^{A\,\prime} \otimes W'^{(n)}_i, \rho^{\otimes n} \right\|_1 \to 0.$$

The definition (8.198) of $C_c(\rho)$ implies that

$$\overline{\lim} \frac{1}{n} \log |\tilde{\Psi}_c^{(n)}| = \overline{\lim} \frac{1}{n} \log M_n \geq C_c(\rho), \tag{10.69}$$

which implies that

$$R_{g,c}(\rho) \geq C_c(\rho).$$

■

Next, to show the direct part of Theorem 10.6, we prepare the following lemma.

Lemma 10.7 *Given M states ρ_i^A on $\mathcal{H}_A$ and M states ρ_i^B on $\mathcal{H}_B$ for $i = 1, \ldots, M$, There exists a code Ψ_c such that*

$$\tilde{\varepsilon}_\rho(\Psi_c) \leq \left\| \frac{1}{M} \sum_{i=1}^{M} \rho_i^A \otimes \rho_i^B - \tilde{W}_p \right\|_1, \tag{10.70}$$

$$|\Psi_c| = M. \tag{10.71}$$

Lemma 10.7 will be shown latter.

Proof of Theorem 10.6 Using Lemma 10.7, we obtain the direct part as follows. Lemma 10.6 and the definition (8.198) of $C_c(\rho)$ yield that

$$R_{V,c}(p, W) \leq C_c(\tilde{W}_p)$$

■

Proof of Lemma 10.7 *Construction of code Ψ_c satisfying* (10.70) *and* (10.71). Firstly, we define the distribution q on $\mathcal{X}$:

$$q_x \stackrel{\text{def}}{=} \sum_{i=1}^{M} \frac{1}{M} \langle e_x^A | \rho_i^A | e_x^A \rangle.$$

Next, we define the encoder Q_x and the decoder W_i' as

$$Q_x(i) := \frac{1}{M} \frac{\langle e_x^A | \rho_i^A | e_x^A \rangle}{q_x}, \quad W_i' := \rho_i^B.$$

Then, Inequality (10.51) follows from this construction. Also, Inequality (10.50) holds under this construction of Ψ_c, as shown below.

Proof of (10.70) The recovered state of the above operation is

$$\tilde{\rho} \stackrel{\text{def}}{=} \sum_{x \in \mathcal{X}} p_x |e_x^A\rangle\langle e_x^A| \otimes \sum_{i=1}^{M} \frac{1}{M} Q_x(i) W_i',$$
$$= \sum_{x \in \mathcal{X}} p_x |e_x^A\rangle\langle e_x^A| \otimes \sum_{i=1}^{M} \frac{1}{M} \frac{\langle e_x^A|\rho_i^A|e_x^A\rangle}{q_x} \rho_i^B.$$

Now, we introduce another state:

$$\tilde{\rho}' \stackrel{\text{def}}{=} \sum_{x \in \mathcal{X}} q_x |e_x^A\rangle\langle e_x^A| \otimes \sum_{i=1}^{M} \frac{1}{M} \frac{\langle e_x^A|\rho_i^A|e_x^A\rangle}{q_x} \rho_i^B.$$

Then, applying the monotonicity of a trace norm to a partial trace and the pinching of PVM $\{|e_x^A\rangle\langle e_x^A|\}$, we have

$$\|q - p\|_1 \leq \left\|\tilde{\rho}' - \tilde{W}_p\right\|_1 \leq \left\|\frac{1}{M}\sum_{i=1}^{M} \rho_i^A \otimes \rho_i^B - \tilde{W}_p\right\|_1.$$

Hence,

$$\left\|\tilde{\rho} - \tilde{W}_p\right\|_1 \leq \|\tilde{\rho} - \tilde{\rho}'\|_1 + \left\|\tilde{\rho}' - \tilde{W}_p\right\|_1 = \|q - p\|_1 + \left\|\tilde{\rho}' - \tilde{W}_p\right\|_1.$$

Thus, (3.48) implies (10.70). ∎

10.8 Compression with Shared Randomness

Next, we consider the case when the sender and decoder share a common random number a priori. In this problem, the **encoder** is given by a c-q channel T_X with the input system $\mathcal{X}$ and the output system $\mathcal{K}$, where T_X depends on the common random number X. The **decoder** is represented by a TP-CP map ν_X from $\mathcal{K}$ to $\mathcal{H}$, which also depends on the common random number X. Hence, our code in this problem is given by the triplet $\Psi_r \stackrel{\text{def}}{=} (\mathcal{K}, T_X, \nu_X)$. The error is defined as

$$\varepsilon_{p,W}(\Psi_r) \stackrel{\text{def}}{=} \mathrm{E}_X \sum_{x \in \mathcal{X}} p_x \left(1 - F^2(W_x, \nu_X \circ T_X(x))\right).$$

Further, when the storage is a classical system, the problem is modified as follows. That is, the **encoder** is given by a stochastic transition matrix Q_X with the input system $\mathcal{X}$ and the output system $\{1, \ldots, M\}$, where Q_X depends on the common random number. The **decoder** is represented by a c-q channel $\{W'_{X,i}\}_{i=1}^M$ with the output system $\mathcal{H}$. The c-q channel W'_X also depends on the common random number. Hence, our code is given by the triplet $\Psi_{c,r} \stackrel{\rm def}{=} (M, Q_X, W'_X)$.[4] Then, these optimal compression rates are defined as

$$R_{V,q,r}(p, W) \stackrel{\rm def}{=} \inf_{\{\Psi_r^{(n)}\}} \left\{ \overline{\lim} \frac{1}{n} \log |\Psi_r^{(n)}| \,\middle|\, \varepsilon_{p^n, W^{(n)}}(\Psi_r^{(n)}) \to 0 \right\}, \tag{10.72}$$

$$R_{V,c,r}(p, W) \stackrel{\rm def}{=} \inf_{\{\Psi_{c,r}^{(n)}\}} \left\{ \overline{\lim} \frac{1}{n} \log |\Psi_{c,r}^{(n)}| \,\middle|\, \varepsilon_{p^n, W^{(n)}}(\Psi_{c,r}^{(n)}) \to 0 \right\}. \tag{10.73}$$

Clearly, we have

$$R_{V,c}(p, W) \ge R_{V,c,r}(p, W) \ge R_{V,q,r}(p, W), \quad R_{V,q}(p, W) \ge R_{V,q,r}(p, W). \tag{10.74}$$

Lemma 10.8

$$R_{V,q,r}(p, W) \ge I(p, W) = I_{\tilde{W}_p}(A : B) = C_d^{A\to B}(\tilde{W}_p). \tag{10.75}$$

Proof Let $\Psi_r^{(n)} \stackrel{\rm def}{=} (\mathcal{K}_n, T_X^{(n)}, \nu_X^{(n)})$ be a sequence of codes achieving the optimal rate $R_{V,q,r}(p, W)$. Defining the bipartite state $\rho_n \stackrel{\rm def}{=} \sum_{x\in\mathcal{X}^n} p_x^n |e_x^A\rangle\langle e_x^A| \otimes (\mathrm{E}_X \nu_X T_X^{(n)}(x))$, we have

$$\begin{aligned}
\log |\Psi_r^{(n)}| \ge & \mathrm{E}_X H\left(\sum_{x\in\mathcal{X}^n} p_x^n T_X^{(n)}(x) \right) - \sum_{x\in\mathcal{X}^n} p_x^n H(T_X^{(n)}(x)) \\
= & \mathrm{E}_X \sum_{x\in\mathcal{X}^n} p_x^n D\Big(T_X^{(n)}(x) \Big\| \sum_{x\in\mathcal{X}^n} p_x^n T_X^{(n)}(x) \Big) \\
\stackrel{(a)}{\ge} & \mathrm{E}_X \sum_{x\in\mathcal{X}^n} p_x^n D\Big(\nu_X T_X^{(n)}(x) \Big\| \sum_{x\in\mathcal{X}^n} p_x^n \nu_X T_X^{(n)}(x) \Big) \\
\stackrel{(b)}{\ge} & \sum_{x\in\mathcal{X}^n} p_x^n D\Big(\mathrm{E}_X \nu_X T_X^{(n)}(x) \Big\| \mathrm{E}_X \sum_{x\in\mathcal{X}^n} p_x^n \nu_X T_X^{(n)}(x) \Big) \\
= & H\left(\sum_{x\in\mathcal{X}^n} p_x^n \mathrm{E}_X \nu_X T_X^{(n)}(x) \right) - \sum_{x\in\mathcal{X}^n} p_x^n H(\mathrm{E}_X \nu_X T_X^{(n)}(x)) = I_{\rho_n}(A : B),
\end{aligned}$$

where (a) and (b) follow from the monotonicity (5.36) and joint convexity (5.38) of quantum relative entropy, respectively. From the choice of our code we can check that $F(\rho_n, \tilde{W}_p^{\otimes n}) \to 0$. Hence, using Fannes inequality (Theorem (5.12), we have

[4] The last subscript r denotes shared "randomness."

$$\liminf_{n\to\infty} \frac{1}{n} \log |\Psi_r^{(n)}| \geq \liminf_{n\to\infty} \frac{1}{n} I_{\rho_n}(A:B) = \liminf_{n\to\infty} \frac{1}{n} I_{\tilde{W}_p^{\otimes n}}(A:B) = I(p, W).$$

In fact, these optimal rates are calculated in the classical case. ■

Theorem 10.8 (Bennett et al. [22], Dür et al. [23]) *If all of states W_x are commutative, we have*

$$R_{V,q,r}(p, W) = R_{V,c,r}(p, W) = I(p, W) = C_d^{A\to B}(\tilde{W}_p). \tag{10.76}$$

Hence, from (10.74) we have

$$R_{V,c}(p, W) \geq R_{V,q}(p, W) \geq R_{V,q,r}(p, W) = R_{V,c,r}(p, W). \tag{10.77}$$

Proof From Theorem 10.6 it is sufficient to show the inequality $R_{V,c,r}(p, W) \leq I(p, W)$. Now we construct a protocol achieving the rate $R = I(p, W) + \epsilon$ for any $\epsilon > 0$ by employing Lemma 10.9 given below. First, we apply Lemma 10.9 to the case with $\mathrm{P}_{Y|X} = W^{(n)}$, $\mathrm{P}_X = p^n$, $M = e^{nR}$, and $\delta = e^{-nr}$. Then, there exists a code $\Psi_{c,r}^{(n)}$ such that

$$\bar{\varepsilon}_{p^n, W^{(n)}}(\Psi_{c,r}^{(n)}) \leq e^{-nr} + \sum_{\boldsymbol{x}} p^n(\boldsymbol{x}) W_{\boldsymbol{x}}^{(n)} \{W_{\boldsymbol{x}}^{(n)} \geq e^{n(R-r)} W_{p^n}^{(n)}\}.$$

In the above discussion, we take the expectation for $\boldsymbol{x}$ under the distribution $\boldsymbol{x}$. Applying (2.168) to $X = \log \frac{W_x(y)}{W_p(y)}$, we have

$$\begin{aligned}\sum_{\boldsymbol{x}} p^n(\boldsymbol{x}) W_{\boldsymbol{x}}^{(n)} \{W_{\boldsymbol{x}}^{(n)} \geq e^{n(R-r)} W_{p^n}^{(n)}\} \leq & e^{ns(D_{1+s}(p\times W \| p\otimes W_p) - R + r)} \\ = & e^{ns(I_{1+s}(p,W) - R + r)}, \quad \text{for } \forall s \geq 0.\end{aligned}$$

Hence,

$$\bar{\varepsilon}_{p^n, W^{(n)}}(\Psi_{c,r}^{(n)}) \leq e^{-nr} + e^{ns(I_{1+s}(p,W) - R + r)}, \quad \text{for } \forall s \geq 0. \tag{10.78}$$

Due to Exercise 10.2, Inequality (10.78) implies that the rate $R = I(p, W) + \epsilon$ for any $\epsilon > 0$ is achievable. Hence, we obtain (10.76). ■

Lemma 10.9 *For any classical source $\{(\mathrm{P}_X(x), \mathrm{P}_{Y|X}(y|x))\}$, there exists a code $\Psi_{c,r} \stackrel{\text{def}}{=} (M, Q_X, Q'_X)$ such that*

$$\frac{1}{2} \sum_y \left| \mathrm{E}_X \sum_{i=1}^M (Q'_X)_y^i (Q_X)_i^x - p(y|x) \right| \leq \delta + \sum_{y: \frac{1}{M} \frac{\mathrm{P}_{Y|X}(y|x)}{\mathrm{P}_Y(y)} \geq \delta} \mathrm{P}_{Y|X}(y|x) \tag{10.79}$$

for $1 > \forall \delta > 0$, where $\mathrm{P}_Y(y) \stackrel{\text{def}}{=} \sum_x \mathrm{P}_X(x) \mathrm{P}_{Y|X}(y|x)$.

Proof of Theorem 10.9 First, the encoder and the decoder prepare the M i.i.d. common random numbers $Y_1, \ldots, Y_M$ subject to $\mathrm{P}_Y(y) \stackrel{\mathrm{def}}{=} \sum_x \mathrm{P}_{Y|X}(y|x)\mathrm{P}_X(x)$. When the encoder receives the original message x, he sends the signal i obeying the distribution $P(i) \stackrel{\mathrm{def}}{=} \frac{\mathrm{P}_{X|Y}(x|Y_i)}{\mathrm{P}_{X|Y}(x|Y_1)+\ldots+\mathrm{P}_{X|Y}(x|Y_M)}$, where $\mathrm{P}_{X|Y}(x|y) \stackrel{\mathrm{def}}{=} \mathrm{P}_{Y|X}(y|x)\frac{\mathrm{P}_X(x)}{\mathrm{P}_Y(y)}$. The receiver recovers $y = Y_i$ when he receives i.

In this protocol, when the original message is x, the probability of $i = 1$ and $y = Y_1$ is given as

$$\mathrm{E}_{Y_1,\ldots,Y_M} \frac{\mathrm{P}_Y(y)\mathrm{P}_{X|Y}(x|y)}{\mathrm{P}_{X|Y}(x|y) + \mathrm{P}_{X|Y}(x|Y_2) + \ldots + \mathrm{P}_{X|Y}(x|Y_M)}.$$

Hence, the recovered signal is equal to y with the probability

$$\mathrm{E}_{Y_1,\ldots,Y_M} \frac{M\mathrm{P}_Y(y)\mathrm{P}_{X|Y}(x|y)}{\mathrm{P}_{X|Y}(x|y) + \mathrm{P}_{X|Y}(x|Y_2) + \ldots + \mathrm{P}_{X|Y}(x|Y_M)}.$$

Thus, since $\frac{1}{x}$ is concave,

$$\begin{aligned}
&\frac{1}{2}\sum_y \left| \mathrm{P}_{Y|X}(y|x) - \mathrm{E}_{Y_1,\ldots,Y_M} \frac{M\mathrm{P}_Y(y)\mathrm{P}_{X|Y}(x|y)}{\mathrm{P}_{X|Y}(x|y) + \sum_{i=2}^M \mathrm{P}_{X|Y}(x|Y_i)} \right| \\
=& \sum_y \left(\mathrm{P}_{Y|X}(y|x) - \mathrm{E}_{Y_1,\ldots,Y_M} \frac{M\mathrm{P}_Y(y)\mathrm{P}_{X|Y}(x|y)}{\mathrm{P}_{X|Y}(x|y) + \sum_{i=2}^M \mathrm{P}_{X|Y}(x|Y_i)} \right)_+ \\
\overset{(a)}{\le}& \sum_y \left(\mathrm{P}_{Y|X}(y|x) - \frac{M\mathrm{P}_Y(y)\mathrm{P}_{X|Y}(x|y)}{\mathrm{P}_{X|Y}(x|y) + (M-1)\mathrm{P}_X(x)} \right)_+ \\
\overset{(b)}{=}& \sum_y \mathrm{P}_{Y|X}(y|x) \left(\frac{\frac{1}{M}\left(\frac{\mathrm{P}_{Y|X}(y|x)}{\mathrm{P}_Y(y)} - 1\right)}{1 + \frac{1}{M}\left(\frac{\mathrm{P}_{Y|X}(y|x)}{\mathrm{P}_Y(y)} - 1\right)} \right)_+ \\
\overset{(c)}{\le}& \sum_{y: 0 \le \frac{1}{M}(\frac{\mathrm{P}_{Y|X}(y|x)}{\mathrm{P}_Y(y)} - 1) \le 1} \mathrm{P}_{Y|X}(y|x) \left(\frac{1}{M}\left(\frac{\mathrm{P}_{Y|X}(y|x)}{\mathrm{P}_Y(y)} - 1\right)\right) \\
&+ \sum_{y: \frac{1}{M}(\frac{\mathrm{P}_{Y|X}(y|x)}{\mathrm{P}_Y(y)} - 1) > 1} \mathrm{P}_{Y|X}(y|x) \\
\le& \delta + \sum_{y: \frac{1}{M}(\frac{\mathrm{P}_{Y|X}(y|x)}{\mathrm{P}_Y(y)} - 1) \ge \delta} \mathrm{P}_{Y|X}(y|x) \\
\le& \delta + \sum_{y: \frac{1}{M}\frac{\mathrm{P}_{Y|X}(y|x)}{\mathrm{P}_Y(y)} \ge \delta} \mathrm{P}_{Y|X}(y|x),
\end{aligned}$$

where (a) and (b) follow from Jensen inequality and Exercise 10.6, respectively and (c) follows from Exercise 10.5 and $\frac{1}{M}(\frac{\mathrm{P}_{Y|X}(y|x)}{\mathrm{P}_Y(y)} - 1) \ge -\frac{1}{M} > -1$. ∎

Exercises

10.5 Prove the inequality $\frac{x}{1+x} \le \min\{x, 1\}$ for any real number $x \ge 0$.

10.6 Show that

$$\mathrm{P}_{Y|X}(y|x) - \frac{M\mathrm{P}_Y(y)\mathrm{P}_{X|Y}(x|y)}{\mathrm{P}_{X|Y}(x|y) + (M-1)\mathrm{P}_X(x)} = \mathrm{P}_{Y|X}(y|x)\frac{\frac{1}{M}\left(\frac{\mathrm{P}_{Y|X}(y|x)}{\mathrm{P}_Y(y)} - 1\right)}{1 + \frac{1}{M}\left(\frac{\mathrm{P}_{Y|X}(y|x)}{\mathrm{P}_Y(y)} - 1\right)}.$$

10.7 Prove (10.40).

10.8 Consider the bipartite state generation via channel with shared randomness. Our code is given by the triplet $\tilde{\Psi}_r \stackrel{\text{def}}{=} (\mathcal{K}, \rho'_X, \nu_X)$, where ρ'_X is the generated state dependently of the common randomness X and ν_X is the decoder dependently of the common randomness X. When the target bipartite state is ρ, we surpass the condition $\mathrm{Tr}_K \rho'_X = \mathrm{Tr}_B \rho$ for any value of the common randomness X. Then, we can define the size $|\tilde{\Psi}_r|$ and the error $\tilde{\varepsilon}_\rho(\tilde{\Psi}_r)$ for the target bipartite state ρ in the same way. Then, these optimal compression rates are defined as

$$R_{g,q,r}(\rho) \stackrel{\text{def}}{=} \inf_{\{\tilde{\Psi}_r^{(n)}\}} \left\{ \overline{\lim} \frac{1}{n} \log |\tilde{\Psi}_r^{(n)}| \,\middle|\, \varepsilon_{\rho^{\otimes n}}(\tilde{\Psi}_r^{(n)}) \to 0 \right\}. \tag{10.80}$$

Show that

$$R_{g,q,r}(\rho) \ge I_\rho(A:B). \tag{10.81}$$

10.9 Relation to Channel Capacities

In the previous section, we have discussed the simulation of c-q channel by the pair of classical noiseless memory and shared randomness. This section discusses the relation between this simulation problem and c-q channel coding.

In the above discussion, we consider the average error under the prior distribution on the input system. Sometimes it is suitable to treat the worst error with respect to the input signal as follows:

$$C^R_{c,r}(W) \stackrel{\text{def}}{=} \inf_{\{\Psi_{c,r}^{(n)}\}} \left\{ \overline{\lim} \frac{1}{n} \log |\Psi_{c,r}^{(n)}| \,\middle|\, \tilde{\varepsilon}_{W^{(n)}}(\Psi_{c,r}^{(n)}) \to 0 \right\}, \tag{10.82}$$

$$\tilde{\varepsilon}_W(\Psi_{c,r}) \stackrel{\text{def}}{=} \max_{x \in \mathcal{X}} 1 - F^2(W_x, \mathrm{E}_X(W'_X)_i(Q_X)^x_i).$$

This problem is called reverse Shannon theorem, and is closely related to c-q channel coding. To discuss this relation, we focus on the c-q channel capacity with shared randomness:

$$C_{c,r}(W) \stackrel{\text{def}}{=} \inf_{\{\Phi_X^{(n)}\}} \left\{ \overline{\lim} \frac{1}{n} \log |\Phi_X^{(n)}| \,\middle|\, \mathrm{E}_X \varepsilon_{W^{(n)}}[\Phi_X^{(n)}] \to 0 \right\}, \tag{10.83}$$

where Φ_X is a c-q channel code randomly chosen by the random number X shared by the sender and the receiver, and is written as the triplet (M, φ_X, Y_X). Here, we assume that the size of codes Φ_X does not depend on the shared random number.

The difference of $C_{c,r}(W)$ from the conventional c-q channel capacity $C_c(W)$ is allowing the use of the shared randomness X. Hence, we have

$$C_c(W) \le C_{c,r}(W). \tag{10.84}$$

However, we can show the equation because for any code Φ_X with shared randomness, there exists a code Φ such that $\varepsilon_W[\Phi] \le \varepsilon_W[\Phi_X]$.

$$C_c(W) = C_{c,r}(W). \tag{10.85}$$

Since the two capacities $C_{c,r}(W)$ and $C_{c,r}^R(W)$ allow use of the shared randomness X, the relation is similar to that between entanglement distillation and dilution, in which, dilution corresponds to $C_{c,r}^R(W)$, distillation corresponds to $C_{c,r}(W)$, and maximally entangled states correspond to noiseless channels. Hence, we can show

$$C_{c,r}(W) \le C_{c,r}^R(W).$$

The first inequality above follows from the comparison between the definitions of $C_c(W)$ and $C_{c,r}(W)$. As shown in Theorem 4.1, we have

$$\max_p I(p, W) = C_c(W).$$

Further, the following theorem holds.

Theorem 10.9 (Bennett et al. [22]) *When all of states W_x are commutative,*

$$C_c(W) = C_{c,r}(W) = C_{c,r}^R(W) = \max_p I(p, W).$$

Proof It is sufficient to show $C_{c,r}^R(W) \le \max_p I(p, W)$ for this theorem. This inequality follows from the proof of Theorem 10.8 with the distribution $p = \operatorname{argmax}_p I(p, W)$. ■

Further, we can define $C_{c,e}(W)$ and $C_{c,e}^R(W)$ by replacing the shared randomness by the shared entanglement.[5] Since the shared randomness can be generated from the shared entanglement,

[5]The last subscript e denotes the shared "entanglement" while the superscript e denotes entangled input.

$$C_{c,r}(W) \le C_{c,e}(W) \le C^R_{c,e}(W) \le C^R_{c,r}(W). \tag{10.86}$$

Hence, when all of states W_x are commutative,

$$C(W) = C_{c,r}(W) = C_{c,e}(W) = C^R_{c,e}(W) = C^R_{c,r}(W) = \max_p I(p, W).$$

When W is replaced by a q-q channel κ, we can consider the simulation of the output states of entangled inputs. Considering this requirement, we can define the capacity $C^{e,R}_{c,e}(\kappa)$ as the reverse capacity of $C^e_{c,e}(\kappa)$ [22, 24]. This capacity can be regarded as the capacity of teleportation through a noisy channel κ. Similarly, we have

$$C^e_{c,e}(\kappa) \le C^{e,R}_{c,e}(\kappa).$$

Recall our treatment of $C^e_{c,e}(\kappa)$ in Sect. 9.3. Originally, the reverse capacities $C^R_{c,e}(W)$ and $C^{e,R}_{c,e}(\kappa)$ were introduced for proving the converse part of $C^e_{c,e}(\kappa)$ by Bennett et al. [22]. They proved the equation $C^e_{c,e}(\kappa^{\mathrm{GP}}_p) = \max_\rho I(\rho, \kappa^{\mathrm{GP}}_p)$ for the generalized Pauli channel κ^{GP}_p by showing the two inequalities

$$\begin{aligned} C^e_{c,e}(\kappa^{\mathrm{GP}}_p) &\ge \max_\rho I(\rho, \kappa^{\mathrm{GP}}_p), \\ C^{e,R}_{c,e}(\kappa^{\mathrm{GP}}_p) &\le \max_\rho I(\rho, \kappa^{\mathrm{GP}}_p) = \log d - H(p), \end{aligned} \tag{10.87}$$

where d is the dimension of the system. They also conjectured [24]

$$C^e_{c,e}(\kappa) = C^{e,R}_{c,e}(\kappa) = \max_\rho I(\rho, \kappa).$$

In addition, when W is a q-c channel, i.e., a POVM, this problem was solved as the compression of POVM by Winter [25] and Massar and Winter [26].

In the same way, we can define $C^e_{c,r}(\kappa)$ and $C^{e,R}_{c,r}(\kappa)$. Then, in the same way as (10.85) and (10.86), we can show that

$$C^e_c(\kappa) = C^e_{c,r}(\kappa) \le C^e_{c,e}(\kappa) \le C^{e,R}_{c,e}(\kappa) \le C^{e,R}_{c,r}(\kappa). \tag{10.88}$$

However, $C^{e,R}_{c,r}(\kappa)$ is infinity when κ is not entangled-breaking because this capacity requires the simulation of quantum channel by the pair of classical memory and classical shared randomness.

Moreover, replacing the classical noiseless channel by the quantum noiseless channel, we can define the capacities $C_{q,e}(\kappa)$, $C_{q,r}(\kappa)$, $C^R_{q,e}(\kappa)$, and $C^R_{q,r}(\kappa)$. Here, we measure the quality of approximation by using the entanglement fidelity. Then, in the same way as (10.88), the relations

$$C_{q,2}(\kappa) = C_{q,r}(\kappa) \le C_{q,e}(\kappa) \le C^R_{q,e}(\kappa) \le C^R_{q,r}(\kappa)$$

hold.

Proof of (10.87) Here, we give a proof for inequality (10.87). Assume that the sender and the receiver share the maximally entangled state $|\Phi_d\rangle\langle\Phi_d|$ on the tensor product $\mathcal{H}_B \otimes \mathcal{H}_C$. When the sender performs the generalized Bell measurement $\{|u_{i,j}^{A,C}\rangle\langle u_{i,j}^{A,C}|\}_{(i,j)}$ on the composite system between the input system $\mathcal{H}_A$ and the sender's local system $\mathcal{H}_C$, he obtains the data (i, j) subject to the uniform distribution p_{mix,d^2}. In this case, the generalized Pauli channel κ_p^{GP} can be written as

$$\kappa_p^{\mathrm{GP}}(\rho) = \sum_{(i,j)} \sum_{(i',j')} p(i'-i, j'-j)\overline{\mathbf{X}_B^{i'}\mathbf{Z}_B^{j'}}$$

$$\cdot \mathrm{Tr}_{A,C}(I \otimes |u_{i,j}^{A,C}\rangle\langle u_{i,j}^{A,C}|)(|\Phi_d\rangle\langle\Phi_d| \otimes \rho)\overline{\mathbf{X}_B^{i'}\mathbf{Z}_B^{j'}}^{*}.$$

Hence, if the classical channel $Q_{(i',j')}^{(i,j)} \stackrel{\mathrm{def}}{=} p(i'-i, j'-j)$ is simulated with the shared randomness, the generalized Pauli channel κ_p^{GP} can be simulated with the shared entanglement. Since $C_{c,r}^{R}(Q_{(i',j')}^{(i,j)}) = \log d - H(\kappa_p^{\mathrm{GP}})$, we have (10.87). ■

10.10 Proof of Lemma 10.3

We first prove the following lemma.

Lemma 10.10 *A visible encoder may be represented by a map from $\mathcal{X}$ to $\mathcal{S}(\mathcal{K})$. Consider the convex combination of codes T and T':*

$$(\lambda T + (1-\lambda)T')(x) \stackrel{\mathrm{def}}{=} \lambda T(x) + (1-\lambda)T'(x), \quad 0 < \forall\lambda < 1.$$

Then, the set of visible encoders is a convex set, and the set of extremal points (see Sect. A.4 for the definition of an extremal point) is equal to

$$\{T\,|T(x) \text{ is a pure state } \forall x \in \mathcal{X}\}. \tag{10.89}$$

Proof When $T(x)$ is a pure state for every input x, T is therefore an extremal point because it is impossible to represent the encoder T as a convex combination of other encoders. Hence, to complete the proof, it is sufficient to show that an arbitrary visible encoder $T(x) = \sum_{j_x} s_{j_x}|\phi_{j_x}\rangle\langle\phi_{j_x}|$ can be represented as a convex combination of encoders satisfying the condition in (10.89). Define a visible encoder $T(j_1, j_2, \ldots, j_n)$ by

$$T(j_1, j_2, \ldots, j_n|i) = |\phi_{j_x}\rangle\langle\phi_{j_x}|.$$

Then, this encoder belongs to the set (10.89). Since $T = \sum_{j_1,j_2,\ldots,j_n} s_{j_1}s_{j_2}\cdots s_{j_n}$ $T(j_1, j_2, \cdots, j_n)$, the proof is completed. ■

We also require the following lemma for the proof of Lemma 10.3. This lemma is equivalent to Theorem 8.3, which was shown from the viewpoint of entanglement in Sect. 8.4.

Lemma 10.11 *Let $\rho \in \mathcal{S}(\mathcal{H}_A \otimes \mathcal{H}_B)$ be separable. Then,*

$$\max\{\operatorname{Tr} P\rho_A | P : \text{Projection in } \mathcal{H}_A \text{ with rank } k\}$$
$$\geq \max\{\operatorname{Tr} P\rho | P : \text{Projection in } \mathcal{H}_A \otimes \mathcal{H}_B \text{ with rank } k\}$$

holds for any integer k.

Proof of Lemma 10.3 According to Lemma 10.10, it is sufficient to show (10.28) for a visible encoder T in (10.89). From Condition ⑥ in Theorem 5.1, there exist a space $\mathcal{H}'$ with the same dimension of $\mathcal{H}$, a pure state ρ_0 in $\mathcal{H}' \otimes \mathcal{H}$, and a unitary matrix U in $\mathcal{K} \otimes \mathcal{H}' \otimes \mathcal{H}$ such that $\nu(\rho) = \operatorname{Tr}_{\mathcal{K},\mathcal{H}'} U(\rho \otimes \rho_0)U^*$, and the state

$$\rho_x \stackrel{\text{def}}{=} \frac{(W_x \otimes I)\, U\, (T(x) \otimes \rho_0)\, U^*\, (W_x \otimes I)}{\operatorname{Tr} U\, (T(x) \otimes \rho_0)\, U^*\, (W_x \otimes I)} \in \mathcal{S}(\mathcal{K} \otimes \mathcal{H} \otimes \mathcal{H}')$$

is a pure state. Since $UT(x) \otimes \rho_0 U^*$ is a pure state and $(W_x \otimes I)$ is a projection, we have

$$\operatorname{Tr} \nu(T(x))W_x = \operatorname{Tr} UT(x) \otimes \rho_0 U^*\, (W_x \otimes I) = \operatorname{Tr} U(T(x) \otimes \rho_0)\, U^* \rho_x. \tag{10.90}$$

Since $\operatorname{Tr}_{\mathcal{K},\mathcal{H}'} \rho_x = W_x$, we may write $\rho_x = W_x \otimes \sigma_x$ by choosing an appropriate pure state $\sigma_x \in \mathcal{S}(\mathcal{K} \otimes \mathcal{H}')$. Hence, the state $\rho_p \stackrel{\text{def}}{=} \sum_{i \in \mathcal{X}} p(x)\rho_x = \sum_{x \in \mathcal{X}} p(x)\, W_x \otimes \sigma_x$ is separable and satisfies $W_p = \operatorname{Tr}_{\mathcal{H},\mathcal{K}'} \rho_p$. Since $I_{\mathcal{K}} \geq T(x)$, we have $U\,(I_{\mathcal{K}} \otimes \rho_0)\, U^* \geq U\,(T(x) \otimes \rho_0)\, U^*$. Thus, from (10.90) we have

$$\begin{aligned}
\sum_{x \in \mathcal{X}} p(x) \operatorname{Tr} \nu(T(x))W_x &= \sum_{x \in \mathcal{X}} p(x) \operatorname{Tr}_{\mathcal{H}} \operatorname{Tr}_{\mathcal{K} \otimes \mathcal{H}'} U\,(T(x) \otimes \rho_0)\, U^* \rho_x \\
&\leq \sum_{x \in \mathcal{X}} p(x) \operatorname{Tr} U\,(I_{\mathcal{K}} \otimes \rho_0)\, U^* \rho_x = \operatorname{Tr} U\,(I_{\mathcal{K}} \otimes \rho_0)\, U^* \rho_p.
\end{aligned} \tag{10.91}$$

According to $I \geq U\,(I_{\mathcal{K}} \otimes \rho_0)\, U^* \geq 0$ and $\operatorname{Tr} U\,(I_{\mathcal{K}} \otimes \rho_0)\, U^* = \operatorname{Tr} I_{\mathcal{K}} = \dim \mathcal{K}$, we obtain

$$\begin{aligned}
&\operatorname{Tr} U\,(I_{\mathcal{K}} \otimes \rho_0)\, U^* \rho_p \leq \max \left\{ \operatorname{Tr} P\rho_p \,\middle|\, \begin{array}{l} P : \text{Projection in } \mathcal{K} \otimes \mathcal{H} \otimes \mathcal{H}', \\ \text{rank } P = \dim \mathcal{K} \end{array} \right\} \\
&\leq \max\{\operatorname{Tr} PW_p | P : \text{Projection in } \mathcal{H}, \text{rank } P = \dim \mathcal{K}\}.
\end{aligned} \tag{10.92}$$

(10.92) may be obtained from Lemma 10.11 and the separability of ρ_p. The projection P on $\mathcal{H}$ satisfies

$$\operatorname{Tr}(W_p - a)P \leq \operatorname{Tr}(W_p - a)\{W_p - a \geq 0\}.$$

If the rank of P is $\dim \mathcal{K}$ (i.e., if $\operatorname{Tr} P = \dim \mathcal{K}$), then

$$\operatorname{Tr} W_p P \leq a \dim \mathcal{K} + \operatorname{Tr} W_p \{W_p - a \geq 0\}. \tag{10.93}$$

From (10.91)–(10.93),

$$\begin{aligned} 1 - \varepsilon(\Psi) &= \sum_{x \in \mathcal{X}} p(x) \operatorname{Tr} \nu(T(x)) W_x \\ &\leq \max\{\operatorname{Tr} P W_p | P : \text{ Projection in } \mathcal{H}, \text{ rank } P = \dim \mathcal{K}\} \\ &\leq a \dim \mathcal{K} + \operatorname{Tr} W_p \{W_p - a \geq 0\}. \end{aligned}$$

We therefore obtain (10.28). ■

10.11 Historical Note

First, we briefly treat the pure-state case. The source coding problem in the quantum case was initiated by Schumacher [3]. In his paper, he formulated the blind scheme and derived the direct part and the strong converse part assuming only unitary coding. Jozsa and Schumacher [4] improved this discussion. Barnum et al. [27] introduced the purification scheme and proved the strong converse part without assuming unitary coding. Further, Horodecki [19] introduced the visible scheme as an arbitrary coding scheme and showed the weak converse part. Further, Barnum et al. [28] pointed out that the previous proof by Barnum et al. [27] could be used as the proof of the strong converse part even in the visible scheme. In this book, Lemma 10.3 plays a central role in the proof of the strong converse part. This lemma was proved by Hayashi [13]. Winter [29] also proved the strong converse part using a related lemma. Using this formula, Hayashi [13] derived the optimal rate with an exponential error constraint. When the probability of the information source is unknown, we cannot use the coding protocol based on the prior distribution p. Using the type method, Jozsa et al. [17] constructed a fixed-length universal code achieving the optimal rate. In addition, in the classical case, Han [14] showed that compressed states that achieve the minimum rate are almost uniformly random in the fixed-length scheme. In this book, a part of the quantum extension of the above Han's result is proved as Theorem 10.3.

In the variable-length scheme, the problem is not so easy. In the classical case, we can compress classical data without any loss. However, Koashi and Imoto [6] proved that if all information sources cannot be diagonalized simultaneously, compression without any loss is impossible. Of course, using Schumacher's [3] compression, we can compress quantum information sources with a small error. Further, using Jozsa et al.'s [17] compression, we can compress quantum information sources with a small error based only on the knowledge of the entropy rate $H(W_p)$ if the information is generated by an independent and identical distribution of the distribution p. Hence, universal variable-length compression with a small error is possible if we can estimate the entropy rate $H(W_p)$ with a negligible state reduction. For this estimation, the estimation method in Sect. 7.4 can be applied. Using this idea, a variable-length universal compression theorem is constructed in this book. This construction is slightly different from the original construction by Hayashi and Matsumoto [9]. The modified construction by Hayashi and Matsumoto [10] is closer to the construction of this book. Further, Hayashi and Matsumoto [9] showed that the average error of variable-length compression does not approach 0 exponentially in the two-level system when the compression scheme has group covariance and achieves the entropy rate $H(W_p)$. Jozsa and Presnell [30] applied this idea to the Lempel–Ziv method. Bennett et al. [31] considered the complexity of universal variable-length compression. Hayashi [32] proposed another formulation of variable-length universal compression, in which there is no state reduction. This formulation cannot decide the coding length to avoid state reduction. Hence, Hayashi [32] considered the average coding length and Kraft inequality.

In the analysis presented in this book, we have only considered probability distributions that satisfy the independent and identical condition for the source. Petz and Mosonyi [33] showed that the optimal compression rate is $\lim_{n\to\infty} \frac{H(W_{p_n})}{n}$ when the information source p_n is stationary. Bjelaković and Szkoła [34] extended this result to the ergodic case. Datta and Suhov [35] treated nonstationary quantum spin systems. Further, Bjelaković et al. [36] extended Bjelaković and Szkoła's result to the quantum lattice system. Nagaoka and Hayashi [37] derived the optimal compression rate without any assumption of the information source based on the quantum information spectrum method. Using Lemma 10.3, they reduced quantum source coding to quantum hypothesis testing. Indeed, it is expected that the above results will be derived based on the asymptotic general formula by Nagaoka and Hayashi [37]. Kaltchenko and Yang [38] showed that this optimal rate can be attained by fixed-length source coding in the ergodic case.

The mixed-state case was firstly discussed by Jozsa's talk [39]. For this case, Horodecki derived the lower bound $I(p, W)$ (10.40) [19] and derived the optimal rate (10.42) [20] in the visible case. However, our optimal rate (10.41) has a slightly different form. Koashi and Imoto also derived the optimal rate in the blind case (10.39).

When the memory is classical, Bennett and Winter [40] pointed out that the compression problem with commutative mixed states is essentially equivalent to Wyner's [41] problem (Theorem 8.13). Theorem 10.6 can be regarded as its quantum extension. Further, Hayden et al. [12] treated the tradeoff between the sizes of classical memory and quantum memory with the visible scheme in the pure-state case.

Next, let us proceed to compression with shared randomness. Bennett et al. [22] introduced a reverse Shannon theorem (Theorem 10.9) and proved Theorem 10.8 as its corollary. Dür et al. [23] also proved Theorem 10.8 independently. In this book, we prove it via Lemma 10.9. Since this lemma has a general form, it can be extended to a general sequence of channels.

Further, we can consider the tradeoff between the sizes of the classical noiseless channel and the shared randomness as an intermediate problem between $R_{V,c,r}(p, W)$

and $R_{V,c}(p, W)$. Bennett and Winter [40] treated this problem in the commutative case.

In the classical case, Slepian and Wolf [42] considered the compression problem when the information source lies in the composite system and has a correlation. In their problem, the encoder in each system is divided into two players, who can only perform local operations. However, the decoder is allowed to use both encoded information. Devetak and Winter [11] treated the quantum extension of this problem in the special case with an ensemble scheme. Ahn et al. [43] treated a more general case with the ensemble scheme. Further, Abeyesinghe et al. [44] treated this problem with a purification scheme. When there is only one encoder and the other system can be accessed by the decoder, the problem is called the source coding with the side information. This problem is slightly easier than Slepian and Wolf coding while it is often confused with Slepian and Wolf coding. When the side information is quantum and the information to be sent is classical, this problem has been discussed in detail in the paper [46].

10.12 Solutions of Exercises

Exercise 10.1 Since $2(1 - F(W_{\boldsymbol{x}}^{(n)}, \nu_n(T(\boldsymbol{x}))) \geq 1 - F(W_{\boldsymbol{x}}^{(n)}, \nu_n(T(\boldsymbol{x})))^2 \geq 1 - F(W_{\boldsymbol{x}}^{(n)}, \nu_n(T(\boldsymbol{x})))$, we have $2(1 - \sum_{\boldsymbol{x} \in \mathcal{X}^n} p_{\boldsymbol{x}}^n F(W_{\boldsymbol{x}}^{(n)}, \nu_n(T(\boldsymbol{x})))) \geq 1 - \sum_{\boldsymbol{x} \in \mathcal{X}^n} p_{\boldsymbol{x}}^n F(W_{\boldsymbol{x}}^{(n)}, \nu_n(T(\boldsymbol{x})))^2 \geq 1 - \sum_{\boldsymbol{x} \in \mathcal{X}^n} p_{\boldsymbol{x}}^n F(W_{\boldsymbol{x}}^{(n)}, \nu_n(T(\boldsymbol{x})))$. Hence, the condition $\sum_{\boldsymbol{x} \in \mathcal{X}^n} p_{\boldsymbol{x}}^n F(W_{\boldsymbol{x}}^{(n)}, \nu_n(T(\boldsymbol{x}))) \to 1$ is equivalent to the condition $\sum_{\boldsymbol{x} \in \mathcal{X}^n} p_{\boldsymbol{x}}^n F(W_{\boldsymbol{x}}^{(n)}, \nu_n(T(\boldsymbol{x})))^2 \to 1$.

Exercise 10.2 Due to (3.48) and (3.52), we have

$$\begin{aligned}
&2d_1(W_x, \nu \circ \tau(W_x)) \geq 2(1 - F(W_x, \nu \circ \tau(W_x))) \\
\geq &1 - F^2(W_x, \nu \circ \tau(W_x)) \geq d_1^2(W_x, \nu \circ \tau(W_x)).
\end{aligned}$$

Taking expectation for x and applying Jensen inequality to $a \mapsto a^2$, we have

$$2\bar{\varepsilon}_{p,W}(\psi) \geq \varepsilon_{p,W}(\psi) \geq \bar{\varepsilon}_{p,W}(\psi)^2.$$

Hence, the condition $\bar{\varepsilon}_{p,W}(\psi) \to 0$ is equivalent with the condition $\varepsilon_{p,W}(\psi) \to 0$. Hence, the optimal rate $R_{B,q}(p, W)$ does not changed by this replacement. We can show the same thing for the other optimal rate $R_{V,q}(p, W)$.

Exercise 10.3 Equation (10.25) implies that $R_{P,q}(\rho) \leq H(\rho)$.

Next, for a given ρ, we choose p and W such that all of W_x are pure and $\rho = W_p$. Then, the second inequality of (10.17) and Theorem 10.1 guarantee that $R_{P,q}^{\dagger}(\rho) \geq R_{B,q}^{\dagger}(p, W) \geq H(\rho)$.

Exercise 10.4 Equation (10.34) can be shown as follows:

$$\dim \Upsilon_n(R,B) \le \sum_{q\in T^n:H(q)\le R} |T_q^n| \overset{(a)}{\le} \sum_{q\in T^n:H(q)\le R} e^{nH(q)}$$
$$\le \sum_{q\in T^n:H(q)\le R} e^{nR} \overset{(b)}{\le} (n+1)^d e^{nR},$$

where (a) and (b) follow from (2.155) and (2.154), respectively.

(10.35) can be shown as follows:

$$\mathrm{Tr}(I - P_{n,R,B})W_p^{\otimes n} \le \sum_{q\in T^n:H(q)\le R} r^n(T_q^n) \overset{(a)}{\le} \sum_{q\in T^n:H(q)\le R} e^{-nD(q\|r)}$$
$$\le \sum_{q\in T^n:H(q)\le R} \exp(-n \inf_{q:H(q)>R} D(q\|r))$$
$$\overset{(b)}{\le}(n+1)^d \exp(-n \inf_{q:H(q)>R} D(q\|r)),$$

where (a) and (b) follow from (2.156) and (2.154) respectively.

Exercise 10.5 When $0 \le x < 1$, we have $x+1 \ge 1$. Hence, we have $\frac{x}{1+x} \le x = \min\{x,1\}$. When $1 \le x$, we have $0 < x \le 1+x$, which implies $\frac{x}{1+x} \le 1$.

Exercise 10.6 Since $\mathrm{P}_{X|y}(x|y) = \frac{\mathrm{P}_X(x)}{\mathrm{P}_Y(y)}\mathrm{P}_{Y|X}(y|x)$, we have

$$\mathrm{P}_{Y|X}(y|x) - \frac{M\mathrm{P}_Y(y)\mathrm{P}_{X|Y}(x|y)}{\mathrm{P}_{X|Y}(x|y) + (M-1)\mathrm{P}_X(x)}$$
$$=\mathrm{P}_{Y|X}(y|x) - \frac{M\mathrm{P}_X(x)\mathrm{P}_{Y|X}(y|x)}{\frac{\mathrm{P}_X(x)}{\mathrm{P}_Y(y)}\mathrm{P}_{Y|X}(y|x) + (M-1)\mathrm{P}_X(x)}$$
$$=\mathrm{P}_{Y|X}(y|x) - \frac{M\mathrm{P}_{Y|X}(y|x)}{\frac{\mathrm{P}_{Y|X}(y|x)}{\mathrm{P}_Y(y)} + (M-1)}$$
$$=\frac{\mathrm{P}_{Y|X}(y|x)\left(\frac{\mathrm{P}_{Y|X}(y|x)}{\mathrm{P}_Y(y)} + (M-1)\right) - M\mathrm{P}_{Y|X}(y|x)}{\frac{\mathrm{P}_{Y|X}(y|x)}{\mathrm{P}_Y(y)} + (M-1)}$$
$$=\frac{\mathrm{P}_{Y|X}(y|x)\left(\frac{\mathrm{P}_{Y|X}(y|x)}{\mathrm{P}_Y(y)} - 1\right)}{\frac{\mathrm{P}_{Y|X}(y|x)}{\mathrm{P}_Y(y)} + (M-1)} = \mathrm{P}_{Y|X}(y|x)\frac{\frac{1}{M}\left(\frac{\mathrm{P}_{Y|X}(y|x)}{\mathrm{P}_Y(y)} - 1\right)}{1 + \frac{1}{M}\left(\frac{\mathrm{P}_{Y|X}(y|x)}{\mathrm{P}_Y(y)} - 1\right)}.$$

Exercise 10.7 Equation (10.40) follows from the combination of Lemma 10.8 and the second inequality of (10.74).

Exercise 10.8 Let $\tilde{\Psi}_r^{(n)} \overset{\text{def}}{=} (\mathcal{K}_n, \rho^{(n)\prime}{}_X, \nu_X^{(n)})$ be a sequence of codes achieving the optimal rate $R_{g,q,r}(\rho)$. We have

$$\log |\Psi_r^{(n)}| \geq \mathrm{E}_X I_{\rho^{(n)'}_X}(A:K) = \mathrm{E}_X D(\rho^{(n)'}_X \| \mathrm{Tr}_K \rho^{(n)'}_X \otimes \mathrm{Tr}_A \rho^{(n)'}_X)$$
$$= \mathrm{E}_X D(\rho^{(n)'}_X \| \mathrm{Tr}_B \rho^{\otimes n} \otimes \mathrm{Tr}_A \rho^{(n)'}_X)$$
$$\overset{(a)}{\geq} \mathrm{E}_X D(\nu_X(\rho^{(n)'}_X) \| \mathrm{Tr}_B \rho^{\otimes n} \otimes \nu_X(\mathrm{Tr}_A \rho^{(n)'}_X))$$
$$\overset{(b)}{\geq} D(\mathrm{E}_X \nu_X(\rho^{(n)'}_X) \| \mathrm{Tr}_B \rho^{\otimes n} \otimes \mathrm{E}_X \nu_X(\mathrm{Tr}_A \rho^{(n)'}_X))$$
$$= I_{\mathrm{E}_X \nu_X(\rho^{(n)'}_X)}(A:B),$$

where (a) and (b) follow from the monotonicity (5.36) and joint convexity (5.38) of quantum relative entropy, respectively. From the choice of our code we can check that $F(\mathrm{E}_X \nu_X(\rho^{(n)'}_X), \rho^{\otimes n}) \to 0$. Hence, using Fannes inequality (Theorem (5.12)), we have

$$\liminf_{n\to\infty} \frac{1}{n} \log |\tilde{\Psi}_r^{(n)}| \geq \liminf_{n\to\infty} \frac{1}{n} I_{\mathrm{E}_X \nu_X(\rho^{(n)'}_X)}(A:B)$$
$$= \liminf_{n\to\infty} \frac{1}{n} I_{\rho^{\otimes n}}(A:B) = I_\rho(A:B).$$

References

1. C.E. Shannon, A mathematical theory of communication. Bell Syst. Tech. J. **27**, 623–656 (1948)
2. T.S. Han, K. Kobayashi, *Mathematics of Information and Encoding* (American Mathematical Society, 2002) (originally appeared in Japanese in 1999)
3. B. Schumacher, Quantum coding. Phys. Rev. A **51**, 2738–2747 (1995)
4. R. Jozsa, B. Schumacher, A new proof of the quantum noiseless coding theorem. J. Mod. Opt. **41**(12), 2343–2349 (1994)
5. Y. Mitsumori, J.A. Vaccaro, S.M. Barnett, E. Andersson, A. Hasegawa, M. Takeoka, M. Sasaki, Experimental demonstration of quantum source coding. Phys. Rev. Lett. **91**, 217902 (2003)
6. M. Koashi, N. Imoto, Quantum information is incompressible without errors. Phys. Rev. Lett. **89**, 097904 (2002)
7. L.D. Davisson, Comments on sequence time coding for data compression. Proc. IEEE **54**, 2010 (1966)
8. T.J. Lynch, Sequence time coding for data compression. Proc. IEEE **54**, 1490–1491 (1966)
9. M. Hayashi, K. Matsumoto, Quantum universal variable-length source coding. Phys. Rev. A **66**, 022311 (2002)
10. M. Hayashi, K. Matsumoto, Simple construction of quantum universal variable-length source coding. Quant. Inf. Comput. **2**, Special Issue, 519–529 (2002)
11. I. Devetak, A. Winter, Classical data compression with quantum side information. Phys. Rev. A **68**, 042301 (2003)
12. P. Hayden, R. Jozsa, A. Winter, Trading quantum for classical resources in quantum data compression. J. Math. Phys. **43**, 4404–4444 (2002)
13. M. Hayashi, Exponents of quantum fixed-length pure state source coding. Phys. Rev. A **66**, 032321 (2002)
14. T.S. Han, Folklore in source coding: information-spectrum approach. IEEE Trans. Inf. Theory **51**(2), 747–753 (2005)
15. M. Hayashi, Second-order asymptotics in fixed-length source coding and intrinsic randomness. IEEE Trans. Inf. Theory **54**, 4619–4637 (2008)

16. I. Csiszár, J. Körner, *Information Theory: Coding Theorems for Discrete Memoryless Systems* (Academic, 1981)
17. R. Jozsa, M. Horodecki, P. Horodecki, R. Horodecki, Universal quantum information compression. Phys. Rev. Lett. **81**, 1714 (1998)
18. M. Koashi, N. Imoto, Compressibility of mixed-state signals. Phys. Rev. Lett. **87**, 017902 (2001)
19. M. Horodecki, Limits for compression of quantum information carried by ensembles of mixed states. Phys. Rev. A **57**, 3364–3369 (1998)
20. M. Horodecki, Optimal compression for mixed signal states. Phys. Rev. A **61**, 052309 (2000)
21. H.-K. Lo, S. Popescu, Concentrating entanglement by local actions: beyond mean values. Phys. Rev. A **63**, 022301 (2001)
22. C.H. Bennett, P.W. Shor, J.A. Smolin, A.V. Thapliyal, Entanglement-assisted classical capacity of noisy quantum channels. Phys. Rev. Lett. **83**, 3081 (1999)
23. W. Dür, G. Vidal, J.I. Cirac, Visible compression of commuting mixed state. Phys. Rev. A **64**, 022308 (2001)
24. C.H. Bennett, P.W. Shor, J.A. Smolin, A.V. Thapliyal, Entanglement-assisted capacity of a quantum channel and the reverse Shannon theorem. IEEE Trans. Inf. Theory **48**(10), 2637–2655 (2002)
25. A. Winter, "Extrinsic" and "intrinsic" data in quantum measurements: asymptotic convex decomposition of positive operator valued measures. Commun. Math. Phys. **244**(1), 157–185 (2004)
26. A. Winter, S. Massar, Compression of quantum measurement operations. Phys. Rev. A **64**, 012311 (2001)
27. H. Barnum, C.A. Fuchs, R. Jozsa, B. Schumacher, A general fidelity limit for quantum channels. Phys. Rev. A **54**, 4707–4711 (1996)
28. H. Barnum, C.M. Caves, C.A. Fuchs, R. Jozsa, B. Schumacher, On quantum coding for ensembles of mixed states. J. Phys. A Math. Gen. **34**, 6767–6785 (2001)
29. A. Winter, Schumacher's quantum coding revisited. Preprint 99–034, Sonder forschungsbereich 343. Diskrete Strukturen in der Mathematik Universität Bielefeld (1999)
30. R. Jozsa, S. Presnell, Universal quantum information compression and degrees of prior knowledge. Proc. R. Soc. Lond. A **459**, 3061–3077 (2003)
31. C.H. Bennett, A.W. Harrow, S. Lloyd, Universal quantum data compression via nondestructive tomography. Phys. Rev. A **73**, 032336 (2006)
32. M. Hayashi, Universal approximation of multi-copy states and universal quantum lossless data compression. Commun. Math. Phys. **293**(1), 171–183 (2010)
33. D. Petz, M. Mosonyi, Stationary quantum source coding. J. Math. Phys. **42**, 48574864 (2001)
34. I. Bjelaković, A. Szkoła, The data compression theorem for ergodic quantum information sources. Quant. Inf. Process. **4**, 49–63 (2005)
35. N. Datta, Y. Suhov, Data compression limit for an information source of interacting qubits. Quant. Inf. Process. **1**(4), 257–281 (2002)
36. I. Bjelaković, T. Kruger, R. Siegmund-Schultze, A. Szkoła, The Shannon-McMillan theorem for ergodic quantum lattice systems. Invent. Math. **155**, 203–222 (2004)
37. H. Nagaoka, M. Hayashi, An information-spectrum approach to classical and quantum hypothesis testing. IEEE Trans. Inf. Theory **53**, 534–549 (2007)
38. A. Kaltchenko, E.-H. Yang, Universal compression of ergodic quantum sources. Quant. Inf. Comput. **3**, 359–375 (2003)
39. R. Jozsa, Quantum noiseless coding of mixed states, in Talk given at *3rd Santa Fe Workshop on Complexity, Entropy, and the Physics of Information*, May 1994
40. C.H. Bennett, A. Winter, Private Communication
41. A.D. Wyner, The common information of two dependent random variables. IEEE Trans. Inf. Theory **21**, 163–179 (1975)
42. D. Slepian, J.K. Wolf, Noiseless coding of correlated information sources. IEEE Trans. Inf. Theory **19**, 471 (1973)

43. C. Ahn, A. Doherty, P. Hayden, A. Winter, On the distributed compression of quantum information. IEEE Trans. Inform. Theory **52**, 4349–4357 (2006)
44. A. Abeyesinghe, I. Devetak, P. Hayden, A. Winter, The mother of all protocols: Restructuring quantum information's family tree. Proceedings of the Royal Society of London A: Mathematical, Physical and Engineering Sciences, rspa20090202 (2009)
45. M. Hayashi, Optimal Visible Compression Rate For Mixed States Is Determined By Entanglement Purification. Phy. Rev. A Rapid Commun. **73**, 060301(R) (2006)
46. M. Tomamochel, M. Hayashi, A Hierarchy of Information Quantities for Finite Block Length Analysis of Quantum Tasks. IEEE Trans. Inf. Theory **59**(11), 7693–7710 (2013)

Erratum to: Quantum Information Theory

Erratum to: M. Hayashi, *Quantum Information Theory*, Graduate Texts in Physics, DOI 10.1007/978-3-662-49725-8

The original version of the book was inadvertently published without including the additional bibliographic information on the copyright page. The erratum has been updated with the changes in this book.

The updated original online version for this book can be found at
DOI 10.1007/978-3-662-49725-8

M. Hayashi (✉)
Graduate School of Mathematics, Nagoya University, Nagoya, Aichi, Japan
e-mail: masahito@math.nagoya-u.ac.jp

M. Hayashi, *Quantum Information Theory*, Graduate Texts in Physics,
DOI 10.1007/978-3-662-49725-8_11

Appendix
Limits and Linear Algebra

A.1 Limits

In this text, frequently we discuss the asymptotic behaviors in several problems when the number n of prepared systems is sufficiently large. In this situation, we often take the limit $n \to \infty$. In this section, we give a brief summary of the fundamental properties of limits. Given a general sequence $\{a_n\}$, the limit $\lim_{n\to\infty} a_n$ does not necessarily exist. For example, a sequence a_n is a counterexample when a_n diverges to $+\infty$ or $-\infty$. In such a case, it is possible to at least denote these limits as $\lim_{n\to\infty} a_n = +\infty$ or $\lim_{n\to\infty} a_n = -\infty$. However, the sequence a_n has no limit as $n \to \infty$, even allowing possibilities such as $+\infty$ or $-\infty$, when a_n is defined to be 0 when n is even and 1 when it is odd. This is caused by its oscillatory behavior. In this case, we can consider the upper limit $\underline{\lim}\, a_n$ and the lower limit $\overline{\lim}\, a_n$, which are given as $\underline{\lim}\, a_n = 0$ and $\overline{\lim}\, a_n = 1$. More precisely, $\underline{\lim}\, a_n$ and $\overline{\lim}\, a_n$ are defined as follows:

$$\underline{\lim}\, a_n \stackrel{\text{def}}{=} \sup\{a | \forall \epsilon > 0, \exists N, \forall n \ge N, a \le a_n + \epsilon\},$$
$$\overline{\lim}\, a_n \stackrel{\text{def}}{=} \inf\{a | \forall \epsilon > 0, \exists N, \forall n \ge N, a \ge a_n - \epsilon\}.$$

When $\underline{\lim}\, a_n = \overline{\lim}\, a_n$, the limit $\lim_{n\to\infty} a_n$ exists and is equal to $\underline{\lim}\, a_n = \overline{\lim}\, a_n$. The following three lemmas hold concerning limits.

Lemma A.1 *Let sequences* $\{a_n\}_{n=1}^{\infty}$ *and* $\{b_n\}_{n=1}^{\infty}$ *satisfy*

$$a_n + a_m \le a_{n+m} + b_{n+m}, \quad \sup_n \frac{a_n}{n} < \infty, \quad \lim_{n\to\infty} \frac{b_n}{n} = 0.$$

Then, the limit $\lim_{n\to\infty} \frac{a_n}{n}$ *exists and satisfies*

$$\lim_{n\to\infty} \frac{a_n}{n} = \overline{\lim} \frac{a_n}{n} = \sup_n \frac{a_n}{n}. \tag{A.1}$$

M. Hayashi, *Quantum Information Theory*, Graduate Texts in Physics,
DOI 10.1007/978-3-662-49725-8

If $a_n + a_m \ge a_{n+m} - b_{n+m}$ *and* $\inf_n \frac{a_n}{n} > \infty$, $b_n \to 0$, *then similarly* $\lim_{n\to\infty} \frac{a_n}{n} = \underline{\lim} \frac{a_n}{n} = \inf_n \frac{a_n}{n}$, *as shown by considering* $-a_n$.

Proof Fix the integer m. Then, for any integer n, there uniquely exist integers l_n and r_n such that $0 \le r_n \le m-1$ and $n = l_n m + r_n$ for each n. Thus, we have

$$\frac{a_n}{n} = \frac{a_{l_n m + r}}{l_n m + r} \ge \frac{a_{l_n m}}{l_n m + r} + \frac{a_r - b_n}{l_n m + r} \ge \frac{l_n a_m}{l_n m + r} + \frac{a_r - b_n - b_{l_n m}}{l_n m + r}.$$

Since $l_n \to \infty$ as $n \to \infty$, taking the limit $n \to \infty$, we have $\underline{\lim} \frac{a_n}{n} \ge \frac{a_m}{m}$ for arbitrary m. Next, taking the limit $m \to \infty$, we have $\underline{\lim} \frac{a_n}{n} \ge \sup_m \frac{a_m}{m} \ge \overline{\lim}_{m\to\infty} \frac{a_m}{m}$. Since $\overline{\lim} \frac{a_n}{n} \ge \underline{\lim} \frac{a_n}{n}$, we obtain (A.1). ∎

Lemma A.2 *Let* $\{a_n\}$ *and* $\{b_n\}$ *be two sequences of positive real numbers. Then,*

$$\overline{\lim} \frac{1}{n} \log(a_n + b_n) = \max \left\{ \overline{\lim} \frac{1}{n} \log a_n, \overline{\lim} \frac{1}{n} \log b_n \right\}.$$

Proof Since $(a_n + b_n) \ge a_n, b_n$ and

$$\overline{\lim} \frac{1}{n} \log(a_n + b_n) \ge \overline{\lim} \frac{1}{n} \log a_n, \overline{\lim} \frac{1}{n} \log b_n,$$

we obtain the $\ge$ part of the proof. Since $2 \max\{a_n, b_n\} \ge (a_n + b_n)$, we have

$$\max \left\{ \overline{\lim} \frac{1}{n} \log a_n, \overline{\lim} \frac{1}{n} \log b_n \right\} = \overline{\lim} \frac{1}{n} \log \max\{a_n, b_n\}$$
$$= \overline{\lim} \frac{1}{n} \log 2 \max\{a_n, b_n\} \ge \overline{\lim} \frac{1}{n} \log(a_n + b_n),$$

which gives the reverse inequality. This completes the proof. ∎

Lemma A.3 *Let* $\{f_n(x)\}$ *be a sequence of functions such that* $f_n(x) \le f_n(y)$ *if* $x \ge y$, *and* $f_n(x) \to 0$ *if* $x > 0$. *There exists a sequence* $\{\epsilon_n\}$ *of positive real numbers converging to zero such that* $f_n(x) \to 0$.

Proof Let N be a positive integer. Choose positive integers $n(N)$ such that $n(N) < n(N+1)$ and $f_n(\frac{1}{N}) \le \frac{1}{N}$ for $n \ge n(N)$. We also define $\epsilon_n \stackrel{\text{def}}{=} \frac{1}{N}$ for $n(N) \le n < n(N+1)$. Then, $\epsilon_n \to 0$. If $n \ge n(N)$, then $f_n(\epsilon_n) \le \frac{1}{N}$. Therefore, $f_n(\epsilon_n) \to 0$. ∎

For any two continuous functions f and g on an open subset $X \subset \mathbb{R}^d$, we define

$$[f, g](a) \stackrel{\text{def}}{=} \min_{x \in V} \{ f(x) | g(x) \le a \}. \tag{A.2}$$

Lemma A.4 *When X is closed and bounded, i.e., compact,*

$$[f, g](a) = \lim_{\epsilon \downarrow 0} [f, g](a + \epsilon). \tag{A.3}$$

Proof From the definition for $\epsilon > 0$, $[f, g](a) \geq [f, g](a + \epsilon)$. Hence, $[f, g](a) \geq \lim_{\epsilon \downarrow 0}[f, g](a + \epsilon)$. From the compactness, for any $\epsilon_1 > 0$ there exists $\epsilon_2 > 0$ such that $\|x - x'\| < \epsilon_2 \Rightarrow |f(x) - f(x')| < \epsilon_1$. Further, from the compactness of X we can choose a small number $\epsilon_3 > 0$ such that $\{x|g(x) \leq a + \epsilon_3\} \subset \cup_{x':g(x) \leq a} U_{x',\epsilon_2}$. Hence,

$$\min_{x|g(x) \leq a+\epsilon_3} f(x) \geq \min_{x \in \cup_{x':g(x) \leq a} U_{x',\epsilon_2}} f(x) \geq \min_{x|g(x) \leq a} f(x) - \epsilon_1, \tag{A.4}$$

which implies (A.3). ■

A.2 Singular Value Decomposition and Polar Decomposition

Any $d \times d'$ complex-valued matrix X has the form

$$X = U_1 X' U_2^* \tag{A.5}$$

with isometric matrices U_1 and U_2 and a diagonal matrix X'. This is called a **singular value decomposition** (the matrix U is an **isometric matrix** if U^*U is the identity matrix; U is a **partially isometric matrix** for the partial space $\mathcal{K}$ if it is a projection onto the partial space $\mathcal{K}$). Choosing a $d \times d'$ partially isometric matrix U in the range $\{X^*Xv | v \in \mathbb{C}^d\}$ of X^*X, we have

$$X = U|X|, \quad |X| \stackrel{\text{def}}{=} \sqrt{X^*X}, \tag{A.6}$$

which is called a **polar decomposition**. If X is Hermitian and is diagonalizable according to $X = \sum_i \lambda_i |u_i\rangle\langle u_i|$, then $|X| = \sum_i |\lambda_i||u_i\rangle\langle u_i|$. Since $X^* = |X|U^*$,

$$XX^* = U|X||X|U^* = UX^*XU^*, \quad \sqrt{XX^*} = U\sqrt{X^*X}U^*, \tag{A.7}$$

$$UX^*U = X. \tag{A.8}$$

Therefore,

$$X = \sqrt{XX^*}U. \tag{A.9}$$

If X is a square matrix (i.e., $d = d'$), then U is unitary. If $d \geq d'$, then U can be chosen as an isometric. If $d \leq d'$, U can be chosen such that U^* is isometric. We now show that these two decompositions exist.

Since X^*X is Hermitian, we may choose a set of mutually orthogonal vectors $u_1, \ldots, u_l$ of norm 1 such that

$$X^*X = \sum_{i=1}^{l} \lambda_i |u_i\rangle\langle u_i|.$$

In the above, we choose $\{\lambda_k\}_{k=1}^l$ such that $\lambda_i \geq \lambda_{i+1} > 0$. Hence, l is not necessarily equal to the dimension of the space because there may exist zero eigenvalues. Defining $v_i \stackrel{\text{def}}{=} \sqrt{\frac{1}{\lambda_i}} X u_i$, we have

$$\begin{aligned}\langle v_i | v_j \rangle &= \sqrt{\frac{1}{\lambda_i}} \sqrt{\frac{1}{\lambda_j}} \langle X u_i | X u_j \rangle = \sqrt{\frac{1}{\lambda_i}} \sqrt{\frac{1}{\lambda_j}} \langle u_i | X^* X | u_j \rangle \\ &= \sqrt{\frac{1}{\lambda_i}} \sqrt{\frac{1}{\lambda_j}} \delta_{i,j} \lambda_j = \delta_{i,j}.\end{aligned}$$

Furthermore, from the relation

$$\begin{aligned}\langle v_i | X | u_j \rangle &= \sqrt{\frac{1}{\lambda_i}} \langle X u_i | X | u_j \rangle = \sqrt{\frac{1}{\lambda_i}} \langle X u_i | X | u_j \rangle \\ &= \sqrt{\frac{1}{\lambda_i}} \langle u_i | X^* X | u_j \rangle = \sqrt{\lambda_i} \delta_{i,j}\end{aligned}$$

we can show that

$$\sum_i \sqrt{\lambda_i} | v_i \rangle \langle u_i | = \sum_i | v_i \rangle \langle v_i | X \sum_j | u_j \rangle \langle u_j | = X. \tag{A.10}$$

One may be concerned about the validity of the second equality if $X^* X$ has some eigenvectors u with zero eigenvalue. However, since $\langle u | X^* X | u \rangle = 0$, we have $Xu = 0$. Hence, the image of vector u is the zero vector in both sides of (A.10). We define $U_2 \stackrel{\text{def}}{=} (u_i^j)$ and $U_1 \stackrel{\text{def}}{=} (v_i^j)$, which are $d \times l$ and $d' \times l$ isometric matrices, respectively. Let X' be an $l \times l$ diagonal matrix $(\sqrt{\lambda_i} \delta_{i,j})$. This gives us (A.5).

Using the above, we obtain the following lemma.

Lemma A.5 *Let a density matrix ρ be written as*

$$\rho = \sum_{j=1}^{d'} | v_j \rangle \langle v_j |, \tag{A.11}$$

where $\{v_i\}$ is a set of vectors that are not necessarily orthogonal. Let its diagonalization be given by $\rho = \sum_{i=1}^l \lambda_i | u_i \rangle \langle u_i |$. Since $\lambda_i > 0$, l is not necessarily equal to the dimension of the space. Then, the vector v_j can be written as $v_j = \sum_{i=1}^l w_{j,i}^* \sqrt{\lambda_i} u_i$ by using an $l \times d'$ isometric matrix $W = (w_{j,i})$ [1].

The set of vectors $\{v_i\}$ satisfying (A.11) is called the **decomposition** of the density matrix ρ.

Proof Let Y be a $d' \times l$ matrix given by (v_i^i). Then,

$$\rho = \sum_{i=1}^{l} \lambda_i |u_i\rangle\langle u_i| = YY^*.$$

Define $w_i \stackrel{\text{def}}{=} \sqrt{\frac{1}{\lambda_i}} Y^* u_i$. Then, $Y^* = \sum_{i=1}^{l} \sqrt{\lambda_i} |w_i\rangle\langle u_i|$. Taking its conjugate, we obtain $Y = \sum_{i=1}^{l} \sqrt{\lambda_i} |u_i\rangle\langle w_i|$. Looking at the jth row, we obtain $|v_j\rangle = \sum_{i=1}^{l} (w_i^j)^* \sqrt{\lambda_i} |u_i\rangle$. Since $\sum_j (w_i^j)^*(w_{i'}^j) = \delta_{i,i'}$, w_i^j is an isometric matrix. The proof is complete. ■

Next, we consider the case where X is a real $d \times d$ matrix. Since a real symmetric matrix can be diagonalized by an orthogonal matrix, the unitary matrices U_1 and U_2 may be replaced by orthogonal matrices O_1 and O_2. In fact, we may further restrict the orthogonal matrices to orthogonal matrices with determinant 1 (these are called special orthogonal matrices). However, the following problem occurs. Assume that the determinant of O_i $(i = 1, 2)$ is -1. Then, O_i may be redefined by multiplying it by a diagonal matrix with diagonal elements $-1, 1, \ldots, 1$. The redefined matrix is then a special orthogonal matrix, and $O_1^* X O_2$ is diagonal. Choosing O_1 and O_2 in a suitable way, all the diagonal elements of $O_1^* X O_2$ will be positive if $\det X > 0$. On the other hand, if $\det X < 0$, then it is not possible to make all the diagonal elements of $O_1^* X O_2$ positive for special orthogonal matrices O_1, O_2.

Exercises

A.1 Define $J_{j,i} \stackrel{\text{def}}{=} \langle u_i | u_j \rangle$ for a set of linearly independent vectors $u_1, \ldots, u_k$ in $\mathcal{H}$. Show that

$$\sum_{i,j} (J^{-1})^{j,i} |u_i\rangle\langle u_j| = \sum_{i,j} ((J^{-1})^*)^{j,i} |u_i\rangle\langle u_j|. \tag{A.12}$$

Show that this is a projection to the subspace of $\mathcal{H}$ spanned by $u_1, \ldots, u_k$.

A.2 Using relation (A.6), show that

$$AA^* f(AA^*) = A f(A^*A) A^*. \tag{A.13}$$

A.3 Norms of Matrices

We often focus on the norm of the difference between two matrices as a measure of the difference between them. There are two types of norms, the matrix norm and the trace norm. The matrix norm $\|A\|$ of a matrix A is defined as

$$\|A\| \stackrel{\text{def}}{=} \max_{\|x\|=1} \|Ax\|. \tag{A.14}$$

Since $\|x\| = \max_{\|y\|=1} |\langle y, x\rangle|$, we have $\|A\| = \max_{\|y\|=\|x\|=1} |\langle y, Ax\rangle|$; therefore, $\|A\| = \|A^*\|$. From the definition we have

$$\|U_1 A U_2\| = \|A\| \tag{A.15}$$

for unitary matrices U_1 and U_2. Defining

$$w(A) \stackrel{\text{def}}{=} \max_{\|x\|=1} |\langle x, Ax\rangle|, \quad \text{spr}(A) \stackrel{\text{def}}{=} \max\{|\lambda| : \lambda \text{ is the eigenvalue of } A\},$$

we obtain

$$\text{spr}(A) \le w(A) \le \|A\|. \tag{A.16}$$

Assume that A is a Hermitian matrix. Then, it may be diagonalized as $A = \sum_{i=1}^{d} \lambda_i |u_i\rangle\langle u_i|$. Thus,

$$|\langle y, Ax\rangle| = \sum_{i=1}^{d} |\lambda_i||\langle y|u_i\rangle\langle u_i|x\rangle| \le \max_i |\lambda_i| \sum_{i=1}^{d} |\langle y|u_i\rangle||\langle u_i|x\rangle|$$
$$\le \max_i |\lambda_i| \sqrt{\sum_{i=1}^{d} |\langle y|u_i\rangle|^2} \sqrt{\sum_{i=1}^{d} |\langle u_i|x\rangle|^2} = \max_i |\lambda_i| = \text{spr}(A).$$

The above inequality implies the equality sign in (A.16). Since $\|A\|^2 = \max_{\|x\|=1} \langle x|A^*A|x\rangle = \text{spr}(A^*A) = (\text{spr}(\sqrt{A^*A}))^2$, then

$$\|A\| = \|\sqrt{A^*A}\| = \|A^*\| = \|\sqrt{AA^*}\|. \tag{A.17}$$

On the other hand, the trace norm $\|X\|_1$ of a matrix X is defined as

$$\|X\|_1 = \max_{U:\text{unitary}} \text{Tr} UX. \tag{A.18}$$

Choosing a unitary matrix U_X such that $X = U_X|X|$ (i.e., a polar decomposition), we obtain[Exe. A.8]

$$\|X\|_1 = \max_{U:\text{unitary}} \text{Tr} UX = \text{Tr} U_X^* X = \text{Tr}|X|. \tag{A.19}$$

Hence, we also have

$$\|X^*\|_1 = \max_{U:\text{unitary}} \text{Tr} U^* X^* = \text{Tr} U_X X^* = \text{Tr}|X^*|.$$

Further, we often focus 2-norm:

$$\|X\|_2 := \sqrt{\text{Tr} XX^*}. \tag{A.20}$$

We have the relation

$$\|X\|_2 \leq \|X\|_1. \tag{A.21}$$

Then, we can show that

$$\|YX\|_i \leq \|Y\| \|X\|_i \tag{A.22}$$

for $i = 1, 2$.

Further, as generalizations of the norms $\|X\|_1$ and $\|X\|_2$, for a real number p, we define p-norm. For a function $f(x)$, the p-norm $\|f\|_p$ is defined as

$$\|f\|_p := \left(\sum_x |f(x)|^p \right)^{1/p}. \tag{A.23}$$

For a square matrix X, the p-norm $\|X\|_p$ is defined as

$$\|X\|_p := (\mathrm{Tr}|X|^p)^{1/p}. \tag{A.24}$$

When $p, q > 0$ satisfy $\frac{1}{p} + \frac{1}{q} = 1$, the **Hölder inequality**

$$\left| \sum_x f(x)g(x) \right| \leq \|f\|_p \|g\|_q \tag{A.25}$$

holds for two functions f and g. The equality holds if and only if there is a constant c such that $f(x)^p = cg(x)^q$. Then, the **matrix Hölder inequality** [2, Theorem 6.21]

$$|\mathrm{Tr}XY| \leq \|X\|_p \|Y\|_q \tag{A.26}$$

holds for two matrices X and Y. Since $|\mathrm{Tr}XY| \leq \mathrm{Tr}|X||Y|$, it is enough to show (A.26) for positive semidefinite matrices X and Y. The inequality (A.26) in this case can be shown from (6.122) of Proof of (6.17) in Sect. 6.7 when $\mathcal{H}_A$ is a one-dimensional space, $\lambda = 1/p$, $\rho^{A,B} = Y^q$, and $\sigma^{A,B} = X^p$.

When $0 < p < 1$ and $q < 0$ satisfy $\frac{1}{p} + \frac{1}{q} = 1$ we can show the **reverse Hölder inequality** [3]

$$\left| \sum_x f(x)g(x) \right| \geq \|f\|_p \|g\|_q \tag{A.27}$$

for two positive-valued functions f and g, and the **reverse matrix Hölder inequality**

$$|\mathrm{Tr}XY| \geq \|X\|_p \|Y\|_q \tag{A.28}$$

for two positive semidefinite matrices X and Y.

Proof of (A.28) Since (A.27) can be regarded the diagonal case of (A.28), we show only (A.28). It is sufficient to show the case when X and Y are invertible. The non-invertible case can be obtained by the limit of the above case. We choose the real number $s := -\frac{1}{q}$ and the two matrices $A := \log X$ and $B := \log Y$. Then, we apply the matrix Hölder inequality (A.26) to the matrices $(e^{A+B})^{\frac{1}{1+s}}$ and $(e^{A})^{-\frac{1}{1+s}}$ with $p' := \frac{1+s}{s}$ and $q' := 1+s$. We obtain

$$
\begin{aligned}
&\|e^{A+B}\|_1^{\frac{1}{1+s}} \|e^{-\frac{a}{s}}\|_1^{\frac{s}{1+s}} = \|(e^{\frac{A+B}{1+s}})^{1+s}\|_1^{\frac{1}{1+s}} \|(e^{-\frac{A}{1+s}})^{\frac{1+s}{s}}\|_1^{\frac{s}{1+s}} \\
\geq &\|e^{\frac{A+B}{1+s}} e^{-\frac{A}{1+s}}\|_1 \geq \mathrm{Tr} e^{\frac{A+B}{1+s}} e^{-\frac{A}{1+s}} \overset{a}{\geq} \mathrm{Tr} e^{\frac{B}{1+s}} = \|e^{\frac{B}{1+s}}\|_1,
\end{aligned}
$$

where a follows from Golden-Thomson trace inequality (5.48). Therefore, we have $\|e^{A+B}\|_1^{\frac{1}{1+s}} \geq \|e^{\frac{B}{1+s}}\|_1 \|e^{-\frac{A}{s}}\|_1^{-\frac{s}{1+s}}$. Since Golden-Thomson trace inequality (5.48) yields $\|e^A e^B\|_1 \geq \|e^{A+B}\|_1$, we have $\|e^A e^B\|_1 \geq \|e^{\frac{B}{1+s}}\|_1^{1+s} \|e^{-\frac{A}{s}}\|_1^{-s}$, which implies (A.28). ∎

Exercises

A.3 Show that the trace norm of a Hermitian matrix $\begin{pmatrix} -a & b \\ \bar{b} & a \end{pmatrix}$ is equal to $2\sqrt{|b|^2 + a^2}$.

A.4 Show that

$$
\|X\|_1 \geq \|\mathrm{Tr}_B X\|_1 \tag{A.29}
$$

for a matrix X in $\mathcal{H}_A \otimes \mathcal{H}_B$.

A.5 Let A and B be square matrices of dimension d. Show that the eigenvalues of BA are the same as the eigenvalues of AB including degeneracies if A or B possesses the inverse.

A.6 Show that $\mathrm{spr}(AB) = \mathrm{spr}(BA)$.

A.7 Show that the function $t \mapsto t^{1/2}$ is a matrix monotone function following the steps below.
(**a**) Show that $\|A^{1/2} B^{-1/2}\| \leq 1$ when the Hermitian matrices B and A satisfy $B \geq A \geq 0$ and B possesses the inverse.
(**b**) Show that $1 \leq \mathrm{spr}(B^{-1/4} A^{1/2} B^{-1/4})$ under the same conditions as (**a**).
(**c**) Show that

$$
B^{1/2} \geq A^{1/2} \tag{A.30}
$$

under the same conditions as (**a**).
(**d**) Show that (A.30) holds even if B does not possess the inverse.

A.8 Prove (A.19) following the steps below.
(**a**) Show that $\max_{v:\|v\|=1}\langle v|\,|X|\,|u_i\rangle = \langle u_i|\,|X|\,|u_i\rangle$ for eigenvectors u_i of $|X|$ of length 1.
(**b**) Show that $\max_{U:\text{unitary}}\langle u_i|UX|u_i\rangle = \langle u_i|U_X^* X|u_i\rangle = \langle u_i||X||u_i\rangle$, where U_X is given by using the polar décomposition $X = U_X|X|$.
(**c**) Prove (A.19).

A.9 Show (A.22).

A.10 (**Poincaré inequality**) Let A be a $d \times d$ Hermitian matrix. Let a_i be the eigenvalues of A ordered from largest to smallest. Show that $\min_{x\in\mathcal{K},\|x\|=1}\langle x|A|x\rangle \le a_k$ for any k-dimensional subspace $\mathcal{K}$.

A.11 Show that $\max_{P:\text{rank}P=k}\min_{x}\frac{\langle x|PAP|x\rangle}{\langle x|P|x\rangle} = a_k$ under the same conditions as above.

A.12 Let A and B be Hermitian matrices, and let a_i and b_i be their ordered eigenvalues from largest to smallest. Show that $a_i \ge b_i$ if $A \ge B$.

A.13 Assume that $\frac{1}{p}+\frac{1}{q}=1$ and $X \ge 0$. Show the following relations by using the matrix Hölder inequality (A.26) and the matrix reverse Hölder inequality (A.28)

$$\max_{Z\ge 0:\text{Tr}Z=1}\text{Tr}Z^{\frac{1}{q}}X = \|X\|_p \text{ for } p > 1 \tag{A.31}$$

$$\min_{Z\ge 0:\text{Tr}Z=1}\text{Tr}Z^{\frac{1}{q}}X = \|X\|_p \text{ for } p < 1. \tag{A.32}$$

A.4 Convex Functions and Matrix Convex Functions

Linear functions are often used in linear algebra. On the other hand, functions such as x^2 and $\exp(x)$ do not satisfy the linearity property. If we denote such functions by f, then they instead satisfy

$$f(\lambda x_1 + (1-\lambda)x_2) \le \lambda f(x_1) + (1-\lambda)f(x_2), \quad 0 \le \forall\lambda \le 1, \forall x_1, x_2 \in \mathbb{R}.$$

A function is called a **convex function** when it satisfies the above inequality. If $-f$ is a convex function, then f is called a **concave function**. In the above, its domain is restricted to real numbers. However, this restriction is not necessary and may be defined in a more general way. For example, for a vector space, we may define the **convex combination** $\lambda v_1 + (1-\lambda)v_2$ for two vectors v_1 and v_2 with $0 < \lambda < 1$. More generally, a set is called a **convex set** when the convex combination of any two elements is defined. Further, a convex set L is called a **convex cone** if $v \in L$ and $\lambda > 0$ imply $\lambda v \in L$. Therefore, it is possible to define convex and concave functions for functions with a vector space domain and a real number range. Similarly, convex and concave functions may be defined with a convex set domain. Examples of convex sets are the set of probability distributions and the set of density matrices. In particular, an element v of the convex set V is called an **extremal point** if $v_i \in V$ and $v = \lambda v_1 + (1-\lambda)v_2$, $(0 < \lambda < 1)$ imply $\lambda = 1$ or 0. For example, a pure state is an extremal point in the set of density matrices. When the convex V is closed, any point $v \in V$ can be written as a convex combination of extremal points. Such an expression of v is called the **extremal point decomposition** of v. The extremal point decomposition characterizes the property of the point v. Further, for a given subset S of a convex set V, we can define the convex subset $\mathrm{Co}(S)$ of V as the set of convex combinations of elements of S. The convex subset $\mathrm{Co}(S)$ is called the **convex hull** of S.

Here, we prepare several important properties of convex functions.

Lemma A.6 *When a convex function f is defined on an open convex subset V of $\mathbb{R}^d$. Then, f is continuous.*

Proof For a point x of V, we choose $d+1$ points y_i in V such that $x = \sum_{i=1}^{d+1} \frac{1}{d+1} y_i$. When a point z in the convex hull of $\{y_i\}$ is close to x, we can choose two positive numbers a_1 and a_2 in $(0, 1)$ that are close to 1 and non-negative numbers $b_{1,i}$ and $b_{2,i}$ such that $z = a_1 x + \sum_i b_{1,i} y_i$ and $x = a_2 z + \sum_i b_{2,i} y_i$. Then, we have $f(z) \le a_1 f(x) + \sum_i b_{1,i} f(y_i)$ and $f(x) \le a_2 f(z) + \sum_i b_{2,i} f(y_i)$. Thus,

$$\frac{1-a_2}{a_2} f(x) - \frac{1}{a_2} \sum_i b_{2,i} f(y_i) \le f(z) - f(x) \le (a_1 - 1) f(x) + \sum_i b_{1,i} f(y_i). \tag{A.33}$$

When $z \to x$, we have $a_1, a_2 \to 1$ and $b_{1,i}, b_{2,i} \to 0$. So, we obtain the continuity of f at x. ■

Lemma A.7 *Let $\bar{V}$ be a compact convex subset set of $\mathbb{R}^d$ and f be a convex function f defined on the inner V of $\bar{V}$. Assume that when a sequence of $\{x_n\}$ in V converges to the boundary of $\bar{V}$, the value $f(x_n)$ goes to $+\infty$. Then, the convex function f has the minimum.*

Proof Assume that the function f does not have the minimum. Since Lemma A.6 guarantees the continuity in V, there exists a sequence $\{x_n\}$ in V such that $\lim_{n\to\infty} f(x_n) = \inf_{x\in V} f(x)$. Since $\bar{V}$ is compact, there exists a subsequence $\{x_{n_k}\}$ converging in $\bar{V}$. However, $\lim_{k\to\infty} f(x_{n_k}) = +\infty$, which contradicts the assumption. ■

Lemma A.8 *Let f be a convex function on the convex set V. For any element v_0 of V, there exists a linear function g such that*

$$g(v_0) - f(v_0) = \max_{v\in V} g(v) - f(v).$$

When f is differentiable, g coincides with the derivative of f at v_0. Further, for any linear function g and a constant $C_0 \ge 0$, there exists the Lagrange multiplier λ such that

$$\max_{v \in V} f(v) + \lambda g(v) = \max_{v \in V : g(v) \leq C_0} f(v) + \lambda g(v).$$

In this case, λg *coincides with the derivative of* f *at* $\operatorname{argmax}_{v \in V : g(v) \leq C_0} f(v)$.

Lemma A.9 ([4, Chap. VI Prop. 2.3]) *Consider two vector spaces* V_1 *and* V_2 *and consider a real-valued function* $f(v_1, v_2)$ *with the domain* $V_1 \times V_2$. *If* f *is convex with respect to* v_2 *and concave with respect to* v_1, *then*[1]

$$\sup_{v_1 \in S_1} \min_{v_2 \in S_2} f(v_1, v_2) = \min_{v_2 \in S_2} \sup_{v_1 \in S_1} f(v_1, v_2),$$

where S_1 and S_2 are convex subsets of V_1 and V_2.

Next, we focus on the set of probability distributions on $\mathcal{S}(\mathcal{H})$ and denote it by $\mathcal{P}(\mathcal{S}(\mathcal{H}))$. In particular, we consider extremal points of the set $\mathcal{S}(\mathcal{H})$:

$$\mathcal{P}(\rho, \mathcal{S}(\mathcal{H})) \stackrel{\text{def}}{=} \left\{ p \in \mathcal{P}(\mathcal{S}(\mathcal{H})) \middle| \sum_i p_i \rho_i = \rho \right\}.$$

Such extremal points of the above set are characterized as follows.

Lemma A.10 (Fujiwara and Nagaoka [5]) *Let* $p \in \mathcal{P}(\rho, \mathcal{S}(\mathcal{H}))$ *be an extremal point and* $\{\rho_1, \ldots, \rho_k\}$ *be the support of* p. *Then,* $\rho_1, \ldots, \rho_k$ *are linearly independent. Hence, the number of supports of* p *is less than* $\dim \mathcal{T}(\mathcal{H}) = (\dim \mathcal{H})^2$.

Note that we obtain the same result when we replace $\mathcal{P}(\rho, \mathcal{S}(\mathcal{H}))$ by $\mathcal{P}(\mathcal{S}(\mathcal{H}))$.

Proof Assume that $\rho_1, \ldots, \rho_k$ are linearly dependent. That is, we choose real numbers $\lambda_1, \ldots, \lambda_k$ such that $\sum_{i=1}^k \lambda_i \rho_i = 0$ and $\sum_i \lambda_i = 0$. Define two distributions q^+ and q^- with the same support by

$$q_i^{\pm} \stackrel{\text{def}}{=} p_i \pm \epsilon \lambda_i. \tag{A.34}$$

Then, we have $p = \frac{1}{2} q^+ + \frac{1}{2} q^-$ and $q^+ \neq q^-$. It is a contradiction. ■

Indeed, applying this lemma to ρ_{mix}, we can see that any extremal POVM has at most $(\dim \mathcal{H})^2$ elements. So, the set of extremal points are compact. Thus, we have the following lemma.

Lemma A.11 *A continuous convex function* f *for a POVM* $\boldsymbol{M}$ *has the minimum* $\min_{\boldsymbol{M}} f(\boldsymbol{M})$.

Further, we focus on the cost functions $f_1, \ldots, f_l$ on $\mathcal{S}(\mathcal{H})$ and treat the following sets:

[1] This relation holds even if V_1 is infinite dimensional, as long as S_2 is a closed and bounded set.

$$\mathcal{P}_{=(\leq)c}(\rho, f, \mathcal{S}(\mathcal{H})) \stackrel{\text{def}}{=} \left\{ p \in \mathcal{P}(\rho, \mathcal{S}(\mathcal{H})) \middle| \sum_i p_i f_j(\rho_i) = (\leq) c \forall j = 1, \ldots, l \right\}$$

$$\mathcal{P}_{=(\leq)c}(f, \mathcal{S}(\mathcal{H})) \stackrel{\text{def}}{=} \left\{ p \in \mathcal{P}(\mathcal{S}(\mathcal{H})) \middle| \sum_i p_i f_j(\rho_i) = (\leq) c \forall j = 1, \ldots, l \right\}.$$

Lemma A.12 (Fujiwara and Nagaoka [5]) *Let p be an extremal point of one of the above sets. Then, the number of supports of p is less than $(l+1)(\dim \mathcal{H})^2$.*

Using the convex function, we can show the following lemma.

Lemma A.13 *When $0 \leq A \leq B$, we have*

$$\mathrm{Tr} A^s \leq \mathrm{Tr} B^s \tag{A.35}$$

for $s > 0$.

Proof Since the function $x \mapsto x^s$ is matrix monotone with $s \in [0, 1]$, (A.35) holds in this case. Assume that $s > 1$. We make the diagonalizations $A = \sum_j a_j |u_j\rangle\langle u_j|$ and $B = \sum_l b_l |v_l\rangle\langle v_l|$. Since $\sum_l \mathrm{Tr} |v_l\rangle\langle v_l | u_j\rangle\langle u_j| = 1$ and the function $x \mapsto x^s$ is convex, we have

$$\begin{aligned}
\mathrm{Tr} B^s &= \sum_l b_l^s \mathrm{Tr} |v_l\rangle\langle v_l| = \sum_j \sum_l b_l^s \mathrm{Tr} |v_l\rangle\langle v_l | u_j\rangle\langle u_j| \\
&\geq \sum_j \left(\sum_l b_l \mathrm{Tr} |v_l\rangle\langle v_l | u_j\rangle\langle u_j| \right)^s = \sum_j \left(\mathrm{Tr} B |u_j\rangle\langle u_j| \right)^s \\
&\geq \sum_j \left(\mathrm{Tr} A |u_j\rangle\langle u_j| \right)^s = \sum_j \left(a_j \right)^s = \mathrm{Tr} A^s.
\end{aligned}$$

■

The concept of "convex function" can be extended to functions of matrices. If a function f with the range $[0, \infty]$ satisfies

$$\lambda f(A) + (1-\lambda) f(B) \geq f(\lambda A + (1-\lambda) B),$$

for arbitrary Hermitian matrices A, B with eigenvalues in $[0, \infty]$, it is called a **matrix convex function**. See Sect. 1.5 for the definition of $f(A)$. Also, the function f is called a **matrix concave function** when the function $-f$ is a matrix convex function. The following equivalences are known for a function from $(0, \infty)$ to $(0, \infty)$ [6]:

① $f(t)$ is matrix monotone.
② $t/f(t)$ is matrix monotone.
③ $f(t)$ is matrix concave.

Furthermore, it is known that if the function f satisfies one of the above conditions, $1/f(t)$ is matrix convex [6]. Hence, since the functions $t^s, -t^{-s} (s \in [0, 1])$, and $\log t$ are matrix monotone, the functions t^s $(s \in [-1, 0] \cup [1, 2])$, $-t^s$ $(s \in [0, 1])$, $-\log t$, and $t \log t$ are matrix convex functions. The following theorem is known.

Theorem A.1 ([2, 6]) *The following conditions are equivalent for a function f.*

① *$f(t)$ is matrix convex on $[0, \infty)$.*
② *When a matrix Z satisfies $Z^*Z = I$, any Hermitian matrix X with eigenvalues in $[0, \infty]$ satisfies $f(Z^*XZ) \le Z^*f(X)Z$.*
③ *When matrices $Z_1, \ldots, Z_k$ satisfy $\sum_i Z_i^*Z_i = I$, any Hermitian matrices $X_1, \ldots, X_k$ with eigenvalues in $[0, \infty]$ satisfy $f(\sum_i Z_i^*X_iZ_i) \le \sum_i Z_i^*$ $f(X_i)Z_i$.*

As its consequences, we have the following corollaries.

Corollary A.1 *$f(t)$ is matrix convex on $[0, \infty)$. Given a Hermitian matrix X on $\mathcal{H}_A \otimes \mathcal{H}_B$ and a state ρ_0 on $\mathcal{H}_B$, we have $f(\mathrm{Tr}_B(I \otimes \rho_0)X) \le \mathrm{Tr}_B(I \otimes \rho_0)f(X)$.*

Proof of Corollary A.1 Assume ③. Consider the spectral decomposition $\rho_0 = \sum_i p_i|u_i\rangle\langle u_i|$. Choose the map $Z_i : |v\rangle \mapsto \sqrt{p_i}|v\rangle \otimes |u_i\rangle$. We have $\mathrm{Tr}_B(I \otimes \rho_0)$ $f(X) = \sum_i Z_i^* f(X)Z_i$ and $\mathrm{Tr}_B(I \otimes \rho_0)X) = \sum_i Z_i^*XZ_i$. So, we obtain the desired argument. ∎

Corollary A.2 *Assume that $f(t)$ is matrix convex on $[0, \infty)$ and that $f(0) = 0$ or $\lim_{t\to\infty} f(t) = 0$, When a matrix C satisfies $C^*C \le I$, any Hermitian matrix A with eigenvalues in $[0, \infty]$ satisfies $f(C^*AC) \le C^*f(A)C$.*

Proof of Corollary A.2 Choose another matrix $B := \sqrt{I - C^*C}$. When $f(0) = 0$, $f(C^*AC) = f(C^*AC) + f(B^*0B) \le C^*f(A)C + B^*f(0)B = C^*f(A)C$.

Similarly, when $\lim_{t\to\infty} f(t) = 0$, $f(C^*AC) + f(B^*tB) \le C^*f(A)C + B^*$ $f(t)B$ for any positive real $t > 0$. Taking limit $t \to \infty$, we obtain the desired argument. ∎

Now, we focus on the equation;

$$\frac{\pi}{\sin p\pi} = \int_0^\infty \frac{t^{p-1}}{1+t} dt \tag{A.36}$$

for $p \in (0, 1)$. For each $x > 0$, substituting u/x into t in the above, we obtain the decomposition of the matrix convex function $-x^p$ as

$$-x^p = -\frac{\sin p\pi}{\pi} \int_0^\infty \frac{u^{p-1}x}{u+x} du = \frac{\sin p\pi}{\pi} \int_0^\infty \left(\frac{u^{p-1}}{u+x} - \frac{u^p}{u+x} \right) du. \tag{A.37}$$

Multiplying $-x$, we also have the decomposition of the matrix convex function x^{p-1} as

$$x^{p-1} = \frac{\sin p\pi}{\pi} \int_0^\infty \frac{u^{p-1}}{u+x} du. \tag{A.38}$$

This relation shows that the function x^s with $s \in [-1, 0]$ can be written as the positive sum of a family of matrix convex functions $\{\frac{1}{x+u}\}_{u>0}$.

Next, we consider the matrix convex function x^{1+p}. Since

$$\frac{x}{u+x} = \frac{1}{u^2+1} + u \left(\frac{u}{u^2+1} - \frac{1}{u+x} \right), \tag{A.39}$$

we obtain

$$\begin{aligned} x^p &= \frac{\sin p\pi}{\pi} \int_0^\infty \frac{u^{p-1}}{u^2+1} du + \frac{\sin p\pi}{\pi} \int_0^\infty u^p \left(\frac{u}{u^2+1} - \frac{1}{u+x} \right) du \\ &= \cos \frac{p\pi}{2} + \frac{\sin p\pi}{\pi} \int_0^\infty u^p \left(\frac{u}{u^2+1} - \frac{1}{u+x} \right) du \end{aligned} \tag{A.40}$$

because the relation $\int_0^\infty \frac{u^{p-1}}{u^2+1} = \frac{\pi}{2 \sin \frac{p\pi}{2}}$ follows from (A.36) with replacing t and p by u^2 and $\frac{p}{2}$, respectively. So,

$$\begin{aligned} x^{1+p} &= \cos \frac{p\pi}{2} x + \frac{\sin p\pi}{\pi} \int_0^\infty u^p \left(\frac{ux}{u^2+1} - \frac{x}{u+x} \right) du \\ &= \cos \frac{p\pi}{2} x + \frac{\sin p\pi}{\pi} \int_0^\infty u^p \left(\frac{ux}{u^2+1} - 1 + \frac{u}{u+x} \right) du. \end{aligned} \tag{A.41}$$

This expression shows the non-linear factor of the function x^{1+p} can be reduced to the functions $\{\frac{1}{x+t}\}_{t\ge 0}$.

As another example, matrix convex functions $-\log x$ and $x \log x$ can be decomposed as

$$-\log x = \int_0^\infty \left(\frac{1}{x+t} - \frac{1}{1+t} \right) dt \tag{A.42}$$

$$x \log x = \int_0^\infty \left(-\frac{x}{x+t} + \frac{x}{1+t} \right) dt = \int_0^\infty \left(\frac{t}{x+t} + \frac{x}{1+t} - 1 \right) dt. \tag{A.43}$$

Generally, we have the following expression.

Theorem A.2 ([7, Theorem 5.1], [6, Problem V.5.5]) *Let f be a matrix convex function defined on $(0, \infty)$. There exists a positive measure μ on $[0, \infty)$ such that*

$$\begin{aligned} f(x) &= f(1) + f'(1)x + b(x-1)^2 + \int_0^\infty \frac{(x-1)^2}{u+x} \mu(du) \\ &= f(1) + f'(1)x + b(x-1)^2 + \int_0^\infty \left(x - (2+u) + \frac{(u+1)^2}{u+x} \right) \mu(du), \end{aligned} \tag{A.44}$$

where $b = \lim_{x\to 0}\frac{f(x)}{x^2} \ge 0$. *When* f *is a sub-linear, i.e.,* $f(x)/x \to 0$ *as* $x \to \infty$, *there exists a positive measure* μ *on* $[0,\infty)$ *such that*

$$f(x) = f(1) + \int_0^\infty \left(\frac{1}{u+x} - \frac{1}{u+1}\right)\mu(du). \tag{A.45}$$

Let f *be a matrix convex function defined on* $[0,\infty)$. *There exist a constant* a *and a positive measure* μ *on* $(0,\infty)$ *such that*

$$\begin{aligned} f(x) &= f(0) + ax + bx^2 + \int_0^\infty \frac{x^2}{u+x}\mu(du) \\ &= f(0) + ax + bx^2 + \int_0^\infty \left(x - u + \frac{u^2}{u+x}\right)\mu(du), \end{aligned} \tag{A.46}$$

where $b = \lim_{x\to 0}\frac{f(x)}{x^2} \ge 0$. *When* f *is a sub-linear, There exist a constant* a *and a positive measure* μ *on* $(0,\infty)$ *such that*

$$f(x) = f(0) + \int_0^\infty \left(\frac{1}{u+x} - \frac{1}{u}\right)\mu(du). \tag{A.47}$$

Therefore, the non-linear factor of a matrix convex function can be reduced to the functions $\{\frac{1}{x+t}\}_{t\ge 0}$ and x^2. That is, the set of matrix convex functions defined on $(0,\infty)$ forms a convex set, and its extremal points are given as the functions $\{\frac{1}{x+\lambda}\}_{\lambda\ge 0}$ and x^2. Theorem A.2 contains the **extremal point decompositions** in four types of matrix convex functions. In particular, the two functions $\frac{1}{x}$ and x^2 play a special role. The sub-linearity corresponds to the absence of the factor x^2, and the extendability of the domain to $x = 0$ does to the absence of the factor $\frac{1}{x}$.

Remark A.1 Nevertheless a matrix convex function is very important mathematical object, no textbook cover it including the extremal point decomposition perfectly. As shown in (A.44) and (A.46), the extremal point decomposition depends on the domain. The paper [7, Theorem 5.1] gives the extremal point decomposition (A.44). The book [6, Problem V.5.5] gives (A.46) when the derivative $f'(0)$ exists. However, the derivative $f'(0)$ does not exist in general [8]. The current form (A.46) was obtained by Hiai [8]. When we impose the sub-linearity, the coefficients of x and x^2 vanish. So, we have the extremal point decomposition (A.45) and (A.47). When the domain is $[-1,1]$, we have another type of extremal point decomposition [2, Theorem 4.40], [9, Theorem 2.7.6].

Exercises

A.14 Show that an extremal point decomposition of an arbitrary density matrix ρ is not unique in the set of density matrices when ρ is not a pure state. That is, give at least two extremal point decomposition of a density matrix ρ.

A.15 Show the concavity of the von Neumann entropy (5.77) using the matrix convexity of $x \log x$.

A.16 Show that the inequality $\phi(s|\rho\|\sigma) \geq \phi(s|\kappa(\rho)\|\kappa(\sigma))$ does not hold in general with the parameter $s \in (-\infty, -1)$.

A.5 Solutions of Exercises

Exercise A.1 Choose a basis $\{|e_i\rangle\}$ of the subspace spanned by $|u_1\rangle, \ldots, |u_k\rangle$. Define the matrix $A = (a_{i,j})$ by $|u_i\rangle = \sum_j a_{i,j}|e_j\rangle$. Then, $J_{j,i} \stackrel{\text{def}}{=} \langle u_i|u_j\rangle = \sum_k \overline{a_{i,k}} a_{j,k}$. That is, $J = AA^\dagger$, which implies $J^{-1} = (A^{-1})^\dagger A^{-1}$. Since $J = J^\dagger$, we have $J^{-1} = (J^{-1})^\dagger$. Thus, (A.12). Hence,

$$\sum_{i,j}(J^{-1})^{j,i}|u_i\rangle\langle u_j| = \sum_{i,j}\sum_{k,l,n}\overline{(A^{-1})}_{n,j}(A^{-1})_{n,i}\overline{a_{j,l}}a_{i,k}|e_k\rangle\langle e_l|$$
$$= \sum_{k,l,n}\delta_{n,l}\delta_{n,k}|e_k\rangle\langle e_l| = \sum_n |e_n\rangle\langle e_n|.$$

Exercise A.2 Using the polar decomposition $A = U|A|$, we have

$$AA^* f(AA^*) = U|A|^2U^* f(U|A|^2U^*) = U|A|^2U^*Uf(|A|^2)U^*$$
$$= U|A|^2 f(|A|^2)U^* = U|A|f(|A|^2)|A|U^* = Af(A^*A)A^*.$$

Exercise A.3 The eigen equation is $(-a-x)(a-x) - |b|^2 = 0$, which is equivalent to $x^2 = |b|^2 + a^2$. Since the eigen values are $\pm 2\sqrt{|b|^2 + a^2}$. So, the trace of $\left|\begin{pmatrix} -a & b \\ \bar{b} & a \end{pmatrix}\right|$ is equal to $2\sqrt{|b|^2 + a^2}$.

Exercise A.4 Using the definition (A.18), we have

$$\|X\|_1 = \max_{U_{AB}} \mathrm{Tr} U_{AB} X \geq \max_{U_A} \mathrm{Tr} U_A \otimes I_B X$$
$$= \max_{U_A} \mathrm{Tr} U_A \mathrm{Tr}_B X = \|\mathrm{Tr}_B X\|_1.$$

Exercise A.5 Since $BA = A^{-1}ABA$, we have $(BA - xI) = A^{-1}(AB - xI)A$. Hence, the kernel of $(BA - xI)$ has the same dimension as that of $(AB - xI)$. Thus, the eigenvalues of BA are the same as the eigenvalues of AB including degeneracies.

Exercise A.6 If A possesses the inverse, Exercise A.5 yields the desired argument. If A does not possess the inverse, we choose an invertible matrix A_ϵ approximating A. Since the eigenvalue is a continuous function of a matrix, we have $\lim_{\epsilon\to 0} \mathrm{spr}(A_\epsilon B) = \mathrm{spr}(AB)$ and $\lim_{\epsilon\to 0} \mathrm{spr}(BA_\epsilon) = \mathrm{spr}(BA)$, which implies the desired argument.

Exercise A.7

(**a**) Since (1.34) implies that

$$I = B^{-1/2} B B^{-1/2} \geq B^{-1/2} A B^{-1/2} = (A^{1/2} B^{-1/2})^\dagger (A^{1/2} B^{-1/2}),$$

(A.17) yields that

$$\|A^{1/2} B^{-1/2}\| = \|\sqrt{(A^{1/2} B^{-1/2})^\dagger (A^{1/2} B^{-1/2})}\| \leq \|I\| = 1.$$

(**b**) Exercise A.6 and (A.16) yield that

$$\operatorname{spr}(B^{-1/4} A^{1/2} B^{-1/4}) = \operatorname{spr}(A^{1/2} B^{-1/2}) \leq \|A^{1/2} B^{-1/2}\| \leq 1.$$

(**c**) The relation $1 \leq \operatorname{spr}(B^{-1/4} A^{1/2} B^{-1/4})$ implies that $B^{-1/4} A^{1/2} B^{-1/4} \leq I$. Hence, we have $A^{1/2} = B^{1/4} B^{-1/4} A^{1/2} B^{-1/4} B^{1/4} \leq B^{1/4} B^{1/4} = B^{1/2}$.
(**d**) Since $B + \epsilon I \geq A$, we have $A^{1/2} \leq (B + \epsilon I)^{1/2}$. Taking the limit $\epsilon \to 0$, we have $A^{1/2} \leq B^{1/2}$.

Exercise A.8

(**a**) Let $\lambda_i \geq 0$ be the eigenvalue of $|X|$ associated with u_i. Then, we have $\max_{v:\|v\|=1} \langle v|\, |X|\, |u_i\rangle = \max_{v:\|v\|=1} \lambda_i \langle v|u_i\rangle = \lambda_i \langle u_i|u_i\rangle = \langle u_i|\, |X|\, |u_i\rangle$.
(**b**) We have

$$\begin{aligned} \max_{U:\text{unitary}} \langle u_i|UX|u_i\rangle &= \max_{U:\text{unitary}} \langle u_i|UU_X|X||u_i\rangle \\ &= \max_{v:\|v\|=1} \langle v|\, |X|\, |u_i\rangle = \langle u_i||X||u_i\rangle. \end{aligned}$$

Since the above maximum can be realized with $U = U_X^\dagger$, we have $\langle u_i||X||u_i\rangle = \langle u_i|U_X^* X|u_i\rangle$.
(**c**) We have

$$\begin{aligned} \max_{U:\text{unitary}} \operatorname{Tr} UX &= \sum_i \max_{U:\text{unitary}} \langle u_i|UX|u_i\rangle \\ &= \sum_i \langle u_i||X||u_i\rangle = \operatorname{Tr}|X| = \operatorname{Tr} U_X^* X. \end{aligned}$$

Exercise A.9 We choose the unitary U_X by using the polar decomposition $X = U_X|X|$, and the basis $|u_i\rangle$ as the eigenvectors of $|A|$. Then,

$$\begin{aligned} \|YX\|_1 &\overset{(a)}{=} \max_{U:\text{unitary}} \operatorname{Tr} UYX = \max_{U:\text{unitary}} \operatorname{Tr} UYU^\dagger UU_X|X| \\ &= \sum_i \max_{U:\text{unitary}} \langle u_i|UYU^\dagger UU_X|X||u_i\rangle \\ &\overset{(b)}{\leq} \sum_i \max_{U:\text{unitary}} \|UYU^\dagger UU_X\| \max_{u:\|u\|=1} \langle u||X||u_i\rangle \\ &\overset{(c)}{=} \sum_i \max_{U:\text{unitary}} \|UYU^\dagger UU_X\| \langle u_i||X||u_i\rangle \overset{(d)}{\leq} \|Y\| \|X\|_1, \end{aligned}$$

where (a), (b), (c), and (d) follow from (A.18), (A.14), (**a**) of Exercise A.8, and (A.15), respectively.

Exercise A.10 Let $\mathcal{K}'$ be the $(d-k+1)$-dimensional subspace spanned by the eigenvectors corresponding to the eigenvalues $a_k, \ldots, a_d$. Then, for any k-dimensional subspace $\mathcal{K}$, the intersection space $\mathcal{K} \cap \mathcal{K}'$ has at least dimension 1. So,

$$\min_{x \in \mathcal{K}, \|x\|=1} \langle x|A|x\rangle \le \min_{x \in \mathcal{K}\cap\mathcal{K}' : \|x\|=1} \langle x|A|x\rangle \le a_k.$$

Exercise A.11 When $\mathcal{K}$ is the image of k-dimensional projection P, Exercise A.10 yields that

$$\min_x \frac{\langle x|PAP|x\rangle}{\langle x|P|x\rangle} = \min_{x \in \mathcal{K}, \|x\|=1} \langle x|A|x\rangle \le a_k.$$

Taking the maximum, we obtain

$$\max_{P:\mathrm{rank} P=k} \min_x \frac{\langle x|PAP|x\rangle}{\langle x|P|x\rangle} \le a_k.$$

The equality holds when P is the projection spanned by the eigenvectors corresponding to the eigenvalues $a_1, \ldots, a_k$.

Exercise A.12 Exercise A.11 yields that

$$a_k = \max_{P:\mathrm{rank} P=k} \min_x \frac{\langle x|PAP|x\rangle}{\langle x|P|x\rangle} \ge \max_{P:\mathrm{rank} P=k} \min_x \frac{\langle x|PBP|x\rangle}{\langle x|P|x\rangle} = b_k.$$

Exercise A.13 The matrix Hölder inequality (A.26) yields that

$$\mathrm{Tr} Z^{\frac{1}{q}} X \le \|Z^{\frac{1}{q}}\|_q \|X\|_p = \|X\|_p$$

for $p > 1$ and a matrix $Z \ge 0$ satisfying $\mathrm{Tr} Z = 1$. When $Z = \frac{X^p}{\mathrm{Tr} X^p}$, we have

$$\mathrm{Tr} Z^{\frac{1}{q}} X = \mathrm{Tr} \frac{X^p}{(\mathrm{Tr} X^p)^{\frac{1}{q}}} = (\mathrm{Tr} X^p)^{\frac{1}{p}}, \tag{A.48}$$

which implies (A.31).

The matrix reverse Hölder inequality (A.28) yields that

$$\mathrm{Tr} Z^{\frac{1}{q}} X \ge \|Z^{\frac{1}{q}}\|_q \|X\|_p = \|X\|_p$$

for $p < 1$ and a matrix $Z \ge 0$ satisfying $\mathrm{Tr} Z = 1$. When $Z = \frac{X^p}{\mathrm{Tr} X^p}$, we have (A.48), which implies (A.32).

Exercise A.14 We make a spectral decomposition of ρ as $\sum_{i=1}^{d} \lambda_i |u_i\rangle\langle u_i|$, which is an extremal point decomposition of the density matrix ρ. To make another extremal point decomposition of the density matrix ρ, we assume that $\lambda_1, \lambda_2 > 0$ without loss of generality. So, using $|u_\pm\rangle := \frac{1}{\sqrt{2}}(|u_1\rangle \pm |u_2\rangle)$, we have

$$\begin{aligned}&\lambda_1|u_1\rangle\langle u_1| + \lambda_2|u_2\rangle\langle u_2| = (\lambda_1 - \lambda_2)|u_1\rangle\langle u_1|\lambda_2(|u_1\rangle\langle u_1| + |u_2\rangle\langle u_2|)\\ =&(\lambda_1 - \lambda_2)|u_1\rangle\langle u_1|\lambda_2(|u_+\rangle\langle u_+| + |u_-\rangle\langle u_-|).\end{aligned}$$

So, $(\lambda_1 - \lambda_2)|u_1\rangle\langle u_1|\lambda_2(|u_+\rangle\langle u_+| + |u_-\rangle\langle u_-|) + \sum_{i=3}^{d} \lambda_i|u_i\rangle\langle u_i|$ is another extremal point decomposition of the density matrix ρ.

Exercise A.15 Since $x \log x$ is matrix convex, two density matrices ρ_1 and ρ_2 and a real number $p \in (0, 1)$ satisfy

$$(p\rho_1 + (1-p)\rho_2) \log p\rho_1 + (1-p)\rho_2 \le p\rho_1 \log \rho_1 + (1-p)\rho_2 \log \rho_2.$$

Taking the trace, we obtain the concavity of the von Neumann entropy (5.77).

Exercise A.16 We show the desired argument by contradiction. Consider that the states $\rho' := p\rho_1 \otimes |1\rangle\langle 1| + (1-p)\rho_2 \otimes |2\rangle\langle 2|, \sigma' := p\sigma\rangle\langle|1\rangle\langle 1| + (1-p)\sigma \otimes |2\rangle\langle 2|$ with an arbitrary state σ.

Apply the assumption of contradiction to the partial trace. Then,

$$\mathrm{Tr}((p\rho_1 + (1-p)\rho_2)^{1-s}\sigma^s) \le p\mathrm{Tr}(\rho_1^{1-s}\sigma^s) + (1-p)\mathrm{Tr}(\rho_2^{1-s}\sigma^s).$$

Since σ is arbitrary, the above inequality is equivalent with $(p\rho_1 + (1-p)\rho_2)^{1-s} \le p\rho_1^{1-s} + (1-p)\rho_2^{1-s}$, which implies the matrix convexity of the map $x \mapsto x^{1-s}$. Since this matrix convexity holds only for $s \in [0, 1]$, we obtain the contradiction.

Postface to Japanese version

My research on quantum information theory started in October of 1994, when I was a first year master's student. At that time, although Shor's paper on factorization had already been published, I was still unaware of his work. Nor was the field of quantum information theory very well known. What follows is a brief summary of how I got started in the field of quantum information theory. This is merely a personal account of my experiences, but I hope that my story will help those considering embarking on graduate or postgraduate studies and pursuing a career in research.

I began my university studies at Kyoto University studying both mathematics and physics, thanks to the university's policy of allowing students to graduate without choosing a major. In my case, I was mainly interested in physics, and I decided to study both physics and mathematics because I was not entirely comfortable with the type of thinking found in physics; I was more naturally inclined toward mathematics. As a result, during my undergraduate years, on the one hand, I had a reasonable understanding of mathematics; on the other hand, I could not understand physics sufficiently. More seriously, I could not grasp the essence of statistical mechanics, in which "physics thinking" appears most prominently. In my fourth year of undergraduate studies, I noticed that, based on my understanding of physics, I probably would not pass the entrance exams for graduate course in physics. Therefore, I decided to apply to a graduate program in mathematics (into which I was just barely accepted). In particular, while I settled on the early universe in the cosmology group as the main focus of research in my undergraduate studies, its outcome was rather hopeless due to my poor knowledge of statistical mechanics. In fact, when I told a professor of physics that I would work the next year as a tutor to help high school students cram for their physics exams, he told me, "I would never let you teach physics to anyone." I managed to graduate, but I could not assimilate physics.

The following April I began my graduate studies in twistor [10] theory[1] under Professor Ueno, a professor of mathematics at Kyoto University. I chose this topic

[1] Professor Richard Jozsa also studied twistor theory as a graduate student.

M. Hayashi, *Quantum Information Theory*, Graduate Texts in Physics,
DOI 10.1007/978-3-662-49725-8

because it is related to relativity theory, which I was interested in at that time. However, as is the case with many topics in mathematical physics, it is rooted in physics, but it was essentially mathematical. I also realized how difficult it was to understand the physics behind the mathematical concepts. Ultimately, I realized that it did not suit my interests. Although I was capable of thinking in a mathematical way, I was not interested in mathematics itself. Therefore, I could not focus on pure mathematics and started to search for another research topic. Meanwhile, teaching high school physics as a tutor to help students cram for exams school during my graduate years led me to the conviction that for the first time I truly understood physics. Until then, I was enslaved by difficult mathematical structures in physics. At this time, I realized, how important it was to understand physics based on fundamental concepts.

While searching for a new research topic, I met Dr. Akio Fujiwara, who came to Osaka University as an assistant professor. He advised me to study Holevo's textbook [11], and I decided that I would start research in quantum information theory. Up until that time, I had mainly studied abstract mathematics with little connection to physics. I was particularly impressed with the quantum-mechanical concepts described by Holevo's textbook without high levels of abstraction. Although Holevo's textbook was not an easy book to read from the current viewpoint, it was not very difficult for me because I had read more difficult books on mathematics. In retrospect, it might be fortunate that I did not proceed to a graduate course in physics because the physics community had an implicit, unwritten rule never to attempt the measurement problem in quantum mechanics due to its philosophical aspect in Japan. Therefore, while I appeared to take a rather indirect path during my years in undergraduate and graduate courses, my career may have been the most direct path. However, I faced a problem upon starting my research. Since I had only studied physics and mathematics until that point, I was wholly ignorant of subjects in information science such as mathematical statistics. In particular, despite having had the opportunity to study these subjects, I had not taken the opportunity. During my undergraduate years, compared with physics, which examines the true nature of reality, I regarded statistics to be a rather lightweight subject. I considered statistics as only a convenient subject, not an essential one. This perception changed as a result of reading Holevo's text. The reason is that it is impossible to quantitatively evaluate the information obtained by an observer without a statistical viewpoint because the measurement data are inherently probabilistic under the mathematical formulation of quantum mechanics. Ultimately, I was forced to study subjects such as mathematical statistics and information theory, which should be studied in an undergraduate program. In the end, the research for my master's thesis would be completed with an insufficient knowledge of mathematical statistics.

Further, as another problem, I had no colleagues in this research area that I could discuss my research with. Hence, I had to arrange opportunities to discuss my work with researchers at distant locations. Moreover, since I was also financially quite unstable during the first half of my doctorate program, I was dividing my research time back then between casual teaching work in high school and at an exam-preparation school. In particular, in the first six months of my doctoral program, my research progress was rather slow due to a lack of opportunities to discuss my research interests. Then, the Quantum Computation Society in Kansai opened in November 1996, and it gave me the chance to talk about topics closely related to my interests. As a result, I could continue my research. During this period, I also had many helpful discussions via telephone with Keiji Matsumoto, who was a research associate at the University of Tokyo. Thus, I was able to learn statistics, and I am deeply indebted to him. I am also grateful to Professor Kenji Ueno, who accepted me as a graduate student until my employment at RIKEN.

In less than 10 years, the situation in Japan with quantum information theory has changed completely. What follows are my thoughts and opinions on the future of quantum information theory. Recently, sophisticated quantum operations have become a reality, and some quantum protocols have been implemented. I believe that it is necessary to propose protocols that are relatively easy to implement. This is important not only to motivate further research, but also to have some feedback on the foundations of physics. I believe that the techniques developed in information theory via quantum information theory will be useful to the foundations of physics.

Thanks to the efforts of many researchers, the field of quantum information theory is now well known. But I feel that many universities in Japan have trouble internalizing quantum information theory in the current organization of disciplines. Scientific study should have no boundaries among the different fields of knowledge. Hence, I take as my point of departure the assumption that it is possible to create a more constructive research and educational environment through the treatment of fields such as quantum information theory that transcend the current framework of disciplines.

My hope is that this book will introduce to quantum information theory people dissatisfied with the existing framework of science as it is currently practiced.

References

1. M.A. Nielsen, I.L. Chuang, *Quantum Computation and Quantum Information* (Cambridge University Press, Cambridge, 2000)
2. F. Hiai, D. Petz, *Introduction to Matrix Analysis and Applications* (Universitext, Springer, 2014)
3. L.P. Kuptsov, Holder inequality, ed. by Hazewinkel, M. *Encyclopaedia of Mathematics* (Springer, 2001)
4. I. Ekeland, R. Téman, *Convex Analysis and Variational Problems* (North-Holland, Amsterdam, 1976); (SIAM, Philadelphia, 1999)
5. A. Fujiwara, H. Nagaoka, Operational capacity and pseudoclassicality of a quantum channel. IEEE Trans. Inf. Theory **44**, 1071–1086 (1998)
6. R. Bhatia, *Matrix Analysis* (Springer, Berlin, 1997)
7. U. Franz, F. Hiai, É. Ricard, Higher order extension of Löwner's theory: operator k-tone functions. Trans. Am. Math. Soc. **366**, 3043–3074 (2014)
8. F. Hiai, private communication (2015)
9. F. Hiai, Matrix analysis: matrix monotone functions, matrix means, and majorization. Interdisc. Inform. Sci. **16**(2), 139–248 (2010)
10. R.S. Ward, R.O. Wells Jr., *Twistor Geometry and Field Theory* (Cambridge University Press, Cambridge, 1991)
11. A.S. Holevo, *Probabilistic and Statistical Aspects of Quantum Theory* (North-Holland, Amsterdam, 1982); originally published in Russian (1980)

Index

M. Hayashi, *Quantum Information Theory*, Graduate Texts in Physics,
DOI 10.1007/978-3-662-49725-8

Printed in the United States
By Bookmasters